Vickery

THE DARWINIAN HERITAGE

Including Proceedings of
The Charles Darwin Centenary Conference,
Florence Center for the History
and Philosophy of Science,
June 1982

THE DARWINIAN HERITAGE

EDITED BY

DAVID KOHN

DREW UNIVERSITY

WITH BIBLIOGRAPHIC ASSISTANCE FROM

MALCOLM J. KOTTLER

PRINCETON UNIVERSITY PRESS

IN ASSOCIATION WITH

NOVA PACIFICA

Published by Princeton University Press, 41 William Street, Princeton, New Jersey 08540

In the United Kingdom:
Princeton University Press, Guildford, Surrey

Library of Congress Cataloging in Publication Data will be found on the last printed page of this book

ISBN 0-691-08356-8

Clothbound editions of Princeton University Press books are printed on acid-free paper, and binding materials are chosen for strength and durability. Paperbacks, although satisfactory for personal collections, are not usually suitable for library rebinding

Designed by Publication Graphics Ltd., Wellington, New Zealand
Typeset by Quickset Platemakers Ltd., Christchurch, New Zealand
Printed in the United States of America by Princeton University Press, Princeton, New Jersey

To

Dov Ospovat, *in memoriam*

and

Sydney Smith, mentor of us all

ACKNOWLEDGEMENTS

It is a great pleasure to thank the following colleagues: The contributors to this volume for their support and unflagging patience. Massimo Piatelli-Palmarini and Paolo Rossi of the Florence Center for the History and Philosophy of Science, which sponsored the 1982 Darwin Centenary Conference. Pietro Corsi, who acted as my liaison during preparation for the conference. Dorian Kottler who copy-edited the manuscript. Anna Rogers, who co-ordinated editorial for Nova Pacifica. J. A. Secord for invaluable assistance in correcting the bibliography. Ray Labone of Nova Pacifica and Judith May of Princeton University Press, who saw this project through to completion. Special thanks to Roger Chapman, who knocked on my door with a very bright idea.

CONTENTS

Contributors

Michele Acanfora, Istituto di Genetica, Biochimica ed Evoluzionistica, Pavia, Italy

John Beatty, Department of Philosophy, Arizona State University, Tempe, Arizona

Gillian Beer, English Faculty, University of Cambridge and Girton College, Cambridge, England

Peter J. Bowler, History and Philosophy of Science, The Queen's University of Belfast, Belfast, Northern Ireland

Janet Browne, The Collected Letters of Charles Darwin, Cambridge University Library, Cambridge, England

Richard W. Burkhardt, Jr., Department of History, University of Illinois at Urbana-Champaign, Urbana, Illinois

I. Bernard Cohen, Department of the History of Science, Harvard University, Cambridge, Massachusetts

Pietro Corsi, Department of the History of Science, Harvard University, Cambridge, Massachusetts

John R. Durant, Department for External Studies, Oxford University, Oxford, England

Howard E. Gruber, Faculté de Psychologie, Université de Genève, Geneva, Switzerland; and Rutgers University, Newark, New Jersey

Sandra Herbert, Department of History, University of Maryland-Baltimore County, Catonsville, Maryland

M. J. S. Hodge, Department of Philosophy, University of Leeds, Leeds, England

David L. Hull, Department of Philosophy, Northwestern University, Evanston, Illinois

David Kohn, Graduate Program in Nineteenth Century Studies, Drew University, Madison, New Jersey

Malcolm J. Kottler, Bell Museum of Natural History, University of Minnesota, Minneapolis, Minnesota

Antonello La Vergata, Dipartimento di Filosofia, Università della Calabria, Cosenza, Italy

Ernst Mayr, Museum of Comparative Zoology, Harvard University, Cambridge, Massachusetts

James R. Moore, Faculty of Arts, The Open University, Milton Keynes, England
Giuliano Pancaldi, Dipartimento di Filosofia, Universita di Bologna, Bologna, Italy
Duncan M. Porter, Department of Biology, Virginia Polytechnic Institute and State University, Blacksburg, Virginia
William B. Provine, Department of History, Cornell University, Ithaca, New York
Stan P. Rachootin, Department of Biology, Mount Holyoke College, South Hadley, Massachusetts
Jacques Roger, Université de Paris I, Paris, France
Martin J. S. Rudwick, Trinity College, Cambridge, England
Silvan S. Schweber, Department of Physics, Brandeis University, Waltham, Massachusetts
Francesco M. Scudo, Istituto di Genetica, Biochimica ed Evoluzionistica, Pavia, Italy
James A. Secord, Churchill College, Cambridge, England
Phillip R. Sloan, Program of Liberal Studies, University of Notre Dame, Notre Dame, Indiana
Elliott Sober, Department of Philosophy, University of Wisconsin-Madison, Madison, Wisconsin
Frank J. Sulloway, Department of Psychology and Social Relations, Harvard University, Cambridge, Massachusetts
Paul J. Weindling, Wellcome Unit for the History of Medicine, Oxford, England
Robert M. Young, Free Association Books, London, England

INTRODUCTION: A HIGH REGARD FOR DARWIN

The Darwinian Heritage represents the present rich state of historical work on Darwin and Darwinism. The common thread of the essays in this volume is a sensitivity to the pressing need to place Darwin in the context of Victorian science. The organization of the work reflects the goal of building bridges between the study of an individual and his place in scientific culture. Part One, *The Evolution of a Theorist*, explores Darwin's growth as a scientific thinker from his student days in Edinburgh to the writing of the *Origin of Species*. Part Two, *Darwin in Victorian Context*, examines both Darwin's social roles and his contributions to specific branches of natural history, including the sciences of man. Part Three, *The Comparative Reception of Darwinism*, considers the development of evolutionary communities in Europe and America. Finally, Part Four, *Perspectives on Darwin and Darwinism*, is devoted principally to philosophical and historiographic studies. This book is the product of a scholarly community that has become increasingly international, institutionalized, and diverse in its historiographic approaches. As strong as this community was in 1982, centenary of the death of Darwin, it simply did not exist in 1959, centenary of the *Origin of Species*. Hence, it is fitting that by way of introduction we should consider how the Darwin community came into existence. A critical examination of its origins is important for understanding its present achievements and future potential.

The 1959 and 1982 centenaries were celebrated in qualitatively different ways. In 1959 the principal subject of interest was not Darwin, but modern evolutionary biology. The celebrations were organized by scientists, not by historians. The major collection of papers, *Evolution after Darwin*, edited by Sol Tax (1960), was concerned with evolution and genetics. A comparable collection of historical papers would have been inconceivable. Indeed, we should remember that in 1959, far from being a passionate subject of historical research, Darwin's significance was largely discounted by historians. Darwin was seen as a second-rate thinker — an observer and compiler of facts, a scientist of no great philosophical sophistication. In short, Darwin was regarded as a sort of inductive innocent. In contrast, the implicit premise for the current community is that Charles Darwin was a thinker of profound intellect and influence. Indeed, what characterizes the present community is a belief in the importance of Darwin. This high regard for Darwin is its central tenet. In an important sense, it is the revaluation of Darwin that has fostered the outpouring of sophisticated *historical* research on Darwin that we have witnessed between 1959 and the present.

To ask what brought this community into existence we need to consider the sources and consequences of the community's high regard for Darwin. It was not historians who rediscovered the eminence of Darwin, but rather biologists. Darwin became a focus of detailed study only after the evolutionary synthesis, which enshrined Darwinian natural selection, was consolidated and widely diffused. In other words, only after biologists legitimated Darwin did historians rush to study him. Indeed the first steps were taken by biologist-historians (G. De Beer, E. Mayr, S. Smith). The professional Darwinists came only later. These professionals have of course profited from the enormous archival resources of the Darwin papers, now principally at Cambridge University. It was not the case, however, that the "rediscovery" of Darwin's archives brought about the revaluation of Darwin. Rather it was the new zeal for Darwin that prompted two biologists, Sir Gavin de Beer and Sydney Smith, to follow Darwin's granddaughter Lady Nora Barlow in rediscovering the archives by making them accessible to scholars. Thus not only the interest in Darwin but also the very access to his papers, which provided the essential raw material for much of the Darwin community, was a foundation gift from a growing and consolidating science — evolutionary biology — to a nascent and rather precarious profession, history of biology. The close links of professional Darwin students to professional Darwinians, particularly in the United States, created both institutional and conceptual opportunities, constraints, and tensions that continue to the present.

Let me sketch briefly how these links have operated. Evolutionary biology provided a devoted audience for the scholarly and semi-popular productions of Darwinists. More broadly, as evolution came to occupy a prominent position in biology curricula, a substantial body of undergraduate and graduate students, primed on Darwin, became available to historians of science. As much as these close links fostered the growth of Darwin studies, they also reinforced a certain dependency relationship. The loyalty of the scientific audience depended on the historians' ability to confirm the scientists' and historians' expectation: to confirm the shared high regard for Darwin. Thus a major thrust of Darwin studies has been an endeavor to come closer and closer to grips with Darwin — to heighten our regard by studying him ever more closely. The struggle has been challenging, rewarding, and as one would expect, carried out with increasing subtlety. But the inevitable result has been to emphasize the particularity and uniqueness of Darwin. This has been its considerable conceptual value, but equally important, this has been its social utility.

Thus for just over two decades we have been allowed to have Darwin in focus. As the essays in this volume show, the gift of evolutionary biology to the historians and philosophers of biology has proved to be a treasure.

But the emphasis on the uniqueness of Darwin has had its costs. Interestingly enough, particularly in recent years, this emphasis has not precluded the use of a variety of historiographic orientations. Particularist Darwin studies have been far from the exclusive domain of "internalists". Of course, work centered on the structure and development of Darwin's scientific ideas has been the hallmark of Darwin studies. But, in addition, we have seen work that stresses the metaphysical and methodological foundations of Darwin's view of nature, and illuminating work has been written from a social history perspective — two pioneering studies are included in the present volume (Secord on Darwin and the breeders, and Moore on the social role Darwin fulfilled in the village of Down). But many potentially contextualizing works have put Darwin in the center of the picture as much as the studies of his theories and their development.

If the most assiduously studied and hence most fruitful area of Darwin research has been concerned with Darwin's intellectual development, perhaps the area that has received the second most concerted interest has been the national receptions of Darwin. *Ipso facto* this work has not been quite so closely focussed on Darwin. But I would suggest that it too has had a defining bias. As the professional interests of Anglo-American Darwinists have been enhanced by giving Anglo-American evolutionary biologists what they want, so too the study of national reception enhances national prestige — even if studies come to the "negative" conclusion that there really was no reception of Darwin in a particular country or that the prevailing conditions of introduction altered Darwin practically beyond recognition. Indeed one wonders whether such conclusions, valid though they may be, themselves serve to enhance national prestige by emphasizing the uniqueness and particularity of national conditions. Notice that I have not used the term comparative reception. As far as I have seen, we are still beginning to move towards the *comparative* reception. For example, whether or not it makes any sense to speak of an international dimension to Darwinism, or of international evolutionary communities in the nineteenth century, or of such communities defined by biological sub-disciplines, remains a series of important and untested hypotheses.

It seems clear that both the studies on Darwin himself and the work on national receptions have been, each in their own way, overly particularist in orientation. It is against this background that *The Darwinian Heritage* should be considered. These essays are attempts to go beyond the particularist past of Darwin studies, but to do so *without* sacrificing the clarity of detailed focus that is their present strength. This is a difficult task since, as I have indicated, our particularist orientations have been neither accidental nor dictated purely by our subject. Rather they have been contingent on the conditions that brought the Darwin community into existence. If we now ask what will be the future maturation of this community, I would suggest that the direction it ought to take is the development of disciplinary studies.

If we are to continue the process of historical contextualization, then we need to reconstruct Darwin's effective scientific community. This would allow us to make greater sense of why Darwin's science took its characteristic shape. To do this, however, we must first understand, in their own right, the conceptual debates and institutional structures of those disciplines in which Darwin participated. Ironically, this means that if we are to truly "find" Darwin we must be prepared to let him go out of focus as the historical evidence requires. Studies of this nature are not only beyond the dimensions of this volume and the present scope of the Darwin community, but they seem also beyond the present ken of historians of nineteenth-century biology. Nevertheless, Herbert's paper in this volume, on Darwin and the geological community, is an important step in this direction. Such disciplinary studies, of course, have national and international dimensions. They would provide the correct level of analysis for studying differential national reception. Our goal should be, at the very least, to grasp such commonalities as existed in late nineteenth-century natural history.

What is the likelihood that the Darwin community will grow along the lines suggested? What contingencies of the current state of the community favor such a future for Darwin studies? I can see three hopeful components of the present situation. First of all the gold mine in Cambridge, while not exhausted, has been well worked. The archival gold that has been the currency of much of our present wealth isn't in the hills any more. At the very least, the developmental studies should be entering a synthetic phase. The only way I can see to accomplish a synthesis is to understand the problematics set by disciplinary debates. Furthermore, the gold that does remain lies in two veins: (1) Darwin's marginalia, which ought to force thorough examination of what Darwin read and (2) Darwin's 14,000 pieces of correspondence, which are gradually being published. The publication of the correspondence ought to radically alter our focus away from Darwin's master theory and onto the execution of his research program. Whether or not we can sustain an interest in Darwin's normal science I do not know. But the way I see to cope with the coming flood of correspondence will be for Darwin scholars to become social historians. Better yet, perhaps social historians will find their way to Darwin. I say this because the mass of Darwin's letters deals with the construction and maintenance of chains and networks of informants, which Darwin manipulated to provide much of the evidentiary substrate for his research program. Thus, to exploit this material we will have to learn who these informants were and we will have to reshape our present, rather elitist, concept of the scientific community in the nineteenth century. I see two further developments. The Darwin community now has the potential to become truly international; hence, there is some prospect that the comparative studies I have called for will be undertaken. Finally, the maturation of history of science as an institutionally autonomous discipline ought to encourage

Darwinists, particularly those who are trained in evolutionary biology and those who have retrained themselves, to consider the relationship between Darwin studies and the central issues of contemporary history of science — such as the structure of scientific communities and the interaction between social and conceptual factors in the production of scientific knowledge. As we create our own audience and our own institutions, we may be able to redirect our orientation.

David Kohn
Drew University
October 1983

PART ONE

The Evolution of a Theorist

1

GOING THE LIMIT: TOWARD THE CONSTRUCTION OF DARWIN'S THEORY (1832–1839)

Howard E. Gruber

As a cognitive psychologist, my forays into the history of science have as their ultimate aim to contribute something to the psychology of thinking and the psychology of creativity. I hoped to learn from historical studies, and enrich my own rather crabbed, often Philistine field. In the course of this effort, my students and I found ourselves developing what we now call, quite provisionally, an "evolving systems approach to creative work" (Gruber 1980a, b).

In this view, creative work is seen as a purposeful growth process. Much work on the psychology of creativity reveals a certain tropism toward monolithicity. In such diverse ideas as: one great insight, one ruling passion, one overarching metaphor — there is a common term, *one*. In contrast, our work has persistently revealed a striking pluralism of events and processes. For Darwin there were many insights, each with a complex inner structure; rather than representing a break with his own past, they reflect the ongoing function of the evolving system of thought (Gruber 1981a). Similarly, there are many influences, several candidates for his "father figure", many metaphors (Gruber 1978), and many enterprises.

In addition to this emphasis on growth and pluralism, we stress the idea of creativity as *purposeful* work. Since it always seems to take a long time, the creative individual must go to some lengths to organize the conditions of life that make possible such continued work. If it were easier, faster, and more straightforward than experience shows to be the case, spontaneity might be enough. But if it were so easy, fast and straightforward, many would accomplish the same thing, and we would not deem it so creative. In the real world, then, purpose is indispensable for creativity.

The person doing creative work exhibits the continuous interplay of three loosely coupled sub-systems: the organizations of knowledge, of purpose, and of feeling. This interplay is displayed with particular clarity when the thinker undertakes to push ideas to their extremes, to abandon cautious middle-of-the-road strategies and instead to test the limits of his innovations. Sailing to the edge of one's intellectual world does not happen by accident: it requires deep knowledge and a sense of direction. It is, moreover, so taxing an effort that it requires intellectual courage and, if not the ability to enjoy life at the edge, at least the resolve to endure it.

The current status of Darwin studies provides an object lesson in the density and complexity of a creative thought process. Instead of being apologetic that we, the collectivity of Darwin scholars, have written so much, we ought to brace ourselves for the probable future. The history and philosophy of science, cognitive science and developmental psychology have reached a promising confluence. The idea that the work of hermeneutic interpretation is a legitimate part of our enterprise has at least taken hold, and description is becoming thicker and thicker. We are, I think, growing more skilled in relating the internal history of science to wider issues in personal psychology and social history. Out of all this will emerge a new generation of Darwin studies, and during its gestation we should all be very patient. Newell and Simon, in their book *Human Problem Solving*, analyze the thinking of one subject solving one problem, thinking aloud while he did it. The subject took twenty minutes. The analysis covers 100 pages (Newell and Simon 1972).

The study of Darwin's thinking is many orders of magnitude more complex. He was solving not one problem but many. The problems were not chosen for him but by him as part of a broader effort to construct a new point of view. He faced a double task. On the one hand, he had to make the best possible use of a wide array of professionally accepted, normalized scientific knowledge. On the other hand, he had to organize his efforts so as to raise and answer questions hardly dreamt of within that conventional framework. To understand Darwin's thinking, we must study the connections between these quite different aspects of his work — his intellectual navigation in well-charted scientific waters and his explorations of the farthest horizons.

I. Networks of Enterprise

If we are to deal with the complexities of a creative life we absolutely must develop some methodical ways of surveying it as a whole. As we go deeper and deeper into detail, we need to avoid losing our sense of direction. One orienting device that I have proposed is the *network of enterprise* (Gruber 1977). This is a diagrammatic way of examining the creative person's organization of purpose by depicting all of the activities of the person as they are connected in time. It permits us to see both the continuity within and the diversity among simultaneous ongoing activities. I use the term *enterprise* to suggest something larger than a problem or project; it has no necessary termination, and the stock of projects within it are usually renewed in order to keep it functional. Of course, at any given time some enterprises are dormant or less active than others.

As it happens, quite independently of my work, Sandra Herbert in her edition of Darwin's *Red Notebook* has published some excellent diagrams that capture the same idea in a simple and illuminating way (Herbert 1980, pp. 14–17). Although a number of colleagues (and I, too) have drawn up networks of enterprise for Darwin, I believe the best reasonably complex diagram currently available was drawn by Martin Rudwick (Rudwick 1982b). This was constructed in such a way as to show that Darwin's network was not only a set of activities, but an agenda. More specifically, it was a plan for the sequence in which his different enterprises would rise from the privacy of Darwin's mind to the level of public disclosure.[1] Needless to say, a network of enterprise has other dynamic properties. For example, one enterprise can steer another, distract attention from another, provide thought-forms and metaphors useful in other contexts.

More broadly still, the network of enterprise represents the organization of purpose for the creative person. As such, since he or she is more or less aware of its structure, it is a fundamental part of the self-concept.

In Darwin's case, as the present essay and for that matter this entire volume show, it is indispensable to see each part of his activity in relation to the others. Ideas or actions which seem ambiguous in a narrow context are clarified as the frame is widened. The point is not so much that we the interpreters clarify Darwin's meaning, but rather that we come to understand how Darwin, over time, disambiguated himself.

II. The Shape and Function of Controversy

As the fund of solid scholarship mounts, disagreements emerge. If the reconstruction of thought processes were an art form, these differences could simply be allowed to stand. As things are, there is an increasing convergence and even collaboration among relevant disciplines concerned with the growth of scientific knowledge: social history, history of science, philosophy of science, cognitive psychology, and the sociology of knowledge. There is even some hope that this confluence is producing a science of science, in which issues can be settled, questions really answered, and knowledge accumulated. So a productive strategy for dealing with differences should be sought.

At present there seem to be two main strategies at work. I want to describe them and propose a third. For want of better terms I will call them *the cave*, *the shadow box*, and *the solution tree*.[2]

The cave strategy is simply the pessimistic subjectivism inherent in believing that we are all looking at mere shadows of the world, and all from the same station point. We can never decide what is really there: if differences arise they can never be resolved. Since we all see the same shadows, any differences must have a subjective origin. The best solution is to accept our fate.

The shadow box strategy. In Plato's cave there is only one source of illumination and only one wall on which shadows are projected. Imagine, instead, a box with an unknown object in it, with two sources and two screens, and hence two station points. Now if two observers begin by disagreeing about what is in the box, they may discover that they are looking at two different shadows of the same thing. They may be able to settle their differences by synthesizing their two perspectives.

In the cave strategy:
"triangle" versus "circle" → disagreement
In the shadow box strategy:
"triangle" x "circle" → "cone!"

I have done this experiment in the laboratory. People can solve quite complex problems fairly soon. But first they must get over the egocentric tendency to discount the other person's report; they must build up trust and a shared descriptive language. For all its merits and its resemblance to some moments in scientific work, the shadow box strategy, or the strategy of multiple perspectives, has two limitations. First, there are really innumerable perspectives and no finite number will tell all. For example, convexities appear nicely in shadows, but to detect concavities other exploratory devices must be introduced. Second, the strategy assumes that there is one unchanging reality, and that a more powerful synthesis will eventually reveal it.

But suppose this is not the case. Suppose, for example, that there is not one Darwin and one sequence of ideas he entertained, waiting to be discovered . . . but many! This thought leads to the third strategy — and beyond.

The solution tree. Investigators of problem solving have for some time been interested in an approach which entails mapping out all of the possible solutions to a problem, separately from observations of actual solutions produced by experimental subjects. Armed with such a set of possible pathways one can then more easily identify the one actually chosen. This approach, like the other two, assumes that there is, for a given thinker, only one pathway. Moreover, it requires that the investigator know more than the experimental subject. This is not a good model for us, for a reason that we all tacitly accept — we may not be able to think about the problem in hand *as well* as Darwin, much less generate all possible solutions. The solution tree strategy may be appropriate for understanding an experimental subject solving a relatively simple, closed problem — where all the rules and conditions are set by the experimenter. But it seems inapplicable for understanding creative scientific thinking, where the limits of the problem and the rules of the game are all constantly changing.

And yet there is a gleam of light in the solution tree approach. It is plausible that a man like Darwin explored many pathways, found partial solutions to numerous problems, and often found several solutions to the same problem. Each successful move would increase his confidence in the general approach that was guiding him. Each *un*successful move, remembered, would increase his knowledge of the intellectual terrain over which he was moving — and by the same token, increase his confidence in his developing point of view.

As lived by Darwin then, there is not a simple pathway to be charted, but a set of them. If we want to know the moves Darwin actually made, knowledge of the set of moves potentially open to him may be enormously helpful.[3]

But how can we get such knowledge? Must we surpass Darwin? I think not. This is where our pooled knowledge and effort are useful. Instead of each rejecting the other's contributions and vaunting our own as better, we can look at each attempt as one of the many moves necessary to fill out the solution tree. Any description of Darwin by a reasonably competent person is a candidate for inclusion in the solution tree. Moreover, descriptions of *anyone* else working in the same domain (Lamarck, Erasmus Darwin, Lyell, Owen, Hooker, etc.) are also plausible candidates. So we can and do generate a greatly expanded solution tree, far exceeding our individual capacities. All we need is respect for each other and the patience to organize our combined efforts in such a fashion.

But our use of the solution tree need not be restricted to finding the *one* pathway Darwin followed. The approach I am proposing is inherently phenomenological. We want to reconstruct Darwin's thinking as *he* experienced it.[4] He had the time, the energy, and the absence of smugness that allowed him to explore widely in the set of possible solutions. He had also the technique of note making, developed in a powerful way, to help him re-explore, retrace the pathway taken.[5] For him, vagrant thoughts were less ephemeral than for most, because he was committed to writing them down. Finally, he believed that "the subjective probability" of an hypothesis increases as the number of partial proofs, following different lines, rises.

III. The Voyage Begins

When Darwin set out in the *Beagle*, it took him a while to get his sea legs and longer still to find his feet as a professional naturalist moving towards the life in science we know him for. Even then he remained vulnerable to *mal de mer* and to a certain *mal d'esprit* reflected in remarks such as

> This multiplication of little means & bringing the mind to grapple with great effect produced is a most laborious & painful effort of the mind . . . (*C*75)

During the voyage, alongside his scientific notes he kept a diary of all sorts of narratives and personal feelings (but nothing too intimate to publish in the *Journal* he may have already been contemplating). There were in the *Diary* many observations pertinent to what would eventually play a major role in his evolutionary theories, and become a distinct enterprise in its own right — his reflections on *homo sapiens*. A few early entries in this *Diary* reveal his state of mind, his plans, and some of his basic orientation at the time (Darwin 1934).

On 13 December 1831, two weeks before the *Beagle* weighed anchor, he wrote a brief sketch of plans for work during the voyage.

> I am often afraid I shall be quite overwhelmed with the number of subjects which I ought to take into hand. It is difficult to mark out any plan & without method on shipboard I am sure little will be done. The principal objects are 1st, collecting, observing & reading in all branches of Natural history that I possibly can manage. Observations in Meteorology, French & Spanish, Mathematics, & a little Classics, perhaps not more than Greek Testament on Sundays. I hope generally to have some one English book in hand for my amusement, exclusive of the above mentioned branches. If I have not energy enough to make myself steadily industrious during the voyage, how great & uncommon an opportunity of improving myself shall I throw away. May this never for one moment escape my mind & then perhaps I may have the same opportunity of drilling my mind that I threw away whilst at Cambridge. (*Diary*, p. 14)

Thomas Huxley, writing his resolutions at a similar stage — the beginning of the voyage of the *Rattlesnake* — was far more specific and more professionally crisp (Huxley 1935, pp. 16–17). Perhaps Darwin's initial looseness and openness was a great asset, when coupled with certain other attributes.

On 11 January 1832, sailing from Tenerife to Cape Verde Islands, he has been working hard with his marine catches.

> *January 11th.* I am quite tired having worked all day at the produce of my net. The number of animals that the net collects is very great & fully explains the manner so many animals of a large size live so far from land. Many of these creatures, so low in the scale of nature, are most exquisite in their forms & rich colours. It creates a feeling of wonder that so much beauty should be apparently created for such little purpose. (*Diary*, p. 23)

Presumably, his nets caught mostly small organisms, so he realizes there is a good food supply for larger ones. Here, then, is Darwin thinking about the food chain, very early. Note also the ease with which he steps back from his own assumption of functional order to enjoy "a feeling of wonder" at the apparent lack of purpose in the beauty of the natural world. On 28 February 1832 he records his early reactions to tropical scenery:

> But these beauties are as nothing compared to the Vegetation; I believe from what I have seen Humboldt's glorious descriptions are & will for ever be unparalleled: but even he with his dark blue skies & the rare union of poetry with science which he so strongly displays when writing on tropical scenery, with all this falls far short of the truth. The delight one experiences in such times bewilders the mind; if the eye attempts to follow the flight of a gaudy butter-fly, it is arrested by some strange tree or fruit; if watching an insect one forgets it in the stranger flower it is crawling over; if turning to admire the splendour of the scenery, the individual character of the foreground fixes the attention. The mind is a chaos of delight, out of which a world of future & more quiet pleasure will arise. I am at present fit only to read Humboldt; he like another sun illumines everything I behold. (*Diary*, p.39)

Darwin sees himself going beyond "the chaos of delight". His cathexis with nature is deepening. He shows his strong sense of connection with Humboldt, whose writings had enthralled him during his student days. But even Humboldt "falls far short of the truth". Again we see Darwin's ability to stand away from the things he admires, and to go beyond the moment. I believe this passage records the moment when Darwin began to construct his metaphor of the "entangled bank", which became the organizing principle of the celebrated closing passage of *The Origin of Species*.

On 18 December 1832 he recorded his first reactions to a primitive group, the Indians of Tierra del Fuego: "I would not have believed how entire the difference between savage and civilized man is. It is greater than between a wild and domesticated animal, in as much as in man there is great power of improvement" (*Diary*, p. 119). In this and other passages Darwin conveyed his vivid sense of the strangeness of these "inconceivably wild" people. These entries also reveal his commitment to the ideal of progress and show him aware of the vast transformations possible within a species.

IV. 1832–1834: Darwin Assimilates Lyell's Principles

Throughout the voyage, Darwin's major activity, by a long margin, was in the field of geology. During the first two years, the main manifest event in Darwin's development was his reading of Lyell's *Principles of Geology* (1830–1833), moving toward the increasingly explicit decision to reject the catastrophist geology he had learned from his teacher Adam Sedgwick, in favor of Lyell's uniformitarianism. Each theory had something to say about physical geology and something to say about the relations among geology, paleontology, and biogeography.

As Hodge has recently emphasized, the theoretical situation in geology then called for systematic search for fossils, with or without benefit of Lyell's *Principles* (Hodge 1982). And we see that Darwin sprang into action on this front early in the voyage. He went out looking for fossils and he made exciting finds. Although there is a clear distinction between the two positions, it is not hard and fast. There are slow processes in Sedgwick's and fast ones, even floods, in Lyell's. There are extinctions in both, and both rely on some mysterious "creation" to replace the lost species.

In the field of physical geology, the matter is clear. Darwin became a uniformitarian, we may even say Lyell's disciple. It took him perhaps two years to accomplish this transition (Gruber and Gruber 1962).

Paleontology played an important role in Lyell's physical geology. From fossil evidence one could reason about the probable course of geological events. Finding beds of seashells on mountain tops suggested the former residence of the sea: either the mountains have been upraised or the sea level has subsided. Further reasoning and evidence of the same kind could decide the matter. An exciting array of issues could be dealt with in this manner.

Matters are much harder to interpret when we see Darwin using the same range of evidence to settle the biological questions of the extinction of some species and the appearance of others. Modern scholars can take the same remark to show that Darwin was coming "to face directly general difficulties in Lyell's account of extinction" (Hodge 1982, p. 35), or "a convinced Lyellian, which means he was committed to (1) the immutability of species; (2) local extinction and local creation as opposed to catastrophism; (3) extinction proceeding gradually by the successive deaths of individuals; (4) the concept of local species distribution" (Kohn 1980, p. 71).

This passage and its alternative interpretations are worth examining. It is a part of his Geological Notes, a few pages written in February 1835 and later removed to be filed with notes on South American geology. In the nearly 1400 pages of geological notes Darwin made during the voyage, this passage may be his first (and almost only) extended discussion of issues mentioned above. Although its interpretation has occasioned some disagreement, a few major points can be summarized.

1. Darwin rejects the idea of a single "diluvial débâcle" as the cause of extinction. He is also skeptical about a series of such events as the likely cause.
2. He is dubious about changes in climate as the cause of extinction.

3. He is interested in the compensatory relationship of regions of elevation and regions of subsidence.
4. He accepts Lyell's metaphor, likening the death of species to the death of individuals, both as natural processes.
5. He extends the metaphor to include both the "gradual birth and death of species". While the phrase, "gradual birth" occurs only once, and almost in passing, it is hard to ignore: Darwin is not only a future evolutionist, he has a past, through contact with the ideas of his grandfather, of Grant, and of Lamarck.
6. After this one *lapsus linguae* he reverts to the more Lyellian formulation, "successive births must repeople the globe". This phrase happens also to echo one of his grandfather's poems (Erasmus Darwin 1803, Canto IV).
7. He probably believes that in the order of nature which "the Author of Nature has now established" the number of species remains approximately constant.

In spite of numerous ambiguities, it seems to me that we can sum up Darwin's most general ideas about extinction at this time as lying within a certain range on a number of issues.

Extinction. Definitely occurs. Sudden débâcles rejected as cause. Possible mechanisms: species senescence, disadaptation due to environmental change.

Approximate constancy in number of species. Accepted as an explicit but unexamined premise.

Replacement of old species by new ones. Follows from the above. Possible mechanisms: "successive births" or "gradual births". Both are vague terms, and it must be noted that the apposition of "gradual" considerably modifies the metaphor of "birth".

In the theories then current, species death could be Sedgwick-sudden, or Lyell-gradual disadaptation, or Brocchian senescent.[6] Do we have Darwin becoming an evolutionist as early as February 1835?

On balance, I think not. All the other evidence points the other way. Kohn would probably accept the interpretation Hodge has now given the passage, as I do. Darwin was dealing with the issue of extinction in a somewhat confused way. He could not interpret his own fossil findings without more expert help, which he received later (see below). The passage does represent the beginning of his rather longstanding commitment to *some version* of the species senescence idea.

There are several versions, and Darwin probably vacillated among them. But to my mind we should not negotiate away these differences of interpretation. They reflect something important — the ambiguities in Darwin's position at *every* point in his development. He was skillful and creative in using ambiguity productively, both to help him get on with what could be settled and to suggest openings. He was capable of living with ambiguity. Also he could sustain ambivalence, entertain several theories during the same period. Closer and closer study of Darwin's thinking should not be aimed at finding the one right pathway that correctly describes his route. He had the time to explore a number of paths. So should we.

I do not say all this in an especially conciliatory spirit, although I see nothing wrong with that. Rather, I wish to underline the value of many eyes, many minds, many station points. The way toward understanding sometimes passes through choice and other times through synthesis.

What can we now say of Darwin's commitment to Lyell? Let us review what we know.

In 1832 his unseen mentor and hero was still undoubtedly Humboldt. By sometime in 1833 he had assimilated enough Lyellian geology to reject, with increasing resolution, throughout 1833–1834, his earlier training in catastrophist ways of thought, especially concerning physical geology.

Sometime after receiving it in April 1834, Darwin began to read and absorb volume III of Lyell's *Principles*. Not long after, Darwin began to think along Lyellian lines with regard to a group of related issues connecting biogeography, and paleontology with uniformitarian geology, all under the aegis of a creationist (albeit multiple creationist) point of view. These commitments are expressed mainly in Darwin's geological notes of February 1835. And it must be noted that this is not a very rich record compared with the documentation we have on other matters. Furthermore, it must be noted that even this commitment was more than a little "iffy".

By December 1835 we have Darwin (a) criticizing Lyell's theory of coral reefs and (b) questioning the immutability of species. It should be noted that even a firm belief in mutability of species would not necessitate espousal of evolution. Although there are still many points of agreement between Darwin and Lyell on biological questions, the atmosphere of discipleship, which lasted between two and three years, has dissipated. When Darwin steps off the *Beagle* in 1836 he is on his own.

Among Darwin scholars, there is good measure of agreement about the theoretical outcome of the voyage for Darwin's progress. To be sure, an older generation of scholars may have believed in a sudden eureka experience in or just beyond the Galapagos experience. But it is now widely recognized that there was during the voyage no grand "Aha!" about the idea of evolution, not to speak of the mechanism of natural selection. In spite of much theoretical and personal growth, Darwin had still a long way to go.

V. Coral Reefs: A Theoretician Upward and Outward Bound

There are two themes that appear and reappear throughout most of Darwin's life, adaptation as both state and process, and continuity through transformation. Both make an early appearance in a surprising place: Darwin's theory of the formation of coral reefs, which he worked out in December 1835, before visiting the coral islands of the Pacific toward the end of the voyage.

Adaptation can be thought of in two ways. On the one hand it refers to a steady state, in which the different parts of a system are so formed that they function in harmony with each other. On the other hand, it refers to a process in which adaptive change in one part of the system compensates for change in some other part. Darwin's coral reef theory argued that a series of local compensatory changes in the growth of coral organisms generates, in the long run, a continuous series of forms of coral reef. The coral organism flourishes within a certain distance of the ocean surface. As the bottom sinks, due to the action of large-scale geological processes, the live coral flourishes at a new level. Meanwhile, a corresponding increment is added to the column of dead coral. As the reef column grows upward and outward, its interaction with the rough and tumble of the sea changes in ways that account for the ultimate shape of the reef. Under different conditions, different types of reef are formed. These are not sharply distinguished but, Darwin argued, grade into each other. Thus, a series of smooth changes in outward physical forces produced a continuous series of forms: fringing reefs, barrier reefs, and coral atolls.

This theory bears a striking *formal* resemblance to the theory of evolution through natural selection. The similarities have been pointed out independently by Gruber and Gruber (1962), and by Ghiselin (1969). First, both theories contain a principle of population growth, e.g. the coral organism does not grow beyond some limiting distance from the ocean surface. In both cases the limiting principle is described by Darwin as a *struggle* — in the case of coral formations, a struggle "between the two nicely balanced powers of land and water". Second, both theories combine this limiting principle with geological ideas to explain the major facts of geographical distribution. Thus the hypothesis that a pattern of regions of subsidence of the Pacific floor (together with other geological factors) determines the places in which the coral organism grows and forms reefs. Third, both theories generate a continuous series of forms where other theories posited only certain classes. Thus for example, ". . . barrier reefs, when encircling islands, are thus converted into atolls, the instant the last pinnacle of land sinks beneath the surface of the ocean."[7]

This coral episode is important for a number of reasons. First, it shows Darwin as a confident theoretician: extrapolating not only from observations but from his own prior theoretical work; formulating the theory before ever seeing a coral reef. It shows Darwin thinking on a global scale: over wide spaces, coordinating the elevation of continental land masses with the subsidence of remote ocean floors; over long periods of time, imaginatively reconstructing the formation of reefs through the interaction of geological and biological processes. It shows Darwin comfortably handling the complexities of a multi-level theory that requires: close knowledge of a small invertebrate organism; clear thinking about the consequences of its colonial mode of life in relation to its environment; working out the reef building effects of periods of elevation and subsidence; connecting all this with a still hypothetical picture of geological processes on a global scale.

Second, it shows Darwin in December of 1835 forming a theory that disagrees with one advanced by Lyell. This did not represent a sharp break with Lyellian thinking, as Lyell was quick to admit, in expressing his admiration for Darwin's idea. Nevertheless it does show that Darwin felt free to criticize his still unseen mentor.

Third, the theory expresses Darwin's interest in a more general theme, the way in which living organisms transform both their own immediate environment, and the earth in general. This "life makes land" theme was made evident in 1837 when Darwin published two papers bearing on it, the May 31st paper on the formation of coral reefs (*CP* 1:46–49, 1837) and the November 1st paper on the formation of vegetable mould through the action of earthworms (*CP* 1:49–53, 1837). The joint occurrence of the two papers, the fact that the earthworm paper seems to come out of nowhere, and the fact that both topics were taken up at later times — all this argues for the idea that the coral theory was not an isolated event, but one related to Darwin's general point of view and embodied in an enduring theme.

Since the term *adaptation* is generally used to refer to morphological and behavioral changes in the organism, the reader may question my use of it to refer to a system of compensatory changes maintaining an invariant. The key point is that Darwin's thinking, from an early date, was permeated with the idea of self-regulating systems. In the eighteenth century there had been a marked increase in the development of self-regulating machines. During the same period the concept of society as a self-regulating system became prominent in the work of Adam Smith and others. The American constitution was constructed as a system of "checks and balances". Although Darwin never used the analogy between natural selection and man-made feedback devices, Alfred Russel Wallace did. In his 1858 paper, presented for him at the Linnaean Society, he wrote of natural selection, "The action of this principle is exactly like that of the centrifugal governor of the steam engine, which checks and corrects any irregularities almost before they become evident . . ." (Wallace 1858b).

How like the "nicely balanced powers" in Darwin's coral reef theory!

Nevertheless, Darwin's first theory of evolution — whether we take Gruber's, Hodge's, or Kohn's version of it (or all of them as there was not necessarily only one at a time . . .) — does not have a formal structure of the kind described above. An adequate account of Darwin's intellectual development should deal with that rather surprising inconsistency.

Darwin's actual visit to the coral islands was a significant event, providing him with the opportunity to make observations supporting his already constructed theory. His increase in self-confidence as a theoretician is reflected in an entry in the *Diary*. As the *Beagle* sailed away from Keeling Island on 12 April 1836, he wrote:

> In the morning we stood out of the Lagoon. I am glad we have visited these Islands: such formations surely rank high amongst the wonderful objects of this world. It is not a wonder which at first strikes the eye of the body, but rather after reflection, the eye of reason. (*Diary*, p.400)

The sense of self Darwin experienced at this time is expressed in a letter to his sister Caroline, written 29 April 1836. He mentions his work on coral formations and remarks, "The idea of a lagoon island, 30 miles in diameter being based on a submarine crater of equal dimensions, has always appeared to me a monstrous hypothesis" (Darwin 1945, pp. 138–139). This was Lyell's idea that he was rejecting. Later on in the letter he writes of his plans to live in London and work as a geologist, "It is a rare piece of good fortune for me, that of the many errant (in ships) Naturalists, there have been few, or rather no, Geologists. I shall enter the field unopposed."

With the theoretical equipment and empirical knowledge we have now described, it might seem as though Darwin was in a good position to move toward a theory of evolution, and that that theory would be one involving an equilibration model of the kind he already knew well, having created it himself. But there were obstacles to be removed. Chief among them were Darwin's belief, although somewhat shaken, in the immutability of species and his inability to interpret his own puzzling biogeographical and paleontological materials. These two kinds of issues were closely related, and their resolution would, it has been argued, make an evolutionist of Darwin. How were they resolved? And did their resolution suffice?

VI. The Self-Construction of a Transformationist

It is now widely agreed among Darwin scholars that when Darwin stepped off the *Beagle* he was not yet an evolutionist. Although our knowledge of the immediately post-voyage period is quite incomplete, Sandra Herbert's publication of the *Red Notebook* is an important landmark in scholarship for this period *(RN)*. And Frank Sulloway (1982a, b, c, 1983) has now done a masterful job of tracking down and organizing the empirical work that moved Darwin toward transmutationism. Sulloway speaks of Darwin's "conversion" but I prefer to think of it as "self-construction" — for three reasons. First, for the whole period from about February 1835 to July 1837 Darwin seems to be moving in a direction, making a set of choices, constructing a point of view and applying it over a wide range of phenomena. Second, at any given time his belief system is assembled out of many components, each with considerable inner structure and all fitted together with some care, albeit not always perfectly coherently. Third, conversions come to an end, constructions do not — and there seems to be no end point in Darwin's activity in any of the enterprises or themes in question. This lack of finish means also that there are always loose ends and ambiguities, continually re-animating the creative process.

The reader may object to my description of movement toward a rather vague goal as purposeful. I grant that Darwin's purposes are not always clear. But remember, we are not speaking of history or of evolution; abstract criticisms of teleology are not at issue here. Human beings do have purposes, and they need to organize their work. The very concept, *work*, is saturated with the idea of purpose. Goal, purpose, plan, aspiration, self-concept, ideal self — these are fundamental human attributes. For years, I have wanted to become a pacifist; I may someday achieve that aim. What is wrong with thinking that Darwin, especially given his family history, may have wanted to become an evolutionist, may have been consciously aware that some intellectual moves took him in that direction and others did not?

During the voyage Darwin collected wonderful material. He later wrote that the relation between fossil and living forms in South America and the facts of geographical distribution, especially the peculiar array of species he found in the Galapagos, were critical in swaying him toward evolution (*Autobiography*, pp. 118–119). But he was not, during the voyage, in a position to use these materials in an evolutionary theory. He was not competent enough in anatomy to make the necessary analyses of his fossils; nor was he enough of a systematist to solve the classificatory problems his far-ranging collections posed. His Galapagos collections were not complete, many specimens were initially misclassified, and the famous tortoises and finches were not adequately labeled to know which island they came from. To some extent these problems were due to Darwin's lack of expertise. But also, he lacked the evolutionary perspective that would have led him to collect and label more assiduously, island by island in the Galapagos. As he put it, "I never dreamed that islands, about fifty or sixty miles apart, and most of them in sight of each other, formed of precisely the same rocks, placed under a quite similar climate, rising to a nearly equal height, would have been differently tenanted" (*Journal of Researches* 1845, p. 394).

To take the next step Darwin needed to fit three ideas together: first, the idea that one species could be transmuted into another; second, the idea that the repetition of such a process could accumulate over geological time to produce large differences; and third, the idea that this scenario, played out on a world scale, with organisms constantly migrating to new environments and becoming isolated from their forebears, could produce the whole system of organic nature.

To establish transmutability, the small differences among related species on the different islands of an archipelago would be ideal material. This step requires that the specimens be differentiated from each other as belonging to different species, and yet classified together as belonging to the same genus. Moreover, if the fundamental biogeographical connection is to be made, the specimens collected must be correctly labelled as to their location. For the birds of the Galapagos Archipelago, the collaboration of the ornithologist John Gould was indispensable, and the work was done between January 4th and early March, 1837. The ornithological findings broke the "species barrier" (Sulloway's phrase): there was no longer an intrinsic limit keeping variation within the boundaries (on which Lyell had insisted) of the species. Other zoologists contributed to the new picture, but Gould's work was the most important.

But establishing the transmutability of species would not lead to a full-scale evolutionary conclusion unless coupled with the more general changes that could only be observed over wider reaches of space and time. Regarding geological time, the paleontological work of Richard Owen was the key collaborative effort. This work began in December 1837. Almost immediately, Owen was able to pronounce that Darwin's fossils included a rodent (*Toxodon*) the size of a rhinoceros and an anteater (*Scelidotherium*) the size of a horse. These and other findings were communicated to Lyell. In his presidential address to the London Geological Society on 17 February 1837, Lyell summarized Owen's findings. He showed how these results dramatically confirmed the law of the succession of types: on large continents, existing species and extinct ones are closely related anatomically. This law really has two parts: first, new species closely resemble the ones they are replacing; and second, the difference between species sufficiently separated in time can become very great.

It should be noted that this law was by no means a new discovery.[8] Why did its confirmation now help move Darwin toward an evolutionist commitment? Perhaps the dramatic confirmation, using his *own* fossil specimens, and the attendant recognition he received, provoked him to think more about it. This highlighting of a known idea took place just as other key results of the voyage were coming into focus, and it was, after all, the integration of such widely different classes of data into a new synthesis that became Darwin's role.

The third class of data growing out of the zoologists' processing of the *Beagle* specimens has to do with the issue of representative species. Darwin revealed some awareness of this idea in his celebrated ornithological notebook in a passage (now dated by Sulloway as written June or July 1836) mainly on the birds of the Galapagos, but also mentioning the foxes of the Falkland Islands. Darwin was struck by the point that organisms "slightly differing in structure and filling the same place in Nature" could be found in different places. But that famous note remains ambiguous, in good part because Darwin injected the phrase, "I must suspect they are only varieties." Only if this suspicion was removed would "such facts . . . undermine the stability of species." The suspicion was not alleviated until early 1837, when the zoological results of the voyage poured in. Extended over a wider scale, Darwin's intuition (as against his prudent "suspicion") was richly confirmed. At a taxonomic level higher than species, there is a broad pattern of resemblances between the forms found in neighboring regions. The greater their isolation from each other — in time, reinforced by space and other barriers — the greater the differences. But islands typically have a general relation of similarity to nearby continents in their flora and fauna.

In the *Red Notebook*, this idea is conveyed in an odd phrase: ". . . new creation affected by Halo of neighboring continent . . ." (*RN* 127, written mid-March, 1837). In one possible reading, Darwin is suggesting that a geographic region somehow imposes a character on its organic productions. In his discussion of this passage, where Darwin wrote "peculiar plants created", Sulloway has added "[by colonization and gradual transmutation]". This is a plausible interpretation of Darwin's meaning, but certainly not the only possibility.

Thus, to assimilate his zoological work of the voyage to his emerging scheme, Darwin had to clarify the relations among three quite different classes of results. No one of them alone required an evolutionary explanation. Even all of them together could be assimilated to other theoretical schemas.

Sulloway has argued convincingly that the new information that Darwin gained from the expert processing of the *Beagle* specimens is not sufficient to account for his turn toward evolutionism; others sharing the same knowledge, indeed responsible for producing it, did not move in the same direction as Darwin. Sulloway attributes the difference to Darwin's "genius". I will not discuss here whether "genius" is an adequate explanatory concept (see Gruber 1982). However that question is decided, we must try, as well as possible, to understand what other moves Darwin was making that would lead him to the turn he took.

The *Red Notebook* may offer some help. Most scholarly attention has been centered on the frankly evolutionary or proto-evolutionary passages in the second half of it, written probably from the end of May 1836 to the close of the voyage. But here I want to draw attention to the first half, which deals mainly with more strictly geological issues.

VII. Going the Limit

What strikes me in the *Red Notebook* is an aspect of his style of thought. He is interested in pushing ideas to their limits, in making global generalizations. He writes of the need to focus on one region (for him, America), then to draw parallels with what is known about Europe, and finally to draw conclusions "applicable to the world" (*RN* 18). Since he knows how marine organisms capture lime, and he believes that this has gone on for a very long time, he asks, "How does it come that all Lime is not accumulated in the Tropical oceans detained by organic powers. We know the waters of the oceans are all mingled" (*RN* 29–30).

He is interested in the relation between very small events and their accumulation to great effects, sometimes not such obvious ones. Thus he tries to explain how gradual processes can lead to coastal steps (*RN* 39–41). He returns to this point a little later: "Mr Lyell . . . considers that successive terraces mark as many distinct elevations; hence it would appear he has not fully considered the subject" (*RN* 60). The more general idea of a qualitative leap emerges in another form in a reference to an experiment by Humphrey Davy showing that a small electric charge on a ship's copper bottom (produced by a bi-metallic contact) prevents fouling: "From Sir H. Davy experiment on the copper bottom, we see a trifling circumstance determines whether an animal will adhere to a certain part" (*RN* 95).

The question of scale occurs over and over in different forms. In writing of the flow of seemingly solid earth, he writes, "Mountains, which in size are grains of sand, in this view sink into their proper insignificance; as fractures, consequent on grand rise, & angular displacement, consequent of injection of fluid rock. — Try on globe, with slip paper a gradually curved enlargement" (*RN* 48). His mind moves eagerly from one scale to another: "Volcanos must be considered as chemical retorts" (*RN* 78). Within a few pages he remarks on "immense time", "immense areas", and "stupendous mass" (*RN* 107–109).

The idea of systems of compensating variables comes up repeatedly. He is fascinated by proposals that the system of volcanic action is a global system of subterranean forces. A line of volcanos in the Cordilleras could have "originated . . . from a fissure in a deep & therefore weak part of the ocean's bottom" (*RN* 10). The system of variables captured in the phrase "deep and therefore weak" deserves reflection.

Thus, while still on the voyage he was perfecting a style of thought in which (a) ideas are pushed to both their limits, such as the very great and the very small; (b) relationships are worked out between these extremes, and are often not obvious; and (c) since the limits in question include time as well as space, matter, and energy, the question of ultimate origins is never very far away.

We do not know just when the note on the inside cover was written, but it was appropriate for Darwin to place it at the front of the *Red Notebook*.

> The living atoms having definite existence, those that have undergone the greatest number of changes towards perfection (namely mammalia) must have a shorter duration, than the more constant: This view supposes the simplest infusoria same since commencement of the world.
> (*RN*, inside front cover)

VIII. The First Notebook on Transmutation

We now turn to the beginning of the B *Notebook*, a momentous step for Darwin. Darwin announces that something is happening. He begins a new notebook. He names it *Zoonomia*, the title of his grandfather's evolutionist essay (Erasmus Darwin 1794-1796). Most important is the change of style. The first thirty pages or so are no longer a miscellany of jottings, but a connected series of reflections. I will take the passage a few pages at a time. On the whole, within the passage, late ideas are added to or combined with earlier ones; revisions and rejections come later.

B 1–5. 1. Adaptive change is necessary. This is nowhere stated but assumed throughout.

2. The function of the life-cycle is to make adaptive change possible. "Generation" is used to refer to the cycle of reproduction, maturation, and death. "There may be unknown difficulty with *full grown* individual with fixed organisation thus being modified, — therefore generation to adapt and alter the race to *changing* world. On other hand, generation destroys the effect of accidental injuries, which if animals lived for ever would be endless . . . Therefore final cause of life" (*B* 4–5).
3. If the young must be born, this is taken to imply the necessity of death. In other words, the population remains approximately constant.
4. Variation is necessary for adaptive change. Two mechanisms are discussed, sexual reproduction and direct response to environmental circumstances. The latter is *not* the Lamarckian idea of inheritance of acquired characteristics. Rather, by some unspecified mechanism, change is induced during reproduction. For example, "seeds of plants sown in rich soil, many kinds are produced . . ." (*B* 3).
5. Variation must be disseminated to a whole population. The theory is not about individual adaptation but about populations and species. This is accomplished by sexual reproduction: "With this tendency to vary by generation, why are species all constant over whole country [?] Beautiful law of intermarriages partaking of characters of both parents and then infinite in number" (*B* 5).
6. There is an explicit denial of the efficacy of asexual reproduction as

an agent in this process of adaptive change: the offspring are uniform. This leaves a question unsolved: Did Darwin think that asexual organisms do not evolve? Did he think that all organisms are at least occasionally sexual? Or was the denial not so absolute, perhaps a rhetorical device to accentuate the value of sexual reproduction? These questions are confused with that of the significance of the opening lines, on pages *B* 1 and *B* 2. Kohn (1980, p. 84) takes them to be a clear and succinct summary of a passage in Erasmus Darwin's *Zoonomia*. I fail to see such a close resemblance, and see the passage as a still rather confused paraphrase and extension of a passage in the *Red Notebook* (*RN* 132), with a reference to *Zoonomia*. But we do not need to settle these questions in order to agree on the others. This opening passage strongly suggests Darwin's aspiration for a theory that would go from monad to *homo sapiens*: from "the original molecule" to "civilized man". Both phrases occur here.

B 6–13. These pages deal with the wider consequences of the initial moves. Darwin begins to discuss the set of resemblances and differences that form a taxonomic system broad and flexible enough to encompass island-to-island differences in an archipelago, representative species in different regions of a continent, and the peculiar pattern of resemblances (which he had earlier called a "halo") between a continent and a nearby island in their flora and fauna. Both geographical and sexual isolating mechanisms are mentioned.

B 14–17. The relation between the extinct and extant animals of a region is cited. Historical geology is brought to bear. "Countries longest separated — greatest differences" (*B* 15).

B 18–23. The issues of the limits of the system, and the direction of evolution come into focus: "Each species changes. Does it progress [?] Man gains ideas. The simplest cannot help becoming more complicated; and if we look to first origin there must be progress" (*B* 18). So far as direction goes, Darwin is cautious but clear: there must be progress.

So far as the first limit of the system, its origin, is concerned, Darwin makes two points about monads, or simplest living forms. First, if monads are constantly formed, there would be lawful similarities among them, due to prevailing worldwide conditions. Second, if monads have a specifiable, finite existence, then their derivatives share this duration in lawful ways.

7. Isolating mechanisms, geographical and sexual, are necessary to stabilize species change.
8. The metaphor likening the life-cycle of a species to that of an individual, which appeared much earlier in his thinking, is reiterated. "There is nothing stranger in death of species, than individuals" (*B* 22).

9. Not only population, but the number of species remains approximately constant.
10. The taxonomic system is a branching one. "Organized beings represent a tree, irregularly branched; some branches far more branched, — hence general. As many terminal buds dying as new ones generated" (*B* 21). Notice that these "buds" must vary, since the intent of the metaphor is to describe the evolution of new species, so they are not the literal buds of a real tree in Erasmus Darwin's *Botanic Garden*.

One of the vexed points in pages 1–23 is the status of extinction. Darwin clearly implies a system of nature in which extinction is both a lawful phenomenon and a formal requirement if new species arise while the species number remains constant. But what is the mechanism of extinction? The phrase, "death of species" states the problem but not the mechanism. There is only a hint of the idea of cumulative disadaptation. The idea of species senescence is not expressed here. Only the idea that I have called "monad life span" — with the rider that the monad includes the things it becomes — is clearly stated. It seems to me that one plausible reading of the passage in question is this: Mammalia have evolved the most from their monadic origins; that is, they have undergone the most change. Species longevity is inversely proportional to amount of change undergone; "Hence shortness of life of mammalia" (*B* 22). Built into this reading is the idea that the monad life span is being shared among its derivatives. So in spite of the copious criticisms Hodge and Kohn have heaped on me, I stand unrepentant on this point. For a brief period Darwin entertained the monad life span idea as a mechanism of extinction. Recognizing this idea is important in order to see the significant change Darwin soon underwent. Whether Darwin at this time relied on monad life span, species senescence, or cumulative disadaptation due to environmental change — or some combination of them — it is clear that he was unsatisfied with his position. And it is reasonably clear that he moved soon to what I have called the idea of "becoming" (Gruber 1981b): unless species change they "die" (*B* 61–63).

Most important of all, the branching model emerged together with these considerations, and it deserves attention. The series of tree (and coral) diagrams in the B *Notebook* evolved over the years into the only diagram in the *Origin*, and the one that was used to explicate the important idea of divergence. At this early time, I believe Darwin saw branching evolution as a good way to describe the empirical facts of taxonomy, biogeography, and paleontology. Moreover, he had some trace of the idea of the exponential growth function implicit in any branching model, and this was soon to become quite explicit. Except for the phrase "*irregularly branched*" (Darwin's italics) and a certain feel of the whole thirty pages, there is little to suggest that Darwin had a clear view of the probablistic view of nature that would eventually justify the branching model.

IX. From Monad to Man

If the theoretical issues at stake for Darwin and his contemporaries could have been contained within the shift from within-species *variability* to between-species *mutability*, their lives would have been much simpler. But it was not hard for them to see that once the "species barrier" was broken, an explosive theoretical change might set in. In the pre-Darwinian debate, the issues of evolution and of the natural origin of life were considered as twin (Farley and Geison 1974). In *Zoonomia*, for example, Erasmus Darwin dealt with them together.[9] In the 1850s, in his Species Notebook, Lyell remarked repeatedly that transformationism could not be contained at either end of the scale. He took some solace in Lamarck's view (as compared with Darwin's) that monads were still being constantly produced by spontaneous generation; this squared with his uniformitarian conscience (Lyell 1970, p. 124–125). Thinking about both limits together was not restricted to the Darwins and Lyell. In 1860, Leonard Jenyns wrote to Darwin, perceptively noting that in the *Origin* Darwin had gone to both extremes. In the conclusion of the *Origin* Darwin wrote plainly and vigorously: "probably all the organic beings which have ever lived on this earth have descended from some one primordial form, into which life was first breathed" (*Origin*, p. 484). Only a few pages later he wrote, far more prudently, "Light will be thrown on the origin of man and his history" (*Origin*, p. 488). Jenyns pointed this out and centered his objections on exactly this issue, the *scope* of Darwin's theoretical aims.[10]

But the shape of these conclusions in the *Origin* is quite different from the shape of Darwin's career as a whole. Faced with the prospect of both "going the whole Monad" and "going the whole Ourang", he made a lop-sided decision. He decisively dropped the issue of the origin of life. It is simply not present in his later work. The trenchant sentence in the *Origin* quoted above represents an abstract conviction, not a program of work. But at the other end of the scale, circumspect as he was in the *Origin*, he labored mightily and took a clear stand, early in the M and N *Notebooks*, and much later in *Descent* and *Expression*. When was this asymmetrical decision made? In the B *Notebook*, both ends of the scale are moderately well represented, although neither was his main preoccupation. In the Spring of 1838 he wrote, "The intimate relation of Life with laws of chemical combination, & the universality of latter render spontaneous generation not improbable" (*C* 102e). Meanwhile, the C *Notebook* was full of remarks about *homo sapiens* and by July 1838 he began the M *Notebook*, on man, mind, and materialism. In several places in the transmutation notebooks Darwin reiterated his mysterious idea, "If all men were dead, then monkeys make men. — Men make angels" (*B* 169). But nowhere do "monads make monkeys". Here again we see Darwin's use of deferral and ambiguity. He put one question firmly aside, and buried the other in his notebooks. And yet, when the time came, anticipating his readers' question "It may be asked how far I extend the doctrine of the modification of species" (*Origin*, p. 483), opening the section quoted above, he answered in his odd mixture of forthrightness and circumspection.

We have seen how Darwin experimented with the idea that the longevity of a species is inversely proportional to its position in the scale of nature: the more evolved species, i.e., mammalia, have the shortest species life span. This idea soon gave way to a quite different formulation.

> ?Law: existence definite without change, superinduced or new species. Therefore animals would perish if there was nothing in country to superinduce a change? (*B* 61)

In this new formulation, *amount* of change is not mentioned as a consideration. On the one hand, the particular change must be in some sense adaptive. On the other hand, change itself is necessary. Fortunate is the species that inhabits a region where something will "superinduce a change". Although stated here between question marks, thc idea is reiterated several times and soon becomes quite definite:

> If *species* generate other *species*, their race is not utterly cut off: — like golden pippins, if produced by seed, go on, — otherwise all die. —the fossil horse generated in S. Africa zebra — and continued, — perished in America. (*B* 72–73)

In the sense that one species is transformed into another, the first is the parent of the second — and in the making of it enjoys a "second life", the phrase Darwin used in his notes on marriage and having children (*Autobiography,* Keegan and Gruber 1983). This does away with any clear meaning that might be assigned to the species life span idea and its variant, monad life span.

Dropping the ideas of species life span, monad life span, and original monads from his thinking was an important step, tantamount to a decision to deal with the system of nature as an ongoing system, and to avoid questions of ultimate origins. But there were numerous vacillations and backslidings, and it was not until May 1839 that he could write unambiguously, "My theory leaves quite untouched the question of spontaneous generation" (*E* 160).

X. Toward Natural Selection

Here the story diverges in a number of ways. Intricate as each path may be, I can only summarize briefly.

First, there is the main line — from the explorations in the B *Notebook* in July 1837 to the moment some fifteen months later when he read Malthus's *Essay on Population* (Malthus 1826) and formulated the principle of evolution through natural selection. Insisting too much on the singular and climactic nature of this moment misses important points. There was the work he had to do to arrive at 28 September 1838. Then there was the work of the moment. As Kohn (1980) has nicely shown, the "moment" of insight had a complex inner structure. Darwin wrote and then, probably immediately, rewrote his ideas. I believe that in the initial version there is a predominant tendency to take species, and in the rewrite to take the individual as the unit of analysis.

The work of the moment also included the task of significantly transforming Malthus's ideas (Keegan and Gruber 1983). The latter anthropocentrically dichotomized the world into a human population tending to increasing geometrically and a food supply increasing arithmetically. For Darwin, the food was also organisms, all with a potential for exponential population growth, unless checked. So generalizing and de-centering went hand in hand. Moreover, Malthus wrote within a context of social theory in which the complex interrelationships among human sexuality, population growth, and social class differences were matters of intense controversy. Darwin abstracted one key idea out of this context and turned it upside down — from the scourge of humanity to the motor of evolution. The first mention of Malthus in the M and N *Notebooks* occurs in an entry made between 4 and 7 October 1838, only a few days after the great moment. It has nothing much to do with the population principle, but deals with Malthus's other preoccupation, sexual continence. The first and probably only suggestion of the principle of natural selection in the M and N *Notebooks* occurs on about 16 March 1839:

> N.B. According to my view marrying late, will make average of life longer. — for short-lived constitutions will then be cut off. (*N* 67)

Second, there is the issue that went underground for so long, the question of divergence. The early B *Notebook* pages strongly suggest the fact of divergence. But why? As Janet Browne has shown (Browne 1980), when Darwin came back to this question in the 1850s, the language he used resembled that of the B *Notebook*. What he did not settle in 1837–1838 was the *reason* for divergence: what makes it necessary? It is widely agreed that it was not until the 1850s that he succeeded in answering that question to his own satisfaction (Browne 1980; Schweber 1980; Ospovat 1981; Kohn this volume).

Third, there is the seeming tangent — the initial exploration of the evolution of mind, recorded in the M and N *Notebooks*. This was not only an effort to extend the theory of evolution to one of its limits, but also to use the limiting case — a "frontier instance", Darwin called it (*N* 49) — to solve problems within the theory of evolution. This is a subject still largely unexplored.

Fourth, there is the disputed issue of artificial selection. Several authors have argued that Darwin came to natural selection via artificial selection. It is true that in the C *Notebook* and the D *Notebook* before Malthus there is much about plant and animal breeding. But it now seems clear that Darwin was investigating the work of breeders in order to find clues to the mechanism of variation: in some way, breeding under artificial conditions was thought to disturb the natural process of sexual reproduction. Nevertheless, this process of steeping himself in the subject was fruitful; when he did arrive at the idea of natural selection, he could then turn around 180° and use artificial selection as a small scale demonstration of the principle. Even this seemingly small step took some months.

While the model of artificial selection may have been a stepping stone on some of the possible paths to natural selection, it was not a necessary way station. As late as 1858, Alfred Russel Wallace arrived at natural selection while explicitly denying the relevance of results of artificial breeding.

Conclusion

I think it is at least tacitly agreed that Darwin's development was a true epigenesis: a series of structures with each phase growing out of the previous, always in interaction with new circumstances provided by a changing scientific and social environment. No one has suggested that when Darwin set out on the voyage he knew exactly where he was going, or that when he began the First Transmutation Notebook the theory of evolution through natural selection was a foregone conclusion. At the same time, Darwin's intellectual activity was far from random exploration. Starting at some early point, he seems to have been moving in a direction. In part this direction was given by certain family traditions, in part by broader historical currents to which he was exposed, and in part by his opportune encounter with Lyell's *Principles*. The voyage itself seems to have evoked in him a strong tendency to be that kind of natural historian who goes beyond local description and explanation to generalize on a world-wide scale. Perhaps we should say that the voyage reinforced a tendency already evident in his pre-*Beagle* admiration for Humboldt's *Personal Narrative*. The combination, tradition × education × circumnavigation, made a global thinker of the young naturalist.

As Darwin's sense of purpose emerged, it rapidly became more and more complex. We have summed up and surveyed this pattern in the "network of enterprise" — a diagrammatic way of showing the simultaneous development of a number of strands of scientific work. One of the themes of this essay has been the need to make sense out of this diversity.

Throughout this early period, we see the emergence and spread of a number of thought-forms. Among the most prominent is the summing of small effects over many iterations to produce large, often surprising results:

"the multiplication of little means" that Darwin found such a "laborious and painful effort of the mind" (*C* 75). This idea involved, for Darwin, the movement from one time-scale to another, from the scale of localized events to the scale of their long-range consequences. So the scale of time and space intellectually available profoundly affects the significance of such summative processes. For Darwin, this scale rapidly became geological in time and global in space.

A second very general thought-form we see emerging in Darwin's work is the equilibration model. Each natural phenomenon hovers around some value governed by a host of factors. Departures from this value provoke an equilibrating process. This is not quite the same as a static "balance of nature" since from an early point Darwin was thinking of a *changing* world, so this re-equilibration was a moving process, as shown dramatically in his theory of coral reef formation.

A third characteristic of Darwin's thought was to think in terms of the whole range of phenomena within whatever domain was in question. Just as geological processes were happily generalized on a world scale, when he saw his first Tierra del Fuegian he immediately thought of the whole range from wild animal to civilized man. When he encountered, in his reading of Lyell, the idea of the "death" of species, he wondered also about their "birth". If one was gradual, why not the other? Moreover, he often thought about the connection between the very small and the very great.

This characterization helps to understand Darwin's evident tendency, at the beginning of his thinking about evolution, to raise questions about the scope of the theory: What is the function of birth and death of individuals? Of species? Can one theory go all the way from simplest living being to most complex, from monad to man?

There has been a valuable trend, in writing about Darwin, to "normalize" his life — to show how he became a true professional, how his work depended on that of other true professionals. This is important if we are to demystify, as far as possible, his achievements. This procedure is likely to accentuate that part of his thinking which was in the solid middle-of-the-range of scientific work.

At the same time, this normalized picture of Darwin de-emphasizes that part of his thinking in which he was testing the limits, exploring the possible scope of his theory. But the scope he achieved was a fundamental part of his contribution. Darwin was a revolutionary thinker. We need to understand what forms of thought he used that permitted him to consider so deeply and so unflinchingly the whole range of possibilities.

Notes

1. Rudwick and I have had a fruitful exchange on the matters covered here. On the relation between public and private science, see also my essay, "The Many Voyages of the *Beagle*" (Gruber 1981b, 259–299).
2. For the cave, see Plato, *The Republic*, Book VII; the shadow box experiments are work in progress; the solution tree, or search tree, is discussed in Newell and Simon (1972).
3. For an approach to playful exploration of micro-worlds as a way of mastering a domain, see Papert (1980).
4. For discussions of the phenomenological approach see Gruber (1981c) and Gruber (1980).
5. He was specifically trained in keeping notebooks by his teachers at Edinburgh University; and Erasmus Darwin's *Commonplace Book* (Ms at Down House) contains a lengthy preface explaining the connection between the practice of recording one's experiences and the empirical philosophy of John Locke; see *Darwin on Man* (Gruber 1981c, 21-22).
6. Both Hodge (1982) and Kohn (1980) concur on the Brocchian source of the species senescence idea. Lyell discussed it and disagreed with it in Vol. III of *Principles*, which Darwin read during the voyage. Lyell learned of it from, and cited the Italian geologist, Giovanni Battista Brocchi. I see no reason to doubt the importance of Brocchi in the story. But I would add that at least one key part of the idea, the gradual deterioration over generations, of grafted apples — an example Darwin alluded to, metaphorically, repeatedly for many years — can be found in Erasmus Darwin's poetry, spelled out in full in a prose note. What is more, the context it occurs in is the poet's celebration of the value and power of sexual love. This attitude was a Darwin family tradition. Erasmus Darwin, *The Temple of Nature or, the Origin of Society: A Poem with Philosophical Notes* (London: Johnson, 1803, posthumous), Canto II, p. 57.
7. In most respects the above description of Darwin's coral reef theory is very close to the version I wrote in *Darwin on Man* (Gruber 1981c).
8. For a brief account of its history see Eiseley (1958, 161–166).
9. Erasmus Darwin, *Zoonomia*, Section XXXIX, "Of Generation". See also the *Temple of Nature*, "Additional Note I", which is an essay on spontaneous generation of simple organisms. The sections of this poem have the following titles: Canto I, "Production of Life"; Canto II, "Reproduction of Life"; Canto III, "Progress of the Mind"; Canto IV, "Of Good and Evil".
10. Jenyns' letter is reprinted in *Lyell's Scientific Journals* (Lyell 1970: 349-351).

Special Note: Writing this essay was completed during a stay at the Institute for Advanced Study, whose hospitality I gratefully acknowledge. I thank Martin Rudwick and Doris Wallace for helpful comments. The idea of *thought-form* is being elaborated in a doctoral dissertation by Robert T. Keegan on Darwin's unpublished "Diary of an Infant".

2

THE WIDER BRITISH CONTEXT IN DARWIN'S THEORIZING

Silvan S. Schweber

Home is where one starts from. As we grow older
The world becomes stranger, the pattern more complicated
Of dead and living.
. . . .
We must be still and still moving
Into another intensity
For a further union, a deeper communion . . .
In my end is my beginning.

T. S. Eliot, *Four Quartets*, 'East Coker' V

Introduction

The *Origin of Species* was the culmination of Darwin's theorizing of the previous twenty years. Its unique role in delineating the subsequent debates over all aspects of evolution account for the enduring interest in the construction of the *Origin* and the intellectual and other factors that helped shape its final form. We know from Darwin's correspondence that he saw himself as constantly engaged in "species-work" during the period from 1840 to 1854. It was "far-distant work" but he did indicate to several of his correspondents that he intended to write a book on the species question, though he would "not publish on the subject for several years" (for example, LL (NY) 1: 392, 394-395). My aim is to trace the development of Darwin's understanding of the divergent pattern of evolutionary history, particularly the mechanism of divergence.

I see the dynamical explanations that Darwin advanced in the *Origin* as the amalgamation of two great insights. The first occurred in the Summer of 1838, and consisted in the apprehension of the Malthusian mechanism. It led to natural selection, and was the high point of Darwin's theorizing following his voyage on the *Beagle*. The second was gleaned in the mid 1850s and resulted in the principle of divergence. The Malthusian principle reflects a deterministic, quantitative, Newtonian mechanistic conceptualization of the world; the principle of divergence is modeled after the Scottish explanations of the social and economic order and, in particular, their conception of the self-regulating, open and progressive character of the market. Viewed this way the *Origin* is the embodiment of Darwin's own intellectual upbringing: the synthesis of the great Cambridge and Edinburgh traditions.

In all of Darwin's theorizing, the dynamics always relied on mechanisms that emphasized gradualism.[1] The reason usually adduced for Darwin's commitment to gradualism is the influence of Lyell and his uniformitarian geology.[2] But that particular answer, though undoubtedly relevant, does not shed light on why a host of eminent continental naturalists were not convinced, nor does it suggest why gradualism should prove so attractive to so many British minds, for example, Hutton, Erasmus Darwin, Lyell, and in particular Charles Darwin. The question to be answered is: "Why were evolutionary theories with a gradualistic, materialistic dynamics so peculiarly British in their conception?" To do so, the pre-existent ideological commitment that such theories reflect must be made explicit. When I refer here to ideology I mean the system of meanings that the members of a social and intellectual community share with one another. Ideology, as I use the term denotes the consensus of values that makes possible intellectual discourse, the set of shared presuppositions, categories, and explanations — which, once accepted, acquire an objective reality for those it informs. In fact, only under the most scrupulous and "persistent examination does the content of ideology reveal itself as a social product rather than as a reflection of universal truth" (Appleby 1978, p. 6).

An evolutionary world view accepts change as a natural feature of the world. The sixteenth century in Western Europe saw the beginnings of the sustained demographic and economic growth that has characterized the modern world to the present. By the beginning of the seventeenth century the changes were so pervasive that the seemingly static equilibrium that had existed between people and land, peasant and lord, lord and king, work and rest, production and consumption was brought to an end. These changes — and the opportunities they created — required the acceptance of new values, the acknowledgement of new occupations, and the imposition of new relationships between individuals (Appleby 1978, p. 3). The typological conceptualization of the social and political world and the view that the aim of science was the inference of ideal types despite their imperfect manifestation in the world were part of the ideology that the new commercial and political system replaced.

It was in the effort to understand and control the new and evolving commercial system that, particularly in England, new interpretative models were elaborated during the seventeenth century which would mold all subsequent economic thought. This enterprise culminated in the writing of the classical economists, in particular James Steuart and Adam Smith. Tawney made the important point that the modern transformation of the British economy took place over so long a period that the categories of thought associated with capitalism appeared to the British mind "as timeless forms imprinted on the very stuff of the human brain" (Appleby 1978, p. 8). Two centuries of agricultural and commercial change preceded the industrialization of England that began in the middle of the eighteenth century. On the continent that time-scale would later be compressed to a single generation. As Appleby has emphasized:

> This dimension of time powerfully affected the perception of the change. For those on the continent industrialization was a radical force that required explanation; for the British, the final stages of capitalism appeared as the end product of what seemed a predictable and wholly natural progression. Modes of behaviour shaped by a commercial society were viewed as characteristics of human nature and of nature in general. Relationships in a modern economy appeared as laws of nature, applicable to all societies and discoverable through empirical investigations. (Appleby 1978, p. 9)

Apposite statements could be made regarding the development of the English political system in contrast to the experience on the continent, particularly France.

The English conceptualization of economic, political, and social life that was elaborated between the sixteenth and eighteenth centuries, and more particularly, the Scottish views that were propounded to understand the evolution and operations of the market deeply influenced Darwin — and for that matter all the nineteenth-century British evolutionists.

Almost all the Scottish Enlightenment inquiries on the nature of the social and economic order were evolutionary in outlook. They described the history of the social and economic order as an *ordered* set of states and ascribed a temporal direction to its evolution. Their history of human culture was described not simply in terms of the change from hunting and gathering to primitive agriculture, from feudal agriculture to commercial industry, but also included a graded scale which reflected the degree of division of labor in the economic sphere (Adam Smith 1937) and the degree of complexity the political and social order had achieved (James Steuart 1966). For Steuart it was "the complicated system of modern economy" that stabilized public affairs. Darwin also believed that an increase in complexity was a feature of the evolutionary history of life on earth and held the view that stability and complexity were connected, in fact the complexity resulted in stability. Complexity for Darwin was the consequence of the strong interactions between the many diverse assemblages of organisms that make up the economy of nature. For Darwin this rich diversity of form and function, strongly interdependent by virtue of strong interactions, were the elements of the "complexity" that stabilized the natural economy. Evolution for Darwin led to greater diversity, complexity and stability much in parallel with the Scottish view of the artificial economy. In fact many of the same mechanisms were invoked: division of labor, divergence of character, competition, etc.

It is my contention that Darwin's evolutionary biology reflects a characteristically British intellectual outlook in its conception. To paraphrase Merz (1904, vol. 2, pp. 395–396, 415): in the *Origin of Species* biology and economics "joined hands" or perhaps more accurately biology joined hands with Scottish political economy, sociology, and historiography, and with English philosophy of science. The political economy was that of Adam Smith and his disciples, the sociology that of Ferguson, the historiography that of Hume and Dugald Stewart. The philosophy of science was woven from more complex and original strands to which Whewell, Herschel, and the Scots contributed importantly.

To understand Darwin's theorizing better I have gone back to some of his first readings in science, in particular, to the chemistry texts he used in the Summer of 1825. Chemistry was Darwin's introduction to a quantitative science, one whose foundations had been bolstered recently by Dalton's work, but also one based on the assumption of the existence of unseen elementary entities. In Section I, I review the models of organisms that were presented in these chemistry texts. These suggest that Darwin's tendency to look at classification as an "essentialist" but to see dynamics "populationally" may have its roots in the then prevalent view of organisms as self-reproducing *machines*. In Section II, I look once again at Herschel's, Whewell's, and Babbage's influence on Darwin and abstract their view of what constituted the essential features of the Newtonian dynamical description. I compare the Newtonian model presented in Herschel, Whewell and Babbage to the view of dynamics expressed in *E* 95–97 based on Darwin's notion that Nature's dynamical equilibrium maximizes the amount of life per unit area, and that diversity is a way of accomplishing this. I inquire into the sources of this approach and point to the literature on scientific agriculture (D. Stewart, Brougham, Davy, and Liebig), to that of the political economists, and more particularly Adam Smith, for similar approaches. Section III looks briefly at the *Essay* to analyze Darwin's conception of the economy of nature and of adaptation in 1844. In Section IV, I review the recent work on the genesis of the principle of divergence by Browne (1980), Schweber (1980), Kohn (1981), and Ospovat (1981) and indicate its relation to the Scottish interpretation of the operation of the market.

It is my contention that from 1839 on, Darwin never wavered from the view that natural selection was a law of nature having *universal* applicability in the organic world in the same sense as Newton's laws of motion and of gravity, and that the quantifiable explanatory models presented by the physical and chemical sciences always loomed in the background and held a certain attraction for Darwin. Although natural selection by itself was not sufficient to explain the appearance and pattern of all evolutionary phenomena — geological, geographic, generational as well as other factors must be brought into the explanation — nevertheless, *all* properties of living organisms are due at least in part to natural selection. The period of 1844–1858 is best understood as a constant confrontation of Darwin the grand theorist, the unifier, whose driving passion is to account for *all* of the features of the organic world in terms of general principles, and Darwin the diversifier, the investigator of individual processes, studying the details of the organic world's diversity, disentangling the complex pattern of *particular* species and individual organisms to see whether his principles do, in fact, explain the observed phenomena. Cannon (1976b) has stressed, I believe correctly, that Darwin could have published his initial version of the theory in 1839 without adverse reaction if he had been prepared to limit its scope to biogeographical and other zoological and botanical questions without raising questions about man. But like Newton, Darwin wanted a theory that had a universal character, applicable to *all* aspects of *all* living organisms including man and his "higher faculties", and that was very likely not acceptable in 1840.

It is no accident that the *Origin* closes with a vision of "this planet cycling according to the fixed laws of gravity" on which "endless forms most beautiful and most wonderful have been, and are being, evolved". Newton had explained the dynamics of the former, Darwin that of the latter.

I. Darwin's Early Chemical Studies

In the present section I want to focus on some very specific early readings and activities that influenced Darwin in important ways in his later craft, namely Darwin's chemical investigations as a teenager. I believe that this subject had particular importance in Darwin's intellectual development. This early exposure to atomism helped shape his characteristically British propensity toward atomic explanations in the physical as well as in the social realm. There were, of course, many other influences.[3] I am merely focussing on one of the strands in the intricate web of his growth in an attempt to obtain further insight into the way he came to think about organisms and their evolution.

Darwin first studied chemistry during the Summer of 1825 with his brother Erasmus. His recollection in the *Autobiography* was that

> Toward the close of my school-life, my brother worked hard at chemistry and made a fair laboratory with proper apparatus in the tool-house in the garden, and I was allowed to aid him as a servant in most of his experiments. He made all the gases and many compounds, and I read

> with care several books on chemistry, such as Henry and Parkes' Ch. Catechism. The subject interested me greatly and we used to go on working till rather late at night. This was the best part of my education at school, for it showed me practically the meaning of experimental science. The fact that we worked at chemistry somehow got known at school, and as it was an unprecedented fact, I was nicknamed "Gas". I was also once publicly rebuked by the head master, Dr. Butler, for thus wasting my time over such useless subjects; and he called me very unjustly a "poco curante", and as I did not understand what he meant it seemed to me a fearful reproach. (*Autobiography*, pp. 45-46)

While a student at Edinburgh he attended and took an interest in the chemistry lectures of Professor Hope. Chemistry was Darwin's first introduction to the physical sciences and their particular emphasis on measurements and quantification. A case can be made that as Darwin matured quantitative measurements assumed an ever greater role in his experimentation. Similarly, the use of statistical data to verify hypotheses became more pronounced as time went on. The *Origin*, since it was addressed to the general public, did not expose or emphasize these Darwinian tendencies, except for the Malthusian statements and the brief summary of his botanic arithmetic (1859, pp. 55-58). This emphasis on quantification was certainly a legacy of Darwin's chemical studies. But this is not the feature I want to explore. The aspects of his chemical studies that I want to look at relate to the kind of understanding and explanation that were presented to account for the constitution of inorganic and "organized" bodies.

All the texts I have looked at — the ones Darwin mentions, Parkes and Henry as well as Thomson on which the latter two were based, and other comparable ones — share a characteristic attitude with respect to the constitution of inorganic bodies and of organisms. The constitution of inorganic objects (crystals, macroscopic objects . . .) can be understood as *structures* made up of atoms and molecules held together by forces and affinities whose precise character it was the job of the natural scientist to discover. Organisms, on the other hand, are "living *machines*". Although they are made up of the same entities (atoms, molecules, . . .) as inorganic ones — some presentations will invoke "vital principles" to account for their unique features — all consider living organisms as *functionally* defined *machines*. It is the consequence of this viewpoint that I want to consider later.

Henry's *The Elements of Experimental Chemistry*, which was dedicated to John Dalton "as a testimony of respect for the zeal, disinterestedness and success with which he has devoted himself to the advancement of chemical philosophy" defined chemistry as the science dealing with those transformations of matter which take place "without apparent motion", in contrast to the "facts of natural philosophy that [were] always attended with sensible motion" (Henry 1819, p. xii).[4] Chemistry, an important science with a large body of empirical data, was based on "elements" that "the eye is unable to see". This fact left its mark on Darwin. Like Buffon, he will later postulate unobservable particles to account for some of the empirical facts of heredity.

In his prefatory essay on the utility of chemistry Henry also noted that

> The animal body may be regarded as a living machine, obeying the same laws of motion as are daily exemplified in the production of human art . . . the living body is a laboratory in which various chemical processes are constantly taking place: conversion of food, production of animal heat . . . (Henry 1819, p. xix)

Parkes' *Chemical Catechism* (1818) is an even more interesting introduction to chemistry. The great importance of chemistry "to the arts and manufactures" suggested to the author that an "initiatory book, in which simplicity was united with perspicuity, would be an acceptable present to a variety of persons" . . . "especially . . . parents who are not qualified by previous acquirements, to instruct their children in the elements of this science." The book is in the form of a catechism — questions and answers — with extensive footnotes that are particularly sensitive to the history of the concepts introduced. The wealth of physics, physiology, natural history and technology that the footnotes (and the Notes) contain is remarkable. Written in a simple, unassuming way designed to spark the imagination of the young reader, the book succeeded admirably as a "first book" for the "chemical student". Indeed it went through numerous editions. The *Catechism* concluded with a series of experiments which could "be performed with ease and safety" that illustrated the chemical principles expounded in the main text. There are experiments on change in temperature, on gases, on the formation and crystallization of salts, on dyeing and inks, on combustion and detonation, and metals. These probably were some of the experiments that Charles and Erasmus performed in the Summer of 1825.

To the second and later editions Parkes added Notes as new information became available. The 1818 edition, which is the one Darwin used, was substantially revised "to be accommodate to the present state of chemical knowledge", and in particular, to the "highly interesting and truly important discoveries of Davy". The Notes to the edition close with an assessment of Davy's work on the alkali. They compare him to Newton as being responsible for a new era in the history of chemistry and eulogize him:

> Immortal Newton thus with eye sublime,
> Mark'd the bright periods of revolving time;

> Explored in Nature's scenes the effect and cause,
> And, charm'd, unravell'd all her latent laws. (Parkes 1818, p. 510)

The young Charles must have found Parkes' *Catechism* particularly stimulating. It may well be the basis for his statement in the *Autobiography* "that there are no advantages and many disadvantages in lectures compared with reading" (*Autobiography*, p. 47). There may also have been an additional reason why Charles found the book memorable: The frequent quotations by Parkes of Erasmus Darwin's *Botanic Garden* could not have escaped his notice.

Parkes also meant his catechism to be "a body of uncontrovertible evidence of the wisdom and beneficence of the Deity, in the establishment and modification of laws of matter which are so infinitely and beautifully varied" (Parkes 1818, Preface p. iii). The stress throughout is constantly on laws. Each chapter ends with an inference regarding the Deity, a Deity who operates through laws. The startling diversity found in the realm of the inorganic world could all be fitted in a framework of lawful behavior designed by a beneficent Creator. The moral that Parkes conveyed was the same as Whewell did in his *Bridgewater Treatise* (1834), in particular that

> with regard to the material world, we can at least go as far as this — we can perceive that events are brought about not by insulated interpositions of divine power exerted in each particular case, but by the establishment of general laws. (Whewell 1834, p. 356)

This last quotation is one of the two that Darwin put on the front page of the *Origin*.

The introductory "Essay on the Utility of the Study of Chemistry" that Parkes wrote for his *Catechism* succinctly summarizes his views of the importance of chemistry:

> The well-informed people of France are so satisfied of the importance of chemical knowledge, that chemistry is already become an essential part of education in their public schools — it shall be my business in this place to endeavor to demonstrate it to be of *equal* importance to the various classes of our countrymen . . .
>
> Is your son born to opulence, — is he heir to an extensive domain; make him an analytic chemist, and you enable him to apprehend the real value of his estate, and to turn every acre of it to the best account. Chemistry will teach him also how to improve the *Culturated* parts of his estate; and by transporting and transposing the different soils, how each may be rendered more productive . . . Should he *occupy* his own estate, and become the cultivator of his own land; he must of necessity be a chemist, before he can be an economical farmer. It will be his

> concern not only to analyze the soils on the different parts of his farm, but the peat, the marle, the lime and the other manures must be subjected to experiment, before he can avail himself of the advantages which might be derived from them . . .
>
> If the profession of medicine be your son's choice, charge him, when he walks the hospitals, to pay particular attention to the lectures on Chemistry, and to make himself master of the chemical affinities which subsists between the various articles of the Materia Medica. This will inspire him with professional confidence; and he will be as sure of producing any particular chemical effect upon his patient, as he would if he were operating in his own laboratory. Besides, the human body is itself a laboratory, in which, by the varied functions of secretion, absorption, composition and decomposition are perpetually going on — how, therefore can he expect to understand the animal economy, if he be unacquainted with the effect of certain causes chemically produced? . . . (Parkes 1818, pp. 1–2)

Note the stress on agricultural chemistry, undoubtedly stemming from the influential lectures on this subject that Humphrey Davy had delivered at the Royal Institution between 1802 and 1812 (Davy 1839, vol. 1, p. 99), which were published in 1813 (Davy 1813, Davy 1840, vols. 7, 8). Once again, note the conception of organisms as laboratories and as machines. Both Henry and Parkes employed these metaphors. So did Thomson in his many books on chemistry. The latter were the sources of much of the chemical information to be found in both Henry and Parkes.

Thomas Thomson's *A System of Chemistry* (1802) went through numerous editions (1807, 1810, 1818) and was the most widely used introductory text in the chemistry courses at the Scottish universities. The various editions grew in size, eventually splitting into volumes dealing separately with inorganic and organic chemistry (Thomson 1838). Already in the 1820s more than half the volumes dealt with "Animal and Vegetable Chemistry". Thomson's *System* was one of the first chemistry texts to be based on Dalton's atomic theory. Indeed, starting with the 1807 edition, Thomson's *System* became a leading vehicle for the dissemination of Daltonian chemistry. The 1818 edition contained a "full-development and illustration of the atomic theory, and the doctrine of definite proportion" (Thomson 1818, p. vi).[5] In 1825, Thomson issued *An Attempt to Establish the First Principles of Chemistry* whose purpose was to present Daltonian chemistry to the medical students at Glasgow. Thomson wrote this text in order to allow him to shorten his discussion and presentation of atomic theory in his lectures during the academic year and to allow him "to enter more into detail respecting those parts of chemistry which are more intimately connected with the theory of medicine" (Thomson 1818, p. vi). The approach of the book was experimental and stressed the role of physical measurement with simple quantitative experiments which were designed so that the students could perform them on their own. Thomson's *First Principles* (1825) was very probably one of the "several books on chemistry" Charles "read with care" during his chemical studies in 1825-1826 (*Autobiography*, p. 46).[6]

In all his presentations of the chemistry of vegetable and animal bodies Thomson represented organisms as machines. The metaphor assumed its most succinct form in Thomson's *Chemistry of Organic Bodies* (1838) in which organisms were conceived as functional machines designed to produce a certain end.

It should not be thought this kind of description of living organisms was unique to the chemists. It is also the model of plants that Henslow presented in the lectures on botany which Darwin attended religiously while at Cambridge. To Henslow botany consisted of the

> investigation of the outward forms and conditions in which plants, whether recent or fossil, are met with, as to the examination of the various functions which they perform whilst in the living state, and to the laws by which their distribution on the earth's surface are regulated. (Henslow 1836, p. 2)

While descriptive botany examined, described and classified

> all circumstances connected with the external configuration and internal structure of plants, which we here consider in much the same light as so many pieces of machinery, more or less complicated in their structure [in physiological botany] we consider these machines as it were in action, to understand their mode of operation, and to appreciate the ends which each was intended to effect. (Henslow 1836, p. 2)

Although Henslow believed "in the presence of a living principle which operating in connection with the two forces of attraction and affinity" was responsible for the structure of plants, the practical consequence of this belief in the actual investigation of phenomena is not apparent.

This conceptualization of organisms as machines fitted naturally into the Natural Theology tradition that conceived of organisms as functional structures that God had designed. External functional relations were emphasized and the utility to the organism of its various parts in relation to a particular mode of life in a particular environment was the "final cause" and the main explanatory tool in this teleological approach. As is well known, adaptation was used as a demonstration of the natural goodness of the Creator and as a certification of his unique talents for perfect design. The notion that organic bodies are machines or automata was of course an old one, which by the 1830s had a long history. Descartes had given the view scientific credence and mechanical models were widely used thereafter to discuss the action of the stomach in digestion, the process of locomotion and the action of muscles.[7] The view received much attention during the Enlightenment and was popularized in the influential works of LaMettrie and d'Holbach. Reaction against their materialistic views was still being felt in Great Britain in the 1830s. And in fact, it was not these models that Thomson and the others were alluding to. The source of their metaphor was the factories sprouting all over the land.

The first half of the nineteenth century witnessed a dramatic expansion of the British economy. Factories, their economic, political and social impact — the machinery question as Berg (1980) has called it — were the constant focus of a wide ranging, never ending inquiry into the causes of economic change. Political economy — the natural science of economy and society — was born of the attempt to understand economic growth and the seemingly "limitless prospects created by technological advance" (Berg 1980, p. 10). Babbage's *On the Economy of Machinery and Manufactures* (1832) and Ure's *Philosophy of Machinery* (1835) were probably the two most influential books on the subject in the early 1830s. But besides these, numerous articles on machinery and their economy appeared in the periodical literature and in the encyclopedias explaining the workings and impact of the new technology. It was saying the obvious when Carlyle characterized the 1830s as the Age of Machinery. For Ure

> The philosophy of manufactures is . . . an exposition of the general principles, on which productive industry should be conducted by self-active machines. The end of a manufacture is to modify the texture, form, or composition of natural objects by mechanical or chemical forces, acting either separately, combined or in succession . . . An indefinite variety of objects may be subjected to each system of action, but they may be all conveniently classified into animal, vegetable, and mineral. (Ure 1835, p. 1-2)

And in a striking paragraph Ure noted

> The term *Factory*, in technology, designates the combined operation of many orders of work-people, adult and young, in tending with assiduous skill a system of productive machines continuously impelled by a central power . . . But I conceive that this title [factory], in its strictest sense, involves the idea of a vast automaton, composed of various mechanical and intellectual organs, acting in uninterrupted concert for the production of a common object, all of them being subordinated to a self-regulated moving force. (Ure 1835, p. 26)

Compare this definition of a factory by Ure with that given by Thomson of an organism:[8]

> It is well known that every vegetable and animal constitutes a machine of greater or lesser complexity, composed of a variety of parts dependent on each other, and acting all of them to produce a certain end. (Thomson 1838, p. 3)

Interestingly the most sophisticated use of the metaphor of organism as machine was made by the French zoologist Henri Milne-Edwards. Milne-Edwards presented his thesis of animals as automata whose parts have become more specialized as a result of a division of labor first in his entries "Nerfs" and "Organization" in the *Dictionnaire classique d'histoire naturelle* (Bory de Saint-Vincent 1822-1831) and thereafter in his influential books on zoology. Darwin undoubtedly read these entries in the *Dictionnaire* while aboard the *Beagle* since they dealt extensively with the invertebrates that were central to Darwin's zoological investigations at the time. From the Transmutation Notebooks we know that Darwin had read Milne-Edwards' *Elémens de Zoologie* (1834). Its zoological philosophy is striking:

> In animals whose faculties are most limited and whose life are simplest, the body presents everywhere the same structure. The parts are all similar; and this identity of organization brings about an analogous mode of action. The interior of these organisms can be compared to a workshop where all the workers are employed in the execution of similar labors, and where consequently their number influence the quantity but not the nature of the products. Every part of the body performs the same functions as the neighboring parts, and the life of the individual is made up of those phenomena which characterize the life of one or the other of these parts.
>
> But as one rises in the series of beings, as one comes nearer to man, one sees organization becoming more complicated; the body of each animal becomes composed of parts which are more and more dissimilar to one another, as much in their morphology, form and structure, as in their functions; and the life of the individual results from the competition of an ever greater number of "instruments" endowed with different faculties. At first it is the same organ that smells, that absorbs from the environment the needed nutrients and that guarantees the conservation of the species; but little-by-little the diverse functions localize themselves, and they all acquire instruments that are proper to themselves. Thus, the more the life of an animal becomes involved in a variety of phenomena, and the more its faculties are delineated, or the higher the degree to which division of labor is carved out in the interior of the organism, the more complicated is its structure.
>
> The principle which seems to have guided nature in the perfectibility of beings, is as one sees, precisely the one which has had the greatest influence on the progress of human technology: *the division of labor.* (Milne-Edwards, 1834, vol. 1:8)

I want to suggest that for Darwin — the theorizer of 1839 after coming to Malthus, and thereafter — one model of an organism is the Ure-Thomson-Milne-Edwards model of an animal as an automaton composed of various organs (external, internal, physiological, intellectual, . . .) "acting in uninterrupted concert for the production of a common object" — its own reproduction — all organisms being subordinated to a "self-regulated" force, — natural selection. What differentiated organisms as automata from purely mechanical devices was the fact that they were self-reproducing. The act of self-reproduction (which was regulated by laws) was however not perfect; the duplicated copies contained slight variations and differed in "trifling ways". Some of these slight variations were heritable. One aim of Darwin's investigation was precisely to determine the interactions of these self-reproducing entities with one another and with the physical environment.

The problem was how to account for the production and reproduction of a diversity of forms that could be classified hierarchically into species, genera, families etc. It is my claim that the Newtonian paradigm deeply influenced Darwin in his initial formulation of a solution and that freeing himself of aspects of that paradigm marked an important transition in the way he thought about the dynamics involved in the economy of Nature.

Let me therefore turn to the relation of Darwin to Newton.

II. Darwin and Newton: Dynamical Descriptions

The singular importance of Newton in the development of the physical sciences and of mathematics was deeply impressed into the history and the practice of British science. That chemistry may have had its Newton in the person of Lavoisier or Davy — and had since then entered a new era — was a widely held belief. That biology was still in need of its Newton was clear to Darwin's generation — the eulogies of Cuvier calling him the Newton of biology notwithstanding. After 1838 Darwin believed he had done for biology what Newton did for physics (Schweber 1979).

To understand the development of Darwin's thought and his theorizing we must try to appreciate what Newton meant to him. I want to separate three strands in the relation of Darwin to Newton. One relates Darwin to Newton the methodologist and philosopher of science. Another is Darwin's understanding of Newton the physical scientist, the theorist who formulated the three laws of dynamics. The third concerns the psychological associations that Newton had for Darwin. Darwin's reading of Herschel's *Preliminary Discourse* (1830), in 1831 and again in the Fall of 1838, of Whewell's *History of the Inductive Sciences* (1837) in 1838 both before and after coming to Malthus, and of Babbage's *Ninth Bridgewater Treatise* (1837) again in 1838, contributed to all three strands.

It may well be that in 1831 when Darwin read Herschel's *Preliminary Discourse* for the first time he only learned "that it would be wonderful to be a scientist" (Cannon 1976b) but by 1838 when Darwin was deeply immersed in the theorizing of his Transmutation Notebooks there were other insights to be gleaned. Hodge (1982) refining Ruse's thesis (Ruse 1975c) has suggested that what Darwin obtained from Herschel was a deep appreciation of the central importance of the Newtonian *vera causa* principle (VCP) in all theorizing.[9] I believe there were other insights which Darwin obtained from the *Preliminary Discourse*. For Herschel, the scientific study of nature consisted in the analysis of complex phenomena into simpler ones. Although it would greatly assist us if we could "by means ascertain what are the ultimate phenomena into which all the composite ones presented by [nature] may be resolved" there is, however, no way which this can be ascertained a priori. No general rules for the analysis of a complex phenomenon into simpler ones can be given. Such rules, could they be discovered, "would include the whole of natural science" (Herschel 1830, p. 97). We must go to nature herself, and be guided by the same kind of rule as the chemist, "who accounts every ingredient an *element* till it can be decompounded and resolved into others" (Herschel 1830, p. 92). I believe Darwin was sensitive to these remarks of Herschel when he pondered what constituted an "element" in his theory of natural selection. Were the individuals the "elements", or were varieties and species "elements"?

Also, Herschel — for all his commitment to the search for causes — advocated adopting a positivistic stragey, as Newton had for gravity:

> Dismissing, then, as beyond our reach, the enquiry into causes, we must be content at present to concentrate our attention to the laws which prevail among phenomena, and which seem to be their immediate results. (p. 91)

This is precisely the attitude adopted by Darwin when dealing with variations after 1838. He was perfectly willing to state his ignorance of the causes of variations, and to look for laws of variation. I shall not here consider other aspects of the Herschel influence: the stress on quantification, the role of quantitative verification, of induction and deduction and of prediction. I want instead to concentrate on the Newtonian model of theorizing that emerged from his reading of Herschel and Whewell.

It is my contention that from 1837 on, Darwin was constantly trying to answer the question: "What constituted an explanation in biology, and more particularly a *dynamical* explanation that identified and referred to causes, *vera causae*?" The models presented by the physical, chemical and geological sciences — as well as their historical development — were important factors in arriving at his own formulation.[10] I would like to suggest that in his answer to the above question, Darwin abandoned the Newtonian model of dynamical explanations in important respects and came to a novel conceptualization of dynamics for biological systems.

Both Herschel in the *Preliminary Discourse* and Whewell in his *History of the Inductive Sciences* expounded at great length the workings of Newton's dynamical explanation of planetary trajectories based on his three laws of motions and his law of universal gravitation. Both stressed that the dynamical laws of motion for composite bodies could be deduced from those of the constituent parts. In a detailed and impressive reconstruction Whewell indicated how the gravitational interaction between the constituents of spherical bodies allowed one to deduce the gravitational interaction between the bodies themselves. He also stressed the linear character of the gravitational interaction: the force on a given mass resulting from the introduction of a new object merely *adds* to the gravitational forces produced.

There are important parallels between Darwin's dynamical scheme after Malthus and the Newtonian model. Just as Herschel and Whewell claimed that Newton's principles of mechanics could answer "every question that can arise respecting the motion and rest of the smallest particles of matter as well as the largest masses", similarly Darwin claimed that natural selection was a law of nature of universal scope in the organic world and applied to all living organisms. As in the Newtonian scheme, one can understand the dynamical response of the organism as a whole by looking at the dynamics of its "elements" ("elements" here understood in Herschel's sense), that is of its parts. Natural selection operating on the parts of any given organism acts like a force (it is a *versa causa*). It selects according to the variation carried by the parts of the organism. Variations determine the response of the system (that is, the eventual survival of the organism); the Newtonian analogue for the response is the acceleration of the system. Variations are the analogue of inertial mass. They are additive and it is this additivity which allows the determination over time of the response of the organism as a whole. The Newtonian construction of the action of the force as occurring incrementally in infinitesimally small steps is also present in Darwin.

But Darwin recognized that the analogy with gravity broke down at the next level of biological organization. One could not describe the dynamics of a community of organisms simply from the pairwise interaction of the organisms and from interaction of the organisms with the environment. Living systems were infinitely more complicated than Newton's planetary system. Biological "elements" had characteristics that were changing in time: they had a history. All the interactions of organisms whether with one another or with the environment were non-additive, non-instantaneous and exhibited memory. It was the ahistorical nature of the objects with which physics dealt that gave the Newtonian scheme the possibility of a simple, mathematical description (Herschel 1830). It was precisely the *historical* character of living objects which gave biological phenomena their unique and complex features.

Certainly one aspect of the staggering entry in the E *Notebook* dated January 1839 is quite explicit about this:

> The enormous *number* of animals in the world depends on their varied structure and complexity — hence as the forms became complicated, they opened fresh means of adding to their complexity — but yet there is no *necessary* tendency in the simple animals to become complicated although all perhaps will have done so from the new relations caused by the advancing complexity of others. It may well be said, why should there not be at any time as many species tending to dis-development . . . my answer is because, if we begin with the simplest forms and suppose them to have changed their very changes tend to give rise to others — . . . I doubt not if the simplest animals could be destroyed, the more highly organized would soon be disorganized to fill their places.
>
> The geologico-geographico changes must tend sometimes to augment and sometimes to simplify structures. Without enormous complexity it is impossible to cover *whole* surface of world with life — for otherwise a frost if killing the vegetable of one quarter of the world would kill all . . . it is quite clear that a large part of the complexity of structure is adaptation . . .
>
> Considering the kingdom of nature as it is now, it would not be possible to simplify the organization of different beings, (all fishes to the state of the Ammocoetus, Crustacea to —? &c) without reducing the number of living beings — but there is the strongest possible [tendency] to increase them, hence the degree of development is either stationary or probably increases. (*E* 95-97 Jan. 1839)

The entry *E* 95–97 presents a view of organisms as entities that have a history and a memory. Only by virtue of having obtained a degree of complexity can organisms add further complexity. Although there is no necessary tendency in the simple animals to become complicated, "new relations caused by the complexity of other" will induce complexity to evolve — and complexity means diversity. Change entrains change. Note further that it is the interaction *among* organisms that gives rise to further complexity, that is novelty and diversity. Darwin is silent here as to any need for external conditions to change in order to induce organisms to vary and produce new adaptations: the interactions between organisms is sufficient. This line of thought does not appear in the *Essay* — for reasons I shall take up in the next section — but emerges again in the mid 1850s.

The last paragraph of the entry is equally impressive. Since there is the strongest possible tendency to increase the number of living beings[11] — since every organism tries to increase in a Malthusian fashion — the development probably increases. Darwin's statement "the strongest tendency to increase [living beings]" marks I believe an important deduction from the Malthusian principle. It becomes stated in *Natural Selection* as follows:

> Every single organism may be said to try its utmost to increase (geometrically) therefore there is the strongest possible power tending to make each site to support as much life as possible.

This maximalization principle — "the strongest tendency to increase" — was an important insight. Darwin suggests that the dynamics of the situation, though very complicated — geometric increases through reproduction, intense inter- and intra-specific competition, interaction with environment, climate etc. — results in a dynamical equilibrium such that the greatest amount of life possible is supported in a given area of the surface of the earth. A comparison with the complicated processes that determine the topography of the surface of the earth is revealing. Lyell had argued that the complicated effects of igneous and aqueous action combine to produce changes such that the net amount of surface area of land masses over the globe is constant. Darwin's maximum amount of life principle, however, represents something novel. What Darwin asserts emerged from the dynamics is not a conservation law but a *maximalization principle.* In 1839 this principle operated under the Lyellian constraint that globally the number of species is conserved: individuals may increase, but the total number of species is constant (Lyell 1832, vol. II, p. 134).[12]

When coupled with the *C* 147e entry that the "quantity of life on planet" depends on "subdivision of stations & diversity" and the *C* 146 one, that reads

> The end of formation of species & genera, is probably to add to quantum of life possible with certain preexisting laws — if only one kind of plant not so many. —

E 95–97 contains many of the insights which will later go into the principle of divergence.

But in 1838–1839 when these entries were written there were indeed many "pre-existing" constraints: conservation of species, conservation of land masses and adaptation to *stations* conceived as geographical localities. But the notion that diversification could increase the "quantum of life" became from that time on part of Darwin's theoretical assumptions. It is given as an explanation for the paucity of species and the diversity of genera in Coral Islets in the Spring of 1844:

> Explanations of fewness of species and diversity of genera, I think must be partly accounted for the plants groups could subsist in greater numbers, and interfere less with each other. This must be explanation of Arctic Regions — How are alpine plants. Several Genera? (DAR 150, unfoliated slip following Hooker's letter of Jan. 1844)

The note is in fact more explicit: in any locality the largest amount of plant life will be supported if there is diversity for then "the plants groups [will] interfere less with each other".

There were several sources responsible for Darwin's approach in characterizing the equilibrium as one which maximalizes the amount of life per unit area. One of these was certainly Brougham's (1839) *Dissertations on Subjects in Science.*

Darwin read Brougham's *Dissertations* in January 1839 after coming to Malthus while working on questions relating to instinct. One of the dissertations is entitled "Observations, demonstrations and experiments upon the structure of the cell of bees". In it Brougham asked what do the requirements "that the greatest possible saving should be made both of space, of wax and labour" imply for the geometry of the cells. If "the form of all [cells is to] give the largest proportions of the walls, and the smallest of rhomboidal base", how should the cells be made so as to place the greatest number in each set or comb? Darwin carefully read and annotated this lengthy article, and followed all the steps in the algebraic calculations. Although he does not refer to these investigations in the *Essay*, Darwin when discussing the cell-making instinct in the hive-bee in the *Origin* does indicate that "we hear from mathematicians that bees have practically solved a recondite problem, and have made their cells of the proper shape to hold the greatest amount of honey, with the least possible consumption of precious wax in their construction" (p. 224).[13]

There is another, more important source for Darwin's usage of maximalization principles: the literature on scientific agriculture and the related writings on political economy.[14] Agriculture was a central concern in all the discussions of political economy, starting with Quesnay and the physiocrats and with Adam Smith. I shall here refer only to the influential lectures on *Political Economy* that Dugald Stewart delivered in the first decade of the nineteenth century. In them, he gave a critical overview of political economy at the turn of the century.[15] As befitted his Scottish training and outlook Stewart's presentation was sensitive to the sociological aspects of the subject matter and offered illuminating comparisons between French and Continental practices. He began his lectures with the role and interrelation of agriculture[16] and manufactures in the economy of a nation. A disciple of Adam Smith, he stressed the *self-regulating* character of the free market:

> In the midst of this conflict of contending interests and prejudices, it is the business of the Political economist to watch over the *concerns of all*, and to point out to the Legislator the danger of listening exclusively to claims found in local or in partial advantages, to remind him that the pressures of a temporary scarcity brings along with it in time its own remedy while an undue depression of prices may sacrifice to a passing abundance years of future prosperity: — above all, to recommend to

> him such a policy, as by securing in ordinary years a regular *surplus*, may restrain the fluctuation of prices within as narrow limits as possible; (Stewart 1855, p. 12)

In Part I of his *Political Economy* lectures Stewart was concerned with the relation of the size of the population of a country to its agricultural practices. The central problem for Stewart was to answer the question: "What kind of agriculture will maximalize output so as to support the largest and best fed population?" (p. 103 ff.). Drawing on Arthur Young's *Political Arithmetic* (1774) Stewart noted that

> in the French system of husbandry . . . much of the farm is arable; — the meadow and pasture being very trifling, except in spots that can not otherwise be applied, and near great towns. Thus very little cattle can be kept except for tillage; in very many farms no other. Here we find manuring cut off at once, almost completely, and consequently the crops must be poor. Besides this, one half or one-third of the land is fallow . . . (Stewart 1855, p. 108)

Stewart went on, by quoting Young approvingly

> It must surely be evident to everyone, that there is great advantage to the *English* farmer, from corn and cattle being in equal demand, since he is thereby enabled to apply all his lands to those productions only to which they are best adapted; while at the same time, the one is constantly the means of increasing the produce of the other. (Stewart 1855, p. 108)

and continued by endorsing Young's conclusion that

> . . . where tillage and pasturing are properly combined, so as to have the farms from one-third to half of meadow and pasture; and the other two-thirds or half thrown into a proper course for the winter support of the cattle, such a farm will be found to feed more men than if it is all ploughed up, and as much wheat as is possible raised upon the French system. (Stewart 1855, p. 108)

A further inquiry of relevance for maximal agricultural output concerned the size of farms. Arthur Young (1794), the famous editor of the *Annals of Agriculture*, had also addressed this question and with him, Stewart concluded that *The best size of a farm is that which affords the greatest proportional produce, for the least proportional expense* (Stewart's italics, p. 128).[17]

The questions raised and the answers given by Dugald Stewart were surely known to Darwin. Josiah Wedgwood the younger ("Uncle Jos") was very much concerned with scientific agriculture and James MacIntosh had an abiding interest in political economy. Darwin greatly respected both of them and had lengthy conversations with them on his visits to Maer. Furthermore, as is the case with many of Charles's investigations, his grandfather Erasmus Darwin had been there first. In 1800 Erasmus had written his *Phytologia* subtitled *The Philosophy of Agriculture and Gardening*. It is a treatise on agricultural chemistry and the political economy of agriculture. Charles had studied it during his Transmutation Notebooks period!

Darwin was undoubtedly also familiar with Humphrey Davy's influential lectures on *Agricultural Chemistry* (1813). These contained many useful insights on the economics, that is to say the dynamics of agriculture. Thus when discussing the yield from pastures Davy indicated that

> Nature has provided in all permanent pastures a mixture of various grasses, the produce of which differs at different seasons. Where pastures are to be made artificially such a mixture ought to be imitated; and perhaps, pastures superior to the natural ones may be made by selecting due proportions of those species of grasses fitted for the soil, which afforded respectively the greatest quantities of spring, summer, later math [mowing], and winter produce; a reference to the details of the Appendix will show that such a plan of cultivation is very practicable. (Davy 1813, p. 324)

The Appendix is a seventy-page account of "the results of the experiments in the produce and nutritive qualities of different grasses and other plants used as food of animals instituted by John Duke of Bedford."

Davy also inquired into the relation between the grasses or plants grown on the pasture, the animals that can be raised on these plants, and the nutritive and regenerative value (in terms of mineral and organic matter) of the manure excreted by these animals to these same fields and plants. The intent was to discover which animals to raise and which grasses to grow in order to maximize the value of the output (Davy 1840, vol. 8, p. 28, pp. 77-79). Davy conceived his approach as a scientific procedure to achieve the steady state, that is equilibrium, that maximized output consistent with nature's constraints.[18] Darwin followed the growing literature on this ecological approach to land management and animal husbandry after he moved to Down. This scientific approach to agriculture probably assumed an even greater importance after Darwin purchased a farm in Lincolnshire in 1845 as an investment.

Agricultural chemistry as a scientific discipline was given a great stimulus by Liebig's *Chemistry in its Applications to Agriculture and Physiology* (1840).[19] Liebig had initially prepared this treatise as a report to the BAAS meeting of 1841. In it he presented the known chemical facts and the scientific means to investigate "the nutrition of vegetables, and the influence of soils and the action of manure on them". A companion volume on *Animal Chemistry or Organic Chemistry in its Application to Physiology and Pathology* (1842) presented Liebig's views of the process of nutrition in animals and in particular "the origin of animal excrements" and "the cause of their beneficial effects on the growth of vegetables".

From his reading notebooks we know that Darwin read Liebig's *Chemistry in its Application to Agriculture* in November 1841: The entry for November 21 states: "Liebig's Agriculture — do." Darwin also read Liebig's *Familiar Letters* (1843) first in 1844 and again in 1851. The tenth letter in Liebig's *Familiar Letters* dealt with the relation between agriculture and the growth of human population. Liebig's views were that

> The cultivation of our crops has ultimately no other object than the production of a maximum of those substances which are adapted for assimilation and respiration, in the smallest space. Cultivation is the economy of force [energy]. Science teaches us the simplest means of obtaining the *greatest* effect with smallest expenditure of power, and with given means to produce a maximum of force [energy]. (Liebig 1843, p. 8)

Finally it should be noted that this maximalization approach had wide currency in Great Britain in Darwin's time. In part it was a legacy of the Benthamite tradition with its felicity calculus and Darwin was exposed to it both at home (Schweber 1980) and at Cambridge in its Paleyan version (Garland 1980). In mathematics isoperimetric problems were widely popularized: most encyclopedias had entries on them; John Herschel for example wrote one for the *Encyclopedia Metropolitana*. An extremum approach was also the basis of William Rowan Hamilton's widely heralded formulation of classical mechanics in the early 1830s (Hankins 1980). It is possible that Darwin would have heard about it from Babbage or Whewell because Hamilton reported on it on several occasions at BAAS meetings during the 1830s.

But undoubtedly the most important reason for the popularity of the approach was the influence of Adam Smith and the school of political economy for which he was responsible (Schweber 1980, 1983). It was Adam Smith's basic tenet that in a *laissez-faire*, politically uncontrolled economy, the efforts of each person to act in his own self-interest, i.e. to better himself, would result in that distribution of capital, labor and land which maximized their respective returns by maximizing the value of the output to the public. Moreover, even though "Each intends only his own gain", in the end "he promotes that of society though this was no part of his intention" (Smith 1937, p. 423).

But the differences between Davy and Adam Smith ought to be stressed. With Davy it is an *externally* imposed maximalization requirement, that the value of the output be greatest, which determines the equilibrium. In Adam Smith's situation, the maximalization is a *self-regulating, self-consistent* one. This is, of course, also the parallelism between the artificial economy (read selection) and the natural economy (read selection) and the reason that Darwin's project mirrors that of Adam Smith.

Although maximalization formulations will play an important role in the construction of the principle of divergence and in its presentation in

Natural Selection and the *Origin*, the *Essay of 1844* does not reflect these insights. Let me turn briefly to the *Essay* to set the stage for these later developments.

III. The *Essay of 1844*

In the *Essay* Darwin's conception of the process by which "admirably adapted" forms are produced is based on the idea that "accidental" variations are differentially adaptive. When individuals of a species are transported or migrate to a new region, or when geological changes alter the conditions under which a species lives, the reproductive system of the individuals is affected, and the structure of the offspring is rendered in some degree "plastic" and variations result. "Every part of the body of the organism would tend to vary from the typical form in sight degrees" but "in no determinate way" (*1844 Essay*, p. 114). These variations are not automatically adaptive, the chances of survival of the better and "less well adapted" to the new conditions, will be different. Natural selection will operate to produce forms that are as well adapted to conditions as their hereditary structure allows. When such organisms have bred for many generations under new conditions, the alterations become fixed in their constitution and they will breed true, and are then "exquisitely", or as Darwin also often says "perfectly adapted". But natural selection only works intermittently, its action being limited by the availability of variations — of which there are few: "Most organic beings in a state of nature vary exceedingly little" (*1844 Essay*, p. 111). Variations are only produced when the organisms cease to be perfectly adapted, that is when they breed (or are bred) under conditions different from the natural ones of the species. Variation under domestication is proof of this. But "organisms in a state of nature must occasionally in the course of ages, be exposed to analogous influences" (p. 113) by being introduced into new regions or by virtue of geological changes in its habitat. However, since

> *without selection* the free crossing of these small variations (together with the tendency to reversion to the original form) would constantly be counteracting the unsettling effect of the extraneous conditions on the reproductive system. (p. 114)

Darwin postulates geographical isolation as the mechanism for speciation. Islands are the paradigms, but geographical barriers, such as continental "islands" resulting from geological change (for example mountain ranges, great plains, rivers, . . .) can likewise be responsible for the allopatry.

Ospovat (1981) in his *The Development of Darwin's Theory 1838-1859*, and in earlier publications has energetically argued that as late as the *Essay* Darwin still believed that organisms are "perfectly not merely relatively well adapted" (Ospovat 1981, p. 77). Moreover, according to Ospovat

> Darwin assumed that the generation of variations is regulated by the perfection of adaptation. As a result of this assumption, in the *Essay* the theory of natural selection is constructed on, and limited by, an essentially natural theological foundation. (Ospovat 1981, p. 77)

My own stress would not be on the "perfection" of the adaptation since Darwin's meaning and use of "perfect" is ambiguous[20] but rather on the slowness, the gradualness of the process of speciation and on the "indeterminate" nature of variations. The equilibrium in nature is dynamic, but it is an equilibrium maintained by a constancy in the total number of species, and the progressive changes which do occur are very slow. This is probably the reason why the view expressed in *E* 95 of evolution creating new evolutionary opportunities does not appear here. The tempo and, in fact the dynamics, are determined by the slow changes in geology and climate. These processes are governed by physics and chemistry, hence are deterministic. The paucity of variations — even though they are indeterminate — suggests a quasi-deterministic, deistic, description of the overall process. My other emphasis would be on Darwin's acceptance of variations that are indeterminate and undirected, that is random, as the "elements" (in Herschel's sense) that enter in the dynamical description of the changes in an organism. It is the Darwin for whom final causes have been banished, for whom man is a "chance" production rather than the "one great object". It is the Darwin for whom *small* contingent effects can be responsible for the large consequences. The *Essay* represents what Darwin thought would constitute an adequate account of the phenomena of natural history, *and a putative theory that met the criteria of, and thus might be acceptable to, his scientific community*. It was, however, a private statement, *not* intended for publication. The *Essay* indicates how far Darwin had gotten with his theory by 1844 to find explanations for the generalizations and various "laws" that had been proposed in the study of morphology, embryology, paleontology, classification and geographical distribution.

In the *Essay* the theory operates at two levels. At one level the individual members of a species are accepted as structural entities which differed very slightly from one another. The origin of variations is not accounted for and they are accepted as phenomenological facts. But there is enough *structural* identity to classify these entities into real, morphological species.[21] Natural selection explains phyletic evolution, that is, the transformation of a population in which variations ranged continuously (without limit) into one in which only a discontinuous range existed. Since Darwin's assumption in the *Essay* is that once the organism is adapted to a station or niche, variations cease, the stability of the clump of individuals that constituted a morphological species was understood.

Classification of group within group follows from common descent and extinction. Classification is linked to biogeography because the higher taxa (family, genera . . .) are formed by the process of migration, isolation, extinction and, of course, natural selection (pp. 215–217). There is an appreciation of divergence in the *Essay* but it is certainly *not* a dominant concern either in classification or in speciation.

The *Essay* operates at a second level when, in dealing with embryology and comparative anatomy, it looks *into* organisms and their development. Here Darwin invokes the stability of the structural *development* (perhaps thought of in analogy with crystallization) to explain Von Baer's generalization of the facts of embryologic development and the insights into classification afforded by the concept of unity of type.

This dichotomy in considering an organism both as a structural and as a functional entity is consistent with considering organisms as self-replicating automata. Natural selection operates on organisms considered as *functional* units — whose structural stability is assumed. Darwin's approach is typological with respect to classification because classification is based on viewing organisms as *structural* units (like molecules, and crystals in chemistry) but it is populational with respect to dynamics because Darwin considered variations as *real* and as the elements driving the process.[22] This tendency persists in the *Origin*. But note that one of the consequences of Darwin's approach in the *Essay* is that it operates at one level. The process of speciation is essentially accounted for at the individual level: It is a few individuals that are the progenitors of new varieties, and thus allopatry plus natural selection (which operates on individuals) can make new species.

IV. Divergence of Character

When in 1838 Darwin conceived of natural selection as the mechanism for adaptation he had simultaneously designed a theoretical structure that generated a constant stream of problems. Every known biological fact, generalization and law would be confronted and the question posed of how to account for it or how to fit it into an explanatory scheme based on descent and natural selection. Every branch of botany and zoology (anatomy, morphology, embryology, physiology etc.) and of natural history (biogeography, habits and instinct of animals, classification etc. . . .) was combed for this purpose. That the facts considered were always of relevance to several enterprises, was the result of Darwin's uncanny ability to pose the right questions, and of his extraordinary communion with nature. The answers or putative answers to questions from one field became transferred to others to raise new questions or as checks for consistency. Thus questions relating to the hardiness and transport of seeds were crucial to Darwin's biogeography but they also became relevant in classification. Similarly the phytogeographic fact that the lower the organization of the plant the more widely distributed the plant is, became a question about "highness" and "lowness", and triggered an inquiry into the temporal sequence of the introduction of plants. Many other such examples could be given.

In several places the theory of natural selection rested on weak foundations. Thus in the *Essay* Darwin did not — and could not — present empirical data to justify the assumptions he had made about variations — their origin, the mechanism of their transmission and their frequency. Observations about variations under domestication together with the premise that variations occur when organisms are taken out of their natural location were used to justify his assumptions about variations in the state of nature. The provisional character of the *Essay* in these matters is made plausible by the fact that very few botanical illustrations are presented in it. Immediately after completing the *Essay* Darwin immersed himself in the botanical literature. Darwin certainly was aware that much more was known about variations in the state of nature in botany than in zoology (Sulloway 1979b). Hybridization in plants suggested a morphological rather than a biological concept of species. It has even been proposed that Darwin's interests in botanical subjects reflected a realization that he might have a more sympathetic audience for his transmutationist ideas among botanists. Some of them, for example, William Herbert (1837), were still taking the genus as "fundamental" and allowed new species to be introduced "naturally" by hybridization (or saltation).

Nor did it escape Darwin's notice when writing the *Essay* that how one classifies, whether one is a splitter or a lumper, makes important assumptions about variations: *species may be made to appear more or less variable according to whether a genus is divided into few or many species*.[23] By and large zoologists were splitters (and admitted but little variations in the definition of their species) whereas botanists were lumpers. Reviewing Waterhouse's book for the *Gardeners' Chronicle* in 1847, Darwin "rejoice[d] to see no sign of that rage to create new species, so prevalent amongst zoologists" (*CP* 1: 214). Hooker's strong stand against "taxonomic splitters" was having its effect.

The *Essay* in fact was unsatisfactory as a *self-consistent* dynamical theory. If one allowed a broader definition of species, hence more variations, the assumption that variations cease when the organism is "perfectly adapted" is called into question. Thus, the very notion of perfect adaptation is called in question.

To lay firmer foundations to his "species work" Darwin embarked on his researches of the Cirripedes in 1846. What is remarkable about this enterprise is not only the skill with which he organized and carried out the in-house anatomical dissections, but also the impressive managerial talents he exhibited in creating the scientific network which made these activities possible. A world-wide zoological community — of amateurs and professionals — was enlisted to provide him with specimens and scientific information. Incidentally, the cooperative spirit that existed in natural history and related scientific communities — a spirit already evident in the writings of White of Selborne — merits further study (Allen 1976). Such activities became the pattern for Darwin for his subsequent work. To obtain the empirical facts necessary for him to complete his species-work he joined various breeders' clubs (Secord 1981b, and this volume). He actively continued organizing and cultivating an extensive scientific network. His queries were sent to numerous correspondents, to various horticultural and gardening journals, and were presented to the participants of the small scientific workshops he organized at his house in Downe. At Down House he set up numerous experiments in his study, in his gardens, in his greenhouse, and in the pastures on his and on the adjoining lands. He raised a variety of animals, including pigeons, rabbits and poultry.

The centrality of these experimental activities to the writing of *Natural Selection* cannot be overemphasized. Nor should Darwin's commitment to (British) empiricism in his philosophical outlook be underestimated. Experiments on the hardiness of seeds, on the relation of fertility of crossing in and between wild and domestic plants, on the embryological developments of different races of pigeons, on the yield of pastures as a function of the genera and families planted were the empirical foundations upon which his various hypotheses rested. Their result critically determined the nature of the "long argument".

Although Darwin had been constantly at work "on species" from 1844 on (as the massive accumulation of notes in the portfolios indicates) the task of preparing a coherent, comprehensive account of his theory, one designed for presentation to the scientific community did not have to be faced until September 1854 at which time he read and studied the *Essay* again and all the notes "on species" he had accumulated. As important as the confirmations of his theory that he had gleaned from 1844 to 1854 were the gaps and difficulties that he discovered in his presentation in the *Essay*.

Darwin's explanation in the *Essay* of the similarity of early embryonic stages as evidence of common descent had acquainted him with the work and views of Von Baer, Muller and Owen. In 1846 upon starting his research on the Cirripedes, Darwin read and was deeply impressed by Milne-Edwards' paper on classification based on embryonic characters (Milne-Edwards 1844). Milne-Edwards' work made him aware that he had to explain the process responsible for the divergence from initially similar embryonic states according to the affinity of the different adult forms. The tendency in the evolutionary process to make embryos diverge from one another according to the degree of affinities of the adult form, was something Darwin had not expected. Milne-Edwards' work on classification indicated to Darwin "that divergence could not be the occasional and accidental result of the diffusion of single species that he had described in the *Essay of 1844* but must be rather a universal tendency among organisms" (Ospovat 1981, p. 173) and thus a general feature of nature.

In July 1847 Darwin wrote himself a note which indicated that the production of the branching had become a problem for him to solve:

> The affinities of organisms are represented by distance — species being called dots by their being placed thus
>
> ●●● ●●●●●● ●●●●●● ●● ●●●
>
> As this arrangement leads to idea of common genetic causation, as we see with chemists in proportion of elements, we are led to compare things thus arranged to branching of tree and may be said to diverge from common stem.
>
> Begin with stating fact — universal at all times and places — overlooked for familiarity, — genus not an entity — Begin with single species vars — action of divergence — the same cause which formed special species will make others and when once formed into larger genus, we know these are the very groups, which do vary most and therefore will give rise to more varieties and thus give divergent character. — Here again we have resemblance to most great forest trees in which 2 of 3 great branches from some accidental advantages have exterminated the others. The affinities are represented down the line of stems. (DAR 205.5: 120)

Darwin's work on the Cirripedes further reinforced his conviction of the universality of the phenomenon of divergence. Ospovat has claimed that it was classification which brought Darwin to confront again the problem of a mechanism for divergence. He suggested that divergence had to be faced at a fundamental level because Darwin had assimilated the work of the morphologists of the second quarter of the century who "by combining unity and diversity in a single scheme, . . . produced a branching conception of organic nature, a developmental conception in which diversity was seen as proceeding out of an initial unity" (Ospovat 1981, p. 116). Ospovat put particular stress on Owen's work (1846, 1848, 1849a, 1855) as an influence on Darwin. Owen accepted Von Baer's claim that in development one sees a gradual change from the general to the special. It was Owen's conception that all organisms started with basic archetypical forms, but thereafter diverged and became progressively adapted and specialized to their particular ecological niches. At one level as E. S. Russell put it, Darwin took Owen's morphology "lock, stock and barrel, to the evolutionary camp" making archetypes into ancestors and explaining homologies and unity of type by descent (Russell 1916, p. 247; Cassirer 1950). But at another, deeper, level it was the *dynamics of the process* that was of concern to Darwin when he read and annotated Owen's and Milne-Edwards' works in the early 1850s. And at the heart of any dynamical explanations were variations.

By giving classificatory and embryological problems pre-eminence, Ospovat has unduly minimized Darwin's confrontation with the problems of the copiousness and ubiquity of variations that his work on the Cirripedes had indicated. From 1847 on, Darwin viewed this copiousness as a universal feature of that part of the biological world which reproduced sexually. Although he had been unable to establish any laws of variations for *individuals*, it had been apparent to Darwin since the early 1840s, and even before, that *empirical* biogeographical *data* could be used to incorporate into the theory the copiousness of variations in large, mundane genera and species without having to make detailed *assumptions* about variations and their heredity in individuals. The July 1847 note quoted above, makes this clear as does his extensive correspondence with Hooker between 1844 and the time Hooker left for India.

In September 1854 Darwin started work on the Big Species book. In November of that year he recorded the following hypothesis:

> To explain why the species of a large genus, will hereafter probably be a Family with several genera, we consider, that the species are widely spread, and therefore exposed to many conditions and several aggregation of species: they will occasionally mingle with a new group, and then on the principle, that the most diverse forms can best succeed, it may be selected to fill some new office, and mere change would determine the origin of a large genus of some new and good modification. (DAR 205.5: 151)

In his *Essay* as well as in a subsequent note of July 1847 on the relation of divergence to classification, Darwin had noted that the process must be linked to the observation that

> The same cause which formed special species will make others when once formed into larger genus, we know these are the very groups, which do vary most and therefore will give use to more varieties and thus give divergent character.

What is novel in November 1854 is that the idea is linked to his observations of January 1844 that "the most diverse forms can best succeed". Recall that in 1844 Darwin's explanation of the flora of the Arctic Regions was that diversification reduced competition. What thus happened in November 1854 was the *simultaneous* linkage of the observation "that large genera vary more and give rise to more species" to both natural selection *and* to the ecological notion that diversity will allow groups to "subsist in greater numbers". I concur with Kohn's strong evidence that November 1854 is the beginning of the genesis of the final form of the principle of divergence (this volume). As Kohn stresses, however, the basic assumptions that the widely spread families present genera with many species and that diversity maximalizes the number of different organisms supported in a given region had to be proven convincingly. If that were indeed the case, then natural selection via the principle "that the most diverse forms can best succeed" will explain why there will be a tendency for these large and successful groups to go on spreading and growing *in the future* and thus diverging ever further from the parent stock.

It was to verify the assumption about the number of species presented by wide ranging genera that Darwin embarked on his botanical arithmetic (*Origin*, p. 53-59; *Natural Selection,* p. 134-164), which has been the focus of Browne's paper (1980). I believe she is right in stressing that the Cirripedes research had made clear to Darwin that variations were much more copious in nature than he had thought in the *Essay* (see also Schweber 1980). This fact would eventually weaken his dependence on geological and geographic changes as being responsible for the cause of variations. Browne's work is of importance not only for Darwin studies — but also more generally because it verifies the increasing importance of statistics in the first half of the nineteenth century, and exemplifies the growing acceptance of *statistical* laws as *explaining* phenomena. Their prominence in *Natural Selection* and *The Effects of Cross and Self-Fertilization* (Darwin 1876) should be noted. Darwin's explorations in botanical arithmetic also illustrate his growing use of quantitative data. The statistical nature of the data reflected the *complexity* of the materials investigated. Darwin's acceptance of the *distribution* of measured values as exhibiting *real* effects in nature was indicative of his positivistic stance.

But only one part of the analysis was still to be completed — natural selection operated on individuals and all the separate arguments thus far referred to different levels of description — the "elements" in Herschel's nomenclature — namely varieties, species and higher taxa (Schweber 1980). That Darwin was aware of the difficulty in January of 1855 is suggested by the following note:

> On Theory of Descent, a divergence is implied and I think diversity of structure supporting more life is thus implied . . . Now in considering amount of life supported in a given area, besides size as an element, as in trees and elephants, besides period of non-action during winter in cold climates, I think such element as amount of chemical change should if possible be taken as measure of life, viz. amount of carbonic acid expired or oxygen in plants. I have been led to this by looking at heath thickly clothed by heather and a fertile meadow, both crowded, yet one cannot doubt more life supported in second than in the first; and hence (in part) more animals are supported. This is the final cause but mere result from struggle (I must think out last proposition). (DAR 205.5: 167)

Increasingly explicit statements of the linkage — how diversity is accomplished in individuals by natural selection — come over the succeeding pre-*Origin* years. One is recorded in a note dated 23 September 1856. It reads:

> The advantage in each group becoming as different as possible, may be compared to the fact that by division of ~~land~~ [sic; Darwin's cancel] labour most people can be supported in each country — Not only do the individuals of each group strive one against the other, but each group itself with all its members, some more numerous, some less, are struggling against all other groups, as indeed follows from each individual struggling. (DAR 205.5: 171)

Although the insight expressed in this note was part of the theoretical baggage that Darwin had carried for many years, and was seemingly a tacit component in the previous argumentation, it is interesting to note that an explicit statement had to be made as late as September 1856. The principle of divergence in the form to be found in Darwin's letter of 5 September 1857 to Asa Gray — the letter[24] that was one of the papers submitted to the Linnean Society meeting at which Wallace's and Darwin's work were presented by Lyell and Hooker — was now essentially in place.

My presentation thus far has focussed on the "internal" aspects of the genesis of the principle of divergence — because that is where the problems arose — and has emphasized the connections that stemmed from the problems Darwin encountered in classification, embryology, biogeography etc. Darwin's solution, as the above suggests, drew on external factors, and reflected the British context. In constructing his principle, Darwin obtained useful analogies for understanding the "natural economy" from the analysis of the marketplace by the new "science of political economy and agricultural chemistry" (Schweber 1980).

I have indicated elsewhere the influence of Adam Smith and the Scottish conjectural historians — Ferguson, Hume, Adam Smith, Dugald Stewart — on Darwin's intellectual development (Schweber 1977, 1980, 1983). Let me here only point to Adam Smith's corollary to the principle of the division of labor — namely that the division of labor *leads to* and is *sustained* by the growth of an economy. Its role as a model for Darwin's principle of divergence should be readily apparent.

The title of Chapter 3 of the *Wealth of Nations* announces "That the division of labour is limited by the extent of the market". Smith in that chapter then develops the following remarkable doctrine: The more the selling of goods and services takes place, the greater can be the division of labour. The subdivision of labour can be made finer and finer as the market extends. That is, there is a mutually reinforcing relation between specialization and the growth of the market economy in which people are buying and selling to each other (Arrow 1979). Stated differently: The division of labour, with its attendant specialization and differentiation of tasks, results in greater efficiency and less competition, which are responsible for the greater profitability of trade. This in turn results in the growth of the market. But the more the selling of goods and services takes place, the greater can be the division of labour and the greater the number of people that can be supported. *So the cycle repeats itself.* Indeed as Darwin put it in September 1856: "by division of labour most people can be supported in each country". In *Natural Selection* he noted

> The view that the greatest number of organic beings (or more strictly the greatest amount of life) can be supported on any area by the greatest amount of their diversifications . . . is in fact that of the "division of labour" so admirably propounded by Milne-Edwards . . .

In the *Origin* the presentation of the principle of divergence of character is based on the following premises (Kohn 1981):

(a) a locality can support more life if occupied by diverse forms partitioning resources.
(b) specialization ("division of labour") in order to maximalize access to these resources is therefore advantageous to an organism as it minimizes competition with other organisms.
(c) hence natural selection which explains all adaptation will favor the evolution of new varieties, hence of the new species.
(d) by reiteration of such forks in the branching phylogeny all classification is generated.

The influence of political economy should be obvious. Kohn has characterized Darwin's principle of divergence as "a set of *nested arguments* comprising the idea of speciation without isolation, and the view that the relations between organisms create new evolutionary situations". Characterizing the arguments as *ecological* would be equally appropriate: It is their self-regulating, self-consistent, interlocking character, that gives the formulation its stability.

A more complete analysis of the final construction of the principle of divergence are presented in Kohn's paper in the present volume; in particular, Darwin's acceptance of the importance of sympatric speciation. Darwin's insight that the ubiquity and copiousness of variations depended but "very little" on the environment — a conclusion he communicated to Hooker in November 1856 (*LL* (NY) 1: 444–445) — and that it was the interaction *among* organisms that was of primary importance had significant consequences for the rate of evolution: It was no longer determined principally by geologic and geographic time scales. Kohn also considers the impact of Darwin giving up the conservation law of species. His article is, in my view, an account of Darwin's *re*-cognition of the importance of niche, of sympatric speciation, and of his coming again to a conception of nature in which evolution creates new evolutionary opportunities, a view first expressed in the Transmutation Notebooks entries *E* 95–97.

I believe that it is the *re*-cognition of this *openness* and *limitlessness* of evolution that is the salient feature of the development of Darwin's theorizing in the 1854–1857 period. Divergence of character is its technical expression, the famous diagram of Chapter IV of the *Origin* its visual representation.

ACKNOWLEDGEMENTS

There is no doubt in my mind that the invitation to write an essay for this volume covering the period from 1844 to 1859 would have been given to Dov Ospovat. His untimely death has taken from the community of Darwin scholars one of its ablest and most gifted historians. There is no better introduction to the subject under discussion that his *The Development of Darwin's Theory*. This book has set standards for scholarship in Darwin studies and will surely be the point of departure of future researches on the genesis of the *Origin*.

This paper was written while I was a fellow of the Edelstein Center for the History and Philosophy of Science at the Hebrew University of Jerusalem during May and June of 1982. I thank the Center for its invitation. I profited from the presence of Martin Rudwick at the Center, from the lectures and seminars he gave as well as from private talks with him. I sincerely thank him. I also benefited from discussions with David Kohn and John Beatty. As always these were helpful, informative and useful. I am indebted to them for these and other encounters. I would also like to express my appreciation to Yehuda Elkana for making available the pleasant and stimulating surroundings of the Van Leer Institute during my visit to Jerusalem.

Notes

1. "Natural selection can act only by the preservation and accumulation of infinitesimally small inherited modifications, each profitable to the preserved being; and as modern geology has almost banished such views as the excavation of a great valley by a single diluvial wave, so will natural selection if it be a true principle, banish the belief of a continued creation of new organic beings, or any great and sudden modification in their structure" (*Origin*, p. 95).
2. "As natural selection acts solely by accumulating slight successive, favourable variations, it can produce no great or sudden modification; it can act only by very short and slow steps. Hence the canon 'Natura non facit saltum', which every fresh addition to our knowledge tends to make more strictly correct, is in this theory simply intelligible" (*Origin*, p. 471).
3. To mention but one: Gilbert White's *Natural History and Antiquities of Selborne* (1774)
4. An autographed copy of Vol. II of the ninth edition (1823) of Henry's *The Elements of Experimental Chemistry* is in the Darwin Library at Cambridge University Library.
5. In that edition the science of chemistry is characterized as follows: "Natural objects present themselves to our view in two different ways; for we may consider them

either as separate individuals, or as connected together and depending on each other. In the first case we contemplate nature in the state of rest, and consider objects merely as they resemble one another, or as they differ from one another: in the second we examine the mutual action of substances on each other, and the changes produced by that action. The first of these views of objects is distinguished by the name of *Natural History*; the second, by that of *Science*.

Natural science then is an account of the *events* which take place in the material world. But every event, or, which is the same thing, every *change* in bodies, indicates motion; for we cannot concern change, unless at the same time we suppose motion. Science then is in fact an account of the different *motions* to which bodies are subjected, in consequence of their mutual action on each other" (Thomson 1825, p. 17). Thomson's commitment to Dalton's atomism was unequivocal and was forcefully stated: "There can be no reasonable doubt about the propriety of adopting practically the opinion, that substances extraneous to us, are the causes and sources of our sensation that these substances are made up principally of particles apparently homogenous; but in fact are composed of particles different in properties and more simple: that all compound bodies are composed ultimately of particles which admit of no further division of analysis; and which are not only with respect to our knowledge, but which are in themselves, and absolutely indivisible, and indecomposable particles, atoms, monads, or molecules (by whatever name they may be designated) whereof, in different proportions, all the other particles and masses of matter, of whatever kind, are formed and composed" (Thomson 1825).

6. Also in the Darwin Library at Cambridge University Library is an autographed, slightly annotated copy of Andrew Ure's *A Dictionary of Chemistry* (1823). Darwin seems to have availed himself of the numerical tables in the Dictionary (pp. 806 ff.) as late as 1880.

7. Recall in this connection Hobbes' Introduction to his *Leviathan*: "NATURE, the art whereby God hath made and governs the world, is by the *art of man*, as in many other things, so in this also imitated, that it can make an artificial animal. For seeing life is but a motion of limbs, the beginning whereof is in some principal part within; why may we not say, that all *automata* (engines that move themselves by springs and wheels as does a watch) have an artificial life? For what is the *heart*, but a *spring*; and the *nerves*, but so many *strings*; and the *joints*, but so many *wheels*, giving motion to the whole body . . . *Art* goes yet further, imitating that rational and most excellent work of nature, *man*. For by art is created that great LEVIATHAN called a COMMONWEALTH . . . which is but an artificial man" (Hobbes *Leviathan* Introduction).

8. The biological metaphors used by Ure should also be noted: "A mechanical manufacture being commonly occupied with one substance, which it conducts through *metamorphoses* in regular succession, may be made nearly automatic (Ure 1835, p. 2).

   It is in a cotton mill, however, that the perfection of automatic industry is to be seen; it is there that the elemental powers have been made to animate millions of complex organs, infusing into forms of wood, iron, and brass an intelligent agency (Ure 1835, p. 2).

   The processes that may be employed, to give to portions of inert matter, precise movement resembling those of organized beings, are innumerable" (Ure 1835, p. 9).

   To describe the main-shafting and wheel-gearing of the machines in a mill Ure characterized them as: "the grand nerves and arteries which transmit vitality and volition, so to speak, which due steadiness, delicacy and speed, to the automatic organs" (Ure 1835, p. 32).

9. Recall a cause is a *vera causa* if it can be shown (1) to be real, that is, to exist in phenomena other than the one under consideration, (2) to be competent to effect the consequences attributed to it and (3) to be responsible for these effects.

10. Incidentally, Darwin's comments in the Transmutation Notebooks indicate that he appreciated already then the social components in the characterization of an explanation as a scientific one: What constitutes an acceptable theory is determined by a scientific community and the latter's religious and political beliefs are reflected in the criteria. From Herschel, as quoted in Babbage's (1838) *Ninth Bridgewater Thesis*, Darwin had concluded that a scientific community whose religious outlook was theistic could accept a putative dynamical theory to account for the origin of species in which the Deity is conceived as operating through secondary laws. The model of astronomy was the constant referent.

11. I have interpreted "beings" as individuals. It is also possible in the context of the entry to understand by "beings" not "individuals" but "kinds", a reading which follows from the remark "(all fishes to the state of the

Ammocoetus . . .)". Darwin in the famous tree of life passages of the B *Notebooks* — (*B* 21-23; *B* 21: "Organized beings represent tree, irregularly branched") — had proposed the splitting of species into branches as a phenomenological fact. The reading of "beings" as "kinds", here would suggest that Darwin is adducing (postulating?) a Malthusian multiplication *mechanism* for species and would constitute an interesting illustration of Darwin's use of analogies in his theoretical models. See in this connection Gruber and Barrett (1974, pp. 130-150), and Gruber (1978).

12. In *Natural Selection* in his discussion of the struggle for existence Darwin refers to Sir C. Lyell's "equilibrium in the number of species" with the caveat that "it expresses far too much quiescence" (p. 187).

13. By 1840 the bee's geometrical talents had been the focus of a good deal of mathematical interest. Maraldi in 1712 evidently was the first to have measured accurately the obtuse angle α of the three bottom rhombs of the cell, and the angle β they form with the prism wall. He found that $\alpha=\beta$, each having the value of about 110°. He posed the geometric question: "What must be the angle α of the rhomb so as to coincide exactly with β?" He found $\alpha=\beta=109° 28'$ and thus suggested that the bees had solved this geometric problem. After minimalization principles were introduced into the study of mechanics by Maupertuis, Réaumur conjectured that the value of α is determined by the most economical use of wax: with every other angle more wax would be used to form cells of the same volume. Réaumur's suggestion was confirmed by the Swiss mathematician, Koenig, but not without some further controversies made famous by Fontenelle in his judgement to the French academy (See Weyl 1952, p. 91).

14. I have elsewhere (Schweber 1977) referred to other possible sources for such an approach namely the philosophical writings of Hume, Smith, and Bentham; that is, the philosophical tradition which based ethics on a pleasure-pain calculus. For this approach in the Scottish circles see Smith (1980) and the essays introducing the texts.

15. Incidentally, Stewart's *Lectures on Political Economy* (1855) are a rich source for a historical overview of demography, agricultural practices and agricultural economics during the eighteenth century. See also C. A. Browne's *A Source Book of Agricultural Chemistry* (1944).

16. "To begin, then, with that science, which, in the judgement of the most enlightened politicians, is the most essential of all to human happiness, — I mean the *Science of Agriculture*; how various and important are the subjects, which belong exclusively to its province! The general principles of vegetation; the chemical analysis of soils; the theory of manures; the principles which regulate the rotation of crops, and which modify the rotation, according to the diversity of soil and climate; the implements of agriculture, both mechanical and animal; . . ." (Stewart 1855, p. 11).

17. It is interesting to note that Stewart then suggested that "In general, it should seem, that in proportion as Agriculture advances, the size of farms should be reduced; or rather, that farms should divide themselves in proportion as the task of superintendences become more difficult". The mechanization of farms had not yet begun by 1800!

18. His "global philosophy" was outlined in his introductory lecture and merits being quoted for his views on the balance of nature: ". . . all varieties of material substances may be resolved into a comparatively small number of bodies, which, as they are not capable of being decompounded, are considered in the present state of chemical knowledge as elements. The bodies incapable of decomposition at present known are forty seven. Of these, . . . the chemical elements acted upon by attractive powers combine in different aggregates. In this simplest combinations, they produce various crystalline substances, distinguished by the regularity of their forms. In more complicated arrangements, they constitute the varieties of vegetable and animal substances, bear the higher character of organization, and are rendered subservient to the purposes of life. And by the influence of heat, light, and electrical powers, there is a constant series of changes; matter assumes new forms, the destruction of one order of beings tends to the conservation of another, solution and consolidation, decay and renovation, are connected, and whilst the parts of the system, continue in a state of fluctuation and change, the order and harmony of the whole remain unalterable" (Davy 1813, pp. 7-8).

19. For a valuable assessment of Liebig and the emergence of agricultural science see Rossiter (1975). See also Krohn and Schäfner (1976). In his *Chemistry* Liebig defined "The GENERAL object of agriculture is to produce in the most advantageous manner certain qualities, of a maximum size, in certain parts or organs of particular plants . . . The rules of a rational system of agriculture should enable us, therefore, to give each plant that which it specially requires for the attainment of the

object in view" (Liebig, p. 108). As his reading lists indicate (Vorzimmer 1977) Darwin read extensively in the literature of scientific agriculture in the 1840-1846 period.

20. The ambiguity is manifest in the *Origin*. Darwin's use of "perfect" is such that he can modify it as follows: "almost perfect", "most perfect", "absolutely perfect", "even as perfect", "equally perfect", "wonderfully perfect", "least perfect" and can refer to an "organic being as perfect as, or slightly more perfect than, the other inhabitants of the same country" (*Origin* 1859, p. 201). Similarly in the *Origin* he will speak of "extreme perfection", "inimitable perfection"; likewise the adverb "perfectly" is often modified into "more perfect". See Barrett et al. *Concordance to Darwin's Origin* (1981). See also Cannon (1976b) for further comments on the meaning of "perfect".

21. One of the features of the *Essay* is Darwin's abandonment of the concept of a biological species which he had in the Notebooks, and his adaptation of the concept of morphological species (Kottler 1978).

22. Cuvier is of course also a source for viewing organisms as functional entities. There is perhaps a consilience of early dispositions and later understanding of the comparative anatomy literature.

23. In the *Essay of 1844* the section entitled "On the Variation of Organic Beings in a Wild State" began as follows: "Most organic beings in a state of nature vary exceedingly little: . . . The amount of hereditary variation is very difficult to ascertain, because naturalists (partly from the want of knowledge, and partly from the inherent difficulty of the subject) do not all agree whether certain forms are species or races"[4] (*Essay*, p. 111). Darwin appended the following note to this passage: "Every naturalist at first when he gets hold of new variable type is *quite puzzled* to know what to think species and what variations . . . among British plants (and I may add land shells) which are probably better known than any in the world, the best naturalists differ very greatly in the relative proportions of what they call species and what varieties. In many genera of insects, and shells, and plants, it seems almost hopeless to establish which are which."

Darwin in his introduction *The Variation of Animals and Plants under Domestication* himself indicated that "In treating the several subjects included in the present and any other works, I have continually been led to ask for information from many zoologists, botanists, geologists, breeders of animals, and horticulturists, and I have invariably received from them the most generous assistance. Without such aid I could have effected little . . . I cannot express too strongly my obligations to the many persons who have assisted me, and who, I am convinced, would be equally willing to assist others in any scientific investigation" (*Variation*, p. 14).

24. "Another principle, which may be called the principle of divergence, plays, I believe, an important part in the origin of species. We see this in the many generic forms in a square yard of turf, and in the plants or insects on any little uniform islet, belonging almost invariably to as many genera and families as species . . . Now, every organic being, by propagating so rapidly, may be said to be striving its utmost to increase in numbers. So it will be with the offspring of any species after it has become diversified into varieties, or subspecies, or true species. And it follows, I think, from the foregoing facts, that the varying offspring of each species will try (only few will succeed) to seize on as many and as diverse places in the economy of nature as possible. Each new variety or species, when formed, will generally take the place of, and thus exterminate its less well fitted parent. This I believe to be the origin of the classification and affinities of organic beings at all times; for organic beings always *seem* to branch and sub branch like the limbs of a tree from a common trunk, the flourishing and diverging twigs destroying the less vigorous — the dead and lost branches representing extinct genera and families" (Gray 1976, 246-247, 458-459, n.23).

3

DARWIN'S INVERTEBRATE PROGRAM, 1826–1836: PRECONDITIONS FOR TRANSFORMISM

Phillip R. Sloan

Introduction

A source of difficulty in demonstrating a fundamental coherence in Darwin's intellectual development has been created by a tendency to read his scientific thought as developing through a linear sequence of historical stages, each tending to replace the previous one. In terms of this familiar sequence, the Edinburgh years stand as an apparently self-contained period, leaving no mark of interest after 1827. Cambridge divinity studies and amateurish "beetle collecting" still seem to define for us the pre-*Beagle* Darwin, a stage which terminated when this somewhat unfocussed Cambridge gentleman was chosen to accompany the *Beagle* as its naturalist. The *Beagle*-Lyellian geologist phase succeeds this, returning Darwin to England, as some have interpreted it, ". . . ignorant of most branches of natural history with the exception of geology and geographic distribution" (Grinnell 1974, p. 273). This period is followed by the transformist theory, the Malthus reading, and the working out of natural selection theory, a sequence then curiously "interrupted" by the barnacle years. These periods in Darwin's development have become almost common knowledge, and are implied in recent studies (Ruse 1979a). The analyses by Howard and Valmai Gruber (Gruber and Gruber 1962), Sandra Herbert (1974), Martin Rudwick (1982b), and Frank J. Sulloway (this volume), have demonstrated the complexity of these developments, without challenging some of the basic assumptions in this historiography, and only Jonathan Hodge (1982, this volume) has attempted to show in some depth the evidence for some of the important continuities in the Darwinian program that I suggest need greater emphasis.

The difficulty in gaining a fully satisfactory picture of Darwin's intellectual development from this standard sequential ordering suggests that the interactions, rather than the successions, of the multiple strands in Darwin's thought require more exploration. An analogy might be useful here. In place of picturing Darwin's intellectual development in terms of linear periodicity, I would suggest that a more synchronous model describes more accurately the complex development of his thought. We can reasonably view Darwin's intellectual development as being played on a complex keyboard instrument with several keyboards and registers, these registers each able to act sometimes in solo, other times contrapuntally, and at times in synchronous harmony. Some themes and registers form dominant melody lines at various times in Darwin's life — geology, animal breeding studies, the natural selection theory, the barnacle investigations. Other themes function more as a *basso-continuo*, often submerged but nevertheless present if one looks closely enough. Here we might see the interests in chemistry, botany, the "methodological" current, the earthworm studies, and of course the summation of the vast and continuous "extraneous" reading in authors like Gibbon, Hume, Lessing, Montaigne, Scott, Martineau, and Goethe that accompanied his more obviously scientific work. At times these lesser lines of inquiry rise to the surface in ways that seem to have little immediate connection with proceeding work. Darwin's work on natural selection was "interrupted" by the intensive barnacle studies that seem to emerge from nowhere. Pursued more deeply, however, the analogy suggests that we should consider the possibility that these are not real discontinuities at all, but continuous themes that are interacting in important ways with the other strands of inquiry.

Analogies can, of course, present difficulties, but this one is at least useful for suggesting that Darwin's intellectual development should be read not so much as a sequence of phases, but as an accumulation of layers, each with a different set of intrinsic problems, but all able to interact in fruitful and creative ways at some point. If one ignores these multiple thematic lines in his work, or gives undue emphasis simply to the more prominent ones, such as the geological interests during the *Beagle* and immediately post-*Beagle* period, the result is an incomplete and distorted image of the full complexity of Darwin's intellectual development.

My intent in this paper is to develop the continuity and significance in simply one strand of Darwin's interests as these extend from the Edinburgh through the *Beagle* years. By selecting Darwin's invertebrate work for specific focus, my suggestion is that alongside, and in parallel to, Darwin's well-known development as a Lyellian geologist in these years, he was also making important theoretical advances on a lesser known set of problems, and that these advances provide a highly important foundation for certain issues leading to his transformist theorizing in 1837. I shall argue that Darwin's biological work in these years was considerably more significant than that of a mere collector of specimens. More controversially, I shall also suggest that Darwin's theoretical consideration of a set of functional questions concerned with the colonial marine invertebrates provided one of the necessary, if not sufficient, conditions for the rapid theoretical developments that were to take place in his thought in the nine months after his return from the voyage in October 1836.

The paper is comprised of four main sections. In the first, Robert Grant's important reflections on the zoophytes will be detailed. The second will analyze Darwin's early intellectual relations with Grant, and the formation of his early interests in reproduction in the zoophytal groups. The third section will explore the continuity of these researches into the *Beagle* zoology inquiries. Finally, the important theoretical synthesis achieved by Darwin in 1836 and the continuity of this with select aspects of the early transformist reflections in 1837 will be discussed.

I. Robert Grant and the Zoophyte Problem

In speaking of an invertebrate "program" in Darwin's work as an underlying thematic enterprise, I intend this in a reasonably technical Lakatosian sense of a historical sequence of theory-guided researches elaborating the implications and research problems contained in some kind of definable "hard core" of theory and assumptions. While acknowledging that several of these conditions are not satisfied in Darwin's case at the commencement of his invertebrate inquiries at Edinburgh, his work can be seen, as Jonathan Hodge has suggested in this volume, in the light of the work of his first professional guide into this area of study, Robert Edmond Grant. I shall argue that Darwin's earliest research was an attempt to work out some of Grant's own enterprise, and for reasons we shall explore, the result of this was to place Darwin and Grant initially on the same path. At the same time, Darwin's failure to embrace key points in Grant's program led to a divergence that allowed his own theoretical program to take form in the *Beagle* period.

As a preliminary, it is necessary to emphasize the need for a substantive revision in the standard historical accounts given of Darwin's development in the Edinburgh years. This period was depicted in the *Autobiography* as theoretically sterile, but almost immediately this characterization was challenged by at least one author directly aware of the documentary evidence to the contrary. Nevertheless, Darwin's depiction has tended to render this period insignificant even in recent discussions.[1]

To analyze the significance of these years for Darwin's later developments in invertebrate zoology, we must first be aware of the unique opportunities for invertebrate study at the University of Edinburgh in the 1820s, opportunities equalled by no other major university in Europe at the time. The University of Edinburgh is situated at what would probably have been less than one hour's carriage ride from the coast of the Firth of Forth, a deep-water marine estuary with large tidal fluctuations, falling within the rich

North Atlantic Boreal faunal zone. In this faunal region, a great luxuriance of invertebrate fauna is characteristic (Ekman 1953, chap. 6). Only a short distance from the shore are the black rocks, providing an excellent collecting site for sessile intertidal forms. In the then unpolluted Firth, oyster trawlers worked the bottom of the estuary, bringing in with their hauls a rich variety of deep-water bottom invertebrates, many known for the first time to science. Not surprisingly, Edinburgh figured prominently in the foundations of modern biological oceanography and marine biology, through the work of Edward Forbes and John Murray.[2]

Edinburgh was also remarkable in the British university system of the 1820s for the presence of several intellectually active student societies, officially sponsored by the University, and usually overseen, at least nominally, by a prominent faculty member.[3] These societies conducted often highly professional and intellectually rich meetings, a fact that is particularly surprising when it is realized that many of the student members were, like Charles Darwin, only in their teens.[4]

Young Charles had travelled to Edinburgh in the company of his elder brother, Erasmus Alvey Darwin, in October 1825 to begin medical schooling at the University. The nature of the intellectual association of the two brothers in this year is elusive to document, but was among other things a kind of active initiation into the study of select areas of science.[5] Fresh from three years of study in anatomy, physiology, chemistry, and mineralogy at Cambridge in the newly reformed medical curriculum, Erasmus had already functioned as Charles's guide in the sciences, introducing him at a technical level into chemistry, about which we shall speak subsequently, and serving as his guide in entomology, natural history, and botany.[6] Darwin's entry into invertebrate zoology was consequently along a path that the two brothers had already begun to pursue during their year as roommates together in 1825–1826. In early 1826, Charles and Erasmus began a series of expeditions to the nearby Firth of Forth, where their primary interest, it seems, was in the rich invertebrate fauna.[7]

For most of March and April 1826, Charles had been left alone in Edinburgh by Erasmus's departure in early March.[8] The continuation of his sea-shore expeditions after this date suggests that he was probably making them in the company of other students. His immediate acceptance into the student Plinian Society in November, and his prompt election to its governing council in December 1826, strongly suggests that he had probably made contact with Plinian Society members in his first year at Edinburgh.[9] It was probably also at this time he first came into personal contact with Robert Edmund Grant, a local physician and part-time lecturer in comparative anatomy, and the Secretary of the Plinian Society in 1825–1826.[10]

The Plinian Society had been founded in January 1823 by Professor Robert Jameson, who held the University chair in Natural History, to provide a society "where young men discuss subjects which they heard in the class of Natural History, or in the other classes where such subjects are considered" (Jameson in: Great Britain 1837, p. 145). Robert Grant's position of leadership in the society indicates, however, that its membership was not confined to undergraduate students, and his greater age and wide experience made Grant the effective leader of the society for at least that group of student members drawn to invertebrate marine biology. After his initiation into the Society, Charles had begun a close and professional collaboration with Grant on his research topics. He had accompanied him on expeditions, and had also accompanied trawlers onto the Firth of Forth to collect live specimens of deep-water invertebrates, which were apparently to be brought to Grant for his research purposes.[11]

To appreciate the importance of Darwin's collaborative inquiries with Grant, it is necessary first to detail the important differences between some of Grant's conclusions on the invertebrates and those of his main authority in invertebrate zoology, Jean Baptiste Lamarck. Although Lamarck's pervasive influence stood beneath most of Grant's invertebrate zoological work in this period, Grant's sources also included another group of zoologists — Etienne Geoffroy Saint-Hilaire, C. G. Carus, Karl Rudolphi, and J. F. Meckel — whose work explored and extended several functional and anatomical issues into a range of questions only touched on, or never dealt with at all, by Lamarck.[12]

Of particular interest are Grant's views on the status of the colonial plant-like invertebrates, falling presently into such groups as the Hydrozoa, Gorgonia, Anthozoa, Bryozoa, and Porifera, but commonly designated until the mid-nineteenth century as the "zoophytes". It is most strikingly in his conclusions on this group of forms that Grant broke with Lamarck and developed his researches in new conceptual territory in the 1820s. To appreciate this, a brief examination of Lamarck's views on these forms is necessary.

When Lamarck assumed the chair of *Vermes* at the newly constituted Muséum national d'Histoire naturelle in 1794, he undertook the great revision of the invertebrate groups that has served to define the understanding of these forms until the present. In proceeding to do this in the years after 1794, he also made two conceptual developments important for understanding the character of Grant's own programmatic enterprise.

The first of these, one with which Grant clearly concurred, was to claim that the "natural" ordering of the animals began with, rather than terminated in, the lowest invertebrate forms. In this move, Lamarck placed himself in opposition to the venerable tradition descending from Aristotle through Linnaeus, Buffon, and his contemporary Georges Cuvier, which had consistently rendered the invertebrates either the final degradation of animal form, or at least an anticlimactic appendage to vertebrate zoology.[13]

This ancient tradition, still implicitly advocated at Edinburgh in the 1820s in Robert Jameson's "Cuvierian" approach to zoology,[14] was in sharp contrast to the Lamarckian approach adopted by Grant. The closeness of Lamarck and Grant on this issue can be discerned by the comparison of the ordering of groups as they were to be dealt with in Grant's Comparative Anatomy lectures at University College, London in 1828 (Grant UCL MS 1179.13), with the ordering found in Lamarck's magnum opus on the invertebrates, the *Histoire naturelle des animaux sans vertèbres*:[15]

Grant, 1828	*Lamarck, 1815–1822*
Mammalia	Infusoires
Aves	Polypes
Reptilia	Radiares
Fishes	Tuniciers
Mollusca	Vers
Conchifera	Insectes
Cirrhipeda	Arachnides
Annelida	Crustacés
Crustacea	Annelides
Arachnida	Cirrhipèdes
Insecta	Mollusques
Vermes	[Poissons]
Radiata	[Reptiles]
Zoophyta	[Oiseaux]
Infusoria	[Mammifères]

Except for the reversal of ordering, the subdivision of the molluscs, and the terminological difference in the designation of the "zoophytes", a group name Lamarck refused to admit for reasons I will examine shortly, the general correspondence is exact; and Grant's reversal of ordering was a concession to practice rather than an indication of theoretical divergence. With Lamarck, Grant accepted even the more controversial historical thesis in the Lamarckian ordering — the claim that a natural ordering from simple to complex also represented the historical order of appearance of life on earth.[16]

More important for formation of the core of Grant's invertebrate enterprise was a theoretical issue that for him, as for Lamarck, was the theoretical justification for giving great emphasis to invertebrates. Representing a tradition articulated proximately by Lamarck and John Hunter, with roots as far back as William Harvey, Grant's theoretical concern with the invertebrates was with their power as *analytic models* for the solution of the major questions of form and function. For those who approached invertebrate zoology with this analytic intent, the invertebrates did not represent "lower" or "degraded" forms, as they tended to remain even for Cuvier and Jameson, but presented instead simplifying paradigmatic models for understanding the complicated biological processes in higher forms.[17] Through a comparative study of forms analyzed by models supplied by the invertebrates, Grant felt that a law-like understanding of structure and function generally was to be attained. As he expressed this in his opening lecture on comparative anatomy at University College, London only a year after ending his association with Darwin:

> We compare the organs of the inferior animals with the similar organs of man, to determine the extent of their deviation, and by watching the result of this change of structure, important light is thrown on the functions of the various parts. (Grant 1828b, p. 4)

Grant's concern with the "analytic" power of invertebrate zoology went even deeper, however, and this led him beyond Lamarck. More in keeping with the approach of Etienne Geoffroy Saint-Hilaire, C. G. Carus, J. F. Meckel, and Etienne Serres, Grant also saw this analytic analysis to be assisted by the study of the embryological development of the invertebrates. By this means one was provided with a key to understanding the development of higher vertebrates in the "recapitulationist" tradition of the 1820s. Again expressing this in the 1828 London lecture:

> And by comparing the adult organs of the inferior classes, with the embryo state of the same organs in the higher orders of animals, many extraordinary analogies are discovered, which throw much light on the functions of the parts, and serve to unravel the most complicated and difficult forms of organization. (1828b, p. 3)

Grant was offering these remarks in 1828 as conclusions built on the foundation of extensive work on the invertebrate organisms carried out in Edinburgh in the 1820s. Exploration of these earlier Grantian investigations at Edinburgh also reveals that the point of these inquiries in the 1820s was to determine by means of analytic insights made possible by invertebrate zoology some kind of underlying "law" governing relationships in these forms, a law that he had hoped to uncover in live studies on the life histories and reproduction of live colonial invertebrates. It is in this set of researches that we find Darwin's own beginnings as a public scientist.

THE ZOOPHYTE PROBLEM

The ingredients of Grant's program I have specified to this point do not form any necessary break with Lamarck. In some respects they might be seen only as elaboration of Lamarckian insights, with the addition of embryological concerns that Lamarck did not articulate. On a second important issue, however — the nature of the relations of the plant and animal kingdoms — Grant and Lamarck sharply diverged. In this respect Grant followed out much more closely the lines of inquiry and reflection expressed particularly by German authors whose work Grant had apparently become closely familiar with during studies in Germany.

Frederick Tiedemann, Professor of Anatomy and Physiology at the University of Heidelberg, whose work on comparative anatomy is frequently referred to in Grant's manuscript lectures in the 1830s, clearly expressed this alternative position. Tiedemann's claim was that there existed a common unifying point of the plant and animal kingdoms achieved through true "zoophytal" forms:

> Organic bodies, divided into two great sections, the vegetable and animal kingdoms, do not come into contact at their boundaries, so that the plants most complicated in structure verge upon the most simply constructed animals, and form the passage from one kingdom to the other. . . . On the contrary, the most simple vegetables, the cryptogamia, particularly the algae, the ulvae, the tremellae, &c., and the most simple animals, the zoophyta, the infusoria, and polypi, approach the nearest of all. . . . The two kingdoms approach so near to each other in their most simple forms, that there are some of these, regarding which it has not been determined, at least hitherto, whether they are plants or animals . . . . One might even be almost tempted to believe that, in certain circumstances, the most simple vegetable and animal forms may pass from one to the other. Confervae are resolved into infusoria, and infusoria produce confervae by their union. (Tiedemann 1834, p. 3)[18]

Lamarck, on the contrary, had consistently rejected such a claim, and had considered the whole concept of a linking "zoophyte" as fallacious. Lamarck's explicit position was that organisms fell clearly into two fully independent series of forms — plants and animals — and that there was no fundamental linkage point possible between them for a variety of theoretical and empirical reasons. A section of the *Philosophie zoologique* of 1809 makes this point quite explicit (pp. 51–54; 195–200). Appealing to the traditional Hallerian functions, he considered animals to possess the two fundamental faculties of sensibility and irritability, both of which he denied to the plants, even to the famous "sensitive" plant *Mimosa* (p. 196). Furthermore, the "life" of plants was viewed by Lamarck as concentrated at a single spot, the root collar, while that of animals must be considered diffused throughout their substance (p. 197). Finally, Lamarck raised what was to be a fundamental objection in the nineteenth century. Chemical analyses of animal and plant materials suggested that the two were made of chemically dissimilar compounds, and as a result there was no form that could combine both of these materials in the way demanded of the "zoophytes".[19]

As a consequence, Lamarck was willing to admit obvious analogies between animals and plants, particularly evident when one compares a colonial invertebrate such as *Gorgonia* to a higher plant. But in direct contrast to the position articulated by German naturalists like Tiedemann, these could be no more than superficial and ultimately misleading disanalogies. As he put this in an important discussion in the introductory essay to the *Histoire naturelle des animaux sans vertèbres:*

> In the same way that nature has formed compound plants, it also forms compound animals, and in this, it has not in the least changed either the animal or plant nature. In considering the compound animals, it would be entirely absurd to say that they are *animal-plants*, just as it would be to say, in the case of compound plants, that they are plant-animals. (Lamarck 1835–1845, 1: 72)[20]

Because there could be no such thing as a genuine "zoophyte", Lamarck overtly rejected the claim that the two kingdoms could be united in some form that could be a common branching point for the origin of the two series:

> It is certain that if the plants could be connected and gradated with the animals by some point on their series, it would be solely by those which are the most imperfect. . . . If there was a gradation at this point, one could not be prevented from acknowledging that in place of forming a single chain, the plants and animals present two distinct branches, united at their base like the two branches of the letter "V". (pp. 75–76)

This possibility was raised, however, only to be firmly rejected. Chemical differences between plant and animal matter assured Lamarck that no such bridging was possible:

> But I am going to make it clear that there is no point of gradation at the point referred to. Each of the branches which I am going to discuss are found in reality separated from the other at its base, and there is a positive character which pertains to the chemical nature of the bodies on which nature has worked, which furnishes a prominent distinction between the creatures embraced by one of these branches, and those which belong to the other. (p. 76)

Grant's break with Lamarck on this issue, and his pursuit of the German alternative, emerged gradually in the 1820s, and was supported, in his view, by strong empirical evidence derived from studies on the structure and reproduction of the colonial invertebrates.

As one of the original members of the Plinian Society, Grant entered this association with a background of several periods of study between 1815 and 1824 with leading zoologists and comparative anatomists on the continent. In 1825 Grant opened his publishing career in marine invertebrate zoology with studies on the reproduction in the sponges, in which he had shown them to reproduce by microscopic "gemmules" (Grant 1825–1826). Several studies on the colonial invertebrates, which he insisted on designating by the group name of "zoophytes", followed in the next three years.

Grant's concern with the importance of the embryonic stages in these forms was quickly to have taxonomic implications. Reporting on studies on the structure and reproduction of the boring sponge *Cliona celata* in April 1826, Grant found this form to constitute a "distinct genus, forming a connecting link between the *Alcyonium* and the *Sponge*. . . ." (1826a, pp. 78–81).[21]

Near the same date Grant delivered publicly a paper of greater theoretical importance that provided the particular context for Darwin's subsequent work with Grant. Following on his observations on the sponges, in which he had noted the presence of motile, free-swimming reproductive gemmules, Grant concluded in May 1826 that this was a wide-spread phenomenon in the zoophytes, possibly revealing a law-like model of zoophyte reproduction generally. Examining forms from several diverse and taxonomically distinct groups, Grant reported finding in each the same common microscopic, free-swimming and ciliated stage. His conclusions from this study are remarkable:

> Polypi, therefore, are not the first formed parts of this zoophyte [*Plumularia*], but are organs which appear long after the formation of the root and stem, as the leaves and flowers of a plant.
>
> From these observations it appears that the so-named *ova* of many zoophytes, when newly detached from the parent, have the power of buoying themselves up in the water, by the rapid motions of ciliae placed on their surface, till they are carried by the waves, or by their own spontaneous efforts, to a place favourable for their growth, where they fix their body in the particular position best suited for the future development of its parts. How far this law is general with zoophytes, must be determined by future observations. (1826b, p. 156)[22]

II. Darwin and Grant: Emerging Tensions

With Darwin having already been engaged in the study of live marine invertebrates even before meeting Grant, it is not surprising that on his initiation into the Plinian Society six months after Grant's report of these researches, Darwin would take up these issues again. In the Plinian as a member of the governing council, Darwin was in regular weekly contact with Grant, and the impact of these contacts is directly reflected in Darwin's exploration of exactly the issues of interest to Grant in his own invertebrate studies. These can be followed out in Darwin's "Zoology Notebook", opened on 16 March 1827 (DAR 118).[23]

Notable about essentially all of the entries in the Edinburgh portion of this Notebook is that they were inquiries into the early reproductive stages of the invertebrates, as if to test the generality of Grant's suspected "law". On specimens of *Purpura, Doris,* various "univalves", the bryozoan *Flustra foliacea,* in egg masses collected on marine algae, and on the deep-water sea pens *Pennatula* and *Virgularia,* Darwin described researches inquiring into the presence of microscopic ciliated "ova" capable of "self-motion" in the life histories of these forms. His novel extension of these findings to the highly anomalous group of the colonial invertebrates — the bryozoans falling in the genus *Flustra* — provided the occasion of his first scientific paper, delivered to the Society in March 1827.

The immediate background to this paper is provided by Grant's own paper on the *Flustra* presented to the *Wernerian* Society on 24 March 1827, only three days before Darwin read his own contribution to the Plinian Society (1827b). Although Grant did not refer to Darwin in his paper, it is transparent that Grant incorporated into his contribution the results of Darwin's researches.[24]

Grant's concern with the "Flustrae" was a result of the anomalous position these forms seemed to occupy with respect to all other colonial invertebrates he had studied previously. These complex creatures — currently placed in a distinct phylum in three main classes, the Phylactolaemata, the Gymnolaemata, and the fossil Stenolaemata — could be collected both intertidally at the Firth, and also in abundance from the deep-water oyster trawls.

As their more common British name "Polyzoa" suggests, these lowly forms seem to be formed, unlike other colonial invertebrates, of virtually autonomous polyps, seemingly unconnected with the other Polyps on the same or adjoining branches. Stimulation of one polyp does not cause reactions of others, and the polyps are able to carry out their own autonomous reproduction almost as if independent creatures.[25] Furthermore, as Grant reported in his observations, the complex life history of these forms involves a highly unusual process in which a senescent polyp, after the release of small ova into the water, then degenerates into a small mass within the capsule, the so-called brown body, which forms into a small oval body that then gives rise to a completely new polyp within the same capsule. Noting the "incomprehensible laws which regulate the formation and growth of the ova, and the whole economy of the zoophyte", Grant had summarized his studies as revealing that "the whole of the old cells are thus never found entirely deserted, [and] the same cells may repeatedly produce ova and polypi, and the whole zoophyte retain its energy for several seasons" (1827b, part II:342).

Darwin's contribution to this study, at least according to his if not Grant's account, was the first discovery of the presence of the ova growing within the capsules on a conveniently studied transparent deep-water form, *Flustra carbasea*, in which he found that these were possessed, like the motile ova observed by Grant on the more common zoophytes with a penumbra of rapidly moving cilia by which they were able to swim when released:

> Having procured some specimens of the / Flustra Carbacea (Lam:) from the dredge / boats at Newhaven; I soon perceived / without the aid of a microscope / small yellow bodies studded in / different directions on it. — They were / of an oval shape & of the colour / of the yolk of an egg, each occupying / one cell. Whilst in their cells / I could perceive no motion; but / when left at rest in a watch / glass, or shaken they glided to & / fro with so rapid a motion, as / at some distance to be distinctly / visible to the naked eye [. . . .] That such / ova had organs of motion does / not appear to have been hitherto / observed either by Lamarck [,] Cuvier [,] / Lamouroux or any other author: — / This fact although at first it/ may appear of little importance/yet by adducing one more to / the already numerous examples / will tend to generalize the / law that the ova of all Zoophytes / enjoy spontaneous motion. (DAR 118: 5–6)[26]

Simply in itself, this would seem to constitute only a minor student contribution to Grant's own research. Darwin evidently saw it as more than this: he became more than routinely fascinated with this issue in the Spring of 1827, to the point that an apparent sense of rivalry began to develop between Darwin and Grant, which may have been responsible for the permanent chilling of the relationship between the two men.[27]

More theoretically interesting is the fact that Grant was also drawing from these researches on the colonial invertebrates conclusions that Darwin would — surprisingly — not follow until 1837. For Grant, all of this research on the reproduction of the invertebrates was part of a larger program to determine a single set of laws governing both the organic and inorganic world. Already in 1826 Grant had stated this point with some clarity in referring to the sponges:

> It is interesting to observe, that the earthy matter of the skeleton of these earliest inhabitants of the ocean, is the same with what we know to have paved the bottom of the vast abyss at the remotest periods we can reach of the earth's history, whether we imagine the silica of the primitive rocks formed by the oxidation of the solid surface, or precipitated from the superincumbent fluid. The appearance of many of their crystalline silicious pointed apicula is the same with that of the slender hexaedral acumenated [sic] prisms which silica naturally assumes in the crystallized state; and the silicious crystals formed by nature contain cavities and fluids like those formed by organic life. The laws, therefore, which regulate

> the forms of the simplest silicious spicula composing the skeleton of the marine sponge, do not appear to differ much from those which regulate the forms of brute matter. (1826c, part 1: 351)

After March 1827, this suggested to Grant a much more ambitious enterprise. In a paper published after his move to London in 1828, and reporting on research carried out in October 1827, Grant extended his generational studies to the subtidal colonial hydroid, *Lobularia (=Alcyonium) digitata*, commonly known as "Dead-men's fingers". Noting that these forms, like all other zoophytes he had now directly or indirectly studied, reproduce by small, ciliated ova, Grant now drew the startling conclusion that all of these forms reproduced by a body directly analogous to the infusoria, and that these same infusorial "monads" formed the elementary units, not only of the zoophytes, but of all organic tissues. His German and — to a degree — French sources for these claims are evident:

> The clusters of ova found in autumn at the base of the polypi of the Lobularia, have no relation to the ovaria of higher animals. They are true gemmules or buds which grow from the sides of the internal canals. . . . Their spontaneous motion establishes the existence of this remarkable property in a tribe of zoophytes with a fleshy axis, where it had not before been observed, and opens to our contemplation a new and singular arrangement for aiding and directing the passage of these delicate reproductive globules, through the complicated bodies of animals where irritability is nearly extinct. . . . The transformation of the ova above described, from their moving, irritable, and free condition of animalcules, to that of fixed and almost inert zoophytes, exhibits a new metamorphosis in the animal kingdom, not less remarkable than that of many reptiles from their first acquatic condition, or that of insects from their larva state. Ulvae and confervae [i.e. marine algae] have been seen to resolve themselves into animalcules, (*Schweigger's Beobacht. auf N.R. s.* 90.), and Professor Aghard has seen these animalcules reunite to construct the plants. Mosses and Equiseta are found to originate from confervae, (*Mém. du Mus.* tom. ix. p. 283), and all the land confervae with radicles appear to pass into the state of more perfect plants. The Oscillatoriae which cover the stones in our fresh water pools with a green and velvety crust, resolve themselves into animalcules and lively moving filaments, whose motions have been described by Saussure, Vaucher, and others. The globules of our blood have been seen to arrange themselves into fibres, (*Phil. Trans.* 1818. p. 172), and the densest fibres have been resolved into their regular component globules. (Grant 1828a, pp. 109–110)[28]

The implication of this claim was directly in opposition to the conclusions drawn by Lamarck. In terms of this universal monadism, Grant by 1828 was arguing that plants and animals were united by means of common monadic bodies, analogous to the animalcules in infusions, and that these were also direct analogues to the units seen as the motile ova in the life history of the zoophytes. As he stated in his inaugural lecture at University College, London on 23 October 1828:

> From numerous experiments, Naturalists have been led to believe that the simplest organized bodies, as *Monads* and *Globulinae*, originate spontaneously from matter in a fluid state, and that these simple bodies, of spontaneous origin, are the same with the gelatinous globules which compose the soft parts of Animals and Plants. Many of the phenomena of Plants . . . , are dependant [sic] on the common laws of inorganic matter; and in the Animal Kingdom the same laws operate in the formation of the silicious cristals [sic], which compose the skeleton of many Zoophytes, and the calcareous crystals of many Radiated Animals. . . . (1828b, p. 18)

While admitting that there were also many differences between plants and animals, Grant then pressed this point to conclude that as we trace their origins down to the lowest infusoria and simple filamentous algae, the two kingdoms did indeed unite:

> The Animal and Vegetable Kingdoms are so intimately blended at their origins, that Naturalists are at present divided in opinion as to the kingdom to which many well-known substances belong — as the *Codium tomentosum, Alcyonium bursa,* the *Corallina officinalis, rubra,* and *opuntia, Dichotomariae, Tremellae, Globulinae,* &c. Several organized bodies, as *Oscillatoriae, Confervae,* and *Monades*, which have neither roots, nor capillary vessels, nor a digestive stomach, nor other distinct organs of plants or animals, connect the Vegetable and Animal Kingdoms by imperceptible gradations. (1828b, p. 20)

Grant's ambitious speculations, connecting him more closely with the tradition of Treviranus, Tiedemann, and Carus than with Lamarck, rendered the "zoophytes" for him true "plant-animals", since these were but compounds, in his view, of the elementary monads. Besides endorsing the view, at least by 1828, that the infusoria were also the first forms to arise on earth, Grant was working toward a larger biological synthesis in the late 1820s that Darwin was surely aware of, at least in part.

In the *Autobiography*, Darwin commented on Grant having spoken one day "in high admiration of Lamarck and his views on evolution", which sixty years later he could recall, somewhat paradoxically, as having made no impression upon him (p. 49). I would suggest that Grant's version of Lamarck had gone considerably beyond Lamarck on some issues, and for good reason was considered untenable on theoretical grounds by Darwin when he heard Grant on this issue in the Spring of 1827. This suggestion requires some comment on Darwin's studies prior to his contact with Grant.

Before he established regular contacts with Grant, Darwin had already completed an intensive 140-lecture course on chemistry under Thomas C. Hope in the 1825–1826 school term; and Darwin was well prepared, even at age seventeen, to appreciate the finer points in this series of lectures, having already been indirectly initiated into university-level chemistry even before arriving in Edinburgh, through his brother Erasmus.[29] Continuing these joint interests, the two brothers had immediately enrolled upon arrival in Thomas C. Hope's chemistry lecture course, reputed to be one of the finest general chemistry lecture courses in Great Britain in the 1820s.[30]

Darwin's enthusiasm for this course midway through the lectures is expressed in a letter home, describing how he "liked both [Dr. Hope] and his lectures very much" (C. Darwin to Caroline Darwin, 6 January 1826, DAR 154). And it was his enthusiasm to attend Hope's concluding lectures on electricity and galvanism that was responsible for his decision to remain in Edinburgh after Erasmus's departure in March (C. Darwin to Caroline Darwin, 8 April 1826, DAR 154).[31] The few pages of notes on these lectures surviving in Darwin's hand attest to both the rigor of the course and Darwin's comprehension of the material.[32]

At least one issue in these lectures is directly relevant to the questions raised by Grant — the possibility of linkage forms between plants and animals. In at least one lecture, Hope had treated this issue, and his conclusions are of some direct interest:

> Almost all animal matters have certain / characters in common, as we have / seen to hold true of veg^{tbl}[.] / It proceeds from the same cause, namely, / that they all consist of the same chemical / elements, which are few in number; & that / the diversity among them arises from a difference / in the proportions of these elements, or in the / manner, the atoms of these Elements are associated / & grouped together — / Three of these are the ordinary constituents of / Veg^{ble} matter & to them Nitrogene [sic] is added / in the Animal Constitution —
>
> It is the addition of this ingredient wc / proves the source of the / peculiar chemical / character of animal substance & the / difference from Veg^{ble} (T. C. Hope MSS, UEL, Gen. 268, Box 1, Item 14.)[33]

We can, however, contrast this with Grant's ambitious claims in 1828, which led him to the conclusion that there was no fundamental distinction to be drawn on chemical grounds between plants and animals, and that all organisms consequently could be understood in terms of the same set of fundamental laws:

> In thus investigating the various functions of the animal economy, in all their different stages of simplicity and complication, [comparative physiology] determines the true nature of animal life, and discovers the characteristic properties which distinguish animals from the beings belong-

> ing to the Vegetable and Mineral Kingdoms. And by the successful applications of the principles of chemical and mechanical science to the explanation of their complicated functions, it shows that, not withstanding the disturbing forces of the animal economy, which have hitherto defied all attempts at generalization, the true solution of all vital phenomena and the laws of organized beings are to be looked for in those magnificent arrangements which embrace the whole system of the visible universe. (Grant 1828b, p. 5)

If we can, therefore, speak of a "Grantian" research program in the 1820s, it included ingredients that Darwin could and, I suggest, did embrace — interest in the "analytic" function of invertebrate zoology; concern with embryological stages as indicative of taxonomic relationship; direct interest in the "zoophytes" as possible keys to more complicated issues in biology. At the same time Darwin's subsequent career demonstrates that he was in a much more ambiguous position in the late 1820s and early 1830s concerning the rest of the "core" of Grant's enterprise — the transformism, the search for a unified single set of laws uniting the organic and inorganic domains, the thesis of universal monadism — which, surprisingly, would be positions to which Darwin would finally return in 1837.

On one issue of relevance, the status of the "zoophytes", Darwin's divergence from Grant would have been reinforced — but on less theoretically firm grounds than provided by Hope's chemical lectures — by J. S. Henslow's probable discussions of the topic in his descriptive and physiological botany course, which Darwin faithfully attended three times during his stay at Cambridge (Henslow MSS, UCL O. XIV, p. 261).

Unfortunately, neither Darwin's own notes nor Henslow's lecture notes from the 1829–1831 years seem to have survived. Consequently, any conclusions drawn must depend at this point on the probability that Henslow's 1836 textbook, *Descriptive and Physiological Botany*, represents the approximate content of his lectures in the period Darwin was a student in his course. There is at least strong warrant for making these assumptions.[34]

Henslow characteristically addressed in the first lecture of his course the question of the relation of animals and plants. As he expressed this in the text of 1836, we read:

> Among the higher tribes of organized bodies, indeed, there is no difficulty in pointing out numerous lines of demarcation between the two kingdoms; but, as we descend in the scale of each, we find an increasing similarity in external characters, and a closer approximation between the analogies existing in many of those functions which mark the presence of the living principle, both in the animal and in the vegetable kingdoms. Perhaps, until the contrary shall be distinctly proved, we may consider the superaddition of "sensibility" to the living principle as the characteristic

> property of animals. . . . But the most constant, if not universal, distinction, — and one which we can readily appreciate, between animals and vegetables, — consists in the presence or absence of those internal sacs or stomachs, with which the former alone are provided. . . . (Henslow 1836, pp. 7–8)

At the same time, Henslow made the issue of the plant-animal relationship a problem to be investigated, rather than one to be resolved solely by chemical dogma. This tentativeness, which I suggest is the attitude with which Darwin resumed his investigations into these problems aboard the *Beagle*, is well expressed in what seems to be the opening lecture of the "physiological" portion of Henslow's course:

> In fact the general phenomena of life and death, are scarcely less striking in the vegetable than in the animal kingdom; and probably the vital principle, considered apart from sensibility, is something of the same kind, if not the very same thing, both in animals and vegetables. This similarity or unity in essence must led us to expect, what experience has shown to be the fact, that a considerable analogy exists between the functions of animal and vegetable life. Although every argument which may be derived from this analogy, cannot be too severely scrutinised before we admit the particular conclusion which it may seem to establish, yet we may confidently reckon upon the certainty of its existence, as one of the best guides which we now possess, towards obtaining a more perfect elucidation of the general laws of physiology. (Henslow 1836, p. 156)

III. The *Beagle* Zoology Researches

The previous discussion has given us a general background for a more intensive discussion of the theoretical developments we may now follow in Darwin's biological thought during the 1831–1836 period. Since Darwin's biological studies in these years were primarily on marine invertebrates, an analysis of his work on this specific group can build directly on the Edinburgh background, and can give us a clear picture of the continuity and theoretical development of these interests. Awareness of this background should also enable us to see the Darwin who embarked on the *Beagle* as much more deeply prepared for this work than the popular image of the gentleman dilettante beetle collector with a copy of Paley under his arm would lead us to expect. We shall appreciate this point of departure by considering briefly Darwin's preparations for the biological work aboard the *Beagle*.

To prepare himself for his biological investigations on the voyage, Darwin consulted his old Plinian friend, John Coldstream, for details on chain-dredge construction and more detailed information on the collection of deep-water invertebrates (Darwin to John Coldstream, 13 September 1831, DAR 204).

Furthermore, he sought the advice of several London experts. For his entomological work, he consulted his friend and entomological authority, the Reverend F. W. Hope, the leader among the London entomologists. The elusive Charles Stokes, interested in conchology and corals, gave him advice on preparations on these groups. Robert Grant, whom Darwin had been advised to contact by John Coldstream, seems to have been the main source of Darwin's information for details on the preservation of both invertebrate and vertebrate animals. Robert Brown had itemized unusual plants Darwin should collect from Tierra del Fuego, and evidently given him substantial instructions on microscopy and microscope design for the voyage, an item I shall discuss below.[35]

Although such instructions alone do not indicate tasks more advanced than those required of a shipboard collector, Darwin's prior preparation in the theoretical questions I have outlined made his work from the beginning much more interesting than the "collector" role would imply. This point is borne out in the highly differential nature of his biological researches aboard the *Beagle*. In that area in which he had already been actively working before 1831 — invertebrate zoology — Darwin's work was deep and specific, and went beyond taxonomy and description to functional and theoretical levels. In areas in which he had been trained and interested for some time, such as ornithology and mammalogy, both of which seemed to form the focus of Robert Jameson's zoology section of his natural history course at Edinburgh (which Darwin attended), Darwin showed detailed and specific interest in classification, distribution and gross-description, if not in functional questions. By contrast, in those areas where he seems to have had only the slightest preparation, such as ichthyology, Darwin rarely acted as more than a mere collector, only occasionally attempting even to place specimens in their proper orders or families, let alone concern himself with more specific identification, distribution, or variation.

Darwin's interest in the invertebrates was also not simply fortuitous, a matter of availability of material. A letter to his cousin and university friend, W. D. Fox in May 1833 makes this quite clear:

> You ask me about Ornithology; my labours in it are very / simple. — I have taught, my servant to shoot & skin birds, / & give him money [. . . .] I collect reptiles, small quadrupeds, & fishes / industriously, especially the first: The invertebrate marine / animals are, however my delight; amongst them I have / examined some, almost disagreeably new; for I can find // no analogy between them & any described families. — / Amongst the Crustacea I have taken many new & curious genera: / The pleasure of working with the microscope ranks second / to geology. — I strongly advice [sic] you to by [sic] from / Bancks in Bond St. a simple microscope such as tht Mr. / Browne recommends. — & then make out insects scientifically / by which I mean separate, examine &

> describe the trophi: it is very easy & exceedingly interesting; I speak from / experience, not in insects, but in most minute crustacea. (Darwin to W. D. Fox, 23 May 1833, Christ's College, Cambridge 46B)

These institutions can be quantified to some degree by a summary of the contents of the Zoology Diary (DAR 30.1, 30.2, 31.1, 31.2) through an analysis of the total lines of discussion devoted to various zoological and biological topics during the cruise. This diary, containing a total of 14, 145 lines of text and added notes, covers the dates of 6 January 1832 to approximately 1 August 1836, and represents a declining effort in zoology as the voyage proceeds, as pointed out by Gruber and Gruber (1962, p. 191).[36] Table 1 summarizes the average lines of discussion in the diary broken down by its archival divisions and presented as an average per-diem rate:

Table 1. Per-Diem Distribution of Zoology Diary Contents in Line Entry / Day

	Zoology Diary			
	DAR 30.1	DAR 30.2	DAR 31.1	DAR 31.2
Dates Covered	6 January–16 September 1832	17 September 1832–22 July 1833	23 July 1833–21 (?) December 1834	28 (?) December 1834–1 August 1836
Total Days	255	289	517	583
Total Lines	4156	3952	3679	2358
Per-Diem Distribution	16.3	13.7	7.2	4.0

At the same time, this should not blind us to two important features of this activity. First, even with this declining total effort — reflecting often no more than the character of the cruise after 1834, and the increasing time spent in inland expeditions — we clearly perceive the differential attention given the different groups of organisms. Table 2 presents a summary of the line count by diary broken down into major groups corresponding to Cuvier's *Embranchements*, a relevant grouping since Cuvier's *Règne animale* (Cuvier 1830) was evidently with Darwin on the cruise, and seems to have been frequently consulted on classification.

In total effort, we can perceive that the discussion of invertebrate groups strongly predominates over that of the vertebrate groups in this diary, constituting approximately 66 percent of the total volume of discussion. The main single vertebrate group forming an exception to this is, predictably, the birds, given a total of 1676 lines of discussion in the diaries, and constituting individually approximately 12 percent of the total discussion. The separate Ornithological Notes and the eleven specimen catalogs have not been

Table 2. Total Number of Zoology Diary Entry Lines by Major Group Arranged According to Cuvier's Embranchements

	Zoology Diary				
Group	DAR 30.1	DAR 30.2	DAR 31.1	DAR 31.2	% TOTAL
Radiata (Zoophyta)	1660	1019	1916	1307	
% Text	39.9	25.8	52.4	55.4	41.7
Articulata	569	1015	90	189	
% Text	13.7	25.7	2.4	8.0	13.2
Mollusca	1159	111	105	124	
% Text	27.9	2.8	2.9	5.3	10.6
Vertebrata	257	1127	1213	605	
% Text	6.2	28.5	33.0	25.7	22.6
Botany	109	421	182	133	
% Text	3.3	10.7	4.6	5.6	6.2
Meteorology, Geology	372	259	173	0	
% Text	9.0	6.6	4.7	0	5.7
Line Totals	4156	3952	3679	2358	100

analyzed, since these have been dated from April to July 1836 (Sulloway 1982).

Within the invertebrate groups themselves, a further interesting discrimination is made possible by these data. Table 3 gives a further breakdown of the invertebrate discussion into separate invertebrate groups. To coincide as closely as possible with Darwin's own conception of these

Table 3. Total Lines of Invertebrate Zoological Entries According to Lamarck's Arrangement

	Zoology Diary				
Group	DAR 30.1	DAR 30.2	DAR 31.1	DAR 31.2	% TOTAL
Infusoria	123	13	73	288	5.4
Polyps ("Zoophytes")	440	869	1389	494	34.4
Radiarians	471	59	236	245	10.9
Tunicata	267	0	0	0	2.9
Vermes	406	51	218	280	10.3
Insecta	341	220	80	39	7.3
Arachnida	0	134	0	0	1.4
Crustacea	228	538	10	40	8.8
Cirrhipedia	0	123	0	110	2.5
Conchifera	154	0	11	0	1.8
Mollusca	1005	111	94	124	14.4
Totals	3435	2118	2111	1620	100

groupings on the voyage, these have been analyzed according to the groups found in Lamarck's *Histoire naturelle des animaux sans vertèbres* (Lamarck 1815–1822), the main reference work on invertebrate zoology Darwin apparently had with him on the cruise. Darwin's "Zoophytes", like Robert Grant's, corresponds in practice to Lamarck's *Polypes* rather than to Cuvier's more comprehensive *Zoophyta* (=*Radiata*).

From this analysis we see the strong preponderance of work on the "zoophytal" forms, precisely the group that had interested Darwin in his Edinburgh studies. And although the content of the DAR 31.2 discussion is almost totally devoted to one group of these forms, the corals, one should not view this as a line of interest separable in principle from the wider interest in this total zoological group, as we shall see subsequently.

A finer discrimination of Darwin's work on the "zoophytes" can be seen in Table 4. In this table, the line totals have been computed on a monthly basis, and correction has been made for notes added later to earlier diary entries, to the degree that these can be dated with some assurance. Because of the important interaction of Darwin's work on the coralline algae in conjunction with this work on the zoophytes, the line entries on this plant group, listed under "Botany" on Table 2, are also shown.

Table 4. Monthly Line Entries in Zoology Diaries Devoted to Groups Falling in Lamarck's Polypes and Coralline Algae, January 1832–August 1836.

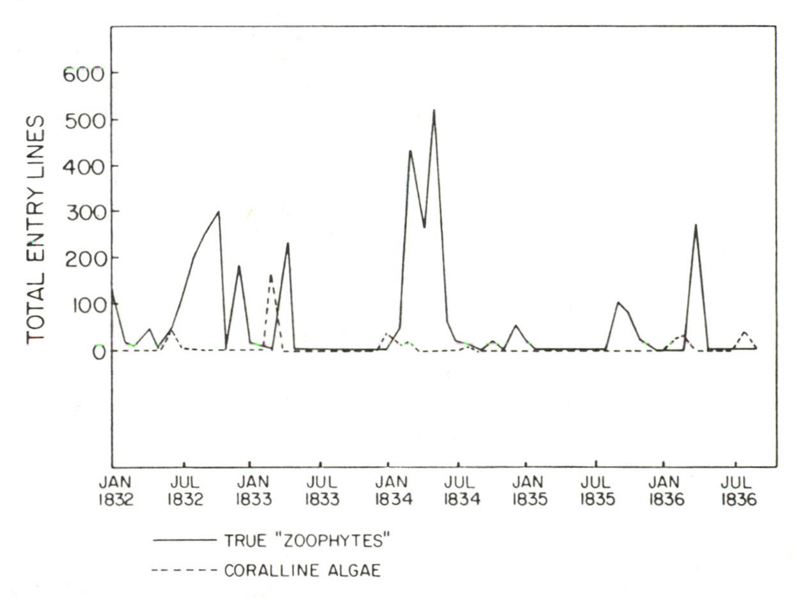

This table reveals that Darwin's work on this group, while often dormant during periods when specimens were unavailable or when Darwin was occupied in inland geology, nevertheless breaks forth at periodic intervals to a degree that would not support a thesis of a loss of zoological interests. Instead, it would appear that Darwin's biological interests, at least in certain groups, were continual, discriminatory, and suggestive of a recurrent set of problems that he was possibly working upon. Analysis of the more specific content of these discussions will disclose this.

THE ATOMS OF LIFE

One of the striking features of the content of Darwin's entries on the lower invertebrate groups in the *Diary* is a continual concern with what he termed "granulated matter". His concern with this issue was, on one level, a direct consequence of a new set of observational issues that were made strikingly evident to Darwin through his use of microscopy on live planktonic and sessile invertebrates during the cruise.

As the letter to W. D. Fox quoted previously documents, Darwin, in preparing for the cruise, had been in direct contact with Robert Brown on the design of the most suitable microscope to use on the voyage. In seeking Brown's advice on this matter and in following his recommendation on the acquisition of a simple microscope from the Banks firm of London, Darwin had in fact managed to take with him on the cruise probably the finest quality microscope available prior to the advent of the achromatic compound microscopes in the 1830s. In company with all the greatest microscopists of the period, Brown had carried out his own work with the simple microscope, since only this possessed the all-important resolving power needed for the finest microscopic studies at high magnifications. Consequently, it is highly significant that Brown had recommended that Darwin obtain a microscope very similar in design to that with which he had first been able to resolve "Brownian"motion in 1827.[37]

The importance of the use by Darwin of a microscope of this design and quality during the *Beagle* voyage was that when the new observational qualities of this instrument were combined with Darwin's pre-existent interests and knowledge of marine invertebrate forms, he was placed in a unique position to prosecute and extend the general questions we have seen him involved with under Robert Grant in his Edinburgh years. One could examine live specimens conveniently with this microscope at high powers by placing them in a small quantity of sea water in a concave watch glass fitted to the object stand. High resolution studies of internal structure by transmitted light could be made with ease even at $^{1}/_{20}$th inch focal lengths (200 X).

The repeated discussion in the Diary of "granulated" matter was a direct result of these new observational possibilities. In numerous forms, Darwin reports finding small, often motile particles, that from his observations would appear most often to be observations of the various microscopical structures generically termed "organelles".[38] Prior to the cell theory, however, which achieved its preliminary formulations by Schwann only in 1839, Darwin had none of the advantage of modern retrospect in determining the nature and function of these ubiquitous particles. His observations revealed them to be almost universally present in the tissue of small invertebrates and marine plants that he was able to study in a live condition, particularly localized in the reproductive structures. His frequent attention to these can be more clearly understood in the context of Robert Brown's theory of the "active molecules", which had been a subject of active debate in the years immediately preceding Darwin's departure on the voyage.

Brown's molecules, like Darwin's granules, were in the main the consequence of the use of a high resolution microscope — in both cases almost the same instrument. Brown had first found these in the interior of pollen grains in 1827. His observations suggested that in all organic materials minute, active particles could be discerned, moving in rapid, vibratory patterns. Quite at odds with the eventual account of these motions as due to random atomic collisions, Brown saw these motions as apparently inherent in the particles themselves. He concluded in his controversial 1828 paper that these were probably the *molécules organiques* of Buffon and Needham:

> Reflecting on all the facts with which I had now become acquainted, I was disposed to believe that the minute spherical particles or molecules of apparently uniform size, first seen in the advanced state of the pollen of Onagrariae, and most other Phaenogamous plants. . . , were in reality the supposed constituent or elementary molecules of organic bodies, first so considered by Buffon and Needham, then by Wrisberg with greater precision, soon after and still more particularly by [Otto] Müller, and very recently, by Dr. Milne Edwards, who has revived the doctrine, and supported it with much interesting detail. I now, therefore, expected to find these molecules in all organic bodies; and, accordingly, on examining the various animal and vegetable tissues, whether living or dead, they were always found to exist. . . . (1828, p. 363)

We know from a note appearing early in the Zoology Diary, by all appearances written about the same time as reports on the observations of microscopic granules in the substance of live invertebrates at Bahia Blanca in 1832, that Darwin was aware of the debate over Brown's "active molecules" by the date of his departure on the *Beagle*.[39] At the same time, there is little direct evidence to suggest that Darwin was setting out explicitly to test Robert Brown's claims on the universal presence of these "active molecules" in his microscopic studies on "granular matter". More plausibly, Brown's controversial claims had drawn his attention to this micro-particular structure he could now observe with clarity, and Brown's association of these with the famous "organic molecules" of Buffon had drawn Darwin's attention to their possible reproductive function. Furthermore, if these microscopic particles, rather than the macroscopic ova studied by Grant, were the true reproductive primordia, a new set of theoretical possibilities was opened up to the thoughtful observer.

Darwin's attention to this question is almost immediate with the opening of the Zoology Diary in early 1832. Initially he considered these particles to be allied with some kind of primitive circulatory scheme in the invertebrates, but he soon concluded that their main function was reproductive. This is described in a second series of observations on live specimens of a deep-water planktonic chaetognath first collected in large numbers in early 1832 whose systematic status was a continual problem to him during the voyage. In 1844 Darwin would place this in D'Orbigny's *Sagitta exaptera* (*CP* 1977, 1: 177–182). His entry in the Zoology Diary is as follows:

> I imagine that the ova (are first formed / in [the structure labeled] D & then pass on into?) F. where they are / perfected & then excluded or <burst forth> by the pap (n) / If (D) had no connection with ova. Why should / the quantity & size of small globules or grains / vary. — [. . . .] I watched one of / the ova after being removed from ovary. — (never taking my eyes from it). — the process / as described [. . . .] [c.o. illeg.] <went> on till. the ova appeared / made up of two equal balls. — they then separated; a capsule remaining; the other / composed of globular mass of pulpy-granular / matter. in which was ~~the~~ a small transparent ball . . . . (DAR 30.1: 74 verso)[40]

Of more direct concern is the involvement of this question in the resumption of Darwin's inquiries aboard the voyage into the relationships and reproduction of the colonial invertebrates. Showing direct continuity with the studies undertaken with Grant in Edinburgh, Darwin was now in a position to prosecute these inquiries much more deeply. In chaetognaths, planarian flatworms, and in several of the colonial "zoophytes", he had found this granular matter. The function of this material, and its relationship to organization and reproduction, were subjected to a sustained series of reflections and shipboard studies between February and July of 1834 (Table 4). Occupying 1208 lines of text, and accompanied by detailed microscopic drawings, Darwin analyzed nineteen distinct forms of colonial invertebrates in this period.

In a series of studies on a gymnolaematid bryozoan identified by Darwin simply as "Flustra (encrusting)",[41] Darwin explored the remarkable process of reproduction by "Brown-Body" formation, which Grant had commented upon in his own foundational studies on the bryozoa in 1826 considered previously (DAR 31.1: 223–224). Darwin's own drawing of the polypary on this form, appropriately labeled for our purposes, clarifies some of the observational issues in this discussion.

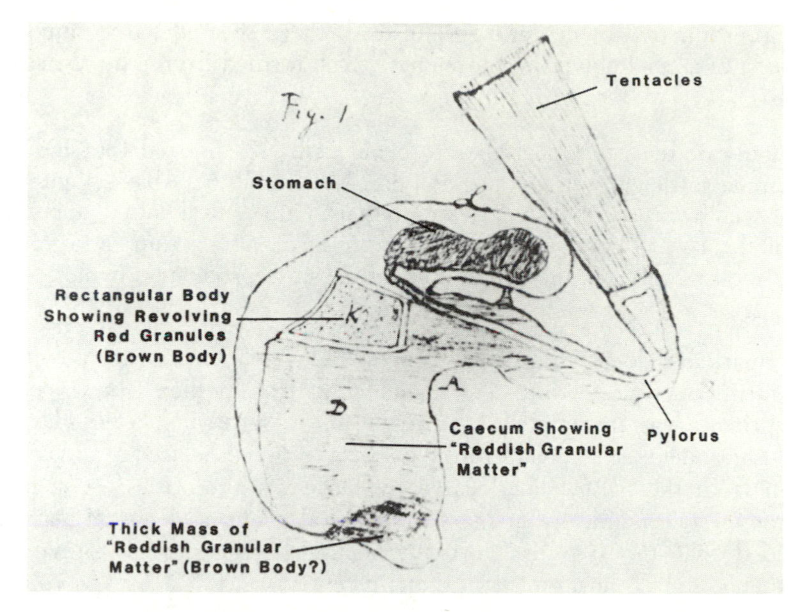

Figure 1. Darwin's drawing of the polypus of the encrusting Bryozoan, identified as Specimen #878, *Zoology Diary* DAR 29.3: 59, Plate IX, fig. 1. Labeled according to Darwin's description at DAR 31.1: 223. Used with permission of the Syndics of Cambridge University Library.

In viewing the conglomerate of "red matter" forming the "Brown body", Darwin could perceive in its interior "a rapid revolution of small red grains", which he concluded must somehow be involved in reproduction (DAR 31.1: 220). Furthermore, this same granular reddish matter was seen as forming a thin connection between the otherwise isolated polyparies, a connection that had escaped Grant in his studies. In an entry at Tierra del Fuego in March 1834, he wrote:

> There is another curious organ; In any row. the base / of one cell. is contracted & cylindrical & unite[s] itself / to the posterior one. beneath the mouth. — [. . . .] The connecting brackets appear / hollow, where two rows of cells diverge, in the / centre of an anterior bracket, a globular enlargemt [sic] / takes place, which afterward forms a cell. so as to/ / fill up the divergence. between the rows. —
>
> In the young & extreme cells, the arms of Polypus / do not reach half its length (Fig. 6). they are / enclosed in a bud, the neck of which is attached to anterior extremity of cell. — [. . . .] The youngest form of cell, is / globular mass with central spot or mark. — (DAR 31.1: 221–222)[42]

Continuing this set of entries in different ink, and presumably at a later time, he reported his lack of success in finding in more mature forms any actual reproductive ovules or eggs. Instead, there seemed to be simply a *replacement* of a dead polyp by a new one, which formed from the red matter at its base:

> In some of the central. & therefore old cells. I / noticed (but did not examine sufficiently. a / young Polypus — as at (F 6). ~~Above~~ <anterior to> which was a shrunken / dark red viscus. with central bulb: it appeared as if the ~~re~~ old Polypus had died (or produced an / ovum) & a young one took its place in the cell. / I could see no reproductive ovules. (DAR 31.1: 222)

This remarkable description of the process by which the mature polyp in these forms degenerates into a compact body from which a new polyp then springs, was interpreted by Darwin in a surprising way. Having previously studied the colonial hydroid *Obelia*, a form now placed in a separate phylum from the *Flustra*, but located by Darwin's authorities — such as Lamarck — taxonomically in the same general group as *Flustra*,[43] Darwin concluded that the two had virtually identical processes of formation. Continuing, after a short omission, the same entry:

> This Polypus. is closely allied to that of Obelia P[ages] 174(a), / there. the vessel. which comes from the base of / arms is elongated passed [sic] a red organ, bends, contains / a revolving mass & ends in a red-gut-shaped / mass. — there is no difference, excepting that in this one, the longitudinal <vessel> joins an oblique one / instead of passing by the Liver & then bending. (DAR 31.1: 222)

In subsequent investigation on a form of "encrusting" *Flustra*, Darwin again described the presence of a central red mass filled with revolving particles, and the same was found in a curious creeping form, propagating itself by horizontal rhizomal shoots that periodically gave rise to vertical vase-like capsules containing the polyp (DAR 31.1: 224).

This "granular matter", which seemed to be involved both in the "Brown Body" formation and also in the interconnection between the seemingly autonomous polyparies of the bryozoans, was to have larger ramifications as these studies progressed. In analyses of the internal structure and functions of the colonial form *Clytia*, presently placed in the Phylum Hydrozoa, and arranged even by Darwin's authorities such as Lamarck, at some taxonomic distance from the "Flustrae",[44] Darwin's studies convinced him that a similar connection of granular matter, only in a more prominent and obvious way, was uniting the polyparies in this form. In an entry dated simply March at East Falkland Island he writes:

> [The base of the polypary capsule] is traversed by central vessel. which / being surrounded by granular matter forms the / ~~cell~~ stem. — this granular matter can be forced / to circulate in ~~the~~ its case. — the <living>

> stem having passed through the two semi-globular / enlargements at base of cell, is much contracted, / & chiefly consists of the central vessel; it is then suddenly enlarged into cylinder almost / filling the cup: which ~~contains~~ <is filled by> granular matter / in which I twice perceived corpuscular motion. (DAR 31.1: 228)

Furthermore, the structures forming the polypus and its organs only seemed to be transformations of this primitive matter:

> We may imagine [structures] E & B to be enlargemts [sic] / of central vessel of stem & the tentacula, the / coat of granular matter in a different form. (DAR 31.1: 228)

This may be clarified by Darwin's accompanying drawing of *Clytia*, showing the central stem of granular matter branching in zig-zag fashion to form polyps:

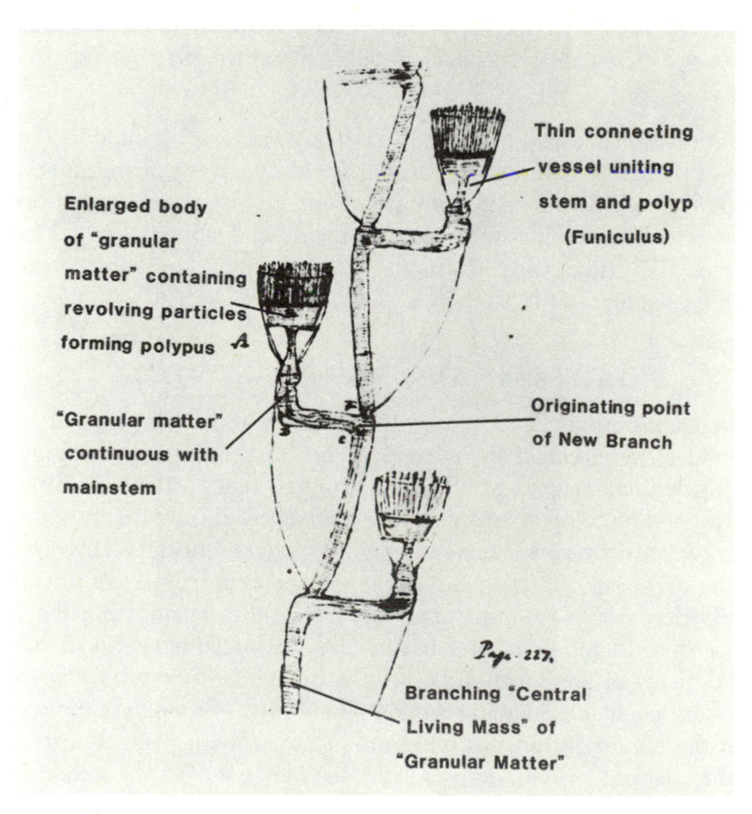

Figure 2. Darwin's drawing of the branching polyparies on bryozoan identified as genus *Clytia*, Specimen #894, *Zoology Diary* 29.3: 61, Plate XI. Labeled according to Darwin's description at DAR 31.1: 227–229. Used with permission of the Syndics of Cambridge University Library.

With what was apparently the same primordial granular matter serving as the basis for unification in the morphologically very different forms, and also as the apparent source of the structure of the polyps themselves, Darwin saw in this a strong reason to doubt distinctions made between these groups in the accepted taxonomies. The degree of polymorphism even possible in a single species, such as *Bugula*, which could have on the same form distinct polyps and also the curious "vulture-head" beaks, literally astonished him. Describing his observations in May 1834 off Santa Cruz on the Patagonian Coast on the bryozoan *Sertularia*, which showed the remarkable property of having two distinctly different polyps arise from the same capsule, he wrote:

> The central living stem (which I believe is / pulpy matter contained in a vessel) / is slightly zig-zig [sic] & comes in contact with the / base of each cell. — When first watching this Coralline, I was astonished at seeing, as / I then thought, 2 different <sorts of> polypi protruding themselves, not only from different cells. but / from the same: I presently saw two / distinct Polypi, each furnished with eight / arms, protrude themselves from a cell . . . . (DAR 31.1: 246)

These observations warranted, in Darwin's eyes, a substantial taxonomic revision of the zoophytes. Before turning to this, we must further investigate the degree to which the same considerations were weakening his original conclusions on the relationships of the plants and animals, which at the opening of the cruise were distinctly at odds with the claims of Robert Grant discussed previously.

THE ZOOPHYTE QUESTION REVISITED

The discussions in the 1820s and 1830s about the possibility of a true "zoophytal" creature had involved workers in invertebrate zoology in a search for defining criteria of plants and animals that particularly concerned the status of the colonial invertebrates and infusoria. Another taxonomic group drawn into this problem was the coralline algae, curious forms occurring primarily in warm waters, and often involved in coral-reef formation. Currently these are placed unambiguously among the plants, but they were a subject of substantial debate in the early nineteenth century.[45] On the one hand, many of these forms occurred in branching tufts with a calcareous skeleton, or even in spreading fungus-like forms, strongly reminiscent of some of the corals, and in this were unlike any known plant. On the other hand, they lacked evident polyps, and showed none of the other animal functions. Generally Darwin's authorities placed the "Corallina" (as distinguished from the animal "corallines") among the colonial animals, but there was clearly much uncertainty on this.[46]

On three occasions before the Summer of 1834, Darwin had made microscopic examinations on the "Corallina", both to study their mode of reproduction, and also to look for signs of their animality. Commenting in the Diary on the form *Amphiroa* in June 1832, Darwin had said he had been unable to find any signs of "irritability" or the presence of a polyp; in a microscopic study of new bud formation, he concluded that these forms, unlike the colonial "zoophytes", reproduced by an "alteration [rather] than continuation of central substance" (DAR 30.1: 56).[47] His general conclusion was that these reproduced by the production of projecting "paps" or "cones" from the surface, each terminated by a small orifice. Within these "cones" were then formed what he considered to be small "ovules", also termed "gemmules", which then protruded out the orifice and by a balloon-like swelling formed a new section. In studies on a different fungoid-shaped form at the Falkland Islands in March 1833, he again found the same "cone" formation, but on closer inspection found that these contained "granular" matter that seemed directly involved in the formation of the new shoot:

> If the cone / is removed in one of the early [generated] ones, the bottom is / concave & on it there is a layer of the pulpy cellular// tissue or granular matter, such as occurs at / the extremities of the branches. — this lies on / the white softer substance of the Corall. — so that / the stony layers are perforated. — At a later / age. the granular matter is collected into / semi-opake. spherical or oval bulbs, with a / transparent case: these are slightly coloured & / between 30 & 40 in number. — in diameter / 1/500th of inch. — They are ovules & the cones / ovaries. — The simplicity of this generative / process is shown by its ~~the~~ similarity to ordinary / growth. — the external border is thickened / composed of precisely a similar substance & enveloped / in a transparent membrane; it may be considered / as formed by a juxtaposition of cones, or / rather the cone & ovules owe their origin / to the creative power acting on a point. / where the growth of extension cannot take place; / hence the granular matter is enveloped in / a spherical case & seeks an exit through / the stony layers, instead of increasing laterally. (DAR 30.2: 161–162)

These investigations, on the one hand, served to reinforce Darwin's conclusion that the "Corallina" were definitely plants, rather than animals. The formation of "ovules" or "gemmules" within these projecting cones seemed decisively demonstrated in early 1834. In a note added possibly in May to a discussion in January 1834 of the articulated, branching coralline algae *Halimeda* he had found at Port Desire, Darwin found the "gemmules" to contain different articulations, and presumably each articulation generated a new branch. This was quite unlike the process he had found in the *Flustra*, where a single globular mass of material generated a single polyp:

> This observation appears to me of considerable / importance in settling

> the long disputed point whether / the genus Corallina belongs to the grand division / of plants. or to that of animals being included / in the Zoophites. — The gemmules containing / several distinct articulations. I believe is entirely / contrary to any analogy drawn from the propagation / of Zoophites . . . . But anyhow, we / should, <certainly> expect that one gemmule would produce / only one young Polypus, & we might as certainly / expect that each <articulation> one (or pair or some definite / number) would contain & be formed by a Polypus [.] / neither of these expectations are realized in the manner of propagation of the Corallina [.] / Therefore I do not believe Corallina to have / any connection with the family of Zoophites. (DAR 31.1: 211 verso)[48]

On the other hand, the observations in 1834 on *Halimeda* reinforced the earlier conclusion he had already drawn concerning the flat, non-articulated alga *Amphiroa* studied at the Falklands in March 1833. At that time he had already postulated that all the "corallina" were true plants, and reproduced by a common model:

> Upon reading over description / of Amphiroa P 56. it will [be] evident how very / close a relationship, in <no [illeg. c.o.]> manner of growth & cones// there exists between that Coralline & this. / The absence of articulations is the chief difference: / I think we may hence expect that the / propagation in the whole family Corallineae / will be somewhat similar to the one / described. — I have never been able / to perceive any Polypus or true cell. & till / I do I must rank these beings as / belonging to the Vegetable rather than animal / world. — the simplicity of the reproduction / would seem rather to favor this idea. — / I suspect the strongest argument against it. / is ~~the~~ <a false> analogy of form, with respect to / Corallines; in this case however there is a / stronger one to Lichens. (DAR 30.2: 162–163)

Through the South American Winter of 1834, Darwin was affirming on the one hand the communality of the reproductive process in the colonial invertebrates, and on the other a common model of reproduction in the "corallina". But he denied any connection between the two. In his long letter to J. S. Henslow, begun in July and finished in November 1834, Darwin summarized the conclusions of his inquiries to that date on the lower plants and animals:

> In T. del Fuego I collected & examined / some Corallines: I have observed one fact which quite// startled me. — it is, that in the genus Sertularia, (taken / in its most restricted form as by Lamouroux) & in < 2 > species / which, excluding comparative expressions, I should find much / difficulty in describing as different. — The Polypi quite / & essentially differed. in <all> their most important & evident / parts of structure. — I have already seen enough to be / convinced that the present families

of Corallines, as arranged by Lamarck, Cuvier &c are highly artificial — It appears they are in the same state which shells were when / Linnaeus left them for Cuvier to rearrange. — [. . . .] I forget, whether I mentioned, having seen / something of the manner of propagation in that most / ambiguous family, the Corallinas: I feel pretty well// convinced if they are not Plants they are not Zoophites: / the "gemmule". of a Halimeda contain~~ing~~ <ed> several articulations / united & ready to burst their envelope & become attached / to some basis. — I believe in Zoophites, universally the / gemmule produces a single Polypus, which afterwards / or at the same time, grows with its cell or single / articulation. (Darwin to Henslow, 24 July 1834, Darwin MSS Royal Botanic Gardens, Kew, fol. 3–4)[49]

The conclusions of the Winter and Spring of 1834 did not, however, represent Darwin's final position, and the move away from the conclusions expressed in this letter discloses for us the nature of the substantial theoretical development in his biological thought in the latter part of the voyage.

In June 1834 the *Beagle* made its first visit to the island of Chiloe, and in an entry in the Zoology Diary in July, made approximately at the same time as the opening of his letter to Henslow, Darwin first reported on the remarkable phenomenon of apple-tree propagation encountered on the island (DAR 31.1: 266)

Asexual propagation in plants in itself was surely nothing new to Darwin by this date, and the topic had apparently been extensively discussed in Henslow's botany course at Cambridge.[50] Henslow's claims on the significance of this form of reproduction, at least as they can be reconstructed from his 1836 text, are highly interesting:

> There are two distinct modes, according to which the propagation of the vegetable species is naturally secured, viz. "subdivision" and "reproduction." In the first the individual plant may be subdivided into several parts, each of which when detached from the parent stock is capable of existing as a separate individual. A familiar example of this mode of propagation may be seen in the common strawberry . . . , elms, poplars, and other trees throwing up suckers from their roots at a distance from the trunk, all of which are capable of becoming so many distinct trees, under favourable circumstances. Man has availed himself of this property, to extend the means which nature has provided for the propagation of the species. (1836, pp. 248–249)

Henslow's interpretation of this phenomenon is of distinct importance. Many horticulturists endorsed the claim that this form of propagation was distinctly "unnatural", and that in time the strain would die out. Henslow, however, was more flexible on this point. While some forms, indeed most higher flowering plants, obeyed this rule, the same was not necessarily true of all plants, as the examples of poplars, elms, and strawberries indicated:

> But although the propagation of many plants may be effected by the means here alluded to, and although some species are most frequently and readily propagated by subdivision . . . , *at least* all those which bear flowers, secure the continuation of their species by a distinct process, of a very different nature. This constitutes the function of "reproduction," properly so called; which consists in the formation of seeds, containing the germs of future individuals. This function of reproduction is to the species, what life is to the individual — a provision made for its continued duration on the earth. (1836, p. 249)

The unusual feature of the Chiloe apple trees that struck Darwin was not simply that they could be reproduced so freely by cuttings, almost to a degree unheard of in his experience. It was more that they seemed to do so even spontaneously. A passage from the published *Journal of Researches* of the voyage describes this:

> I have never seen any country where apple-trees appeared to thrive so well as in the damp part of South America. . . . At the lower part of almost every branch, small, conical, brown wrinkled points project: these are always ready to change into roots, as may sometimes be seen, where any mud has been accidentally splashed against the tree. (p. 363)

By simply cutting a branch containing one of these conical points, growers could have an apple-bearing tree in a season, and within three years an entire orchard bearing typically seedless apples could be created.

Darwin encountered the Chiloe apple trees precisely at the time he had been exploring the propagation of colonial zoophytes and coralline algae, and their theoretical relevance did not escape him. Coralline algae had already been found to reproduce by the formation of small "cones", and the apparently "natural" process being carried on by the apple trees, even independently of human assistance, seems to have suggested to him that the reproduction in lower plant forms was directly relevant to understanding that in higher ones.

Some months later, following directly on the extended second stay of the *Beagle* at the Island of Chiloe from November to February 1834–1835, Darwin turned to this question in a comment on a different set of issues, the extinction of species. Jonathan Hodge has drawn attention to the importance of these reflections that clearly seem to consist of a commentary on Charles Lyell's discussion of species life and death in the *Principles of Geology* (Hodge 1982).[51] Lyell's comments were, however, based on a much stricter and more inflexible position on the question than that adopted by Henslow concerning the relevance of asexual propagation to this question. As Lyell had written:

> The propagation of a plant by buds or grafts, and by cuttings, is obviously a mode which nature does not employ; and this multiplication, as well

> as that produced by roots and layers, seems merely to operate as an extension of the life of an individual, and not as a reproduction of the species, as happens by seed. All plants increased by the former means retain precisely the peculiar qualities of the individual to which they owe their origin, and, like an individual, they have only a determinate existence. (1832, 2: 32–33)

By the date of the composition of his comments on Lyell's passage, however, Darwin's studies on the zoophytes had disclosed the great diversity in form possible within a common material unity. Henslow's more flexible view on these questions, if we can assume Darwin had encountered these, at least indicated that asexual propagation was the normal means of reproduction in several plants. In a note to his so-called Geology Note of February 1835, Darwin added an interesting comment that considered the larger possibilities in this phenomenon. Discussing the possibility raised by Lyell that "Species may perish as well as individuals," Darwin commented:

> The <following> analogy I am aware is a false one; but / when I consider the enormous extension of life of / an individual plant, seen in the grafting of an / Apple tree. & that all these thousands trees are / subject to the duration of life which one bud / contained. I cannot see such difficulty in believing / a similar duration might be propagated ~~by~~ with true generation. (DAR 42.ii: 2 verso)[52]

At least two reasons can be suggested for Darwin's rejection of this analogy in February 1835. Primarily, there was, as we have seen, no clear reason to affirm that the reproductive process in plants was anything like that in animals. His close studies on a wide range of "plant-animals" in which this might seem most likely had in fact revealed a disanalogy. Second, Darwin had apparently not satisfied himself that reproduction by cuttings, known since Trembley's work in 1740 to take place in animals and long recognized by horticulturists in plants, was really a "true" reproduction, from which some larger conclusions might be drawn. The issue is posed simply as a contrary-to-fact option, but one that was at least a possibility in early 1835.

IV. The Two Kingdoms United

After an almost four-month stay at Valparaiso, Darwin departed with the *Beagle* for the return voyage in July 1835, and the record of the stops after this point supports what Darwin described as a pattern of "flying visits" that prevented much sustained geological and zoological work except, of course, at the Galapagos. The Zoology Diary, correspondingly, becomes briefer in content, and many of the most interesting entries take the form of notes added to the rear of earlier Zoology Diary entries, usually with a date to indicate their later insertion, and identifiable by the different ink and pens used.

Prior to his departure, Darwin had been instructed by Charles Stokes to study the propagation of shallow water corals,[53] and his long delay in doing this had undoubtedly been a result of the geographical distribution of the true reef-building corals, which are primarily confined to the Atlantic and Indo-Pacific faunal regions, and are almost entirely absent from all the regions the *Beagle* visited until April 1836.[54] Studies on live solitary corals of the genera *Caryophyllia* and *Actinia* had, however, commenced at St. Jago in February 1832, and continued at Chiloe in July 1834 (DAR 30.1: 9–12; DAR 31.1: 264). Deeper inquiries into the generation of the corals were reserved for studies at the Galapagos. At James Island in October 1835, Darwin reported on generation in the solitary coral genus *Caryophyllia*, and it is clear that his intent was to determine the relationship of the reproduction in solitary lamelliform corals to that he had already observed in the other "zoophytes":

> Having placed a living specimen of this / Corall in Basin of water. whilst at James Isd [...?] in the Galapagos. — soon observed <several> orange coloured ovules swimming in the water. / — When the eye was four feet from the basin a / progressive motion might be very distinctly / seen. — ova generally elongated oval. the narrower / end slightly truncated. — length about 1/3 of / inch — body contractile as ~~as~~ to alter / form. — The motion is progressive. steady / & quick. the obtuse end — being the head. — / [...] The motions of Ovules / noticed in the Sertularia & Flustraceae is now known to exist in the / Lamelliform Coralls. (DAR 30.1: 12, appendix)[55]

This conclusion did not, at this time, have any evident bearing on the relationship of the animals and plants. That issue came to the surface prominently, however, in February 1836 when the *Beagle* stopped at Hobart Town in Tasmania from 5–17 February. The circumstances of this study are revealed in a memorandum written to accompany Darwin's shipment of his *Beagle* zoophyte specimens to William Henry Harvey in April 1847. Referring to a specimen labeled as number 3503 collected at Hobart Town in February 1836 he wrote:

> On lifting up a/ fragment of sandstone which had lately fallen/ into a tidal pool, I found some branches/ of this Nullipora attached to its lower edge./ Their branches had been broken off by some/ violence from their parent tuft, & the terminal/ segments of joints having been pressed by the/ stone had adhered to it and expanded. These/ little foliaceous expansions had exactly the same/ appearance as the first growth of encrusting/ Nulliporae; but from them. little buds were/ springing,

> evidently determined to be branches, & thus/ to form a new tuft. (Memorandum of C. Darwin to W. H. Harvey, 7 [?] April 1847, University of Dublin Archives)[56]

The theoretical importance of these observations for him in February 1836 is revealed by the entry in the Zoology Diary:

> Hence we/ have this novel method of extending the limits of any tuft in the family of Corallines./ It calls to mind the propagation of trees by/ laying; & can hardly be supposed to take place/ in a true Corall where each cell is inhabited/ by its Polypus. — The fact is of interest in/ showing, the close identity in nature of the/ Corallina articulata & inarticulata: & in itself/ is as much as the observation is made/ in that part of the family. where true propagation/ by ovules has not been observed. (DAR 31.1: 279 verso)

In this, Darwin shows he now clearly saw that propagation in the coralline algae was analogous to that he had seen earlier in the apple trees of Chiloe. Superficially, at least, this meant that the coralline algae could in fact propagate in ways similar to at least some of the colonial zoophytes, which also sent out rhizomal shoots. Furthermore, asexual budding was as "natural" for these forms he had always affirmed to be true plants, as he had recently discovered it to be at the Galapagos for the true corals, which were definitely animals in this view.

In March 1836, Darwin commented further on this possibility. In a note apparently added at King George's Sound in South-Western Australia to an earlier March 1833 discussion of Falkland Islands coralline algae, Darwin reported on microscopic studies performed on a specimen of the "articulated" form *Halimeda*, found growing on granitic rocks in tide pools. Noting the presence of the "generative bladders" described in earlier entries, he spoke of finding in their interior microscopic particles that we have seen him describe before. Then in a note in the same ink and pen he added:

> Decandolle & Sprengel Botany P.[1]92 Consider that propagation in Lichens & Conferva is a kind/ of budding & not true generation. In Halimeda/ <& in the Inarticulata> such certainly I think is the process. —/ In the method described in Corallina of/ Hobart town of the extremities of branches./ being "laid" as branches of trees, & when from/ the foliaceous expansion buds appeared. perhaps/ in this method we see the only kind of/ propagation known to this genus in which/ the bladder-formed cones have not been/ discovered. (DAR 30.2: 162 verso)[57]

The precise dating of events in this theoretical development taking place between February and April 1836 is difficult; it depends to some degree on the dating of Darwin's change in the spelling of a single word — from "Corall" to "Coral" — that, as Frank Sulloway has recently indicated, occurred in the latter part of the voyage (1982). At some point before the spelling change occurred — internal evidence suggests a date between the last use of "corall" at Hobart Town in February and the entries made at Keeling (Cocos) Islands from 2–13 April, when Darwin's spelling was exclusively "coral" — an important summary reflection was written. On internal grounds my conclusion is that this reflects the synthesis Darwin was attaining approximately by the dates of the King George's Sound entries in March. I present this text in full:

> That the number of arms in Polypus <of ȳ Flustraceae> varies from/ 8 to 28 & is no more than a Specific character:/ That a proportion is kept up between simplicity of Polypier & number of arms. — that the same essential/ organ[s] are found in very varying forms of Polypier. —/ That the degree of stony nature in Corallines is/ entirely futile as a character. —
>
> That the ~~fo~~ orders of Lamouroux of Cellepora-Cellaria./ & Flustra should be included in one family (probably)/ also some Eschara & Millepora). —
>
> That one Sertularia ~~would~~ <is> also included. —/ That the structure of the Flustracea is most widely/ different from the Clytias. not only in the Polypus, but/ in the generation[.] in the former case each ovule &/ Polypus has some intimate connexion. in the latter it is/ a young Polypus altered. — (manner of growth?)/ General anatomical discussion. — <*[*(Study Hydra &) Actinia my madrepore & Sigillina in Blainville (Sigillina-Polypus)*]*>
>
> That the connexion of the cells although not apparent in the/ true Flustra must exist: from similarity in growth &/ chain of gradation in the Capsule Flustra: & in the/ Flustra of P. 234 [of Zoology Diary] & true Flustra *[*<& Cellaria>*]* having same body. —/ That the Polypier is the essential part in the Corallines, it/ produces the cells & ~~young~~ Polypi <in young> (& after death of Polypus)/ consequent on generation reproduces them?) —/ That the mere possession of arms has grouped very heterogenious [sic] animals. — That Corallina is a plant — *[*Does it not emit in Suns Rays gaz. —*]*//
>
> In Virgularia does the truncate extremity correspond/ to ~~extremity of~~ <root> branch in Corallinum? Examine extremities/ <and the bag to extremity of branch> The relative position of Polypier. with living mass in/ the Lamelliform. —
>
> The structure of transparent extremities of Corallina. —/ Regrowth of Corallines when separated./ In the Capsule Flustra. cells. without Polypi have/ capsules. (moveable)? <yes? I believe> strong proof of disconnection: —//
>
> A close connection and co-sensition [sic] between the/ Polypi of many Corallines is established by the/ co-movements/ of "Capsules Flustras" of the setae in/ Crisia: the flashes of light in Clytia: strongly/ seen in

Virgularia & in Alcyonium[.] an injury in/ stem causing all to collapse.: whilst one <. . . ?> being injured/ did not affect the mass. — on the other/ hand. one point in a Synoicum Blanvi: affected all/ round it for some distance: —/ Have not the Escarae [Escharae?] in the growth of the Polypier an/ analogy with the Celleporaria: where cells appear formed/ in a <cellular> tissue, (or group of hoods, or angular tubes as/ in Favosites) ~~& &~~ stone? —/ A cell reproduces its Polypus/ The stony striae on outside of Lobularia connecting/ link with stony Zoophites. —/ The Lobuted ~~form~~ <[illeg. ...?]> position of tentacula in Chiloe Actinia/ perhaps is an analogy in change between a Caryophyllia &/ Gorgonia or Corallium?. — it shows a passage of this/ arrangement, without material change in animal. —/ It is important to see in Clytia, substance included/ in a young cell appearing equally ready to form/ Polypus or ovules. — The Coralline must produce this matter: not the Polypus the gemmule. —/ I am inclined to think in Corallines, such as Sabularia/ & Flustra, the Polypier is as much a living ~~mo~~ being/ as any Plant, (as a Lichen or Corallina). that it/ communicates with the circumambient fluid either/ simply as in Clytia, or <in> more complicated manner, as in Flustra// How little organization can be seen in Corallina,/ yet even the basal articulation[s] produce paps with/ gemmules. — In the Polypier of the Flustracea/ it seems to make little difference, whether a central/ living axis is clearly living visible or whether it (probably/ forms a thin fold at the base of cells, in the/ encrusting Flustrae. —/ I imagine in the Lamelliform Coralls, the Polypier is only an ~~ex~~ internal secretions[sic], (a bony axis to give/ support) the Polypier being then the mass of living/ matter: we see it thus, in Virgularia. —/ There is an analogy between the Corall-forming Polypi & turf-forming plants. — Hence here the/ soft matter ought to form the gemmules, as/ in the hard matter in the other cases. —/ I think there is much analogy between Zoophites & Plants/ the Polypi being buds; the gemmules the inflorescence/ which forms a bud & young plant. —/ in Sertularia, the Capsules with gemmules appear to/ have no relation with any one Polypus. how/ could it form a totally different sort of capsule/ to its own. & in a place where it, the/ Polypus is never found. — *[In Lamarck good account of Lobularia]* (DAR 5: 98–99)[58]

Read in the context I have provided, we see the remarkable degree of theoretical synthesis that has taken place. First, Darwin has rejected as adequate a taxonomy of the zoophytes based simply on external morphological difference, or on such characteristics as the degree of calcification of the skeleton. On the contrary, his studies have suggested a great lumping of groups, often separated by his authorities by wide taxonomic distance.

Second, Darwin feels that he has, at least theoretically, resolved the problem left unsolved by Grant with regard to the *Flustra* — namely the degree to which they shared an "associated" life. His microscopic studies have revealed that even in these polymorphic forms with ability to respond and reproduce in almost complete independence, there is still a material connection, and their analogy with other zoophytes is strongly affirmed.

His conclusions on the material unity that overrides a remarkable morphological diversity is the opening to the third striking point we see in this document. Not only is there a material connection evident, but also the same material can change into different forms and patterns of organization. The "granular matter" he has noted so often is able to form discrete ovules, individual organized polyps, and give rise to a great diversity of forms that, if his lumping of forms is accepted, would transcend the limits of several recognized orders.

Finally, we see the remarkable theoretical change in Darwin's position on the relations of plants and animals. Although still affirming that "Corallina is a plant", Darwin now considers the process of generation to be strikingly like that in some of the encrusting Flustrae and the "Lamelliform Coralls" he had studied at the Galapagos Islands. The separation between the "corallina" and the true corals has become highly ambiguous, and for the first time Darwin is now willing to affirm a *direct* analogy between *naturally* asexually-propagating plants and the zoophytes. The barriers that were decisive in the February 1835 "Geology Note" have vanished. What was now needed was some test of this conclusion.

With the stop at the Keeling Islands, a small cluster of coral atolls in the eastern Indian Ocean, Darwin was finally in a position to make sustained studies on live forms of reef-building corals, which could easily be collected by wading out into the shallow lagoons. The remarkable impression these islands made upon him is recorded in a passage from the manuscript voyage *Diary* kept through the cruise:

> I am glad we have visited these islands: such formations surely rank high amongst the wonderful objects of this world. It is not a wonder which at first strikes the eye of the body, but rather after reflection, the eye of reason. We feel surprised, when travellers relate accounts of the vast piles & extent of some ancient ruins; but how insignificant are the greatest of these, when compared to the matter here accumulated by various small animals. Throughout the whole group of Islands, every single atom, even from the most minute particle to large fragments of rock, bears the stamp of having been subjected to the power of organic arrangement. (pp. 399–400)

The Zoology Diary records live observations at the Keeling Islands on specimens of the genus of solitary "brain" corals, *Meandria*, on two different forms identified as "millepores", and on specimens of common branching madreporean forms.

Striking to Darwin in these studies, and in contrast to the situation found in the colonial invertebrates he had studied previously, was the great difficulty in finding anything clearly identifiable as a discrete polyp. In *Meandria*, the polyps simply form long, convoluted rows continuous over the surface (Hyman 1940, 1: p. 611), and Darwin reported finding only that "the fleshy matter is united over the whole surface" (DAR 31.2: 353).

Even greater difficulties in identifying the polyps were presented by the form identified simply as "millepore", which was found to grow in "stony vertical plates, which frequently intersect each other & so form a coarse honeycombed mass" (DAR 31.2: 358).[59] In such forms the polyps are usually found only in their expanded form at night, and as far as Darwin could determine, they did not exist at all. His studies revealed only the outer layers of the plates to demonstrate any signs of "animality", presumably meaning any response to stimulus, and "I could not perceive any trace even of a Polypus in the terminal cell" (DAR 31.2: 359).

By subjecting the sections of the specimen to microscopic examination, Darwin was able to determine that the coral grew simply by concentric concretions at the termini of the branches, lacking any evident involvement of a living polyp. All of this suggested to him that the affinities of this form, considered by his main authorities as a genuine animal,[60] were most clearly with the anomalous *Corallina,* which he had continuously affirmed were plants:

> With respect to the nature of these Millepora,/ I cannot help suspecting, that their nature is/ allied to Corallina. rather than to Polypiferous/ Corals. — I am led to this idea from not being/ able to discover any trace of an <organized> Polypus in the cells:/ their position with respect to extremities of branches./ their size varying & their method of grouping:/ All which facts would better agree. with the/ idea that the cell is the seat. where the/ ovum is produced. — Their manner of growth./ & the absence of slime is analogous to the/ Corallineae . . . (DAR 31.2: 360)

Like no other forms he had encountered, these organisms gave empirical warrant for breaking down all the structural bases for distinctions he had previously made between animals and plants at this low level of organization. With this barrier gone, Darwin was at last fully in a position, both theoretically and empirically, to assert a communality in the fundamental bases of plant and animal existence.

V. Transformist Openings

In speaking of an "invertebrate program" in Darwin's *Beagle* work, my intent has not been to claim that he began the voyage with a clearly formulated enterprise. But by the end of the voyage, the lines of this program, in a technical sense, seem to emerge. We can identify the elements of this as involving the "analytic" use of invertebrate zoology, the strong analogy between plants and animals and the possibility of uniting these two kingdoms in a common point. There is a key role given to a common granulated matter in reproduction of both animals and plants: and finally, there is empirical warrant for the universalization of both the process of reproduction, and also the model of "associated" life in a branching network, displayed by the zoophytes, to a much more expanded set of questions.

The key importance of the zoophytal interests to all of this should be clear. In these forms of life, as in no others, Darwin had been presented with a striking and even astonishing model of a material continuity of living substance, represented in a range of degrees, which gave those forms a unity that transcended the great morphological diversity even on the same colony. Ordinary taxonomic boundaries between recognized species, genera, even families of "zoophytes" were made obscure and indecisive when viewed in terms of their reproductive processes. Furthermore, the "false" analogy between animals and plants, suggested by these forms, had become by mid-1836 a genuine identity, suggesting for Darwin a common unifying model that united *all* forms of reproduction.

In identifying these key features of Darwin's invertebrate inquiries up to 1836, I have intentionally avoided talking about "transformism". I would, as a result of this study, defend the conclusions of those who date the transformist theory to early 1837. Until Darwin had achieved the points of theoretical unification I have described, genuine transformism was not really a serious possibility. After he had achieved this, Darwin had satisfied a necessary, but not a sufficient condition for the theoretical unity he would achieve only by July 1837.[61]

We can nevertheless pursue at least two indications of the early post-voyage development of these questions as evidence of the direct continuity of these problems into the transformist reflections. Two key passages in the pre-transformist A and *Red Notebooks* will indicate the way these issues were developing in Darwin's post-voyage thinking.

As the notebook that bridges the close of the *Beagle* voyage in October 1836 and the first half of 1837, the *RN* is of particular interest in following these issues. There we see that Darwin was willing for the first time to *universalize* the fundamental issue emerging from this *Beagle* invertebrate work. From these studies, he had clarified the grounds for claiming that both plants and animals could be understood as generating by a variant of the process he had found in the colonial invertebrates. Furthermore, by this analogy, the relationships in plant and colonial animals were analogous examples of the "associated life" found in the zoophytes.

Sometime shortly after 15 March 1837, Darwin was ready to extend this to include *all* forms in space as well as time:

> Propagation. whether ordinary. hermaphrodite. or by cutting/ an animal

> ~~in~~ two. (gemmiparous. by nature or/ accident). we see an individual divided either at/ one moment or through lapse of ages. — Therefore we are not so much surprised at seeing/ Zoophite producing distinct animals, still partly united. & egg[s] which become quite separate. — / Considering all individuals of species. as/ each one individual <divided> by different methods, associated life only adds one other/ method where the division is not perfect. (*RN* 132)[62]

Contrary to Lyell's claims examined previously, Darwin had now clearly concluded that asexual reproduction by division was "natural" and directly analogous to sexual reproduction. The "associated" life of the zoophyte — already found to consist of distinct degrees ranging from the coordinated response to stimuli, to the almost complete dissociation of relationship in the millepore corals, in which only the tip of the growing branches seemed to be alive, leaving a dead calcareous framework uniting the colony — was now being applied to extension in time as well.

To support such claims required, however, that Darwin do more than speculate. The zoophyte-plant analogy was an issue that needed more empirical investigation, and these investigations would lead into issues occupying his attention into the 1840s.[63] A passage in the A *Notebook* points up the research interest clearly:

> Many interesting experiments might be tried/ by comparing Zoophite to plants. — grafting/ length of life &c &c.(*A* [180])[64]

The first transmutation notebook, begun in July 1837, amplifies on this:

> All animals ~~are~~ of same species are bound/ together just like buds of plants,/ which die at one time. though/ produced either sooner or later. —
>
> ---
>
> Prove animals like plants; — trace/ gradation between associated & non associated animals. — & the story will be complete. (*B* 73)

Darwin's inquiries into invertebrate zoology had begun as a continuation of the researches of Robert Grant, and they retained many marks of this "Grantian" heritage through the *Beagle* period. By 1837, however, these investigations had been transformed into a research program peculiarly Darwin's own, one that was able to interact fruitfully and creatively with his parallel theory-development in geology, biogeography, variation studies, and taxonomy. If other of these themes surely would sound much more prominently during subsequent periods of Darwin's work, the underlying continuity and theoretical importance of his early purely biological interests can scarcely be neglected.

ACKNOWLEDGEMENTS

I wish to express my particular debt in this paper to Professor Sydney Smith of Cambridge University for invaluable criticisms, and for generously sharing with me his extensive knowledge of the Darwin archives. I have also profited in several ways from conversations on aspects of this paper with Frank J. Sulloway, William Montgomery, Michael Ghiselin, Jonathan Hodge, David Kohn, Peter Gautrey, and Stan Rachootin. For valuable archival assistance and comments, I am indebted to Frederick Burkhardt and Peter Gautrey. For fruitful discussions and bibliographical assistance I am always indebted to my colleague Edward Manier. Support for this study was given by the Faculty Development Fund of the University of Notre Dame, and a National Science Foundation Grant No. SES-7925112 as part of a larger study of the genesis of Darwin's species concept. I wish to thank the library staffs of Cambridge University, the University of Edinburgh, the University of Notre Dame, and Yale University for many assistances. Considerable assistance was also given me by the workers on the Darwin Letters Project at Cambridge and the American Philosophical Library. All citations and quotations from the Cambridge archives are with the permission of the Syndics of Cambridge University Library. Other permissions as noted.

Notes

1. See, for example, Ruse (1979a). The beginning point for the study of these years is the unsigned two-part article "Darwin in Edinburgh" (Anon. 1888), issued in response to Darwin's comments in his *Autobiography*. This author alone seems to have had full access to an unmutilated archive at Edinburgh, which included at the time Darwin's check-out slips for library books and other documents on the Edinburgh years. I wish to acknowledge the assistance of Dr. J. D. T. Hall of the Edinburgh University archives for assistance in a search for these materials. The scientific aspects of Darwin's Edinburgh period have been deeply, if not conclusively studied in Ashworth 1935).
2. Edinburgh's importance in biological oceanography extends as far back as John Walker's professorship in the chair of Natural History that commenced in 1779. For useful remarks on this tradition see Deacon (1971) and Mills (1984).
3. Valuable information on the documents surviving from these more prominent societies can be found in Finlayson (1958). By 1826 there was faculty concern over the large number of student societies in which faculty supervision on either the meetings or content was minimal. See the testimony of John Leslie (Great Britain 1837, p. 132). This fundamental document, reporting on faculty interviews in July 1826 and October 1830, includes testimonies from all of Darwin's most important teachers, and contains an extensive appendix giving details of student enrolments, course requirements, and even occasional course syllabi.
4. It was not uncommon to enter Scottish universities in the teen-age years. Thomas Carlyle, for example, entered the University of Edinburgh in 1825 at age fourteen.
5. The author of the *St. James's Gazette* article comments on the association of two brothers as revealed by the documents available to him: "Indeed, so far as can be seen, they did everything together" (Anon. 1888, part 1, p. 5).
6. Erasmus A. Darwin was in Edinburgh to fulfill the external study requirements that were in the process of being instituted at Cambridge as part of John Haviland's reform of Cambridge medical education. These reforms eventually made two years of clinical and external university study required for the degree of

Bachelor of Medicine. These requirements are set forth in Great Britain 1852, p. 35. At Cambridge, Erasmus had completed required courses in chemistry, physiology, anatomy, mineralogy, and geology, and he had studied with both Adam Sedgwick and J. S. Henslow. He was also actively engaged with the Cambridge entomologists under Henslow's leadership. See letters of Erasmus to Charles of 14 November 1822 and 5 March 1823 (DAR 204.1). The joint concern of the two brothers with pure science in Edinburgh is suggested from correspondence and other evidence. In his partial list of the books checked out by Charles from the university library in 1825–1826, Anon. (1888, part I, p. 5) listed Thomas Young's *Introduction to Natural Philosophy*, Newton's *Opticks*, John Fleming's *Philosophy of Zoology*, "two volumes on entomology", a book by Wood on insects, and one by Brooks on conchology.

7. The record of these first excursions is contained in a small pocket diary begun on 1 January 1826 (DAR 129). Record is made in this of the collection of gastropods, a squid, the sea-mouse *Aphrodita*, and various colonial invertebrates.
8. Erasmus departed for Shrewsbury by 9 March, but Charles remained behind to complete T. C. Hope's chemistry lectures. See letter of Erasmus to Charles, 9 March 1826 (DAR 204.1) and quote below, note 31. The pocket diary (DAR 129) records marine biological observations on 13, 17 and 29 March after Erasmus had departed.
9. Charles was proposed for membership in the Plinian at the second meeting of the 1826–1827 session of the society (21 November 1826). His application was sponsored by the current president and disciple of Robert Grant, John Coldstream, by Andrew Fyfe, Jr., a private lecturer in chemistry, and by the surgeon and phrenologist William A. F. Browne. His application was approved unanimously at the next meeting and he was elected to the governing council on 5 December (Plinian Society 1823–1826).
10. In addition to medical practice, Robert Grant was employed, at least until 1825, as a demonstrator in invertebrates for John Barclay's popular comparative anatomy course at his famous proprietary anatomical school in Edinburgh. At thirty-three, Grant would undoubtedly have held a position of commanding authority in a student society composed of many members who were not much more than half his age, and none of whom seem to have had his wide range of experience and contacts with life scientists in Germany, France, and Italy.
11. Darwin reports in his *Autobiography* having "sometimes accompanied [the Newhaven fishermen] when they trawled for oysters, and thus got many specimens" (p. 50). Specimens he collected of the sea-pen *Pennatula* (=*Virgularia*) *mirabilis* and *P. phosphorea* on 15 April 1827, seem to have been those used by Grant — "brought to me alive in sea water" — for his paper on the generation of these forms. See Darwin's report (DAR 118: 9) and Grant (1827a, pp. 330–334).
12. Although Grant drew heavily upon French sources, especially Lamarck, for many of his ideas, he showed also the deep impact of his contacts with German and Italian biologists following his studies on the continent in 1815–1820. In his letter of application for the new chair of comparative anatomy at University College, London in May 1827 he remarked that he had ". . . prosecuted my favourite Anatomical and Zoological pursuits in France, Italy, Germany, and at home" (Grant MSS, UCL, 26 May 1827).Grant's private library contained the main works of Tiedemann, Meckel, Carus, Geoffroy Saint-Hilaire, Rudolphi and Burdach (Grant MSS. Add. 58, UCL, n.d.), and he frequently cited these authors in his comparative anatomy lectures in London.
13. Cuvier's rejection of the assumptions underlying the "chain of being" makes this only partially true, since all *embranchements* theoretically stood on the same taxonomic level. In practice, however, the groups found in the *Radiata*, encompassing the traditional "zoophytes", were treated last in Cuvier's expositions, and typically in scanty detail. Important discussion of Cuvier's *Radiata* and its composition is in Winsor (1976).
14. Robert Jameson's natural history lectures appear to have been pronounced in this respect, beginning with the natural history of man and terminating in the "least perfect" animals. Even though his published syllabus suggests he gave extended treatment of the invertebrates (Great Britain 1837, appendix, p. 117), at least one set of surviving student notes on the lectures around 1827 UEL MSS Dc.7.114) devotes forty-two pages to mammals, twenty-four to birds, and only brief comments to the invertebrates at the end.
15. Summarized from Grant's manuscript (1827?) outline (UCL MSS 1179.3). This is on 1827 watermark paper. Lamarck did not actually treat the vertebrates in the *Histoire naturelle*

except in one schematic table, and I am interpolating here from Lamarck (1809).

16. As Grant summarized the content of one of his concluding lectures intended for his proposed course: "Proofs of the existence of Infusoria before the creation of zoophytes, the oldest known Class of fossil animals" (UCL MSS 1179.3, p. 6). Quoted by permission of the archives of the Watson Library, University College London.
17. This "analytic" use of invertebrate zoology is developed in John Hunter's lectures; in writings by German workers such as C. G. Carus; and especially in the 1830s by Richard Owen in his lectures at the College of Surgeons. I am analyzing this issue in a separate study.
18. This text constitutes Tiedemann's lectures at Heidelberg. I have been unable to determine whether Grant was in contact with Tiedemann during his studies in Germany. However, such ideas were current in German biology from the time of Treviranus' work (1803, 2: 338, 344, 350), and Grant could have been acquainted with them from other sources.
19. Lamarck was drawing heavily in his discussion on the parallel claims of Richerand (1802, 1: xxviii–xxxiv).
20. I will utilize the expanded second edition of Lamarck's work except where noted.
21. Grant presented this paper to the Wernerian Society on 8 April 1826. *Cliona* is a boring parasitic sponge.
22. This was presented to the Wernerian Society Meetings on 27 May 1826.
23. Although begun at Edinburgh, this notebook contains entries on zoological matters from several later dates, and includes an early post-*Beagle* reading list. The Edinburgh entries cover the first seventeen pages of the text.
24. Darwin's name was never mentioned in the paper, but Grant was clearly drawing on Darwin's work. The report of Grant's public presentation of 24 March 1827 (Wernerian Society, Vol. 1, UEL MSS) states: "Dr Grant read a paper regarding the anatomy and mode of generation of Flustrae, illustrated by preparations and drawings 240 times magnified . . ., the ova of which have lately been ascertained by Mr. Charles Darwin." (Quotation by permission of the archives of the University of Edinburgh library.) In his own report on this issue, Darwin noted that his paper "was read both before the Wernerian & Plinian Societies", (DAR 118: 6) which Wernerian society by-laws would not have permitted him to do personally, since he was only a student. There is no separate mention of Darwin's paper having been presented independently from Grant's in the *Minutes Book* of the Wernerian.
25. For details on the Bryozoa I have found particularly useful Harmer (1910), Hyman (1940), Brien (1960a), and Rylands (1970).
26. In his paper, Grant noted the great advantages of using the deep-water form, *Flustra carbasea*, due to its transparent supporting skeleton. In presenting extended manuscript quotations in this paper I will follow the following conventions. All editorial insertions by myself will be indicated by [ ] brackets. Interlineations and insertions by the author will be indicated at the proper points by < > brackets, and crossed-out words will be included as presented. Pencil insertions, or insertions in obviously different pen will be indicated by slant brackets *[ ]*. Lineation is presented marked with / for line breaks, and // for page breaks.
27. In his interesting paper on the Darwin-Grant relationship, Jespersen quotes a comment by Henrietta Darwin Litchfield recalling her father's comments on Grant's response to his *Flustra* studies at Edinburgh: "When he was at Edinburgh he found out that the spermatozoa [ova] of . . . Flustra move. He rushed instantly to Prof. Grant who was working on the subject to tell him, thinking, he wd be delighted with so curious a fact. But was confounded on being told that it was very unfair of him to work at Prof. G's subject and in fact that he shd take it ill if my Father published it" (1948–1949, pp. 164–165). The relationship between Darwin and Grant after 1827 seems surprisingly distant, although Darwin consulted Grant in preparation for the *Beagle* voyage. See below, page 88, and Darwin's notes on preparation for the voyage (DAR 29.3: 78).
28. Grant was in close touch with August Schweigger, professor of medicine and botany at the University of Königsberg, and was concerned to make his work known to British naturalists, considering him the leading contemporary authority on the zoophytes. Grant translated and commented upon a short paper by Schweigger on this group (Schweigger 1826, pp. 220–224).
29. Erasmus had attended as part of his medical curriculum James Cumming's chemistry lectures in 1822–1823, which were apparently lecture-demonstrations without a laboratory component. At the same time he gained practical laboratory training in inorganic chemistry in J. S. Henslow's mineralogy

course. The active concern of the two brothers to acquire necessary chemical apparatus, glassware and books in this period seems to have been part of a conscious attempt on Erasmus's part to carry out with Charles's assistance the actual experiments from Cumming's course, at Shrewsbury. This project is described in a letter of Erasmus to Charles, 14 November 1822 (DAR 204.1).

30 As Joseph Black's successor to the chair of chemistry at Edinburgh, Hope held a preeminent position in British chemistry. Foreign visitors commented favorably on the quality of his lectures (Griscom 1824, 2: 221–222), and his chemistry lectures drew the largest number of students (505) of any listed lecture course at the University in the 1825–1826 year, exceeding by almost twice those attending any other series (Great Britain 1837, appendix, p. 130).

31. "Dr Hope has been giving some very good Lectures on Electricity &c and I am glad I stayed for them. The Classes are beginning to thin. I think I shall stay about nine days or a fortnight longer" (letter of Charles to Caroline Darwin, 8 April 1826, DAR 154).

32. Darwin's surviving chemistry lecture notes (DAR 5.i) consist of eight large sheets, covering only Hope's opening lecture on chemical theory, and treat matter theory, the caloric theory of heat, latent and specific heat, and specific gravity determination techniques. Since the two brothers were attending these lectures together, the incomplete nature of these notes probably indicates that Erasmus, already having completed a university-level chemistry course at Cambridge, was doing the note taking, while Charles was free to devote full attention to the demonstration lectures. The surviving notes are so neat and thorough as to suggest that they are a careful reworking of rougher lecture notes.

33. This lecture, like most of the surviving lectures, seems to have been used, with inserted pencilled changes, for all of Hope's courses from 1817 to 1840. Quoted by permission of the archives of the University of Edinburgh library.

34. Two printed syllabi for the Henslow botany courses for 1828 and 1833 survive in the Cambridge archives (Henslow 1828, 1833). The course consisted of approximately five weeks of lectures, with Tuesday–Thursday lectures on morphology, anatomy and taxonomy, and Monday–Wednesday–Friday lectures on plant physiology, reproduction, and distribution. In describing these lectures, Henslow's first biographer and brother-in-law, Leonard Jenyns, reported that the lectures only varied slightly over the years, but were "much improved in details from year to year, as new matter came to hand. . ." (1862, p. 40). The structuring of Henslow's 1836 text closely follows the plan of the 1833 syllabus, even divided into "periods" in sections that correspond to the periods outlined on the syllabus. For further remarks on Henslow's course see Walters (1981, chap 5).

35. These instructions can be reconstructed from two pages of notes in Darwin's hand (DAR 29.3: 78–79). Although undated without watermark, the document gives such information as the shipping address for the *Beagle* at Buenos Aires, and all other indications place it before the voyage. Queries are listed of specific people, followed by short answers. Perhaps most interesting is the entry: "Humming birds from Juan Fernandez & every thing especially from Gallipagos [sic]. M[r] Stokes." The reference is to Charles Stokes, a minor geologist and zoologist and member of the Linnean Society. He seems to have also been consulted by Darwin on coral reef studies as well (see below, note 53). Stokes published a small number of papers on marine invertebrate zoology, and was especially interested in the corals. He was consulted by Darwin on the genus *Caryophillia* after the return of the *Beagle* (*Coral Reefs* 1889, p. 117n), and his library, was utilized by Darwin in his early London readings (DAR 119, note inside front cover). He should not be confused with the *Beagle* officer John Stokes.

36. Gruber and Gruber (1962) draw from this the conclusion that Darwin's interests shifted markedly from biological to geological issues, but it should be evident that I feel some revision of this interpretation is necessary. I am not, however, discounting the evidence for Darwin's increasing geological interests during the cruise, and I am indebted to Frank J. Sulloway (personal communication) for penetrating comments on an earlier draft of this paper on this issue. Table 1 should make clear, as Sulloway (this volume) argues, that the total effort devoted to biology in lines/diem is definitely declining, and this should be observed in reading Table 2. Although Sulloway (this volume), like Gruber and Gruber (1962), maintains that Darwin's interests shifted markedly from biology to geology during the course of the *Beagle* voyage, he also emphasizes the relative importance of Darwin's researches in invertebrate zoology. Consequently, Sulloway and I are in substantial accord concerning the nature of Darwin's

activities and interests during the voyage.

37. Darwin apparently had two microscopes with him on the *Beagle*. The first, a compound microscope manufactured by George Cary of London, utilized an unusual "Coddington" lens in the objective, and was presented to Darwin as a gift by J. M. Herbert in the Spring of 1831. The second was a "simple" microscope of a modified Ellis-aquatic design, and was manufactured by Robert Bancks (or Banks) and Sons of London on a design very similar to a simple scope designed for Robert Brown by the same firm. The simple microscope, rather than being an inferior instrument, was in fact the preferred instrument for serious microscopic study by the leading authorities in the 1820s, and a high quality simple microscope could attain resolution values superior to those attained by any compound microscopes, with minimal chromatic and spherical aberrations (Van Cittert 1934, pp. 5–12).

    The simple microscope is mainly hindered by the sharp limitation of field and the short working depth at higher magnifications. My personal examination of the Bancks simple scope on display at Down House revealed it to be a remarkably effective instrument to use, except at shortest focal lengths. It was able to resolve classic resolution test objects, such as the striae on butterfly wing scales, with ease. Although Darwin apparently had the compound Coddington scope as well, his reported observations never seem to correspond with those possible on this awkward instrument, which probably explains his preference for the simple scope throughout his scientific career. Darwin had a second simple microscope constructed for him in the 1840s by Smith and Beck, of much larger proportions, and this is the scope recommended by him for shipboard use in his section on microscopy in Herschel (1849, pp. 389–393). Clarifying background on the relation of these two scopes is provided by Darwin's letter to Richard Owen of 26 March 1848 (Archives, New York Botanical Gardens). This letter was drawn to my attention by Professor Frederick Burkhardt of the Darwin Letters Project, who has provided considerable assistance on the question of Darwin's microscopes. I have also been assisted in this by conversations with Sydney Smith, Gerard L. E. Turner of the History of Science Museum, Oxford, and by documents accompanying both of Darwin's scopes at Down House and Robert Brown's microscope at the Linnean Society of London by W. A. S. Burnette of the Royal Society. The helpful assistance of Mr. Gavin Bridson of the Linnean Society and Philip Titheradge of Down House made it possible for me to examine the original instruments.

38. In an attempt to repeat Darwin's observations on live specimens of marine forms, I have been able to obtain similar results. Observing a hydrozoan, probably belonging to *Aglaophenia struthionides* collected intertidally at Pt. Loma, California, I used a compound microscope of 50, 150, 650, and 1450 X. In the main stem were observed numerous corpuscular particles of approximately 0.02 mm. in diameter moving in a dancing pattern, rather than in an obvious circulatory one. These were clearly visible at 150 X, bringing them within the range of Darwin's microscope. A second set of observations on a coralline alga, probably belonging to *Bossea orbigniana*, also revealed a surprisingly similar set of phenomena. Although the main body of this plant was opaque, in clear areas near the site of new bud formation were observed numerous granular particles, located particularly around the articulation points. These particles were again visible at 150 X and again measured approximately 0.02 mm. by the optical micrometer. These lacked all evident motion. I wish to thank Dr. Alberto Zirino of the Naval Oceanographic Laboratory at Pt. Loma, California for providing facilities for making these studies.

39. Darwin's note reads: "Does not the great size of the observed particles entirely separate this fact from the 'molecular movements' of Browne [sic]." This note is added to an entry of *Virgularia* (DAR 20.2: 110 verso). Brown's "active molecules" were much smaller than the particles described by Darwin, which seems to be his point of comment here. In his report on their discovery, Brown (1828) used a simple microscope of 1/32 inch focal length, giving a magnification of 320 X, and seemed to feel this was minimal for their observation. Darwin's scope can be documented to have had only a 1/20 inch focal length lens as a maximum, and would probably not have revealed the Brownian molecules.

40. This accompanies an entry dated August 1832. Darwin was continually puzzled throughout the voyage about the taxonomic position of the chaetognaths. They are first placed among the Pteropod molluscs in the genus *Limancina* of Lamarck, and then simply listed later as "Polype?" Only after his return, when he had access to other literature, was Darwin able to place this form with some certainty. The bulk of his *Beagle* studies on this form, accom-

panied with a redrawing of the original Zoology Diary plate, was published later (Darwin 1844).

41. The detailed description of this form is given at DAR 31.1: 223–224. The exact identifications of the structures Darwin is describing are not always clear, and I have profited most from the discussion and plates in Brien (1960), fig. 1108, p. 1236. The particular specimen Darwin utilized seems to have had more than one "Brown Body" in the process of formation.
42. I have been unable to locate Darwin's drawing of plate 9, fig. 6, in the extant collection in DAR 29.3.
43. Lamarck (1835, vol: 2) placed many of the branching "zoophytes" in his section *Polypiers à réseau*, which also included the bryozoans and many of the hydrozoans. For assistance in his shipboard invertebrate studies, Darwin's notes suggest that he had ready access to the first edition of Lamarck 1819–1822; Lamouroux 1821; Blainville 1816—1830; Audouin et al. 1822—1831, and Cuvier 1829—1830.
44. Lamarck synonomizes Lamouroux's genus *Clytia* under *Campanularia* in his section "vaginiform polyparies", which also included genera such as *Tubularia, Sertularia,* and *Cellularia,* which are often discussed by Darwin in the Zoology Diary.
45. A useful summary of the controversy can be found in Lamouroux's article "Corallinées," in Audouin et al. (1823, 4: 457–460).
46. Lamarck and Lamouroux argued for the animal status of Corallina, and Lamouroux, after surveying all the arguments on either side, decided that the absence of a distinct polyp could not be decisive, since some of the Nullipore corals also lacked one (Audouin et al., 1823, 4: 459).
47. This entry is dated June 1832 at Rio de Janeiro.
48. This is an added note, presumably entered later.
49. Darwin is distinguishing *Coralline* from *Corallina* in this quote, the first referring to the animal forms, such as *Sertularia,* and the second to forms like *Halimeda,* which he considered to be true plants. In the Nova Barlow transcription of this letter (Barlow 1967, pp. 93–94), *Corallinas* has been mistranscribed as *Corallines.* Henslow omitted the second passage in his publication of this letter (*CP* 1: 3–16, 1835). Quotation by permission of the trustees of the Royal Botanic Gardens, Kew. I wish to thank the library staff of the Royal Botanic Gardens, Kew for assistance on this matter.
50. The published syllabus of Henslow's 1833 lecture course devoted five lectures to reproduction, which included a treatment of "propagation by subdivision" (Henslow 1833).
51. In addition to access to the manuscript of his paper, I am also indebted to Jonathan Hodge for extended discussions on these issues.
52. A full text of this document is published in Hodge (1982), and aspects of it have been treated in Kohn (1980) and Herbert (1974), p. 236n.
53. Darwin's preparatory notes for the voyage record the following: "Species of Fungias. ascertain from fleshy parts, & propagation: found in shallow water: Mr. Stokes" (DAR 29.3: 78). See above note 35.
54. Reef-building corals and large solitary coral formations are generally confined to an 18° isotherm. Reefs are absent from the entire western South American coast below the equator, and have no formation at the Galapagos (*Coral Reefs* 1889, pp. 82, 199, 200; Gross 1972, p. 71 and Ekman 1953, pp. 4–6). Darwin had not come in contact with the limited reefs found on the eastern coast of South America (*Coral Reefs* 1889, p. 83). The only entries on corals from the Zoology Diary on the Atlantic side are some brief remarks on the solitary corals *Caryophillia* and *Actinia* at St. Jago in early 1832, and entries on solitary brain corals during a brief stop at the Albrolhos Islands in April 1833. Darwin had, on his own report, formulated his coral reef theory "before I had seen a true coral reef" while still in South America (*Autobiography*, p. 98). No coral reefs were found at the Galapagos (*Coral Reefs* 1889, p. 82).
55. This is inserted in DAR 30.1 headed "1835 Octob. Appendix to p. 12" on a different paper marked with an "M" and Crown watermark.
56. This is an undated memorandum of six pages evidently sent with a list of all the *Beagle* coralline specimens to Harvey for his identification. The dating has been determined by means of a letter to Joseph Hooker on 7 April 1847, which refers to the material Darwin had recently sent to Harvey. For both drawing my attention to this letter and for information on the dating I am indebted to William Montgomery of the Darwin Letters Project. I wish also to thank Prof. David A. Webb of the Botany School, University of Dublin, for supplying a copy of this item. Quotation by permission of the University of Dublin Archives.
57. This note is undated, but is in ink and pen similar to the note added to the rear of the preceding page headed "King George's Sound" (DAR 30.2: 161 verso). The *Beagle* was

at King George's Sound, on the south-east tip of Australia, from 3–14 March 1836.

58. The dating of this document is important for the thesis of this paper, and the following details are relevant. Conventions used in the transcription are as given in note 26. The document is in pen with pencilled insertions, on one large sheet of a crude paper measuring 22.5×32 cm, folded horizontally in half, with a "£" or "ƍ" watermark, but without identifying dates, and matches the paper used in Darwin's undated "Reflections on Reading my Geological Notes", (DAR 42: 49–52, 73). With the assistance of Peter J. Gautrey, Frederick Burkhardt and Frank J. Sulloway (personal communications), this same paper type has been located in three dated specimens. Five sheets, with the last bearing "Port Desire Jan 1833", have been located by Mr. Gautrey in the Geology Diaries at DAR 34.1: 29–34, with the date apparently a mistake for January 1834, when the *Beagle* was at Port Desire. Another scrap, bearing the date "Jan 16th" and "Goree Sound" has also been located in the Geology Diaries at DAR 34.2: 185, which places this paper type on board the *Beagle* as early as January 1833, when the *Beagle* was at Goree Sound. A third important specimen for my conjectured dating, also located by Mr. Gautrey, contains geology notes made at Hobart Town, Van Diemen's Land, and is dated 12 February 1836. This sheet is folded, I am informed, horizontally in half like the sheets used for the "Zoophyte Note", and unlike the folding of the January 1833 specimen. This same folding is found in the "Reflections on Reading my Geological Notes". Other undated specimens located by Mr. Gautrey and Professor Burkhardt consist of two pages of geological notes dealing with South America (DAR 42: 43). Paper type alone, therefore, would provide no firmer date than a January 1833–February 1836 range.

The use of this anomalous paper type for writing these synthetic reflections on the zoophytes does point to a late dating of the document on some other grounds, particularly in light of the use of this paper at Hobart Town in February 1836 in a dated document. The typical paper utilized for writings on corals and related questions in November and December 1835 is uniformly a paper bearing an "RM 1831" watermark, or a crown, on the half sheets. This is found in the Tahiti notes on the coral genus *Fungia* in November 1835 (DAR 31.2: 345–347) and the original draft of the "Coral Islands" essay (DAR 41), which Frank Sulloway (personal communication) dates to December 1835. This paper was also still available to Darwin in January, since it is used, I have determined, in the letter of Darwin to Susan Darwin from Sydney on 28 January 1836 (Down House *Beagle* documents). Darwin seems to have nearly exhausted his supply of this paper by Hobart Town, however, and complained to W. D. Fox in his letter of 15 February 1836 near the end of the *Beagle* stay, that he was running out of paper, and "after touching at King George's Sound/ we proceed to the Isle of France, — It will clearly be necessary/ to procure a small stock of Testament on the occassion [sic]." (Fox-Darwin Correspondence, Christ's College Cambridge, letter no. 48, cited by permission of the Master and Fellows of Christ's College). The exhaustion of the "RM 1831" paper would then be a plausible explanation for the return to the use of the large, somewhat coarsely-made sheets of "£" paper for synthetic reflections. By folding these horizontally in half, these would form sheets that would fit a wallet of the size necessary to hold the Zoology Diary sheets, which measure approximately 25×20 cm. for the full series.

The Zoology Diary itself reflects this paper shortage in an interesting and suggestive way. The entries at Hobart Town begin on the same sheet of "RM 1831" paper used for the November entries at Tahiti, and this same paper folded, forming sheets of the 20×25 cm. size, is used for the next six pages of Zoology Diary notes through the initial April 1836 entries at the Keeling Islands (through DAR 31.2: 352). The entries then continue on sheets of a C. Wilmot 1834 or "ME" bond, which form a remainder of DAR 31.2 through the entries at Bahia, Brazil in August 1836. However, on this same paper, but *out of sequence*, following the Keeling entries, are Hobart Town notes dated February, on experiments on land planaria commenced at Hobart Town, and continued until the early part of April (DAR 31.2: 363–366). This sequence strongly suggests that Darwin recognized at Hobart Town that his supply of paper of the appropriate size for the Zoology Diary should be conserved, and retrospective entries were then made later when a supply of the domestic British "G. Wilmot 1834" paper was obtained either at the Keelings in early April, where a small shipping settlement is reported (*Diary*, p. 397), or else at Port Louis, Mauritius, which was an active trading centre with numerous shops. Darwin records, at the end of the Field Notebook labeled "Sydney-Mauritius," a list of needed writing supplies and other items for purchase (Beagle Notebooks, Down House, see partial list in Darwin 1945, p. 252. Barlow's conjecture (1945) that this is a Cape Town list seems to have little support). If the G. Wilmot paper was obtained only at Port Louis, this would also require that the latter portion of the Keeling Islands Zoology Diary entries were retrospectively written.

This evident paper shortage at Hobart Town would then explain why Darwin would have reverted to the use of the oversize "£" paper for his reflections on the zoophytes, rather than writing them on the "RM 1831" paper he used for the "Coral Islands" essay. This would also explain the linkage in paper with the "Reflections on Reading my Geological Notes" essay. Although I have found there to be no internal barriers that preclude a composition date for this essay of February 1835, as argued by Hodge (this volume), Darwin also reports in his letter to W. D. Fox from Hobart Town that he has been drawing up "very imperfect sketches/ of the Geology of all the places, to which we/ pay flying visits" (Darwin to Fox, 15 February 1836, Christ's College, Cambridge archives letter 48). The appearance of synthetic geological reflections on this paper, folded in the same horizontal fashion, provides a suggestive, if not definitive, linkage of this document with the Hobart Town period.

Further external information on the dating of this manuscript can be obtained from the two uses of the spelling "corall", which Sulloway has identified as a key term in the dating of the *Beagle* manuscripts (1982c, p. 331). Although my data are fully consistent with Sulloway's claim that Darwin had uniformly altered his spelling to "coral" by the latter part of the voyage, the two uses of "corall" in this note do not preclude a composition data as late as February 1836, when the incorrect spelling was still used in a single instance in the Hobart Town note (see above, p. 105), and Sulloway (1983, p. 362) has agreed that this February usage is an authentic, if isolated, exception to his claim that "corall" was last used in December 1835. (Sulloway 1982c, p. 331).

While external means of dating this manuscript do not necessarily narrow the range of its composition to the early months of 1836, internal evidence is more convincing. Darwin's concern with determining the relationship of the solitary lamelliform coral genera *Actinia* and *Caryophillia* is encountered on several occasions in the Zoology Diary between September 1832 (DAR 30.2: 102) and April 1836 (DAR 31.2: 354), but the explicit attempt to draw the colonial invertebrate *Gorgonia* into this relationship would clearly seem to date to after October 1835, when *Gorgonia* was first studied in Galapagos observations, and its connection to *Actinia* and *Caryophillia* commented upon (DAR 31.2: 327–328. See above p. 104).

The clear analogy drawn in this document between "turf-forming plants" and "Corall-forming Polypi" is a novel development that I do not find in any document datable before 1836, and it seems only understandable in light of the Hobart Town reflections. Furthermore, the reference to "*tuft*" and "turf" in this note is itself an important indication of a developed theoretical position. Darwin makes several references to "tuft" formation with reference to the reproduction of colonial invertebrates and coralline algae in the Zoology Diary, referring to the small buds forming at the base of the branches that give rise to new individuals. At Hobart Town, however, he was finally convinced that the coralline algae not only produced these "tufts", but also that these "tufts" gave rise to extensions that were then capable of developing adventitious roots that formed new colonies, something he had seen in colonial invertebrates and in the production of apple trees by tip-layering, but never before in the corallines. The novelty of this observation is revealed more fully from his retrospective comments in his 1847 memorandum to W. H. Harvey in which he described in detail the significance he had seen in the Hobart Town observations on the coralline algae: "I examined some vigorous tufts/ still attached to the rocks, & in them/ I found a few of the lateral branches,/ with their heads drooping outwards & with// the terminal segments attached to the/ surrounding rocky surface & forming little/ expansions, whence new branches were to/ spring. Hence this Nullipora increases like/ a banyan tree". (Memorandum to W. H. Harvey, MS pp. 3–4, quoted with permission of the University of Dublin archives). The explicit analogy drawn with the reproduction in the Banyan tree is of interest, since this is a classic example of a higher plant that propagates by an adventitious root system developing from the contact of the main branches with the ground, and suggests an analogy with the turf-forming grasses that propagate by runners.

The inserted pencilled additions in the document suggest that Darwin had worked this over late in the voyage or even in England.

It is at the Keelings in April that Darwin had first observed true madreporean corals, and his reference to "my madrepore" would be difficult to understand before April 1836. The inserted reference to gas emission of the corallines in sunlight is a relevant issue since this would be a sign of plant, rather than animal status. This query may be the explanation for the experiments undertaken at Bahia, Brazil on the return in August 1836 when Darwin kept several corallines in sunlight to see if they emitted gas (Memorandum to Harvey, MS p. 5). I am deeply indebted to the careful and penetrating comments of Frank Sulloway, and the important assistance and comments of Peter Gautrey and Frederick Burkhardt, for helping me refine these conclusions on the dating of this Zoophyte Note. I am more generally indebted to the landmark scholarship of Sulloway's 1982c paper for drawing my attention to the importance of the spelling changes in the dating of the *Beagle* manuscripts. Responsibility for these claims is, of course, my own.

59. On this see Hyman (1940, 1: 450).
60. Lamarck had placed these forms in a separate section of the group *Polypes*, and separated them from the one form of coralline algae admitted by him into the animals, the genus *Corallina*.
61. To avoid misunderstanding, my point here is that the theoretical unification of plants and animals that Darwin was finally willing to assert by 1836 removed one key conceptual barrier in the way of transformism, since this would allow that all life, and not simply the separate animal and plant series, could be unified in a common source. In this, Darwin had finally returned to key aspects of the thesis of Robert Grant we have explored previously, but now on a much more sophisticated level. This set of "zoophytal" issues provides, in my view, a critical component in the development of Darwin's transformism in the early months after his return to England. The context these give to the "infusorial monad" discussions of the B *Notebook* is being developed in my "Darwin, Vital Matter, and the Transformation of Species" (MS in preparation). I see my conclusions on these issues closely congruent with those of Hodge (1982). This is not to deny the deep, and even crucial importance of the biogeographical, paleontological and taxonomic issues that form additional layers to this question, as developed in Grinnell (1974) and Sulloway (1982c). I am suggesting that there are important interactions between the invertebrate issues I have developed and these better-known dimensions that need greater emphasis.
62. Sulloway has provided a detailed analysis of the dating of the "transformist" passages of the *RN* (1982, pp. 370–386). The passage in question follows in pencil immediately after the long penned discussion dated by Sulloway to 15 March 1837. It is separated from the continuous discussion on *RN* 127–131 by an ink line and one-half page of blank space. I have emended the Herbert transcription slightly, based on my reading of the original.
63. Darwin's concern with these issues can be followed in his annotations to Johannes Müller's *Elements of Physiology* (1838), and especially Henslow's *Descriptive and Physiological Botany* (1837a). According to the DAR 119 reading list, the first volume of Müller was read on 16 January 1840, and Henslow's text on 15 February 1840. Darwin's probably prior awareness of much of the content of the Henslow text by 1831 should be noted.
64. As was his common practice, Darwin began the A *Notebook* from both ends, and the location of these comments at the end of *A* suggests that they were probably entered early in the writing of the notebook.

4

DARWIN'S EARLY INTELLECTUAL DEVELOPMENT: AN OVERVIEW OF THE *BEAGLE* VOYAGE (1831–1836)

Frank J. Sulloway

Introduction

In December 1831 H.M.S. *Beagle* departed England on a five-year circumnavigation of the globe. The principal objectives of the *Beagle*'s voyage were to survey the southern coast of South America and to perform a series of chronometric measurements around the world. On board as ship's naturalist sailed a young man, Charles Robert Darwin, who had yet to pass his twenty-third birthday. Earlier that year Darwin had taken a degree at Cambridge University, without honors, in preparation for becoming a clergyman. His self-described qualifications for the post of ship's naturalist were those of an amateur "hunter of beetles, and pounder of rocks", and he was, in fact, the third person to receive the offer (*LL* 1:254). In the words of Professor John Stevens Henslow at Cambridge University, Darwin was quite simply "the best qualified person I know who is likely to undertake such a situation" (*LL* 1:192).

Despite his seemingly modest accomplishments, Darwin subsequently made a number of observations and discoveries during this five-year voyage that were to revolutionize the science of biology. Although Darwin's theory of evolution by natural selection did not become widely known until the publication of the *Origin of Species* (1859), his theory was developed in many of its most essential features by 1838, within two years of his return to England. Quickly recognized as one of England's foremost men of science, Darwin was elected to the Council of the Geological Society of London in 1837 and became its Secretary in 1838. Even before publication of his *Journal of Researches* (1839), he was known well enough that young Joseph Hooker, upon applying in early 1839 for the post of ship's naturalist aboard H.M.S. *Erebus*, was told by Captain Ross that only such a person as Mr. Darwin would be accepted. And to this Hooker replied, "what was Mr.

D[arwin] before he went out? . . . the voyage with FitzRoy was the making of him . . ." (L. Huxley 1918, 1:41).

The role of voyages in the education of nineteenth-century naturalists has yet to be given sufficient attention, as Mendelsohn has noted (1964, p. 53). Charles Lyell, T. H. Huxley, Alfred Russel Wallace, Henry Walter Bates, and Joseph Hooker — to name just a few naturalists — all took extensive voyages to other parts of the world in the tradition of the great Alexander von Humboldt, whose multi-volume *Personal Narrative* of the places he visited served as a model for Darwin's own *Journal of Researches*.[1] Unfortunately, Darwin's *Journal of Researches*, largely rewritten for publication after his return to England, reveals little of the actual change that the *Beagle* voyage brought about in his life. As one commentator has asserted, both the "before" and "after" are apparent in Darwin's *Journal*, but the transition is nowhere to be seen (Hyman 1962, p. 14).[2] Partly owing to this circumstance, many of Darwin's biographers have pointed to certain changes in Darwin's thinking manifested in the *Journal* and have assumed that he reached these conclusions on, or near, the dates recorded in that work. The chronology of Darwin's intellectual development during the voyage has therefore remained problematic on many important points.[3] In addition, documentation of possible influences on young Darwin, and of various precursors of his ideas, has not succeeded in putting his voyage experiences within a satisfactory perspective.[4] For it is first necessary to know what Darwin was thinking *at the time* in order to determine what effect, if any, such influences may have had on his intellectual development.

Darwin's various autobiographical recollections about the voyage have done little to rectify the historian's problems. Part of the difficulty stems from Darwin's portrayal of the voyage as a source of intellectual discontinuity — indeed, as a distinct and crucial watershed in his life. In his *Autobiography* he asserted, for example: "The voyage of the *Beagle* has been by far the most important event in my life and has determined my whole career; . . . I have always felt that I owe to the voyage the first real training or education of my mind" (pp. 76–77). This autobiographical assessment has provided a congenial model for those who would emphasize the *Beagle* voyage's remarkable transforming influence on Darwin. Indeed, in its most dramatic form, the story of the *Beagle* voyage has often been portrayed as the *Origin of Species* "writ large", a tendency that Himmelfarb and other recent Darwin scholars have sought to counteract with a more realistic reconstruction of the voyage period. As Himmelfarb has commented in this connection: "There is, in fact, no real continuity between the Beagle and the *Origin*. Between the two there intervened an idea" (1959, p. 123). More accurately, there intervened a series of ideas; and the proper dating of these has accordingly played a key role in recent reassessments of the *Beagle* voyage and its role in Darwin's life.

In particular, Darwin's conversion to the theory of evolution — once thought to have been a typical "eureka" experience stemming from his famous visit to the Galapagos Archipelago — is now generally seen as a slow and largely post-voyage development in his scientific thinking (Sulloway 1982c). Deprived of Darwin's conversion (perhaps the most famous symbol of its transforming role in Darwin's life), the *Beagle* voyage remains, more than ever, a seemingly epic event lacking sufficient visible signs of the hero's remarkable transition. Part of the problem is that Darwin's biographers have traditionally sought evidence for Darwin's intellectual development in a fairly restricted domain, namely, in his purely scientific work. There is far more to Darwin's voyage development, however, than his scientific observations and emerging ideas during the five-year period. This development was also closely associated with Darwin as a person; and the key to Darwin's intellectual development lies in unraveling this intricate connection.

The Technique of Content Analysis

In an effort to pinpoint the elusive transition that the *Beagle* voyage represents in Darwin's life, I have employed a somewhat specialized technique known as content analysis. I have applied this technique to a selection of Darwin's voyage letters, primarily those written to Darwin's former professor John Stevens Henslow (Darwin 1967). Both the use of content analysis and the choice of the documents to which it has been applied require some explanation.

Content analysis generally involves a word-by-word analysis of documents in an effort to reveal certain overall themes and patterns. As a technique, content analysis varies considerably in complexity, from the level of simple word counts to far more sophisticated procedures involving multivariate analysis of word co-occurrence patterns within specified units of text. In addition to elucidating potentially significant associations among words, content analysis is also frequently used to analyze relationships among *categories*, or groupings of thematically related words. Because many words are sufficiently synonymous in ordinary usage, they may often be treated, for the purposes of content analysis, as representatives of the same basic category. For example, *I, me,* and *myself* all have in common their reference to a category that might be termed SELF. Similarly, *you, your,* and *yourself* form part of a contrasting category that might be designated OTHER. By formulating a comprehensive series of such categories, it is possible to use them as the basis for comparison of different texts, as well as to analyze changes in texts written over time. Content analysis has been used in this manner to analyze such diverse documents as folktales, political speeches and texts, newspaper editorials, short stories, letters, and autobiographies.[5] The content-analysis procedure is, of course, no substitute for the careful reading of documents. Indeed, detailed scrutiny of a document is a prerequisite for a successful content analysis. Nevertheless, computer-assisted content analysis can sometimes detect significant patterns of co-occurrence and dissociation that go unnoticed even in the most careful reading of a text. A further advantage of this procedure is that it is replicable by other investigators, thus tending to minimize the influence of various biases and expectations that occasionally interfere with the objective reading of texts. Although content analysis in no way guarantees objectivity, it does entail numerous clearly articulated methodological constraints, which can in turn be related directly back to the conclusions that are derived.

In this study I have applied the technique of content analysis to a series of Darwin's voyage letters. My selection of these documents was determined by several considerations. First, unlike Darwin's *Journal of Researches* or other scientific publications stemming from the *Beagle* voyage, the text chosen for analysis ought to offer a contemporary and unrevised account of Darwin's voyage activities. Potentially suitable in this connection are Darwin's voyage *Diary* and his various letters to family members and friends (*LL*; Darwin 1945; 1967). From this sizable choice of materials, a selection was made in order to bring the amount of text within feasible limits of analysis.[6] Darwin's *Diary* was rejected — in part because of its length and in part because of its predominantly non-scientific focus. Like the *Diary*, Darwin's voyage letters, especially those to his family, give only brief summaries of his scientific work. On the other hand, his voyage letters to his teacher John Stevens Henslow provide a nearly ideal text. These letters were intended to keep Henslow up to date on Darwin's scientific activities during the *Beagle* voyage, as well as to convey information concerning the shipment of specimens. The letters also contain numerous personal details about Darwin's life and thoughts during the voyage. Because they offer a regular and detailed series of scientific reports on his work as ship's naturalist, the letters to Henslow were chosen as the primary text for content analysis.[7] Three extremely short notes to Henslow, written in 1833 and 1834, were not included in the analysis.[8] The two longest of these letters, both written in 1834, were replaced instead by two more substantial and informative letters that Darwin wrote about the same time to a sister and to an old school chum. Altogether, the fifteen letters chosen for analysis encompass more than eighteen thousand words (or about seventy-two pages of double-spaced text) and average one letter every four months. The longest gap between letters is six months.

The first step in the content analysis was to enter the entire text of the letters into a computer and then to generate a key-word-in-context index (or concordance) to the correspondence. This 470-page index was then studied carefully and was used as a guide to formulating the various categories — or groups of similar words — that were judged most appropriate for this particular set of documents. I devised forty-two such categories, based in part on categories that have proved useful in previous content-analysis studies and in part on the nature of Darwin's voyage letters. For example, the category SPECIES includes the words *species*, *genus*, *family* and *order*; the category COLLECT encompasses words like *collect*, *collections*, and *specimens*; and the category OVERSTATE, which is common in other content-analysis studies, includes emphatic or exaggerated words like *always*, *every*, *exceedingly*, and *never*. Most categories, like BOTANY, ENTOMOLOGY, GEOLOGY, ZOOLOGY, DELIGHT, DISTRESS, SELF, and WE, are self-explanatory. The forty-two categories are listed in full in the Appendix. Altogether, they encompass more than four hundred different words, including almost every word in the letters used more than four times.[9]

As the next step of the content analysis I scored each letter for the presence of words within each category. For each letter there are accordingly forty-two category scores (the basic unit of the content analysis). Scoring was done with the help of the key-word-in-context index, in order to correct for idiomatic and other non-literal word usages.[10] The resulting category scores were normalized according to the length of the letter, thus eliminating an extraneous source of variation in the category scores from different letters. I then subjected the normalized category scores to factor analysis, a multivariate procedure that tests for the degree of association between variables and attempts to group them into interrelated clusters. More specifically, factor analysis takes all the category scores and tries to group together those categories that simultaneously and consistently have high or low scores within each letter. This process is somewhat like trying to depict the distribution of hundreds of pins in a pin cushion by imagining a very small number of pins that best describe the arrangement of all the others. Four such factors, accounting for 60 percent of the variance in the category scores, were extracted by computer using a principal components analysis.[11] These factors are listed in Table 1 and will be explicated in more detail in the remainder of this study.

Table 1. Category loadings for Factors I-IV, grouped in order of absolute loadings [a]

Factor I: Growing Self-Assurance	I	II	III	IV
DATE	.81	.18	.05	.17
SIGN-STRONG	.80	−.22	−.19	.00
HOME*	.71	−.63	.00	−.04
DELIGHT*	.64	−.48	.00	−.32
THEORIZE*	.61	.43	−.48	−.06
GEOLOGICAL-CAUSE*	.40	.80	.24	−.02
SIGN-WEAK	−.39	−.09	.09	.06
DISTRESS*	−.41	−.49	−.12	.56
FOSSILS*	−.47	.00	−.06	−.50
NON-SPECIFIC	−.48	−.21	−.22	.14
COLLECT	−.49	.14	.50	−.24
RESEMBLANCE*	−.56	.27	−.54	.18
NEW	−.59	−.16	.08	.28
SPECIES	−.65	.03	.02	.15
SIZE-REFERENCE	−.66	.23	.13	.10
ZOOLOGY	−.94	.00	−.05	.03

Table 1. (Cont.)	I	II	III	IV
Factor II: Dependence versus Independence				
GEOLOGY	.28	.88	.01	.09
GEOLOGICAL-TIME	.31	.81	.25	.10
GEOLOGICAL-CAUSE*	.40	.80	.24	–.02
SPACE-REFERENCE	.24	.53	–.16	–.23
QUANTITY-REFERENCE	–.16	.52	.49	.04
BOTANY	–.27	.50	–.09	–.27
THEORIZE*	.61	.43	–.48	–.06
CONTRAST	.15	–.43	–.21	–.34
DELIGHT	.64	–.48	.00	–.32
DISTRESS*	–.41	–.49	–.12	.56
SELF	.16	–.59	.35	–.17
HOME*	.71	–.63	.00	–.04
TIME-REFERENCE	.16	–.66	.35	.26
OTHER*	.30	–.67	.47	–.10
NEGATION	.14	–.67	–.09	–.20
IF	.11	–.69	.00	.34
Factor III: Anxiety versus Involvement				
ANXIETY	–.15	.23	.84	–.11
HOPE	–.05	–.07	.52	.32
FUTURE	.06	–.27	.51	.11
COLLECT*	–.49	.14	.50	–.24
ENTOMOLOGY	–.21	–.09	.49	.15
QUANTITY-REFERENCE*	–.16	.52	.49	.04
OTHER*	.30	–.67	.47	–.10
COMMUNICATE*	.25	–.23	.42	–.59
WE*	–.07	–.34	–.40	.63
THEORIZE*	.61	.43	–.48	–.06
RESEMBLANCE*	–.56	.27	–.54	.18
PLACE-REFERENCE*	.03	.07	–.62	.42
INTERESTING	–.07	–.15	–.72	–.01
OVERSTATE	–.05	–.20	–.76	–.02
CURIOUS	.12	.19	–.77	.11
Factor IV: Group versus Individual Identity				
UNDERSTATE	–.20	.10	–.06	.77
VOYAGE	–.36	–.08	.03	.66
WE*	–.07	–.34	–.40	.63
DISTRESS*	–.41	–.49	–.12	.56
APPEAR	–.03	–.11	.03	.56
PLACE-REFERENCE*	.03	.07	–.62	.42
FOSSILS*	–.47	.00	–.06	–.50
COMMUNICATE*	.25	–.23	.42	–.59
SCIENTISTS	.17	–.38	.05	–.69
MY	–.21	.01	–.34	–.83

[a]Categories with absolute loadings of .40 or greater on more than one factor are marked with an asterisk and are listed multiply. Loadings vary within a maximal range of ±1.0.

I have interpreted and given labels to the four factors extracted in this study by examining those letters, and especially those sentences within each letter, that contribute most heavily to the highest associating categories. In other words, by retrieving numerous passages that contribute most directly to the factor structure of the letters, one begins to understand precisely what that structure means in *literary* terms. In this sense the factors are merely guides to an interpretation of the letters by the more customary procedures of historical scholarship. Thus the text of Darwin's voyage letters, not the categories used in the content analysis, is what ultimately has determined the identification and description of the four factors. Moreover, it is this final step in identifying the factors that allows the historian to interpret Darwin's letters in the context of Darwin's own particular literary style, judged not in terms of single words (or their purely literal meanings) but rather in terms of the many personal and social conventions that Darwin observed throughout the voyage in expressing these words as integral parts of sentences. In short, content analysis does not by any means rob Darwin's letters of their numerous linguistic subtleties. This procedure simply provides a preliminary vehicle for attempting to understand the letters — as Darwin wrote them — within the wider framework of his voyage experience as a whole.

Factor I: Growing Self-Assurance

The first factor extracted, which accounts for 17 percent of the total variance, has as its highest positive loading the DATE of the letter.[12] Indeed, the basic underlying dimension of Factor I is time, with the two polar ends of the factor representing early and late preoccupations in the contents of Darwin's voyage letters. Inasmuch as Factor I accounts for the greatest amount of variation in Darwin's letters, this factor not only makes intuitive sense but it also gives promise of clarifying the elusive "transition" that is absent from Darwin's formal scientific writings about the voyage.

The underlying theme represented by the negative end of Factor I (and hence the early period of the voyage) is Darwin's preoccupation with the description and cataloging of his collections, especially those in zoology. Those categories with the highest loadings (ZOOLOGY, SIZE-REFERENCE, SPECIES, NEW, RESEMBLANCE, and COLLECT) reflect Darwin's initial exuberance at the discovery and collection of numerous biological specimens, many of them apparently new to science. This theme of discovery and enumeration is especially evident in Darwin's old hobby, entomology. From Rio de Janeiro, for example, Darwin reported in his first letter to Henslow:

> I have just returned from a walk & as a specimen [COLLECT] how little the insects [ENTOMOLOGY] are know[n]. Noterus [ENTOMOLOGY], according to the Dic. Class. contains solely 3 European species [SPECIES], I, in one

> hawl of my net took [COLLECT] five distinct species [SPECIES]. — is this not quite extraordinary? (Darwin 1967, p. 56; letter of 18 May 1832)

Repeated references to taking new species and genera account for the high loading that the category NEW has on Factor I. The high loading of SIZE-REFERENCE reflects Darwin's numerous expressions about the size of his specimens, as well as of his collections. In this connection Darwin asserted in his first letter to Henslow:

> if what was told me in London is true viz that there are no small [SIZE-REFERENCE] insects in the collections from the Tropics. — I tell Entomologists to look out & have their pens ready for describing. — I have taken as minute [SIZE-REFERENCE] (if not more so) as in England, Hydropori, Hygroti, Hydrobii, Pselaphi, Staphylini, Curculio, Bembididous insects etc etc. (1967, p. 55)

Similarly, Darwin boasted in his second letter to Henslow, "I made an enormous collection of Arachnidae at Rio" (1967, p. 58; letter of 15 August 1832).

The high score of RESEMBLANCE on Factor I is associated with Darwin's various descriptive comments about his collections, including the identity or resemblance of particular specimens to those described in reference works. Nevertheless the primary concern in Darwin's early voyage letters is not the analysis of systematic relationships per se but rather the problem of what names should be given to his various specimens. Although the category SPECIES (*species, genus, family, order*) appears frequently in the first two years of the voyage, more often than not the word *species* itself is used simply as a synonym for *specimen*. Similarly, Darwin's frequent use of NON-SPECIFIC words early in the voyage (for example, *it, thing, one, ones, some*) reflects his uncertainty about the precise identity of many of his zoological specimens, especially his fossil Mammalia.

Perhaps the most noteworthy aspect of Darwin's early letters to Henslow is Darwin's repeated use of expressions that indicate a lack of self-confidence in his own observations and opinions. In this connection, the categories SIGN-WEAK and DISTRESS have moderately high loadings on the negative end of Factor I. For example, Darwin confessed in his first letter to Henslow: "One great source of perplexity to me is an utter ignorance [SIGN-WEAK] whether I note the right facts & whether they are of sufficient importance to interest others. — In the one thing collecting, I cannot go wrong" (1967, p. 53). But even as a collector Darwin soon found himself faced with a source of considerable anxiety. "All I can say," he informed Henslow during the eighth month of the voyage, "is that when objects are present which I can observe & particularize about, I cannot [SIGN-WEAK] summon resolution to collect where I know nothing [SIGN-WEAK]" (1967, p. 58). Similarly, the high loading of SIZE-REFERENCE on Factor I^{-} is also closely associated with

Darwin's worries about the scientific value of his collections and the poor impression they might be making on Henslow. "And now for an apologetical prose about my collection," Darwin wrote in August 1832. "I am afraid you will say it is very small [SIZE-REFERENCE], — but I have not been idle & you must recollect that in lower tribes, what a very small [SIZE-REFERENCE] show hundreds of species make" (1967, p. 58). And in the same letter Darwin remarked: "It is positively distressing [DISTRESS] to walk in the glorious forest, amidst such treasures, & feel they are all thrown away upon one" (1967, p. 58). During the first year of the voyage, Darwin went so far as to discredit his own eyesight three times (in May, August, and November 1832); and he frequently attributed his success as a collector to "luck" (SIGN-WEAK) and his shortcomings to "ill-luck". In November 1832 he commented to Henslow: "As I have nobody to talk to about my luck [SIGN-WEAK] & ill luck [DISTRESS, SIGN-WEAK] in collecting, I am determined to vent it all upon you. — I have been very lucky with fossil bones; . . . as many of them are teeth I trust, shattered & rolled as they have been, they will be recognised" (1967, p. 61). Yet Darwin continued to be plagued by doubts about the value of his collections; and he wrote, a year later, of "not feeling quite sure [SIGN-WEAK] of the value of such bones as I before sent you" (1967, p. 81).

In short, the first two years of Darwin's voyage correspondence with Henslow reflect his underlying conception of himself as an insufficiently trained naturalist who had been sent out to collect specimens by the real scientists back in England.[13] Somewhat jokingly, Darwin even described himself in August 1832 as "a Baron Munchausen amongst Naturalists", an allusion to Rudolf Erich Raspe's fictional character known for his fabulous and exaggerated adventures.[14] On a more serious level Darwin commented to J. M. Herbert, an old schoolmate at Cambridge, during the second year of the voyage: "By the way, you rank my Natural History labours far too high. I am nothing more than a lions' provider: I do not feel at all sure that they will not growl and finally destroy me" (*LL* 1:248).

If the negative pole of Factor I may be said to center around the description and enumeration of Darwin's voyage collections, the positive end concerns ideas and opinions (THEORIZE). In contrast to the anxious and insecure self-image in the early voyage letters, a confident self-image is increasingly manifested in the later letters and is especially associated with the high loading that SIGN-STRONG has on the positive end of Factor I. This trend may be seen not only in Darwin's scientific work — especially in the field of geology — but also in his general observations about the places and peoples he had recently visited. For instance, in a letter written during the fifth year of the voyage Darwin enthusiastically praised the work of the Tahitian missionaries, and he discussed at length the marvelous development of England's grand colony Australia.[15] Phrases like "I think", "I suspect", "I believe", and "I firmly believe" are peppered throughout these discussions and underscore Darwin's confidence in his own opinions. Nevertheless, it is primarily in connection with Darwin's geological work that we can see most dramatically the transformation in his self-identity from collector to thinker, as reflected by Factor I. This transformation is particularly evident if we consider Factor I in conjunction with the second of the four factors extracted from the correspondence.

Factor II: Dependence versus Independence

Like Factor I, Factor II accounts for 17 percent of the variance in the overall category scores. The clustering of high loadings for IF, NEGATION, and SELF on the negative end of Factor II (Fig. 1) is not uncommon in content-analysis studies and generally indicates a highly defensive style associated with personal uncertainty (Dunphy 1966, p. 331). Examination of the letters with the highest negative loadings on this factor supports this conclusion but suggests, in addition, that Darwin's defensiveness was closely related to his dependence on Henslow. It is this theme of dependence on Henslow that is responsible for the high loading of OTHER (*you, your, yourself*) in this context.

Darwin's defensive style in certain of his letters to Henslow was closely coupled, during the early part of the voyage, with his many self-doubts about his work as a naturalist and collector. For example he remarked in November 1832: "as for one Flustra, if [IF] I [SELF] had not [NEGATION] the specimen to back me up, nobody would believe in its most anomalous structure" (Darwin 1967, p. 63). A similarly cautious and defensive style, involving the co-occurrence of the categories IF, SELF, NEGATION, and SIGN-WEAK, occurs in another early letter. "If I am not mistaken," Darwin asserted in May 1832, "I have already taken some new genera [of spiders]" (1967, p. 55). The high score for TIME-REFERENCE on this same end of Factor II reflects Darwin's repeated need to account for his time and, in this connection, to excuse the poor impression that he believed his collections were making on Henslow. "I have collected during the last month nothing," Darwin confessed in a letter of 15 August 1832 (1967, p. 60). Similarly, he apologized in a letter of 11 April 1833: "And this makes up nearly the poor catalogue of rarities during this cruize" (1967, pp. 72–73).

Above all, these early letters reflect Darwin's feelings of personal responsibility to, and dependence on, Henslow, who had not only secured him the appointment as ship's naturalist on the *Beagle* but who had also agreed to take charge of all Darwin's collections sent home from South America. Henslow therefore assumed in Darwin's letters the simultaneous roles of father confessor, judge, and jury concerning Darwin's activities during the *Beagle* voyage.[17] As Darwin commented in August 1832, toward the end of the first year of the voyage: "I was not fully aware how essential a kindness you offered me, when you undertook to receive my boxes. — I do not know what I should do without such headquarters" (1967, p. 58). Similarly, statements such as "without you I should be utterly undone" are closely tied to Darwin's repeated requests for advice on packing and preserving different items (1967, p. 63).

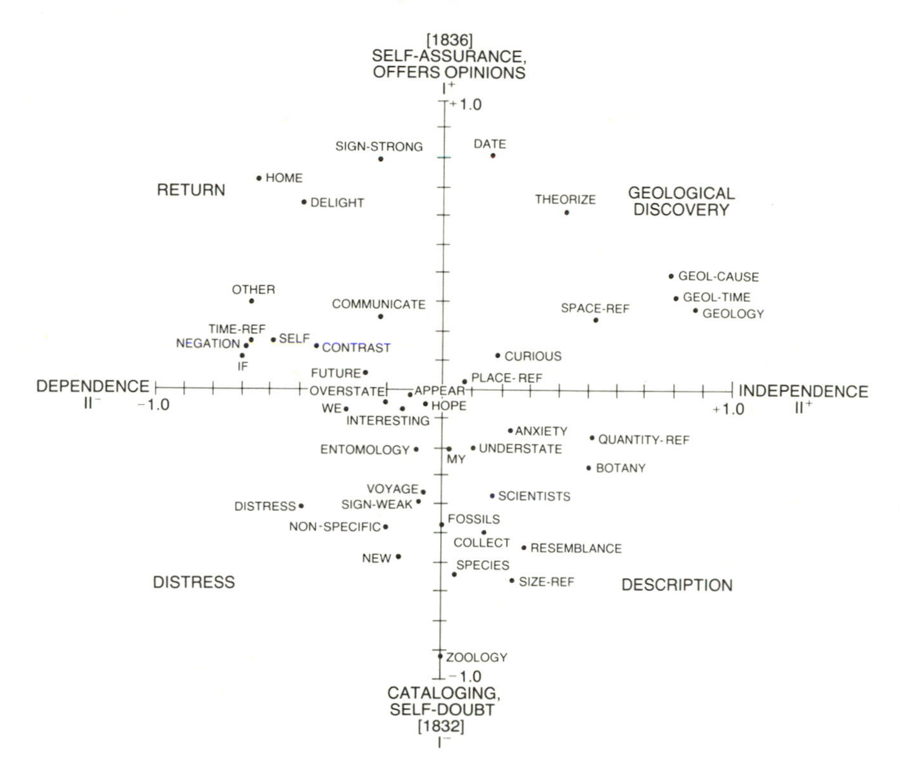

Figure 1. A two-dimensional plotting of the category loadings for Factors I and II (Table 1).[16] Factor I occupies the vertical axis; Factor II, the horizontal axis.

Toward the end of the first year of the voyage Darwin began a recurrent appeal for some reassurance concerning the fate and scientific value of those collections he had already shipped home. Owing to the vagaries of mail shipments to a surveying vessel that was constantly on the move, Darwin heard nothing from Henslow until the third year of the voyage. After waiting for more than a year without hearing from Henslow, Darwin reacted with a large drop in self-confidence, shown in Figure 2 by the document scores of his letters of November 1832 and April 1833.[18] Fearing that silence on Henslow's part signified his teacher's disappointment in his collections, Darwin continued to try to defend himself by emphasizing the amount of time spent by the *Beagle* at sea, and by reiterating his own preference for "the obscure & diminutive tribes of animals" (1967, pp. 64, 75). By mid-1833 Darwin's self-confidence had recovered somewhat, perhaps owing in part to his having hired a servant to help him in his collecting activities. This enabled him to promise Henslow that there would be "a larger proportion of showy specimens" in the future (1967, p. 75; letter of 18 July 1833). Still, Darwin's dependence on Henslow's approval remained unabated during the first two years of the voyage. "I should be so much obliged," he begged Henslow in July 1833, "if you would write to me. — You only know anything about my collections, & I feel as if all future satisfaction after this voyage will depend solely upon your approval" (1967, p. 75). In short, the first half of the *Beagle* voyage was evidently a trying period for young Darwin owing to his nagging self-doubts about the value of his scientific collections, and his continued dependence for advice and encouragement on a strangely non-respondent Henslow.

With Darwin's receipt, in March 1834, of a very supportive letter from Henslow, and with his subsequent receipt in July of two other equally encouraging letters, Darwin's self-confidence was given a substantial boost (Fig. 2). A major change also occurred at this time in Darwin's overall

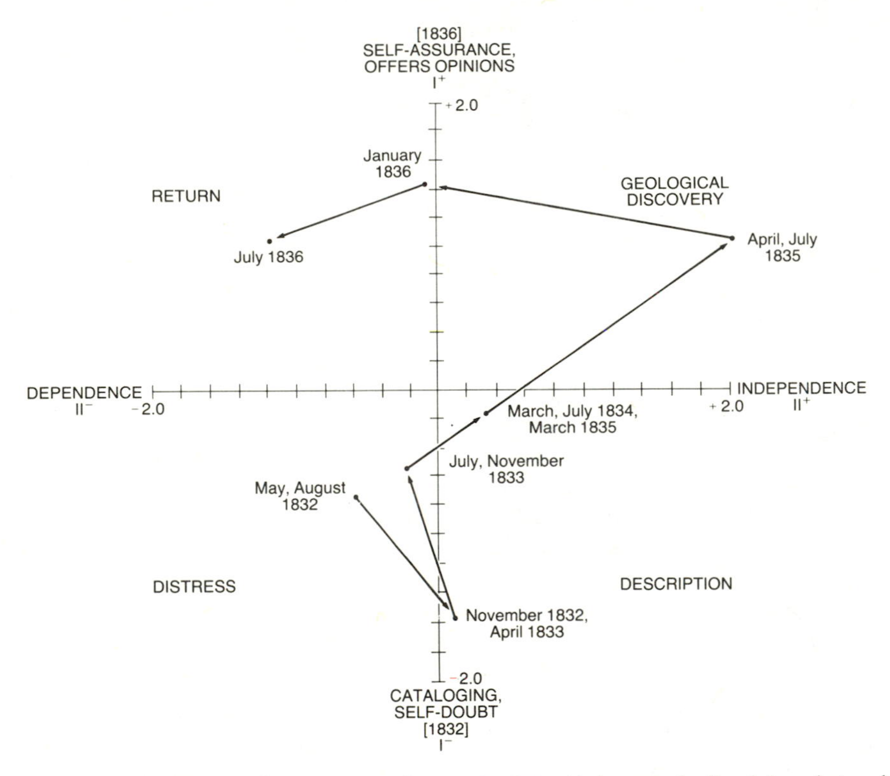

Figure 2. A two-dimensional representation of Factors I and II, with document loadings being substituted for category loadings. Hence the letters themselves, singly or in groups, can be followed temporally through the two-dimensional factor space.

attitude toward the voyage, which he now began to think of in predominantly positive rather than negative terms. At the same time, Darwin's dependence on Henslow — the student-teacher relationship of the earlier letters — began to give way to expressions of growing scientific independence, especially in the sphere of geology. The positive end of Factor II reflects this independent self-image that Darwin increasingly manifested after March 1834. At first his independence emerged in a roundabout way, as reflected by the high loadings that BOTANY and QUANTITY-REFERENCE have on the positive end of Factor II. Having ceased to worry about the value of his collections, Darwin apparently realized that Henslow — who, as a botanist, was naturally pleased with Darwin's plant collections — was now actually dependent on him for further botanical specimens. So the tables were turned! "I am very glad," he replied to Henslow's first letter in March 1834, "[that] the plants give you any pleasure, I do assure you I was so ashamed of them, I had a great mind to throw them away; but if they give you any pleasure I am indeed bound, & will pledge myself to collect whenever we are in parts not often visited by Ships & Collectors" (1967, p. 84). Darwin fulfilled this promise by subsequently sending many seeds and plants (hence the high loadings of BOTANY and QUANTITY-REFERENCE on Factor II) and by discussing those botanical specimens that he thought might especially interest Henslow.

It is in the field of geology, however, that Darwin's scientific independence emerged most clearly in the letters written after March 1834. Whenever Darwin had something of geological interest to write about, he no longer found it necessary to couch his scientific reports in the negative or qualified manner so common in earlier letters. Indeed, by the third year of the voyage Darwin had become sufficiently comfortable with his unresolved geological problems to joke openly about them to Henslow. "I am quite charmed with Geology . . .," he remarked in March 1834. "By the way I have not one clear idea about cleavage, stratification, lines of upheaval. — I have no books, which tell me much & what they do I cannot apply to what I see. In consequence I draw my own conclusions, & most gloriously ridiculous ones they are, I sometimes fancy I shall persuade myself there are no such things as mountains, which would be a very original discovery to make in Tierra del Fuego" (1967, p. 85). Not only did geology increasingly displace biology as Darwin's major preoccupation in the later voyage letters (Fig. 3),[19] but the nature of Darwin's geological discussions underwent a significant change associated with his growing self-confidence and independence of thought.[20] As may be seen in Figure 4, Darwin's references to GEOLOGY (a category composed of purely descriptive terms) actually declined slightly between the second and third years of the voyage. But his references to GEOLOGICAL-CAUSE and GEOLOGICAL-TIME — groups of explanatory terms that are combined and plotted together in Figure 4 as GEOLOGICAL-DYNAMICS — rose sharply over this same one-year period. What this change in geological

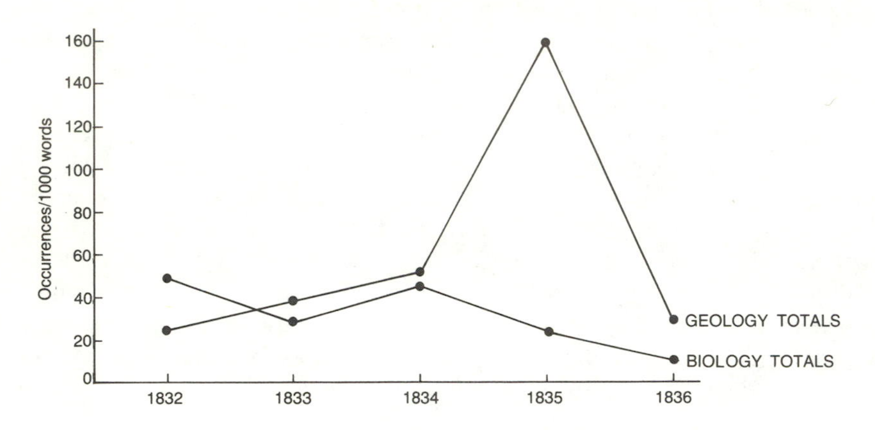

Figure 3. Category scores for BIOLOGY TOTALS and GEOLOGY TOTALS (N=643). BIOLOGY TOTALS=ZOOLOGY, BOTANY, ENTOMOLOGY, and SPECIES. GEOLOGY TOTALS=GEOLOGY, GEOLOGICAL-CAUSE, and GEOLOGICAL-TIME. Reflecting this pattern of changing scientific preoccupations, Darwin wrote to his cousin William Darwin Fox in July 1835: "I am so glad to hear you have some thoughts of beginning Geology. I hope you will; there is so much larger a field for thought than in the other branches of Natural History" (*LL* 1:263).

terminology reflects is Darwin's emergence as a *theoretical* geologist, a new self-image that was associated with three major geological discoveries.

The first of these discoveries was connected with Darwin's researches in southern Patagonia. In March 1834 he excitedly informed Henslow that "the whole of the East coast of South part of S. America has been elevated [GEOLOGICAL-CAUSE] from the ocean, since a period [GEOLOGICAL-TIME], during which Muscles have not lost their blue color" (1967, p. 84). This discovery suggested a very recent time scale for the elevation of the Andes, a conclusion that Darwin knew would greatly interest Charles Lyell, whose controversial uniformitarian views were highly consonant with such facts (1967, p. 93; letter of July–October 1834).

Darwin's second important geological discovery was made the following year, in 1835, and was communicated to Henslow in two letters that have the highest loadings on the positive end of Factor II (Fig. 2). Having seen the effects of the great Concepcion earthquake, and having investigated the geology in the areas of Chile and Valparaiso, Darwin was able to report to Henslow that he could "now prove that both sides of the Andes have risen in the recent period, to a considerable height" (1967, p. 101; letter of March 1835). The following month, when Darwin returned from a trip across the Cordilleras to Mendoza, Argentina, via two different mountain passes, he had even more remarkable confirmation of this view — namely, evidence that the age of the Andes, at a height of roughly 14,000 feet, was no greater than the Tertiary Period. By European standards this was very recent indeed; and Darwin was convinced by other geological

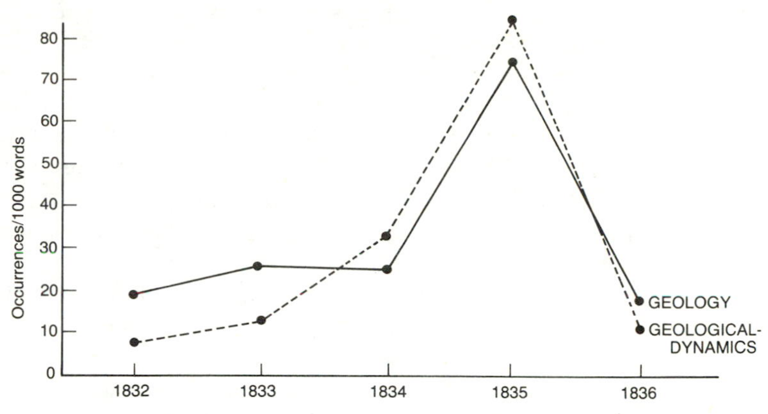

Figure 4. Category scores for GEOLOGY and GEOLOGICAL-DYNAMICS (N=425). The one-year change between 1833 and 1834 is statistically significant (χ^2=6.32, P<.02), as is the combined change between 1832–1833 and 1834–1835 (χ^2=12.43, P<.001).

observations that numerous periods of elevation and subsidence, involving vertical distances of thousands of feet, had all taken place while the Andes themselves were in a gradual process of formation and elevation to their present height (1967, pp. 102–107). As he asserted to his sister Susan in a letter of April 1835, "If this result [the modern age of the Andes] shall be considered as proved, it is a very important fact in the theory of the formation of the world. Because if such wonderful changes have taken place so recently in the crust of the globe, there can be no reason for supposing former epochs of excessive violence" (Darwin 1945, p. 117). By the Spring of 1835 Darwin therefore knew that he would be returning to England with exciting geological findings that would identify him as an active partisan of Lyell's uniformitarian doctrines. Darwin could thus see himself as part of the progressive side of a major revolution in nineteenth-century natural science.[21]

Darwin's third major geological discovery, his theory of coral reef formation, was developed in 1835 while he was still on the shores of South America and had yet to see a coral reef (*Autobiography*, p. 98).[22] In developing this novel theory, Darwin was correcting the views of his geological hero, Charles Lyell; and Lyell subsequently abandoned his own theory in favor of Darwin's. Lyell's theory was that coral reefs are formed on the tops of submerged volcanic calderas, thus explaining their circular form (1830–1833, 3: chap. XVIII). Darwin, in contrast, proposed that coral reefs originate as fringing reefs around volcanic islands, and that the subsequent subsidence of an island causes the coral, which grows upwards, to form a lagoon island in which the original volcanic pinnacle gradually disappears from view (*Coral Reefs*). It is curious that this important theory, which Darwin had a chance to test and confirm over the next year, was never communicated to Henslow. The most plausible explanation for this circumstance is Darwin's having learned, in June 1836, that Henslow had published extracts from Darwin's previous letters to him as a small pamphlet (*CP* 1:3–16; Darwin 1945, pp. 141–142). In choosing to withhold this geological success story in his next (and last) letter to Henslow, Darwin was effectively making sure that he, and not Henslow, would be the first to reveal his new theory to the British geological community. Thus Darwin's assertion of his intellectual independence from Henslow is evident not only from what he reported to Henslow in his later letters but also from what he elected to keep to himself.

Darwin's last two letters to Henslow exhibit a sharp return toward the vocabulary of dependency that was seen in the early letters of the voyage (Fig. 2). His impending return to England evidently reminded him that he was still greatly dependent on Henslow, not only for advice and assistance in connection with his voyage collections, but also for sponsorship within the formal institutional networks of British science. But in asking Henslow, for example, to propose him for membership in the Geological Society of London (July 1836), Darwin was simultaneously exhibiting an ambitious self-assurance that he clearly lacked at the beginning of the voyage. Such self-confidence is evident as well in Darwin's letters to his family. To his sister Caroline he remarked in April 1836, shortly after having buttressed his coral reef theory with researches at the Keeling Islands: "I am in high spirits about my Geology, & even aspire to the hope that my observations will be considered of some utility by real geologists" (1945, p. 138).

Factor III: Anxiety versus Involvement

Just as Factors I and II help to illuminate Darwin's development during the *Beagle* voyage, so the remaining two factors also add to the understanding of his voyage experience, especially when examined in conjunction with Factor I (time). Factor III, which accounts for 14 percent of the variance in the category scores, highlights Darwin's vacillations between uninhibited involvement in his researches, and his repeated anxieties over the merits of his scientific work.

The high loadings of ANXIETY, HOPE, FUTURE, COLLECT, OTHER, and COMMUNICATE on the positive end of Factor III signify Darwin's sense of anxious expectancy concerning his accomplishments as a collector (Fig. 5). Whereas Factors I^- and II underscore Darwin's preoccupation with his activities, and especially his identity, as a collector, Factor III^+ reveals the degree to which Darwin anxiously equated his future in science, at least during the early part of the voyage, with his *success* as a collector. In this connection the high loading of ANXIETY on Factor III is caused by Darwin's constant worry about the "safety" and "worth" of his collections. The high loadings of HOPE and FUTURE reveal his strong feelings of expectancy and apprehension concerning what Henslow (OTHER) will think about his specimens, as well as his concern over the care they might require once they arrived in England. Darwin repeatedly expressed his hope that various specimens would interest Henslow, and he commented frequently about when the next opportunity would arise to send (COMMUNICATE) specimens.

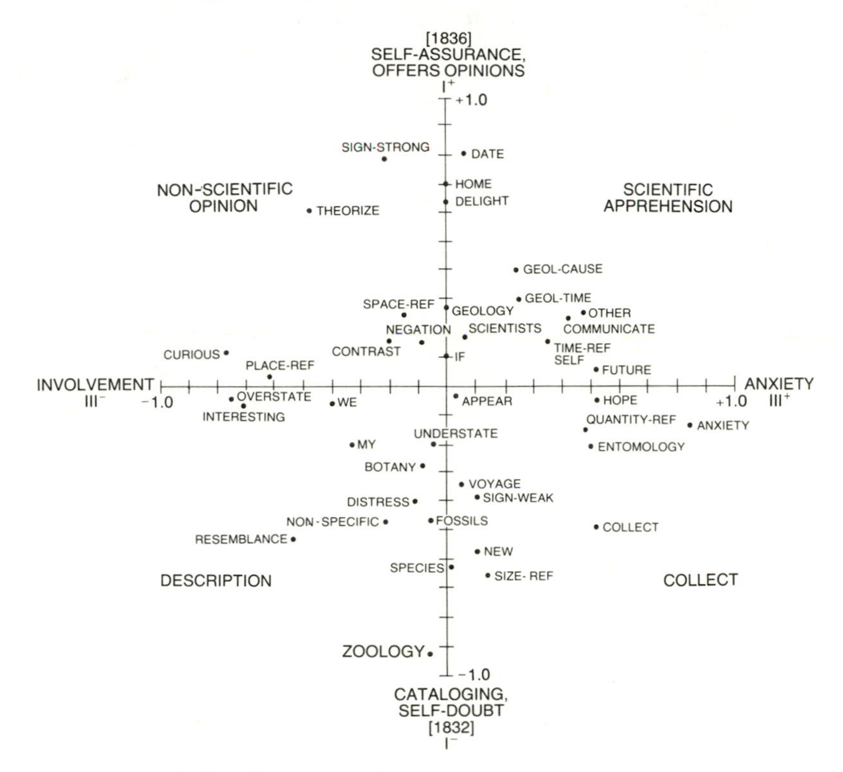

Figure 5. Category loadings for Factors I and III.

After hearing that a French collector had just preceded him around the Horn, Darwin complained of his "ill luck" and exclaimed to Henslow, "I am very selfishly afraid he will [FUTURE] get the cream of all the good things before me" (Darwin 1967, p. 61; letter of 24 November 1832).

Darwin's anxieties reached a peak during the second year of the voyage (Fig. 6), as Henslow's seeming failure to write to Darwin caused him to fear that his teacher was actually too embarrassed to admit how poor his collections really were. With his receipt, at last, of a letter from Henslow (March 1834), Darwin's anxieties temporarily subsided, allowing him to involve himself more freely in reporting the interesting details of his latest voyage findings. This change in the contents of the letters is reflected in the high loadings that the categories CURIOUS, INTERESTING, and PLACE-REFERENCE have on the negative end of Factor III (Fig. 5). When not dominated by feelings of anxiety, as he especially was during the second year of the voyage, Darwin tended to provide Henslow with relatively enthusiastic reports about the various peoples, places, and natural history objects he was seeing. The high loading that the category OVERSTATE has in these letters is caused by Darwin's frequent use of words like "most", "very", and "exceedingly" in describing his interest in what he has observed. These letters are filled with an uninhibited zeal and bring to mind Darwin's father's comment to Henslow, "There is a natural good humoured energy in his letters just like himself" (1967, p. 111).[23]

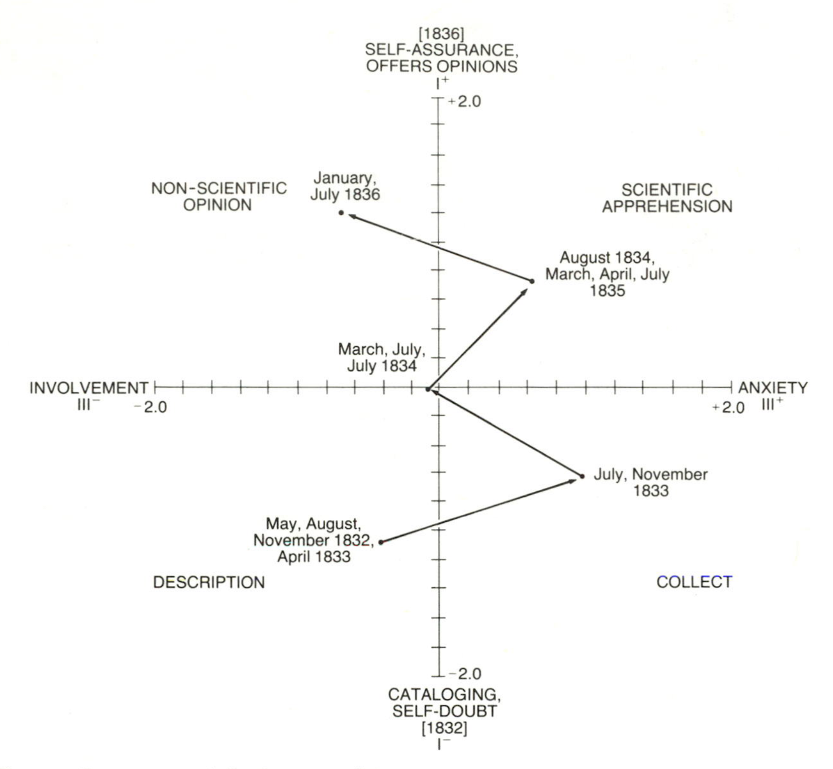

Figure 6. Document scores for Factors I and III.

As Darwin began to develop his geological views more fully during the fourth year of the voyage, the theme of anxiety returned once more to his letters (Fig. 6). This time, however, Darwin's anxieties were connected with ideas and theories rather than with his collections. After propounding a bold theory to Henslow, in July 1835, of how geological changes might be proceeding around the world in an orderly sequence, he added: "I am afraid [ANXIETY] you will tell me to learn my A.B.C. — to know quartz from Feldspar — before I indulge in such speculations" (1967, p. 110). But indulge in such speculations Darwin continued to do, and perhaps the most insightful of these "speculations" led him to his theory of coral reef formation. Toward the end of the voyage, when Darwin's letters became less theoretical owing to the large proportion of time spent at sea, the theme of anxiety once again gave way to a non-anxious involvement in reporting the events of the voyage (Fig. 6).

In examining the overall pattern of the voyage letters as plotted on Factors I and III, it should be noted that Darwin's feelings of anxiety manifested themselves independently of his growing intellectual self-confidence. During the voyage, anxiety was experienced whenever Darwin felt that he was at the limits of his expertise, whether in collecting or in theorizing. As his self-image changed, so then did the problems that were a cause of anxiety to him. Feelings of anxiety were not therefore something that Darwin outgrew during the *Beagle* voyage, like the numerous self-doubts that nagged him in connection with his initial collecting activities. Rather, the potential to experience intense anxiety appears to have been a fixed aspect of Darwin's personality, one that may well have been responsible, at least in part, for his lifelong nervous symptoms after his return to England (Colp 1977a).[24]

Factor IV: Group versus Individual Identity

Factor IV, which accounts for 12 percent of the variance in category scores, depicts Darwin's identification with his shipmates on the *Beagle* (WE) and, alternatively, his identification with his own work (MY) as ship's naturalist (Fig. 7). The positive end of this factor (group identity) centers around the theme of the voyage (VOYAGE), its moments of discomfort (DISTRESS), and its many future uncertainties. Darwin suspected from the very beginning that he would suffer from seasickness, and this suspicion was unfortunately confirmed throughout the entire course of the voyage. Before sailing, Darwin had taken the precaution of having his contract as ship's naturalist altered in order to allow him to leave the ship at any time he should choose (FitzRoy 1839, p. 19). Adding to this constant temptation to desert the voyage were FitzRoy's projections of its increasing length. Darwin had originally been informed that the voyage would last only two years, but by the time the *Beagle* reached South America this estimate had more than doubled (*LL* 1:193). From Rio de Janeiro Darwin wrote to Henslow that he was determined to give the voyage "a fair trial", but he also confessed that "I am sometimes afraid I shall never be able to hold out for the whole voyage. I believe 5 years is the shortest period it will consume" (Darwin 1967, pp. 52, 56; letter of 18 May 1832). Darwin's increasing uncertainty about his ability to stick with the voyage became especially manifest toward the end of the first year (Fig. 8) and is reflected in his frequent use of words like "only" and "nearly" (UNDERSTATE) in describing the future voyage schedule:

the only drawback is the fearful length of time between this & day of our return. — I do not see any limits to it: one year is nearly completed & the second will be so before we even leave the East coast of S America. — And then our voyage may be said really to have commenced. — I know not, how I shall be able to endure it. (1967, p. 63; letter of 24 November 1832)

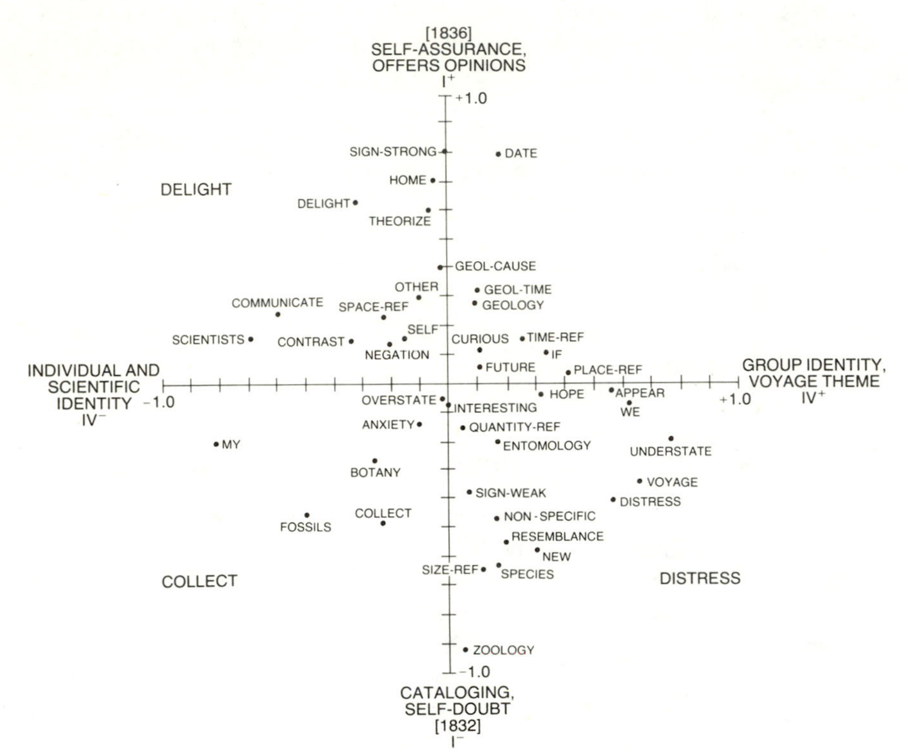

Figure 7. Category scores for Factors I and IV.

During the second year of the voyage, only the prospect of warm Pacific waters and a firsthand look at coral reefs seemed to keep up Darwin's resolution not to resign from his post (1967, pp. 74, 76; letters of 11 April and 18 July 1833).

What is particularly interesting about Darwin's vacillations between group and individual identity is the close association that these contrasting identifications have with Darwin's levels of expressed self-confidence. As may be seen in Figure 8, Darwin's strongest expressions of group identity were always associated with a drop in his level of self-assurance. The first of these confidence-reduction episodes occurred toward the end of the first year of the voyage in connection with worries about completing the voyage, together with increasing doubts about the merits of his scientific work (letters of November 1832 and April 1833). His subsequent acquisition of a servant to assist him in his collecting labors appears to have boosted his self-assurance and, at the same time, to have increased his sense of individual identity (letters of July and November 1833). Then, in March 1834, word finally came that Henslow was greatly impressed with Darwin's collections. This news, which precipitated another round of growing self-confidence coupled

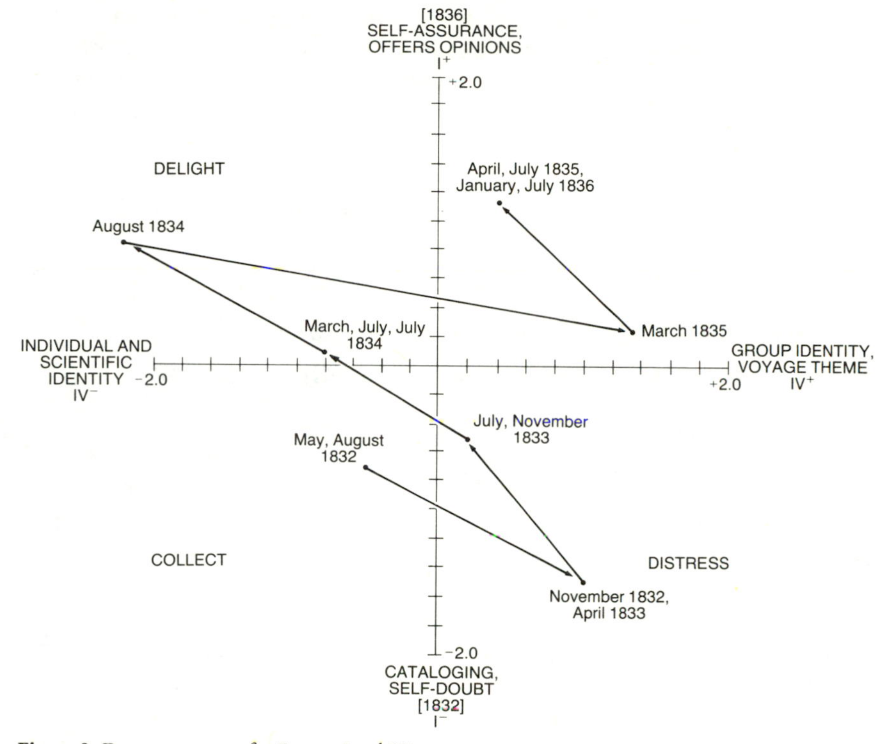

Figure 8. Document scores for Factors I and IV.

with a further increase in Darwin's individual identity, reinforced his resolve to "stick to the voyage", even though, as he quipped, "this may last till we return a fine set of white-headed old gentlemen" (1967, p. 86; letter of March 1834). This trend of growing self-confidence reached a peak in a letter of August 1834, in which Darwin expressed his delight over the reports of eminent scientists regarding the value of his fossil Mammalia (SCIENTISTS, COMMUNICATE). At the same time, he warned that under no conditions should any of the labels be removed from his specimens (MY), lest their scientific value to *him* be destroyed.

Seven months later Darwin once again resumed his identification with his *Beagle* shipmates, jokingly lamenting that the voyage would last "nearly as long as a seven years transportation". No longer concerned about lasting the voyage, Darwin now found his self-confidence temporarily lowered by a new and contrasting realization, namely, that he would have "but little opportunities for Natural History" during the remainder of the voyage (1967, p. 100; letter of March 1835).[25] With his subsequent geological discoveries about the recent formation of the Andes (letters of April and July 1835), Darwin's self-confidence was restored once more, as also was much of his sense of individual identity.

In short, Darwin's identification with the group (his *Beagle* shipmates) appears to have provided a comforting retreat from individual identity in the face of various distressing or confidence-lowering thoughts about the voyage. Like most human beings, it was apparently much easier for Darwin to suffer with others than to suffer alone. Yet any extensive identification with his *Beagle* shipmates was always relatively brief and was easily overridden by his contrasting identity as a scientist whenever his self-confidence was on the rise.

Summary and Conclusion: The Nature of Darwin's Voyage Transformation

Both common sense and the computer-aided content analysis described in this study agree that time (Factor I) is the single most important variable influencing the substance of the letters included in this study. In examining Factor I (growing self-assurance over time) conjointly with three other non-temporal factors, I have followed this representative selection of Darwin's voyage correspondence through a series of thematic patterns that reflect various changes in mood, in preoccupation, and, even more fundamentally, in Darwin's basic personality. As such, these thematic patterns provide a general "study guide" for understanding both the letters and the man who wrote them.

The most important of the *non*-temporal patterns associated with the voyage correspondence are Darwin's alternation between a deferent-defensive dependence on Henslow and his efforts to assert his own independence (Factor II); Darwin's capacity for intense involvement in his work, on the one hand, and his recurring anxieties concerning himself and his researches (Factor III); and, finally, Darwin's identification with his *Beagle* shipmates and, alternatively, with himself as an individual and a scientist (Factor IV). Problems of dependency, anxiety, and identity are basic aspects of human personality; and one is thus tempted to wonder to what extent these same themes continued as important preoccupations in Darwin's later correspondence, not only with Henslow but also with other friends and colleagues.[26] Nevertheless, the question of the generalizability and permanence of these non-temporal themes of the correspondence is one that goes well beyond the scope of this study.[27]

As for Factor I, which highlights the temporal changes in the correspondence, this aspect of the content analysis indeed gives us some insight into the elusive personal and intellectual transformation that Darwin underwent during the *Beagle* voyage. Perhaps no two categories sum up this transformation better than COLLECT and THEORIZE. As may be seen in Figure 9, the early voyage letters are dominated by Darwin's concerns as a collector of specimens, and they reflect his image of himself as an errand boy sent out by the bona fide scientists back in England. With the development of Darwin's identity as a geologist, especially during the third

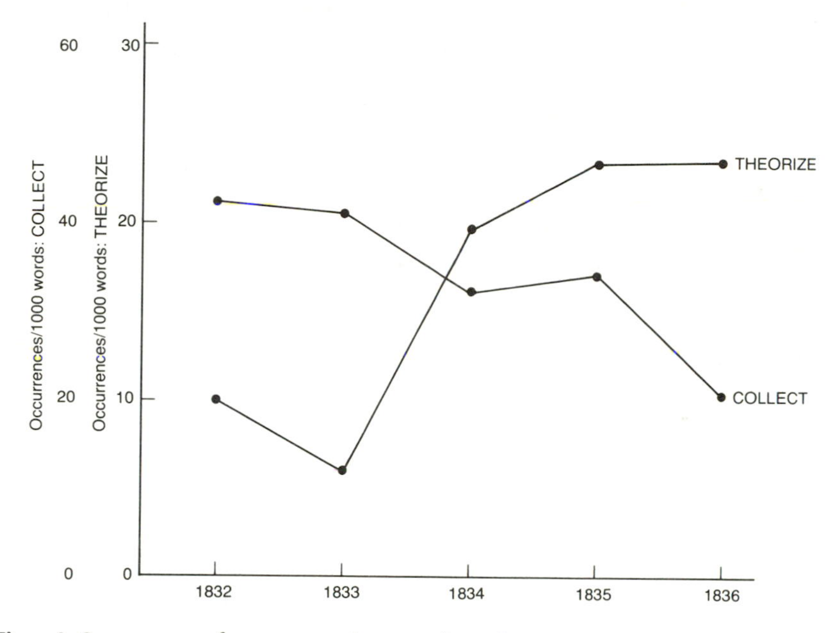

Figure 9. Category scores for THEORIZE and COLLECT (N=359).

year of the voyage, a dramatic change began to take place in Darwin's whole conception of himself. The self-doubting collector became an increasingly confident theorist who could even joke to Henslow about his propensity for drawing "gloriously ridiculous" conclusions. Four decades later, when discussing the voyage in his *Autobiography*, Darwin touched on this aspect of his intellectual development:

> Looking backwards, I can now perceive how my love for science gradually preponderated over every other taste. During the first two years my old passion for shooting survived in nearly full force, and I shot myself all the birds and animals for my collection; but gradually I gave up my gun more and more, and finally altogether to my servant, as shooting interfered with my work, more especially with making out the geological

> structure of a country. I discovered, though unconsciously and insensibly, that the pleasure of observing and reasoning was a much higher one than that of skill and sport. (pp. 78–79)

No matter how "unconsciously and insensibly" this transformation in Darwin's attitude toward science may have occurred, it remains vividly preserved in the text of his voyage letters to Henslow.

Paralleling and, to some extent, building upon this key transformation in Darwin's scientific identity is the related change that occurred in his general level of intellectual self-confidence. This transition is most readily captured by the category scores for SIGN-STRONG and SIGN-WEAK, words that denote self-assurance and self-doubt, respectively (Fig. 10). Like

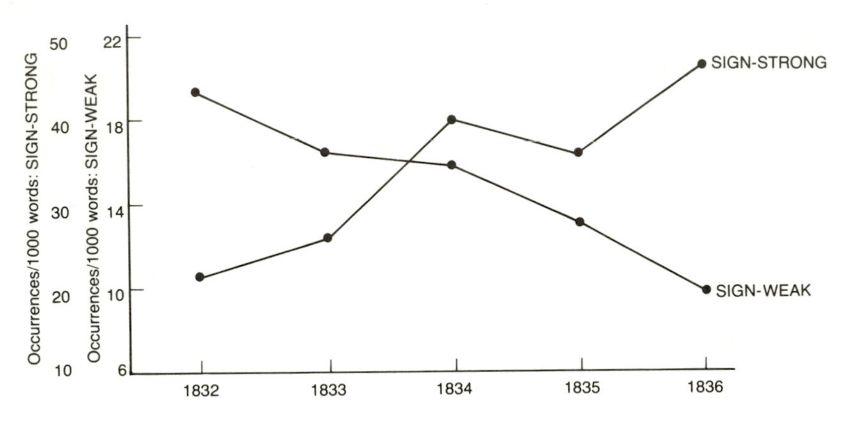

Figure 10. Category scores for SIGN-STRONG and SIGN-WEAK (N=323).

the category scores for COLLECT and THEORIZE, those for SIGN-STRONG and SIGN-WEAK exhibit the same basic crisscrossing pattern when plotted by year of the voyage. They capture a marked transformation in the *tone* of Darwin's voyage letters and exhibit the steadily growing self-esteem that accompanied Darwin's maturation as a scientist and a thinker. Together, the dual transformations that are apparent in Darwin's voyage identity and self-confidence are probably the single most important legacy of the *Beagle* voyage, providing an essential part of the psychological substratum from which Darwin's scientific genius emerged. The man who could write in his last voyage notebook that "Geology of whole world will turn out simple" was clearly not the same person who, four years earlier, had repeatedly doubted the accuracy of his own eyesight.[28] It is particularly this human and personal side of Darwin's intellectual development that later tended to disappear from his formal accounts of the *Beagle* voyage as soon as he began to rewrite his manuscripts for publication. And this is a major reason why the *Beagle* "transformation" has apparently remained so elusive, since even a detailed reconstruction of Darwin's scientific development aboard the *Beagle* is only part of an equally significant and harder-to-document development that occurred in Darwin as a human being.

One last question still remains to be considered in this survey of Darwin's intellectual maturation during the *Beagle* voyage. What about Darwin as a biologist? I have purposely left this question to last, and shall treat it only briefly here, since much of Darwin's development as a biological *theorist* — especially his conversion to the theory of evolution — was a post-voyage episode in his scientific career. This is not to say that Darwin failed to exhibit significant development in his biological views during the *Beagle* voyage. As both Hodge (1982) and Sloan (this volume) have shown, Darwin's considerable interest in problems of marine invertebrate zoology led to important changes in his thinking during the voyage — changes that were subsequently to become closely integrated with his earliest attempts to formulate a general theory of transmutation.[29] In this connection it is worth emphasizing that Darwin's voyage interests and intellectual transformations as a biologist were associated in large part with those fields, like ornithology, entomology, and marine invertebrate zoology, in which he had already deeply immersed himself prior to commencing the voyage.[30] But Darwin's overall intellectual development in these and other biological disciplines consisted primarily in acquiring a greater breadth and depth of knowledge about natural history, not in revolutionizing this field, as has so often been thought. In short, devoted as he was to natural history during the voyage, Darwin simply did not possess sufficient expertise, self-confidence, or theoretical vision as a biologist to develop intellectually in the same way that he did as a geologist. Moreover, surprising as it may seem, the category ZOOLOGY (which includes marine invertebrate zoology) has the highest *negative* association with time (Factor I, –.94) of all the categories included in this content analysis.[31] As reported in his letters to Henslow, Darwin's principal scientific preoccupations clearly lay elsewhere.

In recent years the myth of the *Beagle* conversion, long upheld by Darwin's biographers, has finally been laid to rest by Darwin scholars. It is now known, for example, that Darwin left the Galapagos Archipelago in October 1835 without fully realizing or accepting the evolutionary evidence offered by these famous islands. In fact, Darwin failed to collect specimens of the famous Galapagos tortoises for scientific purposes, mistook certain species of "Darwin's finches" for the forms they appear to mimic, and muddled his ornithological collections so hopelessly by island that he was later forced, after his return to England, to borrow the carefully labeled collections of other shipmates in order to test his newly dawning evolutionary suspicions (Sulloway 1982a, 1982b, 1982c). Even then, what was remarkable about Darwin's conversion to the theory of evolution was that it occurred on the basis of evidence that remained sketchy and ambiguous at best. Although John Gould, Richard Owen, George Waterhouse, and other systematists did much to enlighten Darwin in the Spring of 1837 regarding the full biological significance of his collections, and although these naturalists were able to rectify certain key errors in Darwin's voyage classifications of his specimens, the resulting evidence for evolution was by no means overwhelming. As Darwin himself confessed to Joseph Hooker ten years after his Galapagos visit: "I cannot tell you how delighted and astonished I am at the results of your examination [of the Galapagos plants]; how wonderfully they support my assertion on the differences in the animals of the different islands, about which I have always been fearful" (*LL* 2:22).

Thus Darwin was convinced of the mutability of species based on biological evidence whose complete validity he continued to doubt for nearly a decade! Moreover, of the many naturalists who worked on, or heard detailed scientific discussions about, Darwin's *Beagle* collections, Darwin was the only one compelled to interpret this evidence in terms of the heterodox theory of evolution. How, then, was his conversion possible at all if so many other naturalists, equally or more knowledgeable than he about the science of biology, were unconverted by the *Beagle* "evidence"?

The myth of the voyage conversion has long obscured this interesting historical problem, as well as the nature of Darwin's voyage development more generally. With the myth finally dispelled, along with its "eureka-like" emphasis upon the importance of scientific "facts", one can now see that the key to Darwin's conversion lay as much in Darwin himself as it did in the famous voyage that he undertook. Five years on board the *Beagle* taught Darwin to think for himself and allowed him, especially through his geological work, to envision himself as a theoretician with a penchant for far-reaching explanations and universal laws. Once the anxious collector on the *Beagle* was transformed into an increasingly bold geological theorist, Darwin was able to transfer his developing intellectual talents to many other related fields of science. Thus the influence of the *Beagle* voyage transcended any particular scientific field or discovery on Darwin's part. In the process, the voyage provided Darwin with something much more important, namely, the opportunity to mature intellectually under highly auspicious circumstances and thereby to become the Darwin that history now celebrates.

In concluding, it is appropriate to ask whether the content-analysis procedures employed here have revealed anything that would not otherwise have been apparent from a careful reading of the documents. An answer to this question depends, in part, upon what one means by *apparent*. At one historiographical level, scholars have certainly recognized that Darwin matured intellectually during the *Beagle* voyage and that this development entailed, among other things, a significant increase in his self-confidence. This study, however, suggests something more noteworthy, namely, that Darwin's personal transformation in self-confidence and self-identity — not any specific scientific discovery or his famous Galapagos visit — was actually the *Beagle* voyage's most important contribution to his subsequent success in science. This is not to say that Darwin's scientific work on the voyage is any less significant than historians have generally believed, but rather that its significance *for Darwin* can only be understood within the parallel context of his personal and psychological development on the *Beagle*.

It is for these reasons that, when the technical aspects of this research were first carried out, more than a decade ago, the principal results came to me as a distinct surprise. Having previously chosen the Henslow correspondence as a vehicle for tracing Darwin's *scientific* development, I did not anticipate that a content analysis would underscore the predominance of *psychological* themes in these letters. Indeed, it would seem that I had actually stacked the deck in the opposite direction by including such a high proportion of purely scientific categories in the study. Yet virtually none of these scientific categories ended up "defining" the four factors extracted in the analysis. Ironically, the computer, one of the greatest symbols of dehumanization in present-day society, succeeded in highlighting Darwin's distinctly "human" preoccupations over his strictly scientific ones.

Similarly, even after repeated readings of Darwin's voyage letters, I was not prepared to find a strongly negative association between Darwin's zoological interests and those categories that reflect his emergence as a *thinking* man of science. Like other Darwin scholars before me, I had read these letters with certain dominant interests and expectations in mind; and it was not therefore surprising that I continued to have these expectations fulfilled as long as I was free to concentrate upon those aspects of the letters that I and others considered of greatest importance. In this respect historical scholarship is really no different from science itself; in both fields of research one naturally looks for and tends to find what the current consensus suggests as the expected result.

By altering some of the basic assumptions that historians now share about Darwin, the progress of Darwin studies over the last decade has made some of the findings of this study less novel than they perhaps once were. Nevertheless, Darwin scholarship is still largely preoccupied with the scientific and intellectual, rather than the personal and psychological, aspects of Darwin's life. Within the burgeoning "Darwin industry", the man has become overshadowed by his own increasingly disembodied concepts.[32] In this connection Darwin's manuscripts and published works have tended to take precedence over those "unwritten" aspects of Darwin's life, such as the intricate dynamics that characterized his personality and intellectual style, that can only be reconstructed with great difficulty. Admittedly, making inferences about a great thinker's psyche is a notoriously subjective business, as may be seen in the highly problematic genre of literature that has come to be known as psychohistory.[33] But the absence of a reliable methodology for understanding the minds of people who produce great thoughts should not deter us from seeking interconnections. I hope that this study, which has sought to integrate Darwin's personal development with his scientific work during the *Beagle* voyage, will be seen as a step in that direction.

ACKNOWLEDGEMENTS

I thank the following persons for their advice and criticisms in connection with this study: Frederick Burkhardt, Mary S. Campbell, Ralph Colp, Jr., Stephen J. Gould, Judith V. Grabiner, Jerome Kagan, David Kohn, Ernst Mayr, Daniel M. Ogilvie, Sydney Smith, and Philip J. Stone.

Appendix:

LISTING OF CATEGORIES

ANXIETY: afraid, anxious, fear, safety, trouble, worth.

APPEAR: appear, feel, look, observe, see, seem, think.

BOTANY: plant(s), seeds.

COLLECT: box(es), cask(s), collect, collected, collecting, collection(s), find, finding, found, specimen(s), taken, took.

COMMUNICATE: account, hear, heard, letter(s), say, says, said, send, sending, sent, tell, write, written.

CONTRAST: different, distinct, than.

CURIOUS: curious, peculiar.

DATE: *date of letter.*

DELIGHT: delight, delightful, glad, glorious, pleasant, pleasure.

DISTRESS: bad, difficult, disappoint, distressing, dread, growl, ill, miserable, regret, sad, sick, suffer, weary.

ENTOMOLOGY: insect(s), *proper names of insects.*

FOSSILS: bones, fossil(s), Megatherium.

FUTURE: opportunities, opportunity, shall, till, will.

GEOLOGY: bed(s), chain, coast, cordilleras, formation(s), geological, geology, land, lava(s), rock(s), sandstone, strata, structure.

GEOLOGICAL-CAUSE: action, active, alter, altered, alternate, deposited, depression, elevated, formed, injected, owing, produced, sedimentary, undulations, upheaval, volcanic.

GEOLOGICAL-TIME: age, ancient, epoch, modern, period, *proper names of shells*, recent, shell(s), succession, tertiary.

HOME: Cambridge, England, home, return.

HOPE: hope, want, wish.

IF: although, could, excepting, however, if, may, might, or, whether, would.

INTERESTING: extraordinary, interesting, wonderful.

MY: my.

NEGATION: not.

NEW: new.

NON-SPECIFIC: anything, it, one(s), some, something, somewhat, thing(s).
OTHER: you, your, yourself.
OVERSTATE: all, always, entirely, etc., ever, every, everything, exceedingly, excellent, fine, grand, great, how, least, many, most, much, never, no, nothing, quite, so, such, very, whole.
PLACE-REFERENCE: countries, country, here, place(s), *proper names of places*, where.
QUANTITY-REFERENCE: another, few, half, many, number(s), three, two.
RESEMBLANCE: belonged, identical, like, relation, resemblance, same.
SCIENTISTS: Clift, Henslow, Jenyns, Lyell, Sedgwick, Whewell.
SELF: I, me, myself.
SIGN-STRONG: able, can, certain, clearly, could, good, must, no doubt, ought, really, should, sure, true, unquestionable.
SIGN-WEAK: cannot, doubt, doubtful, ignorance, imperfect, impossible, luck, mistake, nothing, not sure, poor, useless.
SIZE-REFERENCE: enormous, immense, large, little, minute, small.
SPACE-REFERENCE: between, lower, over, near, upper.
SPECIES: family, genus, order, species.
THEORIZE: because, believe, conclusion, consequence, convinced, fact(s), imagine, mind, probably, respecting, suppose, suspect, thus, understood.
TIME-REFERENCE: before, during, long, month(s), now, since, time, year(s).
UNDERSTATE: nearly, only, perhaps, rather.
VOYAGE: Beagle, cruize [*sic*], sail, sailed, voyage.
WE: we, our.
ZOOLOGY: animal(s), *proper names of animals*.

Notes

1. Humboldt (1769–1859), a Prussian, used his small inheritance to finance a five-year expedition to Latin America during the years 1799 to 1804. His thirty-four volume account of his travels, published over a twenty-nine-year period, was partially translated into English under the title *Personal Narrative of Travels to the Equinoctial Regions of the New Continent* (Humboldt and Bonpland 1814–1829). Humboldt's narrative of his explorations subsequently inspired many nineteenth-century naturalists, including Darwin, with a desire to travel. Just a month prior to receiving the offer to sail with the *Beagle*, Darwin wrote to his mentor Henslow: "I hope you continue to fan your Canary ardor: I read & reread Humboldt, do you do the same, & I am sure nothing will prevent us from seeing the great Dragon Tree [of Tenerife]" (Darwin 1967, p. 26). For further information on Humboldt's scientific career, see Biermann (1972).
2. Darwin's *Journal of Researches* was prepared from a diary that he kept during the voyage. The changes made in preparing this diary for publication are significant not only for what was added but also for what was deleted. Only about one-half of the original diary appeared in the 1839 edition of the *Journal*, and even less survived in the revised second edition (1845). As Gruber (1981a) has shown by a careful collation of the changes made in the *Journal* between the first and second editions, this work continued to reflect significant revisions associated with Darwin's intellectual development. Even Darwin's voyage *Diary*, published by Nora Barlow in 1933, provides only a partial record of Darwin's intellectual development on the *Beagle*. This is largely because the *Diary* was intended as a personal rather than a scientific record of his travels and was kept separate from his still unpublished notes on geology and zoology. Only occasionally does the *Diary* offer summaries of the more important scientific observations Darwin was making. Darwin's unpublished scientific notes, which were also kept in diary form, are in the Darwin archive of Cambridge University Library (DAR 31–38).
3. The most valuable attempts to clarify this chronology are those of Gruber and Gruber (1962), Limoges (1970c) and Herbert (1974, 1977, 1980). Elsewhere I have shed new light on Darwin's visit to the Galapagos Archipelago and have provided a detailed account of Darwin's conversion to the theory of evolution (Sulloway 1979b, 1982a, 1982b, 1982c, 1983, 1984). The literature on Darwin's early intellectual development, and the role of the *Beagle* voyage in that connection, also includes contributions from Eiseley (1958), Himmelfarb (1959), Greene (1959a), S. Smith (1960), Wichler (1961), De Beer (1963), Ghiselin (1969), Gruber and Barrett (1974), Grinnell (1974), Keynes (1979), and Kohn (1980).
4. See, for example, Glass, Temkin, and Straus (1959).
5. For a selection of such studies, together with a detailed description of the content-analysis procedure, see Stone et al. (1966). This study differs in one important respect from most other content analyses, in that it deals with the work of a single individual who is also generally acknowledged to have been a genius. The use of computers and content-analysis procedures is increasingly common today in the fields of political and socioeconomic history. Nevertheless, there would appear to be some resistance toward applying such techniques in purely biographical research, especially when the figure involved, like Darwin, is eminent and already well studied. At the Darwin Centennial Conference (Florence, 1982) at which this paper was presented, several colleagues expressed their uneasiness and even antipathies in this regard. One colleague in particular asked me if I was not concerned that this study might inspire other computer analyses of Darwin's work, thus turning Darwin scholarship into a sort of "mindless" activity. Whether the application of computer-assisted research techniques in biography will prove to be more or less limited than in other fields of history remains to be seen; but such techniques are certainly no more inherently "mindless" than various other approaches to historical research. Moreover, computers do not usurp the historian's basic functions; rather, they provide a powerful instrument for advancing historical research in ways that would otherwise be excessively time consuming or virtually impossible.
6. The advent of optical readers and sophisticated word processors, which were not available when the technical aspects of this study were originally done (Sulloway 1969), has greatly transformed the potential use of content analysis by making it much easier (and less expensive) to analyze large volumes of text.
7. So impressed were Henslow and his colleagues

with the scientific caliber of these letters that they arranged for substantial portions of them to be privately printed as a pamphlet by the Cambridge Philosophical Society in December 1835, during the fourth year of the voyage (*CP* 1:3–16). This was Darwin's second scientific publication — his first being records of about thirty insects collected in Cambridge, North Wales, and Shrewsbury (published in Stephens 1829).

Whether Darwin's 1835 pamphlet should actually be regarded as a "publication" has been questioned (Freeman 1977, p. 24), since "publication", in its narrowest twentieth-century sense, generally implies that the item concerned has been offered for public sale (whereas the pamplet was not). The broader essence of "publication", however, consists simply in putting an author's work into general circulation, which Darwin's letters to Henslow certainly were within the Cambridge-London scientific community. The printed circulation of the letters even made Darwin something of a scientific "celebrity" prior to his return to England, and parts of them were subsequently reprinted in the *Entomological Magazine, 3* (1836):457–460. It would therefore seem to be quibbling, especially by nineteenth-century standards, to say that Darwin's 1835 pamphlet was not "published" just because it was issued free and in a limited edition intended for distribution among fellow scientists. Under the 1976 United States copyright law (Public Law 94-553, §101), Darwin's pamphlet *would* constitute a publication.

8. Because the category scores for each letter must be normalized according to the length of the letter, extremely short letters tend to produce unrepresentative results and, for this reason, were omitted from the analysis.
9. Not included in the forty-two categories are approximately fifty words (such as common articles, prepositions, and adverbs) that account for 40 percent of the text of Darwin's letters. Another 35 percent of the text is encompassed by the words included in the forty-two categories. This leaves only 25 percent of the text, composed exclusively of words used four times or less, unrepresented in the chosen categories. At this level, adding more words to the categories (or enlarging the number of categories) would only slightly increase the amount of text encompassed in the study. For example, if every word used more than twice by Darwin had been included, the number of individual *words* encompassed by the categories would have doubled but the amount of *text* covered by the analysis would have increased by only 7 percent. For further information on the relationship between word frequency and text coverage, see Stone et al. (1966, pp. 164–165).
10. The expression *no doubt*, for example, connotes a sense of confident certainty (OVERSTATE, SIGN-STRONG) rather than the opposite (NEGATION, SIGN-WEAK). Similarly, a reference to the Cape of Good Hope would be scored under PLACE-REFERENCE rather than under SIGN-STRONG (*good*) and HOPE (*hope*).
11. The number of factors extracted was limited to four because additional factors accounted for a rapidly decreasing amount of variance. (The average variance accounted for by the first four factors is 15 percent per factor, versus only 6 percent per factor for the next four factors). These first four factors were rotated by the orthogonal varimax method in order to clarify their identity and to maximize their independency.
12. Category *loadings* are the measure of association of each category with the factor. Loadings under .3 in either direction are low, between .3 and .5 are moderate, and over .5 are high.
13. Considering the breadth of Darwin's knowledge in the fields of entomology, ornithology, marine invertebrate zoology, and geology, his self-doubts about his competence as a collector betray his desire to fulfill an unusually high set of standards in his scientific work. Darwin would not have been chosen for the position of ship's naturalist if Henslow or anyone else had doubted his abilities. Thus Darwin's self-doubts are of particular interest in revealing his *own* conception of himself, rather than the conception of his mentors.
14. See Raspe (1786) and Darwin (1967, p. 59; letter of 15 August 1832).
15. Darwin (1967, pp. 112–114; letter of January 1836). FitzRoy, impressed during the last year of the voyage by certain passages from Darwin's personal diary, invited Darwin to collaborate with him on an article discussing the work of the missionaries in the Pacific. This article, which included extracts from Darwin's *Diary*, was completed by June 1836, when the *Beagle* reached Cape Town; and the article was subsequently published in the *South African Christian Recorder* (*CP* 1:19–38). It was during the last year of the voyage that FitzRoy also invited Darwin to collaborate with him in publishing the official account of the *Beagle* voyage. FitzRoy's initial plan of citing extracts from Darwin's *Diary* was fortunately not carried out, and Darwin was allowed to

publish a revised version of his *Diary* as the third and last volume of the official record.

16. If sufficient categories are present to allow an accurate identification, labels are provided in the two-dimensional factor plots for those areas, at a forty-five degree angle from the main axes, that are formed by a fusion of the two factor spaces. In Figure 1, four such "fusion factors" are recognizable and are labeled accordingly.
17. "I look up to you as my father in Natural History", Darwin wrote to Henslow in July 1834, "& a son may talk about himself to his father" (1967, p. 95).
18. Because Factor II delineates Darwin's relationship with Henslow, Darwin's letters to Charles Whitley (July 1834) and Caroline Darwin (August 1834) are not included in the document scores plotted in Figure 2. It should perhaps be mentioned here that Darwin received a now-lost letter from Henslow in April 1832, when he arrived in Rio de Janeiro (1967, p. 55). This letter was not in response, however, to any of Darwin's own voyage letters or shipments of specimens, the first of which were not sent until June and August 1832, respectively. Thus Henslow's lost letter could not have supplied Darwin with any feedback concerning the fate and value of his collections or the merits of his scientific work. Two more years passed before Darwin finally heard again from Henslow.
19. In Figures 3–4 and 9–10, the vertical scale ("occurrences/1000 words") is based on the total number of words included in the forty-two categories (6,389). See further note 9.
20. The relative relationship between BIOLOGY TOTALS (ZOOLOGY, ENTOMOLOGY, BOTANY and SPECIES) and GEOLOGY TOTALS (GEOLOGY, GEOLOGICAL-CAUSE, and GEOLOGICAL-TIME) in Figure 3 is almost identical to the relationship observed by Gruber and Gruber (1962, p. 191) in counting the number of manuscript pages that Darwin devoted to these two subjects during each year of the voyage. Only the thirteen letters to Henslow have been used in compiling Figures 3 and 4. Some portion of the year-to-year differences seen in Figure 3 can be attributed to the different opportunities offered in various fields of natural history by the east and west coasts of South America; but, in general, the totals seem to reflect a genuine shift in Darwin's scientific interests.
21. In the *Origin* Darwin later praised Lyell's *Principles of Geology* (1830–1833) with the comment: "the future historian will recognize [this work] as having produced a revolution in natural science . . ." (p. 282).
22. Darwin's claim in his *Autobiography* (p. 98) that he developed his coral reef theory while still on the South American continent is confirmed by the presence of several brief allusions to that theory in his "Santiago" pocket notebook. The "Santiago" notebook contains entries dating from late 1834 through mid-1836. Because Darwin, in his discussions of the coral reef theory, used the word *Pacific* twice without a terminal *k*, these passages can be dated to the period prior to mid-July 1835, when Darwin began spelling *Pacifick* consistently with a *k* (Sulloway 1983). (Darwin was in Lima, Peru, at this time.) The coral reef passages in the "Santiago" notebook probably date from March or April 1835, about the time of Darwin's transection of the Andes (and his confirmation of the geologically recent elevation of this mountain chain).
23. Robert Darwin's comment was made in reference to the receipt of Henslow's printed pamphlet of extracts from Darwin's voyage letters (*CP* 1:3–16). These extracts tend to lack those passages in the original letters that convey Darwin's repeated anxieties and self-doubts. Hence the extracts represent those portions of the letters producing high category loadings on Factors I^+ (self-assurance), II^+ (independence), and III^- (involvement).
24. Darwin experienced fairly severe somatic symptoms of anxiety (such as upset stomach and palpitations of the heart) on several occasions before he went on the *Beagle* voyage. After the voyage these symptoms became chronic beginning with the Fall of 1837, shortly after Darwin opened his first notebook on the transmutation of species and as he was finishing the proofs for his first book. In later years, Darwin's symptoms were often greatly relieved by hydropathy treatments, which generally involved a cessation of work on Darwin's part. Darwin personally associated his symptoms with hard "mental work" or "excitement", and he would take the water cure or a vacation in order to clear his mind of his scientific thoughts. As soon as he returned to his work, however, his symptoms quickly reappeared. See further Colp (1977a).
25. The precedence that Darwin's scientific interests by this time had gained over the issue of the voyage's length may be seen in the following intention, which Darwin briefly entertained in late 1834. When Captain FitzRoy invalided himself from nervous exhaustion in November of that year (requiring the *Beagle*, by standing orders, to

return immediately to England via the Atlantic), Darwin quickly formulated a plan to remain behind in Chile and Peru for at least a year in order to explore the Cordilleras in detail (Darwin 1945, p. 111; letter of 8 November 1834 to Catherine Darwin). FitzRoy's subsequent reconsideration of his decision to give up command of the *Beagle* was welcomed by all, Darwin included, who was especially anxious to see the coral islands of the Pacific.

26. The remainder of Darwin's correspondence with Henslow, who died in 1861, exhibits many of these same basic patterns of communication. Nothing, for example, could be more defensively deferent (Factor II) than Darwin's 11 November 1859 letter to Henslow, in which he informed Henslow that he was sending him a copy of the *Origin of Species* but did not think that Henslow would at all approve of his book. Like those voyage letters scoring highly on the negative end of Factor II, this letter is studded with words referable to the categories IF, NEGATION, and SELF. Other letters written to Henslow between 1837 and 1860 are also frequently deferent and dependent in tone when requesting scientific information (a literary style that Darwin seems to have perfected over the years for just such requests). On the other hand, many letters, such as Darwin's 1 April 1848 announcement of his discovery of complementary males in cirripedes, score comparatively low in the categories IF, NEGATION, and SELF (1967, pp. 158–161, 200), thus reflecting Darwin's independence from Henslow, especially during periods of important scientific discovery.

27. Inasmuch as the Darwin letters project, which will eventually publish Darwin's known correspondence in full, has been conducted with the aid of computers, it may someday be possible to test this and other questions in a systematic manner. One should not, of course, expect the three non-temporal factors (II–IV) found in this content analysis to be *absolutely* identical in Darwin's later correspondence, either with Henslow or with other correspondents. Although independent of time during the five-year period of the *Beagle* voyage, these factors might well be time-dependent over a longer span. Similarly, Darwin's dependence on other naturalists, including Henslow, for information in connection with his researches would perhaps manifest itself somewhat differently according to the relative age and status of his correspondents, or according to how well Darwin personally knew them. Thus the nature of any "factors" present in Darwin's complete correspondence would inevitably prove somewhat different from the ones manifested in this study. The same point naturally applies to Darwin's voyage correspondence with his family and peers, although Factors I, III and IV are obviously important themes in these other letters, and Factor II (dependence versus independence) is manifested as well, albeit in a somewhat different context (dependence with regard to news, family gossip, and especially money matters). Finally, Darwin's continued development after the voyage, which was associated with many new preoccupations as well as with ongoing changes in his personality, introduced into his correspondence numerous new themes that might well provide the basis for important new "factors".

28. See Darwin's *Red Notebook* (*RN* 50). This sentence was probably written about mid-August 1836. The ambitious, free-flowing speculations put forth in this last voyage notebook provide a record of Darwin's developing thoughts between May or June 1836 and the opening of the first notebook (*B*) on the transmutation of species in July 1837. Regarding the dating of this notebook, which contains Darwin's first evolutionary speculations, written about 15 March 1837, see Herbert (1980) and Sulloway (1982c).

29. I do not believe, however, that Darwin's voyage researches in invertebrate zoology were in any way a *necessary* precondition for his eventual conversion to the theory of transmutation, contrary to Sloan's suggestion (this volume). Rather, it seems clear that the three major classes of facts that Darwin himself later cited as having converted him to a transmutationist position (namely, his South American fossil Mammalia, patterns of geographic distribution among living South American species, and the evidence of the Galapagos Archipelago) were the necessary and nearly sufficient intellectual causes of the conversion (*Autobiography*, pp. 118–119). See further, Sulloway (1982c).

30. In light of the increased attention that has been paid in recent years to Darwin's voyage researches in marine invertebrate zoology, I have reanalyzed the Henslow correspondence in order to distinguish between Darwin's discussions within six distinct biological fields. The six fields are listed here in descending order of their contributions to the BIOLOGY category totals. In addition, the *continuity* of Darwin's interest in these six fields is

reflected, albeit only approximately, by the proportion of discussion occurring during the two halves of the *Beagle* voyage: (1) botany (59%/41%) — clearly botany, hardly one of Darwin's major preoccupations during the voyage, rates highly primarily because of his correspondent's interests in the field; (2) invertebrate zoology (63%/37%); (3) fossil vertebrate paleontology (96%/4%); (4) entomology (58%/42%); (5) vertebrate zoology (94%/6%); and (6) ornithology (71%/29%). Altogether, Darwin devoted more space in his letters to the subject of vertebrate zoology (including ornithology and fossil vertebrate paleontology) than he did to invertebrate zoology (including entomology). Nevertheless, his discussions about invertebrate zoology are more evenly distributed throughout the two halves of the *Beagle* voyage (61%/39%) than are his discussions of vertebrate zoology (91%/9%). At least some of this disparity is undoubtedly owing to the differing opportunities offered for observations and researches in these two general fields as the voyage progressed.

Insofar as these statistics may differ somewhat from those derived from an analysis of Darwin's *Beagle* Zoology Diary (Sloan, this volume), it must be emphasized that Darwin reported to Henslow only what he considered to be of greatest scientific importance, either to himself or to his correspondent. The letters therefore act as a sort of information "filter" — a filter separating out the most significant features of Darwin's voyage thoughts and discoveries, as *he* perceived them at the time, from his researches as a whole.

31. The category ZOOLOGY has as its two most *negatively* correlating categories in this study DATE and SIGN-STRONG (words denoting self-confidence). Both correlations are statistically significant ($P < .01$). The correlation between ZOOLOGY and THEORIZE is $-.49$, which is nearly significant at the level of $P < .05$. In other words, as the voyage progressed, Darwin discussed zoological subjects less and less; whereas his self-confidence and his theoretical interests both increased with time.
32. The work of Gruber (Gruber and Barrett 1974) and Colp (1977) nevertheless represents an exception to this general trend, although even Gruber has favored a "cognitive" over a broadly psychological approach to Darwin's creativity. R. Porter's apt comment about Darwin scholarship is also relevant in this regard: "Whereas the advancement of science used to be the biography of great men, . . . academic history of science has increasingly, in the name of scientific and professional standards, disparaged personal focus. Its goals have become to study problems not people, issues not individuals, ideologies not inspiration. . . . Thus it is striking that no academic historian has written a biography of Darwin over the last twenty years" (1982, p. 18).
33. On the subjectivity of psychohistory, see Stannard's (1980) excellent review of the literature and its pitfalls.

5

OWEN AND DARWIN READING A FOSSIL: *MACRAUCHENIA* IN A BONEY LIGHT

Stan P. Rachootin

She knows the profits of it, but she don't appreciate the art of it, and she objects to it. "I do not wish," she writes in her own handwriting, "to regard myself, nor yet to be regarded, in that boney light."

Charles Dickens, *Our Mutual Friend*[1]

The earliest record we have of Darwin theorizing about the transmutation of species was written in March 1837 in a notebook labeled *RN* (127–131). These few hundred words at first seem impenetrable. By virtue of their conciseness and obscurity, they seem almost like the great thought caught at the moment of its birth. No bit of Darwiniana has inspired so much careful study; it is the Darwinist's shroud of Turin.

Sandra Herbert (1980) has given us the entire text of the notebook and she has used the contents of the rest of the notebook to show that this passage was written after Darwin's return to England, though the notebook was opened on the last leg of the *Beagle* voyage. David Kohn (1980) has ably traced the major themes that twine through this passage and that tie it to the text of the *Journal of Researches*. He gives us a convincing series of transmutation theories based on species senescence and sexuality, that, although they eventually give way to natural selection, sound as undertones in Darwin's later work.

The forest of themes and metaphors in the *RN* transmutation theory has been mapped out. Here, I look at some twigs and leaves of one of the trees. Or, more accurately, the arteries that run closely with the cervical vertebrae to the brains of camels and some other animals. The *RN* transmutation theory, in common with most scientific theories, embeds a series of facts, and it is the transmission, ontogeny, and disintegration of these facts that I examine here. If the facts are understood, we gain a sense of the particulars of science and of genius. They suggest connections of larger theoretical issues in natural history, and they offer a different picture of what Darwin was saying in this early theorizing. If my reading of the internal evidence is correct, the *RN* passage is less the birth of the great idea, than the afterbirth.

In *RN*, Darwin sets out a comparison between two pairs of animals he encountered on the pampas of Patagonia. The common rhea is widely distributed, but a smaller species, which Darwin was the first to collect, is found only in areas where the larger species is absent. Darwin found the bones of an extinct quadruped, and soon after he returned to England, Richard Owen, who described Darwin's fossil mammals, informed him that they belonged to a giant camel. Today the pampas swarm with small camels — the guanaco, and its domesticated form, the llama (then designated *Auchenia*). Darwin considers the replacement in space in the rheas to be comparable to the replacement in time in the guanacos. He considers the possibilities of competition, environmental change, and intrinsic tendencies toward what Kohn (1980) aptly calls species senescence as possible causes of the separations in space and time, which in one of the pairs has led to an extinction. He suggests that in both cases one of the pair gave rise to the other, by a saltation.

This theory, so different from Darwin's mature views, has a considerable difficulty. The fossil "camel", which some ten months later would be described by Owen under the name *Macrauchenia*, was not a camel. The relation between *Auchenia* and *Macrauchenia* partakes of the relation of a duck-billed platypus to a duck. Owen's revised analysis, which I shall argue he communicates to Darwin by the time Darwin writes the relevant passages in *RN*, removes the fossil not only from the family Camelidae, but even from the order Ruminantia (chevrotains, camels, deer, bovids, giraffes, and pronghorns). Owen assigns what he familiarly calls the macrauchene to the order Pachydermata (elephants, hippos, pigs, tapirs, rhinos, horses). As Darwin says in the *Origin*, "Cuvier ranked the Ruminants and Pachyderms, as the two most distinct orders of mammals" (p. 329). Such a chasm makes ideas of species transmutation wildly inappropriate.

Darwin's first theory depends on a certain innocence about the home truths of comparative anatomy. In his case, what he did not see in the way of anatomy allowed him to read stories in bone that were invisible to Owen. Although historical sympathy is a great virtue, it is not clear that a degree of ignorance of the relevant anatomy exceeding Darwin's has led to compensatory insights in Darwin studies.

I. Owen's First Impression

On 23 January 1837, Owen writes to Charles Lyell, who is assembling materials for his anniversary address to the Geological Society, about the fossil mammals

Darwin collected in South America. He writes that the bones have been "again compared in a general manner". His list includes:

> RUMINANTIA
>
> Fam: *Camelidae*
>
> 2 cervical vertebrae, portions of femur, & fragments of a Gigantic Llama! as large as a Camel, but an *Auchenia* (from the plains of Patagonia)
>
> RODENTIA
>
> Great portion of the Cranium of a *Gigantic Rodent*; (size of a Rhinoceros) with some modifications resembling those presented in the cranium of the Wombat. (L. Wilson 1972, p. 437)

The former is to become *Macrauchenia*, and the latter, its foil in many of the arguments presented here, will be named *Toxodon*. Earlier, Owen has told Darwin about his assignment of the cranium to the Rodentia, for Darwin writes to Owen on 19 December 1836 in favor of taking a cast of the "great head of the Rodent" (De Beer, 1959b, p. 49)[2].

It is significant that Owen is so specific about the cervical vertebrae of the first fossil, because it is a single feature of those bones that unites the fossil with camelids. In all other mammals that had been surveyed, the pair of vertebral arteries pass through loops of bone, one on each side of the spinal canal (Fig. 1-A). These loops are perforations at the base of laterally projecting spurs of bone, the transverse processes. An artery threads through a loop on each succeeding cervical vertebra until it reaches the base of the brain, where it joins with its fellow from the other side. Camelids, which in the general case are reputed to be difficult to thread, have in this particular a very different way of routing their vertebral arteries (Fig. 1-B). The vertebral artery enters the posterior face of the vertebra in the spinal canal, adjacent to the spinal cord. About one-third of the way down the body of the vertebra, the artery perforates the pedicle of the neural arch and travels within this canal until it reaches the anterior face of the vertebra. It then enters the spinal canal of the succeeding vertebra and continues in the same manner until it reaches the atlas. The transverse processes are present, but they are not perforated. If we compare a camelid's cervical vertebra to, say, a deer's, we see on the deer's vertebra three openings in both anterior and posterior view — the large central lumen for the spinal cord, and the small foramina at the bases of the transverse processes. In the camelid, we see one opening in posterior view, the lumen that accepts both the spinal cord and the vertebral arteries, but three openings in anterior view — spinal canal, and, laterally to it, in the pedicle of the neural arch, the small foramina of the vertebrarterial canals. The macrauchene cervical vertebrae show exactly the same course of the vertebral arteries.

Could Darwin have figured this out on his own? He excavated the fossils, after all, and he writes of guanacos, "I particularly examined the bones" (*Journal of Researches*, p. 197). That vertebral arteries perforate the transverse processes of cervical vertebrae is not arcane knowledge; the perforated transverse process is the first character you look for, if you are

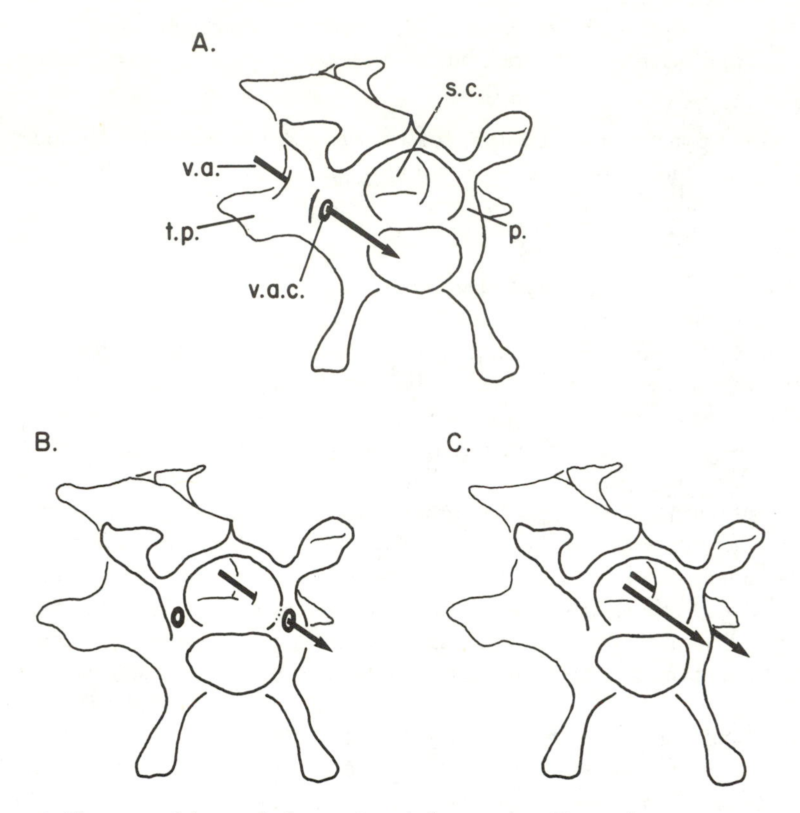

Figure 1. The course of the vertebral artery in typical mammals and in camels.
A. A cervical vertebra, anterior view, showing the typical mammalian arrangement of vertebral arteries. The heavy arrow shows the path of the *right* vertebral artery (v.a.). The vertebrarterial canal (v.a.c.) through which it passes perforates the base of the transverse process (t.p.). The spinal cord runs in the spinal canal (s.c.). The neural arch surrounds the spinal cord; the pedicle (p) is a lateral wall of the arch. (Although this is the typical mammalian pattern, it has, for the sake of comparison, been illustrated on a drawing based on a guanaco cervical vertebra, which at several points departs from the shape of other mammalian cervical vertebrae.)
B. The cervical vertebra of a guanaco, in the same view as (A), showing the altered course of the vertebral artery in camels and *Macrauchenia.* The artery enters in the spinal canal, then perforates the neural arch, and continues in a new vertebrarterial canal in the pedicle of the arch until it emerges on the anterior face of the vertebra. In the normal case (A), the artery is always outside of the main body of the vertebra, while in (B) it is successively in the spinal canal and pedicle. Note that in order to show this difference, the *left* vertebral artery is illustrated in (B).
C. Courses of vertebral artery that do not occur in mammals, but which are consistent with Darwin's description of the course of the vertebral artery in *Macrauchenia.* One arrow shows the left vertebral artery remaining in the spinal canal for the entire length of the vertebra; the other arrow suggests a vertebral artery that remains on the outside of the body of the vertebra but which does not perforate the base of the transverse process.

looking for cervical vertebrae. Knowing the rule is not the same as seeing the exception, however, and noticing the particular course of the vertebral arteries in camelids takes some subtlety. Owen, in his description of *Macrauchenia*, is forced to admit to an oversight in this regard on the part of Cuvier, "who seems not to have been aware of this peculiarity in the *Camelidae*" (*Zoology* 1:43). On the larger issue of the nature of the fossil remains, we can take Darwin at his word when he says, "I had no idea at the time, to what kind of animal these remains belonged" (*Journal of Researches*, p. 208).

Owen's initial view that the remains belong to a giant llama and a giant rodent is in keeping with such generalizations as could then be made about fossil mammals. Two assemblages were at this time well known —the old, generalized mammals, most of medium stature, from the Paris Basin, and the young, often very large mammals from Post-Pliocene alluvium. These latter fossils generally resembled living forms very closely (mammoth, cave bear), or at least had an indisputable family resemblance to extant forms (mastodon, *Megatherium*). Darwin's fossils are very young, mingled as they are with the shells of extant marine molluscs. One of Darwin's fossils shows the vertebrarterial canal that is restricted to camels; the other has the procumbent, open rooted incisors of a rodent. By all previous experience, that should be that.

In addition, both Darwin and Owen are interested in the affinity of giant extinct forms of extant animals in the same region. Fossilized giant sloths are already known, and Darwin recognizes in the field the fossilized dermal pavement of a giant armadillo-like animal. A giant llama and capybara would make the South American nature of the extinct fauna even more pronounced.

II. Owen's Reassessment

Owen describes *Macrauchenia patachonica* in the first fascicle of *Zoology*, published in February 1838.[3] He carefully described the course of the vertebral arteries, the similarity to the condition in camelids, and possible adaptive and developmental explanations of the condition. Yet Owen refers it to the order Pachydermata. Osteological evidence is overwhelming in this regard, although in itself this might not be sufficient to force the conclusion that the macrauchene is not a camel. It is Owen, after all, who will find ways to put man and chimpanzee in separate orders. The finality of the assignment has chiefly to do with the way the logic of classification of the ungulates evolved. Once the characters are stated, the diagnoses of the orders absolutely determine the assignment.

Owen describes in turn each of the bones Darwin collected. For the purposes of this study, the information afforded by the available bones is far more important than our modern interpretation of *Macrauchenia*, based as it is on complete skeletal material, numerous relatives, and a biogeographical analysis that Darwin and Owen would barely recognize. The bones Owen had available are shown in Figure 2; Owen's reading of them is largely dependent on the very different degrees of theoretical significance associated with each. To a lesser extent, their interpretation turns on their condition, and the presence of outstanding peculiarities for which theoretical explanations do not already exist. At this point, one need appreciate that *Macrauchenia* is indeed large, at least the size of living Old World camels, and probably significantly heavier.[4]

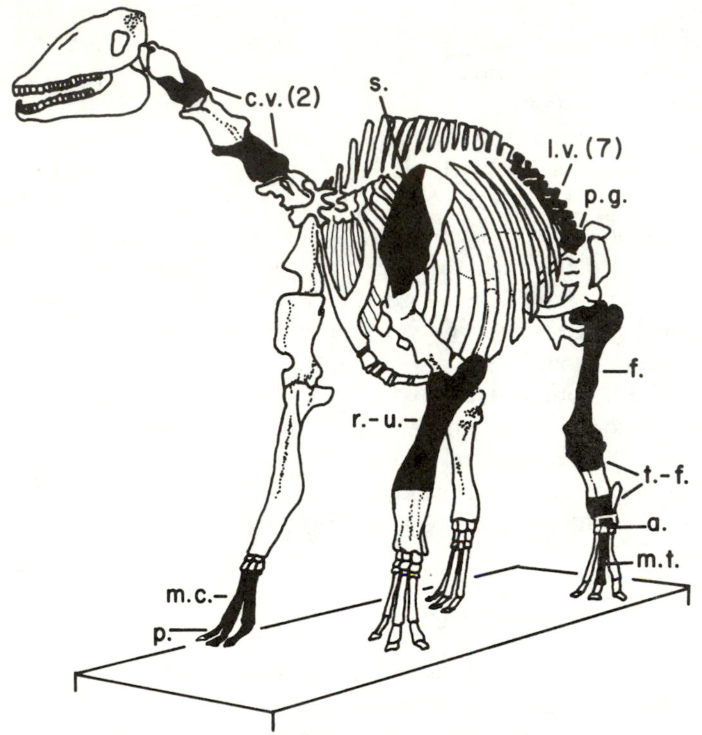

Figure 2. The bones Darwin collected of *Macrauchenia* (black) superimposed on the complete skeleton as we now know it. The skeleton is after the photograph in Herbert's edition of *The Red Notebook of Charles Darwin*, p. 114. The remains consisted of two cervical vertebrae (c.v.), part of the scapula (s), seven lumbar vertebrae (l.v.) and a fragment of the pelvic girdle (p.g.); forelimb: part of the fused radius and ulna (r.-u.), most of the metacarpals (m.c.) and phalanges (p.) of a fore-foot; hindlimb: a femur (f.), proximal and distal portions of the fused tibia and fibula (t.-f.), the astragalus (ankle bone--a.), and one metatarsal (m.t.).

Owen atomizes each bone into features that are then compared to homologous features in a wide array of mammals. As to the acuity of his powers of observation, I have reviewed much of the later osteological work on *Macrauchenia*, and I have found no detail that has subsequently played an important part in our modern picture of the animal that Owen did not also recognize, if he had the appropriate bones. For example, he discusses the calcaneum (heel-bone) of *Macrauchenia*, which he has not got. He does have the distal fibula, however, which has a long facet that "unerringly indicates a corresponding articular projection in the calcaneum", and a conformation that departs from that in *Paleotherium*, the fossil pachyderm from the Paris Basin that, of all genera, most frequently agrees with *Macrauchenia*. Near the end of the nineteenth century, this relation of the calcaneum and fibula, in a considerable number of extinct South American ungulates, served to diagnose the order Litopterna ("smooth heel"), to which *Macrauchenia* is assigned today.

The two cervical vertebrae receive special attention. It is worth noting that the vertebrae do not on the whole resemble the cervical vertebrae of camels. Indeed, with their slightly developed transverse processes, flattened anterior and posterior articulating surfaces, and uniform diameter, they look more as though they belong in the neck of a middling dinosaur than in the neck of a large mammal (P. Olsen, personal communication). As to the origin of the novel course of the vertebral arteries, Owen asserts that in both camelids and the macrauchene it is an adaptation to having a long neck. Owen notes that in fetal mammals the perforation in the transverse process arises as a space between what begin as two transverse processes and a rudimental rib, a condition that is retained into the adult state in the crocodile and platypus. Were the vertebra to be stretched to the great extent that is found in the macrauchene, the perforation of the vertebral arteries would be smeared into a shallow groove:

> This groove would not, however, afford sufficient defence for the important arteries supplying those parts of the brain which are most essential to life; and accordingly the vertebral arteries here deviate from their usual course, in order that adequate protection may be afforded to them in their course along the neck. (*Zoology* 1:37)

An obvious test suggests itself; the giraffe is the one mammal with longer cervical vertebrae than *Macrauchenia*. Owen argues that the case in the giraffe is not exactly typical; the perforation is slightly above and to the front of the poorly developed superior transverse process (*Zoology* 1:37). It is, however, close to typical. (In fact, in 1853, Owen describes the vertebral arteries of the giraffe as perforating "the upper and fore part of the transverse processes.") As Owen remarks at the beginning of his discussion of this question, sometimes we know why a peculiar structure exists, but other times "the connection of the structure with the exigencies of species is by no means obvious, and in this predicament stands the osteological peculiarity, which is immediately connected with our present subject" (*Zoology* 1:36).

Comparisons with camelids remain throughout the description a special interest to him. Wherever camels have an unusual feature, Owen pays special attention in the macrauchene. For instance, the acromial angle of the camelid scapula is particularly acute. Owen struggles with the macrauchenid scapula, which is in fragments, and which is missing the acromion. He concludes from the nature of the fractured edge of the scapular spine that the angle was not as acute as in camelids. Overall, Owen finds some general features shared between camelids and macrauchenes, but only one more point of striking agreement — the macrauchene and camelids are the only ungulates with seven, rather than six, lumbar vertebrae. The form of these vertebrae is definitely pachydermatous, however, and the transverse processes of these vertebrae "deviate remarkably from those of the *Camelidae*" (*Zoology* 1:41). This pattern of argument is not confined to comparisons with camelids; for any bone, Owen will let shape suggest one affinity, and size or number another, and then he will turn to, say, its articular surfaces, and explain how they contradict whatever affinities he has just detected in shape or size. It is the exception rather than the rule for him to consider explicitly a bone as a single functional element, or part of one.

That anything emerges from this welter of detail is the result of certain features being privileged, implicitly for reasons of functional analysis. Georges Cuvier, the father of modern comparative anatomy, had established the primacy of teeth and feet in the classification of mammals. In theory, if one could take as fixed points what an animal ate and how it moved, the rest could be interpolated. Owen has no teeth for *Macrauchenia*; thus the evidence afforded by the bones of the feet will be crucial.

In all ruminants, there are two weight-bearing digits which articulate on a cannon-bone comprised of a pair of fused metacarpals or metatarsals. In the pachyderms, the number of weight-bearing digits varies from one to five, but there is never a cannon-bone of fused origin. Owen described the remains of the macrauchene in light of these distinctions:

> The confirmation of the close affinity of the Macrauchenia to the Pachydermatous Order, which the structure of the cervical vertebrae alone might have rendered very doubtful, is afforded by the bones of the right fore-foot (P1. XI.); these are fortunately in so perfect a condition, as to make it certain that this interesting quadruped had three toes on the fore-feet, and not more; and that the fully developed metacarpal bones are distinct, and correspond in number with the toes, and are not anchylosed into a single cannon-bone, as in the Ruminants. (*Zoology* 1:46)

Owen's discussion of these bones is one of the few places where he departs from purely formal analysis of isolated features; here, theory sanctions comparison of function as well as form:

> The structure of the above described joints proves that the motion of the toe upon the metacarpus was much freer and more extensive than

> in the Rhinoceros, which is the only existing Ungulate mammal which presents the tridactyle structure in the fore-foot . . . . (T)he hog and the horse in this respect approach nearer to the Macrauchene, though the structure of the metacarpo-phalangeal joints in the Hog falls far short of the compactness and strength combined with freedom of play in flexion and extension which distinguish those of the Macrauchene. (*Zoology* 1:47)

Owen turns to the hind-foot, which tells the same story:

> (T)he structure of the distal articular surface of the tibia is attended with particular interest, because we are taught by Cuvier that it reveals to us in the Ungulate animals the didactyle or tridactyle structure of the foot. (*Zoology* 1:51)

The surface resembles that of Cuvier's *Paleotherium*, pachyderm (now perissodactyl) from the Paris Basin.

One bone in particular is especially informative about the organization of the foot. Owen offers this paean to the ankle-bone (the astragalus),

> which, of all the bones in the foot, is the one that an anatomist would have chosen, had his choice been so limited, and which most fortunately has been secured by Mr. Darwin, in a very perfect state, in the present instance. (*Zoology* 1:51)

Since Cuvier, anatomists have determined the affinities of ungulates by the cast of astragali. From the astragalus, Owen could read the leg above, the heel behind, the foot below. In no direction was there any sign of a camel:

> I have compared this astragalus with that of the Giraffe, and other Ruminants, the Camel, the Anoplothere, the Horse, the Hog, the Hippopotamus, Rhinoceros, Tapir, and Palaeothere: it is with the Pachyderms having three toes to the hind-foot, that the Macrauchenia agrees in the main distinguishing characters of this bone . . . . (*Zoology* 1:51–52)

Macrauchenia presents two other important departures from the typical condition in the Ruminantia. The femur of *Macrauchenia* carries a third trochanter, a common condition in the pachyderms but unknown in the ruminants. The fibula is entire (only the chevrotain retains an entire fibula among the ruminants; the rest, including camels, retain only its distal portion, which articulates with the tibia, astragalus and calcaneum).

III. The Logic of Ungulate Classification

It should be clear that the preponderance of characters indicates that the macrauchene is a pachyderm. When we consider the characters of the leg and foot, to which Owen, and most anatomists to follow, accord particular significance, there is no feature that is inconsistent with this assignment. What is peculiar to this case, however, is an asymmetry in the way that the pachyderms and ruminants are classified. It happened to be very easy to show that a particular ungulate is not a ruminant, and very difficult to show that it is not a pachyderm. The only way, in fact, that an ungulate could be shown certainly not to be a pachyderm would be to show that it had the essential features of a ruminant.

Owen knows the ruminants to be united by the following features: two weight-bearing toes, cannon-bone produced by fusion, upper premolars different from upper molars, and a chambered (ruminating) stomach. Most members in addition have a reduced fibula (except in chevrotains), cotyledonary placentas and bony horns or antlers (except chevrotains and camels), and no upper incisors (except camelids). Cuvier wrote of the Ruminantia,

> This order is perhaps the most natural and best determined of the class, for nearly all the animals which compose it have the appearance of being constructed on the same model, the Camels alone presenting some trifling exceptions to the general characters. (1831, 1:182)

The pachyderms are essentially all large ungulates that are not ruminants, or smaller ungulates with clear affinities to one of the large pachyderms. Weight-bearing toes number one (horse), three (rhinoceros), four (hippopotamus), or five (elephant). Teeth are as varied, the third trochanter is present or absent, the fibula usually present. No one character is diagnostic. To quote Owen, the order Pachydermata is "extensive and heterogeneous" (*Zoology* 1:15).

Owen accepts both orders as equally real. Zoologists generally learn the major groups before they are in a position to evaluate the characters that distinguish them. No matter how ambiguously the groups are defined, they are reified into natural assemblages and protected by intellectual inertia. What is surprising is not that Owen accepts the pachyderms as a natural group, but that he perceives their heterogeneity.

In the case of *Macrauchenia*, there is no possibility of assignment to the Ruminantia, because several absolutely dependable characters of that order are violated. The assignment to the Pachydermata is inevitable; as far as the more important characters go, there is not even any increase in the heterogeneity of the group.[5]

Owen's description comes to emphasize not the camelid nature of *Macrauchenia* but rather the approaches that the camelids make to Pachydermata, "without losing the essential characters of Ruminantia" (*Zoology* 1:55). These pachydermatous features are, most notably, the diffuse placenta and presence of upper incisors. The particular features shared by the macrauchene and camelids — the vertebrarterial canal and seven lumbar vertebrae, for instance — are not characteristic of pachyderms any more than of ruminants. Thus the features that ally camels and the macrauchene are not the features that ally camels and pachyderms.

That an unquestionable ruminant might mirror characters that are typically found in a pachyderm, and that a morphologically isolated, extinct pachyderm might present a mixture of characters, some of which are shared with that same ruminant, and others of which recall a wide variety of pachyderms, is in keeping with the expectations of quinarian classification.

The quinarian or, as its proponents preferred, the *natural* system, was set forth between 1819 and 1821 by William Sharpe Macleay, naturalist and foreign service officer.[6] He provided a geometrical and numerological periodic table of the living world. Though it was to be remembered only for its excesses in the later nineteenth century (and forgotten completely until Sydney Smith demonstrated its importance for Darwinian studies in 1965), in the 1830s it was the most advanced and promising system in natural history, and the focus of much theoretical and practical interest. Macleay proposed that all species, if properly recognized, would be found to be arranged five to a genus, and that five genera comprised a family, and so on to the five phyla of the animal kingdom. At each level, the five taxa that comprised the next largest group were found to form a circle, so that the living world consisted of circles within circles. At high taxonomic levels, subordinate circles representing rare and distinctive forms ("osculant groups") alternated with the great circles. These osculant groups partook of some essential characters of the adjacent great groups, while showing their own curious organization. For instance, squid, with its internal shell, eye, cartilaginous skull, and radula, recalls molluscs on one side and vertebrates on the other. Within, between and across the circles at any one level, Macleay discerned reflections, polarities, and analogies that served to confirm the rightness of the circles.

For it has been written:

> These smaller links of the great chain appear to have no very distinct type of peculiar construction. They are all very imperfect beings, and seem in general to be compounded of properties which more peculiarly belong to the two great divisions they link together; [of osculant groups between larger circles]
>
> I have, however, designated the great intervals which sometimes separate two such adjoining groups as chasms or hiatus, rather than as saltus . . . because I cannot help thinking, from analogy, that if they never should be filled by living animals, they may have, at some time or other, been occupied by species now extinct. (MacLeay 1819–1821, pp. 320, 368)

For Owen, the larger significance of Darwin's fossils is to be found in the way they link large natural groups. He writes in the general introduction to the *Zoology*, "of two large and singular aberrant forms, one of which connects the Pachydermatous with the Ruminant Order; the other . . . manifests a close affinity to the Rodent Order" (p. 15).

Ospovat reports that Owen "cribbed extensively from MacLeay's book in preparing his first course of Hunterian lectures" for 1837 (1981, p. 107). In *Zoology*, his sort of quinarianism has been freed from the constraints of geometry and numerology. Owen's penchant for atomized anatomy permits him to find affinities that go in many directions. The macrauchene has camelid affinities, and, more significant, to a series of pachyderms. *Macrauchenia*, however, is homogeneous compared to the osteological salad that comprises *Toxodon*. This pachyderm has particularly close affinities to rodents and "herbivorous Cetacea" (Sirenia), and more than a touch of wombat and armadillo.

Owen concludes the description of *Macrauchenia* with several general observations that would have been of interest to Darwin. He notes the general resemblance of the macrauchene and *Palaeotherium* from the Paris Basin. He notes a general tendency toward ruminants in the "anchylosed and confluent state of the bones of the fore-arm and leg." (Owen is stretching to make this point, given the greater confluence that is seen in the horse, a pachyderm; it means as well that he is equating partial fusion of the macrauchene's tibia and entire, rather robust fibula with the extreme reduction, without fusion, in the fibula of camels and "typical" ruminants.) He reiterates the "singular modifications of the cervical vertebrae" that point to an affinity with camels:

> the evidence which the lost genera, *Macrauchenia* and *Anoplotherium*, bear to a reciprocal transition from the Pachyderms to the Ruminants, through the *Camelidae*, cannot but be viewed with extreme interest by the Zoologist engaged in the study of the natural affinities of the Animal Kingdom. (*Zoology* 1:55)

Finally, he emphasizes the geographically interesting point of a fossil from South America that is "in a remarkable degree" a transitional form connecting two extant animals of that continent, the tapir and llama.

IV. What Does Darwin Make of This?

The affinities of the "extinct llama" are crucial to the validity of Darwin's speculation in *RN* about one species arising from another. If the extinct llama is in reality a long-necked tapir with a soupçon of camel about its cervical vertebrae, then the theory Darwin is proposing is less symmetrical than his words would suggest. "Large northern rhea is to small southern rhea as large fossil tapir is to small living camel" does not have the right ring to it. I shall argue that by the time Darwin was writing *RN* and *Journal of Researches*, Owen had told him that the fossil was not a camel.

In order to reconstruct Darwin's interpretation of the fossil he collected, I shall begin with the way he reads bones. In the *Journal of Researches* he puts the bones of the guanaco into a larger context:

> The guanacoes appear to have favourite spots for dying in. On the banks of the St. Cruz, the ground was actually white with bones, in certain circumscribed spaces, which were generally bushy and all near the river. On one such spot I counted between ten and twenty heads. I particularly examined the bones; they did not appear as some scattered ones which I had seen, gnawed or broken, as if dragged together by beasts of prey. The animals in most cases, must have crawled, before dying, beneath and amongst the bushes . . . I do not at all understand the reason of this, but I may observe, that the wounded guanacoes at the St. Cruz, invariably walked towards the river. At St. Jago in the Cape de Verd islands I remember having seen in a retired ravine a corner under a cliff, where numerous goats' bones were collected: we at the time exclaimed, that it was the burial-ground of all the goats in the island. I mention these trifling circumstances, because in certain cases they might explain the occurrence of a number of uninjured bones in a cave, or buried under alluvial accumulations; and likewise the cause, why certain mammalia are more commonly embedded than others in sedimentary deposits. Any great flood of the St. Cruz, would wash down many bones of the guanaco, but probably not a single one of the puma, ostrich, or fox. I may also observe, that almost every kind of waterfowl when wounded takes to the shore to die; so that the remains of birds, from this cause alone and independently of other reasons, would but rarely be preserved in a fossil state. (*Journal of Researches*, pp. 197–198)

Darwin carefully examines guanaco bones for signs that they may have been dragged to the sites in which he found them. It is not the sort of examination that would turn up a peculiar course in the vertebral artery. What he is looking for will be familiar to anyone who has heard of taphonomy — the science that interprets the distribution and abundance of fossil remains in terms of the natural history of the fossilized organisms, the environment of deposition, and the composition of the remains. The various biases that are introduced in the process of fossilization are studied, in order that the environment in which the organisms lived can be reconstructed. In this passage, Darwin considers the accumulation of bones first by carnivores, and second as a result of the habits of dying animals. He attends to the conditions that would produce uninjured fossils. He takes the desert scene before him and imagines the torrential floods, rare in themselves, that are the common way in which bones would come to be deposited under sediment, and he notes that the preponderance of guanaco bones so buried would be a biased representation of the community in which they lived. In typically

Darwinian fashion, he weaves the goat grave of St. Jago, the perfect skulls fossilized in a cave, the proportions of fossils of different kinds in a deposit, and the dying wishes of guanacos and ducks into a single story. And it is particularly wonderful that there is no precedent for this kind of thinking; indeed, it was not until the 1960s that vertebrate paleontologists paid serious attention to the layers of interpretation that are intercalated between the discovery of a fossil and its description.

Darwin reads the bones as a naturalist, in a synthetic and completely original way. Owen's eyes are those of an anatomist — basically an analytical approach to form, but informed by developmental, functional, and systematic questions as well. The insights that they bring to the same subject exist in different spheres. What one sees clearly, the other may barely perceive at all. Owen writes of teeth and foot-bones as if there is only a theoretical reason for studying them. From the point of view of a taphonomist, the density and compactness of these structures will allow them to persist long after skulls and cervical vertebrae have been smashed to fragments. Cuvier's emphasis on these remains in fossil mammals makes an anatomical virtue of taphonomic necessity. When Owen writes of the astragalus "which most fortunately has been secured in a very perfect state", it is not pure fortune that is at work, and this in turn has helped make the astragalus the one bone "that an anatomist would have chosen, had his choice been so limited". Similarly, Darwin does not quite appreciate the osteological perspective that Owen brings to the bones.

The ecological and geological questions Darwin brings to the guanaco bones also open his discussion of the fossil that his table of contents calls "Fossil gigantic llama":

> In one spot this earthy matter filled up a hollow, or gully, worn quite through the gravel, and in this mass a group of large bones was embedded. The animal to which they belonged, must have lived . . . at a period long subsequent to the existence of the shells now inhabiting the coast . . . . From the small physical change, which the last one hundred feet elevation of the continent could have produced, the climate, as well as the general condition of Patagonia, probably was nearly the same, at the time when the animal was embedded, as it now is. This conclusion is moreover supported by the identity of the shells belonging to the two ages. Then immediately occurred the difficulty, how could any large quadruped have subsisted on these wretched deserts in lat. 49° 15′? (*Journal of Researches*, p. 208)

The environment of deposition is Darwin's major concern; he has evidence that the bones were buried relatively recently (the sediment is unconsolidated and the shells that accompany the bones belong to species that still live on the coast), and he knows by measuring the terrace that about 100 feet of elevation have occurred in the mean time. This is insufficient to make much difference in the climate.

Darwin's ecological question receives an ecological answer:

> The puzzle, however, was soon solved when Mr Owen examined them; for he considers that they formed part of an animal allied to the guanaco or llama, but fully as large as the true camel. As all the existing members of the family of Camelidae are inhabitants of the most sterile countries, so may we suppose was this extinct kind. (*Journal of Researches*, p. 208)

Darwin then gives the independent osteological evidence that supports this interpretation of the fossil, which in turn supports Darwin's reconstruction of the harsh environment of deposition:

> The structure of the cervical vertebrae, the transverse processes not being perforated for the vertebral artery, indicates its affinity . . . . (*Journal of Researches*, p. 209)

Owen writes of the peculiar condition in the camelids, which "consists in the absence of perforations for the vertebral arteries in the transverse processes of the cervical vertebrae" (*Zoology* 1:36). The difference between these two statements is that Owen's is an accurate description of the case in camelids, whose reasonably prominent transverse processes are indeed imperforate, while Darwin's is supposedly a description of the case in the fossil vertebrae, which have little in the way of transverse processes to be perforated. Owen gives a careful description of the route of the artery, its passage in the spinal canal, its perforation of, and emergence from, the base of the neural arch — this is the real nature of the similarity. Darwin says nothing else about the vertebral artery. His description would fit as well Owen's description of the course of the vertebral artery in giraffes, which loops into the wall of the neural arch (though without entering the spinal canal) but misses the transverse process (*Zoology* 1:37). Either of the paths shown in Figure 1-C would be consistent with Darwin's description, but neither would be true of camels or *Macrauchenia*. Were this passage a test of Darwin's grasp of comparative anatomy, a case could be made for withholding full credit.

Having given his version of the vertebrarterial canal, and having said that it indicates the affinity of the animal, Darwin continues, "some other parts, however, of its structure, probably are anomalous". This information could only have come from Owen. Darwin is not an initiate in comparative anatomy; he does not reverence the astragalus. Owen by this point must have commenced his serious study of the remains. He understands the logic of ungulate classification as it is received from Cuvier, and he pays special attention to those observations of Cuvier that are the basis for the thorough reorganization of the orders that he will shortly publish. He knows the difference between two toes and three — once the specimens are unwrapped and cleaned, all he has to do is count. The anomalous features, once seen, require Owen to change his assignment of the fossil to the order Pachydermata. I have no manuscript proof of this, and none is necessary. If one understands the characters, the logical basis of the classification of these orders, and the rudiments of comparative anatomy, this conclusion is inescapable.

It strains credulity to believe that Owen would tell Darwin about the anomalous features, and not tell him that they prohibit the assignment of the remains to the camelids. That Darwin had sufficient motivation to keep seeing a camel when none remained, is suggested by the way Darwin phrased the addendum to the *Journal of Researches* that introduces the scientific name of the "fossil llama" Darwin discusses in the text. He writes:

> For an admirable description . . . of the great fossil (*Macrauchenia Patachonica*), mentioned at p. 208, which in some respects is allied to the Camelidae, I must refer to the first part of the Zoology of the Voyage of the Beagle by Mr Owen. (p. 609)

Darwin is true to the letter of the text Owen has written (and that Darwin had edited), and he has subtly introduced a note of qualification, but the unprepared reader would hardly interpret this sentence to mean that Owen decides that the animal is not a camelid, nor even a ruminant.

Why does Darwin decline this opportunity to correct the several statements in the *Journal of Researches* in which he states that the fossil is that of a gigantic llama? He has a bit of an interest in using a fossil camelid to corroborate his reconstruction of the recent history of Patagonia, but this fact is hardly indispensable. Darwin's use of the "giant Llama" in *RN* provides a more persuasive reason for his reluctance.

The *RN* speculation builds on the ecological conclusion that Darwin works out in the *Journal of Researches*: conditions in Patagonia have not changed; thus extinction of the giant llama must be ascribed to other agencies (*RN* 129). Another ecological explanation, competition, is also explored: "Speculate on neutral ground of 2 ostriches; bigger one encroaches on smaller" (*RN* 127). Darwin considers this theme in connection with the large extinct inhabitants of South America and their small extant relations:

> we can scarcely credit that the armadillo has devoured the food of the immense Megatherium, the capybara of the Toxodon, or the guanaco of the camel-like kind. (*Journal of Researches*, p. 211)

The dramatic difference between *RN* and the *Journal of Researches* is Darwin's suggestion in *RN* that the species arose, one from another, in both the large and small rheas, as well as the large and small llamas. Darwin eschews gradual change or degeneration, in favor of a saltatory origin of the new species:

> not *gradual* change or degeneration. from circumstances: if one species does change into another it must be per saltum — or species may perish. This <inosculation> representation of species important, each its own limit & represented. (*RN* 130)

As Kohn says, "The critical word here is inosculate" (1980, p. 76).[7] Darwin uses it one other time in the *RN* passage — "as in first cases distinct species inosculate, so must we believe ancient ones" (*RN* 130). What does the word mean? Kohn correctly cites MacLeay, and gives an informal definition of "osculant". Herbert notes that inosculation is "a medical term referring to the joining of one blood vessel to another" (1980, p. 7). She does not exploit this anastomotic perspective;[8] nor does she explain why a technical term from the science of classification is out of place in a discussion of the relations of species.

In fact, MacLeay provides a technical definition of inosculant groups — they are the members of major circles that most closely approach the osculant groups between circles, though only if they happen not to be located around the circumference of the large circle comprised of the five main sub-groups and five minor osculants (1819–1821, p. 396 fn). In that camelids are perfectly fine ruminants, and *Macrauchenia* is a splendid osculant to the camelids from the pachyderms, it is barely within the realm of possibility that this is what Darwin intends. The most formulaic aspects of the quinarian system were not pursued, at least in the classification of ungulates, and in the absence of an agreed-upon geometric organization of the hoofed mammals, such a technical use would have been unlikely. By the 1830s, the choice between "osculant" and "inosculant" was determined by euphony. For example, Darwin writes to Henslow from Montevideo:

> there is a poor specimen of a bird, which to my unornithological eyes, appears to be a happy mixture of a lark pigeon & snipe. (1967, p. 62 and fn. 2)[9]

If Darwin knows about the anomalous features of his fossil quadruped, then the word is perfectly appropriate in this context. If he has no reason to doubt that the fossil is a llama, however, then there is no reason to use the word "inosculate", or to insist on saltatory change. He could say instead that the change is gradual (which he explicitly denies) or that the two species are "annectent", to use a word favored by MacLeay and Owen.

Were there justice in the world, there would be no evidence that Darwin ever considered the two rheas as an example of inosculation, in spite of his statement quoted above that as "first cases [that is, the rheas] inosculate, so must we believe ancient ones". The differences Gould finds between the two species — *Rhea darwinii* has skin less bare around the eyes, proportionally larger feet and longer beak, the scales on the leg are reticulated where the common rhea's are banded, and the plumage is tipped with white

(*Zoology* 3:123–124) — are not the sort of differences that suggest profound reorganization. Even Gould, a notorious splitter, kept the new species in the established genus. Nevertheless, Darwin writes of the tarsi,

> they are feathered six inches beneath the knee. In this latter respect, and in the broader feathers of the wing, this bird perhaps shows more affinity with the gallinaceous family than any other of the Struthionidae. (*Journal of Researches*, p. 109)

This bit of invigorated ornithology is Gould, not Darwin (see Gould 1837e); it does, however, provide a feeble reason to consider the smaller rhea to be inosculant. There remains a tremendous difference between Owen's discovery, in which a somewhat quinarian approach serves to put a surprising fossil at the center of basic questions about form, function, development, affinity, geographical distribution, and paleontology, and Gould's, which as a specimen of vulgar ornithological quinarism is chiefly interesting for having made an impression on Darwin.

I conclude this section with a conjecture as to the steps by which Darwin came to set down the speculation about rheas and llamas in *RN*. Initially, Owen tells Darwin that one of his large fossils is a gigantic version of a capybara, the other, a gigantic llama. This fits with what Owen knows about the affinities of recently extinct quadrupeds. It is of interest to both Owen and Darwin for obvious biogeographical reasons. It is of special interest to Darwin because it supports his reconstruction of the recent geological history of Patagonia; it also fits into his inquiry on the ability of apparently sterile regions to support large quadrupeds.

Darwin is unwilling to ascribe the extinction of these giants to competition with their extant vicars, perhaps because he is vaguely predisposed to think that bigger means better (this is explicit in the case of the rheas, and implicit in the llamas — at least, the living guanaco leaves him unimpressed.)[10] There are physiological and geological arguments that lead him to the idea of extinction by species senescence (Kohn 1980); this would suffice to explain extinction without any change in the external environment.

Darwin now comes to one species originating from an allied species in time and space. If the theory that Darwin presents in *RN* really is a precursor of his later theories, as opposed to a flight of quasi-quinarism, then at its genesis, the fossil llama had to be a real llama for Darwin. Since it serves well in some of his other theoretical pursuits at the time, this seems an acceptable supposition. Owen had told him about the cervical vertebrae, and, in all likelihood, he showed Darwin with the actual bones how peculiarly similar the vertebrarterial canals of the llama and the fossil are.

This fact in the web-work of Owen's mind serves as a focal point for the reflections, essences, and a priori correlations that contribute to

Owen's wonderfully complex, abstract, and static view of the organic world. Darwin's web is even at this point differently wrought — his is a dynamic and rather concrete world. A pile of dead bones becomes the result of an instinctual behavior, and the prelude to the torrent that produces a fossil bed. For Darwin, the shared and unique similarity in the course of the vertebral artery is a token of common descent. It is not necessary that Darwin understand the fine points of arterial routing; what is crucial is that the peculiarity is unique to these creatures. In the lingo of cladistics, Darwin has a synapomorphy, a shared derived character that ties two groups together. This seems a perfectly fine way to get a theory of descent off the ground.

The two rheas are generally similar, with differences only an ornithologist could love. Nothing about them is so special, so different from thousands of other pairs of species, that it could in itself suggest descent. This pair serves as a foil for the llamas — associated with Patagonia, differing in size, a pattern in space that recalls a pattern in time, both produced by the same process.

This would be a plausible first Darwinian theory of transmutation, which we have not got. What we have in *RN* and the *Journal of Researches* are Darwin's modifications in light of Owen's discovery that the gigantic fossil llama is merely a gigantic fossil. It is possible that Darwin really does not understand that a three-toed ruminant is chimerical; the vertebrarterial canals are after all not affected. I think it more likely that Darwin knew pachyderms to be very different from ruminants, though without having too clear an idea as to why. He reports that anomalies exist, without being willing to accept the conclusion that must be drawn from them, even after Owen publishes. As an extremely promising theory takes off, it is very difficult to lay to rest a fact that got it started.

Inosculating and saltatory changes are invoked so that the non-camel can still be immediately related to the extant South American camels. Rheas can act again as foils, though not convincing ones. This is not a stable equilibrium; the facts regarding camels have disintegrated, and they can be shored up only by putting weight on aspects of the rhea story that will not bear it. More fundamentally, a naturalistic theory of transmutation cannot sit comfortably atop Owen's idealistic morphology: the symmetries, reflections, formal causes, and webs of affinity of the latter all argue against any of the transmutation theories Darwin explored.

If one comes to *RN* cognizant of the osculant and anomalous features of the "fossil llama", it is possible to sketch an interpretation of the passage very different from what would be suggested if one took the claim for a fossil llama at face value. Indeed, it is almost possible to read the rhea-llama section of *RN* as a completely idealistic, quinarian argument: although species are gradually worn down by the circumstances of their lives, this natural deterioration cannot produce new adaptive types. Rather, these are created at one blow, each new form a representation in some of its details of pre-existing types. Were change limited to the natural running down of species, the result would be extinction:

> not *gradual* change or degeneration. from circumstances: if one species does change into another it must be per saltum — or species may perish. — This <inosculation> representation of species important, each its own limit & represented . . . inosculation alone shows not gradation. (*RN* 130)

I do not find this quite as convincing as the first version I have given, but it is intriguing that the naturalistic and gradual processes are identified as what is *not* happening, and the unexplained patterns of representation and inosculation are taken as premises.

V. Further Adventures of the Macrauchene

"The structure of the pachyderm leg was a favourite with the author." So writes Francis Darwin of his father (*1844 Essay*, p. 157, fn. 2). The story draws not only on the macrauchene, but also on the exciting reorganization of the Ungulata that Owen produces in the 1840s. This reorganization gives Darwin a new use for *Macrauchenia*; by this point, he has very nearly come to terms with it as a pachyderm.

In *Odontography* (1840–1845), Owen revises the Ungulata into three orders (or suborders): the Perissodactyla, or odd-toed ungulates (tapir, rhino, horse, hyrax, paleothere, and macrauchene);[11] the Artiodactyla, or even-toed ungulates (pigs and hippos plus the Cuvierian order Ruminantia); and the Proboscidea (elephants). In 1853, he removes *Toxodon* and some related South American fossils from the Perissodactyla and gives them their own order, Toxodonta (the modern Notoungulata (Owen 1853b)). As one might have expected, the diagnoses turn on teeth and feet — the astragalus in artiodactyls articulates distally with the navicular and cuboid in equal measure (which fuse in the most advanced group), while in perissodactyls the facet for the former is much larger than the facet for the latter. The effect of this reorganization is to make the Perissodactyla a far less heterogeneous group than the old Pachydermata — the new group can be described in terms of characters it possesses, rather than in terms of characters it does not possess. Compared to the Ruminantia, the Artiodactyla is a slightly more heterogeneous assemblage because of the inclusion of the pigs and hippos, which have more teeth than other artiodactyls, and lack a completely fused cannon-bone. It is easy, however, to perceive morphological trends that reach their culmination in the higher ruminants (deer and bovids, for instance) that are foreshadowed in the pigs — there is no question that the new order is still a "natural" one. The reorganization is the most significant advance in ungulate classification since Linnaeus, and it remains to this day. The logic of classification and the characters on which it is based have both stood the test of time.

To anyone raised with a sense of the naturalness of the Pachydermata, Owen's reorganization would have been a considerable surprise. (It is telling that the word "pachyderm" remains in everyday usage almost a century and a half after the zoological justification for such a grouping had been dissolved.) Darwin follows the revision, more or less successfully. Thus, in the *1844 Essay*, Darwin writes of intermediates between "the genus *Sus* and the tapir" (p. 156 — *Sus* is the pig); this is revised in the *Origin* in light of the distinction between artiodactyl and perissodactyl to intermediates between the horse and tapir (*Origin* p. 261). There is a bit of backsliding, however. For instance, in 1857 in *Natural Selection*, he writes,

> in the pig, — which has the snout much developed & which is allied, but, as Owen has shown, not so closely as we formerly thought to the Pig & Elephant, a monstrous trunk is developed oftener than in any other animal. (p. 319)

The reorganization of the ungulates can be interpreted in terms of descent, as Darwin argues in the following passage:

> Cuvier ranked the Ruminants and Pachyderms, as the two most distinct orders of mammals; but Owen has discovered so many fossil links that he has had to alter the whole classification of these two orders; and he has placed certain pachyderms in the same sub-order with ruminants: for example, he dissolves by fine gradations the apparently wide difference between the pig and the camel. (*Origin*, pp. 329–330)

This is not strictly true to either the logic of Owen's discovery (which is not based on fossils) or the logic of justification (it is not that Owen finds links between the natural groups Pachydermata and Ruminantia; rather, he shows that the former is not a natural group). It is close enough so that naturalists, if not anatomists, would take the point.

Macrauchenia serves as a particular example of the same point. In the *1844 Essay*, Darwin writes,

> (Fossils) that fall between our existing groups fall in, according to the manner required by our theory, for they do not directly connect two existing species of different groups, but they connect the groups themselves: thus the Pachydermata and Ruminantia are now separated by several characters, for instance Pachydermata have both a tibia and fibula, whilst Ruminantia have only a tibia; now the fossil *Macrauchenia* has a leg bone exactly intermediate in this respect, and likewise has some other intermediate characters. But the *Macrauchenia* does not connect any one species of Pachydermata with some one other of Ruminantia but it shows that these two groups have at one time been less widely divided. (p. 157)

For the record, the fibula in the horse, a pachyderm, is reduced to a rudiment; the fibula in most ruminants, though typically small, is never rudimentary; and *Macrauchenia's* fibula is neither reduced nor rudimentary, although it is anchylosed to the tibia for much of its length. Darwin's statements indicate an acquaintance with Owen's description of *Macrauchenia*, but not the bones of which he speaks.

By 1845 Darwin is fully aware of the non-ruminant nature of *Macrauchenia*, though his discussion of it in the second edition of the *Journal of Researches* is not without traces of earlier interpretations:

> It belongs to the same division of the Pachydermata with the rhinoceros, tapir, and palaeotherium; but in the structure of the bones of its long neck it shows a clear relation to the camel, or rather to the guanaco and llama . . . I was at first much surprised how a large quadruped could so lately have subsisted, in lat. 49° 15′, on these wretched gravel plains with their stunted vegetation; but the relationship of the Macrauchenia to the guanaco, now an inhabitant of the most sterile parts, partly explains this difficulty. (*Journal of Researches*, 1845, pp. 172–173)

Almost the whole historiography of *Macrauchenia* and the zoological issues that surround it, receive a curious recapitulation a year after the publication of the *Origin*. On 21 November 1860, Thomas Henry Huxley read a description of a new species of *Macrauchenia*, a small animal about the size of a guanaco. The fossil came from the mountains of Bolivia, appropriately; the spaces within the bone had "for the most part been filled with threads of native copper" (Huxley 1861, 2:401). Huxley has an array of bones quite reminiscent of Owen's — a cervical vertebra, a piece of lumbar vertebra, portions of scapula, ulna, tibia, and both astragali. He also has some of the skull and teeth. The fused epiphyses indicate that it is indeed an adult of a small species, rather than a young specimen of *M. patachonica*. Huxley's description follows Owen's for the bones that are available to both. He atomizes the skull, and finds many similarities to the skull and teeth of artiodactyls, some of which recall camelids, and some of which particularly depart from them. He does not contest the assignment of the genus to the perissodactyls, but he remarks that the genus is more deserving of the term "generalized type" than even Cuvier's Eocene quadrupeds (p. 415).

Toward the end of his paper, Huxley draws together the themes of size, representation and extinction in a single biogeographic province. He could not have known that this timely illustration of the principles Darwin used in the *Origin* is almost identical to the very example that got Darwin started on transmutation, twenty-three years before:

> there lived in the highlands of Bolivia a species of *Macrauchenia* not half as large as the Patagonian form, and having proportions nearly as slender as those of the Vicugna, with even a lighter head; and it is very interesting

> to observe that, during that probably post-pleistocene epoch, a small and a large species of more or less Auchenoid Mammal ranged the mountains and the plains of South America respectively, just as at present the small Vicugna is found in the highlands, and the large Guanaco in the plains of the same continent. (Huxley 1861, 2:415)

In fossil and living llamas, Darwin sees a microcosm which reveals the macrocosm. With nearly the same materials, Huxley produces a miniature. He does, however, get his llamas right.

Huxley, whose list of anatomical papers might suggest a plan to re-examine every major study undertaken by Owen, finds in *Macrauchenia* a lesson for a hypothetical anatomist who puts faith in the anatomical method:

> The structure and geological date of the genus *Macrauchenia* may serve, if taken together, to point to an important palaeontological moral. Professor Owen in the able memoir cited above, has clearly pointed out the remarkable combination of Artiodactyle and Perissodactyle characters exhibited by *Macrauchenia*, which unites the eminently characteristic cervical vertebrae of the Artiodactyle *Camelidae* with the three-toed fore-foot and the triply trochantered femur of the *Perissodactyla*; and with an astragalus which, in the apparent entire absence of any facet for the cuboid, is, I may affirm, more Perissodactyle than that of any member of the order, except *Hyrax* . . . (Ibid.)
>
> Again, *Macrauchenia*, alone, affords a sufficient refutation of the doctrine that an extinct animal can be safely and certainly restored if we know a single important bone or tooth. If, up to this time, the cervical vertebrae of *Macrauchenia* only had been known, palaeontologists would have been justified by all the canons of comparative anatomy in concluding that the rest of its organization was Camelidan. With our present knowledge (leaving *Macrauchenia* aside), a cervical vertebra with elongated centrum, flattened articular ends, an internal vertebral canal, and imperforate transverse processes, as definitely characterizes one of the Camel tribe as the marsupial bones do a Marsupial — and, indeed, better; for we know of recent non-marsupial animals with marsupial bones. Had, therefore, a block containing an entire skeleton of *Macrauchenia*, but showing only these portions of one of the cervical vertebrae, been placed before an anatomist, he would have been as fully justified in predicting cannon-bones, bi-trochanterian femora, and astragalia with two, subequal scapho-cuboidal facets, as Cuvier was in reasoning from the inflected angle of the jaw to the marsupial bones of his famous Opossum. But, for all that, our hypothetical anatomist would have been wrong; and instead of finding what he sought, he would have learned a lesson of caution, of great service to his future progress. (Huxley 1861, 2:415–416)[12]

It is unlikely that Huxley would have known that Owen had at first acted just as Huxley's hypothetical anatomist did. Huxley is perhaps being a bit unfair, since the canons of comparative anatomy would emphasize bones other than the cervical vertebrae, and these more dependable bones tell the story that both Owen and Huxley accept.

Huxley's sermon contains one bit of ignorance that would give Owen cause to smile. Huxley says, as Owen had in 1838, that the vertebrarterial canal of camelids occurs otherwise only in *Macrauchenia*. In 1853, however, in the *Descriptive Catalogue of the Osteological Series contained in the Museum of the Royal College of Surgeons of England* (published anonymously, and transparently the work of Owen) the following is described under *Myrmecophaga jubata*:

> The transverse processes of the three succeeding cervicals are imperforate, the vertebral artery entering the neural canal behind, and perforating obliquely the base of the neurapophysis, anteriorly, as in the *Camelidae*. (Owen 1853a p. 426)

Here, in an animal of no conceivable relation to either the macrauchene or the guanaco (except geography), is our peculiarity again. I have not found any mention of this in any of his subsequent papers on the giant ant-eater. Owen's later discussions of camelids and *Macrauchenia* say only that they are the only *ungulates* to show this character.

I would guess that Owen had known of this peculiarity since 1840. In his description of another of Darwin's fossils, *Scelidotherium leptocephalum*, a ground sloth with transverse processes of the cervicals "perforated as usual" (*Zoology* 1:84), Owen embarks on a survey of edentate necks, which includes a discussion of form and function in the neck of the giant ant-eater, and a good figure of the cervical vertebrae (plate 24), though not one that could show anything about vertebral arteries. Although he says nothing about the vertebral arteries, it is difficult to believe that he would have missed a feature for which his eye, normally superb, was already prepared.

I shall not trace *Macrauchenia* to the present day, although there is one final twist to the vertebrarterial canal that I would like to share. In 1886, E. D. Cope asserted that there is no sign of a vertebral artery at all in camel cervicals — he presumed that the artery ran with the spinal cord in the spinal canal (1886). In 1891, Cope reported that in *Macrauchenia*, as in camels, there is no sign of a vertebral artery in the cervical vertebrae. (This paper is decidedly a mixed bag. Cope establishes the Litopterna, the order to which *Macrauchenia* belongs, as an independent group of extinct South American ungulates. He also says that Owen's description was based on skull material, and that Owen was handicapped by not having any bones of the feet (Cope 1891).) S. D. Webb (1965) reports that von Zittel's *Handbuch der Palaeontologie* (1891–1893) makes the same error about the vertebrarterial canals of camels, and that most textbooks in this century repeat it. Webb emphatically states that indeed there is a vertebrarterial canal in camels. Science has reconquered the peak from which Owen commenced.

Conclusion

The history of the vertebrarterial canal, and the relation between camels and *Macrauchenia*, begins with Owen, Cuvier providing a convenient pre-history. The approaches that Owen takes to this fact comprise a catalogue of his wide interests and astonishing genius. They are a window on his way of working, and perhaps, by the time we come to the giant ant-eater's part in the story, even a hint about his scientific character. We are used to thinking of MacLeay having had a temporary effect, largely by way of fools, on the course of speculative zoology, but Owen is able to take the static symmetries of such an approach and find by them interesting relations in the realm of normal comparative anatomy.

In Darwin, we catch sight of one of the triumphs of his method —his ability to misprise in a strong and creative way what experts in other fields have done. Darwin gives little evidence of being aware of either the osteological subtlety that is just beneath the surface in Owen's descriptive work, or the logic that leads Owen to his reorganization of the ungulates. But Darwin takes the vertebrarterial canals as evidence of descent, and he places that descent in a geological, geographical, and ecological context that is his own.

We can also observe, by way of this case study, the tension between the theory of transmutation and mental categories that are non-transmutational. For instance, Darwin remarks in the *Origin*, that under his theory, affinity, rather than being metaphorical, will be a statement of real relation. In this example, we see how different those concepts can be. Owen's multi-dimensional affinities would make a theory of descent impossible. The ability of Darwin and Huxley to accept *Macrauchenia* as simultaneously a perissodactyl and a "more or less auchenoid" creature is no more compatible with genealogy. Could the older concept of affinity be rather more deeply ingrained in both than they would have suspected?

The transmission and modification of facts, from person to person and from field to field, is easier to study than the way in which these facts shape and are shaped by an original theory. While Owen is all that could be asked for in a comparative anatomist, the kinds of questions he asked about form, function, development, and classification can in the main be anticipated. Darwin, on the other hand, practices unrestricted science (Pantin 1968), and, in a case such as the taphonomy of guanaco bones, an unexpected science. Although it is not too difficult to find Darwin foreshadowing modern biological theory — all one must know is the modern theory — it is harder to do justice to theoretical insights that neither modern theory nor close attention to Darwin's contemporaries would predict. The range of meanings in the llama-rhea passages in *RN* point to this difficulty. Knowing biology as Darwin did — and as we do today — at the level of individual facts, not abstract theories, is, however, helpful.

In particular, I suggest that Darwin is brought to salvage a theory of species origins by invoking sudden reorganizations. He takes this step because Owen tells him that his large fossil camel is not a camel. We know that Owen originally said the fossil was a camel, and Darwin's acknowledgement of anomalous features in his discussion of the fossil in the *Journal of Researches* indicates that Owen had re-examined the specimen. Given what the anomalies are, and how ungulates were then classified, it is deducible that Owen knew that he was not describing a camel as soon as he found the anomalies.

The idea of transmutation ought to have developed sometime before Darwin wrote the passage in *RN*, and this precursor probably would not have been based on MacLeay. As Frederic L. Holmes once remarked, the writings in the notebooks are unlikely to be a record of raw Darwinian brain waves. The *RN* passage is perhaps a reaction to the unhappy news that a promising theory had been compromised by a change in the most interesting fact embedded in it. In any event, there are probably more paths through the notebooks than there are Darwinists.

An understanding of the detailed scientific issues Darwin faced does no harm in itself, and allows far more detailed interpretations of what Darwin was doing. It even suggests the sort of evidence that would overturn the view I have offered. For example, it would be fatal to my view if someone discovered an Owen letter from late 1837 in which he wrote of a fossil camel that ran on three toes, and whose femur carried a third trochanter.

The Darwin industry, or a considerable part of it, has produced an achievement that would have been difficult to predict — an internalist history of biology in which the biology is basically irrelevant. Metaphors, philosophical themes, and dental problems are all to be explored, especially if the search can be based on unpublished manuscripts. I would be as depressed as the next person were there only a steady diet of analyses of Darwin's biology, but such is not presently the case. There are very few in the league of, say, Ghiselin and Jaffe (1973) or Sulloway (1982a, b, c), that examine critically the concrete biological problems Darwin deals with.

It might be argued that meeting the biology head on requires more specialist knowledge than anyone trained as a historian of science is going to have. If all the secrets are indeed locked in the cirripede volumes, then few will be able to enter the kingdom. Although the more biology one knows, the easier it is to notice that Darwin is the most brilliant biologist of all time, it is possible to find a good deal without being a practicing anatomist, let alone a practicing transcendental morphologist or quinarian. In retrospect, I realize that all the anatomy I needed for this project is contained in the eleventh edition of the *Encyclopaedia Britannica*.[13]

There is more profit and less art in the boney light than most historians of biology realize. By turning that light on our texts and manuscripts, and by going back to the bones, we can reconstruct what Owen saw, what Darwin saw, what Huxley saw. We can find differences in what they observe and understand, differences that reflect and help to define their theoretical commitments. We can trace the evolving interpretations of a single bone in different perspectives, and we can observe the relationship between a disintegrating fact and the theory it helped to launch. We can look at the bone ourselves, and learn how blurred the boundaries are between observation, interpretation, and authority, and how all contribute to the construction of a fact.

Further, I recommend cautious anachronizing — an awareness of how ruminants are classified today, why litopterns are in their own order, how systematists treat characters for which no function is known, why there is a fuss about paraphyletic groups. Such knowledge can suggest what is being seen by nineteenth-century naturalists, and what is being missed today. Darwin was not a cladist, and Owen was not aware of gradal classifications; but the roots of their differences in 1837, and the roots of the debate in modern systematics, grow in the same soil. In such cases, we may be dealing with perennial, perhaps insoluble, problems in evolutionary biology. If so, this can be of as much interest to the biologist or philosopher as to the historian.

Darwinists have a natural constituency among biologists whom they can instruct and from whom they can learn. Systematists are interested to know that the theory of evolution may well have begun by the recognition of a shared derived character, and many will be just as interested to know that the first shared derived character soon became the first example of homoplasy — a character shared in two groups, but not by descent. That turn of events epitomizes a great deal of current systematic debate. On a more profound level, it is useful for scientists to know that bones are texts — there are inspired readings, such as Owen's, and sloppy ones, such as Cope's. But there are no certain readings except perhaps on trivial matters. Darwin's taphonomic reading and Owen's morphological one are equally strong, and each sees something that is not visible to the other. As Darwin was to learn, he could conduct a dialogue between a naturalist's approach and an anatomist's, but for that reading, richer still, one must turn to the barnacles — or to modern evolutionary biology.

Dialogue or no, to study naturalists without studying bones, to confine ourselves to metaphors, philosophy, and general social currents, is to transmute Charles Darwin into Herbert Spencer. We can do worse than to aspire to the proficiency of Mr. Venus, the articulator of bones, who offered to "sort your wertebrae in a manner that would equally surprise and charm you."

Notes

1. The lament of Mr. Venus, the articulator of bones. I thank Marjorie Grene for making the introduction.
2. De Beer's note adds, "Presumably the Capybara." Presumably not. There is no reason to cast the skull of a common animal, or to discuss it in a paragraph largely devoted to the troubles of dealing with fossil bones.
3. Obviously, the mention of an extinct giant camel in addresses by Lyell or Darwin does not constitute the introduction of this animal to science. Owen does not act as if the published abstract (1837) of his address to the Geological Society of 19 April 1837 on *Toxodon* (in which he introduces that name) has official standing either — he does not bother to address differences between that paper and the later *Zoology* account.
4. Although it is not really germane to this paper, a reviewer reasonably enough asked what *Macrauchenia* and *Toxodon* are. They are representatives of two of the five extinct orders of hoofed mammals that flourished in South America while it was an island continent. Edentates (sloths, ant-eaters, armadillos), endemic rodents (for instance, capybara and agouti) and New World monkeys joined them in a spectacular adaptive radiation.

   They were the last of their orders to disappear; both lived to see the results of the reappearance a few million years ago of the land bridge between North and South America. This bridge allowed such animals as llamas, elephants, horses, tapirs, and cats into a continent that had done rather well inventing convergent forms from whatever was on hand early in the Tertiary. The macrauchene did resemble a massive camel, though with a short trunk (or, according to other reconstructions, nares very high on its head). *Toxodon* was a massive creature, built along the same lines as a hornless rhinoceros.

   Owen and Darwin had no way of knowing that llamas and tapirs were new arrivals in South America, and that the capybara, *Macrauchenia*, and *Toxodon* were unrelated remnants of the old guard.

   For further information on the history and convergences of South American mammals, see George Gaylord Simpson's very accessible *Splendid Isolation* (1980). Though one of the great evolutionists of this century, he converges on historians of biology when he comes to the discovery of the macrauchene. His version of the standard error: Darwin "had the field impression that it was 'a llama or *guanaco*, fully as large as the camel' " (p. 100).
5. The assignment of *Toxodon* is nowhere near as clear cut. Owen has no information on feet, and although the procumbent, open rooted incisors definitely suggest rodentian affinities, the dental formula and some characters of the skull deny it.
6. Recent reviews of the quinarian system include Winsor (1976, pp. 82–87), and Ospovat (1981, pp. 101–111).
7. To this point in the discussion of *RN*, I have been retracing the excellent case made by Kohn (1980). I find Kohn's general argument convincing; specific cavils (for instance, "Owen assigned *Macrauchenia* to the Camellidae [sic]" p. 74) suggested this paper.
8. Darwin can use the word to mean anastomosis; but when he does, the context provides obliging cues — for instance, in the *Origin*, "mountain ranges, which from an early period must have parted river-systems and completely prevented their inosculation . . ." (p. 384).
9. Darwin is referring to a seed-snipe (Thinocoridae in the order Charadriiformes), which is a ground-living seed eater descended from an aquatically inclined ancestor (typical members of the order include the snipe and plover). Both in morphology and behavior, the bird presents a mix of characters that normally are found in very different birds.
10. Darwin may be somewhat influenced in this by his low estimate of the mental equipment of guanacos; he notes that they are attracted to gunfire, "have no idea of defense", and are easily bewildered (*Journal of Researches*, p. 196).
11. In the same work, Owen issues the macrauchene its (taxonomically) "important organs of manducation". The molars resemble in some important ways those of artiodactyls, though with significant differences as well. The dental formula is non-artiodactyl, and closer to the perissodactyls, to whom he assigns it.
12. The story of Cuvier's feat of predictive paleontology is recounted in Coleman (1964, pp. 125–126).
13. In fact, much of what I needed to learn was taught to me by the following comparative anatomists: Paul Olsen, August Pivorunas, V. Louise Roth, and Kathleen Scott. They cheerfully coped with questions that were evenly divided between the shockingly ignorant and the unanswerably arcane. Observing the way

they do their work has proved most useful in trying to understand how Owen did his. I am most grateful to them.

I also wish to acknowledge helpful discussions with Frederic Holmes, Malcolm Kottler, and Michael Ghiselin.

V. Louise Roth helped greatly in the planning of the figures, which are the work of Saran Twombly.

Finally, I give thanks to Sydney Smith for all that he has taught, to me and many others. Were he to write a footnote on the interpretation of *Macrauchenia,* he would put between its lines more research and far more insight than I have managed here. If there is anything to this paper, let it be a crude likeness of what he would leave unsaid.

6

The Immediate Origins of Natural Selection

M. J. S. Hodge and David Kohn

Introduction

Two decades of studies of the early notebooks have illumined the origins of Darwin's theory of natural selection far beyond what was possible at the 1959 centennial when biographers still relied on Darwin's later reminiscences.

The theory was arrived at during the months from Summer 1838 to Spring 1839 as Darwin was filling his *Notebooks* D and E.[1] We offer here a brief narrative of this theoretical discovery, drawing on more detailed analyses given, with fuller reference to the secondary literature, in Kohn (1980) and in Hodge (1982; forthcoming). Of accounts by others, ours agrees most closely with the late Dov Ospovat's (1981).

Any attempt at such a narrative faces at least three kinds of difficulty.

First, there are exegetical uncertainties. Darwin cut from his notebooks many pages, not all located since, nearly one in six of *D* and *E* being still unknown; and the pages we do have contain some entries added later. The notebooks often record only the tips of conceptual icebergs. The more important an insight became the less need was there for further reminders in writing; so the reasoning has to be reconstructed conjecturally by supplying the premises Darwin most likely had in mind but left tacit. Again, an entry may be explicitly related to something Darwin has read or heard, but it normally takes much more than this to show that he was not already thinking along those lines and would not have begun to do so had he not met with that source then. Finally, Darwin marked and annotated many books and articles; but as he usually returned to them later they may contain several layers of responses, none dateable with confidence.

Second, any narrative presupposes interpretative decisions about "periods" or "stages" and the "turning-points" or "watersheds" between them. Hitherto, everyone has found the most consequential moment coming with Darwin's reflections, of 28 September 1838, on Thomas Malthus's *Essay on the Principle of Population*. Even Ospovat (1981), who has gone furthest in spreading the elaboration of natural selection over many weeks rather than a single day, has chapters on "Darwin before Malthus" and "Darwin after Malthus". But Darwin's theorizing underwent complex developments both before (Kohn 1980, Hodge 1982) and after (Ospovat 1981) the meeting with Malthus, and the time has come to insist that these reflections did not start him moving, suddenly or gradually, to an entirely new way of thinking. Our account is hinged on September 1838, but in order to show that no cluster of reflections was uniquely consummatory in seeing "the last piece of the puzzle falling into place" or "the separate elements finally brought together".

All such metaphors are misleading in implying that Darwin's theorizing only became truly coherent when some conclusion, identifiable retrospectively as "natural selection", first occurred to him. For, what the entire B-E *Notebook* sequence, from July 1837 to July 1839, shows is that he was working knowingly with a comprehensive structure or argument throughout these two years, and that the theory of natural selection emerges towards the end of this period in a series of steps such that no one can be picked out as taking Darwin from lacking to possessing the theory.

Third, there are difficulties over the very nature of Darwin's enterprise. Is natural selection a "fact" about the world that he "recognizes", or is it a "vision" that he "expresses"? More generally, are the notebooks building towards a "cosmology", or from an "ideology"?

Although other choices of terms might well be justified, the ones Darwin used have a certain priority. He put his enterprise under the heading at the very opening of *Notebook* B: "*Zoonomia*", the laws of life. It was the title his grandfather had given his best-known book, a work admired by Darwin when at Edinburgh and reread and cited in July 1837. The young Darwin was undertaking his own zoonomical inquiry in order to complement Lyell's new system (1830–1833) of the inorganic terrestrial world, by providing a suitable successor to Lamarck's transmutationist system as carefully presented but rejected by Lyell himself (Hodge 1982).

Only by concentrating on Darwin's self-conscious aims, can we appreciate how he understood the structure of his theorizing. Only then can we discern how the theory of natural selection was eventually reached through successive transformations of a very deliberate configuration of argumentation. It is argumentation that takes off, in 1837, from a quite precise point — the sexual generation of one individual from two others — and moves on to the propagation of one species from another. It is argumentation that integrates generational and geographical considerations. For maturation and crossing distinguish sexual from asexual generation; and, while the changing conditions in any area are inducing adaptive innovations in maturations, the shifting avenues or barriers to migration are permitting or preventing crossings. Geography, given time and motion to make history as geology, thus mediates between the generation of individuals and the propagations of species whereby there grows over eons a whole tree of life.

To emphasize the persistent structuring of the problems that Darwin's notebook theorizing was intended to solve is not to deny that his intellectual and emotional outlook was undergoing major developments bearing directly on his zoonomical program. Far from it, for these were manifestly years when his reading and conversation were often combining with career and family concerns to prompt revision of many beliefs and attitudes — not least longstanding religious ones — and to motivate new inquiries, including those taken up in his *Notebooks* M and N (July 1838–July 1839 approximately) on "metaphysics". Our hope is to clarify those developments in showing that they arose while he continued to pursue explanatory challenges first engaged in the Spring and Summer of 1837.

I. From Individual Generation to Species Propagations

After the first ten pages of *Notebook* B have set out the theory of species propagation, the next seventeen (*B* 11–27) extend it first to "change" (*B* 11–17) and then to "progress" (*B* 18–27) in the widest spans of space and longest stretches of time. (This volume has a fuller discussion in the paper by Hodge.)

Over the next few weeks, Darwin revised much of the argument about "progress" and the "tree of life" in those last ten pages; but his revisions left quite intact the reasoning of the first ten on species propagations and the next seven on change. For, in the revised tree of life progress is made, on a mutable but stable Lyellian earth's surface, in ramifying species propagations analogous to the bud propagations in any tree growth: many buds or shoots ending without branching, in extinctions, while others branch without ending, in species multiplications.

In this arboriform ancestry and descent, groups high in the traditional ranking of organizational types in a scale of complexity, such as reptiles, have arisen by the adaptive improvement for life on dry land of lower groups such as fish. But within any group many lines have diverged from the common ancestral stock, and all lines do not make organizational progress at all times. In adapting to various conditions and ways of life some lines may lose while most are gaining in complexity. So, although progress has been inevitably the norm, there is no developmental law necessitating that life, like a tadpole growing into a frog, must go in one preferred upward direction if it is to go on at all. Any ancestral stock has arisen in adaptive divergence from a still earlier stock, but the resemblances among the many species descended from it are not explicable as a common adaptation to a common set of conditions or way of life. They are rather to be explained as a common legacy from that ancestral stock; while the differences among these descendants are mostly due to differing adaptive divergences from their common ancestors.

For these reasons Darwin's theorizing about species propagations had to provide for change being invariably adaptive and, as a consequence, usually but not invariably progressive in the long run.

The argumentation of *B* 1–10 engages, in this order, the problem of the initiation of hereditary adaptive variation from one sexual generation to the next in changed conditions; the problem of proceeding beyond this variation to the formation of a new variety; and, lastly, the problem of going on from there to species formation, to the formation of a race or form that is no longer merely a variety, because it meets the criteria for specific distinction.

So, we may take adaptation as one theme and species formation as another. For the most instructive thread to follow through 1837 to Spring 1838 is species formation; while adaptation is the best one to follow from that Spring (March) to the next Autumn (November) 1838. Then in tracing the revisions made over the Winter and Spring of 1838–1839, we have to follow both species formation and adaptation. By March 1839 Darwin has the theory of natural selection very much as he presented it in the unpublished *1842 Sketch* and so, too, the *Origin* of 1859. In following these threads on from July 1837, we can understand the steps that took him to natural selection as he understood them at the time, as so many modifications to his thinking at the opening of *Notebook* B.

At *B* 1–10, the argument begins with hereditary adaptive variation accompanying sexual generation in changing conditions, thanks to the impressionability of maturing organization. But how then can any species be constant in character across its entire range? It can because crossing, with the blending of parental characters, keeps the species constant as long as conditions are constant overall and only changing temporarily and locally. Conversely, then, a new variety can be formed and adapted to new conditions, if this conservative action of crossing is circumvented by the reproductive isolation and inbreeding of a few individuals in the new conditions, whether that reproductive isolation arises with or without geographical separation. And how may this variety formation proceed to new species formation? Well, the usual criteria for specific rather than mere varietal distinctions were three: true breeding, demarcation by a character gap that no intermediates fill, and being unable or at least not inclined to cross and produce fertile offspring.

Although jotting down only promissory memoranda at *B* 9–10, Darwin had developed by the end of this notebook, in February 1838, an elaborate analysis of how the origins of character gaps and sterility gaps are related. Between two varieties of a domestic species there is often a marked character gap but no sterility gap. And, Darwin argues, this is because the very conditions of domesticated life disrupt the reproductive system, including any associated instincts in animals; witness the greater variability of domestic species. So, in the wild the divergence leading to such a character gap would be accompanied by an increasing inability, or at least disinclination, to interbreed. Then, with this reproductive isolation, further, unlimited constitutional divergence in differing conditions could proceed, and so there would eventually arise the sharp difference in character and total intersterility that usually separate one species from another.

The incomplete blending of parental characters, and reversion to grandparental characters, Darwin takes as signs of incipient intersterility and so incipient frustration of one object of sexual generation: that is, keeping a species constant in character, when living in conditions not changing overall, by crossing and blending of local variations. For there will be intersterility if the constitutions of the two partners to the cross are too different to be blended irreversibly in a constitutional compromise that their offspring can embody and pass on to their progeny.

These arguments were reinforced, early in *Notebook* C and Spring 1838, with what Darwin calls "Yarrell's law" (after William Yarrell): that is, in any crossing the characters of an older domestic variety dominate rather than blend with those of a younger one. For Darwin this law showed that characters long in the breed are more deeply embedded in the hereditary constitution, and so more resistant to the influences of both crossing and changing external conditions. True breeding, a further mark of species rather than mere varieties, could then be explained as, like a character gap and a sterility gap, a concomitant of the prolongation of adaptive constitutional divergence with reproductive isolation in changed conditions.

This analysis of adaptive species formation in the wild made various comparisons with variety formation in domestic species. But they all drew on a distinction, one Darwin thought confirmed in his consultations with Yarrell and John Sebright, between two ways whereby varieties in domestic species could be formed (*C* 4, *C* 133–7, *D* 20, *D* 107). "Local varieties" were "natural" in being produced without the arts of man, by the influences of the local conditions to which they had adapted, often following isolation or migration or transport to another country; whereas "artificial varieties", not "adaptations" indeed often "monstrosities", were made and maintained by the breeders' art of "picking" (selection). Moreover, although Darwin accepted that some varieties in domestic species were products of both local conditions and careful selective breeding, this judgement explicitly invoked the contrast between the natural and artificial modes of production (*D* 20).

The effectiveness of conditions in making natural varieties depended, Darwin assumed, on all or most individuals in the locality being exposed to the same influences during maturation; for only then would the same new characters be elicited and like be breeding with like without counteraction by blending in crossing. In picking, however, the breeder, choosing which individuals are to be parents of the next generation can also match them as mates for each other; so, by pairing rare, monstrous, minority variants and then discarding any offspring with other characters, he can initiate and perpetuate a new variety starting with variations that could not thrive in the rigorous conditions nor persist with free crossing in the wild.

Following this distinction throughout *Notebooks* C and D (through September 1838), Darwin consistently compared the formation of species, as adaptations, with what nature rather than with what art had wrought in domestic species. The comparison was reinforced by his analysis of adaptation itself. He had done little with this in *Notebook* B beyond what was laid down at its opening in July 1837: that hereditary adaptive change is made possible by the impressionability of the immature organization issuing from a sexual but not an asexual generation. What he did in *Notebook* C reaffirmed rather than replaced that premise.

II. Fertilization, Maturation, Crossing and Adaptation

As Darwin saw it, the adaptations of species had to be traced to adaptive responses made by maturing individuals.

In species, adaptations were those structures and habits that had arisen in adaptive divergences among the descendants from the common ancestry for the group, whether genus, family, order or class; especially divergences giving some species "analogies" with those similarly adapted in other groups, analogies such as aquatic species in various mammal groups often exhibit, seals and otters for instance.

A change arising early in development as a maturing individual adapts to its conditions of life, will be, unlike a mutilation to an adult, hereditarily transmitted. So, consider how an aquatic species of North American feline could arise (*C* 63). If geological changes resulted in habitable muddy areas beside lakes abounding in fish, the jaguars there could be tempted into swimming. Then, Darwin supposes, any structural change such as webbing of the toes, elicited by this new habit and advantageous to the parents, would be duly transmitted in each generation to their offspring; and provided the "whole race", not just a few individuals, took up the new habit, its hereditary effects would not be lost in crossing.

To Darwin, in Spring 1838, this sort of scenario was exemplary for adaptive divergence. Certainly, he had just (*C* 61) asked himself, "Whether species may not be made by a little more vigour being given to the chance offspring who have any slight peculiarity of structure [?]" But this possibility does not contribute to the adaptive divergence problem. Darwin extends it explicitly to one sex only, males, where, he argues, it would make for similarity in structure throughout several orders as they are fitted not for different conditions of life but for what is common to any combative competition for females (*C* 62). He may have reasoned, too, that when limited to one sex such chance peculiarities would be exempt from the usual fate of congenital, heritable peculiarities, such as polydactyly in humans: that is, being "counteracted in crossing" (*C* 83–4).

One and the same variant character may arise as a "monstrosity" or as an "adaptation". When a puppy is born in a warm country with abnormally long fur that is a monstrosity, Darwin says; but were a puppy with a normal coat to grow it longer on being moved to a colder region this would be counted an adaptation. For here the maturational response is presumed to be one always and only elicited by the cold, the external condition to which the longer hair is an adaptation and of which it is an invariable, necessary effect not a rare, accidental, chance concomitant (*C* 65 and *C* 83). All such adaptations are, therefore, what Darwin will later call necessary adaptations.

So, a monstrosity and an adaptation differ in their causes as well as their effects. But could not a congenital deformity be considered as an adaptation, Darwin asks, in that it has presumably arisen in a maturational response to the conditions of pre-natal life in the womb? It could, he decides, but to be a "real adaptation" any variant character must be fitted to the whole life of the organism not merely the foetal phase (*C* 65 and *C* 83).

This *Notebook* C analysis of adaptation in individuals and in species is reaffirmed throughout *Notebook* D and well on into *Notebook* E in October 1838. It is complemented but not corrected by the extensive theorizing about sexual generation that dominates *D* and, equally, by the first Malthusian reflections at the end of *D* and opening of *E*.

According to that generation theorizing (more fully discussed elsewhere in this volume, by Hodge) the unfertilized egg, from which a grandchild will one day grow, is budded off as a constitutional facsimile of its mother when she is herself still immature and a foetus within the grandmother. So, if a father impresses a character on a child he will impress it heritably on the grandchild too, because what is earliest in development is most permanent in the constitution and most transmissible in heritance.

The object of sexual generation is the acquisition and transmission of changes in maturation. So an organism cannot serve that end as a parent if it acquires no new transmissible characters and grows up exactly as its own parents did. The effects of mutilations will not be acquired and transmitted, because they are incurred too suddenly (*D* 18) or too late (*C* 83) in maturation, and because the very process of ontogenetic recapitulation of phylogeny permits only innovations in harmony with all the previous ones, which mutilations are not (*D* 174e).

However, growing up so as to serve the end of sexual generation does not entail acquiring only progressive rather than deteriorational changes for transmission (*D* 174e). Nor are useful variants alone transmitted; witness hereditary diseases and congenital abnormalities such as hare-lip (*D* 169). The maturational necessities inherent in sexual generation as the means of adaptive change ensure that its end can be achieved, although neither progress nor utility, advance and advantage, are ensured in every generation.

Crossing rather than self-fertilization is needed for adaptation to the slow physical changes affecting large areas over long ages. For with selfing the characters everywhere elicited by local and temporary fluctuations are accumulated, not lost through blending. And, even with crossing, if the changes in conditions are solely local and temporary, then only mere "individual differences" will result, differences distinguishing the individuals within a species but not marking off one species from another. For a new species to be formed, conditions must be changing throughout the whole country in the same way (*D* 167 and *D* 174e–175).

Within a week or so of articulating these September conclusions, Darwin was responding to Malthus in the two pages (*D* 134e–135e) now become so well known. (He had already filled *D* 152e–179.) But before fitting his response into our narrative, we should pause to consider another account of how Darwin's thinking has developed to this point.

Like Ospovat (1981), we are not persuaded by several suggestions once made by Schweber (1977). He suggested that reading about Auguste Comte, in August 1838, brought Darwin finally to a fully "evolutionary view of nature", and also to seek the origin of species in an analogy between selective breeding in the wild and by man. But there was nothing in Darwin's general thoughts about nature, relevant to his *Notebook* D theorizing, that had not been there when *Notebook* B was opened the year before; and he has no such positive selective breeding analogy until several months later.

Schweber also suggested that Darwin had learned from reading about Adam Smith, in Summer 1838, a new way to relate chance variation in individuals to the adaptive formation of species. But, at this time, it is not chance or accidental variation but necessary adaptation in individuals (such as the puppy moved to a colder climate) that Darwin thought initiative of adaptive species formation.

We must, then, disagree with Schweber's main conclusion, that before reading Malthus Darwin already had species arising through a natural selective breeding of chance variants consequent on a competitive struggle for existence; and that this reading was to give him principally a quantitative understanding of the population pressure causing that competition.

We would have liked to welcome Schweber's further suggestion that Darwin may have sought out Malthus's book on reading of it in a review of work by the Belgian social statistician Adolphe Quetelet. Certainly, Darwin did resolve, in a list of things to be read (on *C* 268), to find out from the Statistical Society where Quetelet had published on "sexes relative to age of marriage", and the gray ink there dates this resolution most likely to sometime from August to mid-October 1838. However, the note recording his pursuit of the matter is added to *D* 152e in brown ink, dating it very probably to after mid-October.

Darwin's conjecture that all higher animal foetuses go through an hermaphroditic stage surely prompted his interest in Quetelet's findings in the sex ratio at birth. And even if the Quetelet reference did not lead him to Malthus, Darwin may well have gone also to the *Essay on Population* as yet another work on the subject then dominating his zoonomical inquiries: the ontogeny, phylogeny and teleology of sexual generation.

III. Fertilization, Maturation, Crossing, Superfecundity and Adaptive Change

What Darwin concludes in reflecting on Malthus is that superfecundity is always contributing to the adaptation of structure in changing conditions over the long run, where this adaptation is still understood as due to the hereditary embedding of adaptive responses made by maturing individuals in the manner explicated in *Notebook* C. For, the *D* 134e–135e entry records two lines of reasoning.

The first argues that one ought not to wonder at "changes in number of species, from small changes in nature of locality"; because Malthus's amazingly short doubling time — twenty-five years or even much less —for humans with plentiful food, shows that "there is a force like a hundred thousand wedges trying [to] force every kind of adapted structure into the gaps in the oeconomy of nature, or rather forming gaps by forcing out weaker ones." This whole line of reasoning concerns the effects on species population numbers of changes over space and over time in physical conditions.

Only in the second line of reasoning, confined to a closing sentence (inserted interlinearly but apparently without delay, Kohn 1980), is adaptive change in structure considered: "The final cause of all this wedging, must be to sort out proper structure, and adapt it to change — to do that for form, which Malthus shows is the final effect (by means however of volition) of this populousness, on the energy of man."

In the first line of reasoning, Darwin is responding anew to Lyell's argument that the way species are limited in their spatial spreading shows how their duration in time may be ended in extinctions (Lyell, 1830–1833, 2:130–140). A species is often abundant in one locality but surprisingly rare or missing somewhere else that is both accessible and similar enough to be suitable, because, Lyell argued, of De Candolle's ubiquitous war of species against species. For this other locality is already occupied by other species better fitted to the very slightly different climate and soil, and to the distinctive assortment of animal and plant residents there. The range and numbers that any species can maintain are much less, when limited by conditions as influencing occupancy and so competition, than they would be if limited solely and directly by conditions themselves. Likewise, a species achieves no monopoly within its range, because checks and balances among the various species there prevent unlimited increases in any one; a prey species increase causing an increase in the predators that duly brings a return to former prey numbers. So, changes in physical conditions can lead to extinctions, both by facilitating invasions of alien species at the expense of residents and by favoring some residents at the expense of others; for in an area already fully stocked permanent increases in any species must entail permanent decreases in the number of others.

Darwin had long been weighing these proposals with care, especially in Spring 1837 when he was still rejecting them as inadequate for certain South American quadruped extinctions, and developing for these an alternative, species senescence, explanation (see the paper by Hodge in this volume). He had since come to think that these cases were exceptional, and that adaptability in changed conditions determines whether a species ends its life without issue or multiplies itself by propagating descendant species (*B* 38, *D* 72). He had accordingly sought to understand extinction by integrating Yarrellian generational causes of constitutional fixity and Lyellian ecological causes of competitive defeats (*C* 153). He had, then, returned to the Candollean "wars of organic beings" (*C* 73), and to the Lyellian arguments for precision of adaptation and precarious balance in competition, arguments from the rapid increases in species invading new countries, and from the rarity and commonness of species in slightly different localities within their native countries (*C* 73 and *C* 160e).

Consequently, he now saw Malthus as strikingly vindicating Lyell's appeal to interspecific competition, by confirming that very small differences in conditions can make for large differences in the populational representation of species from one place to another. The rapid rates of increase, that even a slow-breeding species such as man can achieve, entail that any species would quickly gain in numbers upon any slight lessening of the checks exerted by the causes of mortality. The wedging simile is to extend this argument to the consequences for all the species living in an area that is presumed to be already fully stocked with life. The greater width of a wedge at its protruding end stands for the great potential numbers realized when any species is able to expand. Thanks to full stocking, every species cannot realize that potential. On all the better adapted species, but on them only, is there a wedging force acting, that is a tendency to increase quickly when any check lessens, at the expense of numbers among the less adapted species.

In the second line of reasoning, Darwin is responding to Malthus's passage in Book One, opening the sixth chapter "Of the checks to Population among the Ancient Inhabitants of the North of Europe". There it is argued that without the law of nature that humans, like other populations, tend to increase beyond the means of subsistence, all the world could never have been stocked with people; for man who is naturally slothful would have had no motivation to spread beyond the original Asian seat of the species into desert and tundra regions. As it was, those inhospitable regions were not only settled once but subsequently too; for new settlers have been moved — by recurrent shortages of land and food consequent on superfecundity — to migrate, to invade and so to struggle with both the earlier settlers and the adverse climate and soil. And the invaders have been victorious because of the enhancement of "the energy of men acting under such powerful motives of exertion". For these "contests" were "struggles for existence" wilfully fought in conscious awareness that "death would be the punishment of defeat and life the prize of victory" (Malthus 1826, 1: 92–95).

The whole argument was obviously consonant with analogies that were already Darwin's own. Not only had Lyell's extinction theorizing drawn parallels with European conquests over American and Australasian native peoples; Darwin's longstanding biogeographical arguments, from colonizing species of alien genera overrunning native residents, had done so no less explicitly.

Moreover, Malthus's providential view of reproductive law conformed well with the teleological interpretation of generation fundamental to Darwin's zoonomical program. Darwin's new argument makes a double comparison: the final cause — the beneficial effect, divinely intended —of population wedging, in animals and plants, is to do for their structure ("form") what such wedging, in man, does for his activity ("energy"). The sorting out of the proper structure is here the expansion of the best adapted varieties at the expense of others, as Malthus has some tribes or races among humans doing likewise. Led by Malthus to consider the eventual adaptive benefits to invading winners rather than the population losses among resident losers, Darwin now has a way round the objection made by Lyell against Lamarck: that transmutation of the species adapting to slow change in any area could never be completed before they were overrun by aliens already fitted to the new conditions.

In Lyell, the manifold interactions constituting the economy of living nature ensured that, just as very small differences in physical conditions sufficed to make one species the winner here and another there, so in any one spot very slight differences in structural character could make one species far more successful than another in maintaining its ground and its numbers. Shown by Malthus how to construe intraspecific competition as analogous to interspecific, Darwin could take very small differences in structural characters to determine the outcome of both.

Darwin is making here no analogy between the contribution of superfecundity to adaptation and "picking", the practice of selective breeding. His talk of sorting is only to indicate the expansion and retention of the adapted and its corollary the contraction and elimination of the others. There is an implicit analogy with all sorting processes whatever, all processes whereby some items tend to be retained and others eliminated, including mechanical siftings (*E* 112), but there is no allusion to the breeders' art.

Darwin's understanding of how superfecundity contributed to the adaptiveness of change in the long run depended on a premise prominent in early October: that the extensive changes in structure occurring when one species is slowly formed from another require only a great extrapolation, to the eons of gradual physical change, of the transmission, sorting and embedding of the variation acquired in maturations from one generation to the next (*E* 4–9e). For Malthusian populational wedging ensures that only those structures arising in maturational responses that count as real adaptations will be retained long enough to be embedded permanently in the hereditary constitution. If a structure is to persist it must be an "adaptation to *whole* life of animal" not only "to womb as in monster" or to "childhood" or "manhood"; otherwise it will be "driven outwards in the grand crush of population" (*E* 9e).

About a month later (*E* 50), he argues that the accumulation of many successive variations, required for major advances in organizational progress, can succeed thanks to the permanent embedding made possible by crossing. And, sometime between November 27th (not the 7th as printed editions have it) and December 2nd (*E* 55e–*E* 59), he goes on to argue that any variation once embedded need not continue to be adaptive to persist. Any structure is "capable of innumerable variations" that can "accumulate" as long as each is "permanently adapted to circumstances of *times*" and persistent due to "slow formation" (*E* 57).

The well-known entry (*E* 58) that follows thus epitomizes Darwin's current, elaborate integration of three conclusions that had all been explicitly in play before the end of September: first, that what is transmitted for one generation normally continues for more, because an ovum owes its initial constitution to the fertilization of the grandmother by the grandfather; second, that small maturational changes, transmissible as large ones often are not, do occur, especially in changed physical conditions; third, that far more offspring are produced than can live on the amount of food and other resources that supported their parents. For Darwin wrote:

> Three principles will account for all
> (1) Grandchildren like grandfathers
> (2) Tendency to small change especially with physical change
> (3) Great fertility in proportion to support of parents

These principles were to remain fundamental. Darwin is still working with them implicitly next May (1839) when considering whether a principle of compensation (one structure developed at the expense of another) may make "a fourth cause or law of change" (*E* 150). We need to remember them as we follow his theorizing into the next months.

IV. Adaptation, Selection, Chance Variation, Varieties, Races and Species

So far, Darwin has made, in the extant pages of *Notebook* E, no explicit appeal to chance variations as contributing to adaptive change (*E* 26e concerns chance conditions not chance variations). Nor has he made any analogy between wild species and domestic races as adapted products of selective breeding.

However, notes already entered on November 27th, in his *Notebook* N on metaphysics (*N* 41–45), contain his earliest known appeal to chance variation. They also show how the selection analogy may have been first perceived at about this time.

Darwin insists there that some instinctive associations in man are only explicable if we have animal ancestors; the association, present in dogs and horses too, of sexual arousal with salivation and biting being his example. He argues that any "habitual action" somehow affects the brain in a heritable way, and that "this is analogous to blacksmith having children with strong arms". But just as this is not the only way for any bodily structure to be changed, so likewise with the brain: "The other principle of those children which chance [with underlining and a "?" added later in pencil] produced with strong arms, outliving the weaker ones, may be applicable to instincts, independently of habits."

We can only guess how Darwin had come to formulate this principle. But he could well have done so by comparing and contrasting success in the struggle to survive among all offspring, the struggle so vividly intensified for him in September, with success in competition for mates by male offspring with some chance peculiarities. For this success of some chance male variants was what was implicitly contrasted, in *Notebook* C, with adaptation through the inherited effects of changed habits. And so he may now have come to reason that, because differential survival among chance variants in the struggle for life depends on all the external conditions of life, it could contribute to adaptation to diverse conditions as the struggle to win females could not. His example here of children (for Darwin presumptively boys) with strong arms outliving weaker ones would certainly fit this guess well. In any case, however he reached this new principle, it implied that wild species had something in common with domestic varieties that they had not had before in Darwin's view. For he had long accepted (*B* 118) that domestic varieties sometimes traced to chance variations.

Now, there are signs that he came in the closing days of November to think of some domestic varieties made by selection as he had not thought of any of them before: as adaptations. And there are signs that it was this new comparison — between wild species and certain picked or selected domestic varieties as adapted products — that led him on to the new analogy between the processes forming these products.

Hence the special significance, in the last days of November and first days of December, of dogs: more precisely sporting dogs, greyhounds in particular, including most particularly the greyhounds of one Lord Orford; greyhounds seen as distinctively fitted in structure and habits for their own peculiar manner of hunting and, in that, analogous to a wild predatory canine species.

Darwin was developing his most general thoughts about instincts on November 27th in *Notebook* N. He had long argued that differences in instincts depend on differences in bodily structure, especially in the brain, so that any accounting for adaptive divergence in bodily structure could also account for the origins of diverse instincts. And he had applied this argument to the special case of the differences in breeding instincts that keep distinct animal species from interbreeding in the wild, holding that the only workable criterion for specific distinction was lack of interbreeding between two forms living in the same region (*B* 212, *E* 24). Accordingly, he had long sought to discredit any notion that the different instincts ensuring this segregation were special immaterial endowments, independent of material organization and somehow superadded to it, rather than arising as a consequence of a gradual divergence in bodily structure.

Now (*N* 41), he takes the acquiring of instincts by dogs to show the influence of habit and so to exemplify the blacksmith principle. On his explicit materialist premise relating "mind and brain", he has, therefore, another confirmation that a brain structure peculiarity is no less heritable than any other. A jackal-fox hybrid had inherited from the jackal grandparent the habit of digging for mice in a distinctively jackal way, so presumably its brain was intermediate like the rest of its body. There is, then, support for the general thesis about divergence in instincts.

Moreover, this cross between these two wild species can support also, Darwin reasons, his view of divergent breeding instincts and species distinctness (*N* 44). For an exact "analogy", even an "identity" with such cases, is found where the cross is between a wild and a domestic species, wolf and dog, or again between two domestic races, greyhound and sheepdog or greyhound and bulldog. When greyhounds are crossed with sheepdogs it is their instinctive tendency to hunt hares, and not the sheepdogs' instinct to round up sheep, that dominates and is transmitted to the offspring. Hence,

Darwin's insistence that facts about "crossing races of dogs" with different instincts are "*most important*" because these facts conform to "the same laws" as the wild species crosses.

Concerning the greyhound and bulldog case he had recorded the details way back that Spring; but he had done so immediately after noting that he and his apparent informant, Yarrell, were agreed in thinking that the picking of varieties was an "unnatural circumstance" (*C* 120). These, the greyhounds of Lord Orford, although the fleetest in the nation, lost courage, Darwin had then noted, and would not run up hills. So a cross was made with a bulldog, and the hybrid descendants were bred for several generations with pure-bred greyhounds; there was thus a "dash" of "bulldog blood" but with the "whole form of greyhound — picking out finest of each litter and crossing them with finest greyhounds".

Greyhounds had, then, long been for Darwin exemplars of the power of picking. So, consider the analogy even identity, now emphasized by him, between the heritability of the wild jackal's and the domestic greyhound's predatory instincts; and recall that predatory habits, and changes in them as initiating changes in structure, had long been among his principal exemplars for adaptive divergence (and notice his consistent later preoccupation with predatory canines). Then it is surely probable that these dogs were the first products of selection to be interpreted by Darwin as adaptations analogous to wild species, and that they were so within only a few days of this 27 November note.

For the earliest known entry to deploy any analogy between nature's and man's selective breeding was written most likely on December 4th or 5th; and it shows that the analogy had by then already been given careful thought. The entry combines the new hybrid, predatory canine lore with Darwin's old concentration (inherited from Lamarck as retailed by Lyell) on swimming, preying and webbed feet (remember the jaguar). It concerns a breed notable for thick, wavy hair obviously advantageous when in water (shades of the puppy moved to a cold climate). Having read recently in Audubon (*N* 45) about duckshooting with large dogs of the Newfoundland and water spaniel cross that had distinctive aquatic retrieving habits, Darwin now asks himself (*E* 63): "Are the feet of water-dogs more webbed than those of others [?]"; and he continues: "if nature had had the picking she would make such a variety far more easily than man, — though *man's practised* judgement even without time can do much. — (yet one cross and the permanence of his breed is destroyed)."

Where Darwin had formerly thought species not picked at all, now he judges them even more picked than the breeder's varieties. But his argument continues to credit greater permanence to greater prolongation of production, even as it concludes that nature uses "precisely the same means" to make her species as man does to make domestic races (*E* 71). What the new argument appeals to is also a difference in degree: the selecting in nature is not only more prolonged, but more discriminating and comprehensive; so that every part of the structure of an aquatic canine, its feet no less than its hair, would be adapted to the water. Species are thus "only ancient and perfectly adapted races" (*E* 72).

The reason why this sameness of the means used by man and nature is such a "beautiful part" of the new theory (*E* 71), in Darwin's eyes, is that one can observe how this means makes domestic races to be adaptations. As Darwin wrote in his notes on John MacCulloch's theological treatise, notes written almost certainly in March 1839 (Hodge, forthcoming): "Get instances of adaptation in varieties — greyhound to hare. — waterdog, hair to water — bulldog to bulls. — primrose to banks — cowslip to fields." These, says Darwin, "are adaptations just as much as Woodpecker. — only here we see means", as we do "not in the other", the wild species, the woodpecker (DAR 71:57; Gruber and Barrett 1974, p. 419).

The new selective breeding analogy was thus cherished as an evidential triumph; because it allowed Darwin to go beyond invoking sexual generation as the observable means of individual adaptive variation from one generation to the next, to identifying selection as the observable means for the formation of a whole race that is an adaptation.

V. Chances, Designs and Selection

The perfect adaptation of species, as compared with the imperfect adaptation of domestic races, is due to nature's selection being so much more discriminating and comprehensive (whole body, inside and out, and whole life before and after birth) than man's. Outward greyhound form might be made by selection by man away from all hares and hunting, but a perfected greyhound race would arise in nature only through the perfecting selection that living by hunting would entail (*E* 71–2, *E* 75).

So, the selection analogy itself implied that the adaptive perfection of species may not be due to any difference between variation in the wild and under domestication. It confirmed, then, the possibility that the adaptations of wild species may be formed from chance or accidental variation just as those of domestic races often are. Darwin accordingly came to favor the contribution of chance variation to adaptation, over the other, blacksmith principle, in extending the new selection analogy to cases, especially seed dispersal structures in plants, where he not only saw no habits having inherited effects but no plausible way for them to arise in any necessary adaptations.

In his March 1839 notes on MacCulloch, he sees his "theory of grain of small advantages" explaining the curling seed pod valves of the broom, on the presumption that these are not "necessary adaptations" but "accidental"; for "they would not be detrimental accidents, and domesticated variations show us accidents may become hereditary . . . if man takes care they are not detrimental" (DAR 71:53; Gruber and Barrett 1974, p. 416). Likewise for acclimatizing plants to harsher conditions, he now identifies "my principle" as "the destruction of all the less hardy ones and the preservation of *accidental* hardy seedlings", adding that "to sift out the weaker ones there ought to be no weeding or encouragement but a vigorous battle between strong and weaker" (*E*112).

The new reliance on chance variation required, therefore, no new understanding of the competitive struggle for existence beyond what the Malthusian vindication of Lyell's appeal to De Candolle had provided in September. It was because the Candollean warring among species had always been a warring among the individuals of the different species that the Malthusian crush of population within each species intensified it so much. The extended senses of "strong" and "weaker", and the figure of "a grain in the balance", could therefore be used of intraspecific and interspecific competition alike. In December, Darwin had reflected that two human races act on meeting like two animal species, struggling to the death: so that Bolivian and Tasmanian natives have sometimes been "exterminated on *principles* strictly applicable to the universe" (*E* 64–65). In March, he acknowledges the difficulty he has in believing "in the dreadful but quiet war of organic beings" when he goes into the "peaceful woods, and smiling fields" (*E* 114). But the difficulty is overcome, he argues, by concentrating on the difference between the actual range of a species and the one it would have were there no others to limit its spread. A rhododendron species, limited to a band a thousand meters high in the Pyrenees, thrives in London gardens, showing that its range is limited not by physical conditions alone but by competition on its borders. There is, he says, "a contest and a grain of sand turns the balance" (*E* 115e). Small differences in the structural character of the various species make one victorious here, another there in conditions only very slightly different.

On going over to chance variations, for many adaptations, Darwin gave himself a further reason for taking adaptive species formations to be very slow and gradual: advantageous variants would arise by chance in only a few individuals initially. In an imaginary canine scenario, in more March notes on MacCulloch, he supposed six puppies born per litter, only two living to breed in the "Malthusian rush for life", and it "so chance that one out of every hundred litters is born with long legs"; then, if circumstances slightly favor that variety, it will "get the upperhand" in "ten thousand years", even with "intermarrying with ordinary race" (DAR 205.5: 28–29; Ospovat 1981, p. 70). The long time, previously needed for embedding them, is now also needed to take variants conferring only a small advantage beyond their initial rarity. Small differences in structure will make for small differences in the chance of survival, but in a long run of generations those small differences in survival chances will determine how the species is changed.

Thus did Darwin come to a two-fold deployment of chance notions. Variations are produced "by chance"; but they are not uncaused, merely not caused as necessary effects of the physical conditions to which the species is adapting. And, among those variations, some will have in these conditions a "better chance" of surviving (*E* 137). Thus, too, did Darwin reach a resolution of the traditional choice between chance and design. An adapted structure, such as the "beautiful seed of a Bull Rush", need be interpreted neither as a "fortuitous" growth (*E* 137) nor as a design impressed on matter in a special creation. For, with chance variations, some of them with slightly better chances of survival, and with many generations, such a structure, Darwin argues in his notes on MacCulloch (DAR 71:58; Gruber and Barrett 1974, p. 419–420), can be formed and perfected slowly in many "trials", as a consequence of general laws enacted by the Creator rather than as a result of a special unknowable act of the Divine will.

The initial rarity of the advantageous chance variants required Darwin to argue that selection, in the struggle for existence, can alter the species despite the effect of crossing with blending. Selection was thus like isolation in so far as it opposed these effects.

Having long understood isolation to facilitate local conditions in producing "natural" varieties in domesticated species, Darwin had previously compared wild species with such local varieties, while contrasting them with the "artificial" ones made by selective breeding. But with the new selection analogy these comparisons and contrasts were reversed. As races made permanent and perfectly adapted by sustained and sensitive selective breeding, species were likened, in December, to what human artifice rather than changed conditions had wrought in domestic species (*E* 71–72). Another well-known March entry (*E* 118) begins accordingly by premising that varieties in domesticated species are "made in two ways". One is when an entire portion of a species is subject to the same influence of conditions as happens on its moving from one country to another. "[B]ut," it is now insisted, "greyhound, race-horse and pouter Pidgeon have not been thus produced, but [the other way] by training, and crossing and keeping breed pure." And in plants likewise "*effectually* the offspring are picked and not allowed to cross." The question is, then, "Has nature any process analogous [?] — if so she can produce great ends"; and, "But how [?] — even if placed on Isld. if etc. etc. . . . Here give my theory. — excellently true theory."

This memorandum epitomizes the main argumentative strategy later followed by Darwin in presenting the theory of natural selection in the opening sections of his manuscript *1842 Sketch.* For what that *Sketch* does is to start just as *Notebook* B did in July 1837, with individuals varying heritably and adaptively in new conditions thanks to sexual as contrasted with asexual generation; then, again as in July 1837, it goes on to the blending out of this variation due to crossing. Now, at this point, the July 1837 argumentation introduced isolation and consequent inbreeding as a cause sometimes effective in counteracting the conservative action of crossing. The *1842 Sketch*, however, introduces a series of "ifs", in considering how human selection might counteract the conservative action of crossing, especially with the aid of some geographical isolation. The conclusion from all the "ifs" is that even if human selection were aided by isolation and enabled to operate to the full extent of its power, it would not produce very much in the way of permanent, adaptive change, because man's judgement is poor, is restricted to external characters and cannot fit a race to all the conditions of its life. The stage is thus set for natural selection whose powers are superior to man's in exactly the ways needed to achieve those results that man's selection cannot. To complete the stage setting, a distinction between direct and indirect adaptation to conditions is made next. And, then, at last, De Candolle, Malthus, the war of nature and chance hereditary variations are called upon, and their implications, for the legs of canine predators and pods full of plant seeds, are elaborated. The discussion can thus move on to explain how the races (that is, species) produced by natural selection, will be permanent, adapted and so on, as domestic varieties are not, all thanks to the greater sensitivity, comprehensiveness and persistence of nature's selection as compared with man's.

Reading the *1842 Sketch* alongside *Notebook* D and *Notebook* E shows, then, that Darwin was not only in possession of the theory of natural selection in the middle of March 1839, but also committed to arguing for it much as he would in 1842, and so, too, in the *Origin*. Six months before, in the middle of September 1838, he had not had the theory, but what he did over those months makes it impossible to name a day, a week or even a month as the moment when he first came to it.

Conclusion

This brief analysis of the Autumn, Winter and Spring months of 1838–1839 will indicate that various interpretative issues may need reformulating.

Consider, first, the topic of adaptation. Limoges (1970) has argued that Darwin's theory originated in an attempt, characteristically English, to undermine the view of adaptation upheld by such natural theologians as William Paley. But, as Ospovat (1981) has insisted, Limoges's analysis founders on a confusion. Darwin, from 1837 on, was explicitly rejecting the adaptational explanations found in Lyell's historical biogeography; and he was soon following other authors in rejecting certain adaptational explanations in comparative anatomy. But to insist that adaptation cannot explain some phenomena did not imply that species are not perfectly adapted; and, as Ospovat emphasizes, the selection analogy, as first developed by Darwin at the end of 1838, was prized by him as explaining how species can be formed as perfectly adapted as ever any Paleyan held them to be. So the theory, far from arising in a rejection of that perfectionist conception of adaptation, presupposed its appropriateness. Nor is that surprising; natural selection was worked out by a theist whose theorizing was explicitly teleological throughout.

The selection analogy itself has also divided interpreters (Kohn 1980), some seeing it as essential, others as peripheral to Darwin's discovery. The truth has to be that it was decisive but not in September, when it was still months away, rather in December, when both the adaptation and species formation theorizing were being revised to conform with new comparisons between domestic races and wild species.

In general, little is gained by asking what prevented Darwin from coming up with natural selection before he did; as if he must have done so in July 1837, if only some inhibition — about God or nature or species —had not gotten in the way of his reasoning. There is then no gain in interpreting the arrival at natural selection as something that happened as soon as certain blocks were removed. For throughout the notebook period, Darwin was consciously and unconsciously favoring some theses and discounting others, and one can usually find an adequate reason for any of the many successive shifts in thought that he makes in these two years. Moreover, the theory of natural selection was the outcome of so many successive shifts, that it makes no sense to insist that it must have arisen when some single insight replaced a formerly blank or prejudiced gaze. It emerges from a continuation of the process begun in 1837, a process of constantly revising the species propagation theorizing that opened *Notebook* B.

In understanding natural selection as emerging from this process, one tempting dichotomy is to be rejected. For nothing is gained by trying to decide whether Darwin's theory owed more to his "biology" or to his "views on man", as if these were separable. As his initial response to Malthus and many other notebook entries show, Darwin was often and deliberately arguing back and forth, between his general zoonomical conclusions and his specific conclusions about man as instantiating those much more general ones. When arguing from man to all species, he allowed for the obvious lack, in lower animals and plants especially, of higher mental faculties, such as "volition" and a "moral sense". When his argument ran the other way, he allowed for the peculiar, or rather pre-eminent, presence of these in man. There is an asymmetry, in that all his conclusions about man were not, as such, conclusions about all species; whereas any general zoonomical conclusion was one about all species including man. But there was no invariable procedural priority one way or the other; no policy that a conclusion about man must always be derived in the application of a general conclusion previously drawn concerning all species; or conversely, that a general conclusion always be derived as an abstraction from the human case. The very structure of Darwin's whole inquiry, from July 1837 on, made any such procedural constraints inappropriate. And sure enough, we find him throughout the whole *B–E* (July 1837–July 1839) period, including the *M–N* period (July 1838 on) theorizing through both kinds of arguments, some running one way, others the other way; so that no sense can be attached to the retrospective question as to which was the primary mode of speculation.

Obviously, we do not accept the view, taken by Shapin and Barnes (1979), that a theorist's intentions can never be reconstructed from documents such as we have in this case. Radical agnosticism about intentions, on behaviorist or any other grounds, is philosophically far too suspect to be a defensible historiographical stance. Nevertheless, we do agree wholeheartedly with Shapin and Barnes that there should be no reluctance in inquiring what continuities and connections there may have been between the origins of Darwinism and the doctrines of Social Darwinism. However, we would suggest that an adequate analysis of the historical sources of natural selection will show that such inquiries have been far too narrowly conceived by most writers, including Shapin and Barnes themselves. There should be vastly more to these inquiries than will ever be engaged by concentrating on the particular issue of whether Darwin was indebted to the political theory no less than the demographic doctrines in Malthus's *Essay*. For, at a minimum, one needs to consider the institutional and ideological context of Darwin's entire notebook zoonomical program, including its "metaphysical" offshoot (see some remarks in the paper by Hodge in this volume). And even on returning to Darwin's many entries on struggles and extinction among men, other animals, and plants, we need to look beyond the doctrine of *laissez-faire* in Britain to the whole theory and practice of the European colonial expansions. Directly and indirectly, through both reading and personal experience, Darwin derived from these worldwide endeavors preoccupations bearing decisively on his response to Malthus on ancient empires; although Smithian division of labor was to be more relevant to this theorizing about divergence in the 1850s (see Schweber's and Kohn's papers in this volume).

By advancing to a broader outlook, the social no less than the intellectual historiography of Darwin's science could move to a new agenda, one no longer dominated by the issues arising from the 1959 centennial.

Notes

1. Darwin's B–E *Notebooks* were published by De Beer, Rowlands & Skramovsky (1960–1967) and again, in computer print-out form, by Barrett (1972). The M-N *Notebooks,* in Barrett's (1974) edition, are available in Gruber and Barrett (1974). All these six notebooks, together with the *Red Notebook* (1836–1837), *Notebook* A (1837–1838), the *Torn*

Apart Notebook (1839–1842), and the *Questions and Experiments Notebook* (1839–1844) have been published by Cambridge University Library and the British Museum (Natural History) in color-film form (1983) and a complete printed text edited by P. H. Barrett, P. J. Gautrey, S. Herbert, D. Kohn and S. Smith is in preparation. The publishers of the printed text are the British Museum (Natural History) and Cornell University Press.

7

Darwin as a Lifelong Generation Theorist

M. J. S. Hodge

Introduction

Darwin as a lifelong generation theorist is a theme that can contribute very directly to any centennial reinterpretation of the Darwinian heritage.

Generation, here, concerns not only the production of an offspring by two parents; or any asexual productions from one parent, in budding say; or any from none at all, as in spontaneous generation. For Darwin extends the term to include the generation or propagation of new species from old; and even beyond that the propagation of the whole tree of life. The full range of Darwin's biological thought is thus confronted.

It is confronted, moreover, in a way that departs fundamentally from what may be called (with no allusion to anything monastic) the Franciscan view of the Darwinian landscape. Most people first meet with Darwin either in his son Francis Darwin's monumental *Life and Letters of Charles Darwin* (1887) or in some book largely based on that work. Along with the father's correspondence and the son's commentary and reminiscences, the three Franciscan volumes include Darwin's *Autobiography* (with omissions) and an essay by T. H. Huxley on the reception given the *Origin* (1859). They provide a pretty comprehensive view of the Darwinian landscape.

Strikingly, the Darwiniana occasioned by the 1959 centennial (notably De Beer 1963) consistently perpetuated the Franciscan view. We should ask, then, why this view can no longer satisfy today. The answers cannot be found solely in Darwin scholarship itself. For the progress made there has often opened up new interpretative prospects through interactions with larger forces at work, as one says, in our society.

Consider how the 1960s and 1970s have come between the 1950s and our collective centennial selves. A few trends can recall many others: recent challenges to neo-Darwinian syntheses; the increasing ideological self-consciousness among biologists; the waning of positivism in all its domains; the historical turn in philosophy of science; the several sociological turns (in both senses) in the new professionalized history of science. Many familiar

Geisten of our lively *Zeiten* provide reasons why the Franciscan view could go unchallenged in 1959 and yet fail altogether to satisfy today. An obvious analogy is with Newton who was also commemorated in monumental Victorian volumes, by the redoubtable David Brewster. Only in the last quarter of a century has Brewster's Newton been replaced.

Consider, then, just one reason why the main thesis of the present paper — that generation theory is a unifying theme throughout Darwin's entire career — must conflict with anything like the Franciscan view.

The Franciscan Darwin is very much the Darwin of the books. In the *Autobiography* Darwin recalls for his family how he came to write the books he did; in his essay, Huxley looks back to the 1830s for sources of the later battles over the *Origin*, and Francis Darwin organizes the correspondence mostly around topical divisions given by the chapters in Darwin's books. The bibliocentricity of the Franciscan view conforms well with the assumption that science is public knowledge, and that a scientist's biography should present the youthful pioneer as the private precursor of the mature, published authority.

A historian, however, cannot give interpretative priority to what came later over what came first; he must, rather, take the Darwin of the famous books and great debates as the successor to the younger man of the earlier private thoughts and manuscript notes.

Again, there is a parallel with Newton. Traditionally his physical science was equated with the *Principia* and the *Opticks*; and the "Queries" appended to the *Opticks* were seen as secondary excursions. So one sought to understand how Newton could come to write the two main texts, and then how subsequent developments could have led to other subsidiary efforts. Likewise, then, with Darwin's hypothesis of "pangenesis", so named because it has a new offspring formed initially from minute buds, "gemmules", supplied by the whole body of a parent. This hypothesis, first published in 1868 at the end of the *Variation of Animals and Plants under Domestication*, has looked like a subsidiary endeavor confined to a tentative addendum to a text that is itself to supplement the *Origin*.

But such impressions have proved as misleading in Darwin's case as in Newton's. Much in Newton's "Queries" pursues long-standing preoccupations, concerning space, time, matter, force and God, that pre-dated the composition of the canonical published texts and provided decisive contexts for the theorizing whose outcomes they expound. Equally, we now know that the generation theorist who we see surfacing in print in 1868, in the pangenesis chapter, was there all along thirty years before in the early notebooks.

So, in understanding the lifelong generation theorist we must reverse the traditional priorities. For this theorizing is not a secondary venture appended to the inquiry consummated in the *Origin*. Rather, both are equally integral to a larger enterprise that was never to surface in print. Moreover, to understand Darwin's larger enterprise, his general generation theorizing, we cannot study him as the man who will one day write the books provoking the controversies of the 1860s. We need rather to make a new start. We have to study him as the student at Edinburgh (1825–1827) consorting with Robert Grant and reading admiringly in his own grandfather, Erasmus Darwin's *Zoonomia* (2 vols., 1794–1796); as the zoologist and geologist who was, consequently, when on the *Beagle* voyage (1831–1836), speculating in distinctive ways about reproduction and colonial life, in invertebrates, and about reproduction and the extinction of species, among mammalian quadrupeds; as the zoonomical theorist who accordingly opened his *Notebook* B, in July 1837, as he would also open both his *1842 Sketch* (on natural selection) and his exposition of pangenesis in 1868: namely, by comparing and contrasting sexual and asexual modes of generation.

In taking generation as a persistent preoccupation throughout Darwin's entire career, we must, therefore, move away from anything like the Franciscan view. For, on that view, any thesis about unifying themes must be a corollary of three assumptions: that the principal components of Darwin's biological theorizing are those "ideas" that we know from his books, "evolution", "natural selection", and "pangenesis"; that before he reached these conclusions he had no clear and distinct thoughts on such subjects; and, finally, that when he did reach them they were related to one another in his thinking as they would be in the books of the 1860s.

Contrary to all such assumptions I shall be proposing that, in the early notebook period (1837–1839), there was a characteristic feature of Darwin's theorizing that, although never explicit in the later books, did leave its mark on everything thereafter; namely, a tendency to try to understand entities above the level of the individual organism — "species" and "trees of life" — as scaled-up analogues to individual organisms; and entities below the level of individual organisms — "buds", "cells", "gemmules", "living atoms" and "monads" — as scaled-down analogues to them.

The work of Olby (1963, 1966), Herbert (1968), Ghiselin (1969, 1975), Gruber and Barrett (1974), Kohn (1980) and Sloan (forthcoming, this volume) has shown how such a new start can and should be made. And it is a privilege to record my great indebtedness to their writings and to discussions with them. Darwin as a lifelong generation theorist (including his changing views on heredity, variation, hybridism, hermaphroditism and so on) is a subject that now cries out for book-length treatment and is a natural for a dissertation (I have no plans to undertake this monographic task and know of no one who has). Whoever takes up the assignment will have to engage all the wider reinterpretative issues involved in transcending once and for all the Franciscan view.

I. From Edinburgh to London (1827–1837)

Although this section and the next cover a decade, they can be presented in summary form. For the relevant documentation can now be found among four recent papers (Kohn, 1980, Hodge 1982, Sloan forthcoming, this volume).

These papers show that there were two decisive sources for Darwin's early biological thinking: his Grantian generational theorizing and his Lyellian geological (including biogeographical and ecological) theorizing. They also show that Darwin the generation theorist should not be offered as a replacement for the better-known Darwin the geological theorist; rather, we have always to ask how he was relating generational and geological considerations to one another.

The roots of both sources often trace to Edinburgh. Darwin as a Cambridge boy is a familiar figure (Francis was a don there). But now Darwin as an Edinburgh lad is an idea whose time has come (Brent 1981).

In the Spring of his last year in Scotland, Darwin studied closely what he took to be "ova" (in fact larvae) in the genus *Flustra* (Bryozoans as classified today). And this work with Grant introduced him to a rich cluster of theoretical issues. First, there were questions about "associated life". In these "zoophytes", a prolific growth could arise from a motile "ovum", by successive buddings, while continuing to act like a single organism. This associated life was, thus, explicitly connected by Grant and others with two issues: whether the similarity between a zoophyte colony and a tree, considered as a colony of buds, indicated that these "polyzoans" were really plants; and how it was that a single principle of life or vital force could pervade the organisation and coordinate the actions of such a colony. Second, there were questions as to the generality, among the *Zoophyta*, of generation by the kind of minute ciliated "ova" or "gemmules" that Grant and Darwin observed in *Flustra*. Particularly, Grant thought he had direct evidence that the large eggs of many species were composed of myriad "gemmules", themselves made from gelatinous parental material (Sloan forthcoming, this volume).

Darwin, and his older brother Erasmus who was at Edinburgh with him, must have been very much aware that they were following in their grandfather's footsteps at this time. The most famous chapter in all the grandparental works was the last one in *Zoonomia* on "Generation". There, conformably with Erasmus Darwin's fascination with Linnaeus' "sexual system" of botany and with Trembley's "polyps", close attention is paid to sexual as contrasted with asexual modes of generation and to colonial animals as compared with plants. Moreover, the production of higher animals and plants is credited to the powers of generation in the simplest.

The young Charles Darwin's move, from Edinburgh and medicine to Cambridge and the prospect of the church, took him from a context where

Grant and his grandfather were natural authorities to one where they could never be so.

Darwin himself may have sensed as much. The *Natural Theology* (1816), of William Paley, that exemplary Cambridge author, featured as its two principal enemies to sound thinking about animals and plants Erasmus Darwin and the Edinburgh sage praised in his *Zoonomia*, David Hume. "*Generation*" (the italics are Paley's) was moreover a decisive issue. For Paley set himself to refute the proposal, mooted playfully by Hume in his *Dialogues on Natural Religion* (1779) and cited enthusiastically by Erasmus Darwin, that when one organism produces another we see adapted organization produced by generation, so that perhaps we should credit order and design throughout the universe to an analogous generation. He also set himself to refute the *Zoonomia* proposal that life can take on new adapted organization in individual embryonic responses to surrounding nutrients and stimuli, and so also, perhaps, in a far vaster progress from simple infusorian origins over millions of years. Accordingly, Paley counters Hume by arguing that generation is no "principle", no productive power, but only a "process", and so not responsible for any product nor therefore for any feature, such as design, in that product; while to the older Darwin, he objects that generation, as we observe it, is a process whereby forms are perpetuated not produced, and that intelligent design remains indispensable in explaining how any such perpetuation could itself have been instituted.

Charles Darwin was apparently quite content with Paley's arguments on reading them as an undergraduate. But his very acquiesence can alert us to a striking ambivalence in his upbringing and education. Diverse issues in British national life were then conducted and perceived as so many challenges to various Tory, Anglican and Oxbridge hegemonies (Halévy 1924–1926). The challenges often came from alliances between English provincial dissenters and metropolitan Scottish reformers. As both an Edinburgh medic manqué and a Cambridge cleric manqué Charles Darwin was eventually heir to both sides of such divisions; but his family cultural inheritance, compounded as it was of Unitarianism, the Lunar Society and Scottish medicine, was a far from neutral legacy. The context of his initiation as a theorist of generation included these divisions and this legacy.

Henslow's teaching at Cambridge may have fostered his Grantian interests. By halfway through his *Beagle* years, in mid-1834, he was theorizing, and reporting in letters to Henslow (Darwin 1967), on his invertebrate findings, all in a thoroughly Grantian manner. Concerning associated life in polyps, such as *Flustra*, his main concern was whether the matter making up the "central living mass" of their growth is continuous and so connecting the several component "cells". As for "gemmules" as a mode of generation, Darwin was now satisfied that these often arise from "granular" matter in the parent, and that this mode of generation is more widely found among diverse invertebrate groups than had been hitherto appreciated (Sloan forthcoming, this volume).

It was theorizing about species extinctions that brought Darwin to first integrate the Grantian and the Lyellian legacies in a single speculative venture.

By mid-1834, he had accepted Lyell's teachings in the *Principles of Geology*. Lyell had new land made and old land destroyed all the time, and new species arising and old ones becoming extinct in a gradual "birth" and "death" of species. Rejecting the transmutation of species, he left unexplained how new species were "created". But he found causes for species extinctions in disturbances to the fine balance of interspecific competition, disturbances initiated by climate, land and sea changes.

Darwin, in February 1835, found this theory inadequate for some South American quadruped extinctions, where he could see no evidence of any appropriate physical changes or ecological disruptions. He, therefore, adopted the theory of Brocchi as presented but respectfully rejected by Lyell. On this theory, the deaths of species depended, like those of individuals, on constitutional peculiarities conferred at birth; and so, in explicating Brocchi, Lyell drew the analogy with plants propagated by cuttings that reportedly died out eventually even when the conditions of life remain unchanged. Elsewhere, Lyell argued that such artificial propagations were like natural asexual propagations by roots and layers; in that they extended the life of a single individual rather than reproducing the species by multiplying individuals; the single life although extended was still limited.

Darwin, in February 1835, considering "the enormous extension of life of an *individual* plant seen in grafting of an Apple tree and that all these thousand trees are subject to the duration of life which one bud contained", did not see "much [or "such"?] difficulty in believing a similar duration might be propagated with true [i.e. sexual] generation" (DAR 42 as quoted in Hodge 1982, where the full text is given).

He is still pursuing this analogy two years later, in Spring 1837, in his *Red Notebook* and in the manuscript of his *Journal of Researches*. He argues now that all modes of generation, sexual or asexual, natural or artificial, are essentially "methods" of "division". And he compares the individuals of a species reproducing sexually to a single individual divided by any other method. In such a sexually reproduced species the division is successive not simultaneous, and the products of the division are thus separated from one another. But Darwin finds his beloved *Flustrae* producing asexual offspring without separation and sexual ova with separation; so they help to establish the unity of all generation as division by showing the continuity between separating and successive division and any other division.

His reflections on associated life, as exemplified by *Flustra*, are apparently now informed by his reading (*RN*) in Erasmus Darwin's *Phytologia* (1800). Its opening section is on "Individuality of the Buds of Vegetables", and the younger Darwin (*Journal of Researches* 1839, p. 262) argues that the organization of a tree, as a union of many buds each with its own "individuality", makes the union of individuals in a "compound animal", such as *Flustra*, more comprehensible.

So, his Brocchian extinction conjectures have now joined up with his invertebrate researches and his renewed study of Erasmus Darwin to take his theorizing about generation into a first phase characterized by a quest for positive analogies among all modes of generation, a quest concluding in favor of a strong thesis that all generation is, in essence, division. This thesis will never be given up. Later we shall distinguish two other phases. In the next one, the second, contrasts between sexual and asexual modes of generation will feature largely. But, notwithstanding these contrasts, the division thesis will only be qualified and reinterpreted. Then, eventually, in a third and final phase, probably beginning in about 1840–1841, Darwin will be moving to new identity theses that once again confirm his divisionism.

II. Generation in a New Zoonomical Program (1837–1838)

By July 1837, Darwin had completed the *Journal of Researches* and was opening his *Notebook* B with the heading "*Zoonomia*". He was going to construct his own system of the laws of life; what is more, he already had a provisional outline for it. For the notebook opens with twenty-seven pages of entries that constitute, as Howard Gruber (1974) first brought out, a sustained and integrated structure of systematic argument evidently thought through before being put down here in writing.

Obviously, Darwin has now moved far away from his 1834 agreement with Lyell's account of the organic world. He had probably done so in two steps. First, in 1836, he had come to favor transmutation as the explanation for various biogeographical findings that seemed inconsistent with Lyell's separate creations of fixed species. Second, in March 1837, being much impressed with new evidence for transmutation, he had apparently decided to go the whole hog and develop a transmutationist system as fully comprehensive as Lamarck's as presented and rejected by Lyell. He had apparently decided, too, to devote himself to this task as soon as the *Journal* was finished. (Although Hodge 1982 dates these two steps roughly correctly, its analysis of Darwin's 1836 views on the Galapagos fauna is quite mistaken, as can be seen from Sulloway 1982c; however, the proposals in Sulloway's paper, as to the first favoring of transmutation are not, I think, established.)

Regardless of when Darwin made his departures from his 1834 position, there can be no question that his generation theorizing is decisive for his July 1837 zoonomical sketch.

The sketch comprises two movements of argumentation. The first starts with sexual generation, as the means of adaptive variation from one generation to the next. It proceeds to "change" in the longer and longer run, all the way, eventually, to divergences as wide as those between Australian and other mammals (*B* 15). Because this movement starts from sexual generation, as distinctive in involving "ovules" (see *Journal of Researches*, p. 262 written about this time), and because it concerns "change" but not "progress", we may call it the ovule and change movement.

A transition (*B* 15–17) introduces the second movement. Here the issue is "progress". And the starting point for progress, as in Lyell's version of Lamarck, is the simplest organisms of all, infusorian "monads". Since this movement concerns the progress that takes life from such simple monads up to the highest animals, including man, we may call it the monad and progress movement.

The two movements are developing quite different arguments that are only confused by talking, as Gruber did, of a "monad theory of evolution", to be replaced one day by natural selection. As Kohn (1980) insisted (and as he and I explain elsewhere in this volume), natural selection, when it came, was to be a revision of the opening steps (*B* 1–10) of the ovule and change movement, not a replacement for the monad and progress arguments.

The argument of the ovule and change movement proceeds through four steps: individual variation in sexual generations; adaptive variety formation; species formation; and, lastly, intergeneric, interfamilial and wider divergences.

In the first two steps appeal is made to the two features distinguishing sexual from all asexual generation: maturation in the offspring (*B* 1–5) and interaction in mating between the parents (*B* 5–7). These two, maturation and mating, are going to mean a great deal to Darwin for the rest of his life, and he has already given much thought to their combined contribution to adaptation and species formation.

In its maturation, the organism repeats the organizational changes made in the progress from a monadic beginning; ontogeny recapitulates phylogeny. Only organization that is undergoing this maturational process is susceptible of acquiring fitting and hereditarily transmissible variations, and so only sexual generation can provide for adaptation to changed conditions. Sexual generation shortens the life of the individual as compared with the extension of it that an asexual generation would allow. The adaptation to change is thus the final cause of the shortness of life entailed by sexual generation. Beginning a new life rather than extending the old one also allows for accidents, maladaptive variations, to be left behind.

If sexual generation provides for the acquisition and transmission of variation, the constancy of a species all across its range requires explaining. Here mating is the answer. The providential law, of sexual offspring partaking of the characters of both parents in crossing with blending, ensures that individual adaptations to local and temporary fluctuations in conditions do not lead to inconstancy of character in the species when conditions are constant overall.

This conservative action can be circumvented, however, if a few individuals, perhaps only a pair, are separated, with consequent inbreeding, in new conditions. A new variety can then be formed; and then, in due course, with greater divergence and intersterility, a species.

This familiar line of argument, at the opening of *Notebook* B, alerts us to a telling parallel between Darwin's new species origin theorizing and his Brocchian species extinction conjecture. His Grantian researches into generation did not lead to his rejection of Lyell's explanation for extinction; fossil finds did. But once the fossil finds had done so, those generation researches could be called on in developing an alternative to Lyell's extinction explanation. Likewise, then (and here I must disagree, I suspect, with Sloan the Grantian researches did not contribute to the adoption of transmutation. For that, biogeography alone was initially decisive. But once Darwin set himself to understand how transmutation was possible, the generational inquiries were the vital resource. In the Brocchian extinction theorizing Darwin had argued for comparisons between sexual and asexual generation. In the transmutation theorizing contrasts were what was needed.

The conclusion from these contrasts is that sexual generation is distinguished by maturation, which is potentially innovative, and mating, which is normally conservative. And these two corollaries of sexual generation jointly determine what happens in the long and the short run, in conditions changing permanently and overall or only locally and temporarily. Small wonder, then, that Darwin will be thinking about the relation between maturation and mating, growth and fertilization, development and crossing, for decades to come.

Moving now to the monad and progress movement, we can see that it only makes sense when read alongside its precedent: the equivalent part of Lyell's presentation of Lamarck. For, there, Lyell insisted that Lamarck needed spontaneous generations of infusorian monads from lifeless matter; and needed them not only once, eons upon eons ago, to start life on the upward progress that has produced the highest organisms now living, but also in every period since, to explain the persistence of simple grades of organization into the present. With Lamarck's presumption of a tendency to progressive advance, these grades must be constantly refilled from below.

Darwin is evidently following Lyell's version of Lamarck (at *B* 18–*B* 22) when he has infusorian monads "constantly formed", and it seems that he has spontaneous generation in mind. Moreover, he would invest them with a scaled-up Brocchian senescence that was never contemplated by Lamarck. For, he can then explain a favorite generalization of Lyell's: namely, that mammal species (not individuals) last less long than mollusk species do. Thus Darwin reasons that if the entire progressive issue from a monad beginning has a vast but limited lifetime, then those lines that have reached the highest grades of organization must have changed in species most rapidly (*B* 22). Hence, the correlation between "shortness of life" in species, and organizational complexity.

The whole argument, although overturned by Darwin himself within a day or a week or so, shows his concern to bring generational theorizing to the explanation of Lyellian laws about the birth and death of species. It also amplifies his teleology of the shortening of life; here, the lives of the species are shortened so that a higher object, progressive advance to high grades of organization, can be realized.

"Shortness of life" is a phrase prominent in Erasmus Darwin's *Temple of Nature*, and it appears faintly inscribed by Charles Darwin on the inside back cover of his copy of volume one of *Zoönomia* (now at Cambridge). Darwin's July 1837 theorizing about generation is often indebted to his rereading in his grandfather's books, where the distinctive powers of sexual generation were held to make possible the origin of new species from old, whether by hybridization of two others or the progressive improvement in changing conditions of one. Charles Darwin is extrapolating from the shortening of life in individuals, as distinctive of sexual generation, to the shortened lives of mammal species as quasi-individuals.

The tree of life metaphor is providing not only analogies for the ramifying form of descent, but also for the births, lives and deaths of species as analogues of the buds by whose successive propagation a tree grows. Darwin can thus invoke, too, albeit briefly, a coral of life metaphor (*B* 23–25); for character gaps between major groups, such as birds and fishes, may be due to the deaths of stems or main branches, as on a coral whose tips are still living.

These analogies are drastically revised, when the limited lifetime for a whole monad issue is rejected because it falsely entails (*B* 29e) that all congeneric species should become extinct simultaneously. Moreover, the constant formation of monads is soon redundant, when Darwin realizes that with ramifying lines of descent and no invariable tendency to progress, the persistence of simple organizational types need not be explained by refilling from below; for some lines may merely have made no progress or even regressed while others have advanced. Again, stem deaths, and the special implications of the coral metaphor, are no longer needed by the end of that Summer (1837), once gaps are credited to many twig deaths (extinctions) accompanying twig splittings (species formations); the longer divergences leading to the larger gaps will entail more extinctions.

What is left after these revisions is, then, a tree of life, but one with no life at any one time beyond the lives of the species extant at that time. Within this tree of life, the births of species remain analogous to the successive bud propagations whereby a tree grows.

The rejection of a limited lifetime for any monad, and so for a whole tree of life growing from it, did not in itself bring rejection of a limited

Brocchian lifetime for each species. But species extinctions did now have to be reinterpreted to integrate them with the propagations of new species from earlier ones. Even in its 1835 formulation, as reaffirmed in the *Journal of Researches* in the Spring of 1837, Darwin's Brocchian theory had obviously allowed for the possibility that many species, unlike the South American quadrupeds, may have succumbed to a change of conditions, sufficing to initiate a Lyellian ecological defeat, before their limited duration of life came to an end. So, there were two ways for species to die without issue. To these two possible fates Darwin has now added a third: a species may end its existence before its lifetime runs out; but it may do so with rather than without issue; because a change in conditions may induce changes in habits and structure, resulting eventually in the propagation of a new species. Hence, then, the explicit analogy now used in interpreting any change of conditions that brings about a new species propagation and so rescues the older species from death without issue: such a change in conditions is, to a species reproducing sexually, as a sexual crossing is to an apple tree succession previously propagated by cuttings (*B* 61–64; *B* 73–74).

For it is a law of life that any individual has a limited duration of life as that individual; so if it is to go on beyond that lifetime it must do so as another individual. Hence, an individual apple tree being extended by grafting can avoid death without issue, by mating and producing another individual. Likewise, then, a species that adapts to changing conditions is avoiding death without issue by shortening its own life and starting the life of another species. In sexual generation a life is shortened, as compared with the possible, albeit limited, asexual extensions of it, in order that adaptive innovations can be perpetuated in the offspring. Likewise, then, Darwin argues, with a species: its changing into another species allows new characteristics to be perpetuated in the new species that could not otherwise be so. And just as some individuals have been more successful in leaving descendants than others, so with species.

With Darwin's tree of life and its branching growth by species propagations settled, by late Summer 1837, into pretty much its final form, the structure of his argument for integrating the long run and the short has also settled into its final form. On the physical side, of land and sea and climate changes, the integration is the one he takes over from Lyell: every spot on the globe sees such changes but the system of igneous and aqueous causation is stable in the indefinitely long run of the past and future. On the organic side, all change traces ultimately to the maturations and matings that distinguish sexual generation and make possible adaptation to a changing world.

It is worth pausing here to consider how much Darwin is taking as given. On the physical world — the stars, the sun and the planets, including the earth — he found no grounds to quarrel over fundamentals with Herschel and Lyell. His theorizing about the organic world was accordingly accommodated to their teachings on such questions as changes in the earth's orbit and changes in its climate. On the organic side, he took as given the "vital powers" — of generation, growth, nutrition and so on — usually thought to distinguish even the simplest living bodies from any lifeless matter. He could and did leave open whether these "powers" or "forces" or "principles" were or were not reducible to the powers of all matter: inertia, gravity and so on. Darwin was a vitalist in that he took such "vitalist" terms to designate real entities (Sloan, forthcoming). But, *pace* Sloan, he was no vitalist if vitalism includes positively rejecting reductionism.

It is worth emphasizing, too, what Rudwick (1982b) has rightly insisted upon, that Darwin's notebook years saw him moving in several scientific circles in London and actively integrating public knowledge and private conjecture. It is also salutary to reflect that on a topic such as generation there are likely to be connections between what is going on in a man's theorizing and what is going on in his psyche and in his society. A man contemplating marrying a Wedgwood knowing that his father did too has his mind wonderfully concentrated on the effects of inbreeding. Darwin in Autumn 1838 will be such a man. Anyone living in that phase of Western civilisation is likely to give the male element the more active role in fertilization, and Darwin is no exception. Again, speculations about colonizing species — and organisms as colonies of cells or tissues — may be reflecting contemporary perceptions of political realities. I have left such psychological and sociological possibilities unexplored. But a fuller treatment of Darwin's generation theorizing could free us from the Franciscan view even more thoroughly by pursuing them wherever they lead.

What will be emphasized, here, is what Herbert (1968) was really the first to insist upon, and what can easily be confirmed from *Notebooks* B, C and D themselves, and from Kohn (1980) and Ospovat (1981): namely, that in the months from July 1837 to September 1838, Darwin elaborated themes, concerning generation in general and sexual generation in particular, that were to influence his thinking for the next four decades.

By distinguishing such themes in a schematic, summary way we can thus prepare ourselves not only to understand Darwin's generation theorizing in a month that was to be decisive for that theorizing, September 1838; we can prepare ourselves also to understand issues that were to be fundamental for his entire career as a biological theorist. For these are themes that will continue to be influential, whether affirmed or revised and replaced subsequently.

First, all generation, sexual or otherwise, arises as excess growth. Like many earlier writers, Darwin has the product of any generation formed from superabundant material being used in growth. If an organism has more than enough material to make more of itself by growing itself, then it makes yet more by growing another. It is because all generation arises as the organism makes more of itself that the norm intrinsic to all generation is perfect reproduction, with no change in inner organizational constitution nor, then, any gains or losses in the characteristics manifested in constant conditions. This theme is obviously fundamental. Many an analyst of scientific thought (one thinks especially of Meyerson) has insisted that any explanatory enterprise is grounded in assumptions about expected states, in principles of conservation, in presuppositions as to what does not need explaining by extrinsic influences because it is presumed to result from the nature of the processes themselves; recall inertial motion in classical physics, the Hardy-Weinberg principle in Mendelian population genetics or the first law in thermodynamics. Well, Darwin, in 1837, has already concluded — what he will never reject — that generation is in, of and by itself perfectly conservative.

Second, sexual and asexual modes of generation are to be distinguished as modes of reproductive growth. The product of a sexual generation matures as it grows, going through a succession of stages that consists of a condensed repetition of the vast series of organizational advances made in the progress from its simplest and remotest ancestors, and it only generates sexually on completing its own advance to adult organization. So, when it does, it produces an offspring that is not initially coeval with it, being, rather, of the simplest organization that the remotest ancestors had. Now, when generating asexually an organism is producing a coeval replica that is, like its own growth, fitting in and for present circumstances; while, by contrast, when a mature adult produces sexually a simple, immature offspring it is not doing so only as a fitting response to these circumstances, but also as an adaptation to the very inconstancy of conditions in a world that is always changing in the long run, as geology has disclosed. For the offspring can accommodate its maturing organization to changed conditions as an adult can not. It is by reason of these presuppositions, then, that a theory of adaptive changes in organization in the long run has to trace all such changes ultimately to the influence of changed conditions acting upon individual maturations. For the effects of such influences on a parent can be transmitted when its eggs or sperms are formed, but they do not arise then; such effects on a male parent can be impressed on an egg when it is fertilized, but they are not arising then.

Third, the conservation in sexual generation, of the constitutional effects of any new maturational responses made by the mother, is ensured because the ovum is formed by excess growth. A young female dog moved to a colder climate, and growing a thickened coat of fur, does not grow a coat of fur, thick or thin, on the unfertilized ovum forming within it. But the organizational change, that is made in response to the cold and results in the thickened fur on the parent, is transmitted to the ovum as a product of the same changed growth. By contrast the effects of mutilations are not transmitted; for they are sudden alterations inflicted on the mutilated tissue, not gradual changes arising in responses made by organization as it is maturing. Likewise, consider the effects of any influences upon an egg at fertilization, effects wrought by the male element because its constitution is not exactly like the egg's own. These effects are transmitted to offspring, the grandchildren, eventually produced by the individual arising from that fertilization; for, as effects of influences exerted at the very start of a new life they are even more transmissible than those arising during the course of life, in maturation. In general, then, any influence on a parent that affects an offspring will be passed on to its offspring, the grandchildren. For the changed constitution, conserved in the growth of an ovum, determines a changed maturation, including in due course growth of ova of that constitution, and so on; any new character being transmitted for a single generation tending to pass indefinitely into future ones. Moreover, since what is transmitted is the constitutional determination for the changed maturation, then the new characters will appear in the descendants at the same stage and age in maturation as they were first acquired, whether by a parent, a grandparent or remoter ancestor.

Fourth, sexual generation brings together at fertilization a male and a female element that are of more or less unlike constitution because produced by parents of a more or less unlike constitution. But too much unlikeness is as unbeneficial as too little; with too much, no compromise between the constitutional difference distinguishing the parents is possible; and so no intermediate offspring can be formed, and one object of sexual generation — ensuring constancy of character in the species by crossing — is frustrated; while, with too little, the very object of sexual as contrasted with asexual generation is frustrated, for the fertilized egg is left like an asexual bud, an exact constitutional facsimile of its mother. Hence, if the offspring does not come out intermediate between the character of the two parents, but more like one than the other, that is a sign that the normal function of crossing, preserving constancy, is close to frustration. So, the inclination and ability of animals in domesticated species to breed successfully, across a wider character gap than wild animals, shows that the conditions of domesticated life have disrupted the reproductive system and its associated instincts.

Fifth, if new influences on maturation are discounted, an offspring from a sexual crossing can be interpreted as the outcome of the departure from the straight line of the exact replication of the mother that would have occurred in the absence of the influence of the male element; it is thus a record of the influence of this male's contribution. Moreover, since the influence of a mating is analogous to the influence of a change in conditions, then crossings offer a short-run analogical means of understanding the influence of changing conditions in the long run. Hence, then, the significance of one of Darwin's favorite notebook laws, "Yarrell's law", after William of that name; the law stating that in any cross a parent of an older breed will have more influence on the character of the offspring than a parent of a younger breed. This law may be taken to imply that older characters are likewise more permanent, less susceptible to the influence of changing conditions. So, there is the further consequence that what is quickly induced by quickly changing conditions will be quickly lost by further changes or in crossing; while, conversely, what is slowly induced by slow changes in conditions is less easily lost. In general, what the organizations of the two mates do to each other, as shown by the character of their offspring, indicates what changing conditions can do to organizational constitutions over many generations. Accordingly, an explanation may be found here for the sterility of interspecific hybrids when they are crossed with one another. Each partner to such a cross has been given in one generation the big difference between itself and either of its two parents. But such a very sudden innovation cannot be impressed on a subsequent generation; so an ovum or male element cannot be formed that would transmit it. Most generally, therefore, such hybrids cannot breed for the same reason that the effects of mutilations are not hereditary. Changes over successive generations and changes in the successive transformations involved in a single individual's maturation conform to the same law: quickly come is quickly gone, so, slowly does it; if a major shift is to be achieved it must be done gradually over many small steps, each taken over plenty of time; for only then can the influences working the earliest steps have time to produce effects sufficiently permanent that they will be added to, not replaced, by later ones.

Such, very briefly, are the leading thoughts about generation that Darwin was working with by the Summer of 1838 and the earlier pages of *Notebook* D filled in July and August. We are thus prepared to take a closer look at his generation theorizing in the month when it came to a remarkably instructive crescendo: September 1838.

III. Generation in September 1838: The Sexes, Buds and Fertilization

Any month in Darwin's life may be hailed, trivially, as initiating the rest of his career from then on. But, no less obviously, there can sometimes be a case for seeing a particular moment as peculiarly decisive in the precedents it is setting for what follows years and decades later. We have already seen more than one moment, notably early 1835 and early 1837, as having such significance for Darwin as a generation theorist. The value of viewing September 1838 in this way derives from two considerations. First, Darwin became at this time even more explicit and self-conscious than before in pursuit of the "theory of generation". Second, by concentrating on September 1838 as a month dominated by generation theorizing, we can liberate ourselves from the powerful but profoundly misleading myth of a uniquely decisive moment of Malthusian revelation. It is not merely that the famous reflections on Malthus came late in the month, on the 28th, and that to pick on them, as what the preceding weeks were building to, is to go into the documents from the wrong end. Beyond that, we would forgo any chance of reconstructing most of what was going on then, as Darwin originally lived it, including those reflections on Malthus when they came. For (as David Kohn and I have emphasized elsewhere in this volume) Darwin met with Malthus as he encountered almost everything else at that time: as a resource in understanding the aetiology, including the teleology, of the ontogeny and phylogeny of sexual generation.

In making a new start on *Notebook* D, then, we do well to observe that the very ordering of its filling was conditioned by the central place that generation was now taking in Darwin's theorizing. In the middle of July, he began on the first page, reaching *D* 70 on the 8th of September. At about this time, however, he went also to *D* 176 near the back of the book and entered there as a special heading: "Proved facts relating to Generation." By September 11th, he had filled *D* 176–179, gone to *D* 174–175 and then to *D* 152e (where the heading "Generation" is entered, as it is likewise at *D* 168), mixing further "facts" about generation with abundant speculation. By then, he had also reached *D* 95e at the front of the book. So, from September 11th it was being filled both from *D* 95e and from *D* 152e onwards. Several parallels, including common references to an Ehrenberg article, suggest that *D* 170–173e were filled at the same time as *D* 127–131, between September 23rd and 28th. If so, that would have left *D* 136e–151e for the remaining days from the 28th to October 2nd, when *Notebook* E was opened.

In the sequence *D* 176–179 and *D* 174–175, Darwin elaborates a complex train of reasoning starting from generational premises now made quite explicit: namely, that the very object of sexual generation is to initiate and perpetuate "differences", departures, that is, from parental characters, that this object is secured when the organism acquires new characters from the influence of external circumstances on the maturation that takes it through the whole series of changes in form previously undergone by its entire ancestry; and that the ovum, from which the organism grows, begins its life, before its fertilization and maturation, as a bud produced by the mother when still herself a foetus in the grandmother (see also *D* 112). For, Darwin now goes on to clarify how an organism, itself produced by sexual generation, can eventually mature and mate and reproduce by that means. In this way, he can clarify how sexual generation serves its object over successive generations. Taking the mammalian case as exemplary, he can clarify all other cases by comparing and contrasting fertilization and maturation in mammals with mollusks or apples.

If the ovum is not to remain a mere bud, an exact constitutional facsimile of that foetus, it must be fertilized by a male that is of unlike constitution, and so capable of impressing a difference on it and making it able to undergo the maturation that will allow it to acquire further differences, additional departures from the parental characters. At least, this influence at fertilization is usually required in every generation, but in aphids fertilizations sometimes suffice for several generations of offspring able to produce still others.

Now, if a pair of organisms come together to mate who have not acquired, in each case, any differences distinguishing them from their parents and remoter ancestors, then the object of sexual generation is frustrated and "desire fails", thanks to the "correlation of structure" throughout the whole body from brain to gonads (*D* 179). Breeding in and in, the repeated pairing of close relatives, does not entail frustration of this object, with that consequence for the "passions", merely because the mates are of similar constitution to one another; but it does when that similarity is due to their not differing, through lack of distinctive influences on maturations, from their recent common ancestors. (It is in reading these entries that one senses Darwin's concern, not to say anxiety, about the consequences physical and otherwise of his own grandfather, father, mother, cousin and prospective wife and children being related as they were to be; corollaries of his theory being, for example, that his grandfather has determined his own character to an extent even greater than usual because his wife was his Wedgwood cousin; and that he and his own cousin, thanks to differences in maturational circumstances, may well be constitutionally as different as more distant relatives).

There is, then, a necessity that, in every generation before an organism can reproduce sexually it must not only mature so as to repeat earlier phylogenetic changes of form, it must undergo some new, additional changes, induced by the present influences of external circumstances. But why? Darwin finds he has no answer. He can reiterate that maturational recapitulation of phylogeny "separates those differences which are in harmony with all its previous changes, which mutilations are not." But as to why "some further change" is required: "At present I can only say the whole object being to acquire differences, indifferently of what kind, either progressive improvement or deter[ioration] . . . that object failing, generation fails." In conclusion, then: "How completely *circumstances* alone make changes or species!!" (*D* 174e). For, "if the circumstances which must be external which induce change are always of one nature species is formed, if not — the changes oscillate backwards and forwards and are individual differences." And, as two parenthetic comments: "(hence every individual is different)", and "(All this agrees well with my view of those forms slightly favoured getting the upper hand and forming species)"; the ink and placing on the page (*D* 175) of this last entry suggesting that it is not a later addition, but rather a recollection of a view, explicit since *Notebook* B: namely, that some generic forms, being slightly better adapted to the changes in conditions occurring, flourish at the expense of others and produce new, descendant species.

Darwin has now not only a foundation in generation theory for securing old insights concerning species formation; he has also extended his insights into sexual generation itself.

Maturation in the higher animals repeats the organizational progress that has led to the separation of the sexes, the confinement to separate individuals of male and female organs. From *Notebook* C on, he had favored the view, associated with John Hunter (whose teachings were sometimes communicated to him in conversation with Richard Owen and with William Broderip), that a higher animal foetus has an early hermaphroditic stage. By late September, he has (*D* 162) an explicit "theory of sexes", that is a complete conjectured series of stages, at once ontogenetic and phylogenetic, wherein an original distinction between two "substances" with no distinct organs, gives way to the hermaphroditic condition of foetal mammals and adult mollusks, before this in turn gives way to the adult higher animal condition.

It is in these weeks, too, that he makes most explicit his speculations as to what is going on in copulation and fertilization, especially in higher animals, and again, by comparative extension, in other organisms including plants. His thinking often starts (*D* 176) from evidence that fertilization is not a one-to-one interaction. To be sure, a single copulation brings together one male and one female; but, Darwin reflects, while it may issue in a single offspring, it often suffices for many births, as in a litter of pups. Moreover, in cases of telegony — notably Lord Morton's famous mare — two matings have contributed to the character of a single offspring. Morton's mare, mated to a quagga (a zebra-like species now extinct) had produced a quagga-like foal; and then when later mated to a horse male subsequently produced foals that also had some quagga features. The case, having fascinated Darwin since *Notebook* B, had been put with one or two others and generalized to become "Morton's law"; and it was to remain, to his dying day, second to no other, in his mind, as an item of the greatest biological significance. In September 1838, it was significant for two main reasons. First (*D* 176; see also *B* 181) it confirmed what the aphid case indicated: namely, that fertilization is not needed in each generation; once in several generations will do for the production from eggs of fertile offspring. And, second (*D* 168), it promised to provide some analogy with the cumulative embedded change wrought in successive generations of crossing with males possessing some particular character. For the influence on the later foals' character of the first mating with the quagga was, Darwin reasoned (*D* 168), comparable to the influence of a grandfather in determining a grandchild's character; in that there has been a contribution to character determination even though another mating has come in between.

Darwin's thinking about insemination is, as a consequence, mostly concerned with understanding how it is possible for a single mating sometimes to achieve multiple fertilizations at the time, and also — as in telegony and in aphid parthenogenesis — later influences upon successive births. One speculation ascribes the amount of paternal influence on the offspring's character to the amount of seminal material (*D* 170–173e). Referring to the possibility (one his grandfather had dwelled on) that the "imagination of the mother" may help to determine the character of the offspring, he asks whether extra time in the womb for one member of a litter would make it more like the mother than the others and so less like the father; while "more semen" going to one child would have the opposite effect; the tacit assumption being apparently that as one source of its nutrition any extra semen can influence an ovum just as uterine sources can; although the power to fertilize an ovum seems to be associated in Darwin's mind with the presence in semen of neural material, entailing vivifying, that is vital and animating, not merely vegetative and nourishing powers (*D* 132, 162, 173e).

To make a bud that, as a fertilized ovum, is to mature, requires, then, two organs, as the making of an asexual bud does not. But could we not suppose the products of the male and female organs to mature before meeting? Hardly, for such a view would be irreconcilable with Mortonian telegony unless, Darwin notes (*D* 173e), the "nervous matter" supplied by the male is supposed to consist "of infinite numbers of globules: generally sufficient for one birth or other."

These speculations, although often inconclusive, only confirm that a prime challenge is to understand the fundamental contrast between, on the one hand, a maturing sexual bud requiring the interaction of products from two organs (whether in separate individuals as in higher animals or in one individual as in many plants) and subject to the effects on its maturation of external influences; and, on the other, a common bud made when an adult individual makes a facsimile of itself with no such interaction, maturation or variability.

Darwin's entries at *D* 128–131 pursue this challenge from the asexual side, by working out further implications of his longstanding view that, from budding to regeneration of severed bodily fragments, asexual generation is essentially a division of a whole into two or more parts, each having in it the power to make all the other parts it needs so as to grow into a new whole.

Commercial varieties of rose can be propagated truly by cuttings as they cannot by seed, even though there is eventually some slow and slight deterioration. And asexual buds likewise show this constancy (*D* 128–129). Now, where Ehrenberg sees artificial division as gemmation or budding, Darwin says that for him it is the other way round: gemmation is essentially division (*D* 130). But, in division, where a fragment regenerates a whole, one must suppose that the fragment, like an asexual bud, contained elements sufficient to determine growth of all other parts and so the whole. The main conclusion is therefore that, by contrast, to make a sexual germ, a fertilized ovum, that will mature to grow into a new whole, requires bringing together two lots of different material present only in the sexual parts; whereas any severed fragment, given appropriate support, nutritional and otherwise, can, like an asexual bud, make a new whole without maturing thanks to elements that are in it as in all parts. When the "whole" is grown from a "part" then "in the separated part every element of the living body is present"; whereas in sexual generation "something is added from one part of the body, (or [if not a selfing hermaphrodite] of other similar, body) to another part of body" (*D* 129). So, on this view "each particle" of an animal "must have structure of whole comprehended in itself, — it must have the knowledge to grow and therefore to repair wounds — but", Darwin emphasizes, "this has nothing to do with generation", meaning sexual generation. He is, then, left with "the two kinds of generation" contrasted as before; the "vast difference" between them being shown, indeed, by their "occurring in the same plant" (*D* 131). For if not vastly different, why would a plant be endowed with both? To be sure, an ovum before fertilization is, for Darwin, a bud, a case of gemmation and so of division. But it is a constitutional facsimile of the immature individual that has produced it. To become a mature whole, it must undergo maturation, which it can only begin when something is added to it that it does not now have; until then it does not have the totipotency, the power to produce a mature whole that any severed part of a mature flatworm has.

Even this quick tour through Darwin's generation theorizing up to September 1838 may show how many aspects of his subsequent thinking it can clarify.

For Darwin, as generational and geographical theorist, a species is both a quasi-individual, a subject of birth, death and propagations, and a population kept more or less constant by crossing with blending of individual differences. So Hull's (1983a) and Mayr's (1982b) concerns with species as individuals and as populations are very relevant here. Consider, too, Gould's (1982) thesis that reformist politics inspired the gradualism in much early Victorian science. In integrating his Yarrellian generational gradualism and his Lyellian geological gradualism, Darwin himself reflected that the rule of constant law "baffles the idea of revolution" in nature as in "government" and "institutions" (*E* 6e).

Many other, more specific topics are clarified too: Darwin's later distinction between the direct and indirect effects of external conditions on variation; his questions to the animal breeders (Vorzimmer 1969); his concern with self-fertilization and hermaphroditism; his relating, in *Notebooks* M and N and the *Descent* (1871), of the moral sense to sociability and so to life with separate sexes; the theorizing about adaptation and species formation that leads him to the theory of natural selection (see Hodge and Kohn's paper in this volume).

Moreover, Darwin will always be inquiring into the powers of generation themselves, in sexual and asexual buds, as mediating between the propagation of species in the tree of life and the vital powers of nutrition and growth in the tiniest gemmules and granules, the very atoms of living matter. In following this line, we are led to pangenesis. For that hypothesis originated in those inquiries; and did so, perhaps, within two or three years after September 1838.

IV. Generation Over Thirty Years: Pangenesis, as Possibly Conceived in 1840–1841 and Published in 1868

Two preconceptions must be countered here. First, Darwin is often thought to have constructed pangenesis to make heredity "soft". But, in fact, we shall see that the theory really involved an extension to asexual generation of the soft heredity he previously held to be distinctive of sexual generation.

Second, when he published pangenesis it was often taken to be his contribution to the new cytological interpretations of reproduction of the 1860s. But, in fact — and notwithstanding that that view is upheld in Geison's (1969a) invaluable article — the roots of pangenesis in the 1840s are on the contrary nowhere more apparent than in its dependence on the old cell theories of the 1830s and its consequent inconsistency with the new cytology of the 1860s.

As Olby (1963, 1966a) first brought out fully, pangenesis had its principal rationale as a theory of the origin in the maturing organism of the same powers — and so, presumptively, the same formative material — in both its sexual germs and its asexual buds. As Ghiselin (1975) argues, it is a theory of generation in the eighteenth-century sense, one reflecting, as he and Olby have both emphasized, Darwin's reading in those authors writing on growth and reproduction, such as Erasmus Darwin, Johannes Müller and Geoffroy Saint-Hilaire (father and son).

That the hypothesis had this rationale should surprise no one familiar with Darwin's notebook generation theorizing. For Darwin explicitly introduces pangenesis, in his manuscript draft of 1865 (Olby, 1963) on behalf of a thesis about germs and buds — the identity of powers and material thesis. And that thesis is directly at odds with his September 1838 contrasts between sexual and asexual generation.

So, it will already be manifest that I will not be arguing that Darwin had anything like pangenesis in 1838; nor that he had the elements of it then, or that it was implicit in his early notebook work. Rather I will be arguing that pangenesis could only have been formulated in a radical rethinking of the 1838 position. I take it to be obvious that to argue for this view is consistent with arguing also that pangenesis, as the outcome of such a radical rethinking, was very much conditioned in its form and content by the notebook theorizing that it replaced.

The whole exposition of 1865 hinges on the introduction, halfway through, of the theory itself. Prior to that Darwin is establishing the need for the hypothesis, after that he is giving the evidence for it.

The need case opens by declaring that there is a continuous series running from sexual or seminal generation through asexual bud reproduction to regeneration, healing and ordinary growth. This continuity thesis is then argued for from the powers and structures observable in all these instances. The initial step is to show that there are no powers or structures peculiar to the sexual germ, especially the unfertilized ovum or ovule, distinguishing it from asexual buds. Accordingly, Darwin argues not only that the unfertilized germ is visibly indistinguishable from bud tissue, but that, as parthenogenesis shows, it has the power to give rise, even if not fertilized, to a whole adult. So, the concurrence of two sexual elements is not an invariable peculiarity of seminal or germinal generation. Nor are the powers of variation, including reversion, confined to seminal generation, as sporting and reversion in buds show; while so-called graft hybrids indicate that two lots of asexual tissue have the power to impress characters on each other. Further, even though no product of budding ever undergoes maturational metamorphoses, this lack does not distinguish all asexual from all seminal generation, for in aphids and higher plants there is germinal generation without maturation and metamorphosis.

The continuity of germs with buds is thus secured by citing the exceptions to any exclusive correlation, with sexual generation, of fertilization, maturation and impressionability by mates or external conditions. The exceptions are to overturn precisely the exclusive correlation Darwin was committed to in 1838. In going on, however, to secure the continuity of budding or gemmation with healing and growth, he can appeal to the very same sorts of evidence he had used to this same end in 1838.

Having duly made that appeal he now adduces indirect evidence for the new thesis of continuity between sexual generation and growth, by considering its principal theoretical consequence: namely, that inheritance is the rule, its lack the exception, even for new characters. Accordingly, with inheritance and the reproductive conservation of any character the rule, reversion is interpreted as a temporary suspension of the normal causation of inheritance, and variation like sterility is similarly credited to a disruption of the proper working of the reproductive system.

So, Darwin has shown what he is going to "connect together by some intelligible bond" (MS p. 55). Accordingly, he moves on from the continuity thesis to what we may call the hylo-dynamic identity thesis. For he may now presume that "the protoplasm or formative matter, included within the germ and male element, and endowed with vital force cause[s] in seminal generation the development of each new being"; and that in this new being, in turn, there are "germs and buds" that agree not only in visible structure but in the powers of "varying inheritance, reversion, and hybridisation" and in the "developed product" that they eventually produce.

The "simplest belief" is, therefore, that "protoplasm, identical in nature with that within the germ, collects at certain points to form buds"; and it is a belief that must be extended to "fissiparous generation" and even "continuous growth". He thereby concludes that "protoplasm of the same nature, must be diffused throughout the whole of each organic being" ready, when "superabundant" in youth or adulthood, for budding, or for healing or for growth. Finally, then, given this view, he must conclude, too, that the reproductive organs do not form all of the "generative protoplasm", that they may not, indeed, form "any of it", and that they may only "select and accumulate it in the proper quantity and make it ready for separate existence"; which would explain why sexual and other modes of reproduction, including growth, seem to proceed at the expense of one another.

Having gone this far on the evidence cited, Darwin now introduces pangenesis itself, as a conjecture as to how the generative protoplasm, always diffused throughout the organism, is produced and how it may then come to be formed into a germ.

He supposes that it is "generated by each different tissue and cell or aggregate of similar cells"; and that "as each tissue or cell becomes developed, a superabundant atom or gemmule as it may be called of the formative matter is thrown off." And he supposes that "these almost infinitely numerous and infinitely minute gemmules unite together in due proportion to form the true germ"; that they "have the power of self-increase or propagation"; and that on uniting and before developing into tissue or cells, they run "through the same course of development" as the germ does whose elements they constitute. This complex of suppositions is the "hypothesis of Pangenesis" (MS pp. 57–58).

Now, Darwin does not claim to have observed these atoms or gemmules nor, therefore, these interactions among them. It is explicitly from "analogies" that he supports his suppositions. They are analogies between known properties of wholes and their putative unobservable parts, analogies like those that Newton had subsumed under a general "analogy of nature", whereby the existence and properties of what is too small to see are inferred by extrapolation down the scale of size among the bodies and their parts that we can see.

Thus the initial supposition, that the cells and tissues are producing such gemmules, is defended on the ground that a whole organism is analogous to a colony of component organisms, so that each of those may be supposed to be throwing off tiny buds, just as organisms themselves do. As for the further suppositions, about the actions of these gemmules that lead to germ or bud formation, these are defended by arguing, again, that what the known generational products of an observable organism do may be presumed to be done also by the gemmular facsimiles produced by every cell or tissue, every quasi-mini-organism that composes it. Thus a maturing embryo conforms to Von Baer's generalizations; so, Darwin argues, any gemmules developing into cells or tissue may do so too. Again, the female tissue of a plant has an elective affinity for pollen of the same species; so, such elective affinities may be supposed possessed by the gemmules and to enable them to arrange themselves in the ways required to reproduce the differentiation of structure in their parent source (MS pp. 59–63).

Even this brief glance at pangenesis indicates how Darwin's generation theorizing could be both strongly continuous with his 1838 theorizing, and yet markedly departing from it. The most obvious departure is that there has been an evening-up of the powers on both sides of the old divide between sexual and asexual generation. The unfertilized ovum now has the totipotency formerly denied to it but credited to a severed flatworm fragment or any healing tissue; while asexual budding is now credited with the impressionability and variability earlier reserved for germinal propagation.

No less plainly, however, this evening-up has been made in accord with premises laid down in the notebook period. The impressionability of asexual parts of the body is, on the new hypothesis, traced to the presence there of immature gemmules, in accord with the principle that impressionability requires immaturity. And it is this extension of that principle that inspires Darwin's intense emphasis on the ability of pollen to affect the skin on citrus fruits, and on Mortonian telegony, both now interpreted as instances of actions that are akin to impregnation but exerted on tissue that is not germinal (MS pp. 68–69). So, just as the totipotency of all asexual parts has now been extended into the ovary, equally the impressionability of sexual material has been extended pan-micro-ovulationally throughout the whole body. Each cell now does what any female foetus did in 1838; it buds off immature, impressionable facsimiles when still immature and impressionable itself.

It is essential to grasp firmly this feature of pangenesis. In *Variation* (1868), Darwin put great emphasis on the significance, for his entire understanding of generation, of those reported facts — especially bud reversion, graft-hybrids and the effects of the male element on female plant tissue other than the ovary — that he took as evidence for the impressionability and so, presumptively, the immaturity and quasi-fertilizability of the formative material all through the body. For this reason, the gemmules of pangenesis could not be equivalent to miniature versions of vegetative buds as Darwin had understood such buds in 1838. They had to be equivalent to miniature versions of the special buds that were ovules in 1838.

This pan-micro-ovulational feature of pangenesis is directly connected with an aspect of it that was to be of great historical importance especially through its influence on De Vries (1889, 1910). For consider Darwin's pangenesis conception of the organism as a colony of component organisms, the cells or tissues. On this conception, the diversity of those cells or tissues is analogous, as De Vries insisted, to the diversity among organisms of different species. Now, each cell or tissue type breeds true by means of the distinctive gemmules it buds off. Hence, for the entire organism — the colony of which these cells and tissues are components — to reproduce itself, whether sexually or asexually, gemmules from all the different cells and tissues have to gather together at one site. As Olby has rightly emphasized (personal communication), it is this requirement that necessitates the assumption that diverse gemmules from all over the body must be transported to any site where a totipotent bud or germ is produced.

Now, we can see how this feature of pangenesis could well have originated, by Darwin's joining a pan-impressionability conclusion with a conception of the higher organism as a differentiated colony of diverse component organisms. And it could have been by such a route that Darwin reached, as early as the 1840s, a theory of generation that, as De Vries emphasized, breaks down the unity of the specific character of any organism into different elements by making intraorganismic differentiation and interspecific diversity analogous to one another.

Our brief glance at the 1865 exposition also suggests what may have been involved in Darwin's earliest movements towards his original formulation of pangenesis. For it shows that the conjecture most likely originated as an attempt to go beyond hylo-dynamic identity to an understanding of how that identity arises. Now, everything we know about Darwin indicates that whenever he concluded in favor of that identity he would have speculated intensely as to the causes of it; and would have done so in accord with the principles laid down in the notebook period.

So, in the absence of evidence to the contrary, our best guess, as to when and how he first reached pangenesis, must be that he did so about the time when he reached the identity conclusion, and did so in speculating on the relevance to it of those principles. And as it happens, there is not only no evidence to the contrary; there is some confirmation. For all the indications are that the identity conclusion was most probably reached around 1840–1841, which is the very period that Darwin himself recalled as seeing the birth of pangenesis.

It has long been noted that Darwin himself, writing to Lyell in August 1867, dated this birth to those years. But there has often been reluctance to accept this testimony as decisive. Most notably, Geison (1969a) took Darwin to mean only that from that time on he had been trying to formulate such an hypothesis; for Geison thought it very improbable that he could have achieved such a result so early. It was to support this view that Geison cited two other letters, one to Wallace and one to Hooker, on consecutive days in February 1868.

However, once one compares the relevant passages in all these three letters (*LL* 3: 70–80), it is plain that the first, to Lyell, contains the only attempt by Darwin at an autobiographical chronology of any accuracy. Nor is that surprising; for the dating given there traces pangenesis to years when Darwin and Lyell (himself a stickler over priorities, and so for chronology, as Darwin knew all too well) were in close association, in London, and before Darwin knew Hooker well or Wallace at all. Darwin, thanking Lyell for his comments on proofs of *Variation under Domestication*, was pleased that Lyell had "noticed Pangenesis", and he went on to ask if Lyell had "ever had the feeling of having thought so much over a subject" that he had "lost all power of judging it." This, he reported, is "my case with Pangenesis (which is 26 or 27 years old), but I am inclined to think that if it be admitted as a probable hypothesis it will be a somewhat important step in Biology."

Here, then, we see pangenesis, the hypothesis itself, identified as the very subject that Darwin says he has thought about for so long as to lose all confidence in his ability to appraise it appropriately. Now, in the letter to Wallace, half a year later, we have a continuation of the same theme, but with a new twist. Delighted that Wallace is like him, in finding it "a relief" to have some hypothesis that explains many, diverse facts until a better one is found, Darwin says that it has "certainly been an immense relief to my mind, for I have been stumbling over the subject for years, dimly seeing that some relation existed between the various classes of facts." And it is this same theme, with the new twist taken even further, that is continued next day when Darwin reports to Hooker the welcome agreement with Wallace, and adds that "perhaps I feel the relief extra strongly from having during many years vainly attempted to form some hypothesis".

Plainly, what has happened in six months of correspondence is that a period of over a quarter of a century, initially recalled as one of growing uncertainty about the value of pangenesis, as formulated at the beginning of that period, has been transformed, in the escalation of Darwin's characteristic, hyperbolic, not to say disingenuous, rhetoric of self-deprecation, to become so many years of stumbling about unenlightened by any hypothesis at all. No doubt, Darwin stood to enhance by this means the hearing given by Wallace and Hooker to a prized conjecture; but this tacit tactic confirms that the letter to Lyell records his only attempt to be accurate, not to say honest, as to when this conjecture had been conceived originally.

So, back to the 1840–1841 period we must go. For, although there is no other direct evidence to lead us back that far, there is also no reason to hesitate, in particular no sign that Darwin's arrival at the conjecture depended on resources — whether concepts or data, doctrines or authorities

— that only became available to him later. Obviously, any conclusion of this kind is subject to revision in the light of new documentary findings; but everything that I have seen would seem to make it quite possible that 1840–1841 was indeed the time when Darwin made departures from his 1838 position that included formulating pangenesis.

It is only fair, however, to emphasize the difficulties arising from the nature of the documentary record for these years. *Notebook* E ends in the middle of 1839. There was a sequel to it that ran into 1841, but Darwin later took it apart and only some three and a half dozen sheets or fragments have been retrieved, presumably less than half of the notebook (Kohn, Smith and Stauffer 1982). In addition, there is the *Questions and Experiments Notebook* (DAR 206) that was used from 1839–1844. And then there are the reading lists of books read (Vorzimmer 1977) as well as the notes on the reading, either in the books themselves or on separate sheets.

Notebook E continues the preoccupation with the relations between sexual and asexual modes of reproduction, and Darwin is often responding there to his readings in Müller, Herbert and Knight and to extensive consultations with Henslow (*E* 77, *E* 83, *E* 142–144, *E* 151–164). He also reads in Blumenbach's and in Spallanzani's writings on generation at this time.

Although the fundamental gulf, established in September 1838, between seminal and all other kinds of generation persists, two possible analogical bridges across it are admitted: Knight's "analogy between grafting and sexual union" (*E* 77) and the "analogy of production by gemmation and by seed — which Henslow is inclined to think very close" (*E* 162). But, admission of these analogies is conjoined with a reaffirmation of the old insistence that sexual generation, with crossing, is unique in allowing adaptive change in the long run. Thus Darwin (*E* 143–164), under Henslow's influence perhaps, now questions Knight's claim that without sexual crossing a species cannot perpetuate itself beyond a limited time, objecting that Knight's conclusion is drawn from studying successions of grafted cuttings, and that the senescence accompanying such propagation may not be entailed by natural rooting or layering. So, plants have not been provided with the means for sexual crossing in order to escape any limitation on the duration of asexual life, but rather to make adaptive change possible. Equally, crossing is still seen as needed to keep the species constant in conditions not changing overall; although those reproducing asexually by buds may be able to stay constant, either because buds allow no change even in changing conditions, or because the conditions of their lives have not changed. In any case, the simplest organisms are less susceptible to the influence of conditions, presumably because they do so little maturing; so they have less need for crossing.

Tulips propagated by buds do not always replicate all their characters truly; they do eventually "break", and such variation being independent of sexual generation is like a variation in a graft succession that is not due to any effect on the graft of the stock. So, such breaking is a "strong case" for the analogy Henslow is urging between gemmation and seminal reproduction (*E* 162).

Darwin's 1838 position could accommodate such an analogy; he had long assumed that all reproduction was division and so gemmation, that is to say budding, including unfertilized ovum production; he had been sure that in polyps one could even see "young buds changing into ovules" (*D 68*); and he had held that all reproduction used the same material, surplus to the requirements of growth, so that any generational production, sexual or not, will have consumed material not then available for use in any other.

These longstanding assumptions combined, accordingly, with the renewed interest in budding and seminal generation in May 1839, to prompt intense fascination with "endless curious facts about every part of plant producing buds, so that Turpin [a writer on the origin of *tissu cellulaire*] says that each cell of plant is individual"; and to prompt Darwin to note that most plants propagating by buds and layers and so on "do not seed freely" (*E* 165; see also *E* 183).

Standing back a moment, then, we can see that, even by the end of *Notebook* E there were three lines of thinking that were potentially leading to reappraisal of the 1838 contrasts between sexual and all other modes of reproduction: the common material all generation draws on; the formation of unfertilized ova as buds; and the variation sometimes accompanying bud propagation, together, conversely with any unexpected trueness of variety propagation by seeds (as at *E* 183). These three are, therefore, all to be kept in mind as we follow Darwin into the 1840–1841 period itself.

The early months of 1840 see generation prominent in his reading lists. In January he returns to the first volume of the English translation of Johannes Müller's *Physiology* (1838); in February he is reading, presumably not for the first time either, Henslow's *Descriptive and Physiological Botany* (1837a) and in April he is returning to Erasmus Darwin's *Phytologia* and also to Spallanzani on generation (unfortunately his copy of this book is apparently no longer with the rest of his library).

However, even if one could be confident in dating them, the annotations on the Darwin *Phytologia* and on the Henslow *Botany* would permit no sure reconstruction of his thoughts at this time. But it is reasonable to suppose that he was finding his grandfather's conjectures suggestive. For Erasmus Darwin elaborates a completely general theory as to the formative matter common to both buds and ovules in plants, and he integrates this theory with conjectures about all animal reproduction too; what is more the theory appeals to analogies with attractive forces and affinities among particles as familiar from physics and chemistry. In Henslow's book, several sections deal with reproduction in plants, and Darwin's marginal comments and markings, if they were made at this reading, show that he paid special attention to the places where Henslow was comparing and contrasting seminal and other forms of reproduction. Certainly, the *Questions and Experiments Notebook*, the *Torn-apart Notebook* and Darwin's letter to Henslow (in Darwin 1967) all show that in the early 1840s he saw Henslow as an authoritative source in his efforts to understand fertilization, hybridization, variation, monstrosities and so on in plants.

On the fundamental issue of how buds and unfertilized ova are to be compared and contrasted, Müller was already the decisive authority for Darwin, as he would still be in the 1860s. In December, 1838 (*E* 83), Darwin had dwelled on the implications, for analogies between ontogeny and phylogeny, of Müller's report (1838, p. 24) of Von Baer's thesis that the "germ" of any vertebrate species is initially indistinguishable from that of any other. The ink and pen here (*E* 83) matches well with ink annotations made by Darwin on this and earlier pages in Müller's thesis (p. 19) that the unfertilized ovum is distinguished only as the one part of the organism that can ever live separately: "There must be," Darwin wrote, "some wider difference between ovum and bud."

In his eventual revision of this dissent and so in his eventual departure from this 1838 position, Darwin will be knowingly coming to agree with Müller on this issue. He will also be coming to draw on Müller as a main source of insights and information concerning theories about the cellular constitution of the organism.

In 1842, in the second volume of the English *Physiology* (1842) Darwin found Müller not only arguing that Planarian regeneration shows that a severed part can act like a germ, but also reinforcing such comparisons with the cytological generalization that all parts of plants and animals are developed from cells, that the germ in animals and in many plants is a cell, and that the essential part of a gemma or bud is either an aggregate of cells or a cell (pp. 1425–1427 and 1448). Darwin was to make many markings on these passages over the next quarter of a century; but some were made presumably on his first reading in 1842. And he may have entered then (p. 1428) a note as to how the product of any asexual generation differs from that of any sexual generation: "differs in duration of life — none [changed to "non"] metamorphoses and less variation." Now, whenever this entry was made it marks agreement with the 1865 rather than the 1838 position; for the 1865 exposition accepts that buds are unlike unfertilized ova only in being able to live longer, in never metamorphosing and in varying less.

Some slight evidence that Darwin has made this shift — to the new comparisons and contrasts between ova and buds — probably before 1849, can perhaps be had from reading a well-known letter to Huxley, most likely written in 1857 (*ML* 1: 102–103), in the light of Darwin's notes in and on his copy of Owen's *Parthenogenesis* (London, 1849b). In the letter he recalls that he "never from the first believed" in Owen's interpretation of parthenogenesis, and he goes on to say: "I cannot but think that the same power is concerned in producing aphides without fertilisation, and producing, for instance, nails on the amputated stump of a man's fingers, or the new tail of a lizard." And this same thought is indeed entered as an objection on a sheet of paper pinned into Owen's book, where Darwin also objects that Owen's theory does not distinguish appropriately between generation and growth.

Overall, then, it seems reasonable to suppose that Darwin may well have moved, even as early as 1841, to credit unfertilized ova with the totipotency that had always impressed him in buds and flatworm fragments.

Consider next, then, the other side of the old 1838 contrast, the side of buds and their powers of variation and so on. For here, too, the breaking down of that contrast has gone far in the early 1840s. These powers were often evidenced in the 1860s by citing findings ascribed by Darwin to Giorgio Gallesio of Florence. And so it is reasonable to conjecture that he saw them as inconsistent with his earlier 1838 views when he studied Gallesio's two books, *Traite du citrus* (Paris, 1811) and *Storia della Riproduzione Vegetale* (Pisa, 1816). Darwin's copy of the *Traité* shows signs of very attentive reading presumably dating mostly to April 1842, when the book appears in his reading list. On the *Storia* he made extensive notes (DAR 71), at the very end of 1841 according to that list. The notes record preoccupation not only with variations of fruit born on one branch, but also with the lemon striping of the peel of the fruits when an orange tree was impregnated with lemon tree pollen. Moreover, we see from the opening pages of the *1844 Essay* that Darwin has considered Gallesio's and others' findings on bud variation and double flowering to confirm that a plant is in some respects not an individual but a colony of individuals; and yet also to show that the same variations and characters that are transmitted by buds are transmitted equally by seminal generation. This bridging of the old gap between budding and seminal generation has, therefore, been deliberately pursued by this time and with the most general issues in mind.

Independently of precise matters of dating, no one can read Darwin's copies of Henslow, Müller, Owen and other works, for instance Steenstrup (as translated by Busk) on *Alternation of Generations* (1845) or Virchow (as translated by Chance) on *Cellular Pathology* (1860) and study his extensive and detailed notes written in them, without appreciating that right through the 1840s and 1850s and on into the 1860s to 1865, the same cluster of issues was pursued as intensely as in the early 1840s.

The 1865 exposition of pangenesis itself draws at every turn on the old cell theories of the 1830s and 1840s as Darwin had learned them then, especially from Müller, but it never appeals to the newer cell theories of the 1850s. These newer theories, as promulgated especially by Virchow, he introduced only in the published 1868 version. Moreover, one consequence of introducing them was, as Darwin shows every sign of sensing, to create a fundamental inconsistency. The old cell theories allowed for a cell to arise other than by division of an existing cell; the new consensus of cytology did not; and so the formation of a germ cell from numerous gemmules was obviously anomalous. By the time, 1875, of the second edition, the inconsistencies were even more extensive. By then (Farley 1982), the cytological consensus was agreed that every cellular organism is either a single cell or a cell colony arising from the divisions of a single cell, and that two cells come together to form one at fertilization, each having arisen by the division of one cell in the respective parent body. But pangenesis conflicted with these generalizations; for if each of the two masses of gemmules coming together at fertilization is taken to be one cell, then it has not arisen in the division of one cell in that parent; while, if each is taken to be a myriad of cells, then far too many are coming together at fertilization.

In 1868 and 1875, when suggesting how to reconcile pangenesis with the prohibition against having new cells arise except by the division of existing cells, Darwin took a very telling line. He conceded that cells do normally reproduce by replicative division in that way; but he insisted that in addition they could also be reproducing by budding off tiny maturing, impressionable gemmules capable of uniting with other cells in interactions analogous to fertilizations. So, he now has for cells very much the distinction between asexual and sexual budding that he had had for most organisms back in 1837. At the end, then, he had not dispensed with his old fundamental contrast between generation that involves maturation and mating and generation that does not; he had merely relocated it so that it holds true of the cells, the organisms composing the colony, rather than holding for the colony, the larger organism itself as a whole.

Such continuities between the notebook generation theorizing of the 1830s and the published hypotheses of the 1860s are but special cases of continuities that run even deeper. No one can read together the Grantian texts of the 1820s and 1830s and the various versions of pangenesis from the 1860s and 1870s and not be struck with the persistence of Darwin's preoccupation with "atoms" and "gemmules" and "cells" and all those general theoretical issues associated with them, issues concerning "individuals" and "colonies" and the "vital forces" and "formative matter" active in their growth. For Darwin, late as well as early, there was no full understanding of anything to do with living beings until it was fitted into a hierarchy of analogies among wholes and their parts that extended from living atoms to a tree of life.

Conclusion

There can be no doubting that a major preoccupation of Darwin's from 1837 to his final years, was in extending analogical inferences, from the comparison of sexual and asexual modes of generation in individual organisms, to entities above and below them in the organizational hierarchy, all the way down to living "atoms" and all the way up to the "tree of life".

It would, however, be a serious mistake to try to reduce Darwin's entire career as a general biological theorist to the pursuit of this preoccupation. I have dwelled elsewhere (Hodge 1982) on the Lyellian sources of his notebook zoonomical explanatory program; and indicated that, as I see it (Hodge 1983) a main challenge is to understand how two inheritances, the generational or Grantian and the geological or Lyellian, could be decisive for the origins and character of that program. Moreover, I would be prepared to suggest that there is a definite asymmetry in the integration of the two. To simplify drastically, it is often true that, in the initial phase from 1835 to 1837, it was Darwin's prosecution of Lyellian geology (including biogeography, ecology and so on) that gave rise to the problems constitutive of the program, while the Grantian inquiries provided resources for their reformulation and solution. Now, I emphasize that this asymmetry is characteristic of the initial phase, because, obviously, any such suggestion has to acknowledge that problems on becoming solved can engender resources while solutions equally can become problematic, as an explanatory program develops further.

Allowing for that complication, however, I would go as far as to propose that these two inheritances were together by far the dominant determinants of Darwin's biological theorizing; so that the burden of proof would seem to lie with anyone who would make a case for any other inheritance being independent of these two and of comparable influence.

A general conclusion of this kind, as to Darwin's own intellectual ancestry, has implications, naturally, for the interpretation of the social history of his science and, no less so, for the interpretation of the intellectual legacy he bequeathed to the century of posterity since.

A brief indication of what these implications may include will provide a chance both to forestall some possible misunderstandings of the analysis given in this paper of Darwin as a lifelong generation theorist, and an opportunity to engage more fully those general issues of reinterpretation introduced at its opening.

As to the first, the social history, it is now apparent, I suggest, that two very different proposals, Walter (later Susan) Cannon's (1961b) and Robert Young's (1973a), cannot take us as far beyond the Franciscan view as it may once have seemed they could. For they both share two common limitations with that older view; because both accept the traditional evolution-plus-natural-selection-as-a-mechanism-for-it formula for Darwin's thought, and both also fail to take institutions seriously. Thus Cannon accepts that Darwin's "worldview" is given by the "ideas" of "evolution" and "natural selection". For, inverting T. H. Huxley's thesis — that "evolution" was necessitated by "uniformitarianism" as an alternative to "Christianity" — Cannon traces "evolution" instead to the liberal Anglican (Sedgwick and Whewell particularly) commitment to "progress" and "development" in the history of the earth and its inhabitants; while assimilating natural selection to that tradition's neo-Paleyan commitment to causal imposition of adaptive design on passive matter. The Cambridge "network", as the home of these commitments, is therefore, according to Cannon, a principal social location for the sources of Darwin's science. But not only is this inversion of Huxley's contribution to the Franciscan view unsatisfactory as intellectual analysis, it is inadequate as social history too. For, like all "network" theses, Cannon's ensures that the institutions, whether Cambridge University, the British Association or the Church of England, only appear as contingent instruments, whereby and wherein particular individuals, taken one at a time or gathered into "networks" or "coteries" or "circles", are acting out their singular or collective ambitions. So, as in the historiography of Lewis Namier, the emigré doyen of all afficionados of English clubbing and jobbing, no account is taken of the wider, shifting group interests and ideologies to which those institutions owe their formation and transformation in the long run of social and economic change.

Marx has often inspired the necessary correctives to Namier, but Young's Lovejoy Marxed proves no more adequate here than Cannon's Lovejoy Namiered. For Young, content to leave Huxley standing on his feet, moves from "evolution", through the more general "ideas" of the uniformity of nature and the continuity of man with nature, to the ideology of "naturalism" and thence, without more ado, to the industrialization of all European society as the economic process giving rise to that ideology. And he moves from "natural selection" to the ideology of utility and competition and so to capitalism as the economic condition for this industrialization. Here, too, then, institutions and their constitutive relations with distinctive interests and conflicting ideologies within society are missing. Where Cannon has associated his Lovejoyan units of intellectual life with social units of extreme narrowness and transience, Young has referred them to social trends permanently affecting everyone and everything.

Moreover, the most recent contributions towards a social history for Darwin's science, notably Manier's (1978) and Rudwick's (1982b), have continued to delineate circles and colleges, visible and invisible, that are comparable to Cannon's networks in how far they extend if not in what they enclose.

Generation, as a preoccupation central to Darwin's intellectual career can, then, help us to overcome the limitations in existing social histories of Darwinian science, both because, as a theme or program or whatever, it is larger than an "idea", and because its institutional life could lead us to consider social changes at a scale between a Cambridge liberal Anglican network on the one hand and industrial capitalism on the other.

Thus we would have to consider, for example, how generation, like the passions of the soul, came within ground disputed between clerical and medical interests for centuries if not millennia; and how social changes as fundamental as those ingredient in the industrial revolution obviously had consequences for all the professions, whether of law, medicine or the church, consequences that transformed the relations arising from their institutional and ideological interactions. The intellectual divergences over generation between Erasmus Darwin and William Paley or between Robert Grant and John Henslow were associated, accordingly, with diverging interests that were being pursued along institutional and ideological lines that had recently begun to run in new directions; think of the Lunar Society or of German universities or of the Edinburgh intelligentsia as sources of agitation for educational reforms, agitation arising from new challenges to established checks and balances among wider group and class interests.

Obviously, these observations present us with the staples of the social historian of the period, whether of left, right or center. And that, I would suggest, is as it should be. For although the social historians obviously have not done their work with any heed to the special historiographical issues raised by Darwinian science, such work has the one great advantage of historical appropriateness. By contrast, although the resources supplied by sociological theories of scientific knowledge may sometimes prove suggestive, they can easily introduce distracting fallacies, as for instance, when a Kuhnian analysis of disciplinary matrices is applied in a context, England in the 1830s, when scientific disciplines had yet to emerge in the form typical of a century later.

The bearing of such observations on our centennial Darwinian business could do with far more extended discussion, needless to say. But I trust that the relevance to them of Darwin's generation theorizing is now plain, and that it is also apparent that I am not offering that theorizing as another Lovejoyan "unit idea" to be duly Namiered or Marxed in the Cannonical or Youngian manner. For I see it able to take us further beyond the limitations of the Franciscan view than either of those responses would permit.

Turning now to the reinterpretation of the Darwinian heritage, in the sense not of what Darwin inherited but of what he bequeathed, my main proposal can be introduced best by considering an objection to all "synthesis" historiographies for evolutionary biology since Darwin.

Any "synthesis" talk inevitably suggests that until what eventually came together did so then some parts were apart or defective or missing altogether. And, indeed, such connotations are present in all the synthesis views familiar from a large and growing literature (Mayr and Provine 1980) on the history of "evolutionary biology"; the view, for example, that a new synthesis of the 1930s brought Darwin on selection together with Mendel on heredity; or that the same period saw a new synthesis of results from many fields as diverse in methods and material as genetics, paleontology, systematics and embryology; or that (E. B. Wilson 1900), in an earlier period, Weismann's theory of germ plasm continuity integrated what had previously developed independently: the theory of evolution and the theory of cells.

Taking this last synthesis view first, it is certainly tempting to think of biology in the middle of the last century as comprising on the one hand natural history, broadly defined so as to cover geology as well as entomology, and, on the other hand, physiology including histology, cytology and the rest. And it is plausible to associate the first with voyages, museums and mining, with observing and collecting and with questions about species and their coming and going in the long run; while associating the second with medicine, dissection, microscopy, experiment and questions about individuals and their functioning, in health and disease, over the short run. Moreover, Darwin may then seem quintessentially exemplary of the first, Virchow and Bernard of the second. And so one would be prompted to ask when this divide was bridged and by whom, and whether the completion of that achievement also awaited a "synthesis" of the twentieth century.

However, historians of science have been learning to be critical of the historiographies, not to say myths, that scientific communities develop and deploy for their own ends, and the time may have come to be skeptical of any myth of synthesis, if only because such a myth will always tend to imply that what went before was incoherent, or incomplete by the new standards set when, as the vernacular would have it, the synthesizers finally got it all together.

That this skepticism is valuable in the present case is evident as soon as we refuse to beg questions about scope and structure with decisions about correctness of content. As any slight acquaintance with the total structure of Darwin's biological theorizing confirms, it ranged right across most of the biological board as one then played upon it. A quick way to appreciate how comprehensive it was, is to take those two vast treatises of the 1830s, Lyell's *Geology* and Müller's *Physiologie*, see how between them they cover all the topics about living things then in play, with several such as hybridization and comparative embryology appearing in both, and then to see that Darwin's thinking, through his generational theorizing, included a response to almost every major biological issue in those two treatises.

How, then, if Darwin's theorizing ranges across so much of both Lyell's and Müller's ground, can he be represented as sending his "evolutionary biology" into the world walking on only one real leg, natural selection, supported if at all by a flimsy physiological (including cytology) crutch, pangenesis, where the other leg, genetics, should be? He can be, obviously, if his cytology and any associated concepts are deemed not to count, historically, as a real limb and a genuine part of the whole body of his "evolutionary biology", on the ground that they have been superseded by the birth and growth of genetics.

Now, put so crudely the fallacy is manifest. But to avoid the more sophisticated versions of it, we have to be wary of all suggestions — deriving, as they do, from retrospective preconceptions — to the effect that Darwin's biological theorizing was such as to raise a need that he could not satisfy, the need for a synthesis of a scope that he could not achieve and that had to come later.

Once we do set aside those suggestions, then our reinterpretative task can be framed in appropriate historical terms. For we may start by thinking of Darwin in the 1860s as someone, like the Haeckel of *Generelle Morphologie* of 1866 or the Spencer of *Principles of Biology* of 1864-1867: a biological theorist responding to a vast complex of issues that had come to prominence in the 1830s and were represented most accessibly in the treatises of Lyell and Müller. Then, having taken that step, we can return to such dominant figures in subsequent decades as Weismann, De Vries and Wilson, who are engaging a range of issues no less broad as they, too, seek to relate germs and cells to species and trees of life. And we can see these biologists within the same succession. On this view, then the De Vries of *Intracelluläre Pangenesis* of 1889, rather than being excluded from the Darwinian heritage because of opposition to gradualistic selectionism, becomes included in a succession of generational theorists of evolution that has included Darwin and that will eventually include Morgan, Muller and Dobzhansky. On this view, then, a De Vries shares in a common heritage with Darwin no less than a Poulton does; for if Darwin is the theorist who comes between the gemmules of Grant and the pangens of De Vries, then De Vries is the theorist who comes between the gemmules of Darwin and the genes of Morgan. Such suggestions are not intended to beg any of those questions, now addressed by specialists in the period (Coleman 1971a) as to what schools, traditions and disciplines were developing in the late nineteenth century. But these suggestions are designed to provoke a reconsideration of conventional preconceptions about the place of Darwin's thinking in the larger successions of biological theory, preconceptions based on misconceptions as to the sort of science he was doing.

Thus, for instance, we need to get away from any notion that Darwin treated, or should have treated, the organism itself as a black box, merely taking variation and heredity as given observationally and independently of all conjectures as to their causes. For it is not simply that such a notion can survive no examination of his writings, even the published ones; it is misleading for the much more general reason that it makes Darwin's thinking about organic diversity and species origins anomalous in comparison with the thinking of nearly all other theorists who have addressed these problems; for they have almost invariably grounded their understanding of those problems in assumptions about the constitution of the animal or plant body: think of Aquinas's souls as the forms of life; Linnaeus's cortex and medulla; Buffon's organic molecules, Erasmus Darwin's fibres and filaments, Lamarck's cellular tissue and so on; although in passing, we may note that Wallace is an exception here, and that this alerts us to the fact that, while he shared Darwin's commitment to Lyellian historical biogeography and, like Darwin, had his own sources of youthful materialism, there was in his thinking nothing equivalent to the Grantian (and Erasmus Darwin) heritage of generation theorizing.

In sum, then, Darwin should be seen as feeding his generation theorizing, as an integrated complex, into the broad stream of biological thought; so that subsequent theorists, such as Weismann or De Vries, were responding to that complex along with other such complexes constructed by the Haeckels and Spencers, revising this bit, rejecting and replacing that and thereby reaching their own integrations. And thus has it been ever since, right on into our century of "evolutionary synthesis". For it could only have been otherwise, been as the myths of synthesis imply, if some parts of Darwin's own synthesis were ignored by everyone. But we all know that none of them were. Of course, only one man, Darwin himself — and he sometimes had doubts — was ever content with all the parts of his whole. But then the same goes for Buffon or Newton or Aristotle. Darwin's legacy was often responded to part by part, and with discrimination, but its historical role was decided by the scope and structure of the whole and by its relations to the main successions of general theory.

8

Darwin's Principle of Divergence as Internal Dialogue

David Kohn

Introduction

However strongly we may see scientific ideas as socially and culturally contingent in their origin and expression, we must acknowledge that they are also the products of individuals. Hence even if we all consider scientific activity to be the reworking of prior scientific activity, the dynamics by which individual scientists develop their theories is a subject integral to the history of science. If we accept the proposition that knowledge grows by public and critical dialogue, we should not ignore the fact that important phases of the dialogue may occur within an individual. Such is the case for Charles Darwin, who over the decades prior to publication of the *Origin* engaged in an extended reworking not only of natural history, but of his own emerging ideas. For a scientist with the scope of Darwin, the internal personal debate is as fierce and as fertile as many a public debate.

The subject of my paper is the internal dialogue that produced Darwin's principle of divergence. Let me begin by defining the principle. The argument Darwin called the principle of divergence runs as follows:

1. First there is an ecological premise. A locality can support more life if occupied by diverse forms partitioning resources. This is the ecological division of labor. Thus specialization is an adaptive advantage to an organism. Hence natural selection, which explains the origin of all adaptation, favors the evolution of new specialized varieties.
2. The making of a new variety occurs sympatrically, that is, with parental and offspring forms inhabiting the same locale. Thus the making of varieties, which Darwin saw as incipient species, occurs by vigorous selection for specialization overcoming the swamping effect of crossing.
3. From this first fork of the branching phylogeny it is a matter of reiteration to generate all of classification. Simply put, niche within niche engenders group within group.

Darwin's principle is itself a set of nested arguments comprising the idea of natural selection, the idea of speciation without isolation, and the view that the relations among organisms create new evolutionary situations. One thing about the argument stands out. It is internally unified by natural selection. That is, explanation at the three classic levels of evolutionary theory — adaptation, species formation, and the hierarchical classification of organic diversity — is portrayed as the application and consequences of natural selection.

One difficulty in studying Darwin one hundred years on is that we all are, or believe we are, Darwinians. Modern evolutionary theory —including post-synthesis versions — comes equipped with its own historiography, which includes a view of Darwin. In particular, Darwin's theory of natural selection has been reduced to an explanation of one aspect of evolutionary process: the origin of adaptation. For Darwin, however, selection explained more than adaptation. It was always intimately bound up in his thought with the multiplication of new species. The dominant modern explanations of species formation do not accept the same role for natural selection in this process that Darwin did. The question is complicated by the often repeated slogan of the modern theory that evolution is the unifying theory of biology. This slogan rests on a number of claims, one of which is that evolution explains a hierarchy ranging from adaptation through species formations to the major trends of organic diversity. Modern explanations of these three levels are, however, not continuous. That is, they do not employ the same mode of argument to explain processes at each level. Thus the modern theory, including post-synthesis versions, is not internally unified. It is a chain of explanations, each more or less appropriate to its own level. This disunification may be one of the keys to its success. Nevertheless the slogan remains, and some other of its claims may be justified. Like his modern "followers", Darwin saw evolution as a unifying theory; unlike them, he sought a theory that was internally unified by natural selection, and he thought he had accomplished this goal through the principle of divergence. One would assume there ought to be a basic disparity in the way that a historian and a biologist would interpret Darwin's principle of divergence. For the historian Darwin's principle is a part of his thought. It requires explanation in context. For the biologist the principle is something Darwin got wrong. Since the biologists have, at least in the past, maintained a claim to be Darwinians, they have wanted to either sweep Darwin's error under the rug or figure out where Darwin made his mistake. This has had its impact on historians. We have tended to follow the biologists' lead and have either ignored divergence or considered it a curiosity.

The past decade has witnessed important efforts — first by Limoges (1970c), and then almost simultaneously by Browne (1980), Ospovat (1981), Schweber (1980), and Sulloway (1979) — to reassess the principle of divergence. This has been stimulated by two factors internal to the recent history of Darwin studies. First of all, our period has witnessed a considerable concentration on Darwin's intellectual development. To get beneath the often enigmatic surface of the Darwin of the *Origin*, the present generation of Darwin scholars has been led to study how Darwin's ideas were formulated. Inevitably, the course of this work has followed the course of Darwin's life. Workers have tried to understand the process of his conversion to transformism, his first formulations of evolutionary explanation, and of course the construction of natural selection. By around 1977, the time had come to tackle the period from 1844 leading up to the writing of the *Origin*. For that phase of Darwin's career, the principle of divergence stood out as a critical intellectual development. It was an idea whose turn to be studied had come. The second factor directing attention to divergence was access to the necessary materials. The publication of *Natural Selection* by Stauffer in 1975 was the primary stimulus. This led scholars, under the guiding hand of Sydney Smith, to attempt to make sense of the great collection of loose notes Darwin had accumulated in the 1840s and 1850s. In these notes, which Darwin organized in topical portfolios, was thought to lie the evidence for reconstructing Darwin's formulation of the principle of divergence. The contributions of Browne (1980) and Ospovat (1981), as well as the present essay in particular, are based on both Stauffer's edition of *Natural Selection* and Smith's efforts to organize and date the portfolio notes.

Suddenly we have had a great deal of light on the principle of divergence. Yet I feel more clarification is needed. We need to have a better understanding of two things: why we are studying the principle of divergence, and how to go about it. My remarks so far suggest an answer to the first issue. Scientists in the construction of their theories not only seek to explain natural phenomena. They also often have methodological goals. The search for unity is one such goal. As historians we ought to be cautious neither to worship nor to ignore the powerful lure of unification. We ought to recognize it as a recurrent tendency in scientific debate that is often internalized as a value in individual scientific practice. In other words, unification can guide the content of science in the same way that ideology has long been known to do. This was the case for Darwin's principle of divergence. He conceived of it as a principle of unification. This explains why he called it "a keystone of my book" (*ML* 1:109). Unification was not a detached value; it was an internal guide that penetrated and directed the development of his research. This is the principal message of the story I am about to tell. The second matter that needs clarification is methodological. The progress achieved since 1959 in understanding Darwin's development has come when we have identified and clearly analyzed concrete episodes in Darwin's career. This is true for Darwin's conversion, his first theory, and natural selection. The principle of divergence, however, had a much longer gestation than those episodes. We still need both a descriptive and a causal embryology for the development of the principle of divergence. I hope to define and analyze one episode in that development.

I. Defining an Episode

Given the importance of analyzing *episodes* in a developmental approach, clearly the prior and primary step is the identification of significant episodes. Let me describe the archival problem briefly. We are dealing with several hundred loose pieces of paper distributed, not in the order in which they were written, but in topical portfolios (DAR 205). Furthermore, within each portfolio there is no reason to assume the notes reflect Darwin's writing order. Given this lack of stratigraphy (a problem that does not exist with Darwin's notebooks), it seemed to me that the safest course is to concentrate on those notes that Darwin took the trouble to date. For the period 1852 through 1857 this came to 124 notes. I was somewhat reassured in this approach when I found few undated notes that I felt troubled about ignoring. When rearranging the dated notes from all the portfolios into chronological order, I had two dominant impressions. First, I was impressed, and to a degree depressed, by the low density of dated note-taking. During the seventy-two months of 1852–1857 there are thirty-three months with no dated notes at all, sixteen months with one note per month, and fifteen months with two — four notes. Thus during over 80 percent of this period, note-taking, at least dated note-taking, can be characterized as only a sporadic activity. My second impression was that one date dominated all others in frequency: November 1854. There are nineteen notes so marked. Thus some 15 percent of the dated notes from the six years of interest were written in that one month of 1854. At last I had at least one episode around which to build a theory. Lest the reconstruction of a period in Darwin's thinking from nineteen scraps of paper appear an implausible object of study, think of the cogent theory, and indeed program of research, that Darwin concentrated into the first twenty or so pages of the B *Notebook*.

II. November 1854 in Perspective

November 1854 was a point in Darwin's career that one would predict to be of interest. In September 1854 Darwin took a critical step towards the eventual writing of the *Origin*. He sorted the loose notes he had accumulated since writing the *1844 Essay* and distributed them into topical portfolios on biogeography, classification, hybridity, variation, embryology, paleontology, and behavior. In November 1854, having no doubt completed a review of his old notes, he in effect initiated the process of writing his species book by writing a spate of new notes.

Before looking at the details of Darwin's position in November 1854, it is well to put the conceptual boundaries of this episode into perspective. It is evident that Darwin had already set himself the problem of showing that divergence was a tendency in nature that required a mechanism or explanatory principle. Ospovat showed that, contrary to his position in 1844, when Darwin rather took divergence for granted, by 1847 he was already inclined to see it as a problem to be solved (Ospovat 1981, pp. 172–173). Ospovat explained this shift as Darwin's response to those comparative anatomists, such as Milne-Edwards following on Von Baer, who held a branching conception of systematic relationship but who maintained a creationist outlook. I think it might be profitable to see Darwin's shift as part of his more general effort to translate the several theories of contemporary systematics into evolutionary terms. This included an evolutionary reinterpretation of Swainson, Owen, and Strickland, as well as Milne-Edwards and Carpenter. But this is an area for future study. At any rate, the problem of explaining divergence was already constitutive to Darwin's thinking before November 1854. In contrast, the nub of Darwin's solution to the problem of divergence, namely ecological division of labor, was formulated shortly after November 1854. So this episode is the original locus of neither the problem nor its solution. Nevertheless, I see it as the turning point of the story, first because here Darwin consolidated his argument and established its characteristic unified structure, and second because in structuring his argument he here discovered the particular line of reasoning that led subsequently and rather quickly to the division of labor.

Turning now to the core of the episode, we find that the focus of Darwin's attention was the use of biogeographic data to draw conclusions on the pattern of divergence. As Janet Browne (1980) has shown, he worked in the botanical arithmetic tradition of Humboldt, Robert Brown, and A. P. de Candolle. In this tradition the evidence of geographic distribution was tabulated and summarized in ratios, typically expressing the number of species per genus in a particular geographic area. By comparing such ratios one drew general conclusions about patterns of distribution. In Darwin's hands this method became a powerful tool to show that the contemporary data of present distribution patterns could be used for two related ends: (1) to portray the stages in the historical process of evolutionary divergence, and (2) to portray the hierarchical classification of natural groups as the product of evolutionary divergence.

In November 1854 Darwin in fact began two processes. The first was the actual tabulation of data from catalogues, monographs, and synoptic works such as the Candollean *Prodromus* (1824–1873). Tabulation became a major project that continued into 1858. The other process was the conceptual one of drawing conclusions from tabulated data and establishing hypotheses to be tested against tabulated data. Out of these two processes the solution to the problem of divergence was formulated. According to Browne (1980) and Ospovat (1981) the twin processes of tabulation and conceptualization went on hand in hand from 1854 to 1858. Thus they contend that, as Darwin made new calculations, he substantially modified his views as to the bearing of his data on the principle of divergence. This led Ospovat (1981, pp. 176–189) to characterize the formulation of the principle as occurring gradually over a period of years and Browne to see it as happening in 1857 near the end-point of the calculations (1980, pp. 82–89). I have to disagree with the basic premise of both Browne and Ospovat. As I see it, in November 1854, when Darwin made his first calculations, which he considered to be a success, he became convinced that the botanical arithmetic approach would allow him to prove certain hypotheses.

Starting from this premise, my thesis is threefold: (1) That Darwin drew his complete set of conclusions from the biogeography of living groups in November 1854 and simply tested these by laborious calculations over the succeeding years. The long series of calculations always confirmed Darwin's hypotheses. In other words, this episode is one more instance of the priority of theory over evidence in Darwin's intellectual development. (2) That also in November 1854, he used a particular form of historical reasoning to transform his biogeographic conclusions into a proof that the history of life was divergent. (3) Finally, that in November 1854 his transformation of biogeographic data into a historical narrative established the framework from which he drew the further critical conclusions necessary to complete his principle of divergence. These conclusions were two: that speciation can occur without any form of isolation. This was a temporary position from which he retreated. It provided, however, the intellectual emancipation that led to the second critical conclusion, namely that species are formed not just by natural selection, as he had long believed, but that species multiply in those ecological conditions that permit vigorous selection. In sum my thesis means that the central structure of Darwin's argument for divergence —with its characteristic unification of natural classification, speciation, and ecology, as well as the defining conditions for the ecological division of labor — was all dashed off in November 1854. The division of labor itself, which completed the principle, was added three months later in January 1855. All the rest, including writing the divergence sections of *Natural Selection*, was calculation, revision, and exposition.

This is my thesis; let me see if I can put some evidentiary flesh on these bare bones.

III. Biogeography as Historical Narrative

Although Darwin's calculations over 1854–1858 were extensive, their basic logic was in place in November 1854 and they can be summarized succinctly. He made four basic calculations; two were focussed on small genera and two on large genera. The starting point was small genera. In November 1854, he had George Robert Waterhouse mark the aberrant or peculiar genera in Schoenherr's catalogue of the Curculionidae (the weevils) (1849). He calculated the number of species per aberrant genus, and found that the aberrant genera had fewer species per genus than the average number of species per genus in the family as a whole. Second, he also examined the geographic range of all small genera in the curculionids, whether they were aberrant or not. He found that the aberrant genera had scattered ranges, and the non-aberrant small genera had predominantly local distributions. Thus Darwin thought he had discriminated between two kinds of small genus. Small genera with local distributions, where the species are morphologically closely allied, he interpreted as rising or nascent genera. On a sheet summarizing his calculations from Schoenherr he wrote:

> All rising genera must be local <(& closely allied)>: . . . (DAR 205.9:290)

In contrast, small genera with morphologically very distinct species — distinct enough to be called peculiar or aberrant — were found to have a scattered distribution. He interpreted these aberrant genera, as he had since the 1830s, as dying genera or living fossils. He continues in the same note:

> . . . all dying genera, <with species very distinct> . . . wd be small, aberrant & <if they had died equally over world, wd be> widely distributed . . . (DAR 205.9:290)

In other words, the biogeographic data could be used to identify nascent genera by their pattern of species fanning out or diverging in the local area of their birth. They could also be used to characterize the end point of divergence as the scattered and peculiar relicts of what might once have been large genera.

These, then, were the hypotheses Darwin tested and the evolutionary inferences he drew from the work on small genera in the curculionids. In November 1854, however, Darwin also saw that the link between small nascent and small dying genera was formed by large genera.

He wrote:

> . . . if extinction has fallen near & around the aberrant genera, then *creation* has fallen on the typical & larger genera. — (DAR 205.5:147)

Darwin in November 1854 saw two hypotheses to be tested with respect to large genera: (1) that the species in large genera tend to be wide ranging and (2) that the species in large genera tend to be polymorphic, that is, broken up into varieties.

He wrote:

> Undoubtedly large genera are partly large because they are widely distributed & have representative species in different countries. (DAR 205.2:111)

This hypothesis has a very definite history in Darwin's thought. From the 1830s Darwin had sought to make some sort of statement about mundane genera and species. He produced a number of indefinite and contradictory statements until 1845, that is after the *1844 Essay* was written, when he found in Swainson the proposition that by typical genera, systematists meant large genera and that large genera were wide ranging. There is a rich correspondence between Darwin and Hooker in 1845 that reflects Darwin's attempts to have these propositions confirmed by Waterhouse and Westwood. From that time the relationship between wide range and large genus became, I think, something of an *idée fixe* that Darwin turned over in his mind.

Hand in hand with wide range, went polymorphism. He wrote:

> to explain why the species of a large <<(& consequently polymorphous)>> genus, will hereafter probably be a Family with several genera, we must consider, that the species are widely spread & therefore exposed to many conditions & several aggregations of species: . . . (DAR 205.5:151)

Implicit in this proposition is Darwin's long-held view that varieties are incipient species. Darwin began the calculations to test these hypotheses in January 1855. Using Hooker's *Flora of New Zealand* (1853–1855), he calculated the number of species that present varieties, and he found more such polymorphic species in large genera than in small genera. He applied calculations of this type to an ever expanding number of botanical works through early 1857. From August 1855 he began calculating the number of species with wide geographic ranges, and found more wide-ranging species in large genera than in small genera. Again he applied this calculation to many botanical works through early 1857. Throughout this undertaking Darwin's calculations confirmed his expectations. In July 1857, John Lubbock informed him that his calculations were in error because he had not precisely defined small and large genera (*LL* 2:104). He then repeated all of his laborious calculations on large genera, which took him well into 1858. He concluded that the new calculations also confirmed his expectations.

I hope that the first point of my threefold thesis is evident from the foregoing: Darwin's calculations only confirmed the hypotheses he held in November 1854. Furthermore, most of these hypotheses were long-held constructs that had implications for divergence. Which brings me to my second point. Out of this well-seasoned timber Darwin built the following description of divergence:

> Hence, <small,> genera will be local <owing to> — their origin <from

> common parent>; & small genera (. . .) certainly, from Schoenherr, are local in proportion of 215:52 (. . .). As to make species is slow work, & [to make] genus increase to <considerable> size much time would be required, hence as Forbes says [they] wd be local in their origin in past time; the species wd extend over continuous spaces in area & time. But it wd generally happen during the time necessary to make a large genus, that geographical mutations & chance accident wd disperse genera & then [the] very fact of the genus having become large in one area, we may suppose wd give it some better chance in another area, & thus the genus wd get bigger & bigger, and certainly most large genera are widely extended. When a genus began to fail & die out, if large, it wd leave probably a few species in distant quarters of the world: Hence this would be another cause of small genera: these wd be aberrant [.] (DAR 205.9:303–304)

It is clear here that Darwin has taken biogeographical patterns, demonstrable in the present, to represent the historical stages of divergence. Biogeography supplied what the inherently imperfect fossil record could not: a coherent historical narrative. Stephen J. Gould (1982) has observed that the argument in most of Darwin's so-called minor books is covertly structured by historical reasoning. He views these lessons in historical reasoning as among Darwin's most lasting contributions. He also recognizes three categories of historical reasoning in Darwin, which are distinguished by the amount and nature of available evidence. Gould describes one form of historical reasoning as follows:

> If rates are too slow or scales too broad for direct observation, then try to render the range of present results as stages of a single historical process. (1982, p. 386, n. 1)

This, I think, is exactly what Darwin did in November 1854. He transformed biogeography into a historical science to "prove" divergence. Moreover, there is evidence that Darwin was methodologically self-aware that he had found a way to read the past from the present. He commented:

> I am inclined to think that it is very curious how similar all laws of relations between organisms separated by time & space; . . . (DAR 205.9:252)

And on one of his undated slips he noted:

> Space & time analogous. (DAR 205.9:360)

This methodological self-awareness in 1854 strongly echoes the early passage in the first transmutation notebook where Darwin wrote:

> . . . as we see them in space, so might they in time — (*B* 17)

Shortly after penning this remark Darwin sketched his first branching diagrams of the tree and the coral of life (Kohn 1980, pp. 94–95). What separated Darwin of July 1837 from Darwin of November 1854 is that, by 1854, he had identified and hoped to quantify those particular patterns in space that showed what species might become in time. It was his specific biogeographic hypotheses, reworked over years and clarified in dialogue with his self-chosen colleagues, that gave substance to the historical analogy between space and time.

But the goal of Darwin's biogeographic work in November 1854 was not just to show that the history of life was divergent. It was to show that this history accounted for the hierarchical, hence tree-like, natural classification. His goal was, as he wrote:

> . . . [to] explain why the species of a large genus will hereafter probably be a Family with several genera . . . (DAR 205.5:151)

It was to show that the theory of descent

> . . . give[s] the diverging tree-like appearance to the natural genealogy of the organised world. (DAR 205.5:149)

It was to show that natural classification was a "natural genealogy".

IV. Speciation and Natural Selection

What further distinguishes the November 1854 episode is Darwin's determination to find an explanatory mechanism for divergence. This brings me to the third and final point of my thesis. On the same paper as the Schoenherr calculations, Darwin wrote a long note under the title "Theoretical Geographic Distribution". Here he addressed the problem of speciation. As I have argued elsewhere, Darwin in July 1837 elaborated two models of species formation: a phyletic model appropriate to a continuous range and a geographic-isolation model appropriate to islands (Kohn 1980, pp. 88–93). The phyletic one came close to sympatric speciation in as much as Darwin recognized that this model would be bedevilled by blending inheritance. As Sulloway (1979) has shown Darwin came to strongly favor the geographic isolation model into the early 1850s.

But in November 1854 Darwin took a fateful step. He wrote:

> When a species breaks & gives rise to another species, the chances seem favourable (. . .) to its giving birth to others. [No doubt here comes in question of how far isolation is necessary, & I shd. have thought more necessary than facts seems to show it is] In fact there never can be isolation for the parent forms must always be present & tend to cross & bring back, to ancestral form; it will *always* be a struggle against crossing, & will require either vigorous selection or some isolation from habit, fewness

> [,] nature of country to separate] Hence, small genera will be local <owing to> their origin <from common parent>; . . . (DAR 205.9:303–304)

Here we have Darwin firmly turning his back on the strict necessity for isolation in favor of pure sympatric speciation. We can feel the tense strain of this movement as he wrote, "I should have thought [isolation] more necessary than facts seems to show it is." In fact he establishes a dichotomous cleft. Species may be multiplied *either* by vigorous selection, with no isolation, or by various indirect but effective isolating barriers: behavioral shifts, low population density, partial topographical barriers. I shall have more to say presently about this dichotomy, to which I attach considerable importance. But our attention should now be focussed on the implications of pure sympatric speciation, where intense selection is seen to be as powerful a force in breaking species as a mountain chain or an ocean. It is clear from the text that the immediate conditions for Darwin's shift away from isolation is his biogeographic work on the proliferating species of small nascent genera. As he comes to see the species of local genera as the primary locus of divergence, he comes to see small locales with no chance of geographic isolation as the primary sites of speciation. Appreciating full well the swamping effect of crossing, he is forced to invoke vigorous selection as the only effective countervailing force. But more important, this line of thinking leads Darwin to look for the local, hence ecological, conditions that favor vigorous selection. The focus of his attention goes to the biotic interactions of assemblages of organisms in small and uniform areas. He writes:

> It is indispensable to show that in small & uniform areas there are many Families & genera. For otherwise we cannot show that there is a tendency to diverge (if it may be so expressed) . . . (DAR 205.5:149)

In other words, his attention goes to the ecology of crowding. It is there that he expects to find the reason for the "tendency to diverge".

This is as far as Darwin went in November 1854. The characteristic three-tiered structure of the principle of divergence is in place. Biogeography allowed him to reconstruct the history of life as divergent, which allowed him to construe branching natural classification as a consequence of divergence. That is the first tier. His biogeographic work on local genera focussed his attention on speciation in a locale without isolation. The result was sympatric speciation by vigorous natural selection. That is the second tier, which led him to look directly for the ecological conditions where vigorous selection would prevail, namely crowded small and uniform areas. That is the third tier. He has conceived an integrated structure that is unified by natural selection. He has yet to complete the structure by the ecological division of labor. But he knows what he wants and he knows where to look for it.

V. Division of Labor

The missing element that breathes life into this structure is found in a note dated 30 Jan 55:

> On theory of Descent, a *divergence* is implied & I think diversity of structures supporting more life is thus implied . . . I have been led to this by looking at a heath Thickly clothed by heath, & a fertile meadow, both crowded, yet one cannot doubt more life supported in <second> than in first; & hence more animals are supported. This is not final cause, but more [a] result from struggle, (I must think out this last proposition) — (DAR 205.3:167)

The idea here is very simple: as a result of struggle, more life can be supported in a meadow with its diverse flora than in the monoculture of a heath. He does not call this idea the ecological division of labor. Instead this is the idea that he later compared to the division of labor. The label was not applied until September 1856. To mistake the labeling for the conception would, I believe, be a misinterpretation of Darwin's developmental process. In my view the principle of divergence was structured in November 1854, including the form of the solution, and by or before January 1855 Darwin had his "keystone".

Conclusion

On this reading the switch to sympatric speciation was a watershed. I will conclude by returning to Darwin's dichotomous views on speciation. We saw that as an alternative to what became species formation by the principle of divergence, Darwin recognized various forms of isolation. In *Natural Selection*, indeed in the section on divergence, Darwin discusses the conditions for speciation and he opposes natural selection to isolation. But it seems to me he has convinced himself that, as Sulloway (1979) concluded, some form of isolation is almost always necessary. He adds complete and partial geographic isolation to the ethological and habitat barriers he recognized in November 1854. He also diagnoses the degrees of isolation required according to the breeding system and mobility of the organism. Animals that are highly mobile and freely crossing require the most intense isolation. Plants that do not cross for each birth and are sessile but may hold a ground by proliferating rapidly require less isolation, but, of course their breeding system and habit are forms of isolation. But even in this case he says:

> I can well believe that a small body of any selected variety might be more quickly formed & hold their own against the ill effects of crossing,

> without being completely isolated. Though in such cases, isolation, at least partial isolation at first, would be favourable to their natural selection. (*Natural Selection* p. 256)

The simplest way of putting the situation is that Darwin wanted to have his cake and eat it. He wanted speciation by natural selection alone, but he was in fact a rather woolly isolationist. The reason he never resolved this internal contradiction is plain. The principle of divergence, which he valued for the unification it gave his theory, was grounded both conceptually and, perhaps more important, developmentally on sympatric multiplication of species by vigorous selection.

It was John Keats who defined that *negative capability* which marks men of achievement as "capable of being in uncertainties, mysteries, doubts, without any irritable reaching after fact and reason." Darwin's scientific achievement owes something to a robust negative capability. But his gift for creative contradiction certainly confused his followers. Mapping the conflict in the reception of Darwin's theory, at least among those English naturalists who considered themselves Darwinians, we find that the lines of demarcation follow the internal lines-of-cleavage formed during the development of Darwin's thought. The late nineteenth century found Romanes and Gulick pitted against Wallace over opposing resolutions of Darwin's contradiction, with Wallace championing selectionism and Romanes and Gulick laying the ground for isolationism, the dominant modern view (Lesch 1975). Ultimately, the issues I have been discussing suggest that the structure of post-Darwinian debate reflects the dichotomous structure of Darwin's thinking. There is no intention here to canonize Darwin. Just as in a Moebius strip there is only one side, so in the history of science there is only reception within scientific communities. But to understand that public critical dialogue through which knowledge grows, the case of Darwin's principle of divergence shows we must attend, with careful scrutiny, to the internal dialogue that is individual intellectual development.

9

DARWIN'S INTELLECTUAL DEVELOPMENT (COMMENTARY)

Giuliano Pancaldi

If I were to give a title to my comments on the papers by Sulloway, Schweber, and Kohn I would suggest "Disciplines to work by". To explain what I mean, I would add that I want to make a plea for a discipline approach to the study of Darwin's intellectual development. "Disciplines to work by" is of course an allusion to David Kohn's well-known essay entitled "Theories to work by" (1980), which alludes in its turn to a crucial sentence in Darwin's *Autobiography* (p. 120). That I have substituted "disciplines" for "theories" is connected with the point I wish to make here. It is my contention that, taking as a vantage point recent scholarly contributions on Darwin's theory as it developed in Darwin's own mind, we can profitably look again at what was going on around Darwin. I suggest that scientific disciplines and discipline boundaries are the appropriate frame of reference for this purpose,[1] and that the work of Dov Ospovat (1981) and the paper by Silvan Schweber in this volume have already shown that the perspective is promising.

When discussing the "modern synthesis" that established the theory of evolution current in our century, it is readily agreed that it resulted from the cooperation of different disciplinary traditions, which we call genetics, systematics, embryology, paleontology, and so on (Mayr and Provine 1980). When considering the "old" synthesis that Darwin himself achieved, it is not sufficiently emphasized that it also derived from the confluence of different disciplinary traditions, indeed from a still wider range of disciplines. I am not implying, of course, that Darwin scholars ignore such obvious facts as the relevance of theoretical debates within taxonomy or embryology for the development of Darwin's thought. I contend that the sorts of stimuli or constraints that disciplines as such exerted on Darwin's intellectual development have not been fully realized, except perhaps in the case of geology.

It is clear and Sulloway's paper confirms this, that Lyell's view of theoretical geology encouraged Darwin to put forward bold speculations in his early geological work. As Sulloway has remarked, the same did *not* happen with zoology or with botany during the *Beagle* voyage (see also

Sulloway 1982). It is also clear, as Sandra Herbert has shown in her introduction to the *Red Notebook* (1980, p. 11), that confrontation with the London zoologists' professional opinion was crucial in Darwin's conversion to transformism in the Spring of 1837. In my attempt to make a plea for a discipline approach to the study of Darwin's intellectual development, I shall say something first about the tradition Darwin was confronted with in systematic zoology in 1837. I shall then add some remarks on a much more controversial discipline, also involved in Darwin's synthesis of the following year: political economy.

Recent research on what was going on around Darwin in those years enables us to say something new on the status of the species question among systematic zoologists. I rely on the evidence offered by the works of Martin Rudwick (1976, p. 207 ff) on Richard Owen, of Toby Appel (1980) on Henri de Blainville, of Pietro Corsi (1978) on the circulation of French transformist ideas, and of myself (Pancaldi 1983) on Italian zoologists. It seems to me that these works should counsel caution in characterizing the leading attitude on species among zoologists in the 1830s and 1840s as "creationist" or "essentialist". Rudwick and Appel have shown, independently of each other, that *a third alternative* to creationism and transformism was represented by the two different but highly representative figures of Richard Owen and Henri de Blainville. Corsi has pointed out the many channels through which Lamarckian transformist ideas circulated in Europe even before 1830. My work on the provincial but receptive Italian scientific community shows that at least three varieties of transformism were well represented around 1840. One was the "moderate" transformism supported by Isidore Geoffroy Saint-Hilaire in his works on variation and monsters. It advocated a reform of the system of classification based on a drastic cut in the number of "true" species, and the identification of other forms as varieties produced according to certain laws. The second sort of transformism was connected with work on classification and embryology by the German and British naturalists, directly or indirectly inspired by *Naturphilosophie*. The third was the "radical" transformism stemming directly from the Lamarckian heritage.

Thus there is growing evidence that a significant part of the zoological community around 1840 adhered either to the "alternative" represented by Owen and Blainville, or to the varieties of transformism just mentioned. Committed transformists were no doubt a small minority, but committed creationists were probably much less numerous than is usually claimed.

This does not amount to re-exhuming the old story that evolution was "in the air" when Darwin entered on the scene, nor to claiming that Darwin's conversion was somewhat "easier" than the psychologists concerned with scientific creativity would admit. On the contrary, a discipline approach shows that Darwin had to overcome still greater, though different obstacles within systematic zoology. These were represented, for example, by the zoologists' firm resolution to avoid too broad questions on species, raised by both creationism and transformism, from becoming the focus of attention of their discipline. A resolution that certainly reflected the creationist preoccupation of some, but also the empirical, professional attitude of many.

In the light of Sulloway's research, one can guess that Darwin was early attracted by geology, while being somewhat repelled by zoology, precisely because of the different attitudes toward broad theorizing in the two disciplines. Remember Darwin's explicit sentence: "there is so much larger a field for thought [in Geology] than in other branches of Natural History" (*LL* 1:263). We may also guess that the hope of changing this situation made it particularly challenging to him to enter biology, where the sort of revolution he attributed to Lyell had not yet been accomplished. Certainly he engaged in a lifelong battle to reconcile his broad evolutionary views with the tradition of empirical, systematic zoology.

I am suggesting that the sorts of questions raised by Darwin's confrontation with disciplinary traditions such as systematic zoology should be given a more prominent place in the study of Darwin's own intellectual development. They might find their place, perhaps, in Gruber's discussion of Darwin's creative work, where Darwin's "evolving organization of purpose" is dealt with, or in Gruber's description of Darwin's "network of enterprise" (Gruber 1981b, p. 312 ff). I am confident that Gruber will be sympathetic to this. He has recently remarked that Darwin's attitude as an "eager and skilled collaborator has not yet been fully told" (1981c, p. xiv). And of course the collaborative side of a scientist's activity is one connecting him with the "disciplines" I am talking about. Hence Sulloway should perhaps stress, even more than he does, that in Darwin's personal contacts with Henslow his entire connection with British, or at least Cambridge natural history, was involved.

In brief, a look at systematic zoology around 1840 suggests that "creationism" or "essentialism" are rather vague labels to be adopted without qualification as the starting-point of Darwin's conversion. It also suggests that established, scientific disciplines are too serious and self-imposing entities to be left on the margin of our historical reconstructions. That systematic zoology was a well-established discipline in the 1830s and 1840s nobody will dispute, I think. More controversial is the status and relevance of another discipline: political economy.

It seems to me that David Kohn (1980) has definitely shown that Malthus's views did play a crucial role in Darwin's path to natural selection. And Silvan Schweber (1980) has convincingly depicted the communications with political economy open to British and French naturalists of the time. As everybody knows, however, the point of Darwin's link with political economy is still one likely to raise controversy, or "grave philosophical problems", as Antonello La Vergata puts it (present volume). Historians from different quarters fear, or else are eager to proclaim, that to acknowledge the connection with political economy is like spreading the clouds of ideology over Darwinian science.

Comparative discipline history may help to disentangle the controversy, by showing, first, that political economy enjoyed in the 1830s a much higher scientific status than we are inclined to accord it today. One recent historian, Maxine Berg, has given a good portrait of the advent of political economy as a well-established discipline, pointing to its wide popularity in Britain during the first half of the nineteenth century (1980, p. 32 ff). If some contemporary, primary evidence is required, consider that John Herschel, an opinion-leader in scientific circles, described Malthus in 1831 as "one of the most profound but at the same time popular writers of our time" (1831, p. 12, n.).

Second, and more relevant, discipline history helps to assess the comparative progress of demography within political economy, and of the study of animal populations within biology. As Kohn aptly remarked, in the 1830s the human species "was the only one for which population data and evidence of the effects of competition were available at the time" (1980, p. 145). Frank Egerton's papers on the study of animal population confirm this (1968, 1970, 1977). And of course there was no reason why evolutionists like Darwin or Wallace should refrain from applying data on human populations to animal populations in general.

The application of human demography to biology actually had dramatic effects on the development of Darwin's theory. But the point I am concerned with here is that with Darwin's theory there was a real transfer of knowledge from political economy to biology. This transfer of knowledge was in fact made easier because of the prestige Malthusian views and political economy enjoyed in Britain for ideological and political reasons. These have long since been well illustrated by John Greene (see the essays collected in Greene 1981a) and Robert Young (1969, 1973a). Recent scholarship, at any rate, carries conviction that there was a scientific as well as an ideological side to Darwin's debt to political economy.

I said earlier that it might prove fruitful to approach Darwin's synthesis with a strategy comparable to the one we adopt when considering the "modern synthesis" of our century. There is, however, one obvious difference between the two that is implicit in David Kohn's paper in this volume: while the modern synthesis was accomplished by a number of scientists, Darwin's synthesis was the work of an individual scientist, namely Charles Darwin. Let me conclude, however, with a risky prediction. I hazard the guess that future scholarship, especially that arising from the forthcoming edition of Darwin's correspondence, will show that this difference between the two syntheses is less sharp than it appears today.

Notes

1. In the 1830s and 1840s scientific disciplines were no doubt bodies of knowledge and cultural behavior less structured than they are today. Historians of scientific institutions and of then recently established fields, such as geology, have shown, however, that organized fields of knowledge and professional expertise already played an important role in scientific activity. On discipline history see Lemaine et al. (1976) and Kohler (1981).

10

SPEAKING OF SPECIES: DARWIN'S STRATEGY

John Beatty

I am often in despair in making the generality of naturalists *even comprehend me. Intelligent men who are not naturalists and have not a bigoted idea of the term species, show more clearness of mind.*

(Darwin to Ansted, 27 Oct. 1860. *ML* 1: 175)

Introduction

There is a wealth of secondary literature on Darwin's species concept, covering many different perspectives of the topic.[1] Of the various accounts available, I have always been particularly intrigued by Frank Sulloway's suggestion that Darwin's choice of species concept was guided by "tactical" considerations. Among those tactical considerations was the decision to employ his fellow naturalists' species concept, in order to speak to them "in their own language" (Sulloway 1979, p. 37). Implicit in the suggestion is that Darwin was a member of a fairly clear-cut community of naturalists. In order to communicate with them about natural history, either to agree or disagree, he had to conform to some extent to their language rules, including their rules for using the term "species".

The suggestion is more perplexing when we consider the respects in which "species" definitions of the time were at odds with evolutionary theory. Darwin apparently faced a dilemma: to communicate his theory of the evolution of species to the community of naturalists, he had to conform to their rules for using the term "species", but his theory undermined their definitions.

We can pursue Sulloway's suggestion down another avenue, however. Perhaps it was possible for Darwin to conform to the language rules of his community without accepting its definitions. Perhaps, in particular, it was possible for Darwin to use the term "species" in a way that agreed with the use of the term by his contemporaries, but not in a way that agreed with his contemporaries' definitions of the term. Another way of asking this question is to consider whether historians might not sometimes more fruitfully approach scientific concepts — like Darwin's species concept

— in ways other than via the definitions of the terms associated with those concepts. An alternative is to try to distinguish "what in the world" scientific terms are used to refer to in practice, from what beliefs about those things serve to define the terms. Agreement of the former sort among members of a scientific community might be conformity enough for purposes of intercommunication.

I shall argue that Darwin indeed perceived the difficulty posed by definitional language rules that undermined the theory he wished to communicate. He tried to get around this difficulty by distinguishing between what his fellow naturalists *called* "species" and the non-evolutionary beliefs in terms of which they *defined* "species". Regardless of their definitions, he argued, what they *called* "species" evolved. His species concept was therefore interestingly minimal: species were, for Darwin, just what expert naturalists *called* "species". By trying to talk about the same things that his contemporaries were talking about, he hoped that his language would conform satisfactorily enough for him to communicate his position to them.

Perhaps it is worth emphasizing from the outset that the so-called community of "naturalists", of which Darwin considered himself a member and to which he addressed his theory, is no mere philosophical construct. It is not my concern to delimit either the members or the membership requirements of the group, but just to show that the group was a real and distinct one *in Darwin's mind*. For now, it suffices to consider Darwin's many references to "naturalists" in the *Origin*. Darwin considered himself a naturalist: for example, "When on board the H.M.S. 'Beagle', as naturalist, . . ." (*Origin*, p. 1). He acknowledged the assistance he received through communication with other "naturalists" (p. 2). He cited majority and minority opinions and cases of dissent among "naturalists": for example, "the view which most naturalists entertain" (p. 6), "the very general opinion of naturalists" (p. 149), "in the eyes of most naturalists" (p. 449), "the protest lately made by some naturalists" (p. 199), and "Let it be observed how naturalists differ . . ." (p. 469).

Moreover, Darwin made clear that it was to "naturalists" that the *Origin* was addressed. For example, he lamented, "I by no means expect to convince experienced naturalists whose minds are stocked with a multitude of facts all viewed, during a long course of years, from a point of view directly opposite to mine" (p. 481). But he hoped that "a few naturalists, endowed with much flexibility of mind, and who have already begun to doubt on the immutability of species, may be influenced by this volume; but I look with confidence to the future, to young and rising naturalists, who will be able to view both sides of the question with impartiality" (p. 482). In short, Darwin recognized a group whose members had common interests, and whose members communicated to each other agreements and disagreements concerning those interests. It was a group to which he also belonged, and the group to which he most wanted to communicate his theory of the evolution of species. But communication was a problem.

I. Definitions, Referents, Examples of Practice, and Theory Change

There are some constraints on theory change that make it a wonder that theory change occurs at all. Theory-laden definition is such a constraint. Consider, for instance, the many pre-Darwinian definitions of "species" in terms of immutability. Those definitions not only reflected non-evolutionary theories of natural history, but also served those theories well, making it difficult to communicate alternative theories in the same terms. How, for example, was one to argue for the mutability of species given Buffon's definition, according to which "we should regard two animals as belonging to the same species if, by means of copulation, they can perpetuate themselves and preserve the likeness of the species; and we should regard them as belonging to different species if they are incapable of producing progeny by the same means" (Buffon 1749, p. 10; quoted in Lovejoy 1959c, p. 93). On such a definition, one-and-the-same species cannot possibly change with regard to its "likeness", and that represents a significant constraint on the communication of a theory of the evolution of species.[2]

Other definitions of "species" were at odds with evolutionary theory in yet other respects. Some were compatible with the mutability of species, but at odds with the transmutation, or lineal descent, of species. Consider, in this regard, Cuvier's definition of "species": "a species comprehends all the individuals which descend from each other, or from a common parentage, and those which resemble them as much as they do each other" (Cuvier 1813, p. 120). Cuvier's definition does not rule out, a priori, unlimited change in a species, but it does rule out the possibility that one species should be descended from another. Thus definitions of "species" in the late eighteenth and early nineteenth centuries were at odds with both the mutability and transmutability of species.

Where scientific terminology is so loaded in behalf of received theory, the proponent of a new, contrary theory apparently faces a dilemma. In order to communicate his alternative in the same terms, he must use those terms differently — he cannot use them in the way they are defined without contradicting himself. But to use the terms differently — to use them contrary to the ways they are defined — is to invite the objection that the difference at issue is purely verbal, a dispute about words rather than about the world. How, in other words, can one respect the language rules to which the other members of one's community are subject *and still* communicate to them an alternative view of the world? How, on the other hand, can one communicate to them an alternative view of the world *without* respecting their language rules? Is it possible, perhaps, to communicate an alternative view of the world by respecting *enough* of the language rules of one's community (or by paying them *enough* respect)?

The answer to the last question is yes — a substantive difference can be communicated even though one respects the previous language rules of one's community only in part. The substance of such differences becomes apparent only when we distinguish between "what in the world" scientific terms are used to refer to in practice, and what beliefs about those things serve to define the terms.

Consider a scientist in the position of questioning a non-evolutionary theory about species, where the definition of "species" is loaded in behalf of the very non-evolutionary theory in question. How, according to that scientist, could his rivals possibly be wrong in any substantive sense? The dissenting scientist has at least two options, depending on what *kinds* of language rules he chooses to respect. First, he might respect his community's *definition* of "species", in the sense that he accepts that the term is to be used only to refer to things that satisfy the non-evolutionary definition of the term. He might, however, still object that there is nothing in the world that actually satisfies the definition of the term. So the theory about species may not be altogether incorrect, in the sense that it is true of whatever satisfies the non-evolutionary definition of "species"; but it is still substantially lacking in the sense that it is not about anything whatever in the world.

The scientist might also choose *not* to respect his community's non-evolutionary definition of "species", in the sense that he *rejects* that the term is to be used only to refer to things that satisfy the non-evolutionary definition. His grounds for not respecting this language rule, however, may be his respect for another of his community's language practices — namely, *examples* of his fellow community members' use of the term "species". By some sort of mistake, he argues, the other members of his community have used the term "species" to refer to things that do not satisfy their non-evolutionary definitions of the term. So, again, the old theory about species may not be altogether incorrect in the sense that it is true of whatever satisfies the non-evolutionary definition of "species". But the old theory about species, and the theory-laden definition of the term, are substantial mischaracterizations of things that the community members have actually *called* "species".

In both cases, the dissenting scientist can communicate substantial disagreement to the other members of his community by satisfying, at least in part, their language rules. In the first case, he accepts their *definitions* as rules governing how their terms are used to refer. He disagrees with their belief that there is anything that satisfies the definitions of their terms — he disagrees, that is, with their belief that their terms refer to anything real. In the second case, he accepts *examples* of the way his community uses its terms as rules governing further such use. He disagrees with their belief that the definitions of their terms correctly characterize what they refer to when they use those terms.

Darwin's ploy was more the latter than the former. Rather than use the term "species" in a way that agreed with his fellow naturalists' definitions of the term, he chose to use it in a way that agreed with his fellow naturalists' actual referential uses of the term. This point is obscured by a couple of factors. First, Darwin did not propose an evolutionary redefinition of the term. This decision not to amend the definitional language rules of his community might be perceived as evidence of his acceptance of those rules. Second, that interpretation is reinforced by his many references to the unreality of species, which suggest that he took non-evolutionary "species" definitions seriously, but denied that anything existed that satisfied those definitions.

As for the first "obscuring" factor, Darwin may not have proposed a redefinition of "species", but he also did not recognize any one definition with which all naturalists, including himself, agreed. And as for the second factor, Darwin did not altogether deny the reality of species. He acknowledged the reality of referents of the term "species", as the term was actually used, but simply denied that there was any one definition that all those referents satisfied.

Definitions of "species" are part of the story of Darwin's species concept, inasmuch as non-evolutionary definitions placed constraints on the communication of his evolutionary alternative. But Darwin's own species concept is not to be found in any definition. Once we appreciate the extent to which Darwin tried to get beyond definitions to referents, we can better understand his conceptions of the reality and unreality of species, and we can better understand the substance of the dispute at the heart of the Darwinian revolution.[3]

II. Darwin's Strategy

Given a non-evolutionary definition of "species", and the assumption that definition *determines* reference, the reality of species goes hand in hand with their immutability and non-transmutability. For, given those assumptions, no single real species can change with respect to its likeness, and there can be no real daughter species separate from their parent species. Thus, defenders of non-evolutionary conceptions of species often presented the evolution issue as a choice between the reality or unreality of species. Charles Lyell is a case in point. Consider first the non-evolutionary definition of "species" that prefaced his discussion of the transformation issue in his *Principles of Geology*:

> The name of species, observes Lamarck, has been usually applied to "every collection of similar individuals produced by other individuals like themselves." This definition, he admits, is correct; because every living

> individual bears a very close resemblance to those from which it springs. But this is not all which is usually implied by the term species; for the majority of naturalists agree with Linnaeus in supposing that all the individuals propagated from one stock have certain distinguishing characters in common, which will never vary, and which have remained the same since the creation of each species. (Lyell 1835, 2: 407)

Assuming this definition determines the reference of "species", a real species cannot be modified with regard to its essential characteristics. Lyell himself made explicit the connection between the reality of species and their immutability. The choice of the matter, as he put it, was "whether species have a real and permanent existence in nature? or whether they are capable, as some naturalists pretend, of being indefinitely modified in the course of a long series of generations?" (Lyell 1835, 2: 405; see also Coleman 1962).[4]

Basically the same choice was offered by other non-evolutionists of the time. As William Hopkins put it, in creationist, non-transmutationist terms, "Every natural species must by definition have had a separate and independent origin, so that all theories — like those of Lamarck and Mr. Darwin —which assert the derivation of all classes of animals from one origin, do, in fact, deny the existence of natural species at all" (Hopkins 1860, p. 747).

Apparently in keeping with this choice, Darwin not only defended the mutation and transmutation of species, but also often seemed to deny the reality of species. In the context of Darwin's evolutionism, and in the context of the sort of choice offered by Lyell and Hopkins, that seems a reasonable interpretation of passages like the following: "In short, we shall have to treat species in the same manner as those naturalists treat genera, who admit that genera are merely artificial combinations made for the sake of convenience" (*Origin*, p. 485).

Louis Agassiz also interpreted Darwin — as a "transmutationist" —to be denying the reality of species, and noted a peculiar consequence of Darwin's having done so: "It seems to me that there is much confusion of ideas in the general statement, of the variability of species, so often repeated of late. If species do not exist at all, as the supporters of the transmutation theory maintain, how can they vary?" (1860, pp. 89–90, n. 1). In other words, Agassiz objected, who cares if species evolve by natural selection or any other means if there are no such things as species?

Darwin's purported confusion, and the denial of the reality of species that apparently occasioned it, make sense in light of the assumption that Darwin accepted the non-evolutionary "species" definitions entertained by his fellow naturalists. And the latter assumption makes sense in light of the assumption that Darwin had to respect the definitions of the community of naturalists in order to communicate with them. But as I suggested earlier, Darwin chose another channel of communication with his fellow naturalists. It is still possible to make sense of his apparent denials of the reality of species, given this alternative manner of communication, though the sense is somewhat different.

Darwin's actual strategy of communication is perhaps more apparent in his never-completed manuscript *Natural Selection* than in the *Origin*. *Natural Selection* — Darwin's detailed account of evolution by natural selection, begun in 1856 — was interrupted by Wallace's independent discovery of the same, and Darwin's scurry to get a complete, if necessarily less detailed version of his theory into print. The version he published was the *Origin*. But for more detailed analysis of problems raised in the *Origin*, it is sometimes helpful to consult *Natural Selection*.

In Chapter 4 of *Natural Selection*, "Variation Under Nature", Darwin dealt more comprehensively than anywhere else with the nature of species and his fellow naturalists' conceptions of species. He made significant use of the work of the British botanist Hewett Cottrell Watson, in whose publications Darwin would have found explicit references to the problems of theory-laden language. For instance, in a two-part review of Robert Chambers's *Vestiges of Creation* — which Darwin made much of — Watson twice pointed out the incompatibility between accepted definitions of "species" and the theory of the transmutation of species. First, he warned: "as to the metamorphosis of one species into another, it must be remembered, that the very definition of 'species' comes in the form of a *petitio principii*; since the widest change ever seen, in the descendants of any plant or animal, would only entitle them to the name of 'variety', according to recognized usage of these terms" (Watson 1845a, p. 111). And in concluding the review, Watson reminded his readers that the transition of one species into another would be "a difficult subject to treat, because the very definition of the term 'species', as usually given, involves an assumption of non-transition; so that any cause of real transition — supposing such a case to be adduced — would be set down simply as evidence to disprove the duality of the species" (p. 147). Thus, Darwin might have been made aware of the constraints of theory-laden language on theory change by Watson (or by some other similarly concerned naturalist), if he was not aware of it already.

Watson might not only have brought the problem of theory-ladenness to Darwin's attention, he might also have suggested to Darwin a means of dealing with that difficulty. At any rate, Watson's treatment of the evolution issue is similar in very important respects to Darwin's. In an earlier article, Watson had drawn a distinction similar to the distinction previously discussed between definitions and referential use in practice. It was important to recognize, he pointed out,

> the necessity of distinguishing two kinds of species; namely, those forms which nature appears to have made permanently distinct, and those which are described in books under a supposition that they are so. The former I shall beg here to designate *natural species*; applying to the latter the

> epithet of *book species*. A book species and a natural species may be strictly identical, or one natural species may be improperly divided into two or more species. (1843, pp. 617–618)

By saying that "natural species" were "permanently distinct", Watson meant that natural species were what actually satisfied the accepted non-evolutionary definition of "species". "Book species", on the other hand, were those entities to which naturalists referred (in books) as "species", on the *supposition* that those entities were permanently distinct. Of course, the entities to which naturalists referred (in books) as "species" might nevertheless turn out *not* to satisfy the accepted definition of "species" — i.e., *not* permanently distinct. This point was made increasingly clear in Watson's later works (1845a; 1845b; and especially 1859, pp. 27—64).

Watson's distinction allows a naturalist to communicate a theory of the evolution of species to his fellow naturalists, even when the latter employ a non-evolutionary definition of the term "species". That is, the dissenting naturalist must certainly acknowledge that whatever satisfies a non-evolutionary definition of "species" does not evolve and is not a product of evolution. But he can also maintain that what non-evolutionists *call* "species" are not only mutable, but also related by descent. In short, he can communicate the point that, while "natural species" do not evolve, "book species" do. Indeed, by way of introducing the distinction, Watson referred to the suggestion that the "alleged" species *Primula veris* (cowslips) and *Primula vulgaris* (primroses) were actually related by descent (Watson 1843, p. 617). In other words, what were *called* species in books were sometimes not natural species according to the non-evolutionary definition of "species".

The genealogical relationship of cowslips to primroses was taken up later in the report of an "experiment" that purportedly showed just that. From seeds of a recognized variety of primrose, namely oxlips (*Primula vulgaris intermedia*), Watson claimed to have grown cowslips as well as true primroses, neither of which, he further claimed, could have cross-fertilized the original oxlips (1845b). What is most important about the report is the manner in which Watson communicated the results of his experiment. First, he pointed out that his materials were *recognized* varieties and species — "book" varieties and species. The publication he invoked for this purpose was the *London Catalogue*. And in reporting his conclusion, he made explicit the reputation of his materials: "The conclusion appears unavoidable to me, that a variety of primrose gave origin at the same time to cowslips, to primroses, and to many varieties of these two *reputed* species" (Watson 1845b, p. 218; my italics).

Watson further argued that his results supported the abandonment of the *definition* of "species" (whatever "the" definition was) as a mischaracterization of what are *called* "species": "If we allow the cowslip and primrose to be two species, and yet allow that one can pass into the other, either directly or through the intermediate oxlip, we abandon the definition of species, as usually given, and fall into the transition-of-species theory, advocated in 'Vestiges' " (1845b, p. 219).

Of the two kinds of rules for using the term "species", Watson thus considered the possibility that examples of referential use in practice might prevail over definitions. He actually made that choice in his review of *Vestiges*, where he proposed "to write of 'species' as commonly understood by botanists, without attempting any rigorous definition of the term, which may hereafter be found to represent only a fiction of the human mind" (1845a, p. 142). He intended, in other words, to discuss the transmutability of what botanists *called* "species", rather than to concern himself with whatever satisfied their *definitions* of "species", which might amount to nothing at all. This strategy was remarkably similar to the one Darwin employed.

Watson and Lyell both prefaced their discussions of the evolution of species with discussions of a preliminary semantic issue, namely their use of the term "species". They differed in their means of settling that issue. Lyell conditioned his use of the term on a definition — a non-evolutionary one at that — while Watson conditioned his use on examples of the use of the term by his fellow naturalists. Darwin also prefaced his discussion of the evolution of species with a discussion of his use of the term —a semantic issue he settled like Watson and unlike Lyell. As he explained in *Natural Selection*, "In the following pages I mean by species, those collections of individuals, which have commonly been so designated by naturalists" (p. 98).

Elsewhere, Darwin explicitly objected to non-transmutationist definitions as rules for using the term "species", in favor of examples of established usage. One such objection was raised, interestingly enough, in the context of the primrose-cowslip issue. Reflecting upon Watson's and others' investigations of primroses, cowslips, and oxlips, and upon the consequences of using the non-transmutationist definition in light of those investigations, Darwin reasoned,

> An able Botanist has remarked that if the primrose and cowslip are proved to be specifically identical, "we may question 20,000 other presumed species." If common descent is to enter into the definition of species, as is almost universally admitted, then I think it is almost impossible to doubt that the primrose and cowslip are one species. But if, in accordance with the views we are examining in this work, all the species of the same genus have a common descent; this case differs from ordinary cases, only in as much as the intermediate forms still exist in a state of nature, and that we are enabled to prove experimentally the common descent. <Hence common practice and common language is right in giving to the primrose and cowslip distinct names.> (*Natural Selection*, p. 133)

In other words, proof of genealogical ties between forms that were commonly *called* "species" would be an *argument ad absurdum* against the use of the non-transmutationist definition. Thus "common practice" was given priority over traditional definition as a guide for using the term "species".

The evolution issue was accordingly, for Darwin, an issue concerning the evolution of *species so designated by naturalists*. As he formulated the transmutation issue in particular, "we have to discuss in this work whether *forms called by all naturalists distinct species* are not lineal descendants of other forms" (*Natural Selection*, p. 97; my italics). This is clearly a different issue from "whether forms that satisfy the definition of 'species' are not lineal descendants of other forms." The former version leaves open the question in a way that the latter does not — and in a way that Darwin clearly wanted that question left open.

But Darwin's decision to talk about what naturalists called "species", rather than to talk about what satisfied their definitions of "species", served more of a purpose than just leaving open the question of whether species evolve. It also allowed Darwin to communicate the position that the term "species" was undefinable. In other words, Darwin not only rejected non-evolutionary definitions of "species", he also rejected the idea that the term could be defined at all. As it turns out, his position concerning the undefinability of "species" was part and parcel of his argument for the evolution of species. It is worth considering, briefly, the connection between Darwin's concern about what his fellow naturalists called "species", his position on the *undefinability* of the term "species", and his position on the *evolution* of species.

The discussion so far has aimed at explaining why Darwin dissociated his use of the term "species" from his fellow naturalists' predominantly non-evolutionary definitions of the term. But Darwin went further in rejecting his fellow naturalists' definitions, denying even that there was one definition upon which they all agreed:

> In this Chapter we have to discuss the variability of species in a state of nature. The first and obvious thing to do would be to give a clear and simple definition of what is meant by a species; but this has been found hopelessly difficult by naturalists, if we may judge by scarcely two having given the same. (*Natural Selection*, p. 95)

There was good reason, Darwin believed, for lack of agreement on a definition of "species". The term was simply "indefinable". As he expressed his scepticism about the term's definability to Joseph Hooker:

> It is really laughable to see what different ideas are prominent in various naturalists' minds, when they speak of "species"; in some, resemblance is everything and descent of little weight — in some, resemblance seems to go for nothing, and Creation the reigning idea — in some, descent is the key, — in some, sterility an unfailing test, with others it is not

> worth a farthing. It all comes, I believe, from trying to define the indefinable. (Darwin to Hooker, 24 December 1856. *LL* 2: 88)

Similarly, in the final pages of the *Origin* Darwin urged that, upon adoption of his views, naturalists would "at least be freed from the vain search for the undiscovered and undiscoverable essence of the term species" (*Origin*, p. 485).

To hold such a position, and still to want to talk about the nature of species, one would have to base one's use of the term "species" on something besides a definition of the term. And as we have seen, Darwin did. But why did he ever defend the undefinability of "species" in the first place?

This question has been taken up already by Michael Ghiselin (1969, pp. 89–102) and Frank Sulloway (1979, pp. 36–39), and I shall rely in part on their analyses. In order to understand the point of maintaining the undefinability of "species", Ghiselin argues, we must first take a closer look at what was being maintained. According to Ghiselin, the crucial issue was that there was no way of defining "species" that distinguished species from varieties — no way of defining the difference. Indeed, that was often the context in which the undefinability position was raised. For instance, in *Natural Selection*, Darwin elaborated upon his remarks to Hooker:

> how various are the ideas, that enter into the minds of naturalists when speaking of species. With some, resemblance is the reigning idea and descent goes for little; with others descent is the infallible criterion; with others resemblance goes for almost nothing, and Creation is everything; with others sterility in crossed forms is an unfailing test, whilst with others it is regarded of no value. At the end of this chapter, it will be seen that according to the views, which we have to discuss in this volume, it is no wonder that there should be difficulty in defining the difference between a species and a variety; — there being no essential, only an arbitrary difference. (p. 98)

Before discussing the point of this position, it is also worth taking a closer look at it in terms of the distinctions raised in this paper. Since the definability of "species" and "variety" is at issue here, a position with regard to that issue cannot take for granted any particular definitions of those terms. Communication concerning that issue must instead be based on some other sort of use of those terms — like examples of their use. And that is precisely how Darwin set up his position on the issue. The passage just quoted immediately preceded the announcement that he would use the term "species" to mean "those collections of individuals, which have commonly been so designated by naturalists". He used "variety" in the same manner. And that allowed him to argue against essential differences between species and varieties on the grounds that what many naturalists *called* "species", many other naturalists *called* "varieties". In other words, if we take as species and varieties what naturalists *call* "species" and "varieties", then we must admit that there is no definition of "species" that excludes all varieties, and no definition of "variety" that excludes all species. Apparently, Watson was of great service to Darwin in this regard, listing for him many of those forms ranked species by some naturalists and varieties by others (*Natural Selection*, pp. 102–103, 159, 168–169; *Origin*, p. 48). From such considerations, Darwin concluded,

> in determining whether a form should be ranked as a species or a variety, the opinion of naturalists having sound judgement and wide experience seems the only guide to follow. We must, however, in many cases, decide by a majority of naturalists, for few well-marked and well-known varieties can be named which have not been ranked as species by at least some competent judges. (*Origin*, p. 47)

The significance of this position, as Ghiselin and Sulloway point out, is that it is intelligible on the assumption of the evolution of species, or more correctly, on the assumption of divergent evolution. In turn it supports that assumption. According to Darwin's notion of divergent evolution, the varieties of a species are incipient species in their own right.[5] More specifically, *what are called "varieties"* of species are, in time, transmuted into *what would be called "species"* in their own right. As Darwin stated that notion, and qualified it at the same time, in those very terms:

> Now comes the question, what is the value of the *varieties recorded in Botanical works*? Am I justified in hypothetically looking at them as incipient species? . . . I may here repeat that I am far from supposing that all varieties become converted into *what are called species*; extinction may equally well annihilate varieties, as it has so infinitely many species. (*Natural Selection*, p. 159; my italics)

If what are called "varieties" are gradually being transmuted into what are called "species", then it is no wonder that there are intermediate stages that are called "varieties" by some naturalists and "species" by others. Divergent evolution thus accounts for the fact that there is no definition of "species" that excludes all of what are called "varieties", and no definition of "variety" that excludes all of what are called "species". As Darwin concluded,

> According to these views it is not surprising that naturalists should have found such extreme difficulty in defining to each other's satisfaction the term species <as distinct from variety>. It ceases to be surprising, indeed it is what might have been expected, that there should exist the finest gradation in the differences between organic beings, from individual differences to quite distinct species; — that there should be often the

gravest difficulty in knowing what to call species and what varieties . . . . (*Natural Selection*, p. 167)

Thus, by formulating his position in terms of the evolution of *what naturalists call "species" and "varieties"*, Darwin was not only able to avoid contradicting himself with regard to predominantly non-evolutionary definitions of those terms, but was also able to communicate and defend a position concerning the undefinability of those terms. The latter position substantiated the evolutionary position. So Darwin's strategy was quite well chosen.

Before concluding, I would like to return briefly to Darwin's views on the reality of species. As I suggested earlier, his references to the "arbitrariness" and "convenience" of species groupings might be interpreted as denials of the reality of species. Such an interpretation makes sense in light of the non-evolutionary "species" definitions that Darwin faced. Had those definitions determined his referential use of the term "species", he would certainly have denied that there was anything to which the term referred (see also Ghiselin 1969, p. 92). But that would have left his position on the evolution of species unclear. For if the non-evolutionary definitions of "species" had determined his referential use of the term, and if he had denied the reality of species accordingly, then his theory of the evolution of species would have amounted either to a contradiction in terms, or as Agassiz noted, to a theory about nothing whatsoever.

As we have seen, however, Darwin did not tailor his use of the term "species" to suit pre-Darwinian, non-evolutionary *definitions* of the term. Instead, he used the term in accordance with *examples* of its referential use by members of his naturalist community. But we still have to contend with all those references to the "arbitrariness" and "convenience" of species groupings, and hence with the possibility that Darwin denied the reality of species after all, on some other grounds. In fact, I have already discussed those other grounds. I just discussed the fact that Darwin not only rejected non-evolutionary definitions of "species" as determining the reference of the term, but also denied that *any* definition determined the reference of the term. And *that* is why he viewed the term "species" "as one arbitrarily given for the sake of convenience . . ." (*Origin*, p. 52). So we are still left with that nagging worry whether Darwin's theory of the evolution of species was a theory of the evolution of anything whatsoever.

Ghiselin has also addressed this problem, and has offered a solution to the apparent confusion. His approach is, moreover, very much in accord with the approach taken in this paper. According to Ghiselin, Darwin's references to the arbitrariness and unreality of species pertained only to the species *category*, not to species *taxa*. In other words, Darwin denied the reality of a species category distinct from a genus category on the one hand, and from a variety category on the other hand. But he did not deny the reality of the various species taxa like the cabbage and the radish (Ghiselin

1969, p. 96; and see *Natural Selection*, p. 98). Darwin's theory of the evolution of species was, of course, about the evolution of species taxa rather than the evolution of the species category. So his denial of the reality of the species category did not render his theory domainless.[6]

Basically the same solution can be constructed in terms of the distinctions employed in this paper. That is, Darwin denied that there was a definition of "species" that excluded all of what were called "varieties", or a definition of "variety" that excluded all of what were called "species". But he affirmed the reality of what naturalists called "species" and of what they called "varieties" — of what were given species and variety names. In other words, Darwin affirmed the reality of recognized taxa. And, as we have seen, it was the evolution of these taxa — what were *called* "species" and "varieties" — that was at issue in Darwin's work.

What, then, *were* called "species" and "varieties" according to Darwin? To what in the world did he believe his fellow naturalists were referring when using their various species and variety names? They were referring, Darwin believed, to chunks with the genealogical nexus of life. They did not refer to *one kind* of chunk with their species names and to *another kind* of chunk with their variety names. That was why there was no definition of "species" that excluded all of what were called "varieties", and so on. Nevertheless, their names referred to real genealogical segments in each case.

This raises the question why Darwin did not at least define a joint "species-variety" category in genealogical terms? The reason might have been that he was concerned to avoid distinguishing them in that manner from the *higher* categories. That is precisely what some of the traditional transmutationist definitions had done, distinguishing the species category from higher categories as *the* category whose taxa were genealogical segments. But this placed constraints upon transmutationist theories of the genealogical relationships of species in a genus, genera in a family, and so on. Understandably, that was a constraint Darwin wanted to avoid:

> On the views here discussed, the idea of common descent of all the individuals of the same species . . . comes into play; but it is not confined, as in the ordinary definition, to the individuals of the same species, but is extended to the species themselves belonging to the same genus and family, or to whatever high group our facts will lead us. (*Natural Selection*, p. 166)

So Darwin was more than just content to do without a definition of "species" that distinguished "species" from "variety" and "genus". He was more than just content to talk about the evolution of what naturalists *called* "species". This means of formulating his position provided him not only with common grounds for discourse with his fellow naturalists, but also common grounds for disagreement. Semantic issues thus settled, he sought to convince his fellow naturalists that the genealogical segments they called "species" evolved over time and were connected to each other genealogically:

> Although much remains obscure, and will long remain obscure, I can entertain no doubt, after the most deliberate study and dispassionate judgement of which I am capable, that the view which most naturalists entertain, and which I formerly entertained — namely, that each species has been independently created — is erroneous. I am fully convinced that species are not immutable; but that *those belonging to what are called the same genera* are lineal descendants of some other and generally extinct species, in the same manner as the *acknowledged* varieties of any one species are the descendants of that species. Furthermore, I am convinced that Natural Selection has been the main but not exclusive means of modification. (*Origin*, p. 6; my italics)

Conclusion

In order to communicate any more than a verbal disagreement with members of one's scientific community, it is necessary to respect their language rules, at least in part. But when the community's theory-laden definitions undermine the rival position being proposed, then those particular language rules cannot be respected — some other language rules of the community must be adopted instead as common grounds for discourse. Those other rules may include actual examples of language use within the community.

For instance, Darwin's theory of the evolution of species was undermined by the non-mutationist and non-transmutationist definitions of "species" to which his fellow naturalists adhered. He clearly could not defend the evolution of species, in any of those senses of "species". He could and did defend, however, the evolution of what his fellow naturalists actually *called* "species" — on the supposition that what they called "species" did not satisfy their non-evolutionary definitions of "species". As Darwin explained his use of the term "species", "In the following pages I mean by species, those collections of individuals, which have been so designated by naturalists" (*Natural Selection*, p. 98). And as he formulated his transmutation position in particular, "we have to discuss in this work whether *forms called by all naturalists distinct species* are not lineal descendants of other forms" (*Natural Selection*, p. 97; my italics).

Darwin's decision to talk about what naturalists *called* "species", rather than about what satisfied naturalists' definitions of "species", served another important function as well. It allowed Darwin to make sense of the position that the term "species" was not definably distinct from the term "variety". What he argued, in effect, was that there was no definition of "species" that excluded all of what were called "varieties", and no definition of "variety" that excluded all of what were called "species". This position was part and parcel of Darwin's notion of divergent evolution, according to which varieties are incipient species.

The suggestion that natural history could really get by without definitions of the categories of classification — especially a definition of "species" — is admittedly hard to swallow. Of course, it should be acknowledged that natural history was only temporarily without a definition of "species". The non-evolutionary definitions rejected by Darwin have since been replaced. Definitions such as Ernst Mayr's "biological species concept" and George Gaylord Simpson's "evolutionary species concept" are already so well entrenched as to be considered traditional. Moreover, just as the old definitions reflected the non-evolutionary theories of natural history in which they were employed, the new definitions reflect the version of evolutionary theory generally accepted at the time they were composed (Beatty 1982).[7]

Following the Darwinian revolution, then, theory-laden definitions of "species" were replaced by theory-laden redefinitions. But the apparent inevitability of theory-laden definition should not be overemphasized —especially not to the point of overlooking possible rationales behind dispensing with definitions at particular periods in the history of science. Far from just "getting by" without a definition of "species", Darwin felt that natural history would be liberated by abandoning the search for one — liberated in particular from the constraints of non-evolutionary thinking built into pre-Darwinian definitions of the term.

ACKNOWLEDGEMENTS

This paper is dedicated to the long life of the Systematics and Biogeography Discussion Group at Harvard. I am also indebted to Michael Ghiselin, David Hull, David Kohn, Ernst Mayr, and Frank Sulloway for sharing their insights into the nature of species and the Darwinian revolution.

Notes

1. The list includes Ghiselin (1969, chap. 4), Hull (1967b, 1976, 1978a, 1980 and forthcoming), Kottler (1978), Mayr (1957, 1964, 1972a, and this volume), Sulloway (1979), Vorzimmer (1970, chap. 7), and Beatty (1982). I am greatly indebted to all these contributions except my own, which raised more problems than it solved. The present paper in some ways takes off from, and in some ways corrects Beatty (1982).
2. I do not mean to overemphasize the non-evolutionary, "likeness-preserving" aspect of Buffon's definition, especially to the point of overlooking other important features of the definition. Phillip Sloan (1979) has argued persuasively that Buffon's definition is part of a tradition of historical-genealogical definitions — a tradition inspired in part by epistemological considerations. However, Sloan seems to me to overemphasize the historical, "perpetuation-through-copulation" aspect of the definition to the point of overlooking the static, "likeness-preserving" aspect.
3. The problem of actually distinguishing between what a term is used to refer to in practice, and what satisfies the definition of the term, is a thorny problem indeed. In fact, it is one of the most central issues in philosophy of science today. A solution to the problem would considerably enlighten this paper. But what is most important for the purposes of this paper is that *Darwin* actually thought such a distinction could be drawn. That fact will, it is hoped, become clear in what follows. The distinction, clues to its solution, and

suggestions as to its use in the history of science are discussed in a very clear and very thoughtful essay by Philip Kitcher (1979).

4. That the dichotomy between the reality and mutability of species actually constrained evolutionary thinking has been pointed out on numerous occasions by Ernst Mayr. In one place Mayr refers to this failure to distinguish reality from constancy as "one of the minor tragedies in the history of biology" (1957, p. 2), and in another place as a "violation of scientific logic" (1972a, p. 987). But these epithets obscure the intrinsic place of such language constraints in theory change.
5. The development of Darwin's theory of divergent evolution has been carefully analyzed by Ospovat (1981), Browne (1980), Kohn (this volume), Schweber (1980), and Sulloway (1979).
6. The distinction between species taxa and the species category has received a good deal of attention lately, in the context of discussions of the notion of "species as individuals". According to this notion, the species category is a spatiotemporally unrestricted class whose member taxa are spatiotemporally restricted individuals. See Ghiselin (1966; 1969, chap. 4; and 1974b), Hull (1976, 1978a, 1980), and Mayr (1976d).
7. The newer definitions reflect in particular the reproductive isolation theory of divergent evolution. According to this theory, breeding groups are the units of evolutionary change. Divergent evolution occurs *between* them, not *within* them. To speak of the evolutionary divergence of species is thus to imply that species are reproductively isolated breeding groups. Mayr defined "species" as "groups of actually or potentially interbreeding natural populations, which are reproductively isolated from other such groups" (1942, p. 120). And Simpson defined "species" as "a phyletic lineage (ancestral-descent sequence of interbreeding populations) evolving independently of others, with its own separate and unitary evolutionary role" (1951, p. 289).

11

THE ASCENT OF NATURE IN DARWIN'S *DESCENT OF MAN*

John R. Durant

What a chance it has been . . . that has made a man.

(*Darwin,* E *Notebook*, 68–69)

It is a fact familiar to all historians of science that Darwin was extremely slow to put his most important ideas into print. Having become a convinced transmutationist in 1837, he made such rapid progress over the next few years that he soon foresaw the prospect of writing a work that would revolutionize natural history. Yet it was not until 1844 that he produced an essay that was suitable for publication by his family in the event of his death; and fourteen years later, the unexpected arrival of Wallace's short paper "On the tendency of varieties to depart indefinitely from the original type" found him still hard at work on the definitive version of his theory. Only when faced with the awful prospect of being pre-empted by the younger man did Darwin finally act with a real sense of urgency to prepare an "Abstract" of his views for immediate publication. It is widely agreed that the reasons for this long delay have to do as much with Darwin's cultural context as with the state of his own opinions. The scientific community in early Victorian Britain was largely hostile to the sort of high-level theorizing in which Darwin indulged, and it was particularly opposed to the idea of the natural transmutation of species. In this situation Darwin was driven to playing a waiting game by the obvious absence of a suitably receptive audience for his views. Opting to pursue his heterodox ideas in secret, he developed a private dialogue with a number of key contemporaries who have been described as constituting the "cultural circle" within which the earliest drafts of the theory of evolution by natural selection were written (Manier 1978). By engaging with the members of this cultural circle in the safety of his study, Darwin was able both to advance a theory and to formulate a strategy for "going public" when the time was right.[1]

One of the first casualties of Darwin's developing strategy for the eventual presentation of his views to the scientific community was the sensitive question of man's place in nature.[2] Privately, Darwin was never in any doubt about where he stood on this question. From the very outset, the transmutation notebooks treated man and mind as part of the natural world. For Darwin, man was at once a fertile source of insights into the rest of the world of life and an important illustration of the process of transmutation. In particular, the different human races were an ideal model for the generation of several varieties or species from a common ancestor; human behavior provided valuable clues to the relationship between habit and instinct, and more generally to the role of an organism's experience in the process of transmutation; and man was a crucial test case for the all-important principle of continuity in nature. Within a year of beginning his systematic investigation of transmutation, Darwin had decided that these and related issues deserved separate treatment in a new series of notebooks. The two notebooks on man that were filled between July 1838 and August 1839 (approximately) are among the most important records we possess of Darwin's early work, and their publication in 1974 was a milestone in the recent history of Darwin studies (Gruber and Barrett 1974, pp. 259–381). Yet the bulk of the material contained in these notebooks was carefully excluded from even the earliest and most tentative of the extended drafts of his theory that Darwin wrote between 1838 and 1858 — his *1842 Sketch*, *1844 Sketch*, and *Natural Selection*. In reply to a query from Wallace in 1857 about the scope of his planned work on transmutation, Darwin clarified the strategy that had shaped these early drafts: "You ask whether I shall discuss 'man'. I think I shall avoid the whole subject, as so surrounded with prejudices; though I fully admit it is the highest and most interesting problem for the naturalist" (*LL* 2:109). Darwin was determined that nothing should stand in the way of his argument for transmutation, and two years later this determination led him to exclude from the *Origin* virtually any reference to "the highest and most interesting" problem of all. But in the end he could not leave man out altogether. In the name of honesty, as he later confessed, Darwin wrote what was to become the most over-quoted understatement in the entire literature of evolutionary theory: "Light will be thrown on the origin of man and his history" (*Origin* 1959, p. 757).

After more than two decades of painstaking preparation, this was a strategic masterstroke. In a single sentence Darwin had hinted at the real extent of his theoretical ambitions while providing almost nothing by way of a visible target for his critics to aim at. Of course, this did not prevent the scientific community from taking up the question of man's place in nature after 1859. As Darwin's opponents had no difficulty in discerning the broader implications of his views, so many of his supporters were keen to carry the battle into fields that the *Origin* had so studiously avoided. In the early 1860s Huxley (1863a), Lyell (1863), Wallace (1864), and others opened up the question of the antiquity and origin of man for public debate. But although Darwin applauded his friends' efforts from the sidelines, once again he chose to play a waiting game. In the decade after the publication of the *Origin* he concentrated almost exclusively on defending his theory of transmutation; and only when this argument had been largely won —and, in the process, Darwin's reputation within the scientific community had been enhanced enormously — did he satisfy professional and public interest in his views on man. With the publication of two closely related works, the *Descent* (1871) and the *Expression* (1872), what Howard Gruber has termed the "two grand detours" of Darwin's career were finally over; at last the full scope of his theoretical vision had been revealed (Gruber and Barrett 1974, p. 24).

The interval between Darwin's earliest speculations and his final publications on the question of man's place in nature spans the greater part of his productive life. Throughout this period, observations and reflections on man were an integral part of a steadily developing program for a revolutionary natural history founded upon the naturalistic principle of transmutation. But at the same time the ideologically explosive issue of man's place in nature was sequestered from the core issue of transmutation within a far-sighted strategy for presenting this program to the scientific community. Darwin's private dialogue with his cultural circle; his increasingly confident command of an argument that embraced the worlds of life, man, mind, and morality; and his organization of this argument within a strategic framework that was eventually encoded in the *Origin* and the *Descent*: these are the central themes with which an adequate account of Darwin's views on man must deal. Accordingly, this paper begins with a review of man's place in Darwin's notebooks, and goes on to consider the way in which the major themes of these notebooks came to be presented to the public. It will be argued that, partly because of the heavy self-restraint that he exercised for most of his professional life, the *Descent* reveals more clearly than any of Darwin's other works the structure and scope of his transmutationist program for natural history.

It was Josiah Wedgwood's opinion that a two-year voyage of discovery would do his nephew no harm at all: "The undertaking would be useless as regards his profession," he wrote to Robert Darwin, "but looking upon him as a man of enlarged curiosity, it affords him such an opportunity of seeing men and things as happens to few" (Brent 1981, pp. 116–117). This perceptive remark captures in essence what the voyage of the *Beagle* did for Charles Darwin. Of men and things he saw enough to last a lifetime; and when in 1836 he came to look back upon his experiences it was the sight of true savages in Tierra del Fuego, as well as the beauty of primeval forests in Brazil and the effects of a violent earthquake at Concepcion, which stood out in his mind (*Diary*, pp. 425–430).

Darwin's first encounter with the inhabitants of Tierra del Fuego in 1832 was to stay with him for the rest of his life. At the close of the *Descent*, almost forty years later, he recounted the experience with a freshness that belied the passage of time:

> The astonishment which I felt on first seeing a party of Fuegians on

> a wild and broken shore will never be forgotten by me, for the reflection at once rushed into my mind — such were our ancestors. These men were absolutely naked and bedaubed with paint, their long hair was tangled, their mouths frothed with excitement, and their expression was wild, startled, and distrustful. They possessed hardly any arts, and like wild animals lived on what they could catch; they had no government, and were merciless to every one not of their own small tribe. He who has seen a savage in his native land will not feel much shame, if forced to acknowledge that the blood of some more humble creature flows in his veins. (2:404)

Darwin had not drawn this transmutationist conclusion from his encounter with the Fuegians at the time. In fact, when he had approached the natives in December 1832, his first thoughts had been of the perfectibility of man. Yet as he tried to grasp the significance of the enormous gulf that separated Fuegians from Englishmen, he resorted to an analogy with the world of nature. "I would not have believed," he wrote in his diary at the time, "how entire the difference between savage and civilized man is. It is greater than between a wild & domesticated animal, in as much as in man there is greater power of improvement" (*Diary* p. 119). This parallel between wild and domesticated animals, on the one hand, and savage and civilized man, on the other, was central to Darwin's subsequent accounts of his experience. It embodied his belief that man was a single species, and that human history was, in the main, a passage from barbarity to civility. In addition, it captured his conviction that on Tierra del Fuego he had seen man in a true state of nature. Visiting the region again in 1834, Darwin was moved to pen a gloomy portrait of a people whose skill, "like the instinct of animals, is not improved by experience" (*Diary* p. 213); and two years later he returned to the contrast between wildness and domestication:

> Of individual objects, perhaps no one is more sure to create astonishment, than the first sight in his native haunt, of a real barbarian, — of man in his lowest & most savage state. One's mind hurries back over past centuries, & then asks, could our progenitors be such as these? Men, — whose very signs & expressions are less intelligible to us than those of the domesticated animals; who do not possess the instinct of those animals, nor yet appear to boast of human reason, or at least of arts consequent on that reason. I do not believe it is possible to describe or paint the difference of savage & civilized man. It is the difference between a wild & tame animal: & part of the interest in beholding a savage is the same which would lead every one to desire to see the lion in his desert, the tiger tearing his prey in the jungle, the rhinoceros on the wide plain, or the hippopotamus wallowing in the mud of some African river. (*Diary*, p. 428)

Reading this passage with the benefit of hindsight, we find it difficult to forget that it was written less than a year before Darwin opened his first transmutation notebook. The vision of man as an animal, with his own nature and habits; the interest in the relationship between instinct and reason; and the employment of the familiar analogy between domestication and civilization: all these were to find their place in the later notebooks. Yet Darwin drew no transmutationist conclusions from his encounter with the Fuegians before the Summer of 1837. Up until then, he appears to have been preoccupied with the relationship between savagery and civilization; and even here he seems to have been certain of little except the fact that the one had given rise to the other.[3] There is no doubt, however, that Darwin's interest in the question of man's place in nature was awakened by his first shocking encounter with the inhabitants of Tierra del Fuego. As his views changed in the months after his return to England, so he came to place a new significance upon these primitive people. Within a year of opening his first transmutation notebook, he had begun to compare them, not with the lion, the tiger, the rhinoceros, and the hippopotamus, but rather with the monkey and the ape. The difference was crucial.

Recent research by Sandra Herbert, David Kohn, and others has helped to clarify man's place in the development of Darwin's theory of transmutation, and nothing more than the briefest of reviews will be attempted here. Although the subject of man was not involved directly in Darwin's conversion to transmutationism, it was an integral part of his subsequent inquiries (Herbert 1974, 1977). In the opening pages of the first transmutation notebook, Darwin was already working with a complex theory according to which sexual generation and geographical isolation played key roles in the development of a number of distinct varieties or species from a common ancestor (Kohn 1980). Man was relevant to this theory in at least two ways. First, as a relatively young species that had become differentiated into a number of geographically distinct races, man provided clear evidence for the reality of transmutation.[4] Second, as the only species whose mental processes (as opposed to mere behavior) could be studied directly, man provided unique insights into the role of habit in the process of transmutation. Sandra Herbert has argued convincingly that it was Darwin's increasing interest in the role of behavior in the generation of adaptation that led him to undertake an expanded program of reading in the Spring and Summer of 1838, and that it was in the course of this reading that he encountered Malthus's *Essay on the Principle of Population*. "Thus," Herbert writes, "because of the enormous effect of Malthus on Darwin's work, biology remains permanently indebted to the field of political economy, as it does to the ability and willingness of certain individuals to transgress the boundaries between fields" (1977, p. 216).[5]

It is clear that, from the very outset, Darwin had no hesitation in using his own species as a source of insights into the rest of the world of life.

In this sense, the idea that man was an animal, to be studied and known in the same way as any other, was not so much a conclusion of the theoretical endeavors of the notebooks as it was a precondition for them. But although Darwin took for granted an essentially naturalistic perspective on man, as his inquiry developed he began to explore its wider implications — if people were animals, then they must be the products of transmutation; if they were to become extinct, then perhaps they would be replaced by other, similar beings; and if they were to survive, presumably they would continue to change (*B* 169, 214–215, 227–232). These and similar ideas in the second half of the first transmutation notebook marked the beginning of Darwin's systematic investigation of man's place in nature. Significantly, this investigation was conducted in the form of a private dialogue between Darwin and his imaginary critics. At the heart of this dialogue lay Darwin's dissent from a prevailing anthropocentrism that virtually deified man by setting him apart as a creature possessed of unique qualities not amenable to scientific analysis. Reason, will, consciousness, morality: these and other similar attributes were widely regarded as the distinguishing marks of man. But for Darwin they constituted a direct challenge to a naturalistic view of the world of life as a single domain, characterized by the possession of common properties and powers, and subject to universal natural laws. It was toward meeting this challenge that virtually the whole of his work on man was ultimately directed.[6]

Darwin soon settled upon his general approach to the question of man's place in nature. In a key passage in the second transmutation notebook, he summed up as follows the position that he was to hold for the rest of his life:

> Once grant that species and genus may pass into each other . . . & whole fabric totters & falls. — Look abroad, study gradation, study unity of type, study geographical distribution, study relation of fossil with recent. The fabric falls! But man — wonderful man "divino ore versum coelum attentior" is an exception. — He is mammalian, — his origin has not been indefinite. — he is not a deity, his end under present form will come, (or how dreadfully we are deceived) then he is no exception. — He possesses some of the same general instincts all & feelings as animals. They on other hand can reason — but man has reasoning powers in excess, instead of definite instincts — this is a replacement in mental machinery so analogous to what we see in bodily, that it does not stagger me. — What circumstances may have been necessary to have made man! Seclusion want &c & perhaps a train of animals of hundred generations of species to produce contingents proper. — Present monkeys might not, — but probably would, — the world now being fit, for such an animal — man (rude, uncivilized man) might not have lived when certain other animals were alive, which have perished. Let man visit Ourang-outang

> in domestication, hear expressive whine, see its passion & rage, sulkiness & very extreme of despair; let him look at savage, roasting his parent, naked, artless, not improving, yet improvable and then let him boast his proud preeminence. — not understanding language of Fuegians puts on par with monkeys. (*C* 76–79)

This passage contains a number of basic Darwinian themes — first, the dissent from anthropocentrism; second, the very tentative reconstruction of the circumstances that may have made man; and third, the re-instatement of the principle of continuity by means of the analogy between apes and savages. These themes were central to Darwin's defense of a transmutationist philosophy of man's place in nature, and it is worth considering each of them in turn.

Darwin's opposition to anthropocentrism took a characteristic form. Looking at man "as a Naturalist would at any other Mammiferous animal" (*OUN* p. 42), he set out to demonstrate that there was no difference in kind between animal nature and human nature. This task necessarily involved the criticism of a whole series of conventional dualisms. For example, faced with the familiar contrast between animal instinct and human reason, Darwin argued that animals possess "some slight dash of reason" while people were "creatures of habit" (*M* 70). Throughout the Summer of 1838, Darwin explored the interrelationships between habit, instinct, memory, and reason, in order to show that even the most advanced mental powers were no more than a smooth extension of capabilities found throughout the animal kingdom. Significantly, it was at this time that he adopted what he termed a "materialist" position. Not pretending that he was able to divine the ultimate nature of reality, Darwin insisted nonetheless on the naturalist's right to presume that matter, when appropriately organized, could think.[7] This denial of the dualism between the mental and the physical underlay his use of animal and human expression as the basis for a natural history of the mind. Darwin's method was quite simple. Having identified particular mental and emotional states with their corresponding expressive signs, he traced their descent through groups of related organisms. Thus human feelings such as love, hate, anger, and fear were treated as discrete organic entities (often called instincts), and shown to have clearly discernible roots in the animal world. "The whole argument of expression," Darwin wrote, "more than any other point of structure takes its value from its connexion with mind (to show *hiatus* in mind not *saltus* between man and Brutes) no one can doubt this connexion" (*M* 151). In such ways, Darwin undermined the anthropocentric position, replacing it with the outlines of a unified framework embracing the whole of animate life under the aegis of natural law.

Darwin complemented his theoretical attack on anthropocentrism with an attempt to demonstrate how, as a matter of historical fact, man had arisen from the animal world. Here, however, he soon encountered difficulties. To begin with, he had toyed with the idea that man was a necessary stage in the progressive development of life on earth, but by the end of the first transmutation notebook (*B*) he had given up this notion in favor of the more radical view that man was caught up in the web of time and chance.[8] On this view, there had been room in the economy of nature for a creature such as man, and (other things being equal) if *Homo sapiens* were to disappear, a similar animal might eventually come to take its place; but the details of the process were neither predictable in advance nor, necessarily, discoverable in retrospect. This was the basis for Darwin's exclamation: "What circumstances may have been necessary to have made man!" (*C* 78). As he reflected on the contingencies of human origins, Darwin recognized the need to prescribe his explanatory task with some care. Returning to this theme in the fourth transmutation notebook, he wrote:

> What a chance it has been, (with what attendant organization, Hand & throat) that has made a man. — any monkey probably might, with such chances be made intellectual, but almost certainly not made into man. — It is one thing to prove that a thing has been so, & another to show how it came to be so. — I speak only of the former proposition. — as in races of Dogs, so in species & in man. (*E* 68–69)

Realizing how very slight was his ability to reconstruct the circumstances that had made man, Darwin became preoccupied with establishing the fact of man's animal ancestry, and with illustrating in very general terms what a historical account of this ancestry, were one available, would be like. In this task the role of analogy was crucial — "as in races of Dogs, so in species & in man" — and this brings us to the final theme in the passage under discussion, namely Darwin's re-instatement of the principle of continuity.

The analogy between apes and savages, enunciated clearly for the first time in the C *Notebook*, but repeated several times thereafter, served Darwin well as a substitute for the historical account of man's place in nature that he could not provide. As part of his investigation of behavior, Darwin had begun watching the primates at the London Zoo. This experience had put a new complexion on his understanding of the lives of primitive people. Rethinking some of his experiences aboard the *Beagle*, Darwin came to see savages as symbolic not only of the state of nature but also of the historical links between animals and man. By placing the Fuegians midway between apes and Englishmen, he gave himself a concrete observational basis for the analogical reconstruction of human origins; and at the same time, he tapped a powerful source of cultural imagery with which to convey his unorthodox views.[9] Darwin's awareness of the heuristic significance of his analogy is evident in the following extract from the first notebook on man:

> Nearly all will exclaim, your arguments are good but look at the immense difference between man, forget the use of language & judge only by what you see. Compare Fuegian & Ourang-outang, & dare to say differences so great . . . 'Ay Sir there is much in analogy we never find out'. (*M* 153)

Darwin's commitment to the principle of continuity was as strong as his conviction that civilized man was the most recent and the most advanced product of a long process of progressive development; and both were embodied in this evocative analogy between apes and savages.

Darwin's opposition to anthropocentrism, and his preoccupation with establishing the principle of continuity, gave to his early writings on man's place in nature their most distinctive aspect. For in his concern to bridge the gulf between nature and man he consistently interpreted animals and people in terms of each other. On the one hand, human thoughts and actions were explained in terms of animal instinct, and on the other, animal behavior was explained in terms of human thoughts and feelings. Of course, the result was a convergence of animal and human nature in accordance with the principle of *natura non facit saltum*. In the notebooks on man, Darwin probed beneath the superficial rationality of human life to expose its irrational and impulsive foundations. He turned not only to savages but also to such groups as children and the insane for evidence of the existence of instincts; and he began to sketch out plausible mechanisms for the development of the more complex mental faculties. For example, he argued that the moral sense or conscience had its origins in the interplay between instinct, memory, and reason. Indeed, by making use of the principles of associationist psychology, he was able to suggest how, from comparatively simple beginnings, even the most elevated precepts of morality and religion might have been produced.[10] This was the reductive side of the argument. But even as he lowered man's mind into nature, Darwin raised the minds of the other animals to meet it. Thus, at the same time that human conscience was resolved into simpler, instinctive elements, animals were endowed with a moral sense. Commenting on the behavior of the Wedgwood family's pet dog, for example, Darwin wrote in the first notebook on man: "I feel sure I have seen a dog doing what he ought not to do, & looking ashamed of himself. — Squib at Maer, used to betray himself by looking ashamed before it was known he had been on the table, — guilty conscience" (*M* 24). A month or two later, he returned to this subject: "What difference is there between Squib after having eaten meat on table, & criminal, who has stolen. neither, or both, may be said to have fear, but both have shame" (*N* 25). Here, Darwin's dependence upon a human standard for his analysis of animal expression was particularly clear, for without having known what it felt and looked like to be ashamed it would have been impossible for him to have recognized the signs of canine conscience. This principle applies to the bulk of the work on animal and human behavior in the notebooks. Setting out to locate in animals the seeds of every major human faculty, Darwin inevitably painted much of nature in human colors. This is evidenced by the very terms in which animal behavior was reported. Apes were described as affectionate, passionate, sulky, and despairing; dogs were said to feel courage, shame, jealousy, and joy; and even the lowly wasp was endowed with intellect (*C* 77–79; *M* 23–24, 63, 84, and 149; *N* 2, 44). This descriptive language was an integral part of Darwin's theoretical enterprise, for it was only by simultaneously demoting man and promoting animals that he achieved his naturalistic synthesis. The following is an example of the two tendencies at work in the second notebook on man:

> The tastes of man, same as in Allied Kingdoms — *food, smell* (ourang-outang), *music*, colours we must suppose Pea-hen admires peacock's tail, as much as we do. — touch apparently, ourang-outang very fond of soft, silk handkerchief — cats & dogs fond of slight tickling sensation. — in savages other tastes few. (*N* 64)

Thus was anthropomorphic zoology combined with zoomorphic anthropology in effecting the unification of animals and man, matter and mind, nature and morality.

Darwin was well aware of the anthropomorphism involved in his analysis of the relationship between animals and man. In the second notebook on man he suggested that "arguing from man to animals is philosophical . . .", since man was "a 'travelling instance' a 'frontier instance', for it can be shown that the life and will of a conferva is not an antagonist quality to the life & mind of man" (*N* 49). Darwin had been reading Sir John Herschel's *Preliminary Discourse on the Study of Natural Philosophy*, and he had come across a reference to Bacon's "travelling instances", in which the nature or quality under investigation "travelled" or varied in degree. In such cases, Herschel had argued, the natural philosopher was able to trace "that general law which seems to pervade all nature — the law, as it is termed, of continuity" (1831, p. 188). In other words, by describing man as a travelling instance in nature Darwin was invoking the principle of continuity to justify his extension to other animals of capabilities more commonly associated with man. Man was a limiting case in nature, an extreme example of laws, properties, and powers common to the whole domain of animate life. This was the central assumption, and from it there followed the methodological principle that prior knowledge of human nature — whether derived from observation or from introspection — was a legitimate source of insights for the naturalist. Thus Darwin's anthropomorphism was the corollary of his rejection of anthropocentrism, and this rejection in turn followed from his meta-theoretical commitment to the principle of continuity in nature.

It has been argued that while Darwin's theoretical program involved the closest integration of man with nature, his strategy for presenting this program to the scientific community dictated that the question of man's place in nature be deferred in favor of the central issue of transmutation. This decision cost Darwin dear. The *Origin* was a work of iron self-discipline, and only in its closing paragraphs did it provide so much as a glimpse of the larger naturalistic vision that was its ultimate inspiration. Moreover, it seems that the long years of concealment caused Darwin to neglect and even to devalue his early work on man. In 1864, for example, he wrote to Wallace, congratulating him on the appearance of an article on human origins and offering him the use of his accumulated materials on the subject: "Do you intend to follow out your views," he wrote, "and if so, would you like at some future time to have my few references and notes? I am sure I hardly know whether they are of any value, and they are at present in a state of chaos" (*ML* 2:33). While the terms of this offer must be interpreted with caution (Darwin's relationship with Wallace over the question of priority was always difficult, and in addition he was often disarmingly modest about his achievements), they are clearly worlds away from the early theoretical notebooks. Even at this comparatively late stage, it appears that Darwin had no definite plans for putting his ideas on man into print. On the contrary, his thoughts were fully taken up with the preparation of the *Variation* and it was only when this work was finished that he turned at last to man.

The *Descent* is an enigmatic book. One of the most eagerly awaited and (in retrospect, at least) significant works in the history of biology, its publication was nevertheless something of an anticlimax. Surveying the literature in 1958, Alvar Ellegård noted "a slight tone of disappointment in many reviews of the book" (1958, p. 296); and a similar tone is detectable in a number of more recent historical assessments.[11] Certainly, the *Descent* provided obvious grounds for criticism. Lacking both the elegance and the authority of the *Origin*, it appeared to struggle with a mass of material that was never quite under complete control. Indeed, to many readers it appeared to consist of two completely different works bound together. Taking up this point in his review, Wallace suggested that for his second edition Darwin might consider separating the material on human evolution from that on sexual selection in animals, bringing out two distinct volumes (Wallace 1871, p. 180). But Darwin did not take this advice, and his second edition appeared in 1874 as a single volume containing almost five hundred pages on the courtship and mating habits of animals.

Perhaps even more surprising than the form of the *Descent* was the fact that much of its content, particularly on the subject of man's place in nature, was unoriginal. Only too well aware of this fact himself, Darwin informed his readers at the outset that, had he known earlier of Ernst Haeckel's *Natürliche Schöpfungsgeschichte* (first edition, 1868), he might never have completed his own work: "Almost all the conclusions at which I have arrived," he wrote, "I find confirmed by this naturalist, whose knowledge on many points is much fuller than mine" (*Descent* 1: 4). In fact, Haeckel was only one of many authors who by 1871 had published their views on the antiquity and origin of man and society, and Darwin drew on a large number of them for his own account. From Boucher de Perthes, Lubbock, and Lyell he took evidence concerning the great antiquity of man (*Descent* 1: 3); from Haeckel, Huxley, and Vogt he obtained support for the descent of man from animals (chap. 1); from Lubbock, Maine, McLennan, Spencer, and Tylor he borrowed ideas on the early evolution of society (chap. 2); and from Bagehot, Galton, Greg, and Wallace he drew ideas relevant to the application of the principle of natural selection to mental and social phenomena (chap. 5). In all these fields, then, the *Descent* covered ground with which its better-informed readers were already quite familiar.

The task of criticizing the *Descent* is easy, but it can be very misleading. Certainly the book was not Darwin's most impressive achievement, but neither was it merely a derivative account of human origins tacked together with an apparently irrelevant and inordinately long-winded discussion of sexual selection in animals. On the contrary, it is best described as the missing half of the *Origin*. Complementing the limited naturalism of Darwin's most famous book, it sought to integrate the realms of life, man, mind, and morality within a single compass. In this sense, the *Descent* was nothing less than the fulfillment of the original program of the early notebooks. Of course there were significant differences between the two. The notebooks were a record of the intellectual adventure of a young man at the peak of his power, whereas the *Descent* was the altogether more cautious and less radical product of late middle age. But for all that it lacked both the fire of the notebooks and the finesse of the *Origin*, the *Descent* bore a closer resemblance to Darwin's early naturalistic vision than anything else that he ever published. Its theme was the unity of man with the rest of the evolving world of animate life on earth, and although it was written in the light of the evolutionary anthropology of the 1860s, it followed fairly closely the arguments that Darwin had rehearsed in private in the late 1830s.

In the late 1860s, far more than in the late 1830s, Darwin found it relatively easy to establish man's physical affinity with the rest of the animal world. Following the notorious dispute between Huxley and Owen over the *hippocampus major* earlier in the decade, there had appeared a succession of works testifying to the close anatomical similarities between apes and man. Drawing on these works in the *Descent*, Darwin dispensed with comparative anatomy in a short introductory chapter. He dwelt far longer on the question of mind, however. On this, one of the most controversial issues since the publication of the *Origin*, there was still nothing like a consensus within the scientific community. Although Spencer and others were moving towards a naturalistic psychology, many even among Darwin's inner circle of supporters withheld their consent from this project. In the later 1860s, for example, not only did the Catholic zoologist St. George Mivart campaign with some success on behalf of a sharp distinction between body and soul, but also Lyell refused to accept a naturalistic account of the human mind. Worst of all, from Darwin's point of view, was the fact that by the end of the decade the cofounder of the theory of natural selection had defected from the ranks of the naturalists over this very issue. Wallace's conversion to spiritualism was a major blow to the program that Darwin had hinted at in the closing paragraphs of the *Origin*, which had called (among other things) for a psychology "based on a new foundation, that of the necessary acquirement of each mental power and capacity by gradation" (*Origin* 1959, p. 757).[12] It was to the task of making good this program that the *Descent* was chiefly devoted.

Darwin introduced his discussion of psychology in the *Descent* by reasserting his commitment to the principle of continuity: "My object . . .," he wrote, "is solely to shew that there is no fundamental difference between man and the higher mammals in their mental faculties" (*Descent* 1:35). In accordance with the method of the notebooks, Darwin rested his case upon a judicious blend of zoomorphic and anthropomorphic arguments. Savages, who were said to possess smaller brains and more prehensile limbs than the higher races, and whose lives were said to be dominated more by instinct and less by reason than those of civilized people, were placed in an intermediate position between nature and man; and Darwin extended this placement by analogy to include not only children and congenital idiots but also women, some of whose powers of intuition, of rapid perception, and perhaps of imitation were "characteristic of the lower races, and therefore of a past and lower state of civilisation" (*Descent* 2:326–327). Conversely, each of the major human mental attributes was located firmly in the animal world. As in the notebooks, so in the *Descent*, Darwin drew no distinction between the observation of behavior in an animal and the ascription to it of the appropriate (that is, human) mental or emotional experience. Animals felt pleasure and pain, happiness and misery; they felt jealousy and pride, and were capable of both magnanimity and revenge (1:39–48). All the higher animals, Darwin suggested, possessed similar senses, emotions, and faculties, "though in very different degrees" (1:48–49); and even the most elevated of human capabilities, such as perfectibility, language, and the moral sense, had their analogues elsewhere in nature.

Darwin's program for a natural history of mind is well illustrated by the way in which he dealt with religion. Faced with the difficult task of providing a naturalistic account of religious belief, he turned first to savages in order to discover the simplest and most primitive forms of the phenomenon. Leaning heavily on the work of the evolutionary anthropologists, he suggested that religious ideas originated in people's earliest attempts to understand the world. McLennan had argued that the simplest hypothesis to occur to savages was "that natural phenomena are ascribable to the presence in animals, plants, and things, and in the forces of nature, of such spirits prompting to action as men are conscious they themselves possess" (*Descent* 1:66). Such animism (whose logic was startlingly similar to that of Darwin's comparative psychology), was seen as the basis for a succession of more sophisticated beliefs — fetishism, polytheism, and monotheism. With this conclusion established, Darwin turned to the animal kingdom for more clues. Now that religion had been defined as little more than the tendency to project subjective experience into nature, might not some animals be fairly described as religious? Darwin continued as follows:

> The tendency in savages to imagine that natural objects and agencies are animated by spiritual or living essences, is perhaps illustrated by a little fact which I once noticed: my dog, a full-grown and very sensible animal, was lying on the lawn during a hot and still day: but at a little distance a slight breeze occasionally moved an open parasol, which would have been wholly disregarded by the dog, had any one stood near it. As it was, every time that the parasol slightly moved, the dog growled fiercely and barked. He must, I think, have reasoned to himself in a rapid and unconscious manner, that movement without any apparent cause indicated the presence of some strange living agent, and no stranger had a right to be on his territory. (*Descent* 1:67)

With reasoning like this it is not difficult to turn one's pets into passable philosophers, and once this has been done the transition from animals to man presents few serious problems. A little further on, Darwin suggested that "the deep love of a dog for his master" was analogous to a person's sense of devotion to his God (1:68). Since the beliefs of savages had already been reduced to the level of crude superstition, and the behavior of domesticated animals was now elevated to that of spirituality, Darwin's readers were left in no doubt as to the true origins of religion.

Having established the principle of continuity in relation to the mental, moral, and social capacities of animals and man, Darwin went on in Chapter 4 to reconstruct the probable course of human development. People varied greatly both in body and mind, and they tended to increase beyond the means of subsistence. Hence they were subject to natural selection. This agency, together with other influences such as the inherited effects of habit, had modified man's ape-like ancestors into progressively more upright, more intelligent, and more social creatures. The outlines of Darwin's argument were clear enough, but the details proved far more difficult to establish. As in the notebooks, so now, Darwin found it impossible to specify with any degree of precision the particular forces that had made man. Indeed, he confessed that natural selection was incapable of accounting for many aspects of the human condition (*Descent* 1:152–153). Unlike Wallace, however, Darwin did not use this confession as a justification for abandoning a naturalistic perspective on man. Instead, he resorted to what had hitherto been a rather minor element within his transmutationism, namely the idea of sexual selection. This idea was the keystone of the *Descent*, for it helped to secure the unity of nature and man in a fashion perfectly consonant with the rest of Darwin's theoretical synthesis.

The idea of sexual selection had its roots in Darwin's early cultural circle. In his famous *Zoonomia*, for example, Erasmus Darwin had suggested that the widespread contest among males for the possession of females had as its final cause, "that the strongest and most active animal should propagate the species, which should thence become improved" (1794–1796, 1:503). The same idea was contained in fragmentary outline in the transmutation notebooks, and it occupied a coherent though subordinate place in the *1842 Sketch*, the *1844 Essay, Natural Selection*, and the *Origin*.[13] Sexual selection involved reproductive competition between individuals of the same species and sex. Darwin recognized two quite distinct forms of such competition — "male combat" and "female choice" — and in the *Descent* he described each of these processes with the help of exactly the same analogy that he had used in his defence of natural selection. In the *Origin* he had made much of the comparison between artificial and natural selection. In the *Descent* he wrote:

> In the same manner as man can improve the breed of his gamecocks by the selection of those birds which are victorious in the cock-pit, so it appears that the strongest and most vigorous males, or those provided with the best weapons, have prevailed under nature, and have led to the improvement of the natural breed or species. (1:258–259)

Moving on to the second aspect of the theory, Darwin simply extended the analogy with domestication one stage further. He continued:

> In the same manner as man can give beauty, according to his standard of taste, to his male poultry . . . so it appears that in a state of nature female birds, by having long selected the more attractive males, have added to their beauty. No doubt this implies powers of discrimination and taste on the part of the female which will at first appear extremely improbable; but I hope hereafter to shew that this is not the case. (1:258–259)

Darwin applied the theory of male combat to those species in which the male was larger and more aggressive than the female, or in which he possessed distinctive armor or weaponry; and he invoked the theory of female choice wherever the male was distinguished by color, ornamentation, or song. The difference between these theories was rather great, for while the one was simply an analogical extension of the idea of natural selection to cover the case of intra-specific male competition, the other was a far more literal transfer of the idea of artificial selection into the natural world. According to Darwin, many female animals literally chose their sexual partners on grounds of subjective preference; and what is more, they did so with sufficient single-mindedness to stamp their likes and dislikes upon future generations. He wrote: "It would even appear that mere novelty, or change for the sake of change, has sometimes acted like a charm on female birds, in the same manner as changes of fashion with us" (*Descent* 2:230).

Darwin's preoccupation with the theory of sexual selection in the *Descent* undoubtedly disconcerted many of his readers, yet the theory occupied a central place within his larger program. First, it bolstered the claims of evolutionary naturalism in the face of widespread criticism; second, it reinforced the all-important principle of continuity between animals and man; and third, it provided a versatile, not to say protean, source of solutions to some of the more problematic aspects of man's place in nature. In connection with evolutionary naturalism, sexual selection provided Darwin with a welcome explanation for many phenomena with which natural selection was unable to cope (Kottler 1980, pp. 205–206). In particular, it answered the arguments of those natural theologians who continued to cite the existence of beauty in nature as an objection to Darwinism. For example, the Duke of Argyll made much of the argument that, in the case of the humming birds, "Mere ornament and variety of form, and these for their own sake, is the only principle or rule with reference to which the Creative Power seems to have worked in these wonderful and beautiful Birds"; and he suggested that Darwinian natural selection provided no explanation whatever for such phenomena (1867, pp. 232–234). Darwin's reply illustrates very clearly the importance of the theory of sexual selection within his naturalistic synthesis:

> The Duke of Argyll says, — and I am glad to have the unusual satisfaction of following for even a short distance in his footsteps — 'I am more and more convinced that variety, mere variety, must be admitted to be an object and an aim in Nature.' I wish the Duke had explained what he here means by Nature. Is it meant that the Creator of the universe ordained diversified results for His own satisfaction, or for that of man? The former notion seems to me as much wanting in due reverence as the latter in probability. Capriciousness of taste in the birds themselves appears a more fitting explanation. (*Descent* 2:230)

Here the logic of Darwin's position was completely consistent. Just as the idea of God as cosmic craftsman had been replaced in the *Origin* by the selecting power of nature, so the idea of God as cosmic artist was replaced in the *Descent* by the selecting power of animals. Interestingly, the confusion that had been created in the minds of many readers by the one metaphor extended also to the other (Young 1971). For example, one reviewer of the *Descent* found in the theory of female choice clear evidence of "a cause which will seem to most men more needful of explanation and more worthy of it, than the effect itself." Seeing in the aesthetic instincts of animals clear evidence of the handiwork of God, the reviewer decided that the *Descent* was "a far more wonderful vindication of Theism than Paley's *Natural Theology*" (Anon. 1871, pp. 319–320). This extraordinary conclusion was made possible because the natural theologian used an anthropomorphic analogy to interpret the phenomena of nature in terms of the transcendent will of God, while Darwin used the same device to interpret the same phenomena in terms of immanent powers and laws. It was around this rather tricky distinction that a great many of the Darwinian debates of the late nineteenth century ultimately revolved (Durant 1977, pp. 84–96).

Sexual selection not only answered the natural theologians but also strengthened the links between animals and man. Indeed, the account of female choice in the *Descent* constituted an almost endless series of variations on the theme that animals were far more like people than they might appear. "With respect to animals very low in the scale," wrote Darwin in his discussion of the natural history of mind, "I shall have to give some additional facts under Sexual Selection, shewing that their mental powers are higher than might have been expected" (*Descent* 1:35–36). Darwin consistently described animal reproduction in terms drawn from the world of Victorian courtship and marriage; he compared the sexual ornamentation of many males with the gaudy appearance of savages; and he used a human standard in order to establish the nature of the mental processes that operated throughout the animate world. In a passage on sexual selection among birds, for example, Darwin explained the logic of his position in the following words:

> We can judge . . . of choice being exerted, only from the analogy of our own minds; and the mental powers of birds, if reason be excluded, do not differ fundamentally from ours . . . If this be admitted, there is not much difficulty in understanding how male birds have gradually acquired their ornamental characters. (*Descent* 2:124)

If this be admitted . . . In fact, many of Darwin's contemporaries were unwilling to grant as much as this, and as a result the theory of sexual selection met with considerable skepticism.[14] It is important to notice, however, that the idea of female choice was of a piece with the rest of Darwin's comparative psychology, and indeed with the whole of his evolutionary thought. For in the end it was simply the most overtly voluntaristic interpretation of a fundamentally anthropomorphic analogy between nature and human artifice.

After an enormously detailed survey of sexual selection in the animal kingdom, Darwin turned once again in the closing chapters of the *Descent* to man. Armed with a new and powerful explanatory device, he suggested that many distinctive human characteristics were the combined result of male battle and both male and female choice. Darwin was a master of the art of story-telling; in his hands the theory of sexual selection was made to deliver a series of plausible explanations of what were assumed to be the distinct natures of men and women. Thus, by suggesting that brain as well as brawn had been important in male combat, he accounted for man's superiority over woman in both physical and mental powers (*Descent* 2:328–329); and by simply reversing the roles of the sexes when it came to mate selection, he reconciled the theory of "female choice" (*sic*) with the obvious and widespread subjection of women (2:371–372). Men's superior strength, he argued, had given them the power of sexual choice, and as a result women had become progressively more beautiful. But at the same time, women had retained a degree of sexual choice as well — as was evidenced, for example, by the existence of male adornments such as the beard. In equally ingenious ways, Darwin accounted for temperamental differences between the sexes; for racial variations in hair distribution and color, skin color, and so on; and even for a number of universal human attributes, such as musical abilities and language. To illustrate the enormous power of sexual selection, he compared it with the unconscious human selection that was responsible for the continual transformation of domesticated animals. "Each breeder," he wrote, "has impressed . . . the character of his own mind — his own taste and judgment — on his animals" (*Descent* 2:370). In just the same way, and just as unconsciously, the human race had molded itself down the generations in conformity to its own changing inclinations and ideals.

Darwin had now come full circle. Having begun by applying the model of artificial selection to nature, he had returned to man, rediscovering in human history the very process with which he had commenced his investigation. In the long chain of this argument, the theory of sexual selection was a key link not only between animals and man, but also between past and future. For if it was true that people had brought themselves to their present position — if it was true, in other words, that mankind was quite literally self-domesticated — then the question arose as to what might yet be accomplished by way of further improvements in human nature. Significantly, therefore, it was immediately after Darwin had terminated his lengthy account of sexual selection with the "remarkable conclusion" that mind had played an important part in the progressive development of the higher animals that he went on to consider the implications of his theory for the future of mankind. He wrote:

> Man scans with scrupulous care the character and pedigree of his horses, cattle, and dogs before he matches them; but when he comes to his own marriage he rarely, or never, takes any such care. He is impelled by nearly the same motives as are the lower animals when left to their own free choice, though he is in so far superior to them that he highly values mental charms and virtues. On the other hand he is strongly attracted

> by mere wealth or rank. Yet he might by selection do something not only for the bodily constitution and frame of his offspring, but for their intellectual and moral qualities. Both sexes ought to refrain from marriage if in any marked degree inferior in body or mind: but such hopes are Utopian and will never be even partially realised until the laws of inheritance are thoroughly known. All do good service who aid towards this end. When the principles of breeding and of inheritance are better understood, we shall not hear ignorant members of our legislature rejecting with scorn a plan for ascertaining by an easy method whether or not consanguineous marriages are injurious to man. (*Descent* 2:402–403)

Far from having been idle speculations tacked on to the end of the *Descent*, these ideas were implicit in the very structure of Darwin's thought. They were the final, and perhaps the most obvious, application of the model of artificial selection to the natural world. Unlike his cousin Francis Galton, Darwin saw no practical means whereby to translate the conclusions of evolutionary theory into effective social policy.[15] This disagreement should not be taken as a sign of any fundamental difference between the social philosophies of the two men, however. Darwin believed that the English aristocracy had been made handsomer than the middle classes by means of sexual selection (*ML* 2:34); he was a fierce critic of primogeniture, which he regarded as a disruptive influence in the selective process (*Descent* 1:170); he was a somewhat reluctant advocate of the struggle for existence as a necessary precondition of human progress (1:167–184); and he took comfort from the thought that, having risen to "the very summit of the organic scale", man might go on to "a still higher destiny in the future" (2:405).[17] Thus Darwin's position was perfectly clear. Man was an animal, like any other, and his past development, present condition, and future prospects were alike dependent upon those natural laws that governed the entire domain of earthly life.

From the time of his first encounter with the Fuegians aboard the *Beagle* to his final flirtation with eugenics in the closing paragraphs of the *Descent*, Darwin elaborated his views on nature and human nature within a larger vision of a world continuously active in the generation of new forms of life and mind. This was materialism, and Darwin knew it; but it was a materialism that humanized nature every bit as much as it naturalized man. Far more committed to the principle of continuity than he was to any particular doctrines concerning the ultimate constituents of the universe, Darwin developed his case by moving freely between the domains of nature and human affairs, seeing in each the reflected image of the other. The ideas of artificial, natural, and sexual selection were an integral part of this process of mutual illumination; but so too was Darwin's comparative psychology, which depended upon a characteristic combination of zoomorphic and anthropomorphic analogies. To Benjamin Disraeli's famous question:

is man an ape or an angel? Darwin's reply was clear. But he defended it, not by discrediting the angels, but rather by painting the apes in such human colors that further dispute was made to seem almost futile. As the title of this chapter suggests, Darwin's program involved the ascent of nature in the descent of man.

In his important study of Darwin as a creative thinker, Howard Gruber notes that "to a striking extent, Darwin's thinking about nature seems marked by images drawn from human experience and conduct" (Gruber and Barrett 1974, p. 10). It is extremely difficult to deny the truth of this assessment, but scholars have differed over its interpretation. Gruber himself acknowledges that it is tempting to see the profusion of human imagery in Darwin's theoretical writings as evidence that "human life is the fundamental source of our creative imagery", and that therefore "the general forms of scientific thought are directly constrained by existing social relations which govern the limits of our images of man"; but he rejects this conclusion as being "both anthropomorphic and un-Darwinian", suggesting instead that the lesson to be learned from Darwin's work is "the value of abundant and varied images", and the continual effort to transcend them (Gruber and Barrett 1974, pp. 12–13). This interpretation lies somewhere between two rather more extreme views that have been taken by Darwin scholars. On the one hand, some historians have sought to minimize the significance of anthropomorphic imagery in Darwin's work. For example, Michael Ghiselin has attempted to distinguish between the "misleading language" in which Darwin couched his theories and the content of those theories themselves. "What matters," he has written, "is ideas, not the language in which they are expressed" (1969, p. 240). On the other hand, Robert Young (1971) has pointed to the broader cultural and ideological significance of the metaphor of natural selection; and Edward Manier (1978) has argued that anthropomorphism was deeply embedded in Darwin's metaphysics. Where Ghiselin sees Darwin's rich descriptive and theoretical language as little more than an irrelevant encumbrance — added, as he puts it, for "literary effect" — Young and Manier see it as constitutive of his scientific enterprise. The interpretation offered here is intended to provide a way out of this dilemma. Anthropomorphic imagery occupied a coherent place in Darwin's philosophy of nature, man, and society, and at the same time it served him well as an effective literary device by which to present this philosophy as persuasively as possible. In the terms employed in the opening section of this chapter, anthropomorphism was part and parcel of both Darwin's program and his strategy.

So far as Darwin's program was concerned, it has been argued that the use of anthropomorphic imagery was closely related to the rejection of anthropocentrism. Darwin's commitment to the principle of continuity led him to treat man as a "travelling instance" in nature, and this in turn allowed him to project into nature as immanent properties and powers many of the complex human attributes whose origins he sought. Similarly, anthropomorphism played an important part in Darwin's strategy for presenting his program to the scientific community. Of significance here is the fact that, as he prepared for the public presentation of his views, Darwin placed more and more emphasis upon argument from analogy. The case for natural selection in the *Origin* was organized around the analogy with artificial selection, as was that for sexual selection in the *Descent*. In the end, as Manier has noted, Darwin adopted a story-telling mode of discourse that was well-suited to the task of persuasion, particularly in those cases in which the direct evidence for evolution was rather weak (Manier 1978, pp. 110–111). The *Origin* and the *Descent* were both, among other things, highly effective pieces of naturalistic propaganda, and each depended in their different ways upon the vocabulary of evocative analogies and metaphors to convey their central message.

The problem with describing Darwin's thinking as constitutively anthropomorphic is that this description risks being interpreted as stern criticism. Interestingly, this problem is not so great in the case of zoomorphism, which was an equally important part of Darwin's work. The reason for this discrepancy is surely that twentieth-century biology recognizes a legitimate place for zoomorphism but not, on the whole, for anthropomorphism. In recent years, a number of biologists have published highly successful zoomorphic accounts of man and society, whereas anthropomorphism has been almost universally abjured as what the American biologist William Morton Wheeler once called "a very terrible eighth mortal sin" (1939, p. 47). It need hardly be said, however, that it is unhelpful to employ the conventions of our own day if what we seek is a better understanding of nineteenth-century science. If we attempt to rewrite Darwin's theories in language other than their own, stripping them of all "extraneous" analogies and "unfortunate" metaphors, we stand to lose at least as much in historical perspective as we gain in supposed philosophical clarity. To separate Darwin's ideas from their distinctive terms of reference is not merely to sacrifice context for content but ultimately to distort both in the interests of some ulterior view of science. For Darwin's language reflects some of his most fundamental assumptions about nature, man, and society; it embodies the particular meaning that he attached to his theoretical synthesis; and it points beyond this synthesis to the wider culture in which it was constructed and to which, after so many delays, it was eventually directed. In the last analysis, to say that Darwin's theories were constitutively anthropomorphic is not to criticize them, but it is to recognize that they were constitutively social as well.

Notes

1. From the moment when he opened his first notebook on transmutation in the Summer of 1837, Darwin displayed a sensitivity toward the views held by his scientific colleagues that amounted at times to real fear of persecution. This emerged from time to time in phrases such as "Opponent will say . . ." (*B* 217), "Mention persecution of early Astronomers . . ." (*C* 123), and "I fear great evil from vast opposition . . ." (*C* 202). For good discussions of this subject, see Gruber and Barrett (1974, pp. 35–45), and Herbert (1977, pp. 157–178).
2. Throughout this chapter, the term "man" will be used in the same way that Darwin used it, namely, to stand for the whole of humankind. This principle will also be adopted for other key terms, such as "savage", whose place in contemporary English has become contentious.
3. It is difficult to arrive at a clear conception of Darwin's views on the status of savages during the voyage of the *Beagle*. For example, he appears to have been torn between pity for the unimproved condition of the Fuegians and admiration for the way that "Nature, by making habit omnipotent, has fitted the Fuegian to the climate and productions of his country" (*Diary* p. 213). Similarly, he was ambivalent about the exact nature of the gap that separated savage from civilized people. Sandra Herbert has pointed out that, although in a sense the *Beagle's* return of several Fuegians to their native environment after several years in the company of Englishmen amounted to "an experiment in acculturation", it was one whose outcome provided no simple understanding of the difference between the two peoples (1974, p. 227).
4. The first transmutation notebook contains many entries on the origins of and interrelationships between the different human races. Darwin appears to have envisaged a threefold analogy between natural species, human races, and domesticated varieties. This analogy is never stated very explicitly, but the entries move back and forth between these categories in such a way as to make the thrust of the argument quite plain. Man is a young species, and consequently both he and his domesticated animals have had time to diversify only to a very limited extent; nevertheless, this diversification indicates the general way in which larger-scale transmutation occurs over longer periods of time in nature (see *B* 3–4, 32–34, 93, 119–120, 147–148, 169, 217, and 244). The important point here is not that man and his works were the sole source of insights into transmutation, but rather that they were part and parcel of the larger inquiry. Of course, it should be emphasized that at this early stage of the investigation the analogy with domestication carried none of the selectionist overtones that it was to acquire after September 1838.
5. It is well known that Darwin himself claimed for Malthus an important role in the genesis of the theory of natural selection (see *Autobiography*, p. 120). The question of the relationship between Malthus and Darwin has remained controversial, however, not least because of the wider questions that it raises concerning the relationship between science and its social context. Although Darwin's claim was disputed by De Beer (1960, p. 121), on the basis of a reading of the (incomplete) third transmutation notebook, the recovery and subsequent publication of several missing pages from this notebook (*D* 162–163) has revealed very clearly the essential accuracy of his recollection, and Herbert's judgement may be taken as representative of the current consensus. However, her comment concerning Darwin's willingness to "transgress the boundaries between fields" rather begs the question (to which she is extremely sensitive in other parts of her essay) of the nature of disciplinary boundaries in the early nineteenth century. For the argument that, in turning to Malthus in the late 1830s for insights into the natural world, Darwin transgressed no perceived disciplinary boundary whatever, see Young (1969). For more recent accounts of the nature of Malthus's influence on Darwin see Limoges (1978), Herbert (1971), Bowler (1976b), and Kohn (1980).
6. It is impossible to arrive at an adequate understanding of Darwin's early work on man without taking into account its oppositional form. By making his primary objective the overthrow of the conventional view of human nature, Darwin was led to consider a very particular set of qualities that, in a very real sense, was not of his own choosing. As Greta Jones has pointed out, the result was that he naturalized, not every conceivably significant aspect of human life, but rather just those aspects that were of most concern to his contemporaries (1978, pp. 6–7). This is another

of the ways in which Darwin's social context left an enduring imprint on his scientific work.

7. On this point, Darwin frequently returned to an analogy with the world of physics. For example, in an important passage on habit, in the second transmutation notebook, he wrote: "Thought (or desires more properly) being hereditary it is difficult to imagine it anything but structure of brain hereditary, analogy points to this. — Love of the deity effect of organization, oh you materialist! — Read Barclay on organization!! . . . . Why is thought being a secretion of brain, more wonderful than gravity a property of matter?" (*C* 166). See also *OUN*:37, and 39–41. In the first notebook on man, Darwin again described himself as a materialist (*M* 57).
8. The change in Darwin's thinking on this subject is clear from a comparison of the following entries in the first two transmutation notebooks: "Man in *savage* state may be called species in *domesticated* races. — If all men were dead, then monkeys make men. — Man makes angels" (*B* 169). "Without *two* species will generate common kind, which is not probable, then monkeys will never produce man, but both monkeys and man may produce other species" (*B* 214–215) and "The believing that monkey would breed (if mankind destroyed) some intellectual being though not MAN — is as difficult to understand as Lyells [sic] doctrine of slow movements &c &c" (*C* 74).
9. The notion of savagery was of great ideological as well as scientific significance throughout the eighteenth and nineteenth centuries. In the late eighteenth century animals and man were commonly assigned their places on a continuous scale of increasing complexity, the so-called great chain of being (Lovejoy 1936). This may appear to have been a radical step, but in practice it often had highly conservative implications, since "savages" and "primitive" people generally, and Negroes in particular, were commonly placed between the "civilized" races of mankind and the rest of the natural world (see Jordan 1968, and Bynum 1974a, chap. 1 for further discussion of this point). In the early nineteenth century the idea of the chain of being fell into disrepute, but assumptions about the biological and social inferiority of "savages" persisted. Polygenist anthropologists assigned separate origins to the "inferior" races of mankind, and even the monogenist James Cowles Prichard argued that "Civilized life holds the same relation to the condition of savages in the human race, which the domesticated state holds to the natural or wild condition among the inferior animals (1813, p. 209). This was the image of savagery that had struck Darwin so forcibly during the voyage of the *Beagle*, and that he now brought to the defence of a transmutationist philosophy of man's place in nature. Its broad appeal throughout the nineteenth century is illustrated by the fact that in due course it became incorporated into virtually the entire literature of evolutionary anthropology (see Stocking 1968, chap. 6, and Weber 1974, pp. 260–283).
10. In the first notebook on man, Darwin wrote: "May not moral sense arise from our enlarged capacity yet being obscurely guided . . . or strong instinctive sexual, parental, & social instincts, giving rise 'do unto others as yourself.' 'love thy neighbour as thyself.' Analyze this out, bearing in mind many new relations from language . . . May not idea of God arise from our confused idea of 'ought', joined with necessary notion of 'causation', in reference to this 'ought', as well as the works of the whole world" (*M* 150–151). Associationism served Darwin well as a way of accounting for the development of complex ideas by small steps.
11. In his book on the reception of the *Origin*, Peter Vorzimmer commented that "the *Descent* is undoubtedly Darwin at a disadvantage" (1970, p. 233). A year later, in a volume published to mark the centenary of the *Descent*, Loren Eiseley referred rather apologetically to Darwin's having written it with "tired and shakey" hands (B. Campbell 1972, p. 2). Finally, the most damning assessment of recent years is probably that of Greta Jones. The *Descent*, she writes, "as well as having a derivative character, is confused, self-contradictory and obscure in places" (1978, p. 16).
12. Mivart's most effective work was *On the Genesis of Species* (1871a). Mivart's biographer wrote that, by 1871, he "could maintain with some vehemence that man differed more from the gorilla than the latter did from the dust of the earth" (J. Gruber 1960, p. 40). As for Lyell, he never accepted a completely naturalistic view of transmutation. On 5 May 1869, for example, he wrote to tell Darwin that, "as I feel that progressive development or evolution cannot be entirely explained by natural selection, I rather hail Wallace's suggestion that there may be a Supreme Will and Power which may not abdicate its function of interference, but may guide the forces and laws of nature" (K. Lyell 1881, 2:442). For the argument that it was the questions of man's place in nature that lay

at the heart of Lyell's rejection of evolutionary naturalism, see Bartholomew 1973. "Wallace's suggestion" was first made in a review of the tenth edition of Lyell's *Principles of Geology*. Wallace wrote: "While admitting to the full extent the agency of the same great laws of organic development in the origin of the human race as in the origin of all organized beings, there yet seems to be evidence of a Power which has guided the action of those laws in definite directions and for special ends . . . We must . . . admit the possibility, that in the development of the human race a Higher Intelligence has guided the same laws for nobler ends" (1869, p. 393). This change of heart came as a great disappointment to Darwin (*ML* 2:39–40). For more detailed analysis of the reasons for Wallace's defection from the ranks of the evolutionary naturalists, see R. Smith (1972), Kottler (1974, and this volume), Turner (1974a, chap. 5); and Durant (1979).

13. For a more detailed exposition of Darwin's path to the theory of sexual selection, see Richard Burkhardt (this volume).
14. The most detailed argument of all on the subject of sexual selection was conducted between Darwin and Wallace (see Kottler 1980).
15. The *Origin* had an inspirational effect on Darwin's cousin, and it led him to undertake a life-long study of human nature and of the ways in which it could be improved. (For futher details, see Cowan 1977, and Durant 1977, chap. 5.) Darwin, in turn, admired greatly the first major product of this study, Galton's book *Hereditary Genius* (*ML* 2:41), but he was never a wholehearted supporter of Galtonian eugenics. For example, early in 1873 Galton sent his cousin a paper outlining a eugenic program designed to encourage "a sentiment of caste among those who are naturally gifted" (1873, p. 126). Darwin wrote back, thanking Galton for the paper, but pointing out many practical difficulties. He concluded: "Though I see so much difficulty, the object seems to me a grand one; and you have pointed out the sole feasible, yet I fear utopian, plan of procedure in improving the human race. I should be inclined to trust more (and this is part of your plan) to disseminating and insisting on the importance of the all-important principle of inheritance" (*ML* 2:44).
16. For an excellent discussion of Darwin as a social evolutionist, see Greene (1981a, pp. 95–127).

12

Darwin and the Expression of the Emotions

Janet Browne

In a letter to Thomas Henry Huxley about an advance copy of the *Variation*, Charles Darwin added a short note to Mrs Huxley, the long suffering and redoubtable Henrietta: "Give Mrs Huxley the enclosed," he suggested, "and ask her to look out when one of her children is struggling and just going to burst out crying." What, he wanted to know for instance, did Leonard Huxley's eyebrows do? "A dear young lady near here plagued a very young child for my sake, till it cried, and saw the eyebrows for a second or two beautifully oblique, just before the torrent of tears began" (*ML* 1:287).

Darwin was off on his latest hobby horse, the physical expression of the emotions, and was busy collecting information from as many different sources as possible. Mothers — particularly those like Henrietta accustomed to a life of science — were as knowledgeable in their own field as geologists or horticulturists in theirs, and could be relied upon to provide accurate observations of the faces that their children made. Darwin therefore lost little time in making sure that the Huxleys and other friends received a copy of his latest printed work, a single sheet of questions about expression (Freeman and Gautrey 1972), so that "definite descriptions" of infant behavior could be recorded for his eventual use.

Darwin did not, of course, depend exclusively on the observations of Victorian mothers, although he does seem to be one of the few natural philosophers to have made use of this relatively unexploited area of expertise. The object of his exercise was to publish yet another defense of evolutionary theory, a study designed to show that human behavior and emotional expressions were derived from those of animals. As always, he conducted his own thorough investigation into infant expression, taking his immediate family as experimental subjects. Indeed, it was the birth of his eldest child William in 1839 that first prompted him to make a study of expression, for Darwin responded to fatherhood in the same distinctive way that he reacted to all new phenomena — by sitting down and recording the baby's development as if it were a barnacle or a primrose, turning his private life into a scientific essay, his family into facts. From late 1839 Darwin regularly recorded the behavior, emotions, voluntary and involuntary actions of his growing children, filling the famous M and N *Notebooks* with speculation about the psychology of behavior, instinct, and the will. Later, to these "metaphysical" observations he added anecdotes and descriptions from his correspondents, notes from his own extensive reading, and a series of sketches and photographs illustrating the range of human passions — in all, one of the nineteenth century's most comprehensive surveys of infantile development. Towards the end of his life, much of this material was published as "A Biographical Sketch of an Infant", a charming yet authoritative account of William's early life and development (*CP* 2: 191–200).

Nor did Darwin depend exclusively on observations of children, although these were obviously important to his project, for there were many other interesting sources of expression. Children were significant because their faces displayed the smallest emotional change in dramatic emphasis, and different moods were expressed in an apparently pure, uncomplicated way. But it was also important to know if expressive gestures and facial movements were identical throughout the different races of man, if they were the same in deranged people as in sane, and if they matched the behavior and feelings of animals. Only when all these avenues had been thoroughly explored could Darwin make a general statement about expression and its relation to evolutionary theory. Like all good hobby horses, Darwin's had a long road to travel.

Since the central ideas of Darwin's study of expression were more or less fixed at the outset and remained relatively unchanged throughout the twenty or so years of his active research (Swisher 1967), there is perhaps little need for historians to rehearse the development of his views within this specific topic. Excellent studies of Darwin's early theories of behavior and instinct have already been made by Gruber and Barrett (1974), Swisher (1967), and Richards (1981). The finished, published work is a genuine account of what he thought — and had thought — important, the product of many years' consideration and mature reflection on the problem. Nevertheless the book itself, the *Expression of the Emotions in Man and Animals* (1872), was written rapidly even by Darwin's standards, being composed in four months during 1871 in the gap between reading proofs of the *Descent* (1871) and compiling the sixth and last edition of the *Origin* (*LL* 3: 133–134, 171).[1] Indeed, the *Expression* had originally been planned as a single chapter of the *Descent*, until the mass of material persuaded Darwin that a survey of expression in its own right would dramatically advance his arguments for human evolution. It would confirm — though to a limited extent only, as he fully admitted — that man was derived from some lower animal form, and all human races had descended from a single "parent" stock. By showing in considerable detail that all the chief expressions exhibited by man were the same throughout the world, he could effectively argue that so much similarity could hardly have been acquired by independent means. It was "far more probable that the many points of close similarity in the various races were due to inheritance from a single parent-form, which had already assumed a human character" (*Expression*, p. 361). This single parent would have inherited a full range of emotional expressions from its own animal progenitors, and was capable of gradually acquiring others, which afterwards became instinctive. Studies of children and the mentally ill, coupled with a general survey of the animal kingdom, demonstrated the possible route that this evolutionary history might have taken, and indicated, to Darwin at least, the physiological basis of the majority of expressions. The whole was intended to show that even the most "human" of human characteristics were, at root, derived from animals.

So the key to Darwin's *Expression* is to see it as a sequel to the *Descent*. Knowing that publication of the *Descent* in 1871 would provoke severe criticism, much of which would center on the impossibility of linking man's higher attributes with the animal kingdom, Darwin had already prepared the *Expression* in reply.[2] Indeed, it is essential to read one volume after the other to get to the real meat of his arguments, for much that appears in the *Expression* is only summarized in the *Descent*, and vice versa. The aims of the two works were virtually identical. In both, Darwin intended to demonstrate that "man was derived from some lower animal form" (*Expression*, p. 367), or rather, "man, like every other species, was descended from some pre-existing form" (*Descent* 1:3). The *Descent* was, of course, undoubtedly the more important study, being his magnum opus on man, an answer to the crucial question left untouched by the *Origin*. The *Expression* is clearly a secondary book. It is easy to identify, for example, the places in *Descent* where Darwin had intended to insert sections on expression. In Chapters 3 and 4 he ran through a list of emotions conventionally ascribed to human beings, explaining how they were also displayed by the lower animals. Curiosity, terror, suspicion, courage, rage, and revenge were only a few of his examples, backed up by many anecdotal details from friends and relatives, as well as travel books, and correspondence with Mr Sutton, a keeper at the Zoological Gardens in London, among others. Nor did Darwin ignore the higher attributes such as memory, imagination, attention, and reason. He claimed that animals possessed these gifts in a nebulous form, not always as fully developed as in man, but present all the same. He also insisted that many animals could be self-conscious, could even feel guilt, two faculties that were more usually believed peculiar to humans alone. Even morality, the last great "human" characteristic of them all, could be found in a rudimentary way in animals, particularly in those with social or gregarious instincts. After all, argued Darwin, morality was only social instinct overlain with the effects of long-continued habit and man's own intelligence (*Descent* 1: 148–192, 195–224).[3] No mental or moral faculty was thus special to man alone; the difference was only one of degree, not kind.

Having worked through these important ideas in the *Descent*, it is clear that Darwin intended at first to buttress his case with examples drawn from human and animal expression. Stereotyped facial and body movements displaying rage, fear, or sorrow, for example, were to interlock with Chapter 3, and blushing — being a special case — to exemplify the discussions in Chapter 4 on the evolution of the moral attributes of man.

As the book progressed, however, it must have become increasingly obvious that the *Descent* was taking another, rather different direction. It became a book uneasily divided, containing more about animals than people, and certainly more about sexual selection than evolutionary anthropology or "descent". The text completely belied its authoritative title. The bulk of the second volume thus contained a minutely detailed account of sexual selection in the animal kingdom, an extremely long prelude to the supposed theme of the work. Selection in man, and his origin in some ancestral form, provided the final coda. So the facial expression of human and animal species was an inappropriate digression, and Darwin thought it "better to reserve my essay for separate publication" (*Descent* 1: 5).

In this case he was probably right, for in his book on expression he had the space to ruminate on questions left unasked in the earlier work and the freedom to speculate at greater length on the emergence of human behavior patterns. Much more than the *Descent*, this was Darwin's book on man, his most explicit account of human origins.

In the *Expression*, Darwin arranged his material around three explanatory principles of his own invention. Many gestures and facial expressions were due, he thought, to the force of habit, in that certain behavior patterns that were repeated generation after generation could become "fixed" in the population. Rather like the effects of use and disuse on the physical form of a body, continued habitual movements could become incorporated into the heritable make-up of an animal.[4] Indeed, the more useful or serviceable the habit to the life-style of an individual, the more likely it was, he thought, to become inherited: "movements which are serviceable in gratifying some desire, or in relieving some sensation, if often repeated, become so habitual that they are performed, whether or not of any service, whenever the same desire or sensation is felt, even in a very weak degree" (*Expression* p. 348). One classic example, well known to Darwin and other country gentlemen, was the tendency of young, untrained gundogs to "point" or "set" as if the trait were in some way natural to the breed. After many eons of time, such "serviceable associated habits" would be indistinguishable from genuinely primary instincts.

The second kind of expression was based loosely on the first, although considerably more complicated. Darwin believed that some behavior patterns — particularly those that were instantly recognizable, yet not directly useful to the organism — had arisen in opposition to other, more fundamental experiences. Expressions of delight, with the face wreathed in smiles, clapping the hands, "laughing, talking, kissing", had developed as exact opposites to the behavior of people under great distress. Another such expression, a more clear-cut example in Darwin's opinion, was the external appearance of submission, particularly in the case of domestic cats and dogs. These lowered their bodies, ears and tails down, and turned the head to present a vulnerable part of the neck in order to assure their opponents that they had no intention of being aggressive. Every submissive gesture was the opposite of some typical aggressive response: a flat back instead of an arched one, depressed ears for pricked, the side of the head offered up to be touched instead of being protected. Foreshadowing those like Robert Ardrey who assumed that violent, defensive behavior was more basic to animals than the niceties of social intercourse, Darwin believed that attacking behavior was a primary instinct, and that when the need for appeasement became crucial in their social development, animals simply reversed the symbols to get the new message across, responding with a complete set of opposite expressions (*Expression*, pp. 50–57). He called this the principle of antithesis:

> Our second principle is that of antithesis. The habit of voluntarily performing opposite movements under opposite impulses has become firmly established in us by the practice of our whole lives. Hence, if certain actions have been regularly performed, in accordance with our first principle, under a certain frame of mind, there will be a strong and involuntary tendency to the performance of directly opposite actions, whether or not these are of any use, under the excitement of an opposite frame of mind. (*Expression*, p. 348)

For his third principle of expression Darwin invoked the direct action of the nervous system, believing like many of his contemporaries that there was a finite quantity of nervous fluid or "nerve-force", which, when generated in excess, tended to overflow in certain predetermined directions.[5] When the brain or "sensorium" was strongly excited, Darwin thought that extra nerve-force was transmitted in definite directions, depending partly on the connections of the nerve cells and partly on habit, and effects were produced in the body that we recognize as expressive. The crucial point was that these effects were judged independent of the will. Under this heading Darwin included several purely neuro-muscular responses, such as trembling, but conceded that most nervous reactions were combined with one or more of his serviceable habits. For example, a person in mental agony will writhe, gnash his teeth, stare wildly, and so forth, and the circulation and respiration are generally much affected, as if the distress was actually physical. To Darwin these struggles were literally an attempt to escape from the cause of the suffering, to move away from the "pain" within. Long-continued attempts to escape would establish the habit of exerting all the body's muscles whenever great suffering was experienced, and excess nervous fluids would overflow into the face and chest[6] to be relieved only by grimaces and contortions, perspiration and screams:

> The frantic and senseless actions of an enraged man may be attributed in part to the undirected flow of nerve-force, and in part to the effects of habit, for these actions often vaguely represent the act of striking. They thus pass into gestures included under our first principle; as when an indignant man unconsciously throws himself into a fitting attitude for attacking his opponent, though without any intention of making an actual attack. We see also the influence of habit in all the emotions and sensations which are called exciting; for they have assumed this character from having habitually led to energetic action; and action affects, in an indirect manner, the respiratory and circulatory system; and the latter reacts on the brain. Whenever these emotions or sensations are even slightly felt by us, though they may not at the time lead to any exertion, our whole system is nevertheless disturbed through force of habit and association. (*Expression*, p. 349)

Darwin found that the majority of human expressions could be explained by a suitable combination of these three basic principles, with additional references to the behavior of other animals, the various races of man, children, and the insane. Like many Victorians, he believed there was a sort of scale of humanity proceeding from the animal kingdom to man by way of monkeys and apes, with primitive races, savages, and idiots occupying some intermediate position.[7] While many scholars and the public at large understood this to be the centerpiece of Darwinism itself, thinking that natural selection actually made the scale a physical reality, Darwin was more ambivalent and took care not to call one man closer to the beast than another. He merely used each set of individuals as an example of what *might* have been the case at some earlier, less sophisticated time. So his studies on expression, complete as they may seem, are not a blow by blow account of the evolution of every twitch and grimace, but rather a reconstruction, a hypothetical story of the stages that may have been passed through on the way to a smile.

Darwin grouped human expressions into six broad categories based on their obvious relationships with each other, describing all the different manifestations of, say, grief, in a single chapter. Repudiating the classification schemes of earlier writers such as Le Brun (who based his work on the movements of the eyebrows)[8] or Sir Charles Bell (the respiratory movements of the chest),[9] Darwin tried to reduce every expressive gesture to the stark outlines of neuro-muscular physiology. The wide staring eyes of astonishment were, he thought, the natural result of attempting to see more of the thing that caused surprise, the curled lip of a sneer a side effect of originally wrinkling the nose at unpleasant odors. It was only by attributing such functional purposes to every nuance of the human face that Darwin could hope to show these expressions as animal in origin; only by describing the raw physiology of expression that he could claim continuity between animals and men. Charles Bell, for example, had insisted there were muscles in the human face without analogue in the animal kingdom designed by the Creator for the display of specifically "human" emotions like morality, shame, or spirituality.[10] If Darwin could demonstrate another, more practical purpose for such facial muscles, a purpose that clearly linked us to the rest of nature, he would weaken Bell's and all other natural theologians' arguments, and strengthen the case for expression as a product of evolution.

Two expressions in particular carried the weight of Darwin's reductionism and set the tone for the rest of his long exposition, expressions that probably interest us, as people, more than any others. Homing in on the very qualities that make human beings feel human rather than like Cartesian machines, he analyzed first the face of sorrow and then the face of joy. Boldly, he set out to explore the neurophysiology, as it were, of Hamlet.

Eyes and eyelids were the key, he thought, to all expressions of joy and suffering. In order to prevent the eyeballs from becoming dangerously engorged with blood under extremes of emotion, particularly with the rapid intake of breath that accompanies a cry of pain or laughter, various sets of muscles were brought into play, the contraction of one causing the contraction of another, each producing part of the overall appearance of either facial expression. For example, a complex train of events lay behind the face of suffering. Distress began with the simultaneous contraction of the orbiculars (muscles surrounding the eyes) and the muscles of the eyebrows, causing wrinkles all round the eyes and a heavy frown. When these were strongly contracted, muscles running to the upper lip also contracted and raised it, producing a well defined fold or furrow on each cheek, characteristic, for instance, of the crumpled expression of a crying baby. As these muscles were drawn up, others pulled down on the corners of the mouth, and the effect of such opposing tension, above and below, tended to give the mouth an oblong, almost squarish, outline. This was the frantic gape of Laocoon reduced to a paroxysm of despair.

Yet Darwin realized that no amount of written description could adequately convey the complexities of any such expression, for faces and their feelings are quintessentially visual — expression, of course, being the prime means of non-verbal communication between individuals.[11] His subject positively demanded illustration. Darwin used more than 200 photographs during his researches, material from Rejlander, Kindermann, Duchenne, and others, and found that this recently improved technique provided him with faces more or less frozen into whichever expression he needed. Although many were actually posed, and the expressions simulated rather than spontaneous (*Expression*, pp. 14, 23), he nevertheless managed to put the photographs to good use, arranging them in sequence (when the subject allowed) to illustrate the stages of muscular contraction. Thirteen heliotype reproductions (Jay 1980) of grief, despair, and weeping, for instance, argued his case far more effectively than any amount of written material. Three of these are reproduced here.

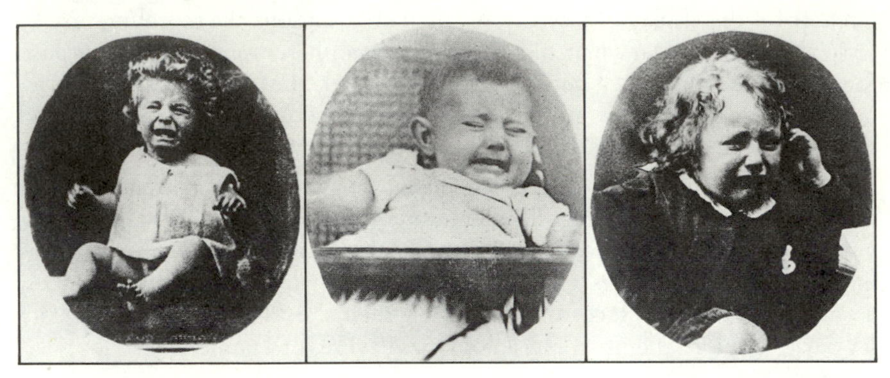

Figure 1. (left) A child screaming without the production of tears, photographed by Oscar Rejlander. The firm closing of the eyelids protects the eyes from becoming gorged with blood, and causes the forehead to contract into a frown. The lips are open and retracted, which makes the mouth characteristically square in outline.
Figure 2. (centre) Another illustration of the same phenomenon, photographed by Herr Kindermann of Hamburg.
Figure 3. (right) Darwin included this photograph by Rejlander in order to show moderate crying in an older child.

To suffer is not always to weep, and Darwin correctly recorded that the screaming of children and adults was not necessarily accompanied by tears. The secretion of tears, he explained, was a purely physiological response to the pressure of contracted muscles around the eyes, and occurred under the most opposite of emotions, and under no emotion at all (*Expression* p. 163). Since weeping was also a trait that emerged only when a child was three or four months old, Darwin suggested it was a learned response, associated with distress solely through habit: prolonged screaming would lead to the gorging of the blood vessels in the eye and this would have led, at first consciously and then habitually, to the contraction of the muscles around the eyes to protect them; at the same time spasmodic pressure on the surface of the eye and distension of the vessels within, would have affected, through reflex action, the lacrymal glands (pp. 169–174). "Finally, through the three principles of nerve-force readily passing along accustomed channels — of association, which is so widely extended in its power — and of certain actions, being more under the control of the will than others — it has come to pass that suffering readily causes the secretion of tears, without being necessarily accompanied by any other action" (p. 176).

The outward appearance of grief was therefore explicable as an amalgam of physiology, habit, and inherited responses, whereas weeping, the most obvious manifestation of sorrow, was more or less a learned reaction, an incidental result as purposeless as the secretion of tears from a blow to the eyes.

With joy, however, the situation was reversed. Laughter was as fundamental to the origin of happy expressions as weeping was superfluous to sorrow. Expressions of delight were founded on man's desire to laugh — to emit sounds, at least — when pleased, and out of this evolved a smile. The physical mechanism was much the same as that of grief, for once again the orbiculars and mouth muscles were called into play, although here the cheek muscles also contracted and lifted the corners of the mouth up and apart. Darwin guessed that this might be caused by a need to have the mouth wide open in order to utter sounds loudly and distinctly (*Expression*, pp. 207–208). Laughter itself was the opposite of a shriek or cry of distress, having short and broken expirations of breath interspersed with long inspirations, where sobbing was based on long, continuous expirations with sharp intakes of breath. It was possible, thought Darwin, that these two very different patterns of sound had evolved in tandem, through his principle of antithesis.

But happiness was not simply the reverse of sorrow. Unlike grief, the face of joy had one peculiar characteristic. It could not be imitated in any realistic manner — a false smile being instantly recognizable as a false smile, no matter what the surrounding circumstances. Darwin discovered this more by accident than design, through his use of G. B. Duchenne's photographs of experimentally induced expressions. Duchenne had studied this topic by galvanizing muscles in the face of an old man whose skin was relatively insensitive, thereby producing various expressions that were photographed on a large scale and published in his *Mécanisme de la physionomie humaine* (1862). Greatly impressed by these illustrations and his meticulous analysis of the contraction of each facial muscle, Darwin had a large number of Duchenne's photographs copied for his own research purposes, many of which were further reproduced in the *Expression*. He showed several of the best plates without a word of explanation to twenty or so people of various ages and sexes, asking them what emotion was being expressed and recording their answers. As might perhaps have been expected, some of the expressions were quickly recognized and described in more or less the same terms, whereas others were more perplexing. In particular, Duchenne's photograph of a smile induced by electricity confused almost all of Darwin's helpers, most of whom, although they identified it as an unnatural smile, inferred that malice or surprise was the stimulating cause. A genuine smile from the same old man was easily recognized as such (*Expression*, pp. 203–204). Duchenne's plates are reproduced overleaf.

From this and other instances of confusion or misidentification, Darwin was able to draw several valuable conclusions. It was a useful demonstration of the fine degree of discrimination possessed by the human eye, for his survey showed that people only recognized expressions if *all* the muscular details were as they should be; without crinkled eyelids, a laughing mouth meant nothing. So the eye learns to "read" faces and stumbles over errors in their syntax. Yet this faculty of instant recognition can be easily misled by our imagination, for as Darwin and many others have discovered (Ekman 1973), the caption or the context promotes a "correct" reading of the situation whereas the absence of any external clues, or the presence of those that actively misrepresent particular expressions (Montague 1959), can often hamper identification. Darwin confessed that if he had examined his test photographs without any explanation, he would have been as perplexed, in some cases, as others had been (*Expression*, p. 14). He was anxious therefore to stress that faces were not everything in the expression of the emotions, and that gestures and general behavior could be equally significant.

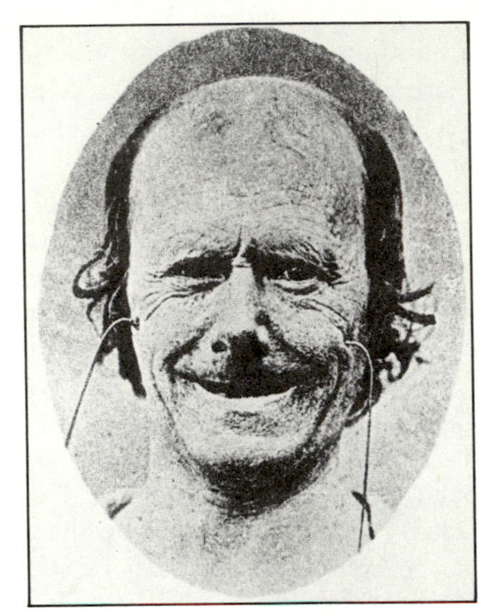

Figures 4 & 5. Left: During laughter the mouth is curved upwards and the eyelids are crinkled. G. B. Duchenne photographed this old man naturally smiling, and Darwin reproduced it in his *Expression of the emotions*. Right: The same old man, this time galvanised by Duchenne's electrical equipment. The stimulated muscles were exactly the same as those used in a normal smile, but the result was obviously unnatural — a "false smile" according to Darwin.

Here Darwin was lending scientific weight to an idea well known to artists and orators over the centuries. No painter of history scenes would have made a face tell its story without a full supporting cast of objects and gestures, unless he wanted the deliberate ambiguity of a solitary portrait. Certainly Charles Le Brun, who provided seventeenth-century Academicians with a pattern book of expressions that were widely copied at least until the time of Bell, created huge, minutely detailed canvasses, complete to the last gesture and body posture, in order to make his message clearly understood.[12] A century after Le Brun, Théodore Géricault, famous for his sensitive portraits of the inmates of the Salpêtrière, where the interest lies in the faces alone, nevertheless made full use of the power of a caption and labeled his works "Monomanie de l'envie", "Monomanie du vol", and so on, encouraging the viewer to discover insanity in his passive figures (Cadinouche 1929; M. Miller 1940–1941). Even Darwin, who obtained a collection of portrait photographs from the well-known psychiatrist James Crichton Browne, fell into the trap of "seeing" insanity in the faces because he knew they were patients in the Wakefield Asylum (Gilman 1979). Like Browne, Darwin thought he could distinguish particular expressions for each special category of mania — aided no doubt by the copious notes Browne had supplied (E. J. Browne 1985).

But Darwin seemed relatively uninterested in exploring the way human beings learned to read and recognize expressions, perhaps because this depended to such a marked degree on cultural, developmental factors (*Expression*, pp. 358–360). The aim of the *Expression* was, after all, to concentrate on the evolution of expressions themselves, not the psychology of their identification; on the physical attributes, rather than mental perception and conventions. In the light of his case for evolution, Darwin was more interested in the way man's body actually worked, than in the theory of perception: real phenomena were more useful in the fight to establish continuity between humans and other species.

The core of Darwin's book, and possibly the most difficult section of his argument, was a discussion of the origins of the higher — more "human" — emotions. All the higher qualities of man, all the Victorian virtues of modesty, innocence, and sensibility, were dramatically revealed by the flustered self-consciousness of an outright blush. Blushing showed the spiritual and moral side of human nature more clearly than any other facial display. Only blushing, it was thought, could prove that men and women had a conscience, that they could tell right from wrong and feel guilty when they overstepped the boundaries of convention. Only the hot flush of embarrassment could establish that people had the ability to reflect upon themselves and their behavior, were alive to their situation, could feel *self-conscious* in the literal sense of that term. Completely by-passing the will, and surfacing despite all our efforts to subdue it, blushing was seen to be a direct manifestation of our innermost thoughts and feelings. It was a window to the soul far more revealing than those other windows, the eyes. Christopher Ricks's elegant discussion of *Keats and Embarrassment* (1976) provides a stimulating account of the role that theories of blushing could play in literary and philosophical circles.

For Darwin, blushing was the quintessential human expression, far more so than smiling or laughing: it was "the most peculiar and the most human of all expressions. Monkeys redden from passion, but it would require an overwhelming amount of evidence to make us believe that any animal could blush" (*Expression*, p. 310).

That an animal cannot blush might seem, at first glance, an awkward problem for Darwin, in that all other facial expressions could be explained by man's evolutionary inheritance. Darwin had to contest the idea that blushing was specially designed by the Creator in order to display the moral feelings, that it acted as a check on ourselves and as a sign to others that certain sacred rules had been violated. Thomas Burgess, for instance, taking up the whole issue in 1839, believed the physical phenomena of embarrassment were intimately connected with the soul. A change of color signified shame and, because involuntary, indicated an innate sense of guilt. Such genuine "moral instinct", asserted Burgess, was an attribute given to men and women by God to prevent them "deviating from their allotted path" (1839, p. 24). It separated the world of man and his morals from that of the beasts. This view was also advocated by Charles Bell, who claimed blushing "is not acquired; it is from the beginning" (1844, p. 95).

To evolutionists, on the other hand, blushing could mean no such thing. In a crucial concluding chapter to the *Expression*, Darwin set out to destroy the claims of Burgess and others by demonstrating that the outward appearance of shame, guilt, modesty, and embarrassment could be explained fully by physical, mechanistic devices. Although the internal feelings might, by all means, be the result of some intrinsic moral quality (Darwin was careful to hide the true extent of his heterodoxy), the actual expression of these emotions was due to the way the human body was constructed.[13] Evolution could explain it all.

True to his usual style of writing, Darwin began by describing the natural history of a blush. Young people reddened more frequently than the old, but not at all during infancy when the mental powers were insufficiently developed to allow a sense of shame or self-attention. Women blushed more than men, but all human beings shared a tendency to blush whenever the attention of others made them uncomfortable. Dark-skinned people were not spared.[14] Extremes and waves of blushing were inherited, so that whole families were prone to go scarlet at the slightest provocation. As Samuel Coleridge well knew, it was sufficient to stare hard at some people to make them blush. Even the way a blush began — coming up from the chest, spreading from one cheek to another, starting as a glow or a single splash of color — did not escape Darwin's attention. He observed the blushes of his family and friends, the awkwardness of adolescent sons, the agonizing shyness of daughters, with a cool detachment that makes the less scientific reader begin to tingle with embarrassment on his behalf. Perhaps this was the only way Darwin could cope. One blush will inevitably stimulate another, and Darwin ruefully acknowledged that it was impossible even to write or speak about the phenomenon without feeling self-conscious oneself.

Darwin's theory of blushing was in essence simple. He believed all the phenomena of heightened color, embarrassed gestures, averted eyes, and so forth, were different manifestations of a solitary cause, described by him as self-attention. According to various medical authorities, attention directed to any part of the body tended to interfere with the ordinary muscular tone of the surface capillaries, making the vessels relax and fill with bright red arterial blood. Of all the bodily parts, the face received the most attention and so became more and more susceptible to blushing. Through force of association, added Darwin, the same effect would tend to follow whenever we thought that others were considering our personal appearance or behavior. *We* think people are looking at our faces and our own self-attention brings on the blush.

Always the first to see the flaws in any theory, Darwin admitted his hypothesis might seem ill-considered. It rested on the claim that mental attention had some power to influence the capillary circulation, an unabashed declaration of the power of the mind to influence the physiology of the body. Darwin could not call on any clear-cut examples of this particular vascular phenomenon, except, that is, the state of sexual erection, which was obviously a topic unsuited for a general readership. Darwin nevertheless had this in mind when first considering blushing (*N* 52). But he hoped to reinforce his case by describing many similar instances of mind-body relations, the most telling of which came from the work of Pierre Gratiolet, a distinguished French anatomist who gave a course of lectures on expression at the Sorbonne, published in 1865 as *De la physionomie et des mouvements d'expression*. Darwin found Gratiolet's study a valuable stimulus to his own researches and full of interesting conclusions. From it he took a case of a man who could, at will, make his heart skip a beat, adding that his father, Robert Waring Darwin, had also known someone with control over the involuntary muscles of the heart. Other involuntary muscles (the unstriped muscles of the body) including the intestines, salivary and mammary glands, and bladder could on occasion be stimulated or controlled by mere thought (*Expression*, pp. 339–344). Further corroboration was taken from prominent medical authors of the day, Henry Maudsley, Sir James Paget, and James Crichton Browne all having written on the mind-body problem, or "mental physiology", as Sir Henry Holland called it.[15] But although Darwin managed to establish that such things did indeed happen, he was unable to explain them. He deferred to the opinions of physicians and philosophers such as Johannes Müller. Following Müller, Darwin therefore suggested that sensory and motor nerve cells were so intimately connected that when we voluntarily concentrate on any part of the body, the first will stimulate the second and cause an appropriate reaction. In the case of blushing, the recollection of, say, heat on our cheeks, or the mere act of thinking about our faces, will create the sensation of actual heat and the brain initiates suitable body responses:

> Here, again, it seems not improbable that if we were repeatedly to concentrate with great earnestness our attention on the recollection of our heated faces, the same part of the sensorium which gives us the consciousness of actual heat would be in some slight degree stimulated,

> and would in consequence tend to transmit some nerve-force to the vaso-motor centres, so as to relax the capillaries of the face. Now as men during endless generations have had their attention often and earnestly directed to their personal appearance, and especially to their faces, any incipient tendency in the facial capillaries to be thus affected will have become in the course of time greatly strengthened through the principles just referred to, namely, nerve-force passing readily along accustomed channels, and inherited habit. Thus, as it appears to me, a plausible explanation is afforded of the leading phenomena connected with the act of blushing. (*Expression*, p. 345)

With these words, Darwin firmly closed his discussion of the biological aspects of blushing, and moved on to the realm of mind and morals.

The belief that blushing was specially designed by the Creator was, of course, opposed to the general theory of evolution. Yet up to this point Darwin had not directly confronted Burgess's and Bell's claims. The nub of the natural theological argument was that blushing — and hence morality — was peculiar to man alone, and that this set him apart from the world of mere animality. Darwin, for his part, fully agreed that it was found only in the human species. To further his own interpretation of man's ancient ancestry, he now had to establish that blushing could have evolved from animal behavior. When did blushing emerge in our evolutionary history? At what stage in mankind's mental development did self-attention and embarrassment come into existence, creating the circumstances for physiological responses to become first habitual and then instinctive?

For an answer to these questions Darwin turned, as often, to the penumbra of Victorian society, to the insane. Like Bell and Burgess before him, and like many of the physicians with whom he corresponded, he examined the question whether the insane could truly blush. Crichton Browne proved Darwin's most valuable contact in this respect, supplying him with carefully detailed observations on blushing among idiots and the deranged. On 16 April 1871, Browne sent Darwin a rough copy of his study, concluding that the severely impaired did not blush, even under the most provoking circumstances. "Dr Crichton Browne," wrote Darwin, "observed for me those under his care, but never saw a genuine blush, though he has seen their faces flush, apparently from joy, when food was placed before them, and from anger. Nevertheless some, if not utterly degraded, are capable of blushing" (*Expression*, p. 311). It all depended on the degree of retardation. Indeed some of the inmates of Browne's asylum in Yorkshire were particularly liable to blush. But the majority of these, Browne assured Darwin, were actually not blushing, but flushing — a state induced by inflammation of the membranes of the brain, which then led to the engorgement of the face, ears, and perhaps the eyes, with blood. A flush such as this could be artificially stimulated by the administration of nitrate of amyl (*Expression*,

p. 325; J. C. Browne 1871). Another manifestation of the "insane flush" documented by Browne was the scattered red blotches and mottlings on the chests of epileptic patients, which Darwin also took as evidence of a close connection between capillary circulation and the brain.

Along with most contemporary physicians, Crichton Browne labored under the belief that madness stemmed from a biological cause (Hunter and Macalpine 1963; Scull 1979; Foucault 1965). Insanity was seen essentially as a disease of the will, to which anyone could potentially succumb. This prostration of the mind was best treated by appeals to the will and conscience of the patient, so that the power of self-control could be nurtured and the art of self-government furthered.[16] In this context, the connection between blushing, conscience, and the diseased working of the will was particularly important (Skultans 1977). A true blush, it was thought, emanated from the conscience. Hence idiots were deemed unable to blush, for they suffered from congenital disturbances and, as it were, possessed no will at all. Other mental patients might blush from a morbid sensibility, displaying the complete failure of their inner self-control. Either way, the behavior of these inmates revealed what was thought to be severe moral derangement.

Darwin, on the other hand, refused to believe blushing depended on a *moral* sense, preferring to explain it as a result of self-attention, as a product of consciousness rather than conscience. He was interested in the blushes of the insane not because they revealed a lack of mental restraint, but because they displayed a breakdown in the way people perceived themselves. In Darwin's opinion, idiots were not aware of their behavior. They lacked an idea of self and were unable to become self-conscious; their behavior was like that of a child, completely ignorant of the effect it may have on others, and like a child, they did not blush. These two categories of insanity and childhood emphasized the malleability of the will, its growth, development, and possible distortion, and provided a plethora of metaphors for investigations in either field.

Analogy with children therefore gave Darwin a key to understanding the blushes of the insane: neither were fully conscious of themselves; blushing could not occur without self-attention. The same argument served to explain the origin of blushing itself. It emerged in association with consciousness. As human beings evolved from their anthropoid ancestors, slowly perceiving and understanding, so the foundations for blushing were laid. Human beings alone had learned to reflect on what they were doing, so humans alone could feel self-aware. This recently developed mental attribute was, however, expressed via the ordinary resources of the body. Physically, blushing was no different from the rest of the emotions covered in his survey, being rooted in unexceptional biological responses of a kind that could be found in any order of animals.

Here then was the crux of Darwin's argument and the reason why he believed expressions would substantiate evolutionary theory. As species changed and evolved, as their intellects developed, habits altered, and instincts varied, so the outward appearance of these mental properties also changed. Man had more complex intellectual machinery than his nearest relatives, the primates, and accordingly exhibited a wider and more varied range of facial communication. Yet there was continuity between the mind and emotions of animals and human beings, a continuum closely reflected in external display. The muscles, after all, were much the same, and the physiology consistent within the major classes of nature. "I felt convinced," wrote Darwin at the end of his life, "that the most complex and fine shades of expression must all have had a gradual and natural origin" (*Autobiography* pp. 131–132).

The point was obviously crucial to Darwin. He therefore broke over twenty years of silence and explicitly described the early history of mankind, explaining how we emerged from a pre-human parent stock. Step by step he conducted his readers through their evolutionary heritage, describing the various stages at which pleasure, fear, and all the other emotions were acquired, and drawing close analogies with "our nearest allies, the anthropomorphous apes" (*Expression*, p. 362). Here was the concluding shot in the fight for evolution, his last book in the anthropological cycle that began with the *Origin* and culminated in the *Descent*. Here, in the *Expression of the Emotions*, was the real descent of man.

ACKNOWLEDGEMENTS

I would like to thank all those colleagues who have directly and indirectly helped me with this work, particularly Bill Bynum, Roy Porter, David Kohn, Peter Gautrey, Christopher Lawrence, and Nick Browne. I also owe a special debt to Michael Neve for his comments on the paper and expression in general. The Wellcome Trustees have generously supported me throughout, and William Schupbach of the Wellcome Institute for the History of Medicine kindly supplied the illustrations.

Notes

1. According to Darwin's "Journal" (Darwin 1959), the *Expression* was begun on 17 January 1871, the last proof of the *Descent* having been finished on 15 January. The rough copy was finished by 27 April and in November and December the proofs were taken in hand, occupying him until August 1872. Seven thousand copies of the book were printed for publication on 26 November 1872 (Freeman 1977). There were favorable reviews in the *Athenaeum* (9 November 1872) and *Times* (13 December 1872), and abuse from the *Edinburgh Review* in April 1873 (Baynes 1873). The most interesting commentary was by Alfred Russel Wallace (1873), who considered it "far-fetched" in places. Alexander Bain, whose *Senses and the Intellect* (1864) and *Emotions and the Will* (1865) were carefully read and annotated by Darwin, criticized Darwin for not making use of evolution to explain more complex feelings or intellectual powers: he made it all sound too simple. Darwin had "involved himself in speculations that pass beyond our grasp"; it was impossible to say whether conscious actions or emotional expressions came first in our evolutionary history, and equally impossible to claim that one produced the other (Bain 1873; 1904, p. 320).

Darwin acknowledged the fairness of his criticism (*LL* 3: 172).

2. As Howard Gruber puts it, "Darwin sensed that some would object to seeing rudiments of human mentality in animals, while others would recoil at the idea of remnants of animality in man" (Gruber and Barrett 1974, p. 202). The best account of Darwin's intentions for the *Descent* is still J. C. Greene (1959a, chap. 10) but see also Gruber and Barrett (1974, pp. 175–257). Darwin had, of course, always believed that human expressions and behavior were intimately connected with man's mental, moral, and physical development. Since July 1838 he had been speculating freely on man, mind, and materialism in his M and N *Notebooks,* which were "full of metaphysics on morals and speculations on expression", and "metaphysics and expression".
3. Greene describes the origin of Darwin's ideas on morality (1959a, pp. 322–332), while Gruber mentions what appears to be Darwin's twofold concept: on the one hand, Darwin argued that social behavior that has long been valuable to a species becomes habitual and then instinctive; and on the other, he seems to have accepted the notion that the idea of God arises as a first explanatory principle when man confronts the unknown and the inexplicable (Gruber and Barrett 1974, pp. 324–325; see also pp. 15, 398–401). Darwin expanded on the latter in *Descent* 1: 142–147.
4. Richards (1981) has dealt with the problem of instinct more thoroughly than any other scholar, describing how Darwin never entirely relinquished the use-disuse model. Only a few instincts, such as those of neuter insects, were not habitual in origin, and these are explained by natural selection. By the early 1840s Darwin had become convinced that instincts, like anatomical structures, could vary and that natural selection could preserve and continually accumulate profitable variations to produce the most complex kinds of innate behavior patterns. This is the explanation offered in *Natural Selection* (pp. 466–527) and the *Origin* (pp. 207–244). There was much in common between Darwin's studies of behavior and the work of natural theologians such as Kirby, Brougham, and Wells (Richards 1981, pp. 199–209). Richards (1979) has also examined earlier, mainly eighteenth-century theories of behavior and, in particular, ideas about instinct and will, and discusses the impact of sensationalist epistemology on Darwin's arguments.
5. As the eighteenth century drew to a close, electrical forces, rather than mechanical forces, came under consideration for the explanation of neurophysiological matters; arguments were put forward for the possibility that electricity might be involved in nerve conduction (Home 1970; Jackson 1970). Galvani's electrophysiological experiments seemed to confirm such ideas, and throughout the major portion of the nineteenth century "nerve-force" was a popular expression, used to convey the electrical metaphors on which ideas about nervous conduction were based. Towards the end of the century it became apparent that electrical conduction would require *continuous* nerves, providing the background to Sir Charles Sherrington's proposal of the synapse concept (French 1970b).
6. Here Darwin echoed Sir Charles Bell's (1774–1842) emphasis on the nervous system of the chest in the expression of emotion. In the second and third editions of his *Anatomy of Expression* (1824, 1844), Bell had insisted there was an intimate connection between the brain, respiration, and human facial expression. Four pairs of nerves from the *medulla oblongata* went to regions of the body already supplied by the central nervous system. Bell proposed these "must be sent to bestow properties which the spinal nerves and fifth are incapable of giving" (1844, p. 242). Animals, he claimed, did not have them, so they must be associated with organs and faculties specific to humans: they supplied the chest, where emotions were felt, all the organs of speech, and those muscles of the human face that were specially adapted for and associated with language. The mind, he asserted, could therefore act on the face only via the torso. Almost by mistake, Bell discovered the afferent-efferent (sensory-motor) arrangement of the spinal nerve roots while substantiating this belief (1824, p. 3; 1844, p. 211). Paul Cranefield (1974) provides a full account of these events, reprinting the relevant materials especially Bell's first published description of the nerves. Bell's discoveries were later put forward in two papers, "On the nerves; giving an account of some experiments on their structure and functions, which lead to a new arrangement of the system" (Bell 1821), and "Of the nerves which associate the muscles of the chest in the actions of breathing, speaking and expression, being a continuation of the paper on the structure and functions of the nerves" (Bell 1822). For Bell, if not for Darwin (for example, Darwin's marginalia in Bell's *Anatomy of Expression*, 3rd edn. 1844), such ideas served to explain the obvious connection between the physiological

responses to emotion and the appropriate facial expression: "In the preceding essays, it has been shown, that the powerful passions influence the same class of nerves and muscles which are affected in highly excited or anxious breathing; and it was inferred that the apparatus of respiration is the instrument by which the emotions are manifested. In fear or in grief, the movements of the nostrils, the uncontrollable tremor of the lips, the convulsions of the neck and chest, and the audible sobbing, prove that the influence of the mind extends over the organs of respiration; so that the difference is slight between the action of the frame in a paroxysm of the passions and in the agony of a drowning man" (1844, p. 189).

7. This physical manifestation of the great chain of being (its "temporalization") is most thoroughly described by Lovejoy (1936, pp. 242–287), and Greene (1959). Lovejoy's arguments are reassessed and extended to cover anthropology in more detail in Bynum (1975).
8. Charles Le Brun (1619–1690) dominated the artistic world of seventeenth-century France, dispensing patronage and proclaiming the rules that made the art of his time so thoroughly academic, intellectual, and disciplined. His *Conférence sur l'expression générale et particulière* (1698) is both evidence and source of the fascination with the passions that characterize his era. Le Brun relied on Descartes' system; where Descartes had described the internal movements of each passion or emotion, Le Brun demonstrated how these appeared on the surface. The face should show as on a "dial" the movements of the hidden mechanism, and the eyebrows worked like the hands of a clock to reflect the conditions of the soul (situated close behind in the pineal gland), going up for good feelings, down for bad. The mouth, betraying its animal connotations, followed the wishes of the heart, not the soul, but it also went up and down at the corners. With these two simple aids, Le Brun confidently mapped out the infinite variety of human facial expressions. His task was made considerably easier by the fact that Descartes had already divided the emotions into classes. Le Brun had to identify only six basic expressions, arranged in pairs of opposites and ranked according to the distance or intangibility of the object that stimulated the feeling: thus admiration and contempt were inspired by relatively unattainable objects; love and hatred by those likely to be obtained; and joy and sadness by things actually possessed. All other expressions (fear, desire, anger, and so forth) were mixtures or extremes of these "simple" emotions. Le Brun's work, constantly edited, translated, pirated and bastardized through the following century (Montague 1959), was the single most important text in the study of physiognomy before the work of Lavater (1741–1801).
9. Charles Bell's work on the *Anatomy of Expression* (1806) illustrates the rich diversity of ideas about the face to be found at the beginning of the nineteenth century. His was the first scientific study of the physical manifestation of emotions. He also tried to help artists by drawing attention to the gestures that accompany deep feeling, and by advocating a greater knowledge of anatomy. It is not generally known that the first edition of his book was intended to win him a chair at the Royal Academy (Royal Academy MS Lawrence 1: 199–202, and 4: 310). Only later, having failed to secure a chair, did he turn to his more famous work on the nerves. Subsequent editions of the book were in effect manifestos for his particular interpretation of the nervous system, significant in the controversy with François Magendie over the discovery of the different functions of anterior and posterior spinal nerve roots (Shaw 1839, 1860, 1868; Cranefield 1974). Bell's theory of expression was inspired by natural theology and a belief that the muscles were there for a purpose: the human face had more muscles and was the most mobile of the animal kingdom because it had to express a greater range of thoughts and feelings (Bell 1844, p. 58). He attempted to arrange the passions in a system based on pain and pleasure, exertion and relaxation. All expressions were a combination of these. Such general opinions were widely shared by Bell's contemporaries (Jeffrey 1806), and for this reason Darwin used the 1844 edition as an authoritative source against which he could evaluate his own researches (first edition 1806 read at Maer sometime between 10 June and 14 November 1840, DAR 119; but third edition 1844 heavily annotated and in the Darwin Library). Bell, of course, held a view of expression directly opposite to that of Darwin: "It is obvious thus to observe how the muscles, by producing distinct impressions, afford a new occasion of distinguishing the tribes of animals; and, as signs of superior intelligence, become proofs of the higher endowments of man, and a demonstration of the peculiar frame and excellence of his nature" (Bell 1806, p. 101).
10. The principal muscles peculiar to man were those running from the top of the nose to

the eyebrows (*corrugator supercillii*), and from the base of the lower jaw to the corner of the mouth (*depressor anguli oris* and *triangular oris*). The *corrugator supercillii* "knits the eyebrows with a peculiar and energetic meaning which unaccountably, but irresistably, conveys the idea of mind and sentiment" (Bell 1806, p. 95); the others provide a "combination of muscular actions of which animals are incapable" (p. 97). Darwin's marginalia in his copy of the 1844 edition argue against this view (Darwin's copy of Bell 1844, pp. 95–99, 120–121, and particularly p. 138: "I suspect he never dissected monkey").

11. From the first systematic treatise on physiognomy, attributed to Aristotle (Evans 1969), to the time of Darwin and beyond, the problem of interpreting human feelings without the aid of language has occupied many scholars and has surfaced in many different areas of thought. As a quasi-science, physiognomy always bore a close relationship to the study of medicine and the methods of diagnosis; as an art, to the practice of rhetoric and the theater. The best-known actors and actresses studied the so-called language of gesture, Henry Siddons, for example, publishing *Practical Illustrations of Rhetorical Gesture* in 1822 (see also Foote c.1750). The tradition continued at least until the end of the nineteenth century, as illustrated by J. E. Foster's *Art of Expression* (1885). Physiognomic ideas were also utilized in caricature (Grose 1788) and drawing (Hamilton 1812). Medieval symbolism in the fine arts (Barasch 1976) was transformed into a complete set of rules for the depiction of expressions by Le Brun (Montague 1959), further modified by Lavater, who drew up a system by which specific emotions and characters could (supposedly) be identified. Such systems permeated late eighteenth- and early nineteenth-century art — as in Rowlandson's plates for Ackermann's *Le Brun Traversed* (1800), Hogarth (especially his *Analysis of Beauty*, 1753), and the works of Fuseli — and literature (Graham 1961; Tytler 1982). Modern studies of expression as a visual medium have been made by Gombrich (1960 and 1963), and as a biological property by Blake (1933), Ekman, Friesen, and Ellsworth (1972), and Ekman (1973). It is clear that Darwin recognized expression as a nonverbal, solely visual phenomenon through his careful use of photographs. Such photographic reproductions were extremely rare in the 1860s and 1870s when Darwin was concluding his research and were, at first, considered more as an extension of the arts than as aids to science. Oscar Rejlander, for example, one of the key figures in nineteenth-century photography, was a painter in Rome before settling in Wolverhampton where he began his photographic work (E. Y. Jones 1973). By 1872 he had gained an enviable reputation as a portraitist, genre worker, and photographic artist; not content with photographing other people (mainly actors) at Darwin's suggestion, he made several self-portraits illustrating various expressions, including disgust, defiance, surprise, shrugging, and helplessness, that appear in the *Expression* (for example, facing p. 255). Apart from Duchenne, the other photographic artists who supplied Darwin with material are little known. Herr Kindermann of Hamburg, like Rejlander, ran a flourishing *carte-de-visite* studio specializing in the photography of babies and infants. Dr George Charles Wallich, a field surgeon in India for many years, natural historian, and author of geological texts, was renowned for his portraits of *Eminent Men of the Day* (1870). Duchenne's work is more fully discussed in Bikaplan (1948), and his photographic studies are housed at the Ecole des Beaux Arts in Paris. The largest group of photographs in Darwin's collection, other than the seventy-three from Rejlander, comprise portraits of the insane from the West Riding Asylum at Wakefield, Yorkshire. Although only one appears in the book, they were obviously important sources for Darwin (Gilman 1979). It is possible that either James Crichton Browne, the medical director, or Hugh Diamond, the eminent medical photographer, was responsible for these. Photographing the insane was an extension of a typically nineteenth-century obsession with the outward signs of madness, an attempt to find physical attributes in the faces of patients that would facilitate diagnosis (as in Morison 1840, and Conolly 1858–1859; see also Gilman 1976). By far the most intriguing print is by Rev. C. L. Dodgson.

12. Ernst Gombrich (1978) chronicles the rise and fall of this kind of history painting, and J. A. Leith (1965) describes its ramification into the idea of art as propaganda during the second half of the eighteenth century in France. Diderot's critiques of the Parisian *Salons* were taken to heart, for instance, by David. The new social realism depended to a large extent on conventional expressions and gestures to make the meaning clear.

13. Darwin never actually asked himself, what *is* an emotion? Even Charles Bell had tussled with this question, concluding it was some-

thing felt in the heart and lungs stimulated by a special set of nerves, the "respiratory nerves". A succession of philosophers since Robert Burton had inquired into the nature of emotion, analyzing the relations between passions and the will, the link between emotion and reason. From these beginnings there emerged, in England at least, a school of moralist thinkers anxious to separate environmentally determined passions and emotions from those that were supposed to be innate. Samuel Burgess (1825) studied the bestial nature of man and believed that the advance of civilization would lead to a preponderance of "higher" emotions; in contrast, the phrenologist J. G. Spurzheim argued (1815) that all faculties were innate, that the beast in man would emerge regardless of the state of civilization (Giustino 1975). Both scholars believed that man had been given "higher" emotions in order to control his animal passions. These, and other British empiricists, agreed that emotions originated in sensation, developing as simple ideas about the causes and effects of pleasure and pain. The laws of association thus explained how connections were made between the stimulus and the feeling experienced in the mind; see, for example, Alexander Bain (1864, 1865). David Monro (Wright-St. Clair 1956), Benjamin Brodie (1854), and Samuel Coleridge (Bostetter 1970) attempted to define the passions. See also Rather (1965) and Mischel (1966).

14. The question whether dark-skinned people could blush preoccupied many early nineteenth-century physicians and philosophers, for it bore on the larger issue of the status of Negroes and other races in the hierarchy of nature. For a time it was believed that "primitive" races lacked the sensibilities of more civilized, predominantly European nations and therefore could not produce any signs of sensibility — the blush. Alexander von Humboldt, for example, quoted without protest the sneer of the Spaniard in South America, "how can those be trusted, who know not how to blush?" (*Expression*, p. 319). Charles Bell, who believed blushing a God-given characteristic of the human race, nevertheless doubted if a blush would be seen in a Negro: "in this we perceive an advantage possessed by the fair family of mankind, and which must be lost to the dark; for I can hardly believe that a blush may be seen in the Negro" (1844, p. 96). In 1839, however, Thomas Burgess took considerable pains to counteract this prejudice, claiming that Africans did indeed blush, but that the phenomenon was masked by the dark color of their skin; studies of unpigmented skin surrounding scar tissue, and of albino Negroes, substantiated his point. Even though Burgess, like Bell, believed that blushing was "an evidence of Design", he was not prepared to deny salvation to non-European races. Blushing, he asserted, was caused by moral issues, and it was the *intellectual*, not moral, faculties that distinguished Negroes from the European (Burgess 1839, pp. 30–42). Darwin, who found Burgess's work invaluable, took many of his examples of blushing in the various races of man from it. He also used Burgess extensively in his account of the physiology of blushing.
15. Since the close of the nineteenth century more familiarly known as neurophysiology. French (1970b) and Poynter (1958) provide accounts of the milestones in this field, and Figlio (1975) assesses the physiology of mind in the late eighteenth century.
16. There is a wide range of literature on changing definitions of insanity through the nineteenth century, but see particularly Bynum (1974b) on the rationales for therapy in British psychiatry, and Carlson and Dain (1962) on the meaning of moral insanity. Insanity as the loss of self is discussed in Fullinwider (1975). General texts on nineteenth-century ideas about insanity are Scull (1979), Skultans (1975), and Foucault (1965). Moral management — the art of nurturing self-control — is best described by one of those who developed the technique, Daniel Hack Tuke (Bucknill and Tuke 1858).

13

Darwin on Animal Behavior and Evolution

Richard W. Burkhardt, Jr.

In an obituary notice of 1882 examining the causes of Darwin's success and the importance of Darwin's works, the Genevan botanist and pioneer of the history of science Alphonse de Candolle identified two characteristics in particular that had made Darwin such an exceptional thinker. One was Darwin's ability to occupy himself simultaneously with both the smallest details and the broadest theoretical considerations. The other was the extraordinary *range* of Darwin's researches and the way that each of Darwin's separate studies, however specialized, contributed to the whole of Darwin's *oeuvre* (Candolle 1882).

Although de Candolle did not elaborate upon how the various aspects of Darwin's research were related to each other or to the major intellectual concerns of Darwin's predecessors and contemporaries, historians of science of the present generation have been engaged in precisely this task, and Darwin's contributions to a number of different intellectual domains have now been subjected to careful scrutiny (as the articles in the present volume amply testify). Still relatively unexplored, however, are Darwin's thoughts on behaviour. This is the case even though Darwin was fascinated with behavioral phenomena throughout his life, from his boyhood enchantment with the activities of dogs and birds to his last book, which was on earthworms (1881), and his last communication to a scientific society, which was on sexual selection in a race of Syrian street-dogs (1882, *CP* 2: 278–280). The sweep of Darwin's behavioral concerns encompassed the motions of plants, the instinctive and intelligent acts of animals, and the evolution of the higher mental and moral faculties of humans. The special characteristics that de Candolle found in Darwin's work as a whole — the wide range of interests and the ability to move between factual details and broad theoretical considerations — were manifested in microcosm in Darwin's studies of behavior. Yet the nature of Darwin's work on behavior and the relation of this to his work as a whole remain to be examined.[1]

There are, it seems, a variety of reasons why Darwin's work on behavior has not received a great deal of attention from historians. The first is that Darwin scholarship has concentrated primarily on Darwin's early intellectual development, particularly with reference to problems associated with "the origin of the *Origin*". Although behavioral concerns played an important role in Darwin's early thinking, the extent of these behavioral concerns was not reflected in the *Origin*, where the discussion of behavior was limited to a single chapter on instinct. It was not until twelve years later, in the *Descent of Man (1871)* followed by the *Expression of the Emotions in Man and Animals (1872)* that Darwin revealed to the public a broader — if still not entirely comprehensive — picture of what his behavioral concerns included. Darwin scholars recognize that the chronological ordering of Darwin's publications is not a reliable indication of the order in which Darwin's views actually developed. It does appear, nonetheless, that the orientation of Darwin studies toward Darwin's early years and the development of the *Origin* has resulted in a certain inattention to Darwin's work on behavior.

A second factor that has contributed to the relative neglect of Darwin's work on behavior has been the way in which the study of behavioral evolution itself has developed since Darwin's time. Historians of science insist, quite properly, that historical studies of science should be undertaken for their own interest and not be guided by a "Whiggish" concern with the concepts and issues of modern science. It remains the case, however, that contemporary developments do, for better or worse, stimulate historical studies. For example, present-day debates over group selection and the origins of altruistic behavior have occasioned a discussion of what Darwin himself had to say about these subjects (see Ruse 1980b). But this particular incentive for studying Darwin's behavioral work is a new one, which has emerged only in the last few years as the efforts of ethologists and sociobiologists have made the study of behavioral evolution a vigorous — and controversial — area of investigation. Prior to these recent developments in the study of behavioral evolution, if the "Whig" approach to history had any significant influence on the study of Darwin's thoughts on behavior, it was probably to downplay the interest of Darwin's behavioral studies. The reason is that the anecdotal and anthropomorphic cast of many of Darwin's discussions of behavior, plus Darwin's frequent endorsement, in discussing behavioral phenomena, of the now-discredited idea of the inheritance of acquired characters, tend to give modern readers the impression that Darwin was not at his best on behavioral subjects.

Perhaps the foremost reason for the relative neglect of Darwin's work on behavior, however, is that behavioral considerations, though constituting an *integral* part of Darwin's thinking, did not constitute a *privileged* part of his thinking. By this I mean that Darwin did not take it upon himself, either independently of or in conjunction with his other goals, to establish an evolutionary science of behavior in its own right. Not that he lacked the breadth of vision to do so; his early manuscript notebooks testify to the scope of his behavioral concerns, and in the works that he eventually published he did in effect provide the conceptual foundations on which an evolutionary science of behavior could be erected. But the study of behavior did not exist as a distinct scientific field in Darwin's time, waiting, as it were, for someone to restructure it along evolutionary lines, and Darwin did not approach behavioral phenomena as if they represented a specialized domain. To say that behavioral studies did not represent a specialized domain in Darwin's time is not to say that Darwin's work on behavior was conducted in an intellectual vacuum. Naturalists such as Réaumur, Gilbert White, and the Hubers, among others, had already demonstrated the interest of behavior study as a part of natural history. Philosophers had discussed at length the nature of instinct and intelligence and the different capabilities of animals and man. Natural theologians had found instinctive behavior to provide choice grist for their mills. The context of Darwin's thinking on behavior was thus, to use Robert Young's phrase, a *common* context (Young 1969), not a context defined by and accessible only to specialists. Perhaps not surprisingly, then, Darwin's engagement with behavioral considerations was generally in the service of his broader explanatory goals — such as confuting the creationists or demonstrating the continuity between animals and man — and not for the purpose of dealing with behavioral phenomena in and of themselves.

Young (1973b) has argued that psychological issues operated at different levels in "the great debate on man's place in nature". He has shown how psychological conceptions were central to Malthus's work, which in turn influenced Darwin, Wallace, and Spencer, and he has suggested that psychological conceptions were also fundamental to competing philosophies of nature in the nineteenth century. He has found surprisingly little evidence of the role of psychological issues on the surface of the nineteenth-century debates over evolution, however, and with respect to Darwin in particular he has written: "whatever the *implications* of his work for psychology, it is clear that the main sources of Darwin's theory were derived from the studies of a field naturalist and from geology. This was where his real interests lay . . . ." (1973b, p. 184). In support of this view, Young notes that in the last years of his life, Darwin turned his notes on instinct and comparative psychology over to George John Romanes (1883) instead of bothering to publish these notes himself.

Especially valuable in Young's analysis of the broader context of the evolutionary debates of the nineteenth century is his observation that the thinkers of Darwin's day did not carve up their intellectual endeavors in ways that correspond to the disciplinary divisions of the present. Young in fact argues this both ways, not only identifying a "common context" for these evolutionary debates but also observing that "in any given period, intellectual life is fragmented in ways that appear bizarre to those who have the benefit of hindsight" (1973b, p. 191). Young has not, however, provided much guidance with respect to the way psychological or behavioral issues impinged on the fine structure of Darwin's theorizing. Indeed, with the important exception of Malthus's influence on Darwin, Young has denied a significant role for psychological concerns in Darwin's thinking. If one considers Darwin's activities as a naturalist, however — and recognizes that the word "naturalist" encompassed a considerable breadth of concerns in the nineteenth century — one finds that these activities continually brought Darwin face to face not only with interesting behavioral phenomena but also with theoretical issues that he proceeded to resolve, at least in part, by appealing to behavior. To say that Darwin's "real interests" lay in his studies as a naturalist thus does not allow one to conclude that psychological considerations were of minimal significance for Darwin's work. Whether or not psychological considerations were important for Darwin in his theorizing during the course of his career remains to be investigated.

I wish to survey in this paper the relations between Darwin's thoughts on behavior and the development of his evolutionary theory. These relations have typically been regarded as distinctly one-sided. Darwin's thinking on behavioral evolution has been represented as an extension to behavioral and mental phenomena of evolutionary ideas Darwin formulated initially in addressing other features of living things, such as their geographical distribution, their relations in geological time, and their morphological affinities.[2] What I hope to show here is that Darwin did not simply *apply* his evolutionary theory, once he constructed it, to behavior. On the contrary, at various stages in the development of his thinking, Darwin's attention to behavioral phenomena was of considerable importance for his deepening appreciation of the means by which organic change takes place. His understanding of behavior thus both reflected *and reflected back upon* his understanding of the evolutionary process. His comprehension of each was informed and enriched through his interaction.

Attention will be given here to Darwin's interests in behavior prior to his coming to believe in organic mutability; his behavioral concerns as exhibited in his early manuscript notebooks, in his *1842 Sketch* and in his *1844 Essay*; his treatment of instinct in the *Origin*; the behavioral phenomena he addressed in the *Descent*; and one particular aspect of the argument he presented in the *Expression of the Emotions*. The primary emphasis here will be on Darwin's ideas about *animal* behavior as these related to his evolutionary theorizing. Darwin's views on human behavior will not be explored in much detail, not because Darwin felt there was any fundamental difference between the behavior of animals and the behavior of humans, but because Darwin's views on man are treated elsewhere in this volume by Durant (see also Herbert 1974, 1977; and Gruber and Barrett 1974). For Darwin's ideas on the behavior of plants, Ghiselin (1969) provides the best introduction, although this is another aspect of Darwin's work that remains largely unexplored. The comparison of Darwin's views on the nature of instinct and intelligence with the views of other thinkers of Darwin's century and earlier has been undertaken by Richards (1981, 1982).

I. Behavioral Concerns Through the Voyage of the *Beagle*

Darwin's autobiographical recollections, appropriately enough, reveal more about Darwin's own behavior than they do about his observations on the behavior of animals. These recollections do suggest, nonetheless, that beyond "a strong taste for collecting" it was Darwin's joy in observing wildlife — living animals in their natural environments — that led him to feel, as he put it at the age of thirty, that he had been "born a naturalist" (*ML* 1: 4). This statement does not appear on the face of it to be especially revealing. It may serve, however, to highlight an important difference between Darwin and some of the leading scientists of the beginning and the end of the nineteenth century.

Late in the century, the English naturalist E. B. Poulton remarked that the reason Darwin's theory of sexual selection was still a matter of debate was that there were "comparatively few true naturalists — men who would devote much time and the closest study to watching living animals in their natural surroundings, and who would value a fresh observation more than a beautiful dissection of a rare specimen." It was, Poulton said,

> a very remarkable fact that the great impetus given to biological inquiry by the teachings of Darwin has chiefly manifested itself in the domain of Comparative Anatomy, and especially that of Embryology, rather than in questions which concern the living animal as a whole and its relations to the organic world. And yet these were the questions in which Darwin himself was principally interested. (1890, pp. 286–287)

Poulton was by no means a disinterested evaluator of what Darwin's principal interests had been, and Poulton's description of Darwin's interests perhaps corresponded more closely to Poulton's own interests than they did to Darwin's. There was, nonetheless, an element of truth in Poulton's lament about the disappearance of "true naturalists" from the ranks of science. The face of the biological sciences was changing, and the leaders of the generation of biologists after Darwin were for the most part cast in a different mold than Darwin, Alfred Russel Wallace, and Henry Walter Bates had been.

This is not to suggest, however, that Darwin was necessarily more like his predecessors than he was like the comparative anatomists and embryologists who came after him. This is apparent when one compares him, for example, with the French zoologist J. B. Lamarck. Lamarck's own theory of organic change addressed behavioral phenomena in two fundamental ways. In the first place, Lamarck identified behavior as a key agent in the process of organic change. In the second place, Lamarck sought to explain how behavioral capacities emerged as the complexity of organisms had gradually developed over time. Lamarck is rarely numbered among the precursors of modern ethology, however, and the reason for this seems clear. Although Lamarck is famous for advocating the idea that habits precede structures in the course of organic change, his actual practice as a naturalist gave precedence to structures over habits. The focus of Lamarck's research was not the behavioral repertoires but rather the last remains of living things. His efforts centered on the study of specimens in a museum, not animals in the field. When he attempted to comprehend the phenomena of life, it was through analyzing the structures of organisms that were long since dead (R. Burkhardt 1977, 1981).

Darwin, in contrast, displayed, as Poulton said, an abiding interest in "the living animal as a whole and its relations to the organic world". In his youth, Darwin read Gilbert White's *Natural History of Selborne*, and he thereafter "took much pleasure in watching the habits of birds, and even made notes on the subject", an activity that left him wondering "why every gentleman did not become an ornithologist". Darwin also became an avid hunter. To rationalize his love of hunting, he told himself that being able "to judge where to find most game and to hunt the dogs well" made hunting "almost an intellectual employment" (*Autobiography*, pp. 45, 55).

Nora Barlow writes of Darwin — her grandfather — that "already, as a boy . . ., his powers of perception of more than the formal attributes [of species] can be noticed, and his sympathetic participation in the lives of the creatures he observed helped him to understand their habits; form, function, adaptation and behaviour are all brought to bear on the living of each species in its own surroundings" (Darwin 1963, p 206).[3] When Darwin embarked on his *Beagle* voyage, it may have been true, as Henslow told him, that he was not yet a "*finished* naturalist", but this did not prevent him from appreciating fully that his role as a naturalist-voyager involved describing not only the physical characteristics but also the *habits* of the different animals and peoples he would encounter. And if he later testified that what eventually struck him most among his *Beagle* voyage observations, at least as far as the origin of his evolutionary views was concerned, were facts regarding the geographical distribution of animals and the relations between fossil and living forms, it is also evident that he was profoundly impressed by the behavior of some of the animals and peoples he saw.

Darwin was amazed by the remarkable lack of fear of man displayed by the birds of the Galapagos (*Diary*, p. 334). He was also struck by birds that appeared, by their habits, to represent aberrant or transitional forms with respect to the types of birds with which he was already familiar. One finds, for example, in the same letter in which he wrote to Henslow about his exciting fossil finds in Patagonia, Darwin also reporting on "a poor specimen of a bird, which to my unornithological eyes, appears to be a happy mixture of a lark pidgeon & snipe . . . . — Mr. MacLeay himself never imagined such an inosculating creature. —" (Darwin 1967, p. 62)[4].

While this bird appeared exceptional because of the mixed character of both its habits and its structure, other birds impressed Darwin by the seeming incongruity between their habits on the one hand and their structures on the other. The tyrant flycatcher *Saurophagus sulphuratus*, for one, closely resembled the true shrikes in its structure, but in its hunting habits it often hovered in the air like a hawk or remained stationary near the water's edge like a kingfisher (1963, pp. 216–217). The frigate bird had webbed feet, but as Darwin described it in his Ornithological Notes:

> The bird never touches the water with its wings, or even with its feet; indeed, I have never seen one swimming on the sea; one is led to believe that the deeply indented web between its toes is of no more use to it than are mammae . . . in the male sex of certain animals; or the shrivelled wings beneath the wing-cases firmly soldered together of some coleopterous beetles. (1963, p. 267)

Other apparent incongruities between habits and structures attracted Darwin's attention, such as those displayed by *Pelecanoides Berardi*, a bird from "the deep and quiet creeks and inland seas of Tierra del Fuego". "No one seeing this bird for the first time," Darwin wrote, "diving like a grebe and flying in a straight line by the rapid movement of its short wings like an auk, would be willing to believe that it was a member of the family of petrels; — the greater number of which are eminently pelagic in their habits, do not dive, and whose flight is usually most graceful and continuous" (*Zoology* 3: 138–139). The case from the animal kingdom that seems to have astonished Darwin most, however, was a ground-feeding woodpecker of the pampas. Darwin's entry in his Ornithological Notes reads:

> Picus, not uncommon: frequents stony places & seems to feed exclusively on the ground; the bill of this specimen was muddy to the base: in the stomach nothing but ants. — cry loud, resembling the English manner of the bird; tail seems very little used, although I have seen one, with it a good deal worn: alights horizontally. like any common bird, on the branch of a tree: but occassionally [sic] I have seen it clinging to a post vertically. — are rather wild, frequent the open plains. generally three or four together. (1963, p. 219)

To these examples from the animal kingdom must be added Darwin's reaction to seeing the "savages" of Tierra del Fuego. He found it impossible, he said, "to describe or paint the difference of savage and civilized man." The Fuegians, in his words, were "men whose very signs & expressions are less intelligible to us than those of the domesticated animals; who do not possess the instinct of those animals, nor yet appear to boast of human reason, or at least of arts consequent on that reason" (*Diary*, p. 428).

It is evident that Darwin was jolted by the way the behavior and appearance of the Fuegians contradicted his expectations and by the way the structures and behavior patterns of certain animals likewise failed to correspond to his past experiences. What Darwin initially made of his *Beagle* voyage observations and the significance that he eventually attached to them, however, are of course not necessarily the same thing. It is clear, though, that once Darwin returned to England, and once the idea of the mutability of species took shape in his mind, he treated the example of the ground-feeding woodpecker, among others, as a capital rejoinder to those who denied that organisms could ever change their behavior in such a way as to be successful in circumstances other than those to which they were clearly adapted, and he treated the Feugians as clear support for the view that the differences between man and the higher animals are differences of degree rather than differences of kind.

In going over the dead specimens that he had collected during his travels, Darwin had good reason to reflect upon the behavior patterns these animals had displayed while living. His appreciation that behavioral characters as well as morphological characters were of use in determining species was particularly enhanced for him early in 1837 when his ornithological specimens were analyzed by the distinguished ornithologist John Gould. In South America, Darwin had distinguished from each other — solely on the basis of their differences in habits — two kinds of mocking-birds, one that he found near Maldonado and the other that he found on the plains of Patagonia. Upon shooting specimens of these birds and comparing them, however, he concluded that the birds were actually indistinguishable. Back in England, Gould confirmed that Darwin's original impression — the impression based upon the birds' behavior — was correct and that the birds were in fact different species (*Zoology* 3: 61; Darwin 1963, p. 216). It was in part on the basis of differences in behavior that Darwin had distinguished the common "ostrich" (Rhea) of Northern Patagonia from the much rarer bird that the natives called the Avestruz Petise. This again was a distinction that Gould confirmed (*Journal of Researches*, p. 109). Likewise, it was in part on the basis of what Darwin told Gould about the habits of certain carrion-feeding hawks of the Galapagos that Gould instituted in the subfamily *Buteoninae* a new genus, *Craxirex*, to encompass the new species Darwin had found (*Zoology* 3: 22).

More details could easily be provided here regarding behavioral phenomena that Darwin witnessed during his travels as naturalist of the *Beagle*. Enough has been said, however, to demonstrate Darwin's early concern for the activities of living organisms, the interest of some of the particular observations he made on his *Beagle* voyage, and the way certain of his behavioral observations were given added significance by Gould's professional assessment of Darwin's ornithological specimens. In light of this background, it does not appear surprising that when Darwin opened his "first notebook on 'Transmutation of Species' " (*B*) in the Summer of 1837, he was prepared to regard behavioral and mental changes, as well as structural changes, as part of the immense subject that was before him.[5]

II. Behavioral Concerns in Darwin's Early Evolution Theorizing

"Even mind and instinct," Darwin assured himself as he began his first transmutation notebook, are "influenced in [the] course of generation" (*B* 3). Citing Lamarck's claim that the distinctions between species become less and less clear as collections become more and more complete, Darwin observed that this was "truer even than in Lamarck's time" and that it applied to "every character" that one might consider, instincts included (*B* 9). Later in the same notebook Darwin wrote that as one was led to attempt to discover the causes of change and adaptation, instinct and structure became focal points for further speculation and observation. "My theory," he acknowledged, would lead to the study of many things, including, in his words, "study of instincts, heredity and mind heredity, whole metaphysics" (*B* 227–229).

David Kohn (1980) has explained more persuasively than anyone else what Darwin meant late in 1837 or early in 1838 by "my theory". Kohn indicates that, at that stage in Darwin's intellectual development, Darwin was not simply casting about for "a theory by which to work", but instead had adopted a theory in which sexual reproduction automatically generated adaptive variation. The *purpose* of generation, as Darwin put it at the beginning of his B *Notebook*, was "to adapt and alter the race to *changing* world" (*B* 4). Darwin continued to develop this particular theory in his C *Notebook*, begun in late February or early March 1838. In his C *Notebook* he also introduced a major new idea — habit precedes structure. This idea promoted behavior to a central place in Darwin's explanation of organic change.

The source of Darwin's idea that habit precedes structure has often been assumed to have been the French zoologist and evolutionary theorist Jean-Baptiste Lamarck, although others, such as Frédéric Cuvier (Richards 1981) and Sir John Sebright (Kohn, personal communication), have also been suggested for this role. Identifying a precursor from whom Darwin's idea might have come, however, is perhaps not as important as identifying the *function* that the idea played in Darwin's thinking when he first employed it. The fact is that Darwin initially used the idea that habit precedes structure in a very un-Lamarckian context, a context shaped more by the demands of Darwin's own early theorizing than by any of the problems that had been crucial to Lamarck.[6]

Kohn is certainly correct in stating that "Darwin's view that habit precedes structure had deep roots in his own emergent thought" and in indicating that when Darwin began considering behavioral adaptation, especially in birds and man, it was natural for him to see if his sexual theory could handle behavioral phenomena. In Kohn's view, "the principle that habit precedes structure is the logical parallel to and chronological transformation of Darwin's separation of adaptation and heredity, applied to habit" (1980, p. 131). Indeed it does appear that Darwin did not treat the two hypotheses — adaptive change produced by generation and adaptive change produced by habit — as being in conflict with one another. In each case, in response to changes in external conditions, an adjustment of the organism (or its offspring) to the new external conditions was automatically produced. Nonetheless, the historical connection between Darwin's sexual reproduction theory of change and his introduction of the idea that habit precedes structure was more intimate than even Kohn has suggested. Habit and instinct, for Darwin, did not simply constitute a class of phenomena that any evolutionary theory would have to handle. They did not simply represent an area to which his sexual reproduction theory of change had to be applied. Instead they appeared to Darwin as *the solution*, or at least *a solution*, to one of the key difficulties he perceived with his theory.

Shortly after beginning his C *Notebook*, Darwin allowed that "the most hypoth:[etical] part of my theory, [is] that two varieties of many ages standing, will not readily breed together" (*C* 30). How was it, in other words, that varieties, once developed, were not then swamped by interbreeding? The answer, Darwin decided, was *instinct*. In domesticated animals, Darwin readily acknowledged, varieties did breed with one another, but this was not the case, he maintained, with respect to animals in the wild, thanks to their instincts:

> in wild state (where instinct not interfered with, or generative organs affected as with plants) no animals *very* different will breed together, so when two great (which can be shown probable) varieties may be made in wild state, there will be presumption that they will not breed together. (*C* 30)

Later in the same notebook Darwin underlined the importance of this argument by incorporating it into his definition of the word "species". As he put it, "My definition of species has nothing to do with hybridity, [it] is simply an instinctive impulse to keep separate. . . ." (*C* 161e).

It thus appears that animal instincts, instead of constituting a class of phenomena to which Darwin *applied* his sexual reproduction theory of change once he formulated it, played a key role in that theory itself. Instincts discouraged related varieties from interbreeding before structural differences developed to make interbreeding physically impossible. The point to be emphasized here is that it was in this way, *through* the problem of reproductive isolation (which Darwin identified as crucial for his theorizing), that the idea of habit preceding structure entered Darwin's thinking. The first statement of this idea in Darwin's notes reads:

> Instinct goes before structure (habits of ducklings & chickens young water ouzels) hence aversion to generation, before great difficulty in propagation. (*C* 51)

The logical structure of the above sentence suggests that Darwin did not *derive* the idea of instinct or habit preceding structure from the problem of reproductive isolation, but rather that he called upon the idea to solve the problem. Having once called upon the idea, he then warmed quickly to it, and the scope of its usefulness to him expanded. A few pages after he first introduced the idea he told himself:

> The circumstances of ground woodpeckers, — birds that cannot fly &c. &c. seem clearly to indicate those very changes which at first it might be doubted were possible, — it has been asked how did the otter live before it had its web-feet. All nature answers to the possibility. — (*C* 57)

A few pages later, Darwin offered the forceful generalization: "All structures either direct effect of habit, or hereditary & combined effect of habit, — perhaps in process of change" (*C* 63).

Through the Spring and Summer of 1838 Darwin proceeded to explore at length the themes of the relation between habits and structure, the nature of instinct, the relations between instinct and reason, the expression of the emotions in animals and man, the continuity of the mental faculties in animals and man, and the importance of habits for classificatory purposes. The sources of his reflections included not only his observations from his *Beagle* voyage, but also new observations he was making on visits to the London Zoological Gardens, plus facts and opinions he was finding in articles and books.

Noting in his C *Notebook* that "Gould seems to doubt how far structure and habits go together", Darwin told himself, "This must be profoundly considered" (*C* 81–82). Bringing the subject up again, he stated:

> It is of the utmost importance to show that habits sometime go before structure. — the only argument can be a bird practising imperfectly some habit, which the whole rest of other family practise with a peculiar structure, thus Tyrannus sulphureus if compelled solely to fish, structure would alter. — (*C* 124)

He then asked himself whether or not it was the case that "when two very close species inhabit same country are not habits different [?]" (*C* 125). Further on in his C *Notebook* he wrote: "According to my views, habits give structure, [therefore] habits precede structure, [therefore] habitual instincts precede structure. — duckling runs to water before it is conscious of web-feet. —" (*C* 199).

When Darwin revealed his evolutionary views to the public in 1859 in the *Origin*, he did not discuss the organic basis of instincts but simply argued that instinctive behavior, like corporeal structures, varied among individuals and thus was subject to natural selection. Two decades earlier, however, in his private notebooks, he confronted directly the issue of the organic basis of instincts and ideas. For ideas as well as instincts, he decided, what was involved was "mental machinery". As he put it in the lengthiest of his early comments on man's place in nature:

> [Man] possesses some of the same general instincts all & feelings as animals. They on the other hand can reason — but man has reasoning powers in excess, instead of definite instincts — this is a replacement in mental machinery so analogous to what we see in bodily, that it does not stagger me. — What circumstances may have been necessary to have made man! . . . . Let man visit Ourang-outang in domestication, hear expressive whine, see its intelligence when spoken [to], as if it understood every word said — see its affection to those it knows, — see its passion & rage, sulkiness & very extreme of despair; let him look at savage, roasting his parent, naked, artless, not improving, yet improvable and then let him dare to boast of his proud preeminence. — (*C* 77–79)

Thus, as Darwin saw it, man's instincts, emotions, and intelligence were continuous with those of the higher animals, and all these mental faculties had an organic basis. Habits altered the "mental machinery" of the organism, and this altered machinery was then transmitted through reproduction to successive generations. The idea of mental traits having become hereditary was what led Darwin to charge himself with materialism:

> Thought (or desires more properly) being hereditary it is difficult to imagine it anything but structure of brain hereditary, analogy points out to this. — Love of the deity effect of organization, oh you materialist! (*C* 166)

This was the kind of thinking that led Darwin to begin in 1838 a separate series of notebooks on "Metaphysics and Expression of Emotions". This was also the kind of thinking that led him to write in his third transmutation notebook: "Mine is a bold theory, which attempts to explain, or asserts to be explicable every instinct in animals" (*D* 26).

Bold as his theory was by the Summer of 1838, Darwin had not resolved every question he had raised for himself in his transmutation notebooks. He continued to turn over in his mind, for example, the peculiar case of generation reported by Lord Morton (R. Burkhardt 1979). Furthermore, as will be discussed presently, though he had already given some thought to various puzzling "secondary" sexual differences between males and females of the same species, he had yet to articulate his theory of sexual selection.

Nonetheless, he had constructed a theory of organic change with which he was evidently well pleased. One could take a view of the world which was, as he saw it, "magnificent". Broad physical influences on the earth caused geographic and climatic changes, calling forth in turn adaptive changes in the organic world: "—instincts alter, reason is formed & the world peopled with myriads of distinct forms from a period short of eternity to the present time, to the future. —" This, Darwin told himself, involved a far grander view of God's laws than did the creationist view that God had occupied himself by creating "a long succession of vile molluscous animals". "How beneath the dignity of him," Darwin wrote, "who is supposed to have said let there be light & there was light" (*D* 36–37).

This broad view of organic change was not altered by Darwin's reading of Malthus's *Essay on the Principle of Population*. What reading Malthus changed for Darwin was Darwin's understanding of the means by which organic change takes place. Malthus's book focussed Darwin's attention on intra-specific struggle and the importance of individual differences in this struggle. This is what brought Darwin to the idea of natural selection. The impact on his thinking was immediate. The full repercussions of the impact, however, had to be worked out by Darwin over time.[7]

Before examining how Darwin's ideas on animal behavior developed following his reading of Malthus, it is worth considering Darwin's autobiographical account of what made it possible for him, upon reading Malthus, to realize "how selection could be applied to organisms living in a state of nature". Although the importance of behavioral considerations here may not have been as clear cut as they were when he appealed to instinctive aversions to assure reproductive isolation, once again it appears that behavioral considerations did have an important bearing on the development of his evolutionary theorizing.

Darwin explained in his *Autobiography* that to see, upon reading Malthus, how selection worked in nature, it was not sufficient that he had already "perceived that selection was the keystone of man's success in making useful races of animals and plants." It was also critical that he was "well prepared to appreciate the struggle for existence which everywhere goes on." The source of this appreciation, he said, was his "long-continued observation of the *habits* of animals and plants" (*Autobiography*, pp. 119–120; my italics).[8]

The word "habits", to be sure, is a word that his diverse meanings, and it is evident that Darwin in the sentence just quoted meant something more general than the meaning commonly associated with the word "habits" today. But if he was not referring to the particular behavior patterns that organisms acquire through experience in their lifetimes, he also evidently had something more specific in mind than the broad definition of habit as an organism's "mode of growth or general appearance". It would appear that by "habits" Darwin meant "ways of life", that is, ways of dealing with other organisms and the physical environment. That such dealings necessarily involved *struggle*, however, was, as Darwin acknowledged to himself, not something that was immediately apparent to every observer of nature. In one of the most poetic sentences of his transmutation notebooks, written in March 1839, Darwin stated: "It is difficult to believe in the dreadful but quiet war of organic beings going on [in] the peaceful woods & smiling fields" (*E* 114). The evidence for this "dreadful but quiet war", Darwin suggested, could be seen in "the multitude of plants introduced into our gardens. . . which are propagated with very little care, — & which might spread themselves as well as our wild plants, we see how full nature [is], how finely each holds its place. —" (*E* 114).

Although Darwin referred in the above passage only to plants, the passage is reminiscent of an earlier one from his B *Notebook* involving, among other examples, the ground woodpecker;

> There certainly appears attempt in each dominant structure to accommodate itself to as many situations as possible. — Why should we have in open country a ground woodpecker. — do. parrot. — a desert Kingfisher. — mountain tringas. — upland goose. — water chionis, water rat with land structures; carrion eagles. This is but carrying on attempt at adaptation of each element. — (*B* 55e).

This passage is reminiscent of the now famous passage Darwin wrote in his D *Notebook* in September 1838 upon reading Malthus: "One may say there is a force like a hundred thousand wedges trying [to] force every kind of adapted structure into the gaps in the oeconomy of nature, or rather forming gaps by thrusting out weaker ones. —" (*D* 135e). Darwin did not go on at this point in his D *Notebook* to provide examples of what he was talking about. It is evident, nonetheless, that the ground woodpecker of the pampas and the other examples mentioned in his B *Notebook* represented to him cases of adaptive structures thrust into "the gaps in the oeconomy of nature". This appears, at least, to be one kind of observation that made him "well prepared to appreciate the struggle for existence which everywhere goes on" in nature.[9]

Another sort of struggle that Darwin had considered before he read Malthus was the struggle between males for females. Darwin's grandfather, Erasmus Darwin, had commented on this struggle in his *Zoonomia*, noting that this combat had been the occasion for the development of special weapons on the part of the males, and that organic change was the consequence of the overall process. In Erasmus's words:

> The final cause of this contest amongst the males seems to be, that the strongest and most active animal should propagate the species, which should thence become improved. (1794–1796, 1: 507)

Despite his grandfather's discussion of male combat and its effects on the improvement of the species, Charles Darwin does not appear to have articulated his concept of sexual selection until after the concepts of natural selection and artificial selection were well established in his own mind.[10] He did consider the fights of males for females, however, and he worried whether animals have a sense of beauty. Just two weeks and a day before he read Malthus, Darwin wrote in his D *Notebook*: "The passion of the doe to the victorious stag, who rubs the skin of [f] horns to fight, is analogous to the love of women . . . to brave men. —" (*D* 99). Eleven days before he read Malthus Darwin was wondering whether females fight for males, and he noted: 'Singing best sign of most vigorous males . . . other birds display beauty of plumage" (*D* 114). This sexual struggle too may have thus been in Darwin's mind when he later recalled that the observation of "habits" had led him to some appreciation of struggle in nature before he read Malthus. As we shall see later in considering how Darwin developed his idea of sexual selection, he did distinguish rather carefully between the struggle for existence and the struggle for mates.

Coming to the idea of natural selection did not cause Darwin to discard entirely his previous understanding of the evolution of behavior. It did not require him to give up the idea that habits precede structures, and for the remainder of his career, in fact, he believed that at least some kinds of instinctive behavior were best explained in terms of habits or experiences that had become hereditary. The extraordinary, instinctive stinging behavior of the solitary wasps described by J. H. Fabre seemed to represent such a case (*LL* (NY) 2: 420–421). Nonetheless, while in his E *Notebook* he still called upon instincts to guarantee the reproductive isolation of varieties (*E* 143–144), the general role of habit and instinct in his new understanding of organic change was less than it had been previously. According to his new view, habits and instincts were *tested* and developed by the struggle for existence. As he put it:

> When two races of men meet, they act precisely like two species of animals. — they fight, eat each other, bring diseases to each other &c., but then comes the most deadly struggle, namely which have the best fitted organization, or instincts (i.e. intellect in man) to gain the day. (*E* 63–64)

In considering the strange plumage of some pigeons and the birds of paradise, Darwin stated:

> All that we can say in such cases is that the plumage has not been so injurious to bird as to allow any other kind of animal to usurp its place — & therefore the degree of injuriousness must have been exceedingly small. — This is more probable way of explaining, much structure, than attempting anything about habits. — (*E* 147)

Habit had by no means become inconsequential for Darwin's theorizing. While writing his E *Notebook* on transmutation he was also writing his

N *Notebook* on metaphysics and expression, and in the latter, referring to Sir John Sebright's notion of "hereditary habits", Darwin stated with evident enthusiasm: "Let the proof of hereditariness in habits be considered as grand step if it can be generalized" (*N* 63). But Darwin concluded soon enough that certain instincts were probably not best explained as habits become hereditary. This at least seemed to him to be the case for the instincts by which the solitary wasp provided grubs for larvae she would never see — a case insisted upon by Lord Brougham in his impressive analysis of instincts (Brougham 1845, pp. 14–15, 60–61; Richards 1981, pp. 210–211) — and it was probably also the case, Darwin decided, for the instinctive behavior of the tumbler pigeon and perhaps even for the instinctive behavior of pointer-dogs, although the latter on first glance seemed easily attributable to inherited experience (*1844 Essay*, p. 139).

In the *1842 Sketch* and in his longer manuscript, the *1844 Essay*, Darwin portrayed habits as being of great consequence for the evolutionary process, but not as being directly responsible for adaptive change. Habits generated variations upon which natural selection could act. Selection was thus the key for behavioral as well as structural evolution. In 1842 Darwin phrased it thus:

> It must I think be admitted that habits whether congenital or acquired by practice often become inherited; instincts, influence, equally with structure, the preservation of animals; therefore selection must, with changing conditions tend to modify the inherited habits of animals. If this be admitted it will be found *possible* that many of the strangest instincts may be thus acquired. (*1842 Sketch*, p. 55)

III. From the *1842 Sketch* to the *Origin of Species*

In his *1842 Sketch* and his *1844 Essay*, one can see Darwin not just developing a theory — a theory in which, whatever the source of variation, selection was the primary agent of change — but also working out a strategy of presentation.[11] In presenting his ideas to his contemporaries, Darwin's minimal goal with respect to behavior was the demonstration that complex instincts could be understood as the result of evolution by natural selection and did not have to be interpreted as the result of wise design on the part of the Creator. Knowing just how well complex instincts had served the purposes of natural theologians like William Kirby, Darwin wrote: "I want only to show that [the] whole theory ought not at once to be rejected on this score" (*1842 Sketch*, p. 56). To make his point, Darwin proceeded to address himself to some of the most problematic examples of instinct known. He did not, however, attempt to explain the origin of instincts, memory, attention, or any other faculty of the mind, any more than he attempted to explain the origin of life or even the origin of the different classes of animals. What he did argue was that given the existence of a species with particular instincts, there would be individual variations in these instincts, arising either congenitally or through habits, and these might serve as the source of further adaptive change through natural selection. The "smallest differences" in habits or instincts, he said, just like the smallest differences in structure, health, and so forth, told in the struggle for existence (*1842 Sketch*, p. 47).

Darwin repeated and expanded his brief claims of the *1842 Sketch* in his *1844 Essay*. After dealing in the first chapter of the *Essay* with "the variation of organic beings in a wild state" and "the natural means of selection", he turned immediately in his third chapter to the subject of instincts and other mental attributes. He offered a variety of facts to support the claim that in animals under domestication, "almost infinitely numerous shades of disposition, of tastes, of peculiar movements, and even of individual actions, can be modified or acquired by one individual and transmitted to its offspring" (*1844 Essay*, p. 138). He then argued that "the mental qualities modified or recently improved during domestication" resembled in a number of important respects the instincts of animals in the wild: each frequently required "a certain degree of education . . . to be perfectly developed"; each often involved an action performed without any apparent knowledge of the *purpose* of the action (a crucial point on which Lord Brougham had laid great stress); and each was nonetheless associated with "some degree of reason". Furthermore, the mixed nature of the instincts and habits of mongrels produced by crossing two different breeds of the same domesticated species corresponded to the mixed nature of the instincts and habits of the hybrids produced by crossing different species (pp. 139–141). As for variation in the mental powers of wild animals, this, Darwin said, was appreciated "by all those who have had the charge of animals in a menagerie" (p. 142). Additional evidence of the existence of slight individual differences in instinctive behavior, Darwin maintained, could be seen in the nests constructed by birds of the same species. These facts, Darwin felt, entitled him to claim that instincts as well as morphological characters were subject to the power of natural selection; and that "a series of small changes may, as in the case of corporeal structure, work great changes in the mental powers, habits and instincts of any species" (p. 143).

The problem with this explanation, Darwin recognized, was that the mechanism of change he was proposing — and perhaps any theory he might propose — was likely to strike the skeptical reader as altogether insufficient to account for "many of the more complicated and wonderful instincts" (*1844 Essay*, p. 143). He thus felt it necessary to address the problem of instincts before going further with the discussion of the various ways in which evidence from other sources either supported or opposed the idea of descent with modification. Among the "complicated and wonderful" instincts that he proceeded to treat were insects feigning death, animal migration and navigation, the nest-building of birds, the ability of hive-bees to build hexagonal cells, and parents bringing to their young food "which they themselves neither like nor partake of" (pp. 143–149).

With regard to each of the instincts mentioned, Darwin's strategy was the same. He maintained that by studying gradations in the instinctive behavior patterns of related species, the most singular instincts proved to be less extraordinary than they at first glance appeared. The Australian bush-turkey, for example, displayed the remarkable instinct of heaping together huge piles of fermenting materials, which then generated heat sufficient to hatch the bird's eggs. Darwin pointed to a related species from the tropics that buried its eggs, "apparently for concealment, under a lesser heap of rubbish"; and he suggested that under the right circumstances, the instinct of the bird from the tropics could have developed into the instincts of the Australian bird (*1844 Essay*, p. 144).

Having argued that it was indeed possible to conceive of the kinds of transitional stages that could have led up to the more remarkable instincts exhibited by present-day species, Darwin proceeded to argue that behavioral evidence was also relevant to the problem of the transitional stages between complex and well-adapted *corporeal structures*. To counter the objection that "in its transitional state [an organism's] habits would not be adapted to any proper conditions of life", Darwin returned to examples from his *Beagle* voyage and his transmutation notebooks, notably the ground-feeding woodpecker and the fish-eating jaguar. Of the latter he wrote:

> will it be said that it is *impossible* that the conditions of its country might become such that the jaguar should be driven to feed more on fish than they now do; and in that case is it impossible, is it not probable, that any the slightest deviation in its instincts, its form of body, in the width of its feet, and in the extension of the skin (which already unites the base of its toes) would give such individuals a better *chance* of surviving and propagating young with similar, barely perceptible (though thoroughly exercised), deviations? Who will say what could thus be effected in the course of ten thousand generations? (*1844 Essay*, pp. 152–153)

Darwin was still prepared, in other words, to suppose that changes in habits preceded changes in structures, even though he no longer argued that changes in habits produced changes in structure directly.

The explanation of instincts that Darwin advanced in the *Origin* was very much the same that he set out earlier in his *1842 Sketch* and *1844 Essay*. By 1859, though, his theory had been modified in two important respects, and instincts were involved in or at least affected by each of these changes. The first of these changes was that by 1859 Darwin was no longer inclined to believe that species were "perfectly" adapted to the conditions of their existence. The second was that Darwin's estimation of the importance of habits in the acquisition of new characters had decreased even further by 1859 than it had by 1844.

The more fundamental of these shifts in Darwin's thinking was the shift from a belief in perfect adaptation and limited variability in nature to a belief in relative adaptation and considerable variability in nature, a shift identified and admirably analyzed by Dov Ospovat (1979, 1981). In Ospovat's words:

> In the *Origin* Darwin said that there was much variability in nature, and he supposed that no change in conditions is necessary for variation and transmutation to occur. In 1844 he believed that there is little variation and that changes in conditions are necessary; and this was because in 1844 he still believed that in the absence of change organisms are perfectly adapted. (1979, p. 228)

It does not seem to be the case that Darwin's consideration of habits and instincts was especially instrumental in bringing him to the enlarged view of intraspecific variability that he had come to hold by 1859.[12] It may be the case, however, that thinking about "mistakes in instinctive behaviour" and continuing to think about how certain species, in adopting new habits, were able to fill "gaps in the economy of nature" did contribute significantly to his appreciation of the imperfection of adaptation.

Although the instincts of animals were commonly characterized by their *perfection* with respect to the needs of the individual or species, the most critical natural theologians, such as Lord Brougham, had acknowledged that there could be "mistakes" of instinct: "Mules begotten; flies deceived by the smell of the stapelia to lay their eggs where they cannot breed the maggots, supposing the vegetable an animal substance putrefying; and many others." Brougham, however, dismissed this problem optimistically, saying that such anomalies were no doubt only apparent and that further knowledge would probably explain them, reducing "every thing to order" and demonstrating the consistency of all with the "perfect wisdom and skill" of the Creator (Brougham 1845, p. 64).

Darwin read Brougham carefully, and in his *1844 Essay* he considered mistaken instincts briefly in citing Brougham's claim that animals are unaware of the purposes of their instinctive actions (*1844 Essay*, pp. 140–141). But Darwin did not at that time develop the idea that mistakes in instincts represented *imperfections* in adaptiveness. As of 1857, he was prepared to acknowledge that adaptation did not need to be perfect. "Natural selection," he told himself, "will not necessarily produce absolute perfection. . . ." All that it could do was assure that "each organism . . . be sufficiently perfect in all its parts to struggle with all its competitors in the same country. . . ." (*Natural Selection*, p. 380). In 1859 in the *Origin* he related this view particularly to instincts: "On the view of instincts having been slowly acquired through natural selection we need not marvel at some instincts being apparently not perfect and liable to mistakes, and at many instincts causing other animals to suffer" (p. 475).

In addition to laying more stress on the imperfection of instincts — and of adaptation in general — than he had previously, Darwin in 1859 also attributed less importance to the role of habit in the origin of instincts. In the years immediately after he read Malthus, as has been indicated above, Darwin still considered the inheritance of acquired habits to be a major source of the variations upon which natural selection acted in producing new instincts. By the time he wrote the *Origin*, though, he was prepared to state: "I believe that the effects of habit are of quite subordinate importance to the effects of natural selection of what might be called accidental variations of instincts. . . ." (*Origin*, p. 209). This demotion of the importance of habit in producing heritable variations was not specific to Darwin's treatment of instincts. It was crystallized, however, in the particular triumph for natural selection that Darwin found in confronting the instincts of neuter castes of social insects.

The problem of the specialized characters of neuter insect castes was one that Darwin identified in 1848 as "the greatest *special* difficulty I have met with" (Richards 1981, p. 221). Later, once he had solved it, he happily identified the difficulty as one "which at first appeared to me insuperable, and actually fatal to my whole theory" (*Origin*, p. 236). A certain amount of poetic license should perhaps be granted to Darwin with respect to this statement. In 1838, when he first noted the "wonderful" instinct by which worker bees take a neuter grub and turn it into a queen, he expressed no anxiety that the case might be "fatal" to his theory (*C* 221e). Furthermore, although in the 1840s he did come to see the characters of neuter insect castes as posing a serious special difficulty for him, it is not apparent that he felt unable to proceed with the public presentation of his theory of natural selection until he solved this difficulty in the way he solved it in the *Origin*, or that this difficulty was a crucial factor in the "delay" of Darwin's publication of his views, contrary to Richards's (1981) suggestion.[13] As Richards has pointed out, it appears to have been Lord Brougham's work that led Darwin initially to downplay the role of use-inheritance in the evolutionary process. Darwin himself wrote in 1857:

> For my own part, though I do not doubt that use & disuse may affect structures & be inherited, yet long before thinking of this case of neuter-insects I had concluded that the effects of habit were of quite subordinate importance. (*Natural Selection*, p. 365)[14]

To say that the problem of the characters of neuter castes of insects was not what initially guided Darwin's thinking on the importance of use-inheritance is not to deny the considerable importance of this case for the refinement of his thinking and the way he presented his views to the public. In the *Origin*, the explanation of the instincts and structures of the neuter castes was one of Darwin's real showpieces. He offered the special characters of the neuters as a prime example of "the power of natural selection", not only in showing "that natural selection could have been efficient in so high a degree" but also in providing a case in which "the most wonderful instincts" and structures could not possibly have been acquired through any amount "of exercise, or habit, or volition. . . ." (*Origin*, p. 242).

The steps by which Darwin arrived at his explanation of the evolution of the instincts and structures of neuter castes of insects have been described by Richards (1981). The particular case that Darwin chose to explicate in the *Origin* was that of the development of the characters of worker ants. The problem, as Darwin explained it, was that the worker ant differed greatly in its characters from its parents, but it was itself "absolutely sterile; so that it could never have transmitted successively acquired modifications of structure or instinct to its progeny" (*Origin*, p. 237). The solution to the problem, Darwin indicated, was that "selection may be applied to the family, as well as to the individual, and may thus gain the desired end" (p. 237). This was what enabled cattle breeders, having found in a slaughtered animal that the meat was well marbled, to breed with confidence from the same family as the slaughtered animal, and thereby develop the desired character even though the slaughtered animal never had any offspring. Natural selection, Darwin supposed, produced an analogous result in the social insects:

> Thus I believe it has been with social insects: a slight modification of structure, or instinct, correlated with the sterile condition of certain members of the community, has been advantageous to the community: consequently the fertile male and females of the same community flourished, and transmitted to their fertile offspring a tendency to produce sterile members having the same modification. And I believe that this process has been repeated, until that prodigious amount of difference between the fertile and sterile females of the same species has been produced, which we see in many social insects. (*Origin*, p. 238)

This could apply, Darwin felt, even to species in which the neuters were divided into more than one caste. Gradations of structure within the castes of the same species showed that there were differences upon which natural selection could act, that is, "by the long-continued selection of the fertile parents which produced most neuters with the profitable modification, all the neuters ultimately came to have the desired character" (*Origin*, p. 239). Noting the value of the division of labor for civilized man, Darwin allowed that a division of labor could also be of benefit to a community of social insects, and that this division could be effected best if the workers were sterile. This, he suggested, was what natural selection had accomplished, though as he acknowledged: "I am bound to confess, that, with all my faith in this principle, I should never have anticipated that natural selection could have been efficient in so high a degree, had not the case of these neuter insects convinced me of the fact" (p. 242). The conclusion of Darwin's chapter on instincts in the *Origin* was a powerful one:

> to my imagination it is far more satisfactory to look at such instincts as the young cuckoo ejecting its foster-brothers, — ants making slaves, — the larvae of ichneumonidae feeding within the live bodies of caterpillars, — not as specially endowed or created instincts, but as small consequences of one general law, leading to the advancement of all organic beings, namely, multiply, vary, let the strongest live and the weakest die. (pp. 243–244)

IV. Behavior and the *Descent of Man*

Darwin scholars regard the period from October 1836 — when Darwin returned from his *Beagle* voyage — to September 1842 — when Darwin left London to reside at Down — as the most creative period of Darwin's life. This is consistent both with Darwin's own assessment of his activities (*Autobiography*) and with the general evidence from the history of science, which indicates that scientific creativity tends to exhibit itself more frequently in the earlier rather than the later stages of a scientist's career. Darwin's thinking on behavior would appear to be no exception to this, for the main themes regarding behavior that are to be found in Darwin's later works are also to be found, in one form or another, in Darwin's manuscript notebooks from the late 1830s and the early 1840s. But it does not follow that Darwin's scientific activity after 1844 (or 1859) consisted of little more than developing in print the insights he first reached as a young man. Ospovat (1981) has identified various critical ways in which Darwin's thinking developed between 1844 and 1859, and Darwin's publications following the *Origin* give little reason to believe that Darwin's intellectual abilities faded once the *Origin* appeared. It is true that Darwin's later years have at times been represented as a period in which he occupied himself with relatively innocent and intellectually non-taxing subjects — for example the movements of plants or the habits of earthworms — and in which he lost confidence in natural selection as an explanatory mechanism (see Vorzimmer 1970). But the examination of Darwin's ideas on sexual selection, the evolution of the higher mental faculties of humans, and the expression of the emotions fails to support the idea that Darwin either faltered in intellectual power in his later years or retreated to a position of equivocation and confusion concerning how evolution works. Indeed, both in the *Descent* and the *Expression* Darwin offered special insights regarding the adaptiveness and non-adaptiveness of characters, insights related to his understanding of what natural selection could do and could not do. It does not detract from these insights to observe that Darwin reached certain of them at the same time he was endorsing ideas that are no longer considered correct (such as the inheritance of acquired characters). All this does is remind the historian of the necessity of understanding historical developments in their own historical setting.

Darwin's book *The Descent of Man, and Selection in Relation to Sex* is, as its full title indicates, a book with two parts. The parts were more closely related, both historically and conceptually, than is sometimes realized. In 1864, when Darwin wrote to Alfred Russel Wallace that he had "collected a few notes on man" (but doubted that he would ever use them), he registered his belief that "a sort of sexual selection has been the most powerful means of changing the races of man. I can show," Darwin claimed, "that the different races have a widely different standard of beauty" (*LL* (NY) 2: 272). In February 1867, after Darwin sent the manuscript of the *Variation* to the printer, he set to work on the subjects of man and sexual selection. He wrote to Wallace on 22 February stating:

> I am hard at work on sexual selection, and am driven half mad by the number of collateral points which require investigation, such as the relative number of the two sexes, and especially on polygamy. Can you aid me with respect to birds which have strongly marked secondary sexual characters. . . .? (*LL* (NY) 2: 274)

Four days later in another letter Darwin admitted to Wallace:

> The reason of my being so much interested just at present about sexual selection is, that I have almost resolved to publish a little essay on the origin of Mankind, and I still strongly think (though I failed to convince you, and this, to me, is the heaviest blow possible) that sexual selection has been the main agent in forming the races of man. (*LL* (NY) 2: 276)

Attending to the proofs of the *Variation* stalled Darwin's work on sexual selection and man for the rest of the year, but on 30 January 1868 the book was published and Darwin took up the subjects of sexual selection and the evolution of man again in earnest.

In conducting his researches on sexual selection Darwin did not work in isolation. He exchanged letters with dozens of naturalists, including Wallace, J. Jenner Weir, Fritz Müller, Edward Blyth, J. Blackwall, and many more.[15] With Wallace he had an extended, detailed, and extremely challenging exchange on the causes of sexual dimorphism, an exchange in which Wallace pressed Darwin on — among other things — the possibility that the plumage differences between the sexes in birds were due primarily to the females' need to be protectively colored (Kottler 1980, this volume). Darwin also had an extensive and immensely fruitful correspondence with Jenner Weir. Darwin suggested to Weir, as he had already suggested to W. B. Tegetmeier, the experiment of altering the appearance of a male bird to see what effects the change would have on the bird's success in courting (*ML* 2: 64–65). Other topics Darwin discussed with Weir included the number of birds in the wild that went unpaired, whether birds and butterflies exercise choice in mating, the way birds display the most beautiful parts of their plumage during courtship, Weir's experiments demonstrating the correctness of Wallace's idea that certain caterpillars are conspicuously colored for warning purposes, plumage relationships, the nest-building instinct, bird song, and so forth. With Fritz Müller, Darwin discussed, among other things, "how low in the scale sexual differences occur which require some degree of self-consciousness in the males, as weapons by which they fight for the female, or ornaments which attract the opposite sex" (*LL* (NY) 2:292-293). From Edward Blyth, Darwin received facts on the pugnacity of male birds and on the plumage relations of the young and the adult females and males of birds. From J. Blackwall, Darwin received information on sexual dimorphism in spiders.

On 21 May 1868 Darwin wrote to J. D. Hooker saying: "I have been working very hard — too hard of late — on Sexual Selection, which turns out a gigantic subject; and almost every day new subjects turn up requiring investigation and leading to endless letters and searches through books" (*ML* 1: 303). In November of the following year, in another letter to Hooker, Darwin complained: "I am sick of the work, and, as the subject is all on sexual selection, I am weary of everlasting males and females, cocks and hens" (*ML* 1: 316). When Darwin finally published the *Descent* in 1871, thirteen of the twenty-one chapters of the book — the thirteen chapters that he apparently wrote first[16] — were devoted to the subject of sexual selection. Sexual selection was not an ad hoc hypothesis added to the *Descent* when Darwin decided natural selection was unable by itself to account for human racial differences. Darwin had already discussed sexual selection in the *Origin*, though briefly. Once he decided to publish on human evolution, he treated sexual selection as an integral part of the project.

Darwin's self-proclaimed goal in the *Descent* was to consider explicitly "whether man, like every other species, is descended from some pre-existing form" (*Descent* 1: 2). Among the points Darwin attempted to establish in arguing that man is indeed descended from a pre-existing form, two were especially significant for the further development of animal behavior studies. The first was the claim that "the difference in mind between man and the higher animals, great as it is, is certainly one of degree and not of kind" (1: 105). The second was the claim that the differences in the secondary sexual characters of humans, like those of the higher animals, could be explained as the result of sexual selection, typically involving either male combat or female choice. These two points will be considered here in the order Darwin presented them.

In arguing for the continuity between the mental faculties of man and the higher animals, Darwin was not hesitant in maintaining that "man and the higher animals, especially the Primates, have some few instincts in common" (*Descent* 1:48). In his words:

> All have the same senses, intuitions and sensations — similar passions, affections, and emotions, even the more complex ones; they feel wonder and curiosity; they possess the same faculties of imitation, attention, memory, imagination, and reason, though in very different degrees. The individuals of the same species graduate in intellect from absolute imbecility to high excellence. (1: 48–49)

In preparing the second edition of the *Descent*, Darwin saw fit to expand this list, adding to the attributes of the higher animals "jealousy, suspicion, emulation, gratitude, and magnanimity", and allowing further that these animals "practice deceit and are revengeful; they are sometimes susceptible to ridicule, and even have a sense of humour" (*Descent* 1874, p. 79). Much of the evidence Darwin offered to support his claims about the mental continuity between man and the higher animals was anecdotal, and his interpretation of his animal evidence was decidedly anthropomorphic. These features necessarily make the *Descent* appear quite dated to the modern reader, for anecdotalism and anthropomorphism were in large measure drummed out of scientific discussions of animal behavior at the turn of the century. The conceptual power of Darwin's book thus stands somewhat disguised.[17]

Darwin did not base his argument for the evolution of the higher mental faculties solely on evidence for the continuity between the mental faculties of man and those of the higher animals. He also offered an explanation of the means by which the higher mental faculties could have evolved. In doing so, he paid particular attention to "the moral sense or conscience", because this, he said, was the feature in which man seemed most different from the animals. Many other writers, he indicated, had discussed the moral sense, but no one, as far as he knew, had "approached it exclusively from the side of natural history (*Descent* 1: 71).

Darwin proceeded to argue that the social instinct, like most of the other instincts of animals, had been developed chiefly through natural selection:

> For with those animals which are benefited by living in close association, the individuals which took the greatest pleasure in society would escape various dangers; whilst those that cared least for their comrades and lived solitary would perish in greater numbers. (*Descent* 1: 80)

Darwin also supposed that some kinds of behavior, such as that of bees destroying their nearest relations, had probably been selected because it was of "service to the community" (1: 81). Darwin did not, however, equate "the good of the community" with "the good of the species". Furthermore, he recognized difficulties with "the good of the community" argument.

Darwin's sensitivity to problems regarding the level at which selection operates is evidenced in his discussion of the origins of altruistic behavior in humans. It was clear enough, Darwin felt, that a tribe possessing individuals with well-developed moral qualities would have an advantage over a tribe whose members lacked these qualities. Alfred Russel Wallace had made this point earlier in 1864. As Darwin put it in 1871:

> When two tribes of primeval man, living in the same country, came into competition, if the one tribe included (other circumstances being equal) a greater number of courageous, sympathetic, and faithful members, who were always ready to warn each other of danger, to aid and defend each other, this tribe would without doubt succeed best and conquer the other. (*Descent* 1: 162)

The difficulty with the situation described above, Darwin acknowledged, was in accounting for how altruistic behavior could have arisen and been developed among the members of a single tribe in the first place, since those individuals most likely to sacrifice themselves for others would have been the ones least likely to leave offspring. In his words:

> It is extremely doubtful whether the offspring of the more sympathetic and benevolent parents, or of those who were the most faithful to their comrades, would be reared in greater numbers than the children of selfish and treacherous parents of the same tribe. He who was ready to sacrifice his life, as many a savage has been, rather than betray his comrades, would often leave no offspring to inherit his noble nature. The bravest men, who were always willing to come to the front in war, and who freely risked their lives for others, would on an average perish in larger numbers than other men. Therefore it seems scarcely possible (bearing in mind that we are not here speaking of one tribe being victorious over another) that the number of men gifted with such virtues, or that the standard of their excellence, could be increased through natural selection, that is, by the survival of the fittest. (*Descent* 1: 163)

Just how the social virtues could have been developed within a single tribe, Darwin decided, was too complex a process to attempt to reconstruct with any confidence. He did suggest, nonetheless, that an individual of a tribe might initially acquire the habit of helping his fellows from the "low motive" of perceiving that this might help him receive aid in return, and that the habit of performing benevolent acts, carried out over many generations, would probably come to be inherited. Darwin was not, however, fully satisfied with inherited habit as the explanation of the origin and development of the social virtues. A much more powerful stimulus to the development of the social virtues, he maintained, was the love of praise and the dread of blame. The two of these had developed, he supposed, in conjunction with the instinct of sympathy — an instinct that itself "no doubt was originally acquired, like all the other social instincts, through natural selection" (*Descent* 1: 164). In discussing the evolution of the higher intellectual faculties, Darwin also suggested a mechanism very much like what is now called kin selection when he noted that even if certain specially-endowed individuals failed to leave offspring of their own, "the tribe would still contain their blood relations", and this would allow for the further development of the characters in question. The explanatory model he used in this instance was the same model from animal breeding he had used to illuminate the evolution of the structures and instincts of castes of neuter insects: "by preserving and breeding from the family of an animal, which when slaughtered was found to be valuable, the desired character has been obtained" (1: 16).

If the problem of the evolution of altruistic behavior led Darwin to address, if only briefly, potential conflicts between selection acting at the level of the individual and that of the community, the problem of the evolution of secondary sexual characters led him to reflect further on the powers and limitations of natural selection. Indeed, Darwin's discussion of sexual selection in the *Descent* affords a special, if not unique, opportunity to consider what Darwin, in his later years, believed natural selection was capable of accomplishing.

As Darwin explained in the *Descent*, the kinds of differences between the sexes he had in mind when talking about "secondary sexual characters" were not the differences that resulted when the males and the females of the same species had markedly different habits of life, for such differences could be accounted for by the natural selection of those individuals best fitted to survive in the struggle for existence. The differences Darwin had in mind — the differences produced by sexual selection — included such features as

> The weapons of offence and the means of defence possessed by the males for fighting with and driving away their rivals — their courage and pugnacity — their ornaments of many kinds — their organs for producing vocal or instrumental music — and their glands for emitting odours; most of these latter structures serving only to allure or excite the female.
> (*Descent* 1: 257–258)

Unlike natural selection, Darwin explained, sexual selection "depends on the advantage which certain individuals have over other individuals of the same sex and species, in exclusive relation to reproduction" (1: 256). The issue, in other words, was not one of survival, but rather one of securing a mate and leaving progeny, for it was clear, Darwin said, that "unarmed, unornamented, or unattractive males would succeed equally well in the battle for life and in leaving a numerous progeny, if better endowed males were not present" (1: 258). Darwin's ideas on sexual selection are thus extremely important for understanding Darwin's evolutionary thought, for they illuminate Darwin's understanding of natural selection, relative adaptation, and fitness in a way that can scarcely be appreciated if one defines these words simply in terms of "differential reproduction".[18]

In the *Descent*, Darwin portrayed sexual selection as "an extremely complex affair, depending as it does on ardour in love, courage and the rivalry of the males, and on the powers of perception, taste, and will of the female" (1: 296). He acknowledged that sexual selection would be "dominated by natural selection for the general welfare of the species" (1: 296). Males, he explained, would not acquire characters that "would be injurious to them in any high degree" by causing them to expend "too much of their vital powers, or by exposing them to any great danger" (1: 278–279). At the same time, however, he admitted:

> The development. . .of certain structures — of the horns, for instance, in certain stages — has been carried to a wonderful extreme; and in some instances to an extreme which, as far as the general conditions of life are concerned, must be slightly injurious to the male. From this fact we learn that the advantages which favoured males derive from conquering other males in battle or courtship, and thus leaving a numerous progeny, have been in the long run greater than those derived from rather more perfect adaptation to the external conditions of life. (*Descent* 1: 279)

Here was an extremely important insight on Darwin's part. It illustrates that Darwin did not define natural selection simply in terms of differential reproduction. As Darwin explained in the *Descent*, what natural selection did was produce adaptations to the conditions of life; it made organisms "better fitted to survive in the struggle for existence" (1: 257). But organisms less well adapted for the struggle for existence could still leave the most offspring. Fitness for Darwin was thus not defined in terms of reproductive success, although the two of these often went together.

Significantly, this disparity between fitness and reproductive success was something that Darwin acknowledged was also to be found in modern human populations. Darwin noted, for example, that certain social institutions permitted the reproduction of "the weak members of civilized societies", and he cited with evident approval W. R. Greg's (1868) observation that if one contrasted the numbers of progeny of "the careless, squalid, unaspiring Irishman" with "the frugal, foreseeing, self-respecting, ambitious Scot", one would find that "in the eternal 'struggle for existence', it would be the inferior and *less* favoured race that had prevailed — and prevailed by virtue not of its good qualities but of its faults" (*Descent* 1: 174).

In contrast to what happened because of human social institutions, sexual selection, as Darwin portrayed it, was a process that operated in nature. Sexual selection explained how certain imperfections with respect to the general struggle for existence could be maintained and even fostered in a state of nature. To be precise, these imperfections could be maintained and fostered if they were of use in the struggle for mates. For example, the bright colors that made a male bird more conspicuous to its enemies also made it more attractive to the females of its species. To paraphrase

W. R. Greg — although Darwin himself did not identify the parallel between Greg's words on man and his own words with respect to sexual selection — the brightly colored bird thus prevailed in a reproductive sense by virtue of characters which, with respect to the general struggle for existence, could only be identified as faults.[19]

This conclusion of Darwin's regarding sexual selection was one that Alfred Russel Wallace, significantly enough, was unwilling to accept. Where Darwin emphasized success in mating and *relative* adaptation, Wallace emphasized individual vigor and *perfect* adaptation. This can be seen, for example, in Wallace's book *Darwinism* (1889), in which he discussed "the enormously lengthened plumes of the bird of paradise and of the peacock". Although Wallace acknowledged that these feathers "must be rather injurious than beneficial in the bird's ordinary life", both the specific explanation he offered for these feathers and the general thrust of his argument were markedly different from Darwin's. As Wallace put it:

> The fact that [these feathers] have been developed to so great an extent in a few species is an indication of such perfect adaptation to the conditions of existence, such complete success in the battle for life, that there is, in the adult male at all events, a surplus of strength, vitality, and growth-power which is able to expend itself in this way without injury. (1889, pp. 292–293)

If the major significance of Darwin's concept of sexual selection for understanding his theorizing as a whole is the illumination it provides concerning Darwin's thoughts on adaptiveness, reproductive success, and what natural selection can and cannot do, the major impetus Darwin's theory of sexual selection had for animal behavior studies per se was the attention it drew to the function of *displays* in animal behavior. The most controversial part of the theory of sexual selection in regard to these displays was Darwin's notion of "female choice".

Wallace had at first granted to Darwin that "female choice" might have played a role in developing the secondary sexual characters of some animals, but he eventually came to reject this idea entirely (Kottler 1980). Darwin himself acknowledged that the idea seemed improbable at first glance. "It could never have been anticipated," he wrote, "that the power to charm the female has been in some few instances more important than the power to conquer other males in battle" (*Descent* 1: 279). But Darwin went on to argue that what seemed improbable was in fact the case. Inasmuch as Darwin's concept of female choice — or the male "charming" the female — has been the subject of certain misunderstandings, it is worth paying close attention to his comments on the subject.

Sexual selection, Darwin maintained, was not something that should be expected to apply to all the classes of animals, since the representatives of the lowest animal classes lacked the mental power either to feel rivalry, which was necessary for male combat, or to appreciate beauty, which was necessary for female choice. "Considerable perceptive powers and. . .strong passions", Darwin wrote, were a prerequisite for sexual selection's operation (*Descent* 1: 377). The results of sexual selection's operations, therefore, were not visible in the animal scale until the higher invertebrates (crustaceans, spiders, and insects). These results proceeded then to manifest themselves all the way up to man, where, in Darwin's view, sexual selection had been the chief agent in establishing the distinguishing features of the different races.

That "female choice" had been involved in producing these different secondary sexual characters, Darwin maintained, was a necessary conclusion when one considered animal displays. To make his case for the existence of female choice in birds, Darwin offered the analogy of what a hypothetical visitor from another planet, observing the behavior of "young rustics at a fair", would have to conclude:

> If an inhabitant of another planet were to behold a number of young rustics at a fair, courting and quarrelling over a pretty girl, like birds at one of their places of assemblage, he would be able to infer that she had the power of choice only by observing the eagerness of the wooers to please her, and to display their finery. Now with the birds, the evidence stands thus; they have acute powers of observation, and they seem to have some taste for the beautiful both in colour and sound. It is certain that the females occasionally exhibit, from unknown causes, the strongest antipathies and preferences for particular males. When the sexes differ in colour or in other ornaments, the males with rare exceptions are the more decorated, either permanently or temporarily during the breeding-season. They sedulously display their various ornaments, exert their voices, and perform strange antics in the presence of the females. (*Descent* 2: 122–123)

Darwin insisted that it was inconceivable that such displays were without a purpose, that "all the labour and anxiety exhibited by [the males] in displaying their charms before the females" was to no avail (*Descent* 1: 64). One was therefore justified, he believed, in concluding "that the female exerts a choice, and that she receives the addresses of the male who pleases her most (2: 123). Darwin did not, however, suppose that the exertion of a "choice" necessarily involved conscious deliberation on the part of the female. Instead, he maintained, all he really meant by "choice" was that the female was "most excited or attracted by the most beautiful, or melodious, or gallant males" (2: 123). He reiterated this point in the scientific notice that was read at the Zoological Society of London the day before he died:

> may naturalists doubt, or deny, that female animals ever exert any choice,

so as to select certain males in preference to others. It would, however, be more correct to speak of the females as being excited or attracted in an especial degree by the appearance, voice, &c. of certain males, rather than of deliberately selecting them. *CP* 2: 278)

V. Expression, Evolution and Adaptiveness

Darwin made sense out of secondary sexual characters and display behavior by identifying their usefulness in the competition for mates rather than in the general struggle for existence. He did appreciate, however, that some structures and behavior patterns might not be particularly useful either in the struggle for existence or in the struggle for mates. He sounded this theme in a number of places in his writings. With respect to behavior, the area in which he sounded it most emphatically was in his discussion of the expression of the emotions in man and animals.

Since Darwin's book the *Expression* is described in detail in J. Browne's paper in this volume, it is unnecessary here to survey the contents or general argument of that book. In the present paper the particular aspect of Darwin's work on the emotions that will be considered is the argument identified above: the argument that many forms of emotional expression are without adaptive value. This was not the main argument of the book. It constitutes another important example, nonetheless, of the way in which the consideration of behavioral phenomena was the occasion of the elaboration by Darwin of a significant insight concerning the evolutionary process. It illustrates, furthermore, that Darwin's evaluation of behavioral phenomena did not take place independently of interpretive contexts provided by his predecessors.

In introducing the topic of Darwin's views on the non-adaptiveness of emotional expression, it is worth recalling certain comments on the non-adaptiveness of structures that Darwin offered in the *Descent*. There Darwin allowed that in the early editions of the *Origin* — that is, prior to the fifth edition — he had "probably attributed too much to the action of natural selection or the survival of the fittest". As he put it: "I had not formerly sufficiently considered the existence of many structures which appear to be, as far as we can judge, neither beneficial nor injurious; and this I believe to be one of the greatest oversights as yet detected in my work" (*Descent* 1: 152).

Whether or not Darwin was overstating here the extent of his earlier inattention to the non-adaptiveness of characters is perhaps debatable. From the very first edition of the *Origin*, for example, he had commented upon "the importance of the laws of correlation in modifying important structures, independently of utility and, therefore, of natural selection" (*Origin* p. 144). Be that as it may, the *excuse* Darwin offered in 1871 for what he called "one of the greatest oversights" of his earlier work deserves mention here. In writing the *Origin*, Darwin said, he had "had two distinct objects in view, firstly, to shew that species had not been separately created, and secondly, that natural selection had been the chief agent of change, though largely aided by the inherited effects of habit, and slightly by the direct action of the surrounding conditions" (*Descent* 1: 152–153). His excuse for his oversight was the following:

> Nevertheless, I was not able to annul the influence of my former belief, then widely prevalent, that each species had been purposely created; and this led to my tacitly assuming that every detail of structure, except rudiments, was of some special, though unrecognized service. Any one with this assumption in mind would naturally extend the action of natural selection, either during past or present times, too far. (*Descent* 1: 153)

Darwin allowed, in other words, that in retreating from and reacting against a particular position — the dogma of special creations — he at least temporarily overlooked an important point, namely that certain characters might not be adaptive. Clearly, as the quote from the *Origin* cited above indicates, he had not overlooked this point altogether. Nonetheless, he does seem to have been more prepared to develop this point in 1872 than he had been thirteen years earlier. This was in part due to certain criticism that the *Origin* had received. An additional and crucial stimulus to his thinking on the non-adaptiveness of emotional expressions, however, was his antagonism to the creationist interpretation that had informed the studies on emotional expression in his day.

Darwin acknowledged that what got him started on the subject of emotional expression was, in 1838, reading Sir Charles Bell's *Anatomy and Philosophy of Expression* (1824) (*Descent* 1: 5). Writing to Alfred Russel Wallace in March 1867 on the subject of the expression of the emotions, Darwin stated:

> The subject is, I think, more curious and more amenable to scientific treatment than you seem willing to allow. I want, anyhow, to upset Sir C. Bell's view . . . that certain muscles have been given to man solely that he may reveal to other men his feelings. (*LL* (NY) 2: 278)

Darwin's representation of Bell's position was accurate. Bell had written:

> in man there seems to be a special apparatus, for the purpose of enabling him to communicate with his fellow creatures, by that natural language which is read in the changes of his countenance. There exist in his face, not only all those parts which by their action produce expression in the several classes of quadrupeds, but there is added a peculiar set of muscles to which no other office can be assigned than to serve for expression. (1872, p. 121)

Darwin saw Bell's view as an obvious threat to his own idea that man had descended from a lower form, and Darwin's book, the *Expression*, constituted a refutation of Bell's position.

Darwin cited the work of anatomists to show that the corrugator muscles, which make frowning possible and which Bell had claimed were unique to the human race, were also to be found in orangutans and chimpanzees, although in these lower forms the corrugators were not nearly so well developed (Bell 1872, pp. 137, 139; *Expression*, p. 222). Darwin denied, furthermore, Bell's general contention that there was a sharp discontinuity between the expression of the emotions in animals and the expression of the emotions in man (Bell 1872, p. 141; *Expression*, pp. 146, 367). As for the view of Bell and others that blushing was a special provision for expression, Darwin argued that blushing was of no service either to the blusher or to the beholder, and he pointed out that blushing occurred even in dark-coloured races, where it was "scarcely or not at all visible" (Bell 1872, pp. 95–96; *Expression*, p. 338). The special purpose that Bell had assigned to blushing and other expressive actions in humans, Darwin insisted, was neither essential to these actions, nor were these actions sufficient to set humans off from the rest of the animal kingdom.

On the basis of his studies, Darwin concluded that most expressive actions in humans were *instinctive*. He did not regard these actions, however, as necessarily *adaptive*. He acknowledged that certain expressive actions reveal the state of the mind, but he maintained that "this result was not at first either intended or expected" (*Expression*, p. 357). He indicated further that if humans were able to *recognize* expressive actions *instinctively* — which seemed likely to him but not certain — this was probably an instance of experience having become hereditary, that is, an instance of the inheritance of acquired characteristics. Darwin decided that the expression of the emotions could be explained in terms of the three principles of "serviceable associated habits", "antithesis", and "actions due to the constitution of the nervous system" (*Expression*, pp. 28–29). He refrained from adding a fourth principle involving the natural selection of expressive gestures of communicative value. As he put it, "there are no grounds, as far as I can discover, for believing that any muscle has been developed or even modified exclusively for the sake of expression" (p. 355).

In making the above claim Darwin did not deny that certain organs had been developed for sexual signalling, that is, so that "one sex might call or charm the other" (*Expression*, p. 355). He also acknowledged, if only briefly, that certain species had apparently developed through natural selection certain warning sounds or threat postures. By and large, though, he did not call upon selection to explain major features of emotional expression. In this area, as much as in any other, he showed an appreciation that imagining how a character might be useful does not necessarily constitute an adequate explanation of why it exists or how it originated.

Darwin had, to be sure, advanced a similar argument a decade earlier in his book *Orchids* (1862; second edition 1877) (Ghiselin 1969). But the thrust of the earlier argument had been different, for there he had still been focussing on what he considered to be *adaptations*, and his point was that parts that serve a special purpose at the present may have developed from parts which served a very different purpose in the past. As he put it: "throughout nature almost every part of each living being has probably served, in a slightly modified condition, for diverse purposes, and has acted in the living machinery of many ancient and distinct specific forms" (*Orchids* 1877, p. 284). His emphasis then was still on "the use of each trifling detail of structure" (p. 286).

The thrust of Darwin's argument in the *Expression* was not that the expression of the emotions in man and animals exhibited the modification of actions and structures for new purposes, but rather that these actions and structures could not be understood in terms of purpose at all: they had been neither specially created nor naturally selected for a communicative function. This can only be interpreted as a retreat on Darwin's part if one assumes that Darwin was more committed to arguing for the all-sufficiency of natural selection than to demonstrating the reality of evolution, and this was not the case. In the *Origin* Darwin had felt compelled to combat the creationist view by showing that the most difficult cases of adaptation — such as the instincts of neuter castes of insects — could be explained by natural selection. In the *Expression*, in contrast, Darwin was able to refute Bell's position by pointing out continuities in the structures and expressions of animals and humans and by denying that emotional expression in humans could be understood only in terms of the communicative function that Bell had ascribed to it. Interestingly enough, in constructing his argument against the idea that special structures in man had been designed by the Creator for the purpose of non-verbal communication, Darwin appears to have overreacted, thereby leaving himself ill-disposed to develop an idea that would later be advanced by the ethologists of the twentieth century — the idea that certain expressive actions, whatever their primary origin, had been developed over time by natural selection.[20]

Conclusion

The comments that have been offered here on the development of Darwin's thoughts on the evolution of behavior and the role of behavior in the evolutionary process by no means exhaust these subjects. These comments may suffice, however, to indicate something of the extent of Darwin's interest in behavior and to demonstrate that behavior was not simply something to which Darwin sought to apply his evolutionary theory as it developed. Instead behavior was a source of a variety of considerations that informed Darwin's continuing analysis of organic change.

The main points set forth in this paper can be summarized briefly. From the beginning of his career, Darwin regarded the behavior of animals as one of the key interests of natural history. On his *Beagle* voyage, and afterwards, when he reflected on his *Beagle* observations, among the observations that particularly impressed him were those having to do with the way the *habits* of particular species related both to their particular structures and to the stations they occupied in their respective geographical locations. This evidence, Darwin soon decided, bore directly on the possibility of the transmutation of species. In developing his pre-selectionist understanding of organic change, Darwin called upon behavior — specifically instincts — to assure the reproductive isolation that he identified as central to the evolutionary process. He then proceeded to develop the idea that habits *precede* structures. When he read Malthus in September 1838, it was not just Darwin's thinking about domesticated forms and the practices of breeders that enabled him to derive from Malthus the concept of natural selection. It was also that his observations of the "habits" of organisms had given him already a sense of "struggle" in nature. The competition of individuals for mates and the way species moved into "gaps in the economy of nature" have been advanced here as likely candidates for what Darwin may have had in mind in this reference to "habits".

As Darwin proceeded to develop his idea of natural selection after 1838, he came to attribute less of a role to behavior as a direct generator of adaptive change than he had previously. In other ways, however, behavioral considerations continued to inform his theory. Thinking about mistakes in instincts may, for example, have been one source of his emerging realization that adaptation was only relative and not perfect. In the *Origin*, Darwin offered the instincts and structures of the neuter castes of insects as providing his finest demonstration of the power and the unique ability of natural selection to explain complex adaptations. In the *Descent*, he attributed a special role to behavior — through sexual selection — in shaping future generations in ways that might even be inconsistent with "the survival of the fittest". Indeed, in the course of his discussion of sexual selection he provided some of his clearest statements regarding what natural selection could and could not do. In the same book, his consideration of the evolution of altruistic behavior in humans led him to reflect on potential conflicts between selection at the level of the individual and selection at the level of the community. The following year, his book on the expression of the emotions proved to be an occasion for him to demonstrate at length that he did not believe that all characters needed to be understood primarily in terms of their adaptiveness. In short, behavioral phenomena provided Darwin not simply with a mine of facts requiring explanation but also with a source of issues and answers that were of fundamental importance to him as he pursued the whole subject of the process and products of organic evolution.

My intention here has not been to claim that behavior was the *key* to Darwin's evolutionary theorizing, only that it was an integral part of his thinking (the importance to Darwin of other subjects is detailed in the other studies in this volume). Darwin's work on behavior does seem to tell us some interesting things about Darwin's thinking that have not always been acknowledged, for example that Darwin's theoretical acuity continued well after the publication of the *Origin* and that Darwin was not an uncritical exponent of the ideas of adaptation and natural selection. As Darwin's discussions of sexual selection and the expression of the emotions suggest, his views about adaptiveness were rich and subtle rather than dogmatic. And as the structure of his argument in the *Expression* indicates, Darwin's main effort even in his later years continued to be more the refutation of the creationists and the establishment of the reality of evolution than the promotion of the explanatory adequacy of natural selection. Furthermore, as an *integral* part of his thinking, Darwin's concerns with behavior and psychology provided connections between his social views and his evolutionary theory of a more direct nature than writers such as Young (1973b), who have focussed on broader, underlying ideas, have been concerned to identify. The topics of instinctive aversion to intervarietal mating, altruistic behavior, female choice, and the survival of the *less* fit all figured, as has been shown here, in Darwin's evaluation of the means by which evolution takes place. These are all topics through which ideological factors may have had a bearing on Darwin's theorizing. This subject has not been developed in this paper, but the identification here of the points of contact between these issues and Darwin's evaluation of the mechanisms of evolution might serve as a means of beginning to explore this subject.

In conclusion, it is worth restating that behavioral concerns formed only a part of what de Candolle called Darwin's *oeuvre*, and Darwin never took it upon himself to establish an evolutionary science of behavior in its own right. Furthermore, Darwin's writings did not give rise immediately to a new science of behavior. This is evidenced by Poulton's comments (cited earlier), and it is also suggested by the examples of major observers of animal behavior of the latter half of the nineteenth century who did not conceive of their work in evolutionary terms. There were, to be sure, a number of naturalists and psychologists in this period who did proceed to analyze behavior in evolutionary terms (C. Lloyd Morgan is a prominent example). Furthermore, behavioral questions continued to be of interest in certain turn-of-the-century debates concerning the mechanisms of evolution. Nonetheless, it was not until the twentieth century that the biological study of behavior emerged as a field in its own right, and even then the development of a strong evolutionary base for this field did not come easily (Burkhardt 1983). However much Darwin may have provided the conceptual foundations on which an evolutionary science of behavior could have been erected, he did not create the speciality himself. He remained at one and the same time the naturalist who had written enthusiastically in his early manuscript notebooks about the material basis of mind and the continuity between the higher animals and man, but who had also told himself:

> When we talk of higher orders we should always say intellectually higher. — But who with the face of the earth covered with the most beautiful savannahs and forests [will] dare to say that intellectuality is [the] only aim in this world. (*B* 252)

Notes

1. There is no comprehensive treatment of Darwin's ideas on behavior, although a number of works deal with aspects of the subject. Ghiselin (1969) provides the best sense of the range and theoretical significance of Darwin's behavior studies. Gruber describes in more detail Darwin's thoughts on instinct and the higher mental faculties and the development of Darwin's thoughts on these subjects over time (Gruber and Barrett 1974), but Gruber has less to say than Ghiselin about the particular theoretical insights afforded Darwin by his behavioral studies, and he scarcely mentions the topic of sexual selection. Other works dealing with one or another part of Darwin's thinking on behavioral subjects include Herbert (1974, 1977), Kottler (1980), Mayr (1972b), Richards (1981) and Swisher (1967).
2. Swisher writes of Darwin's arriving at his theory of natural selection and then applying natural selection to the study of behavior, although, contrary to Swisher, Richards notes that Darwin took some time before he did so (1981, pp. 199, 205). Kohn indicates that before Darwin came to the idea of natural selection he applied his earlier, sexual theory to the explanation of the behavior of birds and man (1980, p. 131). The emphasis of the present paper is on how Darwin's thoughts on behavior contributed to, as well as reflected, his evolutionary theorizing throughout his career.
3. Barlow's statement is based on her reading of Darwin's "Diary for 1826" (DAR 129) and the small pocketbooks in which Darwin recorded the observations he made on his expeditions while on the *Beagle* voyage.
4. William Sharp MacLeay proposed a system of classification in which the plant and animal kingdoms were arranged as systems of "circular reticulations". Continuities between forms were represented by the touching or "inosculating" of the circles, and particular organisms were said to be "osculant" if they fell between the circles by combining the characteristics of two quite distinct groups. On MacLeay's system see Ospovat (1981, pp. 101–113) and Rachootin (this volume).
5. For the dating of Darwin's notebooks on transmutation and metaphysics see Herbert (1977).
6. For Lamarck's evolutionary theory and understanding of behavior see R. Burkhardt (1977 and 1981, respectively).
7. This topic has been treated with special care in Ospovat (1981).
8. Kohn (1980) deserves the credit for calling attention to Darwin's recollection that it was thinking about not only artificial selection but also the habits of plants and animals that prepared him to derive the concept of natural selection from his reading of Malthus.
9. One of the things Darwin learned in South America was that during the great drought that occurred between 1827 and 1830 large numbers of many kinds of animals perished, but there was a tremendous increase in the population of mice (*Natural Selection*, p. 178). In his *Zoology* he comments on the surprising numbers of mice that turned up in his traps (3: 31). The mice of South America may thus have provided him with another example of the struggle for existence going on in nature, and a special case of how populations might increase dramatically in size with changed environmental conditions.
10. Ghiselin (1969) has noted that Darwin did not elaborate his theory of sexual selection until after his idea of artificial selection was in place, and that the structure of the argument for sexual selection was comparable to the structure of the argument for artificial selection.

    That Darwin had not developed the notion of "female choice" at the time he began his notebooks on the transmutation of species is evidenced by his early denial that animals have a sense of beauty. Animals, he said, are prevented from mating with dissimilar types by their instincts, not by a sense of beauty. He came to doubt, however, that there was a sharp distinction to be drawn between the instincts of animals, on the one hand, and the sense of beauty in humans on the other (see *C* 178), and in mid-July 1838, he announced to himself in his M *Notebook*: "Beauty is instinctive feeling, & this cuts the knot." He went on to note that Sir Joshua Reynolds's (1778) views on artistic taste might apply to whites aquiring one instinctive notion of beauty and blacks another (*M* 32).

11. Darwin's strategy of presentation has been dealt with in an especially insightful fashion in Herbert (1974, 1977).
12. Ospovat indicates that it was Darwin's work on barnacles and on variation in large genera that apparently led him to conclude there was a great deal of individual variability in organisms in a state of nature (1981, pp. 200–205).
13. Richards's (1981) description of Darwin's thinking on neuter insects is the best treatment of the subject. Richards probably exaggerates, however, the role of neuter insects in "delaying" Darwin's publication of his evolutionary theory. His analysis does not distinguish clearly between the idea of habits as a direct cause of adaptive change and the idea of habits as a source of inheritable variations on which selection could act.
14. This claim is supported by Darwin's annotations of Kirby and Spence (1818–1826). On the last page of Darwin's copy of the first volume of Kirby and Spence's *Introduction to Entomology*, now preserved in the Cambridge University Library, Darwin wrote: "As neuters are sometimes converted into Queens & then breed my argument against instinct arising from habit, is not perfect." Here Darwin's anxiety seems not to have been the difficulty of accounting for the instincts of a caste that could not reproduce its kind, but rather that the example did not provide as air-tight an argument *against* the inheritance of acquired characters as one might suppose. Darwin's annotations of Kirby and Spence indicate that at the time he read that work, he appreciated both that the only instincts neuter insects could have acquired through habit were those actually acquired earlier by the females and that some instinctive tasks performed by the workers could not have been originally performed by the females. As he wrote concerning page 148 of volume two of Kirby and Spence's work: "it is difficult to believe the workers could have acquired this instinct when they were females before their neutrality was gained."
15. See *Handlist of Darwin Papers at the University Library, Cambridge* (Cambridge, 1960), pp. 20–23, for the names of correspondents who provided Darwin with "Materials for the 1st edition of "The Descent of Man 1871" (DAR 80–86).
16. "Darwin's Journal" (Darwin 1959, p. 18) and the comments in Darwin's published correspondence for the period 1868–1871 indicate that Darwin worked on the sexual selection sections of the *Descent* before he took up the parts of the book dealing with the structural and mental continuities between animals and man. Richards (personal communication) first called this to my attention.
17. Ghiselin has made the same point with respect to the assessment of Darwin's *Expression*.
18. Ghiselin writes: "sexual selection is Darwin's most brilliant argument in favor of natural selection, of which it is a corollary" (1969, p. 215). Ghiselin's insights on the power of Darwin's theory of sexual selection are extremely valuable, but he confuses the issue elsewhere in his book by defining natural selection as "differential reproduction with its causes, nothing more" (1969, p. 74). Mayr handles Darwin's distinction between sexual selection and natural selection with welcome clarity, noting: "A separation of sexual and natural selection makes sense only if one adopts the same definition of fitness as Darwin, who employed the term in an uncompliated, everyday sense" (1972b, p. 88).
19. The notion that sexual selection might run to some degree counter to natural selection was first expressed clearly by Darwin in the *Descent*. As early as his *1844 Essay*, however, he was prepared to admit that the struggle for mates was "less rigorous" than natural selection and that "the effect chiefly produced would be the alteration of sexual characters, and the selection of individual forms, no way related to their power of obtaining food, or of defending themselves from their natural enemies, but of fighting one with another" (*1844 Essay*, p. 121). The results of sexual struggle, therefore, would evidence the same kind of imperfection as did the results produced by "those agriculturalists who pay less attention to the careful selection of all the young animals which they breed and more to the occasional use of a choice male" (p. 121). But if Darwin thus hinted in 1844 that sexual selection might produce imperfectly adapted forms, this was still not the basic thrust of his argument. On the contrary, as he said at the end of his *Essay*, in the case of "sexual struggle" among animals, "the most vigorous, and consequently the best

adapted, will oftener procreate their kind" (p. 243).

20. As indicated, Darwin did appreciate that displays involved in mating behavior could be developed through sexual selection, and he also acknowledged the importance of selection in developing certain interspecific warning signals. The selectionist interpretation of intraspecific behavioral and structural releasers later developed by Lorenz (1935) and Tinbergen (1952), however, is not to be found in Darwin's *Expression*.

14

CHARLES DARWIN AND ALFRED RUSSEL WALLACE: TWO DECADES OF DEBATE OVER NATURAL SELECTION

Malcolm Jay Kottler

Introduction

Much of Darwin scholarship in recent years has focussed on the very private Charles Darwin, talking and thinking *to himself* in his *Transmutation,* M, and N *Notebooks* as well as other strictly personal writings. Less attention has been given to the large number of close intellectual relationships Darwin formed with other naturalists, such as John Henslow, Charles Lyell, Joseph Hooker, Asa Gray, T. H. Huxley, and Alfred Russel Wallace. Since these dialogues — conducted primarily through correspondence and thus accessible to historical analysis — often resulted in the clarification or even modification of Darwin's views, they certainly require thorough study if we are to gain as full an understanding as possible of the development of Darwin's scientific thought. In this essay I will analyze the intellectual relationship between Darwin and Wallace, emphasizing their extensive discussions concerning the nature and scope of natural selection.

The story has been told many times how Darwin first formulated his concept of natural selection in 1838, but then for twenty years kept it to himself, telling only a very few other scientists (Hooker in 1844, Lyell in 1856, Gray in 1857); how, in 1858, Wallace independently formulated his own concept of natural selection, immediately wrote it up, and sent his manuscript "On the Tendency of Varieties to Depart Indefinitely from the Original Type" to Darwin; and how Wallace's manuscript precipitated the first public presentation of Darwin's concept, along with Wallace's, at the celebrated meeting of the Linnean Society of London on 1 July 1858, and induced Darwin to write the *Origin* (between 20 July 1858, and 1 October 1859) (see Fig. 1).

It is almost always asserted or simply assumed that Wallace's concept of natural selection as presented in his 1858 paper was identical to Darwin's concept in the first edition (1859) of the *Origin*. Darwin certainly acted in 1858 as if the two concepts were equivalent; and much later both Darwin and Wallace explicitly stated that they were. It has been argued, however, that the two concepts in 1858 were significantly different, since Darwin's natural selection was "competitive selection", whereas Wallace's was "environmental selection" (Nicholson 1960); and since Darwin was concerned primarily with selection acting on individual differences, whereas Wallace's "main theme was the differential survival of varieties rather than individuals" (Bowler 1976c). J. L. Brooks has made the very different claim that Darwin "appropriated, without any acknowledgment, the concept of 'divergence' as it appears in the *Origin of Species* from Wallace's 1855 paper ["On the Law which has Regulated the Introduction of New Species"] and the manuscript that Wallace sent to Darwin . . . early in 1858" (Brooks 1969, 1972). Thus in Part I of this essay ("Darwin and Wallace in 1858") I will re-examine Wallace's 1858 paper in order to evaluate these claims concerning Wallace's concepts of natural selection and divergence in 1858, in relation to those of Darwin.

The dialogue between Darwin and Wallace took place almost entirely through their correspondence and published writings. That correspondence, initiated by Wallace on 10 October 1856, continued for twenty-five years through 1881. Only a small number of letters passed between Darwin and Wallace from 1856 to 1862, while Wallace was still in the Malay archipelago (see Beddall 1968, pp. 319–323). Their extensive interaction over natural selection began, therefore, after Wallace's return to England in 1862, and was especially active from 1864 to 1872. During this period it became abundantly clear to the two men that there were substantial differences between them as to the nature and scope of natural selection. They engaged in three major debates over the role of natural selection in the origin of (1) cross- and hybrid sterility, (2) sexual dimorphism, and (3) man.[1] These debates were truly monumental intellectual confrontations. They continue to be especially interesting and indeed awe-inspiring, because during their debates Darwin and Wallace were the first to raise several fundamental issues concerning the nature of natural selection that are still subjects of controversy among evolutionary theorists, while the opposing positions they upheld over one hundred years ago remain prominent alternatives in the present day. These issues included: (i) "adaptationism" (Gould and Lewontin 1979; Gould 1982), (ii) constraints on natural selection, and (iii) levels of selection (G. C. Williams 1966).

These three debates take on an added fascination when one also considers the very human side of the interaction between Darwin and Wallace. During the 1860s each man developed an enormous admiration for the intellectual vigor and genius, as well as personal character, of the other.[2] Because of these mutual feelings of respect, it mattered a great deal to each to convince the other; and, at the same time, it became rather frustrating when these efforts to convince repeatedly failed and the differences of opinion remained unresolved. Indeed to Darwin, for whom the opinions of those few he considered his intellectual peers were particularly important, the seemingly endless disagreements with Wallace over natural selection proved to be quite distressing. At the climax of the sterility debate for instance, Darwin remarked: "Life is too short for so long a discussion. We shall, I *greatly* fear, never agree" (Darwin's italics; Wallace 1916, p. 172). And five months later at the climax of the debate over sexual dimorphism, he echoed that concern: "I grieve to differ from you, and it actually terrifies me, and makes me constantly distrust myself. I fear we shall never quite understand each other" (Wallace 1916, p. 189). In his autobiography, Wallace observed: "It is quite really pathetic how much he felt difference of opinion from his friends. I, of course, should have liked to have been able to convert him to my views, but I did not feel it so much as he seemed to do" (Wallace 1905, 2: 14).

The debate over the origin of cross- and hybrid sterility has not been the subject of any comprehensive study published to date, so I will devote most of Part II of this essay ("Darwin and Wallace after 1859: The Three Great Debates") to a detailed narrative and analysis of this debate. I will then conclude this part with brief summaries of the other two debates, in order to demonstrate that all three debates were phases of a single extended dialogue.

I. Darwin and Wallace in 1858

Alfred Russel Wallace became an evolutionist in 1845 after reading Robert Chambers's *Vestiges of the Natural History of Creation.* Thirteen years later in February 1858, while suffering from a severe attack of malaria in the Malay archipelago, the idea of natural selection "suddenly flashed" upon him: "In the two hours that elapsed before my ague fit was over I had thought out almost the whole of the theory and the same evening I sketched the draft of my paper, and in the two succeeding evenings wrote it out in full, and sent it by the next post to Mr. Darwin" (Wallace 1891, p. 20). Two questions naturally arise concerning Wallace's formulation of a concept of natural selection in 1858. Since Darwin had formulated his own concept of natural selection in September 1838, twenty years earlier, was Wallace's formulation completely independent of Darwin's? Second, was Wallace's concept of natural selection in 1858 identical to that of Darwin in 1858–1859?

Wallace himself was very definite about his independence from Darwin with respect to the formulation of natural selection: "[natural selection] was conceived by me before I had the least notion of the scope and nature of Mr. Darwin's labours" (Wallace 1870a, p. iv; see 1891, p. 21, and 1905, 1: 359). Almost all historians and evolutionary biologists have accepted

Wallace's statements of independence, because, prior to February 1858, Darwin had not published his idea and had told it to only Joseph Hooker, Charles Lyell, and Asa Gray, none of whom was in communication with Wallace. Furthermore, although Wallace had received one letter from Darwin prior to February 1858, Darwin had not divulged anything to Wallace about the nature of his theory: "This summer will make the twentieth year (!) since I opened my first note-book on the question how and in what way do species and varieties differ from each other . . . it is really *impossible* to explain my views in the compass of a letter as to causes and means of variation in a state of nature; but I have slowly adopted a distinct and tangible idea" (Darwin's italics; Wallace 1916, pp. 107–108).[3] It is incorrect, however, to conclude from these facts that, prior to 1858, Darwin had absolutely no influence on Wallace's thinking.

Wallace entered Darwin's intellectual life in 1856, or possibly a year or two earlier. But Darwin's influence on Wallace began in 1842 when Wallace first read, with evident excitement, Darwin's *Journal of Researches* and then re-read it in 1846 (the second edition of 1845?). In a letter of 1846 to his new friend and fellow enthusiast in natural history Henry Walter Bates, Wallace praised both Darwin and Alexander von Humboldt; and in his autobiography Wallace looked back to Darwin's *Journal of Researches* and Humboldt's *Personal Narrative* of his travels in South America as "the two works to whose inspiration I owe my determination to visit the tropics as a collector" (Wallace 1905, 1: 256). Within just two years (1848), Wallace left England for the Amazon region of South America, in the company of Bates, and his career as a scientist had begun.

It is difficult to determine how much more than "inspiration" Wallace gained from his reading of Darwin's *Journal of Researches*. Wallace attributed his formulation of a concept of natural selection to the combined influence of Malthus's *Essay on Population* and Lyell's *Principles of Geology* (Wallace 1908a, pp. 111–118; McKinney 1972, pp. 160–163). In all probability Wallace first learned of Lyell's views from Darwin's *Journal of Researches*. Since Darwin did not mention Malthus and his *Essay* by name in the *Journal*, Wallace could not have been led to Malthus initially by Darwin. But in his new treatment of the causes of extinction, in the second edition (1845) of the *Journal of Researches*, Darwin did discuss the Malthusian doctrine that "the supply of food, on the average remains constant; yet the tendency in every animal to increase by propagation is geometrical." Recollecting many years later the circumstances surrounding his formulation of natural selection, Wallace noted: "Something led me to think of the 'positive checks' described by Malthus in his 'Essay on Population' " (1891, p. 20). While in the Malay archipelago Wallace did have with him the second edition of Darwin's *Journal of Researches*, so possibly the "something" that reminded him of Malthus's views was Darwin's discussion in the *Journal* (cf. Ghiselin 1980). Natural selection is, however, much more than the combined views of Lyell and

Malthus; so whatever Wallace's debt may have been to Darwin for bringing these views to his attention, I believe that, independently of Darwin, Wallace did put these and other views together to produce a concept of natural selection.

Perhaps the main warrant for the widely-held view that Wallace's natural selection (in 1858) was identical to Darwin's comes from what the two men themselves had to say. On 18 June 1858 — supposedly the very day on which he received Wallace's 1858 paper — Darwin wrote to Lyell: "I never saw a more striking coincidence; if Wallace had my manuscript sketch written out in 1842, he could not have made a better short abstract! Even his terms now stand as heads of my chapters . . . so all my originality, whatever it may amount to, will be smashed . . ." (*LL* (NY) 1: 473). Everything in this famous passage suggests that Darwin regarded Wallace's idea as very similar, if not identical, to his own. In the Introduction to the *Origin*, Darwin remarked that the two men had reached "almost exactly the same general conclusions" (*Origin*, p. 2), while in his autobiography Darwin wrote that Wallace's paper "contained *exactly the same* theory as mine" (my italics; *Autobiography*, p. 121). Finally there are a number of statements in his correspondence in which Darwin treated the two views as identical. For instance, in 1859 Darwin told Wallace "you have thought . . . in so nearly the same channel with myself" (Wallace 1916, p. 115). In 1864 he said "the theory . . . is just as much yours as mine" (p. 127); and in 1869, in a well-known remark, Darwin commented, "I hope you have not murdered too completely your own and my child" (p. 197). Wallace wrote, in an introductory note added in 1891 to his 1858 paper: "This [paper] sets forth the main features of a theory *identical* with that discovered by Mr. Darwin many years before" (1891, p. 20). The most recent discussion of Wallace's concept in relation to Darwin's supports the common view that, although there are "subtle differences", Wallace and Darwin independently arrived at "essentially the same theory" (Mayr 1982b, pp. 494–497). But were there more than just subtle differences?

ENVIRONMENTAL VERSUS COMPETITIVE SELECTION

A. J. Nicholson argued that: "Although Darwin and Wallace regarded their theories as being essentially the same, they emphasized different parts of its mechanism. Darwin's arguments were concerned dominantly with competitive selection, which causes the less fit forms to be displaced as a secondary effect of the preservation of fitter forms, whereas Wallace . . . referred almost exclusively to what may be called 'environmental selection', the active principle of which is the direct elimination of the unfit" (1960, p. 491).

In environmental selection the elimination of the *unfit* is primary, whereas in competitive selection the elimination of the *less fit* is secondary. Since, according to both concepts of selection, there is an elimination of some individuals, the distinction being drawn might appear to be purely semantic. But it is not. The important difference between the two concepts is clearly illustrated by the ideas of "hard" and "soft" selection introduced by Bruce Wallace:

> I use the terms "hard" and "soft" in describing the basis by which natural selection determines which individuals are to be excluded from the ranks of successful breeders. One possibility ["hard selection"] is that the "cutoff point" is determined on an invariate fitness scale or by unconditional selective factors. Consequently, as the distribution of fitnesses that characterize a population fluctuates relative to the constant cutoff point, the number of individuals leaving progeny also fluctuates. Another possibility ["soft selection"], though, is that the cutoff point is not a constant determined according to some fixed fitness scale. Under this possibility, the number of parents may remain relatively constant from generation to generation despite fluctuations in the distribution of fitnesses within the population. (B. Wallace 1968, pp. 427–428; 1975)

Environmental selection is an example of hard selection. The conditions in the external environment establish an absolute standard that must be met if an individual is to survive and then reproduce. The individual organism, under such circumstances, is "struggling" primarily against the external environment, not the other individuals of the species, since its survival is dependent primarily upon its own characteristics in direct relation to the demands of the external environment, not in relation to the characteristics of conspecific individuals. In the extreme, the fate of the individual would be independent of that of every other individual of the species. All those individuals that fail to meet the absolute standard are automatically eliminated, while all those (if any) meeting the standard survive. Under environmental (hard) selection the individuals that do not survive can properly be called unfit. Competitive selection, on the other hand, is an example of soft selection. The individual organism, in these situations, is struggling primarily against the other individuals of the species, since its survival is dependent largely upon its characteristics in relation to the characteristics of conspecifics. An individual might be perfectly able to survive and reproduce in the absence of certain other individuals of the species; but in the presence of those individuals it cannot survive and reproduce. Thus, under competitive (soft) selection, the individuals that do not survive are not really unfit at all. They are simply *less fit* than the survivors and are eliminated only because of the presence and preservation of the more fit individuals.

The main evidence offered by Nicholson that Wallace concentrated on environmental selection is this key passage from his 1858 paper:

> Now, let some alteration of physical conditions occur in the district —

> a long period of drought, a destruction of vegetation by locusts, the irruption of some fresh carnivorous animal seeking 'pastures new' — any change in fact tending to render existence more difficult to the species in question, and tasking its utmost powers to avoid complete extermination — it is evident that, of all the individuals composing the species, those forming the least numerous and most feebly organized variety would suffer *first*, and, were the pressure severe, must soon become extinct [automatic elimination of the unfit]. The same causes continuing in action, the parent species would *next* suffer, would gradually diminish in numbers, and with a recurrence of similar unfavourable conditions might also become extinct. The superior variety would then alone *remain*, and on a return to favourable circumstances would rapidly increase in numbers *and occupy the place of* the extinct species and variety. (my italics; 1858b, p. 274)

Later in the paper Wallace discussed the evolution of the giraffe's neck by means of selection as opposed to "volition" (what he, mistakenly, took to be Lamarck's theory); and here also he described a process of environmental selection: "any varieties which occurred . . . with a longer neck than usual at once secured a fresh range of pasture over the same ground as their shorter-necked companions, and *on the first scarcity of food* were thereby enabled to outlive them" (my italics; pp. 277–278).

The example of the giraffe's neck points up an important difference between the two types of selection in that "with competitive selection . . . the standard of selection automatically rises as a result of the biological improvement already selected, thus causing evolutionary advancement to continue, even in a constant environment, just so long as superior genotypes continue to appear. By contrast, with environmental selection . . . the maximum degree of evolutionary advancement that can be produced . . . is that at which all individuals have a barely sufficient defense to enable them to survive under the prevailing intensity of the selective factor" (Nicholson 1960, pp. 492 and 513). Thus, as Wallace described the process, initially the individuals with longer necks were no more fit than others because all were able to obtain the minimum amount of food necessary for survival; only after adverse environmental circumstances arose, did those shorter-necked individuals that could no longer obtain enough food become unfit. At this point they were directly eliminated, while the longer-necked individuals that could obtain enough food remained. With strict environmental selection, even if individuals with yet longer necks now appeared, no further evolution would take place since all individuals could obtain sufficient food. Another change in the external environment would be necessary before more selective elimination would occur.

Further evidence that Wallace thought primarily in terms of environmental selection is contained in a letter he wrote to Darwin in 1866, in which he proposed that in future editions of the *Origin* Darwin adopt Herbert Spencer's phrase "survival of the fittest". Wallace remarked: "This term is the plain expression of the fact, natural selection is a metaphorical expression of it, and to a certain degree indirect and incorrect, since, even personifying Nature, *she does not so much select special variations as exterminate the most unfavourable ones*" (my italics; Wallace 1916, pp. 140–142).[4] Such a statement clearly reflects the view that the primary type of selection is the direct elimination of the unfit, or environmental selection.

In support of his view that Darwin emphasized competitive selection, Nicholson cited this passage from the sixth edition of the *Origin*: "Owing to the high geometrical rate of increase of all organic beings, each area is already fully stocked with inhabitants; and it follows from this, that *as the favoured forms increase in number, so,* generally, *will the less favoured decrease and become rare*" (my italics; 1960, p. 478; the corresponding passage in the first edition is on p. 109). However, a clearer illustration of Darwin's grasp of the concept of competitive or soft selection, which is based on direct interactions among conspecific individuals, is his analysis of *sexual* selection, the quintessential example of competitive selection. Having enumerated a number of types of sexual dimorphism that he thought were due to sexual selection, Darwin remarked in the *Descent*: "That these characters are the result of sexual selection is clear, as unarmed, unornamented, or unattractive males *would succeed equally well* in the battle for life and in leaving a numerous progeny, *if better endowed males were not present*" (my italics; 1:258).

Darwin certainly recognized that change in the external (physical or biotic) environment was not absolutely necessary for "ordinary or natural selection" to produce evolutionary change (*Origin*, pp. 82 and 91, for instance). But his clearest statement of the difference between competitive and environmental selection with respect to the possibility of continued evolution can be found, again, in his treatment of *sexual* selection. "In regard to structures acquired through ordinary or natural selection, there is in most cases, as long as the conditions of life remain the same, a limit to the amount of advantageous modification in relation to certain special ends; but in regard to structures adapted to make one male victorious over another, either in fighting or in charming the female, there is no definite limit to the amount of advantageous modification; so that as long as the proper variations arise the work of sexual selection will go on" (*Descent* 1: 278).

In conclusion I believe Nicholson was correct that in 1858, and perhaps later, Wallace's major focus was on environmental selection; and it seems that in 1858 Wallace "did not clearly realise that competition plays an important part in selection" (Nicholson 1960, p. 491).[5] At the same time it is apparent that Darwin appreciated both competitive and environmental selection, and in his discussion of *sexual* selection demonstrated an especially clear understanding of the differences between them.

THE MEANING OF "VARIETY": VARIANT INDIVIDUAL OR POPULATION?

A different issue is involved in P. J. Bowler's suggestion that "Wallace's initial concept of selection differed considerably from Darwin's (1976c, p. 18). Bowler's argument starts from the claim that the two concepts were presented "in very different terms". According to Bowler, whereas Darwin described selection as acting, primarily, on differences among individuals to form varieties, Wallace described selection as acting, primarily, on these permanent varieties after they had already been formed. In Bowler's view, Wallace's "considerably" different method of *describing* the action of selection raises some important questions: "need we regard this as anything more than an alternative method of presenting the same basic idea [as Darwin]? May we assume that Wallace was aware from the beginning that selection acted upon individual differences to form varieties, but that he preferred to discuss competition among varieties because he was more familiar with this [secondary] level of variation? Or might it be argued that at first he failed to appreciate the primary Darwinian mechanism of selection acting upon individual differences?" (p. 21). Bowler contends Wallace said so little in his 1858 paper about the formation of varieties by the action of selection on differences among individuals that we can "at best only infer" that he understood this process. But in light of what Bowler thinks Wallace did say about the initial formation of varieties, he is inclined to the view that in 1858 Wallace did not "fully" understand the "primary Darwinian mechanism of selection".

Much of Bowler's argument rests on Wallace's extensive use of the term "variety". Starting with his title, "On the Tendency of Varieties to Depart Indefinitely from the Original Type", Wallace referred to "varieties" at least thirty times in his paper. Consequently a key question is: What did Wallace mean by "variety"? Mayr has pointed out that in the nineteenth century the term "variety" was often applied quite indiscriminately to two very different kinds of variation: variant individuals and variant populations (for example, subspecies) (1959c, p. 222; 1982b, p. 415). With regard to the section of Wallace's 1858 paper devoted to selection, three very definite but at the same time very different opinions have been expressed as to what he meant by "variety". My own opinion is that the matter is hardly as clearcut as Mayr, Bowler, or Brooks has presented it. Mayr has stated that "Wallace applies the term 'variety' to variant individuals, that is individuals *within a population* that do not share the same properties (my italics; 1982b, p. 496). In marked contrast, Bowler has claimed the very opposite: "it is clear from the context that he is referring to 'permanent true varieties', not individual differences. Wallace had shifted the concept of variation to a new, and by Darwinian standards, secondary level, where the unit of discussion was a distinct group within the species, not an individual" (1976c, p. 20). J. Brooks has also held that Wallace's "varieties" were variant populations, not individuals; but Brooks has claimed these varieties were geographically isolated from each other, whereas Bowler argues they coexisted. Brooks has written: "In the 1858 essay . . . Wallace observed that most species are represented *in different geographical areas* by populations in which the individuals all exhibit constant though often slight differences from the individuals of other populations of other areas. These locally distinct populations of a species were called 'varieties' . . . Wallace then postulated that these differences *between populations* must entail differences in the ability of these populations to reproduce themselves" (my italics; Brooks 1972, pp. 50–51). Although Wallace did have an understanding of geographic variation (see below footnotes 7, 37), I can find no place in his 1858 paper where he discussed this phenomenon. It seems very clear that the entities (whether individuals or populations) that Wallace was considering in his paper co-existed with each other. For example, he wrote: "let some alteration . . . occur in *the district*" (my italics; 1858b, p. 274) and "any varieties which occurred . . . with a longer neck . . . secured a fresh range of pasture *over the same ground* as their shorter-necked *companions*" (my italics; p. 278).

In conjunction with the term "variety", Wallace used two other terms that must be considered: "race" and "variation". In several places Wallace used the word "race" as synonymous with "variety", from which we must conclude that sometimes he did mean variant population by the term "variety". Wallace also used the term "variation" at least six times in the original 1858 version of his paper (see below for three important additional uses of "variation" in the 1870 slightly revised reprint of the paper); and, in at least some of these instances, I think he meant, by "variation", variant individuals within a population. Since some of Wallace's references to such individual "variation" occur in the same passages as his description of the action of selection on "varieties", it is by no means obvious that the varieties subjected to selection were supposed to be different populations, rather than different individuals within one population.

The first juxtaposition of "variety" and "variation" occurs in the introductory part of the paper: "it is the object of the present paper to show that . . . there is a general principle in nature which will cause many *varieties* [Wallace's italics] to survive the parent species, and to give rise to successive *variations* [my italics] departing further and further from the original type" (1858b, p. 269). This statement was followed immediately by Wallace's discussion of the "struggle for existence", which concluded: "The numbers that die annually must be immense; and as the individual existence of each animal depends upon itself, those that die must be the weakest . . . while those that prolong their existence can only be the most perfect in health and vigour" (p. 272). Bowler has taken note of these passages and acknowledged they suggest an awareness of the action of selection on individuals. He minimizes these remarks, however, since, in his view, Wallace then "changed the subject" from the struggle among individuals to the struggle among permanent varieties. Yet the very first words in Wallace's discussion of "varieties" concern individual variation: "Most or perhaps all the *variations* from the typical form of a species must have some definite effect, however slight, on the habits or capacities of the *individuals*" (my italics; p. 273). Wallace proceeded to describe briefly the effect on survival of individual differences such as the greater vulnerability to predation of an antelope with shorter or weaker legs, or the greater likelihood of starvation in a passenger pigeon with less powerful wings. He continued: "If, on the other hand, any species should produce a *variety* having slightly increased powers of preserving existence, that *variety* must inevitably in time acquire a superiority in numbers . . . All *varieties* will therefore fall into two classes — those which under the same conditions would never reach the population of the parent species, and those which would in time obtain and keep a numerical superiority" (my italics; pp. 273–274). At this point the key passage on the action of selection — "Now, let some alteration . . ." (see above) — followed. Throughout this passage Wallace continued to use the term "variety". Since Wallace began this section with explicit reference to individual variations, and since his change in terminology in the middle of the section from "variations" to "varieties" does not seem to correspond to a conceptual change, there is a basis for Mayr's view that Wallace meant individual variant by "variety"; if so, he was concerned here with selection acting on differences among individuals.

In the next paragraph of the paper Wallace again moved freely back-and-forth between "variation" and "variety": "*variations* in unimportant parts might also occur, having no perceptible effect on the life-preserving powers; and the *varieties* so furnished [variant individuals or variant populations?] might run a course parallel with the parent species, either giving rise to further *variations* or returning to the former type" (my italics; p. 275). Lastly, in his discussion of domesticated "varieties", Wallace also used the term "variation" — in the sense of individual variation — as virtually synonymous with "variety". First he wrote, "in the domesticated animal all *variations* have an equal chance of continuance; and those which would decidedly render a wild animal unable to compete with its fellows and continue its existence are no disadvantage whatever in a state of domesticity" (my italics). But then he wrote, "Domestic animals are abnormal, irregular, artificial; they are subject to *varieties* which never occur, and never can occur, in a state of nature" (my italics; pp. 276–277). These examples of Wallace's almost interchangeable use of "variety" and "variation" demonstrate the great difficulty in arriving at a proper interpretation of his meaning in this paper.

It is clear that Wallace himself later recognized the existence of ambiguity in his paper. When the paper was reprinted for the first time in 1870, Wallace added subheadings and a few footnotes (although he stated the paper had been reprinted "without alteration of the text, except one or two grammatical emendations" (1870a, p. vi)). The part of the paper containing Wallace's main description of selection, including that key passage "Now, let some alteration . . .", was given the subheading "Useful *variations* will tend to increase; useless or hurtful *variations* to diminish" (my italics; 1870a p. 34).[6] Similarly, to the discussion of why domesticated "varieties" reverted when turned wild, Wallace added this footnote: "That is, they will vary, and the *variations* which tend to adapt them to the wild state, and therefore approximate them to wild animals, will be preserved. Those *individuals* which do not vary sufficiently will perish" (my italics; p. 40). He also changed the word "varieties" to "variations" in the passage on "domestic animals" (p. 41) quoted above.

These additions could be interpreted as a significant conceptual change, in line with Bowler's position — that is, only after 1858, when Wallace became familiar with Darwin's views, did he come to fully appreciate the primary role of selection acting on differences among individuals, and then with these additions attempt to recast his initially quite different concept of 1858 in more Darwinian terms. On the other hand, they could also be interpreted as a clarification by Wallace of an idea present from the beginning but not stated initially in an unambiguous fashion. This is what Wallace himself argued nearly forty years later.

H. F. Osborn in 1894 was, I believe, the first to contend that there was a "wide gap" between Wallace's concept of natural selection in 1858 and that of Darwin:

> Remarkable as this parallelism is, it is not complete. The line of argument is the same, but the *point d'appui* is different. Darwin dwells upon *variations in single characters*, as taken hold of by Selection; Wallace mentions variations, but dwells upon *full-formed varieties*, as favourably or unfavourably adapted. It is perfectly clear that with Darwin the struggle is so intense that the chance of survival of each individual turns upon a single and even slight variation. With Wallace, Varieties are already presupposed by causes which he does not discuss, a change in the environment occurs, and those varieties which happen to be adapted to it survive. There is really a wide gap between these two statements and applications of the theory. (Osborn's italics; Osborn 1894, p. 245)

E. B. Poulton disagreed with Osborn's (= Bowler's) view, and asked Wallace to comment on it. Poulton wrote:

> Further consideration tends to obliterate the supposed distinction. Although Wallace used the term "variety" as contrasted with "species", the whole context proves that he, equally with Darwin, recognised the importance of individual variations and of variations in single characters. This becomes clear when we remember his argument about the neck of the giraffe,

> the changes of colour and hairiness, the shorter legs of the antelope, and the less powerful wings of the passenger pigeon. Wallace has kindly written to me (May 12th, 1896) stating the case as I have given it, and he further explains — "I used the term 'varieties' because 'varieties' were alone recognised at that time, individual variability being ignored or thought of *no importance*. My 'varieties' therefore included 'individual variations'." (Wallace's italics; Poulton 1896, p. 80)

Bowler has also claimed that Wallace "wrote as though the production of permanent varieties were a random process, in the sense that groups might appear with both useful and harmful features" (1976c, p. 22). Clearly if this were so, then Wallace could not have understood the role of selection in forming varieties since natural selection could not produce a variety (subspecies) with characteristics harmful to itself. But I see no evidence that Wallace actually held such a view. The example Bowler has given — an antelope species giving rise to a variety with longer than normal legs (useful) and one with shorter than normal legs (harmful) — is not anything Wallace himself presented. Wallace did state that if "an antelope" had shorter or weaker legs it would suffer more from predation. But Wallace did not refer explicitly to a permanent variety of such shorter-legged antelopes, nor to any antelopes at all with longer legs. In the key passage on selection, he did refer to "the least numerous and most feebly organised variety" (and elsewhere to "inferior" varieties). Of course if Wallace meant, by "variety", individual variant, then there is no issue. But even if he did mean permanent variety, surely such language was meant to convey an idea of *relative* inferiority, whereas Bowler has suggested Wallace believed absolutely harmful traits had somehow evolved in permanent varieties. Furthermore I find it impossible to reconcile such a position with remarks by Wallace at the end of his paper: "no unbalanced deficiency in the animal kingdom can ever reach any conspicuous magnitude, because it would make itself felt *at the very first step*, by rendering existence difficult and extinction almost sure soon to follow" (my italics; p. 278).[7]

In conclusion, my main point is that we should openly acknowledge that in 1858 Wallace was not absolutely clear in his manner of expression, so that we cannot be certain as to what he meant by "variety" in the key passage on selection. Furthermore it is important to keep in mind that aside from his 1858 paper there are almost no other documents by Wallace from the late 1850s that can be consulted for additional information on this question. Whereas we are fortunate to possess an enormous amount of manuscript material concerning the development of Darwin's concept of natural selection in 1837–1839 (and after), we have, by comparison, regrettably little material of this kind illuminating the development of Wallace's concept.

THE MEANING AND CAUSE OF DIVERGENCE

J. L. Brooks was the first to raise the issue of the "derivative nature of Darwin's concept of 'divergence' ". In a two-page-long report, Brooks asserted, without supplying his evidence, that Darwin "appropriated, without any acknowledgement, the concept of 'divergence' as it appears in the *Origin of Species* from Wallace's 1855 paper and the manuscript that Wallace sent to Darwin from the Dutch East Indies early in 1858." Brooks did not explain exactly what he, Wallace, or Darwin meant by "divergence"; but he did state: "Wallace was the only person to have conceived of the manner in which observed patterns of the affinity and distribution of organisms arise through natural processes" (Brooks 1969). This brief report was followed three years later by an article, in which, at the very end in a two-page section headed "Darwin's use of Wallace's Hypothesis", Brooks claimed that "Darwin's statement in Chapter Four [of the *Origin*] on the role of extinction in species formation is different from Wallace's, even though much evidence[9] indicates that the treatment in the *Origin* is based on Wallace's essays . . . [Darwin] incorporated (without acknowledgement) the essence and details of most of Wallace's hypothesis as presented in his two essays . . ." Footnote 9 read, in part: "Recitation of the details of the evidence is too lengthy for this paper. An indication of some salient evidence can be found in Brooks, 1969" (Brooks 1972, pp. 52–54).

H. L. McKinney, unlike Brooks, was tentative in his own brief discussion, but he, too, raised the question whether Darwin formulated his principle of divergence independently of Wallace: "the problem of divergence was the one problem which, by [Darwin's] own admission, he had not worked out satisfactorily in his earlier sketch of 1842 and his essay of 1844 . . . Did Wallace's [1858] paper provide any special insights for Darwin into this or any other problem? Wallace may simply have reinforced past fleeting ideas or illuminated some obscure point . . . on the other hand, we may find that still another important chapter needs to be written about the Darwin-Wallace relationship" (McKinney 1972, pp. 141–142, 144, 153–154). Is there any evidence that Darwin did derive some important ideas from Wallace's 1855 and 1858 papers, which then enabled him to formulate his principle of divergence? D. Kohn has conducted the most complete study of this issue to date and I concur with his conclusions:

> Wallace and Darwin derived two fundamentally different principles . . . There was no principle of divergence in the 1855 paper, nothing to be influenced by, nothing to steal . . . The arrival of Wallace's [1858] paper is high drama. Nevertheless, it was an intellectual non-event. For Darwin learned nothing about the mechanisms of evolution from Wallace's paper. Indeed . . . with regard to divergence, there was a great deal Darwin could have told Wallace. (1981, p. 1106)

The issue hinges, first and foremost, on what each man meant by the phenomenon of divergence. Although both used the word "divergence", I believe they meant two quite different things, or to be more precise Darwin's concept was much broader than that of Wallace, including Wallace's but then going well beyond it. Hence the two principles were necessarily different because they were intended to explain different phenomena. One can think of divergence, in the sense of separation from some starting point, in the context of either linear (phyletic) or branching evolution. Darwin included a diagram in the *Origin* to accompany his discussion of divergence; and by reference to this diagram, the distinction I want to make can be clearly seen. Divergent phyletic evolution is represented by the single lineage A-a^1-a^2-a^3-a^4-a^5-. . .-a^{10}. If "difference" is represented by the horizontal dimension, then a^{10}, well to the left of A, would be quite different from A, that is, it has "diverged" considerably from the starting point A. But as the result of such divergence there is still just one form (a^{10}) at the end of the process. Divergent branching evolution, on the other hand, is represented, in the simplest case, by the lineage A-a^1-a^2. . .-a^{10} *in combination with* the lineage A-m^1-m^2-m^3-m^4-m^5-. . .-m^{10}. As the result of this kind of divergence, there is a multiplication of forms and furthermore the small initial difference *between* the forms, a^1, m^1 is increased substantially over time (a^2,m^2; a^3,m^3;. . .a^{10},m^{10}). Thus not only have the a's and the m's diverged from the common starting point A, but also they have diverged from each other (*Origin*, diagram opposite p. 117).[8] Simply put, Wallace in his 1858 paper considered only divergent linear evolution; whereas Darwin, in *Natural Selection* and then in the *Origin*, sought to account for both types of divergence.

Wallace was certainly aware of the phenomenon of branching divergence. In his 1855 paper he described both linear and branching divergence: "So long as each species has had but one new species formed on its model, the line of affinities will be simple, and may be represented by placing the several species in direct succession *in a straight line*. But if two or more species have been independently formed on the plan of a common antitype, then the series of affinities will be compound, and can only be represented *by a forked or many-branched line* . . ." (my italics; 1855, pp. 6–7). Wallace even called these latter "divergent series" and used the analogy of a tree to represent the natural system of classification. The purpose of Wallace's 1855 paper was to introduce his law that "Every species has come into existence coincident both in space and time with a pre-existing closely allied species" and to demonstrate how it "explained and illustrated" a wide range of phenomena. But, even though Wallace *described* divergence, he did not completely *explain* it in this paper. His law is a *necessary* condition for any evolutionary explanation for both linear and branching divergence, but it is not *sufficient*, by itself, to account for them.

Brooks has especially stressed Wallace's understanding of the role of extinction "in the genesis of observed patterns of diversity", and in particular Wallace's remark that "it is an article of our zoological faith, that all *gaps* between species, genera, or larger groups are the result of extinction of species" (my italics; Wallace 1856b, p. 206). I would argue again that extinction is a necessary but not sufficient condition for an evolutionary explanation of divergence. Extinction eliminates intermediate forms and thereby produces gaps, either in the classification or the geographical distribution of organisms; but it does not produce the extreme (or intermediate) forms in the first place. Hence Wallace, prior to 1858, had grasped two of the necessary components for a principle of divergence, but lacking others — such as concepts of selection and of "character displacement" due to the advantage of diversity — he could not possibly have formulated a complete explanation for divergence.

Even though Wallace's 1855 paper could not have supplied to Darwin a complete, ready-made theory of divergence, could Darwin have derived some key component for his theory from this paper? Darwin had long accepted the idea of gradual evolution, which implies Wallace's law. Already in his B *Notebook* (1837) Darwin had used the analogy of a tree (or coral) of life (*B*: 21, 25, 36–37). Indeed Darwin annotated this part of Wallace's paper, "Uses my simile of tree" (cf. *Natural Selection*, pp. 249–250; *Origin*, pp. 128–130). And in the section "Origin of genera and families", of the *1844 Essay*, Darwin had written: "the arrangement of species in groups is due to partial extinction (p. 217; cf. Ospovat 1981, pp. 171–173; fn. 7, p. 265). Darwin's overall comment to himself on Wallace's paper was: "Laws of Geograph. Distrib. Nothing very new."[9] And, in fact, there was nothing new *to Darwin* of relevance to the problem of divergence. Nevertheless, the 1855 paper played an important role in the origin of the *Origin*. On 16 April 1856 Lyell raised with Darwin the question of the explanation of Wallace's law. Lyell had known since the mid-1840s that Darwin was an evolutionist (*LL* (NY) 1: 312–313; cf. 1:393 and *ML* 1: 50); but he first learned about Darwin's theory of natural selection on 16 April 1856 (Lyell 1970, pp. 54–55; McKinney 1972, chap. 7). Within a month of his meeting with Lyell, Darwin had begun to write for publication because of Lyell's "insistent advice". It seems clear that Lyell gave that advice, because he thought Darwin might be "forestalled" by Wallace (*LL* (NY) 1: 426–430; *Autobiography*, p. 121). (See Fig. 1.)

By 1857 at the very latest Darwin had formulated his principle of divergence as it appeared in both *Natural Selection* and in the *Origin* (Ospovat 1981, chap. 7; Kohn, this volume; Browne 1980). Consequently nothing in Wallace's 1858 paper about divergence could possibly have influenced Darwin. Nevertheless it is still worthwhile to examine just what Wallace did say about divergence in his paper. As Wallace described the process

Figure 1. (opposite) Darwin and Wallace prior to 1859.
This figure summarizes some of the key events in the intellectual relationship between Darwin and Wallace prior to 1859. A solid arrow indicates a direct causal relationship between two events, a dotted arrow a possible causal relationship.

Charles Darwin (b. 1809)		Alfred Wallace (b. 1823)
Mar. 1837 Becomes an evolutionist		
Sept. 1838 Formulates natural selection		
		1842 Reads Darwin's *Journal*
1844 Writes *Essay*, shows it to Hooker		
		1845 Reads *Vestiges*, becomes an evolutionist
		1848–1852 In South America (with H.W. Bates)
		1854–1862 In Malay Archipelago
		Sept. 1855 "On the law which has regulated the introduction of new species"
	Nov. 1855 Lyell reads "On the law", begins first Species Journal	
Apr. 1856 Lyell visits Darwin, asks him about Wallace's "law", and Darwin tells Lyell about natural selection		
May 1856 Darwin begins to write for publication, on advice of Lyell		
		Oct. 1856 Wallace writes first letter to Darwin
Apr.–May 1857 Darwin receives, replies to Wallace's first letter		
July 1857 Darwin tells Gray about his belief in evolution		
Sept. 1857 Darwin tells Gray about natural selection, principle of divergence		Sept. 1857 Wallace writes second letter to Darwin
Dec. 1857 Darwin receives, replies to Wallace's second letter		
		Feb. 1858 Wallace formulates natural selection, writes "On the tendency of varieties to depart indefinitely from the original type", sends MS. to Darwin
June 1858 Darwin receives Wallace's MS., writes to Lyell		
	1 July 1858 Linnean Society of London presentation	
20 July 1858 Darwin begins *Origin*		

of selection, under adverse environmental circumstances those individuals forming "the least numerous and most feebly organised variety" were eliminated first, then those forming the "parent species", until only those forming the "superior variety" remained. Thus there was no increase in the number of forms. Wallace then went on: "this new, improved, and populous race might itself, in course of time, give rise to new varieties, exhibiting several *diverging* [my italics] modifications of form, any of which, tending to increase the facilities for preserving existence, must, by the same general law, in their turn become predominant. Here, then, we have *progression and continued divergence*" (Wallace's italics; 1858b, p. 274). This "divergence" was what I have termed divergent linear or phyletic evolution. Even though at the very end of his paper Wallace referred briefly to "the many lines of divergence from a central type" (p. 278), he made no attempt to explain divergent branching evolution. The explanation for branching divergence (what would now be termed adaptive radiation) rests upon the idea that there is an advantage to diversity, because diversity reduces competition between co-existing forms. That this key idea was absent from Wallace's thinking prior to 1859 is not too surprising in light of his emphasis upon environmental, as opposed to competitive, selection.

What if Wallace had *not* formulated a concept of natural selection in 1858 and then sent his manuscript to Darwin? In June 1858 Darwin was more than half-done with *Natural Selection*, which he had begun in May 1856. It is hard to resist the conclusion that, barring some unusual interruption, Darwin would have completed and published it in the early 1860s, and never written the *Origin*. But the arrival of Wallace's manuscript was an unusual interruption. After the Linnean Society presentation, Darwin shelved *Natural Selection* (temporarily, he then thought), and instead, over the next year, prepared an abridgement, the *Origin*. Both Darwin and Wallace thought that the impact of the *Origin* was much greater than that of *Natural Selection* would have been (*Autobiography*, p. 124; Wallace 1891, p. 21; cf. *LL* (NY) 1: 493–494; Wallace 1916, p. 111). I believe they were right; so even though the arrival of Wallace's manuscript was an "intellectual non-event" for Darwin, it nevertheless had the most important result of leading Darwin to write the *Origin* (see Fig. 1).

II. Darwin and Wallace after 1859: The Three Great Debates

"In regard to the paper in the *Annals* [Wallace 1855], I agree to the truth of almost every word of your paper; and I daresay that you will agree with me that it is very rare to find oneself agreeing pretty closely with any theoretical paper; for it is lamentable how each man draws his own different conclusions from the very same fact" (Wallace 1916, p. 107; Darwin's first letter to Wallace, 1 May 1857).

"Nothing is so humiliating to me as to agree with a man like you (or Hooker) on the premises and disagree about the result. . . . Life is too short for so long a discussion. We shall, I *greatly* fear, never agree" (Darwin's italics; Wallace 1916, pp. 170, 172; Darwin to Wallace, 6 April 1868).

"I am delighted to see that we really differ very little — not more than two men almost always will" (Wallace 1916, p. 175; Darwin to Wallace, 15 April 1868).

"In truth, it has vexed me much to find that the further I get on, the more I differ from you . . ." (Wallace 1916, p. 181; Darwin to Wallace, 19 August 1868).

"You will be pleased to hear that I am undergoing severe distress . . . this morning I oscillated with joy towards you; this evening I have swung back to the old position, out of which I fear I shall never get" (Wallace 1916, p. 183; Darwin to Wallace, 16 September 1868).

"I grieve to differ from you, and it actually terrifies me, and makes me constantly distrust myself. I fear we shall never quite understand each other" (Wallace 1916, p. 189, Darwin to Wallace, 23 September 1868).

"I am sorry to find that our difference of opinion on this point is a source of anxiety to you. Pray do not let it be so. The truth will come out at last, and our difference may be the means of setting others to work who may set us both right. After all, this question is only an episode (though an important one) in the great question of the origin of species, and whether you or I are right will not at all affect the main doctrine — that is one comfort" (Wallace 1916, p. 189; Wallace to Darwin, 4 October 1868).

"As for our not quite agreeing, really in such complex subjects it is almost impossible for two men who arrive independently at their conclusions to agree fully — it would be unnatural for them to do so" (Wallace 1916, p. 212; Darwin to Wallace, 30 January 1871).

"It is a great pleasure to receive a letter from you sometimes — especially when we do not differ very much" (Wallace 1916, p. 250; Wallace to Darwin, 9 January 1880).

"How lamentable it is that two men should take such widely different views, with the same facts before them; but this seems to be almost regularly our case, and much do I regret it" (Wallace 1916, p. 256; Darwin to Wallace, 2 January 1881).

Wallace returned to England from the Malay archipelago in the Spring of 1862. He and Darwin met for the first time during the Summer of 1862 and thereafter, for about the next ten years, they met only about once a year when Darwin came to London to visit his brother Erasmus. Consequently the two men interacted primarily through their correspondence and published writings. This correspondence became "very extensive" (Wallace 1905, 2: 1) during the period 1864–1872 as Darwin and Wallace became increasingly aware of the fact that there were significant differences between them as to the nature and scope of natural selection. These differences emerged in the course of three debates over the role of natural selection in the origin of (1) cross- and hybrid sterility, (2) sexual dimorphism, and (3) man (see Fig. 2). At the heart of these differences were the fundamental and persistent issues of (i) "adaptationism" — is every trait useful ("the problem of utility"), and has every useful trait evolved directly by means

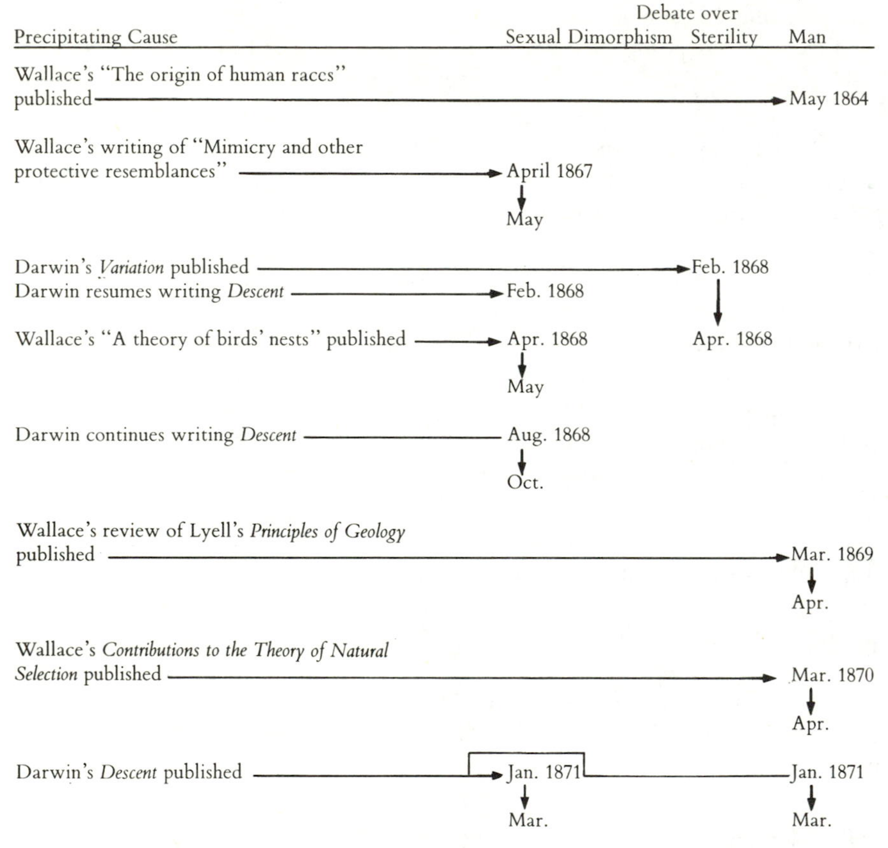

Figure 2. The Three Debates between Darwin and Wallace.
Between 1867 and 1871 Darwin and Wallace engaged in three great debates concerning the nature and scope of natural selection. The major exchanges in these debates are illustrated in this figure, and the precipitating cause for each of these exchanges is indicated on the left.

of natural selection to perform its current function(s)? — (ii) constraints on natural selection — is the action of natural selection limited by modes of inheritance and development? — and (iii) levels of selection — does selection act on groups of individuals, as well as on individual organisms? Adaptationism was a central issue in all three debates. The matter of levels of selection also played a fundamental role in the dispute over sterility, while the question of constraints on natural selection was critical to the disagreement over sexual dimorphism.

Wallace throughout adopted the more consistently adaptationist position. He believed that virtually all traits were now or formerly had been useful; and that any useful trait must have been the direct result of some process of selection. Although Darwin firmly held that natural selection was the "main" mechanism of evolutionary change, he recognized several other important factors of evolution, including some that would result in non-adaptive changes. Consequently he was much more willing than Wallace to grant the existence of non-adaptive traits; and at the same time — and of particular relevance to their debates — he was fully prepared to accept the origin of certain useful traits "by chance", for example as incidental by-products, and thus without the *direct* action of natural selection. Hence to Darwin the utility of a trait did not establish, by itself, its selective origin. Furthermore, whereas Wallace thought there were very few limits to the power of natural selection, Darwin believed natural selection was constrained in its action, especially by modes of inheritance and development. Natural selection, in Darwin's view, could not do everything. Finally Darwin adopted the more consistently individualist position on the question of levels of selection. Wallace did not hesitate to advance the possibility of group selection.

In Part II of this essay I will describe and analyze in detail the debate over the origin of cross- and hybrid sterility and then summarize the other two great debates, in order to reveal the important differences of opinion between Darwin and Wallace concerning the way in which natural selection could and did operate.

THE DEBATE OVER THE ORIGIN OF CROSS- AND HYBRID STERILITY

The debate between Darwin and Wallace over the role of natural selection in the origin of cross- and hybrid sterility took place in 1868 (see Mayr 1959c, pp. 227–228; Ruse 1980b, pp. 619–625; Sober, this volume). Darwin had been concerned since 1837 with the nature and significance of the phenomena of sterility; and from (at least) 1857 — when he wrote the chapter on "Hybridism" for *Natural Selection* — he had been considering the possibility of their origin by means of natural selection. In the early 1860s Darwin became especially interested in this problem because of T. H. Huxley's repeated statements (1860–1863) that the proof of Darwin's theory of natural selection as the solution to the problem of the origin of species was incomplete until the selective origin of cross- and hybrid sterility had been demonstrated, by *artificial* selection experiments for example. Huxley's argument reflected his view that complete proof of any theory required direct evidence from empirical observations. The process of artificial selection — Darwin's own analogy for natural selection — could provide just such direct evidence. At this very same time (1860–1864), Darwin was engaged in performing a series of experiments to elucidate the meaning of heterostyly in di- and trimorphic plants; and he became particularly excited about this work when it appeared, for a time, to provide just the sort of proof Huxley had been demanding. Thus Darwin's debate with Wallace in 1868 was the culmination of a long and thoughtful analysis of the problem of sterility.

This debate is particularly fascinating, since one of its two central issues was: at what levels does the process of selection operate? Over the past two decades, this very same question of the levels or units of selection has once again been the subject of heated debate among evolutionary theorists. Natural selection clearly operates at the level of individual organisms. Traits that are advantageous to the individual become established in a population as the result of the increased reproductive success ("fitness") of those individuals possessing them. But are there also traits that, although disadvantageous to the individual, are beneficial to the group (population or species) to which the individual belongs? Such traits are sometimes called "altruistic", because the individuals possessing them sacrifice their own good "for the good of the species". Genuinely altruistic traits — that is, those that do not benefit the individual either directly *or indirectly* — cannot become established by natural selection acting on individuals, because, by definition, they reduce the fitness of the individuals possessing them. Therefore any such traits, if they do indeed exist, must have become established either "by chance", or by a process of "group selection" since they are good for the group but not the individual. Those populations whose members possess such a trait must be more successful than, and eventually replace, other populations whose members lack the trait. In the great debate between Darwin and Wallace over the origin of sterility, Wallace supported, while Darwin rejected the possibility of group selection.[10]

The second major issue in the debate involved what has recently been called "adaptationism". Indeed it was in large part because of their different positions on this second issue that Darwin and Wallace took different positions on the question of levels of selection. By 1868 Wallace had become quite an ardent adaptationist. Since cross- and hybrid sterility were useful, at least to the species (especially incipient species), if not to the individual, then they *must* have been produced by selection. Even though Darwin agreed completely with Wallace about the utility of sterility to the species, he did not feel compelled by the mere fact of utility to invoke natural selection as the explanation for its origin. Darwin proposed, instead, that cross- and hybrid sterility had originated as incidental by-products, that is "by chance."

Cross- and Hybrid Sterility in Natural Selection *and the* Origin

The fact that Darwin decided to devote an entire chapter in the *Origin* to sterility demonstrates how important he believed the problem to be. The "Hybridism" chapter is part of the "Difficulties" section of the book (comprising Chapters 6–9 in the first five editions, and Chapters 6–10 in the sixth edition). Darwin clearly recognized that most naturalists regarded cross- and hybrid sterility as special, designed qualities with which each species had been endowed at its creation in order to guarantee its distinctness from all other species. Furthermore the distinctness or "reality" of species, as established by cross- and hybrid sterility, was widely held to be equivalent to their "permanence". Thus when Lyell, in his refutation of Lamarck, asked "whether species have a real *and* permanent existence in nature" (my italics), he was asking just one question (Lyell 1830–1833, 2:1; Kottler 1978, pp. 277–278). So one of Darwin's main purposes in his chapter on "Hybridism" was to demonstrate that sterility was not a special endowment and therefore not an insurmountable objection (or, as he put it in *Natural Selection*, a "fatal difficulty") to the origin of species by descent. The table overleaf (Fig. 3) enumerates Darwin's main arguments in the *Origin* against the special endowment hypothesis (column I), and also demonstrates how Darwin had been thinking about many of these matters from as early as 1837–1839 (columns III, IV).

Even though Darwin argued that sterility was not universal, hence not designed, he nevertheless acknowledged that it was very general and consequently required an explanation. Darwin proposed that cross- and hybrid sterility were the "incidental" results of the differences, chiefly in the "reproductive system", that had arisen during the multiplication and divergence of species. In proposing the incidental origin of sterility, Darwin rejected one other explanation besides the special endowment hypothesis. In the first edition of the *Origin* Darwin dismissed in just a single sentence the possibility of the origin of sterility by selection: "On the theory of natural selection the case is especially important, inasmuch as the sterility of hybrids could not possibly be of any advantage to them, and therefore could not have been acquired by the continued preservation of successive profitable degrees of sterility" (p. 245).[11] In the fourth edition of the *Origin*, published in 1866, Darwin added nineteen sentences that gave his full argument in support of this negative conclusion. He began by noting: "At one time it appeared to me probable . . . that this sterility might have been acquired through natural selection slowly acting on a slightly lessened degree of fertility, which at first spontaneously appeared, like any other variation in certain individuals of one variety when crossed with another variety" (*Origin* 1959, p. 443).

I *Origin of Species* (1859)	II *Natural Selection* (1856–1858)	III *1844 Essay*	IV *Transmutation Notebooks* (1837–1839)
1. "The degree of fertility, both of first crosses and of hybrids, graduates from zero to perfect fertility" (p. 255) 250–253, 255–256	pp. 398–401, 409–410, 426–441	pp. 103–104, 124–126	"perfect series, from physical impossibility to unite to perfect prolifickness" (*E*:107) *B*:30, 139–141; *C*:184; *D*: 15–16, 25–26, 87, 105–106
2. "The parallelism between the difficulty of making a first cross, and the sterility of the hybrid thus produced . . . is by no means strict" (p. 256)	415		"remarkable law, that first cross plentiful, second absolutely sterile" (*D*:10) *D*:16
3. "The degree of fertility is . . . innately variable" (p. 256)	411		
4. "The correspondence between systematic affinity and the facility of crossing is by no means strict" (p. 257)	411–412	124	
5. "There is often the widest possible difference in the facility of making reciprocal crosses . . . Hybrids raised from reciprocal crosses . . . generally differ in fertility" (p. 258)	413–415		
6. "The fertility of hybrids is not related to the degree in which they resemble in external appearance either parent" (p. 259)	416–417		
7. "Mere external dissimilarity between two species does not determine their greater or lesser degree of sterility when crossed" (p. 269) 259–260	412, 417, 432	"Mere difference of structure no guide to what will or will not cross" (p. 129, n. 1) 126, n. 1	"It does not bear any practise relation to structure" (*B*:212) *B*: 198, 211, 241; *C*:135
8. "Why . . . has the production of hybrids been permitted?" (p. 260)	418		"My views which would even lead to anticipate mules is very important for Lyell said to me the fact of existence of mules appeared to him most strange . . my theory thus explains a grand apparent anomaly in nature" (*B*:135) *B*:122
9. "A long course of domestication tends to eliminate sterility" (p. 269)	440–441	123n., 127, 130	*B*: 120; *D*: 66, 75
10. Existence of cross-sterile varieties (pp. 269–271)	405–408		*B*: 123
11. Forms, previously ranked as different species, have been reclassified as different varieties when found to be cross-fertile (pp. 246–247)	391, 394, 402	"begging the question" (125)	"arĝument in circle" (*B*: 240) "as long as opponents are not able to tie themselves down, they can find loopholes" (*D*: 66)
12. Forms, previously ranked as different varieties, have been reclassified as different species when found to be cross-sterile (pp. 268, 277)			

Figure 3. Darwin's Arguments against the Origin of Cross- and Hybrid Sterility by Special Creation. Darwin devoted an entire chapter ("Hybridism") in the *Origin* to the phenomena of cross- and hybrid sterility. In the first column (I) are enumerated his main arguments in the *Origin* against the special creation, or design, hypothesis as an explanation for these phenomena. The second column (II) cites the places where these same arguments appeared in the chapter on "Hybridism" in *Natural Selection*. The last two columns demonstrate that Darwin had been thinking about these phenomena since 1837–1839. In the third (III, *1844 Essay*) and fourth (IV, *Transmutation Notebooks*) columns are enumerated the places where Darwin first presented his various arguments against the special creation hypothesis. A blank space in column III or IV indicates that Darwin did not state in 1844 or 1837–1839 the particular argument (for example, no. 5) found in the *Origin* (column I).

In fact Darwin had been inclined to accept the origin of sterility by selection at least two times prior to 1866. In *Natural Selection* Darwin had written: "What we have to show in order to render the facts here treated of, not utterly subversive of our theory, is nearly the same as in the case of any peculiar organ, namely to show how sterility could first arise, to show that it is variable in degree & that there is a gradation in different species from a lesser to greater degree of sterility. And all this, I think, can be done."[12] From this passage it seems probable that, for a time in 1857, Darwin thought sterility could have evolved by means of natural selection. But Darwin eventually cancelled this passage, and his final position in *Natural Selection* came very close to that of the first edition of the *Origin*: "By our theory this sterility . . . must be looked at as an incidental concomitant . . . This must be so, for sterility cannot have been produced, at least in the case of the hybrids themselves, by natural selection, as sterility obviously could not be favourable to them." Darwin did go on to raise the theoretical possibility that cross-sterility, as opposed to hybrid sterility, might have originated by selection but he was not inclined to accept it: "In the case of sterility between species & species, in as much as this is favourable to them by keeping their characters pure & unmixed, it is just possible that the tendency might have been acquired through natural selection; but I know of no fact leading to this conclusion . . ." (*Natural Selection*, p. 390). The second, and more important, time Darwin seriously considered the selection hypothesis was in the early 1860s, in conjunction with his debate with Huxley and his simultaneous study of heterostyly.

The Darwin-Huxley Debate

Thomas Henry Huxley is well remembered as "Darwin's bulldog" because of his vigorous defense of the theory of descent after the publication of the *Origin*.[13] But it is not very well known that Huxley, from 1859 to his death in 1895, remained doubtful about the theory of natural selection. The one and only basis for these doubts that he stated in his published writings was the problem of the origin of sterility. From early in 1860 Huxley frequently expressed his views, to the point where, by 1862, he was referring to "my *old* line about the infertility difficulty" (my italics). Darwin was quite concerned, and at times more than a little exasperated, by Huxley's position; and from 1860 to 1864 (with a few later comments up to 1868) Darwin made a concerted effort to convince Huxley, whom he ultimately called the "Objector-General" on this matter, that the sterility difficulty was not overwhelming.[14]

It is unclear precisely when Huxley "discovered" the sterility difficulty. In his famous letter to Darwin of 23 November 1859 — written immediately after he had completed reading the *Origin* for the first time — Huxley specifically mentioned Chapters 1–5 and 9–13, but not Chapter 8 "Hybridism". He did raise two "objections", the first of which has been cited many times to the present day, but did not refer to sterility: "1st, that you have loaded yourself with an unnecessary difficulty in adopting *Natura non facit Saltum* so unreservedly; and 2nd, it is not clear to me why, if continual physical conditions are of so little moment as you suppose, variation should occur at all" (L. Huxley 1900, 1: 176). Just four days later, in response to another letter from Huxley (which has not been found), Darwin enumerated what Huxley should read about hybridism and offered to send him the manuscript of the chapter from *Natural Selection*. Perhaps Huxley had raised the sterility difficulty by now. In any case, he said nothing about it in his first public pronouncement on the *Origin*, his famous review in *The Times* (26 December 1859). Huxley did refer to the phenomena of sterility, but not as a difficulty for Darwin's theory. He pointed to the facts of cross-fertile species and cross-sterile varieties as evidence against the criterion of cross- (or hybrid) sterility for species, a concept extending back over one hundred years to Buffon (Huxley 1859, pp. 3–4).[15] Huxley did, however, indicate clearly in the review his own qualified acceptance of natural selection and recommended to others the state of mind of "active doubt". Darwin's theory explained a great deal and was not contradicted by the main phenomena of life. Natural selection was a "*vera causa*" (true cause); but since it would take many years to demonstrate its competence to produce everything Darwin had ascribed to it, Huxley was not yet prepared to affirm its truth "absolutely" (pp. 19–20).

Whatever Huxley was thinking in November and December 1859, by January 1860 he had discovered the sterility difficulty and raised it with Darwin. In his reply of 11 January 1860, Darwin commented: "I fully agree that the [sterility] difficulty is great, and might be made much of by a mere advocate." But he concluded: "The whole case seems to me far too mysterious [on which] to rest a valid attack on the theory of modification of species . . ." (*ML* 1: 137). Little did Darwin realize that Huxley was soon to become such an "advocate".

Huxley first raised the difficulty in public in his lecture "On Species and Races, and their Origin" delivered at the Royal Institution on 10 February 1860 (Huxley 1860a). He argued, as had Darwin himself in the *Origin*, that even if sterility between species were not universal, it was common and therefore had to be accounted for by any theory of their origin. A complete solution of the problem of the origin of species required "the experimental determination of the conditions under which bodies having the characters of species are producible." This was the domain of *artificial* selection; and so Darwin's case would be complete "if it can be shown that these breeds [produced by artificial selection] have all the characters which are ever found in species . . ." But so far cross-sterile varieties had not been produced by artificial selection. Huxley believed that "well conducted" experiments "probably" would produce such varieties. But, as the facts now stood, Darwin's theory fell "short of being a satisfactory theory".

Darwin, in letters to both Lyell and Hooker, reacted promptly to Huxley's lecture. He did not feel that Huxley had given a "just idea" of natural selection, so — on this count — the lecture seemed to Darwin an "entire failure". Furthermore, he "remonstrated" against the impression Huxley had conveyed that sterility was a "universal and infallible criterion of species" (*LL* (NY) 2: 74). Darwin remarked to Hooker that if natural selection explained "several large classes of facts" then it "would deserve to be ranked as a theory deserving acceptance". Huxley, in contrast, "rates higher than I do the necessity of Natural Selection being shown to be a *vera causa* always in action" (*ML* 1: 139–140). Ruse has pointed out that in the nineteenth century there were two different conceptions of a *vera causa* (1979a, pp. 235–236). Huxley subscribed to the view that a *vera causa* was directly observable, hence "always in action". Thus, for Huxley, the results of *artificial* selection, which could be observed directly, constituted the best evidence of what *natural* selection could or could not effect. The other view was that a *vera causa* was able to explain "several large classes of facts". Darwin was certainly correct that Huxley considered his "empiricist" *vera causa* to be the more important in the evaluation of scientific theories; and this difference of opinion proved to be the single most important factor in the debate between Darwin and Huxley.

Huxley's next public statement concerning the sterility difficulty came in his second review of the *Origin*, published in the *Westminster Review* (April 1860). Darwin had complained, after the Royal Institution lecture, that Huxley had not even "alluded to the more important parts of the subject" (*ML* 1: 139). Consequently Huxley now presented Darwin's views at length. But he went on to say that even if only two species were cross-sterile, an important problem remained to be solved, and any theory that did not solve it was "imperfect". Huxley repeated his own belief that it should be possible "in a comparative few years" to produce cross-sterile varieties. But until that had been accomplished, there was a "little rift within the lute" that should be neither "disguised nor overlooked". Huxley then observed that he had been unable to "pick holes of any great importance" in the rest of Darwin's argument. Thus the sterility difficulty stood as the only serious objection to Darwin's theories (Huxley 1860b, pp. 43–50, 74–75). Darwin saw Huxley very soon after the review was published and then wrote to Lyell: "I *think* I have convinced him that he has hardly allowed weight enough to the case of varieties of plants being in some degrees sterile" (Darwin's italics; *LL* (NY) 2: 94). The existence of cross-sterile varieties in plants, as originally demonstrated by Gärtner and then confirmed at Darwin's urging by the young horticulturist John Scott, became thereafter a major point in Darwin's many responses to Huxley's repeated critiques.

Darwin next discussed the issue with Huxley in December 1860, one year after the publication of the *Origin*. Darwin "entirely" agreed with Huxley about the "terrific" difficulties raised by his views. Darwin was, however, confident about the eventual acceptance of those notions, because "some who went half an inch with me now go further, and some who were bitterly opposed are now less bitterly opposed." Yet Huxley himself, on account of the sterility question, was a disturbing exception to this trend. Darwin wrote: "And this makes me feel a little disappointed that you are not inclined to think the general view in some slight degree more probable than you did at first. This I consider rather ominous" (*LL* (NY) 2: 147).

Although Huxley had nothing to say in public about sterility during 1861, Darwin clearly did not forget about his objection. In September 1861 Darwin asked Hooker for white and yellow varieties of *Verbascum*, so that he could test Gärtner's "wonderful and repeated statement" (cited by Darwin in the *Origin*) about the cross-sterility of these varieties. "I do not think any experiment can be more important on the origin of species; for if he is correct we certainly have what Huxley calls new physiological species arising" (*ML* 2: 271). By this time, Darwin was well along in his experimental investigation of heterostyly; on 21 November 1861, he read his first paper on the subject to the Linnean Society of London.[16] On the basis of his experimental results, Darwin changed his mind, for a time, about natural selection and the origin of sterility; and thus for the next few years, the phenomena of heterostyly played a central role in the Darwin-Huxley debate. But then the continuation of his study of heterostyly, now in the trimorphic species *Lythrum salicaria*, eventually led Darwin back to his previous position that sterility had not been produced by selection (Whitehouse 1959; Ghiselin 1969, pp. 141–153; Ford 1964, pp. 172–185).

Darwin, Heterostyly, and the Origin of Sterility

Darwin had long been familiar with the phenomenon of dimorphism in plants, which he first regarded as "merely a case of unmeaning variability". In the species of the genus *Primula*, such as the primrose and the cowslip,[17] the flowers are of two forms (dimorphism). In one form ("pin" or long-styled), the style is long and the stamens are short, while in the other form ("thrum" or short-styled) the style is short and the stamens long (see Fig. 4). Closer examination of the two forms convinced Darwin that they were "much too regular and constant" for the dimorphism to be due to mere variability.

Darwin's first hypothesis was that dimorphic species were in a transitional state ("on the high road") from hermaphroditism to dioecy. In the short-

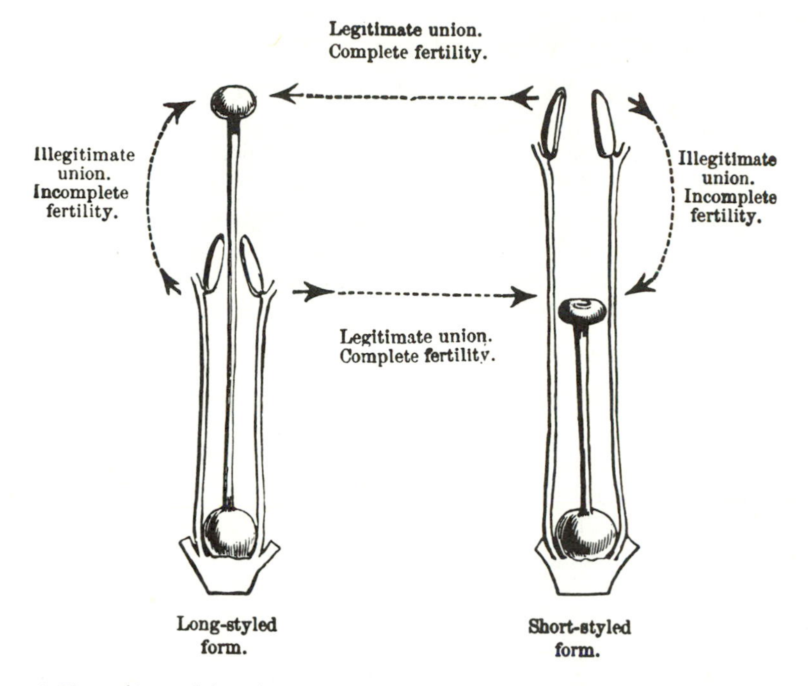

Figure 4. Dimorphism and the Selective Origin of Sterility (Darwin 1877, fig. 2).
Initially, in 1861, after the completion of his first experiments on fertilization in the dimorphic species *Primula vulgaris* (primrose), Darwin was inclined to the view that natural selection had produced self-sterility in order to prevent self-fertilization. The four possible crosses are illustrated in this figure. Since the "illegitimate union" in the short-styled form was both the more likely to occur and the more sterile of the two illegitimate crosses, Darwin was led to propose the selection hypothesis (see text for further details).

styled form, the style was degenerating and the flower, with its long stamens, was becoming male. Similarly, in the long-styled form, the (short) stamens were degenerating and the flower, with its long style, was becoming female. This hypothesis appeared to gain support from the fact that pollen grains in the short-styled form (incipient male) were larger than those in the other form. Thus Darwin predicted that the long-styled form (incipient female) should produce more seed. In June 1860, Darwin remarked to A. Gray: "If it should prove that the so-called male plants [short-styled] produce less seed than the so-called female plants [long-styled], what a beautiful case of gradation from hermaphrodite to unisexual condition it will be! If they produce about equal number of seed, how perplexing it will be" (*LL* (NY) 2: 472). As M. Ghiselin has pointed out, Darwin's hypothesis fit in well with his previous work on barnacles in which he had found just such transitional forms (Ghiselin 1969, p. 144).

Darwin proceeded to test his hypothesis and found, to his considerable amazement, that his rather light-hearted comment (to Hooker) — "it may turn out all a blunder" — had been right on the mark. His hypothesis had been "knocked on the head". Now he wrote to Hooker:

> By Jove the plants of primroses and cowslip with short pistils and large grained pollen [the hypothesized incipient male] are rather more fertile than those with long pistil, and small-grained pollen . . . I never will believe that these differences are without some meaning. Some of my experiments lead me to suspect that the long-grained pollen [from the long stamens of the short-styled form] suits the long pistils [,] and the small-grained pollen [from the short stamens of the long-styled form] suits the short pistils, but I am determined to see if I cannot make out the mystery next spring [1861]. (*LL* (NY) 2: 472–473)

To test his new hypothesis Darwin performed all four possible crosses between the two forms — long × long, short × short, long (♀) × short (♂), short ( ) × long (♂)— always taking care to avoid complicating effects from inbreeding. The results, obtained in 1861, showed that the first two crosses involving like forms ("homomorphic" unions) were almost completely sterile (hence "illegitimate" unions); whereas the last two crosses involving unlike forms ("heteromorphic" unions) were perfectly fertile ("legitimate" unions), confirming Darwin's new hypothesis. At this point Darwin concluded heterostyly was an adaptation to ensure the crossing of different plants: "The meaning or use of the existence in *Primula* of the two forms in about equal numbers, with their pollen adapted for reciprocal union, is tolerably plain; namely, to favour the intercrossing of distinct individuals. With plants there are innumerable contrivances for this end; and no one will understand the final cause of the structure of many flowers without attending to this point" (*CP* 2: 59). Darwin had long believed that perpetual self-fertilization in hermaphrodites did not occur, and that at least an occasional cross with a different individual was a "law of nature": "Nature . . . abhors perpetual self-fertilisation" (*Orchids* 1877, p. 293). This law arose from the fact that cross-fertilization increased vigor and fertility in the offspring, while self-fertilization diminished them. Heterostyly favored cross-fertilization in the following manner. Since all the flowers of a plant were of one form, a fertile cross necessarily involved two different plants of unlike forms. The fertile unions between unlike forms were guaranteed by the reciprocal adjustments of style and stamen lengths. The long style of the long-styled form was the same length as the long stamens of the short-styled form. And the short style of the short-styled form was the same length as the short stamens of the long-styled form. Darwin's own experiments established that insects were necessary for fertilization in heterostyled plants. Thus, for example, when a humble-bee visited first a short-styled plant with its long stamen, pollen adhered to the base of the bee's proboscis, which was inserted all the way down to the bottom of the flower to obtain the nectar; this pollen would then be transferred to the stigma of the long style of a long-styled plant that the bee visited later. In this way, the short × long cross was effected (see Fig. 4).

In his first paper on dimorphism, Darwin suggested that the sterility of homomorphic unions had been produced by selection. One particular experimental result was critical to Darwin's reasoning. Although both homomorphic unions were largely sterile, they were not equally so. The short × short cross was more sterile than the long × long cross. Why? Darwin believed that the proboscis of a bee inserted into a short-styled flower would almost certainly carry some pollen down from the stamens above to the style below, thus effecting self-fertilization, which was manifestly harmful. "On this view we can at once understand the good of the pollen of the short-styled form, relatively to its own stigma, being the most sterile, for this sterility would be the most requisite to check self-fertilization, or to favour intercrossing" (*CP* 2: 60). In other words, because the self-fertilization of the short-styled form was both the more likely to occur and the more sterile of the two types of self-fertilization, Darwin was led to suggest that self-sterility was a special adaptation, produced by natural selection, to counteract self-fertilization, and thereby favor cross-fertilization; and then this self-sterility was transferred to sterility with all other individuals of the same form (see Fig. 4). The reasonableness of Darwin's suggestion is illustrated by the fact that Walter Bodmer proposed to E. B. Ford exactly the same idea one hundred years later, apparently without either's knowledge of Darwin's priority: "Dr. Bodmer points out to me that one reason for the evolution of greater self-sterility in thrum compared with pin is that thrum pollen will naturally fall down on to the thrum stigma, giving greater opportunities for self-fertility if thrums were self-compatible" (Ford 1964, p. 174).

Having suggested the possibility of the selective origin of self-sterility in heterostyled plants, Darwin went on in his 1861 paper to ask whether cross-sterility in general might not also have originated by selection:

> Seeing that we thus have a ground work of variability in sexual power, and seeing that sterility of a peculiar kind has been acquired by the species of *Primula* to favour intercrossing, those who believe in the slow modification of specific forms will naturally ask themselves whether sterility may not have been slowly acquired for a distinct object, namely, to prevent two forms, whilst being fitted for distinct lines of life, becoming blended by marriage, and thus less well adapted for their new habits of life. But many great difficulties would remain, even if this view could be maintained. (*CP* 2: 61)[18]

Darwin's tone in this published remark was very tentative. But in his correspondence during 1862 he was very much more confident.

In January 1862 Huxley delivered two lectures in Edinburgh on "The Relation of Man to the Lower Animals", the substance of which was published in his book *Man's Place in Nature* (1863a). As he put it to Darwin, he took his "old line" on sterility. Huxley stated that a "true physical cause" had to account for *all* the phenomena. He carefully distinguished between a theory that was clearly "inconsistent" with a phenomenon — in which case it had to be rejected — and a theory that so far had been unable to account for all phenomena. Darwin's theory was not inconsistent with any phenomena, but it had not yet accounted for everything. Even though increasing knowledge led him to regard the "hiatus" in Darwin's evidence as "of less and less importance", "one link in the chain of evidence" was still missing. Until "physiological species", that is cross-sterile forms, had been produced by "selective breeding", natural selection would not be "proved to be competent to do all that is required of it to produce natural species". Huxley remarked that he emphasized this difficulty, because he did not want to be regarded as an "advocate" *for* Darwin's theory, someone who sought to "smooth over real difficulties, and to persuade where he cannot convince" (Huxley 1863a, pp. 106–108).

Darwin responded quickly to Huxley's lectures: "I must say one word on the Hybrid question. No doubt you are right that here is a great hiatus in the argument; yet I think you overrate it — you never allude to the excellent evidence of *varieties* of Verbascum and Nicotiana being partially sterile together. It is curious to me to read (as I have to-day) the greatest crossing *Gardener* [Gärtner] . . . insisting how frequently crossed *varieties* produce sterile offspring" (Darwin's italics). Darwin then directed Huxley's attention to his *Primula* research, "for it leads me to suspect that sterility will hereafter have to be largely viewed as an acquired or *selected* character — a view which I wish I had had facts to maintain in the 'Origin' " (Darwin's italics; *LL* (NY) 2: 176).

In his reply Huxley tried his best to reassure Darwin; he had no doubt that twenty years of artificial selection experiments by a "skilled physiologist" would produce cross-sterile forms; "and in this, if I mistake not, I go further than you do yourself . . . when these experiments have been performed I shall consider your views to have a complete physical basis, and to stand on as firm ground as any physiological theory whatever. . . . I am constitutionally slow of adopting any theory that I must needs stick by when I have once gone in for it; but for these two years I have been gravitating towards your doctrines, and since the publication of your primula paper with accelerated velocity. By about this time next year I expect to have shot past you, and to find you pitching into me for being more Darwinian than yourself. However, you have set me going, and must just take the consequences, for I warn you I will stop at no point so long as clear reasoning will carry me further" (L. Huxley 1900, 1: 196). But, as a matter of fact, one year later (1863) Huxley was as unconvinced as he had been in 1860.

During the 1862 growing season, Darwin continued his experiments on the dimorphic species *Linum grandiflorum*, and became even more inclined to accept the origin of cross-sterility by selection. In *Linum* the long-styled form also had long stamens; thus, of the two forms, the long-styled was more likely to be self-fertilized. If self-sterility had been produced by natural selection, then the long-styled form should be the more self-sterile; and Darwin's experiments showed this to be the case. Consequently Darwin wrote to Hooker in December 1862: "my notions on hybridity are becoming considerably altered by my dimorphic work. I am now *strongly* inclined to believe that sterility is at first a selected quality to keep incipient species distinct. . . . It is this which makes me so much interested with dimorphism" (my italics; *ML* 1: 222). Darwin's intense interest at this time in the problem of sterility is especially well demonstrated by his correspondence with W. B. Tegetmeier and John Scott about experiments he hoped each would perform.

W. B. Tegetmeier, a poultry breeder and bee master, had been in correspondence with Darwin since 1855. On 27 December 1862, Darwin wrote to Tegetmeier: "I have been led lately from experiments on dimorphism to reflect much on sterility from hybridism, and partially to change the opinion given in *Origin*." Darwin proceeded to describe an elaborate selection experiment he had designed, "which seems to me well worth trying, but too laborious ever to be attempted." If a cock (A) and a hen (B) of the same breed could be found that happened to be sterile with each other but were otherwise fertile, then each should be crossed with a near relation (see Fig. 5). The offspring of the cock (a, b, c, d, e) should then be crossed with the offspring of the hen (f, g, h, i, j). All fertile pairs should be destroyed, while any (partially) infertile pairs (for example, a♀×i♂) should be preserved. Then a (♀ offspring of A) should be crossed with A, and i (♂ offspring of B) should be crossed with B, "so as to try and get two families which would not unite together; but the members *within* each family being fertile together. This would probably be quite hopeless; but he who could effect this would, I believe, solve the problem of sterility from hybridism" (*ML* 1: 224).[19] Clearly if such an experiment were a success, Huxley's demand would be met. As far as anyone knows, the experiment was never carried out.

Darwin and the young horticulturist John Scott became acquainted in November 1862, when Scott wrote to correct an error in Darwin's recently published *Orchids*. In his second letter to Scott (19 November) Darwin asked whether he had ever tested the "relative fertility of varieties of plants (like those I quote from Gärtner on the varieties of *Verbascum*). I much want information on this head . . ." (*ML* 2: 309). And in his next letter to Scott, Darwin remarked: "To the best of my judgement no subject is so important in relation to theoretical natural science . . . as the effects of changed or unnatural conditions, or of changed structure on the reproductive system. Under this point of view the relation of well-marked but undoubted varieties in fertilising each other requires far more experiments than have been tried" (*ML* 1: 218). Darwin was hopeful that Scott would be the one to do these further experiments. In his letter of 11 December 1862, Darwin repeated his view that a repetition of Gärtner's experiments would be "pre-eminently important". Scott did, in fact, repeat Gärtner's experiments; and Darwin reported Scott's confirmation of Gärtner's findings on cross-sterile varieties, in the fifth edition of the *Origin* (*Origin* 1959, p. 465).

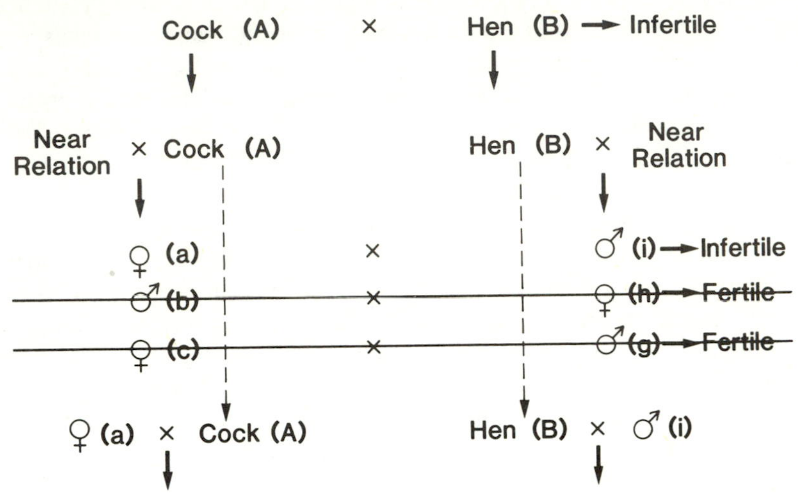

Figure 5. Artificial Selection and the Origin of Sterility (see *ML* 1: 224).
In 1862, stimulated by T. H. Huxley's statements that the complete proof of the theory of natural selection required the experimental demonstration of the selective origin of sterility, Darwin designed the experiment illustrated in this figure. Its objective was the production of cross-sterility between two different families by means of artificial selection. As far as anyone knows, this experiment was never performed (see text for further details).

Scott also conducted a number of experiments with *Primula*. Darwin became very enthusiastic about this work when he learned in 1863 that Scott had found primrose varieties that appeared to be cross-sterile. Scott reported his results in a paper read to the Linnean Society on 4 February 1864. He stressed the "remarkable" finding of "absolute zero of fertility, apparently, attained between undoubted varieties of a species!" Scott concluded: "In view of such evidence, I think I am fully justified in adding that this . . . form [a variety of the cowslip *Primula veris*] is, in fact, judged by the physiological test so much insisted on by Professor Huxley, *a new and distinct species*" (Scott's italics; Scott 1864, pp. 97–108). In the light of these results, it is easy to understand what Darwin wrote to Hooker in April 1864 about Scott's abilities and his experiments: "I believe years may pass before another man appears fitted to investigate certain difficult and tedious points — viz. relative fertility of varieties of plants . . . (already Scott has done excellent work on this head) . . ." (*ML* 2: 328–329). Darwin was "fully convinced" Scott's work would have "permanent value". He was therefore willing to pay Scott for a year or two to enable him to continue his experiments. After Scott's paper on his *Primula* experiments was published, Darwin wrote to both Hooker and Gray on 13 September 1864. The cross-sterile varieties were the "new and, as they seem to me, important points." The reason was clear enough: "Here we have a new 'physiological species' " (*ML* 2: pp. 327–328; cf. Darwin 1877, pp. 224–225).

At the same time (November and December 1862) that Darwin solicited the aid of Tegetmeier and Scott to do experiments in order to meet Huxley's demand for cross-sterile varieties produced by selection, Darwin and Huxley engaged in their last major interchange on the sterility difficulty. Once again the precipitating cause was a series of public lectures by Huxley, which were published as the book *On Our Knowledge of the Causes of the Phenomena of Organic Nature* (1863b). In the fifth of the six lectures, Huxley raised the sterility difficulty and asserted, as before, that there was a "physiological contrast" between species and varieties. Species were frequently cross-sterile, or at least produced sterile hybrids. "Can we find an approximation to this in the different races known to be produced by selective breeding from a common stock? Up to the present time the answer to that question is absolutely a negative one. As far as we know at present, there is nothing approximating to this check." There appeared to be a "physiological limitation" to the amount of divergence producible by selection (Huxley 1863b, pp. 110–117).

Huxley's statements brought an immediate and exasperated reply from Darwin:

> You say the answer to varieties when crossed being at all sterile is "absolutely a negative." Do you mean to say that Gärtner lied, after experiments by the hundred (and he a hostile witness), when he showed that this was the case with *Verbascum* and with maize (and here you have selected races): does Kölreuter lie when he speaks about the varieties of tobacco. My God, is not the case difficult enough, without its being, as I must think, falsely made more difficult? I believe it is my own fault — my damned candor: I ought to have made ten times more fuss about these most careful experiments. I did put it stronger in the third edition of the *Origin*. (*ML* 1: 230; cf. *Origin* 1959, p. 464 (217, 217:c))

In his sixth and final lecture Huxley returned again to the "sterility case", since he regarded it as the only objection to Darwin's theory "of any great value". Natural selection was a *vera causa*. It could explain all the phenomena of races, all the morphological phenomena of species, and most of the physiological phenomena of species. But so far it had not been "wholly competent" to explain the phenomena of hybridism. Therefore, "to place his views beyond the reach of all possible assault" Darwin had to supply evidence of cross-sterile varieties produced by selection. Huxley acknowledged the existence of cross-sterile varieties; however he went on to say: "but if any objector urges that we cannot prove that they have been produced by artificial or natural selection, the objection must be admitted — ultra-sceptical as it is. But in science, scepticism is a duty." Huxley remarked that in view of how little was known concerning sterility, there was no reason to suppose the "crucial result" could not be obtained by experiment. If it could be shown, however, that artificial selection could not possibly produce cross-sterile varieties, then Darwin's theory would be "utterly shattered" (Huxley 1863b, pp. 135–136, 145–149).

After this lecture, Darwin had little more to say to Huxley: "We differ so much that it is no use arguing. To get the degree of sterility you expect in recently formed varieties seems to me simply hopeless. It seems to me almost like those naturalists who declare they will never believe that one species turns into another till they see every stage [in] the process" (*ML* 1:225). D. Hull has noted that one of the methodological objections commonly raised by Darwin's early critics was that he had not supplied *direct* evidence for his theory, since he had not demonstrated the actual transmutation of one species into another (1973b, pp. 49–51). Though these critics were anti-evolutionists, Huxley the evolutionist was raising very much the same objection against natural selection, as Darwin himself perceived. Two weeks later Darwin wrote once more: "Of course I do not wholly agree about sterility. I hate beyond all things finding myself in disagreement with any capable judge, when the premises are the same . . . Thinking over my former letter to you, I fancied (but I now doubt) that I had partly found out the cause of our disagreement, and I attributed it to your naturally thinking most about animals, with which the sterility of the hybrids is much more conspicuous than the lessened fertility of the first cross . . . In plants the test of first cross seems as fair as test of sterility of hybrids. And this latter test applies, *I will maintain to the death*, to the crossing of varieties of *Verbascum*, and varieties, *selected varieties*, of *Zea*. You will say Go to the Devil and hold your tongue. No, I will not hold my tongue" (my italics; *ML* 1: 231–232). Darwin then repeated an argument from the *Origin* and concluded: "Now I will hold my tongue." And hold his tongue he did; for Darwin did not raise the subject of sterility again with Huxley for four years.

*Trimorphism (*Lythrum*) and the Incidental Origin of Sterility*

Darwin's experiments in 1861–1862 with the dimorphic species of *Primula* and *Linum* had led him to seriously entertain the possibility of the origin of sterility by selection. But further experiments in 1862–1863 with the trimorphic species of *Lythrum* convinced Darwin that he had been right in the first place about the incidental origin of sterility. In *Lythrum* there were three different forms of flowers. Each form possessed two sets of stamens and one style. In the long-styled form, the stamens were short and mid length. In the mid-styled form, the stamens were short and long; while in the short-styled form, they were long and mid length. At first

Darwin was driven "almost stark staring mad" by the puzzle posed by *Lythrum*, "a real odd case . . . [which] interests me extremely, and seems to me the strangest case of propagation recorded among plants or animals" (*LL* (NY) 2: 475–476). After he did find the solution to *Lythrum* — "a necessary triple alliance between three hermaphrodites" — Darwin wrote to Hooker, "I have done nothing which has interested me so much as *Lythrum*, since making out the complemental males of Cirripedes" (*LL* (NY) 2: 480; cf. *ML* 2: 419).

Darwin performed all eighteen possible crosses (each of the three styles × each of the six stamens). He found six fully fertile crosses, which resulted when the style of a given length was fertilized by pollen from a stamen of the same length in either of the two other forms (see Fig. 6). The remaining twelve unions were sterile but to different degrees. The degree of sterility followed a distance rule: "with the several illegitimate unions it will be found that the greater the inequality in length between the pistil and stamens, the greater the sterility of the result. There is no exception to this [distance] rule." Thus, for instance, a cross involving a long style was more sterile if the pollen had been derived from a short stamen than from a midlength stamen. Darwin concluded from this rule that sterility in *Lythrum* had to be "an incidental and useless result of the gradational changes through which this species has passed in arriving at its present condition" (*CP* 2: 120). Darwin reasoned as follows. If the sterility had originated by selection, then those illegitimate crosses that were more likely to occur should have been more sterile. Which crosses were more likely to occur? Clearly those in which the style and stamen were closer together. Thus, on the selection hypothesis, the rule should have been: the *smaller* the inequality in length between style and stamen, the greater the sterility of the result. Just the reverse was actually the case.

In both editions of his *Variation* Darwin discussed self-sterility in plants at length, and continued to entertain the possibility of its origin by selection: "With respect to the species which, whilst living under their natural conditions, have their reproductive organs in this peculiar state, we may conclude that it has been naturally acquired for the sake of effectually preventing self-fertilisation" (*Variation* 2: 140; *Variation* 1875, 2: 122). But when he rethought the entire matter anew while writing *The Effects of Cross and Self Fertilisation in the Vegetable Kingdom*, Darwin came to a new conclusion. He began by stating, "it seems at first sight highly probable that self-sterility has been gradually acquired through natural selection in order to prevent self-fertilisation . . . Nevertheless the belief that self-sterility is a quality which has been gradually acquired for the purpose of preventing self-fertilisation must, I believe, be rejected" (Darwin 1878, p. 345).

Darwin offered three arguments in support of his conclusion: First, "there is no close correspondence in degree between the sterility of the parent-plants when self-fertilised, and the extent to which their offspring suffer

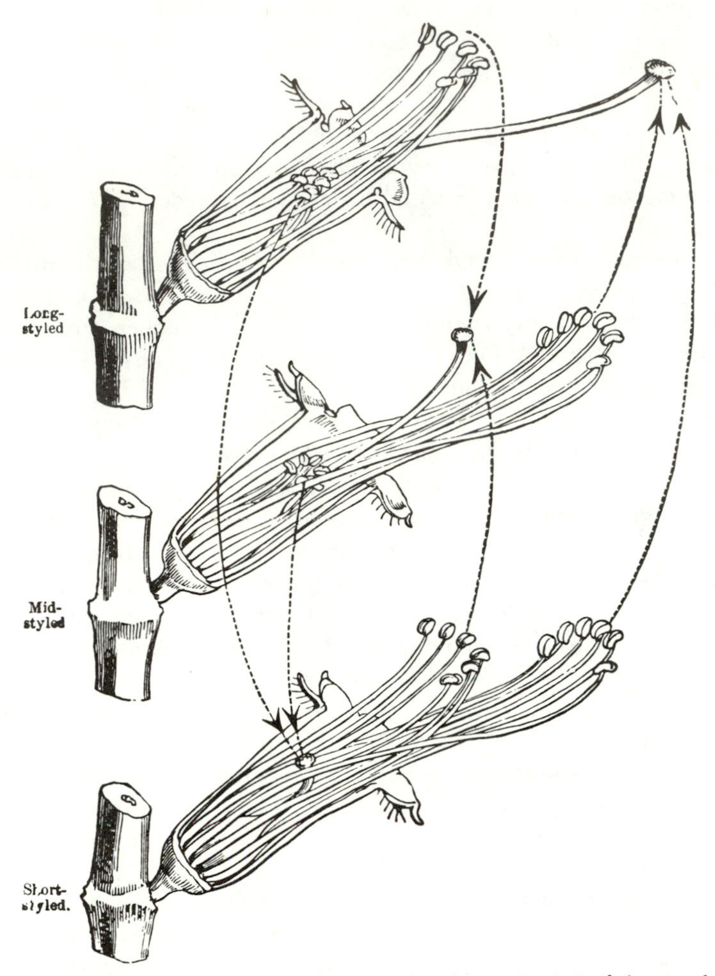

Figure 6. Trimorphism and the Incidental Origin of Sterility (Darwin 1877, fig. 10).
From his experimental study of fertilization in the trimorphic species *Lythrum salicaria* (purple loosestrife), Darwin concluded in 1864 that natural selection could not have produced self-sterility in order to prevent self-fertilization, but rather the self-sterility must have originated incidentally. The six fertile crosses are illustrated in this figure by the dotted lines. The remaining twelve possible crosses were sterile to different degrees, in accordance with a distance rule — "the greater the inequality in length between the pistil and stamens, the greater the sterility of the result" — that was just the reverse of what should have been the case on the selection hypothesis (see text for further details).

in vigour by this process, and some such correspondence might have been expected if self-sterility had been acquired on account of the injury caused by self-fertilisation." Darwin's reasoning here was parallel to that in his analysis of the implications of the distance rule in *Lythrum*. On the selection hypothesis, self-sterility should have been greatest when self-fertilization was (i) most likely to occur and (ii) had the most severe effects on offspring. Second, the degree of self-sterility varied between individuals "of the same parentage". The only way out of this difficulty was to "suppose that certain individuals have been rendered self-sterile to favour intercrossing, whilst other individuals have been rendered self-fertile to ensure the propagation of the species." This was, in fact, the position Darwin had taken just the year before in the second edition of the *Variation*. Lastly, Darwin pointed to the "immediate and powerful effect of changed conditions in either causing or in removing self-sterility." Thus Darwin concluded self-sterility, like other forms of sterility, had been incidentally, not selectively, acquired (Darwin 1878, pp. 345–346).

In his initial analysis of sterility in dimorphic species, Darwin had concentrated on the evils of self-fertilization. Sterility, which prevented such self-fertilization and thereby necessitated beneficial cross-fertilization, appeared to be advantageous. Thus Darwin thought selection for this sterility had come first in the evolution of heterostyly. But in his book on heterostyly, *The Different Forms of Flowers* (1877), written one year after *Cross and Self Fertilisation*, Darwin observed: "Another view seems at first sight probable, namely, that an incapacity to be fertilised in certain ways has been specially acquired by heterostyled plants . . . . It is, however, incredible that so peculiar a form of mutual infertility should have been specially acquired unless it were highly beneficial to the species; and although it may be beneficial to an individual plant to be sterile with its own pollen [self-sterility], cross-fertilisation being thus ensured, how can it be any advantage to a plant to be sterile with half its brethren, that is, with all the individuals belonging to the same form?" (1877, p. 264). Darwin now recognized that sterility could only be advantageous *after* the plant had already become adapted for cross-fertilization, so he now focussed primarily on the benefits of cross-fertilization as opposed to the evils of self-fertilization. "The means for favouring cross-fertilisation *must* have been acquired *before* those which prevent self-fertilisation, as it would manifestly be injurious to a plant that its stigma should fail to receive its own pollen, unless it had already become well adapted for receiving pollen from another individual" (my italics; 1878, p. 383). This theoretical conclusion was supported by the fact that many species that are well adapted for cross-fertilization remained capable of self-fertilization. Thus heterostyled species had first become adapted for cross-fertilization between unlike forms; the sterility between two individuals of the same form was merely the incidental by-product of this prior adaptation (1877, pp. 264–268; Ghiselin 1969, pp. 149–151; Whitehouse 1959, p. 213).

The Fourth Edition of the Origin *and Sterility*

The first three editions of the *Origin* (1859, 1860, 1861) appeared over the short period of eighteen months. Five years then passed before the fourth edition was published in 1866, so it was the most extensively revised edition to date. Darwin added to his "Hybridism" chapter a section of forty-six sentences on heterostyly, as well as a section of nineteen sentences in which he elaborated on his single-sentence rejection, in the earlier editions, of the selective origin of sterility. He presented both theoretical and empirical reasons in support of his view. First, natural selection acting *directly* on individual organisms could not increase sterility:

> He who will take the trouble to reflect on the steps by which this first degree of sterility could be increased through natural selection to that high degree which is common with so many species . . . will find the subject extraordinarily complex. After mature reflection it seems to me that this could not have been *of any direct advantage to an individual animal* to breed poorly with another individual of a different variety, and thus to leave few offspring; consequently such individuals could not have been preserved or selected. (my italics)[20]

Second, natural selection acting *indirectly* on individual organisms could not increase sterility either:

> With sterile neuter insects we have reason to believe that modifications in their structure have been slowly accumulated by natural selection, from an advantage having been thus indirectly given to the community to which they belonged over other communities of the same species; but *an individual organism*, if rendered slightly sterile when crossed with some other variety, *would not thus indirectly give any advantage to its nearest relatives* or to any other individuals of the same variety, thus leading to their preservation. (my italics)

Darwin did believe that cross-sterility "would profit an *incipient species* . . . for thus fewer bastardised and deteriorated offspring would be produced to commingle their blood with the newly-forming variety." But since he focussed here exclusively on natural selection acting, directly or indirectly, *on individuals, not on groups*, he concluded "the various degrees of lessened fertility which occur with species when crossed cannot have been slowly accumulated by means of natural selection" (*Origin* 1959, pp. 443–445).

Besides these theoretical considerations, Darwin pointed to two empirical grounds for the same conclusion. Geographically isolated (allopatric) species were often cross-sterile to some degree, even though they never intercrossed in nature; so "it could clearly have been of no advantage to such separated species to have been rendered mutually sterile, and consequently this could not have been effected through natural selection."[21] Furthermore natural selection could not account for the fact that in some reciprocal crosses one was fertile while the other was sterile, because (as Darwin added in the fifth edition) "this peculiar state of the reproductive system could not possibly be advantageous to either species" (*Origin* 1959, pp. 443–444).[22]

Immediately after the fourth edition of the *Origin* was published Darwin wrote to Huxley (22 December 1866) and raised with him the sterility question for the first time since January 1863. He asked Huxley to read the "Hybridism" chapter in the fourth edition of the *Origin*, "for I am very anxious to make you think less seriously on that difficulty . . . you have been Objector-General on this head" (*ML* 1: 274). Huxley's reply has not been found; but he must have restated his "old line", because in his 7 January 1867 reply to Huxley Darwin added this postscript: "Nature never made species mutually sterile by selection, nor will men" (*ML* 1: 277).

By early 1867 Darwin had also completed the *Variation*. Chapter 19 contained "remarks on hybridism" that were almost identical to the additions in the fourth edition of the *Origin* (*Variation* 2: 185–189). Once again Darwin wrote to Huxley concerning sterility: "You are so terribly sharp-sighted and so confoundedly honest! But to the day of my death I will always maintain that you have been too sharp-sighted on hybridism; and the chapter on the subject in my book I should like you to read: not that, as I fear, it will produce any good effect, and be hanged to you" (*ML* 1: 287). This letter, as far as I know, was Darwin's last comment to Huxley about sterility. Huxley never changed his mind. In his essay "On the reception of the 'Origin of Species' ", written for the *Life and Letters of Charles Darwin*, he repeated that the "logical foundation" of Darwin's theory was still "insecure", because artificial selection had not yet resulted in cross-sterile varieties (Huxley 1887a, p. 551). And in the Preface to *Darwiniana*, published just two years before his death, Huxley stated:

> Those who take the trouble to read the first two essays, published in 1859 and 1860 [reviews of the *Origin* in *The Times* and the *Westminster Review*], will, I think, do me the justice to admit that my zeal to secure fair play for Mr. Darwin, did not drive me into the position of a mere advocate; and that . . . I did not fail to indicate its weak points. I have never seen any reason for departing from the position which I took up in these two essays . . . I remain of the opinion expressed in the second, that until selective breeding is definitely proved to give rise to varieties infertile with one another, the logical foundation of the theory of natural selection is incomplete. (1893b, pp. v–vi)

This was Huxley's final statement on the subject.

The Darwin-Wallace Debate

The great debate in 1868 between Darwin and Wallace over the origin of sterility was precipitated by Darwin's "remarks on hybridism" in the *Variation*. In February 1868, within a month of the publication of Darwin's book, Wallace wrote to Darwin, raising for the first time the matter of selection and sterility.[23] Wallace agreed with Darwin that cross-sterility would not be advantageous to the cross-sterile *individual*; but as long as it would be a "positive advantage" to the two cross-sterile *forms*, then, he argued, it could be increased by natural selection.

> I do not see your objection to *sterility* between allied species having been aided by Natural Selection. It appears to me that, given a differentiation of a species into two forms each of which was adapted to a special sphere of existence, every slight degree of sterility would be a positive advantage not to the *individuals* who were sterile, but to *each form*. If you work it out, and suppose the two incipient species, A, B, to be divided into two groups, one of which contains those which are fertile when the two are crossed [A_f B_f, Figure 7], the other being slightly sterile [A_s B_s, Figure 7], you will find that the latter will certainly supplant the former in the struggle for existence, remembering that you have shown that in such a cross the offspring would be *more vigorous* than the pure breed, and would therefore certainly soon supplant them, and as these would not be so well adapted to any special sphere of existence as the pure species A and B, they would certainly in their turn give way to A and B. (Wallace's italics; Wallace 1916, p. 162)

Darwin replied right away: "I feel sure that I am right about sterility and natural selection. Two of my grown-up children . . . have two or three times at intervals tried to prove me wrong, and when your letter came they had another try, but ended up coming back to my side." Darwin then proceeded to restate his own view. As in the *Origin* and the *Variation*, his conception of natural selection involved the level of the individual only; since increased cross-sterility did not enhance the reproductive success of the more cross-sterile individual, Darwin could not see how selection could increase sterility:

> I do not quite understand your case . . . If sterility is caused or accumulated through Natural Selection then . . . Natural Selection must have the power of increasing it. Now take two species, A and B, and assume that they are . . . half-sterile, i.e. produce half the full number of offspring. Now try and make (by Natural Selection) A and B absolutely sterile when crossed, and you will find how difficult it is. I grant, indeed it is certain, that the degree of sterility of the individuals of A and B will vary, but *any such extra-sterile individuals* of, we will say, A, if they should hereafter breed with other individuals of A, *will bequeath no advantage to their progeny*, by which these families will tend to increase in number over other families of A, which are not more sterile when crossed with B . . . it is a most difficult bit of reasoning, which I have gone over and over again on paper with diagrams. (my italics; Wallace 1916, p. 163)

Two days later (1 March) Wallace accepted Darwin's challenge to "try and make (by Natural Selection) A and B absolutely sterile", sending him a nineteen-point "demonstration, *on your own principles*, that Natural Selection *could* produce *sterility of hybrids*" (Wallace's italics). Wallace began by stating, "I think you will . . . admit that if I demonstrate that a considerable amount of sterility would be advantageous to a *variety*, that is *sufficient proof* that the slightest variation in that direction would be useful also, *and* would go on accumulating" (my italics; Wallace 1916, p. 164). Simply stated, the main point under dispute was that "and". Darwin did not deny that sterility would be useful to a variety (= incipient species). But if this sterility were useful *only* to the variety, would it necessarily be selected? Wallace thought it would; Darwin was extremely doubtful.

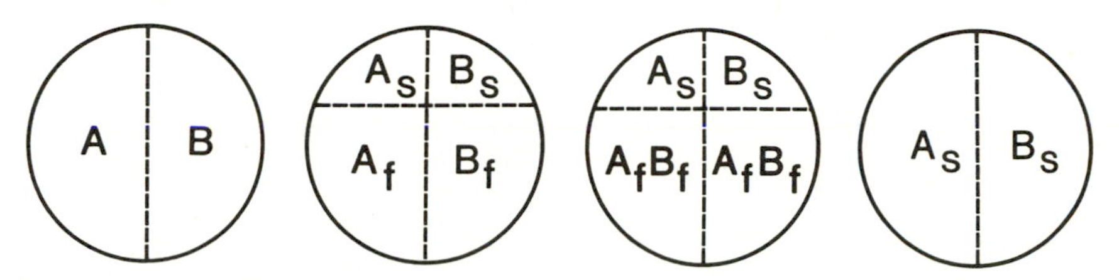

Figure 7. Wallace's Hypothesis of the Selective Origin of Cross-Sterility (see *ML* 1: 290–292, and Wallace 1889, pp. 179–180).
This schematic diagram illustrates Wallace's conception of the origin and increase of cross-sterility by means of natural selection, as he stated it in a letter to Darwin (1 March 1868) and then reprinted it in his book *Darwinism* (1889). Wallace sought to show how two cross-fertile forms A and B would be supplanted by two cross-sterile forms A_s and B_s, as the result of a process of selection (see text for complete details).

In Figure 7, I have constructed a schematic diagram of my understanding of Wallace's conception of the increase of cross-sterility by natural selection. Wallace supposed that a species had begun to divide into two forms A and B, each adapted to somewhat different conditions in the *same* general region (for example, one to woods, the other to open grounds). Wallace further supposed that each form became divided into two varieties A_f A_s and B_f B_s. A_f and B_f were cross-fertile and produced fertile hybrids. Indeed, because of the advantages of a cross between somewhat different forms — such as A_f and B_f — the hybrids, A_fB_f, were *more* vigorous and fertile than A_f and B_f. Hence, Wallace argued, in the area of A_f and B_f, the hybrids of various degrees would "certainly" come to outnumber the two pure varieties A_f and B_f, since the increased vigor and fertility of the hybrids more than compensated for their inferior adaptation to the environment. In contrast, A_s and B_s were partially cross-sterile (and/or produced partially sterile hybrids). Therefore, in their area, the "great majority" of individuals would belong to the pure varieties A_s and B_s, since the hybrids, A_sB_s, were not only less adapted but also less numerous and increased less rapidly than A_s and B_s. Then, when the struggle for existence became severe, A_s and

B_s would "certainly" supplant the A_fB_f hybrids, just as they had supplanted the A_sB_s hybrids. Thus, by means of natural selection between groups, the partially cross-sterile varieties A_s and B_s of the two forms A and B became established as the only ones. By the repetition of this process, increasingly cross-sterile varieties would replace less cross-sterile varieties (*ML* 1: 290–292).[24]

Darwin received Wallace's nineteen-point argument while he was in London, and so his initial response was brief: "I do not feel that I shall grapple with the sterility argument till my return home; I have tried once or twice and it has made my stomach feel as if it had been placed in a vise. Your paper [nineteen-point argument] has driven three of my children half mad" (Wallace 1916, p. 166). In his reply to Darwin, Wallace made the very revealing comment: "I am deeply interested in all that concerns the powers of Natural Selection, but though I admit there are a few things it cannot do I do not yet believe sterility to be one of them" (Wallace 1916, p. 167). This remark reflects the adaptationist thinking that had come to dominate Wallace's approach in the 1860s.

"Adaptationism": Darwin versus Wallace

Wallace had not always been a strict adaptationist. In *Natural Selection*, in the course of his remarks about those (including Huxley) who had "protested against the utilitarian doctrine that every part of every organic being is of use to it", Darwin specifically commented on the non-adaptationist position Wallace had taken in his book *A Narrative of Travels on the Amazon and Rio Negro* (1853): "Mr. Wallace . . . seems to doubt the strict adaptation even of very differently constructed birds; for he lays much stress on the fact of having repeatedly seen the ibis, spoon-bill & heron feeding together on precisely the same food . . . But until it can be shown that these birds feed throughout the year on exactly the same food, & are throughout their lives from the nest upwards exposed to the same dangers . . . the fact of their feeding together for a time or even a whole year, seems to me to tell as nothing against the strictest adaptation of their whole structure to their conditions of existence" (*Natural Selection*, pp. 379–380; Kottler 1980, p. 205). Furthermore, in his 1858 paper Wallace referred twice to the fixation of "unimportant", that is useless, parts: "Variations in unimportant parts might also occur, having no perceptible effect on the life-preserving powers; and the varieties so furnished might run a course parallel with the parent species . . ." (1858b, p. 275). These "unimportant parts" included "colour, texture of plumage and hair, form of horns or crests" (p. 278).

Finally, Wallace himself made an unequivocal declaration of his non-adaptationist position in a truly remarkable passage that seems to have gone entirely unnoticed by historians. In his 1856 paper "On the habits of the orang-utan in Borneo", after arguing that the huge canines of the orang were useless to it, Wallace declared:

> Do you mean to assert, then, some of my readers will indignantly ask, that this animal, or any animal, is provided with organs which are of no use to it? Yes, we reply, we do mean to assert that many animals are provided with organs and appendages which serve no material or physical purpose. The extraordinary excrescences of many insects, the fantastic and many-coloured plumes which adorn certain birds, the excessively developed horns in some of the antelopes, the colours and infinitely modified forms of many flower-petals, are all cases, for an explanation of which we must look to some general principle far more recondite than a simple relation to the necessities of the individual. We conceive it to be a most erroneous, a most contracted view of the organic world, to believe that every part of an animal or of a plant exists solely for some material and physical use to the individual, — to believe that all the beauty, all the infinite combinations and changes of form and structure should have the sole purpose and end of enabling each animal to support its existence, — to believe, in fact, that we know the one sole end and purpose of every modification that exists in organic beings, and to refuse to recognize the possibility of there being any other. Naturalists are too apt to *imagine*, when they cannot *discover*, a use for everything in nature; they are not even content to let "beauty" be a sufficient use, but hunt after some purpose to which even *that* can be applied by the animal itself, as if one of the noblest and most refining parts of man's nature, the love of beauty for its own sake, would not be perceptible also in the works of a Supreme Creator. The separate species of which the organic world consists being parts of a whole, we must suppose some dependence of each upon all; some general design which has determined the details, quite independently of individual necessities. We look upon the anomalies, the eccentricities, the exaggerated or diminished development of certain parts, as indications of a general system of nature, by a careful study of which we may learn much that is at present hidden from us; and we believe that the constant practice of imputing, right or wrong, some use to the individual, of every part of its structure, and even of inculcating the doctrine that every modification exists solely for some such use, is an error fatal to our complete appreciation of all the variety, the beauty, and the harmony of the organic world. (Wallace's italics; 1856a, pp. 30–31)[25]

During the 1860s, however, Wallace adopted just such an extreme adaptationist doctrine, which he never relinquished. In his important paper, "Mimicry and other Protective Resemblances among Animals", Wallace enunciated a principle, which he claimed to be a "necessary deduction from the theory of Natural Selection, namely — that none of the definite facts of organic nature, no special organ, no characteristic form or marking, no peculiarities of instinct or of habit, no relations between species [such as cross-sterility] or between groups of species — can exist, but which must now be or once have been *useful* to the individuals or the races which possess them" (my italics; 1870a, p. 47). (Wallace later observed that in this passage he had laid down the principle of utility "perhaps a little too absolutely"! (1896, p. 379)) But such a principle is a "necessary deduction" from natural selection *only if* one holds the view that all evolutionary change has resulted from the action of natural selection. This was indeed the view that Wallace adopted in the 1860s, and thereafter it became very difficult for him to grant any traits were non-adaptive, and essentially impossible for him to accept that any useful traits — such as cross- and hybrid sterility — had originated "by chance", rather than by a process of selection, acting on individuals or groups. If, in theory, selection could produce a certain change, then it necessarily had produced it. Twenty years later (1889) in his book *Darwinism*, Wallace discussed "useless or non-adaptive characters", concluding that "the assertion of 'inutility' in the case of any organ or peculiarity which is not a rudiment or a correlation, is not, and can never be, the statement of a fact, but merely an expression of our ignorance of its purpose or origin" (1889, p. 137). Wallace's adaptationism is very apparent in this passage. And even though he did acknowledge here Darwin's principle of correlation, he invoked that principle only rarely.

In contrast, Darwin subscribed to a much more "pluralist" view with regard to evolutionary mechanisms. The best known aspect of Darwin's pluralism is his so-called Lamarckism. Darwin held that acquired characters — produced either by the direct action of the environment on the organism or by the use or disuse of parts — could be inherited, thereby leading to evolutionary change without the action of natural selection. When Darwin responded to criticisms that he had attempted to explain everything by natural selection, he pointed to these views (*Origin* 1959, pp. 747–748, (183: f/y and 183.0.0.1–4:f); *CP* 2:223). But in his debates with Wallace, Darwin's "Lamarckism" was not a factor; rather the critical aspect of his pluralism was his emphasis on the laws of growth and, in particular, the principle of correlation (or correlated variation).

Darwin treated organisms as integrated systems; and thus he recognized that a change in one part might very well bring about changes (useless or accidentally useful) in other correlated parts. Darwin, unlike Wallace, made considerable use of this principle. On the one hand he invoked correlation of growth to account for a number of traits that he regarded as genuinely useless. Darwin believed that, in some extreme instances, the laws of correlation might explain a "large part" of the entire morphology: "probably in the case of those insects which live only for a few hours, and which never feed, a *large part* of their structure is merely the correlated result of successive changes in the structure of their larvae" (my italics; *Origin*, p. 86, pp. 143–146). On the other hand, Darwin surmised that a number of *useful* traits might also have originated as the result of the laws of growth

— what Wallace would have called "by chance" — so that "sometimes we may [mistakenly] place to the account of natural selection that which is wholly due to the laws of growth" (*Natural Selection*, p. 377). In the *Origin*, he pointed to, for example, the sutures in the skulls of young mammals, which "no doubt . . . facilitate, or may be indispensable" for, parturition. Nevertheless, despite their great utility or even absolute necessity, Darwin attributed these sutures to the laws of growth, not natural selection, since they were also present in the skulls of young reptiles and birds (*Origin*, p. 197; Gould and Vrba 1982, p. 5).[26]

M. Ghiselin has emphasized quite rightly the extensive use Darwin made of the principle of correlation in his explanations for the origin of many forms of behavior involved in the expression of the emotions (1969, pp. 203–208; see also Burkhardt, this volume). Darwin employed three "principles" to account for such behavior: "serviceable associated habits", "antithesis", and "direct action of the nervous system". Only in the case of the first of these was the resulting behavior necessarily useful: "Certain states of the mind lead to certain habitual actions, which are *of service*" (my italics). But the behavior resulting from the other two principles was initially nothing more than the correlated by-product of other neuromuscular changes. Thus in many instances it was completely useless to the organism. With respect to the principle of antithesis Darwin remarked: "Now when a directly opposite state of mind is induced, there is a strong and involuntary tendency to the performance of movements of a directly opposite nature, *though these are of no use*" (my italics; *Expression*, p. 28). Darwin was aware that some expressive behavior was useful in visual social communication. He explicitly denied, however, that the present communicative function of expressive behavior accounted for its origin. In Darwin's view, all such behavior had originated initially for other reasons, often having nothing at all to do with utility, and then — only secondarily — had been taken advantage of for purposes of communication: "Although they [expressive movements] often reveal the state of the mind, this result was not at first either intended or expected. Even such words as that 'certain movements serve as a means of expression' are apt to mislead, as they imply that this was their primary purpose or object. This, however, seems rarely or never to have been the case; the movements having been at first either of some direct use, or the indirect effect of the excited state of the sensorium" (*Expression*, p. 357).[27] In the light of such views on the role of correlation, it is easy to see why, for Darwin, the fact alone of the utility of cross- and hybrid sterility did not necessarily imply selective origin.

Still in London and hence unable to reply at length to Wallace's nineteen-point argument, Darwin reiterated (19–24 March) his position briefly: "The sterility is a most puzzling problem . . . I am hardly willing to admit all your assumptions, and even if they were all admitted, the process is so complex and the sterility . . . so universal that I cannot persuade myself that it has been gained by Natural Selection . . ." (Wallace 1916, p. 168). Wallace now (24 March) presented to Darwin what he believed to be "a strong general argument". Wallace had contended (point 3) that if two incipient species in the same general region were cross-fertile and produced fertile hybrids, "then the formation of the two distinct races or species will be retarded or perhaps entirely prevented." Now he asked Darwin, if natural selection could not produce sterility between coexisting forms, "how do species ever arise, except when a variety is [geographically] isolated?" The production of sterility by selection was not only useful, it appeared to be absolutely essential unless all speciation were geographic, something Wallace knew Darwin was not prepared to admit. Wallace concluded that sterility had been "generally produced by Natural Selection *for the good of the species*" (my italics; Wallace 1916, p. 169). Still in London, Darwin replied again only briefly (27 March): "I dread beginning to think over this fearful problem . . .; but I will sometime. I foresee, however, that there are so many doubtful points that we shall never agree. . . . Heaven protect my stomach whenever I attempt following your argument" (Wallace 1916, p. 170).

At the beginning of April Darwin returned to Down from London; so finally on 6 April Darwin sent Wallace his complete critique.

> I have been considering the terrible problem. Let me first say that no man could have more earnestly wished for the success of Natural Selection in regard to sterility than I did, and when I considered a general statement (as in your last note) I always felt sure it could be worked out, but always failed in detail, the cause being, as I believe, that *Natural Selection cannot effect what is not good for the individual*, including in this term a social community. It would take a volume to discuss all the points; and nothing is so humiliating to me as to agree with a man like you . . . on the premises and disagree about the result. (my italics; Wallace 1916, p. 170)

Darwin's exclusive focus upon the individual level of selection is strikingly apparent in his critique. He pointed to the case of the cowslip and primrose, which were moderately sterile, but occasionally produced hybrids. Darwin was willing to grant that from the production of these hybrids the two species did "suffer" to a small extent. This did not mean, however, that selection would increase the cross-sterility between them. "Can you conceive that any *individual plants* of the primrose and cowslip, which happened to be mutually rather more sterile . . . than usual, would profit to such a degree as to increase in number to the ultimate exclusion of the present primrose and cowslip? I cannot" (my italics).

Darwin's conception of natural selection acting, directly or indirectly, only for the good of the individual is further illuminated by his recognition in this letter that one form of what is now termed reproductive isolation could be acquired by means of (individual) selection. Reproductive isolation consists of two basic types, pre-mating and post-mating. Cross- and hybrid sterility are forms of post-mating reproductive isolation because they function only after mating has occurred. According to current theory, in agreement with Darwin and in opposition to Wallace, since cross- and hybrid sterility do not increase the reproductive success of the individual they cannot be the direct product of natural selection. But pre-mating reproductive isolation, which prevents sterile crosses and the formation of less fit hybrids in the first place, is advantageous to the individual, so it can be produced by natural selection. In setting forth his argument Wallace had supposed (point 6) that "as soon as any sterility appears under natural conditions, it will be accompanied by some disinclination to cross-unions". Such "disinclination to cross", the mutual aversion so commonly observed between species, is a form of pre-mating reproductive isolation. Darwin responded that he knew of "no ghost of a fact supporting the belief that disinclination to cross accompanies sterility. . . . I saw clearly what an immense aid this would be, but gave it up." Instead Darwin proposed that "disinclination to cross seems to have been independently acquired, probably by Natural Selection; and I do not see why it would not have sufficed to have prevented incipient species from blending to have simply increased sexual disinclination to cross" (Wallace 1916, p. 171). This was Darwin's clearest statement of the selective origin of pre-mating reproductive isolation.[29]

Darwin's last words in this letter were essentially his last words of the debate: "Life is too short for so long a discussion. We shall, I *greatly* fear, never agree" (Darwin's italics; Wallace 1916, p. 172). Wallace answered two days later. In the face of his failure to convince Darwin, Wallace partially retreated: "I am sorry you should have given yourself the trouble to answer my ideas on Sterility. If you are not convinced, I have little doubt but that I am wrong; and in fact I was only *half convinced* by my own arguments, and I now think there is about an even chance that Natural Selection may or [may] not be able to accumulate sterility . . . I will say no more but leave the problem as insoluble, only fearing that it will become a formidable weapon in the hands of the enemies of Natural Selection" (Wallace's italics; Wallace 1916, pp. 172–173).

Epilogue to the Debate: Wallace's Darwinism

After this intense interchange in February-April 1868, Darwin and Wallace said little more to each other about the role of natural selection in the origin of sterility. But Wallace clearly did not forget about the problem (Wallace 1916, p. 243). In his book *Darwinism* published seven years after Darwin's death, Wallace returned one last time to sterility, devoting an entire chapter to the subject, as had Darwin in the *Origin*. Wallace began by observing: "One of the greatest, or perhaps we may say the greatest, of all the difficulties in the way of accepting the theory of natural selection as a complete explanation of the origin of species, has been the remarkable difference between varieties and species in respect of fertility when crossed" (1889, p. 152). Whatever Wallace's doubts may have been in April 1868 when he told Darwin he was "only half convinced" by his own arguments, those doubts vanished in 1888 when he re-examined that argument. While writing *Darwinism* he exclaimed to Raphael Meldola (20 March 1888):

> I have been working away at my hybridity chapters, and am almost disposed to cry "Eureka!" for I have got light on the problem. When almost in despair of making it clear that Natural Selection could act one way or the other, I luckily routed out an old paper that I wrote twenty years ago, giving a demonstration of the action of Natural Selection. It did not convince Darwin then, but it has convinced me now. . . . I really think I have overcome the fundamental difficulties of the question and made it a good deal clearer than Darwin left it. (Wallace 1916, p. 296; cf. 1889, p. 174; *ML* 1: 299)

Thus Wallace repeated his 1868 argument with remarkably little change in *Darwinism*. Indeed he concluded his section on "The Influence of Natural Selection upon Sterility and Fertility" with a lengthy footnote: "As this argument is a rather difficult one to follow, while its theoretical importance is very great, I add here the following briefer exposition of it, in a series of propositions; being, with a few verbal alterations, a copy of what I wrote on the subject about twenty years back" (1889, pp. 179–180). The eleven propositions that followed were the first eleven propositions, repeated almost verbatim, from Wallace's nineteen-point argument sent to Darwin on 1 March 1868.[30]

In his discussion of the debate between Darwin and Wallace, Mayr has claimed, "they used the term 'sterility' where we would use the term 'isolating mechanisms' " (1959c, p. 227). If this were the case, then Darwin advocated the incidental origin of reproductive isolation mechanisms, Wallace their origin by natural selection. Grant has gone on to suggest that it would be "fitting and desirable" to call the selective origin of reproductive isolation mechanisms the "Wallace effect" (1966, p. 99; 1981, pp. 188–189). There can be no question that some late nineteenth-century naturalists did use the word "sterility" where evolutionists now use "reproductive isolation mechanisms". But I would argue that in their debate Darwin and Wallace meant what we do by "sterility". The distinction Wallace drew in point 6 of his 1 March 1868 letter between "disinclination to cross-unions" and "sterility" certainly supports his view. Consequently Wallace was not proposing the selective origin of reproductive isolation mechanisms in general, but rather the selective origin of the particular *post*-mating mechanisms of cross- and hybrid sterility. Since, according to current theory, these forms of sterility are precisely the types of reproductive isolation that *cannot* be produced by selection, the Darwin-Wallace debate provides little historical justification for the term "Wallace effect". The present view on the origin of sterility is essentially Darwin's view of an incidental origin (cf. Littlejohn 1981, pp. 299, 320). Furthermore, during the debate it was Darwin not Wallace who recognized the possibility of the selective origin of pre-mating reproductive isolation ("disinclination to cross"), while rejecting the selective origin of cross- and hybrid sterility.[31]

THE DEBATE OVER THE ORIGIN OF SEXUAL DIMORPHISM

The second, and most extended, of the debates between Darwin and Wallace involved the role of natural selection in the origin of sexual dimorphism, especially the differences in coloration between the two sexes. Since I have recently published a comprehensive account of this debate (Kottler 1980), I will present only a brief summary here, emphasizing the two major issues of adaptationism and constraints on natural selection, which were at the heart of the disagreement.

Wallace's theory of natural selection for the protection of the sex in greater danger grew out of his great interest in protective coloration, and in particular those instances in which protective coloration was restricted to just one of the two sexes. In 1864 Wallace reported his discovery of the phenomenon of sex-limited mimicry in butterflies, and offered an explanation for it. Sexual selection, in the form of female choice, had produced the non-mimetic coloration of the male, while natural selection had acted on the female to produce the protective mimicry:

> The reason why the females are more subject to this kind of modification than the males is, probably, that their slower flight, when laden with eggs, and their exposure to attack while in the act of depositing their eggs upon leaves, render it especially advantageous for them to have some additional protection. (Wallace 1865, p. 22)

This was Wallace's first published statement of his theory. The debate between Darwin and Wallace was initiated by Wallace's letter of 26 April 1867, in which he applied a generalized version of his theory to account for sexual dimorphism in coloration in birds. He had discovered a correlation between the presence or absence of sexual dimorphism in coloration and the mode of nesting. When both sexes in a species were conspicuous, the nest was concealed; whereas when the female was less conspicuous than the male, the nest was out in the open. Wallace proposed that the mode of nesting had determined whether or not there was sexual dimorphism in coloration. When the nest was concealed, sexual selection had acted on both sexes, producing conspicuous males and females. But when the nest was in the open, sexual selection had acted on the male only, while natural selection had acted on the female to produce her less conspicuous coloration:

> I impute the absence of brilliant or conspicuous tints in the female of birds (when it exists in the male,) almost entirely to this *protective* adaptation

> because in birds, the female while *sitting* is much more exposed to attack than the male . . . the *direct cause* of the prevalent dull colours in the female is solely [their?] *dangers*. (Wallace's italics)

The debate between Darwin and Wallace went through three phases. During the first phase (April 1867–March 1868), Darwin stated his substantial, although not complete, agreement with Wallace several times (Wallace 1916, pp. 153, 157, 164). During the second phase (April–May 1868), Darwin began to emphasize an alternative explanation for the origin of sexual dimorphism: sexual selection, in combination with the sex-limited inheritance from the beginning of the selected variations, without the direct action of natural selection. On 6 April 1868, at the climax of the debate over sterility, Darwin had remarked to Wallace: "We shall, I *greatly* fear, never agree" (Darwin's italics; Wallace 1916, p. 172). Now, three weeks later, in the second phase of the debate over sexual dimorphism, Darwin observed (30 April 1868): "But we shall never convince each other" (Wallace 1916, p. 177).

Darwin rejected Wallace's theory almost completely during the third and final phase of their debate (August–October 1868). This phase began when Darwin wrote to Wallace on 19 August 1868: "In truth, it has vexed me much to find that the further I get on, the more I differ from you about the females being dull-coloured for protection . . . This has *much decreased* the pleasure of my work" (Darwin's italics; Wallace 1916, p. 181; cf. p. 183). Darwin's letter "surprised" Wallace, who was now stimulated to make one last effort to convince Darwin. In the style of his letter of 1 March 1868, six months earlier in the sterility debate, Wallace presented to Darwin a fifteen-point argument in support of his theory.[32] At the conclusion Wallace remarked:

> Your view appears to me to be opposed to your own laws of Nat. Select[n] & to deny its power & wide range of action. Unless you deny that the general dull hues of female birds and insects are of *any use to them*, I do not see how you can deny that Nat. Select. must tend to increase such hues, and to eliminate brighter ones. I could almost as soon believe that the *structural adaptations* of animals & plants were produced by "laws of variation & inheritance" alone, as that what seem to me equally beautiful & varied adaptations of *colour* sh[d] be so produced. (Wallace's italics)

Here Wallace stated very clearly and forcibly one of the two central issues in this debate. For Wallace the adaptationist, the fact that a character was useful was sufficient proof of its origin by natural selection; otherwise useful characters had originated "by chance", a view he simply could not accept. Therefore the indisputable utility of its less conspicuous coloration to the sex in greater danger meant to Wallace that such coloration *must* have been produced by natural selection.

In his reply (23 September 1868) Darwin finally broke with Wallace:

"I think we start with different fundamental notions on inheritance. I find it most difficult, but not, I think, impossible, to see how, for instance, a few red feathers appearing on the head of a male bird, and which *are at first transmitted to both sexes*, could come to be transmitted to males alone" (Darwin's italics; Wallace 1916, p. 185). Here Darwin raised the second fundamental issue in the debate, the question of constraints imposed on natural selection by modes of inheritance and development. Wallace believed that variations that first appeared in one sex — and were then selected — were as a rule inherited equally by both sexes. Such a process of selection in combination with equal inheritance led to monomorphism. Consequently the evolution of sexual dimorphism required the conversion of equal inheritance into sex-limited inheritance. Wallace was absolutely convinced that natural selection could alter modes of inheritance and, in fact, *must* have done so frequently. During the first phase of their debate, Darwin had agreed with Wallace. But now he had come to the conclusion that natural selection could convert equal into sex-limited inheritance only with great difficulty and therefore had done so rarely, if ever. As beneficial as such a modification might be, it nevertheless would not occur. There were constraints upon the power and range of action of natural selection.

Wallace's response epitomized his adaptationist position. On Darwin's view, "the colours of one or other sex will be always (in relation to their environment) *a matter of chance*. I cannot think this. I think Selection more powerful than laws of inheritance of which it makes use" (Wallace's italics; Wallace 1916, p. 187). Wallace asked Darwin: "Why should the colour of so many female birds seem to be protective, if it has not been made protective by selection?" (*ML* 2: 87).

Darwin's position did not change after October 1868. In the *Descent* he attempted to show, with an "imaginary illustration" involving artificial selection, why the conversion of equal into sex-limited inheritance by natural selection could occur only with "extreme difficulty". He concluded: "I am unwilling to admit that this has often been effected with natural species" (*Descent* 2: 156–160). At Darwin's special request, Wallace reviewed the *Descent*. In his review he tried one last time to convince Darwin that on the crucial point of the conversion of equal into sex-limited inheritance, natural selection was up to the task: "[Mr. Darwin] appears to be unnecessarily depreciating the efficacy of his own first principle when he places limited sexual transmission beyond the range of its power" (1871, p. 181). Darwin's reply to Wallace's review was, for all practical purposes, the last of this great debate: "I will keep your objections to my views in my mind, but I fear that the latter are almost stereotyped in my mind. I thought for long weeks about the inheritance and selection difficulty, and covered quires of paper with notes, in trying to get out of it, but could not, though clearly seeing that it would be a great relief if I could" (Wallace 1916, pp. 213–214).

THE DEBATE OVER THE ORIGIN OF MAN

The last, best known, and most studied of the three great debates between Darwin and Wallace concerned the adequacy of natural selection — and indeed natural causation in general — as an explanation for the origin of man. Initially there was no disagreement at all between the two men. In 1864, in his paper "The origin of human races and the antiquity of man deduced from the theory of 'natural selection' " (1864), Wallace became the first "Darwinian" to propose in public a completely naturalistic explanation for the origin of man in terms of natural selection. Darwin had very high praise for Wallace's paper. He wrote to Hooker (22 May 1864) that it was "*most* striking and original and forcible. I wish he had written Lyell's chapters on Man [in the *Antiquity of Man*]" (Darwin's italics). Darwin went on to say there was "remarkable genius shown by the paper. I agree . . . to the main new leading idea" (*ML* 2: 31–32; cf. Wallace 1916, p. 127).[33] Wallace's "leading idea" was that in the earliest stages of the development of man natural selection had acted upon both the physical body as well as the mind. But once the social and sympathetic feelings, on the one hand, and the intellectual and moral faculties, on the other, had become "fairly developed", the body had ceased to be subject to natural selection as man became adapted to the changing environment through the action of his mind. Thereafter natural selection had continued to operate on the mind alone to produce modern man.[34]

Five years later, in his letter of 24 March 1869, Wallace gave Darwin the first hint of a change in his views:

> In my forthcoming article in the "Quarterly" [Wallace 1869], I venture for the *first time* on some limitations to the power of natural selection. I am afraid that Huxley and perhaps yourself will think them weak & unphilosophical. I merely wish you to know that they are in no way put in to please the Quarterly readers, — you will hardly suspect me of that, — but are the expression of a deep conviction founded on evidence which I have not alluded to in the article but which is to me absolutely unassailable. (Wallace's italics)

In his reply Darwin remarked: "I shall be intensely curious to read the *Quarterly*: I hope you have not murdered too completely your own and my child" (Wallace 1916, p. 197). But Wallace had done just that. He had come to the conclusion that natural selection could not account for certain purely physical characteristics, as well as the higher intellectual and moral faculties, of man. Wallace proposed that a "Higher Intelligence" had guided the laws of organic development "in definite directions and for special ends" so as to produce man. Darwin's response to Wallace's new view was rapid and incredulous: "I presume that your remarks on Man are those to which you alluded in your note. If you had not told me I should have thought that they had been added by someone else. As you expected, I differ grievously from you, and I am very sorry for it. I can see no necessity for calling in an additional and proximate cause in regard to Man" (Wallace 1916, p. 199). He marked his copy of Wallace's article with a "triply underlined 'No' and with a shower of notes of exclamation" (*ML* 2:40); cf. *LL* (NY) 2: 298; Wallace 1916, p. 206).

Discussion of the debate between Darwin and Wallace over the origin of man is complicated by the fact that Wallace was not concerned about this problem for scientific reasons alone. Indeed he was motivated primarily, if not exclusively, by his underlying belief in Spiritualism to which he became converted in 1866. He alluded to the influence of Spiritualism in his letter of 24 March 1869 (quoted above), and he expanded on this in his letter to Darwin written immediately after Darwin had expressed his surprise and disbelief upon learning of Wallace's new view (18 April 1869):

> I can quite comprehend your feelings with regard to my "unscientific" opinions as to Man, because a few years back I should myself have looked at them as equally wild and uncalled for. . . . My opinions on the subject have been modified *solely* by the consideration of a series of remarkable phenomena, physical and mental, which I have now had every opportunity of fully testing, and which demonstrate the existence of forces and influences not yet recognised by science. (my italics; Wallace 1916, p. 200)[35]

In what follows I will not treat Wallace's belief in Spiritualism, primarily because it has been discussed in great detail elsewhere (Kottler 1974; Turner 1974a; Durant 1979), but also because — having tried and failed to persuade other scientists of the validity and meaning of psychical phenomena — Wallace omitted this evidence concerning the nature of man from his scientific publications. I will summarize and analyze the main scientific arguments presented by Wallace and Darwin in their debate (cf. Gould 1980b).

The logic of Wallace's scientific argument, as presented in 1869–1870, is best illustrated by his analysis of the origin of the human brain. That argument rested upon several premises concerning natural selection, evolution, and the brain and mind. Wallace regarded natural selection as a principle of *present* utility and *relative* perfection. Furthermore, since he was a strict adaptationist by this time, he considered natural selection to be the only *natural* cause of important evolutionary change. Finally, in Wallace's view, the brain was the organ of mind, while brain size was a reliable indicator of mental capacities; in addition, since he accepted phrenology, Wallace regarded the mind as a composite entity, subdivided into many faculties that had developed independently of each other. Wallace noted that living savages and prehistoric humans possessed brains of nearly the same size as civilized men; consequently they had the same mental capacities, at least in latent form. But, Wallace claimed, these capacities — especially the higher ones such as mathematical ability, the ability to form abstract conceptions, and the ability to perform complex trains of reasoning — were utterly useless, if not somewhat harmful, to prehistoric and savage men in their natural environments, and hence the brain was much more highly developed than necessary in the struggle for existence. Being both unneeded and unused, the large human brain could not possibly have evolved by natural selection alone:[36]

> The mental requirements of the lowest savages . . . are very little above those of many animals. The higher moral faculties and those of pure intellect and refined emotion are useless to them, are rarely if ever manifested, and have no relation to their wants, desires, or well-being . . . Natural selection could only have endowed the savage with a brain a little superior to that of an ape. (Wallace 1869, pp. 391–392; 1870b, p. 356)

Since Wallace was a strict adaptationist, the insufficiency of natural selection implied the insufficiency of natural causation in general. As Gould has already observed: "Wallace's error on human intellect arose from the inadequacy of his rigid selectionism, not from a failure to apply it" (1980b, p. 54).

Darwin was completely unable to accept Wallace's conclusion and sought to refute it in the *Descent*. On the one hand, Darwin endorsed Huxley's critique of Wallace. In his review "Mr. Darwin's critics" (1871), Huxley challenged Wallace's opinion that the mental requirements of savages were very little above those of animals. He called particular attention to Wallace's own paper "On instinct in man and animals", published in the same volume as Wallace's "The limits of natural selection as applied to man", in which Wallace himself had given an "admirable" description of the considerable mental challenges facing the savage (Wallace 1870a, pp. 207–209). In the light of these facts the large brain could hardly be considered useless to the savage. Darwin was "delighted" by Huxley's review. After praising Huxley's critique of St. G. Mivart, Darwin went on: "I am mounting climax on climax, for after all there is nothing, I think, better in your whole review than your arguments *v.* Wallace on the intellect of savages" (*LL* (NY) 2: 329). In the *Descent* Darwin enumerated the many intellectual abilities and achievements of man in the "rudest state" and then remarked: "I cannot, therefore, understand how it is that Mr. Wallace maintains, that 'natural selection could only have endowed the savage with a brain a little superior to that of an ape' " (*Descent* 1: 137–138).

Darwin conceded, however, that a number of physical and mental attributes of man had never possessed any selective value in the general struggle for existence, and therefore had not evolved directly by natural selection. But since he was not a strict adaptationist, he was not as a result compelled to invoke the supernatural to explain what natural selection could not. Instead Darwin proposed several additional *natural* explanatory factors, including the inherited effects of habit; sexual selection, which could produce features (such as the naked skin) that might be "in a slight degree injurious" in the general struggle for existence; and the important principle of "correlation", according to which attributes without any immediate selective value might arise as incidental by-products of the action of natural selection.

In stressing the importance of correlation, Darwin endorsed C. Wright's critique of Wallace. Wright agreed with Wallace that the savage possessed mental faculties that were unneeded and unused; but he rejected Wallace's conclusion that the large brain was useless to the savage, because he rejected Wallace's view that the different mental faculties were independent of each other. Wright contended that the large brain had evolved by natural selection as the physical basis for language, and that all other mental faculties of this brain had originated as the incidental by-products of the acquisition of language:

> Why may it not be that all that he [the savage] can do with his brains beyond his needs is only incidental to the powers which are directly serviceable? . . . The philosopher's mental powers are not necessarily different in their elements from those which the savage has and needs in his struggle for existence . . . The philosopher's powers are not, it is true, the direct results of Natural Selection, or of utility; but may they not result by the elementary laws of mental natures and external circumstances, from faculties that *are* useful? . . . Are they not rather implied and virtually acquired in the powers that the savage has and needs — his powers of inventing and using even the concrete terms of his simple language? (1870, pp. 109, 111–112)

In the *Descent* Darwin presented the same counter-argument, citing Wright's critique prominently. He noted that such powers as self-consciousness and abstraction might well be "the incidental results of other highly-advanced intellectual faculties; and these again are mainly the result of the continued use of a highly-developed language" (1: 105). In a footnote added in proof to the first edition of the *Descent*, Darwin quoted Wright's statement on correlation that "there are many consequences of the ultimate laws or uniformities of Nature through which the acquisition of one useful power will bring with it many resulting advantages as well as limiting disadvantages, actual or possible, which the principle of utility may not have comprehended in its action" (Wright 1870, p. 107). Then Darwin added: "This principle has an important bearing . . . on the acquisition by man of some of his mental characteristics" (*Descent* 2:335). Darwin returned to the important role of this principle in the evolution of the human brain, in the last chapter of the *Descent*, and again cited Wright: "The large size of the brain in man . . . may be attributed in chief part, as Mr. Chauncey Wright has well remarked, to the early use of some simple form of language. . . . The higher intellectual powers of man . . . will have followed from the continued improvement of other mental faculties" (*Descent* 2:391).[37] Thus Darwin and Wright, who had fully incorporated the principle of correlation into their evolutionary thinking, argued that, in order to establish the adequacy of natural selection, it was not necessary, *contra* Wallace (cf. Wallace 1905, 1: 427–428), to show that every mental faculty made possible by the large brain had been needed and used by prehistoric and savage men.

Conclusion

In his very first letter to Wallace — written 1 May 1857, one year before his receipt of Wallace's manuscript on natural selection — Darwin observed: "In regard to the paper in the *Annals* [Wallace 1855], I agree to the truth of almost every word of your paper; and I daresay that you will agree with me that it is very rare to find oneself agreeing pretty closely with any theoretical paper; for it is lamentable how each man draws his own different conclusions from the very same fact" (Wallace 1916, p. 107). On the other hand, nearly twenty-five years later and about a year before his death, Darwin wrote to Wallace: "How lamentable it is that two men should take such widely different views, with the same facts before them; but this seems to be almost regularly our case, and much do I regret it" (Wallace 1916, p. 256). This latter statement reflected the fact that between 1867 and 1871 Darwin and Wallace engaged in three great debates over the nature and scope of natural selection. Underlying all three debates was the fundamental issue of "adaptationism": are all traits useful, and have all useful traits evolved directly by natural selection? M. Ghiselin was, I believe, the first in recent times to call attention to the important difference of opinion between Darwin and Wallace on this issue.[38] Commenting specifically about their debate over the origin of sterility, he remarked: "Wallace went to extreme lengths to construct a model of selection for sterility, but Darwin was unconvinced by these efforts, and rightly so, for they were exceedingly far fetched. The dispute is particularly revealing, for it casts much light on their respective attitudes toward methodology and metaphysics. Wallace, who was always seeking plausible reasons, held, erroneously, that the inability to account for sterility argues against natural selection. Darwin . . . by contrast . . . held that it was by no means essential that natural selection explain the origin of every adaptive trait. One should expect an occasional biologically advantageous feature to arise by accident" (Ghiselin 1969, pp. 150–151).

During their debates Wallace advocated strict adaptationism: virtually all traits were, or formerly had been, useful, and all useful traits had originated by natural selection, acting on either individual organisms or groups. His adaptationist position is very clear in the two debates over sterility and sexual dimorphism. Wallace was completely convinced of, on the one hand, the utility of cross- and hybrid sterility to (incipient) species, and, on the other hand, the utility of her less conspicuous coloration to the female bird sitting on an exposed nest. The fact of utility was, by itself, "sufficient proof" to Wallace of the selective origin of these traits. Furthermore, as S. J. Gould has emphasized (Gould 1980b, pp. 53–54), even in the debate over man, in which Wallace *denied* the utility to prehistoric and savage men of the mental faculties made possible by their large brains, Wallace's adaptationism is nevertheless apparent. Wallace held that natural selection was the only *natural* cause of major evolutionary change. The uselessness of the large brain meant it could not have evolved by natural selection; so, on the principle of "all or nothing", the inadequacy of natural selection meant the inadequacy of natural causation. Hence, ironic as it may seem, Wallace felt *scientifically* justified in invoking the supernatural to account for the origin of man.

Since Darwin believed natural selection, a principle of utility, was the "main" mechanism of evolution, he certainly had adaptationist inclinations of his own. At several high points in their debates, Darwin expressed to Wallace his strong desire to agree with Wallace's views. For example, at the climax of the sterility debate, he stressed: "Let me say that no man could have more earnestly wished for the success of Natural Selection in regard to sterility than I did" (Wallace 1916, p. 170). Similarly in the sexual dimorphism debate he informed Wallace: "You will be pleased to hear that I am undergoing severe distress about protection and sexual selection: this morning I oscillated *with joy* towards you" (my italics; Wallace 1916, p. 183). Finally, at the very end of this debate, he commented: "I thought for long weeks about the inheritance and selection difficulty, and covered quires of paper with notes, in trying to get out of it . . . clearly seeing that it would be *a great relief* if I could" (my italics; Wallace 1916, pp. 213–214). But, ultimately, Darwin could not bring himself to adopt Wallace's strict adaptationism. Instead he defended a more "pluralist" position, emphasizing the important role of the principle of correlation, according to which *useful* as well as useless traits had evolved "by chance", as the incidental by-products of natural selection. In Darwin's view, the mere fact of utility to the species of cross-sterility did not prove its selective origin; whereas the uselessness in the struggle for existence of some of man's higher mental faculties did not prove the insufficiency of natural selection and the necessity for divine intervention in human evolution. Consistent with this pluralism, Darwin maintained that natural selection was not all-powerful; it was constrained by modes of inheritance and development.

The other major issue dividing Darwin from Wallace involved the levels of selection: does natural selection operate on groups as well as on individual organisms? Darwin thought a great deal about this question, as is perhaps best illustrated by his formulation of a concept of "family selection" to solve the problem of the origin of the sterile castes in social insects. He did not rule out completely the possibility of selection at the level of groups.

Indeed in the one case of the evolution of the moral sense in man he apparently accepted it. But Darwin was definitely reluctant to invoke group selection. For a time in the early 1860s he had been quite strongly inclined toward the origin of cross- and hybrid sterility by group selection. Eventually, however — after "mature reflection" as he put it in the *Origin* (1959, pp. 443–445) — Darwin rejected this hypothesis; and throughout his debate in 1868 with Wallace he focussed exclusively upon the individual level of selection. Since cross- and hybrid sterility were obviously disadvantageous to the individual and "its nearest relatives", Darwin contended they had not evolved directly by natural selection. Wallace, in marked contrast, never regarded group selection as a problematic process, and accordingly did not display any reluctance to postulate it. The explanation for this difference seems to be, again, Wallace's adaptationism. If utility implied origin by selection, then utility to the group rather than the individual implied origin by *group* selection. But, not being such a strict adaptationist, Darwin did not find this reasoning at all compelling.

At this point the question naturally arises: why did Wallace become such a strict adaptationist? Unfortunately I do not have a good answer to this good question. One way to approach it is through the converse question: Why was Darwin more pluralist in his thinking? A critical empirical factor was Darwin's knowledge of the process of embryological development, gained primarily from his exhaustive study of variation in domesticated animals and plants. M. Ghiselin has led the way in pointing to the significance of this factor: "The contingencies of developmental mechanisms vastly complicate the problems of evolutionary theory. But Darwin foresaw the implications of such phenomena, realizing that an understanding of the laws of variation would provide invaluable explanations for the facts of evolution, and that it could serve as a guide to the further elaboration of his theory. . . . The most fundamental of Darwin's ideas on the relationships between development and evolution is perhaps that of 'correlated variation' " (Ghiselin 1969, pp. 165–166). Wallace, on the other hand, had no first-hand knowledge about the phenomena of development. As a result the principle of correlation had little real meaning to him; although he acknowledged it (Wallace 1896, p. 380, fn. 1), he never really made any significant use of it.

S. J. Gould has argued that "the essential questions of a discipline are usually specified by the first competent thinkers to enter it. The intense professional activity of later centuries can often be identified as so many variations on a set of themes. The arrow of history specifies a sequence of changing contexts within which the same old questions are endlessly debated" (1977c, p. 1). In their debates Darwin and Wallace — the first two competent thinkers, indeed the founders, of the new discipline of evolutionary biology — certainly did raise "essential questions" that have persisted to the present day (cf. Gould 1980b, p. 49). Thus the interaction between Darwin and Wallace concerning natural selection may well rank as the most intellectually and emotionally dramatic episode of its kind in the entire history of evolutionary biology.

ACKNOWLEDGEMENTS

Even though they did not read the manuscript of this paper, I nevertheless want to acknowledge E. Mayr, M. Ghiselin, and S. J. Gould for the inspiration I have received from them in my own work on the history of evolutionary biology.

Notes

1. Besides their debates over natural selection, Darwin and Wallace also had two debates concerning biogeography. One centered on the existence in the past of land bridges, such as extensions of continents to oceanic islands. Darwin was very much opposed to postulating such bridges in order to account for puzzling phenomena of geographical distribution. Initially, Wallace was inclined to invoke certain land bridges, but during the 1860s he became converted to Darwin's position and thereafter became its great champion (Fichman 1977; 1981, chap. 3). The other disagreement concerned the presence of north-temperate and arctic species of plants in the *southern* hemisphere and on mountain tops in the tropics. Both Wallace and H. W. Bates, Wallace's fellow traveler in the South American tropics, rejected Darwin's explanation, which involved a significant cooling of the tropics during the recent glacial period. Wallace favored "aerial transmission of seeds, either by birds or by gales and storms"; (Wallace 1905, 2: 19–21; see also *Origin*, pp. 372–382; Wallace 1916, pp. 252–256; Stecher 1969, pp. 8, 10–12, 16, 27–29, 31–32). Mayr has noted that Darwin "grossly underrated" the power of wind to disperse seeds (1982b, p. 447). In his autobiography Wallace also listed the inheritance of acquired characters as another "chief difference of opinion" between himself and Darwin (Wallace 1905, 2: 21–22). Darwin always accepted such inheritance. At first Wallace accepted it, too, although he, unlike Darwin, hardly ever invoked it to account for the characteristics of organisms. Eventually, after becoming familiar with A. Weismann's writings in the 1880s, Wallace rejected the inheritance of acquired characters totally (1889, chap. 14). Consequently, this "difference" did not arise until after Darwin's death.
2. For Darwin's admiration of: (i) specific writings by Wallace, see Wallace (1916, pp. 107, 111–112, 114, 127, 132, 137–138, 151, 155, 158, 194–196, 204, 208, 213, 234–235, 237, 252; (ii) Wallace's intellectual abilities, see Wallace 1916, pp. 115, 132, 148, 153–154, 164, 198–199, 217, 220, 222, 229, 249); (iii) Wallace's character, see Wallace (1916, pp. 111, 113, 117, 136, 191, 193, 196, 199, 235). For Wallace's admiration of Darwin, see Wallace (1916, pp. 59, 61–63, 128, 131, 190). Wallace dedicated *The Malay Archipelago* (1869) to Darwin; and in the prefaces to his *Contributions to the Theory of Natural Selection* (1870a, pp. iv–v) and *Geographical Distribution of Animals* (1876, 1: xv) expressed his respect for Darwin's accomplishments. After reading the preface to Wallace's *Contributions*, Darwin remarked: "There never has been passed on me, or indeed on anyone, a higher eulogium than yours . . . Your modesty and candour are very far from new to me. I hope that it is a satisfaction to you to reflect — and very few things in my life have been more satisfactory to me — that we have never felt any jealousy towards each other, though in one sense rivals. I believe that I can say this of myself with truth, and I am absolutely sure that it is true of you" (Wallace 1916, pp. 206–207). In his autobiography Wallace commented on this: "This friendly feeling was retained by him to the last, and to have thus inspired and retained it, notwithstanding our many differences of opinion, I feel to be one of the greatest honours of my life" (Wallace 1905, 2: 15–16).
3. Writing thirty years later, Wallace thought he had received two letters from Darwin prior to February 1858 (1891, p. 20); but since Darwin's second letter was dated 22 December 1857, this is unlikely (cf. Beddall 1968, p. 293). Even if he had, however, this letter was as uninformative about Darwin's views as the first one. Darwin again wrote: "it is too long a subject to enter on my speculative notions" (Wallace 1916, p. 109).

4. Darwin followed Wallace's advice, adding "survival of the fittest" to the fifth edition of the *Origin* (*Origin* 1959, pp. 163 (2:e), 164 (13:e), although he continued to use "natural selection" as well.
5. It should be noted that Wallace did use the words "compete" and "competition" in his 1858 paper (I count three uses, pp. 274–276).
6. The wording of this subheading probably reflects the more adaptationist position Wallace adopted in the 1860s (see below p. 410). Although in the original paper Wallace clearly allowed for the establishment of useless variations, the subheading states that "useless" as well as "harmful" variations will diminish.
7. Bowler has also claimed that "There is nothing to suggest that [Wallace] was aware of the role played by factors such as geographical isolation in the formation of varieties and species" (1976c, p. 22). It is true that in this particular paper Wallace did not discuss geographical varieties; but at this time he certainly was well aware of their existence. In a very interesting "Note" published in 1858, Wallace explicitly referred to "permanent and geographical varieties" (1858a; Kottler 1978, pp. 294–295).
8. In Diagram I of *Natural Selection*, the lineage M-m^{10} would represent divergent phyletic evolution, whereas the lineages A-a^{10} and A-l^{10} would be divergent branching evolution (p. 236).
9. In his annotations, Darwin also wrote: "It seems all creation with him." Even though Wallace never explicitly stated in 1855 that evolution was the explanation for his law, the immediate reactions to his paper by the two creationists Lyell and E. Blyth demonstrate clearly that evolution was very much "between the lines" (Lyell 1970, p. 66; Beddall 1972). So why did Darwin write what he did? Wallace's frequent use of the term "creation" might provide at least a partial answer. It is certainly true that at this time Darwin occasionally used the term "creation" simply to mean "origin", without any implication of divine intervention. But Wallace used the term on almost every page of his paper (about fifteen times in all), thereby possibly misleading Darwin.
10. In the *Descent* Darwin did adopt group selection to account for the origin of the moral sense in humans (1: 163, 166; see Burkhardt, Sober, this volume). In 1886 G. J. Romanes wrote to Francis Darwin: "Whether natural selection could in any case act on a type is a question which your father has told me he could never quite make up his mind about, except in the case of social hymenoptera and moral sense of man" (1896, pp. 172–173). As a matter of fact, in order to account for the sterile castes in social insects, Darwin invoked what he called "family selection", which would not be regarded today as genuine group selection (*Origin* 1959, p. 417 (248.x-y:f); *Natural Selection*, pp. 369–370).
11. In the fifth edition of the *Origin* (1869), Darwin rewrote this sentence, stating that cross-sterility, as well as hybrid sterility, could not have been produced by selection (*Origin* 1959, pp. 424–425).
12. It is important to establish what Darwin meant by "our theory". In the first edition of the *Origin* Darwin used the corresponding phrase "my theory" nearly sixty times (Barrett et al. 1981, pp. 747–749), sometimes to mean natural selection, other times descent with modification. There is a very close parallelism between this passage and Darwin's discussion of the origin of "organs of extreme perfection and complication" (*Natural Selection*, pp. 350–352; *Origin*, p. 186). Since he was clearly referring to natural selection in the latter, I don't believe there can be any doubt that in this particular passage he meant natural selection by "our theory".
13. Apparently it was Huxley himself who coined this famous epithet (L. Huxley 1900, 1: 363).
14. Bartholomew has pointed to Huxley's lack of enthusiasm for natural selection, but he never discusses the sterility question (1975, p. 529). Lovejoy, on the other hand, did take note of Huxley's concern about sterility, but mistook it for an objection to the theory of descent, rather than the theory of natural selection (1959c, p. 113; 1959b, pp. 395–396). For previous discussions of the relationship between Darwin and Huxley, see Poulton (1896, pp. 119–143; 1904, pp. 77–83) and Ruse (1979a, pp. 235–236).
15. In 1749, Buffon first enunciated his celebrated species definition: "We should regard two animals as belonging to the same species if, by means of copulation, they can perpetuate themselves and preserve the likeness of the species; and we should regard them as belonging to different species if they are incapable of producing progeny by the same means" (Lovejoy 1959c, pp. 93–94). For Buffon, the transmutation of species was rendered highly improbable by the existence of cross- and hybrid sterility. "In order that two individuals cannot reproduce together it is only necessary that there be some slight dissimilarity in temperament or some accid-

ental defect in the reproductive organs of one or the other of the two individuals . . . but what an immense and perhaps infinite number of combinations would be necessary before one could conceive that two animals, male and female, had not only so far departed from their original type as to belong no longer to the same species — that is to say, to be no longer able to reproduce by mating with those animals which they formerly resembled — but had also both diverged to exactly the same degree, and to just that degree necessary to make it possible for them to produce only by mating with one another" (Lovejoy 1959c, p. 99; Buffon 1753, pp. 389–390). Although Lovejoy found Buffon's logic in this argument a "trifle obscure", it is actually a very sound objection to a particular type of speciation. Buffon envisaged speciation as resulting from a single pair of variant individuals, of opposite sex, fertile with each other, but sterile with all individuals of the parent species. It was easy to understand how either individual might vary in such a way that it could no longer reproduce with any individuals of the parent species. But since variation occurred independently in the two individuals (♂ and ♀) and, furthermore, since so many different kinds of variation could render each individual cross-sterile with the parent species, Buffon thought it highly unlikely that both individuals would have varied in precisely the same way so that, while each could not reproduce with the parent species, they could reproduce with each other (cf. Ghiselin 1969, pp. 147–148).

16. After reading this paper, Darwin wrote to Hooker: "I by no means thought that I produced a 'tremendous effect' in the Linn. Soc. but by Jove the Linn. Soc. produced a tremendous effect on me, for I could not get out of bed till late next evening, so that I just crawled home. I fear I must give up trying to read any paper or speak; it is a horrid bore, I can do nothing like other people" (*LL* (NY) 2: 473). It is worth noting that, at this very same meeting of the Linnean Society, Henry W. Bates read his famous paper on mimicry in butterflies (Bates 1862).
17. The primrose (*Primula*) of interest to Darwin should not be confused with the evening primrose (*Oenothera*) later studied by Hugo de Vries.
18. I believe Ruse is mistaken in his claim that in this passage, "insofar as sterility was being generalized from the individual to the group, it was accidental, in the sense of not being of selective value. There was no question of selection for the group . . ." (1980b, p. 622). The greater self-sterility of the short-styled form of *Primula* could be regarded as advantageous to the individual; but the "two forms" in this passage are two incipient species, and their cross-sterility (or the sterility of any hybrids produced) would be advantageous to the two *groups* only.
19. The day after writing to Tegetmeier, Darwin wrote to Huxley: "I . . . have told him how alone I think the experiment could be tried with the faintest hope of success . . . but the difficulty . . . would be beyond calculation" (*ML* 1: 225–226).
20. In the fifth edition of the *Origin* (1869), Darwin added two sentences on the problem posed by two species that were already largely but not totally cross-sterile: "What is there which could favour the survival *of those individuals* which happened to be endowed in a slightly higher degree with mutual infertility, and which thus approached by one small step towards absolute sterility? Yet an advance of this kind, if the theory of natural selection be brought to bear, must have incessantly occurred with many species, for a multitude are mutually quite barren" (my italics; *Origin* 1959, p. 444; this passage first appeared in *Variation* 2: 186). Here, as in the rest of his discussion, natural selection meant selection of individual organisms.
21. Darwin did concede that "if a species were rendered sterile with some one compatriot, sterility with other [=allopatric] species would probably follow as a necessary contingency". That is, given three species A, B, C — A and B sympatric, A and C allopatric — then *if* selection had rendered A and B cross-sterile because such cross-sterility had been advantageous to species A, the (useless) cross-sterility between A and C might well have followed as an incidental consequence. But Darwin's theoretical argument concerning natural selection acting *on individuals* showed that the cross-sterility between A and B could not have been produced by selection.
22. For a new "conclusive" argument added to the sixth edition of the *Origin*, see *Origin* (1959, p. 447); cf. *Variation* (1875, 2: 171).
23. At least as early as February 1866, Wallace had begun to think about sterility, because he wrote then to Darwin: "If you 'know varieties that will not blend or intermix, but produce offspring quite like either parent', is not that the very physiological test of a species which is wanting for the *complete proof* of the origin of species?" (Wallace's italics; Wallace 1916, p. 140).
24. Advocates of group selection have sometimes

taken for granted the existence of a "variety", the members of which possess the trait that is not advantageous to the individual. They have concentrated on selection between a group with and a group without the trait, but have not dealt with the problem of how the trait became established within a group in the first place. Wallace did not totally neglect this problem; but his two "solutions" left rather little for natural selection to do, as Darwin quickly pointed out. Wallace suggested that the cross-sterility between varieties A_s and B_s arose, on the one hand, from the direct effect on the reproductive system of the different conditions to which each was exposed (point 4) and, on the other hand, from the overall divergence of A_s from B_s as they became adapted to their different conditions of existence (point 11). Wallace's second suggestion was actually nothing more than Darwin's own view of the incidental origin of sterility. As for the first suggestion, Darwin expressed doubt that cross-sterility between A_s and B_s could arise and increase from their continued exposure to different conditions; but if Wallace was correct, then, Darwin observed, "there would be no need of Natural Selection" (Wallace 1916, p. 172). Mayr has also noted that "in retrospect it is clear that Wallace made so many assumptions . . . that he started out with virtually reproductively isolated species" (1959c, p. 228).

25. The view of life so clearly expressed in this passage may help explain why, in his 1855 paper, Wallace interpreted rudimentary organs as nascent, rather than vestigial, and as serving to unite the entire organic world into "an unbroken and harmonious system" (1855, p. 18).
26. Darwin also discussed climbing bamboo plants that possessed "exquisitely constructed hooks . . . no doubt . . . of the highest service to the plant." Yet, since similar hooks were to be found as well in many non-climbing plants, Darwin proposed, at first, that they may have arisen from the laws of growth. But in the last two editions of the *Origin* Darwin suggested instead that the hooks had indeed originated by selection as a defense against herbivores, and subsequently had changed function in those plants becoming climbers. Change of function was Darwin's main explanation for the emergence of novelty (*Origin* 1959, p. 365).
27. If one likes the new terminology proposed by Gould and Vrba, these secondarily useful behaviors would be exaptations that were initially nonaptations, and they illustrate the process of cooptation (Gould and Vrba 1982, table 1, p. 5).
28. For Darwin's views on the role of geographic isolation in speciation, see *Natural Selection* (pp. 256–257, 266, 269), *Origin* (pp. 103, 174), *Origin* (1959, pp. 227 (382.39.4:d), 230 (382.56:c)), Kottler (1976, chap. 4), Sulloway (1979b). For Wallace's views, see Wallace (1855, pp. 5, 10; 1865, pp. 10–14; 1889, pp. 119–120, 144–146, 150; 1896, 382).
29. Compare Darwin's discussion of pollen prepotency in the section on sterility and selection added to the fourth edition of the *Origin*. Darwin noted that whereas selection could not increase sterility, it could increase prepotency, which is now classified as a form of pre-mating reproductive isolation. He went on to suggest that selection for pre-potency might have produced cross-sterility as a correlated result (*Origin* 1959, pp. 445–446; *Variation* 2: 187).
30. It is worth pointing out that in *Darwinism* Wallace made a suggestion that led the way to a fruitful new approach to the problem of sterility. In the debates between Darwin and Huxley, and between Darwin and Wallace, fertility within a species had largely been taken for granted; attention had focussed on the problem of the origin of cross-sterility. Wallace now observed that the maintenance of fertility within a species required constant selection, and that once such selection ceased to act, cross-sterility would almost automatically result. Not long afterwards Poulton developed Wallace's idea and related it directly to Huxley's contention that the complete proof of *natural* selection required the production of cross-sterile forms by *artificial* selection: "If, then, mutual fertility be the result of unceasing selection, and mutual sterility the inevitable, even if long-postponed, consequence of its cessation, it is obvious that Huxley's difficulty is solved while his suggested experimental creation of sterility by selection would not reproduce any natural operation: it would afford a picture of a natural result but would be produced in an unnatural way" (Poulton 1904, pp. 80–82; 1898; Kottler 1976, pp. 398–402).
31. In *Darwinism* Wallace did propose the selective origin of one form of *pre*-mating reproductive isolation. He thought that, during the initial differentiation of a species into two (or more) incipient species, slight differences in coloration would usually arise. These differences would then serve as the basis for "selective association" or preferential mating between

individuals of an incipient species, and natural selection would intensify these differences, which Wallace called recognition marks, "for the purpose of checking the intercrossing of closely allied forms" (1889, pp. 119–120, 171–173, 217–218, 226–228; cf. 318–319 for insect pollination of plants). Although one might wish to justify the term "Wallace effect" on the basis of these remarks, it should be pointed out that Wallace's treatment of preferential mating as a form of pre-mating reproductive isolation was incomplete. Both Wallace and Darwin simply assumed that similarly-colored individuals would automatically prefer to mate with each other. But homogamy will not be automatic. There must be selection for the mating preference, as well as for the recognition marks that permit the expression of such a preference, if those marks are to function in reproductive isolation. Wallace only discussed selection for the marks themselves.

32. This letter has not yet been published, but a draft of it was published in both Wallace (1905, 2: 18–20) and Wallace (1916, pp. 183–185). The important concluding paragraph quoted in the text is not in the draft.
33. In the *Descent*, Darwin presented Wallace's "leading idea" in detail, calling his 1864 paper "celebrated" and "admirable" (1: 137, 158).
34. In his autobiography Wallace remarked that his later view had been "first intimated in the last sentence" of his 1864 paper, and that Darwin in 1864 had been "quite distressed at my conclusion that natural selection could not have done it all" (Wallace 1905, 1: 418; 2: 17). Wallace was, however, definitely mistaken on both counts. He did refer in 1864 to man as "in some degree a new and distinct order of being", "a being who was in some degree superior to nature", and "a being apart". But this uniqueness of man was due solely to the nature of his mind, and throughout his 1864 paper Wallace attributed the development of that mind entirely to the purely natural process of natural selection. Darwin's correspondence proves that in 1864 he had been very enthusiastic, not "quite distressed", by Wallace's "great leading idea". Wallace must have had in mind the *new* last sentence of the *revised* version of his paper, which did indeed express the new view he had first stated the year before (1870a, pp. 330–331). Darwin was quite distressed by that *new* view.
35. Wallace said much the same thing twenty years later, while writing *Darwinism*, to E. B. Poulton: "I (think I) *know* that non-human intelligences exist — that there are *minds* disconnected from a physical brain — that there *is*, therefore, *a spiritual world*. This is not, for me, a *belief* merely, but *knowledge* founded on the long-continued observation of facts — and such *knowledge* must modify my views as to the origin and nature of human faculty" (Wallace's italics; Poulton 1924, p. xxviii).
36. In connection with Wallace's reasoning, it is interesting to compare Mayr's recent remark: "Why primitive man should have been selected for a brain of such perfection that 100,000 years later it permitted the achievements of a Descartes, Darwin, or Kant, or the invention of the computer and the visits to the moon, or the literary accomplishments of a Shakespeare or a Goethe, is hard to understand" (1982b, pp. 622–623; cf. p. 600).
37. For recent statements of the same view that most of the brain's capacities originated incidentally as by-products of the action of natural selection, see Gould (1980b, p. 57; 1982, p. 384), Gould and Vrba (1982, p. 13).
38. G. Romanes, at the end of the nineteenth century, was the first to contrast Wallace's strict selectionism to Darwin's pluralism and, furthermore, to argue that "from this great or radical difference of opinion . . . all their other differences of opinion arise" (1895, p. 5). He coined the terms "neo-Darwinian" and "ultra-Darwinian" in order to distinguish the views of Wallace and August Weismann, who were "seeking to out-Darwin Darwin by assigning an exclusive prerogative to natural selection", from those of Darwin himself (pp. 12–13). Romanes's critique of Wallace included several important points. He noted that, although in *Darwinism* (Wallace 1889, p. 444) Wallace had quoted with approval Darwin's statement that natural selection had been "the most important, but not the exclusive means of modification" (*Origin* 1959, p. 75 (50:e)), Wallace, "to all intents and purposes", had not recognized any cause of evolution other than natural selection (pp. 20–21; cf. Gould and Lewontin 1979, p. 586). Romanes also pointed out that Wallace's statement in 1867 that the utility of all traits was a "necessary deduction" from the theory of natural selection presupposed strict selectionism (pp. 180–185). Lastly he called attention to Wallace's equation of natural selection and natural causation as crucial to his argument concerning the origin of man (p. 28). Given the great similarity between Romanes's and recent analyses of Wallace's views, it should be noted that in his discussion of Darwin's pluralism Romanes, like Darwin himself,

stressed the Lamarckism. Furthermore, he treated the principle of correlation, not so much as an important feature of Darwin's pluralism but rather as a "loophole" through which adaptationists like Wallace could escape (p. 269). (For a summary of Romanes' debate in the late 1880s with the British neo-Darwinians, see Kottler 1976, pp. 273–281.)

PART TWO

Darwin in Victorian Context

15

Darwin of Down: The Evolutionist as Squarson-Naturalist

James R. Moore

> *In my room, at the head of my bed there was a crucifix and under it the words: "Greater love hath no man than this, that he lay down his life for his friends." At the side of my bed was a photograph of Darwin, and under it the text: "Labour, Art Worship, Love, these make men's lives." Darwin was never presented to me as a great scholar, but always as a great lover of nature, and because of nature, of God. He was the high-priest of my mother's religion, and it was not till years after that I discovered him for myself as a great scientist.*
>
> (E. L. Grant Watson, *But to What Purpose? The Autobiography of a Contemporary, 1946*)

Ever since the English establishment appropriated the body of Charles Darwin and buried it in Westminster Abbey, the interpretation of Darwin's religious life has been controversial. Right from the start partisan opinion was divided; explanations had to be dredged up pro and con. No sooner had the coffin sunk ironically beneath the Abbey pavement than the flotsam of Darwin's religious life began to surface in the press. On the weekend the evangelical *Record* reported how the Lord Bishop of Derry had told a crowd of cheering clergymen about Darwin's support for Church of England missions. Some months later readers of freethought literature were gratified to learn from Karl Marx's son-in-law, Edward Aveling, about a conversation in which Darwin admitted giving up Christianity at the age of forty (South American Missionary Society 1882; Aveling 1882b, 1883; cf. *LL* 1:317n). In 1885 the Duke of Argyll graced the godly pages of *Good Words* with an account of a conversation in which Darwin admitted sometimes glimpsing design in nature; and in 1889 G. W. Foote and the Progressive Publishing Company repeated at second-hand how Darwin often escorted his family to church but did not himself "go through the mockery" of attending (Argyll 1885; Foote 1889; cf. Litchfield 1904, 2:315).

Meanwhile, as recollection vied with recollection, other writers competed to fix the true nature of Darwin's religious outlook by publishing his private letters to them. In the week of the Abbey funeral a Devonshire rector offered evidence in a church newspaper that Darwin had "virtually" admitted to him that human evolution was "only an hypothesis" (Savile 1882a, 1882b; letters in DAR 177). A few months later Ernst Haeckel, the Liberal *Pall Mall Gazette*, and the Secularist *National Reformer* printed the letter to a young German student in which Darwin avowed disbelief in any divine revelation.[1] In 1883 John Fordyce and the conservative *Evening Standard* transcribed a letter from the same period, in which, however, Darwin both denied that he had ever been an atheist and affirmed that one might certainly be "an ardent theist and an Evolutionist" (De Beer 1958d, p. 88; Fordyce 1883, pp. 189–190; *LL* 1:304; DAR 139.12). The Nonconformist *British Weekly* followed in 1888 with a letter by Darwin that made out the problem of God's existence to be merely "insoluble" (Darwin 1888).

So there was no disinterested interpretation of Darwin's religious life in the early years after his death. Nor has there been since. The main reason for this, it seems to me, is the deference shown to the judgement of "able men" in religious matters, a deference that was shared "to a certain extent" by Darwin himself (*LL* 1:306; DAR 139.12). Whether or not one agrees with all the particulars of his theories is immaterial; Darwin was by any estimate an "able" scientist, and in Western culture this creates the presumption in many minds that his religious views should be taken very seriously. Because Darwin was a devout and objective student of nature, according to one line of reasoning, his metaphysical understanding is more trustworthy than if, say, he had been an engineer, a barrister, a politician, or a philosopher. Alternatively, because Darwin was such a devout and objective researcher (or is widely thought to have been), any errors in his metaphysics are both salutary to point out and essential to correct in view of their likely pernicious influence. So, there is no disinterested interpretation of Darwin's religious life because, given the general deference shown him as a great scientist, his religious views are considered to have obvious and important implications for what people should or should not believe.

Such reasoning underlies most of the dreary minor literature on Darwin's religious life. Whether the writers praise or blame, adopt Darwin or disown him, it is the measure in which his *religious views* confirm or conform to their own that preoccupies them. Their interest, in other words, lies predominantly in the doctrinal aspect of Darwin's historic religious experience. A liberal Congregational minister finds that, despite the atrophy of Darwin's religious sense, he was a "great soul" who epitomized the race and directed its eyes "toward an ideal" whose outlines are "lost in the bright faith of possible perfectibility" (Cadman 1911, pp. 41, 43). An English Non-conformist divine vindicates Darwin's "spirit of reverence for the Great Creator" and devotes a chapter of his *Typical Christian Leaders* to explaining the coincidences between Darwinism and evangelical doctrines (Clifford 1898, p. 216). A contributor to *The Open Court*, a journal "devoted to the

Science of Religion, the Religion of Science, and the Extension of the Religious Parliament Idea", interprets Darwin as a religious pragmatist who "found in Love all the sustaining strength that others assumed to be the monopoly of Faith" (Nash 1928, p. 463).

On the other hand, a theistic freethinker explains that Darwin never shook off the conception of Paley's extramundane God, and this "prevented him from ever entertaining the notion that God may be Himself the supreme Law and Life of the universe". An orthodox Presbyterian rationalist points out that the "formulated helplessness" of Darwin's inability to trust the mind's religious intuitions was "a logic which strips the very logic on which we are depending for all our conclusions, of all its validity, and leaves us shiveringly naked of all belief and of all trustworthy faculty of thought." A camp-follower of the Moral Re-armament movement offers Christian sympathy to "poor lovable Darwin", a great man who paid insufficient attention to prayer and divine worship but whose shade may still haunt the Sandwalk with its discovery, not of the missing link, but of "the golden cord that joins men with God" (Symonds 1893, p. 428; Warfield 1889, p. 16; Warfield 1932, p. 576; A. J. Russell 1934, pp. 289, 296–297; cf. R. B. Freeman 1978, p. 88).

There are few exceptions to this doctrinal preoccupation among the filio-pietistic and apologetic interpreters of Darwin's religious life. And I have found only one obscure account in which Darwin is expressly deemed a "poor authority on theological questions", although here the reason seems to be that he was not enough of a Calvinist after the manner of the Cumberland Presbyterian Church (Hinds 1900, p. 40). Even in the most distinguished, but now dated, literature on Darwin's religious life, it is his mature "religious views" or "philosophical and religious ideas" that receive attention. "The enormous impact of Darwin's theory on the theological discussions of his contemporaries," explains Maurice Mandelbaum, "makes it of interest to see how Darwin himself faced the implications of his theory" (Luzzatti 1901; Mandelbaum 1958, p. 363).

The *terminus ad quem* of a century's writing on Darwin's religious views has certainly now been reached in Neal Gillespie's *Charles Darwin and the Problem of Creation* (1979). The book is not wholly satisfactory (Moore 1981b). Starting from a loose interpretation of Foucault's concept of "episteme", Gillespie attempts to show the inexorable triumph of positivist over creationist accounts of origins by reviewing how Darwin dealt with theological problems along the way. He declines to launder the metaphysics from Darwin's language, but, in the interests of tidiness, he nevertheless ends up with "two Darwins", Darwin the theist and Darwin the positivist, the historic Darwin in epistemic self-contradiction. This interpretation, though more resolute in its dualism, is not unlike other recent assessments of the religious or ideological content of Darwin's theories (cf. Conry 1974; Ruse 1979a; G. Jones 1980; and Oldroyd 1980 with Moore 1977a, 1980a and 1980b). Its unique

Its unique importance, however, lies in having demonstrated from published sources alone not only the centrality of theological problems to the course of Darwin's research but the ineradicable religious beliefs that undergirded it. No longer should it be possible to write of the "quiet, undemonstrative atheism" at which Darwin "eventually arrived" (Brent 1981, p. 315; cf. Schweber 1977, pp. 233–234). Darwin indeed gave up Christianity long before he wrote the *Origin* but he remained a muddled theist to the end.

Having with Gillespie's help got over this minor — and in my view merely apologetic — crisis in Darwin historiography, the way is clear to entertain interpretations that explore not just Darwin's religious views but the *religious filiation* of his theories. Beginning some twenty years ago with a few leading remarks by Charles Gillispie and a seminal essay by Walter Cannon, historians have looked more seriously at the references in Darwin's autobiography to the Reverends Paley and Malthus (Gillispie 1960, pp. 311ff; W. Cannon 1961b). We now know that in adopting or adapting their ideas about nature Darwin was a participant both intellectually and socially in the world of orthodox natural theology. His theories expressed and exploited its assumptions. And we now know that Darwin acquired even more from this world than Paley's problematic of adaptation or Malthusian population pressure according to a quasi-mathematical law. His reasoning a fortiori, his concept of chance, his logic of possibility, his understanding of instinct and intelligence, and the theodicy that reappears even in the *Origin* — a whole ensemble of beliefs and values pointing towards a new Naturalistic Theology were available to Darwin in the orthodox scientific literature he read early in his career (Altner 1966, Limoges 1970a; Yokoyama 1971; Herbert 1974, p. 219; W. Cannon 1976b; Schweber 1977; Bartov 1977; Manier 1978, pp. 73ff, 165–166; Loades 1979; Ospovat 1979, 1980, and 1981, chaps. 2–3; Moore 1979, chap. 12; Richards, 1981). So clearly and abundantly has this been demonstrated in the two most substantial studies of Darwin's earlier intellectual life, *The Young Darwin and his Cultural Circle* (1978) by Edward Manier and *The Development of Darwin's Theory* (1981) by the late Dov Ospovat, that Charles Gillispie might well take pride in having once found it "inconceivable" that Darwin's individualistic language of struggle and moral improvement "could have been written by any Frenchman or German or any Englishman of any other generation" (Gillispie 1974, p. 224).[2]

Here, however, there is no disinterested interpretation either. Leaving aside the debatable religious implications of suggesting that the theory of natural selection as the legitimate offspring of an orthodox theology of nature, there is the real historiographic problem raised by Gillispie's canny *aperçu*: Did Darwin's account of nature become "true" only in Victorian England? Or can the religious filiation of Darwin's theories affect their validity? Gillispie states his case elsewhere, and many still agree that historians should take their cue from successful practitioners of science and keep the context of discovery entirely separate from the logic of verification. The origins of theories have nothing whatever to do with their truth. This of course looks rather like the deference that has led writers to be preoccupied with Darwin's religious views. The possibility at least ought to be recognized. I myself find the approach unhelpful partly because it tends to discourage or precondition the search for the intellectual filiations of scientific theories, but mainly because it almost precludes looking beyond the intellectual filiations to the material, economic and social context that they mediate. In other words, studying Darwin's theories as bare theories rather than in their natural theological dimension, as we now know, would be historically nearsighted. But studying natural theology as bare theology rather than in its ideological dimension would also, by common admission, be nearsighted (Gillispie 1951, chap. 7). This inevitably suggests that studying Darwin's theories as bare theories or simply in relation to natural theology, rather than in their full and complex relations with the economic structures and social institutions of early Victorian Britain, would offer only a very limited understanding of Darwin's work. A medical researcher might just as well seek the aetiology of color-blindness without recourse to biochemistry, or investigate hallucinatory experiences without the aid of pharmacology and neurophysiology.

My concern, then, is not just with the religious views of a great scientist or the religious filiation of his theories but with the part of Darwin's material culture that may be called his *religious context*. And this concern finds me in good company. In recent years the narrow internalist and intellectualist approaches to Darwin's religious life have been superseded; a new historiography of Victorian science and religion has emerged under the rubric "social-intellectual history" or the "social history of ideas". Beginning with original studies by Robert Young and Frank Turner, and extending latterly to the work of John Durant and Leon Jacyna, historians have begun to trace the transition from theistic to naturalistic cosmologies in Victorian Britain against the backdrop of the professionalization of science or, more generally, as a function of the changing patterns of subordination within an industrializing and imperial social order (Young 1969, 1970b, 1971a, 1972a, 1973a, 1980; Turner 1974a, 1974b, 1974–1975, 1978, 1981; Durant 1977, 1979, 1982; Jacyna 1980a, 1980b, 1981; cf. Moore 1981a, pp. 33–41). These developments, moreover, are reflected in some of the latest and most acute of the studies dealing with Darwin's religious life. Although Manier's *Young Darwin and his Cultural Circle*, with its analysis of audiences, linguistic communities, rhetorical strategies, and the uses of metaphor, appears to be a study in applied communications theory, there is much in the book that looks beyond Darwin's cognitive culture to its material context. As Manier indicates elsewhere, without "locating Darwin's logic in its broader social and rhetorical context" the history and philosophy of biology are "likely to misdefine the logical and empirical aspects" of his work (Manier 1980a, p. 322; cf. 1980b, pp. 19–21). In Ospovat's *Development of Darwin's Theory* the outlook toward society is both more explicit and more hopeful.

The book is exclusively an account of Darwin's intellectual development, but Ospovat remains sensitive to the "subtle ways in which scientific ideas are socially constructed" through the long-term establishment in a culture of "ways of seeing nature that are themselves constructed in response to social, political, and religious, as well as scientific, interests". He recommends that we should view the change in Darwin's thinking "as a microcosm of the more general development from a philosophy of nature and man appropriate for an agrarian and aristocratic world to one suitable for the age of industrial capitalism" (1981, pp. 230, 233).

I find this the most helpful and promising perspective in which to view Darwin's religious life. It is controversial, but no more so than other historiographies, which have their own interests to defend. It is challenging to the researcher, but only insofar as conventional interpretations, established readings of texts, and familiar historical sources tend to concentrate the mind. In this essay, having indicated the value and limitations of some conventional interpretations of Darwin's religious life, I shall proceed to suggest new readings for old texts and to introduce fresh historical sources alongside familiar ones. My aim is to contribute toward a social history of Darwin's thought by analyzing its religious context, as defined above. My argument will run as follows. On going to Cambridge in 1828 Darwin was destined to be a country clergyman. Neither he nor his friends and advisers thought this comfortable and privileged vocation would interfere with the devout study of natural history, whether at university or on a voyage before taking orders, or in a parish afterwards. On the *Beagle* voyage Darwin found deep emotional satisfaction in contemplating the natural world. This, together with uncertain marital prospects, diminished, although it did not extinguish, his happy vision of a country parsonage. After returning to England in 1836 Darwin had to decide in what context his devout and dangerously disturbing work on transmutation could best be pursued. Faced with the conflicting sets of life-options represented, on the one hand, by the married clerical naturalists of Cambridge, and on the other by his single brother Erasmus and Charles Lyell in London, Darwin elected to fulfill his old vision and retire *en famille* to a country parish. On removing to the parish of Down in 1842 Darwin increasingly exercised the prerogatives of squire and parson, culminating in the 1860s when there was no resident vicar. So prominent became his role as "squarson", to use a contemporary term, that it precipitated a conflict when an incumbent arrived who held traditional views on the role of the parish priest. The family ceased attending the church and Darwin relinquished most of his parish responsibilities. Thus, while Darwin's intention of seeking ordination was never "formally given up, but died a natural death", to use his own words, it is possible to trace the relation between what was never "formally given up" and what emerged, Phoenix-like, from the ashes of a clerical career. This continuity or transmutation of Darwin's vocational plans, I shall suggest in conclusion, supplies a context in which to interpret the meaning of his theoretical achievement in Victorian culture.

I. The Vocation

What was Dr. Robert Darwin to do? He had educated his son at Shrewsbury School; he had set him a fair example in his medical practice, and had even taken him on his rounds and taught him to administer prescriptions. Then he had sent Charles off to the famous medical faculty at Edinburgh where he might study in the company (or under the aegis) of his elder brother Erasmus. But all to no avail. The young man had spent his father's money, wasted his professors' time, and at the age of eighteen seemed set to become a professional dilettante. Was this perhaps because he reckoned his future was secured with hereditary wealth? Would vocational indirection be the price of his habitual hob-nobbing with the Shropshire squirearchy and hunting with the Staffordshire bourgeoisie?

Dr. Darwin thought not. And, further, the lad would have to give an account of himself in an established profession before he could have the security of the family exchequer. If not medicine, then, which one should it be? Lawyers and military men were not unprecedented among the Darwins, but Charles had yet to show the self-discipline that such vocations required. The safest bet would be the Church. Charles's first cousin, John Allen Wedgwood (1796–1882), was already in orders, and both Dr. Darwin's uncle, John Darwin II (1730–1805), and his half-brother, John Darwin III (1787–1818), had been rector of Elston in Nottinghamshire, where the senior branch of the Darwin family had been squires since the end of the seventeenth century. Also Dr. Darwin's uncle by marriage, Thomas Hall, had served for many years as rector of Westborough in Lincolnshire (R. B. Freeman 1978). Here were country clergymen with respectable careers whom a wayward son might emulate. Send him to an English university, to Cambridge where his grandfather had attended, and to Christ's College where his brother and now his second cousin, William Darwin Fox (1805–1880), had gone before. Let Charles be educated for three years as a gentleman, and spend another year, if he liked, attending voluntary lectures in theology. Let him marry soon afterwards while serving a comfortable curacy; then present him to a quiet rural benefice, which a man of Dr. Darwin's means might easily afford (Chadwick 1966–1970, 2:169; Heeney 1976, pp. 98–99; Brent 1981, pp. 71–72).

Was this not after all the course pursued by a great many fathers of gentlemanly means? The seeming nonchalance with which Dr. Darwin, a confirmed freethinker, settled upon a clerical career as his second choice for Charles may appear almost cynical a century and a half later. But when over 60 percent of England's population lived outside the towns, while there were few professions among which to choose, and before the mid-century church renewals had quickened priestly commitment, the decision to send an aimless son into a country living made eminent good sense. He need have little theological knowledge, still less any pastoral training, and no intimation of a divine "calling" whatever. In the Christian ministry he might, on the other hand, have "as much desire for increased emolument, reputation, and advancement, as in any other calling", according to a career handbook published as late as 1857. That is, he could look forward to ample rewards: "opportunities for early independence", "comparative security of position", "opportunity of leisure", "absence of any risk of total failure", "easy work compared to the struggles of other callings", "ready admission into society", and lastly, a "satisfactory sphere of usefulness" (quoted in Heeney 1976, p. 94).

In the rural parishes of Georgian England, moreover, the rewards were greatly enhanced. The typical country parson, having graduated from Oxbridge, married well and started a large family. He settled into a commodious rectory, complete with well-laid lawns and a gardener, a housekeeper, and a groom. He might be the pluralist non-resident sportsman of popular caricature or the quiet philanthropic scholar. He was at all events a gentleman, the preceptor of the village and the crony of the squire. He had charge of local charities and he helped mediate local disputes. As a magistrate he shared with the squire the duty of keeping order in the district. If, as sometimes occurred, he was the principal landowner or the chief patron of local trade, he might combine the attributes of squire and parson in a hybrid creature whose concerted social, economic, and spiritual power had vast possibilities for good or evil: the "squarson", as Samuel Wilberforce and Sydney Smith called him (Heeney 1976, pp. 2, 23, 95–96; A. T. Hart and Carpenter 1954, chap. 1; Addison 1947, chaps. 17–18, 20; Keppel 1887).

All this was commonplace to Dr. Darwin when he packed young Charles off to Cambridge in January 1828. His son would have a prominent social role, a steady income, and, all going well, a handsome legacy. He might even resume the hunting and hob-nobbing that had thus far unsettled him for vocational pursuits.

Charles, for his part, "liked the thought of being a country clergyman", although he felt no divine call to the ministry and at first had scruples about subscribing the Thirty-Nine Articles of the Church of England (*Autobiography*, p. 57). And he was in good company. To Cambridge ordinands outside the orbit of Charles Simeon, the evangelical vicar of Holy Trinity Church, becoming a country clergyman was a matter of practical common sense. It was a vocation with modest demands that left room for much else besides — a little shooting, a little drinking, a little doubt, and, if one liked, a good deal of natural history.

Such things also went well with a Cambridge education. Those who were intended for the Church intended to enjoy themselves en route, as this was much the best training for pleasures and privileges to come. Charles accordingly fell in with a merry lot of fellows who seemed set to form the collective image of the big-bottomed, red-faced rector of partisan legend. There was the Gourmet (or "Glutton") Club, for example, most of whose members took orders — holy ones as well. Charles Whitley (1808–1895), a schoolmate from Shrewsbury, became vicar of Bedlington in Northumberland; Frederick Watkins (1805–1888) finished his career as archdeacon of York; J. W. L. Heaviside (1808–1897) became a canon of Norwich; and J. H. Lovett Cameron (1807–1888) spent his last three decades as rector of Shoreham in Kent (*Autobiography*, p. 60; *LL* 1:28, 168–170; R. B. Freeman 1978; De Beer 1958d, p. 111; De Beer 1968, pp. 83–85; DAR 112.4). Had it not been for his cousin Fox and some dedicated clerical naturalists, Charles might well have followed his earlier inclinations in the company of these men and lived out his years, like George Eliot's Parson Gilfil, smoking very long pipes, preaching very short sermons, and hunting contentedly with the squire.

It is questionable, as I shall explain, whether Dr. Darwin thought that country parsons had much business studying flowers, insects, and rocks. How or why he should have reached this conclusion is difficult to imagine. Stephen Hales, Gilbert White, William Kirby, and a host of lesser parish scholars were not held in ill repute (A. T. Hart, 1959, chap. 4; D. Allen, 1976, pp. 21–23). That Charles would have differed strongly with his father on this account, even before going to Cambridge, is undeniable. At this time perhaps up to half the attraction of a country living to him was its recreational prospects. But the other half was surely the opportunity for continuing open-air studies such as he had begun at Edinburgh under the supervision of Dr. Robert Grant. In any case, however, when Charles arrived in Cambridge the balance of his interests swung decisively in favor of natural history. At Christ's College, whose boast was Milton and Paley, his boyhood fascination with bugs was rekindled by Fox, who had decided to seek ordination. Together they tramped the fens in search of the elusive *Panagaeus crux-major*. Before Fox left Cambridge to seek a curacy at the end of 1828 he introduced Charles to the circle of clerically-minded naturalists who surrounded the professor of botany, the Reverend John Stevens Henslow (1796–1861). The group met weekly at Henslow's open house, and members accompanied Henslow and his botanical class on the occasional country excursion. Among them were Richard Dawes (1793–1867), later dean of Hereford, William Allport Leighton (1805–1899), a schoolfellow of Charles's who would become an Anglican lichenologist, and the Reverend Leonard Jenyns (1800–1893), Henslow's brother-in-law, whose father was the squarson of Bottisham Hall near Cambridge. Jenyns had gone from the university to the parish of Swaffham Bulbeck, close to his father's property, where he would remain as vicar for twenty years. Like Charles he was afflicted with beetle-mania, and when his friend visited at the vicarage they would sometimes stalk their wiry prey in the woods at Bottisham Hall (*Autobiography*, pp. 62–67; *CP* 2:72–73; DAR 112 [part 2], 94; *LL* 1:181, 364n; Blomefield 1889; Teidman 1963).

Many years later Jenyns claimed to be unaware that his fellow-collector "ever read for the Church, or had any thought of entering it" (DAR 112. 67–68). This is understandable to some extent because Charles followed an ordinary degree course, and his theological study would have taken place in an additional year, at the end. Decisions and discussions about it could be postponed. Also Charles was much closer to his mates who lived in Cambridge, and to other friends, like Fox, from whom indeed he sought advice on what theological books to read. This occurred in March 1829, and although Charles seems subsequently to have had doubts about his calling, he wrote Fox in May 1830 about reading divinity with Henslow. In the Autumn Henslow had become his tutor and intimate friend, so certainly he knew of Charles's intentions by then. Besides, Fox was off the scene, undertaking parish responsibilities for the first time. Henslow, who served as assistant curate of St. Mary-the-Less in Cambridge, was well placed to advise Charles from day to day on both professional and scientific matters. They walked together, they dined together, they botanized together, and they talked deeply and openly about religion. If anyone would help direct a divinity student and naturalist like Charles into a parish that suited his interests, the man would be Henslow (*LL* 1:171, 182–183; *CP* 2:73–74; *Autobiography,* pp. 64–66; A. T. Hart and Carpenter 1954, p. 61; Russell-Gebbett 1977, chap. 1).

Eighteen-thirty-one was the year Charles would prepare himself for ordination. Fox had just passed the required examination and Charles, now keeping two terms before taking his degree, asked him "about the state of your nerves; what books you got up, and how perfect". "I take an interest about that sort of thing," he explained, "as the time will come when I must suffer" (*LL* 1: 184). But the time was a long way off. Charles spent the early months of the year reading, communing with Henslow, and planning an excursion to Tenerife. With his academic course behind him he could twice savor Humboldt's *Personal Narrative* and take inspiration from Herschel's lately published *Preliminary Discourse on the Study of Natural Philosophy*. Perhaps it was Henslow who recommended these works, while Fox had set his cousin to reading Paley's *Natural Theology* and Sumner's *Evidence of Christianity Derived from its Nature and Reception*. Perhaps it was the other way around. That Charles had read them all with care before the summer seems certain, for by then he had quit Cambridge to diversify his training for a country parish (Darwin 1959, p. 6; *Autobiography,* pp. 59, 67–68).[3] Henslow had interested him in geology, and in August he would travel with the Reverend Adam Sedgwick (1785–1873), the Woodwardian professor at Cambridge, to measure the inclination of hills in North Wales. The autumn would bring another season of hunting in Staffordshire, followed by further preparations for the Canaries expedition that would take place perhaps early in 1832. After all, Fox had spent two years finding a curacy and taking holy orders. Why shouldn't he? (Darwin 1967, pp. 25–26; *Autobiography,* pp. 68–71; Clark and Hughes 1890, 1:380–81; Barrett 1974).

II. The Voyage

What happened next, when Charles returned from his geological tour, is too familiar to bear repeating without a change of emphasis. The Reverend George Peacock (1791–1858), a lecturer in Trinity College and later dean of Ely, showed a degree of inaccuracy unbecoming in a mathematician by soliciting from his colleague, the Reverend Henslow, the name of a man qualified to serve as a naturalist aboard the *Beagle*, which was soon to sail under the command of the young Tory aristocrat, Robert FitzRoy (1805–1865). In fact the *Beagle* already had a naturalist, a man of a lower class than the captain would find congenial in a cabin companion. And this above all was what FitzRoy desired: a social equal with whom he might dine and converse, thereby relieving the personal isolation that had driven his predecessor to suicide (Darwin 1967, pp. 38–41; *LL* 1:207–211; J. W. Gruber 1969; Burstyn 1975). Such a man could certainly be found at Cambridge, even if Peacock did interpolate that he should be qualified as a naturalist; and there was every chance that he would have an ecclesiastical affiliation if he were recruited through the clerical old-boy network. In his letter to Henslow, Peacock suggested the well-connected Reverend Jenyns for FitzRoy's partner, but parish responsibilities prevented his acceptance. Henslow, indeed, declined to put himself forward only out of deference to his wife. There remained Charles Darwin, unbeneficed, unmarried, as yet unordained. And he had proved himself a delightful companion. The captain, Henslow urged, "would not take anyone however good a Naturalist who was not recommended to him likewise as a *gentleman*. . . . I assure you I think you are the very man they are in search of" (Darwin 1967, pp. 28–30; *LL* 1:200).

So the country parish could wait. Henslow and Charles agreed. Dr. Darwin did not. Here was further evidence of his son's aimless preoccupation with enjoying himself. The voyage would be a useless, dangerous lark. And afterward — what then? The experience would, first of all, be "disreputable" to Charles's "character as a clergyman"; it would also unfit him for a "steady life". Not only, in other words, could the voyage lead to yet another change of profession; it might well result in no profession at all. Charles, however, had the wholehearted support of his Wedgwood relatives with whom he was staying, as was his custom, for the start of the hunting season on 1 September. Dr. Darwin regarded Josiah Wedgwood as a man of great common sense and Charles found that his uncle did not think the voyage would be "in any degree disreputable to his character as a Clergyman". He held, on the contrary, that "the pursuit of Natural History, though certainly not professional, is very suitable to a clergyman." Further, Uncle Josiah thought that Charles was not so "absorbed in professional studies" at present that his career could not be interrupted, whereas after a protracted voyage he might in fact be more inclined to settle down (Darwin 1967, pp. 34–37; *Autobiography*, pp. 71–72, 77).

When Charles and Uncle Josiah confronted Dr. Darwin with these counterarguments he relented. In December 1831 Charles therefore embarked on the "most important event" of his life, knowing full well that none of those he most trusted and admired, neither his father nor his uncle nor Henslow, thought the voyage incompatible with a future career in the Church (*Autobiography,* p. 76; *Diary* pp. 3–4).

Five years aboard a ten-gun brig, backtracking its way around the world, furnished enough strange and wonderful experiences to unsettle the most resolute of professional intentions. Charles's underwent no upheaval because, in the first place, they had not originally been his own, and second, because they were none too strong. The third and, in my view, the most important reason why no crisis occurred is that Charles had received the assurance of his closest friends and advisers, by precept and example, that nothing would happen to him on the voyage inconsonant with being a country clergyman thereafter. The career, as presented to him, was not so stringently defined or demanding that his activities as a naturalist need ever interfere. Thus in his *Autobiography* Charles wrote that his intention to be a clergyman was never "formally given up, but died a natural death" (p. 57).

During the *Beagle* voyage, when Charles might have been expected to drop his intention of seeking ordination, his practices and beliefs were conventionally religious. He attended Sunday worship aboard ship; he joined a mate in asking the chaplaincy at Buenos Aires to administer them holy communion before they began the tortuous passage to Tierra del Fuego; and he contributed to the vast sum of £50 that FitzRoy and the officers raised among themselves to help build a church (or chapel) at the Bay of Islands, New Zealand.[4] Charles continued to believe in the Bible as a moral authority — references to "antediluvian" animals did not survive the voyage — and, with the evangelical FitzRoy, he had little but praise for the missionaries who brought its message to native islanders in the South Pacific. Unlike FitzRoy, however, Charles detested slavery. He could only be "a sort of Christian" in South America because Europeans habitually used the word to designate slave-owning Spanish Catholics (*Autobiography,* p. 85; Darwin 1945, p. 76; G. Darwin in DAR 112 (ser. 2): 7–50; *Diary*, pp. 244, 372; *CP* 1:19–38). Theologically there was also conformity and uniformity, if not much profundity, in Charles's views. His hope that "Providence" would spare the *Beagle* further storms like the one at Cape Horn the previous day seems, under the circumstances, not to have been disingenuous. Nor is it surprising to find Charles noting, in Lyellian fashion, the "fitness' with which the "Author of Nature" had established the number of species in a region. In Australia he toyed with the rationalist view that "two distinct Creators" must have worked in the world, one of them on the extraordinary fauna of that country; but he dismissed the notion when a lion-ant beside him furnished a footnote to Kirby's Bridgewater Treatise. "The one hand has surely worked throughout the universe," Charles concluded (*Diary*, pp. 128, 323; Herbert 1974, pp. 233; Gruber and Gruber 1962).

What was changing on the voyage had little or nothing to do with religious doctrines and observances. As I read the evidence, it was rather Charles's attitude toward nature and his assessment of his social prospects on returning to England. The one conditioned the other and both were emotionally fraught. A worshipful attitude toward nature emerged and developed following Charles's solemn encounter with the Brazilian forests in 1832. At almost the same time he began to realize his emotional distance from friends and family in England, whose lives went on without him. He developed a complex longing for home and hearth, for Cambridge camaraderie and female affection, all of which seemed increasingly remote and unattainable in the measure he had previously known, or had anticipated when a country parsonage loomed large. In proportion as this longing, compounded of personal isolation, sexual deprivation, and vocational uncertainty, took hold of Charles's emotional life, his devotion to natural history increased. Was it not the "sublime solitude" of his forest reverie in June 1832 that, he admitted, turned his thoughts toward home, and elicited the confession to his younger sister, "If I gain no other end, I shall never want an object of employment and amusement for the rest of my life"? (Darwin 1945, p. 70; *LL* 1:241; cf. Darwin 1945, pp. 162–165 with *Diary*, pp. 56, 70–71). His future happiness and success might thus reasonably depend on what he could accomplish at the time as a naturalist, not on what he or his father had intended that he should do when the voyage was over. But in proportion as Charles devoted himself to natural history, he felt limited as to the personal and social expectations he could fulfill on returning to England. How, for example, could he enter the Church in his state of unpreparedness? Or how could he make of natural history a profession that would meet with his father's approval? How to do the former, indeed, without a wife, or the latter while being married?

Marriage and career were part of the same problem for Charles. The average Anglican clergyman was expected to have a wife. Early in 1832 Charles learned that his former sweetheart, Fanny Owen, had become a Mrs. Biddulph (Brent 1981, pp. 55–70). Then came news from his cousin Charlotte, as he put it, "of parsonages in pretty countries and other celestial views." She and her husband, the Reverend Charles Langton (1801–1886), were removing that year to the country parish of Onibury near Ludlow in Shropshire. It was all too much. "By the fates, at this pace I have no chance for the parsonage," Charles despaired. Nevertheless a few weeks later he wrote, "I find I steadily have a distant prospect of a very quiet parsonage, and I can see it even through a grove of palms" (1945, pp. 62, 66). But in November came a letter from his bachelor brother Erasmus, then living in London:

> I am sorry to see in your last letter that you still look forward to the horrid little parsonage in the desert. I was beginning to hope I should have you set up in London in lodgings somewhere near the British Museum or some other learned place. My only chance is the Established Church being abolished. (quoted in Herbert 1974, p. 222)

His Whig sympathies and the Reform Bill notwithstanding, Charles ignored the taunt. "I am becoming quite devoted to Nat. History," was his reply (Darwin 1945, pp. 76, 78).

In June 1833 Charles wrote his Gourmet Club companion, J. M. Herbert, wishing him "a dear little lady to take care of you and your house". "Such a delightful vision," he added, "makes me quite envious" (*LL* 1:248). Herbert, however, had not read divinity at Cambridge; it was Herbert's cousin, Whitley, and his own cousin Fox, who were establishing themselves as clergymen. Toward the end of the voyage Charles wrote each of these old friends a franker and more revealing letter.

While at sea off the coast of Chile, on 23 July 1834, Charles inquired about Whitley's marital status, then exclaimed:

> Eheu! Eheu! this puts me in mind of former visions of glimpses into futurity, where I fancied I saw retirement, green cottages, and white petticoats. What will become of me hereafter I know not; I feel like a ruined man, who does not see or care how to extricate himself. That this voyage must come to a conclusion my reason tells me, but otherwise I see no end of it. It is impossible not bitterly to regret the friends and other sources of pleasure one leaves behind in England; in place of it there is much solid enjoyment, some present, but more in anticipation when the ideas gained during the voyage can be compared to fresh ones. I find in Geology a never-failing interest, as it has been remarked, it creates the same grand ideas respecting this world which Astronomy does for the universe. (*LL* 1:255)

Everything in this passage, up to the mention of geology and its "grand ideas", might be understood to have a sexual reference: that is, to marriage ("former visions"), fornication ("ruined man"; "much solid enjoyment, some present"), celibacy ("I see no end to it"), former sexual experiences ("the friends and other sources of pleasure one leaves behind in England"), erotic fantasies ("ideas gained during the voyage"), and future sexual encounters ("fresh ones"). I doubt whether this interpretation could be entirely valid, for the presumed references to celibacy and fornication are not mutually consistent. It is easier to believe that Charles saw his marital prospects being "ruined" by his absence from England and by his singleminded devotion to natural history. Thus he hardly knew what would become of him, so far as "retirement" to a country parish, with its "green cottages", was concerned. But of one thing he was certain: geology with its "grand ideas" had become a "never-failing interest".

A year later, in 1835, Charles felt much the same. He was at Lima, facing up to the loneliness of the transpacific passage, when he received from Fox a long account of wedded life as a clergyman. "How very strange it will be thus finding all my friends old married men with families," Charles told his elder sister soon afterwards (1945, p. 125). To Fox, however, he confessed,

> This voyage is terribly long. I do so earnestly desire to return, yet I dare hardly look forward to the future, for I do not know what will become of me. Your situation is above envy: I do not even venture to frame such happy visions. To a person fit to take the office, the life of a clergyman is a type of all that is respectable and happy. You tempt me by talking of your fireside, whereas it is a sort of scene I never ought to think about. I saw the other day a vessel sail for England; it was quite dangerous to know how easily I might turn deserter. As for an English lady, I have almost forgotten what she is — something very angelic and good. As for the women in these countries, they wear caps and petticoats, and a very few have pretty faces, and then all is said. But if we are not wrecked on some unlucky reef, I will sit by that same fireside in Vale Cottage and tell some of the wonderful stories which you seem to anticipate and, I presume, are not very ready to believe. *Gracias a dios*, the prospect of such times is rather shorter than formerly. (*LL* 1:262–263)

"Celestial views" in 1832, "delightful visions" in 1833; a year later, "former visions of glimpses into futurity". Now, in 1835, the "happy visions" dare not arise. I find no evidence here that Charles believed his devotion to natural history unfitted him directly for a clerical career. Otherwise Fox's account would have lacked its evident appeal. It was rather the absence of an "English lady, . . . very angelic and good", that Charles thought unfitted him for the office, and this was due to a prolonged and distant voyage during which the love of nature had come in to fill up his emotional void.

III. The Venture

Before there was a "clear & not so distant prospect" of returning to England, Charles found in natural history "a prospect to keep up the most flagging spirit". It had become his "favourite pursuit", he told his sister, and "I am sure will remain so for the rest of my life" (1945, pp. 85–86, 110). This did not of course rule out a clerical career. On the contrary, it was his clerical naturalist colleagues to whom Charles immediately looked for validation of his work. If Henslow "shakes his head in a disapproving manner," he remarked in April 1836, "I shall then know that I had better give up science, for science will have given up me." But when he learned a few months later that Sedgwick had spoken highly of his fossil collections, now arrived safely back in England, he found it "deeply gratifying" and a spur to future attainments. "I trust . . . that I shall act — as I now think, — that a man who dares to waste one hour of time, has not discovered the value of life." And there would be no time to waste in sorting through and analyzing the amassed debris of five years' research. Six months before the *Beagle* landed at Falmouth Charles knew he would have to live a year in London, though often he thought that Cambridge and the "real country" would be better. He asked Erasmus to put his name forward for a club and to look out "for some lodgings with good big rooms in some vulgar part of London". There he would put his notes in order, arrange to place his collections in expert hands, and prepare his journal of the voyage for publication (Darwin 1945, pp. 138–139, 141, 145).

When Charles reached London in October 1836 he entrusted himself and his affairs to Henslow. To facilitate their collaboration on the natural history specimens it was decided that Charles should live a few months in Cambridge. Thereafter he would reluctantly follow his first plan and reside "for some time in . . . dirty, odious London". Communicating with FitzRoy and fellow-naturalists, identifying and classifying specimens, liaising with publishers — such things could be done more readily from there. This indeed was the advice of a new mentor named Charles Lyell (1797–1875), whose *Principles of Geology* Charles had devoured on the voyage. In March 1837 Lyell welcomed his admiring student to London, and Erasmus was delighted to have his brother for a neighbor when Charles decided to take lodgings in Great Marlborough Street (Darwin 1967, pp. 118–124; Darwin 1959, p. 7).

Five years earlier Henslow had offered Charles entrée into a Cambridge circle that consisted largely of clerical naturalists like Fox, Jenyns, and Sedgwick. Now, while Henslow and Cambridge continued to represent one set of life-options — Anglican, naturalist-clerical, pastoral, and married — Lyell and Erasmus in London offered him another. Both were gentlemen of independent means and both regarded urban life as intellectually more bracing than rural. Erasmus, a bachelor and a rationalist, had become a node in the London literary network, and through him Charles met all sorts of advanced intellectuals. Lyell, a liberal Unitarian, had, while married, become a professional geologist, and through his good offices Charles was elected a Fellow of the Geological Society in 1836 and a member of the Athenaeum Club two years later (*LL* 1:293–294, 297, 298; Darwin 1967, p. 122; *Autobiography*, pp. 100–101). By paving his path into scientific and professional London, Lyell contributed substantially to the making of Charles's reputation as a naturalist. He exemplified, with Erasmus, a set of life-options — freethinking, scientific-professional, urban, and unmarried — that differed widely from the one represented by Henslow and Cambridge. Faced with these sets of options, Charles had to contemplate for many months the direction his life should take.

Publicly Charles was ambivalent. He enjoyed pleasant evenings with Erasmus and lengthy geological sessions with Lyell. He made the rounds of the London "dons of science", worked diligently at his collections, and found time to publish several scientific papers. He even agreed somewhat reluctantly to accept in 1838 the honor of serving as a secretary to the Geological Society. In the course of these activities Charles also kept closely in touch with the Cambridge circle of clerical naturalists. Henslow remained his father-confessor and literary executor, although after July 1837 he had, in addition to his teaching duties, an increasing responsibility as rector of the parish of Hitcham in Suffolk. At this time Charles paid him a warm tribute summed up in the words, "You have been the making of me, from the first" (Darwin 1967, p. 135; A. T. Hart and Carpenter 1954, pp. 61–64; Russell-Gebbett 1977). Jenyns at Swaffham Bulbeck had kept up his reputation as a precise and fastidious observer. Charles paid him tribute, in turn, by entrusting him with the description of fishes in the part of the *Zoology* that bears his name (*LL* 1:281). To Henslow and Jenyns, as well as Fox in the Isle of Wight, Charles confessed at intervals how much he longed to "escape" from "vile smoky" London, which to him seemed like a "prison". He would visit them when he could — Fox, as he promised late in 1837, on the eve of his departure for the parish of Delamare in Cheshire; Jenyns and Henslow again in the Spring of 1838, when he heard Jenyns "bitterly complaining of his solitude" — but never without a backward glance at the work in London that awaited completion. "My life is a very busy one, . . ." Charles told Fox, "and I hope may ever remain so; though Heaven knows there are many serious drawbacks to such a life, and chief amongst them is the little time it allows for seeing one's natural friends" (Darwin 1967, pp. 124, 131, 142; *LL* 1: 281–282, 289; Darwin 1959, p. 7).

So far the public face of an ambitious young professionalizing geologist. He was working sacrificially for the present to finish a definite task and earn the respect of his scientific elders. Privately, however, in the contemplative world he admitted having inhabited for the previous five years, the world of "air-castles" and "delightful visions", of "sublime devotion" and "grand ideas", there was growing turmoil (*ML* 1:29). Charles still took much pleasure in brooding over the enigmas of natural history, but his thoughts were becoming dangerous, his brooding masochistic. The turmoil in his mind played a subtle and complex counterpoint to the public turmoil in urban Britain during those years. Of this, I believe, Charles must have been at least dimly aware. For what he dared only hint at, even to his closest scientific friends, was that since coming to London in

March 1837 he had secretly espoused, to himself, in a series of pocket notebooks, doctrines that were scientifically disreputable, legally actionable, and politically subversive (*LL* 1:298; Herbert 1980, pp. 10ff). He had begun to articulate a vision of the world in which all living things were linked together through natural generation over vast eons of time; and not only their physical structures, but their habits, instincts, reasoning power, emotions, morality, and even belief in God. From echinoderms to Englishmen, all had arisen through the lawful redistribution of living matter in response to an orderly changing geological environment.

This was rank materialism, and Charles knew it. It was all that Lyell rejected in Lamarck, all that England had feared in the ideology of revolutionary France. There were blasphemy laws and sedition acts to curb it, courts to expose and prosecute it, and severe punishments to deter intellectual offenders, like him, from publicizing their views. While living in London Charles presumably formed his latter-day habit of reading a daily newspaper. Perhaps it was also at this time that he became habitually fascinated by "any curious trial" and found the law report "about the most interesting part of the paper".[5] Not of course as if he need otherwise have been ignorant of the convulsions in Britain's cities between 1837 and 1842, or the notorious activities of Owenites, Chartists, phrenologists, and other radical freethinkers who preached materialist doctrines in the cause of social reform. The streets of London were running sores — full of "dirt, noise vice & misery", Charles complained to Fox in Malthusian terms — and Great Marlborough Street had become particularly abhorrent to him within two years of his removing there from Cambridge (Colp 1977a, p. 28; *LL* 1:297; Moore 1982a). Doubtless such conditions, with all their social antecedents and political repercussions, were high on the agenda at soirées with Erasmus and his friends, such as Thomas Carlyle and Harriet Martineau. Alternatively, as a source of intellectual discomfort Charles need only have recalled the suppressed minute of a Plinian Society meeting at Edinburgh in 1827, when a fellow student had ventured to express materialistic views; or the prosecution two years later of a local proprietor in Cambridge for lodging the infidel missionaries Robert Taylor and Richard Carlile, who had publicly challenged the heads of all the colleges to a debate; or Lyell's account in his *Principles* of the violent struggles between tradition and innovation in the progress of science, including the persecution of early astronomers and the forced recantation of the French naturalist Buffon.[6]

Yet it was Lyell, the believer in scientific progress and the opponent of transmutation, who "strongly advised" Charles, probably during this period, "never to get entangled in a controversy, as it rarely did any good and caused a miserable loss of time and temper." This was in reference to his geological work, but the admonition had greater relevance at the time than Lyell knew (*Autobiography*, p. 126). Neither professionally nor politically, then, was it prudent for Charles to disclose his thoughts. If in the future the "whole fabric" of people's inherited beliefs, and with it their culture and social institutions, "totters & falls", it would not be because the young naturalist had played the young radical. Self-destruction was unbecoming in a man of his pedigree and ambitions, not to say unacceptable to his father. Rather, if such a man were to move the world, he would first require a strategy and a place to stand. These things could not be obtained overnight, especially with responsibilities pressing around. Privacy therefore had to be cultivated in the meantime as best one could, and perhaps a modicum of the discreetest confidential advice.

IV. The Valetudinarian

The excuses for privacy and discretion were mounting day by day. Compounding his inner turmoil, feeding on it and feeding it in turn, Charles felt himself becoming sick. The hard work on his journal, and on the zoology and geology of the voyage, combined with duties at the Geological Society, no doubt took its toll. But Charles had performed prodigious mental and manual feats during the voyage without obvious ill effects. It was only in the Autumn of 1837, six months after returning to London and less than three months after opening his first notebook on transmutation, that serious dyspepsia, headaches, and cardiac palpitations set in (Colp 1977a, pp. 14ff). Three days at Cambridge in May 1838 did Charles "wonderful good", as he relived old times with Henslow, Jenyns, and Sedgwick, and on the Sunday evening heard "The heavens are telling the glory of God" rendered magnificently in Trinity Chapel. Back in London, however, he complained to Fox, there was again "smoke, ill-health and hard work". "Lost very much time by being unwell," Charles later wrote in his diary (*LL* 1:289, 290; Darwin 1959, p. 8). And little wonder. This was the period when in his notebooks he was reproaching himself for being a materialist, recalling the persecution of early astronomers, and defiantly drawing out the implications of his views for the origin of humankind. "Man in his arrogance thinks himself a great work worthy the interposition of a deity. more humble & I believe truer to consider him created from animals" (*C* 196–197). And, "I will never allow that because there is a chasm between man . . . and animals that man has different origin" (*C* 223).[7]

It was during this period, I believe, in the later Spring of 1838, that Charles began to move decisively toward a resolution of the dissonant life-options that faced him. His journal was soon to be published; the zoology of the voyage was well under way, and he had completed part of the geology. In the foreseeable future he would finish the primary task that had brought him to live in London. What then? Charles deliberated to himself on the back of an old letter in two columns, each headed "Work finished". On the left he sketched what his life would be like if he remained single — "travel", "work at transmission of Species", "live in London"

— and on the right, the consequences of marriage. Here "London life, nothing but Society, no country no tours" is rejected. "Could I live in London like a prisoner?" Charles asked himself poignantly. Besides, he could scarcely afford to live there if he had children (unless he took remunerative employment). Cambridge would be "better" but, not being a professor, he would live in "poverty" and feel "like a fish out of water". A "Cambridge Professorship" like Henslow's would offer an income and perhaps the most agreeable life-style, except, said Charles, "I couldn't systematize zoologically so well." For this it would be necessary to have independent means and reside in the country, which would be ideal provided that he did not live there "indolently". At last Charles reached a conclusion on the reverse side of the page:

> I have so much more pleasure in direct observation, that I could not go on as Lyell does, correcting and adding up new information to old train, and I do not see what line can be followed by man tied down to London. — In country — experiment and observations on lower animals, — more space — (quoted in *Autobiography*, pp. 231–232)[8]

Charles had a "line" to follow, a threatening new theory to pursue, not an old book like Lyell's *Principles of Geology* to go on safely correcting and revising. Accordingly, if he married he would follow Henslow, Jenyns, and Fox into the country and there continue his research.

The renewed illness of June 1838 drove Charles into the country, unmarried, to geologize delightedly for the first extended interval since the voyage. At Glen Roy in Scotland, where he examined the famous "parallel roads", the weather was "most beautiful . . . with gorgeous sunsets, and all nature looking as happy as I felt", he later reported to Lyell. "My Scotch expedition answered brilliantly" (*LL* 1:293; Rudwick 1974a). From Scotland Charles made his way back home to Shrewsbury for a fortnight, from 13–19 July. There he was "very idle" — a Darwinian euphemism for the brain working overtime with little tangible results. For at home, safe and secluded, Charles merely seized the occasion to open his third notebook on transmutation (*D*) and his first on metaphysical subjects (*M*). Sixty-odd pages of the latter were filled in the fortnight, and the opening words, "My father says, . . ." set the prevailing tone. Evidently Charles had lengthy and earnest conversations with Dr. Darwin about subjects on which it would have been impossible, even if the young man had desired it, to conceal his heterodox views (Darwin 1959, p. 8; Gruber and Barrett 1974, p. 266). "Some notes from my father," his diary innocently records. Toward the end of these notes Charles wrote of waking unwell at night, beset with irrational fears. He went on to describe his symptoms alongside a strategy for concealing their implications:

> The sensation of fear is accompanied by/troubled/beating of heart, sweat,

> trembling of muscles, are not these effects of violent running away, & must not <this>/running away/ have been usual effects of fear. — the state of collapse may be imitation of death, which many animals put on. — The flush which accompanies passion, & not sweat, is the stated [?] effect of short, but violent action. —
>
> To avoid stating how far, I believe, in Materialism, say only that emotions, instincts degrees of talent, which are hereditary are so because brain of child resembles parent stock. — (& phrenologists state that brain alters). (*M* 53)

Whether the strategy for concealing "how far, I believe, in Materialism", as Charles put it, was Dr. Darwin's (the "parent stock"?) is debatable. But I think there can be little question that Charles received from his father other strategic advice during this time.

"Before I was engaged to be married," Charles recalled many years later in his *Autobiography*,

> my father advised me to conceal carefully my doubts, for he said that he had known extreme misery thus caused with married persons. Things went on pretty well until the wife or husband became out of health, and then some women suffered miserably by doubting about the salvation of their husbands, thus making them likewise to suffer. (p. 95)[9]

So, according to Dr. Darwin, it is the wife who suffers directly because of her husband's ill-health, if he fails to conceal his doubts. Now in July 1838 Charles was full of doubts, his health was poor, and his father was giving him much grave advice. What's more, he was visiting Shrewsbury for the last time, so far as is known, before his engagement. The conclusion seems inescapable that in this period Charles and his father finally resolved the problem of marriage and career. Dr. Darwin agreed to continue supporting Charles to the extent of about £400 per annum, and to increase the amount appropriately when he had taken a suitable wife, provided that Charles would continue to work diligently and make a reputation in the field of his first love, natural history (Darwin 1959, pp. 8–9; Keith 1955, chap. 18; Atkins 1976, chap. 10). The Church was not renounced. Rather, having been delayed, postponed, then ignored, a clerical career had been overtaken by events. The break was not "formal", to use Charles's later term, but it was nevertheless final.

Perhaps the best evidence that an arrangement such as this was reached in July 1838 is another utilitarian calculation by Charles, probably dating from the very month. Its heading, "This is the Question", suggests that "Work finished" in London, and possibly the acquisition of remunerative employment, was no longer a precondition for thinking about marriage. For the rubrics beneath read simply "MARRY" and "Not MARRY". Under the latter, ironically, Charles sketched chiefly the drawbacks, under the former the advantages. Despite losing time and forgoing "Choice of Society, and *little of it*", he found marriage an attractive proposition. The companionship — "better than a dog" — and the "charms of music and female chit-chat" would be "good for one's health". Domestic privacy, even while living in London, might be assured, assuming his wife would tolerate the city and was "an angel and made one keep industrious". The alternatives were "fighting about no Society" and perhaps her "banishment and degradation" with him in a premature retirement to the countryside, where he might well become an "indolent idle fool". But Charles would "trust to chance". There was still a beautiful old vision that seemed to foreshadow its own fulfillment:

> Imagine living all one's day solitarily in smoky dirty London House. — Only picture to yourself a nice soft wife on a sofa with good fire, and books and music perhaps — compare this vision with the dingy reality of Grt Marlboro' St. Marry — Marry — Marry Q.E.D.

On the reverse side of the page Charles asked himself the obvious question, "When?" His father had counselled "soon", and he was inclined to agree. The bonds of marriage had advantages, after all. "There is many a happy slave" (quoted in *Autobiography*, pp. 232–234).[10]

From this time onward Charles's course was set. He had become "conscious of himself as a being in time, that is, as an individual freed for certain kinds of action by opportunities present to him in the moment." Immediately, on 29 July, he visited Emma Wedgwood (1808–1896), his first cousin and future wife, at Maer in Staffordshire nearby. In London a few days later he started a diary and wrote a 1700-word account of his early life. He also began for the first time to date the entries in his transmutation and metaphysical notebooks, which went on filling apace (Herbert 1977, pp. 208–211; Darwin 1959, p. 8; *ML* 1:1–5).

On 12 August he recorded the headache he got the previous week from reading a review of Auguste Comte's *Cours de philosophie positive*. This may have come from contemplating the threat, perceived by the reviewer, that Comte's opinions were capable of "poisoning the springs" of morality and religion (*ML* 81; Schweber 1977, pp. 241ff). It may equally have arisen because reading about Comte's views on natural law and scientific progress served to stimulate and intensify those "grand ideas respecting the world" that Charles had first entertained on the voyage, ideas that had recurred to him the previous year when thinking of the laws of transmutation by analogy with those of astronomy (*B* 101–102). For now, on 16 August, he wrote:

> What a magnificent view one can take of the world [.] Astronomical causes modified by unknown ones, cause changes in geography & changes of climate suspended to change of climate from physical causes, — then

> suspended changes of form in the organic world, as adaptation, & these changing affect each other, & their bodies by certain laws of harmony keep perfect in these themselves. — instincts alter, reason is formed & the world peopled with myriads of distinct forms from a period short of eternity to the present time, to the future. — How far grander than idea from cramped imagination that God created (warring against those very laws he established in all organic nature) the Rhinoceros of Java & Sumatra, that since the time of the Silurian he has made a long succession of vile molluscous animals. How beneath the dignity of him, who is supposed to have said let there be light & there was light. (*D* 36–37)

On the same day Charles noted that "Metaphysics must flourish", and accordingly he "thought much about religion" in September and October 1838 (*M* 84; Darwin 1959, p. 8). This was accompanied by a vivid dream of capital punishment in which a brave man was hung or decapitated. Perhaps the "prisoner" of London had waited too long to plead that his views were not atheistic — although, he admitted in his notebooks, they might "tend to" be. Or perhaps even the apologist did not escape who claimed that it was "grander" to view the Creator's "magnificent laws" as the agents of an evolving material creation (*M* 69, 74, 136e, 154).

In this period Charles also filled his notebooks with observations of his own and other animals' sexual behavior. He was now courting Emma, and this gave rise to reactions that, like his fear and guilt, repaid careful analysis. Their engagement commenced on 11 November 1838; on the 14th there were great debates at Shrewsbury over "suburbs vs. central London" as the matrimonial abode (Charles, a self-confessed "solitary brute", favoring "retired places"); and on the 23rd Charles, full of connubial anticipation, wrote his Cambridge friend Whitley that he and Emma intended "living in London for at least some years, until I have wearied the geological public with my newly acquired cacoethese scribendi". This was, again, Dr. Darwin's advice. As Charles searched around Regent's Park and the central London squares for a suitable residence, he reconciled himself to forgoing "pleasures of the country (gardens, walks, etc.)" for a little while longer (Litchfield 1904 1:416; De Beer 1968, p. 84; Darwin 1945, p. 257).

On 29 January 1839 Charles and Emma were married in St. Peter's Church, Maer, by the perpetual curate, their mutual first cousin, the Reverend John Allen Wedgwood. They returned to London directly, settled in a big vulgar Georgian terrace in Upper Gower Street, and there began a slow withdrawal from society. Emma, pregnant within three months, entered a confinement not unlike the self-imposed one of her spouse. Snug together in their urban "cottage", they revelled in the "grandeur" of London's Autumn fogs. "We are living a life of extreme quietness," Charles wrote Fox in October; "Delamare itself, which you describe as so secluded a spot, is, I will answer for it, quite dissipated compared with Gower Street. We have given up all parties, for they agree with neither of us; and if one is quiet in London, there is nothing like its quietness". Nothing like it, that is, if one had the countryside to look forward to. Charles could "glory in thoughts that I shall be here for the next six months" because, having just spent five weeks in Maer and Shrewsbury, he was well apprised of the delights that the Spring held in store, when, after Emma's delivery, he would return there alone (*LL* 1:299; Darwin 1959, p. 9).

Meanwhile Charles's health deteriorated, giving further cause for escape. In these days Emma had other things on her mind besides himself, and to this Charles was unaccustomed. Then there were the usual anxieties over childbirth and, afterward, the added responsibility of a family. Finally, however, what made these matters so frustrating was the obsessive theorizing from which they detracted, the reading and note-taking that itself portended consequences, personally and professionally, that Charles had long since feared (Colp 1977a, pp. 26–27). A naturalist who could not leave off making notes, even on his wedding day, was bound to find his wife's first confinement at least moderately distressing. And a Malthusian who had departed London the previous Summer in the wake of Chartist riots might well have balked at fatherhood in the Winter while reading essays by Carlyle. "The dreadful but quiet war of organic beings" went on not only among the starving poor of London's snowy streets; it touched the soul of one who would make this the prevailing image of nature in late Victorian Britain (*E* 98, 114; Vorzimmer 1977, pp. 122–123).

So the "fighting about no Society" that Charles anticipated in his latter marriage memo went on chiefly inside himself. He won the battle with Emma, so to speak, by impregnating her, not just once, but a second time within six months of the birth of William, and a third time a year after the birth of Anne. As the family grew, in turn, domestic responsibilities made an increasingly plausible excuse for avoiding society. But Charles did not cease fighting about it; he devoted himself the more to his proper work, the zoology and geology of the voyage, which was detaining them in the city. Visits to the country were a sharp reminder, as he told Fox in July 1840, that "we shall never be able to stick all our lives in London". A year later, while visiting at Shrewsbury, Charles felt sufficiently roused to talk Dr. Darwin into helping him purchase a house. Then in the Autumn of 1841 he and Emma began seriously "taking steps to leave London, and live about twenty miles from it on some railway" (Colp 1977a, pp. 26–27; Atkins 1976, p. 19; *ML* 1:31; *LL* 1:302). There they would enjoy both proximity and seclusion. Charles need no longer offer excuses for avoiding society; he might see people on his own terms, just when and as he pleased. Also his neighbors there would be less likely to judge him for what he thought and wrote than for the kind of man he was. In a quiet country village, indeed, Charles could develop with impunity what he sketched out at Maer and Shrewsbury in the Summer of 1842, on the eve of purchasing his new estate. For it was there that he first allowed himself to give a full account of his theory in a pencil abstract of thirty-five pages (Darwin 1959, p. 10; *Autobiography*, p. 120).

V. The Village

The significance of Darwin's removal to the village of Down in Kent can scarcely be overestimated. It was both a strategic move and the fulfillment of a vision that had inspired the young man since his professional intentions took shape at Cambridge. The village, sixteen miles from St. Paul's, was a secure place to stand for a nervous young intellectual with a growing family, an Archimedean point, outside the fearsome flux of London, from which, given enough time and energy, he would move the world. Equally the village, eight miles from a train station, was a quiet country parish like Henslow's at Hitcham, like Jenyns's near Cambridge, like Fox's in Cheshire, like the one Darwin himself fondly imagined aboard the *Beagle*, the one in which his father may have proposed to install him, now fifteen years ago. This dual attraction of the village was not, however, paradoxical. In removing to Down, Darwin neither renounced his theorizing nor resisted the role of a country clergyman — "a type of all that is respectable and happy", as he had told Fox. For him these things were not incompatible, however heterodox at times his ideas might seem. Five years he had spent reconstituting the knowledge that in England was very largely the special preserve of clerical naturalists and natural theologians. He had been forging a "far grander" and more "magnificent view" of creation. Now, in the hungriest year of the "hungry forties", he had ventured a new theodicy, justifying the divine laws that lead to "death, famine, rapine, and the concealed war of nature" on the grounds that they produce "the highest good, which we can conceive, the creation of the higher animals". "The existence of such laws," Darwin declared in the pencil sketch of his theory, "should exalt our notion of the power of the omniscient Creator." What more appropriate place, then, to get on with this transfiguration of the conventional view of nature than a pleasant rural parish, a parish, Darwin rejoiced to Fox, situated "absolutely at the extreme verge of the world" (*1842 Sketch*, p. 87; *LL* 1: 321–322).

The parsonage, for one, might seem a more appropriate place. And that of course is the residence Darwin bought. Down House, an ugly flat-fronted property just one-third mile south of the village center, had been inhabited for a number of years by the late incumbent, the Reverend James Drummond (1800–1882). Drummond, the son of an officer in the Indian army, took charge of the parish in 1828. He lived first in a large house called Petleys near the church; then in 1837 he purchased Down House and, while improving the property considerably, farmed its fifteen acres in a desultory way. By the beginning of 1841 Drummond had left the parish, probably for a curacy at Highgate in north London, and his replacement, the Reverend John Willott (1813–1846), took a house elsewhere in the village. Having failed at auction, Down House remained empty until the Darwins moved in on 14 September 1842. Drummond received £2020 for the property, about 10 percent less than he had asked (Foster 1888, 1890; *Clergy List*, 1841; KAO P123/2/5; Howarth and Howarth 1933, p. 77; Atkins 1976, pp. 15, 17, 20–21). It was not purchased by Darwin *as* the ex-parsonage, to be sure. But the sum and substance of the transaction nonetheless signifies his social identification with country clergymen of a certain class. The villagers, at any rate, would soon have perceived the affinity, for within three weeks of the family's arrival Emma was delivered of her second daughter, Mary Eleanor, and the child was baptized by Reverend Willott in the parish Church of St. Mary the Virgin. A few days later the infant died. On 19 October Darwin, Emma, and the Reverend Willott gathered in the churchyard to inter the body in accordance with the rite of the Church of England (KAO P123/1/10,14).

Privately Darwin continued his work on species, and when the new Spring arrived he combined this unselfconsciously with the familiar role of parson-naturalist. Down without Darwin, we now realize, would have been as bereft as Selborne without the Reverend Gilbert White. But it is only seldom understood how largely the converse held true. Parsons in rural parishes often found themselves stranded for many years without personal contact with fellow minds. For those with scholarly interests, a substantial library and copious correspondence were generally their only solaces — until, that is, Reverend White showed them classically in 1789 how an incumbent could triumph over isolation by chronicling the natural history of his parish. White's *Natural History of Selborne* has been credited as the chief inspiration of field naturalists in the early nineteenth century and a main reason for the growing interest in natural history among the British upper classes during the same period (D. Allen 1976, pp. 21–22; L. Barber 1980, pp. 15, 41–44). Darwin read the book as a young man and learned from it to take delight and care in observing birds, a practice that paid off handsomely after he visited the Galapagos islands. His library contained a two-volume edition of White's *Natural History*, published in 1825 — it was no doubt the one he had read in his youth. But after coming to Down he acquired the new edition annotated by his friend Leonard Jenyns and published in 1843. Both copies contain Darwin's own annotations, and it is not unreasonable to think that the latter one gave rise to the "Account of Down" that Darwin began in May 1843 and continued for a full cycle of seasons. The manuscript left his son with the impression, at least, that he had "intended to write a natural history diary after the manner of Gilbert White" (*Autobiography*, p. 45; *ML* 1:33–36).

But playing the parson-naturalist and living in the former parsonage did not assuage Darwin's fears. Nor did his health improve. Was he not still a dangerous radical seeking asylum, though now taking a role for which he had been amply prepared? To the south a great chalk escarpment cut him off from the low country of Kent; to the north thick woods, a steep valley, and four hundred villagers stood between him and the road that led to London. On the west, across his field, was a large flat-bottomed valley, and on the east the village lane and open country. Gradually Darwin was detaching himself from the outside world, fearful of the emotional distress and ill health that would result from unplanned social contacts. Only the eastward approach to his inner sanctum remained unguarded. So he fixed a mirror outside the window of his study in order that visitors could be seen as they drew near the house (*ML* 1:31–33; *LL* 1:321–322; *Autobiography*, p. 115).[12]

Safely ensconced at Down, what Darwin feared most of all was censure and persecution by fellow naturalists. Only occasionally in subsequent years did his remarks evoke a larger social context. To his trusted friend J. D. Hooker, later Henslow's son-in-law, Darwin intimated in 1844 a fear of capital punishment: "I am almost convinced (quite contrary to the opinion I started with) that species are not (it is like confessing a murder) immutable" (*LL* 2:23). Again writing to Hooker, twelve years later, Darwin cast himself as a flagrant social offender: "What a book a devil's chaplain might write on the clumsy, blundering, low and horribly cruel works of nature!" (*ML* 1:94). *The* "Devil's Chaplain" was none other than the Reverend Robert Taylor (1784–1844), the Anglican clergyman turned infidel missionary who had outraged Cambridge in May 1829 and afterwards lectured notoriously in London until his second imprisonment (Aldred 1942; Royle 1974, pp. 38–40). Although the analogy between the career of a renegade parson and his own may not have escaped Darwin, what prompted his exclamation was surely the fear that his own identity would be mistaken. It was not Darwin of Down, after all, but a devil's chaplain who might write an infidel work on dysteleology. Similarly, what lay behind his earlier admission to Hooker was the fear that he would be punished unjustly. Darwin of Down was not, after all, a murderer. In the context of a wider Victorian public Darwin felt acutely the incongruity that arguments for the mutability of species, with all their possible implications, should come from a respectable parish scholar like him.

But Darwin had effectively isolated himself from the wider public. His intercourse was now selective, his public the villagers of Down and just so many correspondents as he ventured to make privy to his views. In the context of this restricted public he could act more readily to minimize any incongruity between his self-perception and the expectations of others. Darwin craved acceptance from his neighbors, he feared rejection from fellow naturalists; and his manner toward one and all was thus inordinately modest, upright, and sincere.

Some evidence of this appears, firstly, in the elaborate deferences and convoluted self-abnegations of the letters in which Darwin disclosed his work on species to colleagues. Although such sentiments occurred in letters before 1845, as the one "confessing a murder" indicates, it was only in that year that Darwin read with "fear and trembling" an eighty-five page review by his old professor Sedgwick, cataloging the moral, ideological, and scientific objections to the "development hypothesis" as set forth in the anonymous *Vestiges of the Natural History of Creation* (1844). The advantages of literary anonymity were never more painfully enforced (*LL* 1:344; Egerton 1970–1971).[13] In sharing his views with Jenyns later in the year Darwin therefore feared they would appear "absurdly presumptuous". "I know how much I open myself to reproach," he despaired. "I am a bold man to lay myself open to being thought a complete fool, and a most deliberate one." My "heterodox conclusion" that species are mutable, he told Asa Gray, "will make you despise me." "I always expect my views to be received with contempt." Fox's reaction to the proposal that he might publish these "wild and foolish . . . views" gave Darwin "another fit of the wibber-jibbers". And when the *Origin* finally appeared he warned Richard Owen, "I fear that it will be abominable in your eyes." To Hugh Falconer Darwin wrote at the same time, "Lord, how savage you will be, if you read it, and how you will long to crucify me alive. I fear it will produce no other effect on you" (*LL* 2:32, 34, 71, 79, 120–121, 216; De Beer 1968, pp. 77–78). Even Henslow had to be forewarned, "I fear . . . that you will not approve of your pupil in this case" (Darwin 1967, p. 200). A few years later, when another clerical naturalist, Charles Kingsley, canvassed the possibility of a work on the ancestry of humankind, Darwin still foresaw "how I sh^{d} be abused if I were to publish such an essay". "I shall meet with universal disapprobation, if not execution," he told St. George Mivart shortly before the *Descent* appeared.[14]

Respectability, then, was the paramount thing. Darwin had prepared for it all his life. Slowly, diffidently, and at a distance, he would convince a few able men and thus satisfy his need for professional approval. Then he would cower forebodingly, awaiting the reaction of the rest. On no account would he be a maverick like Chambers, the author of *Vestiges*, or an eccentric like Buckland, Lyell's mentor at Oxford, or a firebrand like his own latter-day defender, T. H. Huxley. His stomach could not bear it, for one. But his stomach could not bear it because, since the time in London when radical thoughts began polarizing in his mind, Darwin had failed to relieve the discrepancy between the respectable man he felt himself to be and the man he feared others would mistake him for if they knew or suspected his thoughts. His marriage and removal to Down were attempts to relieve the discrepancy by privatization, by reducing society to manageable proportions in conformity with the set of life-options represented by Henslow and the clerical naturalists of Cambridge. This strategy failed in part because the scientific world constantly leaked into

Down House through a steady trickle of periodicals and disconcerting letters. And sometimes Darwin ventured ill-advisedly afield. But I would suggest that Darwin also failed to consolidate his real or imagined identities because in removing to an idyllic country parish he took with him a major source of his problem.

Emma, his faithful and long-suffering nurse, the mother of his parson-sized family of ten — Emma was also the permanent household representative of a world full of critics whom Darwin feared. In one respect alone he had disregarded Dr. Darwin's advice: he had neglected to conceal his doubts. Apart from Erasmus and his father, Emma was probably the first to know of his secret work on species. He had told her before their marriage, and her devout, evangelical Unitarianism, so unlike the deistic brand of the Darwins', responded just as the Doctor had said (Litchfield 1904, 2:187; Chadwick 1966–1970, 1:396–397; McLachlan 1934). Although Emma would have dreaded "the feeling", she explained soon after their engagement, "that you were concealing your opinion from fear of giving me pain", it was nonetheless "melancholy" for her to think that "our opinions on the most important subjects should differ widely. . . . I feel it would be a painful void between us" (quoted in Brent 1981, pp. 255–256). That indeed is what it became. Not a year passed before Emma had written again to question whether those "honest & conscientious doubts" did not perhaps arise instead from absorption in one's work, the example of a free-thinking elder brother, and the scientific habit of "believing nothing till it is proved". Excruciatingly she concluded, "I should be most unhappy if I thought we did not belong to each other forever." This was a distinctly evangelical form of torture, the sabre-rattling of the soul. Sweetly, discreetly, obliquely, Emma evoked the perdition to which Christian England consigned its infidels when their views became publicly known. Darwin winced. "When I am dead, know that many times, I have kissed and cryed over this," he scrawled miserably at the bottom of the note (quoted in *Autobiography*, pp. 235–237).[15]

It is tempting to suppose that Emma was disappointed in her husband. They were first cousins, she a year older, and their friendship had begun in childhood. For ten years before they married she had known he was intended for the Church. She remembered well that breathless August day in 1831 when her father and brothers and sisters in one accord declared that the *Beagle* voyage would not be disreputable to his character as a clergyman. Then, after the voyage, she had seen him following up his discoveries and, as she put it in her note, "casting out as interruptions other sorts of thoughts which have no relation to what you are pursuing." This was indeed disreputable, and her betrothed confessed his doubts. But, fearing and hoping, she married him. They would go to a country parish like that of her elder sister Charlotte, whose husband was the incumbent. And there perhaps . . . But then came word in 1841 that the Reverend Langton had resigned his living, having lost his faith, and the family were removing to Maer. A precedent had vanished, or rather another one was set; and Emma, too, must resign herself to a lifetime's ministry without the noble title of parson's wife (R. B. Freeman 1978; M. Watt 1943).

It is tempting, I say, to suppose that Emma was disappointed in her husband's choice of career, but this account is too conjectural. In the absence of additional evidence I think it would be fairer simply to conclude that Emma's deep-seated faith silently impressed on her husband the need to "work out his own salvation" within her immediate precinct, the household and the parish. This was not of course any less his own intention than not writing atheistically or otherwise appearing disrespectable to his colleagues in the outside world. The parsonage, the parish, and their social appurtenances had long since become agreeably familiar to him in the company of the Henslows, the Jenyns, and the Foxs. Being a country clergyman had in fact seemed quite attractive at a still earlier time. Darwin took to the role of squarson like he took to placing epigraphs from Francis Bacon, Bishop Butler, and the Reverend William Whewell at the beginning of the *Origin*, or to introducing references to "the Creator" in the text (*Origin* 1959, pp. 753, 758–759): because it was a respectable thing to do; because he felt himself to be a respectable man and sincerely wished others, including Emma, to think so; and because the deed struck a responsive chord deep within his past.

VI. The Vicarious Naturalist

In applying to Darwin the semi-jocular term "squarson" I do not intend it in the strict sense of a clergyman who holds the position of squire in his parish. The appellation becomes increasingly anachronistic in the later nineteenth century, when the traditional social structure of country parishes was breaking down; and in any event Darwin did not take orders and so never performed the priestly duties of preaching, administering the sacraments, and home visitation. Nor, on the other hand, was he the local squire. There does not seem to have been a squire in the parish for many years. Sir John Lubbock, the principal landowner, had only recently come to live in the area, and his newly rebuilt residence, High Elms, stood a considerable distance outside the village. The original manor in fact lay just opposite the east side of Darwin's estate. Its name, Down Court, and its location suggest that it may have lent a certain prestige to the neighbor across the lane who had succeeded the Reverend Drummond.

And by contemporary standards this would have been a good thing. Any parish gained from having at least one "resident of education and public spirit", a landowning gentleman, usually called the squire, who would raise the moral tone of all whom he patronized and employed (Chadwick 1966–1970, 2:152; A. T. Hart and Carpenter 1954, p. 28). Sometimes this gentleman was the parish priest. But Down did not attract priests of this caliber. It was not a wealthy benefice: it had neither vicarage nor glebe. It enjoyed no squire as its patron, only the rector of Orpington nearby; and its value as a perpetual curacy in 1841 was a very modest £105 per annum (*Clergy List*, 1841). In these respects Down resembled the parish of Seathwaite in the Lake District, whose curate for sixty-six years, the scholar, small farmer, and omnicompetent Reverend Robert Walker (1709–1802), was the original of Wordsworth's "Pastor" in "The Excursion".

> In this one Man is shown a temperance — proof
> Against all trials; industry severe
> And constant as the motion of the day;
> Stern self-denial round him spread, with shade
> That might be deemed forbidding, did not there
> All generous feelings flourish and rejoice;
> Forbearance, charity in deed and thought,
> And resolution competent to take
> Out of the bosom of simplicity
> All that her holy customs recommend,
> And the best ages of the world prescribe.

Wordsworth, a close friend of Sedgwick, declined to follow a family tradition and enter the Church. Whether this knowledge prompted Darwin twice to read "The Excursion" while living in London can only be surmised (Addison 1947, pp. 124–128; Clark and Hughes 1890, 1:247–248; *Autobiography*, p. 85). But the Pastor he encountered there, like the "Solitary" and the "Wanderer", had much in common with the gentleman "farmer", as a local directory called him, who had ceased his global wandering and settled into a life solitary and pastoral at the old parsonage in Down (Manier 1978, pp. 89ff; Howarth and Howarth 1933, p. 81). He was the squarson of the parish because no other word so aptly captures the character of one who for forty years there exercised both genteel and clerical prerogatives.

By the mid-nineteenth century the typical country parson in the south of England was drawn increasingly from the middle classes. He was the best informed and most widely experienced man in the parish. He had acquired an Oxbridge degree, a wife, numerous offspring, and a carriage. His wife managed the vicarage, employing a staff of servants with whom the family gathered for prayers at regular intervals. She also did numerous "good works" about the parish such as ministering to the sick and delivering food and clothing to the poor. The daughters helped their mother, taught in the Sunday School, and occasionally married the curate. The sons followed their father to Oxbridge and thence into professions. The incumbent himself performed the sacred offices and remained on cordial terms with the squire. Together they "upheld the social pattern as it was and their own position in it" (A. T. Hart and Carpenter 1954, p. 31; G. Clark 1973, pp. 145ff; M. Watt 1943).

During this period, with rural depopulation, a new poor law, and the lingering specter of agrarian unrest, country parsons switched from hunting in their leisure to social administration. Many continued to mediate the forgiveness of God and the punishment of the state by serving as county magistrates — more than one in eight English and Welsh J.P.s in 1842 (Zangerl 1971–1972, p. 118). But typically the mid-century parson became deeply involved in conducting the wide variety of social services and financial institutions that would later be absorbed into state bureaucracies, municipal corporations, and large private firms. He took the initiative in the Sunday School, offering sacred and secular instruction cheaply to the children of the poor. He had responsibility for the weekday infants school and for the grant it received from the government until the advent of rates-supported elementary education after 1870. He provided uplifting recreational facilities, such as a parish library and a village reading room to combat not only ignorance but the drunkenness and immorality associated with public houses and the unwholesome literature available there. He encouraged self-help by starting a penny bank or by forming a provident society that would supply his working-class parishioners with things like bulk-price coal and clothing in exchange for regular savings. He was foremost among those who established friendly societies to help working men lay up for the remoter future and thus gain a stake in economic and political stability (Heeney 1976, chap. 4; Colloms 1977, pp. 26ff; Chadwick 1966–1970, 2:186ff; Clark 1973, chap. 4, pp. 182ff; Gosden 1961, pp. 88ff, 169).

These tasks the country parson undertook, sometimes with the squire's assistance, but always with yeoman's service from his own wife and family. The textbook example of the socially enterprising mid-century parson is John Stevens Henslow, the Cambridge professor who had begun reforming the parish of Hitcham in Suffolk even as his most famous pupil was settling at Down in Kent. Henslow started out with fireworks on the rectory lawn. He proceeded to establish a parish school at his own expense, to let allotments to laborers on the glebe land, and to organize a bi-annual horticultural show, complete with exotic exhibits under a marquee — "The Professor's Museum" — and lectures there by the Professor himself. The usual coal and clothing clubs, a public library, and a benefit society also originated under Henslow's energetic administration (A. T. Hart and Carpenter 1954, pp. 62–63; Clark 1973, pp. 173–175; Russell-Gebbett 1977, chap. 3). All this Darwin admired from afar, aiding, abetting, and even emulating where he could, "What good you must do to the present & all succeeding generations," he concluded a letter to Henslow in 1843. "Farewell, my dear patron." The day after moving into Down House, Darwin despatched a collection of Fuegian body paints for his "patron" to exhibit in the Professor's Museum. In 1845, the year he gave the name George to his second son, evidently after Henslow's first, Darwin purchased a large farm at Beesby in Lincolnshire and determined to have allotments arranged there, also after the example he had been set (Darwin 1967, pp. 151, 154–155).[16] A few years later he was asking directions about pyrotechnics. "The most important of my queries on Fire Works," Darwin wrote, is "*What sum of money will procure a fair village display*?" Most of his queries, however, dealt with botanical subjects. Under Henslow's tuition the school children of Hitcham were becoming experts in the local flora, and in the Summer of 1855 a party of little girls collected and counted seeds for Darwin's experiments. Later the "great growers of hollyhocks" among their elders received a list of questions, headed by a note to Henslow indicating, as Darwin put it, "that I am intimate with you & therefore a respectable person" (1967, pp. 157, 171, 176–177, 189).

In his own rural parish Darwin had no more need to prove his respectability than Henslow had in his. Darwin of Down and Henslow of Hitcham made a symmetrical pair. The Darwin infants were christened in the parish church;[17] Henrietta, the only daughter to wed, was married there in 1871;[18] and those who died in childhood, together with other relatives, were buried in the churchyard.[19] As the children grew up family prayers were held on Sunday, a ritual that Emma later discontinued on finding that the servants took little interest in it. But the family attended church regularly. They sat at the front in a large pew of their own, witnessing to their Non-conformist heritage by facing forward when the rest of the congregation turned toward the altar to say the Creed (F. Darwin 1920, pp. 52–53). Although Darwin himself eventually ceased to attend, there is evidence that his continuous support for the church and its activities began early on. About 1845 a list of subscribers was drawn up in the parish to pay £50 for embellishments of the church, including paint and repairs to the interior and relocating the organ. The list contains the signature of Darwin and records his pledge of £5. The heading of the list may also be in his hand, suggesting that he was responsible for soliciting the subscriptions.[20]

Outside the parish Darwin also maintained an image evincing a respectable and responsible station. In 1857 he was made a county magistrate. He served thus until the time of his death, and in 1929 his name was still to be seen on the tablet above the chairman's seat in the Police Court at Bromley, where the names of J.P.s since 1820 were inscribed.[21] About the same time, in the mid-1850s, Darwin undertook to have his younger sons groomed for higher education. William, the eldest, had attended the Bruce Castle School at Tottenham near London before completing preparations at Rugby to follow his father at Christ's College (*LL* 1:385; Colp 1977a, p. 49). The others — George, Francis, Leonard, and Horace — were entrusted to the tutorial care of various Anglican clergymen. Between 1856 and 1865 all of them attended at the vicarage of the neighboring parish of Hayes, where the Reverend George Varenne Reed (1816–1886) gave them lessons in Latin and arithmetic. George and Francis went on to the proprietary grammar school run by the Reverend Charles Pritchard (1808–1893) at Clapham in south London, an institution that emphasized mathematics and scientific subjects. Leonard joined the Royal Military Academy at Woolwich, where science was also stressed. Horace followed his elder brothers to Clapham, when the school had been taken over by the Reverend Alfred Wrigley (1817–1898), and afterwards he studied briefly under the Reverend R. C. M. Rouse (1832–1904), rector of Woodbridge in Suffolk, before again following George and Francis to Trinity College, Cambridge. Darwin made elaborate and costly arrangements to have his sons properly educated as gentlemen. Not least among his advisers were the Reverends Fox and Henslow, and the Reverend John Brodie Innes (1817–1894), vicar of Down (Moore 1977b, pp. 52–53, 67 n.10; *LL* 1:380–387; Darwin 1967, p. 198; Stecher 1961, p. 216).

An odder couple than Innes and Darwin does not appear in the annals of Darwin's long and variegated career. Innes, a Tory, a High Churchman, and an old-fashioned Bible-believing creationist; Darwin a Whig, a nominal Anglican of Unitarian stock, and a Bible-rejecting evolutionist. Innes, an intimate of Bishop Wilberforce; Darwin, the master of an unleashed bulldog named Huxley. Innes, who believed that "a man was made a man though developed into niggers who must be made to work and better able to make them, if those radicals did not interfere with the salutary chastisement needful, neglecting the lesson taught by the black ant slaves to the white"; Darwin, whose blood boiled at the deeds of slavery, "done and palliated by men, who profess to love their neighbours as themselves, who believe in God, and pray that his Will be done on earth!" (Stecher 1961, pp. 235, 256; *Journal of Researches* 1889, p. 500). Innes, the purveyor of animal anecdotes, natural history nonsense, and a design argument from the comb-making instincts of bees; Darwin, who graciously sifted the wheat from the chaff, encouraged Innes to observe the behavior of bees more critically, and sent him a copy of the *Origin* when it was first published.[22] Once before Darwin had lived at close quarters with a man of Innes's convictions, Captain Robert FitzRoy, and from that experience he had understandably recoiled. For FitzRoy possessed a brittle and arrogant temperament to match the fragility and presumption of his beliefs. But with Innes it was different. He was "one of those rare mortals", to use Darwin's own words, "from which [sic] one can differ & yet feel no shade of animosity" (Stecher 1961, p. 232).

It was not, however, merely Innes's gentility that made the relationship work. Darwin and Innes were united through mutual forbearance and admiration in a commitment to the well-being of the parish. Innes, the son of an army officer, came there from Trinity College, Oxford, via curacies in Northamptonshire and in the neighboring parish of Farnborough, where in 1842 he and Darwin first became acquainted. After the death of Willott in 1846 Innes became perpetual curate of Down. He inherited a Sunday

School and a Coal and Clothing Club, to which in 1850 or 1851 was added, thanks to Darwin's good offices, a benefit society (Foster 1888; *Clergy List* 1889; KAO P123/2/5; Stecher 1961, p. 203). Again Henslow seems to have been the chief inspiration. "I . . . believe I have succeeded in persuading our Clodhoppers to be enrolled in a Club," Darwin informed his old friend and adviser. Henslow replied with a long letter, full of suggestions, and Darwin passed it on to Innes, who was greatly obliged. In due course the Down Friendly Club was born, with Darwin as treasurer and guardian. He continued in this capacity until his death, retaining the confidence of members through the difficult negotiations following the Friendly Societies' Acts of the mid-1870s (Darwin 1967, pp. 166–168, 190–191; R. B. Freeman 1977, p. 157; R. B. Freeman, 1978, p. 128; Gosden, 1973, chap. 4).

Innes, meanwhile, was experiencing difficulties of his own. The Archbishop of Canterbury, J. B. Sumner, had assumed the patronage of Down from the rector of Orpington, and Sumner persuaded Innes to acquire it from him, perhaps in return for an elevation to the office of vicar. Innes took possession of the advowson at about the time, toward 1860, when he added "Brodie" to his name and became a Scottish landowner through inheritance. For some years, with a wife, a sickly son, and a menagerie of pets, Innes had sought in vain to purchase a house in the parish, or even a plot on which to build. Now, as he was obliged to retire to his property in Scotland, the matter became urgent. If he did not supply a vicarage it would be difficult to get a priest of appropriate caliber to replace him. By 1862 a house had not been found. Reluctantly Innes left the parish in nominal charge of the curate, a Cornishman named Thomas Sellwood Stephens. And he left Darwin as his virtual agent and administrator of temporal affairs (*Clergy List* 1854 etc; Stecher 1961, pp. 223–224; KAO P123/2/2).

Darwin's "years of controversy", the decade or more when his name, his books, and his reputation were constantly before the public eye, the period that saw perhaps his severest and most prolonged bouts of ill health — these also were the years when Darwin bore his heaviest responsibilities in the parish. Besides serving the Friendly Club he became treasurer of the Coal and Clothing Club, and he agreed with Innes to take over the accounts of the National School. The Sunday School accounts also appear to have fallen to Darwin, for there exists a small paperbound notebook with records from 1866 to the end of 1867 in his inimitable hand.[23] In Scotland Innes kept in touch with his curate, Stephens, and corresponded with Darwin rather more frankly, one may infer, about illnesses, deaths, romances, and pressing parish matters. The National School lacked proper supervision. Innes was delighted by the interest and involvement of the Darwins' elder daughter, Henrietta, whom he considered "my Minister of Education (non political)". There was also the perennial problem of a vicarage. When Innes learned that Emma's eldest sister, Sarah Wedgwood, intended purchasing property in the village, he had the Darwins intercede with her to see if she would sell a building plot.[24]

Most serious of all parish matters was the problem of unreliable curates (Appendix). Stephens, a competent man at best, had neglected the village school. After his departure in 1867 a Mr. Samuel Horsman arrived on the scene and took charge of rather more than his due. He ran up a string of bills and made off with the schools' money after Darwin had mistakenly shared treasurer's duties with him. Darwin thought Horsman "more an utter fool than knave"; Innes replied, "I really think he is mad."[25] But no sooner had he departed in 1868, vaporing threats of litigation, than his replacement, John Robinson, whom Innes had thought "little less than a saint", proceeded to disgrace himself by "walking with girls at night". Or so it was rumored, according to Darwin:

> I did not mention this before, because I had not even moderately good authority; but my wife found Mrs. Allen very indignant about Mr. R.s conduct with one of her maids. I do not believe that there is any evidence of actual criminality. As I repeat only second hand my name must not be mentioned. — Our maids tell my wife that they do not believe that hardly anyone will go to church now. (Stecher 1961, p. 226)

Robinson was soon despatched to a curacy at nearby Brasted, where, it seems, he may have teamed up with Horsman, who was on the point of taking Darwin to court. "It is too bad," wrote Darwin, who thought a cross-examination in court would "half-kill" him,

> for I remember explaining to Mr. R. why people would not pay subscriptions for School etc to him, only to me, and why persons did not call on him, as his predecessor had been a mere swindler. Poor little Mr. R. will look like a fool, if asked in court why he left Down. . . . Certainly we have been unfortunate in Mr. H. & Mr. R. Mr. Powell is, I think, a thoroughly good man & gentleman. Does good work of all kinds in the Parish, but preaches, I hear, very dull sermons. (Stecher 1961, p. 232)

In 1869 the Reverend Henry Powell (1840–1892) brought peace to the parish for a time and relieved Darwin of the Coal and Clothing Club and the National School. He was succeeded in 1871 by the Reverend George Sketchley Ffinden, who remained vicar of Down for forty years.[26]

Here my account of Darwin as squarson-naturalist of Down reaches its consummation. For in Ffinden Darwin encountered his FitzRoy a second time, a Tory High Churchman, ordained in 1861 by Bishop Wilberforce himself, a vicar who took charge of the parish as if it were his command. Ffinden's "clerical ability", Innes admitted, was not rated very highly, but his wife had "capital testimonials" as to her qualifications (Foster 1888; Stecher 1961, p. 46). This may explain in part why Emma was the first to draw the lightning. Over many years, while ministering to a "large clientele" of poor people in outlying areas, Emma had bestowed "wise and good forms of kindness" in her own parish: a lending library for the children, she herself giving out the books every Sunday afternoon; small pensions for the old; dainties for the ailing; and medical comforts and simple medicines in case of illness. "A deep respect and regard were felt for her in the village," her daughter Henrietta declared (Litchfield 1904, 2:181–184; Atkins 1976, p. 104; Raverat 1952, pp. 148ff). But in 1873, when Emma tried to perpetuate a recent initiative under the new régime, an evening reading room for working men held in the village schoolroom, Ffinden tried to persuade the School Committee, of which he was now chairman and treasurer, to withhold its permission to use the room on the grounds that it would be left in a condition "by no means salutary for the scholars". Darwin addressed the Committee, of which he had ceased to be a member under Ffinden, with a memo to the opposite effect, attaching a favorable report on Emma's proposal, which he had solicited from the office of the Privy Council in London. The Committee ruled in favor of the Darwins and Ffinden communicated the result to Emma (24 December 1873) in the curtest of notes.

> I learnt at the Meeting that Mr Darwin had addressed the Education Department on the subject. As I am the only recognized correspondent of the School according to rule 15. Code 1871, I deem such a proceeding quite out of order, especially as I myself had undertaken to communicate with the Office. For your information I now append a Copy of the answer I received, which is, in my opinion, by no means favourable to your view of the case. (KAO P123/25/3)[27]

Innes was still Ffinden's patron, and Emma encouraged him to help ease the vicar out. "This would certainly be a great blessing to this place, as Mr. Ffinden has no influence here & has excited general dislike. . . . You will not think me an impartial person perhaps as he cuts every member of our family when we meet" (Stecher 1961, p. 239).

Emma drew the lightning; her husband caught the thunder. Early in 1875, when Ffinden and the Darwins were no longer on speaking terms, John Lubbock asked on their behalf for the loan of the schoolroom for two evening lectures to the villagers. Lubbock in fact owned the schoolroom and rented it to the Committee, but Ffinden did not allow this knowledge to temper his reply.

> Allow me to observe that I have been so grossly insulted on more than one occasion by the Darwin family that I cannot be expected to take any interest in anything that concerns them. But as the request you make is really one for the decision of the Schl Committee, I have referred to them & finding they see no particular obstacle to the use of the room

> I notify their consent but for what are good & sufficient reasons, you will excuse me if I withhold my own.

Lubbock answered with a plea for reconciliation with the Darwins and an offer to mediate. Ffinden countered with complaints that his authority had been impugned, for he was "the only recognized correspondent at the Council office" in respect of school affairs. He added,

> I had been long aware of the harmful tendencies to the cause of revealed religion of Mr. Darwin's views but on coming into this parish I had fully determined, as far as lay within my power, not to let my difference of opinion interfere with a friendly feeling as neighbours, trusting that God's Grace might in time bring one so highly gifted intellectually & morally to a better mind. For allow me to say: of one thing I am convinced that neither Socinians nor Infidels can feel *quite sure* of their ground.

Lubbock ignored the special pleading and continued to proffer Darwin's reasons for his conduct and the family's desire to "return to the ordinary courtesies of life". But Ffinden's churchmanship remained implacable toward "uncalled for interference".[28] Darwin meanwhile told Innes, "We never cease to wish you had not left us" (Stecher 1961, p. 242).

The controversy threatened to break out again in 1879 when there was a dispute over the extent of the land Ffinden had purchased from Sarah Wedgwood in 1872 for the erection of the long-awaited vicarage. This time the principals, Ffinden and Darwin, exchanged polite letters.[29] But a year later Francis Darwin embroiled himself over the election of a new chairman of the School Committee, on which he now served. Ffinden tendered his resignation from the Committee, not to Francis but to his father. Only on hearing from Francis that the Committee wished him to reconsider did Ffinden state his objections to the election of a Mr. George E. Forrest as chairman:

> You must be aware that, as Vicar of this Parish I am its *legally* constituted head: it would therefore be utterly incongruous for me to accept a position at any meeting here subordinate to Mr. Forrest, neither do I intend to do so. . . . It seems to me that it is certainly no essential, & perhaps hardly a becoming part of a Priest's Office "to serve tables," if you can understand what I mean as a High Churchman.[30]

Ffinden apparently did not withdraw his resignation, and his name disappeared from the School's annual reports (KAO P123/25/5). Nor was this his only defeat. On Christmas Eve 1880 Emma told Innes triumphantly, "The great event last week was the opening of a reading room. . . . We have also a Band of Hope which is of course prosperous at present. . . . Both these undertakings are thorns in Mr. Ffinden's side & he has not been content with holding aloof from them; but has used all his influence to prevent their succeeding" (Stecher 1961, p. 248).

Perhaps this explains why Darwin was so anxious within a few weeks to lend the reading room — the hotly contested old schoolroom, which he himself now rented from Lubbock — to the young evangelist J. W. C. Fegan (1852–1925), a Plymouth Brother, for his revival services. Darwin wrote to Fegan:

> You ought not to have to write to me for permission to use the Reading Room. You have far more right to it than we have, for your services have done more for the village in a few months than all our efforts for many years. We have never been able to reclaim a drunkard, but through your services I do not know that there is a drunkard left in the village.
>
> Now may I have the pleasure of handing the Reading Room over to you? Perhaps, if we should want it some night for a special purpose, you will be good enough to let us use it. (quoted in Fullerton 1930, p. 30)[31]

Members of the Darwin family were sometimes present at Fegan's services, and the family altered their dinner hour so that other members of the household could attend. Parslow, the old family butler, and Mrs. Sales, the housekeeper, were converted; and when news reached Emma in February 1881 that a notable old drunkard "made as nice a prayer as ever you heard in your life," she exclaimed to her daughter Henrietta, "Hurrah for Mr. Fegan!" (Litchfield 1904, 2:313).[32] Surely the continuance of Non-conformist church services in that reading room — latterly called the "Gospel Room" — for more than half a century afterward is a strange and ironic witness to the parish churchmanship, though hardly the religious beliefs, of the squarson-naturalist of Down.

VII. The Valedictory

Within a few years of Ffinden's coming to Down, Emma and the other members of the family who attended church transferred their allegiance to the parish church of Keston, two miles away (Atkins 1976, p. 48).[33] No doubt Emma returned to Down at least once, when Ffinden buried her sister Sarah in 1880; and in 1896 Ffinden may have obtained a certain satisfaction in conducting the funeral of Emma herself. Otherwise the only evidence of the Darwins' participation in Ffinden's ministry at Down lies in accounts of their contributions to the church: £50 in 1872, dropping to £10 the next year, and dwindling to £7-5-0 from the head of the family in 1878, with appropriately smaller donations from the rest. The extant accounts are incomplete, but other records show that Darwin and the family also gave modestly toward a clock for the church, toward restoration of windows, and toward new surplices for the choir. In 1876 Darwin matched the £25 contributed by Sir John Lubbock to the Downe Vicarage Endowment Fund, an act compatible with support for Innes and the institution as much as the vicar.[34] For although Innes had offered the advowson to Darwin in 1868, he remained the patron of the parish;[35] and there is no evidence that Darwin ever deviated from the conviction for which Innes praised him, "that where there was no really important objection, his assistance should be given to the Clergyman, who ought to know the circumstances best and was chiefly responsible" for the life of the community (Innes in DAR 112 (ser. 2): 85–92; cf. Stecher 1961, p. 255).

It is hard to imagine an attitude farther removed from the crusading anti-clericalism of Darwin's followers after 1859. Men of Huxley's mettle had not enjoyed the comforts of hereditary wealth, social seclusion, and a respectable life as squarson-naturalist in a small agricultural village. On the contrary, they had known poverty and discrimination at the hands of the very society into which Darwin fitted so well, the clerically dominated old order that resisted the rise of a new élite of professional interpreters of nature. Darwin belonged to the old order so far as his personal life was concerned. Once removed from the professional London of Lyell and Erasmus, he had more in common with the Reverend Charles Kingsley (1819–1875), rector of Eversley in Hampshire, whose support for evolution is quoted in the second and subsequent editions of the *Origin* (1959, p. 748), or with the Reverend Octavius Pickard-Cambridge (1828–1917), squarson of Bloxworth in Dorset for forty-nine years, whose researches on spiders figure prominently in the *Descent* (1874, pp. 255n, 273; Darwin 1967, pp. 213–214; Colloms 1977, chap. 4), or with his cousin Fox, or with Henslow, or Jenyns or any number of other clerical naturalists whose compendious knowledge of nature Darwin assimilated and reconstituted in the old parsonage at Down. It was not the least ironic that a man such as he should introduce his greatest work with quotations from Bacon, Butler, and Whewell. Indeed, he might almost have said that the *Origin* was "like a Bridgewater Treatise", as he described his next book, *Orchids*, to the publisher (*LL* 3:266).

"Almost", I say, because Darwin's theories, like his life, did not belong exclusively to the old and vanishing pre-industrial order, where a static natural theology justified the ways of man to man. The squarson-naturalist of Down stood between the times. He was neither a clergyman *manqué* nor a professional scientist in the manner of his later-Victorian followers, but a sort of transformed "vicar" in the root sense of the word, the mediator of a struggling, improving, but law-bound nature to a struggling, improving, but law-abiding society. His half-first cousin, Francis Galton, thought him the "hereditary genius" *par excellence*, the father-superior of a new "scientific priesthood" whose ministrations in society would elicit the "religious significance of the doctrine of evolution" (Galton 1874, p. 260; Galton 1883, p. 220). Yet in completing the questionnaire for Galton's *English Men of*

Science (1874) Darwin stated that he was affiliated "nominally to Church of England" and that he had given up "common religious belief almost independently from my own reflections" (Hilts 1975, pp. 10–11). The *un*common religious belief that remained was quite compatible with nominal Anglicanism and it fully justified Galton's vision of a secular priesthood of scientific experts. Its essence was the numinous sense of Nature as the source and ground of being, the arbiter of human nature, human conduct, and human destiny. Nature had long been the natural theologians' locution for God. Darwin merely exemplified their piety in framing his "grander" idea of creation. As the author of a Bridgewater Treatise put it,

> In order to avoid the too frequent, and consequently irreverent, introduction of the Great Name of the SUPREME BEING into familiar discourse on the operations of his power, I have, throughout this Treatise, followed the common usage of employing the term *Nature* as a synonym, expressive of the same power, but veiling from our feeble sight the too dazzling splendour of its glory. (Roget 1840, 1:11)

Only Darwin went a step farther. Nature to him was something in itself, not just a synonym for transcendent divine power, and he attributed the creation solely to the action of God's immanent natural laws. This belief was not necessarily irreverent, but it could be irreligious according to the standards of the time. It certainly gave the nod to those who would rule by the Book of Nature over those who ruled from Scripture and tradition.

Darwin would not have seen the controversies occasioned by the publication of his secret research as an ideological struggle, perhaps not even as a struggle between established and ascendent social élites. Nevertheless he would have wished the controversies to proceed with the decorum and equanimity of his relations within the parish. At first, when his private thoughts threatened to make him a social offender, it was the values of English middle-class Non-conformity expressed in the desires of Dr. Darwin and in his own professional ambitions that counselled respectability. Latterly it was the same sanctions within the framework of naturalistic religion. The recollection of Darwin by his eldest son, William (4 Jan. 1883), has such vividness and authenticity as to bear repeating in full:

> A very strong characteristic was his deep respect for authority of all kinds and for the laws of Nature. He could not endure the feeling of breaking any law of the most trivial kind, even the most harmless form of trespassing made him uncomfortable and he avoided it.[36] He both felt and liked to shew his respect for the position and title of others, and was very careful in addressing them by letter in the proper form. . . . This feeling partly explains his great respect for a title; but in addition to this he had an instinctive admiration for an old title and old family, and used laughingly to say how deeply he admired a Lord as regard[s]

> his respect for the laws of Nature[.] It might be called reverence if not a religious feeling. No man could feel more intensely the vastness or the inviolability of the laws of nature, and especially the helplessness of mankind except so far as the laws were obeyed. He had almost a terror of any infringement however slight of the laws of health, & he would laugh at one as being illogical for such a remark as "just one glass of port", can do no harm. . . Though obeyance to the natural laws and a deep sense of the power of nature may be called in his case a religious feeling, [In one sense *del*] he had no religious sentiment. (DAR 112 (ser. 2): 3c–d)[37]

Here, then, is a chief reason why Darwin got on so well with clergymen that he could even, to some extent, assume their social role. Whatever their theological, or terminological, differences over the ground of moral conduct — some called it Nature, others called it God — Darwin upheld the natural, the existing, order of society. One who served as a county magistrate, founded a Friendly Club, and became treasurer of a parish Sunday School was an agent of the very mechanisms of subordination that were tended by country parsons in mid-Victorian England. Together, increasingly, they could find Samuel Smiles' *Self-Help* (1859) to be a "goodish" statement of the dominant ideology they subserved (Vorzimmer 1977, p. 153; cf. R. Gray 1977 and J. Hart 1977).

But theology counted for something. The Reverend Ffindens of Victorian Britain made the point notoriously, and their arguments and conduct should not be lightly dismissed. Darwin was a threat to Ffinden because, although the parish naturalist had a "deep respect for authority", he openly challenged its traditional rationale. The *Descent* was published, after all, not nine months before Ffinden came to Down. Darwin's views ran counter to "revealed religion"; his participation in parish life — on the School Committee and through sponsorship of a reading room and public lectures — promised to erode or usurp the leadership of its recognized interpreter. So, in this respect at least, Ffinden's fears were not groundless: the parish of Down was a microcosm of society at large. And in April 1882 Ffinden no doubt found it especially galling to be passed over for Innes as the family's choice to perform Darwin's last rites,[38] then to see the body transported to the Church's noblest shrine at the behest of scientists, churchmen, and politicians, the intellectual aristocracy of the land, who rejoiced in the new rationale that Darwin had afforded their growing professional leadership. For in Westminster Abbey the Church not only reclaimed its erstwhile son; it acknowledged Nature, as Darwin had, to be an authority above itself, and interpreters of Nature to be the mediators of God's will to humankind (Annan 1955; Moore 1982b).

Whether more than Darwin's body was interred that fateful day, a century ago, remains an open question. It will not be answered satisfactorily by spurious appeals to authority and tradition, as Ffinden might have made. Nor, perhaps, will its relevance even be appreciated by those who devote themselves to Nature after the manner of Darwin. Only those can tell who transform and transcend the natural theology of the day, not as Darwin did in his, in parochial isolation, but amid the struggles and commerce of the collective life engaged.

Appendix

Clergymen serving the Parish of St. Mary the Virgin Church, Downe, Kent, from 1828 to 1911

The following biographical details have been compiled from: *The Clergy List*, 1841ff; *Crockford's Clerical Directory*, 1858ff; Foster 1888 and 1890; Venn 1940–1954; Stecher 1961; and sources in the Kent Archives Office (Maidstone), P123/2/1–5.

JAMES DRUMMOND b. 1800?; Christ Church College, Oxford, 1818; B.A. 1823; M.A. 1825; perpetual curate of Down, 6 Feb. 1828; curate of Highgate, London?, 1840; curate of Achurch nr Oundle, Northants, 1845; rector of Thorpe-Achurch nr Oundle &c., 1852; hon. canon of Peterborough and rector of Thorpe-Achurch &c., 1854; rector of Gaulby, Leics., 1859–77; d. 18 Nov. 1882.

JOHN WILLOTT b. 1813?; St John's College, Cambridge, 1831; B.A. 1835; M.A. 1838; ord. deacon 1836; ord. priest 1837; perpetual curate of Down, 13 Jan. 1841; d. 8 Mar. 1846.

WILLIAM HICKEY b. 1788?; Trinity College, Cambridge, 1803; Trinity College, Dublin, 1804; St. John's College, Cambridge, 1806; B.A. 1809; M.A. (Dublin) 1832; ord. deacon 1811; rector of Dunleckny, Leighlin, 1811; vicar of Bannow, Ferns, 1820; rector of Kilcormick, 1826; rector of Wexford, 1831; rector of Mulrankin, 1834–75; curate of Down (Kent?), 1841–(48?); agriculturist, author &c.; d. 24 Oct. 1875.

JOHN (BRODIE) INNES b. 1817?; Trinity College, Oxford, 1835; B.A. 1839; M.A. 1842; ord. priest 1840; curate of Corby nr Wansford, and Stanion, Northants, 1841?; curate of Farnborough, Kent, 1842; perpetual curate of Down, 1846; vicar of Downe, c.1860–69; chaplain to bishop of Moray, Ross, and Argyle, 1840–46 (Bishop Low), 1861–80 (Bishop Eden); curate-in-charge of Milton-Brodie mission, diocese of Moray, c.1879ff; chaplain to the primus of Scotland, the bishop of Moray, Ross, and Caithness, c.1886ff; d. 19 Oct. 1894.

JOSEPH OLDHAM b. 1821?; St. John's College, Cambridge, 1845; B.D. 1855; ord. deacon 1845; ord. priest 1846; curate of Walthamstow, Essex, 1845; curate of Down, 1848; vicar of Clay Cross, Derbys, 1851; rector of North Wingfield, Derbys, 1888; d. 2 Aug. 1896.

THOMAS SELLWOOD STEPHENS b. 1825?; Worcester College, Oxford; B.A. 1847; M.A. 1850; ord. deacon 1848; ord. priest 1849; curate of Wanstead, Essex, 1853; curate of Downe, 1859; rector of St. Erme, Cornwall, 1867; d. 1904?

SAMUEL JAMES O'HARA HORSMAN Trinity College, Dublin; B.A. 1857; M.A. and LL.B. 1864; admitted *comitatis causâ*, Oxford, 1864; ord. deacon 1858; ord. priest 1860; curate of All Saints, Northants, 1858; curate of St. Matthew's, Rugby, 1860; asst minister and acting chaplain to the forces, Stirling Castle, 1862; curate of St. Philip's, Liverpool, 1864; curate of Acton Trussell, Staffs, 1865; curate of Downe, 2 Aug.–10 Nov. 1867; curate of St. Luke's, Marylebone, London, 1868; curate of St. George the Martyr, Southwark, London, 1880; curate of St. Mark's, Regent's Park, London, 1883; rector of Condicote, Gloucs, 1884; d. 1887?

JOHN WARBURTON ROBINSON Trinity College, Dublin; B.A. 1859; M.A. 1863; admitted *comitatis causâ*, Oxford, 1864; ord. deacon 1860; ord. priest 1864; curate of St. Mary's, Haggerston, London, 1864; curate of Downe, 30 Aug. 1868–4 Feb. 1869; curate of Brasted, Kent, 1869; curate of Lynsted nr Sittingbourne, Kent, 1872–74; curate of Blisworth, Northants, 1876; emigrated to Melbourne ?, Australia, c.1882.

HENRY POWELL b. 1840?; Clare College, Cambridge, 1857; B.A. 1861; M.A. 1865; ord. deacon 1862; ord. priest 1864; curate of Chertsey, Middx, 1862; curate of St. Luke's, King's Cross, London, 1864–65; curate of Mortlake, Surrey, 1866; vicar of Downe, 12 April 1869; vicar of Oatlands,

Surrey, 1871–72; rector of Lavendon with Cold Brayfield, Bucks, 1874; rector of East Horndon, Essex, 1875–80; rector of Stanningfield, Suffolk, 1882–91; d. 20 Mar. 1892.

GEORGE SKETCHLEY FFINDEN King's College, London; Assoc. (1st class) 1859; ord. deacon 1860; ord. priest 1861; curate of Monks Risborough, Bucks 1860; curate of Newport Pagnell, Bucks, 1861–62; curate of Moulsoe, Bucks, 1863–69?; vicar of Downe, 2 Nov. 1871–1911; domestic chaplain to Lord Carington, 1873(?)ff; d. 1911?

? HOOLE curate of Downe, 1877(?)ff.

Notes

APS=Library of the American Philosophical Society, Philadelphia
CMLA=Cleveland Medical Library Association, Cleveland
DAR=Darwin Archive, The University Library, Cambridge
KAO=Kent Archives Office, Maidstone, Kent

1. Haeckel translated the letter to Nicolas Baron Mengden (5 June 1879) in a note to a published lecture and the *Deutsche Rundschau* reprinted the translation. Katherine Macmillan sent a re-translation into English to the *Pall Mall Gazette* and Charles Bradlaugh's *National Reformer* reprinted the re-translation (Darwin, 1882). Other periodicals also joined in the fray (cf. Lewins 1882; Aveling 1882a; and McCrie 1891). There was some correspondence in *The Academy* about whether the translation had been accurate, and Francis Darwin, in preparing his father's *Life and Letters* (*LL* 1:307), obtained a transcription of the English original by Richard Hodgson of St. John's College, Cambridge. Its authenticity was certified by Mengden and Haeckel in their own handwriting (DAR 139.12).

2. Cf. De Beer's assessment of Darwin's career: "Brought up as a country gentleman to be thoroughly familiar with cultivated plants in spacious gardens and with domestic animals in kennels and stables, passionately fond of riding and shooting, he lived in a country where academic failure did not ruin his career, speed of work counted for nothing, private means could support him without his having to earn his living, and liberal institutions protected him from persecution for his unorthodox ideas. Nowhere but in England would such an environment have been possible in the first half of the nineteenth century, and Darwin played the part perfectly" (1963, pp. 275–276).

3. Darwin's lengthy *précis* of J. Sumner's argument in *Evidence of Christianity* (1824) is written on ordinary paper watermarked SLADE 1822 (DAR 91:114–117) and on tracing paper watermarked I ANNANDALE 1815 (DAR 91:118). The SLADE paper was also used for Darwin's "Early notes on guns and shooting" (DAR 91:1). These notes could be dated as early as 1826 (cf. *Autobiography*, p. 54) or as late as 1830. What suggests a date of 1831 for the notes on Sumner's book is the series of backwards (but not inverted) question marks in Darwin's hand opposite one paragraph (DAR 91:115). These may have been an attempt at the Spanish form of punctuation. Darwin studied Spanish in the Spring of 1831 in preparation for the trip to Tenerife. I am grateful to Sydney Smith for some suggestions on this point. On Darwin's reading of Sumner (not Paley, *pace* Herbert 1974, pp. 218–219 and Manier 1978, p. 212 n.3), see Gruber and Barrett (1974, pp. 125–126).

4. *Diary*, p. 75; R. Hamond to F. Darwin, 19 Sept. 1882, in DAR 112 (ser. 1): 54–55 (cf. *Diary*, pp. 110–111 and Darwin 1945, p. 78, where together Darwin and Hamond also ogle the señoritas); information on the New Zealand church (or chapel — cf. *Diary*, pp. 364ff) supplied by R. B. Freeman in a private communication.

5. Recollections of F. Darwin in DAR 140.3:23. See also the recollections of William Darwin, 4 Jan. 1883, in DAR 112 (ser. 2) 3a–e; the evidence of Darwin's avid reading of *The Times* and the *Daily News* in Colp (1978); and Darwin to Emma Darwin, April [1858], in *LL* 2:114 where Darwin reads at leisure the Chief Justice's summing up in the trial of Simon Bernard as an accessory to Orsini's attempt on the life of the French emperor. Darwin thought Bernard should be convicted.

6. On Darwin's fear of persecution, I agree in general with the analysis in Gruber and

Barrett (1974, chap. 2), which could be carried much farther. I agree with Schweber (1977) that Darwin understood the threat posed by his materialistic speculations, but I doubt whether this initially had much to do with Emma, and I deny that this was because Darwin ever thought of himself as an atheist or agnostic at this stage in his life. I agree substantially with Manier (1978, 1980a) in analyzing the perceived social and cultural implications of Darwin's work from 1837 to 1842, but I deny that the historian should rest comfortably in an analysis that omits to connect Darwin's work with the period's economic and social context (which Manier (1980b) and Brent (1981, p. 321) understand but find somewhat irrelevant). I agree with Richards (1981) that Darwin's materialism was not incompatible with all kinds of theism, but I seriously doubt whether the chief sources of Darwin's anxiety were the "conceptual obstacles" he had to overcome if his theory was to be made scientifically respectable. No amount of analysis, however, can improve on Richards' statement, "That Darwin should not have feared suspicions of materialism, of course, does not mean that he did not" (p. 229).

7. I date these pages shortly before the Summer because of their proximity to *C* 224ff, which from internal evidence could not have been written before 2 June 1838. Cf. Darwin (1959, p. 8).
8. Here Barlow transcribes the memo in DAR 210.10. She places the conclusion in the right-hand column, neglecting to point out that it appears on the reverse side of the original. The *terminus a quo* for the memo is 7 April 1838. This has been established from references in the old letter on which it is written: viz., to the following Monday when the government would bring the Factory Amendment Act into the House of Commons, according to the writer, Leonard Horner, Lyell's father-in-law.
9. It is noteworthy that in the autobiography Darwin placed the section on "Religious Belief" chronologically in the middle of his London period, after an account of the Glen Roy expedition but before any mention of his marriage. The section on religious belief finishes with a paragraph about Dr. Darwin's advice that includes the passage quoted.
10. In her transcription of the memo (in DAR 210), Barlow makes this the "second" of the two memos on marriage, and I believe this is chronologically correct. Marriage was contingent on "Work finished" in the first memo. In the second, marriage alone is "The Question". When his work was finished Darwin would still remove to the country, but in the meantime marriage was not incompatible with city life, presumably because of an arrangement with Dr. Darwin.
11. Atkins (1976, p. 24) gives Darwin's manuscript the title "The General Aspect". Darwin's copies of White's *Natural History of Selborne* are among the volumes of his library received in the University Library, Cambridge, in 1961. See the 1826 diary in DAR 129 for evidence of White's inspiration.
12. The mirror is now on display in The Charles Darwin Room in Down House. To begin with, at least, Darwin seems to have been bothered by public footpaths intersecting his property (*ML* 1:31–33).
13. Darwin read *Vestiges* in November 1844 and the sixth edition in August 1847 (Vorzimmer 1977, pp. 132, 138). He found much objectionable in the book but, in retrospect, he thought it had "produced a good effect on the public mind" by "removing prejudices" (De Beer 1959b, pp. 53, 54).
14. Darwin to C. Kingsley, 6 Feb. [1862], CMLA; Carroll 1976, p. 137 (no. 375). See also Carroll (1976, p. 39 (no. 103); *LL* 2:138, 152) and De Beer (1958, p. 104).
15. The manuscript of Emma's note (DAR 210.10) is written on paper watermarked "W. Warren 1837", the same paper used by Darwin for his second marriage memo, probably written in July 1838. Emma's note may therefore have been written very early in the marriage.
16. Darwin's first boy, William, may have been named after his cousin William Darwin Fox. Leonard, the fourth boy, seems to have been named with Leonard Jenyns in mind (Darwin 1967, p. 166).
17. "Register of baptisms in the parish of Down . . ." (KAO P123/1/10) gives the following names, dates, and officiating clergymen:
    Mary Eleanor Darwin 2 Oct. 1842 Willott
    Elizabeth Darwin 10 Oct. 1847 Innes
    Leonard Darwin 5 Oct. 1850 Oldham
    Horace Darwin 28 Sept. 1851 Edwin Day [?]
    Charles Waring Darwin 21 May 1857 offic. minister
    William was probably christened in London, where he was born, Henrietta and George perhaps at Shrewsbury. Francis, who was christened at Malvern, mentioned that the godfathers and godmothers were "usually uncles and aunts, but this tepid relationship was deprived of any conceivable interest" because, as rumor had it, "the uncles were usually represented by the parish clerk"

(F. Darwin 1920, p. 51). Darwin himself was godfather to T. H. Huxley's eldest son, Leonard, born in 1860 (T. H. Huxley to G. H. Darwin, 22 April 1882, in DAR 215).

18. "Register of marriages" (KAO P123/1/13) gives the date of 31 August 1871, with William and Leonard Darwin as witnesses. Their father attended with some difficulty (Litchfield 1910, p. 124) and the officiating clergyman was Vernon Edlin [?] (see Stecher 1961, p. 220 where a Vernon "Salin" or "Palin" is mentioned).

19. "Register of burials in the parish of Down . . ." (KAO P123/1/14) gives the following names, dates, ages, and officiating clergymen:
Mary Eleanor Darwin 19 Oct. 1842 3 weeks Willott
Charles Waring Darwin 1 July 1858 18 months Innes
Sarah Elizabeth Wedgwood 11 Nov. 1880 86 years Ffinden
Erasmus Alvey Darwin 1 Sept. 1881 76 years A. Wedgwood
Emma Darwin 7 Oct. 1896 88 years Ffinden
Elizabeth Darwin died on 19 August 1926, aged seventy-four, and her ashes were placed in the churchyard at Down. Henrietta died on 17 December 1927, aged eighty-four, and her ashes were likewise buried, with Gibson as the officiating clergyman.

20. "Subscription towards embellishments of the church according to the estimate furnished by Mr. Hardwick", in KAO P123/6/1–2. The date is ascertained by noting that one of the subscribers, the incumbent John Willott, died on 8 March 1846 (KAO P123/2/5) and that the paper itself has an 1845 watermark. Francis Darwin could only remember his father attending church at the christening of Charles Waring Darwin in 1857 and at his brother Erasmus's funeral in 1881. His father's routine on Sunday was exactly the same as on any other day (recollections of F. Darwin in DAR 140.3:79).

21. R. B. Freeman (1978, p. 178); De Beer (1959b, p. 37); Darwin anniversary, *Bromley and District Times*, 15 Feb. 1929, in Baxter newscuttings IX/48, Bromley Central Library (Bromley, Kent). See *LL* 2:225–226 and F. Darwin (1920, p. 58).

22. On Innes's interest in bees, which probably dates from the period 1854–1861, when Darwin carried out field observations on the insects with help from his children (R. B. Freeman 1968), see Carroll (1976, p. 55 (no. 149); recollections of J. B. Innes, in DAR 112 (ser. 2): 89; and Innes's copy of the first edition of the *Origin* in the Sterling Library, Senate House, University of London. Judging from the condition of the pages, all of which have been cut, Innes may have read the entire book. His only annotations are on page 94 ("I have observed humble bees making holes and hive bees using them") and pages 232–233, where he copied out a passage on the architecture of the cells of the hive bee from Samuel Kinns's *Moses and Geology: or the Harmony of the Bible with Science* (1881; 7th ed. 1884), p. 330. For Darwin's views on the subject, see De Beer (1958d, p. 110). On the back endpaper Innes pasted a cutting from *Good Words* in which the Duke of Argyll recounts his conversation with Darwin about design (Argyll 1885).

23. Stecher (1961, p. 217); Darwin to J. B. Innes, 2 Sept. 1868, in APS B/D.25.m (Getz Collection); "Sunday School Account, Downe, 1863 to 1873", in KAO P123/5/1.

24. Stecher (1961, pp. 217–218, 219); Darwin to J. B. Innes, 20. Jan. and 16 Dec. [1868], in APS B/D.25.m (Getz Collection).

25. Darwin to J. B. Innes, 15 June [1868], in APS B/D.25.m (Getz Collection); Stecher (1961, p. 220).

26. "Incumbents of the parish of S. Mary the Virgin Church, Downe" and G[eorge] S[ketchley] F[finden], "Downe, Kent", reprinted from the *Canterbury Diocesan Gazette*, in KAO P/123/2/5. In "Vouchers and accounts for Downe Coal and Clothing Club" (KAO P123/5/3) the audited accounts for 1868–1869 are the latest inscribed in Darwin's hand.

27. The other items of the uncalendared correspondence are in the same location: Emma Darwin to G. S. Ffinden, Saturday [prob. Nov. 1873]; Darwin to "Gentlemen" [Downe School Board], n.d. [prob. Nov. 1873]; G. S. Ffinden to "Downe N.S.", 1 Dec. 1873; and Darwin to Downe School Board, 19 Dec. 1873. Ffinden's letters are draft copies. Darwin's are in Emma's hand but signed by him. On Emma's falling out with Ffinden, see Atkins (1976, p. 48).

28. The full uncalendared correspondence, consisting of draft copies of Ffinden's letters and the autographs of John Lubbock, is in KAO P123/25/3: G. S. Ffinden to J. Lubbock, 30 Jan., 8 Feb., 29 Mar., 3 April, and 23 June 1875; and J. Lubbock to G. S. Ffinden, 4 Feb., 9 Feb., 31 Mar., and 12 April 1875. See also Darwin's draft letter to John Lubbock, 8 [5 deleted] April [1873], in DAR 97 (ser. 3): 15–17. In a private interview with me on 9 February 1982 Lord Avebury, one of Lubbock's grandsons, reported the family tradition that the Darwins fell out with Ffinden

over an anti-Darwinian sermon. Lubbock's daughter-in-law stated that Darwin and Lubbock had both been present in the parish church for the sermon (Grant Duff 1924, p. 15).

29. The uncalendared correspondence, consisting of autograph letters to Ffinden from Darwin and Sarah Wedgwood, and draft copies of letters from Ffinden, is in KAO P123/3/4: S. E. Wedgwood to G. S. Ffinden, 14 Mar. 1873; G. S. Ffinden to White, Barrett & Co., 14 May 1879; Darwin to G. S. Ffinden, 31 May 1879; and G. S. Ffinden to Darwin, 31 May 1879.
30. The full uncalendared correspondence, consisting of autograph letters to Ffinden from Francis Darwin et al. and draft copies of Ffinden's replies, is in KAO P123/25/2: G. E. Forrest, C. Harris, and F. Darwin to G. S. Ffinden, 31 Jan. 1880; F. Darwin to G. S. Ffinden, 8 Mar., 29 Mar., and n.d. [April?] 1880; G.S. Ffinden to Darwin, 19 Mar. 1880; and G. S. Ffinden to F. Darwin, 30 Mar. and 6 April 1880.
31. The Lubbocks had made the schoolroom available to the Darwins on occasion for many years, perhaps as early as 1848 (Carroll 1976, p. 3 [no. 77]).
32. Fegan, the founder of "Mr Fegan's Homes" for children, was almost certainly assisted by Elizabeth Reid Hope (*née* Cotton) between 1880 and 1882. She was the Lady Hope who in 1915 published an account of an interview with Darwin that subsequently gave rise to the legend of his "death-bed conversion".
33. The Lubbocks also left the church (according to Lord Avebury, n. 28 above) and Sir John was buried in the parish churchyard at Farnborough nearby (Howarth and Howarth 1933, p. 74).
34. "Parish of Downe. Kent. Contributions Received, . . ." in KAO P123/3/7; "Downe Church Windows, &c., Account" and "Downe Church Choir Surplice Fund," in KAO P123/5/26; "Downe Vicarage Endowment Fund", in KAO P123/3/6. Neither the Darwins nor the Lubbocks nor Sarah Wedgwood augmented the "Down Parsonage House Building Fund" (KAO P123/3/5), whose well-connected contributors look like the old guard Ffinden might have met while serving as domestic chaplain to Lord Carington (Appendix).
35. Stecher (1961, p. 219); Darwin to J. B. Innes, 15 June [1868], in APS B/D.25.m (Getz Collection). Darwin replied: "I much hope that you may succeed soon in arranging that some clergyman sh^{d} have permanent charge of this parish. I am much obliged for your offer with respect to the purchase of the advowson, but it w^{d} not be in my way."
36. Compare the childhood recollections of a sixty-nine-year-old farmworker, who first encountered Darwin thus in 1871: "I had been bird's nesting in some woods not far from Down House when suddenly I found myself confronted by a very tall man with a grey beard, wearing a black cloak and a flat, wide-brimmed hat rather like the ones clergymen wore. I was terrified at first, but he spoke very kindly to me and showed me a place where a wren had built her nest. There were three eggs in it and I wanted to take one of them, but he told me I mustn't. In spite of this, he went on to tell me how to blow eggs so as to preserve them, which I thought was very funny. When I returned home to Cudham my father told me who the old man was and said he probably knew more about birds' eggs than any person living" (transcribed in Bunting 1974, pp. 106–107).
37. The passage continues: "I remember after Tyndall's Belfast Address my father told me that he asked Tyndall whether he really was conscious of the same sentiment towards Nature as towards a divine power (I forget the exact words) [.]My father told me with a smile that Tyndall hemmed & hawed & said something about the glory of sunsets &c" (DAR 112 (ser. 2): 3c–d).
38. "Under the circumstances of the parish" Innes was willing to take the funeral service. "I will put in a line to Ffinden," he wrote Francis Darwin, "which you can send to him or burn" (22 April 1882, in DAR 215). The "line" does not survive.

16

DARWIN THE YOUNG GEOLOGIST

Sandra Herbert

I. Darwin in the Context of a Field

The question I propose to answer in this paper is this: was the work Charles Darwin did as a young geologist compatible with the development of the field of geology in England in the 1830s, and, if so, how? I shall begin by outlining what I see as the major features of the development of geology as a field in England during the period. First, however, I should like to quote a statement on the subject by Martin Rudwick:

> The dominant cognitive goal of geologists at this period was to discover the true order of succession of the strata and their fossils. Stratigraphical description occupied the centre of the stage. But behind this enterprise lay the higher theoretical goal of reconstructing the main outlines of the history of the earth and the history of life on its surface. (1979b pp. 10–11)

There is much in this concise statement that is true. In my view, however, it is too Whiggish; that is, it stresses what were the successful features of English geology in the 1830s at the expense of its actual features, which were more varied, and not always so successful. It has the additional difficulty that its characterization of English geology during the 1830s leaves little room for Darwin, and, behind him, for Charles Lyell, for neither Lyell's nor Darwin's work during this period had as its dominant focus the identification of specific strata, or the reconstruction of the main outlines of the history of the earth and the history of life on its surface. What I should like to do, then, is to offer a more pluralistic description of English geology during the 1830s, which will modify Rudwick's characterization.

My source for defining the state of the field of geology in England during the 1830s is the series of annual addresses of the presidents of the Geological Society of London. I have chosen to define the decade of the 1830s broadly; I shall include in my discussion sixteen addresses by eight presidents. (Each president served two years and made two addresses.) The first address I shall consider was that given by William Fitton in 1828. It was the first presidential address to be printed in the *Proceedings* of the Society. The last address I shall discuss was that given by Roderick Murchison in 1843. The year 1843 seems a proper point at which to tie off the period since the addresses of Henry Warburton in 1844 and 1845 were not printed.

It is of course always difficult to know how to characterize an entire field, even in retrospect, but I believe the annual addresses of the presidents of the Geological Society of London to be a reasonably good guide to the development of English geology during the period for several reasons. First, the Geological Society of London can fairly be taken to be the chief institutional embodiment of the field as it existed in England during the period. While provincial societies did exist, the London Society outshone them on all counts. Second, those delivering the presidential addresses were, by virtue of their office and their individual competencies, in a position to speak authoritatively on behalf of the Society. The list of the presidents, their terms of service, and a brief characterization of them or their work is as follows:

William Henry Fitton, 1827–1829, the old Huttonian.
The Reverend Adam Sedgwick, 1829–1831, founder of the Cambrian system.
Roderick Impey Murchison, 1831–1833, founder of the Silurian system.
George Bellas Greenough, 1833–1835, founder of the Society and in these years in his third presidency.
Charles Lyell, 1835–1837, author of the *Principles of Geology* (Lyell 1830–1833).
The Reverend William Whewell, 1837–1839, the philosopher of science.
The Reverend William Buckland, 1839–1841, the Oxford geologist, in his second presidency.
Roderick Impey Murchison, 1841–1843, in his second presidency.

This is an impressive list of geologists. One misses a few names of potential presidents — Henry De La Beche for one — but, for the most part, the leading figures of the Society for the period do appear on the list. The third reason that these addresses are a good guide to the field is that they were intended to be reviews of the progress of geology in England during the preceding year. The speakers varied in the manner in which they interpreted its mandate: Murchison, for example, included more discussion of the work of foreign geologists than did the others, while Whewell favored grand philosophical summary over discussion of individual papers. Yet all the authors clearly understood that their remarks were to reflect the entire field rather than simply their own particular interests. Fourth, and probably of most importance, the presidential addresses of the Geological Society of London catch what G. M. Young called "the conversation of the people who counted"; they give one a sense of "not what happened, but what people felt about it when it was happening" (1977, p. 18). One gathers from these addresses which papers were discussed and attended to and which were not. One also gains from the addresses a sense of the momentum of the Society, where its leaders felt it had been and where it was going. Less personal than correspondence, yet more impressionistic and informal than papers, the presidential addresses allow one to listen in on the conversation of English geologists in the 1830s.

The most immediately striking feature of the conversation of English geologists in the 1830s is that it was optimistic. These were boom times for geologists, and every president of the Society referred with obvious satisfaction to the Society's current prosperity and prospects for growth. Membership in the Society, excluding foreign and honorary members, climbed from approximately five hundred members in 1830 to approximately eight hundred members in 1840. But the growth in knowledge during this period is even more striking. It required only twelve pages to print William Fitton's presidential address in 1828; it required eighty-six pages to print Roderick Murchison's presidential address in 1843. Even allowing for Murchison's expansive rhetoric, that is a dramatic increase. In addition, during the period, the science of geology was crowned with increasing official recognition. On the recommendation of the Royal Society, the government sought the counsel of the Geological Society in such endeavors as establishing mapping conventions and standards of coloration for ordnance maps. These last developments contributed to the public standing of the Society and to its authority to make certain kinds of publicly honored recommendations.

In addition to being optimistic, conversation within the Geological Society of London was focussed. On the evidence of the presidential addresses, two ideas dominated the minds of English geologists during the 1830s. First was the notion that the primary work of geologists was to determine the true order of succession of strata. Second was the notion that geology should be a comprehensive science. I shall treat these two notions in turn. I should also like to point out in passing, however, that these two notions were sometimes at odds with each other, and that in cases of conflict the tension between them was resolved by insuring that the first goal — determining the true order of strata — took precedence over the second goal — making geology a comprehensive science. Hierarchy preserved harmony within the field.

The dominance of strata-determination in the minds of English geologists is shown by several features of the presidential addresses. Most obviously, the notion of strata was the central principle around which presidents of the Geological Society organized their presentations. This was true for William Fitton, who used De La Beche's table of English strata to organize his material (Fitton 1829, pp. 115–116; De La Beche 1828). It was also true, with variations, for all of the other presidential authors. What Whewell called "Descriptive" geology, Buckland called "Positive" geology. Nevertheless, both of these headings referred to the same thing: the classification of facts based, as Whewell said, on the belief that the "key of all our geological knowledge of our country, — the doctrine that there is a fixed order of strata . . . ." (1838, p. 636). What is more, these discussions took pride of place: whatever else presidents chose to discuss in their addresses, they first discussed contributions to knowledge concerning the order and character of strata. Another sign of the pre-eminence of interest in strata is the manner in which geologists assessed the lasting worth of their own and their colleagues' achievements. Roderick Murchison in his address of 1842 spoke to the point: "The perpetuity of a name affixed to any group of rocks through his original research, is the highest distinction to which any working geologist can aspire" (1842, p. 649). In this vein it is also interesting to note that in his presidential address of 1835 George Greenough chose to repeat without correction or qualification the grounds on which the Council of the Royal Society had awarded a medal to Charles Lyell as author of the *Principles of Geology*:

> The Council of the Royal Society, premising that they decline to express any opinion on the controverted positions contained in Mr. Lyell's work, entitled "Principles of Geology," state the following as the grounds of their award. 1. The comprehensive view which the author has taken of his subject, and the philosophical spirit and dignity with which he has treated it. 2. The important service he has rendered to science by especially directing the attention of geologists to effects produced by existing causes. 3. His admirable description of many tertiary deposits, several of these descriptions being drawn from original observations. Lastly, The new mode of investigating tertiary deposits, which his labours have greatly contributed to introduce; namely, that of determining the relative proportions of extinct and still existing species, with a view to discover the relative ages of distant and unconnected tertiary deposits.
>
> (Greenough 1835, p. 170)

Of the four grounds given by the Royal Society, two had to do with Lyell's work on tertiary deposits. A modern reader of the *Principles* innocent of the concerns of Lyell's contemporaries would surely be surprised to see the work described so much as a work on Tertiary geology.

Whatever bending it required to make a work like Lyell's *Principles* fit the mold, the preoccupation of English geologists of the 1830s with strata paid handsome dividends, for their achievements rival those of any group of similar size working over a comparable period in any other scientific field. As one might expect, achievement came in stages. At the beginning of the decade the task at hand was seen as filling in knowledge of English strata. With Secondary strata reasonably well worked out by a previous generation of workers, including William Smith, and Tertiary strata less present physically in England than on the continent, an important group of English geologists set out to study the oldest fossil-bearing strata in England. While the utility of fossils for identifying strata was well established (although constantly re-emphasized in presidential addresses), the utility of mineralogical characteristics in identifying strata was also noted,[1] and fossils were attended to primarily for their value as markers of strata, rather than for their biological significance. As is well known, the search for the oldest English strata was successful, and from 1830 on one reads in the presidential addresses of the work of Sedgwick, Murchison, and De La Beche in identifying what came to be called the Cambrian, Silurian, and Devonian formations. The second stage of achievement came as English geologists extended the scope of their results to Europe. By the middle of the decade English geologists were finding what they termed "geological equivalents" of their own strata on the continent. In 1838 William Whewell spoke of "Home" geology meaning England and Northern Europe (1838, p. 633). A year later Whewell was pleased to report that the classification applicable to England and Northern Europe might be extended beyond the Alps:

> In the survey of the progress of our labours which I offered to your notice last year, I stated, that in proceeding beyond the Alps, and I might have added the Pyrenees, we no longer find that multiplied series of strata, so remarkably continuous and similar, when their identity is properly traced, with which we have been familiar in our home circuit. Yet the investigations of Mr. Hamilton and Mr. Strickland appear to show, that we may recognise, even in Asia Minor, the great formations, occupying the lowest and highest positions of the series, which are well marked by fossils, namely the Silurian and Tertiary formations; and also an intermediate formation corresponding in general with the Secondary rocks of the north, but not as yet reduced to any parallelism with them in the order of its members. (1839, pp. 83–84)

This extension of the terms of "Home" geology prompted Whewell to crow: "As if Nature wished to imitate our geological maps, she has placed in the corner of Europe our island, containing an *Index Series* of European formations in full detail" (Whewell 1839, p. 80). The third and final stage of achievement in stratigraphy came as English geologists extended their results worldwide. By the 1840s English strata were seen to be an index series of universal applicability. As Murchison exclaimed with deserved satisfaction:

> The chief aim of this Society has been to gather sound data for classification; and, following out this principle, I have endeavoured to show, how the order of succession established in our own isles, is now extended eastwards to the confines of Asia, and westwards to the back-woods of America. From such researches, and by contributions from our widely spread colonies, we have at last reached nearly all the great terms of general comparison. (1843, p. 149)

To borrow William Whewell's metaphor, the rocks of the world had been reduced to the same geological alphabet (Whewell 1838, p. 633).

Once the succession of strata identified in England was found to be universally applicable, English geologists were quick to draw the conclusion that in their strata they had the "records of creation" (Murchison 1843, p. 150). Fossils were now valued as remnants of past forms of life, rather than simply as markers of strata. Geology had yielded paleontology. In his address of 1843 Murchison presented a capsule history of life on the planet:

> Besides ascertaining where the great masses of combustible matter lie, we can now affirm, that during the earliest period of life, conditions prevailed, indicating a prevalence over enormous spaces — if not almost universally — of the same climate, involving a very wide diffusion of similar inhabitants of the ocean. We have learned, that in the earliest of these stages of animal life, no vestige of the vertebrata has yet been found, whilst in the succeeding epochs of the Palæozoic age singular fishes appear, which, in proportion to their antiquity, are more removed from all modern analogies. In each of these early long-continued periods, the shells preserving on the whole a community of character, differ from each other in each division — and in that later formation, where a very few only of the same types are visible, they are linked on to a new class of beings, the first created of those Saurians, whose existence is prolonged throughout the whole Secondary period; whilst we have this year seen reason to admit that even birds (some of them of gigantic size) may have been the cotemporaries of the first great lizards. With the close of the Palæozoic era we have also observed a gradual change in the plants of the older lands, and that the rank and tropical vegetation of the Carboniferous epoch is succeeded by a peculiar flora. In the next, or Triassic period, we have another flora, whilst new forms of fishes and mollusks indicate an approach to that period when the seas were tenanted by Belemnites and Ammonites, marking so broadly these secondary deposits with which British geologists have long been familiar, and which, commencing with the Lias, terminate with the Chalk. And lastly, from the dawn of existing races, we ascend through successive deposits gradually becoming more analogous to those of the present day, until at length we reach the bottoms of oceans so recently desiccated, that their shelly remains are undistinguishable from those now associated with Man, the last created in this long chain of animal life in which scarcely a link is wanting! — all bespeaking a perfection and grandeur of design, in contemplating which we are lost in admiration of creative power. (Murchison 1843, pp. 149–150)

As the sweep of this passage suggests, English geology had expanded its sights beyond stratigraphy. It had set for itself the new goal of reconstructing the history of life on the planet.[2] This new goal came into focus only when English geologists had achieved their original goal of determining the true succession of English — or, more precisely, British — strata.

But stratigraphy, and its corollary paleontology, did not exhaust the interests of English geologists during the 1830s. The earth is more than its strata, and English geologists of the period insisted on theirs being a comprehensive science. This ideal had two dimensions. First, geologists wanted their science to continue to have interlocking ties with other sciences. Geologists regarded their science as rather like a house with many empty or half-filled rooms. The earth was, after all, a planet, and geologists must reserve room in their science for the insights of physical astronomy. A similar argument was made for chemistry. While most geologists might prefer field work to mathematical computation or laboratory analysis, geology as a science required these techniques. Moreover, as geologists were pleased to point out, they had much to offer related fields: certain kinds of measurements to astronomers, rock specimens to chemists (one recalls here the work of Humphry Davy and J. J. Berzelius), and, ever increasingly, fossils to zoologists and botanists. The second way in which geology was intended to be a comprehensive science was that it was to be a science of causes, or, as William Whewell put it, of "dynamics". As is well known, debates over geological causation were heated during the 1830s and centered on the work of Charles Lyell. While Lyell's impact on geology has usually been discussed in terms of his contribution to the long-running uniformitarian-catastrophist debate, the presidential addresses suggest that Lyell's work had, in addition, a more immediate and consensus-producing impact on the collective mind of the Geological Society. With publication of the first volume of Lyell's *Principles* in 1830, three successive presidents announced to their colleagues in the Society their commitment to Lyell's claim that the Noachian flood be discounted as an agent for explaining the origin of what were then termed "diluvial" deposits. For two of these presidents, Adam Sedgwick and George Greenough, these announcements required them to abjure previously held views. Here is George Greenough on the subject in 1834:

> The vast mass of evidence which [Lyell] has brought together, in illustration of what may be called *Diurnal Geology*, convinces me that if, five thousand years ago, a Deluge did sweep over the entire globe, its traces can no longer be distinguished from more modern and local disturbances. (1834, p. 70)

The practical result of Greenough's conclusion, now shared by nearly all his colleagues in the Society, was that it freed succeeding generations of geologists from the attempt to harmonize Biblical and geological history.

In addition to historical causes, there were two other classes of causes that interested English geologists in the 1830s. The first referred to those that Whewell termed "ulterior" causes. Chief among these causes were those pertaining to the interior of the earth. Despite the nebular hypothesis and the fact that a majority of geologists believed the interior of the earth to be molten, this class of questions received little attention in the presidential addresses, being, in Whewell's words, "obscure" (Whewell 1838, p. 645). The second class of causes, which Whewell termed "proximate" causes, were more amenable to the investigation of geologists working with existing methods (Whewell 1838, p. 644). The proximate causes receiving the greatest attention in the 1830s were the opposing forces of elevation and subsidence. (Elevation refers to the raising of the earth's crust, subsidence to its lowering.) While Lyell in the *Principles* was chiefly responsible for focussing the attention of geologists on the action of these forces, he was not original in invoking them. As Roderick Murchison said in his 1832 address: ". . . whatever discrepancies of opinion may . . . exist, . . . all inquirers agree in this fundamental opinion, that the earth's surface has been mainly brought into its present condition by numerous changes of the relative level between the land and the sea . . . ." (1832, p. 376). Or, as George Greenough said in his 1834 address, even as he prepared to attack Lyell's views on the subject, "Among the subjects which have for some years past engaged the thoughts of geologists none perhaps has excited so general and intense an interest as the Theory of Elevation" (1834, p. 54). Confirming Greenough's comment, Lyell's presidential addresses of 1836–1837 and Whewell's addresses of 1838–1839 were devoted in large part to the subject of the opposing forces of elevation and subsidence.

The next year saw a change. In his presidential address of 1840 William Buckland introduced a new contender in the class of proximate causes: Louis Agassiz's glacial hypothesis. Soon the explanatory possibilities of glaciers attracted the interest of English geologists. Moreover, since Agassiz's glacial hypothesis explained the origin of superficial deposits of detritus (the former "diluvial" deposits) as well or better than Lyell's drifting iceberg hypothesis (itself based on the forces of elevation and subsidence), the victories of Agassiz's hypothesis, even if partial, drew attention away from elevation and subsidence as proximate causes.

The final aspect of geological dynamics in the 1830s that requires notice is the question of species. As Whewell pointed out in his address of 1838, the laws of change regarding species fall under the province of geology by virtue of bearing on the "history of the globe" (1838, p. 645). What is equally interesting is that Whewell made this point in order to discuss some of Darwin's work arising from the *Beagle* voyage.

By way of summary I should now like to offer an amended version of Rudwick's characterization of English geology during the 1830s:

> The dominant cognitive goal of English geology at this period was to discover, using fossils, the true order of succession of the strata. Success in achieving this goal carried with it the possibility of reconstructing the history of the earth and the history of life on its surface. The secondary goal of English geology at this period was to make geology a comprehensive

science by insisting that it complement and draw from other sciences and that it be, in principle, a science of causes as well as of description and classification.

We are now prepared to see where Charles Darwin's own goals as a young geologist fit into this scheme.

Charles Darwin can appear either central or peripheral to the field of geology during the 1830s. What matters is how one defines the field. If one does so in such a way as to equate geology with stratigraphy, Darwin appears as a peripheral figure. Darwin did not work on English strata, the primary interest of English geologists, and so could not hope to win the highest prize of having his name permanently associated with the name of a geological system, as Sedgwick's name was associated with the Cambrian or Murchison's with the Silurian. At best Darwin's early contributions to geology, considered as stratigraphy, appear as those of a first-class colonial agent, one member of the small army of English collectors working abroad whose labors allowed geologists working at home to establish a world-wide system of geological equivalents. If, however, one allows a broader definition of the field of English geology, one that leaves room for other goals in addition to stratigraphy, Darwin appears in a different light. He appears as someone who was very properly welcomed into the inner circle of the Geological Society at a young age. He appears, in short, as someone who was indeed central to the field. To argue for that point of view I should now like to provide a very brief description of Darwin's activities as a geologist during the 1830s.[3]

Darwin's activities as a geologist began in 1831. At the beginning of that year he was relatively untutored in geology; at the end of the year, when the H.M.S. *Beagle* left port, he had as good an education in geology as any man in England his age. In the short span of a year he had become a geologist. This transformation was initiated and directed by J. S. Henslow, Professor of Botany at Cambridge University and, as Darwin termed him, his "Master" in natural history (Darwin 1967, p. 114). Henslow, a Fellow of the Geological Society of London since 1819, and a published author on geological subjects, regarded it as within his province to advise his charge on the subject of geology. Darwin's *Autobiography* is explicit on the point:

> As I had at first come up to Cambridge at Christmas, I was forced to keep two terms after passing my final examination, at the commencement of 1831; and Henslow then persuaded me to begin the study of geology. Therefore on my return to Shropshire I examined sections and coloured a map of parts round Shrewsbury. Professor Sedgwick intended to visit N. Wales in the beginning of August to pursue his famous geological investigation amongst the older rocks, and Henslow asked him to allow

> me to accompany him. Accordingly he came and slept at my Father's house. (pp. 68–69)

A letter dated 28 April 1831 from Darwin to his sister Caroline makes the same point:

> All the while I am writing now my head is running about the Tropics: in the morning I go and gaze at Palm trees in the hot-house, and come home and read Humboldt: my enthusiasm is so great that I cannot hardly sit still on my chair. Henslow & other Dons give us great credit for our plan: Henslow promises to cram me in Geology. — I never will be easy till I see the peak of Teneriffe and the great Dragon tree; sandy dazzling plains, and gloomy silent forest are alternately uppermost in my mind. (DAR 154)[4]

Henslow's "cramming" took several forms: direct instruction, suggestions for reading, and field assignments. Perhaps most important, however, was Henslow's placing of his student in promising situations. Clearly the capstone of Darwin's geological education was the period in the Summer of 1831 he spent working in Wales with the Reverend Adam Sedgwick, professor of geology at Cambridge University (Barrett 1974). Henslow had arranged this field trip. Moreover, it was he who provided Darwin with the opportunity of a lifetime: the offer to serve as a naturalist aboard the H.M.S. *Beagle*.

In his letter to Darwin of 24 August 1831 regarding the *Beagle* position Henslow described what the position entailed: "collecting, observing, & noting anything new to be noted in Natural History" (Darwin 1967, p. 30). "Collecting", as the entire correspondence pertaining to the offer makes clear, would be done for the purpose of enriching already existing holdings, as, for example, those at the British Museum or at Cambridge; Darwin was not being offered the opportunity of creating a private collection. Yet otherwise Darwin was left free to interpret his mission as he saw fit. That he chose to emphasize geology was undoubtedly a result of Henslow's tutelage and their common perception that there was more "new to be noted" in geology than in any other area of South American natural history.

The first half of the voyage saw Darwin working diligently, though with considerable anxiety over whether he was collecting the right kind and number of specimens. When he finally received a letter from Henslow in March 1834, over two years since the *Beagle* had left port, he was overjoyed. On 24 July 1834 Darwin wrote back to Henslow: "Not having heard from you until March of this year; I really began to think my collections were so poor, that you were puzzled what to say: the case is now quite on the opposite track; for you are *guilty* of exciting all my vain feelings to a most comfortable pitch . . . ." (1967, p. 91). Henslow had not only praised Darwin's collections but had communicated to him the news that a choice specimen from his collection of fossil remains of extinct South American mammals had been exhibited at the Geological Section of the British Association for the Advancement of Science at its meeting in Cambridge in 1833 (Darwin 1967, p. 77). Darwin's work as a collector was already attracting the attention of geologists. Similarly, recognition of Darwin's acuity as a geological observer also preceded his return to England. In his presidential address of 1836 Charles Lyell took note of the value of Darwin's geological observations as communicated in his letters to Henslow (1836, p. 367).[5] Since Darwin's observations supported Lyell's classification of Tertiary strata and his general conclusions respecting the gradual nature of the action of elevatory forces, Lyell could take personal satisfaction from Darwin's remarks. In sum, then, even before the end of the *Beagle* voyage, Darwin's work as a collector and observer was receiving favorable attention from English geologists. Given the additional facts that Darwin came with the personal credentials most acceptable to members of the Geological Society of London — that he was a gentleman by birth and education, personally known to several members of the Society, and had distinguished himself by his fieldwork — it is no wonder that when Darwin arrived back in England in the Autumn of 1836 his desire to become a Fellow of the Society was immediately honored. Not surprisingly it was Henslow who put Darwin forward for election to the Society.

Yet on his return to England Darwin had more to offer the field of geology than raw materials. He intended to contribute to that aspect of the field concerned with dynamics. In particular he was interested in contributing to geologists' understanding of the action of elevation and subsidence. Here Darwin took Charles Lyell's work as his point of departure. The major change he made in Lyell's notion of elevation and subsidence was to imagine these forces working on a larger scale — continents rather than patches, and long-term changes in level rather than temporary oscillations. As they derived from Lyell's work, Darwin's views were easily grasped by his colleagues at the Geological Society.[6] Moreover, his views were treated respectfully, even when, as happened in the case of his theory of erratic boulders, they were challenged by rival interpretations (Davies 1969). The institutional sign of the high regard in which Darwin was held by his colleagues in the Geological Society was the forcefulness with which they insisted in 1837 that he serve as one of the secretaries of the Society (Darwin 1967, pp. 138–140). In the end he could not refuse them, for he knew that while he had contributed his intelligence and zeal to their enterprise, they had benefitted him even more by providing him a field in which to work.

In sum, in the short space of seven years Darwin had become a member of the inner circle of the Geological Society of London. He had been able to achieve this position because the goals of the Society, broadly conceived, matched his own. While his initial contributions to the search for the true succession of strata were not of first-order importance, in the sense of being equivalent to the contributions of a Sedgwick or a Murchison, his collections and observations did aid geologists working in England to establish the universality of their system of classification of strata. A good example of Darwin's contribution in this regard is his collection of Silurian specimens from the Falkland Islands.[7] Even more important to colleagues at the Geological Society were Darwin's collections of South American fossil mammals, for the fossils revealed interesting patterns of succession in animal life.[8] Such collections were essential, as geologists at home recognized, to their goal, then emerging, of reconstructing the history of life on the planet. It hardly requires stating that, as a theorist, Darwin would eventually make the major contribution of his career towards achieving that goal. In the 1830s, however, it was to geological dynamics that Darwin made his most significant contributions. The fact that, aside from his coral reef theory, his major contributions to dynamics were ultimately rejected does not detract from their importance at the time. Darwin's ideas of elevation and subsidence, forgotten now, formed part of the conversation of the people who counted in the geology of his day.

II. Historiographical Afterword

What I should like to do in this historiographical afterword is to engage the literature, not by reviewing it, but by suggesting how the conclusions I have drawn from my reading of the presidential addresses confirm or call into question statements made by other authors writing on the history of geology. I have organized my remarks around five topics: (1) the characterization of English geology as a field; (2) the place of Lyell in English geology; (3) the rise of paleontology; (4) the relationship between geology and society; and (5) Darwin himself. I have settled on the first two points because they are presently the most controverted topics in the literature (Lyell being the single most problematic figure for the historian) and because one's view of Darwin as a geologist is ultimately dependent on one's views on these two points. The third and fourth points have drawn less attention in the literature, but I believe them sufficiently germane to the period not to be overlooked. The inclusion of the fifth topic needs no explanation. As a final caveat, what I have to say pertains only to English geology in the 1830s as it was represented in the Geological Society of London. How far discussions within the Society can be taken to be characteristic of geological discussion outside London in the British Isles or on the continent I do not know. Members of the Society travelled extensively, however, and met frequently with other geologists (occasionally, as with Louis Agassiz, attracting foreign geologists to their company), so that in writing of the Geological Society of London one is writing of a conversational circle that was joined to others.

THE FIELD OF ENGLISH GEOLOGY IN THE 1830s

Judging by the presidential addresses, the essential fact about English geology in the 1830s, is that it was multipolar in emphasis rather than unipolar. Put simply, English geology was composed of a paradigm and a list. The paradigm was mapping strata. The list included other questions that geologists wanted answered and all other topics that appeared to them geological. To an extent, the division between "paradigm" and "list" corresponds to the division William Whewell drew in his presidential addresses between "descriptive geology" and "geological dynamics" (1838, 1839). Certainly the desire to recognize division within the field stems from a common perception of its plenitude. Whewell's preference for causal issues biased his naming of the divisions, however. As measured by level of activity and achievement, what Whewell termed the "descriptive" side of geology was the more important of the two; hence I prefer to label it "paradigmatic". Moreover, not all non-stratigraphical topics fall easily under the rubric of "dynamics". Descriptive mineralogy, for example, did not fit easily into either of Whewell's categories. It was neither a stratigraphical nor a dynamical topic, yet it was clearly pertinent to the subject matter of both. Similarly, paleontology did not fall naturally into either of his two categories, although Whewell tried at first to place it under "dynamics". Clearly two divisions were insufficient to capture the richness of the field. Substituting "list" for Whewell's more euphonious "dynamics" acknowledges the multiplicity of interests of English geologists, while, as has been stated, substituting "paradigm" for "descriptive" recognizes the dominance of stratigraphy over other concerns.

In writing on English geology in the 1830s a number of authors have recognized the paradigmatic role played by stratigraphy. Martin Rudwick has already been quoted to that effect in the main body of this paper. In 1960 Walter Cannon, in writing on the uniformitarian-catastrophist dispute, opened his discussion with a disclaimer that acknowledged the pre-eminence of stratigraphy over any other topic in geology of the 1830s, including the dispute in question: "First, most geological monographs of the period, and, we may assume, most scientific energies, were devoted to matters of exact stratification and classification. What follows is by no means a history of geology in the period, but only a consideration of one theoretical question in geology" (W. Cannon 1960b, p. 39). John Challinor, in his article "The Progress of British Geology During the Early Part of the Nineteenth Century", showed by sheer force of evidence how concerned British geologists were with British strata in the first two decades of the century. He did not discuss the shape of geology as a field in any explicit way, although he referred to the fact that "various subjects" were treated under the rubric of geology in the 1830s. Yet his thumbnail sketches of the works of prominent geological authors, and a table he compiled of the successive classifications of British strata (he ended with Lyell's classification of 1833), testify to the central role he assigned to stratigraphy in British geology of the period (Challinor 1970).

Another way of looking at geology as a field is through the daily activities of its practitioners. As one reads through the presidential addresses one is impressed that geologists — real geologists — were expected to be men of the hammer as well as men of the pen. Those who worked in a laboratory setting only, such as mineralogists or conchologists, might be useful to geologists; but, however useful, they were something less, or at least other, than geologists. No one has caught this better than Roy Porter, whose evocation of the geologist as field worker is so beautifully expressed that it must be quoted in full:

> The nineteenth-century amateur ethos did not merely drive researchers out of doors: it made them construct a romance of the field. Earlier investigators such as Catcott and Hutton had certainly forged a practical, fruitful, field tradition. But their fieldwork was sober and instrumental. Thus Hutton could remind his readers that careful armchair pondering upon a specimen would yield more than indiscriminate observations. By contrast — spurred by Romanticism and muscular Christianity — nineteenth-century geologists celebrated 'doing geology on your feet', as the hard-core activity of their science. It was not merely, as Archibald Geikie was to write a little later, that fieldwork 'evidently underlies all solid research in geology'. Or that, gentlemen geologists being for the most part scientifically untrained, fieldwork was one mode of science where their qualities of stamina, shrewd perception and native intelligence could particularly succeed. Rather, fieldwork became a cult, an obsession. Lyell's three pieces of geological advice were: Travel, travel, travel. In his essay, 'To the field!', David Page crooned, 'A day well spent in the field is worth a dozen of reading at the fireside . . . . "To the field on every fitting occasion" should be the guiding maxim of the young geologist.' And, not least, fieldwork became an explicitly religious experience, a spiritual re-creation, elevating geologists above the mundane world of utility and everyday duties to an arena where they could test their moral fibre in the pure presence of Nature. Humphry Davy wrote, 'That part of Almighty God which resides in the rocks and woods, in the blue and tranquil sea, in the clouds and moonbeams of the sky, is calling upon thee with a loud voice: religiously obey its commands and come and worship with me on the ancient altars of Cornwall.' A century later, the Cambridge geologist Newell Arber could still echo Davy's strain: 'I thought I would take a holiday for the rest of the evening and indulge in a fit of "field-fever" or "field-dreams" . . . . A perfect day when one is in the field is one of the greatest things on earth.' (Porter 1978, pp. 820–821)

Porter was interested in field work primarily as it contributed to what he termed the "amateur ethos" of nineteenth-century British geology. But it is not contrary to his point to say that emphasis on field work as an activity contributed to the dominance of stratigraphy in British geology during the 1830s. Even John Herschel, as recently discovered notes show, did some field work (1831); and those men, such as William Whewell, whose interests did not include field work of a traditional sort, could not help but feel like onlookers (though, in Whewell's case, a gifted onlooker) to the subject.

What, then, of the "list"? One recent collection published by the British Society for the History of Science under the title *Images of the Earth* has addressed itself in an enterprising and fresh way to the study of what the editors term "the earth broadly considered as an object of scientific investigation" (Jordanova and Porter 1979, p. vi). Several of the essays in the volume, particularly those by W. H. Brock and D. E. Allen, consider the nature of the disciplinary boundaries of geology in the nineteenth century (Brock 1979; D. Allen 1979). Overall, however, the authors who have contributed the most to our understanding of geology's "list" are those exploring the history of the physical sciences in general rather than geology per se. A conference organized by Harold Burstyn on "The Place of the Geophysical Sciences in Nineteenth Century Natural Philosophy" is a case in point (Gillmor 1975). The interest of historians of the physical sciences in geology has a straightforward explanation, for the field of geology provided some of the materials from which the field of physics — itself a nineteenth-century invention — was constructed (Brush 1978; W. Cannon 1978; Kuhn 1976). Among these materials was a concern with the application of mathematics to certain kinds of problems. In geology the representative figure was William Hopkins, who treated the question of the fluidity of the interior of the earth mathematically. Since Hopkins was tutor to William Thomson, later Lord Kelvin, at Cambridge University, the link between geology and physics was personal as well as intellectual. In treating such problems as the nature of the interior of the earth, geology required exactly those ingredients that Kuhn has characterized as being required for physics — "the establishment of a firm bridge across the classical [that is, mathematical]-Baconian divide" (1976, p. 30). The two prominent issues in nineteenth-century geology that had a mathematical aspect were the question of the nature of the earth's interior, which Stephen Brush (1979) has described in his paper "Nineteenth-Century Debates about the Inside of the Earth: Solid, Liquid or Gas?" and the question of the age of the earth, which Joe Burchfield has treated in his now-classic study, *Lord Kelvin and the Age of the Earth* (1975). As Brush and Burchfield have shown, resolution of these issues brought conflicting scientific communities as well as conflicting points of view into contention. By the end of the century, as all have agreed, physics outranked geology in prestige and had laid claim successfully to certain issues that were once the province of geology alone. In a sense, geology had been stripped of much of its list. Still, in considering English geology one has to remember that questions of the physical nature of the globe, the chemistry of its interior, and its behavior as a planet were the province of the science during the 1830s. To revert to Charles Darwin for a moment, I should like to add that his early notebook on geology, labeled *A*, which dates from this period, fully corroborates this point. Finally, in speaking of geology's "list" I do not wish to leave the impression that all the questions on it lay in the province of what today would be physics or chemistry. One major question on the list had to do with the origin of species, and it, of course, was resolved without recourse to physics or chemistry. But it is interesting that this question, too, was struck from the list in the sense that, while first raised in the context of the field of geology, it soon found its primary residence elsewhere.

Thus far I have characterized the structure of English geology in the 1830s as being composed of a paradigm and a list. To this characterization I should like to add a third point: that the excitement of the field in this period stemmed from the inherent tension beween the goals represented by the paradigm and the goals represented by the list. On the one hand the stratigraphic accomplishments of English geologists were such as to call forth universal admiration; on the other hand the far-reaching nature of the goals on the list — understanding the origin of species and defining the structure of the earth's interior, to name but two — stimulated the theoretically-minded natural philosopher of whatever training. As I have pointed out elsewhere, geology was unique among the branches of English natural history in its toleration of the theoretical (Herbert 1977). If I may expand on that point a moment, I should like to add that it seems to me that the *varied* nature of the goals that geologists espoused (some drawn from natural history, some from natural philosophy, as Rachel Laudan (forthcoming) has suggested) encouraged a freer notion of the definition of the field and gave greater encouragement to the elaboration of theory. Those fields such as zoology and botany that guarded their notion of what it was their practitioners were to study, were inherently less theoretical, and, as Darwin was fond of saying, inherently less exciting. It was impossible for geologists to hold on to this excitement forever; the difficulties in keeping together such a disparate group of goals and disparate group of people was too much of a strain. Yet in the 1830s the tension between geology's "paradigm" and its "list" provided a high level of intellectual excitement. As W. Cannon put it over twenty years ago,

> The Geological Society of London was at the height of its prestige in the 1830's. It was the outstanding center of scientific activity in England; its geologists were thought to be the best in the world, its monographs the most important. Members of the Royal Society yearned for stimulating debates which closed its meetings; other societies imitated its *Proceedings*. Its yearly presidential address gave authoritative and exhaustive summaries

> of the progress of geology, so exhaustive in some cases that to a modern reader it seems incredible that they were ever actually delivered. They were enlivened, however, by the perfect freedom the presidents enjoyed to state their own theories and criticize those of their opponents. Their opponents were in most cases their close friends, and the debates were conducted for the most part by a circle of men who, whether Uniformitarian or Catastrophist, represented the progressives in the Society . . . . All in all, it was a close-knit high-spirited group, and the period of the 1830's appears in the history of the Society as a Golden Age. (1960b, p. 40)

Thus far in characterizing the literature on the topic of English geology as a field I have made only positive comments. I should now like to be more critical. My comments were directed not so much at specific errors of interpretation as at a certain pattern of perception, and, even more, of writing, that has in the past distorted some treatments of the subject. The distortion has arisen from the conflict between the natural desire of the historian to tell a good story and the intractability of the subject matter. In adopting a field as one's unit of analysis, one is faced with what is, for the historian, the painful loss of an obvious narrative thread on which to construct a story. There is too much happening in a field at any given moment to capture it all in a single coherent story line. It is not that there is no drama to a field — Sedgwick and Murchison's attacks in their presidential addresses on scriptural geologists were rich in emotion and must have been spectacular when enacted in person — but that the usual dramatic conventions, such as organizing action into a beginning, a middle, and an end, or introducing a conflict and showing its resolution, are inadequate to portray a field whose life extends beyond that of any individual or controversy. How this problem can be addressed I am not certain. Abandoning the narrative approach in favor of an analytical approach in writing (and thinking) solves the immediate difficulty, while creating others. One longs for a device corresponding to the film technique of montage to capture the richness of events in a field.

One source of distortion derives from the understandable preference of historians for coherent narrative. The inherent difficulty in providing such a narrative for English geology in the 1830s is that the field itself was split between what I have termed its "paradigm" and its "list". Thus in constructing a narrative for the period, one has to take care to specify which tradition of research within the field one is discussing. Otherwise one is likely to mistake the part for the whole. For an example of the kind of error I am describing I should like to go to Roy Porter's excellent book *The Making of Geology*. The example is not from the 1830s but is, I think, a fair one. In the culminating chapter of the book, Porter argues against the tradition of interpretation that makes James Hutton out to be the founder of modern geology. In Porter's view, Hutton was rather "an outsider, who did not spring from the mainstream, and whose impact upon emerging traditions of Earth science was essentially oblique" (Porter 1977, p. 186). I share Porter's hesitation to make Hutton *the* founder of modern geology. However, Porter's use of the metaphor "mainstream" seems to me to go too far the other way. The image of "mainstream" calls to mind a river into which other streams flow. This is not, I believe, an appropriate image to use in representing late eighteenth- and early nineteenth-century British geology. The widest and deepest stream in geology was indeed stratigraphy, and Hutton was not primarily a stratigrapher. But other streams of thought, which did not empty into the mainstream, also ran through the field of geology, and Hutton's work came from, even established, some of these other streams. Among these other streams, or traditions, were physical geology, studies of the interior of the earth, and inquiries concerning the rate and kind of forces molding the earth's surface. In short, "mainstream" seems the wrong word. Why then was it chosen? Because, I suspect, it satisfies the preference of the historian for cogent narrative.

A second source of distortion arises from the desire of the historian to tell an interesting story. This impulse, which is clearly a matter of importance to the reader, can lead the historian to emphasize those elements in a story that are inherently dramatic: the struggle of protagonist and antagonist (witness the emphasis on the FitzRoy-Darwin conflict in the recent British television series, "The Voyage of Charles Darwin"), triumph and defeat, and conflict of all kinds. Thus in 1909, by way of introducing a discussion of Charles Darwin's work as a geologist, Sir Archibald Geikie condensed the entire history of geology into a succession of conflicts: first that between the "neptunists or Champions of Water" and the "Vulcanists, or Plutonists, with Fire as their watchword" which latter struggle was in turn broken down into a struggle between the "Catastrophic or Convulsionist school" and the "Uniformitarian" (Geikie 1909, pp. 3–4). A later work, which has also served as one of the models of scholarship in the history of science, is Charles Gillispie's *Genesis and Geology* (1951). Gillispie's approach in this work is narrative, and he has succeeded brilliantly in capturing the flow of events within a defined area of discourse. The narrative impulse in the book is so strong, however, that it does on occasion highlight the dramatic aspects in the field at the expense of other more structural, and less controversial, developments. To be fair, one must keep in mind what story the author intended to tell. Gillispie did not set about to write a general history of English geology; he was offering, in the words of his subtitle, "a study in the relations of scientific thought, natural theology, and social opinion in Great Britain." Readers as well as authors can create distortions.

LYELL

As Philip Lawrence has put it, rather mildly, "Historians of science have never seriously doubted that Charles Lyell is a figure of major significance.

Of late, however, there has been increased interest in the precise nature of his significance" (1978, p. 101). The question has arisen because Charles Lyell has had the good fortune to have had his work examined by a series of unusually able historians, who have arrived at differing, though in some sense still tentative, conclusions. In looking at Lyell through the glass of the presidential addresses, I hoped to arrive at some conclusions regarding Lyell's place in the history of geology that would assuage my own uncertainties and perhaps aid in a small way in the eventual resolution of the question.

As a source the presidential addresses are useful in telling one about the ways in which Lyell's *Principles* was received by his colleagues. The addresses are a poor source, however, for learning much about the private intentions of Lyell or any of his peers. I make this point because a large part of the historical literature centers on such questions. I have in mind here such important articles as Leonard Wilson's study of the origins of Lyell's uniformitarian ideas and Martin Rudwick's reconstruction of Lyell's strategy in writing the *Principles* (Wilson 1964; Rudwick 1970). What follows, then, are some comments not on Lyell as he was to himself but on his historical persona.

My first observation is that Lyell's colleagues read the *Principles* in a different spirit from that of many modern historians of science, including, until recently, myself. A modern reader, at least a conscientious one, tends to read the *Principles* straight through, as a whole work, to look for interrelations among the parts of the book, and to seek to understand the author through the book. Lyell's contemporaries read the book with less thought to seeing the work as a whole, and more thought to seeing in what ways it advanced or retarded their science. One can imagine them mentally ripping the volumes from their covers, saving some chapters, discarding others. Lyell's contemporary readers were thus serious in their approach to his work, but not necessarily respectful.

This selectivity of response shows up very clearly in the manner in which the Royal Society chose to word its award to Lyell. They did not praise the work indiscriminately as a whole but drew up a list of reasons for their award (see above p. 486). Everything on the list is in Lyell's book, but one could hardly go from this list to a reconstruction of the book. The list is significant, however, in that it suggests how Lyell's contemporaries went about assimilating the *Principles* to their field. In its announcement the Society explicitly "declined to express any opinion on the controverted positions" in the work. A number of topics thus do not appear on the list, notably Lyell's views of earth history (what Martin Rudwick has referred to as his "steady state" position), on species, and on cosmology in relation to geology. What do appear on the list are those features of the *Principles* that the majority of geologists could readily assimilate: his "comprehensive" view of geology, his attention to existing causes, and his work on Tertiary strata. (Rather surprisingly, Lyell's uniformitarian method of estimating ages of Tertiary strata is also credited.)

With regard to the total impact of the *Principles* on Lyell's contemporaries, controversy not excluded, the testimony of the presidential addresses underscores the point made by Gillispie three decades ago: "One thing the *Principles of Geology* unquestionably accomplished. The book administered the *coup de grâce*, to the deluge" (Gillispie 1951, p. 140). As one reads through the presidential addresses for the early 1830s, one cannot help but be impressed as the great men of the Society — Greenough, Sedgwick, Murchison — used the forum of their addresses to announce their abandonment of traditional beliefs. Lyell was credited with, in Greenough's words, "having awakened us to a sense of our error . . . ." (Greenough 1834, p. 70). Remarkably, the presidents of the Society confessed their "error" publicly. By the end of the decade, the deluge had been eliminated from the useful scientific vocabulary of the most eminent men of the Society. The decade in question is the 1830s; the presidential addresses do not support Philip Lawrence's assertion that English geology had rid itself of non-naturalistic causes by the second decade of the century (Lawrence 1978, p. 102).

Where then does that leave one with respect to Lyell's "precise significance" as viewed by his colleagues? First, the presidential addresses do support the notion that Lyell was an epoch-making or revolutionary figure to his contemporaries in the narrow sense that he was the one who forced the disengagement of geological history from sacred history. The geological community in England was of a different mind on the subject of the Noachian flood after publication of the *Principles*. In the presidential addresses, Lyell still appears the great secularizer. On broader methodological issues, however, Lyell does not seem to have been a revolutionary figure to his colleagues. Thus, for example, the Royal Society citation credits him with "directing the attention" of geologists to existing causes; it does not credit him with establishing a new opinion. Further, and as others have pointed out in other contexts, there is no evidence in the presidential addresses that Lyell's strong uniformitarian position was well received. Second, the presidential addresses indicated that Lyell's colleagues in the Geological Society were inclined on publication of the *Principles* to take from it what they could use, and to hold in abeyance or to discard what suited them less. The systematic or synthetic character of the *Principles* does not seem to have been as important to them as other aspects of the work.

The additional factor that emerges from the presidential addresses as important to understanding Lyell's significance to his peers is a matter of dignity and position. Lyell was valued in the 1830s as a geologist whose work advanced the standing of geology as a science. To quote the Royal Society citation once again: Lyell's *Principles* was to be praised for "the comprehensive view which the author has taken of his subject, and the philosophical spirit and dignity in which he has treated it." "Comprehensive" refers, in context, to the fact that Lyell broadened the scope of geology.

His work did deal with stratigraphy, but it also dealt with much more. To return to my earlier terminology, I would say that Lyell was a man of the paradigm and of the list. Lyell's significance to his colleagues was thus not so much that he was the central figure to the field (he was certainly less central than either Sedgwick or Murchison to the paradigmatic aspect of the field), but that he was the spanning figure. He reached out to make a great many subjects, including species, the subject matter of geology. This accomplishment is an element that must be taken into account in judging his significance to the history of geology.

Valuing Lyell for the reasons that his contemporaries did has one further advantage that must be mentioned. As historians have shown, Lyell turned out to be wrong on several important issues. He backed non-progression (Bartholomew 1976); he disengaged geology from the nebular hypothesis (Lawrence 1977); his ideas on geotectonics were not the most useful ones in the long run (M. T. Greene 1982). How then could he have been so important a figure in the history of geology? To this question the perception of his peers offers an answer. As Michael Bartholomew has suggested, Lyell was a "singular figure", rather than the head of a school, whose significance to the field lay largely in the effect his book had upon others (1979, p. 289). Not least among the others was, of course, Charles Darwin.

PALEONTOLOGY

Much of what appears in the presidential addresses on paleontology will be easily recognizable to those familiar with the writings of Martin Rudwick. But I would like to offer two addenda to his conclusions. The first of these pertains to dates. In reading the presidential addresses I was struck that, for the English geologists, the "paleontological synthesis" of which Rudwick has written occurred about a decade later than he suggests. If one substitutes the date 1840 for the date 1830, the following paragraph from *The Meaning of Fossils* fits very well:

> By about 1830, therefore, the spectacular success of some three or four decades of research on fossils had transformed Cuvier's early demonstration of a single recent organic revolution into a palaeontological synthesis of very wide scope and explanatory power. The geological time-scale was firmly established as almost unimaginably lengthy by the standards of human history, yet documented by an immensely thick succession of slowly deposited strata. The successive formations of strata, and even in some cases individual strata, were clearly characterised by distinctive assemblages of fossil species, which enabled them to be identified and correlated over very wide areas. This correlation proved that in its broader outlines the history of life had been the same in all parts of the world. (Rudwick 1972, p. 156)

I have no explanation for this lag on the part of English geologists. It may simply have been a matter of time being required for new ideas to cross the Channel. It may have been — as my geologist colleagues remind me — simply a matter of the inherent difficulty of the earlier materials. In any case, it is interesting that, on the question of dating the paleontological synthesis Peter Bowler also favors a date of "about 1840" for English geological opinion (1976a, p. 34).

The second addendum I should like to make to Rudwick's characterization of the paleontological synthesis is that as of 1830 English geologists were not willing to claim that their stratigraphic identifications had universal applicability. Whether they were merely mouthing an anti-Wernerian and traditionally British empiricist viewpoint, contrary to their own private beliefs, I do not know. Skepticism towards the idea of universal formations is evident in the presidential addresses, however. D. R. Oldroyd has recently noted the same phenomenon:

> The doctrine of universal formations, useful though it had been in the late eighteenth century, was fairly quickly rejected in the nineteenth, and we find, then, the first steps in the stratigraphers' seemingly endless task of piecing together a multiplicity of discretely observed stratigraphical sections, with the constantly associated problem of weeding out stratigraphical synonyms. (1979, p. 243)

While reasons for the skepticism may not be clear, its importance is clear, for until English geologists were confident that the English strata they were describing had foreign counterparts, recognizable by the identity of their fossil assemblages, they could not use their knowledge of stratigraphy to reconstruct the history of life on earth.

More positively, what the presidential addresses for the 1830s do reveal is the emergence in well-marked stages of paleontology as a subfield of geology. At the opening of the decade fossils were treated as tools useful for the identification of strata. The notion of using fossils to describe the history of life was not the dominant theme in discussions. On occasion one senses in reading the stratigraphical summaries offered by the presidents that if some other aspect of nature, such as, say, chemical elements, had proved useful for the purpose of marking strata, the interests of geologists would have been equally well served. During this period the primacy of fossils in strata identification may even have been the subject of some propagandizing, as Rachel Laudan (1976) has suggested. Certainly older methods of strata identification were not abandoned. In his study of the Cambrian-Silurian dispute James Secord (1981a) has suggested that Adam Sedgwick continued to rely more on lithology than on fossils in marking out the Cambrian. In addition, during the late 1820s and early 1830s the focus of geologists was often quite local; they referred to "beds" as easily as to "strata". Such parochialism did not lend itself to generalization concerning universal formations.

During the mid-1830s, however, the climate of opinion changed rapidly. Excitement within the Society was palpable as English geologists reached out first to northern Europe, then to the Continent, and finally to the entire world to apply their system of stratigraphical classification. By the end of the decade the geological column had been set in its major outlines. With stratigraphy in hand, geologists then turned to read the geological record as a record of the history of life. This development occurred in the late 1830s and early 1840s. At this point paleontology had clearly established itself as a subfield within geology, a subfield so powerful in its implications that only a good measure of flat-out traditionalism preserved the dominance of stratigraphy within the field.

What, then, does this mean for historical interpretation? Its major effect would seem to be, again, on our understanding of Lyell's *Principles*. It has always been something of a mystery why Lyell took such a fierce stand against the notion of progression in the second volume of the *Principles*. Michael Bartholomew's suggestion that Lyell was motivated by religious and philosophical convictions seems sound (1973). But even so, one wonders how Lyell could have been so bold as to challenge the common opinion with little more than well-argued skepticism on his side. The history of paleontological opinion sketched above provides a possible way out. If the paleontological component of what Martin Rudwick has called the directionalist synthesis was not articulated until the *late* 1830s, at least within the confines of the Geological Society of London, then Lyell surely would have felt freer to propose his own interpretation of the fossil record in the *early* 1830s. Certainly timing had something to do with it. It is hard to imagine that the second volume of the *Principles* would have been the same book if its initial publication had taken place in 1842 rather than in 1832. In 1832 Lyell could hope to be convincing on the subject of non-progression; a decade later one wonders whether he could have entertained such hopes.

Was Lyell's interpretation of the fossil record, then, a complete failure? On this point I should like to go beyond the presidential addresses to suggest that Lyell's skepticism concerning the reliability of the fossil record as a basis for reconstructing the history of life may have been the factor that dissuaded Darwin from relying on the concrete details of paleontology in building a theory of species change. While fossil finds were important in stimulating Darwin to develop a transmutationist theory (on this point see the *Red Notebook* (*RN*)) he was thereafter extremely cautious in describing the descent of species in concrete terms. His characteristically abstract approach to the question of descent is nowhere better illustrated than in the sole diagram that accompanied the first edition of the *Origin*. In contrast to Darwin's abstract approach to the fossil record, one might consider the approach taken by his great contemporary and sometime rival, the French explorer and paleontologist Alcide Dessalines d'Orbigny. D'Orbigny traveled in South America from 1826 to 1834 as a representative of the Muséum d'Histoire Naturelle. Later in his career d'Orbigny wrote on the subject of species succession. His conclusions are contained in his *Prodrome de paléontologie stratigraphique universelle* (1850–1852). Of this work Heinz Tobien has written:

> It consisted of critical lists of all the fossil mollusks and of other invertebrate groups, which were arranged according to their stratigraphic distribution. D'Orbigny made consistent use of this novel approach and divided the sediments and their fossil contents into twenty-seven stages (étages). The stages were named for localities or regions and all were spelled with the same *-ian* ending (*-ien* in French) — Silurian, Callovian, Aptian, Cenomanian, and so forth. Furthermore, the stages were designated by characteristic fossils, and the 18,000 species under consideration were divided into twenty-seven stages. In this manner d'Orbigny obtained twenty-seven successive extinct faunas. He examined the faunas and ascertained that most species in any given stage no longer appeared in the next younger one; rather, they were replaced by new species. He therefore arrived at a conception of successive destructions and creations of animals in the course of the earth's history. (1974, p. 222)

D'Orbigny's empirical paleontological approach to the subject of species succession relied more on the completeness of the fossil record. It was a route toward understanding the historical relationship of species that was not taken by Darwin. That circumstance may perhaps be traced to Lyell's extreme skepticism with regard to the adequacy of the fossil record for reconstructing the history of life on earth.

GEOLOGY AND SOCIETY

As Martin Rudwick and Roy Porter have shown, the pursuit of geology in nineteenth-century England was shaped by the gentlemanly origins and predilections of the membership of the Geological Society of London (Rudwick 1963; Porter 1978). Thus, for example, when English geologists first became professional, they did so by way of employment in the Geological Survey, an institution that permitted the continuation, under another guise, of their gentlemanly preference for field work as against other forms of geological labor.

Everything in the presidential addresses corroborates Rudwick's and Porter's depictions of the gentlemanly habits of English geologists. The one addition I should like to make to their assessment is by way of comment on the nature of the partnerships formed in the 1830s between geologists and representatives of other spheres of English life. The most natural link between geologists and society was obviously economic, but, as Porter and others have shown, English geologists did not generally rise from the mining and manufacturing classes or solicit contacts with them (1973b). Yet in the 1830s, the decade following the period on which Porter focusses, the presidential addresses show geologists very much interested in forwarding the scientific aspects of mining. The role geologists sought to play in this development was not straightforward. They did not wish to descend into mines to advise mine owners; those they sought to advise were elsewhere. What geologists wanted was a partnership between geology and mining through the mediation of the state. In their opinion the interests of mine owners, and thereby of the nation, would best be served if the government were to establish schools of mining on the continental model. The geologists who pressed this goal most ardently were William Buckland and Roderick Murchison, the most establishment-oriented individuals among the Society's presidents. As is well known, geologists did not see their goal realized since no national system of mining schools was established in England during the period. It was, as Porter has said, a case of the "politics of British *laissez-faire*" that "ruled out this excessively obvious solution" (1973b, p. 334).

While English geology and the English economy remained distinct entities during the 1830s, greater integration of interests did occur in the area of greatest concern to geologists: mapping. During the 1830s the Ordnance survey of Great Britain began to provide financial support for the geological mapping of Great Britain. The story is too detailed to be told here, but, in passing, it is interesting to note that it was Henry De La Beche, the geologist who initiated contacts between the Geological Society of London and the Ordnance survey, who himself eventually became the first head of the Geological Survey. It was thus in the nexus of relationships among De La Beche, the Geological Society, and the Ordnance survey that English geology became professionalized. The presidential addresses are full of information on this point. In sum, English geologists during the 1830s sought to integrate their science into the work of the larger society by serving as scientific advisors to the state. In the area of economic interests, where the English state traditionally sought little role, they were not successful. In the military and political arena, however, where the state was willing to meet them halfway, they had greater success. Both geological surveying and geomagnetic research were supported by state funds (Cawood 1977). Foreign geological exploration was also aided by the state, with the *Beagle* voyage being a case in point. It was, after all, at the instigation of the Hydrographic Office of the Navy that Charles Darwin became a member of the ship's company. His presence aboard ship was testimony to effective work of geologists, among others, in their role as scientific advisors to the state.

DARWIN

In considering the literature on Darwin and geology it is instructive to compare the two major contributions to the 1909 commemorative celebrations with the present-day literature. On 24 June 1909 Sir Archibald Geikie gave the Rede Lecture entitled "Charles Darwin as Geologist" at the Darwin Centennial Commemoration at Cambridge University (1909). In the same year J. W. Judd contributed an essay entitled "Darwin and Geology" to the volume *Darwin and Modern Science* published at the suggestion of the Cambridge Philosophical Society (1909). Both Geikie and Judd were eminent geologists and well prepared to comment on Darwin's place in the history of geology. Geikie was the foremost British geologist of his generation in the sense of representing the science in an official capacity. Judd had known Darwin intimately, having served, after Lyell's death, as Darwin's primary contact with the field of geology.

Geikie and Judd viewed Darwin from the perspective of being his younger colleagues. Geikie was born in 1835, Judd in 1840. In writing on Darwin's career they emphasized the event of which they had greatest personal knowledge: the impact of the publication of the *Origin* on geological science. In addition they emphasized Darwin's links with Charles Lyell, whom they also knew personally. On both these topics Geikie and Judd wrote superbly. In writing on decades of which they had no personal knowledge, however, they are less reliable. Thus, for example, in describing the background to Darwin's achievements, Geikie collapsed the developments of the "Golden Age" of geology into a simple story of the triumph of Plutonism over Neptunism and of uniformitarianism over catastrophism.

In contrast, present-day historians write more convincingly on Darwin's geological work before the *Origin*. They have the advantage, which Geikie and Judd did not, of having access to Darwin's manuscripts. To cite only two examples, Wilson's depiction of Darwin's relation to Lyell and Stoddart's analysis of Darwin's coral reef theory have both benefitted from study of manuscripts unavailable to previous generations of scholars (Wilson 1972; Stoddart 1976). In addition, present-day authors understand more of Darwin's early geological work than did Geikie or Judd because they have a more detailed and sympathetic understanding of early nineteenth-century geological science in general. Thus, for example, whereas Geikie in his 1909 address discreetly avoided discussing Darwin's Glen Roy theory — it was after all erroneous and hence an embarrassment — Rudwick has used the same material to illuminate Darwin's intellectual development (1974a). Of course questions remain to be answered with respect to Darwin's early geology. Darwin's method in working out his coral reef theory has been carefully and sympathetically described by Michael Ghiselin (1969); other portions of Darwin's geological work await similar treatment. In addition, Darwin's geological specimens from the *Beagle* voyage have yet to be identified (see D. Porter, this volume) and his work on South American strata evaluated. But, overall, historians presently active have gone far towards placing Darwin's geology in context.

Conclusion

From the abundance of citations in the Historiographical Afterword it should be clear that whatever novelty exists in my interpretation of the situation of English geology in the 1830s lies not in the introduction of wholly new views but rather in the adjustment and rearrangement of the elements of numerous interpretations. This adjusting and rearranging was undertaken with the intention of providing a more unified description than presently exists of the state of English geology in the 1830s, and of locating Charles Darwin against that background.

Notes

1. Consider William Fitton's remarks in his address of 1828: "The whole series indeed, of the phaenomena developed by recent examination in Scotland and the north of England, gives rise to the most interesting speculations on questions of geological identity, and of the relative value in geology of mineralogical and zoological characters, — which has been so ably treated by Brongniart and other continental writers: —" This passage is interesting as it shows Fitton wanting to take a "balanced" view on the question of the relative merits of lithology and fossils in identifying strata (Fitton 1828, p. 59).
2. "New" is meant here in the sense of "newly emerging" or "newly articulated." By asserting the newness of the goal I wish to question Rudwick's characterization of the intentions of geologists in the sentence: "But behind this enterprise [stratigraphy] lay the higher theoretical goal of reconstructing the main outlines of the history of the earth and the history of life on its surface" (1979b, pp. 10–11). It is the word "behind" in this statement that strikes me as wrong, for it suggests an ulterior motive in the minds of geologists, who could then be presumed to have been interested in stratigraphy only, or primarily, as a way of getting at "higher" theoretical goals. The presidential addresses do not appear to me to support so strong an inference, especially for the early years in the period. Indeed it is the emergence of the goal in this period that is interesting to trace.
3. I am presently engaged in a longer study of Darwin's geology to be published by Cornell University Press. I have completed two shorter pieces on the same subject (Herbert 1982; Herbert, 1983). The first of these compares, in greater detail than in the present essay, current views of Darwin's geology with those held by speakers at the 1909 centenary celebration of his birth; the second compares Darwin's and Lyell's views on elevation and on coral reefs. The point of departure for all of these papers has been the chapter in my dissertation entitled "Darwin as a uniformitarian geologist: his attempt to construct a general theory of elevation and subsidence" (Herbert 1968).
4. I am indebted for this reference to Nancy Mautner.
5. Lyell began his treatment of Darwin's correspondence with the enthusiastic remark that "few communications have excited more interest in the Society than the letters on South America addressed by Mr. Charles Darwin to Professor Henslow", and then went on to summarize the contents of the letters, taking care to emphasize the points where Darwin's observations complemented his own.
6. The major papers in which Darwin expressed his ideas of elevation and subsidence were as follows: "Observations of Proofs of Recent Elevation on the Coast of Chili, Made during the Survey of His Majesty's Ship Beagle, Commanded by Capt. Fitzroy, R.N." [Read 4 January 1837] (*CP* 1: 41–43), "On Certain Areas of Elevation and Subsidence in the Pacific and Indian Oceans, as Deduced from the Study of Coral Formations" [Read 31 May 1837] (*CP* 1: 46–49), "On the Connexion of Certain Volcanic Phenomena in South America; and on the Formation of Mountain Chains and Volcanos, as the Effect of the Same Power by which Continents are Elevated" [Read 7 March 1838] (*CP* 1: 53–86), and "Observations on the Parallel Roads of Glen Roy, and of Other Parts of Lochaber in

Scotland, with an Attempt to Prove That They Are of Marine Origin" [Read 7 February 1839] (*CP* 1: 89–137).

7. Roderick Murchison used some of Darwin's specimens from the Falkland Islands to help establish his claim that Silurian strata were of worldwide distribution. He wrote in his *Silurian System*: "There is . . . a phenomenon of the highest importance, connected with the distribution of organic remains in the older strata, which has not been adverted to; namely, that the same forms of crustaceans, mollusks and corals, are said to be found in rocks of the same age, not only in England, Norway, Russia, and various parts of Europe, but also in Southern Africa, and even at the Falkland Islands, the very antipodes of Britain." To this passage Murchison added the following footnote: "The fossils from the Falkland Islands were discovered by Mr. C. Darwin, and they appear to me to belong to the Lower Silurian Rocks" (1839, p. 583).
8. After discussing Darwin's collection of extinct South American mammals, Lyell made the following remark in his presidential address of 1837: "These facts elucidate a general law previously deduced from the relations ascertained to exist between the recent and extinct quadrupeds of Australia; for you are aware that to the westward of Sydney on the Macquarie River, the bones of a large fossil kangaroo and other lost marsupial species have been met with in the ossiferous breccias of caves and fissures" (1837, p. 511). On Lyell and the law of succession, also see Corsi (1978).

17

Darwin and the World of Geology (Commentary)

Martin J. S. Rudwick

This note is a brief comment on Herbert's interpretation of Darwin's first chosen field of serious scientific research, the field in which he first earned respect as a highly competent "gentleman of science". Herbert takes as her text a published comment of mine about what I termed the "dominant cognitive goal" of geologists at the period when Darwin joined their company (1979, pp. 10–11). I want to explain why this did not in fact imply a "narrow definition" of geology, and why there is therefore no paradox in identifying Darwin (and of course his older mentor Lyell) as central figures in London geology at the time, although they were relatively marginal to the particular cognitive enterprise I mentioned. In short, her disagreement with me is more apparent than real.

Herbert takes the successive anniversary addresses of the presidents of the Geological Society of London, and justifiably treats them as sensitive indicators of what the leading geologists in that circle regarded at the time as the most significant current research. She might have added that one reason why these addresses are such a good barometer of opinion is that the presidency was not controlled by any one faction, and that those who filled the position in the relevant years therefore expressed a wide representative range of viewpoints on both general and specific issues. Analyzing these addresses, Herbert is also right, I think, to stress that what was later termed "stratigraphy" — the correct ordering of the sequence of strata in the Earth's crust — emerges as the kind of geology most consistently emphasized. This is what I termed the "dominant cognitive goal", not only of the London society but of geology as an international enterprise. But I deliberately termed it dominant, not exclusive. Before relating it to one of its important but subordinate counterparts, I should first clear up a confusion about the character of the dominant stratigraphical enterprise itself.

I claimed that "behind" straightforward stratigraphy lay "the higher theoretical goal" of reconstructing the *history* of the earth and of life. I readily accept Herbert's criticism that this historical approach did not lie — even implicitly — behind *all* the work that was being done on strata in the 1830s. On the contrary, the pervasive language of "classification" in discussions of strata, and the common use of the terms "higher" and "lower" in preference to "younger" and "older", are sure signs that many practitioners were *not* going beyond questions of structure and sequence to questions of history (Rudwick 1982a). Nonetheless, there was a continuing tradition of historical reconstruction in geology. Its most influential exemplar, and not only in Britain, was none other than William Buckland's work on the cave faunas from what he interpreted as the immediately pre-"diluvial" period (1823). Only the bad press that Buckland has had in the "Genesis and geology" school of historiography has obscured the significance of this work for the broader historical ambitions of his Oxford student Lyell. I do not think Lyell's contemporaries were surprised to learn that he had been praised so highly by the Royal Society for the contribution of the *Principles of Geology* to Tertiary stratigraphy. Admittedly the micro-politics of the two societies were involved here, for those who were critical of Lyell's broader speculations in geology were anxious to ensure that his work was also publicly approved on more solidly empirical grounds. Nonetheless, Lyell himself did present his work on the Tertiary strata as an exemplar of his whole approach and as the culmination of his three-volume magnum opus (Rudwick 1970). Many of his contemporaries agreed with that assessment, precisely because Lyell's work on the Tertiary strata suggested to them how the whole business of stratigraphy could be enriched by being transformed from a primarily structural matter of sequences of strata into a fully historical matter of sequences of events.

I think Herbert is right to see this transformation of stratigraphy into earth-history as a *growing* feature of the dominant cognitive enterprise in geology during the 1830s, rather than one that was fully fledged already; and I agree that my earlier statement was misleading in this respect. My own impression now — although this needs much more thorough study — is that the spread of the new approach was due particularly to the initiative of a *younger* set of geologists following Lyell's lead. While an older set of men such as Sedgwick (b. 1785) and Murchison (b. 1792) was rehabilitating William Smith as British geology's pioneer hero-figure, Smith's own nephew John Phillips (b. 1800), and younger geologists such as Robert Austen (b. 1808), were arguing for a much more historical and biological use of fossils in stratigraphy. It was their work that gradually brought men like Murchison to the fully historical viewpoint expressed in his public statements in the 1840s.

This growing emphasis on the historical reconstruction of ancient environments and their organisms brought the traditionally dominant enterprise of stratigraphy much closer to that other enterprise in geology that Herbert rightly emphasizes, namely the search for at least "proximate causes" for the "dynamics" of the earth, to use Whewell's terms. This kind of geology, however, was already a well-established tradition within the London society, to look no further afield; but admittedly it received much less attention than stratigraphy, from most of those who regarded themselves as the most competent practitioners of the science. Like the historical component of the stratigraphical tradition, this dynamic or causal tradition had its immediate roots in the diluvial debates of the 1820s; for those debates had largely been concerned with the proximate causes of what all parties agreed were some very peculiar phenomena. It is important to note that by 1830 Lyell was not the only major figure to exploit the causal approach in a more generalized and ambitious manner. For example, the enthusiasm widely expressed for Léonce Élie de Beaumont's theory of mountain ranges was not just a devious way of doing down Lyell. It reflected a genuine — but far from uncritical — admiration for a bold causal theory with wide explanatory power, firmly grounded in the highly plausible physical model of a contracting earth (Rudwick 1971). Lyell's particular approach, with its distinctive emphasis on slow, gradual processes, and its much more controversial rejection of any linear direction to earth-history, must therefore be regarded as a project that was seen at the time to be competing on equal terms with others, within a well-established tradition of causal interpretation in geology.

To summarize my interpretation of the cognitive structure of geology at this period, I still contend that the stratigraphical tradition was dominant. But it was being modified from within by the infusion of more fully "historical" cognitive goals. That modification was effectively narrowing the gap between stratigraphy and the subordinate but nonetheless important enterprise of dynamic and causal interpretation. (Other aspects of geological science, such as mineralogy and paleontology, were in effect treated at this period as *auxiliary* subfields of geology proper.)

This outline description of major and minor enterprises within geology must now be related to the social and cognitive "topography" of practitioners of the science. Here the visible landscape of formal institutions such as the Geological Society of London and the Société géologique de France must be regarded as superimposed on a much more important, invisible topography: the informal and tacit landscape of "*ascribed competence*" (Rudwick 1982b, Fig. 1). At the center of this landscape, there was a small international "élite" of those who, even when they disagreed strongly on specific issues, regarded each other in practice as competent to pronounce on the most fundamental matters of theory and method in geology. Of lesser ascribed competence were those I have termed "*accredited*" geologists — men whose expertise and interpretations were accepted and valued within the limits of their more local or specialized first-hand experience. Still further down the gradient of ascribed competence were "*amateurs*" (of both sexes) whose factual observations and collections were regarded as trustworthy, but whose interpretations (if any) were ignored. The general *public* was not even trusted that much.

I believe that this kind of graded, weak-boundaried "topography" is a much more faithful representation of the world of geology in the 1830s than anything conveyed by the anachronistic use of terms such as "geological community", "professional geologists", or "discipline of geology". And it allows us to resolve the apparent paradox that Lyell could be an important figure in the central zone of élite geologists, even though he contributed much more to the subordinate research tradition of dynamic and causal interpretation than to the dominant tradition of stratigraphical description and correlation. For as a recognized member of the international élite of the science, he had earned the tacit authority to enlarge the cognitive goals of both enterprises, and so to move them both towards what Herbert rightly notes as the universally acknowledged goal: the goal of a comprehensive and unified science of geology.

Having sketched the cognitive and social landscape of geology in the 1830s, as I understand it, I can now turn to Darwin himself. Herbert rightly emphasizes the importance of Darwin's induction into geological practice on the eve of the *Beagle* voyage, particularly through his brief field trip to Wales with Adam Sedgwick. This should help to suggest the implausibility of any suggestion that Darwin chose Lyell's definition of geology in opposition to Sedgwick's, as if they were incompatible. Of course Darwin was deeply influenced by Lyell's *Principles* once the voyage had begun, and particularly by Lyell's emphasis on the explanatory power of currently observable processes ("actual causes"). But Darwin could never have made the kind of observations — or the kind of notes — that he did make in South America, if he had not first been initiated by Sedgwick into the tacit knowledge of field geologists and their routine practice of structural and stratigraphical geology. Without that initiation he might still have been, in my terms, a competent "amateur" collector of geological specimens (cf. Sulloway, this volume); but as it was, Sedgwick transformed him implicitly and potentially into an "accredited" geologist. In doing so, he ensured that Darwin's local interpretations, based on informed first-hand observation — of strata, among other things — and the collection of relevant specimens, would be treated as authoritative and trustworthy.

The new status that Darwin enjoyed is clearly shown by John Henslow's treatment of the letters that Darwin sent him later from South America, and by the exposure of those reports in expert semi-public arenas in Cambridge and London even before his return. Darwin's letters were of immediate interest, because they bore directly on the current "focal problem" of recent continental elevation, and also on the broader issue of the origin of mountain ranges. On both levels his evidence clearly favored Lyell's views, as Herbert mentions; but we should not infer from this that other élite geologists such as Sedgwick felt that Darwin had deserted them. For they would have realized that, whatever Darwin's theoretical inclinations, he was likely in any case to become a valuable addition to the circle of those who contributed actively to the lively and fundamental debates that characterized Geological Society meetings.

Meanwhile, on the other side of the world, Darwin was fitting his geological observations with increasing consistency into the causal explanatory framework provided by Lyell. On a few points, the most famous being his interpretation of coral islands, Darwin improved on Lyell in Lyellian ways (Stoddart 1976). But I do not think it is true, as Herbert suggests, that he was enlarging the *scale* of Lyellian explanations. Lyell himself had certainly been seen by his earliest readers as having explanatory ambitions on a global scale: for example, Henry De La Beche's draft caricature showing Father Time holding Europe and America in a huge balance, the one continent subsiding as the other rose (Rudwick 1975b, Fig. 7), is a clear indication that Lyell's own model of crustal oscillation was perceived as being just as large in scale as Darwin's was to be somewhat later. What Darwin did do, I think, was to go beyond Lyell in attempting a more directly *causal* explanation for such large-scale crustal movements.

This, however, brings us forward to Darwin's early London years. The importance of that brief creative period is hardly controversial; but I think it should be pointed out that a fully non-retrospective analysis of Darwin's career would give far more attention to the creativity displayed in Darwin's *public* science at this time than has been customary in recent scholarship. I have argued elsewhere that what we can see reflected in the succession of Darwin's early geological papers is his rapid trajectory from "accredited" to "élite" status within the Geological Society (Rudwick 1982b). His 1838 paper on elevation (*CP* 1:53–86) marks his tacit acceptance into the élite zone, because it presented a tentative theoretical explanation of crustal elevation *in general*. It was no longer tied to the regions of which he had authoritative first-hand experience, but was in principle of global validity. This theory was sufficiently important to Darwin for him to interrupt his other work at this point, and to undertake the *only* major field trip of his London years (or thereafter). The research he then did around Glen Roy was important to him — at the very least — as an elegant exemplar of his theory that major segments of the earth's crust rise (and fall) in a highly "equable" manner because they are floating on a deeper fluid layer —in his own revealing phrase, "like the sea beneath the polar ice" (Rudwick 1974a, p. 139).

In public science such as this, Darwin was openly pursuing the dynamic causal enterprise in geology in its distinctively Lyellian form. He did so with great success, as perceived by other élite members of the world of geology. He was not only tacitly accepted into the élite zone of geologists generally, but also more specifically into the "core-set" (Collins 1981) that was centrally concerned with the focal problem of recent elevation and with the wider issue of the nature of crustal mobility in general. Stated another way, Darwin joined Lyell as a younger but unquestionably up-

and-coming figure in the Geological Society of London, and as a highly promising advocate of the subordinate but significant enterprise of dynamic and causal interpretation. There is no paradox in pointing out that at the same time his contributions to the dominant enterprise of stratigraphy were even more marginal than Lyell's. The allegedly Silurian fossils that he collected from the Falkland Islands, for example, were eagerly exploited by Murchison to promote his view of a world-wide Silurian System (Murchison 1839, p. 8); but Darwin's role in that focal problem was little more than that of a competent collector. He played a similarly marginal role in the Devonian controversy (Rudwick 1979b, 1985), and only became involved at all because Austen argued that some of the limestones on which that focal problem hinged were ancient coral reefs. Since the equally young Darwin had already made himself the Geological Society's unquestioned expert on modern coral reefs, he was brought in to evaluate the claim.

Such minor involvements are less important, however, than the fact that Darwin in these years used the Geological Society as the primary locus of his sense of identity as a man of science. When asked to serve under Whewell as one of the Society's secretaries, he was at first reluctant to accept; but it was an appropriate appointment, and he must have gained as much from it in personal interaction with other geologists as he lost in research time. Although speculating privately on issues that reached far beyond the Society's self-set terms of reference, Darwin presented himself publicly as a geologist, as Herbert rightly emphasized some years ago (1977). I would go further than she does, however, in claiming the positive importance of Darwin's public career in the Geological Society as an arena in which he learned to be a competent *theorist* in science — or at the very least, learned to have *confidence* in his powers as a theorist (Rudwick 1982b). Ghiselin pointed several years ago to the formal analogies between Darwin's innovative work on coral islands and his later theorizing on the species problem (1969); and I argued similarly for the parallels in structure and presentation between his first major scientific paper, on Glen Roy, and the *1842 Sketch* of a species theory that he composed only four years later (Rudwick 1974a). But such parallels and analogies are only the structural or methodological counterparts of a more substantive linkage between Darwin's geology and his biology — a linkage more widely recognized now than it was at the time of the 1959 centennial. For although geology and geologists already had a clear sense of "disciplinary" identity, it was one that allowed the world of life a very large role in the definition of its activities. In the stratigraphical tradition, this role was expressed in what in these years was just beginning to be called "paleontology". In the dynamic and causal tradition, it was expressed in what was later to be called historical biogeography. Darwin was not breaking any new ground, of course, but merely following in Lyell's footsteps, when he extended his speculations from strictly geological issues to questions involving historical biogeography, even though in the longer run those speculations took him further and further away from the active arena of geological debate.

Darwin's slow trajectory out of the élite zone of geology, and back into the status of a merely "accredited" geologist, belongs to his post-London years. (Of course, he moved simultaneously up into the élite of natural historians.) His illness, family commitments, and consequent inability to travel extensively, were certainly contributory factors in this shift. Furthermore, the writing up of the *Beagle* geology became a burdensome chore to him, because the required format of detailed monographs gave him little scope for the clear development of his earlier theoretical ideas. What he termed, much later, his "great failure" (*Autobiography*, p. 84) or "gigantic blunder" (*ML* 2:188) over Glen Roy was *not*, I think, so important; for the plausibility of Agassiz's Ice Age explanation was still low in the 1840s, and that of Darwin's work correspondingly high. What seems to me to be by far the most important factor in Darwin's effective decision to move out of geology — apart from the obvious fact that by this time he was doing highly creative private work in a related but separate field — is his perception of the direction of his career. For even by 1840 he must have sensed that, although he had gone beyond Lyell in some respects, he was unlikely to be able to make enough of an innovative break *within* geology (in the narrower sense) to get out of Lyell's shadow and thereby fulfill his legitimate ambitions as a man of science. It is at this point, therefore, that we can make sense of his growing involvement on the private level with the crucial but refractory biological problem of the origin of species, not only in terms of an evolving "network" of personal research enterprises but also in terms of the construction of a scientific career.

I think, however, that what is implied by Herbert's paper is that Darwin's species work *emerged* from a pre-existing *matrix* of his geological work. This conclusion does not demote his geology into a merely preparatory role: on the contrary, in any non-retrospective analysis Darwin's geology must always be treated as highly important in its own right, since it was for a time *his* central concern. Nor, on the other hand, does my formulation underplay the importance, in this process of emergence, of inputs from right outside geology, particularly Darwin's long-standing concern with the traditional problems of "generation" (see Hodge, Sloan, this volume). The relation between these various concerns is, I think, most accurately reflected in the pattern of filiation of Darwin's early notebooks, as Sandra Herbert summarized it in visual form a few years ago (1977, Fig. 3). When Darwin "opened" the first of his famous transmutation notebooks (*B*), he also began simultaneously another new notebook, to which he gave the *first* letter of the alphabet, to receive his continuing notes and speculations on geology. That surely marks the point at which he himself became fully aware that his "species-origins" enterprise had taken on a life of its own, having emerged fully from the primarily geological matrix which we can trace in the earlier

Red Notebook (*RN*) and which continued in its own right in the A *Notebook* (Rudwick 1982b, Fig. 2).

Whether detected at the level of notebook analysis, therefore, or on the broader level of the pattern of Darwin's public activities as a young man of science, I conclude, with Herbert, that we should see the young Darwin first and foremost as a geologist. But I would add that his concern with the species problem emerged from that pre-existent matrix. Only some such formulation will satisfy the requirements of a non-retrospective description and analysis of Darwin's career. That his species work would eventually overshadow his geological work was not an outcome that could be known in those early years — even, with any certainty, to Darwin himself. As historians we owe it to Darwin's memory not to distort the shape of his career in science as he himself forged and experienced it.

18

DARWIN AND THE BREEDERS: A SOCIAL HISTORY

James A. Secord

In 1898, sixteen years after the death of Charles Darwin and two years after that of his wife Emma, the fate of their famous residence at Down was highly uncertain. The aging botanist Joseph Hooker, writing to Darwin's son George, suggested that the historic house might well be saved for future generations by turning it to practical use as an experimental station for the study of animal breeding (Atkins 1974, p. 101). Although never taken up, the idea was an appropriate one. Almost from the very beginning of his career as a transmutationist, Darwin looked to the work of animal and plant breeders for clues to the mysterious processes underlying reproduction, and as is well known, he founded the argument of the *Origin* upon an extended analogy between selection by man and selection by nature. From their factual grounding to particular innovations in theory, from their underlying metaphysics to their argumentative structure, the *Origin* and its offshoots reflect in a variety of ways Darwin's immersion in the world of the Victorian plant and animal breeders.

The present essay is a preliminary examination of Darwin's studies of domesticated animals and plants from the standpoint of the social history of ideas.[1] Over the centuries, generations of horticulturalists, beekeepers, cattle and sheep breeders, pigeon fanciers, and nurserymen had accumulated vast stores of practical experience about variation, inheritance, generation, and selection. "Man," as Darwin put it in 1868, ". . . may be said to have been trying an experiment on a gigantic scale; and it is an experiment which nature during the long lapse of time has incessantly tried" (*Variation*, 1:3). In Darwin's view, the very essence of his theory of transmutation demanded that he transgress the boundaries of his own scientific community, so that the lessons of the breeders' "experiment" might be applied to the vexed question of the origin of species in nature. Previous theorists, he believed, had been led sadly astray through inattention to the wealth of knowledge possessed by the breeders. Darwin determined not to repeat their mistake. His conviction of the importance of the breeding community as a source of information is immediately apparent from the species notebooks, the *1842 Sketch* and *1844 Essay*, *Natural Selection*, the *Origin* itself, and above all the two volume *Variation of Animals and Plants under Domestication* of 1868. Virtually the only parts of his species project that received what Darwin viewed as a full and satisfactory depth of treatment were those for which the breeders' aid was absolutely essential.[2]

Recent students of the 1837–1839 notebooks on transmutation have convincingly demonstrated that the notion of artificial selection, far from leading Darwin to his discovery as he later claimed, initially functioned as a barrier to it. As practiced by the breeders, selection seemed limited in extent and productive of maladapted oddities (Kohn 1980). But it would be highly unfortunate if the lack of a role for the breeders on 28 September 1838 — the day of the Malthus reading — led to a discounting of the importance of their work for the rest of Darwin's career. As Gruber has pointed out (Gruber and Barrett 1974; Gruber 1980a), his theory is best viewed as a series of structures of thought, molded in important ways from his early childhood to the very process of correcting the proofs of his books. Darwin's publications, reading lists, and correspondence leave no room for doubt that the literature and personnel of the breeding community had crucial roles to play in this complex story. Many months before discovering the relevance of artificial selection in the Winter of 1838, Darwin fully appreciated the potential importance of the breeders for his inquiries into generation, variation, and inheritance, and over two decades later the *Origin* reaffirmed his belief that domestication "afforded the best and safest clue" to a host of particular problems bearing on transmutation (p. 4).

Considered as a problem in scientific communication, as they will be here, Darwin's links with the breeders of Victorian England are of special interest in that they necessitated attention to fields largely outside the traditional boundaries of natural history. To place his contacts in context, this essay thus begins by outlining the established relationship between the natural sciences and the practical arts of breeding plants and animals; it then characterizes Darwin's own position with respect to these two communities. With the social geography of the Darwinian enterprise established, the second section of the paper examines the particular methods Darwin used to reach the breeders — the letters, reading, experimentation, personal contacts, and questionnaires through which he gained acquaintance with a body of knowledge unfamiliar to most contemporary men of science. In the two concluding sections of the essay, special attention will be focussed on a pair of issues raised by these attempts to reach the breeders. First, from Darwin's own perspective as an expert naturalist, information from a group outside natural history could only be used after suitable reinterpretation. His role as a translator between widely different contexts of evaluation will accordingly be considered. Finally, the tables will be turned and the entire subject of "Darwin and the Breeders" will be briefly examined from the opposite point of view, that of the breeders themselves. How did they perceive Darwin's species project, and why did they help so unstintingly in his search for information about domesticated animals and plants?

I. A Pattern of Interchange

Darwin's systematic attempt to associate with breeders of plants and animals affords a revealing example of a scientist ranging outside the limits of his particular speciality, seeking out and actively using resources available in the wider cultural milieu. Historians have recently focussed attention upon a number of such instances, the most famous of all being Darwin's own extensive reading in political economy (Young 1969; Schweber 1977; Manier 1978; Shapin and Barnes 1979). Such ideologically charged instances of cross-disciplinary links are justly famous and worth continued scrutiny. One suspects, however, that these spectacular cases have sometimes been allowed to obscure the more mundane connections that have typically joined science with other areas of culture. For example, men of science from the seventeenth century onward have drawn upon travel narratives as a source of information, although these works have often been composed by non-naturalists writing with entirely different aims in mind. An analogous case is provided by the Victorian geological community's use of the knowledge of miners and quarrymen, where practical expertise was drawn upon for the advancement of the earth sciences. Such non-controversial borrowings from the wider milieu of science, like Darwin's contacts with the breeders, have often been taken for granted and viewed as unproblematical. Although Darwin's efforts in this sphere are often referred to in passing, they have never been studied systematically or in detail.

Let us look, then, at the intellectual and social orientation of natural history with respect to animal and plant breeding in the mid-nineteenth century. By doing this, even in a tentative fashion, we should be in a position to assess the actual availability of the resources of the breeding community to an individual like Darwin. Moreover, it should also be possible to gain insights into the reciprocal relations of the breeders and naturalists more generally; in this sense, the study of Darwin's contacts opens up the larger historical problem of the nature of the boundary between science and an important part of agricultural practice in the Victorian era.

One point bears emphasis from the very beginning. Information on plant and animal breeding was very widely diffused in Darwin's England, much more so than the major doctrines of Malthusian political economy or Scottish moral philosophy. For all the upheavals of the Industrial Revolution, England remained in many respects a largely agricultural nation with an important proportion of the populace engaged in the production of food, and most wealth and power securely anchored in land (F. Thompson 1963; Orwin and Whetham 1964). Given this situation, it seems rather paradoxical that historians of science have often referred to Darwin's readings in the literature on domestication as obscure. Although his sources on this subject may be difficult to find in modern libraries, for the most part Darwin utilized standard handbooks and journals on the shelves of country houses and farms throughout the kingdom. The *Gardeners' Chronicle* and the *Cottage Gardener* — his most regular readings in this field — were popular newspapers that circulated in thousands of households virtually indistinguishable from the establishment at Down. In this contemporary context it is thus Darwin's reading in *science* that would have appeared esoteric and specialized. By almost any standard the numbers of those concerned with practical plant and animal breeding completely dwarfed those participating in any aspect of the sciences. Nearly all of the major agricultural districts in Great Britain held an annual show of produce and prize animals, while the peripatetic British Association for the Advancement of Science had to muster all its forces merely to hold one meeting each year (Hudson 1972; Trow-Smith 1959, pp. 224–232; Morrell and Thackray 1981). On the metropolitan scene the membership of the popular Geological Society totalled less than a third that of the Royal Agricultural Society. Figures for the circulation of scientific and agricultural periodicals show a like disparity (Woodward 1907; Watson 1939).

Given the critical place of animal and plant improvement in the Victorian economy, the natural sciences could not conceivably have remained untouched by the activities of the breeders: agriculture, livestock husbandry, horticulture, and so on were much too pervasive to be ignored or set aside as irrelevant to scientific inquiry. In consequence, as we shall see, important links often tied practice and science, and Darwin used these connections in elaborating his transmutation theory. But for the most part domestic breeding and the sciences of life were pursued by separate individuals, in separate organizations, and in separate publications. Those occupying the borderland between the two spheres frequently lamented the lack of interchange between them. "In my opinion," wrote a correspondent from Glasgow in an early number of the *Gardeners' Chronicle*, "Botany should be studied by all Gardeners; but I am sorry to say that I have found that five out of six know no more of the classification of flowers than they do of steering a ship" (Towers 1842). Many scientific men were equally ignorant of cereal plants and garden vegetables. "Botanists have generally neglected cultivated varieties," Darwin once remarked, "as beneath their notice" (*Variation* 1:305). An even greater distance appears to have separated zoologists from animal breeders. The Reverend Edmund S. Dixon, himself an amateur naturalist and an enthusiastic breeder of pigeons and poultry, decried the mutual distrust that often kept these two activities distinct. In an article published in the *Quarterly Review* he even spoke of them as occupying different worlds:

> Everybody knows that there is a fashionable world, a literary world, a sporting world, and a scientific world; but everybody does *not* know that there is a poultry world, with its jealousies, excitements, pre-eminences, and interests, just like any of the other worlds that revolve

> "cycle on epicycle, orb on orb" in the midst of the great universal world itself. The grand evil is that the poultry world has hitherto been kept to a great degree distinct from the scientific world, to the disadvantage of both these respectable spheres. Not a few renowned naturalists have disdained *in toto* the scrutiny of domesticated animals. (1851, p. 324)

Writing in 1864, the prominent agriculturist Harry S. Thompson recalled in a presidential address to the Royal Agricultural Society the situation that had prevailed at the time of its foundation a quarter century earlier. "It is scarcely an exaggeration to say," he wrote, "that the thoroughbred British farmer of that day despised science as much as he feared Free-trade, and that the only things which commanded his entire confidence were his father's experience and his own skill. . . . The first attempts of the farmer and the philosopher to run in couples were certainly not encouraging. They conversed with one another in unknown tongues, and many of the early specimens of scientific practice . . . were decided failures" (1864, p. 51). Even when Thompson wrote, the principal links between science and agriculture involved chemistry, where the experimental results of Justus von Liebig had been incorporated into a significant segment of agricultural practice. In comparison, the direct connections with the biological sciences remained relatively circumscribed. Typically, agriculture possessed no section of its own at the British Association, and its existence was recognized only grudgingly in the 1840s by classing it under chemistry (Morrell and Thackray 1981, p. 456); the formation of an agricultural section had to await the new century. In suggesting the conversion of Down House into an experimental farm for studies in animal breeding, Hooker could point to the long-established tradition of research at Rothampsted Station — whose main achievements were in soil science and agricultural chemistry (Atkins 1974, p. 101; E. J. Russell 1966).

The reasons for the apparent divergence of interests are not difficult to find. As with any economically important endeavor, the production of new varieties of plants and animals engendered trade secrets and an unwillingness to broadcast new methods and results. "The men who live by the propagation and sale of valuable beasts and birds," wrote E. S. Dixon, "have had their lips sealed by the dread, that while they were communicating some natural fact, they might betray some precious secret . . ." (1851, p. 324). From the naturalist's perspective, the study of domesticated creatures necessitated ties with commerce and the world of business, an environment many of them had endeavored with great diligence to escape. The dismissal of the world of the breeders had important intellectual roots as well, for men of science typically believed that the products of man's selection were "monstrosities" altered from a naturally occurring progenitor and rendered incapable of surviving or reproducing in the wild.

Despite these fundamental differences in approach, men of science and the breeders came into contact along a broad front of common concerns. These intersections of interest are of special importance for Darwin's work, for they provided the foundations for his own elaborate network of investigation and inquiry. But for these linkages to be fully understood, the complexity and diversity of the contemporary community of breeders must be appreciated, even if only in outline. In viewing all those concerned with domesticated plants and animals as a single homogeneous group, "the breeders", I have thus far been approaching them just as Darwin and the members of his circle did: from the outside. It is much more accurate to speak of a single scientific community in Victorian England, embracing the extremes of astronomical physics and taxonomic botany, than it is to lump together pigeon fanciers with sheep breeders, rose growers with rabbit raisers, or beekeepers with veterinarians. In reality Darwin had to contact not just one group of practitioners and a single body of literature on domestication, but rather a great many, each with its special institutions, publications, and practices. Moreover, each of these segments of the breeding community possessed a different relation to the appropriate field of research in the world of natural history, thus producing many variations on the overall pattern of interchange that I have already sketched. In the end the picture is an extremely complex one. A more complete understanding will have to await careful historical investigation involving particular examples, using more sensitive tools for charting the diffusion of scientific and practical information among various social groups in Victorian England.

The extent and character of the conjunctions of interest between breeders and naturalists were by no means uniform across the range of domesticated plants and animals. Most significantly, botanists generally appear to have possessed a much deeper concern with horticulture and gardening than did zoologists in the corresponding practical arts of animal husbandry. Men like James E. Smith, William and Joseph Hooker, and John S. Henslow cooperated with prominent landowners by introducing new plants from abroad and improving those already cultivated (Brockway 1979). The Honourable and Very Reverend William Herbert was only the most prominent of a host of practical botanists in England who experimented with plant hybridization on a systematic basis (Guimond 1972). Scientific horticulture was pursued even more vigorously on the continent, as exemplified by the outstanding work of Joseph Kölreuter and Karl Friederich von Gärtner in Germany. Insofar as posthumous fame is concerned, this long-standing tradition culminated in Gregor Mendel's studies of the pea (H. F. Roberts 1929; Olby 1966a, 1966b; C. D. Darlington 1937). In comparison most writers on zoological topics referred only fleetingly to the relevant domesticated varieties, locating them in the appropriate place in a classificatory scheme, but without detailing the numerous peculiarities so prized by the fancier or practical breeder. Thus even ornithological authors like William Yarrell or Hugh Strickland who did converse with breeders would typically do no more than note that the common fowls of the farmyard belonged to the species *Gallus bankiva*, thus stopping at precisely the point where a serious poultry fancier picked up the subject. Citations of scientific books by poultry and livestock breeders were equally cursory, usually appearing only in an introductory chapter or two and limited to treatises of the most general kind, such as the English translation of Georges Cuvier's *Règne Animal*. Lengthy practical manuals like those of the veterinarian William Youatt on horses (1831), cattle (1834), sheep (1837), and pigs (1847) were useful and systematic, but their overall perspective differed radically from that of contemporary treatises in scientific natural history. Only at the very end of the nineteenth century, and explicitly in connection with the acceptance of Darwinian evolution theory, were skeletons and skins of domesticated animals placed on show in the Natural History Museum in London (Stearn 1981, p. 185).

Among all the naturalists of his generation, Darwin was especially well placed to bridge any gaps between naturalists and breeders, to take advantage of the wealth of experience and empirical data possessed by horticulturalists, livestock men, gardeners, and poultry fanciers. Growing up in the middle of one of the richest farming districts in the kingdom, from his childhood Darwin would have been familiar with the raising of horses and dogs, the breeding of pigeons (which his mother had kept), and the rudiments of kitchen gardening. Appropriately, one of his earliest memories concerned a cow dashing across the garden of the family house in Shrewsbury (*ML* 1:1). Darwin's uncle Josiah Wedgwood, as Michael Ruse has pointed out, was an eminent breeder of sheep and had introduced three hundred Merinos in an effort to improve his extensive flocks (1979a, p. 178). John Wedgwood, another maternal uncle, was an amateur gardener and a prime agent in the foundation of the Royal Horticultural Society in 1804 (Fletcher 1969). Further aid in Darwin's species work came from neighboring landowners, men like George Tollet, an agriculturalist of considerable renown who participated in many famous schemes of livestock improvement. In his transmutation notebooks and in other early manuscript notes, Darwin records information obtained from many of these men, and his eight-page questionnaire of 1839 on animal breeding appears to have been composed with them in mind. At any rate, the two responses to these queries that have been discovered — one from Tollet, the other from Richard Ford, agent to the Fitzherbert estate at Swynnerton — probably date from a trip that Darwin made early in 1839 to family and friends in Shropshire and Staffordshire (Freeman and Gautrey 1969; Vorzimmer 1969b; Darwin 1968). Darwin's provincial background and his contacts with the landed gentry in these two counties almost certainly helped to lay the foundation for his lifelong habit of viewing the natural world through the spectacles of domestication.[3]

After leaving London for Down House in 1842, Darwin once again entered an environment well situated for utilizing the resources of the breeding community. A variety of immediate neighbors, and even his vicar, presented him with numerous "facts" about rabbits, pigs, horses, and other animals (Atkins 1974; Stecher 1961). But Down's rural character was even more helpful for the space it afforded for Darwin's own extensive experiments in breeding and raising plants and animals. For studies of poultry, pigeons, and rabbits, he had special hutches and a dovecot built in the back garden; for his work in crossing orchids and other plants, Darwin borrowed the services of Horwood, the professional gardener of his neighbor Sir John Lubbock (*LL* 3:269; Atkins 1974, pp. 29–30). The spacious gardens at Down also facilitated the cultivation of the numerous varieties of peas, beans, cabbages, and other cultivated plants that Darwin raised during the course of his researches. The entry in the local directory for 1847 of a "Darwin, Charles, farmer", was not so far off the mark after all (Atkins 1974, p. 24).

The property at Down, of course, provided not only the Arcadian benefits of rural life, but also the signal advantage of nearness to London (Ospovat 1981; Rudwick 1982b). For Darwin, the advantages of the metropolis centered chiefly in its many scientific societies, libraries, museums, and expert colleagues. During the first two decades of his residence at Down, Darwin frequently attended meetings of the Geological, Royal, and Linnean societies; he used the unparalleled natural history collections of the British Museum and the Zoological Gardens; and he often queried scientific colleagues such as Richard Owen, Joseph Hooker, and Charles Lyell. Only in the 1860s and 1870s and during particularly severe bouts of illness did he become tied almost exclusively to a reclusive existence at Down House. But besides these scientific resources, London also contained the largest single concentration of expertise on plant and animal breeding in Great Britain. Many eminent and progressive landowners wintered in London, attending meetings of the Royal Agricultural Society and the Royal Horticultural Society. At the Crystal Palace and at numerous other localities on the outskirts of the city, huge agricultural, poultry, and floral shows were held throughout the year. Even on the relatively scanty evidence as yet available, Darwin is known to have made a point of viewing these exhibitions on several occasions. Moreover, many activities concerning domesticated animals and plants were pursued with special enthusiasm in the urban environment. In the mid-1850s, for example, when Darwin devoted special attention to the breeding of fancy pigeons, he joined two of the metropolitan fanciers' clubs, and his attendance at several of their grand shows is recorded in the *Gardeners' Chronicle* (Secord 1981b, pp. 176–178). London also held pride of place as a horticultural center. Here Darwin could find the rare specimens required for his studies of orchid fertilization and his investigations into plant movement; here he could converse with skilled gardeners and nurserymen like Thomas Rivers (Allan 1977, pp. 224–225; *Orchids*; *Variation*).

From his boyhood in Shropshire to his final years at Down, Darwin was thus unusually well situated to undertake researches into domestication and to contact those with expertise on such subjects. Nevertheless, these opportunities would have counted for little without precedents among Darwin's own scientific peers for utilizing them. Although an appreciation of the Salopian background and residence at Down aids in understanding his interest in the work of the breeders, Darwin always identified himself primarily as a naturalist. Without some sanction from the past practice of his mentors and predecessors in science, it is inconceivable that he would have turned to an outside group like the breeders for help in solving that "mystery of mysteries, the origin of new species". Darwin recognized that good scientific precedents for a concern with domesticated animals and plants could be found, even if no one in England had ever focussed so carefully on the subject. At the same time, he argued that earlier evolutionary theorists had ignored the issue entirely. "I believe all these absurd views arise," he told Hooker in 1844, "from no one having, as far as I know, approached the subject on the side of variation under domestication, and having studied all that is known about domestication" (*LL* 2:29). Within Darwin's immediate scientific circle, on the other hand, several individuals had manifested varying degrees of interest in the work of the breeders.

The earliest of these role models were perhaps in the long term the most important as well. William Darwin Fox, Darwin's second cousin and close contemporary at Cambridge, was three years his senior and already keenly interested in natural history. Throughout his life, Fox maintained an unusually wide-ranging and practical acquaintance with the phenomena presented by domesticated animals. Darwin's correspondence with him is substantial, and shows that Fox aided the species project on numerous occasions (*LL* 1:301; 2:111–112). John Henslow, another formative figure in Darwin's years at Cambridge, became increasingly involved with horticultural and agricultural issues in the 1830s and 1840s, as indeed one would expect from the director of the Botanic Gardens in Cambridge and an active parish priest in an important farming district (Jenyns 1862; Russell-Gebbet 1977; Darwin 1967). Henslow organized agricultural shows in his parish at Hitcham in Suffolk, arranged botanical lessons for the local schoolchildren, participated in the local farmers' club, and (like Darwin at a later date) published many notices in the *Gardeners' Chronicle*. On one occasion in the 1840s, he published a brief extract from Darwin's *Beagle* voyage specimen notes in hopes of shedding light on a contemporary rust blight affecting the wheat crop in England (Henslow 1844). Although much less involved with such practical matters than Henslow or Fox, Charles Lyell exerted part of his profound influence on the young Darwin through his insistence, in good actualistic fashion, upon the study of observable events in the farmyard and greenhouse as a key to understanding the extent of variability in nature. In arguing from the evidence of domestication for the reality and stability of species (like most authors before Darwin), Lyell referred to publications of the Royal Horticultural Society and several important agricultural societies (Lyell 1830–1833, 2:passim). Some of these same papers, particularly those of Herbert, Gärtner, and Kölreuter, later proved of essential aid to Darwin in formulating his distinctly unLyellian conclusions concerning the plausibility of transmutation. The examples of Fox, Henslow, and Lyell could be multiplied, but the instances of these three men indicate that Darwin could draw upon a tradition (albeit a limited one) of links between breeders and men of science.

As I have indicated, in assessing the overall availability of knowledge about domesticated plants and animals to an individual like Darwin, attention to the various contexts in which he lived and worked is essential. His personal and family background brought him into close proximity with men and women actively engaged in improving domesticated varieties. In addition, important precedents existed within his own scientific group for bringing this practical knowledge to bear upon a great variety of problems in natural history, from anti-transmutationist theorizing to particular difficulties in taxonomy. But despite these established interactions, naturalists and breeders viewed the world with different eyes and from very different perspectives. Thus for Darwin to apply the results of human artifice to the study of nature in a truly comprehensive fashion, he had first of all to build bridges between these two distinct social groups; approaching the issue as a naturalist, Darwin would have to extend lines of communication far into the separate world of the breeders.

II. Research Networks

By the end of his life, Darwin was an exceedingly wealthy man in more ways than one, for as he told Hooker in 1864, he had become "a complete millionaire in odd and curious little facts" about inheritance and selection (*LL* 3:27). In some respects, the same attentive perseverance that served Darwin so well in his financial investments had also led to an equally enviable accumulation of natural history information. This scientific fortune, as the following section of this essay suggests, was gathered by methods similar to those used by other expert naturalists. Thus the program of research into domestication, while unusual in relying so heavily on informants from outside natural history, from another perspective exemplifies a set of techniques routinely employed in the most advanced Victorian scientific investigations.

More than any other single technique, the construction of a large network of informants and qualified experts links Darwin's investigations among the breeders with earlier work in natural history. All the best contemporary procedure in science encouraged the collection and explanation of all possible information bearing upon a topic, whatever its source. Lyell, for example, had assembled a range of active correspondents (including Darwin himself), each providing materials for the successive editions of the *Principles* and the *Elements of Geology* (Lyell 1830–1833; 1838; Wilson 1972). In composing his big book on the geology of Wales and the Welsh Borders, the *Silurian System* of 1839, Roderick Murchison built up an analogous network that ranged from illiterate quarrymen to the Earl of Powis (Murchison 1839; J. Thackray 1977). Botanists compiling local floras, zoologists undertaking a monograph of a genus or a family, tidologists like William Whewell studying the vertical movements of the sea: all labored with great diligence to collect a host of scattered observations (W. Cannon 1978; Morrell and Thackray 1981). Darwin, in writing his book on coral reefs during the early 1840s, had already obtained practical experience of his own in consulting a typically miscellaneous assemblage of sources, from travel books and nautical charts to local observers in Australia and the South Seas (*Coral Reefs*). The precedents for such a procedure extend back at least as far as the seventeenth century (D. Allen 1976).

Although the main focus of the present essay is on Darwin's contacts with the breeding community within the British Isles, it is particularly important to emphasize that his quest for information extended much further afield as well, both geographically and historically. Darwin's efforts to render his work comprehensive on the widest possible international scale would well repay further study. For civilized or thoroughly explored nations Darwin usually relied on an authoritative published natural history, such as Johann Matthäus Bechstein's *Gemeinnützige Naturgeschichte Deutschlands* (1789-1795?) or Philip Henry Gosse's *A Naturalist's Sojourn in Jamaica* (1851). For information from areas less frequented by Europeans, he used travel books concerned with natural history to a greater or lesser degree, from his own *Journal of Researches* and the *Himalayan Journals* of Joseph Hooker (1854) to works like David Livingstone's *Missionary Travels and Researches in South Africa* (1857). In many cases Darwin went to great lengths to obtain local residents — frequently missionaries, medical men, or consular officials — as expert correspondents. Such men could provide specimens unavailable in the London museums, or clarify points not covered by the available published literature. Thus Hugh Falconer and Edward Blyth gave details of the domesticated animals of the Indian subcontinent, Dr. Laurence Edmondstone wrote from the Shetland Islands on several occasions, and Henry Layard sent an impression of a Mesopotamian cylinder seal (De Beer 1959b; *Variation* 1:246, 301; 2:161; Beddall 1973).[4] The development of the British colonial empire during the course of the nineteenth century facilitated Darwin's efforts immensely, as did the cosmopolitan ideals traditionally associated with the sciences (Pancaldi 1981; Worboys 1981). By the 1850s his reach extended around the world. A relevant fact involving a foreign country could be retrieved at relatively short notice, either through reading or through a query to the appropriate correspondent. Once established, the geographical network proved useful for a variety of Darwinian projects: individuals who had sent barnacle specimens or data on coral reefs could easily aid in securing materials for the *Variation* and the *Descent*.

Not satisfied with information on breeding from the present era only, however international in scope, Darwin also pushed his questioning backwards in time. In an extensive program of historical research, he read a host of works dealing with domesticated animals from the earliest ages. The *Variation* and the *Origin* cite the Bible, wall paintings in the tombs of the Pharaohs, medieval chronicles, and the works of Homer, Pliny, and Plato (*Origin* and *Variation*, indexes). As in the case of his geographical network, Darwin queried special experts, antiquaries such as Dr. Samuel Birch, Keeper of Antiquities at the British Museum (*Variation*, 1:208). He placed brief notes in the gardening magazines in hope of a response from some historically-minded reader. "I should be very much obliged," Darwin asked in the columns of an 1864 issue of the *Gardeners' Chronicle*, "if any one who possesses a treatise on gardening or even an Almanac one or two centuries old would have the kindness to look what date is given as the proper period for sowing Scarlet Runners or dwarf French Beans" (*CP* 2:93). Here he wished to discover if the modern improvement in the bean involved a shift of its sowing time to an earlier point in the year. As a result of such minute inquiries multiplied many times over, Darwin labored to obtain something akin to a domesticated version of a geological record, illustrative of the remarkable changes produced by man in the plants and animals under his care.

The wide-ranging character of Darwin's search for information, although remarkable to modern eyes, is in large part nothing more than a function of the canons of explanation in contemporary natural history. As Lyell remarked in 1863, "We usually test the value of a scientific hypothesis by the number and variety of the phenomena of which it offers a fair or plausible explanation" (1863, p. 395). Thus if inheritance, generation, and variation were to be comprehensively understood, it could only be through what William Whewell called a consilience of inductions (L. Laudan 1971), by bringing disparate lines of investigation to bear upon the question and uniting them all under a tentative hypothesis. In consequence, broad theoretical goals like those of Darwin required an equally comprehensive search for information. This quest for breadth of coverage is especially evident in Darwin's marshalling of the knowledge of the breeders: in fact, his extensive network of contacts with a motley assemblage of animal husbandry men, pigeon fanciers, and rose growers was in one sense nothing more than the *social* concomitant of the central methodological canon of Victorian natural history.

Having assimilated the standard techniques of contemporary science, Darwin naturally proceeded to work his way letter by letter and book by book to an acquaintance with many practitioners of the breeder's art, and a familiarity with their standards of judgement. The Darwin manuscripts at Cambridge, and the even greater wealth of evidence preserved in the footnotes to the *Variation*, make it possible to trace Darwin's path from the community of naturalists to that of the breeders with unexampled fullness. A knowledge of his precise position with respect to these distinct groups as outlined earlier in this essay is essential to the task of reconstructing these lines of communication. I have already shown that Darwin gradually widened his circle of informants by exploiting the links already established through his family circumstances and by fellow naturalists like Henslow. The process is illustrated with particular clarity in the case of Darwin's work with poultry and pigeons, a subject he studied with special care in the years immediately preceding the *Origin* (Secord 1981). William Yarrell, Fellow of the Zoological Society and author of bulky compilations on the birds and fish of Great Britain,[5] first met Darwin just before the departure of the *Beagle*. Yarrell had long been familiar with leading poultry and pigeon fanciers, and when the young naturalist returned and began to manifest an interest in subjects like variation and inheritance, Yarrell promptly put him in touch with these men. Just before Yarrell's death in 1855, he introduced him to William Tegetmeier, a prolific journalist who rapidly emerged as Darwin's principal expert on domesticated birds (E. Richardson 1916). Although on the periphery of the London scientific scene, Tegetmeier occupied an important — if not exactly central — position in the pigeon and poultry fancying community. During the mid-1850s he fed Darwin references, took him to meetings of the leading pigeon fancying clubs, and introduced him to the outstanding fanciers of the day. From Yarrell, to Tegetmeier, to prize-winning eccentrics like Mr. Bult and John Eaton, Darwin traversed step-by-step the social boundaries that separated different ways of perceiving the natural world.

Although more detailed and systematic work on the subject is necessary, it appears that Darwin followed a similar procedure in his reading.[6] Incidental references in scientific works, in the regularly read *Gardeners' Chronicle*, or from friends, usually initiated forays into the various elements of the breeding literature. Then in his "list of books to be read" or in pages pinned at the back of a publication, Darwin noted those works that promised further enlightenment. By thus tracking down chains of references he penetrated very far indeed into the published sources relating to domestication. "When I see the list of books of all kinds which I read and abstracted," Darwin wrote in his autobiography, "including whole series of Journals and Transactions, I am surprised at my industry" (*Autobiography*, p. 119). By the early 1840s he kept continually abreast of the latest publications on breeding, reading books and periodicals as they appeared from the press. Needless to say, this program of reading was always much more straightforward than putting together a network of correspondents and personal contacts, a task that required personal introductions and quantities of time-consuming letter-writing. As a result Darwin chose his special queries with a care that increased with experience.

How did Darwin proceed once he had decided that the published literature alone had failed to answer a question of importance for his theory? An interesting instance of the methods he could use in such circumstances is provided by his attempt in 1862 to discover the extent of variation in the common honey bee, *Apis mellifica*. More than any other domesticated creatures, bees remained close to their original wild state, for they fed themselves and followed their usual habits of life even under the care of the most assiduous beekeeper. If variability, however slight, could be shown for bees, a closer connection between variation under domestication and in nature might be forged. Moreover, bees raised a critical problem for the entire theory of natural selection, in that the great majority of individuals in any hive were neuters and consequently did not produce offspring. Selection could in such cases operate only indirectly, through the role neuters played in aiding the chances of survival for the hive as a whole. As Darwin wrote in the *Origin*, "this is by far the most serious special difficulty, which my theory has encountered" (p. 242; also Richards 1981).

These two points made an inquiry into the variability of the honey bee well worth pursuing. By early June 1862 Darwin had reached the brief section on bees in the process of writing his *Variation* (1:297–299; Darwin 1959a, p. 16). His files at that time appear to have contained at least three or four notes on the variability of the bee, taken either from his reading or from his usual informants. A Mr. J. Lowe of Edinburgh, writing in the 15 May 1860 issue of the *Cottage Gardener*, had described a new light-colored variety of bee, while D. A. Godron, in his *De l'Espèce* (read by Darwin soon after its appearance in 1859), had stated that bees in the south of France were larger than those further to the north (p. 459). Still unsatisfied with this information, and especially curious about Lowe's discovery, Darwin sent a query to the periodical — by this date retitled the *Journal of Horticulture* — in which the description of the new variety had originally appeared. (Because this brief paper was not included in the *Collected Papers*, the full text is printed in the notes to the present essay.[7]) Darwin hoped in particular to hear from expert apiarians, men who had extensive experience in the practical art of keeping and raising bees. "I should feel much obliged," Darwin began, "if the 'Devonshire Beekeeper' or any of your experienced correspondents would have the kindness to state whether there is any sensible difference between the bees kept in different parts of Great Britain" (1862a). The query was one of Darwin's most successful, for it produced a spirited exchange in the pages of the *Journal of Horticulture*. A few beekeepers claimed that bees did in fact vary, and they pointed to several appropriate instances ("Surrey Highlander" 1862; Newman 1862). But the weight of authority soon rested against this view. S. Bevan Fox, a famous English beekeeper of twenty-one years' experience, had no doubts in the matter: "that there is any difference in the ordinary English hive bee," he concluded, "further than may be seen in any apiary or single colony, I do not for a moment believe" (Fox 1862). The man responsible for introducing movable-frame beekeeping into Britain, T. W. Woodbury — writing incognito as "a Devonshire Beekeeper" — not only answered Darwin's query in the *Journal of Horticulture* but also arranged for it to be inserted (with suitable alterations) in the German apiarian journal *Die Bienenzeitung* (Darwin 1862b). Replies in translation soon came from two German experts, including the pre-eminent Johannes Dzierzon (1862; also Kleine 1862); both men agreed that bees varied only to an insignificant degree. Meanwhile J. Lowe of Edinburgh, whose strange bees had prompted Darwin's initial interest, wrote to the *Journal of Horticulture* with information on the further history of his aberrant variety (1860; 1862). Lowe's testimony was especially useful, for it clearly pointed to the possibility of variation in bees, without which even the most careful selection could not succeed (*Variation* 1:299).

The sheer scale of Darwin's research network among the breeding community comes into perspective when this single series of exchanges is multiplied, as it must be, hundreds or even thousands of times over. Although Darwin found a use for each of the responses to his query about the variability of honey bees, the subject occupied no more than a minor place in the grand design of the species project. Out of a total of over twelve hundred pages in the *Variation* alone, only two or three were devoted to the variability of bees; by way of contrast, the discussions of rabbits and dogs each took up about thirty pages, while nearly one hundred were devoted to pigeons. And for Darwin, the quest for information never ended. For example, he began almost immediately to use the contacts established by his 1862 query among the beekeepers to gather facts for yet another interest, the variability of bees in nature. Woodbury, the ever-helpful "Devonshire Beekeeper", undertook at Darwin's special request measurements of the waxen cells of different species of foreign bees; the cells proved identical in diameter, despite substantial variations in the sizes of the bees themselves ([Woodbury] 1863). This information would presumably have found its place in Darwin's never-completed book on variation in nature, originally planned as a sequel to the *Variation*. By thus having Woodbury deal with combs and bees brought from distant lands Darwin brought the specialized knowledge of an expert to bear on objects outside the usual concern of a beekeeper: as so often in dealing with the breeders, he juxtaposed individuals, approaches, and materials usually kept separate.

III. Contexts of Evaluation

Given the willingness of the breeders to respond to Darwin's requests, the greatest problem in contacting them involved not so much getting information in the first place, but rather in determining its reliability. "*The difficulty*," he emphasized in a letter to Huxley, "*is to know what to trust*" (*LL* 2:281). A project that depended so heavily on materials derived from outside the natural history community raised this problem with special force. As a naturalist, Darwin used one set of criteria for evaluation, while the breeders customarily employed another; in bringing the knowledge of domesticated animals and plants to bear upon his theories, Darwin had first of all to scrutinize, compare, and judge. As a result of his contacts with the breeders, Darwin quickly realized that the meaning of a "fact" depended very much upon the circumstances in which it was originally enunciated. Only with the validity of his information from the breeders established in scientific terms, could he proceed to use their data within a scientific context. This third section will focus especially upon the problems Darwin faced in assessing materials derived from a group of men outside science.

Darwin's procedure in dealing with the breeders differs in several respects from his treatment of his own natural history community. Dov Ospovat has shown how Darwin regularly searched out the major generalizations of contemporary naturalists and then attempted to explain or reinterpret them in evolutionary terms. In Darwin's view, as in that of William Whewell, any well-attested phenomenon could gain the status of a "fact", from the law of unity of type to the length of a particular bird's beak (Ospovat 1981, pp. 95–96; see also Ruse 1975c). In most instances Darwin placed considerable confidence in those "facts" which derived from his expert colleagues in science, such as Henri Milne-Edwards, Karl Ernst von Baer, or Edward Forbes. The situation with the breeders was very different. Although they provided a vast quantity of "facts", these were almost invariably of a lower order than those accepted without question from members of his own natural history community. Indeed, the highest accolade a husbandry man or horticulturalist could receive from Darwin was to be called "scientific". An example is provided by the discussion in the *Variation* of the number of crossings required to produce a true race. Darwin noted that some practical gardeners thought that twelve or even twenty crossings were needed, but Gärtner had said six or seven were enough, and Kölreuter had settled for an approximation of eight. Darwin accepted the latter estimates as the probable ones. "The conclusions of such accurate observers as Gärtner and Kölreuter," he concluded, "are of far higher worth than those made without scientific aim by breeders" (*Variation* 2:88).

The case of Gärtner and Kölreuter also illustrates the principal means that Darwin used to establish standards of evaluation: as in his scientific work, he cultivated one or two expert authorities for every subject with which he had to deal. Just as Hooker advised on botany and Huxley on invertebrate zoology, so too did the works of the two German horticulturalists generally serve as a court of final appeal. In a similar fashion advice came from Tegetmeier on poultry, Mrs. Whitby on silkworms, Thomas Rivers on roses and fruit trees, and Woodbury on bees. These standard authorities

— available either in print, by personal contact, or through the various other means already described — could then be supplemented by a host of lesser sources. Clearly the degree to which Darwin could depend upon a few trustworthy standards differed radically from field to field, for complete reliability in Darwin's eyes was almost a direct function of the nearness of a subject to the world of science. Thus nothing better illustrates the relatively large gap separating zoologists and animal breeders than Darwin's inability to discover any satisfactory treatise on domesticated animals as a whole. When asked by Huxley how to "get up facts about breeding", he could recommend several works (including Gärtner and Kölreuter) for domesticated plants. For animals, however, Darwin lamented that the situation was very different: "no resume to be trusted at all; facts are to be collected from all original sources" (*LL* 2: 281).

But science could not be the sole means of evaluating the expertise of individuals so unconcerned with things scientific. Experience counted too, and so Darwin went to some lengths to master and apply the breeders' own standards to their work. A great many of the facts on domestication in his books are tagged with a certification of their origin; one needed to know not only what had been said, but also who had said it, and on what basis in practical experience. In consequence, Darwin frequently characterizes his sources of information in terms that would do justice to the catalogue of an agricultural exhibition: we learn from "a great and successful breeder of the Improved Oxfordshires", "a great breeder", an "excellent gardener", "a great winner of prizes at the Pigeon-shows", and "an excellent judge of pigs" (*Origin*, *Variation*, passim). By thus evaluating the breeders partly in their own terms, Darwin could obtain at least some means of deciding which elements of practical *experience* to account for in constructing his theory even if that experience was ultimately to be "explained away" from his own perspective in science.

In connection with the problem of reliability, the history of Darwin's *Questions on the Breeding of Animals* of 1839 is of special interest (Vorzimmer 1969b; Darwin 1968). Although of obvious utility in suggesting the state of Darwin's ideas at an important juncture in his career, this brief pamphlet is best viewed as a failed experiment in scientific communication. Remarkably little direct reference to it is found in his notebooks or correspondence, and neither of the two extant replies provided information used in any of Darwin's later works.[8] By the early 1840s Darwin seems to have realized that queries must be closely tailored to the individual respondent if he was to obtain any dependable testimony not available in print. No one breeder, no matter how experienced, could accurately answer forty-eight different questions concerning domesticated animals in general. The "Queries about Expression" of 1867 are noticeably more focussed, for all seventeen questions deal with a single species, man. Moreover, that list closes with cautionary remarks on techniques of observation and reporting (*CP* 2:136–137). In the years following the publication of his 1839 pamphlet, Darwin drew up further queries for breeders, but these were always on a specific topic or for a single individual. The questions about bees and the history of runner beans, both mentioned earlier, are entirely typical in this regard.

If historians and biologists have often criticized Darwin's work on heredity and variation for excessive credulity, the sheer number and extent of his queries demonstrate that he in fact made extraordinary efforts to obtain the most accurate information available. "No one or two statements are worth a farthing," he told Huxley shortly after the publication of the *Origin*; "the facts are so complicated. I hope and think I have been really cautious in what I state on this subject, although all that I have given, as yet, is *far* too briefly" (*LL* 2:281). Thus when the former vicar of Down reported a case of a cross between a Highland cow and a red deer, Darwin rejected it as "too wonderful and opposed to analogy" (Stecher 1961, pp. 225–226). On the other hand, he did allow for the now-discredited phenomenon of telegony — the impress of previous crossings by a female appearing in her subsequent offspring — but on excellent authority and after discovering many cases (R. Burkhardt 1979; *Variation* 2:403–404). Certainly Darwin saw his task in canvassing the breeders as that of a scientific judge, responsible for picking out all those points that a comprehensive theory would have to explain. His book on domesticated animals and plants can in this light be seen as a gigantic catalogue of facts on breeding, explicitly constructed with an eye to the ever-present problem of reliability.

Notably, the *Variation* was the only one of the fully documented expansions of the *Origin* that ever appeared. In Darwin's view the subjects discussed in this work particularly required full-length treatment, much more so than embryology, biogeography, and other topics more fully elaborated in the scientific literature. The general consequences of his theory for these latter fields were perhaps evident enough from the abstracted discussions already provided in the *Origin*; once Darwin pointed the way, others could follow. With heredity, variation, and reproduction, this was by no means the case, for the number of well-established "facts" was remarkably small, in part because the appropriate expertise derived so largely from men outside science. He thus chose to work up his notes on domestication not just as a necessary prelude to a whole series of projected expansions of other elements of the theory. The decision to write the *Variation* resulted from a more positive factor as well, for Darwin had good reason to believe that he could aid in putting the difficult questions raised in the *Origin*'s opening chapter upon a proper foundation for the first time.

IV. A System of Intellectual Paternalism

Up to this point I have considered Darwin's construction of an information-gathering network among the breeders largely from Darwin's own point of view. But how did the breeders perceive his work? More particularly, why did men like Woodbury respond with such marked enthusiasm to Darwin's self-confessedly "troublesome" requests for quantities of seemingly unconnected information? Darwin certainly recognized the distance separating his work from the practical concerns of the breeders: in 1848, for example, he asked the pioneering English silkworm raiser Mrs. Mary Anne Whitby to try breeding a race of silkworms with cocoons destitute of silk. "In the eyes of all silk-growers," he admitted, "this assuredly would appear the most useless of experiments ever tried" (Colp 1972, pp. 873–874). From Darwin's position, the importance of each of these multifarious requests in a larger plan was clear, but for his respondents this must often have not been the case. How did Darwin manage to instigate complicated trains of research among people he had usually never met?

Both before and after the *Origin*, Darwin appeared to contemporary husbandrymen, gardeners, and poultry fanciers above all as a leading man of science engaged upon important and prestigious investigations. In this role he could lend status to these men, who in their turn provided him with information and aid. In short, Darwin and the breeders were mutual beneficiaries of what might well be termed a system of "intellectual paternalism" in which both sides gained through participation in a major theoretical enterprise. The acknowledgements cluttering the footnotes to the *Variation* testify eloquently to the operation of such a system. As a perceptive reviewer wrote of Darwin's two bulky volumes, "his book will make many men happy" (*LL* 2:76–77).

In capitalizing upon the intellectual status of science in his dealings with the breeders, Darwin rather paradoxically benefitted from the general separation of the breeding community from his own scientific circles. Many of the practical men, whose aims and motives contrasted so sharply with those of the naturalists, desperately wished to raise their subjects to the intellectual level of the sciences. Participation in a project like Darwin's, even if only to the extent of measuring the cells of honeycombs, offered one way in which this might be accomplished. Just as the natural history sciences looked longingly up a hierarchy to astronomy and physics, so too did those concerned with domesticated animals and plants dream of transforming their enterprise into an inductive science on a par at least with zoology, botany, or geology. As Philip Pusey, first president of the Royal Agricultural Society, explained in an inaugural address of 1840, these latter three subjects had achieved scientific status only recently, largely by forming associations of like-minded enthusiasts. In his view, agriculture — similarly dependent upon a host of isolated observations — could by the same means attain a similar end (Pusey 1840, pp. 19–21). Not surprisingly, Pusey enrolled a number of leading natural scientists on the Society's list of honorary members, and "Practice with Science" became its motto. Such men saw scientific knowledge not only in terms of its utility, but also as a potential source of social mobility (A. Thackray 1974). One correspondent in the *Gardeners' Chronicle*, although fully willing to admit "that a man may be a good gardener without any knowledge of Botany", nonetheless found "something so pleasing in it, independent of its utility in determining the families of plants", that he recommended it "to all gardeners" (Towers 1842). At the same time, the invocation of science was especially striking among the higher reaches of the social scale. The call to science provided the gentry and aristocracy with one of their principal claims to the leadership of England's agricultural revolution, and served to remove any remaining taint of manual labor from their efforts. Even as these men pursued changes in their show animals that were uneconomical — such as excessive weight gain that led to constitutional delicacy or reductions in milking yields — gentleman breeders like Sir John Sebright and Robert Bakewell brought supposedly scientific method and persistence into their improvements (Trow-Smith 1959).

It goes without saying that invocations of the social status and intellectual cachet of science should not be misconstrued as evidence for direct links between naturalists and breeders. But when combined with the substantive contacts that I have previously outlined, the scientific aspirations of many leading fanciers and breeders unquestionably enhanced their utility for a scientific enterprise like Darwin's. As the *Cottage Gardener* commented after a large exhibition of fancy birds, "It relieves Pigeon fancying from all charge of triviality, when savants of such reputation as Messrs. Darwin and Waterhouse show, by their attendance and interest, that the changes capable of being produced in any species by domestication, are worthy of the deep attention of scientific inquirers . . ." (Anon. 1858). Eager to serve the cause of science as well as their own, the breeders of Victorian England were primed to aid in Darwin's search for information on generation, variation, and selection.

On occasion, of course, the paternalistic system of rewards did not work, and the practical men openly expressed disappointment at being accorded less than what they saw as their due. Thus Patrick Matthew, whose *On Naval Timber and Aboriculture* (1831) had featured a brief appendix anticipating natural selection by almost three decades, would not have been placated to learn that Darwin had mentioned his work in the *Variation* (2:237). Although the list of precursors prefacing the later editions of the *Origin* did include Matthew, Darwin had found *On Naval Timber* really useful for its facts on the variability of forest trees. Matthew, on the other hand, wanted full credit for high theory (Wells 1973). Fortunately for Darwin such objections to the system of scientific paternalism were relatively rare, and his work fitted comfortably into an established pattern of exchange in which both sides benefitted, each in its separate way.

We have seen that Darwin brought the knowledge of the breeders to bear upon questions of science by assembling a remarkable network that ranged from horticulturalists to animal husbandrymen, from pigeon fanciers to beekeepers. While necessitating extensive forays outside the social and intellectual boundaries of the world of science, such an all-encompassing method stemmed directly from Darwin's indoctrination in the research traditions of his own natural history community. That background led him to consider all manner of sources, however distant from science, as potential grist for his theoretical mill. Within the terms of reference employed in Victorian natural history, information from experienced breeders could usefully form part of a satisfactory explanation of the origin of species; for Darwin, man was indeed "trying an experiment on a gigantic scale". In contrast, scientists of the early decades of the twentieth century eventually followed an approach drastically reduced in scope, with Thomas Hunt Morgan and his colleagues studying fruit flies in a single room at Columbia University. Many of their aims remained unchanged, but the relevant conditions for the grand "experiment" had been radically transformed.

Locating Darwin and the breeders within their respective contexts opens several further avenues for historical inquiry. In concluding, I wish first of all to suggest a few ways in which the concerns of the present essay may be linked more directly to the contents of Darwin's theories and arguments. From the outset I have emphasized that his interest in domestication originated in studies of generation, variation, and inheritance. But the breeders became much more than another source of facts on these subjects, for Darwin eventually used them as the foundation for an extensive analogy between selection by man and selection in nature. This analogy of selection, which developed gradually in the months and years after Darwin read Malthus (Cornell 1984; L. Evans 1984; Kohn 1980, pp. 136–139; Limoges 1970c), depended critically upon the existence of a group of men engaged in competitive struggle for prizes and individual success. In Darwin's exposition of the structure and ethos of the breeding community, the "common context" of Victorian social and biological theory characterized by Robert Young (1969) is made manifest and concrete.

A second point meriting additional study concerns Darwin's use of the distinction between breeders and naturalists. I have suggested that this very separation assisted Darwin in his efforts to gather information from practical men, but the existence of these distinctive communities was also helpful in more direct ways as well. For example, the disdain of many naturalists for the unscientific character of the breeders' approach could be turned to positive advantage. In such cases Darwin could direct the consensus of his fellow naturalists away from the special creation of each species by paralleling that view with the "uneducated" opinions of breeders about their favorite fancy varieties. This argumentative device does not appear in the *1844 Essay* (pp. 71–74) and exemplifies the rhetorical elaboration of Darwin's argument in the years preceding the publication of the *Origin*.

The perception of the content of Darwin's theories by various elements of the breeding community also deserves further investigation, particularly as part of the larger and almost totally unexplored problem of his reputation during the post-*Beagle*, pre-*Origin* period. Clearly Darwin was seen at this time as a naturalist of great promise, known especially to the general reader as the author of a book of travels. The *Journal of Researches* must have opened as many doors to the public at large as his volumes on the zoology and geology of the voyage did to the specialist scientific community. To all but a select few, Darwin posed as an agnostic on the subject of transmutation and presented his studies as contributions to the wholly respectable problem of distinguishing species from varieties. By keeping the full extent of his theoretical views prudently under wraps, he hoped to avoid being classed with Lamarck, Robert Grant, or "Mr. *Vestiges*" (Herbert 1977; Desmond 1984); moreover, such a strategy allowed Darwin a chance to gather all possible information from outsiders before the greatly feared storm of execration. Needless to say, this storm never broke with such intensity, and there is no evidence that Darwin's sources among the breeders dried up after 1859. Far from feeling betrayed or incensed, the practical men seem to have been more eager than ever to help. Why this should be the case is not yet entirely clear, although it corroborates recent historical work suggesting that the polarizing effect of the Darwinian debates has been exaggerated (Moore 1979).

In discussing the period of Darwin's career dealt with in the present essay, historians and biologists alike have often lamented a very perceptible reduction in the rate of his theoretical inventiveness. Although a number of important conceptual innovations took place after the completion of the *1844 Essay*, most notably the principle of divergence (Ospovat 1981; Browne 1980; Kohn, this volume), these shrink into relative insignificance when compared to the momentous insights of the 1837–1842 period. By the end of that most creative phase of his career, there is every evidence that Darwin no longer searched actively for ideas that would fundamentally alter the underlying structure of his theory. In this sense the very comprehensiveness of his articulated views served a conservative function by blocking their radical reconstruction, just as the coherence of his pre-selectionist theory had hampered Darwin in the months before the discovery of natural selection (Kohn 1980, pp. 149–154). But rather than regretting the demise of the young biological revolutionary or forcing the Darwin of the 1840s and 1850s into the speculative mold of his younger self, it would seem more appropriate to develop different perspectives for approaching his later research. The examination of Darwin's work during the post-1842 period has the potential for shedding light on methods of communication and information-gathering in science, on ways of reading, writing, experimenting, and collecting in pursuit of a long term scientific objective, and on the structure of scientific communities and their relationships with other social groups. If these years lack the sustained intellectual ferment of the late 1830s, they also show that Darwin remained remarkably fertile of smaller-scale hypotheses and creative extensions to his theory. The great mass of manuscript and printed Darwinian sources thus provide not only a means for investigating a creative naturalist in his moments of insight, but also can aid in understanding the equally vital (if less dramatic) phase of scientific work that grew out of those moments. The immense task that Darwin faced from the early 1840s to the end of his life has in its turn left an equally important one for students of his work.

Notes

1. In particular, the preliminary character of this essay is evident from the fact that I have not made explicit reference here to manuscripts concerning these topics available both in Cambridge and elsewhere. Besides Darwin's annotated library, notebooks, and incoming correspondence, the most important of these include the following: letters to William B. Tegetmeier at the New York Botanical Garden; letters to William Darwin Fox, now mostly in the library of Christ's College Cambridge; a heavily annotated run of the *Gardeners' Chronicle* in the Cory Library of the University Botanic Garden; various items in DAR 205, especially 205.6 (notes on embryology), 205.7 (notes on hybridism); DAR 206, especially the "Questions & Experiments" notebook; and several lengthy abstracts of works on crossing and variation. However, an unusually large percentage of Darwin's manuscript notes, drafts, and slips relating to the subjects dealt with in this essay appear to have been thrown away, presumably because they had been published fully in the *Variation*.

   I wish to thank Dr. Janet Browne for her helpful suggestions and comments. For theoretical perspectives on the approach taken in this essay, see Shapin (1974), Dolby (1977), and Crane (1972); references to much of the relevant literature can be found in Shapin (1982) and Secord (1985).

2. See *Autobiography*, p. 131, where Darwin notes that the subject of sexual selection, "and that of the variation of our domestic productions together with the causes and laws of variation, inheritance, &c., and the intercrossing of Plants are the sole subjects which I have been able to write about in full, so as to use all the materials which I had collected." Even the lengthy work *Natural Selection* did not represent the full and perfected scale on which Darwin had originally hoped to publish.

3. For an additional set of queries on breeding, see "Questions for Mr. Wynne", published in Gruber and Barrett (1974, pp. 423–425). Needless to say, I do not in any sense wish to claim that the effect of growing up in a rural environment *determined* the extent or character of Darwin's interest in domestication; the example of A. R. Wallace, who spent the first years of his life in rural Wales and generally dismissed the relevance of the work of the breeders, is instructive in this regard. A penetrating analysis of links between the landed élite and the zoological community in the metropolitan context is available in Desmond (1985).

4. Although the Layard who sent Darwin the cylinder seal (either directly or indirectly) is presumably Austen Henry Layard (1817–1894) the archaeologist and politician, it was his brother — the amateur naturalist Edgar Leopold Layard (1824–1900) of Ceylon — who actively corresponded with Darwin and sent him pigeon skins in the mid-1850s. The two men are confused in Secord (1981b, p. 178).

5. Despite the comment in Vorzimmer (1975, p. 201), William Yarrell cannot be usefully categorized as a breeder; see for example the biography provided by T. Forbes (1962), which does not, however, mention Yarrell's contacts with Darwin.

6. Serious study of Darwin's reading habits began with S. Smith (1960), and has continued in a variety of works, most notably Ospovat (1981, especially pp. 95–98). For some interesting comments on his reading in natural history journals, see Sheets-Pyenson (1981). A partial list of books read by Darwin is conveniently available (Vorzimmer 1977), but

must be used with considerable care as the transcription is not always accurate.

7. The text of Darwin's 1862 query on bees is as follows: "I should feel much obliged if the 'Devonshire Beekeeper' or any of your experienced correspondents would have the kindness to state whether there is any sensible difference between the bees kept in different parts of Great Britain. Several years ago an observant naturalist and clergyman, as well as a gardener, who kept bees, asserted positively that there were certain breeds of bees which were smaller than others, and differed in their tempers. The clergyman also said that the wild bees of certain forests in Nottinghamshire were smaller than the common tame bees. M. Godson [*sic* for Godron], a learned French naturalist, also says that in the south of France the bees are larger than elsewhere, and that in comparing different stocks slight differences in the colour of their hairs may be detected. I have also seen it stated that the bees in Normandy are smaller than in other parts of France. I hope that some experienced observers who have seen the bees in different parts of Britain will state how far there is any truth in the foregoing remarks. In the Number of your Journal published May 15, 1860, Mr. Lowe gives a curious account of a new grey or light-coloured bee which he procured from a cottager. If this note should meet his eye I hope he will be so good as to report whether his new variety is still propagated by him. — Charles Darwin. (We insert this without expressing any opinion, because we wish to have answers from as many of our readers as have paid attention to the subject. We, as well as the well-known writer of this inquiry, will be greatly obliged by any observations upon the subject. — Eds. J[ournal] of H[orticulture].)"

   I would like to thank Dr. Eva Crane of the International Bee Research Association for providing a photocopy of the German version of Darwin's query (1862b). For some of Darwin's other uses of his contacts with beekeepers, see Brian and Crane (1959). I thank S. V. Pocock for this reference.

8. Darwin does of course allude to the *Questions* ("printed enquiries") in the *Autobiography* (p. 119), and it is always possible that further study of his correspondence and early notebooks will show that important data were obtained through the printed pamphlet.

19

Darwin's Reading and the Fictions of Development

Gillian Beer

We have no knowledge, that is, no general principles drawn from the contemplation of particular facts, but what has been built up by pleasure, and exists in us by pleasure alone . . . The knowledge both of the Poet and the Man of Science is pleasure.

William Wordsworth, *Preface to the Lyrical Ballads*

Darwin's writing profoundly unsettled the received relationships between fiction, metaphor, and the material world. That power of his was nurtured by his omnivorous reading. None of Darwin's reading seems to have been in vain. It was all useable, and used, though relatively little of it was undertaken in a utilitarian spirit. We might apply the remarks of one of his favorite authors, Sir Thomas Browne, who wrote in the *Religio Medici* (1642): "*Natura nihil agit frustra*, is the only indisputable axiom in Philosophy; there are no *Grotesques* in nature; nor any thing framed to fill up empty cantons, and unnecessary spaces" (Browne: Martin ed. 1964). In the first part of this essay I shall discuss some of Darwin's literary reading and suggest ways in which it conditioned his insights, particularly in relation to problems of creation, succcession, and development. In the second part of the essay I shall argue that in the *Origin* he used narrative sequences and functional metaphor to control any over-simple developmental patterns. In the conclusion I shall examine briefly ways in which Darwin's book sharpened contradictions and so created fresh space for development-fictions in literature. These fictions bring to light unresolved implications in Darwin's writing. Here my examples will be George Eliot's novel *The Mill on the Floss* and Robert Browning's poem *Caliban upon Setebos*, both written shortly after the publication of the *Origin*.

Most work on Darwin and literature has been concerned to tabulate and analyze Darwin's influence on specific writers (Stevenson 1932; Henkin 1940; A. E. Jones 1950; Roppen 1956; Willey 1960; Ong 1966). Much of it has effectively discussed evolutionary ideas in general (Peckham 1965). Some major work has also been undertaken on Darwin as an imaginative writer, particularly in relation to other nineteenth-century thinkers (Barzun 1958; S. Hyman 1962;

Culler 1968; Peckham 1965; Eiseley 1958, 1965; Gruber and Barrett 1974; Gruber 1978; Manier 1978, 1980b).

My particular concern in this essay is to study the processes of cultural interchange. An analysis of the interpenetration of ideas and discourses is crucial to an understanding of Darwin's place in the milieu he so fully shared with his non-scientific contemporaries. And more than that, such study is an essential pre-condition to an understanding of Darwin's place in our own culture (Foucault 1970).

Darwin's struggle to realize the theoretical potentiality of his work was a struggle also with the particular language he inherited and with the multiple readerships implied in that language. His insights were to some extent determined by the narrative patterns already taken for granted in his culture. However, his wide reading in a range of older and contemporary literature should not be seen simply as pre-empting or determining his perceptions. His reading gave him access to a range of alternative understandings; these ranging alternatives were particularly needed by an imagination that thrived on abundance and diversity, and which was to make abundance and diversity essential constituents of his theory.

Language itself poses certain conditions, some of which bore particularly hard on Darwin and his theoretical enterprise: (1) Language is intrinsic to ideas; (2) Language is historically and culturally determined; (3) Language is anthropocentric; (4) Language is never neutral; (5) Language is multivocal (it potentiates simultaneously diverse meanings). At the same time, not all potential significations are active. The terms of agreement between writer and implied reader can select and exclude significations (e.g., to give a simple example, "This is natural history: here race refers to cabbages"). This "contract" is not, however, permanent. Signification may be controlled and focussed within a like-minded group (particularly any professional group), but the excluded or left-over significations of words remain potential and can be brought to the surface and put to use by those outside the professional agreement as well as by those future readers for whom new historical sequences have intervened. Furthermore, any radical new theory will itself have the effect of disturbing the "taken for granted" elements in the language it employs. All these factors made Darwin's language an area of debate, reappropriation, and neologism, even when the vocabulary preserved the forms of older discourse.

Darwin's own later comments emphasized the loss of his aesthetic powers. In his *Autobiography*, written for his family towards the end of his life, he summarizes in a pained and self-denigrating passage his loss of affective response.

> I have said that in one respect my mind has changed during the last 20 or 30 years. Up to the age of thirty, or beyond it, poetry of many kinds, such as the works of Milton, Gray, Byron, Wordsworth, Coleridge

> and Shelley gave me great pleasure, and even as a schoolboy I took intense delight in Shakespeare especially in the historical plays. I have also said that formerly Pictures gave me considerable, and music very great delight. But now for many years I cannot endure to read a line of poetry: I have tried lately to read Shakespeare and found it so intolerably dull that it nauseated me. I have also almost lost any taste for pictures or music. — Music generally sets me thinking too energetically on what I have been at work on, instead of giving me pleasure. I retain some taste for fine scenery, but it does not cause me the exquisite delight which it formerly did . . . My mind seems to have become a kind of machine for grinding general laws out of large collections of facts, but why this should have caused the atrophy of that part of the brain alone, on which the higher tastes depend, I cannot conceive. (pp. 138–139)

This later clouding of his affective powers has been read back by many commentators into far too early a period of his life. A similar argument is usual concerning his reading of fiction.

Darwin was an omnivorous reader of fiction even after he had lost his taste for other forms of literature. Indeed, his ingenuous pleasure during middle and later life in having novels read aloud to him and his rejection of "unhappy endings" has contributed to the picture of a simple reader, naive about literary conditions and achievements. This view is itself a naively retrospective reading that masks the significance of Darwin's enthusiasm for literary experience up to and including the major period of theory-formation and theoretical writing.

In the passage quoted above, Darwin's puzzlement at the loss of aesthetic pleasure is registered. In the succeeding passage of the *Autobiography* he records with a greater intensity of regret the waning of his delight in aesthetic experience and his belief that it has implied emotional, intellectual, sensuous, and moral loss. If we seek further evidence of his early pleasure in an extraordinary range of writing, it is to be found in the notebooks and in his reading lists from 1838 to 1860, now in Cambridge University Library (DAR 119, 120, 128) (Vorzimmer 1977). For example, one entry in 1840 runs: "Midsummer N. Dream. Hamlet. Othello. Mansfield Park. Sense and S. Richd. 2nd (Poor) Henry IV. Northanger Abbey. Simple Story" (DAR 119). Another makes it clear that he was reading Wordsworth alongside Erasmus Darwin, and the notebooks have allusions to Scott and references to writers of other periods such as Edmund Spenser and Thomas Browne. The reading lists supplement such references and make it clear how vigorous his reading was. It is likely that he consumed rather than analyzed, but it would be an error to assume that what he read in literature therefore had less effect on him.[1]

Let us first of all take an example from his reading of Wordsworth and see how the context of the allusion throws light on the processes of his thought. One can watch the working of Wordsworth in his imaginative life.

> Pleasures of imagination, which correspond to those awakened during music — connection with poetry, abundance, fertility, rustic life, virtuous happiness — recall scraps of poetry . . . I a geologist, have ill-defined notion of land covered with ocean, former animals, slow force cracking surface etc. truly poetical. (V. Wordsworth about sciences being sufficiently habitual to become poetical) (*M* 39–41)

Wordsworth wrote in the preface to the *Lyrical Ballads*:

> If the labours of Men of Science should ever create any material revolution, direct or indirect, in our condition, and in the impressions which we habitually receive, the Poet will sleep no more than at present, but he will be ready to follow the steps of the Man of Science, not only in those general indirect effects, but he will be at his side, carrying sensation into the midst of the objects of the Science itself. . . . If the time should ever come when what is now called Science, thus familiarized to men, shall be ready to put on, as it were, a form of flesh and blood, the Poet will lend his divine spirit to aid the transfiguration, and will welcome the Being thus produced, as a dear and genuine inmate of the household of man. (Wordsworth: Sampson ed. 1940)

In the passage quoted above, Darwin condenses the pleasures of imagination and of creative scientific thinking by means of a geological image, a reference to Wordsworth, and a half-memory of Milton's description of the creation in *Paradise Lost*.[2] That ready series of associations should alert us to the fullness with which literature permeated his earlier thinking, "carrying sensation into the midst of the objects of the Science itself". One of Darwin's major enterprises was for "scientific and common language (to) come into accordance" just as for Wordsworth poetic discourse was to "be a selection of language really used by men". Darwin hoped to substantize metaphor into "plain signification", as we see in his discussion of metamorphosis, which in his view "may be used literally" (*Origin*, pp. 438–439): "Science, thus familiarized to men, shall be ready to put on, as it were, a form of flesh and blood." Wordsworth's hope that there can be a congruity between the sciences and poetry, mediated by the figure of man, is an imaginative inspiration to Darwin. But it is an ideal whose anthropocentrism also makes problems for Darwin as he seeks to find a form in language for his insights.

I. Methodology: Reading and Writing

The effects of reading can never be proved beyond doubt, since they are multiple and since they depend upon the style of reading (Iser 1978). Darwin recorded the different intensities of reading for some of his texts ("skimmed", "read thoroughly", "failed in reading"). But it would be rash to assume a steady relation between the thoroughness of reading and the stimulus provided. Darwin from time to time scrupulously records that he has failed in reading something: 15 April 1840: "Failed in reading Dryden's Poems except Absalom and Ach. wh. I rather liked"; 20 September: "Failed in reading Niebuhr's Rome" (DAR 119).

It would be an ingenious and confident commentator who could tell precisely what, if anything, Darwin retained from his failed reading of Dryden — though clearly something teased him and drew him back, since already on 15 March, a month earlier, he had recorded, "Skimmed Pope and Dryden Poems — need not try them again". Yet three years later, on 1 October 1843, he records "Scott's Life of Dryden" (DAR 119). The recurrence of Dryden's name in the reading lists is intriguing, but the kinds of pleasure and recalcitrance he possessed for Darwin remain obscure. Niebuhr, however, has been claimed as an influence on Darwin, and Darwin's comment that he "failed in reading him" by no means necessarily rules out that claim. Indeed, difficulty, distaste, and even boredom, particularly when stimulus and pleasure have been anticipated by repute, create a powerful difficulty that may lead to a more sustained brooding on the problems raised by an author than does an enthusiastic, complete, and therefore resolved reading. Darwin's arguments with the canonical are for him an important habit of mind. His own style is full of questions and exclamations, enthusiastic rebuttals and problem-raising queries. So one style of reading which we can document in Darwin's work is dialectical reading, an alert skepticism in the face of the text's "authority". He is a determined reader, and an omnivorous one, his most hostile comments being reserved for the "intolerably prolix". Whereas some works he sweeps aside simply with the comment "poor", it is to be noted that in both the examples cited above he feels *himself* to have failed: "failed in reading". The dialectic is interrupted, but not set aside with the book. It is extended in the reader's own thoughts and may therefore move on at a tangent from the initiating problem.

There is another style of reading in which Darwin engaged with pleasure: that unguarded reading which looks for relaxation — what we call leisure-reading. Here the expectation of debate, the oppositional mode of reception, is in abeyance — though it may of course occur. But such reading, because it is not *expected* to challenge, may quietly describe shapes for experience and establish expectations that are never brought into conscious scrutiny. Its powers may be found as much at the level of the reader's inattention as of his attention. I should make it clear that I am here describing styles of reading rather than categories of subject matter. The domain of maximum scrutiny and skepticism will be wherever a particular reader's professional expertise is engaged. There is no gulf fixed between the two styles of skeptical and assimilative reading — no such crude disjunction ever takes place in reading-process. Nevertheless, there is a difference of emphasis between the different kinds of pleasurable expectation.

Fiction, because of its implicit contractual release of the reader from the need to demur when it deviates from expectations, is particularly protected from any appraisal of its assumptions. The social assumptions of a particular work may happen to irritate a particular reader. We may dislike what happens in a story, or feel uneasy with a style of characterization. But few readers not trained to observe such levels of the text become aware of the extent to which formal ordering and sequence of telling may persuade and may help to fix our expectations beyond the reading of that work. This absolving of fiction from skeptical scrutiny was also historically related to its marginalization. In later life Darwin was ashamed of his reading habits, and of the pleasure he continued to feel in novels even after his taste for other literature had withered. Because it was read as entertainment, it was not read combatively. Its implications were therefore not tested or teased out.

Darwin himself was interested in the psychology of reading and response. On 12 August 1838 he read two items, both of which engrossed him completely. One was Brewster's review of Comte in the July issue of the *Edinburgh Review*. The other was one of the *Sketches by Boz*, an early Dickens work. The Comte gave him a headache. The Boz cured it. Why? he wondered. The Comte "made me endeavour to remember, and to think deeply"; as for the Boz — "Now in this I was interested as was I in the other, and read so intensely as to be unconscious of all around, yet there was no strain on the intellectual powers — the difference is of a man wagging his foot, and working with his toe to perform some difficult task" (*M* 81). That is to say, the difference was one of complexity in reading performance. The task of understanding Comte required more delicate and precise adjustments.

My argument is that we should not relate "influence" solely to alert, combative reading, but should recognize the extent to which patterns for observing experience are learned by means of the unanalytical reception of books read. And these patterns may pre-empt what can later be observed or registered.

Books read do not stay inside their covers. Once in the head they mingle. The miscegenation of texts is a powerful and uncontrollable force. Commentators on Darwin all note that he read Malthus in late September and October 1838. He records his reading in DAR 119 on 3 October. From that statement one might imagine a pure act of reading, sustained and uninterrupted, a virginal encounter. During October 1838, however, Darwin also read at least sixteen other works and in the preceding months very many more.[3] What we have here is a network. Reading has related these random texts so that they are interactive. No one of them is quite the same as if it had been read without the others.

We privilege Darwin's reading of Malthus, and I think rightly so, since it released and disturbed him creatively. But we should not isolate it. Darwin's rereading of Malthus — that is, his combative appraisal — relied on the intellectual conditions of his reading, on what he had read before, often without analytical scrutiny, and on what he was reading alongside it. The rereading of Malthus (the changes of emphasis, the rebuttals, the new emotional emphases that his own theory gave to the material) was a protracted process, taking in reading of the months and years before and immediately *after*.

This essay analyzes mainly works we habitually think of as "literary" rather than scientific, and considers ways in which, read alongside scientific works, they provided the conditions of language, helped to polarize arguments, and sustained his independence of mind. The discussion of his reading will focus in turn on three different groups of issues related to development: (1) creation, production, and succession; (2) biography and history as an analogy to ontogeny and phylogeny; (3) native inhabitants and colonization.

In my first extended example we see how some of Darwin's most vivid early reading shaped his understanding of the productions of nature and husbandry, and his reception of Malthus.

II. Shakespeare, Montaigne, and Milton: Creation and Succession

If we are to achieve an insight into the importance of Darwin's early reading for the imaginative development of his ideas, we need to remember the power of that primary reading which preceded even his young adult leisure-reading. Particularly striking are his boyhood enthusiasm for Shakespeare and his constant reading of Milton during the voyage of the *Beagle*. His constant reading of Wordsworth and its significance for his formative adult years have been well analyzed by Edward Manier (1978) and Marilyn Gaull (1979).

In discussions of reading and of the influence of books it is usual to dwell on works read, analyzed, and responded to in adult life, during the period of controversy and theory formation. Another period and mode of reading precedes this, however. It is that uncritical and absorbed immersive reading we experience in childhood and youth. This unguarded reading is less controlled in its reception, less capable of being held at bay than any later appreciation. It creates shapes for experience, and those shapes endure into the experience we undergo in adult life.

This gives a particular value to Darwin's boyhood enthusiasm for Shakespeare, particularly the history plays. The intimacy and solitariness of his contact with Milton, the one book he never left behind when he set out on his isolated land-journeys from the *Beagle*, places Milton also in a special position. The sustenance he drew from such sources has its bearing on the formation of his ideas and on their mythopoeic powers. His literary resources affect, too, his reception of the *implications* of Malthus's ideas. Let us examine briefly ways in which some of his early reading may have contributed to his own imaginative intellectual development.

He describes himself as a young boy sitting for hours avidly reading the history plays, generally in an old window in the thick walls of the school. The plays emphasize the need for stable succession in order to preserve order and government, to preserve, indeed, the idea of the nation and the race. They presented Darwin with one genetic pattern for interpreting the relationship between race and time. The blood succession becomes a means of stemming the tide of time — replication is emphasized and change is accommodated — the dead king is replaced by a live king whose blood succession ensures that no radical alteration has taken place. Each produces "after his kind". In kingship the aspect of *restoration* is intensified, and succession becomes not a means of change but a way of standing still. No usurper can thrive, however good his individual talents may be, as Bolingbroke demonstrates. The imagery of stock and of engrafting, which is so powerfully used throughout the history plays, lies somewhere between metaphor and substantiality. "The corruption of a blemished stock" brings about downfall. The fortunes of families, like plants, will be affected and can to some extent be controlled by conscious breeding and by mingling the qualities of specified stock.

> This noble isle doth want her proper limbs;
> Her face defaced with scars of infamy,
> Her royal stock graft with ignoble plants . . .
>
> (*Richard III*, III, vii, 134–136)

Darwin's argument in the *Origin* was based from the outset on the same analogy of husbandry. But in Darwin's argument husbandry is always insufficient. Man breeds plants and animals to serve *man's* ends — not particularly to benefit the plants or animals. In contrast, Darwin asserted, natural processes breed always for the good of the individuals of the race concerned. This is a crucial distinction in his argument and points to the benevolence implicit in his view of nature.

In the M *Notebook* Darwin relates the pleasures of the imagination to that release of images often experienced while listening to music. He has earlier been discussing the relations of music and poetry — music he relates more immediately to instinct, and poetry to thought. He goes on to consider daydreaming and then to analyze the pleasures of scenery: his example of the combined pleasures of rhythm and symmetry is the shape of a tree, an image crucial in his later mythography (*M* 33–39). He then turns to "Pleasures of imagination, which correspond to those awakened during music — connection with poetry, abundance, fertility, rustic life, virtuous happiness

— recall scraps of poetry; former thoughts . . ." (*M* 39–40).

"Poetry, abundance, fertility, and virtuous happiness" — the description is apt for that scene in *The Winter's Tale* in which King Polixenes debates with Perdita, the wise shepherdess so soon to be discovered as herself the lost scion of a royal house. They discuss with serious courtesy the propriety of grafting and selecting to produce hybrids — "artificial selection", which is the necessary contrary contained in Darwin's term "natural selection". Perdita will not plant "carnations and streak'd gillyvors,/Which some call nature's bastards." Her reason for rejecting them is:

> For I have heard it said
> There is an art which in their piedness shares
> With great creating nature.
>
> (*IV*, iv, 102–104)

They are not wholly natural; art has intervened, and they are in part products of that art rather than of "great creating nature". But Polixenes affirms the supremacy of nature's powers even in such artful use:

> Say there be;
> Yet nature is made better by no mean
> But nature makes that mean: so, over that art
> Which you say adds to nature, is an art
> That nature makes. You see, sweet maid, we marry
> A gentler scion to the wildest stock,
> And make conceive a bark of baser kind
> By bud of nobler race: this is an art
> Which does mend nature, change it rather, but
> The art itself is nature.
>
> (*IV*, iv, 105–114)

In Shakespeare's late plays there is no longer so strong an insistence as in the history plays on the act of replacement from generation to generation. Instead the emphasis is upon replenishment, growth, and transformation. The new generation represents fresh possibilities rather than simple restitution of continuity. "The art that nature makes" mends and changes nature and is itself nature. In August 1838 and again in October 1843 Darwin read Montaigne's essays,[4] each time specifying Volume I, in which is collected the famous essay "Of the Cannibals" that informed the writing of both *The Winter's Tale* and *The Tempest* (DAR 128: 89; DAR 119). Montaigne teases out the paradoxical relations of "artificial" and "natural" in his extended discussion of the concept "savage". Of the cannibals he writes:

> They are even savage, as we call those fruits wild which nature of herself and of her ordinary progress hath produced; whereas indeed they are those which ourselves have altered by our artificial devices and diverted

> from their common order, we should rather term savage. In those are the true and most profitable virtues and natural properties most lively and vigorous, which in these we have bastardized, applying them to the pleasure of our corrupted taste. . . . there is no reason art should gain the point of honour of our great and puissant mother nature. (tr. Florio 1603)

His cultural relativism is expressed through a metaphor from natural history. The argument between Polixenes and Perdita in Shakespeare's play draws on Montaigne's metaphor. Montaigne continues by quoting Plato: "all things are produced either by nature, by fortune, or by art: the greatest and fairest by one or other of the two first, the least and imperfect by the last." Nature and fortune are set over against art here, but in Sir Thomas Browne's *Religio Medici*, which Darwin was reading within two weeks of Montaigne in August 1838, and from which he cites arguments on chance and providence, a rather different bent is given to the problem:

> Now nature is not at variance with art, nor art with nature; they being both the servants of his providence: Art is the perfection of Nature: Were the world now as it was in the sixth day, there were yet a Chaos: Nature hath made one world, and Art Another. In brief, all things are artificial, for nature is the Art of God. (Browne: Martin ed. 1964)

This debate, with its alternative suggestions that either "Art" or husbandry is creating "Nature's bastards" and "applying them to the pleasure of our corrupted taste", or that "Art is the perfection of Nature", provides a pair of terms for Darwin to think with, terms that lie behind Artificial Selection and Natural Selection. Browne's assertion that nature is the art of God is part of a tradition that allowed Darwin in the *1842 Sketch* to write: "Who seeing how plants vary in garden, what blind foolish man has done in a few years, will deny an all-seeing being in thousands of years could effect (if the Creator chose to do so) . . ." (p. 45). In the *Origin* "Nature" is substituted for the Creator in the parallel passage. This shift splits apart the contraries: they are no longer two aspects of one providence.

The language in which Darwin contrasts the powers of man and of nature in their productions is freighted with the elevated discourse he had learned from the sixteenth- and seventeenth-century writers he relished.

> How fleeting are the wishes and efforts of man! how short his time: and consequently how poor will his products be, compared with those accumulated by nature during whole geological periods. Can we wonder, then, that nature's productions should be far "truer" in character than man's productions: that they should be infinitely better adapted to the most complex conditions of life, and should plainly bear the stamp of far higher workmanship? (*Origin*, p. 84)

The serious pleasure that Shakespeare, Montaigne, and Browne offered had provided him with terms for contemplation at a crucial stage in the precipitation of his theory.

Such reading created an expressive habit of language that allowed Darwin unselfconsciously to register his gradualism by his recourse to an older high discourse. But, more importantly, it increased the imaginative potential of Darwin's divergence from Malthus. And here his reading of Milton becomes crucial.

Almost all commentators stress the importance of reading Malthus for the precipitation in Darwin's imagination of his already half-formed notion of natural selection (e.g. Vorzimmer 1969; Bowler 1969a; Young 1969; Herbert 1971; Ospovat 1979; Kohn 1980). In doing so, they follow Darwin himself. What has not always been sufficiently recognized, however, is the extent to which Darwin transformed the imaginative tone and emotional balance and hence the intellectual potentialities of Malthus's concept. Malthus opens his essay *On Population* with a passage in which celebration and alarm are finely balanced as he describes the energy of fecundity.

> It is observed by Dr. Franklin, that there is no bound to the prolific nature of plants or animals, but what is made by their crowding and interfering with each others means of subsistence. Were the face of the earth, he says, vacant of other plants, it might be gradually sowed and overspread with one kind only, as for instance with fennel, and were it empty of other inhabitants, it might in a few ages be replenished from one nation only, as for instance with Englishmen. This is incontrovertibly true. Through the animal and vegetable kingdoms Nature has scattered the seeds of life abroad with the most profuse and liberal hand; but has been comparatively sparing in the room and the nourishment necessary to rear them. The germs of existence contained in this earth, if they could freely develop themselves, would fill millions of worlds in the course of a few 1000 years. Necessity, that imperious, all pervading law of nature, restrains them within the prescribed bounds. The race of plants and the race of animals shrink under this great restrictive law; and man cannot by any efforts of reason escape from it. (1826, the edition used by Darwin)

Any single species of plant or animal whose propagation went unchecked could rapidly colonize and take over the entire world, leaving no place for any other. Malthus goes on from this natural historical example to a further phase of economic argument in which he proposes that the reproductive energies of man, if not curtailed, must always outstrip the means of providing him with food.

To Malthus fecundity was a danger to be suppressed — particularly by draconian measures among the human poor. To Darwin fecundity was a liberating and creative principle, leading to increased potential for change and development. Because of the myriad superproductiveness of natural generative process, the range of individuality and of possible mutation is immense. And here it becomes important to remember the two books that accompanied him on the voyage of the *Beagle* at the time that he was imaginatively at his most responsive. One of them was Lyell's *Principles of Geology*. The other, which he says in his *Autobiography* was the one book that he never left behind, taking it with him on the long land expeditions from the *Beagle*, was Milton's poems: "in my excursions during the voyage of the *Beagle*, when I could take only a single volume, I always chose Milton" (p. 85). He continued to read Milton on his return to England: for example we find an entry in his reading list for 1840: "March 13th, Minor Poems of Milton and first volume of Wordsworth."

What kinds of imaginative sustenance did Milton offer to Darwin at this intensely formative period of his life? One of the crucial discoveries that came to Darwin as a result of the voyage was that the green English landscape, with its many man-induced harmonies and its sober beauties, could not be considered normative. Beyond England lay other natural landscapes full of tumultuous color and life.

The discovery of diversity and of profusion were of equal importance. The rich, even ecstatic, descriptions that Darwin gives of his travels allow some glimpse of the happiness his experiences engendered in him.

It must have seemed that the natural world came close to justifying Comus's earlier (and very un-Malthusian) view of natural superabundance, and the prodigal productivity, of the earth. Comus, voluptuary and bacchic villain, interprets the abundance of the world as all being provided for the pleasuring of man:

> Wherefore did Nature powre her bounties forth,
> With such a full and unwithdrawing hand,
> Covering the earth with odours, fruits, and flocks,
> Thronging the seas with spawn innumerable.
> But all to please, and sate the curious taste? (ll. 709–713)

Comus claims that man has not only the right to indulge his luxurious appetites, but the duty to do so. Else Nature would be "quite surcharged with her own weight,"

> And strangl'd with her waste fertility;
> Th'earth cumber'd, and the wing'd air dark't with plumes,
> The herds would over-multitude their Lords . . . (ll. 728–730)

Comus's speciously libertarian arguments are countered by the Lady he has imprisoned; she insists that the appearance of over-plenty comes from the imbalance of want and superfluity among men. Instead of a few men engrossing all natural wealth, what is needed is a more even distribution of plenty:

If every just man that now pines with want
Had but a moderate and beseeming share
Of that which lewdly-pamper'd Luxury
Now heaps upon some few with vast excess,
Nature's full blessings would be well dispenc't
In unsuperfluous even proportion,
And she no whit encomber'd with her store . . . (ll. 767–773)

Darwin's preoccupations at this time were with fertility, the mechanisms of increase and generation, and the significance of these for the development of nature through time. The debate in *Comus* on the relation of man's consumption to the fertility of nature creates a precedent, and a counter-reading, which allowed Darwin a vantage point from which to survey Malthus's arguments.

Darwin walked the tropical forests with Milton. His intense sense-arousal took him beyond his own power of language:

It is, when the sun has attained its greatest height, that such scenes should be beheld: then the dense splendid foliage of the mango hides the ground with its darkest shade; whilst the upper branches are rendered from the profusion of light of the most brilliant green. . . . When quietly walking along the shady pathways, and admiring each successive view, one wishes to find language to express one's ideas. Epithet after epithet is found too weak to convey to those, who have not visited the intertropical regions, the sensation of delight which the mind experiences . . . to every one in Europe, it may be truly said, that at the distance of a few degrees from his native soil, the glories of another world are open to him.
(*Journal of Researches*, p. 591)[5]

Furthermore, Milton's descriptions of creation accompanied Darwin's discoveries of virgin lands. We have seen already the informing presence of Wordsworth in Darwin's notebooks. Let us return to the passage analyzed earlier and bring to the surface this time Milton's possible contribution to Darwin's imaginative imagery. In July 1838 Darwin records a train of thought, of day-dream, of creation and growth, which seems to him truly poetical: "I, a geologist, have ill defined notion of land covered with ocean, former animals, slow force cracking surface etc., truly poetical" (*M* 40). What is "the slow force cracking the surface", which to him seems so "truly poetical"? He is discovering a metaphor for the process of creative thought itself, as well as drawing on Lyell's imagery of the immense extent of time in uniformitarian geology with its repudiation of catastrophe and its insistence upon ineluctable process, slippage, congestion, slow force.

Milton's account of the third day of creation in the seventh book of *Paradise Lost* describes the parting of the earth and the water:

Over all the face of Earth
Main ocean flow'd, not idle, but with warme
Prolific humour soft'ning all her Globe,
Fermented the great Mother to conceave,
Satiate with genial moisture, when God said,
Be gather'd now ye waters under Heav'n
Into one place, and let dry Land appear.
Immediately the mountains huge appear
Emergent, and their broad bare backs upheave. (VII: 278–286)

In every line of Milton's description of creation there is superabundance, variety, and plenty, "the Sounds and Seas each Creek and Bay with Frie innumerable swarme"; the fish

part single or with mate
Graze the Sea weed thir pasture, and through Groves
Of Coral stray, or sporting with quick glance
Show to the Sun thir wav'd coats dropt with Gold,
Or in thir Pearlie shells at ease, attend
Moist nutriment, or under Rocks thir food
In jointed Armour watch; (VII: 403–409)

"And" and "or" link the overrunning lines in a sinuous dance of anticipation ("attend Moist nutriment"), free play, and satisfaction. On the sixth day the Earth

Op'ning her fertil Womb teem'd at a Birth
Innumerous living Creatures, perfet formes,
Limb'd and full grown. (VII: 453–455)

In *Paradise Lost* Darwin met the full poetic expression of "separate creation", of fully formed, full-grown species. Sexuality there expresses itself as lyrical union, rather than as generation, descent, or development. Milton emphasizes the direct birth of life from sea and earth: "the Ounce, the Libbard, and the Tyger", all emerge out of the earth:

The Grassie Clods now Calv'd, now half appeer'd
The Tawnie Lion, pawing to get free
His hinder parts . . . (VII: 463–465)

The surreal completeness of this issue from primary matter is also the supreme compression of time:

Aire, Water, Earth,
By Fowl, Fish, Beast, was flown, was swum, was walkt
Frequent: and of the Sixt day yet remain't.

Milton's account extends the dreamlike qualities of Genesis — replacing its assurance of plenitude with a fantastically articulated display of specific life.

What Milton affords here is not agreement nor proto-indications of Darwin's ideas, but rather a stringent tactile joyousness in his imaging of the earth and its inhabitants, the supreme expression of a creationist view of the world. His language made manifest for Darwin in its concurrence with his own sense of profusion, density, and the articulation of the particular, how much could *survive*, how much could be held in common and in continuity from the past. This sense of *continuity* of culture and insight had an emotional and indeed theoretical importance for Darwin. It accorded with the uniformitarianism he had derived from Lyell, *Natura non facit saltum* — and neither, it seems, does mind. Darwin was at pains to emphasize the congruity of his images with those previous myth-systems rather than iconoclastically to throw them aside.

III. The Imagination and Individual Development

Darwin particularly enjoyed reading "lives", those generally optimistic fictions of development. Biographical and autobiographical accounts provided psychological evidence of the growth of creativity. They provided too an opportunity for "voyages imaginaires" through other people's experience. They are thus simultaneously studies of ontogeny and expeditions into new territories. "Gibbon's Life of himself", "Hume's life of himself with corres: with Rousseau", "Several of W. S. Landor's Imaginary Conversations — very poor", "Lockhart's life of Walter Scott", "Boswell's life of Johnson 4 vols.", "Life of Haydn and Mozart", "Lockhart's life of Napoleon", "Cowper's Life & several volumes of letters", "Moore's Life of Byron 6 volumes — poor" (DAR 119); all these works are in the lists for 1838 and 1839. There is then a lull in his recorded reading of biography, though in 1840 he records "Three vols. of Swift's letters" and "The Hour & the Man H. Martineau" — this latter was Harriet Martineau's historical novel on the life of Toussaint l'Ouverture, who led the uprising of the slaves against their masters. (Martineau was one of Darwin's favorite writers if one is to judge by the speed with which he bought her books as they appeared.) Later in the 1840s he turns to more radical lives, the working class autobiographies of Samuel Bamford "Passages in the Life of a Radical" (20 July 1844) and in 1848, the year of revolutions, "Autobiography of a Working man A. Somerville (excellent)" (12 November 1848) and a "Life of M. Wollstonecraft and Rights of Women" along with that other fictional life that *Blackwood's* considered so incendiary, Charlotte Bronte's *Jane Eyre* (8 August 1848) (DAR 119). His studies of individual lives begin to be more insistently accompanied by the reading of history, perhaps in part owing to the influence of Carlyle whose works "Sartor Resartus — excellent", *Chartism, Heroes and Hero-Worship* — "moderate" and *Past and Present* are all read as they appear (DAR 119). Among the histories he records in 1840 to 1844 are "Hume Hist. Engl. Vol. 5 and 6 (March 15th)", "April 4th Hallam's Hist. Lit. 4 vols. good", 18 April "7th & 8th Vol. of Hume's England — admirable", "one (volume) of Dr. Cooke's History of England from Anno VIV." This 1842 entry for 6 May notes also "several plays of Shakespeare": "Midsummer N. Dream, Hamlet, Othello, Richard 2d. poor, Henry IV." "Shakespeare Sonnets" appear among the entries for 1840; *Julius Caesar, Coriolanus* and *Lear* are grouped on 30 July 1841 (All entries DAR 119). With the growing intensity of interest in history he turns again to his earliest literary pleasure, the plays of Shakespeare, especially the history plays.

His very thorough reading of general histories begins in 1843 and 1844 to include rather more theoretical work, such as "Arnold's Lectures on History" and the "History of Civilization by Guizot" as well as works on race such as "Smith Varieties of the Human Race", "White Regular Gradations of Man", and accounts of conquest and the invasion of one race and its land by conquerors, such as "Prescott's Hist. of Mexico" (DAR 119).

In 1838 his preoccupation with psychological development was at a height and accompanied his struggle with the idea of species development and transmutation. Ontogeny thus authenticates in a very general way (and perhaps as assumption, not argued position) the inquiry into phylogeny.

In the midst of his reading of individual lives and his discussion of instincts, memory, and unconsciousness, he composed an autobiographical sketch recalling his own earliest years. At the time of writing it he was still only twenty-nine. The purpose of the fragment is to explore and fix recollection, and it implies no arc of public achievement as was commonly the case with published biography at that time. It is a wholly private work. If anything, it has affinities with Wordsworth's poetic recollections of his childhood in its insistence on remembered particulars. But it makes no claims to metaphysics. It is comic, affectionate, and passionate, in its summoning up of the young child's experience. He remembers horror, astonishment, pleasure, fear. He fails much to remember his mother who died when he was eight and a half (*ML* 1:2–3). That obliteration perhaps provided emotional and intellectual drive for his plumbing back into a more universal past.

The memories that particularly intrigue the young adult Darwin are of the pleasures of collecting and naming — and of the pleasures and dangers of storytelling. Or lying. The stories he invented in his childhood were designed to impress and astonish himself and others. His passion for fabulation sought power, the power to make things be and to control the paradoxes by which he was surrounded in the natural world. At the same time he was exhilarated by the intensity of paradox. What he made up became substantial and vivid to him.

> I was in those days a very great story-teller — for the pure pleasure of exciting attention and surprise. I stole fruit and hid for these same motives, and injured trees by barking them for similar ends. I scarcely ever went out walking without saying I had seen a pheasant or some strange bird (natural history taste); these lies, when not detected, I presume excited my attention, as I recollect them vividly, not connected with shame, though some I do, but as something which by having produced a great effect on my mind, gave pleasure like a tragedy. I recollect when I was at Mr. Case's inventing a whole fabric to show how fond I was of speaking the truth! My invention is still so vivid in my mind, that I could almost fancy it was true, did not memory of former shame tell me it was false. (*ML* 1:3–4)

The prowess of invention gives him "pleasure like a tragedy". This arresting description exactly conveys the *fullness* and the density of his imaginative life. The power of lying, of invention, of telling and not telling, fuels his passion for discovery: "I distinctly recollect the desire I had of being able to know something about every pebble in front of the hall door"; "I was very fond of gardening, and invented some great falsehoods about being able to colour crocuses as I liked" (*ML* 1:3). In his account he realizes the child's obdurate sense of the reality of these inventions —a sense that almost survives: "My invention is still so vivid in my mind, that I could almost fancy it was true, did not memory of former shame tell me it was false." The delight in lies and inventions — their urgency as a form of hoped-for truth — is both wonderfully comic and wonderfully full of insight.

When Darwin was disbelieved as a boy and had to acknowledge his claims false, he felt shame. Only by means of shame did he thoroughly disbelieve his own claims. When a couple of months after writing this sketch in 1838 he formulated his theory of natural selection, he kept quiet about it. This brief autobiographical account was written not only at the height of his imaginative powers but at the height of his *study* of imaginative powers. Simultaneously his mind and his notebooks were thronging with the as yet unorganized and uncommunicated story of metamorphosis, transmutation, and selection. The impulse to avoid the challenge of utterance which may wreck the thought-work is strong (is it, after all, to be treated by others as a castle in the air?) It may be that the length of his account of story-telling or lying, compared with his other memories, registers an elation and a creative disturbance felt anew by the young Darwin in the midst of formulating his theory and akin to that which he had experienced as a ten-year-old. He maintained a powerful and long-continued secrecy in which to relish and develop his own imagined story of a past for the life of our planet. The "pleasure like a tragedy" thrives still in the invention that developed from the early notebooks, through the *1842 Sketch* and *1844 Essay*, to the incomplete *Natural Selection*, and the complete *Origin*.

Alongside the other intensity of his reading he particularly recalls the "Wonders of the World" "which I often read and disputed with other boys about the veracity of some of the statements; and I believe this book first gave me a wish to travel in remote countries which was ultimately fulfilled by the voyage of the *Beagle*" (*Autobiography*, p. 44). Just as the doubtful marvels of the *Wonders of the World* impelled him at last on his journey of scientific discovery, so the history plays of Shakespeare had offered him an initiating language for apprehending patterns of descent and kin.

In the same month as the composition of his autobiographical fragment Darwin thought much about the nature of reading and imagining and tried to analyze the processes of creative labor. He speculates that "Perhaps one cause of the intense labour of *original inventive* thought is that none of the ideas are habitual nor recalled by obvious associations, as by reading a book. — Consider this —." The key to the fatigue of thought and reading is "the comparison with past ideas": "The mind thinks with extraordinary rapidity — We may conclude that neither number, vividness, rapidity, novelty of separate ideas cause fatigue to the mind, it is solely the comparison with past ideas which makes consciousness" (*M* 103). He compares his experience of reading Brewster's account of Comte and his reading of *Sketches by Boz* (*M* 81). What he calls "castles in the air" he says, are work as hard as any train of geological reasoning.

> In castles of air the trouble [I well recollect] is in making things somewhat probable, in comparing every step, & inventing new means — therefore works of imagination *hard* work. (*M* 115)

As he ponders the psychological problems of imagination, instinct, "double consciousness", and expression during that month, he draws on recollections of Spenser's *The Faerie Queene*; Coleridge's drama *Zapolya*; Burke's idea of Sympathy; Montaigne's *Essays*; Lonsdale; Mayo's *Philosophy of Living*; Hume, Adam Smith, and Dugald Stewart; Thomas Browne's *Religio Medici* and Lockhart's *Life of Scott*. Some of these he was reading at the time, as DAR 119 also testifies; some were part of his meditative store. All are set in immediate relation to his current life experience. The idea of descent from kindred forms is succinctly formulated in the iconoclastic wit of his entry for 4 September:

> Plato [Erasmus] says in Phaedo that our "imaginary ideas" arise from the preexistence of the soul, are not derivable from experience — read monkeys for preexistence. (*M* 128)

To him "imaginary ideas" throng with experience, those of the individual, of his culture, and of long patterns of evolutionary descent. The high spirited delight in the "number, vividness, rapidity, novelty" of separate ideas informs all his writing at this time. His long endeavor was to establish connections between all these vigorously disparate ideas and works, for as he summarizes the problem on 21 September 1838: "Believing consists in the comparison of ideas, connected with judgment" (*M* 144).

IV. Language and Selection

One of Darwin's enterprises in the *Origin* was to resolve scientific and common discourse as thoroughly as possible. He attempted, in Wordsworth's phrase, to use "a selection of the language really spoken by men". The will towards plainness can be seen in his amusing 1855 gloss on the title of a reprint. The reprint was entitled: "On the Power of Icebergs to Make Rectilinear, Uniformly-directed Grooves Across a Submarine Undulating Surface." In Darwin's hand on the cover are the words "On Iceberg Scratching Rocks" (DAR 135/11). The accommodations between technical and vernacular usage, however, sometimes produce ambiguity. The hard enterprise of revising the *Origin* to control its multivocality brought home the problem keenly to him. But he seems to have become aware of multivocality *as* problem only as a result of the reception of the published work rather than before the first edition. The attempt to make room for the technical and the vernacular was, moreover, no mere stylistic quirk but part of his desire that his theory should be equivalent to the evidence of the natural world in all its diversity: "we shall at least be freed from the vain search for the undiscovered and undiscoverable essence of the term species" (*Origin*, p. 485). Etymology is an important metaphor for him because it reveals each word as process, history, and development.[6]

His discourse is of the kind that George Eliot characterized as expressing "life" in contrast to that "patent de-odorized, non-resonant language, which effects the purpose of communication as perfectly and rapidly as algebraic signs." His language has not driven out "uncertainty", "whims of idiom", "cumbrous forms", "the fitful stammer of many-hued significance" (Pinney 1963). Darwin's was not an austere Descartian style. He felt the problems of obscurity and of richness of association: the over-rapid condensation of argument and insight that may bury the deep connections obvious to the writer or allow minor terms to unfurl new significations.[7] Some of the problems of "naturalizing" scientific or abstract discourse are suggested in a letter he wrote with the taxing experience of revising the *Origin* behind him. Writing to his protegé John Scott in 1863 he said: "I never study style, all that I do is to try to get the subject as clear as I can in my own head, and express it in the commonest language that occurs to me. But I generally have to think a good deal before the simplest arrangement and words occur to me" (*ML* 2: 322). The apparent easiness of "I never study style" is undermined by the admission of how much thought-work must precede simplicity.

The multivocality of Darwin's language reaches its furthest extent in the first edition of the *Origin*. His language is expressive rather than rigorous. He accepts the variability within words, their tendency to dilate and contract across related senses, or to oscillate between significations. He is interested less in singleness than in mobility. In his use of words he is more preoccupied with relations and transformations than with limits. Thus his language practice and his scientific theory coincide.

Once the *Origin* was published, Darwin became far more aware of the range of implications carried by this generous semantic practice. It was brought home to him, by the criticisms of his contemporaries, that many of his terms could have meanings different from or beyond what he intended (Vorzimmer 1970). When he had uttered the willful fictions of his childhood they had been challenged, and he felt shame. After the publication of the *Origin* he defended his theory by paring away multiple significations, trying at points of difficulty to make his key terms mean one thing and one thing only, as in the case of natural selection. Such labor came hard to him. The exuberantly metaphorical drive of the language of the *Origin* was proper to its topic. The need to establish more parsimonious definitions and to combat misunderstanding may have been in his mind when he later wrote that his mind "seems to have become a kind of machine for grinding general laws out of large collections of facts" (*Autobiography*, p. 139). The need to contain meaning and to retrench potentiality, which underpins much of the enterprise of revision from edition to edition, may help to account for the dimming of his imaginative powers, which he so deeply regretted.

One of the major questions raised by the discourse of the *Origin* is the extent to which it is possible to delimit words and to insist on univocality in a work addressed to a range of general readers as well as to a scientific confraternity (Shapin and Barnes 1979). A related problem is how to control the metamorphic extension of concepts. Metaphors may overturn the bounds of meaning assigned, thereby destabilizing the argument being developed. Even seemingly stable concepts may come to operate as generative metaphors, and in doing so they may reveal inherent heterogeneity of meaning and of ideology. Darwin's use of the concept "struggle" is a well-known example. But there are others, less remarked, such as *generation*, which yields the tree, the great family, the lost parent, the "changing dialect" of life. Each of these consequent ideas extends some element in the initiating one of generation, and itself establishes a further range of incipient meanings.

Sometimes we can watch Darwin seeking to contain the implications of a word, as in the discussions of "slaves" and "masters". He is skeptical of "the slave-making instinct"; "any one may well be excused for doubting the truth of so extraordinary and odious an instinct as that of making slaves" (*Origin*, p. 220). Yet he finds himself obliged to acknowledge the practice as natural to ants from his observations of slave-making ants, while attempting to hold off any naturalization of human slavery. He makes it clear that human behavior functions simply as the second term in the metaphor, providing a vocabulary by means of which to describe the behavior of ants, without allowing the behavior of ants to justify the practices of men. Perhaps there is an edge of covert humor in the miniaturized social analogy, but it is hard to be sure: "In England the masters alone usually leave the nest to collect building materials and food for themselves, their slaves and larvae. So that the masters in this country receive much less service from their slaves than they do in Switzerland" (*Origin*, p. 223). The tones of Gulliver among the Lilliputians come through here.[8]

Related to this problem of colonization and enforcement is his use of the terms of "native inhabitant" and "foreigner". And this pair of terms bears directly on problems in the idea of development as improvement. The relating of development to improvement causes Darwin much uneasiness. Before looking at the literary context of these concepts and their bearing on Darwin's own experience, it is worth taking into account dictionary definitions from Darwin's time, not because he would have gone and looked up the meaning of familiar words but because they can legitimately be taken as expressing shared assumptions about the meaning of words. Edward Manier has made much of Hensleigh Wedgwood's *A Dictionary of English Etymology* in his discussion of Darwin's use of the term "struggle". But Wedgwood's dictionary is really too late to be helpful, even allowing for the fact that Darwin might have had access to it before publication. It appeared in three volumes between 1859 and 1867. Moreover, since struggle begins with "s" Wedgwood was probably working on its definition well after the publication of the *Origin* and so may in his turn have been affected by Darwin's usage!

It is better to turn to the more modest dictionary in Darwin's own library, the highly traditional Dr. Johnson, abridged and revised in 1826. There we find the definition: "To Struggle: To labour, to act with effort, to strive, to contend, to contest . . . To labour in difficulties; to be in agonies or distress." Here, as in Wedgwood, the signification of laboring and effort is the first sense, while that of actual contest comes fifth. It seems clear that in turning away from the concept of "war" to that of "struggle" Darwin was deliberately giving predominance to effort over conflict. These problems of stipulation have their bearing on his discussions of invasion and development.

In the light of Darwin's repeated use of the image of "entanglement" to express the ecological appearances of nature, it is worth noting that the primary meaning given to *evolve* in his dictionary is "To Unfold; to disentangle". Perhaps, with his culture's more lively sense of the Latin origin of words, there is a punning cross-play in his contemplation of "an entangled bank" at the conclusion of the work which has unfolded "the laws acting around us". In his emphasis through metaphor on ecological interdependence and "the inextricable web of affinities" Darwin draws attention to what is fresh in his own work and what distinguishes it from "Development Hypothesis" as elaborated by writers such as Chambers, Spencer, and even Von Baer. Far from being superimposed fine writing, as W. Cannon suggests (1968), the metaphor of entanglement enacts what often remains latent in his argument: the extent to which evolution is a lateral rather than simply an onward movement, whose power lies in multiple relationships as much as in selecting out.

Later in his life Darwin emphasized the speciousness of easy connections between "development" and "improvement". He commented sardonically that "the white man is 'improving off the face of the earth' even races nearly his equal". His quotation marks frame his distaste (D. Freeman 1974; Greene, 1977).

V. Native Inhabitants and Improvement: A Problem of Multivocality

During the 1830s and 1840s, as I have already indicated, Darwin read widely in history and race-theory. There were often conflicting models in the presentation of processes of invasion and colonization (Levin 1959; H. White 1973). Darwin had, in addition, two very important sources for his appraisal of the relations between "native inhabitant" and "intruder". These sources were of two kinds: the first was his own experiences when on the *Beagle* voyage and the understanding of those experiences that he reached in writing them down. The second was his enjoyment of Walter Scott, one of the few writers whom he continued to read throughout his life.

Darwin had seen the establishment of dominance by "intruders" in the name of development and civilization. In the *Journal of Researches* he gives a vivid account of the warfare between Spaniards and Indians. Throughout his description there is a tone of poignant admiration for the heroic (or barbaric) qualities of the Indians — "when overtaken, like wild animals, they fight against any number to the last moment", "they were remarkably fine men, very fair, above six feet high, and all under thirty years of age" (p. 120).

Despite his own fear of ambush during his travels, his account often gives an epic stature to Indian behavior: "The old Indian father and his son escaped, and were free. What a fine picture one can form in one's mind — the naked bronze-like figure of the old man with his little boy, riding like a Mazeppa on the white horse, thus leaving far behind him the host of his pursuers" (p. 123). The Indian becomes a heroic statue "bronze-like" "like a Mazeppa". Darwin was shocked most of all by the genocidal aspect of "this war of extermination" and by the massacre of all women over twenty years of age. "When I exclaimed that this appeared rather inhuman, he replied 'Why, what is to be done? They breed so!' " "This is a dark picture," Darwin comments; "Everyone here is fully convinced that this is the most just war, because it is against barbarians" (p. 20). Darwin saw the ruthlessness of struggle as conflict cast in human terms. He saw also that the destruction of "native inhabitants" resulted in their cultural retrogression: "Not only have whole tribes been exterminated, but the remaining Indians have become more barbarous. Instead of living in large villages, and being employed in the arts of fishing, as well as of the chase, they now wander about the open plains, without home or fixed occupation" (p. 122). Once their relationship to their habitual environment was disturbed by invasion from without, they became less fitted to survive.

Darwin in the *Journal of Researches* grasped how far the designation of "barbarism" or "civilization" is a matter of cultural prejudice. This recognition calls into question the absoluteness of any notion of cultural development. For example, in an expedition from Rio de Janeiro, they passed a spot where some runaway slaves had for a long time established a little settlement at the top of a granite cliff.

> At length they were discovered, and a party of soldiers being sent, the whole were seized, with the exception of one old woman who, sooner than again be led into slavery, dashed herself to pieces from the summit of the mountain. In a Roman matron this would have been called the noble love of freedom: in a poor negress it is mere brutal obstinacy. (p. 22)

His travels had created a continuing interest not only in South America and its indigenous people but in the conquering Spaniards who were naturalized there. We find him, for example, reading "Dublado's letters on Spain — excellent" in 1840, "Robertson's America" and "Don Quixote" in 1841; in 1842 and 1843, "travels in W. America" and "Stephen's Central America". In April 1844 he read two books about travels in Spain, Borrow's "Bible in Spain", and "Townshends Journey through Spain", and then in August and October 1844 "Aug 30th Prescott's Hist. of Mexico/ Oct 1 2d and 3d vols" (DAR 119).

William H. Prescott's *History of the Conquest of Mexico with a Preliminary View of the Ancient Mexican Civilisation and the Life of the Conqueror Hernando Cortes* appeared in 1843. In the first volume, and again in the appendix, Prescott dwells on the character of Aztec civilization, and seeks to give "the reader a just idea of the true nature and extent of the civilization to which the Mexicans had attained." His account (like Montaigne's essay "Of the Cannibals", which Darwin read again in 1843) shows a strong cultural relativism. His vivid account of Aztec civilization, occupying as it does nearly the first two hundred pages of his work, gives the reader a sense of how much was lost through the invasion of the Spaniards, even while in the later part of the work he sets out sympathetically the exploits of Cortes and his followers. Moreover, he makes it clear that the Spaniards altered not only the culture but also the land in their own image. The Spaniards "made indiscriminate war on the forest":

> In the time of the Aztecs, the table land was thickly covered with larch, oak, cypress, and other forest trees, the extraordinary dimensions of which, remaining to the present day, show that the curse of barrenness in later times is chargeable more on man than on nature . . . This spoliation of the ground, however, is said to have been pleasing to their imaginations, as it reminded them of the plains of their own Castile — the table land of Europe; where the nakedness of the landscape forms the burden of every traveller's lament, who visits that country. (Prescott 1843, 1:9)

The willfulness of the imaginative Spaniards may well have fed into Darwin's representation of artificial selection, which is not only "husbandry" but bears a sense of the grotesque and extreme in opposition to wise natural selection. Artificial selection has an exploitative rather than a nurturing function. Darwin's own experiences and his continued reading made him aware of the dark shadows of genocide in any conquering intrusion.

Like Darwin in the *Journal of Researches*, Prescott emphasizes the subsequent degradation of the conquered race. Aztec people could not adapt themselves, and the tone is one of admiration for that proper obduracy: "Their civilization was of the hardy character which belongs to the wilderness. They refused to submit to European culture — to be engrafted on a foreign stock" (Prescott 1843, 1:46). Such reading helps explain the rather curious uncertainty of tone in Darwin's argument at the beginning of the chapter in the *Origin* on Natural Selection where he contrasts the development of countries with open borders with that of islands. Although his example is of a change in climate, the terms he chooses do not exclude human application and this disturbs the sequence of his argument. The immigration of new forms "would seriously disturb the relations of some of the former inhabitants":

> We may conclude, from what we have seen of the intimate and complex manner in which the inhabitants of each country are bound together, that any change in the numerical proportions of some of the inhabitants, independently of the change of climate itself, would most seriously affect many of the others. (*Origin*, p. 81)

He then counter-proposes a sequestered country "into which new and better adapted forms could not freely enter". Here there might be modification of the original inhabitants: "for, had the area been open to immigration, these same places would have been seized on by intruders." In the enclosed model betterment and adaptation would proceed by "slight modification" and "natural selection would thus have free scope for the work of improvement".

The argument at this stage sets natural selection over against the immigration of new forms, with the implicit suggestion that such immigration is a part of its contrary, artificial selection. The problem is never fully spelled out, but the haunting awareness of the range of meaning held in words like "intruders" and "native inhabitants" beset Darwin: "for in all countries, the natives have been so far conquered by naturalised productions, that they have allowed foreigners to take firm possession of the land" (*Origin*, p. 83).

Walter Scott's novels and poems studied in a variety of ways the decay and absorption of ancient Scottish culture by the English. In *Waverley* (1814), for example, Baron Bradwardine grieves "for the blackened walls of the house of my ancestors" and for the failure of the Jacobite rebellion:

> To be sure we may say with Virgilius Maro, *Fuimus Troes* — and there's the end of an auld song. But houses and families and man have a' stood till they fall with honour, and now I hae gotten a house that is not unlike a *domus ultima*' — they were now standing below a steep rock. "We poor Jacobites", continued the Baron, looking up, "are now like the conies in Holy Scripture (which the great traveller Pococke called Jerboa), a feeble people, that make our abode in the rocks."

Donald Davie has well analysed the import of the work:

> It shows the victory of the un-heroic (the English Waverley) over the heroic (the Scottish MacIvor); it shows that this was inevitable and on the whole welcome, yet also sad . . . "Heroic" and "un-heroic" may both be misunderstood, unless we admit that for "heroic" we may substitute "barbarian" for "unheroic", "civilized." The second pair of terms tilt the scales of approval towards the English, as the first pair towards the Scots; the novelist's achievement is in tilting neither way, but holding the balance scrupulously steady." (Davie 1961)

The magnanimous and scrupulous appraisal of the relationship between older and newer, which Walter Scott offered, was of immense worth to Darwin. Scott showed the problems of the border between two countries, as well as between past and present. Darwin's own experiences in South America had led him to distrust any easy claims for the superiority of "advanced" races. Yet he needed to make room in his theory for the possibility of improvement by means of the entry of "foreigners" or "immigrants": "If the country were open on its borders, new forms would certainly immigrate, and this also would seriously disturb the relations of some of the former inhabitants. Let it be remembered how powerful the influence of a single introduced tree or mammal has been shown to be" (*Origin*, p. 81).

Darwin knew that the concept of environment must include that of the invader. A being may be in accord with its environment until the environment is invaded from without, as the Incas, the Indians, and the Scots had been, as well as the trees and mammals he specifies. The terms "original inhabitants" and "intruders" ("had the area been open to immigration, these places would have been seized on by intruders") are not species-specific. Darwin sometimes directs a term like "native" within natural historical terms: "Man keeps the natives of many climates in the same country; . . . he feeds a long and a short beaked pigeon on the same food." But in the preceding paragraph, which first introduces the distinction between Nature's practices and Man's, the non-technical range of senses for "inhabitant", "native", and "foreigner" is allowed to thrive.

> No country can be named in which all the native inhabitants are now so perfectly adapted to each other and to the physical conditions under which they live, that none of them could anyhow be improved: for in all countries, the natives have been so far conquered by naturalised productions, that they have allowed foreigners to take firm possession of the land. And as foreigners have thus everywhere beaten some of the natives, we may safely conclude that the natives might have been modified with advantage, so as to have better resisted such intruders. (*Origin*, pp. 82–83)

Darwin's argument allows room for different readings, and these readings are related to the problem of whether "naturalization" is usurpation or improvement. He is chary of affirming either definitively. So the argument faces two ways: native inhabitants are not fully developed and thus will inevitably be beaten by colonizers; native inhabitants lack perfection only *in that* they do not have the means to resist foreign invaders (see *Origin*, pp. 103–109). W. F. Cannon (1968) held that Darwin formed a language central to British Imperialism; it could quite as well be held that he formed a language appropriate to excluding immigrants.

As ideological conditions change, so readers appropriate diverse elements of Darwin's writing and turn to their own advantage one element in his complicated discourse. That this can be done is in part due to Darwin's own acceptance of broad terms, which he habitually thought about in a natural historical context. (In Johnson's *Dictionary*, "Inhabitant" is simply: "Dweller; one that resides in a place." The human is suggested, but not exclusively: "one that" not "one who".) The gender distinction he establishes between Nature and Man suggests that Darwin himself felt disquiet about the possibly exploitative implications of his argument here. The passages I have analyzed lead straight into the major statement of natural selection as a benign principle, non-exploitative, concerned only for the good of each individual: "Man selects only for his own good — Nature only for that of the being which she tends" (*Origin*, p. 83).

The imaginative trouble implicit in the fate of the American Indians in relation to ideas of improvement haunted Darwin. His emphasis on the "web of complex relations" that net all species together is expressed in the previous chapter on "The Struggle for Existence" in a metaphor that makes a very rare allusion to ways in which this struggle affects human beings as well as all other species. "The ancient Indian mounds" and "the old Indian ruins" in the following famous passage quietly set the lost tribes among the general world of struggling nature:

> When we look at the plants and bushes clothing an entangled bank, we are tempted to attribute their proportional numbers and kinds to what we call chance. But how false a view is this! Every one has heard that when an American forest is cut down, a very different vegetation springs up: but it has been observed that the trees now growing on the ancient Indian mounds, in the Southern United States, display the same beautiful diversity and proportion of kinds as in the surrounding virgin forests. What a struggle between the several kinds of trees must here have gone on during long centuries, each annually scattering its seeds by the thousand; what war between insect and insect — between insects, snails, and other animals with birds and beasts of prey — all striving to increase, and all feeding on each other or on the trees or their seeds and seedlings, or on the other plants which first clothed the ground and thus checked the growth of the trees! Throw up a handful of feathers, and all must fall to the ground according to definite laws; but how simple is this problem compared to the action and reaction of the innumerable plants and animals which have determined, in the course of centuries, the proportional numbers and kinds of trees now growing on the old Indian ruins! (*Origin*, pp. 74–75)

In this passage struggle in "the entangled bank" appears in its sense both of effort and of conflict. The urgency of evolutionary struggle here calls up in Darwin's imagination the related ideas of "Indians" and "war". The poignant intensity of the passage condenses many kinds of reading and experience.

VI. Some Functions for Metaphor in the *Origin*

Colin Turbayne in *The Myth of Metaphor* (1970) remarks that "the sciences are riddled with metaphors, but the scientists who use them, for example, Descartes and Newton, do not always admit to their use" (Black 1962; Hesse 1966, 1974). Darwin, on that scale of awareness, was probably unusually conscious of a spectrum of fictiveness in his use of metaphor. For example, he insists upon the purely metaphorical status of the concept of "the struggle for existence" while wishing to prove the factual basis of the apparently metaphoric and mythic concept of metamorphosis:

> I should premise that I use the term Struggle for Existence in a large and metaphorical sense, including dependence of one being on another, and including (which is more important) not only the life of the individual, but success in leaving progeny. Two canine animals in a time of dearth,

> may be truly said to struggle with each other which shall get food and live. But a plant on the edge of a desert is said to struggle for life against the drought, though more properly it should be said to be dependent on the moisture. A plant which annually produces a thousand seeds, of which on an average only one comes to maturity, may be more truly said to struggle with the plants of the same and other kinds which already clothe the ground. The mistletoe is dependent on the apple and a few other trees, but can only in a far-fetched sense be said to struggle with these trees, for if too many of these parasites grow on the same tree, it will languish and die. But several seedling mistletoes, growing close together on the same branch, may more truly be said to struggle with each other. As the mistletoe is disseminated by birds, its existence depends on birds; and it may metaphorically be said to struggle with other fruit-bearing plants, in order to tempt birds to devour and thus disseminate its seeds rather than those of other plants. In these several senses which pass into each other, I use for convenience sake the general term of struggle for existence. (*Origin*, pp. 62–63)

"May be truly said"; "more properly it should be said"; "may be more truly said"; "can only in a far-fetched sense be said to struggle"; "may metaphorically be said to struggle"; "In these several senses, which pass into each other, I use for convenience sake the general term of struggle for existence": the precise articulation of degrees of distance and congruity makes it clear that in the case of this term at least, Darwin was well aware of what he was doing. But he was also aware that the "several senses . . . pass into each other". Although they can be analyzed and separated out, they cannot be kept rigidly apart. The confluence of the general term makes for a persisting, half-realized image. Manier (1978) makes the point that "In Darwin's own stipulation, 'struggle' was an inherently equivocal term, with no fewer than three meanings: interdependence, chance, and contest, which grade into each other. The tension introduced by this stipulative combination of three meanings transformed the meaning of each of them taken singly."

The deliberately guarded and consciously metaphoric status that he gives to the phrase "struggle for existence", which he sometimes varies as "struggle for life" and even, in one instance, "the great battle for life", also expresses his unwillingness to give dominance to a militant or combative order of nature. He interprets it as interdependence or endurance as much as battle. The essentially egalitarian, horizontal ordering of his view of the natural world means that he eschews the simplicity of hierarchy. Neither the ladder nor the pyramid are useful models for him. When he uses the term "the scale of nature" it is not to sort and distinguish in a vertical order. In nature relations can never be simple. There is no single line of ascent and descent, no straightforward development, but rather an abstruse lateral range of interconnections. "I am tempted to give one more instance showing how plants and animals, most remote in the scale of nature, are bound together by a web of complex relations." "The dependence of one organic being on another, as of a parasite on its prey, lies generally between beings remote in the scale of nature." Metaphor is a counterorganization helping to control any sequential notion of development.

The complexity of interrelation is another reason why he needs the metaphoric. He needs to emphasize the transposed, metaphorical status of his description — its imprecise innumerate relation and application to the phenomenological order it represents. The representation is deliberately limited to that of "convenience" and does not attempt to present itself as a just, or full, equivalent. In another related passage one can see this desire to *specify* complexity without appearing to *simplify* that complexity. He begins with precise instances and then moves into a deliberately vague speculation:

> The recent increase of the missel-thrush in part of Scotland has caused the decrease of the song-thrush. One species of charlock will supplant another, and so in other cases. We can dimly see why the competition should be most severe between allied forms which fill nearly the same place in the economy of nature; but probably in no one case could we precisely say why one species has been victorious over another in the great battle of life. (*Origin*, p. 76)

"Dimly", "nearly", "probably" — the tentative, blurred, half-glimpsed reasons for happening are momentarily stabilized in the vivid martial image of "why one species has been victorious over another in the great battle of life". *Of* life, not *for* life — the preposition harks back to another sense of struggle: the struggle to survive, not to conquer. Or, if we take another statement full of qualifiers, we read: "If our reason leads us to admire with enthusiasm a multitude of inimitable contrivances in nature, this same reason tells us, though we may easily err on both sides, that some other contrivances are less perfect" (*Origin*, p. 202).

That insistent demurring at our powers of judgement is once more parenthetically introduced "though we may easily err on both sides", and again reason twins admiration and analysis in a way very close to that usually accorded to the imagination in opposition to reason: "our reason leads us to admire with enthusiasm" as well as telling us "that some other contrivances are less perfect".

A generous sense of profusion and of the illimitable powers of wonder is as crucial to Darwin's argument here as are the words "reason" and "contrivance". Words like "admire", "enthusiasm", "multitude", "inimitable" create a constant effect of space, aspiration, and uncontainable profusion. The effect is not simply hyperbolic, or the product of an overblown rhetoric; though it is exuberant and multivocal, straining its own seams.

In its insistent extending and surpassing of our powers of analysis it creates within the text a scale that is to express the scope and extension, the complexity of the natural world.

In passages such as this Darwin deliberately sets off against each other the wayward and the iconic elements in metaphor. That is, he gives room for mystery, for exploration, and insists upon the dark space behind the summary formulation of "the struggle for life". The chapter ends by encouraging us to try an experiment "in our imagination" and then proving to us that we cannot sufficiently imagine the complexities of relation in nature or of its causes and effects to succeed in our experiment:

> It is good thus to try in our imagination to give any form some advantages over another. Probably in no single instance should we know what to do, so as to succeed. It will convince us of our ignorance on the mutual relations of all organic beings; a conviction as necessary, as it seems to be difficult to acquire. (*Origin*, p. 78)

Again he places value on the expansion rather than on the stabilization of our sense of the world that surrounds us. This sense of ignorance, of our partial knowledge and our imaginative desuetude, provides a countercurrent to the final sentence of the chapter, which urges itself towards a meliorist belief that it yet never fully shares:

> *When we reflect* on this struggle, *we may console ourselves with the full belief*, that the war of nature is not incessant, that no fear is felt, that death is generally prompt, and that the vigorous, the healthy, and the happy survive and multiply. (*Origin*, p. 78; my italics)

The form of the sentence is optative, "we may", not absolute. It is also urgently assertive, its confidence sagging momentarily with the word "generally" in "death is generally prompt" (but particularly?), and it comes to rest in the Old Testament word "multiply".

The will to believe in a happy world and the dark flood of insight into suffering that accompanies it, is a frequent movement in Darwin's prose. It would be easy to make either an optimistic or pessimistic selection from the *Origin*. This poignant tension between happiness and pain — a sense simultaneously of the natural world as exquisite and gross, rank and sensitive — constantly subverts the poise of any moralized description of it. (Can we, for example, specify the organization of the material world in terms of justice and injustice? Darwin wavers, as we have seen in passages previously analyzed.) The problem was one to which Hardy most powerfully responded.

Schon (1967) has excellently analyzed the emotion of discovery — its dangerousness, its playfulness, the "oscillations between wrenching pain and unexpected joy". "In our culture," he writes, " 'novelty', 'the new', 'innovation', 'creativity', have taken on highly positive emotive meanings

But we are easier in our minds talking about the new than actually experiencing it."

Metaphor is a means both of initiating and of controlling novelty. Schon's description of the reader's participation could be aptly extended from scientific discovery to participation in narrative: a sense of powerlessness is generated by a text that will not permit us to "build up", or select, or fulfill expectation. Too great a freedom for the creator will mean oppression for the reader. Metaphor, both in its residuum of the known and in its heuristic powers, offers a means to recognizable discovery.

Darwin needed the conceptual space offered by metaphor for his arguments to work. He could not afford too sharply to discriminate usages in words such as "struggle" and "selection", or in extended images such as "the entangled bank". He was seeking ways of expressing natural orders that did not center in man or in man's language. He was exploring the complexity of interrelations beyond the domain of the human. He could neither drive out the human nor accord it supremacy therefore; nor, on the other hand, could he have recourse to an elegantly simple, mathematicized description of the natural world. Elegance was not the most striking property of the world as he conceived it. He was drawn more to the umbelliferous than the ellipse.

Schon, in a strikingly evolutionary metaphor, attempts to express the shifting process of concept transference thus:

> The metaphors in language are to be explained as signs of concepts at various stages of displacement, just as fossils are to be explained as signs of living things in various stages of evolution. (1967, p. 51)

Darwin's prose permits those evasive, probing movements of mind by which we explore metaphoric potential and establish the limits of usefulness — limits that may need persistently to be re-assessed as the work proceeds. Darwin's copious use of surface metaphor (as well as implicit model) and of analogy encourages the acts of recognition by which we scan for elements suppressed as well as expressed in argument and by which we glimpse the disanalogous within the activity of analogy. This is, I believe, a major reason why his text became so useable to other writers who read him. But there is also a further reason that makes the *Origin* an open text, at no point sharply delimited.

Despite the metaphoric density of the *Origin* Darwin seems never fully to have raised into consciousness the mythic and sociological implications of his theories. He presents them — and seems to have succeeded in casting them for his own satisfaction — as problems of biology rather than of philosophy or sociology. Yet we know that in his notebooks he saw a good way into the human implications engendered by his theories and that he wished to avoid naturalizing current social organizations. He saw some of the dangers of "authorization".

It was the element of obscurity, of metaphor whose peripheries remain undescribed, that made the *Origin* so incendiary — and that allowed it to be appropriated by thinkers of so many diverse political persuasions. It encouraged onward thought: it offered itself for metaphorical application and its multiple discourses encouraged further acts of interpretation. The presence of *latent meaning* made the *Origin* suggestive — perhaps unstoppably so — in its action upon minds.

VII. Narrative Organization in the *Origin*

Darwin's theories draw on radical narrative themes and organizations. Descent is one of these, development another. To reverse descent is to break taboo, as we see in the Oedipus story. Narrative sets much store by succession, whether of events or of people. It has its own progenitive laws, though these need not serve cause and effect. The theme of descent has frequently been used to demonstrate the nobility of the protagonist and to set him within a pattern either of development or degeneration. Rabelais, for example, in the sixteenth century, satirizes such assumptions in *Gargantua and Pantagruel* (1533). The first chapter of the first book is entitled "Gargantua: His Genealogy and Antiquity" and claims direct descent for Gargantua from the "race of giants". He proceeds to query any such ballasting of the pride of the individual with notions of pedigree.

> Would to God every one could be as certain of his pedigree from the days of Noah's celebrated ark down to the present. To my way of thinking, many a man sprung from a race of sham relic-peddlers and journeyman-carriers walks the earth today an emperor, king, duke, prince or pope. Similarly, not a few of our sorriest, most miserable tramps are sprung from the blood and lineage of proud kings and emperors.

Here, hierarchy is false because topsy-turvy. In Darwin's recasting of the myths of pedigree, all beings are interconnected because sprung from common stock — a stock that goes back far beyond Noah and has no human form (compare Beer, 1983a).

In the Prologue to the Fifth Book Rabelais goes further, and mocks the implications of genetic imaginative orders that extend the self backward through history and make the past the servant and forerunner of the present. He brings into question the notions of improvement, development, and the authority of now:

> Here is my question, then. Would you maintain, by perfectly logical inference, that the world was formerly muddle-headed, but has now acquired wisdom? How many and precisely what conditions made men simpletons; how many and what circumstances developed men into sages? Why were men ever idiots; how should men now be intelligent? . . . Whence came this pristine giddiness; whence comes this present

> equilibrium? Why did hallowed ignorance cease abruptly today, and not a century hence; why did modern authority begin today and not ages ago?

The developmental conclusions of the *Origin* clearly opened the text to this well-established form of hubris. But Darwin used a counterorganization in the narrative, both in total ordering and in local sentence-structure. He evades any suggestion that the world is now accomplished and has reached its final and highest condition, though he does present the movement of evolution as one of proliferation and enhancement: "whilst this planet has gone cycling on according to the fixed law of gravity, from so simple a beginning endless forms most beautiful and most wonderful have been, and are being, evolved" (*Origin*, p. 490).

Cycling and fixed, simple and endless, "have been and are being": he alternates in the sentence the principles of stasis and of motion, of completion and continuity, and sets them spinning and growing on into the silence which succeeds the conclusion of the *Origin*. Darwin's final statement feeds our imaginative sense of continuance and change.

This imaginative release into a continuing and undescribed future is remarkable when it is set alongside the positivistic emphasis on *finality* we find in Comte, a suggestion that the positive and scientific have now achieved mastery and that the world may fully and definitively be described for ever. Darwin persistently emphasizes physical process, not completed idea.

Darwin's work is not a search for an originator nor for a true beginning. It is rather the description of a process of becoming, and such process does not move constantly in a single direction. The title of Darwin's book signals that this is a work where the narrative and the descriptive are inextricably mixed. The usual shortened form, *The Origin of Species*, disguises the element of narrative in the title and changes "origin" from a process into a place, or substantive. The full title reads, "On the Origin of Species by means of Natural Selection, or the Preservation of Favoured Races in the Struggle for Life." The title is in polemical contrast with Chambers's insistence on *Vestiges of the Natural History of Creation*. Vestiges are remnants, surviving fragments of a primordial creative act. Darwin's enterprise is history, not cosmogony. "I must premise, that I have nothing to do with the origin of the primary mental powers, any more than I have with that of life itself" (*Origin*, p. 207). In his book *Beginnings*, Edward Said (1975) emphasizes that beginning includes "the intention to continue". Darwin is concerned with this particular property of beginning. He is interested in initiation, but he is interested in it not as completed ceremony, rather as indefatigable process. So the emphasis in his title is on *means*: "By Means of Natural Selection."

"On the Origin of Species by means of Natural Selection" is in a very precise sense a narrative, because what it describes cannot be correctly described except through the medium of time. Neither analysis nor exposition would in themselves suffice for what was new in Darwin's ideas. Categorization, classification, description, must all be understood to be implicated in movement, process, and time. Darwin rejected the idea of a stable or static world, and would not accept equilibrium as a sufficient description of the relationship between the forces of change and continuance. He thus avoided the pattern by means of which many Victorian writers set limits to change and asserted moderation as an essential natural order: a pattern we perceive equally in a novelist like Trollope and in this passage from Edward Daubeney:

> We seem to catch a glimpse of a general law of nature, not limited to one of her kingdoms, but extending everywhere throughout her jurisdiction, — a law, the aim of which may be inferred to be, that of maintaining the existing order of the universe, within any material or permanent alteration, throughout all time, until the fiat of Omnipotence has gone forth for its destruction. The will, which confines the variations in the vegetable structure within a certain range, lest the order of creation should be disturbed by the introduction of an indefinite number of intermediate forms, is apparently the same in its motive, as that which brings back the celestial Luminaries to their original orbits, after the completion of a cycle of changes induced by their mutual perturbances. (Basalla et al, 1970, p. 308)

Darwin came to see that his own subtitle suggested too inert a procedure: he changed "the preservation of favoured races in the struggle for life" to "the survival of favoured races in the struggle for life" in later editions. The new form of the title obliterates the vestiges of a "preserver". "Survival" suggests a continuing struggle whereas "preservation" may suggest repose: the passive becomes active.

The organization of his narrative emphasizes variability rather than development. The narrative time of the *Origin* is not one that begins at the beginning but rather in the moment of observation. The first words are "When we look", and the first two chapters are concerned with variation: variation under domestication and variation under nature. The ordering reinforces the argument. It suggests two crucial insights. Originating is an activity, not an authority. And deviation, not truth to type, is the creative principle.

Darwin's account of the origin of species ranges to and fro through time in a way that disturbs any simple sequence, chain, or development. Both the introduction and the first chapter open with specific experience: "When on board H.M.S. *Beagle*, as naturalist, I was much struck with certain facts." This is broadened into a shareable repeated present in Chapter 1: "When we look to the individuals of the same variety . . . When we reflect on the vast diversity." This present is not simply at the mercy of the contingent: it includes a sense of repetition and meditation. Its participatory "when we . . . when we" culminates in "I think we are driven to conclude". The movement towards discovery and decision is dramatized.

The emphasis in these first chapters is on individuation and diversity. The range and profusion of the world and the freight of examples threatens to disguise any drive of argument. Moreover Chapters 2 and 3 take further the weight of retardation. Both begin with the word "before", but this is no reference to origins, rather to accumulation: "Before entering on the subject of this chapter, I must make a few preliminary remarks . . ." The movement throughout the first six chapters is into increasing doubt and difficulty, even while they represent with growing complexity the substance of his ideas.

Darwin gives room to difficulty, and with increasing humility sets out the problems of his argument. So he opens Chapter 5 with revision: "I have hitherto sometimes spoken as if the variations . . . had been due to chance. This, of course, is a wholly incorrect expression." Chapter 6, "Difficulties on Theory", opens: "Long before having arrived at this part of my work, a crowd of difficulties will have occurred to the reader." The maximum point of copiousness and confusion having been reached, that chapter ends by establishing the relations between "two great laws — Unity of Type, and the Conditions of Existence". The succeeding chapters analyze specific difficulties, such as Instinct and Hybridism and "the Imperfection of the Geological Record". It is in Chapter 10 that he finally concentrates on the question of the succession of organic beings. Chapters 11 and 12 consider geographical distribution and in Chapter 13 he reaches the problems of classification: "Mutual Affinities of Organic Beings: Morphology: Embryology: Rudimentary Organs." This, together with the recapitulation and conclusion, is probably the most confidently and passionately written section. It begins by relating time to classification and thus produces story:

> From the first dawn of life, all organic beings are found to resemble each other in descending degrees, so that they can be classed in groups under groups. This classification is evidently not arbitrary like the grouping of the stars in constellations. The existence of groups would have been of simple signification, if one group has been exclusively fitted to inhabit the land, and another the water; one to feed on flesh, another on vegetable matter, and so on; but the case is widely different in nature; for it is notorious how commonly members of even the same sub-group have different habits. (*Origin*, p. 411)

There would have been no story, only "simple signification", if each group had been "exclusively fitted" for one milieu. As it is, difference and difficulty provoke plot.

The *final* section in this powerful chapter is on embryology and rudimentary organs. Whereas a simple developmental narrative based on the model of the single life-span might have placed the embryo at the beginning of the story, or a narrative preoccupied with origins and cosmogony might have started with the geological record, Darwin places the initiating emphasis in his narrative on the profusion of individuals, their variability, the diversity of species. Only gradually do laws emerge from the welter of particularity. Even then the law of "Unity of Type" is seen to be secondary to that of "Conditions of Existence". So change, environment, the conditional nature of existence, is reinforced by the ordering as well as the argument of his narrative. Even the stabilizing Recapitulation and Conclusion refuses to let us repose in an accomplished order.

The one permanence in which Darwin concurred with other scientists of his time was that of the possibility of achieved and immovable truth, the tracking of "fixed laws", though these laws primarily described change and motion. He added a quotation from Butler's *Analogy* in the second and subsequent editions of the *Origin*: "The only distinct meaning of the word 'natural' is *stated, fixed,* or *settled*" (Peckham 1959, p. 40, note 13).

The ornamented title page of *Nature* (4 November 1869) shows a globe surrounded by clouds and bears an epigraph from Wordsworth:

> To the solid ground
> Of Nature trusts the mind which builds for aye.

That condensation of the meanings of "logic" and "earth" in the word "ground" was part of the comforting inheritance of Romantic thought in Victorian science, which seemed to assure a continuance of natural truth through the action of permanent discoverable laws — what Whewell in Darwin's first epigraph to the *Origin* called "the establishment of general laws". Darwin invoked the same idea in his final sentence with its implicitly validating reference to Newton's "fixed law of gravity" set alongside his own newly discovered laws of development and change (*Origin*, pp. 489–490).

VIII. Fictions of Development: 1

Darwin's writing drew upon the whole range of resources present in his culture. We shall better understand the fruitfulness of his writing, and better specify the unresolved problems within it, if we take account of the ways in which his creative contemporaries responded to his work. Such responses are not a matter simply of Darwin's "influence", but of his power to disturb the taken-for-granted and to create fresh contradictions. Other writers may assimilate or resist his work, and, as I have argued elsewhere, it is those concepts that provoke most anxiety in a culture that shift most rapidly from field, often as much to control as to solve the problems they raise (Beer 1983b). I have chosen two works to analyze, which very fully share with Darwin a system of reference to Renaissance and later literature and which use his work to probe difficulties in the idea of development.

Browning's poem "Caliban upon Setebos; or, Natural Theology in the Island" appropriates Caliban, the vindictive, monstrous and exploited inhabitant of Prospero's island in *The Tempest*. In Shakespeare's play Caliban claims to be the indigenous dweller usurped by Prospero, whose dominance over him is achieved by a mingling of reason and magical powers. *The Tempest* raises many of the same issues of colonization that I have already analyzed in Darwin's discourse, and also draws on Montaigne's essay "Of the Cannibals". Writing of man's claim to uniqueness and denial of kinship with other creatures, Darwin comments: "Has not the white man, who has debased his nature and violated every such instinctive feeling by making slave of his fellow Black, often wished to consider him as another animal. — it is the way of mankind" (*C* 154). Prospero has taught Caliban language in return for knowledge of the island: how —

> To name the bigger light, and how the less,
> That burn by day and night: and then I lov'd thee,
> And show'd thee all the qualities o'th'isle,
> The fresh springs, brine-pits, barren place and fertile:
> Curs'd be I that did so! (I:ii: 336–341)

In his dramatic monologue, written in the person of Caliban, Browning articulates the processes by which Caliban creates out of his own experience a cosmogony, and a theology. Both are based on that same anthropomorphism that Darwin had mocked in the *Notebooks*: "Mayo (Philosophy of Living) quotes Whewell as profound because he says length of days adapted to duration of sleep in man!! Whole universe so adapted!!! and not man to Planets — instance of arrogance!!" (*D* 49).

In the case of the oppressed and vicious Caliban in the poem, anthropomorphism produces a god (Setebos), captious, meager, and envious of his creation. Strength gives power, which obliterates questions of right or wrong; Caliban himself articulates his insights without a first person. He reaches first person as an assertion of the analogy between himself and Setebos:

> 'Thinketh, such shows nor right nor wrong in Him,
> Nor kind, nor cruel: He is strong and Lord.
> 'Am strong myself compared to yonder crabs
> That march now from the mountain to the sea;
> 'Let twenty pass, and stone the twenty-first,
> Loving not, hating not, just choosing so.
> 'Say, the first straggler that boasts purple spots
> Shall join the file, one pincer twisted off;
> 'Say, this bruised fellow shall receive a worm,

And two worms he whose nippers end in red;
As it likes me each time, I do: so He.

"Caliban" challenges the natural theological assumption that design and benign are more or less the same word and substitutes the pair, design and destroy. He looks at the material of the world about him and, using evidence of depradation such as Darwin also had observed — "the young cuckoo ejecting its foster-brothers, — ants making slaves, — the larvae of ichneumonidae feeding within the live bodies of caterpillars" (*Origin*, p. 244) — Caliban organizes a godhead thriving on random survival and spoliation. He lives, therefore, in terror of Setebos, who may only fitfully be placated. But beyond Setebos he imagines a further level of godhead, more tenuously and disinterestedly associated with his own experience:

There may be something quiet o'er His head,
Out of His reach, that feels nor joy nor grief,
Since both derive from weakness in some way.
I joy because the quails come; would not joy
Could I bring quails here when I have a mind:
This Quiet, all it has a mind to, doth.

The quality of quietness becomes the metaphysical power, Quiet. The poem explores the powers of argument by means of which Caliban rationalizes the physical and metaphysical order into his own image. He is primitive man seeking meaning, but equally he is contemporary man arguing in positivistic style from the evidence of the material world alone.

What part, if any, did the publication of the *Origin* play in creating this powerful work? I have already indicated some of the texts that Darwin and Browning shared, the extent to which Shakespeare and Montaigne were part of a common cultural interchange for them both, just as Paley quite crucially was. The figure of Robinson Crusoe, the rationalistic improver of his island, turning everything to use, is a counter-fable of the island ("Some Arabian Nights. Gullivers Travels. Robinson Crusoe." 1840. DAR 119). Browning, however, in his subtitle "Natural Theology in the Island", turns our attention to the tradition of discovering evidence for God in the material world. When the poem first appeared, reviewers saw it either as a satire on theology (*Athenaeum* 1864: 767) or as an attack on the false gods of scientific reasoning: "a most edifying chapter to innumerable gentlemen of our acquaintance, Darwinians, believers in force and matter, and other such divine and worshipful deities" (*The Eclectic Review* 1864: 70). Huxley read it as an anthropological analysis: "a truly scientific representation of the development of religious ideas in primitive man" (Symons 1886, p. 125). We do not know for certain whether Browning was reading Darwin between 1859 and 1864 (the dates between which the poem was written).

Much later in his life — in a letter to Furnival, 11 October 1881 —

Browning said that he had been "Darwinized", that he believed in evolution, and that he had put it forward in "Paracelsus" well before the publication of Darwin's work (Hood 1933, p. 199).

It is perhaps worth recording that during the 1860s he formed a close friendship with Julia Wedgwood, Darwin's niece, one of the few who Darwin said had "understood my book perfectly" (Irvine and Honan 1975). Browning's close friend the American transcendentalist, Theodore Parker, with whom he was in daily contact at the time, read and was immensely enthusiastic about the *Origin* when it appeared in November 1859. In the following month Parker began writing "A Bumblebee's Thoughts on the Plan and Purpose of the Universe", which uses the critique of anthropocentrism implicit in all Darwin's work, his sense that human reason is an imperfect instrument, and his scorn of organizing the universe in conformity with man's image of his own nobility (Tracy 1938).

Darwin's work interrupted the continuity of natural theological explanation. The term "natural selection" read like a satire on natural theology, and may indeed have been meant to do so. The clash of terms certainly led to efforts to reconcile them, such as Asa Gray's 1861 pamphlet "Natural Selection Not Inconsistent with Natural Theology". But the polemical choice implicit in their semantic closeness, natural selection *or* natural theology, created disturbance. So we do not need, in this instance, to insist on Browning's attentive reading of Darwin for his poem to have issued creatively out of the contradictions set up by the *Origin* and discussed in Browning's circle.

Darwin's exclusion of direct discussion of man from the argument of the *Origin* was as much polemical as tactical. It denied humankind any special place at the summit of hierarchy; it made us one species among many. An act of will by the reader was required to restore man to the center of the text and in itself this transaction problematized his assumed centrality. "Man's place in nature" became a matter of general debate. Browning's Caliban can more easily imagine fickleness than purposive change (the random interventions of the pincer twisted off or two worms given to the one "whose nippers end in red"). Towards the end of the poem, however, Caliban struggles to imagine deep change, using the favorite model of transformation within the life cycle:

> 'Conceiveth all things will continue thus,
> And we shall have to live in fear of Him
> So long as He lives, keeps his strength: no change,
> If He have done His best, make no new world
> To please Him more, so leave off watching this, —
> If He surprise not even the Quiet's self
> Some strange day, — or, suppose, grow into it

As grubs grow butterflies: else, here we are,
And there is He, and nowhere help at all.

But the form of change here is a change in the nature of godhead, an alienated reading of self-change, and the poem ends in panic-stricken and masochistic propitiation of the angered Setebos.

Caliban's terror-ridden existence, interlaced with sensuous brilliance, registers the form of a world in which the old idea of design can no longer work. Paley's *Evidences* and *Natural Theology* (books that had profoundly impressed the young Darwin) cannot contain a mind-order where manifold, slight, unwilled events, are the only means to profound change. Darwin's own imagining of the activities of natural selection is a good deal more melioristic, less willful, than Browning's creation.

> It may be said that natural selection is daily and hourly scrutinising, throughout the world, every variation, even the slightest; rejecting that which is bad, preserving and adding up all that is good; silently and insensibly working, whenever and wherever opportunity offers, at the improvement of each organic being in relation to its organic and inorganic conditions of life. (*Origin*, p. 84)

In Browning's poem "the survival of favoured races in the struggle for life" is as much the result of vagrant interventions as of natural law.

The poem springs out of the tormenting break-up of anthropocentrism: there are no longer sufficient or just equivalences between perception and universe. Any suggestions that we are at the centre of significance or of signification must, as in this poem, take the form of satire. At the end of Shakespeare's play, Prospero sailed away, leaving the island to its original inhabitants. Browning's work makes no mention of the master Prospero, but at the end of his poem Caliban is still in the toils of subjugation, at the mercy of his own malign conception of natural order.

IX. Fictions of Development: 2

George Eliot began to read the *Origin* immediately after its publication, in company with her life-companion, the philosopher, scientist, and writer George Henry Lewes (Levine 1980). She commented in her Journal: "We began Darwin's work on The Origin of Species tonight. It seems not to be well written: though full of interesting material, it is not impressive, from want of luminous and orderly presentation" (Eliot: Haight ed. 1954, 3:214). In a letter two days later she described it as "an elaborate exposition of the evidence in favour of the Development Theory, and so, makes an epoch." She perceived its importance immediately — "it makes an epoch" — but not yet its originality. At this early stage of reception she took it as a summation and scientific authentication of Chambers's *Vestiges* and of essays such as Herbert Spencer's "The Development Hypothesis" written in the early 1850s. She did not yet see what distinguished it from general Comtist accounts of development. I have analyzed elsewhere the complexity with which in her late works *Middlemarch* and *Daniel Deronda* she responded to the difficult implications of Darwin's theories (Beer 1983a). Although in the first phase of response she did not discriminate sharply between Darwin's and Spencer's views on development, the publication of the *Origin* and the controversy surrounding it disrupted all previous assumptions about the relations between individual, social, and species development, to such an extent that it became possible to articulate what had earlier been taken for granted (and therefore was not capable of being recounted). In *The Mill on the Floss* (1860) she views as a problem what might previously have been accepted as normal.[10]

> It is one of those old, old towns which impress one as a continuation and outgrowth of nature, as much as the nests of the bower-birds or the winding galleries of the white ants: a town which carries the traces of its long growth and history like a millenial tree, and has sprung up and developed in the same spot between the river and the low hill from the time when the Roman legions turned their backs on it from the camp on the hill-side, and the long-haired sea-kings came up the river and looked with fierce eager eyes at the fatness of the land. It is a town "familiar with forgotten years." (Eliot: Haight ed. 1980, I: xii: 101)

The parallels by means of which she naturalizes in Comteian style the growth of the town to the process of nature are close to Darwin's natural historical examples. Darwin, moreover, discovered in his diagram of descent the "millenial tree", changing it from a purely formal to an experiential metaphor, appropriating the form of the tree not simply as a diagram of organization but as an expression of communality and of the processes of growth. So close is the connection between representation and actuality that he can claim "truth" for it: "The affinities of all the beings of the same class have sometimes been represented by a great tree. I believe this simile largely speaks the truth" (*Origin*, p. 129). In George Eliot's citation, "familiar with forgotten years", we recognize again the community of reading culture that nets together Darwin and his contemporaries. The creativity of Darwin, who read through the whole of *The Excursion* twice, and of George Eliot equally, was stirred by Wordsworth's poem. The quotation is from the first book (*The Excursion*, I, 276) (Wordsworth: de Selincourt and Darbishire eds. 1949, p. 17). The passage describes the emotional and intellectual development of the child under the influences of Milton, science, and nature.

The problem of development as George Eliot perceives it is that a culture oppresses as well as sustains its inhabitants, and that those best fitted to survive in a meager environment may precisely not be those exceptional spirits whose imaginations range beyond the alternatives provided by their society. She suggests this with peculiar force in her heroine, Maggie Tulliver, as well as to a lesser extent in the intelligent, crippled Philip Wakem. Maggie's old-fashioned miller father ponders the mysteries of breeding that have inconveniently produced a clever daughter and a slow son:

> "Did you ever hear the like on't?" said Mr. Tulliver, as Maggie retired. "It's a pity but what she'd been the lad — she'd ha' been a match for the lawyers, she would. It's the wonderful'st thing" — here he lowered his voice — "as I picked the mother because she wasn't o'er'cute — bein' a good-looking woman too, an' come of a rare family for managing; but I picked her from her sisters o' purpose, 'cause she was a bit weak, like; for I wasn't agoin' to be told the rights o' things by my own fireside. But you see when a man's got brains himself, there's no knowing where they'll run to; an' a pleasant sort o' soft woman may go on breeding you stupid lads and 'cute wenches, till it's like as if the world was turned topsy-turvy . . ." (Eliot: Haight ed. 1980, I: iii: 17)

George Eliot deeply distrusts the terms in which society conceives competition and improvement. Aptness to the environment may require conformity and even obtuseness, which will breed specific forms of competition and selection. Improvement will be cast only in the restricted image of society's own values. Darwin had attempted a way out of this dilemma by proposing three selective forces, artificial selection, sexual selection, and natural selection. But his readers domineeringly insisted on the elements of competition and improvement in his argument. George Eliot reserves her most sardonic comments for "those severely regulated minds who are free from the weakness of any attachment that does not rest on a demonstrable superiority of qualities" (p. 152). Her most direct allusion to Darwin's theories in the work is caustically facetious. Maggie's brother Tom, in a childhood scene, gives evidence of his usual dogged lack of empathy. In a fit of coldness to Maggie, Tom ignores her and finds amusement elsewhere:

> Tom took no notice of her, but took, instead, two or three hard peas out of his pocket, and shot them with his thumb-nail against the window — vaguely at first, but presently with the distinct aim of hitting a superannuated blue-bottle which was exposing its imbecility in the spring sunshine, clearly against the views of Nature, who had provided Tom and the peas for the speedy destruction of this weak individual. (Haight, ed. 1980, I: ix: 76)

The harsh competitiveness that destroys "the weak individual" (whether it be blue bottle or Philip Wakem) repels George Eliot. But she is drawn to all that reminds us of relations and of the pains of any onward progress that may seem to repudiate the past. The "oppressive narrowness" of their environment acts on them "as it has acted on young natures in many generations, that in the onward tendency of human things have risen above the mental level of the generation before them, to which they have nevertheless been tied by the strongest fibres of their hearts." Fibres here is a word that carries the sense not only of nerve pathways but of rooted plants.

The sense of intelligible binding together, of "a large vision of relations" and "a vast sum of conditions", is close to that vision of ecological entanglement, both constricting and sustaining, which distinguishes Darwin's imagination. His emphasis is on "horizontality" as opposed to the vertical hierarchy that places man at the top.

> The suffering, whether of martyr or victim, which belongs to every historical advance of mankind, is represented in this way in every town, and by hundreds of obscure hearths; and we need not shrink from this comparison of small things with great; for does not science tell us that its highest striving is after the ascertainment of a unity which shall bind the smallest things with the greatest? In natural science, I have understood, there is nothing petty to the mind that has a large vision of relations, and to which every single object suggests a vast sum of conditions. It is surely the same with the observation of human life. (Eliot: Haight ed. 1980, IV: 1: 238)

George Eliot seems to have recognized the difficult fit between ideas of development, selection, and variability, even before she had fully articulated the surface problems of Darwin's ideas. It was a critical imaginative leap, which took her straight into the central difficulties that continued to disturb Darwin creatively. She refers at times to writing that related to Darwin's and preceded it. Maggie first notices Stephen Guest with whom she will fall disastrously in love when he describes Buckland's *Geology and Mineralogy Considered with Reference to Natural Theology* (1837), the last of the Bridgewater Treatises. The narrator alludes humorously to Bob's hand, "a singularly broad specimen of that difference between the man and the monkey" (IV: iii: 248) (284). Darwin's later adversary Richard Owen, in his book *On the Gorilla*, sees the hand as contrasting man with other species: "Man's perfect hand is one of his peculiar physical characters" (1859, p. 9). Darwin's parallel reference to the hand draws on Charles Bell's *The Hand Its Mechanism And Vital Endowments As Evincing Design*, a Bridgewater Treatise published in 1833. But for Darwin the hand is not evidence of man's separation but of his *community* with all other living forms, "this element of descent is the hidden bond of connection which naturalists have sought under the term of the Natural System" (*Origin*, p. 433).

> What can be more curious than that the hand of a man, formed for grasping, that of a mole for digging, the leg of a horse, the paddle of the porpoise, and the wing of the bat, should all be constructed on the

> same pattern, and should include the same bones, in the same relative positions? (*Origin*, p. 434)

It is in this sense that morphology may be said to be "the very soul" of natural history. Darwin's work greatly intensified the sense of kinship, of inextricable interconnection, and the dénoument of his work revealed "the hidden bond" so long searched for to be "community of descent". Variation is the creative principle. Individualism carries the full potentiality of change, and of development. Yet at the same time his emphasis on selection and on the need for the organism to conform to its medium or environment created a fierce counterpressure that told against any individualistic concept of freedom and against any idealistic expectation that the most distinguished will necessarily be fit for survival.

In *The Mill on the Floss* George Eliot participated in the irreconcileable dilemmas of consciousness that Darwin aroused. To express her disquiet she adapted the form of *Bildungsroman*, in which a young man comes eventually through vicissitudes into accord with his society. The self is chastened but accommodated. George Eliot substituted a young woman for a young man, more imaginative and intelligent than was apt to her society, with a further-reaching sense of community and connection. What her work shows is that Maggie's very exceptionalness, the degree of human development she represents, simultaneously unfits her to survive in St. Ogg's and gives her a profound and in the end overmastering sympathy with that "hidden bond" of community of descent. She cannot break away and rupture the continuities that stand in the way of her love for Stephen Guest. She cannot marry him even though she half-unwittingly elopes with him. She must go back to her family and renounce extent.

The end of the novel is catastrophist against the profound gradualism of its earlier picture of development. Ontogeny cannot properly correspond to phylogeny. The single life span cannot carry the weight of the accumulated fictions of development: "No wonder, when there is this contrast between the outward and the inward, that painful collisions come of it."[11] In a manuscript passage that originally concluded that chapter, George Eliot wrote:

> A girl of no startling appearance, who will never be a Sappho or a Madame Roland or anything else that the world takes wide note of, may still hold forces within her as the living plant-seed does, which will make a way for themselves, often in a shattering, violent manner. (Eliot: Haight ed. 1980, III: v: 206)

Maggie dies in a flood that sweeps her back into the arms of her brother, Tom, from whom she has been alienated.

Like Darwin, George Eliot brought into consciousness the dismaying sense of hidden violence and irremediable waste associated with development. Her own life experience of alienation opened her to the necessary contradictions of which his work offered an analysis. She transposed Darwin's concerns into the area of the human organism and its medium, and brooded on his conflicts. Darwin thought *Adam Bede* "excellent" when it appeared in 1859 (DAR 128) but according to Francis Darwin found her later books, other than *Silas Marner*, too painful to return to. This tribute should not surprise us, for in her novels George Eliot contemplated with peculiar intensity those unresolved dilemmas — of the relation between individual and species, of struggle, survival, extinction and development — which troubled Darwin and which he continued to contemplate without seeking to resolve them: "each lives by a struggle at some period of its life . . . heavy destruction inevitably falls either on the young or old, during each generation or at recurrent intervals" (*Origin*, p. 66). After the first edition he removed the sentence that draws too disturbingly close the senses of visage and surface:

> The face of Nature may be compared to a yielding surface, with ten thousand sharp wedges packed close together and driven inwards by incessant blows, sometimes one wedge being struck, and then another with greater force. (*Origin*, p. 67)

Darwin excluded humankind from his discussion in the *Origin*. George Eliot made such life her field of study: "how complex and unexpected are the checks and relations between organic beings, which have to struggle together in the same country" (*Origin*, p. 71).

Their cultural community of language and their diverse intellectual modes allowed George Eliot to test the problems raised by Darwin's theories through the medium of fiction, particularly the problem of the female in evolutionary patterns of development. Speaking directly of this in 1867, she significantly turns to that same passage of *The Winter's Tale* in which Montaigne and Shakespeare had helped to compose a Darwinian language for the central problems of variation, selection and development:

> Yet nature is made better by no mean
> But nature makes that mean: so, over that art
> Which you say adds to nature, is an art
> That nature makes.
> . . . .: this is an art
> Which does mend nature, change it rather, but
> The art itself is nature.

For her, the same language proposes a new difficulty and a new aspiration:

> As a fact of mere zoological evolution, women seem to me to have the worse share in existence. But for that very reason I would the more contend that in moral evolution we have 'an art which does mend nature.' (Eliot: Haight ed. 1956, IV: 364)

Notes

1. Most of my examples are from the notebook DAR 119, which Darwin titled "Books read/ Books to be read" and which covers the period 1838 to 1851. The notebook DAR 120 is an alphabetical catalogue of books listed in DAR 119 and DAR 128. The notebook DAR 128 covers the years 1852 to 1860. Peter J. Vorzimmer gives a useful account of the notebooks, but his transcription is unfortunately unreliable, particularly for literary texts (1977). For example DAR 119:21 should read "Giaor" [Byron]; p. 23 should read "Barnaby Rudge" [Dickens], not "Band of Rudge"; DAR 128, 1854 should read "Comte Philosophie Positive by Lewes (curious)," not "Politics by Lewes"; 1857 should read "Thackeray English Humourists", not "Humanists", etc.
2. For discussion see below pp. 17–22. Brief reference to some of these examples appears in my book *Darwin's Plots* (1983a). That work also includes a more developed discussion of some of the arguments concerning metaphor and analogy advanced in this essay.
3. "All September read a good deal on many subjects: thought much upon religion. Beginning of October ditto" (cited Gruber and Barrett 1974: 328).
4. It is not clear whether Darwin read Florio's or Cotton's translation.
5. In the second edition of the *Journal of Researches*, Darwin makes the writing more personal, for example, "I wished to find language to express my ideas" (*Journal of Researches* 1845, p. 496).
6. Compare Rudwick (1979a) for discussion of the etymological image in Lyell.
7. Paul Ricoeur (1976) provides a useful starting-point for discussions of "surplus meaning".
8. Darwin read *Gulliver's Travels* and *Robinson Crusoe* in 1840 among many other narratives, fictitious and non-fictitious, of travel. The figure of the traveler visiting strange societies raised questions about neutrality and intrusion.
9. An excellent discussion of G. H. Lewes and George Eliot is to be found in George Levine (1980) and in Sally Shuttleworth (1984).
10. George Eliot had already written the first draft of volume I of her novel when she read Darwin. She wrote the rest of the work in the following four and a half months.
11. Rosemary Ashton (1980) discusses the relation of organism and medium in George Eliot's work.

20

Three Notes on the Reception of Darwin's Ideas on Natural Selection (Henry Baker Tristram, Alfred Newton, Samuel Wilberforce)

I. Bernard Cohen

Introduction

It is well known that Darwin's theory of evolution was founded in the first instance on two kinds of observations: the occurrence of variations in animals and plants and the inheritability of such variations. In the opening chapter of the *Origin*, Darwin refers specifically to the existence of variation among "individuals of the same variety or sub-variety" and to the "endless" "number and diversity of inheritable deviations of structure, both those of slight and those of considerable physiological importance" (pp. 7, 12). Since far more individuals are regularly produced than can possibly survive, there is a consequent "struggle for life" or "struggle for existence" among them, and those individuals with variations that are "profitable" to them in their "infinitely complex relations to other organic beings and to external nature" will tend to have a better chance of survival and of handing down those variations to their offspring. The offspring will thus be favored in the contest for survival or "will . . . have a better chance of surviving". Taking his cue from the practice of selection by breeders (artificial selection), Darwin called the "principle, by which each slight variation, if useful, is preserved, by the term of Natural Selection" (p. 61). It was a decisive moment in the development of Darwin's thought when he transformed the concept of interspecies competition, as developed by Lyell, into that of intraspecies competition or competition among individuals.[1]

Variation, inheritability, competition, and natural selection are thus the key words in Darwinian evolution. Of the four, natural selection may be the most significant, since Darwinian evolution is evolution "by means of natural selection".[2] It is the concept of natural selection that distinguishes Darwinian evolution from evolution in general (Poulton 1896; Limoges 1970c) and from pre-Darwinian evolutionary theories. Furthermore, although many post-Darwinian "zoologists, botanists, and paleontologists eventually accepted gradual evolution through natural causes", the majority of them did not accept "natural selection as the prime cause of evolutionary change . . . until the 1930's" (Mayr 1976a, p. 294).

The present notes call attention to some unfamiliar aspects of the first application of natural selection in a scientific publication and to the discussion of natural selection by Samuel Wilberforce, Lord Bishop of Oxford.

I. The First Scientific Publication (Tristram 1859) in Which Natural Selection is Applied as a Working Scientific Principle

The doctrine of evolution by natural selection was publicly announced at a meeting of the Linnean Society of London on 1 July 1858, and papers by Darwin and Wallace were published in the *Proceedings* of the Society on the following 20 August, with the general title, "On the Tendency of Species to form Varieties; and on the Perpetuation of Varieties and Species by Natural Means of Selection" (Loewenberg 1957, 1959d; De Beer 1958a; Appleman 1970). Darwin later wrote of their "joint productions", that they "excited very little attention, and the only published notice of them which I can remember was by Professor Haughton of Dublin, whose verdict was that all that was new in them was false, and what was true was old" (*Autobiography*, p. 122). In his biography of Darwin, Sir Gavin de Beer added that the "only person who noticed what had happened was Alfred Newton, who read the joint paper and found 'a perfectly simple solution of all the difficulties that had been troubling me for months past' " (1963, p. 150).[3] Haughton's address was delivered to the Geological Society in Dublin on 9 February 1859,[4] more than nine months before the publication of the *Origin* in November 1859.

A second reference to the Darwin-Wallace communication occurred in the presidential address given by Richard Owen to the British Association for the Advancement of Science at the Leeds meeting in September 1858 (Owen 1858). There is no evidence that Owen read to the BAAS the complete version of his address as published, which runs to sixty-two pages; so we do not know whether Owen actually mentioned Darwin and Wallace in his talk in the same manner in which he did in print.[5] Owen introduced the subject by observing that "the healthiest specimens of Orang or Chimpanzee, brought over in the vigour of youth, perish within a period never exceeding three years . . . in our climate", despite all the care in choice of "food, clothing, and contrivances for artificially maintaining the chief physical conditions of their existence." This led him to ask: "By what metamorphoses . . . has the alleged humanized Chimpanzee or Orang been brought to endure all climates? The advocates of 'transmutation' have failed to explain them . . ." (1858, pp. lxxxiv-lxxxv).

Six pages later, Owen turned to "subjects regarding which we have not, at present, the basis of true assertion" (1858, p. xci). Foremost among these was "the probable cause of the extinction of species". Owen advanced here an idea he had expressed earlier concerning the "gradual changes in the conditions of a country affecting the due supply in sustenance to animals in a state of nature." This explains why "many of the larger species of particular groups of animals" have "become extinct, whilst smaller species of equal antiquity have remained." Large animals suffer from a drought sooner than small ones, he argued, and an alteration of the climate that reduces "the quantity of vegetable food" will affect first "the bulky Herbivore". New enemies will destroy "the large and conspicuous quadruped or bird", while "the smaller species conceal themselves and escape." Hence small animals may exist in countries where there had formerly been large animals, but this does not imply "any gradual diminution of the size of such species" so much as the fact that "the smaller and feebler animals have bent and accommodated themselves to changes which have destroyed the larger species." Then Owen said:

> Accepting this explanation of the extirpation of species as true, Mr. Wallace has recently applied it to the extirpation of varieties; and, assuming, as is probable, that varieties do arise in a wild species, he shows how such deviations from type may either tend to the destruction of a variety, or to adapt a variety to some changes in surrounding conditions, under which it is better calculated to exist, than the type-form from which it deviated. (1858, p. xci)

Owen referred in a footnote to Wallace's paper in the *Proceedings* of the Linnean Society for August 1858.

"No doubt," said Owen,

> the type-form of any species is that which is best adapted to the conditions under which such species at the time exists; and as long as those conditions remain unchanged, so long will the type remain; all varieties departing therefrom being in the same ratio less adapted to the environing conditions of existence. But, if those conditions change, then the variety of the species at an antecedent date and state of things will become the type-form of the species at the later date, and in an altered state of things. (1858, pp. xci-xcii)

He then observed: "Mr. Charles Darwin had previously to Mr. Wallace illustrated this principle by ingenious suppositions." He quoted Darwin's "imaginary example" of the "canine animal which preyed chiefly on rabbits, but sometimes on hares." Owen's only comment was that "Observation of animals in a state of nature is required to show their degree of plasticity, or the extent to which varieties do arise"; and that "Observation of fossil remains is also still needed to make known the ante-types, in which varieties, analogous to the observed ones in existing species, might have occurred, so as to give rise ultimately to such extreme forms as the Giraffe, for example" (1858; p. xcii).

Owen's fair and generous treatment of Wallace and Darwin in his address of 1858 may be contrasted with his review of the *Origin* in the *Edinburgh Review*, April 1860 (Owen 1860). Darwin said of this review that it was "extremely malignant, clever, and . . . very damaging" (*LL* (NY) 2: 94); he was "astonished at the misrepresentations" (*LL* (NY) 2: 96); Owen was "very bitter" — he "scandalously misrepresents" many things and "misquotes some passages, altering words within inverted commas"; "It is painful to be hated in the intense degree with which [Owen] hates me" (*LL* (NY) 2: 94; cf. *LL* (NY) 2: 36-37, 106-107).[6]

Any list of scientists who referred to the Darwin-Wallace papers prior to the publication of the *Origin* must include Joseph Dalton Hooker, whose *Flora of Australia* appeared in 1859. In a postscript at the end of the "Introductory Essay", dated 4 November 1859, Hooker referred to a paper ("which I have not seen") by Wallace read at a meeting of "The Linnaean [sic] Society . . . on the 3rd of November", in which Wallace had independently reached the same conclusion as Hooker "regarding the permanence of vegetable as compared with animal forms". Then, he wrote, "I would further observe here, to avoid ambiguity, that my friend Mr. Darwin's just completed work 'On the Origin of Species by Natural Selection,' from the perusal of much of which in MS. I have profited so largely, had not appeared during the printing of this Essay, or I should have largely quoted it." Throughout the Essay, beginning on the very first page, Hooker referred with evident approval to "the recently published hypotheses of Mr. Darwin and Mr. Wallace" (Hooker 1859a, p. i) and made use of the concepts of variation and natural selection.

But there was one instance in which the Darwin-Wallace papers gave rise to a public acceptance and application of the theory of natural selection. This event is of notable interest because it demonstrates in a particularly striking manner the way in which natural selection solved knotty problems that had seemingly defied enodation. Furthermore, the complete story reveals a conversion followed by a subsequent deconversion that in itself illustrates the power of the received anti-evolutionist opinion, and that incidentally shows how misleading are the usual Whiggish presentations of Huxley's debate with Wilberforce at Oxford in 1860.

The one naturalist publicly to accept and to apply the new concept of natural selection before the publication of the *Origin* was Henry Baker Tristram, a distinguished British ornithologist and later residentiary canon of Durham, but then rector of Castle Eden. Tristram had been studying birds of the Sahara (notably larks and chats). In his article "On the ornithology of Northern Africa" published in October 1859, Tristram openly and plainly declared:

> Writing with a series of about 100 Larks of various species from the Sahara before me, I cannot help feeling convinced of the truth of the views set forth by Messrs. Darwin and Wallace in their communications to the Linnean Society, to which my friend Mr. A. Newton last year directed my attention, "On the Tendency of Species to form Varieties, and on the Perpetuation of Varieties and Species by natural means of selection." It is hardly possible, I should think, to illustrate this theory better than by the Larks and Chats of North Africa. (Tristram 1859, p. 429)

Tristram then proceeded to discuss variations in coloration and anatomical structure, concluding that these differences "doubtless have a very direct bearing on the ease or difficulty with which the animal contrives to maintain its existence" (1859, pp. 429–430). Survival in the Sahara Desert required a coloration similar to "the surrounding country". As he explained:

> There are individual varieties in depth of hue among all creatures. In the struggle for life which we know to be going on among all species, a very slight change for the better, such as improved means of escaping from its natural enemies (which would be the effect of an alteration from a conspicuous color to one resembling the hue of the surrounding objects), would give the variety that possessed it a decided advantage over the typical or other forms of the species. Now in all creatures, from Man downwards, we find a tendency to transmit individual varieties or peculiarities to the descendants. A peculiarity either of colour or form soon becomes hereditary when there are no counteracting causes, either from change of climate or admixture of other blood. Suppose this transmitted peculiarity to continue for some generations, especially when manifest advantages arise from its possession, and the variety becomes not only a race, with its variations still more strongly imprinted upon it, but it becomes the typical form of that country. (1859, p. 430)

The application of these ideas to the larks of the Sahara was simple and straightforward:

> If the Algerian Desert were colonized by a few pairs of Crested Larks, — putting aside the ascertained fact of the tendency of an arid, hot climate to bleach all dark colours, — we know that the probability is, that one or two pairs would be likely to be of a darker complexion than the others. These, and such of their offspring as most resembled them, would become more liable to capture by their natural enemies, hawks and carnivorous beasts. The lighter-coloured ones would enjoy more or less immunity from such attacks. Let this state of things continue for a few hundred years, and the dark-coloured individuals would be

> exterminated, the light-coloured remain and inhabit the land. This process, aided by the above-mentioned tendency of the climate to blanch the coloration still more, would in a few centuries produce the *Galerida abyssinica* as the typical form. And it must be noted, that between it and the European *G. cristata* there is no distinction but that of colour. (1859, pp. 430–431)

The papers published by the Linnean Society, unlike the later *Origin* and other writings of Darwin, did not contain the geological evidence for a time scale of evolution, which may account for Tristram's optimistic guess of "a few hundred years".

Tristram also applied the theory of natural selection "to *Galerida isabellina, G. arenicola,* and *G. macrorhyncha* [in which cases] we have differences not only of colour but of structure." One of these *(G. arenicola)* had a "very long bill", while another *(G. isabellina)* had a "very short one". Since the first of these lived in a region where the food had to be sought in deep sand, its long bill gave it "a great advantage", whereas the second fed "among stones and rocks" where what was required in a bill was "strength rather than length" (1859, p. 431).

Tristram was a deeply religious man, and he accordingly concluded with a reference to God's method of working. He noted that "it is contrary alike to sound philosophy and to Christian faith to doubt the creation of many species by the simple exercise of Almighty volition." Even so, he insisted, "knowing that God ordinarily works by natural means, it might be the presumption of an unnecessary miracle to assume a distinct and separate origin for many of those which we term species." However one might speculate on this question, and might do so for a life-time, "this conclusion alone so far is certain, — that every peculiarity or difference in the living inhabitants of each country is admirably adapted by the wisdom of their beneficent Creator for the support and preservation of the species" (1859, p. 433).

The basic outlines of the story, as I have presented it here, are to be found in Edward B. Poulton's admirable book *Charles Darwin and the Theory of Natural Selection* (1896).[7] Some additional information is available in an article by Alfred Newton, the scientist who brought the papers of Darwin and Wallace to Tristram's attention (Newton 1888a), and in the Tristram-Newton correspondence (Wollaston 1921, pp. 111–122 et al.).[8] Newton, the inaugural professor of zoology at Cambridge University, was one of the major ornithologists of his age. He and Tristram were close friends for over half a century of scientific companionship.[9] A measure of Newton's esteem for Tristram is given by the fact that Newton withdrew his own name from candidacy for Fellowship in the Royal Society so that Tristram could be elected.[10]

In the first part of 1858 Newton had been in Iceland with John Wolley on an ornithological expedition.[11] On his return to England, he stopped for a visit with fellow-ornithologist Tristram, then rector of Castle Eden.[12] Recalling the events thirty years later, Newton said he had been particularly struck by Tristram's findings that "the inhabiters of the desert took a dull drab, but occasionally a warm or sand-coloured hue, while those which did not dwell in the desert wore a suit of much more decided and variegated tint" (1888a, p. 243). He was at once reminded of what seemed to be an instance of "a similar general law", which had been brought to his attention a year earlier in Washington when he had visited Spencer Fullerton Baird.[13] Newton relates:

> Not many days after my return home there reached me the part of the *Journal of the Linnean Society* which bears on its cover the date, 20th August, 1858, and contains the papers by Mr. Darwin and Mr. Wallace, which were communicated to that Society at its special meeting on the first of July preceding, by Sir Charles Lyell, and Dr. (now Sir Joseph) Hooker. I think I had been away from home the day this publication arrived, and I found it when I came back in the evening. At all events, I know that I sat up late that night to read it; and never shall I forget the impression it made upon me. Herein was contained a perfectly simple solution of all the difficulties which had been troubling me for months past. I hardly know whether I at first felt more vexed at the solution not having occurred to me, than pleased that it had been found at all. However, after reading these papers more than once, I went to bed satisfied that a solution had been found. All personal feeling apart, it came to me like the direct revelation of a higher power; and I awoke next morning with the consciousness that there was an end of all the mystery in the simple phrase, "Natural Selection." (1888a, p. 244)

Newton at once tried to interest various correspondents in "the discovery of Mr. Darwin and Mr. Wallace", but was generally disappointed. The only convert he made was Tristram.[14]

On 24 August 1858, Newton wrote to Tristram about the applications of natural selection. This letter was written only four days after the date of publication of the Darwin-Wallace papers and the day after Newton had received it.

> I have been very much pleased with a paper in the last number of the Linnean Society's *Proc.* on "the tendency of Species to form Varieties and on the Perpetuation of Varieties and Species by Natural means of selection," by Darwin and Wallace. I am not quite sure that I altogether agree with them, but there is very much in it that is very good, and most of the ideas propounded are original. I think there is a hint in it on which you might speak, on the subject I suggested to you when at Castle Eden as being a likely one for a paper before the Linnean Society, the variations induced by desert climate, as exemplified in

Northern African Larks and Wheatears. (Wollaston 1921, p. 115)

Then, after pointing out that the "idea is perhaps not new, *i.e.* many naturalists [e.g., Baird and Gould] know perfectly well that birds from desert localities do not exactly resemble individuals of the same species (*i.e.* good species, not those of bird-namers) from more favoured districts", Newton said that he did "not suppose any one has connected these facts with the theory (though it is more than theory) of Darwin and Wallace, nor has any one practically applied their ideas." He concluded:

> It seems to me that they can be connected and should be connected thus: any modification of the structure (using the word in its widest sense, even to comprehend a mere change of colour) of an animal must in some way or other affect the ease or difficulty with which it contrives to maintain its existence. In the struggle for life which we know to be going on among all species, a very slight change for the better, such as improved means of escaping from its natural enemies (which would be the effect of an alteration in colour from one differing much to one closely resembling the hue of surrounding objects), would give that variety a great advantage over the typical or other forms of the species. Allow the advantage to be continued for a considerable period, and the variety becomes not only a race with its variations still more strongly imprinted upon it, but the typical form or varieties having experienced changes not advantageous to their life may even become extinct. Thus to apply the case, suppose an Algerian desert to become colonised by a few pairs of Crested Lark; we know that the probability is that of them one or two pairs would be likely to be of a darker complexion than the others, these and such of their offspring as most resembled them would become more liable to capture by their natural enemies, hawks, carnivorous beasts, etc.; the lighter coloured ones would enjoy more or less immunity from such attacks; let the state of things continue a few hundred years, the dark-coloured individuals would be exterminated, the lighter-coloured remain and inhabit the land.
>
> Again, smaller or shorter-billed varieties would undergo comparative difficulty in finding food when food was not abundant, and had to be picked out from crevices among stones, these would be in comparatively reduced condition, in the breeding season they would not feel their capabilities were such as inclined them to matrimony, the consequences would be in a few hundred years the longer-billed varieties would be the most numerous, they would become a race, in a few hundred years more they would be the sole possessors of the land, the shorter-billed fellows dying out of their way until that race was extinct. Here are only two cases enumerated which might serve to create, as it were, a new species from an old one, yet they are perfectly natural ones, and

such as I think must occur, have occurred, and possibly be occurring still. (Wollaston 1921, pp. 115–117)

It is, I believe, clear from these documents that the application of natural selection to the birds of the Sahara was conceived by Newton, and that Tristram had basically rewritten Newton's private communication to him and had incorporated it into his own article. But there can be no doubt that Tristram was an early and easy convert to the new doctrine of natural selection.

Since Tristram was the first scientist in print to accept and to apply the concept of natural selection, it might be supposed that he would have been a staunch and foremost advocate of Darwinism in the early years of controversy after the publication of the *Origin*. And this is just the impression that is given by Poulton (1896), the only historian of Darwin's work and its reception to discuss Tristram's use of natural selection. But the fact of the matter is that within a year after the *Origin* appeared, Tristram joined the ranks of the rabid anti-Darwinians.

II. The 1860 Oxford Debate and Tristram's "Re-conversion" to the "Old Faith"[15]

The event that triggered Tristram's re-conversion to orthodoxy was the Huxley-Wilberforce debate. Some accounts of this confrontation would have Huxley the clean victor, confounding his bigoted religious opponent in the name of science, progress, and truth. Thus we are told that after the debate, "the opponents [of Darwin] were left crushed on the field", that Huxley's "triumph was complete".[16] Alfred Newton did report that the "feeling of the audience was much against the Bp" (Wollaston 1921, p. 119). But, as Newton also recorded, "On the whole it seemed to be a drawn battle, for both sides stuck to their guns" (1888a, p. 249).[17] According to Newton, his friend Tristram "waxed exceedingly wroth as the discussion went on, and declared himself more and more anti-Darwinian (Wollaston 1921, p. 119). Newton, of course, tried to get Tristram back into the fold of Darwinians, but without success.

I very much doubt that Tristram's "re-conversion at Oxford to the old faith" could have been "inspired [only or primarily by] a feeling of loyalty to the Bishop" (Wollaston 1921, p. 120), but there can be no doubt that Newton was very disappointed in his friend and convert and accordingly "sought (unavailingly) to show him the error of his ways" (Wollaston 1921, p. 120). In a letter of 30 July 1860, written to Tristram just one month after the debate (which took place on Saturday, 30 June 1860), Newton referred to "the original of the speech spoken to the British Asses" (i.e., the members of the British Association for the Advancement of Science,[18] which had just been published in the *Quarterly Review*. "I am," he wrote, "quite converted. I was (I confess it) in a 'state of transition,' but Darwin*oid* I might have remained for a whole geological aeon." However, as he noted, "The Bishop's speech and article have caused me by a process of 'natural selection' to become something better. I am developed into pure and unmitigated Darwinism" (Wollaston 1921, p. 121).

Tristram's reply shows not only an orthodox state of mind following his re-conversion, but indicates that everyone did not share Huxley's and Newton's low opinion of Wilberforce's article (see next section). "How they can answer the *Quarterly*," Tristram wrote to Newton on 31 July 1860, "I cannot tell except by the argument of noise and sneers with which they tried to put down" Bishop Wilberforce "and every one else who did not subscribe to the infallibility of the God Darwin and his prophet Huxley." Tristram hoped that his friend Newton's "monomania" was only temporary, and he concluded by telling Newton that Darwin's work (which he may not even have read in book form) was no more than a mere "renovation of Lamarck". Like the clergyman he was, rather than scientist, he saw the new doctrine of evolution by natural selection as "one blind plunge into the gulph of atheism and the coarsest materialism" (Wollaston 1921, p. 122).

From Tristram's remarks after the Oxford debate one would hardly have guessed that he had been the first scientist to publish an application of the new concept of natural selection to a scientific problem, and indeed the first scientist to give this concept any public recognition and approval. But his approbation was short-lived, lasting barely a year. In the end the clergyman in him overcame the scientist, providing a striking illustration of the power of the received opinion or current orthodoxy to inhibit the acceptance of the new idea of Darwinian evolution by natural selection.

III. Wilberforce and Natural Selection

Although it is sometimes thought that Wilberforce received an intellectual hiding from Huxley in the Oxford debate of 1860,[19] and slunk away to regret his temerity, nothing could be further from the truth. For the fact of the matter is that he proudly published his attack on Darwin's *Origin* in *The Quarterly Review* for July and October 1860 (Wilberforce 1860). This lengthy essay, published anonymously, was revised during the course of printing. The Harvard College Library copy has an errata slip pasted in, indicating that "Part of our impression was printed off without the following corrections . . . ." Three of the four errata were simple press errors, the fourth was a minor intellectual fault.[20] The Harvard copy also shows pages 251–252 to be a cancel, that is, the original leaf containing these two pages has been cut away and replaced by a substitute leaf pasted on to the stub. In a letter to J. D. Hooker (July 1860), Darwin referred to this fact by observing: "I can see there has been some queer tampering with the Review, for a page has been cut out and reprinted" (*LL* (NY) 2: 118); I have not found a copy with the uncancelled page and so I do not know what changes were introduced at the last minute.

Wilberforce began his essay by lauding Darwin's "scientific attainments, his insight and carefulness as an observer, blended with no scanty measure of imaginative sagacity, and his clear and lively style." As to the *Origin* itself, Wilberforce declares it to be "a most readable book" — "full of facts in natural history, old and new, of his collecting and of his observing." Darwin's book, according to Wilberforce, will attract the attention of naturalists and other scientists, but also will appeal "to every one who is interested in the history of man and of the relations of nature around him to the history and plan of creation." Although Wilberforce announces *in limine* (to use a Latinism much favoured by Wilberforce in this essay) that he "shall have much and grave fault to find" with "Darwin's 'argument' ", he professes himself "disposed to admire the singular excellences of his work". He gives three extracts from the *Origin* (pp. 74, 210–211, 219–223) as examples of Darwin's great merit as an observer and experimenter. But it is the theme of the book, the "argument", that Wilberforce admits sets him "immediately at variance" with Darwin: Darwin's inference — as quoted by Wilberforce — " 'from analogy that probably all the organic beings which have ever lived on this earth' (man therefore of course included) 'have descended from some one primordial form into which life was first breathed [by the Creator]' — p. 484." It is obvious that Wilberforce's wrath was kindled by the implications of Darwin's theory for man, a topic that Darwin unsuccessfully tried to keep so latent that it would not obtrude and startle the reader.[21] But Wilberforce so constantly introduced the origin of human beings into the discussion that the otherwise uninformed reader would have concluded that the discussion of this topic had been the primary aim or subject of Darwin's *Origin*.[22]

In order to rebut and demolish Darwin's thesis, Wilberforce set forth four "leading propositions which he [Darwin] must establish in order to make good his final inference":

1. That observed and admitted variations spring up in the course of descents from a common progenitor.
2. That many of these variations tend to an improvement upon the parent stock.
3. That, by a continued selection of these improved specimens as the progenitors of future stock, its improvement may be unlimitedly increased.
4. And, lastly, that there is in nature a power continually and universally working out this selection, and so fixing and augmenting these improvements. (1860, pp. 225, 226, 230, 231).

The first of these propositions is in fact genuinely Darwinian — the continual variation that occurs in offspring — although it is puzzling to know exactly what Wilberforce intended by the adjectives "observed and admitted". The second, however, like the allied third proposition, introduces the non-Darwinian concept of "improvement", with its direct overtones of teleology, whereas Darwin had restricted himself to inheritable variations that give individuals a better chance for survival (and for reproducing) in the Darwinian "struggle for life", the contest to be won not by being an "improvement" on one's forebears but by being better adapted to one's environment than is the case for other individuals exhibiting different variations. No doubt Wilberforce was misled by an over-concentration on the selective practices of breeders whose aim has always been to "improve" their stock.

What I find most interesting about this set of four propositions is that Wilberforce not only was perceptive enough to recognize the basic role of natural selection in the Darwinian conception of evolution,[23] but was completely convinced that natural selection had been proved by Darwin to be a principle or process fundamental to all the operations of nature. Since neither Huxley nor Lyell ever fully accepted natural selection despite the degrees of warmth of their embrace of Darwinian evolution (Poulton 1896, chaps. 16, 18), it is more than ordinarily worthy of note that one of the first vocal opponents to Darwin's theory should have accepted natural selection without qualm. Thus, referring to the fourth of his propositions, "the last in our series", Wilberforce said that it is in his opinion "the newest and the most ingenious part of Mr. Darwin's whole argument" (1860, p. 232). Of course, Wilberforce must "absolutely deny the mode in which he [Darwin] seeks to apply the existence of the power to help him in his argument"; yet he admitted that Darwin had convincingly shown "that such a self-acting power does actively and continuously work in all creation around us." Wilberforce went so far as to say that one of the most "interesting" parts of Darwin's book was that in which he set forth "the principle of 'Natural Selection' " and "establishes this law".

While Wilberforce had "no doubt of the existence or of the importance of the law itself", he had to go on record as differing from Darwin "totally in the limits which he would assign to its action". Wilberforce believed (as did Cuvier, Lyell, and others) in an "eliminative" principle that was natural, and hence a lawful principle of "selection" that was part of a divine plan. Hence, Wilberforce's version of natural selection accorded with the concept of extinction of species, according to scientific and ecclesiastical canons, but not the production or creation of new species. Thus Wilberforce could laud Darwin for advancing a principle that Wilberforce interpreted in an essentially non-Darwinian manner.

Wilberforce then quoted extracts from Darwin to prove the extraordinary rapid increase that would occur in any unchecked natural populations, and commended Darwin's observations and conclusions. Wilberforce "readily" admitted that "a struggle for life then actually exists",[24] and that "it tends continually to lead the strong to exterminate the weak". This struggle he saw as a mechanism mercifully provided by the Creator in order to ensure that there would be no deterioration, "in a world apt to deteriorate", of "the works of the Creator's hands". And it is to be noted especially that although Wilberforce apparently understood that Darwinian natural selection acted "through the struggle of individuals" (1860, p. 233), he was also committed to inter-species rivalry (as Lyell had taught) (see n. 1). He thus believed that "this law" of natural selection not only "maintains . . . the high type of the family" as a result of "the struggle of individuals", but also acts "through a similar struggle of species, to lead the stronger species to supplant the weaker." And it could almost have been Darwin himself writing, or one of his most fervid disciples, when Wilberforce gave an example of the "clear and indisputable" mode of "action of such a law" as follows:

> Hardier or more prolific plants, or plants better suited to the soil or conditions of climate, continually tend to supplant others less hardy, less prolific, or less suited to the conditions of vegetable life in those special districts.

But although Wilberforce appeared to be so fully committed to Darwinian natural selection, he did not go along with the notion that such a process could produce new species. He had two arguments to support his position. One: we must be shown that there exists in nature ("co-ordinate with the law of competition and with the existence of . . . favourable variations") a "power of accumulating such favourable variation through successive descents". The other:

> we must be shown first that this law of competition has in nature to deal with such favourable variations in the individuals of any species, as truly to exalt those individuals above the highest type of perfection to which their least imperfect predecessors attained — above, that is to say, the normal level of the species.

Unless these two points could be established (and Wilberforce proceeded to show why he believed that they could not be), it was not possible — he held — to argue from the existence and operation of the law of natural selection to "a perpetual improvement in natural types".

Wilberforce's attack on Darwin in *The Quarterly Review* for July 1860 was published anonymously. Presumably his arguments were so similar to those he expressed in the Oxford debate that his authorship was obvious.[25] Huxley characterized this article as "the insolence of a shallow pretender to a Master in Science" (1887, 2: 183). He condemned Wilberforce for his gross display of ignorance and his "want of intelligence, or of conscience, or of both", which he held to be all the more egregious in that Wilberforce "held [Darwin] up to scorn as a 'flighty' person, who endeavours 'to prop up his utterly rotten fabric of guess and speculation,' and whose 'mode of dealing with nature' is reprobated as 'utterly dishonourable to Natural Science.' "[26] And, although Wilberforce said that he could not "consent to test the truth of Natural Science by the word of Revelation", he nevertheless — as Huxley remarked — "devotes pages to the exposition of his conviction that Mr. Darwin's theory 'contradicts the revealed relation of the creation to its Creator,' and is 'inconsistent with the fulness of his glory' " (1887b, 2: 183).

Wilberforce's style of attack was to mingle coarse sarcasm with high-minded statments about the radical quality of new ideas that are established by science. In the same paragraph in which Wilberforce poked fun at Darwin by a reference to "our unsuspected cousinship with the mushrooms", he declared that "we are too loyal pupils of inductive philosophy to start back from any conclusion by reason of its strangeness". As an example he cited the way in which "[Isaac] Newton's patient philosophy taught him to find in the falling apple the law which governs the silent movements of the stars in their courses." And he asked whether "Mr. Darwin can with the same correctness of reasoning demonstrate to us our fungular descent" (1860, p. 231).

Wilberforce attacked Darwin on the grounds of his not having been "a loyal disciple of the true Baconian philosophy". This was manifest, according to Wilberforce, in Darwin's use of such expressions as "It seems to me unlikely" — "I do not doubt" — "I can conceive" — "It is not incredible" — "It is conceivable — "I venture confidently" — which led Wilberforce to conclude:

> In the name of all true philosophy we protest against such a mode of dealing with nature, as utterly dishonourable to all natural science, as reducing it from its present lofty level of being one of the noblest trainers of man's intellect and instructors of his mind, to being a mere idle play of the fancy, without the basis of fact or the discipline of observation.

And then, in opposition to Darwin, he quoted from a great authority on the "true spirit of philosophy", Adam Sedgwick, as follows:

> 'Analysis,' says Professor Sedgwick, 'consists in making experiments and observations, and in drawing general conclusions from them by induction, and admitting of no objections against the conclusions but such as are taken from experiments or other certain truths; for *hypotheses are not to be regarded in experimental philosophy.*'

The reference is to " 'A Discourse on the Studies of the University,' by A. Sedgwick, p. 102" — a work from which Wilberforce quoted extracts again and again. Wilberforce was not even aware that this whole paragraph, word for word, was being quoted by Sedgwick from the penultimate paragraph of Isaac Newton's *Opticks*, and was not an original expression of Sedgwick's own philosophy. I find it astonishing that, despite the overt reverence of Isaac Newton by Wilberforce, he did not even know one of Newton's most famous statements about the methods of proceeding in natural (or experimental) philosophy.

Conclusion

The three episodes concerning natural selection are significant for a number of reasons. The general lack of reaction to the publication of the short Darwin-Wallace communications by the Linnean Society in 1858 and the strong reaction to the longer and more fully documented *Origin* a year later show that what we tend to call the Darwinian Revolution is more owing to Darwin's *Origin* than to Wallace's essay. The contrast in Owen's reaction to the initial communications in the Linnean Society and to the *Origin* underlines this point, especially in view of the fact that it was Owen who was Wilberforce's primary coach in the preparation of his article in *The Quarterly Review*, which was the basis of his speech at Oxford.[27] The evidence thus indicates the truth of Ernst Mayr's opinion (personal communication, December 1980) that if Darwin had merely sent on Wallace's paper for publication by the Linnean Society without further ado and had then published his own *Origin* we would think today of Wallace merely as one of the precursors of evolution by natural selection (along with Patrick Matthew and William Charles Wells), rather than a "co-discoverer" with Darwin. In fact, as Wallace himself pointed out, the developed theory of evolution by natural selection and the revolution that it inaugurated in biology was due primarily to Darwin:

> As to the theory of Natural Selection itself, I shall always maintain it to be actually yours and yours only. You had worked it out in details I had never thought of, years before I had a ray of light on the subject, and my paper would never have convinced anybody or been noticed as more than an ingenious speculation, whereas your book has revolutionised the study of natural history, and carried away captive the best men of the present age. All the merit I claim is the having been the means of inducing *you* to write and publish at once. (Wallace 1916, p. 131)

Wallace quite properly acknowledged that it was the *Origin* and not the papers published by the Linnean Society that had "revolutionised the study of natural history".

The analysis of these episodes also shows the shallowness of those who would denigrate Darwin's great and original contribution simply because it was supposed to have been "in the air", apparently floating around and waiting for anyone who was handy to pull it down to earth.[28] Had this really been the case, then there would have been a quite different reaction to that first publication by Darwin and Wallace. Indeed, the president of the Linnean Society, T. Bell, gives us further evidence for this in his annual report for 1858. There he said that the biological world was waiting for one of those great "revolutions" which would, in a single stroke, change the whole aspect of the science of living things. Yet — despite the fact that it was the Linnean Society that had been the forum for the presentation of the Darwin and Wallace communications and that it was the Linnean Society that had published their communications — the president announced that "The year . . . has not, indeed, been marked by any of those striking discoveries which at once revolutionize, so to speak, the department of science on which they bear" (quoted in De Beer 1963, p. 150). The revolution required not merely a statement concerning evolution and natural selection but a treatise such as the *Origin*, which would, by its mass of evidence of all sorts, so forcibly bring the new doctrine to the attention of scientists that it could not be ignored and hence would challenge the existing concepts. The preliminary statements in the Darwin-Wallace communications of 1858 could hardly have produced what Darwin foresaw as "a considerable revolution in natural history".[29]

Additionally, the fact that the only scientist publicly to use the concept of natural selection, as presented in the Darwin-Wallace communications,[30] recanted after the Huxley-Wilberforce debate in Oxford gives us a real insight into the extraordinary power of the received opinion that, far from being ready to embrace a kind of evolutionism that was "in the air", was strong enough to cause Canon Tristram to recant his heresy and return to the fold of orthodoxy. Wilberforce's own published attack on the *Origin* gives further evidence of the newness and power of the concept of natural selection. At the same time that Wilberforce praised Darwin for introducing natural selection, he rejected the notion that there could be an evolution of species according to natural selection. His own view of natural selection, as we have seen, was (unlike Darwin's) part of a limited natural process that eliminates the unfit but never creates new species. Thus this example attests to the force and originality of Darwinian natural selection, of which the soundness was to a degree convincing, even to one who was bitterly opposed to the consequences that Darwin drew from it in relation to evolution and the origin of species.

ACKNOWLEDGEMENTS

This article is based on research supported by a grant from the Alfred Sloan Foundation. I am grateful to Ernst Mayr for many discussions on this topic and others relating to evolution; it was he who first brought to my attention the article by Tristram on the birds of the Sahara.

Notes

1. It was in this transformation that Malthus's ideas were of great importance (see Herbert 1971). The general theory of the development of scientific ideas by such transformations is developed in I. B. Cohen (1980, part 2); the example of Darwin is explored in §4.3.

2. It is to be remembered that Darwin himself made this point explicitly in the title of his book.
3. De Beer's remarks are incomplete and insofar misleading, since Newton did not himself at once make any public announcement of his acceptance of Darwin's ideas, although he did bring those ideas to the attention of Tristram — as shall be seen below — who did use them in a publication.
4. Samuel Haughton's address to the Dublin Geological Society read in part: "This speculation of Messrs. Darwin and Wallace would not be worthy of notice were it not for the weight of authority of the names [i.e., Lyell's and Hooker's], under whose auspices it has been brought forward. If it means what it says, it is a truism; if it means anything more, it is contrary to fact" (*LL* (NY) 1: 512; quoted by Darwin in a letter to Hooker, April or May 1859). Haughton later wrote up a full attack on the *Origin* (1860; reprinted with omissions in Haughton 1863).
5. No information concerning this topic is to be found in the biography of Owen by his grandson (R. S. Owen 1894).
6. Although Huxley had combatted Owen's arguments against evolution (in relation to man and the gorilla) (Bibby 1972, pp. 46–49), he contributed an essay on "Owen's position in the history of anatomical science" to Owen's biography (R. S. Owen 1894, 2: 273–332).
7. The name of Tristram does not appear in the index of Carter (1957), Himmelfarb (1959), Hull (1973b), Irvine (1955), Ruse (1979a). This episode is not mentioned in either Darwin's or Wallace's autobiography.
8. See also the historical summary in Newton (1888b). Newton also wrote a lengthy anonymous review of *LL* (1888c).
9. Neither Newton nor Tristram appears in the *D.S.B.*, but accounts of both are to be found in the *D.N.B.* The index to Wollaston's biography (1921) is very complete and gives ready access to the relations between Newton and Tristram.
10. In a letter to Newton (25 Feb. 1906) written just ten days before Tristram's death, the latter said: "It is utterly impossible to get out of your debt epistolary, as I have found ever since that unparalleled act of friendship many years ago, when you took off your name from the Royal Society in order to secure my election. When one looks back through the long vista of years there is nothing I have found to equal it for self-sacrifice and generosity." To which Newton replied (26 Feb. 1906): "I can never forget the steady, friendly, I may say, brotherly support I have invariably received from you, and if it were my good fortune to have done you a good turn in the matter of the Royal Society, a circumstance that had wholly passed from my mind, it was but a slight return for the aid you rendered in starting the B.O.U. and the *Ibis*, and again at the critical moment when our first Editor threw up the job, and (with one or two more) would not have been sorry had it come to an end" (Wollaston 1921, pp. 67–68). Tristram was elected F.R.S. in 1868, Newton in 1870.
11. On Wolley, see Wollaston (1921, pp. 12–17); a "Memoir of the late John Wolley" appeared in *The Ibis*, (1860, 2: 98). On the trip to Iceland, and the discussions between Wolley and Newton concerning the questions — "What is a species?" and "How did a species begin?" — during the expedition, just before the publication of the Darwin-Wallace papers in 1858, see Newton (1888a).
12. "Arrived in England, I, on my way home, stopped to visit another friend (then rector of Castle Eden, and now a canon of Durham), who had but lately returned from the first of those journeys of exploration whereby so much light has been thrown on the Natural History of the Holy Land. Before making his pilgrimage thither, Canon Tristram, to give him his present title, had passed two winters and springs in Algeria or Tunis, and had diligently collected specimens in those countries. The consequence was that he had amassed such a series as had never before been seen. Among those that most interested me were the so-called Desert-Forms of various animals, especially reptiles, birds, and mammals. In several groups of each of these classes examples were to be seen of individuals from the desert which differed chiefly or only in coloration from those inhabiting the surrounding country, or the oases which the desert itself surrounded; but then this difference was constant" (Newton 1888a, p. 243).
13. "I was . . . reminded of what, in a less degree, I had been shown and told the year before at Washington by the late Professor Baird, who pointed out to me the variations exhibited by examples of the same species of several groups of North-American birds, according as they came from woodland, prairie, or elevated plain-country, of which there was a very considerable series in the Museum of the Smithsonian Institution.

"Among all these there were indications of a similar general law. The woodland examples were the most highly coloured. Those from the prairies were less deeply tinted; while those from the high plains — districts which, from what I heard, seemed to approach in some degree the condition of a desert such as is found in the Old World (Mauritania or Palestine) — exhibited a fainter coloration. Here then was a sign that like causes produced like effects even at the enormous distance which separated the several localities. The effects were plainly visible to the eye; what were the causes? The only explanation offered to me by Professor Baird, so far as I remembered, was that the chemical action of light, uninterrupted by any kind of shade, produced the effect that was patent. With this explanation, though it hardly seemed satisfactory, one was fain to be content" (Newton 1888a, p. 243).

14. Newton was "convinced a *vera causa* had been found, and that by its aid one of the greatest secrets of creation was going to be unlocked." He writes: "I lost no time in drawing the attention of some of my friends, with whom I happened to be at the time in correspondence, to the discovery of Mr. Darwin and Mr. Wallace; and I must acknowledge that I was somewhat disappointed to find that they did not so readily as I had hoped approve of the new theory. In some quarters I failed to attract notice: in others my efforts received only a qualified approval. But I am sure I was not discouraged in consequence; and I never doubted for one moment, then nor since, that here we had one of the grandest discoveries of the age — a discovery all the more grand because it was so simple" (1888a, pp. 244–245). Newton added a note that "at this time I had no acquaintance personally or by correspondence with either of the discoverers" (p. 245).
15. Wollaston (1921, p. 120).
16. In the recent B.B.C. TV series "The Voyage of Charles Darwin", Huxley's alleged complete triumph was shown with particularly dramatic intensity.
17. Newton penned an almost contemporaneous account of the debate in a letter to his brother Edward on 25 July 1860. In part, this reads: "the Bp. of Oxford . . . made of course, a wonderfully good speech if the facts had been correct. Referring to what Huxley had said two days before, about after all its not signifying to him whether he was descended from a Gorilla or not, the Bp. chaffed him and asked whether he had a preference for the descent being on the father's or the mother's side? This gave Huxley the opportunity of saying that he would sooner claim kindred with an Ape than with a man like the Bp. who made so ill an use of his wonderful speaking powers to try and burke, by a display of authority, a free discussion on what was, or what was not, a matter of truth, and reminded him that on questions of physical science 'authority' had always been bowled out by investigation, as witness astronomy and geology. He then caught hold of the Bp.'s assertions and showed how contrary they were to facts, and how he knew nothing about what he had been discoursing on. . . . The discussion was adjourned until the Monday, but it was then thought by the leaders on both sides that it had better be dropped, and so the matter rests" (Wollaston 1921, pp. 118–120).
18. James Clerk Maxwell wrote a number of "serio-comic" verses about papers read to the British "Ass." In "Notes of the President's Address" (1874), he referred to "spherical small British Asses in infinitesimal state" (this poem was published in *Blackwood's Magazine*, together with a translation of it into Greek by Richard Shilleto). His "Report on Tait's Lectures on Force: — B.A., 1876" began: "Ye British Asses, who expect to hear / Ever some new thing." See Campbell and Garnett (1882, pp. 373, 639, 646).
19. The usual sources concerning this event may be supplemented by a lengthy letter written by Alfred Newton to his brother Edward on 25 July 1860, less than a month after the debate (which took place on 30 June); see Wollaston (1921, pp. 118–120); extracts from the letter are printed above in note 17. Bibby (1972, p. 41), gives an extract from Huxley's letter to his friend Frederick Daniel Dyster concerning this affair.
20. E.g., "reservant" for "reservans" (p. 233), "tired out" for "tried out" (p. 236), "powers" for "improvements" (p. 231), and an omitted quotation mark.
21. In the antepenultimate paragraph of the *Origin*, Darwin did refer to two "open fields for far more important researches" in "the distant future": "Psychology will be based on a new foundation" and "Light will be thrown on the origin of man and his history." (In the sixth edition, he altered this latter sentence to "Much light . . ." *Origin* 1959, p. 757). Wilberforce quoted this paragraph as evidence of "the flighty anticipations of the future in which Mr. Darwin indulges."

    Of course, Darwin was fully aware of

the implications of the theory of evolution by natural selection for humans, but he wanted to avoid unnecessary sources of controversy in the *Origin*. As he wrote to Wallace: "I think I shall avoid the whole subject, as so surrounded with prejudices; though I fully admit that it is the highest and most interesting problem for the naturalist" (*LL* (NY) 1: 467).

22. For example, Wilberforce said that "the theory which really pervades the whole volume" was that "Man, beast, creeping thing, and plant of the earth, are all the lineal and direct descendants of some one individual *ens*, whose various progeny have been simply modified by the action of natural and ascertainable conditions into the multiform aspect of life which we see around us."

    He referred to the unexpected conclusion "that mosses, grasses, turnips, oaks, worms, and flies, mites and elephants, infusoria and whales, tadpoles of to-day and venerable saurians, truffles and men, are all equally the lineal descendants of the same aboriginal common ancestor, perhaps of the nucleated cell of some primaeval fungus, which alone possessed the distinguishing honour of being the 'one primordial form into which life was first breathed by the Creator' " (1860, p. 231).

23. This in itself was hardly a great achievement, since the expression "natural selection" appears so prominently on the title page of Darwin's book. Additionally, Wilberforce had leaned heavily on Owen in composing the article.
24. The expression "struggle for life", like "natural selection", is part of the full title of the *Origin*.
25. In 1874, Wilberforce's authorship was openly acknowledged when this essay was included in Wilberforce 1874. Darwin knew who the author was, since — in a letter to his neighbor, Brodie Innes — he said (in a postscript): "If you have not seen the last 'Quarterly,' do get it; the Bishop of Oxford has made such capital fun of me and my grandfather" (*LL* (NY) 2: 119).
26. Huxley's examples are as follows: "And all this high and mighty talk, which would have been indecent in one of Mr. Darwin's equals, proceeds from a writer whose want of intelligence, or of conscience, or of both, is so great, that, by way of an objection to Mr. Darwin's views, he can ask, 'Is it credible that all favourable varieties of turnips are tending to become men'; who is so ignorant of paleontology, that he can talk of the 'flowers and fruits' of the plants of the carboniferous epoch; of comparative anatomy, that he can gravely affirm the poison apparatus of the venomous snakes to be 'entirely separate from the ordinary laws of animal life, and peculiar to themselves'; of the rudiments of physiology, that he can ask, 'what advantage of life could alter the shape of the corpuscles into which the blood can be evaporated?'" (1887b, 2: 183).
27. On Owen as the coach of Wilberforce, see L. Huxley (1900, 1: 180–189). Owen had personally taken his stand against evolution on Thursday, 28 June, two days before the Wilberforce-Huxley debate, citing evidence on the difference between the brain of man and the brain of the gorilla; on this topic and the subsequent argument between Huxley and Owen, see Chapter 15 of the above-mentioned biography of Huxley and also the account given by Bibby (1972, pp. 46–48).
28. An extreme view occurs in C. D. Darlington (1959a).
29. In the conclusion to the first edition of the *Origin*, Darwin wrote: "When the views entertained in this volume on the origin of species, or when analogous views on the origin of species are generally admitted, we can dimly foresee that there will be a considerable revolution in natural history" (*Origin*, p. 484). In the second edition, Darwin added a phrase so that this sentence would read: "When the views advanced by me in this volume, and by Mr. Wallace, or when . . ." (*Origin* 1959, p. 754).
30. Tristram was the only scientist, so far as is known, to have used the new concept of natural selection in a publication; he was not the only scientist to accept this concept, of course, since Newton was an earlier convert who brought Tristram into the fold, however temporarily.

21

DARWINISM *IS* SOCIAL

Robert M. Young

It strikes me that there should be little need for this paper. Only positivists believe that scientific facts and theories are separate from human meanings and values, and even they, inconsistently, set out to extrapolate human and social conclusions from putatively decontextualized facts. Only religious fundamentalists believe that a belief in God cannot be reconciled with science, and that true religion is based on the literal truth of Scripture. This is a sort of religious positivism, as is the notion of creation science, which the ultra-right is currently deploying in opposition to a vulnerable, neo-Darwinian scientific orthodoxy, as part of an attack on the role that science plays in giving legitimacy to a liberal vision of capitalism.

Except for scientific positivists and religious fundamentalists, then, the connection between Darwinism and society is acknowledged. Indeed, the nineteenth-century Comtean positivist historical progression of stages — from theological to metaphysical to positive science — is now more likely to be seen as conceptual layers, rather than stages, with the social totality as the most basic level below those three.

But, of course, I've already made the situation far too simple. There is no such thing as "the" connection between Darwinism and society. And we must pick our way carefully among the various versions of the connection that are now on offer and that were on offer in the nineteenth century.

I have heart for this task, but I want to begin by registering a certain weariness, even impatience, that it's still necessary to argue that: first, the intellectual origins of the theory of evolution by natural selection are inseparable from social, economic and ideological issues in nineteenth-century Britain (I nearly wrote "Victorian", but that would beg the question of what happened for over a third of that century); second, the substance of the theory was, and remains, part of the wider philosophy of nature, God, and society, where the conceptions of nature and God are themselves changing in complex ways which are integral to the changing social order; third, the extrapolations from Darwinism to either humanity or society are not separable from Darwin's own views, nor are they chronologically subsequent. They are integral.

Let me reiterate that it still seems odd to have to argue that the great nineteenth-century debate was about "man's place in nature". Yet I well recall a seminar in which the historian of biology Jonathon Hodge said, in a barbed aside, that not everything in the nineteenth-century evolutionary debate was about man's place in nature. And yet Darwin called it the highest and most interesting problem for the naturalist. Another way of making the point is to say that it's very implausible indeed that the great nineteenth-century quarterlies — rooted in history, literature, and social questions — would have devoted so much space and would have gone into such detail about geology and natural history, unless something rather important was implicitly, and often explicitly, at issue, which was itself centrally concerned with the natural order as the symbolic basis for the social order.

Efforts are still made to separate the origins of the theory of evolution by natural selection from the substance of the theory and from extrapolations to society. Efforts are also made to separate the origins and substance of the theory from social and economic contexts and debates. There is something of a supermarket of distinctions on offer. It's argued that contemporary social and economic conditions and theories played no part; or that since Malthus didn't say what Darwin took him to say, the connection is thereby weakened — as if there were a neutral "what X said", any more than there is a neutral observation language in science; or that we can legitimately trace the geological and/or theological connections while being silent about the social origins and resonances of ideas of the earth or of the deity; or that we can separate out the positivist Darwin from the ideologue, either within the *Origin* and other relatively strictly scientific works, or between what he said about species and what he went on to say about humanity and society; or that we can separate the Darwin and Wallace scientific theory from the wider debate embracing, most notably, Chambers, Spencer, and (Lyell's version of) Lamarck. Then there is the attempt to privilege natural selection as the mechanism of evolution and deny the real strength of other mechanisms in the *Origin,* and their growing prominence in subsequent editions and in Darwin's other writings.

I don't deny that there are meaningful distinctions to be made among all these issues, disciplines, figures, and periods, but none of those distinctions is ultimately important. I'm not arguing for a concept of the evolutionary totality so Leibnitzean that every monad reflects all the others with equal intensity. Rather, I'm suggesting that, then as now, the issues are all related to changing notions of humanity and society, and that the points at which the distinctions of issue, discipline, or level are made are themselves of socioeconomic and ideological interest. Once it is granted that natural and theological conceptions are, in significant ways, projections of social ones, then important aspects of all of the Darwinian debate are social ones, and the distinction between Darwinism and Social Darwinism is one of level and scope, not of what is social and what is asocial.

Nature is a societal category, and so is God. The ideological process that, it seems to me, underlies these developments is one that must be seen as arising in nineteenth-century secularization and that culminates in twentieth-century functionalism and sociobiology. That process is the naturalization of value systems. If we look at the debate about man's place in nature in those terms, we have to look much more widely: that is, look backward into a wider process of biological naturalization in the nineteenth-century movement that embraced the work of St. Simon, of Comte, of Gall. Looking forward, we have to consider much more carefully the biologization of human sciences, which is most prominently displayed in the present in ethology and sociobiology.

Once we have begun to consider the process of the naturalization of value systems in broader terms and see the debate on man's place in nature as a part of that wider set of issues, we must also consider our historical explanations as calling for a more comprehensive set of determinations. Just as we are interested in the findings, the data, the ruminations and the thought processes of Darwin — the notebooks, the scraps of paper and the crossings-out — we should also be interested in the large-scale forces and their resolutions, and the prevailing compromises of the period, as well as the issues that frame the inquiries of disciplines and the figures in them. We can, for example, trace these determinations for Paley, for Malthus, and for Lyell. Respectively, they help us to grasp the meaning of a utilitarian natural theologian, a Newtonian concept of progress through struggle, and a religious uniformitarian. Each is arguing a case in relation to particular prevailing views and traditions.

Let's develop these points. Lyell's opposition to a caricatured catastrophism was on behalf of a less hide-bound theology and conception of nature. His exposition of Lamarck made evolution so plausible that it convinced Spencer, among others, and Lyell finally reached a point where he had to put in what amounted to a scholium, to preclude the impending conclusion that he finally came to as late as 1869. He said in the *Principles* that even if other animals came to be by transmutation, to extend this view to man would "strain analogy beyond all reasonable bounds" (Lyell 1830–1833, 1: 156).

Paley, in his *Natural Theology,* expressed a certain dying pastoral order. Paley managed to absorb the issues that became the motor of Darwinian progress into a balanced order of nature. Here is the flavor of his world:

> But, to do justice to the question, the system of animal *destruction* must always be considered in strict connection with another property of animal nature, viz, *superfecundity.* They are countervailing qualities. One subsists by the correction of the other. (Paley 1816, p. 408)

He comments on how this attribute keeps the world full and in balance. But what happens when fruitfulness gets out of hand?

> But then this *superfecundity*, though of great occasional use and importance, exceeds the ordinary capacity of nature to receive or support its progeny. All superabundance supposes destruction, or must destroy itself. Perhaps there is no species of terrestrial animals whatever, which would not overrun the earth if it were permitted to multiply in perfect safety; or a fish, which would not fill the ocean: at least, if any single species were left to their natural increase without disturbance or restraint, the food of other species would be exhausted by their maintenance. It is necessary, therefore, that the effects of such prolific faculties be curtailed. In conjunction with other checks and limits, all subservient to the same purpose, are the *thinnings* which take place among animals, by their action upon one another. In some instances we ourselves experience, very directly, the use of these hostilities. One species of insects rids us of another species; or reduces their ranks. A third species perhaps keeps the second within bounds; and birds or lizards are a fence against the inordinate increase by which even these last might infest us. In other more numerous and possibly more important instances, this disposition of things, although less necessary or useful to us, and of course less observed by us, may be necessary and useful to certain other species; or even for the preventing of the loss of certain species from the universe; a misfortune which seems to be studiously guarded against. Though there may be the appearance of failure in some of the details of Nature's works, in her great purposes there never are. (1816, pp. 411–412)

He concludes reassuringly:

> We have dwelt the longer on these considerations because the subject to which they apply, namely, that of animals *devouring* one another, forms the chief, if not the only, instance in the works of the Deity of an economy stamped by marks of design, in which the character of utility can be called in question. (1816, p. 413)

Paley's pastoral order was being challenged by an urban industrializing order, in which progress was not the smooth process of the pleasure/pain principle of utility. It was, rather, progress through a more disruptive and rapacious version of pain, evil, suffering, famine, war, and death. Paley tried to accommodate this Malthusianism with a gentle rendering of God's superfecundity: the necessity of "thinnings". Malthus's Law of Change was more brutal: not pruning shears, but unremitting pressure, the "thousand wedges" we find in Darwin's D *Notebook*. These were the same wedges that prevented the huge gap between arithmetic increase of food and geometric increase of population from ever opening up.

Malthus's order of society and nature had a very different flavor from Paley's:

> The history of the early migrations and settlements of mankind, with

> the motives which prompted them, would illustrate in a striking manner the constant tendency in the human race to increase beyond the means of subsistence. Without some general law of this nature, it would seem as if the world could never have been peopled. A state of sloth, and not of restlessness and activity, seems evidently to be the natural state of man; and this latter disposition could not have been generated but by the strong goad of necessity, though it might afterwards be continued by habit, and the new associations that were formed from it, the spirit of enterprise, and the thirst of martial glory. (Malthus 1826, 1: p. 92)

He then reflects on the consequences of a population in a congenial environment.

> These combined causes soon produce their natural and invariable effect, an extended population. A more frequent and rapid change of place then becomes necessary. A wider and more extensive territory is successively occupied. A broader desolation extends all around them. Want pinches the less fortunate members of society: and at length the impossibility of supporting such a number together becomes too evident to be resisted. Young scions are then pushed out from the parent stock, and instructed to explore fresh regions, and to gain happier seats for themselves by their swords. 'The world is all before them where to choose.' Restless from present distress, flushed with the hope of fairer prospects, and animated with the spirit of hardy enterprise, these daring adventurers are likely to become formidable adversaries to all who oppose them. The inhabitants of countries long settled, engaged in the peaceful occupations of trade and agriculture, would not often be able to resist the energy of men acting under such powerful motives of exertion. And the frequent contests with tribes in the same circumstances with themselves, would be so many struggles for existence, and would be fought with a desperate courage, inspired by the reflection, that death would be the punishment of defeat, and life the prize of victory.
>
> In these savage contests, many tribes must have been utterly exterminated. Many probably perished by hardship and famine. Others whose leading star had given them a happier direction, became great and powerful tribes, and in their turn sent off fresh adventurers in search of other seats. (1826, 1: 94–95)

Paley and Malthus described different social orders, with very different moods and sanctions of God and nature, producing very different conceptions of biological stability and change — both theistic and both orderly. Paley extolled being content with your lot, while Malthus offered upward social mobility in return for moral restraint. Very different mechanisms were at work. The Malthusian law of progress was not inescapably pessimistic. Indeed, Panglossian renderings of it were expressed by Spencer, whereby progress to perfection was the consequence of the law of organic change, and the law of population was its proximate cause (cf. Young 1969, pp. 130–137).

Spencer said in 1851,

> Progress, therefore, is not an accident, but a necessity. Instead of civilization being artificial, it is a part of nature; all of a piece with the development of the embryo or the unfolding of a flower. The modifications mankind have undergone, and are still undergoing, result from a law underlying the whole organic creation; and provided the human race continues, and the constitution of things remains the same, those modifications must end in completeness. As surely as the tree becomes bulky when it stands alone, and slender if one of a group; as surely as the same creature assumes different forms of cart-horse and race-horse, according as its habits demand strength or speed; as surely . . . so surely must the things we call evil and immorality disappear; so surely must man become perfect. (1851, p. 65)

In 1857 Spencer put forward a comprehensive law of progress:

> This is the history of all organisms whatever. It is settled beyond dispute that organic progress consists in a change from the homogeneous to the heterogeneous.
>
> Now, we propose in the first place to show that this law of organic progress is the law of all progress. Whether it be the development of the Earth, in the development of Life upon its surface, in the development of Society, of Government, of Manufactures, of Commerce, of Language, Literature, Science, Art, this same evolution of the simple into the complex, through successive differentiations, holds throughout. (1901, p. 10)

He then develops the law of progress throughout many manifestations and passes from individual humanity to society, where he concludes: "The authority of the strongest or the most cunning makes itself felt among a body of savages as in a herd of animals, or a posse of schoolboys" (1901, p. 19).

Once again he draws the most general conclusion:

> It will be seen that as in each event of to-day, so from the beginning, the decomposition of every expended force into several forces has been perpetually producing a higher complication; that the increase of heterogeneity so brought about is still going on and must continue to go on; and that, thus progress is not an accident, not a thing within human control, but a beneficent necessity. (1901, p. 60)

Nor was Spencer the only person to put a dramatically optimistic interpretation on Malthusianism. Consider, in this light, the passage that ends:

> Thus, from the war of nature, from famine and death, the most exalted

> object which we are capable of conceiving, namely, the production of higher animals, directly follows. There is grandeur in this view of life with its several powers, having been originally breathed into a few forms or into one; and that, whilst this planet has gone cycling on according to the fixed laws of gravity, from so simple a beginning endless forms most beautiful and most wonderful have been, and are being, evolved. (*Origin*, p. 490)

If it is thought that the last paragraph of the *Origin* was merely a rhetorical flourish, one also has to explain away the last sentence of the chapter on Instinct:

> Finally, it may not be a logical deduction, but to my imagination it is far more satisfactory to look at such instincts as the young cuckoo ejecting its foster-brothers, — ants making slaves, — the larvae of Ichneumonidae feeding within the live bodies of caterpillars, — not as specially endowed or created instincts, but as small consequences of one general law, leading to the advancement of all organic beings, namely, multiply, vary, let the strongest live and the weakest die. (*Origin*, pp. 243–244)

Recent scholarship has confirmed the close link between Darwin's work and both social theories and theological theories, which were in turn closely linked with changing conceptions of the order of nature and society. There is also a growing consensus that Darwinism was a subtle accommodation within natural theology, rather than a clean break with it. Anyone wishing to take Darwin's mature views outside the context of natural theology has a lot of explaining to do, from the frontispiece quotations to many of the forms of reasoning and rhetoric in the *Origin*. Darwin was meticulous in his revision, as is obvious from Peckham's variorum edition (*Origin* 1959). Why would Darwin fail to remove forms of address and reasoning if they had become odious to him?

To the extent that the conclusion is gaining credence that Darwinism was a subtle accommodation within natural theology, then Darwin takes a place within the history of Victorian theology, on the one hand; on the other, given the direction taken by middle-brow theology with respect to science, it also points to an increasing embedding of value systems in conceptions of living nature.

Recall, for example, R. H. Hutton's rendering of the deliberations of the Metaphysical Society, where he wrote that "The uniformity of nature is the veil behind which, in these latter days, God is hidden from us" (Hutton 1885, p. 180). It is a changed and very watered-down natural theology within which one finds Darwin's mature work, but it is natural theology nonetheless.

An analogous change is that as nature, not God, bore the weight of the laws of life and mind, fundamentalism — itself born in opposition to the presumptuous reductionism of science — gave the believer a much more personal, intimate, and ethical God. Science could not carry the role of the transcendent. This was clear to Darwin, who on the whole avoided such questions. It was true of the spiritualist Wallace, who invoked a rather pale deity to account for the origin of important (and to him otherwise inexplicable) intellectual and moral faculties of man. He invoked socialist politics to deal with the need for a just and ethical society. Huxley was as clear as Wallace in believing that evolution itself could not bring about the millennium. He argued against Spencer, saying contra inevitable millenniarianism that ethics has to be brought in against the results of biology.

Here we come upon another curious set of distinctions. Edward Thompson has treated Darwin as the careful empiricist and Huxley as the ideologue. In his controversy with Perry Anderson, his side of which has recently been published in the collection of essays, *The Poverty of Theory,* Thompson tries to make Darwin an empiricist of the first order and to draw a very sharp separation between Darwin on the one hand as an inductive scientist, and Huxley on the other as a political and ideologically tainted publicist (1978, pp. 60–62; cf. pp. 255–256). (Is there a whispered parallel between himself and Darwin on the one hand, and Perry Anderson and Huxley on the other?)

What is striking about Thompson's position in this matter is the shocking isolation of his writing as a social historian from the mainstream of debate then and now about these matters. That is, he very surprisingly argues that there should have been much more of a furor, much more manifesto writing, much more debate within the periodicals. And in saying so he ignores just what ubiquitous debate there was throughout the literature of the period. It was not confined to the periodicals; but were one to consider only that sector, it takes Ellegård fifteen pages just to list the periodicals that were involved in his research about that debate. That is, Thompson, has simply ignored the breadth and texture of the debate in which both Darwin and Huxley were embedded, a debate I should add, in which science and ideology were inextricably intertwined (cf. *Rad. Sci. J. Collective* 1981, pp. 25–26).

In a related set of distinctions, Greta Jones (1980) has also set about separating the scientist Darwin from the ideologue, and both of those from Social Darwinism. As I see it, both Huxley and Darwin were expressing commonly held positions that were relatively progressive for their time, but relatively shocking to our eyes. I'm thinking, for example, of what Huxley had to say about blacks and women. I shall quote this, as well as passages from Darwin, in some detail, in the hope that these striking examples will destroy once and for all the notion that it's possible to distinguish sharply the scientist from the ideologue.

Huxley's essay is called "Emancipation — Black and White". First blacks:

It may be quite true that some negroes are better than some white men;

> but no rational man, cognizant of the facts, believes that the average negro is the equal, still less the superior, of the average white man. And, if this be true, it is simply incredible that, when all his disabilities are removed, and our prognathous relative has a fair field and no favor, as well as no oppressor, he will be able to compete successfully with his bigger-brained and smaller-jawed rival, in a contest which is to be carried on by thoughts and not by bites. The highest places in the hierarchy of civilization will assuredly not be within the reach of our dusky cousins, though it is by no means necessary that they should be restricted to the lowest. But whatever the position of stable equilibrium into which the laws of social gravitation may bring the negro, all responsibility for the result will henceforward lie between nature and him. The white man may wash his hands of it, and the Caucasian conscience be void of reproach for evermore. And this, if we look to the bottom of the matter, is the real justification for the abolition policy. (Huxley 1865, pp. 17–18)

Notice that he's opposing slavery and saying that blacks are biologically inferior, but that we shouldn't make it worse by adding social oppression. He continues:

> The like considerations apply to all the other questions of emancipation which are at present stirring the world — the multifarious demands that classes of mankind shall be relieved from restrictions imposed by the artifice of man, and not by the necessities of Nature. (1865, p. 18)

On the question of women, he is equally enlightened for his time.

> For our parts, though loth to prophesy, we believe it will be [like the result] of other emancipations. Women will find their place, and it will neither be that in which they have been held, nor that to which some of them aspire. Nature's old salique law will not be repealed, and no change of dynasty will be effected. The big chests, the massive brains, the vigorous muscles and stout frames, of the best men will carry the day, whenever it is worth their while to contest the prizes of life with the best women. . . . The most Darwinian of theorists will not venture to propound the doctrine that the physical disabilities under which women have hitherto labored, in the struggle for existence with men, are likely to be removed by even the most skilfully conducted process of educational selection. (1865, p. 22)

And he concludes: "The duty of man is to see that not a grain is piled upon that load beyond what Nature imposes; that injustice is not added to inequality" (p. 23).

This essay illustrates the principle that the science/ideology distinction is at any point a contingent resolution of historical forces, playing its own ideological role. The more ostensibly pure the science, the deeper one often has to look in order to demonstrate this principle. It is therefore easier in the case of, say, a Spencer or a Chambers than a Darwin or a Lyell. The evaluative conceptions that constitute the problems and parameters of a discipline, however, apply to a Newton and an Einstein just as much as they do to a Voltaire or a Velikovsky. Here, for example, is Darwin in the *Descent*:

> With savages, the weak in body or mind are soon eliminated; and those that survive commonly exhibit a vigorous state of health. We civilised men, on the other hand, do our utmost to check the process of elimination; we build asylums for the imbecile, the maimed, and the sick; we institute poor-laws; and our medical men exert their utmost skill to save the life of everyone to the last moment. There is reason to believe that vaccination has preserved thousands, who from a weak constitution would formerly have succumbed to small-pox. Thus the weak members of civilised societies propagate their kind. No one who has attended to the breeding of domestic animals will doubt that this must be highly injurious to the race of man. It is surprising how soon a want of care, or care wrongly directed, leads to the degeneration of a domestic race; but excepting in the case of man himself, hardly anyone is so ignorant as to allow his worst animals to breed. . . . (*Descent* 1874, pp. 133–134)

He goes on:

> We must therefore bear the undoubtedly bad effects of the weak surviving and propagating their kind; but there appears to be at least one check in steady action; namely that the weaker and inferior members of society do not marry so freely as the sound, and this check might be indefinitely increased by the weak in body or mind refraining from marriage, though this is more to be hoped for than expected. (*Descent* 1874, p. 134)

A little bit later:

> But the inheritance of property by itself is very far from an evil; for without the accumulation of capital the arts could not progress; and it is chiefly through their power that the civilised races have extended and are now everywhere extending their range, so as to take the place of the lower races. (*Descent* 1874, p. 135)

Further on:

> The presence of a body of well-instructed men, who have not to labour for their daily bread, is important to a degree which cannot be overestimated. As all high intellectual work is carried on by them, and on such work, material progress of all kinds mainly depends, not to mention other and higher advantages. (*Descent* 1874, p. 135)

American Social Darwinism could take comfort from the following:

> There is apparently much truth in the belief that the wonderful progress of the United States, as well as the character of the people, are the results of natural selection; for the more energetic, restless, and courageous men from all parts of Europe have emigrated during the last ten or twelve generations to that great country, and have there succeeded best. (*Descent* 1874, p. 142)

He carries on:

> Obscure as is the problem of the advance of civilisation, we can at least see that a nation which produced during a lengthened period the greatest number of highly intellectual, energetic, brave, patriotic, and benevolent men, would generally prevail over less favored nations. (*Descent* 1874, p. 142)

And further on:

> Nevertheless the more intelligent members within the same community will succeed better in the long run than the inferior, and leave a more numerous progeny, and this is a form of natural selection. The more efficient causes of progress seem to consist of a good education during youth whilst the brain is impressible, and of a high standard of excellence, inculcated by the ablest and best men, embodied in the laws, customs and traditions of the nation, and enforced by public opinion. (*Descent* 1874, p. 143)

I skip now to the general summary where Darwin reprises the quasi-imperialist views in the above passages.

> The advancement of the welfare of mankind is a most intricate problem: all ought to refrain from marriage who cannot avoid abject poverty for their children; for poverty is not only a great evil, but tends to its own increase by leading to recklessness in marriage. (*Descent* 1874, p. 618)

Who says, by the way, that Darwin didn't take in what Malthus said? He goes on:

> On the other hand, as Mr. Galton has remarked, if the prudent avoid marriage, whilst the reckless marry, the inferior members tend to supplant the better members of society. Man, like every other animal, has no doubt advanced to his present high condition through a struggle for existence consequent upon his rapid multiplication; and if he is to advance still higher, it is to be feared that he must remain subject to a severe struggle. Otherwise he would sink into indolence, and the more gifted men would not be more successful in the battle of life than the less gifted. Hence our natural rate of increase, though leading to many and obvious evils, must not be greatly diminished by any means. There should

> be open competition for all men; and the most able should not be prevented by laws or customs from succeeding best and rearing the largest number of offspring. Important as the struggle for existence has been and even still is, yet as far as the highest part of man's nature is concerned there are other agencies more important. For the moral qualities are advanced, either directly or indirectly, much more through the effects of habit, the reasoning powers, instruction, religion, &c, than through natural selection; though to this latter agency may be safely attributed the social instincts, which afforded the basis for the development of the moral sense. (*Descent* 1874, p. 618)

I have quoted at length these passages from Huxley and Darwin to show the inseparability of so-called Darwinism from so-called Social Darwinism and, congruent with that, between science and ideology. Anyone wishing to separate the scientific from the social from the theological will have to contend with these passages in these men's work. And anyone wishing to confine Darwin's Social Darwinism to his post-*Origin* work will have to contend with Silvan Schweber's claim: "To the best of my knowledge the M and N notebooks contain the first presentation of an evolutionary view of society based on an evolutionary view of nature" (1977, p. 232).

Would-be separators of Darwin the biological scientist from Darwin the Social Darwinist would also be likely to stumble over passages from the E *Notebook*; the projected Chapter 6 of *Natural Selection* ("Theory Applied to the Races of Man"); the marginal annotations in Darwin's own books on the races of man; a letter to Lyell in 1859 that applied natural selection and the effects of inherited mental exercise as follows: "I look at this process as now going on with the races of man; the less intellectual races being exterminated" (*LL* 2: 211). These evidences of continuity, along with many more, have been set forth in John Greene's convincing essay on "Darwin as a Social Evolutionist" (1981a, pp. 95–127). This complements his earlier essay on "Biology and Social Theory in the 19th Century" (1981, pp. 60–94), and both invite us to broaden and deepen our views on the mutual constitutiveness of scientific and social thought.

Turning now to Social Darwinism per se, my first point is that there is no such clearly separable thing. There was, however, a movement that was concerned chiefly with the interpretation of evolutionary ideas in the social context. It was a Malthusianism buttressed by the law of the history of life. It was based on a conception of the imbalance between human instincts and needs on the one hand, and human industry and nature's bounty on the other. It was not always pessimistic, but it was never very pleasant. Moreover, it was almost always associated with concepts of social hierarchy and mobility via competition.

My own conception of Social Darwinism is that it was an attitude toward nature with common elements, usually including Malthusianism, a belief in the science of social laws, and a belief that nature decreed extreme inequalities that most thought would lead to progress. Social Darwinists usually invoked some version of the survival of the fittest, although there were differing views about what the fittest were fit for. For more on conceptions of Social Darwinism, we can look at some passages from Robert Bannister's interesting monograph, *Social Darwinism: Science and Myth in Anglo-American Social Thought.* One definition presents Social Darwinism as "the type of theory that attempts to describe and explain social phenomena chiefly in terms of competition and conflict, especially the competition of group with group and the equilibrium and adjustment that ensues upon such struggles" (Bannister 1979, p. 4). Another described it as "the name loosely given to the application to society of the doctrine of the struggle for existence and survival of the fittest" (p. 5). Another definition said that it's "the more general adaptation of Darwinian, and related biological concepts to social ideologies" (p. 5). A last example was: "a ruthless form of laissez-faire that it has become fashionable to call 'Social Darwinism' " (p. 6).

Moving away from definitions to the question of how broadly this attitude towards nature was represented, it's important to remember that it was very widespread and not confined to post-Darwinian writings. For example, Lyell wrote that "In the universal struggle for existence, the right of the strongest eventually prevails" (Young 1969, p. 129). The concept of struggle is very common in the *Principles* (see Young 1969, p. 129, n. 76).

In a way that is echoed in the last passage from the *Descent,* Malthus himself said: "Had population and food increased in the same ratio, it is probable that man might never have emerged from the savage state" (Malthus 1798, p. 364). Malthus remarked:

> The first great awakeners of the mind seem to be the wants of the body. They are the first stimulants that rouse the brain of infant man into sentient activity, and such seems to be the sluggishness of original matter that unless, by a peculiar course of excitements, other wants, equally powerful, are generated, these stimulants seem, even afterwards, to be necessary, to continue that activity which they first awakened. The savage would slumber for ever under his tree unless he were roused from his torpor by the cravings of hunger, or the pinchings of cold; and the exertions that he makes to avoid these evils, by procuring food, and building himself a covering, are the exercises which form and keep in motion his faculties, which otherwise would sink into listless inactivity. From all that experience has taught us concerning the structure of the human mind, if those stimulants to exertion, which arise from the wants of the body, were removed from the mass of mankind, we have much more reason to think, that they would be sunk to the level of brutes, from a deficiency of excitements, than that they would be raised to the rank of philosophers by the possession of leisure. In those countries, where nature is the most redundant in

> spontaneous produce, the inhabitants will not be found the most remarkable for acuteness of intellect. Necessity has been with great truth called the mother of invention. (pp. 356–358)

Of course, Social Darwinism was, one might say, a broad church. A very optimistic interpretation was put on it in the writings of Darwin; the same was true of Spencer's rendering of it, just as was the use to which it was put in Social Darwinist ideas of the American robber barons. There is perpetually an undertone, however, as there was in Malthus, another note, a sense of pessimism. In Malthus's case it was a pessimism that allowed scope for striving and moral restraint. But there were worse forms of pessimism. Henry George — one of the passionate followers of Spencer — became disillusioned and played in turn an important role in inspiring the social ideas of Wallace (see Young 1968b). George once recalled a conversation between himself and Spencer's great American publicist, E. L. Youmans, concerning the state of American society. "What do you propose to do about it?" George had asked. To this Youmans responded "with something like a sigh": "Nothing! You and I can do nothing at all. It's all a matter of evolution. We can only wait for evolution. Perhaps in four or five thousand years evolution may have carried men beyond this state of affairs. But we can do nothing" (quoted in Bannister 1979, p. 75).

The doctrine of this broad church, then, conveyed both optimism and pessimism — both a concept of progress and a fatalism about its parameters and its pace. But more important of all, it rooted social ideas in biological ideas. The point I'm making is that biological ideas have to be seen as constituted by, evoked by, and following an agenda set by, larger social forces that determine the tempo, the mode, the mood, and the meaning of nature. In particular, evolutionary concepts of society changed quite fundamentally in the three decades from 1880 to 1910, from those of Social Darwinism to those of functionalism. The change coincides with the shift in the epochs of capitalism from that of primitive accumulation to that of managerialism; from a conception of the frontier and of moving ever onward, exploiting new areas of nature, to a conception of ordering and managing the space in society that is already occupied; from a doctrine of pure competition to one of competition within meritocracy, or consensus; from brawling to achievement; from survival to careers; from omnivorous trusts à la Rockefeller's Standard Oil to philanthropic trusts à la the Rockefeller Foundation; from conquest to organization. And of course some of the most vehement defenders of rampant capitalism set up some of the most philanthropic foundations — not just Rockefeller, but also the steel magnate Andrew Carnegie. In the succeeding epoch, the great defender of managerial capitalism was the patron of yet another foundation, Henry Ford. The Ford Foundation was concerned with the next generation of philanthropy in the social sciences.

Although it was fed by many streams, the managerial order of capitalism and its theoretical representation, functionalism, used the rhetoric of evolutionary biology. Concepts of structure, function, adaptation, evolution, were fundamental to the whole conceptual vocabulary of both nature and society for two, and probably three, generations.

This framework of ideas, about which I've written in detail elsewhere, became pervasive throughout the human and social sciences and, in particular, in psychology, sociology, and anthropology (Young 1981). The most succinct summary of the assumptions of functionalism, the one in which the density of biologistic concepts is the greatest, is the address "On the Concept of Function in the Social Sciences", which A. R. Radcliffe-Brown gave to the American Anthropological Association in 1935, where he began:

> The concept of function applied to human societies is based on an analogy between social life and organic life. The recognition of the analogy and of some of its implications is not new. In the 19th century the analogy, the concept of function, and the word itself appear frequently in social philosophy and sociology. (Radcliffe-Brown 1935, p. 178)

He goes on to develop the analogy between social life and organic life, and to dwell on the concept of function. He says,

> As the word 'function' is here being used the life of an organism is conceived as the *functioning* of its structure. It is through and by the continuity of the functioning that the continuity of the structure is preserved. If we consider any recurrent part of the life-process, such as respiration, digestion, etc., its *function* is the part it plays in, the contribution it makes to, the life of the organism as a whole. (1935, p. 179)

By analogy, the function of a particular social usage

> is the contribution it makes to the total social life as the functioning of the total social system. Such a view implies that a social system (the total social structure of a society together with the totality of social usages in which that structure appears and on which it depends for its continued existence) has a certain kind of unity, which we may speak of as a functional unity. We may define it as a condition in which all parts of the social system work together with a sufficient degree of harmony or internal consistency, i.e. without producing persistent conflicts which can neither be resolved nor regulated. (1935, p. 181)

Radcliffe-Brown acknowledges that the idea of the functional unity of the social system is hypothesis (p. 181). In the functionalist tradition, however, this hypothesis has become the model according to which societies are viewed. It is a model that therefore systematically places second and treats secondarily concepts of conflict, and a model within which concepts of contradiction simply do not arise. In particular, irreconcilable class conflicts are unthinkable, as is the notion of a mode of production as a contradictory unity of forces and relations of production (Clarke 1977). Put another way, the language of functionalism has a profound ideological role to play in the way that social theories operate as lenses through which to view societies. Using concepts like morphology, organic unity, physiology, structure, sets of relations, continuity, adaptation, etc., the functionalist tradition leads us to think in certain ways and systematically diverts our gaze from other directions (Russett 1966).

The applications of this model are not confined to the sociology of "primitive" peoples but are orthodox in the sociology of our own society, and extend to the ruling conceptual framework for the social interpretation of science itself. For, as Barbara Heyl showed in a brilliant paper on "The Harvard 'Pareto Circle' " (1968), it was within a certain social group centered around Harvard in the 1930s and 1940s that models were taken up from physiology, and particularly the circulatory physiology of L. J. Henderson. These models were applied to society, and extrapolated from the society to units within society, including the sociology of science itself. Robert Merton, the doyen of the sociology of science, was a member of this self-same circle, which has provided the reigning model of the interpretation of science within society (Young 1982a). Looking more broadly, it is a model that has not been confined to the social sciences, but has been applied to epistemology itself, in the work of Karl Popper, Stephen Toulmin, and David Hull. The provocative epigram "Darwinism *is* Social" is meant to evoke the role of biological and organic analogies, which move freely from biology to society to the theory of knowledge itself, and lead us to think in certain ways about the most abstract levels of nature and society, from the thermodynamics of particles to systems theory, itself a metafunctionalism (Emery 1969; Beishon and Peters 1972; Haraway 1981–1982).

In attempting to understand the ways in which these conceptions operate in the society itself, it's important to realize that in general culture, and in upper-middle-brow culture, the sharp distinctions we might choose to make between pure science, applied science, extrapolation, ideology, and popularization, simply do not apply. Examine the illustration from the *Sunday Times* magazine of 24 July 1977. We have here a gestalt: a picture of the double helix of DNA and portraits with potted biographies of Charles Darwin, Konrad Lorenz, Robert Ardrey, Edward Wilson, and — very beautifully portrayed — Richard Dawkins. People see the chemistry of life and *On the Origin of Species, On Aggression, The Social Contract, Sociobiology,* and *The Selfish Gene* as of a piece.

It's worth adding that in the same issue of the magazine there are pictures of women dressed up as leopards in very provocative poses. That is, concepts of biologism and animality are not only intermingled in the gestalt of the illustration, but are also adjacent to the gestalt of feline conceptions of woman, femininity, and sexuality.

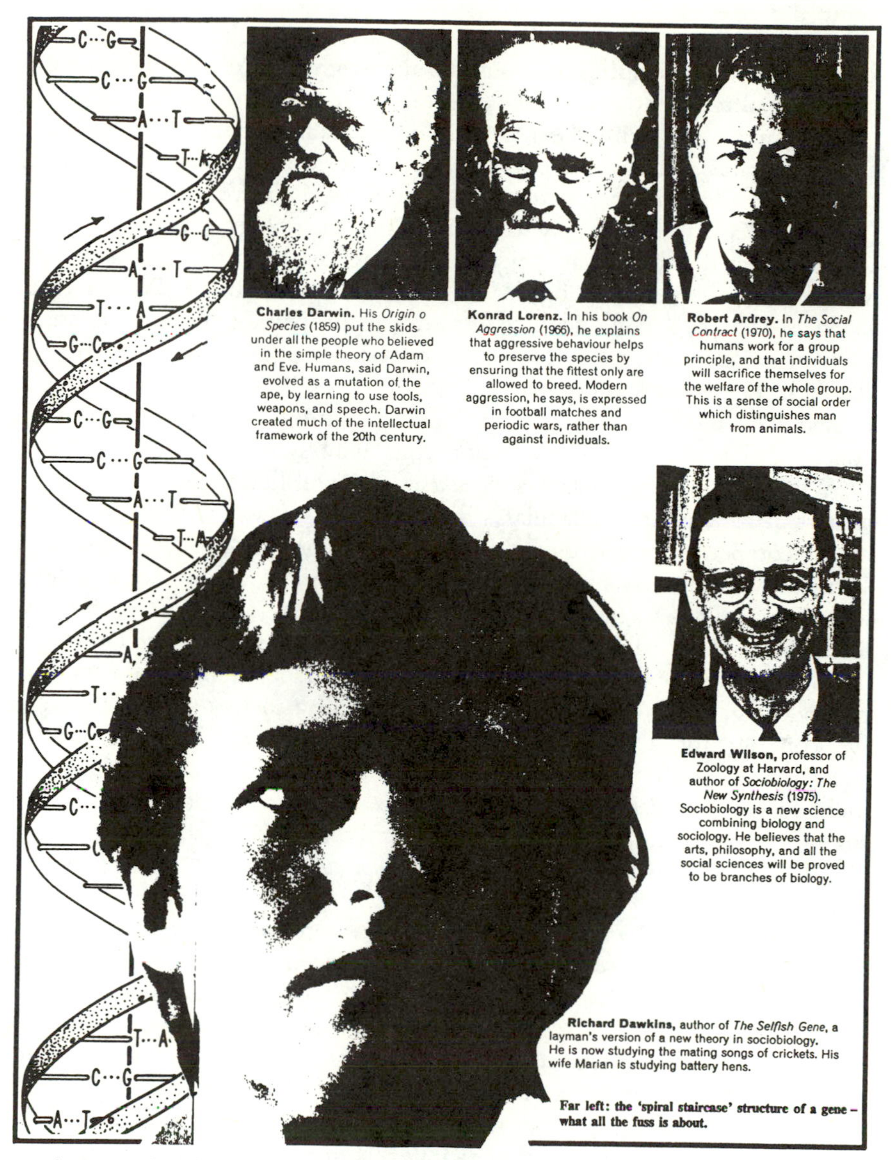

Sunday Times (London, 1977): Where is the line between science and ideology?

The different elements of that gestalt are really those of a right-wing liberal consensus. In the present, of course, that is under attack. Darwinism and forms represented by ethology, sociobiology, behavioral genetics, are seen as an appropriate target for people who are in opposition to the liberal consensus and feel that Keynesianism, the United Nations, Trilateralism, meritocracy, and expertocracy are undermining traditional values and threatening the moral fabric of society. (Spare a thought for the poor ole liberal-capitalist-scientific consensus: the ultra-right attacks it for being liberal, while the ultra-left attacks it for being capitalist.)

I do not agree with the Moral Majority/radical right about abortion and the nuclear family, for example, but I do see their point in wanting to maintain a basis for values that is not caught up in instrumental rationality (Young 1982b). I also agree with them that sociobiology is pernicious, and utterly reject the thesis that ethics should be given for a time to the people who will — here's a neologism — "biologicize" it (E. O. Wilson 1975, p. 562). That is, I don't look to genetics, neurophysiology and the study of ants, troops and prides for my conception of society, nor do I accept the thesis that biology is destiny.

And here's where we come up against quite a profound truth about conceptions in science. If you work your way systematically through E. O. Wilson's *Sociobiology: The New Synthesis,* you will find the following terms as part of his working vocabulary: division of labor (sexual and task), hierarchy, competitiveness, domination and submission, peck order, aggression, selfishness, altruism, rank, caste, role, worker, slave, soldier, queen, host, harem, promiscuous, mob, combat, spite, bachelor, jealousy, territoriality, leadership, indoctrinability, élites. If we look a bit wider to Richard Dawkins's *The Selfish Gene*, we find cheat, sucker, grudger; wider still, we find nepotism, philandering, rape. What possible source except a society such as ours can we consider for a conceptual vocabulary such as that? What possible significance except the scientific underpinning of a competitive, fatalistic, individualistic, élitist, patriarchal, sexist society can be attached to the following titles that have appeared recently around these questions: *On Aggression, The Naked Ape, The Territorial Imperative, The Imperial Animal, The Dominant Man, The Inevitability of Patriarchy, The Biological Imperative,* and, once again, *The Selfish Gene?* Only two of those were written by people who were not professional, academic biological or social scientists. And of course we have *On Human Nature* itself, the Pulitzer Prize-winning work of E. O. Wilson.

These texts provide the current analogy to the nineteenth-century debate, more evidence that the relationship between so-called purely biological and so-called purely ideological ideas, books, and concepts is one that can't be sorted out at all easily. We find that levels and concepts intermingle and that it is from society that we derive our conceptions of nature. These conceptions are in turn inextricably intermingled with our conceptions of human nature. It is, after all, the Darwinian theory of evolution by natural selection that made conceptions of nature, living nature, and human nature part of a single framework of ideas. It is also in the age of Darwinism that we live in our attempts to formulate a single science.

What is problematic about that attempted formulation is that our conception of living nature may be so paltry as to give us a pessimistic and fatalistic notion of humanity. Some people say that modern science's concepts of matter are not rich enough to give us mind (Young 1967a, 1967b). It can also be argued that conceptions of animality are not rich enough to give us humanity in the required sense. By that I mean a conception of humanity that envisages a society worth living in and a world worth changing.

Here are some examples of recent Social Darwinism, which are so pessimistic that they leave us with conceptions of humanity not worth bothering about. The first is from the work of the Nobel Laureate in ethology, Konrad Lorenz, who wrote during the Nazi era:

> This high valuation of our species — specific and innate social behavior patterns — is of the greatest biological importance. In it as in nothing else lies directly the backbone of all racial health and power. Nothing is so important for the health of a whole Volk as the elimination of 'invirent types': those which, in the most dangerous, virulent increase, like the cells of a malignant tumor, threaten to penetrate the body of a Volk. This justified high valuation, one of our most important hereditary treasures, must however not hinder us from recognizing and admitting its direct relation with Nature. It must above all not hinder us from descending to investigate our fellow creatures, which are easier and simpler to understand, in order to discover facts which strengthen the basis for the care of our holiest racial, volkish, and human hereditary values. (quoted in Kalikow 1978, pp. 174–175)

Lorenz felt that if it turned out that in conditions of civilization where natural selection was inoperative, there was an increase in mutants leading to "imbalance of the race, then race-care must consider an even more stringent elimination of the ethically less valuable than is done today, because it would, in this case, literally have to replace all selection factors that operate in the natural environment" (quoted in Kalikow 1978, p. 176).

These avowedly fascistic views are, of course, the extreme. But, they are by no means extinct. Sociobiology is used to argue for analogously Social Darwinist and racist views in the present. The National Front magazine *Spearhead* draws routinely on biological reductionism, for example, in the article "Sociobiology: the Instincts in our Genes" by Richard Verrall, who points out that "Genetically determined instinctual behaviour lies at the root of social organisation and even ethical and altruistic impulses" (1979, p.10). He goes on to review recent sociobiological and biologistic literature and draws the expected racist conclusions.

More individualistic forms of Social Darwinism are not hard to find in the mass media. Consider the most gripping scene in *The Third Man,* when Harry Lyme meets his friend Harley on the top of post-World War II Vienna's ferris wheel. When Harley confronts his old friend with the consequences of his selling diluted penicillin — horribly brain-damaged children — the conversation continues as follows:

> Harley: Have you ever seen any of your victims?
> Harry: (moving as if to push his friend out) You know I don't feel comfortable on these sorts of things. Victims? Don't be melodramatic. Look down there [at pedestrians far below]. Would you really feel any pity if one of those dots stopped moving forever? If I offered you £20,000 for every dot that stopped, would you really, old man, tell me to keep my money, or would you calculate how many dots you could afford to spare? Free of income tax. Free of income tax: it's the only way you can save money nowadays.

After an interchange in which it becomes apparent that killing Harley won't eliminate the only evidence against him, Harry chuckles and says that he still believes in God and mercy, but that the dead are better off dead since there's not much to miss on earth. As he walks away, he says,

> Don't be so gloomy. After all, it's not that awful. You know what the fellow says. In Italy for 30 years under the Borgias they had warfare, terror, murders, bloodshed, but they produced Michelangelo, Leonardo da Vinci, and the Renaissance. In Switzerland they had brotherly love; they had 500 years of democracy and peace. And what did that produce? The cuckoo clock.

As to corporate Social Darwinism, we have the following example from an editorial in *Computer Weekly* commenting on the dominant role of IBM in the industry:

> The problem of trying to regulate IBM is that what is good for IBM is in general good for the user, and as the company becomes more innovative and more competitive that becomes increasingly true. IBM is an inevitable product of the capitalist system in which survival of the fittest must always tend toward monopoly. (Anon. 1980)

These examples — drawn from ethology, fascism, film, and business — might be dismissed as on the wrong side of the science/ideology divide. Then try this:

> The evolution of society fits the Darwinian paradigm in its most individualistic form. Nothing in it cries out to be otherwise explained. The economy of nature is competitive from the beginning to end. Understand that economy, and how it works, and the underlying reasons for social phenomena are manifest. They are the means by which one organism gains some advantage to the detriment of another. No hint of genuine charity ameliorates our vision of society, once sentimentalism has been laid aside. What passes for cooperation turns out to be a mixture of opportunism and exploitation. The impulses that lead one animal to

> sacrifice himself for another turn out to have their ultimate rationale in gaining advantage over a third; and acts 'for the good' of one society turn out to be performed to the detriment of the rest. Where it is in his own interest, every organism may reasonably be expected to aid his fellows. Where he has no alternative, he submits to the yoke of communal servitude. Yet given a full chance to act in his own interest, nothing but expediency will restrain him from brutalizing, from maiming, from murdering — his brother, his mate, his parent, or his child. Scratch an 'altruist,' and watch a 'hypocrite' bleed. (Ghiselin 1974a, p. 247)

The author of that charming integration of biological and social thought is a distinguished academic biologist and holds a MacArthur Fellowship, one of America's most prestigious research awards. It would be wrong to claim that this is not a minority, even eccentric, view among scientists. But holders of it and of closely related views — for example, the "lifeboat" theory of scarcity, Friedmanite economics, and the biological, social, and historical synthesis of C. D. Darlington and that of Sir Hans Krebs (cf. Young 1973c, esp. pp. 247–249) — are all well within the scientific and social scientific cultures of eminent professors at the Universities of California, Chicago, and Oxford as well as Fellows of the Royal Society and Nobel Laureates (see Kalikow 1978; Hirshleifer 1977; and Anon. 1978 for further extremes of conservative biologism).

The point of this portion of the argument is to reject Social Darwinism in an extremely important sense, while embracing it in another. Having rejected crude Social Darwinist extrapolations from other animals to humankind, it's therefore legitimate to ask: What next? Do we turn altogether away from extrapolations? Or do we choose others?

The first option, it seems to me, isn't open, since ideas of nature and humanity, as I've said repeatedly, are mutually constitutive. There is nature apart from human values, priorities, and perceptions, to be sure. But as far as we know it — as far as we characterize it, have research programs, put questions to nature, and have criteria of acceptable answers — we do so in inescapably anthropocentric and anthropomorphic terms (Young 1973c, 1982c). So my rejection of Social Darwinists' characterization of nature is not in the service of avoiding illegitimate extrapolations in favor of nature "as it is". Nature is as we characterize it, and extrapolations are as inescapable as the humanocentric relationship with nature in the first place.

The issue is how we characterize and work out the humanity and/as nature — that is, the humanity *and* nature, and the humanity *as* nature — relationship. I want to treat it as a transformative process of human labor. This is as true of knowledge as it is of any other form of human industry.

If we take that point seriously, we have to take seriously that, as Marx put it,

> *Industry* is the *actual*, historical relation of nature, and therefore of natural

> science, to man. If, therefore, industry is conceived as the exoteric revelation of man's *essential powers,* we also gain an understanding of the *human* essence of nature, or the *natural* essence of man. In consequence, natural science will lose its abstractly material — or rather, its idealistic — tendency, and will become the basis of *human* science, as it has already become the basis of actual human life, albeit in an estranged form. *One* basis for life and another basis for *science,* is *a priori* a lie. The nature which comes to be in human history — the genesis of human society — is man's *real* nature. Hence nature as it comes to be through industry, even though in an *estranged* form, is true *anthropological* nature. (Marx 1961, pp. 110–111)

The point of that quotation is to reinforce once again that nature, knowledge, and human industry are part of a single mode of relating, conceiving, doing.

If we try to look at Darwinism and Darwinism-as-social in this way, the basis for humanity is not biology, genes, instincts, the givenness of our species in an evolutionary sense: not body, not mind, but the concept of person, and that concept is ontologically primitive. There's a parallel ontologically primitive concept that promises to resolve the nature/culture dualism: labor. Labor is neither nature nor culture, but their matrix.

It is at this point that my historiographic argument about how we should think about Darwin, Darwinism, and the debate about the place of humanity in nature — as the nineteenth century called it, the debate on "man's place in nature" — has to be recontextualized and connected up with the points I've been making here about the concepts of industry, the concept of a person, and of labor. Historiography has to be reintegrated into a new conception of what we mean by humanity, a conception that is not based on nature/culture, body/mind, animal/human dualisms.

This would give us a notion of humanity — and secondarily of biology — that is not fatalistic, pessimistic, reifying, and scientistic. I would like to think that it is a progressive and optimistic historiography, one without blinkers, as opposed to the historiography of much of what I've come to think of as the Darwin industry, which is very much a historiography whose distinctions and whose narrowness of perspective makes it a historiography of the status quo.

I'd like to drive this point home by recalling a letter that Engels wrote to the sociologist P. L. Lavrov in 1875, because I think it captures quite a lot of the points I've been making. It came as something of a surprise to me after I'd completed the argument. I thought I would include a bit from this letter and found that it really said, for all my reservations about some of Engels's ideas, quite a lot of what I've been trying to say here. The first passage will be very familiar, but I shall go on to three others that I think are quite helpful. Engels says,

> The whole Darwinist teaching of the struggle for existence is simply

> a transference from society to living nature, of Hobbes' doctrine *bellum omnium contra omnes* [that is, the war of all against all] and of the bourgeois-economic doctrine of competition, together with Malthus's theory of population. When this conjuror's trick has been performed (and I question its absolute permissibility, as I've indicated . . . particularly as far as the Malthusian theory is concerned), the same theories are transferred back again from organic nature into history and it is now claimed that their validity as eternal laws of human society has been proved. The puerility of this procedure is so obvious that not a word need be said about it. (Marx and Engels 1965, p. 302)

But of course he does go on. The key to the above is, of course, the sentence, "When this conjuror's trick has been performed, the same theories are transferred back from nature to history and claimed as eternal laws of society", which I think is a fair summary of a great deal of what goes on in ethology, sociobiology, and the human sciences, vis-à-vis their relationship with biology, in particular, genetics, behavioral genetics, and the study of the nervous system, Engels says,

> When therefore a self-styled natural scientist takes the liberty of reducing the whole of historical development with all its wealth and variety to the one-sided and meager phrase 'struggle for existence', a phrase which even in the sphere of nature can be accepted only *cum grano salis,* such a procedure really contains its own condemnation. (Marx and Engels 1965, p. 302)

Once again, a fair summary of much of what I've been saying. But here's the point that moves us on the relationship between animal and human. He says,

> The essential difference between human and animal society consists in the fact that animals at most *collect,* while men *produce.* This sole but cardinal difference alone makes it impossible simply to transfer laws of animal societies to human societies. It makes it possible, as you properly remark, 'for man to struggle not only for existence but also for pleasures and *for the increase of his pleasures* — to be ready to renounce his lower pleasures for the highest pleasure.' Without disputing your further conclusions from this I would, proceeding from my premises, make the following inferences: At a certain stage the production of man thus attains such a high level that not only necessaries but also luxuries, at first, true enough, only for a minority, are produced. The struggle for existence — if we permit this category for the moment to be valid — is thus transformed into a struggle for pleasures, no longer for mere means of *subsistence,* but for means of *development, socially produced* means of development, and to this stage the categories derived from the animal kingdom are no longer applicable. (p. 303)

And then later he says: "The struggle for existence can then consist only in this: that the producing class takes over the management of production and distribution from the class that was hitherto entrusted with it but has now become incompetent to handle it, and there you have the socialist revolution" (p. 303).

Now what Engels did in this letter was make a critique with which I absolutely agree, about the interplay between social conceptions, their biologization and then extrapolation back to being laws of nature. He then pointed out the limitations of that interplay when applied to humanity. Instead, at a certain point the conceptual and historical framework moves away from mere animality to production, which entails human industry and the concept of labor. The concept of labor is not one which we find inside biology in its narrow, Darwinian sense. But we do find social concepts at every level inside the Darwinian theory. In that sense Darwinism is social.

What I have attempted to convey in this essay is the need for a greater sense of different scope and, more importantly, different levels, of the mutual constitutiveness of conceptions of nature, science, and society, including the deity. We're not forced to choose between Darwin's falsely conscious claims to have nothing to do with politics, on the one hand, and scholars' fears, on the other, that to invoke politics in the broader sense is to pollute natural science. The "pollution" is inherent in the labor of knowing.

The historiographic and political issue is how societies constitute their knowledge, and why they conceive of that process as they do. And that's as important a question about the Darwinian debate as it is about our own time. It entails a historiography of then and a historiography of now and their relationship to whether we're trying to keep knowledge and society as they are or put them in the service of a better world. A better world would be one in which the struggle for existence doesn't have the resonances that it had in the period of Social Darwinism and that it now has in the renewed period of *laissez-faire*, Friedmanism, Thatcherism, and Reaganism.

Historiographic Afterthoughts on the Science of History

Just as I have argued that the history of science must find its place within history, it is important to realize that the history of science, as an academic discipline that reflects on the history of science per se, must also find its place in history. That is, this volume and the historians writing in it are doing so within a cultural, politico-economic and ideological framework. Putting it yet another way, it can be argued that just as Darwinism is social, so is Darwinian scholarship.

I should like to illustrate this point of view with some reflections on the conference at the Villa di Mondeggi near Florence, where the papers for this volume were presented and discussed. I came to the conference in the capacity of a Rip Van Winkle. In the period between 1968 and 1973 I had written a dozen or so papers and a monograph containing a series of hypotheses about the role of certain factors in the nineteenth-century debate on man's place in nature, in particular, natural theology, Malthusianism, political economy, the concept of progress, and the relations between the history of the earth, biology, psychology, neurology, and wider political and economic issues in the period, around the general theme of "The Naturalization of Value Systems in the Human Sciences". This research, and historical and personal issues in the period around 1968, led me to move away from Darwinian scholarship. Indeed, my subject was never Darwin. It was always the nineteenth-century debate on man's place in nature, and my effort was always to combine a detailed reading of texts with a wider perspective on the issues. However, the definition of "texts" is itself a contentious issue in Darwinian scholarship, since my texts have been primarily published ones; in particular, the debate in the Victorian periodicals, which provided the intellectual milieu for the debate on man's place in nature, into which the major papers and monographs were received and which, in turn, provided the context for their own views.

Returning to these questions after a decade in order to prepare a paper for this volume and to collect my essays, I felt strongly that there has been a restriction of framework and perspective. In particular, the general issue of the nineteenth-century debate on man's place in nature seems to have fallen off the agenda. The zeal with which current scientists-historians seek to separate Darwin's genius and achievements from the work, ideas, and influences of Spencer, Chambers, and Wallace seems to be to betray a pathetic, sycophantic hagiography — Great Man history — which I had thought was waning in the history of science, as historians of science thought of their discipline in terms of the history of ideas, the history of culture, and the history of society. Indeed, one distinguished biologist-historian concluded his comments by saying that Darwin was the author of "the greatest and most universal revolution ever experienced in the history of human thought". I found myself asking, why do we defer to great men? Why do we defer to working scientists who are part-time historians? Why do we defer to great men in the history of science? Why do we not consider the social processes of scientific change in their broadest contexts? Where have these questions gone in the past decade?

In place of these issues, we find that scholars are looking deeper and deeper and in greater and greater detail into the minutiae of Darwin's notes and thought processes. What is it that we wish to find there? Is it the key to genius? Why is a higher and higher power microscope applied to rethinking the thoughts of the "great"?

Turning to a more particular issue, I would argue that the influence of Malthus or of other social and political issues on Darwin's thinking is not an empirical question in the sense that, for example, De Beer, Mayr, Schweber, or Greene pose it. The dichotomy between so-called internal and so-called external factors (which I have been advocating transcending since 1969) neglects two points. First, history — including intellectual history — is overdetermined. Of course we can find sufficient factors to explain the origin and development of Darwin's theory of evolution by natural selection without appealing to Malthus. Indeed, my own reading of what Darwin "got" from Malthus is remarkably close to interpretations of so-called internalists.

As I re-examine my own claims and compare them to those of Mayr and Schweber, I am at a loss to understand the difference between us. I wrote,

> It appears, then, that it was the removal of Malthus's idea of 'moral restraint', and an emphasis on the concept of 'population pressure' which left a natural law about plants and animals, that characterized Darwin's interpretation. He was, in effect, reverting to the purity of the inescapable dilemma of Malthus's first edition. It is 'the strong law of necessity'; which Malthus emphasizes repeatedly in both editions, even though in the second it lies side by side with the partial palliative of 'moral restraint'. References with this deterministic basis appear in tens of places in both editions and might themselves have influenced Darwin's application of the principle to man . . . (Young 1969, p. 129)

I go on to point out that Lyell's *Principles of Geology* was the work that most influenced Darwin and that there are innumerable references there to the struggle for existence.

I have no quarrel with Mayr's claim that "the role of Malthus was very much that of a crystal tossed into a saturated fluid" (1982b, p. 493), nor with Schweber, who says, "It seems to me that the Notebooks support the view that Darwin was struck with the numerical and deterministic aspects of the Malthusian statement" (1977, p. 296). Schweber also says,

> How much we attribute to the Malthusian insight is to a certain extent a reflection of our proclivities. My own reading is that the Malthusian statement gave Darwin the *quantitative* element he needed to make the theory meet the standards of theories in the natural sciences. (p. 303)

I find it ironic that the work of David Kohn, which Mayr acidly contrasts with my own, concludes, "The work of one recent commentator, Robert M. Young, stands out as nearly definitive" (Mayr 1982, p. 492; Kohn 1980, p. 142). Kohn proceeds to characterize the relationship in terms with which I wholly agree (pp. 142 sqq.). This agreement relates to my opposition to attempts to demarcate Darwin's thinking sharply from ideological connections with his age. De Beer and Schweber are also at pains to stress that "internal factors" are sufficient to account for Darwin's concept of natural selection. In varying degrees, they are keen to separate Darwin's originality and thinking from the age — dramatically so in the cases of De Beer and Mayr, less so in that of Schweber, and not at all in the case of Kohn. The wider and deeper claim, which some are rejecting, is that the history of science is part of history; that science is part of culture, not above it, or an alternative to it; that science is the embodiment of the values of the epoch. It is ever so strange. Scholarship about ancient Greece and Rome, Islamic science, the Middle Ages, the Renaissance, the seventeenth and even the eighteenth centuries — all take as a premise the thesis that the intellectual life of the age, including and especially its conceptions of rationality and of science, were part and parcel of its social and economic structure and value system. The systems of knowledge are part of the culture of the age. Yet at the same time there are scholars who know this, but put on blinkers and burrow deeper and deeper into the minutiae of papers, early drafts, unpublished manuscripts, correspondence and minor works as if these were not occurring in an age and in a context whose role is not contextual but constitutive.

In Darwin studies a trend has become dominant that would be welcome if it were not becoming a near orthodoxy. It has been made possible by the publication of more and more notebooks and early works, especially the M and N *Notebooks, Natural Selection*, and other manuscript material. Some, but by no means all of its practitioners are people who did graduate work in biology. They are doing important and interesting research in reconstituting the processes of intellectual development of Darwin and his colleagues. But this work does not interdigitate or articulate with the wider issues in the nineteenth century.

It appears to me that we are, in the late 1970s and 1980s, in a period in which scholars are attempting to cultivate their own gardens as a withdrawal from the social activism of the preceding period. They look at the past in the same terms as they approach the present. One aspect of this orthodoxy was expressed in some waspish remarks at the Florence conference. Reference was made to "primary Whiggism", in which it is claimed that the past leads to the present without any space for the contemporary context of issues, without any consideration of the "losers in history". The concept of "secondary Whiggism" was mentioned and also criticized. It is the belief that a scientist's immature views lead only to his or her mature views. In both cases the retrospect wipes out the integrity and the texture of the prospect. Whiggism also implies that people don't hold clear views until they hold the views we remember them for. Secondary Whiggism, on the other hand, has a tendency to underemphasize people's mature work and can succumb to the temptation of disappearing without trace into the minutiae of someone's "immature" thought processes.

We can go on with this sort of thinking and produce a notion of "tertiary

Whiggism", which ignores other figures in the period and our hero's real situation vis-à-vis fame and fortune. A tertiary Whig could leave out the eminence of a Buckland in the geology of the 1830s and could fail to take seriously the Bridgewater Treatises. Carrying on, a "quarternary Whiggism" could privilege topics and issues we consider important and ignore, for example, the role of phrenology in the debate on man's place in nature.

If we look for a way forward in these matters, it would appear important that people define and defend their reference group. Mine has always been Victorian culture and the debate on man's place in nature: man, God, nature, society. The relevant reference group is the debate in the Victorian periodicals. I suspect that the relevant reference group for some current Darwin scholars is either the peer group of the scientific community at the time or the peer group of present knowledge.

With Whiggism goes positivism. Primary positivism treats facts as decontextualized from their matrix of meanings and values. Secondary positivism does not take our hero and his theory out of the context of the scientific community of his time. No, these are meticulously considered, as are all nuances and contemporary meanings of theories and concepts, no matter how they have been treated by subsequent history. But, the secondary positivist draws a sharp boundary around the professional community of contemporary scientists. The secondary positivist also treats all connections as contextual and ignores immanent, structural or epochal causality. Therefore, for example, if there is sufficient evidence in the texts to explain an influence, no consideration is given to the possibility that other ambient forces might be at work in the intellectual formation of a scientist.

Darwin and Darwinism are important because humanity is part of the history of life at the same time that human history is an open prospect. Or is it? Does biology set the limits to destiny? Shall we await the verdicts of the biologists — even the Darwin scholars — to set and pursue our social, cultural and political goals? Marx once said, "We know only a single science: the science of history" (Marx and Engels 1968, p. 28n). I think the science of history was and should be much richer than the history of science seems to be making it.

When I say Darwinism *is* social, I mean it in two senses. First, in Darwin's own work there was never a clear separation of his biological research and thinking on the one hand, and its origins in and extrapolation to social evolution or Social Darwinism on the other. I don't find that conclusion very interesting, except as a stick with which to beat positivists and Whigs of the higher orders. Second, science is social. Of course we can disappear into the texts, but we must ask ourselves what counts as a text. These were people who read and contributed to Victorian periodicals and who lived in places that must be for us texts, for example, Shrewsbury, Edinburgh, Cambridge, the *Beagle*, London, Down. In the same way that a machine and Victorian Manchester are "texts" for the social and economic historian, these locations are texts for a Darwin scholar. These determinations are efficacious, and no amount of reading Darwin's reading lists and marginal annotations will get us exhaustively through the determinations of Darwin's thinking, however much we might welcome the interpretation of marginal notes done by, for example, Gillian Beer, Jim Moore, John Greene, and the mentor of us all, Sydney Smith.

Darwinism is social because science is. And of all science the theory that links humanity to the history of nature is likely to be most so. Those who wish to find sciences furthest from society should go to the haven of mathematics and physics, but alas, even there, there are polluters such as Hodgkin and Forman to show the social constitution of the issues in those esoteric disciplines.

Why not instead join up scholarly traditions and make contact with political, cultural, literary, and ideological studies of the period? In failing to do so the orthodoxies of the left and right meet. The scientific left celebrates science and tries to show that socialism is scientific. The right attempts to defend science and its autonomy in a way that guarantees that ruling ideas of the prevailing ruling class are scientific. The history of science, is of course, one battleground in this struggle. At the moment it appears to me that the right is winning hands down.

The connection between these two points is very important. It is because science is not above history that no clear separation can be made between Darwin's Darwinism and Darwin's Social Darwinism. That Darwin was a Social Darwinist is not news, however often it is conveniently forgotten. The point about that is a deeper one: the search for the neat, isolable influence or cleavage plane is a search for a will o' the wisp. It is a positivist search, and positivism was a historical movement in the nineteenth century just as physicalism in the philosophy of science was in the 1940s–1960s, with its search for a decontextualized neutral observation language. I fear that Darwin studies are lapsing into a positivism about the origins, originality, and unequivocalness of Darwin's theory.

I have no quarrel with people who wish to pursue the most detailed studies of Darwinian texts. I wish only to challenge their doing so in a way that fails to connect with other dimensions of the determination of scientific, intellectual, and cultural phenomena. It is important to point out which questions a given social formation wants — through its science — to pursue. This broader question extends from the most general features of its philosophy of nature and society to its most mundane facts. At the most general level a given socioeconomic order — a mode of production — constitutes and is constituted by a world view, which includes a framework of assumptions and methods about what is known, what is discoverable, what it wants to discover, and how to set about discovering it. At an intermediate level certain sorts of issues preoccupy investigators at a given phase in the development of the mode of production, reflecting, in more or less mediated ways, the contradictions of that period. In the eighteenth century it was classification. In the mid-nineteenth century it was origins — the historicity of genesis of earth, life, mind, and society. In the late nineteenth and early twentieth centuries it was structures and functions in the psychological and social sciences with particular emphasis on stability, systems, and equilibria. In our own time it is mechanisms and abilities — the least elements and their recombination to suit specified needs.

These intellectual preoccupations are closely linked (in ways we need to explicate further) with the development of machinofacture, the division of labor, de-skilling, and the call for general ability in the society — abstract ability for abstract labor. Scientific research is seeking a secure foundation in our own epoch for gradations of ability, for élitism (usually at least formally meritocratic), for hierarchy, for a growing split between mental and manual labor, for dominance and patriarchy. It seeks to root these social relations in biological givens — to naturalize them. These preoccupations can be seen as our era's analogy to the nineteenth and early twentieth centuries' attempt to rebase its socioeconomic order on biological, evolutionary, and physiological equilibria rather than the deistic principle. Competitive individualism and functionalist views of the social order, cohesion and progress, were more consistent with an urban industrialism and mobility of labor than with the rural pastoral order that suited a deistic age of fixed, classified social stasis — the world of Paley. Looking at the issues and attempting to conduct the ideological battle on this terrain makes it completely unsurprising that investigators whose disciplines — however unself-consciously — favor the existing socioeconomic order will propose and defend certain inquiries, and that radicals and some liberals will not. This is not just to prove them wrong at the empirical or even the conceptual level but to say it is wrong to ask such questions in isolation and to pursue whole areas of inquiry in a blinkered way. Nor is it because one group is right and the other wrong, but because they have starkly conflicting visions of the social order that throw up starkly different issues for scientists and historians to pursue. The debate, therefore, becomes one between competing ideologies and interest groups. My own perception of it is that it is a conflict between those scholars concerned with the struggle for socialism and those concerned with the struggle for existence.

Part Three

Towards The Comparative Reception of Darwinism

22

SCIENTIFIC ATTITUDES TO DARWINISM IN BRITAIN AND AMERICA

Peter J. Bowler

Certain images spring immediately to mind whenever the scientific reaction to Darwinism is mentioned. For many, Thomas Henry Huxley's response to Bishop Samuel Wilberforce at the Oxford BAAS meeting in 1860 symbolizes the scientists' refusal to bow to outside pressure. Huxley's debate with Richard Owen over man's relationship to the apes illustrates the clash between the radical and conservative responses within science, as do the efforts of Asa Gray and William Barton Rogers to defend Darwinism against the attacks of Louis Agassiz in the United States. Many laymen no doubt assume that once the initial opposition was overcome, Darwinism soon rose to the dominant position it still occupies in modern biology. The historian, however, realizes that these simple images conceal the true complexity of both the scientific reaction to Darwinism and the continuing developments in evolution theory in the last hundred years.

Detailed accounts of the reaction to Darwinism can be found in the standard histories of evolution theory by writers such as Loren Eiseley (1958), John C. Greene (1959), Gertrude Himmelfarb (1959), and William Irvine (1955). Michael Ruse's more recent *Darwinian Revolution* (1979a) is valuable because of its careful exploration of the intellectual climate into which Darwinism was injected. Among the works devoted specifically to the reception of the theory, Alvar Ellegård's survey of the periodical press also throws light on some aspects of the scientific debate (1958). James R. Moore's recent discussion of the religious controversy sparked off by Darwinism helps to undermine the simplistic image of a conflict between science and theology (1979). On the details of the scientific response, David Hull's valuable collection of reviews of the *Origin* is prefaced with an account of the methodological arguments used against the theory (1973b). Other aspects of the opposition have been studied by Peter Vorzimmer (1970) and Joe Burchfield (1975). Most directly relevant to my purpose, though, is the volume edited by Thomas F. Glick (1974b), in which M. J. S. Hodge and Frederick H. Burkhardt covered the reception in Britain, Edward J. Pfeifer in the United States.

Can a single article do any more than survey this mass of existing literature? I hope it can, but in any case there is much to be gained by simply putting together a comprehensive account of the scientific arguments used for and against Darwinism in the late nineteenth century. Many of the existing discussions deliberately limit themselves to particular issues or time-periods, and it is probably fair to say that we still do not have a properly balanced analysis of anything but the earliest phase of the debate. Since we are now commemorating Darwin's death in 1882, this would seem an appropriate time to extend the analysis of the reception of his theory into the later part of the century. The debate over the adequacy of the selection theory was certainly not over in 1882 — indeed, it intensified during the following decades. The main purpose of this article is to go beyond the debate over the *Origin* and present a balanced account of the reception of Darwinism in the late nineteenth century. The first section will merely provide a brief survey of the scientific arguments, depending upon published accounts to supply the details. It will be as comprehensive as possible, pointing out the various levels of disagreement over Darwinism, and trying to show how some of the arguments changed in character as the debate moved into the later decades of the century. This survey will then provide the basis for an analysis of the general impact of Darwinism, and of the different responses in the various biological disciplines. Finally, I shall offer my own thoughts on the forces that shaped the changing perception of Darwinism in the course of the late nineteenth century.

I. For and Against Darwin

Some of the less flattering accounts of Darwin's efforts to promote his theory describe the opposing arguments so forcefully that one is left wondering why anyone took him seriously at the time (Eiseley 1958; Himmelfarb 1959; Vorzimmer 1970). Darwin is pictured as someone who built a basically unsound structure and then tried desperately to shore up one part after another threatened with imminent collapse. More positive accounts go to the opposite extreme, picturing Darwin as a hero of the scientific method, who created the foundation of modern evolutionism and presented the advantages of his theory so that it was able to transcend the limitations imposed on it by the common misconceptions of his time (De Beer 1963; Ghiselin 1969). Our final goal must be to establish just how successful Darwin and his followers were, but before we can do this, we must have some idea of his theory's strengths and weaknesses as judged from various scientific perspectives. A simple list of pros and cons will not allow us to understand what happened, since the really interesting questions center on which arguments appealed to which scientists, and why. But in a general survey such as this, it may be worthwhile beginning with a straightforward account of the arguments that were used, particularly if this can be done in a way that will reveal the various levels of debate. The mechanism of natural selection was built into a radically new way of looking at the organic world, so that "Darwinism" could be taken to mean either the detailed mechanism or the more basic principles of the new world view. If we are to understand the impact of Darwinism we must begin with the fundamentals, and then move on to the arguments for and against selection. At each level we shall then be able to see how the Darwinian approach challenged the traditional interpretation of nature, and how the supporters of the traditional view could both criticize the new theory and try to adapt their own ideas to serve as alternative explanations of organic development.

The most revolutionary aspect of Darwin's approach was its strict policy of scientific naturalism, that is, his determination to explain the development of life solely in terms of natural processes that can be seen operating in the world today.[1] A few younger naturalists, including T. H. Huxley and J. D. Hooker, were on the lookout for a natural explanation of the origin of species during the 1850s, and welcomed Darwin's theory as the first plausible solution to the problem. It may seem obvious that a policy of naturalism is essential if the question is to be opened up to scientific investigation, but at the time there were many who still believed that certain aspects of the creation of new forms did indeed lie outside the scope of science. This is not to say they accepted the miraculous creation of species in a purely Biblical sense, but as Neal C. Gillespie has argued (1979), there was a widespread conviction that — whatever the details of the process — certain aspects of it represented a direct expression of the Creator's will that could never be explained in terms of ordinary natural law. The classic "argument from design" was still accepted as part of the scientists' intellectual framework, allowing certain kinds of phenomena to be put on one side as expressions of a higher purpose. Biologists who thought in this way could never accept a purely naturalistic theory, least of all one based on the selection of random variations. The reactions of a host of conservative thinkers from Adam Sedgwick and Richard Owen to more sympathetic figures such as Charles Lyell and Asa Gray reveal the extent to which the question of design preyed upon their minds. To accept Darwinism was to accept that there was no way of *proving* a higher purpose in nature, although as Gray pointed out it did not prevent one from *believing* that the laws of nature were designed to achieve certain goals.[2]

How could those who wished to retain the argument from design respond to Darwin's challenge? They could, of course, simply express the general belief that no naturalistic theory would ever prove adequate to explain the development of life, reinforcing this by pointing out various detailed problems with Darwin's own mechanism of natural selection. On a slightly different tack, they could argue that Darwin's methodology was unsatisfactory, thereby excluding his theory from the ranks of those truly scientific studies of nature made in the Newtonian tradition. David Hull (1973b) has shown how this charge arose out of the Victorians' limited understanding of how science functions, and how it was deliberately exploited as a means of defending the traditional world view. By appealing to an inflated idea of the certainty achieved in the physical sciences, these opponents were in effect trying to ensure that *no* theory based on the quite different foundations needed to investigate the past would ever be accepted as scientific. Even some of Darwin's supporters fell into this trap, as when Huxley agreed with John Stuart Mill that the theory had not been "proved" (Huxley 1860b, p. 74; Mill 1874, p. 328; see Kottler, this volume). Darwin himself realized that a theory can never be "proved" except in the sense that it can be shown to explain a whole range of phenomena. Huxley accepted Darwinism as a working hypothesis because of his desire to open up the origin of species to scientific investigation, but the majority of Darwin's opponents proclaimed the inadequacy of his methodology in order to defend their own intuitive sense that species were real (that is, fixed) entities within a purposeful divine plan. Whether one labeled Darwinism as inadequate or unscientific, the end result was the same: one could go on to insist that natural processes would have to be boosted by a supernatural power to bring about the creation of new living forms. There was, however, an increasing tendency for the supporters of this view to concede that the supernatural agency might operate through the transmutation of existing forms — theistic evolution rather than pure creation.

By accepting that design worked through evolution rather than by creation *ex nihilo*, theistic evolutionists such as St. George Jackson Mivart hoped to accommodate the evidence suggesting a genetic relationship between the succession of forms in a particular area. Yet their concession exposed the essential weakness of a position based on the claim that no naturalistic theory would ever be devised to satisfy their requirements. Once it had been shown that some of the evidence could be explained in natural terms, it became more plausible to suppose that further refinements would produce a natural explanation of the whole process. Darwin pointed the way in this direction, and his most important success was in establishing the basic point that science could *hope* to explain the evolution of life, even if his own theory was not completely satisfactory. Theistic evolution was a popular alternative for a while, but by the end of the century scientists had accepted that it was no longer part of their job to appeal to the supernatural whenever they ran into difficulties. Many alternatives to natural selection were suggested in the late nineteenth century, but they were all natural mechanisms of evolution that left no room for an explicit element of design. If nothing else, Darwinism had established the biologist's right to assume that the whole of organic nature was open to his investigation.

At first there did seem to be one other way in which natural theology might be extended into an anti-Darwinian argument. What if there were actual phenomena observable in nature that could be explained only as a result of divine forethought? Here we must recognize that design had traditionally been interpreted in two quite different ways: it had been sought both in the wonderful adaptation of each organism's structure to the functions it must perform, and in the overall pattern of nature displayed by the relationships between different forms. I have called these the "utilitarian" and the "idealist" versions of the argument from design (Bowler 1977a). Natural selection was, of course, a mechanism of adaptation (a point I shall return to below), and it thus challenged directly the whole idea that utility could prove design. Almost all Darwin's opponents conceded that his theory had undermined this version of the argument. One might continue to believe that adaptation was a sign of divine benevolence, but one could no longer hope to prove it. The only exception was the Duke of Argyll, who argued that rudimentary organs were not the relics of once useful structures now diminishing in size, but new structures being prepared for future use (Argyll 1867, p. 213; 1888). Such an anticipation of future needs would indeed have been teleological in a way that was incompatible with Darwinian naturalism, but it was never taken seriously. Several of Darwin's opponents did, however, suggest that there were orderly patterns in the development of living things that could only be explained as the result of a divine purpose. Mivart pointed to the parallel development of structures such as the eye in various phyla (1871a, pp. 84–87), while William B. Carpenter saw regular sequences in the evolution of the Foraminifera (1888). Some later naturalists agreed that there were regularities in nature that could not have resulted from the selection of random variation, but it is typical of the growing naturalism of the late nineteenth century that the patterns were now explained in terms of non-Darwinian mechanisms of natural evolution such as orthogenesis (Bowler 1979; 1983).

The fact that the supposed linearity of evolution could be reinterpreted in this way shows that it was an issue that could be treated independently of the argument from design. This leads us to the next level of Darwin's challenge to the traditional view of nature, which is precisely his insistence that there is no predetermined pattern of development. For Darwin, it was the forces acting on the individual organism during its daily life that shaped evolution, and hence there could be no long-range trends apart from those related to the environment (for instance, specialization for a particular way of life). The fact that evolution is a causal, historical process inevitably imposes limits on its activity. Each form is the result of compromise between the pressure of adaptation and the structures inherited from ancestral types.

Nevertheless, in Darwin's theory the constraints imposed by ancestry did not completely predetermine the future. In general, evolution was an irregular, constantly branching process, with developments in one branch having no relationship to those elsewhere on the "tree of life". In particular, there could be no single goal of evolution, no return to the old idea of a linear hierarchy of organization aimed at man — the basis of Agassiz's vision of natural order. This vision had already begun to break down before Darwin published (Bowler 1976a; Ospovat 1976, 1981). Yet within individual branches of evolution it was still possible to look for patterns that would be inconsistent with any theory in which evolution was shaped only by pressures from the local environment. It was believed that related species advanced in parallel through a predetermined line of development toward a non-adaptive goal. This was the essence of the theory of orthogenetic evolution, in which linearity of development was thought to be imposed by internal forces predisposing living forms to vary in particular directions — thereby breaking the link with design. This was a fundamentally anti-Darwinian view of evolution, not only because it denied the role of adaptation, but because it asserted the regularity, and indeed the predictability, of the process. The paleontologists of the "American school" were among the first to develop this view into a coherent alternative to Darwinism.

Although paleontology was later thought to provide the best evidence for orthogenesis, it was in this field that Darwinism won some of its first triumphs. These were derived not from the selection theory itself, however, but from the broader vision of branching evolution. Indeed, some of the leading figures involved were Darwinians only by the loosest definition. Ernst Haeckel, whose work was admired by Darwin and became immensely influential in the English-speaking world, openly linked Darwin with Lamarck and Goethe (Haeckel 1876), while even T. H. Huxley was by no means a true selectionist (Bartholomew 1975; Kottler, this volume). In this loosely defined form, Darwinism merely predicted that distinct modern forms must have evolved from a common ancestor, the point of divergence lying further back in time the greater the difference between the modern descendants. There was also a more detailed prediction that highly specialized modern forms must have evolved from generalized ancestors. Darwin pointed out that although most of the details were missing, the general outline of the fossil record was consistent with this view. Although he himself was reluctant to speculate on the detailed course of evolution, many of his followers were only too willing to reconstruct the history of life using indirect evidence from morphology and embryology. Haeckel was the boldest speculator along these lines, drawing up complete evolutionary trees based on his "biogenetic law" that ontogeny recapitulated phylogeny. Stephen Jay Gould (1977b) has shown that the recapitulation theory was by no means identified solely with Darwinism, although in general Haeckel's speculations were consistent with the Darwinian picture of branching evolution. Only his tendency to assume that earlier steps in the ascent of life must have survived through to the present unchanged betrayed Haeckel's inability to escape from the last vestiges of the old chain of being.

The crucial question was whether or not the predictions of the evolutionists would be fulfilled by new fossil discoveries, at least in enough cases to make their general argument seem plausible. M. J. S. Rudwick (1972) has shown how a number of important cases studied in the decades after 1859 helped to support the evolutionary interpretation. The discovery of *Archaeopteryx* provided a fossil with characters intermediate between those of two modern classes, the birds and the reptiles. Although this did not prove that the birds had evolved from the reptiles, it did confound the creationists who had insisted that no link between the classes would ever be found. On a smaller scale, Othniel C. Marsh's discovery of a sequence of American fossils leading toward the modern horse was hailed by Huxley as "demonstrative proof of evolution" (Huxley 1877, p. 90). Already, however, there were problems, since it was only in their most general outlines that such sequences confirmed the Darwinian view. Marsh's great rival Edward Drinker Cope saw an exaggerated linearity in the evolution of the horse, which to him seemed evidence of a more direct mode of adaptation than was possible with the selection of random variations (Cope 1868, pp. 146–150; see Bowler 1977b, 1983). His friend Alpheus Hyatt uncovered vast regular trends among the fossil cephalopods that seemed to defy explanation in Darwinian terms (1866, 1889; see Gould 1977b, chap. 4). Adrian Desmond (1982) points out that Huxley did not appreciate the significance of the mammal-like reptiles discovered during the late nineteenth century, because they were described by Owen and his followers, who attributed the evolution of the mammals to a purposeful trend. Thus after the first flush of new discoveries, the fossil record seemed to turn against Darwinism — although not quite in the way that Darwin himself had anticipated.

Darwin's own fears about the fossil record had centered on quite a different problem, that of continuity. He had built his theory on the belief that only the relatively small variations observable in modern populations were accumulated over a long period of time to give a gradual process of evolution. Yet the fossil record showed not a gradual development of life but a series of distinct forms. There were many substantial discontinuities where totally new forms appeared suddenly in the record with no sign of an evolutionary process whereby they had been generated. Following Lyell, Darwin proclaimed the "imperfection of the geological record" — yet the discontinuities have remained a fertile source of opposition to his gradualism to the present day. Eventually most paleontologists sided with Darwin on this issue, although even Huxley felt that Darwin had tied himself too closely to the principle of continuity.[3] If there were "steps" in evolution, they were small enough to be accounted for in natural terms. It is worth noting, however, that the continuing lack of evidence for links between the classes became increasingly puzzling to those who had expected more discoveries along the lines of *Archaeopteryx*. By the end of the century, the paleontologists' disappointment was being exploited by the new generation of experimental biologists as an argument to discredit the whole technique based on the speculative reconstruction of the history of life.[4]

The debate over continuity was not just a technical disagreement over the state of the fossil record. It also symbolized Darwin's assault upon yet another citadel of the traditional world view, what Ernst Mayr (1959a) has called the "typological" concept of species. According to this more or less Platonic interpretation, the species was defined by a typical form or structure existing at a deeper level of reality than the mere individual organisms that make up the population at any one time. Individual variation was thus by definition only a trivial effect that could not alter the basic form of the species. The theory of gradual evolution necessarily broke down this concept of a permanently fixed morphological type. Modern Darwinists no longer define the species by its morphological structure; instead the species is regarded as equivalent to the breeding population. Darwin himself was never able to free himself entirely from the tendency to define species in morphological terms, but his theory made it clear that there could be no guarantee of stability once selection began to act upon a variable population. Small wonder that many conservative thinkers refused to accept the destruction of an idea that had fitted in so neatly with the theological view of nature. But there were practical reasons too why naturalists concerned with description and classification should resent the new trend. If accepted, it would undermine the logic of their neat pigeon-holing, making it impossible for them to justify their arbitrary distinctions between true species and mere local varieties. The degree of morphological difference between two specimens would no longer be an adequate criterion for classification. To some extent, Darwinism flourished precisely because such naturalists were constantly falling out over which forms were entitled to species status, and the theory explained why this kind of uncertainty was inevitable (see Beatty, this volume). On the other hand, the instinctive tendency of many naturalists to think of species as distinct units meant that even after evolution had become generally accepted, they were constantly tempted to look for non-Darwinian mechanisms based on large saltations.

Darwin refused to treat species as units in an abstract pattern of development because he was convinced that evolution took place only in response to external pressures. His theory was thus one of uncompromising utilitarianism: it suggested that all evolutionary changes take place because they are useful to the individual organisms in their struggle to cope with an ever-changing environment (or because they confer reproductive success). But this emphasis on adaptation was a controversial point, since many naturalists believed that the trivial characters used to distinguish between closely related species were *too* trivial to be of any use. Karl Nägeli was one of the first to stress the widespread existence of non-adaptive characters as an argument against the theory, and Darwin eventually conceded that he had not sufficiently allowed for this in some of his earlier discussions (see Vorzimmer 1970, chap. 9). His own solution was to invoke the "correlation of growth" to explain how non-adaptive characters might have been developed as a by-product of natural selection. Even his supporters paid little attention to this, however, and in later years those naturalists who continued to doubt the utilitarian interpretation of nature turned to non-Darwinian mechanisms such as orthogenesis and mutations to explain the production of non-adaptive characters. The fact that such non-utilitarian versions of evolution were still flourishing in the late nineteenth century shows the limited success achieved by this aspect of the Darwinian approach. Those who *were* converted, cheerfully accepted that an adaptive explanation would eventually be found for the origin of every character, but those who harbored doubts about Darwinism were constantly tempted to think that life evolved by a process *not* totally subservient to external influences.

Darwin's supporters had at first placed a great deal of emphasis on Henry Walter Bates's studies of mimicry in insects. Protective coloration in general, and Bates's particular discovery of insects mimicking inedible forms, were seen as evidence for Darwinism because selection seemed the only possible way of explaining this kind of character (Bates 1862; see Beddall 1969 and Woodcock 1969). Lamarckian use-inheritance would be ineffective since insects could not control their color, and if the similarities were to be given any adaptive significance at all, they would have to result from the selective elimination of those individuals in which the resemblance was less obvious. It is significant that some of the more extreme anti-Darwinists did, in fact, deny the adaptive significance of "protective" coloration, attributing the resemblances to parallel variation-trends affecting widely different species. In this respect they rejected not only the selection theory, but also the whole Darwinian emphasis on adaptation. By the end of the century, Darwinians such as Edward B. Poulton (1908) could point to several decades of field research in which the concept of protective coloration had proved its value, but this did not prevent their opponents challenging the validity of the whole enterprise.

Although it gave rise to many objections, Darwin's insistence that evolution was guided solely by external factors was certainly the source of his theory's most promising applications. His viewpoint had been shaped by his own early studies of biogeography, and it was here that it offered his followers their best chance to show off its advantages. Darwinism explained the distribution of modern species in terms of their ancestors' ability to migrate around the world, under the limitation of geographical barriers, coupled with the tendency of each form to adapt to any new region it colonized. Alfred Russel Wallace soon showed that the complexity of geographical variation among Malayan butterflies was entirely consistent with the theory (1864). He also showed that many problems of geographical distribution on a wider scale could be explained. In particular he defined "Wallace's line", which separated the Asian and Australian faunas of the Malay archipelago (modern Indonesia), and explained it in terms of the strait between the islands of Bali and Lombok being too deep to have been bridged even if the sea-level had been lower in the geological past (Wallace 1876, 1880; see Mayr 1954b; Beddall 1969; Fichman 1977, 1981, chap. 3). Botanists used the theory in a similar way. Joseph Dalton Hooker's account of the flora of Tasmania was one of the earliest pro-Darwinian arguments (1860b), while Asa Gray (1876) applied the same principles to explain the geographical distribution of American plants.

By postulating that species could adapt as well as migrate, the Darwinians gave a new impetus to the study of geographical distribution. On the more detailed question of the precise role of geographical factors in speciation, however, there were disagreements that would eventually lead the theory into major difficulties. Was geographical isolation essential for separating a homogeneous population into distinct groups that would no longer interbreed and were thus potentially distinct species? Darwin's experiences in the Galapagos islands showed him that isolation was important in this respect, but before he wrote the *Origin* he had become convinced that speciation could occur by adaptive specialization *without* geographical separation. Many of his followers adopted the same view, openly repudiating Moritz Wagner's claim that isolation was essential for speciation (1868). Ernst Mayr has described the confusion that resulted from this decision, as the Darwinians' attempts to find a satisfactory mechanism of sympatric speciation proved futile (1959b; also Lesch 1975; Sulloway 1979b). Only at the end of the century did Karl Jordan (1905) and others succeed in demonstrating that Darwin's populational definition of species required geographical isolation as the only way of building up an effective barrier to interbreeding between two groups derived from a single original population (Mayr 1955). In the meantime, opponents of gradualistic evolution exploited the confusion to argue for discontinuous steps or mutations as the sole cause of speciation.

The Darwinian approach to geographical factors could be exploited without making a commitment to the selection mechanism itself. Mayr (1955) points out that Karl Jordan was not a dogmatic selectionist, while the American naturalist Alpheus Packard regarded geographical isolation as a vital component of his neo-Lamarckian view of evolution (1901, pp. 404–405). We have seen, in fact, that many aspects of the Darwinian viewpoint can be specified without defining the details of natural selection, and hence could be challenged on grounds other than any supposed ineffectiveness of the selection mechanism. Darwinism meant much more than natural selection; it referred to a whole complex of ideas that challenged the traditional view of nature on fundamental grounds concerning the overall pattern of life's evolution. Selection was Darwin's particular way of explaining how his vision of evolution was supposed to unfold, but other mechanisms of adaptation such as the Lamarckian inheritance of acquired characters could, in theory, produce the same effect. Thus the revival of interest in Lamarckism toward the end of the nineteenth century did not necessarily challenge some of Darwin's most productive insights, as Packard (1901) was eager to point out. A far more basic challenge to everything that Darwin stood for came from those naturalists who tried to link their Lamarckism with orthogenesis. If evolution did indeed proceed through internal forces driving groups of species along parallel lines of development, sometimes toward non-adaptive goals, then the whole Darwinian image of life's history was in error.

All this is not meant to imply that the well-known arguments against natural selection were irrelevant. On the contrary, picking out supposed flaws in Darwin's reasoning was a favorite pastime of his opponents, and certainly helped to boost the search for alternative mechanisms. But we must recognize that these criticisms could be employed with two very different purposes in mind. Some naturalists merely wished to replace natural selection with an alternative mechanism of adaptation within a generally "Darwinian" view of evolution — this was Packard's position, for instance. But others, including some of Packard's colleagues in the American school, saw their attacks on selection as part of a much wider-ranging campaign against the whole Darwinian perspective on the history of life. It should also be noted that most of the arguments *for* Darwinism were derived from its broader applications to paleontology and geographical distribution. There was no demonstration of natural selection's efficacy — although W. F. R. Weldon's biometrical experiments in the last decade of the century were meant to plug this gap (1894–1895, 1898, 1901; see Provine 1971; Norton 1973). In effect, the argument over the plausibility of natural selection was a debate within a debate. There was always opposition to selection, but it took some time for the more extreme opponents to work out a completely anti-Darwinian philosophy of evolution. Bearing this point in mind, we can now pass on to look at the various arguments that were meant to demonstrate the weakness of Darwin's mechanism.

One objection that was in principle applicable to any theory of slow evolution — but in practice was aimed directly at natural selection — centered on the question of geological time. Joe D. Burchfield (1975) has described Lord Kelvin's efforts to show that geologists such as Lyell had greatly exaggerated the amount of time available in the earth's history (Kelvin 1863, 1871, 1891–1894, vol. 2). Darwin had relied upon Lyell's concept of an almost indefinite period of time during which a very slow process such as natural selection could have achieved the major evolutionary results we observe. His theory was certainly one of Kelvin's targets, and there can be little doubt that the reduced age of the earth, backed up by the authority of a leading physicist, became a major stumbling block holding back acceptance of natural selection. Although some of Darwin's followers tried to argue that selection could work faster than he had supposed, Darwin himself refused to abandon his position and insisted that something must be wrong with Kelvin's calculations. Only in the first decade of the twentieth century did it become apparent that the radioactivity of materials deep in the earth might account for the discrepancy. In the meantime, many of the alternative mechanisms of evolution had been hailed as improvements over selection precisely because they seemed to imply that the history of life could be compressed into a much shorter time-scale.

In biology, the most obvious problems arose from the limitations of mid-nineteenth century ideas on heredity and variation, or from what were perceived as fallacies in Darwin's reasoning. Such problems were pointed out gleefully by numerous opponents, and some are still used today by modern anti-Darwinists. But there were also weaknesses exposed by the Darwinians themselves, as they tried to extend the theory in what at first seemed promising directions. Malcolm Kottler (1980 and this volume) has discussed the debate between Darwin and Wallace over the origins of sexual dimorphism, noting that in the end neither was able to convince the other. The two men disagreed completely over the relevance of Darwin's notion of sexual selection, as they did over a number of other topics. The inconclusive nature of this debate illustrates an important limitation of the original selection theory: it opened up the prospect of finding natural explanations of various phenomena that had hitherto been seen merely as objects to be described, but it contained serious flaws that prevented it from generating unambiguous solutions to the problems. At first it was hoped that the disagreements would be cleared up by further research, but when the ambiguities proved impossible to resolve, an atmosphere of frustration built up and stimulated the search for alternative mechanisms.

A problem that has received considerable attention from some modern historians is centered on Darwin's views on the origin and inheritance of variations. Because his ideas have been replaced so drastically by Mendelian genetics, they seem to indicate a fundamental weakness in the original selection theory, which must surely have limited its impact on nineteenth-century science. Eiseley wrote of Darwin's slide into Lamarckism, as the inadequacies of his original ideas were pointed out to him (1958, p. 217, 240); a similarly pessimistic image of Darwin's ability to resist criticism on this score has been presented in Vorzimmer's account (1970, chaps. 2–6). Darwin was certainly persuaded that some of his original views on selection were untenable, but there is no reason to suppose that he was forced to give up the selection theory altogether. Even within the commonly accepted notion of "blending heredity", selection was still a workable mechanism (Bowler 1974a). Fleeming Jenkin's famous review (1867) of the *Origin* pointed out the incompatibility between blending and the selection of *single* variations, but this argument did not apply to a range of variation seen as an integral characteristic of the population. Jenkin himself conceded that selection could work at this level, but pointed to the common experience that it was constrained by a fixed limit of variability apparently built into the species. The Darwinians had to assume that this "variation barrier" could be broken, if enough time were allowed — a view that was to be challenged again by many early geneticists. Before the advent of rigid particulate theories of heredity, though, the Darwinians' assumption did not seem too implausible. There is thus no reason to be surprised that Darwin was able to continue promoting selection, despite the weakness of his views on heredity.

It is true, of course, that the continued study of evolutionary problems helped to reveal the inadequacy of existing views on heredity, particularly Darwin's theory of pangenesis. But many new initiatives were attempted in response to this problem, not all of them hostile to Darwinism. The biometrical school under Karl Pearson and W. F. R. Weldon exploited Francis Galton's "law of ancestral heredity" to create a workable mathematical model of selection based, in effect, upon blending heredity (Froggatt and Nevin 1971b; Provine 1971; Norton 1973). For them, blending helped to explain how selection could exert a continuous effect upon the range of variation existing within a population. August Weismann's germ plasm theory (1891–1892, 1893a) was a theoretical initiative based on an incomplete notion of particulate heredity, intended as a means of upholding natural selection as the only mechanisms of evolution. On the other hand, the confusion in heredity theory also allowed the Lamarckians to move in the opposite direction, invoking the gradual assimilation of individual experiences into the character of the species. The most radical alternative was that demanding a strictly experimental study of heredity, purged of the previous speculations on the physical nature of the process. This approach eventually gave rise to the new science of Mendelian genetics, whose supporters at first thought that their views on particulate inheritance were incompatible with the Darwinians' gradualistic form of natural selection (Cock 1973; De Marrais 1974). The late nineteenth-century crisis in the study of heredity was a genuine one, precipitated by the unsatisfactory nature of the Darwinists' attempts to deal with the problem. But it is a gross oversimplification of the situation to claim that the original form of selection was rendered unworkable by its assumption of blending heredity.

In the end, perhaps the most damaging charges that could be made against selection take us back to the more basic levels of opposition that we have seen emerging out of the traditional world view. These focussed on what were perceived as weaknesses in the logic of Darwin's argument for selection. Curiously, the claim that the "survival of the fittest" is merely a tautology, which has become the stock in trade of modern "philosophical" opponents of Darwinism, was not a significant factor in the nineteenth-century debates. More serious then were a number of arguments that were meant to show the impossibility of selection ever having a significant effect.[5]

Some naturalists including the young T. H. Morgan (1903), doubted that the struggle for existence could affect adult characters, on the grounds that the elimination of excess population generally took place among the young, on a more or less random basis. Most crucial of all was the claim that natural selection could never be a "constructive" mechanism, because it could only evaluate those characters presented to it by variation. The origin of *variation* was thus the real problem of evolution, and many refused to accept that variation was random in the sense demanded by Darwinism. For many naturalists trained in the old tradition, it was only a short step from the claim that selection was only a negative mechanism for eliminating the unfit to the assumption that variation must be a positive force guiding the production of new characters. Originally, such an assumption was used to defend the old belief in divine guidance. But as the demand for naturalistic explanations of evolution grew, the opponents of Darwinism were forced to develop more realistic theories of how variation was guided in certain directions. Lamarckism, orthogenesis, and the mutation theory were all the results of efforts to undermine Darwinism's central reliance on random variation, thereby de-emphasizing the selective role of the environment.

II. The Pattern of Debate

My all too brief survey of the scientific arguments has revealed that Darwinism could be evaluated on many different levels. This means that any attempt to describe the impact of the theory upon science must be subdivided into separate discussions of the various levels of debate. We cannot simply ask, "What was the success of evolutionism, or of Darwinism?" (cf. Mayr, this volume). Instead, there are at least four questions that need to be answered. What was the success of (a) the basic idea of transmutation, (b) the principle of naturalism, (c) the concept of branching adaptive evolution, and (d) natural selection itself? The problem is further complicated by the fact that attitudes toward all these issues changed through time, not necessarily in a consistent manner, and that the various scientific disciplines did not react in the same way. Our next goal must be to provide an outline description of the success or failure of Darwinism at these various levels throughout the late nineteenth century.

I begin with the general concept of transmutation. When Robert Chambers's anonymously published *Vestiges of the Natural History of Creation* (1844) suggested that transmutation might be the means whereby the divine plan of living development had been unfolded, the idea was met with a barrage of criticism despite Chambers's effort to compromise with the traditional view of design (Gillispie 1951; Hodge 1972; Millhauser 1959). By 1859, some of this hostility had evaporated, and an increasing number of naturalists were beginning to recognize the existence of trends within the fossil record that seemed to indicate some genetic relationship between the successive forms (Bowler 1976a; Ospovat 1981). Left to itself, this movement might have gradually matured into a general consensus based on theistic evolutionism. Even so, Darwin's bold statement of gradualistic transmutation went far beyond what most of his contemporaries had in mind. Yet the general impression given by those who lived through the debate, and accepted by most historians, is that Darwin was successful in converting the scientific community to evolutionism within a decade or so of the *Origin's* publication. A recent attempt to test this belief suggests that it may be a little extravagant, but even this study concedes that by 1869 approximately three quarters of the scientists investigated had accepted evolution (Hull, Tessner, and Diamond 1978). Ellegård's survey of the periodical press (1958) showed that even outside science there was a widespread acceptance of evolution by the early 1870s. Ruse notes that in the course of the 1860s the evolutionists were so successful in their takeover of Cambridge University that the examination questions switched from design to natural selection (1979a, p. 262). Looking back from a period when Darwinism was not so highly regarded, Eric Nordenskiöld's classic *History of Biology* concluded that the theory "reached its zenith in the seventies and eighties" (1928, p. 528). Although a few scientists continued to oppose evolution to the end of the century — Sir John William Dawson of Montreal is perhaps the best example — they were increasingly perceived as cultural fossils and lost much of whatever prestige they had once enjoyed (Dawson 1890; see C. O'Brien 1971).

It may be true that by the 1870s there were few scientists still opposing evolution altogether, but we should not exaggerate the theory's immediate impact, particularly on some areas of biology. F. Burkhardt notes that in the British learned societies, theoretical discussion of evolution was at first kept to a minimum, even by Darwin's supporters (1974, p. 72). Michael Bartholomew (1975) has shown that even T. H. Huxley's conversion to evolutionism had little real effect on his anatomical and paleontological work. Philosophically, there was a difference between looking to common ancestors rather than Platonic archetypes to explain the similarities between related forms, but in terms of how one named and described the forms themselves it often made little difference. To a surprisingly large extent, Darwin had adapted his theory to exploit the morphological techniques and concepts developed earlier in the century (Ospovat 1981). Only in paleontology did the search for evolutionary sequences actually direct the interests of some researchers. Once the theory of evolution was firmly established, it became common practice to discuss the evolutionary significance of fossils even in the most academic literature. In this area, the theory's triumph came not so much through a revolution of technique, but through a direction of interest toward those fossils that might help the cause of evolutionism, and a growing willingness to regard evolutionary relationships as a legitimate area in which hypotheses could be proposed.

The popularity of evolutionary morphology during the 1870s and 1880s is perhaps the best illustration of how the general idea of common descent had gained scientific respectability. Comparative anatomy and embryology were no longer merely descriptive sciences, but were explicitly used as means of providing indirect evidence for the reconstruction of key steps in the history of life, where the fossil record was blank. Morphological relationships, especially during the early stages of growth, were appealed to as evidence of evolutionary connection, a technique pioneered in Germany, brought to Britain by students of Huxley such as E. Ray Lankester, and refined by Francis M. Balfour at Cambridge. A veritable hive of industry centered on such important questions as the origin of the vertebrates (E. S. Russell 1916, chap. 15). However, as we have already seen, the fossil record remained stubbornly incomplete in some of these areas, and this generated a good deal of skepticism as time went on. By the end of the century, exponents of the experimental method such as Bateson were openly repudiating Darwinism because of the speculative nature of these reconstructions. In fact, such criticisms should have been levelled against the whole idea of evolution, since the American school, for instance, had been equally involved in the movement. One member of the American school accused Bateson of expressing his doubts so forcefully that he *appeared* to be rejecting evolution altogether (Osborn 1922). In Britain, though, Darwinism was associated in everyone's mind with the technique of phylogenetic speculation, and rejection of this technique led inevitably to a repudiation of other aspects of Darwinism, especially the postulation off dubious adaptive explanations of every phase of evolution.

In the case of evolutionary morphology, there was a distinct cycle of optimism and development, followed by skepticism and rejection. In other areas of science, evolutionism never broke through the barriers of indifference or hostility erected against it. So general a theory was seen by some as at best an irrelevance to their detailed work, at worst a threat to the established framework of research. This was the case in entomology, for instance. Wallace and Bates might promote the evolutionary explanation of mimicry, but to those entomologists whose chief concern was description and classification, such hypothetical accounts of the origin of characters that had only just been discovered were of little interest. The internationally renowned Belfast entomologist Alexander Haliday lived through the whole Darwinian debate without even commenting on it.[6] Other entomologists were openly hostile to Darwinism, a reaction that Darwin and his followers had anticipated and were prepared to ignore. So great is the variety of insects waiting to be described that, even today, many entomologists are still so deeply involved in this kind of work that they can see little point in evolutionary speculations. The fact that some sciences were simply not in a position to benefit from evolution, while others experienced a cycle of interest and suspicion, suggests that the triumph of Darwinism represented far more than a simple recognition of the theory's superiority in dealing with the facts.

One possibility to be explored is that Darwinism succeeded in part because it became a symbol of the new spirit of scientific naturalism, which demanded that science must have access to all the questions that had hitherto been ruled off-limits to it. This is consistent with the surprisingly brief career of theistic or designed evolutionism. Such a position was a blatant compromise, accepting transmutation to explain the regularities in the development of life, but retaining design by assuming that something more than natural causes was required to account for the development. Violently rejected by the opponents of *Vestiges* in the 1840s, this position was soon adopted as the most obvious means of salvaging something of the old way of thought from the Darwinian challenge. Yet within a couple of decades the whole project had been virtually abandoned. A few diehards such as Argyll (1898) continued to resist any natural explanation of evolution, but very little of the vast body of anti-Darwinian literature produced at the end of the century was intended to demonstrate the inadequacy of all natural causes. Alternative natural mechanisms such as Lamarckism and orthogenesis were now upheld as preferable explanations of the facts that had once been used as evidence of design. Cope, a founder of the American school, had at first appealed openly to the Creator's will as an explanation of the regularities he saw in the fossil record, but soon he went on to argue the same phenomena represented the best evidence for Lamarckism.[7] Even Mivart (1884) eventually accepted a more or less Lamarckian position.

The increasing isolation of creationists such as Dawson and theistic evolutionists such as Argyll represented a triumph for the principle of scientific naturalism. Even those who remained opposed to the Darwinian view of evolution had now conceded that it was not legitimate for them to invoke the supernatural to account for the phenomena they regarded as incompatible with Darwinism. Yet the triumph was in some respects a hollow one, since many aspects of the old tradition soon began to re-emerge in a new guise, superficially adapted to the now dominant philosophy of naturalism. The supporters of Lamarckism and orthogenesis were eventually successful in convincing the scientific community that the postulation of evolutionary trends was acceptable, provided the apparently goal-directed nature of the trends was ascribed to some hypothetical ordering principle in the behavior or growth-pattern of the individual organisms. This raises a very general question about the impact of Darwinism, since it implies only a limited acceptance of the concept of constantly branching evolution upon which the selection theory was based. By the 1850s, the general idea that the history of life was represented by a branching tree had been fairly widely established, and some naturalists were beginning to realize that the individual branches represented specializations for different ways of life (Bowler 1976a; Ospovat 1981). Darwin reinforced this trend by supplying a mechanism to explain both the branching and the specialization. Yet the popularity of theories based on directed evolution during the late nineteenth century suggests that it was only at the broadest level that the principle of branching development was universally accepted. Many naturalists refused to give up the belief that within each branch, there was some force ensuring a degree of orderly development in a particular direction. To this extent they remained committed to the idealist concept of directed evolution modeled on embryological growth.

Two general points need to be made concerning this retention of a quasi-teleological concept of orderly development within late-nineteenth-century evolutionism. The first concerns the complex relationship between evolution and the idea of progress. Earlier in the century, the notion of a general trend in the history of life was almost universally linked with the ascent of a linear hierarchy toward man. This view of biological progression was now ostensibly abandoned as naturalists turned instead to the concept of branching development. It is obvious, however, that most evolutionists continued to incorporate the idea of progress into their thinking in one form or another (Greene 1981b). The link with progress was a good selling-point at the non-scientific level for all theories of evolution, Darwinian and non-Darwinian alike. Nevertheless, it must be noted that in their most successful scientific applications, most theories either ignored or threatened to undermine the hierarchical view of organic relationships upon which biological progressionism was based. The processes of migration and adaptation invoked to explain geographical distribution sidestepped the issue altogether, while the recognition of evolutionary specialization forced naturalists into a reinterpretation of what progress might mean. Although most American Lamarckians were progressionists, the linear patterns of development they postulated within each branch went in many different directions, and in the case of racial senility ended up in degeneration and extinction. The debate over whether or not evolution was an orderly process on a small scale was thus fought out independently of the wider concept of progress, to which evolutionists of many different backgrounds appealed as a source of cultural influence.

The second point concerns the link between orderly patterns of development and the utilitarian view that adaptation represents a sign of divine purpose. Once they had realized that Lamarckian use-inheritance guided by habit could explain at least the adaptive trends in the fossil record, the paleontologists of the American school were quite willing to see the process as an indication of design. Lamarckism did not allow one to claim proof of divine benevolence, but the belief that animals' efforts to adapt to their environment became cumulative through use-inheritance certainly seemed more compatible with the existence of a benevolent Creator than did natural selection (Pfeifer 1965; Moore 1979, pp. 146–151). The element of design was merely transferred from a supernatural deity into the forces of life itself, and was thus internalized within nature. This was also the point made by Samuel Butler in his long campaign against natural selection (1877, 1879, 1880, 1887; see Willey 1960). Butler too saw use-inheritance as a means of preserving the natural theologians' concern for purpose within a system of adaptive evolution, providing a morally acceptable alternative to the Darwinian nightmare of trial-and-error variation. On the other hand, the willingness of some American neo-Lamarckians to invoke non-adaptive trends in evolution suggests that we should be careful not to exaggerate the link between the concepts of directed and purposeful evolution. If use-inheritance could be seen as the salvation of Paley's utilitarian concept of design, directed evolution was more a product of the idealists' vision of orderly development. The two could be linked in some cases, but they could also be defended separately.

The link between the theological defense of Lamarckism by Butler and the American school brings out an important difference between the fortunes of Darwinism on the two sides of the Atlantic. American "neo-Lamarckism" began to develop in the 1870s and had become a fully fledged alternative to Darwinism by the time it was given its name by Packard in 1885. Butler, on the other hand, was virtually ignored in Britain until the last decade of the century, when his ideas at last began to be taken seriously by at least some biologists.[8] The concept of "self-adaptation" was also developed independently by other British naturalists with a theological axe to grind (G. Henslow 1888, 1895). As in America, this somewhat delayed outburst of support for Lamarckism was also linked with the idea of orthogenetic evolution. When the marine biologist Joseph T. Cunningham turned to Lamarckism in the late 1880s, one of his first actions was to translate a German work by Theodor Eimer, the founder of the theory of orthogenesis.[9] Yet the link between Lamarckism and orthogenesis was never as strong in Britain as it was in America. Even Butler accepted that on the small scale, Lamarckian evolution would be haphazard and irregular. The continued influence of Herbert Spencer also ensured that the inheritance of acquired characters could be accepted as a mere assistant to natural selection within a generally Darwinian framework of utilitarian evolution, and without Butler's natural theology.

The earlier development of American neo-Lamarckism, and its much closer link with the idea of regular, non-adaptive evolution, suggest that here there was a wholesale opposition to the Darwinian view of evolution that prevented the selection theory from ever reaching a dominant position in nineteenth-century American science. In Britain, on the other hand, the anti-Darwinian forces were driven underground rather more effectively — although they were not eradicated and were able to renew the challenge once Darwinism began to show signs of weakness at the end of the century. The temporary success of Darwinism was ensured by the ability of Huxley and his circle to shape the emergence of a new generation of British naturalists.

Men such as Lankester and Poulton were raised on a diet of branching, adaptive evolution and never doubted that selection was the chief, if not the only, cause. As they rose to dominate the scientific establishment, they became the high priests of a Darwinian orthodoxy that had become firmly entrenched by the 1880s. The ability of this group to prevent Lamarckism from being taken seriously as a major alternative to Darwinism can be judged from the total lack of success enjoyed by Butler's writings on the topic, despite his use of the theological argument that was so popular in America. Herbert Spencer was perhaps the only influential writer to accept a more prominent role for Lamarckism than most orthodox Darwinians, but since he also acknowledged the role of selection, his view did not at first generate significant controversy. Only in the late 1880s did Spencer begin to speak out openly against Darwinism, charging that natural selection had become an inflexible dogma imposed as a matter of faith on the scientific community (1887, 1893; see also G. Henslow 1898). This was Butler's point too — although the two men's reasons for accepting Lamarckism were quite different — and it became a standard complaint of the new Lamarckism.

Spencer's reaction was provoked by developments occurring within Darwinism itself, particularly the increasing dogmatism that came to be associated with the name of August Weismann (1893b). Toward the end of the century, the term "neo-Darwinism" came into use, denoting a more rigid adherence to the selection mechanism. To the early Darwinians, the question of Lamarckism as an alternative to selection had not seemed crucial. Darwin himself had always accepted a subordinate role for the inheritance of acquired characters; later in his career he became even more willing to admit that selection needed this supplement. The claim that Darwin was driven into Lamarckism by his failure to cope with the problem of blending inheritance is an exaggeration, but it *was* the problem of heredity and the failure of Darwin's own theory of pangenesis that led Weismann to develop his theory of the germ plasm in the 1880s (see Churchill 1968). The belief that the material substance responsible for heredity was completely isolated from the rest of the body made Lamarckism a theoretical impossibility, a view that Weismann backed up with his famous experiment proving the non-inheritance of mutilations in mice. His works were rapidly translated into English (1891–1892, 1893a) and gained wide support among Darwinists. Wallace had already committed himself to natural selection as the only mechanism of animal evolution, and Weismann's theory helped to boost support for this position in Britain. The attempt to purify Darwinism backfired, however, since there were many who found the germ plasm theory unconvincing, and who saw neo-Darwinism as a threat to the broader view of evolutionism they preferred. It was in response to Weismann that Spencer took up his pen in defense of the inheritance of acquired characters, and thus for the first time allowed himself to be perceived as an opponent of Darwinism (Churchill 1978).

American neo-Lamarckism received some extra support from opponents of Weismann's theory, but the emergence of a self-conscious Lamarckian movement within British science was largely a direct response to the dogmatic selectionism of neo-Darwinism. Several quite different kinds of support for the inheritance of acquired characters were now integrated to make a common front against Weismann. Spencer was turned from a conventional Darwinian into an opponent of Weismann — and his influence guaranteed that the case for the inheritance of acquired characters would receive a serious hearing among scientists.[10] The analogy between heredity and memory, the foundation of Butler's theological Lamarckism, was now accepted by some as a plausible alternative to Weismann's theory of fixed hereditary particles (E. S. Russell 1916, chap. 19; Gould 1977b, pp. 96–100). Cunningham (1895) could raise the inheritance of acquired non-adaptive characters as further evidence against the isolation of the germ plasm (Bowler 1979). British Lamarckism was a diverse movement comprising various lines of scientific and philosophical support for the theory, but its supporters were united by this hostility to neo-Darwinism (R. Burkhardt 1980). Whatever the final outcome of this first effort to base Darwinism on a theory of absolutely "hard" heredity, its initial result was to polarize opinions to such an extent that the 1890s saw the first major attempt to provide an alternative to Darwinism within the framework of scientific naturalism.

The revival of interest in Lamarckism during the last decade of the century was one component of the more general reaction against the selection theory that Julian Huxley would later call the "eclipse of Darwinism" (1942, pp. 22–28). Yet for all its increased popularity, Lamarckism did not represent a new threat. In both its American and its European forms, it rested upon conceptual foundations put together in the early years of the Darwinian debate and derived from a much older tradition in natural history. With hindsight, one can easily see that its success could only be temporary. Its alternative view of heredity seemed plausible only because the germ plasm theory fell so far short of substantiating Weismann's claims. The germ plasm clarified the notion of hard heredity, but Weismann's views on its internal structure were not fruitful, and left him open to the charge that he was speculating far beyond the available evidence. A new initiative was needed to put this approach on a firmer footing, and this came with the experimental techniques that led to the rediscovery of Mendel and his laws in 1900. As Mendelian genetics established itself, Lamarckism soon showed itself incapable of modernizing its traditional concept of heredity. Whatever the experimental successes claimed by its supporters in the early twentieth century — and these were, to say the least, highly controversial[11] — the lack of any theoretical alternative to genetics ensured that Lamarckism was reduced to a peripheral role within the new science of heredity (Bowler 1983).

The second component of the eclipse of Darwinism was precisely the emergence of an experimental study of heredity. Under Pearson and Weldon, the biometrical school of Darwinism tried to refine the measurement of variation and to develop a theoretical model of heredity and selection within large populations. Weldon even provided experimental evidence of selection (1894–1895, 1898, 1901). But this approach did not satisfy those critics who looked for a more radical break with the old tradition of natural history within which both Darwinism and Lamarckism had flourished. Apart from the problem of heredity, frustrations had built up because the fossil record persistently refused to yield evidence that would either support or refute the speculations of the evolutionary morphologists. The Darwinists' failure to solve the problem of speciation also left them open to the charge that some unknown, possibly saltative, factor in heredity might be responsible for the sudden production of new forms. Dissatisfied with the vagueness of much Darwinian speculation, a new generation of biologists turned instead to the experimental method, which had already proved its worth in other fields. By applying its techniques to the problems of heredity and variation, they hoped to transcend the now evident limitations of natural history. For someone like W. Bateson (1894), neo-Darwinism was far too speculative, and continued reliance on selection still left room for unverifiable hypotheses on the adaptive value of each evolutionary development. Further progress could only be ensured by abandoning all the preconceptions of existing theories to concentrate on those factors open to direct observation.

The subsequent development of Mendelism and the mutation theory lies outside the scope of the present study. Here we need only note the extent to which these new initiatives were shaped by the frustrations that had built up within late nineteenth-century Darwinism. Convinced that the artificial world of the laboratory would yield insights superior to those of the earlier generation of naturalists, the Mendelians inevitably rejected the theoretical principles of Darwinism along with its techniques. Since discontinuous variations were easier to study than continuous ones, they argued that the Darwinians had been wrong to suppose that evolution must be a gradual process. Furthermore, since new characters survived in the laboratory whatever their adaptive value, it was easy to believe that the selectionists had exaggerated the role of utility. The instinctive defense of both these positions by the biometrical school ensured that the split between Darwinism and Mendelism would be complete. Yet the new experimentalism continued one central theme of neo-Darwinism: its insistence on hard heredity. Thus despite its rejection of the selection mechanism, genetics became the implacable enemy of Lamarckism and soon succeeded in destroying the credibility of the experimental evidence for the inheritance of acquired characters. The emergence of the "modern synthesis" in the 1930s represented far more than a recognition that their common reliance on hard heredity allowed Mendelism and Darwinism to be combined. It also signalled the bridging of a major split that had emerged within the ranks of scientific biology, a split between the old traditions of field study and paleontology and the new experimentalism.

III. Tempo and Mode in the Darwinian Revolution

By extending the study of the reaction to Darwinism into the late nineteenth century, we are forced to take note of a complex pattern of developments. It is no longer enough merely to ask how the general idea of evolution, or the particular mechanism of natural selection, was received by the various biological disciplines. Accepting that the principle of naturalism was rapidly adopted in most areas, we need to know why the opposition to Darwin's particular interpretation of the process unfolded in the way it did. Why was Darwinism able to gain so strong a hold in Britain, but was forced to grapple with an alternative view of evolution almost from the very beginning in America? Why did support for Lamarckism intensify in the 1890s, even in Britain, and what was the link between this movement and the earlier opposition to Darwinism? Why did the climax of Lamarckism coincide with parallel rejection of Darwinism by the new generation of biologists committed to the experimental method? Questions such as these must be answered if we are to have a comprehensive understanding of the role played by Darwinism in nineteenth-century biology. My survey of the debate itself has already hinted at some possible answers, but we must now look more closely at the factors that may have shaped the rise and fall of the original form of Darwinism.

The eclipse of Darwinism at the end of the nineteenth century introduces a major complication for the historian trying to fit the Darwinian revolution into any of the existing schemes of scientific development. Several writers have already suggested that T. S. Kuhn's notion of revolutions as paradigm changes (1962b) is not adequate to deal with the rise of evolutionism, or at least will have to be extensively modified (for instance Greene 1971; Mayr 1972a). These opinions are based mostly on the study of events leading up to the introduction of Darwinism, but we must now deal with further complications in the process by which the theory eventually came to serve as the basis of modern evolutionism. Kuhn's scheme certainly does not suppose that a new theoretical initiative is turned into a paradigm overnight — in the case of the Copernican revolution it took over a century before Newton realized the full potential of the sun-centered cosmology. Yet there was no "eclipse of Copernicanism" corresponding to the rise and fall of nineteenth-century Darwinism. The Copernican revolution saw a number of parallel developments that were not synthesized for some considerable time, but there was no crucial part of Copernicus's original theory that was first accepted, then widely rejected, and finally revived by Newton. The fact that Darwinism did undergo such a fluctuation in its popularity points to an even more complex system of factors influencing scientists' views of evolution in the late nineteenth century.

The starting point for the analysis that follows is the hypothesis that the eclipse of Darwinism was essentially the coincidence in time of two quite different reactions against the theory. What we now perceive as the "Darwinian revolution" occurred in two phases; the complexity of its structure results from the fact that a backlash against the first phase reached its climax just as the second phase was getting underway. The first phase is the revolution in natural history whereby the subject was converted to a naturalistic viewpoint that posited the legitimacy of a scientific investigation of the origin of species. Darwinism clearly played a major role in precipitating this revolution, and its supporters cleverly exploited the initiative of this pioneering role to create the theory's first wave of popularity. From the beginning there was opposition from conservative thinkers within science, but at first these thinkers had no naturalistic alternative to offer, and it took some time for them to put together a system based on Lamarckism and orthogenesis that would salvage as much as possible of the pre-Darwinian world view. The Americans were quicker off the mark in this respect; for this reason Darwinism did not gain so dominant a status in American science. In Britain, on the other hand, it was only when the limitations of the original form of Darwinism became apparent that the more conservative alternative was able to make a serious bid for scientific respectability. At the same time, however, the dissatisfaction with Darwinism was breeding an even more radical challenge directed not only against Darwinism but against the whole tradition of natural history. Where the Lamarckians were trying to turn the clock back, the supporters of the new experimentalism wished to transcend the limitations of the techniques upon which both Darwinism and Lamarckism were based. This was the second phase of the revolution, in which the insights of Darwinism would be revived only when the study of heredity had provided them with more secure foundations.

The second phase of the revolution lies outside the scope of the present article, except in the sense that we are able to pinpoint the source of the frustrations that led the experimentalists to reject the original form of Darwinism. Our real concern must be the first revolution, the introduction of naturalism. The chief aim of the following discussion is to suggest that the eclipse of Darwinism must force us to look again at the circumstances that allowed the theory to become so successful in the first place. Once we realize that Lamarckism and orthogenesis were compromises whereby certain aspects of an earlier world view were adapted to the new climate of naturalism, it becomes necessary to ask why it was Darwinism, rather than these less radical alternatives, that played the crucial role in converting the scientific world to evolutionism. Why were these compromises not developed before, or at least at the same time as Darwinism, thereby smoothing the path of the general trend toward naturalism? Why was it necessary to go through the trauma of having evolutionism presented in its most radical form in order that the revolution could be initiated? If less radical alternatives were not only conceivable, but able to attract considerable support later in the century, should they not have led the way — or at least gained wide support as an immediate reaction to the Darwinian challenge? We know that theistic evolutionism was indeed suggested first, in Chambers's *Vestiges*, and would probably have gained in popularity even without the stimulus of the Darwinian challenge. But there is no evidence that any significant move was underway to develop natural mechanisms that would still preserve the elements of order and purpose so basic to the traditional interpretation of nature. Thus Darwinism gained a head start and was able to capitalize upon the fact that it was the first "scientific" mechanism of natural evolution. So great was the confusion in the conservative camp, that it took some time for the details of a satisfactory alternative to be worked out, even in America, while in Britain the traditional viewpoint was virtually driven underground after the collapse of theistic evolutionism.

These events must be explained partly in terms of the unique circumstances surrounding the development of natural selection and the eventual publication of the theory. But we also seem to be dealing with a failure of initiative on the part of those supporters of the traditional view of nature who were so slow in responding to the challenge of Darwinism. The Darwinian revolution was so much more a *revolution* (in the sense of a relatively sudden change) because the most obvious line of conceptual development leading toward a naturalistic theory of evolution was circumvented. This situation must be explained not only in terms of the success of Darwin and his followers in expounding the positive applications of their theory, but also in terms of the failure by the opposition to come up with anything more than merely negative arguments against selection during the early years of the debate. If we can identify the factors that helped to slow the development of the Lamarckian alternative, we shall have added a new level to our comprehension of the process by which Darwinism gained its initial success. This is not, of course, a conventional approach to the study of Darwinism, and it might be argued that we should not be spending so much time dealing with the opposition to the selection theory. It is more usual to explain the triumph of Darwinism by showing how the theory's successes confounded those who had insisted that no naturalistic mechanism of evolution would ever prove adequate to cope with the problems of natural history. This aspect of the debate is clearly important, and has been outlined above, but its details have already been explored at great length by a host of competent historians. It is time that the role played by the failure of the opposition is taken into account in the preparation of a more balanced understanding of the Darwinian revolution.

One advantage of this approach for those with more orthodox concerns is that it should force us to define more clearly the vague notion that

Darwinism was somehow "in the air" by the late 1850s. It is certainly true that a number of naturalists were becoming interested in the process of specialization revealed by the trends in the fossil record. But so long as their thinking on this topic was confined within the limits of natural theology, the trends were still looked upon as manifestations of divine purpose rather than the subject for scientific investigation. Even when converted into a form of theistic evolutionism, this approach did not touch on the Darwinian question of the origin of species. If we can show why there was no chance of the traditional world view converting itself spontaneously to naturalistic evolutionism, we shall have shown precisely why the new initiative of Darwinism was so essential. At the same time, we shall have to ask once again why Darwinism was able to gain so much headway during its early years. The only writers who have expressed surprise that Darwinism was ever able to get off the ground in the first place have been those who started from a conviction that the selection theory was fatally handicapped either by flaws in its basic logic or by the weakness of components such as the available theory of heredity. It does not make sense to imply that Darwinism was able to flourish despite these limitations, as though scientists were somehow blinded by their materialistic preconceptions into accepting a totally unworkable theory. But it is legitimate to ask why the opposition to the selection theory found it so difficult to create a less radical form of evolutionism, since it seems intrinsically plausible to suppose that had the Lamarckian alternative been in the running at the very beginning of the debate, the Darwinists might have had a much more difficult job to persuade their contemporaries that the selection theory represented the best introduction to naturalistic evolutionism.

Implicit in all this is the assumption that the changing attitudes toward Darwinism and its alternatives were not conditioned solely by new scientific discoveries. All of the arguments used against Darwinism during its eclipse had already been formulated during the first decade of the debate. The original opponents of natural selection had identified its chief weaknesses, but they were unable to prevent the theory from dominating scientific thought on the origin of species, at least in Britain. Much of the later opposition was generated not by the discovery of new, contradictory facts, but by the growth of a feeling that Darwinism had not fulfilled its original promise. Similarly, there was no dramatic improvement in the evidence for Lamarckism toward the end of the century that would explain that theory's rise to popularity. Perhaps the American paleontologists had been able to put together more examples of supposedly linear developments in the fossil record, yet the essence of their argument for Lamarckism and orthogenesis had already been established by the early 1870s. The best experimental evidence for the inheritance of acquired characters — C. E. Brown-Séquard's work on epilepsy in guinea pigs — was already available in 1860, and the rise in the popularity of Lamarckism came *after* the publication of Weismann's experimental disproof.[12] These points all suggest that the changing fortunes of the competing evolutionary theories were conditioned by other than purely rational factors.

At the same time it should be obvious that the rise of Darwinism cannot be explained as an inevitable consequence of the growing tide of scientific naturalism. If that were the case, then, being the most radical theory, it ought to have been the last to gain popularity, not the first. In part, the success of Darwinism in Britain can be explained by the skill with which Huxley and his circle gained control of the means of publication and the system of scientific education. This is hardly a new suggestion about the tactics of a scientific revolution, and it does not go far enough in this particular case. Lamarckism revived at the end of the century *despite* the weight of academic orthodoxy ranged against it. The implication of this is that the popularity of scientific theories is controlled by even less tangible factors. David Hull (1978b) has suggested that the success or failure of a new theory may be determined in part by the "image" its supporters present to the scientific world. They must be adept at twisting the theory to meet every objection, while always presenting any changes as modifications of, rather than replacements for, their original ideas. They must present a united front that stresses the most successful areas of application and ensures that the difficulties keep a low profile. They must not fall out in public on matters of principle; they must impose their own terminology on the language of scientific discourse. If they are more able than their opponents in the employment of these debating skills, the theory will take on a life of its own and will establish itself in the scientific imagination. The success of Darwinism may certainly be explained by these factors, although it may be equally important to show why its opponents failed for so long to create an adequate image for their alternative ideas.

There are some who will no doubt object to the claim that good public relations are important for the promotion of a new theory. But all this is not meant to imply that a good PR team can "sell" a bad theory, only that in a case in which the theory cannot cope with all the problems confronting it, good tactics at this level may play an important role in ensuring its success. In any case, the opponents of Darwinism have always claimed that the theory represents an entrenched dogma that has somehow mesmerized its supporters so that they cannot appreciate its weaknesses. This view was expressed during the eclipse of Darwinism and it is still the opinion of many anti-Darwinists both inside and outside science. The fact that an increasing number of professional biologists have become prepared to challenge the dogma in recent years should make it all the more easy for us to explore the non-scientific factors that helped to create Darwinism's original image.

To begin with the Darwinians themselves, we can note a number of points about their tactics. Although small in number at first, they successfully influenced the systems of both scientific publication and education so as to create a favorable climate for the theory. Huxley's role in this process was just as important as his public defense of Darwinism, although he was aided by an ever-increasing number of allies and disciples. Ruse points out that Darwin had deliberately cultivated a set of friends and colleagues who would be in a position to ensure his theory a smooth entry into the scientific community (1979a, pp. 253–254). This group was aided by the emergence of a new generation of scientists with no formal ties to religion, who were now taking control of teaching at both the old universities and newer institutions such as the Royal School of Mines. There was a snowball effect here, since the more influence the Darwinians gained in the appointment of new professors and lecturers, the greater became their ability to control future developments within the profession. There was also a good deal of behind-the-scenes activity centered on the "X Club", to which Huxley, Hooker, and Spencer belonged (MacLeod 1970). By deliberately avoiding too much open propaganda for natural selection in the more orthodox scientific societies and their publications, the Darwinists ensured that less enthusiastic naturalists would not be alienated during the early phases of the debate. As their numbers increased, they were able to permit more open discussion through their control of the editorial process. Darwinians were deeply involved in the creation of the new, semi-popular journal *Nature* in 1870, to which Huxley contributed many articles during its early years (MacLeod 1969). Significantly, it was the appearance of frequent references to the inheritance of acquired characters in *Nature* around 1890 that Samuel Butler saw as evidence of a swing at last toward Lamarckism (1908, p. 309).

In both their scientific and more popular writings, the Darwinians defended their theory vigorously, but not dogmatically. The original form of Darwinism was based on a coherent central theme but was extremely flexible in its details. None of the original supporters was an all-or-nothing selectionist, and several additional mechanisms of evolution were suggested. Huxley opted for saltations, while Darwin admitted the inheritance of acquired characters and Spencer included a major role for Lamarckism. But the Darwinists were careful not to criticize one another when they disagreed over the most likely supplementary mechanisms, and all were presented as *additions* to selection, compatible with the overall Darwinian view of evolution. Thus, far from undermining the credibility of the basic theory, these concessions helped to disarm the opposition by showing that criticisms of natural selection were not necessarily fatal. One of the many reasons why Lamarckism was not at first seen as an alternative to selection was that it was successfully incorporated as a minor element *within* Darwinism. It was precisely the breakdown of this open-minded, flexible policy under the influence of Weismann's concept of the germ plasm that helped to create the dogmatic image of neo-Darwinism, thereby forcing anyone with an interest in non-Darwinian mechanisms into the opposing camp. Weismann's views may have been developed in response to genuine problems within the original form of Darwinism, but their sheer inflexibility was guaranteed to boost the level of opposition by forcing many biologists to oppose a theory with which they had once been willing to compromise.

Turning now to the opposition, it seems plausible to suppose that its failure to stem the tide of Darwinism in Britain may be attributed at least in part to the weakness of its tactics. Desmond (1982) points out that Owen and his followers had some success in describing goal-directed trends in the fossil record, but concedes that their work was too diffuse to form a coherent alternative. Owen's abrasive personality ensured that he would be unable to prevent Huxley's group from dominating the scientific community. Mivart was even more inept, allowing the Darwinians to maneuver him into a position in which he appeared socially unacceptable.[13] By contrast, the fairly rapid appearance of a coherent neo-Lamarckian opposition in the United States can be linked to the influence of Louis Agassiz upon the educational system and the scientific community in general. Although Agassiz's opposition to evolution left him out on a limb, his followers rapidly adapted certain aspects of his world view to the new situation and were able to exploit their position within the establishment to promote their alternative interpretation of evolution. Agassiz's son, Alexander, and innumerable students and disciples continued to dominate the scene at Harvard and many other universities, and in prominent groups such as the Boston Society of Natural History.[14] A whole range of American journals allowed their ideas to be printed at considerable length as they formulated the Lamarckian alternative around 1870. They were also able to found their own journal, the *American Naturalist*, dedicated to preserving the link between science and natural theology.[15] Agassiz may have lost the original debate with Gray, just as Owen lost to Huxley, but here the resemblance ends. Gray was unable to prevent an anti-Darwinian school of evolution from flourishing within the scientific community that Agassiz had done so much to shape.

The ability of the American school to promote its alternative within the existing academic framework gave them a head start in the fight against Darwinism, but it cannot explain the relative ease with which they adapted Agassiz's view of nature to the new current of scientific naturalism. There was no shortage of conservative naturalists in Britain, but none of them took up Butler's suggestion that Lamarckism could be used to preserve the spirit of natural theology. This can be explained only by supposing that there were intellectual forces at work in America, but not at first in Britain, facilitating the emergence of the Lamarckian alternative. To understand this, we must take into account the fact that the inheritance of acquired characters can be built into a number of quite different conceptual systems. Lamarckians from Butler to Arthur Koestler have insisted that their thesis is morally preferable to the nightmare of Darwinian trial-and-error, but the fact that Spencer was also a Lamarckian shows that this moral connection is not a necessary one. For Spencer, the inheritance of acquired characters was an obvious component of a philosophy of natural evolution, quite compatible with natural selection and a generally "Darwinian" picture of how life developed. For Butler, natural selection had to be *replaced* by the inheritance of acquired characters for the world to retain its moral purpose — yet evolution was still supposed to be a shortsighted and haphazard affair. The Americans appreciated the moral possibilities of Lamarckism, but it was their belief that evolution is *not* irregular that led them to take a stand against Darwinism. Thanks to the influence of Agassiz's idealist philosophy, their conception of evolution was modeled on embryological growth, and the recapitulation theory became the foundation of their philosophy of development (Bowler 1977b; Gould 1977b). Cope and Hyatt both proposed their "law of acceleration of growth" to explain the parallel between ontogeny and phylogeny, without at first offering any explanation of why new stages were added on to growth in a regular sequence in the course of a species' evolution. Only in the 1870s did they begin to realize that, in the case of adaptive characters, the regularity of the additions could be accounted for by assuming the inheritance of characters acquired in response to a consistently applied behavior pattern. For the paleontologists of the American school, Lamarckism was a secondary product of their search for an orderly pattern of development in evolution, not an alternative to selection within a generally Darwinian framework. The gulf between their Lamarckism and that of Spencer — or even Butler — is clearly illustrated by the fact that Darwin himself, who had no problems with the inheritance of acquired characters, found the writings of Cope and Hyatt quite unintelligible.[16]

It will be necessary for us to identify why the peculiar origins of American neo-Lamarckism allowed it to serve as a means of rallying conservative opposition to Darwinism. But first we must ask why Lamarckism in its most simple form (that is, a direct reliance upon the inheritance of acquired characters) was unable to play the same role. Indeed, it may be worth asking why the theory did not play a greater role in the original conversion of the scientific world to acceptance of adaptive evolution. By the end of the 1850s, we know that a small group of younger naturalists had become impatient with the restrictions imposed by the old philosophy of design and were on the lookout for a naturalistic theory to explain the origin of species. The reaction of people such as Huxley and Hooker to Darwin's proposal confirms this — and yet there is no evidence of these naturalists exploring the possibility that Lamarckism might break the deadlock in which they found themselves. To put the question another way: what would have happened if Wallace had not forced Darwin's hand, so that the selection alternative had remained concealed for a few more years? There can be no doubt that Spencer would have promoted a naturalistic approach to evolutionism based on Lamarckism, and would have tried to interest the scientific community in the possibility. He would have written his *Principles of Biology* anyway, and even in its post-Darwinian form (1864–1867) this work gives a greater role to Lamarckism than to natural selection. In the absence of Darwin's theory, could Spencer have convinced the younger naturalists that Lamarckism offered a plausible foundation for a theory of naturalistic evolution? The question is not as ridiculous as it seems, since we know that later in the century students of paleontology and geographical distribution did take Lamarckism seriously. If Spencer could have started the evolutionary ball rolling in this way, Darwin would have had to publish his new mechanism in a climate in which the alternative explanation of adaptation had already seized the initiative, and the subsequent history of evolutionism would have been very different.

In fact, it seems likely that Spencer would have found it very difficult to convince even the most radical naturalists that Lamarckism was an adequate basis on which to found a new science of evolution. Since he did not think of natural selection himself, Spencer had from the beginning of his career assumed that Lamarckism was the only available solution to the conceptual problem of the origin of species. The evidence suggests that no working scientist at the time was prepared to take the inheritance of acquired characters this seriously. Men such as Huxley and Hooker never seem to have given the mechanism a second thought, and Huxley, at least, remained profoundly opposed to it throughout the rest of his career. Whether Spencer's philosophical arguments could have forced them to take another look at the whole issue of evolution is a question we cannot answer. The only purpose of proposing such a hypothetical scenario is to bring home to us the extraordinary impact that Darwin did in fact have on the scientific world. His new theory came as such a revelation precisely because the only other possible mechanism of adaptive evolution was not taken seriously at the time.

Since Darwin did publish before Spencer had time to develop a detailed case for the inheritance of acquired characters, Spencer was forced to incorporate natural selection into his own system. He was thus perceived by most scientists as a somewhat anomalous Darwinian. But what of those who opposed the wholesale slide into scientific naturalism? Lamarckism could be used as a device for going along with this trend while still retaining some element of the old concept of benevolent design. Yet there is little to suggest that the opponents of Darwinism were anxious to take up this solution to their problems; even Butler's appeals fell upon deaf ears. Lamarckism simply did not have the power to command attention even from those who might have made the same use of it as Butler and the American school. There are occasional references to Lamarck's views in the British anti-Darwinian literature, but the inheritance of acquired characters was always treated at best as a subsidiary mechanism, of little real importance. The later editions of Chambers's *Vestiges*, still promoting the mysterious law of progress, referred in addition to Lamarck's mechanism of use-inheritance to explain the production of adaptive characters (1846, pp. 235–236; 1860, pp. 160–161). We have already noted Mivart's eventual acknowledgement of the evolutionary significance of the environment's ability to affect the growth of the organism (1884), and Carpenter also wrote of the inheritance of acquired mental characters (1873). Yet none of these theistic evolutionists saw Lamarckism as a sufficient alternative to the Darwinian philosophy of trial-and-error. In some basic way, Lamarckism did not satisfy their requirements for an anti-Darwinian weapon, and it was allowed to remain on the sidelines of the debate.

Two reasons can be suggested to explain the failure of Lamarckism to play a significant role in the British reaction to Darwinism. The most obvious is that the theory's image was still tarnished as a result of its rejection by Lyell and other writers earlier in the century. As Chambers recorded (1846, p. 234), no one at the time considered Lamarck's concept of use-inheritance to be a plausible mechanism to explain the evolution of life. Darwin always wrote slightingly of Lamarck in private (for instance *LL* 2:23, 39, 215), although the paragraph in the "Historical Sketch" prefaced to later editions of the *Origin* is rather more respectful (*Origin* 1959, p. 60). In his analysis of the reception of the *Origin*, Huxley dismissed Lamarck's views as without influence and referred to Lyell's "trenchant and effective criticism" (1887b, 2: 189). Later Lamarckians have complained about the misrepresentation and oversimplification of Lamarck's views by the early Darwinians and have openly hinted at a conspiracy to play down the role of Lamarckism in order to exaggerate the significance of Darwin's work (H. G. Cannon 1959, chap. 2). The evidence suggests, however, that the lack of faith in Lamarck was not confined to the Darwinians, but was a general feeling shared by the majority of biologists in the mid-nineteenth century. Butler noted that Lamarck had been "so systematically laughed at that it amounts to little less than philosophical suicide for anyone to stand up on his behalf" (1879, p. 61). Although he believed that this in part explained the refusal of the scientific community to take his own anti-Darwinian views seriously, it is significant that Butler himself devoted as much space to Buffon and Erasmus Darwin as to Lamarck in his book *Evolution, Old and New* (1879).

It is significant that most of the Americans developed their ideas without reference to Lamarck's name or actual writings. The possibility that the inheritance of acquired characters might be linked to the acceleration of growth only occurred to them after they had committed themselves to the recapitulation theory. Cope later admitted that he had not read Lamarck when he first proposed his "Lamarckian" views (1887, p. 423). Even Packard, who did know of Lamarck's writings and actually coined the term "neo-Lamarckism" in 1885, started with an interest in the law of acceleration rather than the inheritance of acquired characters (1870, 1872). The creation of a self-proclaimed school of "Lamarckism" was the result of conscious decision by Packard and others to promote Lamarck's name as a label for an anti-Darwinian philosophy of evolution that was already fully developed. The movement succeeded in reviving Lamarck's reputation where Butler had failed, to the extent that even a Darwinist such as Lankester (1888–1889) could concede that Lamarckism was a "reputable denomination" for those opposed to natural selection. Yet many Lamarckians remained indifferent to Lamarck's own writings, and the circumstances in which his name was grafted on to a movement that was *not* inspired by his own ideas suggests that far more than a spirit of historical objectivity was at work. Lamarck's name was seized upon as a convenient symbol by those who saw the inheritance of acquired characters as a bulwark against neo-Darwinism, and who quite correctly perceived that the dogmatism of Weismann and his followers had brought Darwinism itself into disrepute. On its own, Lamarckism could not have precipitated the revolution in biology. Under that name, at least, it could be revived only as an alternative explanation of evolution once Darwinism had shown that it too was by no means without fundamental weaknesses.

The fact that the more successful American form of neo-Lamarckism drew its real inspiration from Agassiz rather than from Lamarck's own writings suggests that it was the idealist philosophy of nature that lay at the heart of the most sustained anti-Darwinian feeling. Many naturalists at first opposed natural selection as an adequate explanation of adaptive evolution, but this issue was not crucial enough to require the creation of an alternative philosophy of evolution. To the extent that evolution *was* adaptive, many were prepared to concede that natural mechanisms might be involved. Lamarckism was no doubt morally preferable to selection, as Butler pointed out; but even the more conservative naturalists were not prepared to follow Butler in seizing upon this issue as the basis for the construction of a complete anti-Darwinian system. The real source of their opposition to Darwin was not his particular mechanism of adaptation, but his basic utilitarian assumption that all evolution can be explained in terms of adaptation. The emphasis on the role of linear trends that links theistic evolution, American neo-Lamarckism, and the theory of orthogenesis, was associated with a belief that nature is an orderly system whose development cannot be reduced to the trivial and haphazard requirements of adaptation. To replace selection with Lamarckism was not enough, if it was still admitted that adaptation was the sole driving force of evolution. As Chambers put it, the weakness of simple Lamarckism lay in its "giving this adaptive principle too much to do" (1860, p. 161). For most of Darwin's opponents, there had to be a non-utilitarian factor ensuring a level of orderliness in the development of life. This was the real point of Sir J. F. W. Herschel's famous complaint that natural selection represented the "law of higgeldy-piggeldy".[17] It was the reason why Mivart was so interested in parallel evolution, and why Carpenter and the paleontologists of the American school all stressed the regularity of evolutionary trends and the lack of adaptive purpose in some of them. What became known as "Lamarckism" acquired the status of an alternative to Darwinism only when the Americans incorporated it into this more general level of opposition on the basis of the preconception that development must be a linear orthogenetic process.

For the American school, Butler's point about the moral superiority of use-inheritance was merely a bonus gained by adopting Lamarckism as an explanation of the regular extension of growth required by their interpretation of the recapitulation theory. Their system could resist the tide of Darwinism that overwhelmed the idealist philosophy of nature in Britain because it successfully adapted the concern for natural order to the new tide of naturalism. The great weakness of theistic evolutionism, as expounded by Mivart, Carpenter, Argyll, Herschel, and others, was that it continued to present the "laws" of evolution as the embodiment of a preconceived plan of development aimed at a definite goal. Theistic evolutionism thus retained the supernatural as a (scientifically unverifiable) explanatory factor. The Americans realized correctly that this was no longer acceptable in the climate of post-Darwinian naturalism, and they sought instead to explain the regular trends in evolution as the result of biologically determined "laws of growth", influenced sometimes by the organisms' response to the environment. That these mysterious processes might themselves prove inexplicable in terms of pure mechanism was not the point. There was no explicit appeal to the supernatural in their system, nor did they invoke purposeful goals for every trend; their belief in the orderliness of natural development could thus be promoted as an article of scientific rather than religious faith. But this still leaves one question unanswered: why was it only in America that the traditional view of nature was successfully "modernized" in this way? There was no shortage of naturalists in Britain who shared the belief in natural order, yet none of them was able to create a coherent alternative to Darwinism in which prior beliefs were purged of their overtly metaphysical overtones.

The answer to this must lie in the particular character of Agassiz's idealist view of development, which retained elements of a much earlier concept of embryological growth that had been abandoned by most British idealists. In its earliest form, the recapitulation theory had been based on the "law of parallelism", in which the growth of the embryo and the history of life on earth both represent the ascent of a linear hierarchy toward man (E. S. Russell 1916; Gould 1977b). Agassiz had gone far beyond the notion of a simple "chain of being" linking all forms of life, but he still tended to think in terms of a basic hierarchy of the vertebrate classes through which life had ascended toward the perfection of the human form. It was this aspect of his thought that his students adapted to evolution by postulating linear patterns of growth-extension leading toward apparently predetermined goals. The success of the American school derived from their acceptance of a "Darwinian" image of branching evolution on a *large* scale, meanwhile retaining Agassiz's concept of parallel, linear development on a smaller scale within each branch of the tree of life (Bowler 1976a). By limiting their search for order to the trends within a single group, they could accept that each branch of evolution had gone off in its own direction and that there was no final goal for the whole of life. Agassiz's simple progressionist vision was thus purged of its more obviously teleological implications. The American paleontologists could then search for linear trends within each branch without apparently having to postulate goal-directed forces. Indeed, in their willingness to accept that the "goal" of some trends was non-adaptive, or even fatal, they eliminated the normal interpretation of purpose from their system. Hyatt's concept of racial senility is the ultimate extension of the idealists' belief that individual growth contains the pattern of all development, into a realm where the concept of a "goal" or "purpose" has a purely formal significance. Nothing could illustrate more clearly how limited was the success of the movement to introduce a concept of branching evolution, of which Darwinism was merely the most radical manifestation. For the Americans, the concept of branching was the means not of breaking down the traditional image of orderly development, but of preserving it in a new, more flexible, and far less obviously teleological form. Their success is a measure of the extent to which the naturalists of the late nineteenth century were not yet ready to abandon completely the old, hierarchical view of nature.

In Britain there was nothing to parallel this subtle transformation of the traditional view of orderly development. The closest approach to it was made by Chambers, who had long been advocating a law of progress modeled on embryological growth. But his vision of development was still largely confined to the old, unilinear approach, and it presented the upward ascent of life as the unfolding of a divine plan. Even when he postulated parallel lines of evolution, Chambers merely arranged modern forms into progressive hierarchies that no naturalist could take seriously as evolutionary genealogies (Hodge 1972; Bowler 1976a, pp. 54–62). Among the trained naturalists, two of the leading exponents of the idealist view of order had modified their interpretation so as to abandon altogether the notion of linear hierarchies. Under the influence of Karl Ernst von Baer's much earlier refutation of the law of parallelism, both Owen and Carpenter had recognized that the most basic trend to be observed in the fossil record was a multifaceted process of divergence into an ever-increasing number of evolutionary avenues (Ospovat 1976, 1981). Owen's concept of the archetype as the central theme linking all of these various manifestations still served the explicit purpose of satisfying his idealist desire for order, but it was profoundly at variance with Agassiz's hierarchical approach. In a sense, Owen and

Carpenter were prevented from developing a coherent alternative to Darwinism precisely because they had advanced beyond the linear viewpoint of the old recapitulation theory. Independently of Darwin, they had spearheaded the breakdown of the old hierarchical interpretation of natural order. Although they could not accept Darwin's radical explanation of branching development, their view of the basic pattern of life's history was so close to his that it was difficult for them to identify a means of demonstrating the falsity of his mechanism. Desmond (1982) has shown that Owen and his followers did make use of the concept of purposeful evolutionary trends to create a theory of the polyphyletic origin of the mammals, but even some of the Darwinists accepted the possibility of such trends. Under the influence of Agassiz, however, the Americans hoped to demonstrate the existence of linear, and at least partly non-adaptive, trends at a much more precisely defined level. To the extent that they were able to provide apparently valid evidence for these trends, they could provide an effective scientific argument against natural selection. By modern standards, Owen was justified in his rejection of the linear viewpoint upon which the Americans built. But in the late nineteenth-century context, the very sophistication of his approach prevented his creating a coherent alternative to Darwinism. It was the Americans, with their more traditional view of development, who were able to replace the vague idea of a divine order with an apparently naturalistic, but very un-Darwinian theory of evolution by extension to growth.

In a paradoxical sense, the unique character of American neo-Lamarckism illustrates the crucial impact of Darwinism upon the scientific world. I have already suggested that simple Lamarckism did not have the power to convert the more radical naturalists to evolution. Yet it may also be argued that, by itself, the idealist philosophy of nature could not have promoted a fully fledged evolutionary system. In their very different ways, both Agassiz and Owen had helped to generate an interest in the trends that could be observed within the fossil record. In addition, Owen had correctly recognized the significance of specialization as a corollary of the thesis that the development of life consists of a series of ever-diverging branches. It could perhaps be argued that naturalists such as Owen were inevitably converting the traditional world view into an evolutionary one, by coming ever closer to the idea that the succession of related forms follows a pattern most easily explained by assuming the adaptive transmutation of one form into the next. Yet the furthest that Owen would go, even under the stimulus of Darwinism, was to concede a form of theistic evolutionism in which the whole question of a mechanism of change was side-stepped by retaining the appeal to design (1866–1868, 3: chap. 40). If we accept that there was a trend toward theistic evolutionism in the mid-nineteenth century, we must still recognize the extent to which the concept of design limited the scientific value of this approach. It was Darwin who broke the deadlock by showing that the naturalist could go beyond the level of mere description to inquire into the cause of the developments he observed. It is also clear that the Agassiz school would never have developed their idealist version of evolution without the stimulus of Darwinism. Agassiz himself was too strongly committed to the belief that each species represented a fixed element in the divine plan ever to accept transmutation (Mayr 1959a). It took the impact of Darwin's mechanism of evolution to convince Agassiz's students that creationism was no longer a viable position, thus forcing them to break with their teacher on this issue (Dexter 1965). For all the anti-Darwinian implications of the theory they eventually devised, it is probable that without the prompting of Darwinism they would never have expanded the embryological analogy into anything like a naturalistic mechanism of evolution.

It was no accident that the leading proponents of the more orthogenetic form of American neo-Lamarckism were paleontologists. By concentrating on the fossil record for particular orders and families, Cope and Hyatt were able to make out an apparently plausible case for linear and even non-adaptive trends. Many later paleontologists dismissed this aspect of their work as a product of their imagination, imposed upon evidence so inadequate that it concealed the true complexity of the evolutionary process (Simpson 1944; Jepsen 1949). Yet it is worth noting that pupils of Cope such as William Berryman Scott and Henry Fairfield Osborn continued to advocate the idea of linear evolution in the early twentieth century, long after they had been forced to abandon a Lamarckian interpretation of the trends (W. B. Scott 1929; Osborn 1929; see Rainger 1981). Here again the fascination of orthogenetic development reveals itself as the truly significant anti-Darwinian factor. On the other hand, those members of the American school who worked in other areas, particularly geographical distribution, were drawn away from the orthogenetic approach and sometimes adopted a more "Darwinian" interpretation of evolution. This is the case with Packard, who began in the 1870s with an interest in the law of acceleration of growth, but then went on to suggest a synthesis of the Darwinian and Lamarckian approaches (1901; see Dexter 1979). In those areas in which the evidence would not support the concept of linear development, the inheritance of acquired characters could be recognized as an evolutionary mechanism in its own right. American neo-Lamarckism thus moved in two rather different directions, with the paleontologists' concept of linear evolution being only loosely correlated with the field naturalists' growing belief that the character of the individual, and hence of the species, was shaped by the changing nature of the environment. The appearance of this more conventionally "Lamarckian" interpretation of the theory was a natural product of the American school's extension into areas other than paleontology. Without the head start given to it by Agassiz's influence, however, even this form of Lamarckism would not have been able to flourish several decades before it began to get a hearing in Britain (Bowler 1983).

By studying the alternatives to natural selection, I have been able to throw some light on the role played by Darwinism in late nineteenth-century science. We have a better understanding of the forces that were ranged against it and of the degree of its success in overcoming these forces. Although certain aspects of the Darwinian world view, particularly its use of branching development, were pioneered independently by naturalists such as Owen and Carpenter, there is little evidence that these more conservative thinkers could ever have engineered the transition to a naturalistic view of evolution. The Lamarckian alternative was too heavily discredited by earlier criticisms to serve as a stimulus for either conservative or radical naturalists to think again about the origin of species. Darwinism thus succeeded in converting the scientific world to evolutionism because it broke the deadlock that had built up within conventional nineteenth-century thought on the issue. Those who were already predisposed toward a naturalistic, utilitarian viewpoint seized upon the theory of natural selection, in its original, undogmatic form, as the key that would unlock the door to the whole new world of scientific investigation. The more conservative naturalists did not approve of natural selection, but they found it difficult to resist the general trend toward evolutionism. They turned instead to theistic evolutionism, and it is a measure of Darwinism's success that this effort to resist the trend toward naturalism was comparatively short-lived.

There can be little doubt that the idealist concept of orderly development was the most powerful source of conservative opposition to Darwinism and lay at the heart of most versions of theistic evolutionism. The idealists' options for converting their viewpoint into an apparently naturalistic alternative were limited, however. Only the Americans succeeded in this endeavor, because Agassiz's rather outdated notion of linear development could be exploited on a smaller scale by appealing to the paleontological evidence for directed evolution. The growth of a genuinely "Lamarckian" interpretation of this approach came later, as the paleontologists began to realize that use-inheritance could account at least for those trends with an adaptive goal. A form of Lamarckism that placed more stress on the direct influence of the environment was also developed by those students of Agassiz who moved into those areas stressing geographical studies. In Britain, the idealism of Owen — although in some respects more sophisticated — could not form the basis for a theory of directed evolution. The Lamarckian alternative remained largely dormant until growing dissatisfaction with Darwinism allowed naturalists to take a more sympathetic look at the only utilitarian alternative. Here, at least, the sudden flourishing of Lamarckism at the end of the century was the result of changes within Darwinism itself, changes that were intended as a response to the problems that had arisen, but that were perceived as unjustified dogmatism by those more aware of the difficulties.

Although Lamarckism gained in popularity toward the end of the century, it did not offer any really new initiative to solve the problems that had become apparent within evolution theory. By taking embryological growth as its model for evolution, one could make it appear naturalistic; but this approach proved to be of little value in solving the problem of heredity that had been exposed by the weakness of Darwinism. The Lamarckian movement was essentially a reaction against the mechanistic image of Darwinism within the established tradition of natural history, drawing most of its support from those who wished to make one last effort in defence of an older view of natural development. The simultaneous rejection of Darwinism by the exponents of the new experimental study of heredity had an entirely different motivation. Here there was no turning back to the past, but a radical call to abandon the foundations of natural history so that the study of evolution could be placed on firmer foundations. The frustrations underlying this movement may have been the same as those encouraging Lamarckism: a feeling that the neo-Darwinians had failed to deal with the problems inherent in their whole approach. But instead of using this as an excuse to revive an older view of nature, the experimentalists intended to precipitate a second revolution in the field, equivalent in scope (though not at first in direction) to the first Darwinian revolution.

The repudiation of natural selection by Bateson and the early Mendelians was only one manifestation of their conviction that the techniques of natural history were played out. In the end, their hostility to Lamarckism was far more profound, and the Lamarckians were unable to modernize their concept of orderly growth to cope with the theory of particulate inheritance. The reliance of Darwinism upon hard heredity would eventually allow it to be reconciled with the experimental movement, once the techniques of population genetics had been worked out. The Mendelians' second revolution thus turned out in the end to be a second *Darwinian* revolution, with genetics completing the destruction of the idealist view of orderly development in both embryology and evolution theory. The "modern synthesis" of the mid-twentieth century was able to finish the job that Darwin had begun because it was based on foundations that transcended the limitations of nineteenth-century natural history. In this sense, the original form of Darwinism can be seen as a half-way house, opening up a whole range of topics to scientific investigations, yet unable by itself to break down the framework of natural history within which a more traditional view of nature could still flourish.

Anyone familiar with the eclipse of Darwinism around 1900 cannot escape a feeling of *déjà vu* when confronted with the modern debate over the adequacy of natural selection. Many of the arguments are the same; there is even a superficial similarity between the alternatives offered. Darwinism is rejected because its attempts to reconstruct the adaptive processes by which the various forms of life have evolved seem too speculative and are incompatible with the observed state of the fossil record. The modern alternatives include discontinuous evolution by "punctuated equilibria"

(Gould and Eldredge 1977) and even a new form of Lamarckism (Steele 1981; but see Dawkins 1982). Yet one thing is clear: none of the alternatives — not even the revival of use-inheritance — involves a return to the kind of pre-Darwinian philosophy of nature against which Darwin himself struggled with only limited success. The question of continuity, on the other hand, seems to be a recurrent one. The fossil record is incomplete and likely to remain so; therefore any attempt to fill in the details of evolution will involve an element of speculation. Unless we are to follow creationists in putting the whole question once again off-limits to scientific inquiry, a level of uncertainty must be tolerated. The history of Darwinism tells us that the willingness of scientists to propose hypotheses that "fill in the gaps" depends upon the intensity of their faith in the theory available to them. In periods of confidence, there is a tendency for the supporters of a theory to extend it beyond the level of the available evidence, but in a time of crisis the theory's critics dismiss all efforts to make it more flexible as a sign of weakness. The rise and temporary fall of the original form of Darwinism seems to be repeating itself in the twentieth century. Whether the outcome will be equally favorable to the principles of Darwinism this time around, only time will tell.

Notes

1. Darwin's opponents called this a philosophy of "materialism", while one modern commentator has preferred "positivism" (Gillespie 1979). Both of these terms carry ideological overtones, however, and "scientific naturalism" seems more appropriate in a discussion of the scientific debate.
2. Gray (1876) tried to reconcile the selection theory with the argument from design, although to Darwin's dismay he did eventually concede that variation might somehow be led in beneficial directions; see Dupree (1959). Despite this concession, Gray was never attracted to the American school's attempt to link design and evolution by means of Lamarckism — a clear indication that the utilitarian view of design was *not* the real inspiration of American neo-Lamarckism.
3. See Huxley's letter to Darwin, 23 November 1859 (*LL* 2:230–231); also his 1860 review of the *Origin*, reprinted in Huxley (1893b), especially p. 38.
4. Karl von Zittel was the most open exponent of the view that paleontology had not confirmed the evolutionary links between the classes; see the discussion of his views of Depéret (1909, chap. 12) and E. S. Russell (1916, pp. 357–358). William Bateson's revolt against evolutionary morphology was clearly influenced by the recognition that his own speculations on the origin of the Chordata were never likely to be confirmed; see W. Bateson (1894).
5. Probably the most comprehensive barrage of anti-selection arguments can be found in Thomas Hunt Morgan's early work (1903; see G. Allen 1968 and Bowler 1978). The best survey of the anti-Darwinian literature is Kellogg (1907).
6. I am indebted to Robert Nash of the Ulster Museum for my information on Haliday.
7. See Cope's early paper (1868, pp. 243–244, 269). This paper is reprinted along with his later Lamarckian writings in Cope (1887). See Bowler (1977b).
8. Butler was never inclined to overestimate his impact on the scientific community, but note the more optimistic tone of his 1890 essay "The Deadlock in Darwinism" (reprinted in Butler 1908, pp. 234–340). For examples of scientific interest in Butler's views see Francis Darwin 1908, pp. 15–16n; H. F. Jones 1911; E. S. Russell 1916, chap. 19; and Marcus Hartog's introduction to the 2nd edition of Butler's *Unconscious Memory* (reprinted in Butler 1920).
9. The translation is Eimer (1890). Note the reference to the "laws of growth" in the

subtitle of this work, anticipating Eimer's later interest in directed evolution or orthogenesis (see Bowler 1979). Francis Darwin supported Lamarckism because he believed it to be more consistent with the rhythmical nature of individual growth (1908, p. 14).

10. J. Arthur Thomson, a strong opponent of Lamarckism, nevertheless conceded that it would have to be taken seriously because of Spencer's support; see Thomson (1912, p. 166).
11. The best-known account of the controversy over the experimental "proof" of Lamarckism is Arthur Koestler's description of the "midwife toad affair" (1971), although this is hardly a reliable source. For a more general description, see for instance Fothergill (1952).
12. The guinea-pig experiments were first described in English in Brown-Séquard (1860); see Olmsted (1946). Weismann's classic experiment to prove the non-inheritance of mutilations in mice is described in Weismann (1891–1892, 1:433–461).
13. Owen's reaction to evolution was far more complex than might be imagined from his clashes with Huxley; see MacLeod (1965). As Ospovat (1981) pointed out, Owen was a leading figure in the movement to discover trends of gradual specialization in the fossil record before Darwin published (see also Bowler 1976a). Ruse notes that by 1869 Owen was writing favorable referee's reports on Darwinian papers (1979a, p. 260). On Mivart's career, see Jacob W. Gruber (1960).
14. Space prevents me from describing in detail the extent to which Louis Agassiz's students dominated American biology. His son, Alexander, worked at the Museum of Comparative Zoology from 1860 onward. Nathaniel Shaler was Dean of Harvard's Lawrence Scientific School. Hyatt was curator of the Boston Society of Natural History, Professor of Zoology and Paleontology at the Massachusetts Institute of Technology and Professor of Zoology at Boston University. For more details see Lurie (1960).
15. The first issue of the *American Naturalist* in March 1867 was prefaced by a statement indicating the journal's intention to support natural theology. The original editors were Hyatt, Packard, E. S. Morse, and F. W. Putnam. Cope became an editor in 1878.
16. See Darwin's correspondence with Hyatt (*LL* 2: 154, *ML* 1: 338–348). In a letter to Morse in 1877, Darwin complained that he had given up in despair the attempt to understand Cope and Hyatt (*LL* 3: 233).
17. Herschel's objection was noted by Darwin in a letter to Lyell, 12 December 1859 (*LL* 2: 241). Herschel later noted that evolution must act "according to law (that is, a preconceived and definite plan)" (1861, p. 129n).

23

Darwinism in Germany, France, and Italy

Pietro Corsi and Paul J. Weindling

Introduction

The eleven-year interval since the conference on the reception of Darwinism organized by Thomas F. Glick (1974) has witnessed important changes in research on evolutionary ideas in Europe. More case studies on Germany have appeared; Professor Yvette Conry has published her large volume on French non-reactions to Darwin; and a new generation of Italian historians of science has undertaken to explore the immense and immensely under-researched territory of Italian reactions to Darwin. Yet at present, as in 1972, the task of offering a balanced comparative assessment of Darwinian debates within the major European countries proves daunting. For the most part, the historical literature concerning the three cultures is still fragmentary. Most studies completed in recent years have emphasized the complexity of the problem and the need for further systematic investigation.

There is a rapidly growing literature on many aspects of biology in nineteenth-century German culture, but no immediate prospect of a comprehensive synthesis of German Darwinism. Although German scientists are seen as having made fundamental discoveries on development and heredity, the effect of Darwin's writings remains controversial. From a series of excellent case studies by such authors as G. Allen, Churchill, Coleman, Hoppe, Mann, Mocek, Querner, and Uschmann, it has emerged that the Weltanschauung guiding the work of German biologists was complex and highly individuated in its intellectual ingredients. Views for and against Darwin represented only one element of their scientific concerns. Extra-scientific considerations often influenced opinions on evolutionary theories.

The wealth of information available is still largely unexplored from the point of view of the social history of scientific culture. Contradictory theses have been supported by appropriately selected material. In a stimulating study on the teaching of natural history in schools, Scheele (1981) has reached the conclusion that the introduction of Darwinian biology into the school curricula was delayed for many decades. Yet studies of the development of German nationalism show student demands for lectures on Darwin, and they provide suggestive but impressionistic evidence for the popularity of the new evolutionary Weltanschauung.

Certain key areas like anthropology remain virtually unresearched. Proliferation of case studies, changing historiographical perspectives, and the many controversies in need of assessment, make it inappropriate to offer here a comprehensive analysis of German Darwinism. Montgomery (1974) has succinctly summarized some of the major contemporary writings. Instead the focus is on the formation of Haeckel's Darwinism and on the influence of Spencer, since they are often regarded as the twin pillars of evolutionary ethics and Social Darwinism. Concentration on Haeckel permits analysis of key factors in the transformation of Darwin's views in German culture.

Historians of Darwinism in France and Italy are becoming increasingly aware of the considerable influence exercised by Haeckel's philosophy and science in these countries. Indeed, the defense of Darwinism and evolutionism was often carried on in Haeckelian terms. To those who regarded evolutionary thought as a powerful weapon against Catholic and other Christian views of nature and society, the works of Haeckel provided convincing answers to problems Darwin was rather reluctant to tackle. It could indeed be argued that the powerful rhetoric of Haeckel's writings, which to certain more pragmatic Darwinists appeared mystical, did in fact contribute to their diffusion throughout Europe and the world.

As far as France and Italy are concerned, the small number of studies published on national reactions to Darwinism has made it possible to approach the topic through the format of the essay-review. Areas of research that promise to improve understanding of the complex dynamic of continuity and change within French natural history disciplines have been indicated. Conry's work undoubtedly represents the major single effort as yet undertaken to offer a comprehensive picture of debates on the scientific dimension of Darwinism, and it provides precious bibliographical information on the French literature of the period. R. E. Stebbins's 1965 thesis, abstracted in his contribution to the Glick conference (1974), is still useful. In the last few years, the *Revue de Synthèse* has published several papers dealing with French neo-Lamarckism. Among young historians of science, there appears to be a growing concern for French debates over transformism before and after the French translation of the *Origin* in 1862. A dissertation recently written by Claude Blanckaert (1981), dealing with the controversy over polygenism and monogenism, as well as work in progress by Ruth Harris, highlight new dimensions of the relationship between physical and social anthropology, and of the discussion of the concept of species and species degeneration amongst anthropologists and political thinkers before and after 1859.

There is no study of Italian reactions to Darwin comparable to Conry's book. Moreover, with the exception of a few paragraphs on Catholic attitudes toward evolutionism, the Glick volume failed to account for the variety and extent of Italian debates on Darwin as well as on Spencer and Haeckel. This gap in the coverage of European diffusion of Darwinism reflected the state of Italian studies on the subject. In recent years, a group of young Italian historians of science has contributed several case studies of selected features of the controversies over evolutionism. Research is hampered by a deplorable lack of basic bibliographical information, and of appropriate financing of projects on the scale required by the task.

It has been decided to offer a panorama of studies on Italian Darwinism, and to provide a case study of the close links between some of the major supporters of Darwin and leading representatives of Italian Lamarckism active during the early decades of the nineteenth century. Italian natural history of the time had lost the high status it had achieved in previous centuries, and was heavily reliant on research undertaken in France, Germany, and to a lesser extent in England. Even though little original research was completed by Italian evolutionary biologists, it is of considerable interest to explore the way in which scientific and broadly philosophical problems discussed within the European context found original recombinations in Italy.

In view of the lack of bibliographical aids on the subject of Italian Darwinism, a sample of the range of topics touched upon by participants in the evolutionary controversies has been offered. The major focus of the review is the scientific reactions, although appropriate mention is made of religious and ideological dimensions of the debate. As in the cases of France and Germany, research on the impact in Italy of biological categories on anthropology, and on the work of Cesare Lombroso in particular, will contribute to an appreciation of the interpenetration between broad ideological commitments and scientific research.

I. Darwinism in Germany

P. J. Weindling

Charles Darwin was astonished at the divergences from his views and at the ferocity of German debates on *Darwinismus*. He remarked that nationality had a curious effect on scientific opinion, as German supporters often put an exaggerated value on his work, whereas the French appeared uninterested (*LL* 3: 68–69, 118). But it is misleading to dismiss as simply exaggeration the scientific originality and the widespread acclaim for a new Weltanschauung, characterizing *Darwinismus*. It is more appropriate to discuss the response in terms of differences between the English and German scientific traditions (Mullen 1964, p. 3), although this approach neglects the distinctive social and cultural circumstances of the 1860s and 1870s.

A dominant concern among German scientists was the laws of organization, particularly of animal morphology. This was the theme of a treatise published in 1858 by the paleontologist Heinrich Georg Bronn (1800–1862). His belief in an immanent developmental force accounted for the bowdlerization of his prompt translations of the second and third editions of the *Origin* (1860, 1863). Until Ernst Haeckel (1834–1919) launched his enthusiastic campaign for Darwin's natural selection, along with a host of other adaptive mechanisms, the response to the *Origin* had been cautious. Thomas Huxley (1825–1895) remarked that "Germany took time to consider" (1887b, 2: 186), because there were a number of divergent classificatory and developmental theories, which were not immediately reconcilable with Darwin's views. Huxley's own differences with Darwin were influenced by Von Baer's morphology (Querner 1978). How it came about that prior German debates on morphology and development became subsumed under the banner of Darwinism will be examined here.

Since Germany was politically fragmented and culturally diversified, there are difficulties in generalizing about the impact of Darwinism. Even after unification in 1870, the exclusion of Austria and the continuing control of universities and schools by the constituent states ensured variations in the arts and sciences. It was at the University of Jena, which was administered by four Saxon Grand Duchies, that Darwinism initially found a niche in the 1860s. Heterogeneity in science was offset by virtue of certain common areas of research, and indeed science could be used for assertion of nationalist ideals.

Darwin admired German superiority in cytology and embryology. Johannes Müller (1808–1858) harnessed the rapid advances being made in cytology and embryology to explain the mysteries of animal organization. His comparative anatomy was a major influence in the 1840s and 1850s. That Haeckel was a pupil of Müller, who was renowned as "the German Cuvier", and that Haeckel was to achieve distinction as "the German Darwin" suggest that a necessary preliminary was reinterpretation of cell theory and embryology in terms of Darwinism. Haeckel combined his adoption of transformism with growing nationalist and anti-clerical convictions. These resulted in the distinctive character of his influential formulation of Darwinism.

Particularly important for Haeckel's transformation of Darwinism was the reform of the *Staatsgrundsgesetz* of the cell initiated by the protoplasmic theory of Max Schultze (1825–1874) (1861). Mechanisms of development were investigated in cytological terms, and embryological researches were used by comparative anatomists to reconstruct the genealogy of life. Concerns with the history and organization of life were related to intensified political aspirations for national unity. Many Darwinists considered that they were bringing biology into line with the standards of the historical — rather than physical — sciences, and could thereby prescribe laws for national development.

Haeckel's role in the transformation and dissemination of Darwinism was substantial, but subject to unceasing controversy. After a cautious mention of Darwin in his monograph on *Radiolaria* (1862), he lectured in Jena on

Darwinii theoriam de organismorum affinitate. In 1863 he began to publicly discuss Darwin's theories as the basis for reinterpretation of biological classification and of ethical and social thought. He relentlessly constructed ever more comprehensive evolutionary syntheses, and popularized a monist creed of the unity of man and nature. Huxley's verdict was that Haeckel was over-zealous and provocative in his scientific procedures and polemics. It meant that whereas Haeckel became the target for bigoted prejudices against evolution, Darwin was regarded as the epitome of "forethought and moderation" (*LL* 3: 68). Haeckel's evolutionary syntheses combined beguiling titles like *The Riddle of the Universe* (*Welträthsel*) with vibrant imagery. Their presentation of the facts of evolution of organic, mental, and social life provided a compelling picture of universal progress. Haeckel's influence was acknowledged by revolutionary thinkers like Freud (1856–1939) (Sulloway 1979a, pp. 258–263) and Lenin (1870–1924), who applauded the effect of Haeckel's writings as undermining professorial philosophy and theology (Lenin 1948, p. 362). They were avidly read throughout the world, far exceeding the *Origin* in numbers of cheap editions (McCabe in Boelsche 1906, pp. 294–324).

The immense variety of German responses to Darwin makes it appropriate to focus on the development of Haeckel's views in order to evaluate factors in the transformation of Darwinism in German culture. Because "Darwinism" and "Germany" are problematic categories, the existing literature has many flaws. Historians of science have appreciated the multiplicity of reactions to Darwinism, which was not directly dependent on any one physiological or anatomical approach; but they have often neglected the implications of political fragmentation, university expansion, and ideological issues. By way of contrast, general historians are prone to see Darwinism as a monolithic unity (D. Thomson 1977, pp. 101–109). So distinct have the two veins of literature been that Altner begins his anthology with the paradox *Darwinismus und Darwinismus* (1981, p. 1).

Although commentators often plead that the complexity of evolutionary writings extenuates their fragmentary analyses, they have not hesitated to deliver definitive verdicts on the life and death of Darwinism. Gregory has argued that the battle for Darwinism was won before 1859 since the temper of German materialism had already been established (1977b, p. 175). In the Glick volume, which pays scant attention to the implications of nations as emergent categories, Montgomery suggests that a younger generation of German Darwinists forged historical modes of explanation in the 1860s and 1870s (1974a, pp. 80–116). Kelly's innovative study of *The Popularization of Darwinism in Germany* observes, "by 1875 Darwinism had carried the day at least among the scientific community" (1981, p. 21). Coleman, with an eye on Germany, contends that "by 1900 biological Darwinism had prevailed" (1971a, p. 84). The discrepancies between these dates of clear-cut victory can be matched by dates marking the defeat of Darwinism, in an older generation of histories of biology, which were critical of Darwinism (Ràdl 1930, p. 42, and Nordenskiöld 1928, pp. 478, 528), and in more recent interpretations, as G. Allen's view of a "revolt from morphology" in the 1880s (1979, p. 21).

The discrepancies need to be resolved by more precise definition as to whether "Darwinism" meant such criteria as transmutation of species (accepted by many prior to 1859), descent from a common ancestor (much contested by polyphyletic opponents of Haeckel's monophyletic genealogical tree), or natural selection, in which even staunch Darwinians like Haeckel placed only limited faith. Mullen sees Darwin's theories as mechanistic and utilitarian, whereas Haeckel may be seen as employing a differing historical interpretation of the concept of a causal mechanism (Mullen 1964, p. 96). Analysis of the contemporary meaning of Darwinism has yet to be undertaken, despite a case study of the popular Catholic press (Dörpinghaus 1969), and of research on the records of particular institutions at Jena (Uschmann 1959), Munich (Hoppe 1972), and Berlin (Weindling 1981). There has also been philological analysis of the range of meanings of the term "evolution" from military maneuvers to biological preformationism (Briegel 1963). But there is no comprehensive bibliography of *Darwinismus*, despite early attempts (Seidlitz 1871; Spengel 1872) and May's survey of the 1909 anniversary literature (Altner 1981, pp. 454–471).

The importance of achieving a balanced interpretation of Haeckel is that this enables reconciliation of studies concerned with ideological and scientific aspects of *Darwinismus*. Coleman's study of Haeckel's fellow Darwinian, Carl Gegenbaur (1826–1903) at Jena, traces the transition in comparative anatomy from type concept to descent theory. In contrast to Haeckel, Gegenbaur shrank from the problem of the forces determining the changes in organisms (Coleman 1976, p. 172). Not only were there important conceptual differences between Haeckel and Gegenbaur over cell theory; Haeckel set himself the task of unravelling the causal factors in descent. Another leading Darwinian, Fritz Müller (1822–1897), examined the relations between descent theory and embryology. His *Für Darwin* paid special attention to adaptation, but did not offer a causal theory of recapitulation (1864; D. Peters 1980, p. 61). The services of Victor Carus (1823–1903) as a translator of all Darwin's publications from 1866 meant minor adaptations to Darwin's concepts, such as Bronn's *Züchtung* becoming Carus's Züchtwahl (Darwin 1860; 1863; 1867; Zirnstein 1977).

Although Darwin drew on anthropological writings of Haeckel and of German materialists like Carl Vogt (1817–1895) for his *Descent*, there were significant differences between Haeckel and the materialists (Gregory 1977b, pp. x, 76, 180). Indeed the *Origin* did not mark a fundamental turning point for the materialists, as it did for Haeckel and Gegenbaur. Haeckel's monism differed from materialist reductionism of life to physics and chemistry, and it resulted in bitter scientific disputes, as with the embryologist Wilhelm

His (1831–1904) over the application of physics to embryology. Major scientific and philosophical differences between Vogt and Haeckel make Kelly's view — that the two men were allies in the campaign to popularize materialism — questionable. Regarding monism's social impact, Lenin observed in 1908 the irony of how Haeckel furthered the materialism to which he was opposed (1948, p. 363).

Controversies over Darwinism are reflected in conflicting views of Haeckel. His scientific work has often been dismissed as speculative. For example, Oppenheimer has viewed Haeckel's biogenetic law of recapitulation and his theory of the germ layers as primitive organs, as obstructing advances in embryology (1940). Biographers have also concentrated on philosophical aspects of Haeckel's writings, particularly his monism (Boelsche 1906; H. Schmidt 1926). The 1920s editions of Haeckel's travel descriptions (1923), and his letters to his parents (1921a), to his betrothed (1921b), and to philosophers (Jodl 1922) and admirers concentrated on Haeckel's spiritual struggles and naturalistic ethics. With the exception of the love letters (Werner 1930) to Franziska von Altenhausen (the disguised name for Frieda von Uslar Gleichen, the 335 letters now being located in the *Staatsbibliothek Preussischer Kulturbesitz*, West Berlin), these drew on the archive of Haeckel's *Villa Medusa*, a remarkable *fin de siècle* villa, which Haeckel ensured would enshrine his achievements for posterity.

Although publication of Haeckel's correspondence continued during the Third Reich with letters to the poet Hermann Allmers (1821–1902) (Koop 1941), and to his outstanding students Oscar Hertwig (1849–1922) and Richard Hertwig (1850–1937) (Franz 1941, pp. 9–72), an *Ernst-Haeckel-Gesellschaft* was established to replace the then discredited Monist League (Franz 1941, pp. 157–159; Franz 1944, pp. 205–206). A number of post-War studies gave a hostile twist to the view of Haeckel as a key figure in the romantic and authoritarian "volkish tradition" (Gasman 1971, p. xi). The case rests on many distortions and disregard for discontinuities, such as the strong connections between monism and workers' free-thinking organizations, and between monism and free-masonry (Breitenbach 1913; Bolle 1981). Greater historical sensitivity has been shown in a variety of other approaches to Haeckel. Among Uschmann's many outstanding books and articles are an edition of Haeckel's letters and travel diaries (1954), and a masterly study of zoology at Jena, using institutional archives to assess Haeckel's relations with colleagues in the "citadel of Darwinism" (1959). Work also emanating from the German Democratic Republic has highlighted Haeckel's impact on popular materialism (Dorber and Plesse 1968) and the persistence of Lamarckism (Mocek 1982); G. Schmidt (1974) has assessed Haeckel's role in the literary reception of Darwinism. Kelly (1981) has also emphasized the materialism and radicalism of popular Darwinism, whereas others have emphasized Haeckel's idealistic commitments and pantheistic religion (DeGrood 1965; Holt 1971). In the Federal Republic of Germany the symposium on *Biologismus* (Mann 1973a), resurrecting a term from 1911, of the philosopher Heinrich Rickert (1863–1936), marked a broadening of interest by historians of biology and medicine in cultural and social issues. It has furthered understanding of Haeckel's historical and ideological concerns (Mann 1980; Winau 1981). From this overview it can be seen that apart from Uschmann's balance between scientific and biographical issues, there has been a tendency to stress social and cultural aspects of Haeckel's contribution to *Darwinismus*. It is the purpose of the rest of this paper to assess Haeckel's role in the dissemination of Darwinism with special regard to scientific premises.

Darwin had a modest reputation in Germany prior to 1860. His *Journal of Researches* had been translated (1844), and his studies on barnacles had received critical attention (Gegenbaur 1912, pp. 425–426). On 1 October 1857 he was elected to the Leopoldina academy of naturalists (Leopoldina 1982). The initial response to the *Origin* was sluggish. Haeckel read it during May 1860 in Berlin, but he could discuss it only with a still skeptical Alexander von Braun (1805–1877) (Hoppe 1971), owing to the hostility of other Berlin naturalists. He subsequently came to regard Berlin as a center of hostility to Darwinism (Franz 1941, p. 54). Like Bronn, whose translation of the *Origin* reached a second edition in 1863, Braun saw evolution in terms of an inner formative force or *Bildungstrieb*. Only when Haeckel visited the comparative anatomist and marine zoologist Gegenbaur in Jena in June 1860 did Haeckel become convinced of the truth of Darwinism and transformism (Boelsche 1906, p. 133). In this early phase he was still cautious about Darwin's theory when constructing a genealogy of the *Radiolaria* (Boelsche 1906, pp. 140–143; Haeckel 1862, pp. 231–232).

Temkin — echoing Braun in 1862 — has observed that "the theory of descent did not reach an unprepared science" (1959, p. 324). For example, the *Vestiges of Creation* of Robert Chambers (1802–1871) was translated in 1851 and achieved considerable influence, although its translator, Vogt, opposed transmutation of species (Gregory 1977b, p. 176). Temkin has suggested that German biology during 1848–1858 was dominated by materialist and mechanistic approaches reacting against a *Naturphilosophie* tainted with transformist speculation (1959, pp. 336–337). But neither the materialism of Vogt and Ludwig Büchner (1824–1899) nor the "1847 biophysics program" formulated by the physiologists Ernst Brücke (1819–1892), Emil Du Bois Reymond (1818–1896), and Hermann von Helmholtz (1821–1894), nurtured Haeckel's views. On reading Vogt in 1853, Haeckel admired his comparative zoology as reminiscent of Müller, but he deplored Vogt's radicalism and attack on Christianity (Haeckel 1921a, pp. 52–53).

The major influence on Haeckel — and a critic of materialism and naive physiological experimentalism — was the comparative anatomist Johannes Müller. His interest in animal organization led to appreciation of how certain organisms were fundamental to particular phyla, like the

Amphioxus for the vertebrates. Haeckel's deduction of primitive ancestors, whose forms were recapitulated in developmental processes, shows traces of Müller's influence. Müller had encouraged Haeckel's interest in the simple forms, which could be observed on marine zoological expeditions. Haeckel accompanied Müller on an expedition to Heligoland in 1854, and, meeting again at Nice in 1858, Müller urged Haeckel to work on *Radiolaria* (Haeckel 1862, pp. 231–232; Boelsche 1906, pp. 69, 97). It was to Müller's memory that Haeckel's monograph on *Radiolaria* of 1862 was dedicated. Müller, although resolutely opposed to transformism until his death in 1858, inspired a generation of Darwinists like Haeckel, Max Schultze, and Gegenbaur.

Haeckel was also influenced by another Müller protégé, Rudolf Virchow (1821–1902), who advocated cell division and the theory of the organism as a cell state. Haeckel was at Würzburg in 1852–1854 and 1855–1856, when Virchow's views were moving away from physiological reductionism in the direction of Müller's conviction of the distinctiveness of vital organization (Rather 1959). Boelsche justifiably stressed how the theory of the cell state motivated Haeckel to search for the origins of social organization. In 1859 at Messina, Haeckel became enraptured by the beauty of the social *Radiolaria*, united into colonies by a network of protoplasm (1862, pp. 116–127). He hoped that study of their radially symmetrical forms, which were reminiscent of crystals, would offer insight into the elemental forms of life.

Haeckel's studies on *Radiolaria* coincided not only with his reading of Darwin's *Origin*, but also with a growing debate on the material basis of life, unleashed by Max Schultze's reform of the constitutional theory of the cell. In 1860 Schultze redefined the cell as nucleated protoplasm. He thereby shifted attention away from the membrane to the cellular protoplasm, which, being common to all plants and animals, could suggest descent from a common ancestor (Lücker 1977). Protoplasm exhibited the basic properties of life: of contractility and movement for Schultze, and of reproduction, motility, and irritability for Haeckel (1862, pp. 92–106; Weindling 1982, pp. 39–44). Haeckel elaborated a cellular hierarchy from a non-nucleated plasmatic *Moner* (the elemental form of life) to nucleated multi-cellular organisms (1866, pp. 269–326).

In the transition from Müller's static comparative anatomy to strictly historical criteria of descent, evolutionary understanding of the cell was a prerequisite for establishing the genealogy of life. Histological research was undertaken by Gegenbaur, Haeckel, and Schultze, so that understanding of the cellular processes of development could yield insight into the causal mechanisms of evolution. Haeckel's first published mention of Darwin was in his monograph on radiolarians of 1862, an extended version of his 1861 *Habilitation* qualifying him to teach at Jena. He stressed the importance of Darwin's theories — even naming a species *Coccodiscus darwinii* (1862, pp. 482–488) — and highlighted the importance of transitional species. But he pointed out that the chief defect of the Darwinian theory was that it threw no light on the origin of the earliest organism, which was probably a simple cell (1862, pp. 231–232).

Darwin's *Origin* provided scant guidance in the reformulation of comparative anatomy. Nowhere did cell theory figure in the *Origin* (Hughes 1959, p. 77; Oppenheimer 1967, p. 216). When privately musing on natural selection in 1856–1857 Darwin recognized that common cellular structure indicated "by analogy that all living beings descended not from 4 or 5 animal types, but from *one* single created protoplasm" (Ospovat 1980, p. 174). He failed to develop this line of reasoning, in contrast to the keen German interest in histology, which ultimately revealed the cellular mechanisms of reproduction and heredity.

Schultze and Haeckel were among the first to consider the implications of the *Origin* for cell theory. They have been seen as distant in their personal relations, the speculative temperament of Haeckel contrasting with the cautious skepticism of Schultze, who far outshone Haeckel as an innovator in histological techniques (Geison 1975). Hitherto unnoticed among the approximately 36,000 letters in the *Haeckel-Haus* are fifty-two letters between Schultze and Haeckel (Best. A-/Abt. 1 No. 1006). In response to Haeckel's preliminary communication on *Rhizopoda*, Schultze explained why the term *sarcode* (introduced by Félix Dujardin (1801–1860), and referring to animal ground substance) should be replaced by *Protoplasma*, as the substance accounting for all organic tissue formation (17 June 1860; 10 December 1860). Although Haeckel continued to speak of *sarcode*, a substantial section of his Radiolarian monograph was devoted to proof that sarcode and protoplasm were equivalent (1862, p. 96). Adopting Schultze's criterion of a nucleus, Haeckel recognized that protoplasm that was undivided into cells gave the *Radiolaria* their unicellular characteristic. Schultze was pleased that Haeckel endorsed his views on the organization of *Protozoa*, on which their great mentor Müller had worked (21 October 1862; 14 January 1863). Schultze was well aware of Darwinism, having visited England in 1862, and he encouraged Haeckel's Darwin studies, exclaiming, "Die Sache muss fleissig erhalten werden" (29 January 1865; 11 March 1865). He regarded Darwin as of major importance to his research on the retina (7 June 1866), and he commended Haeckel's Darwinian treatise, *Generelle Morphologie*, as a major foundation upon which to establish Darwinism (5 October 1867; 9 May 1868).

The attack on protoplasmic theory and Darwinism by the Berlin anatomist Carl Reichert (1811–1883) meant that Haeckel and Schultze had common foes in Berlin. Schultze also deplored Du Bois Reymond's opposition to Darwinism (9 May 1868). There resulted a major rift among the disciples of Müller. Schultze was among those who, because of dislike of Prussian illiberalism, did not wish to take up Müller's mantle in Berlin, and Du Bois Reymond and Reichert had been appointed. Reichert excluded Schultze from the major journal *Müllers Archiv*, so that Schultze established the *Archiv für mikroskopische Anatomie* in 1865 as the organ for Darwinian histology. Its opening article commended Haeckel's researches on protoplasm (M. Schultze 1865a; 1865b, p. 17). Haeckel and Schultze had both become vehemently anticlerical, a further indication of their common outlook.

Haeckel scrawled emotive annotations over Schultze's final letter of 16 December 1871. Although Schultze supported Haeckel's theory of a primal *Gastraea* form, accounting for fundamental differences between *Protozoa* and *Metazoa*, he criticized Haeckel's concept of the cell — part of a hierarchy of elemental forms — as too uniform. Haeckel established that the sponges were the lowest tissue-forming animals. Drawing on earlier work on germ layers, as by Aleksandr Kovalevski (1840–1901) on invertebrate germ layers, Haeckel suggested that they were to be regarded as identical throughout the animal kingdom. Important in the *Gastraea* theory was the historical approach to the causes of morphology. He emphasized that the methods of embryology — *Entwicklungsgeschichte* — were those of the archeologist or cultural historian (1877, p. 16). Phylogeny was the historical cause of ontogeny. Growth occurred by division of cellular labor, being "dependent on the physiological functions of inheritance and adaptability" (1874; Lankester 1876, p. 145).

Haeckel's synthesis of comparative anatomy, cytology, and embryology provided a powerful conceptual framework for further research in the 1860s and 1870s. Major discoveries by Oscar and Richard Hertwig on fertilization and development exemplify how Haeckel was able to inspire students to research on the cellular mechanisms of growth and inheritance. Although they learned advanced cytological techniques from Max Schultze, it was Haeckel who initiated the Hertwigs into Darwinism. Haeckel encouraged their nationalist enthusiasms, as when on a marine biological expedition to Dalmatia their evenings were spent in discussion of Darwinism and politics (Uschmann 1954, p. 84). Much of the embryological work of the Hertwigs related to their aim of providing an explanation of the growth of the middle germ layers, and producing a *Coelom* theory to complement Haeckel's *Gastraea* theory. Serious differences of opinion quickly became apparent in the 1870s among Darwinians, however. Some interpreted fertilization in terms of the formative powers of protoplasm, which produced the nucleus in the fertilized egg as a new formation. In contrast, Oscar Hertwig (1875–1877) deduced the morphological continuity of sperm and egg nuclei in a process of cellular fusion. This refuted Haeckel's theory that fertilization had to recapitulate the primal *Monera* form of life, passing through a non-nucleated stage (Weindling 1982, pp. 71–108). These differences of opinion came at a time when Haeckel's views on Darwinism were subjected to a major attack by Virchow.

The keen attention paid to these debates resulted from the expectation that Darwinism was to solve a range of social and metaphysical questions. Darwin and Huxley were dismayed at how Haeckel mixed religion and politics with science. Huxley advised Haeckel that one "public war-dance against all sorts of humbug and imposture" was enough, and that any translation of the *Generelle Morphologie* ought to exclude "the aggressions" (Uschmann and Jahn 1959–1960, pp. 13, 19). Particularly important for the development of Haeckel's convictions on the social significance of Darwinism was that while he was in Italy in 1859 researching on the "social Radiolaria", he experienced a new patriotic ardor for freedom and fatherland (Koop 1941, pp. 38–40, 46–47, 52–54). These events preceded his reading of the *Origin* and stimulated his interest in the origins of social and mental life.

Political considerations were prominent in debates on Darwinism at the *Gesellschaft Deutscher Naturforscher und Aerzte.* Since its foundation by the radical *Naturphilosoph* Lorenz Oken (1779–1851) in 1822, its discussions show how nationalist aspirations were channelled into science. In 1860 the *Origin* was commended in teleological terms by the Bonn philosopher Jürgen Meyer (1829–1897). On 19 September 1863 at Stettin (Sczezin) Haeckel sparked off major discussion by introducing Darwin's *Origin* as a history of creation, which modified "personal, scientific and social views". Darwinism represented "development and progress" as opposed to "creation and species" (1863, p. 18; Querner 1975, p. 440). The laws of progress were to be understood by examination of the simplest stages of organization, and by resolution of the question of whether life had originated in the form of a simple cell of plasma particle. Progress resulted from struggle and selection in nature as in society, but also from cooperative interaction between organisms. Priests and despots blocked progress to national unification (Haeckel 1862, pp. 23-26). Virchow spoke of organisms as federal unities (Virchow 1864, p. 41). Virchow's liberal principles meant that he was opposed to any concept of centralization, and Haeckel therefore had to look elsewhere for integrating principles.

Haeckel's *Generelle Morphologie* particularly drew on the theories of Bronn (1858) to explain the formative laws of organization. Divergence and progress were not always identical. Higher forms of organization were achieved by reduction of numbers of organs, the concentration of functions and their respective organs leading to centralization under a *Central-Organe.* This process was accompanied by increasing internalization of physiological powers (Haeckel 1866, 2: 251, 258). In a discussion of the relations between individuality and the cell state, Haeckel suggested that higher individuals were composed of a community of lower individuals, just as cells composed human organisms, who in turn could unite in the higher organism of the state (1866, 1: 269–372).

It was at this point in Haeckel's intellectual development that he encountered the evolutionary laws of progress of Herbert Spencer (1820–1903). While on Tenerife in the Winter of 1866–1867, Haeckel became interested in the *Staatsqualle* or social medusa, the *Siphonophora.* Soon after he returned from Tenerife he received Spencer's *First Principles* and *Principles of Biology*. Spencer's article "The Social Organism" described the processes of mutual dependence and organic integration by the nervous system, analogous to the coordinating function of the telegraph in the modern state (1860, pp. 399, 427–428). These principles of mutual dependence and of physiological integration resulting from specialization of functions were substituted by Haeckel for Bronn's principles of centralization and internalization (Spencer 1864-1867, 2: 372-376). The principles were used to interpret how the *Siphonophora* could form colonies based on the principle of division of labor. Haeckel traced how this medusa developed from the ovum and then budded to form a community (1869a; Boelsche 1906, pp. 246–249). In popular lectures Haeckel used *Siphonophora* to exemplify the state-forming instinct, and stressed how the brain and nervous system — analogous to the telegraph — achieved organic integration in higher organisms (1896b; 1923).

Haeckel's incorporation of Spencerian concepts coincided with intensification of debates on the political constitution of Prussia, and particularly with the war of 1866 between Prussia and the Hapsburg lands. The *Generelle Morphologie* contrasted progressive and conservative inheritance, accounting for the degenerative tendencies in the Prussian aristocracy (Haeckel 1866, 2: 170–190). Haeckel debated the role of Prussia in German nationalism with August Weismann (1834–1914), who favored Prussian expansion as furthering national unity (Churchill 1968; Montgomery 1974b, pp. 201–202; Uschmann and Hassenstein 1965, p. 18). There were similar concerns in Austria with the question of the reconstitution of the Empire after the defeat by Prussia. The pathologist Carl Rokitansky (1804–1878) began a preliminary debate on this question by raising the issue "whether Charles Darwin is right or no" (Geikie 1924, p. 129). Rokitansky (1869) believed that although aggression was rooted in protoplasmic hunger, this was offset by integrating mechanisms in higher organisms.

Offers of chairs in Vienna in 1872 and at the reconstituted university of Strassburg (Strasbourg), where there was a deliberate policy to promote German cultural values, show how Haeckel's reputation was rising in the years around the unification of Germany. Haeckel's admiration for Otto von Bismarck (1815–1898) steadily grew, culminating in Haeckel proclaiming him doctor of phylogeny (Franz 1941, pp. 82–86; Franz 1944, p. 119). But the division that occurred between Virchow's Progressive Liberal Party and the pro-Bismarck National Liberals was reflected in controversies between Virchow and Haeckel. At the 1877 *Naturforscher-Versammlung* these differences erupted in a dispute over whether Darwinism was proven law or only a hypothesis, and over the suitability of teaching Darwinism in schools. Haeckel and Virchow argued in terms of contrasting theories of the cell state. Haeckel's speech exemplifies how his Darwinism was based on embryology, as providing proof of the history of descent and cell theory, each cell being the active citizen of an organism (1877, p. 17). Unicellular organisms showed that the cell was the unit of mental life, and could be termed a *Seelenzelle*. As protoplasm was composed of molecules, termed by Haeckel "plastids", so the lowest psychological unit was the *Plastidulseele*, which itself was a unity of inorganic substances. The evolution of social instinct made it possible for individuals to cooperate in forming higher organisms. Examples drawn from animal societies proved the existence of a natural religion based on duty, division of labor and the subordination of egoism to the good of the society. Haeckel therefore argued that evolutionary theory should be the basis of education in the new nation (1877, pp. 17–20).

Virchow refuted Haeckel by taking his hierarchical principles to absurd extremes. Whereas discoveries like cell division were incontrovertible facts, Virchow ridiculed cosmologies seeking to explain laws of atoms and astronomy. If the theory of descent were made the basis of social and religious principles, this would be a dangerous distortion based on ignorance. The theory of the plastidule soul was a possibility, but it could not be proved as fact, and as such it was inappropriate to teach it in schools. Such dangerous distortions bred from half knowledge opened the way for socialism (Virchow 1877, pp. 68–69). Haeckel denied this charge by invoking the principle of organic integration: as greater centralization was the result of evolution, it could not lead to the disappearance of the state demanded by socialists (Haeckel 1878).

The debate marked a waning of Haeckel's academic reputation. Although he continued research on *Protozoa* from the Challenger expedition, Haeckel's credibility was further weakened by his inability to keep up with advances in histology, leading to discoveries of the chromosomal mechanisms of reproduction, and to the cytological and embryological experimentation of the 1880s pioneered by his pupils the Hertwigs, Hans Driesch (1867–1941) and Wilhelm Roux (1850–1924) (Churchill 1970; Coleman 1965; Mocek 1974; Weindling 1982). Controversies over fraudulent illustrations of recapitulation — with the use of the same illustration for a number of species — also tarnished his reputation (Gursch 1980). Haeckel's social evolutionism increased in influence, however, as its emphasis on corporate integration was better suited to prevailing social problems than Virchow's individualism. Haeckel can be seen as blending the social views of two staunch individualists by fusing Virchow's concept of the cell state with Spencer's principle of organic integration. Despite fundamental differences between Haeckel and Spencer over state centralization, they recognized common classificatory aims, with Haeckel's account of the progress of life from the crystal forms of *Radiolaria* to man, and Spencer's attempt to classify human social formations. Spencer's individualism came into conflict with the corporatism and belief in state intervention of Patrick Geddes (1845–1932), the Edinburgh botanist, but Geddes and Haeckel corresponded on the unity of the biological and social sciences (*Haeckel-Haus* Best. A — Nr. 1461/1–11).

Haeckel emphasized that his philosophical approach continued Jena traditions of Goethe's morphology, of the developmental *Naturphilosophie* of Oken, Friedrich Schelling (1775–1854), and of Schleiden's cell theory. Schleiden had formulated the theory of the cell as an elementary individual, and although critical of materialism and Darwinism, he admired Haeckel's writings on the *Seelenzellen* (*Haeckel-Haus*. Best. A-Abt. 1 Nr. n006, 5 July 1878). Haeckel's indebtedness to these intellectual traditions at Jena also points to the differences between his monist pantheism and materialist reductionism (1866, 2: 448–452). The four Saxon duchies that administered the university were enlightened patrons of the arts and sciences. Saxe-Weimar aspired to be a center of German culture. Its Grand Duke was a benign protector of Haeckel; he ensured that monuments like the Kyffhäuser and Wartburg should be used for festivals symbolizing national unity. It was a deliberate policy to appoint adventurous young scientists like Gegenbaur, Haeckel, and Thierry William Preyer (1841–1897) in the 1860s, because pioneering of Darwinism enhanced the university's reputation. Students, renowned for their nationalist enthusiasms, petitioned for lectures on Darwinism (Museum of the *Ernst-Haeckel-Haus* Jena).

After his dispute with Virchow, Haeckel's position at the university weakened. The town was undergoing rapid industrialization, particularly due to the improvement of precision optics by the physicist Ernst Abbé (1840–1905), and the establishment of the *Zeiss-Stiftung*, which considerably benefited the university. Haeckel lamented that he understood little of the new Jena, although admirers of his work like the merchants Paul von Ritter and Albert von Samson provided funds making it possible to further Darwinian zoology and research on the natural basis of ethics. Haeckel was not the only Jena academic to recast his scientific discipline for popularization. The Jena philosopher Rudolf Eucken (1846–1926) promulgated idealist *Lebensphilosophie*, and the historian Dietrich Schäfer (until 1885 at Jena) was prominent in the ultra-nationalist *Alldeutsche Verband*. He was succeeded by a disciple of the Aryan ideologue of Gobineau, Alexander Cartelleri (1867–1955). A setback came when the Jena publisher Gustav Fischer (1845–1910) refused to publish Haeckel's *Welträthsel*; although Fischer recognized it would be a bestseller, he considered it to be unscientific. Rational and empirical features in evolutionary theory as in the historical sciences gave way to irrationalist mysticism and nationalism.

These developments provide insight into the scientific, philosophical, and social reasons as to why Haeckel's Darwinism diverged from Darwin's own views. Kelly has concluded that the *Welträthsel* has little trace of social Darwinism, "so Haeckel exerted no mass influence as a social Darwinian" (1981, p. 120). This is an important charge, as I have endeavored to show social factors in the propagation of Darwinism. Holt has observed that the *Welträthsel*, an exposition of the evolutionary foundations of monist ethics, belongs to a later phase of Haeckel's intellectual development, coinciding with an era of disenchantment with Darwinism, and — one might add — hostility to Weismann's neo-Darwinism. That Darwin tried to excise political passages from the planned translation of the *Generelle Morphologie* suggests discontinuities between Haeckel and Darwin. Despite such differences of opinion Haeckel's indebtedness to Darwin in the formulation of his views in the 1860s was immense. Darwin's *Origin* was a major stimulus in the establishing of an evolutionary understanding of embryology — about which Darwin was enthusiastic — and cytology — on which Darwin was silent. It is important not to judge Haeckel only in the light of later work, but to recognize the effect of the reading of the *Origin* in the context of Haeckel's researches on *Radiolaria*.

Similarly, it is important to realize how reading of Spencer in 1867 came at a critical stage in the researches on the social medusae. Like Spencer, Haeckel made only limited use of natural selection. Bannister has suggested that the epithet "Social Darwinism" in Spencer's case is misleading in that Spencer drew on earlier evolutionary writings of such people as Chambers and developmental theories of Carl Ernst von Baer (1792–1876) (1973, p. 43). Unlike the case of Spencer, it was in the wake of the *Origin* that Haeckel provided an evolutionary interpretation of comparative anatomy and cell theory. Darwin's *Origin* was thus important as a catalyst.

It is artificial to separate biological from social concepts, as some commentators have done (Montgomery 1974b, p. 214). Historical concerns intensified by nationalism were applied to biology, from which social laws were derived. This suggests that the success of Darwinism in Germany was due to the *Sehnsucht* for liberty enjoyed by a constitutional nation. Darwin's *Origin* not only indicated a number of important directions for systematic investigation and raised fundamental questions; it also had a symbolic value. After unification, critics of Darwinism attacked the theory of natural selection as deriving from inhumane *Manchestertum* (O. Hertwig 1916, pp. 634–640). Attention to the formulation of Haeckel's Darwinism reveals the importance of a wide range of concerns, as with the origins of vital organization, sensibility, and the evolution of coordinating organs in higher organisms. Darwin's *Origin* inspired Haeckel to fundamentally reinterpret comparative anatomy in the evolutionary terms of the genealogy of organisms. Haeckel's evolutionary comparative anatomy established a predominant trend in *Darwinismus* with its attempt to demonstrate the descent of species by cytological and embryological investigations, and it thereby made a substantial contribution to biological, ethical, and social thought.

II. Recent Studies on French Reactions to Darwin

P. Corsi

Even though German *Darwinismus* was bound to displease Darwin in the course of the 1860s, he had no reason to complain about the circulation of his book, his name, and his ideas in the German states. With all the spade work yet to be done in order to gain a fairly accurate picture of the German debates on Darwin and on evolution theories, it is nonetheless clear that the publication of the *Origin* produced significant and lasting controversy in that country. In February 1863, three years after the publication of his book, and one year after the first translation of the *Origin* in French, Darwin wrote to Camille Dareste: "as far as I know, my book has produced no effect whatever in France" (*LL* (NY) 2: 192).

Early attempts to arrange for a translation had produced no result. Major publishers, such as Baillière, Masson, and Hachette, turned the book down (*LL* (NY) 2: pp. 30–31). In September 1861 arrangements were finally made with the publisher Guillaumin, and a copy of the third English edition of the *Origin* was sent to Clémence-Auguste Royer (1830–1902), a woman who had been teaching philosophically refurbished Lamarckism in Lausanne a few years previously. When the translation appeared in 1862, it was clear that Mlle Royer had strong views on the significance of Darwin's work for contemporary culture. The title of the work, with no authorization from Darwin, read *De l'origine des espèces ou des lois du progrès chez les êtres organisés*. Moreover, a fifty-page-long preface contained repeated and unequivocal declarations of faith in progress and a secular picture of nature:

> Yes, I believe in revelation, but in a personal revelation of man to himself by himself, in a revelation which is only the result of the progress of science. . . The doctrine of M. Darwin is the rational revelation of progress, putting itself in its logical antagonism with the irrational revelation of the fall. (Darwin 1866, p. xx)

Disagreement with Mlle Royer over the criteria of translation finally convinced Darwin to look for alternatives, and in 1873 a new translation by Jean-Jacques Moulinié (1830–1873) appeared (R. Stebbins 1965, p. 45, chap. 3, "The Translation of Darwin"; Farley 1974, pp. 286–287; Conry 1974, pp. 19, 262–266).

As Robert Stebbins has pointed out, it would be wrong to make Mlle Royer responsible for what Huxley called "the conspiracy of silence" that surrounded the publication of *De l'origine des espèces*. Indeed, no major French naturalist spoke on Darwin's side; his theories were consistently misrepresented, reduced to a re-enactment of older transformist credos, or rejected on a priori epistemological assumptions, especially by supporters of positivism. Between 1870 and 1878 Darwin's name was placed six times in the nomination list for the zoological section of the Académie des Sciences. When he was finally elected, in August 1878, it was to the botanical section — a result not taken by Darwin as a compliment (Camerano 1896, pp. 331–332; R. Stebbins 1965, pp. 215–331).

The sparse and unfavorable reaction to Darwin in the years following the French edition of his work has been echoed by the lack of historiographical enthusiasm for, or even interest in, the issue. French historians of *transformisme* of the past and the present century have shown little concern for the actual state of affairs of debates on Darwin in the decades following the publication of the *Origin*. Indeed, the few major systematic studies of French reactions to Darwin were not produced until the 1970s, 1972 being the date of publication of Professor Conry's important preface to her edition of the correspondence between Gaston de Saporta (1823–1895) and Charles Darwin, and 1974 the date of publication of *The Comparative Reception of Darwinism*, edited by Thomas F. Glick. In the *Comparative Reception*, Robert Stebbins devoted approximately fifty pages to France, and concluded: "there was discussion of transformism, and there were many transformists in France from 1859 to 1882, but little Darwinism and fewer Darwinists" (1974a, p. 117). Harry Paul (1974) contributed to the same volume a paper on "Religion and Darwinism", devoted mainly to reactions of French Catholics to the diffusion of broadly evolutionary theories. I shall consider this topic when discussing the much improved version of Professor Paul's paper, published in 1979 as the first part of a volume devoted to French Catholic reactions to science in the last decades of the nineteenth century.

Stebbins's compact essay does not constitute a completely fair representation of the work he had done on French reactions to Darwin. Indeed, Stebbins's 1965 dissertation, "French reactions to Darwin, 1859–1882", did contain an attempt at applying the methodology elaborated by Alvar Ellegård (1958) in his famous study on the reception of the *Origin* by the British press. The attempt proved unrewarding, even though of considerable interest. The patient perusal of several contemporary French periodicals produced the meager and frustrating result of thirty-four papers published between 1859 and 1862 relating to broadly transformist issues. Of these, only ten touched directly on Darwin's work. The five reviews of the *Origin* that the author was able to trace set the tone for all future reactions to Darwin (Stebbins 1965, chap. 2, pp. 23–38).

With the exception of Edouard Claparède (1832–1871), a Swiss naturalist who wrote a favorable account in the *Revue Germanique* (Claparède 1861), friends and foes alike showed limited awareness of the originality of Darwin's theory, and of the specific problems and topics he touched upon. All appeared convinced that the cause of transformism was the issue at stake, and that the theory of natural selection represented only one of the possible solutions. In December 1860, Jean-Louis-Armand de Breau De Quatrefages (1810–1892), the life-long but just critic of Darwin, took notice of the theory put forward by the English naturalist in the "Histoire naturelle de l'homme", a series of lectures given in 1856, and revised for publication in the prestigious *Revue des deux mondes*. As Stebbins has noted,

> The Darwin publication was not sufficient to suggest a radically new

> and forceful approach. For De Quatrefages, and probably for his readers, Darwin did not represent something totally new or intrinsically important, but instead represented another chapter in a question which had been debated before 1859 and would have continued to be a matter for consideration after that date, even without the stimulus of the *Origin of Species*. (1965, p. 36; cf. Sillard 1979)

To many contemporary French naturalists, Darwin's book was nothing more than yet another attempt to reformulate theories put forward by Jean-Baptiste Lamarck (1744–1829) and Etienne Geoffroy Saint-Hilaire (1772–1844). In 1864, Pierre-Jean-Marie Flourens (1794–1867) published his *Examen du livre de M. Darwin sur l'origine des espèces*. Flourens, a pupil of Geoffroy Saint-Hilaire, went over to Cuvier, and was nominated by the dying baron to be his successor as Perpetual Secretary of the Académie des Sciences. Flourens distrusted theories and generalizations in science, and he found Darwin deficient in the basic requirements of his discipline. Darwin had not offered a definition of species, and yet he claimed it was variable; he was clearly unaware of the limits of variability; his language was far from clear and not up to scientific standards. The arguments used against Lamarck proved immediately applicable to Darwin (Flourens 1864, pp. 1–2).

The polemical issues that most concerned French naturalists around 1860 were the debate between Pasteur and Pouchet over spontaneous generation, and the discussions among anthropologists concerning a polygenist or a monogenist theory of the origin of the human races (Farley 1974, Farley and Geison 1974; Farley 1977). Significantly, Flourens himself devoted part of his book on Darwin to refuting the doctrine of spontaneous generation, and favored a strict creationist approach to the origin of life. As might be expected, polygenist anthropologists were attracted by Darwin's theory, and there was no lack of discussion on transformism within the Anthropological Society of Paris, established in 1859 by Paul Broca (1824–1880). As Stebbins has shown, during the 1860s the *Bulletin* published by the Society devoted more and more space to papers discussing transformism. In 1870, a peak of 350 pages devoted to transformism was reached, in a volume of less than twice that length. Yet even a supporter of transformism like Eugène Dally (1833–1887), the translator of Huxley's *Man's Place in Nature,* made it clear that

> it is important to separate the cause of transformism from that of Darwinism. . . . His [Darwin's] views are only one of the explanations that can support the undeniable fact of variation. (Stebbins 1965, p. 192; Dally 1868, p. 710; Conry 1974, p. 68; Schiller 1979)

Paul Broca was equally convinced of the need to separate the "fact" of transformism from the "hypothesis" put forward by Darwin (Conry 1974, pp. 51–64). In 1871 Darwin was duly elected Foreign Associate of the Anthropological Society, an honor that could hardly have consoled him for his second defeat, that same year, at the Académie des Sciences. From 1871 to 1882, the Society seemed to lose interest in discussing transformism; only six papers on the issue appeared in the *Bulletin*. Thus, even for the society in which polygenists and transformists came to dominate, Darwinism was hardly a crucial issue, a problem, or a conviction:

> The discussion was virtually never for or against Darwin, but always for or against transformism. Favor of transformism was ultimately preponderant. None claimed to be a Darwinist, most were transformists. (Stebbins 1965, pp. 202–203)

If this was the reaction of the Anthropological Society, it is not surprising to note that more conservative societies, from the Académie des Sciences to the geological and botanical ones, or the Association Française pour l'Avancement des Sciences established in 1871, took little or no notice of Darwin (cf. F. Burkhardt 1974 for English scientific societies). What was known of the debates over Darwin's nomination to the Académie revealed that opponents had obviously strong views against Darwin, but that his supporters were not converted either. De Quatrefages and Henri Milne-Edwards (1800–1885), who were prominent amongst Darwin's friends, emphasized that their votes in no way implied approval of the "theoretical" aspects of his work (Camerano 1896, p. 332; Stebbins 1965, pp. 215–231).

Stebbins also noted the silence of Claude Bernard (1813–1878) and Louis Pasteur (1822–1895), the cool skepticism of Isidore Geoffroy Saint-Hilaire (1805–1861), and the opposition of positivist philosophers to Darwin's speculations. Even the naturalists who in the last three decades of the century favored a broadly evolutionary interpretation of the history of life — as did Jean-Octave-Edmond Perrier (1844–1921), who in 1879 publicly announced his conversion to transformism; Jean-Albert Gaudry (1827–1908); or Alfred-Mathieu Giard (1840–1944) — were clearly defending a view of nature different, to say the least, from the one put forward by Darwin. During the 1880s and the 1890s, when supporters of transformism gained prestigious academic positions, broadly Lamarckian allegiances prevailed over Darwinism (Stebbins 1965, pp. 113–226, 147–152, 157–165; Tetry 1974; Blanckaert 1979; Viré 1979; Gohau 1979; Laurent 1980).

Stebbins attempted an explanation for the reasons behind the opposition to, or non-communication with, Darwin's theory in French naturalist circles. The weight of the anti-transformist Cuvierian tradition, the anti-uniformitarianism and anti-gradualism of the majority of French geologists and paleontologists, the religious convictions deeply embedded in the basic teleological approach shared by a considerable majority of contemporary naturalists — these are regarded by Stebbins as the main factors responsible for the unsympathetic French reaction to Darwin. Toward the end of the century, nationalism, Roman-Catholic spiritualism, and a variety of vitalistic interpretations of organic evolution favored conciliatory moves between opponents and supporters of transformism, who were united in denying the validity of Darwin's solution to the problem of species variability. Even hard-line supporters of evolution tended, not surprisingly, to find Herbert Spencer (1820–1903) and Ernst Haeckel (1834–1919) philosophically and cosmologically more acceptable than Darwin (Stebbins 1965, chap. 8, pp. 266–312; Dougherty 1979; Roger 1982).

Yvette Conry, the major historian of French reactions to Darwin — or rather non-reactions — published in 1972 the correspondence between Gaston de Saporta and Charles Darwin. In a long and dense preface devoted to reconstructing the conceptual foundation and development of modern paleobotany, she poses a direct and crucial question: "was de Saporta the Darwinist he claimed to be?" (Conry 1972a, p. 9). If Stebbins notes the discrepancy between French transformist thought and the Darwinian view of evolution, Conry questions the existence of any significant impact of the theory of natural selection even upon naturalists who openly sided with Darwin. De Saporta was certainly convinced by Darwin of the truth of transformism, but sought to explain its mechanism in non-Darwinian terms. He found "the process of differentiation which led organic beings from simplicity to complexity, from homogeneity to heterogeneity", as described by Karl Ernst von Baer (1792–1876) and Wilhelm Friedrich Hofmeister (1824–1877), a better explanatory tool than the principle of natural selection: "it was not phylogeny which provided the key to understanding ontogeny, but the opposite" (Conry 1972a, pp. 61, 57). Thus, even though Darwin's theory represented a "radical turning point" in the transformist tradition, "it is equally true that after Darwin, pre-Darwinism did not disappear. Indeed, even though Darwin's successors could not ignore him. . . , nevertheless they took their explanatory models from a tradition which preceded the *Origin of Species*" (Conry 1972a, p. 81).

In her lengthy book *L'introduction du Darwinisme en France au XIXe siècle* (1974), which constitutes the first major study published on the topic, Conry pursues her analysis of the scientific and epistemological doctrines dominating contemporary natural sciences, both within and outside the transformist camp. The theme underlying and guiding her sophisticated reconstruction of concepts relevant to the debate on Darwin is a strict definition of the terms "introduction" and "influence". Professor Conry stresses the need for carefully defining such historiographic tools. According to her definition, it is possible to speak of the "introduction" of a theory only when its concepts and assumptions become integrated parts of the relevant disciplines and are capable of reshaping their boundaries, objects, and goals. The actual priorities and research traditions within the French scientific community prevented any significant communication with Darwin's theoretical proposals and shaped even the answers of authors who were prepared to view the history of life in evolutionary terms (Conry 1974, pp. 15–28).

Critics have pointed out that Conry's definition of "introduction" would lead to the conclusion that Darwin's theory was not accepted, or introduced, anywhere in Europe or the United States, and possibly found no supporters even in Britain. Indeed, the standpoints defended by Spencer or Haeckel cannot be equated with the logic and contents of the explanatory strategy developed by Darwin. H. Paul, who is one of the most outspoken critics of Conry's thesis, argues that

> since scientific groups do not convert to new paradigms quickly, and it is usually the new generation that embraces the new paradigm, a time lag of twenty years or more for the general acceptance of Darwinism was normal for the French situation. (H. Paul 1979, pp. 22–23)

He does not, however, enter into any detailed definition of the "Darwinian paradigm", and he avoids naming the naturalists who accepted it (H. Paul 1979; Moore 1977a; Roger 1976. Cf. Guilhot 1976; Marquez-Breton 1977).

After Conry's analysis of the interpretation of evolution by De Saporta, Gaudry, and Giard, or the qualified approval of transformism by Broca and other representatives of the Anthropological Society, it would be difficult to argue that Darwin's key doctrine, the theory of natural selection, found supporters in France. Authors maintaining this would be forced to agree with those nineteenth-century French commentators who, having reduced Darwin's theory to a "long argument" in favor of evolution, claimed that there was nothing new in the *Origin*. To many contemporary naturalists, the theory of natural selection was unacceptable on many different grounds, and the "facts" quoted by Darwin proved only that life had evolved according to some natural law, either supernaturally preordained or hidden amongst the properties of matter. It was thus possible to search for alternative models purporting to explain those problems and phenomena — such as the cause of individual variation, the origin of life, or the philosophical interpretation of the evolutionary pattern — for which Darwin had no solution to offer, or which were excluded from or by his theory.

In the chapter devoted to examining the epistemological and research priorities within paleontology, Conry argues that naturalists active within the discipline — those who attacked as well as those who approved of Darwin's theory — "read it as a doctrine of progress" (1974, pp. 195–227). Gaudry, the son-in-law of the arch-Cuvierian and strict creationist Alcide Dessalines d'Orbigny (1802–1857), became the earliest paleontologist to be "converted" to a transformist interpretation of the fossil record: the first, of course, in the "reaction" period. Yet, in Conry's words, "to the Darwinian theme of specific differentiation through ecological struggle, Gaudry opposed a philosophy of diversity" (1974, pp. 222, 221–227). Darwin rejected the concept of progress, of plan, of necessary development toward "higher" forms of life — whatever the meaning of "high" and "low" as applied to organisms. On the contrary, Gaudry, like De Saporta, found inspiration in the embryological doctrines put forward by Von Baer, even though he also consistently applied the concept of division of physiological labor restated by Milne-Edwards in 1867 (cf. Ospovat 1981). The development of life on earth, to Gaudry, revealed the unfolding of the divine plan for the living creation (De Stefano 1907). As will be noted below, if he accepted a "paradigm" or followed an approach in preference to others, it was not Darwin's theory of natural selection, but the creationist evolutionism defended since the early 1830s by the Belgian geologist Jean-Baptiste-Julien D'Omalius D'Halloy (1783–1875). Analogous considerations apply to Alfred Giard, a paleontologist placed by his birthdate (1840) into the "new" generation of French naturalists. He had no a priori objection to transformism or Darwinism, but regarded his version of Lamarckism as the best explanatory model available (Viré 1979; Roger 1982).

The implicit corollary of Conry's remark on the concept of introduction is the warning that the historian must not reduce the issue of Darwinism in France to a listing of naturalists favoring or opposing transformism. The close examination of the theses put forward by authors who were regarded or who regarded themselves as Darwinists or transformists, reveals the complex articulation of French concepts about nature and its operations, and a variety of standpoints on the goals and priorities of the natural sciences. The second part of *L'introduction du Darwinisme en France* is thus devoted to reconstructing the "conditions of impossibility" that prevented the "translation" of Darwin's explanatory strategy and its presuppositions into French naturalistic thought. Indeed, as in the case of crucial terms such as "natural" and "artificial" selection, the actual translation by Royer helped to create the impression that Darwin personalized nature, a misunderstanding common among readers of the *Origin* and not only in France (Conry 1974, pp. 263–269, 290; Claparède 1861, p. 531).

More basic obstacles did, however, prevent the acceptance of concepts that played a crucial role in Darwin's work and understanding of nature. Thus, for instance, biogeographical considerations were at the foundation — both historically and conceptually — of Darwin's theory. Yet French biogeography stressed the physical side of geographical investigation. "The historical exegesis of the distribution of organic forms" became dominated by studies on climate, and tended to join forces with neo-Lamarckism (Conry 1974, p. 293). Moreover, organic beings were regarded as strictly dependent upon their "locality", "endemic" within their region. The model of a static and providential "economy of nature" underlined biogeographical research. Equally static was the concept of organic economy prevailing within physiological and anatomical disciplines, through the permanence of Cuvierian "conditions of existence", the application of Milne-Edwards's division of physiological labor, or Claude Bernard's formulation of the concept of "internal milieu". Thus the study of debates on Darwin or evolution in general cannot dispense with carefully evaluating the actual concepts and traditions that organized scientific disciplines and research within each naturalist culture, in France or elsewhere.

Before turning to a consideration of the major religious and ideological features of French reactions to Darwin, it is appropriate to consider briefly the suggestions for further research that emerged from the studies here reviewed. Stebbins's contribution has provided ample evidence of the non-reaction to Darwin by French naturalists. His assessment of publications relating to evolutionary and Darwinian topics requires improvement and further research. Conry's conceptually challenging book has investigated the reasons why French science proved epistemologically and theoretically unreceptive to the methodology of, and the conclusions put forward in, the *Origin*. She argued that the weight of the naturalistic tradition of the country played a crucial role in the rejection, or the particular interpretations, of Darwin's work. This same conclusion was reached by John Farley who stated that "in many ways the Darwinian debate was a reenactment of the Cuvier-Lamarck debate of earlier years" (Farley 1974, p. 275).

Several trends of evidence support the conclusions of Conry and Farley. It is therefore appropriate to suggest further inquiries into the French natural history debates of the period 1830–1860. Supporters of various brands of transformism were certainly isolated, but they kept alive a tradition of broad transformist thought, which stood in opposition to the fixity of species defended by Cuvier and Flourens. Thus, for instance, it would be of some interest to assess the influence of theories by Etienne Geoffroy Saint-Hilaire on the interpretation of paleontological findings, and to determine whether his model of transformation of organic beings through viable monstrosities was discussed by contemporaries. It is indeed possible to establish a connection between Geoffroy Saint-Hilaire's doctrines and the early formulation of a transformist model for the interpretation of the paleontological record by D'Omalius D'Halloy. Of equal interest would be whether there was any reaction to the synthesis of Lamarckism and the principle of embryological recapitulation put forward by Bory de Saint-Vincent in his *Dictionnaire Classique d'Histoire Naturelle*. Bory's dictionary was published between 1822 and 1831, and was reissued in Belgium in 1853 (Drapiez 1853). The work was translated into Italian, and it was well known in Britain too. The articles on geographical distribution and on "Creation" were of particular interest; they have never been systematically studied. Julien Joseph Virey (1775–1846), whose *Nouveau Dictionnaire d'Histoire Naturelle* (1803–1804) enjoyed European circulation early in the century, put forward a synthesis of embryological and transformist thought, probably the earliest attempted in France. Virey's doctrine was creationist and anti-Lamarckian; it was indeed an answer to the publication in 1802 of Lamarck's *Recherches sur l'organisation dès corps vivans*. According to Virey, each step in the process of unfolding of higher forms of life was achieved through an act of creative intervention.

Virey's dictionary went through a second edition between 1816 and

1819. His theories on the history of life, and on the physical history of man — Virey was a convinced polygenist — were discussed in works published during the period 1820–1850. His treatises included discussions on the relationship between animal and human societies. His impact was probably slight on professional naturalists, but considerable among amateur naturalists, doctors in particular. I should like to suggest that an investigation of the cultural background of the doctors taking part in the establishment of the Société d'Anthropologie, who were ready to accept transformism, though not Darwinism, would reveal interesting continuities within a little known and less studied sector of the community of naturalists. To quote a further instance of the relevance of debates between 1830 and 1860 to the understanding of French reactions to Darwin, it is appropriate to mention that since 1831 D'Omalius D'Halloy defended a broad transformist interpretation of the fossil record and had no difficulty in keeping his Catholic faith (Omalius D'Halloy 1831). He certainly played a significant role in alerting colleagues to the problem of transformism. As Albert Gaudry himself acknowledged in a letter to D'Omalius, "the chapter on the appearance and succession of living beings in your Abrégé of geology, has contributed to inclining me towards transformism" (Omalius D'Halloy 1868).

The four volumes of Isidore Geoffroy Saint-Hilaire's *Histoire naturelle générale des règnes organiques* (1854–1862) contributed powerfully to reminding colleagues and the public of the state of affairs in the controversy over the definition of species. Chapter 6 of volume 2 (1859) was devoted to discussing the history of the debate on species and on transformism in French naturalist circles. Isidore did not accept transformism, but he severely criticized the concept of the fixity and creation of species. He defended the theory of the limited variability of organic forms, and the "simple and rational" doctrine of "paleontological filiation" (2: 365–446, 434). Dominique Alexandre Godron (1807–1880) published in 1859 his book *De l'espèce et des races dans les êtres organisés, et specialement de l'unité de l'espèce humaine*. He too rejected transformism, although his summary of ideas by Bory de Saint-Vincent, Lamarck, and Etienne Geoffroy Saint-Hilaire was fair and respectful. The question of the limits of specific variability was a perfectly legitimate issue in natural sciences (1859, p. 13).

As is shown by the recurrent Lamarckism and temptations to ontogeny of several representatives of nineteenth-century French and European evolutionism, the understanding of the reception of Darwin's ideas cannot dispense with thoroughly investigating alternative naturalistic traditions and interpretations of the history of life. The debate on evolutionism — to use the word in its wider connotation — did not start in 1859, and the controversy over the Darwinian theories was deeply influenced by the discussion of theories put forward during the previous decades (Fischer 1981; Blanckaert 1981).

As far as the broad cultural dimensions of French reactions to Darwin are concerned, it should be pointed out, with Conry, that Social Darwinism, usually regarded as an important feature of the impact of the *Origin* on western culture, was never dominant or popular in France. Due to the obvious and well-known associations with Germany, Social Darwinism was not appealing to French intellectuals, who felt that in the struggle among nations their country had been on the losing side. There were of course attempts to apply biological models to the interpretation of colonial expansion, class division, and class struggle, especially toward the end of the century. Linda Loeb Clark (1968) has written an interesting dissertation discussing some of these attempts and the debates over the relationship between contemporary society and the development of natural sciences (see Lagarde 1979). It should be pointed out that in France, Germany, and Italy, as well as in many other countries, the phenomenon of Social Darwinism was complex and diffuse. In Conry's words, Darwinism "was only an excuse, as is shown, for instance, by the various interpretations of the same element taken from the theory by representatives of diverging systems" of Social Darwinism (1974, pp. 397–406, 404).

All the studies of French reactions to Darwin reviewed here examined the religious dimension of the debate. Stebbins argued that the period 1859–1880 was characterized by the silence of the Roman Catholic hierarchy, and by individual attempts to refute Darwinian and transformist theories. French Roman Catholics were clearly more worried by debates on the age of the earth, than by Darwinism. Indeed, if Clémence Royer stressed the rationalist import of Darwin's work, and other supporters of transformism stated the utter incompatibility of transformism and religion, the contemptuous silence of the official scientific bodies, the outspoken criticism by Flourens, or the calm but severe refutations by De Quatrefages provided religious critics of evolution with all the ammunition they needed. In a famous letter to Constantin James (1813–1898), author of a violent and inelegant attack *Du Darwinisme ou l'homme-singe* (1877), Pope Pius IX attacked Darwinism as an absurd, dangerous corruption of morality (James 1882, pp. 84–85; Conry 1974, p. 230). But he too appeared convinced that the scientific opposition to the theory was sufficient to curb the heresy. The Pontificate of Pius IX (1792–1878) ended in 1878, and Leo XIII (1810–1903) was the head of the Church from 1878 to 1903. The period was characterized by enormous political and ideological difficulties for the Church. The reaction to pressure, initially at least, was harsh and deeply reactionary. In September 1870 the Papacy lost its millennial temporal power. Toward the end of the century, liberal movements within the Church sought ways to find a conciliation with moderate and conservative sectors of the socialist interpretation of the Christian message.

Theoretical difficulties, debates over the mechanism of heredity, and a variety of evolutionary models made it possible to find basically non-Darwinian formulations of the theory of evolution. Thus, even though the diffusion and popularity of evolutionary interpretations of the history of life on earth in the last two decades of the century made it difficult for Catholics to maintain that the theory had no followers, the philosophical and cosmological implications drawn from it offered ample possibility of conciliation.

Conry touched upon the issue of French Roman Catholic reactions to Darwin when discussing the work of Nicolas Boulay (1837–1905), a botanist who taught at the University of Lille. The different ideological, political, and broadly cultural commitments of Pius IX and Leo XIII were reflected in the reactions to Darwin and evolution by Catholics active in Lille. The rejection of compromise that characterized the period 1860–1880, was followed by a period of conciliatory effort. This was partly due to pressure concerning the updating of the curriculum of Catholic schools and institutes for higher education. The Abbé Boulay took an active part in the Congresses of Catholic Scientists, which held their first session in Paris in 1888. He accepted the evolutionary interpretation of the history of life, but he rejected the mechanism of natural selection offered by Darwin. The opposition to the concept of natural selection, and of struggle, was also linked to the effort of social reconciliation made by Pope Leo XIII in the Encyclical *Rerum Novarum* (15 May 1891). The worker had to obey the master; those who claimed the right to freedom of expression and of organization, let alone the socialists, were condemned to eternal death. The masters were firmly invited to exercise paternal care over their dependents, however. The ideal *Pax Christiana* was based on a society free from struggle, organized according to traditional ideologies of benevolent paternalism. In this context, reconciliation with the scientific community, possibly led by Catholic scientists, appeared as a reasonable project, but the theory of natural selection proved impossible to integrate (Conry 1974, pp. 228–237).

More moderate — and more willing to compromise — were the representatives of the minority Protestant community. Stebbins examined contributions to the *Revue Germanique*, which was published under changing titles from 1859 to 1869, and to the *Revue Chrétienne* (1859–1882). He found no trace of full support for Darwin, but he did find a growing sympathy for him, because of the moderation of his position, as opposed to the more radical views of some of his English and German supporters. Conry evaluated the response to Darwin by Armand Sabatier (1834–1910), professor of anatomy at Montpellier, who was noted for his Protestant zeal. Sabatier had no difficulty in accepting evolutionism, but he viewed it as a process pre-ordained by a powerful Creator, the true organizer of the succession of organic beings throughout the history of the earth (Stebbins 1965, pp. 308–315; Conry 1974, p. 246).

The major study of French Catholic reaction to Darwin and to evolutionism in general was published by H. Paul in 1979. The thesis Paul sets out to prove is that "Catholicism cannot be automatically equated with hostility towards evolution or even Darwinism". The first phase of Catholic reaction to Darwinism, up to the early 1880s, was characterized by a strong defensive stand, provoked by the rationalist if not openly atheistic overtones of some supporters of evolution, and reinforced by the unfavorable reaction of the French scientific community (H. Paul 1979, pp. 24, 53, 64; Farley 1974). Toward the 1880s, however,

> Catholics opposed to evolution had . . . to face the fact that most of the scientific community and an increasing number of intellectuals were accepting evolution, in spite of the so-called irrefutable case against it. (H. Paul 1974, p. 78)

Paul closely scrutinizes the debate over transformism and Darwinism at the five congresses of the Catholic scientists held between 1888 and 1900, two in France, one in Brussels (1894), one in Freiburg (1897), and the last one in Munich (1900). The discussion showed a certain degree of acceptance of evolutionism, appropriately purged of obnoxious overtones: "a small if vocal and persuasive minority of voices was raised in defence of a restricted type of evolution, including some scientific aspects of Darwinism" (H. Paul 1974, p. 87). Paul's work does undoubtedly represent a firm starting point for further research on Roman Catholic reactions to Darwinism and evolutionism in France. Yet the reader is left with the impression that the author has tried to revise Whiggish views on the relationship between science and religion in the nineteenth century, but could prove only that few Catholic scientists and fewer Catholic theologians showed a sincere desire to come to terms with contemporary developments in biology, and with Darwin's ideas in particular. Not surprisingly, therefore, the opening statement that Catholicism cannot be equated with opposition to Darwinism and evolutionism is followed by repeated and significant qualifications. Thus Paul acknowledges that "Catholic journals opposed nearly all forms of evolution from the appearance of the *Origin of Species* well into the first decade of the twentieth century" (1979, p. 40). Moreover, "the hard line defence of the fixity of species remained an obsession of a substantial part of the Catholic community" (1979, p. 78). Finally, the listing of Catholic authors who suitably revised various evolutionary models discussed at the end of the century — the theory of the natural selection of chance variations being rigorously excluded — in order to preserve plan, finality, and providence in nature, cannot avoid the fact that the Catholic Church was a theocratic power, and the ultimate word rested with the hierarchy. As Paul acknowledges,

> The integrists seem to have been powerful enough to keep Rome on the side hostile to evolution . . . . Unable to find enough support from the hierarchy, French Catholic scientists organized themselves to fight the evil effects of the integrists on Vatican opinion in scientific matters. (1979, p. 104; cf. pp. 40 and 78)

The interpretation implicit in the sentence quoted above suggests that the Roman Catholic hierarchy opposed evolution because it was under the devious influence of French reactionary integrists. It could indeed be argued that Paul's attempt at revisionist historiography fell victim to "justificationist" temptations.

The reconstruction of French Catholic reactions to Darwinism and evolutionism provided by Stebbins and by Paul contains significant reference to a variety of standpoints within the Catholic community, and to serious and at times dramatic infighting between Jesuits, secular clergy, laymen and scientists. Yet there is no study available relating political, ideological, and theological differences in attitudes toward the variety of evolutionary models available toward the end of the nineteenth century, and in the first two decades of the twentieth century.

Notwithstanding clear differences of emphasis, of historiographical, epistemological, and theological orientation, the available studies of French reactions to Darwin do contribute to widening the scope of the history of science. Indeed, it could be argued that the general conclusion to be reached is that the focus of attention is narrowed by the concentration on French Darwinism, or the lack of it; that this has tended to preclude a deeper understanding of scientific debates at the time, and of their impact upon, or integration with, broader dimensions of French intellectual and social life. There was undoubtedly a great debate over French traditional transformism, over European evolutionism in general, and the theory of Charles Darwin in particular — a debate that often ended with the acceptance of a broad evolutionary model for the interpretation of the history of life on earth, and the rejection of the specific evolutionary mechanism put forward by Darwin. Thus representatives of the French scientific, philosophical, and theological communities undoubtedly felt the weight of Darwin's "long argument" in favor of a naturalistic interpretation of the succession of organisms on the surface of the earth and during its history, but found alternative explanations philosophically, cosmologically, and theologically more rewarding. As far as the broader social, political, and theological implications and dimensions of the debate were concerned, it is clear that only the application of strict historical methodologies, helped by the specific epistemological tools required by the topic, will contribute to clarify this crucial and — with the few exceptions here discussed — rather neglected episode of modern French culture.

III. Recent Studies on Italian Reactions to Darwin

P. Corsi

The volume on the comparative reception of Darwinism edited by Glick (1974) was characterized by a revealing gap: no chapter was devoted to

Italy, even though H. Paul briefly mentioned Italian debates on evolution, with particular references to their religious dimension (1974, pp. 408–413). The omission of Italy reflected the state of studies on nineteenth-century natural sciences in general. Indeed, naturalist-historians such as Lorenzo Camerano (1856–1917), Carlo Fenizia, Giovanni Canestrini (1835–1900), or Michele Lessona (1834–1894), who at the end of the nineteenth century assessed the Italian reaction to Darwinism, left no significant legacy to Italian historiography of the twentieth century. The idealistic and spiritualistic philosophies prevailing in the early decades of the twentieth century, coupled with exacerbated nationalism and political despotism, created a climate unfavorable to the development of an independent scientific culture, and little concerned with the history of science (Camerano 1902, 1904, 1905–1909; Canestrini 1894; Lessona 1883, 1884; Fenizia 1901).

The philosophy of Benedetto Croce (1866–1952) and Giovanni Gentile (1875–1944) did not acknowledge the speculative and cognitive value of science, which they regarded as merely an efficient and useful technique. The history of science suffered from this lack of appreciation of the role of scientific theories and methods in modern culture.

The history of philosophy, of idealistic inspiration became, and to some extent still is today, the predominant trend in Italian intellectual history. Aldo Mieli (1879–1950), the founder of *Archeion*, and Federigo Enriques (1871–1946), the well-known mathematician and historian of science who opposed the idealistic ascendancy, were forced to leave Italy well before the racial laws forced more intellectuals of Jewish origin to emigrate. The initiatives of Mieli and Enriques were defeated at the end of the 1920s, despite the sporadic interest of the fascist regime in celebrating the scientific glories of the country.

Diverse developments within intellectual circles of the opposition favored the preservation of a fringe group of scholars sensitive to the history of science. Moreover, after World War II, representative historians of philosophy made significant concessions to the role of science in the making of modern culture. The main concern, in Italy as well as elsewhere, was with the history of mathematical and physical sciences. Studies on such figures as Galileo Galilei by Ludovico Geymonat (1957), or on Francis Bacon by Paolo Rossi (1957), exemplified the growing attention toward the history of science. The physical sciences of the Renaissance period attracted the largest share of contributions to the field. A pioneering attempt by Pietro Omodeo to call attention to an important collection of manuscript notes taken by Giosuè Sangiovanni (1775–1849) at courses given by Lamarck, and probably used by Lamarck himself in the writing of the *Histoire Naturelle des Animaux sans Vertèbres*, failed to alert colleagues to the importance of Italian sources for the history of nineteenth-century evolutionary ideas. For many years, Omodeo, a zoologist, and Giuseppe Montalenti, the distinguished Italian biologist, were among the few Italian contributors to the history of the biological sciences (Omodeo 1949a, 1949b, 1959, 1969; Montalenti 1958).

The translation into Italian of works by Alexandre Koyré, a historian of science particularly sympathetic to the philosophical approach to the field; the diffusion of ideas put forward by Gaston Bachelard and Michael Foucault; and the translation of Bernal's works, should be mentioned as further evidence of the concern for the history of science in various quarters, and particularly among philosophers. The creation of the first six university chairs in Italy for the history of science in 1981, has provided official acknowledgement for a discipline that for a long time occupied a peripheral position in university teaching and curricula.

It was probably the growing popularity of the history of biology in France, Britain, and the United States, as well as the shift from the study of seventeenth-century science to the investigation of scientific debates of the nineteenth century, that produced in the early 1970s a noticeable impact upon the younger generation of Italian historians of science. It is significant that five of the eleven studies on Italian reactions to Darwin here reviewed were published in the years 1976 and 1977.

In 1977 Giovanni Landucci produced the first monograph on the impact of Darwinism on the culture of Florence, a town, needless to say, of particular importance for the intellectual life of the country. In the previous year Gian Battista Benasso completed the first part of a study on the history of Italian evolutionism, devoted to assessing the shortcomings and leading features of Italian zoological investigations in the first part of the nineteenth century. In 1977 Rossi published a preface to the *Ascent of Man* by Antonio Fogazzaro (1849–1911), an Italian novelist who attempted a compromise between Roman Catholic theology and his own spiritualistic interpretation of evolution. The book *Charles Darwin. 'Economy' and 'History' of Nature* by Giuliano Pancaldi, which also appeared in 1977, contained a dense forty-page chapter on "Darwinism in Italy, 1860–1900". The physical anthropologist Giacomo Giacobini published a short paper on the debates in Turin from 1864 to 1900 on the origin of man (1977).

All the authors listed above expressed full awareness of the difficulty of the task they undertook and noted the almost total neglect of the topic in histories of Italian philosophy and science of the period. The sources for the history of evolutionary debates in Italy are numerous and are printed in all the provincial centers of the time, from Milan, Venice, and Turin to Naples, Messina, and Palermo. It is important to emphasize that there is no single library in the country containing a comprehensive selection of such literature. A very conservative estimate of printed sources for the period 1860–1920, listed in an avowedly incomplete and often unreliable bibliography of Italian books, amounted to about 450 pamphlets and books directly relating to evolutionary debates (Pagliaini 1903–1928). A summary survey of the daily and periodical press reveals hundreds of contributions on the variety of topics — scientific, philosophical, and political — that were seen as relevant to the debate on evolution. In Italy, as in France and England, journals of scientific societies or local scientific academies proved reluctant to embark upon a scientific discussion of the doctrines put forward by Darwin. Literary, philosophical, and religious journals, on the other hand, were prominent participants in the debate.

Pancaldi attempted a broad survey of the major trends of the Italian debate on Darwinism and evolutionism. The earliest reviews of the *Origin*, which appeared in the *Civiltà Cattolicà*, the intellectual quarterly of the Jesuits, and in *Il Politecnico*, failed to notice the novelty of the approach and the solutions put forward by Darwin. Giovan Battista Pianciani (1784–1862), who was inspired by a review of Darwin's work by the Swiss naturalist François-Jules Pictet (1809–1872), wrote in the *Civiltà Cattolicà* that the new theory was nothing but a restatement of old transformist hypotheses (Pianciani 1860b, 1862; Pictet 1860). The Italian public and Italian naturalists initially appeared little responsive to the *Origin*. The excitement caused by the War of Independence of 1859, and the successful operations conducted by Garibaldi in the south during the year 1860, crowned four decades of social, political, and intellectual agitation. The temporal power of the Church retreated within the Roman Walls, surrounded by a state professing scarce sympathy for, if not open hostility against, the Papacy. Rome was regarded as the natural capital of Italy, as the scientists convened for one of the congresses of Italian naturalists strongly indicated.

As in France, naturalists and intellectuals attentive to developments within natural sciences were more concerned with debating the issues of spontaneous generation and the physical history of man, than the many questions Darwin touched upon in his book. Not surprisingly, therefore, the official starting point of the controversy over evolutionism was a lecture given by the zoologist Filippo de Filippi (1814–1867) in Turin on 11 January 1864, on the subject of "L'uomo e le scimie", ("Man and the monkeys"). In view of Darwin's restraint on the subject of man, De Filippi found in Huxley's *Man's Place in Nature* and in the controversy between Huxley and Richard Owen over the anatomical differences between human and ape brains, the basic material for his approach to the subject. De Filippi was a respected naturalist, of known Catholic sentiments. He had fought a long personal battle over the problem of reconciling the antiquity of man with his own religious beliefs. De Filippi accepted all the basic arguments put forward by Huxley and Darwin, but maintained that moral evolution was not comparable with physical evolution. The kingdom of man — characterized by such exclusive prerogatives as philosophic doubt, or moral and religious sentiments — could not be equated with the animal kingdom (Pancaldi 1977, pp. 167–168; Benasso 1976, pp. 59–64; Giacobini 1977; Lessona 1883, pp. 161–206, 194–196).

De Filippi's proposed reconciliation between evolutionism and traditional religious beliefs did not please those who saw scientific naturalism as the final stage of a long struggle against religious superstition, nor those who looked with horror at the progress of materialism. Michele Lessona, who translated many of Darwin's works, and who divided his allegiances between Darwin and Lamarck, was a close friend of De Filippi, to whom he was related by marriage. He told the story that when De Filippi died in 1867, two preachers in Turin pointed out from the pulpit that on the brink of death the impious naturalist had asked for religious consolation, whereas a rationalist magazine reported that the story was a fabrication of the clergy (Lessona 1883, p. 196; Fenizia 1901, pp. 325–327).

The first phase of the debate on evolutionism was thus concerned mainly with anthropological, philosophical, and theological issues. Indeed, the first complete Italian edition of the *Origin* appeared only in 1865 and was translated by Giovanni Canestrini, at the time teaching in Modena, and by Luigi Salimbeni. It has, however, escaped the attention of students of Italian Darwinism that the first partial translation of Darwin's work was completed and published in 1864 (R. Freeman 1977). The publisher Zanichelli issued the first part of the *Sull'origine delle specie* containing the translation of Chapters 1–3 as a publicity installment. The pamphlet was sent to readers — and it would be interesting to know who were the chosen ones — with the invitation to subscribe to the entire work, or to return it to the publisher. Italian translations of Darwin, a list of which was compiled by Conry in an appendix to her book, were usually late, even though the circulation was satisfactory (Conry 1974, p. 438; R. Freeman 1977). The harsh polemics of the years 1864–1865, which I shall discuss in some detail below when reviewing Landucci's book on Darwinism in Florence, were followed during the 1870s by a different kind of reaction. In Pancaldi's words, there was "a subtle work of assimilation of evolutionary problematics within various intellectual trends of contemporary Italian culture" (Pancaldi 1977, p. 177). If French positivists opposed Darwin, Italian positivists eagerly sought to appropriate suitably adapted features of evolutionary doctrines, and they soon found the cosmic systems of Spencer and Haeckel more rewarding and consoling. Even a few representatives of the Neapolitan Hegelian tradition accepted a philosophic, strongly finalistic interpretation of evolutionary processes. This was the standpoint defended by Pietro Siciliani (1830–1885), whereas the idealist physiologist Angelo Camillo de Meis (1817–1891) proposed to view evolution as guided by the unfolding of a logical process (Oldrini 1973; Benasso 1978, pp. 110–111; Pancaldi 1977, pp. 179–182).

As in France, but on a much larger scale, evolution was kept distinct from Darwinism, and preference was accorded to Spencerism, Haeckelism, and variously refurbished versions of Lamarckism (Morselli 1887; Salvadori 1900; Bulgarini 1887–1888). In Italy the debate over evolution was general; it concerned a variety of social, political, philosophical, and theological topics, plus a few specifically related to natural sciences. As a consequence, what was gained in breadth was clearly lost in depth. In Pancaldi's words, "the distance between the philosophical reflection and the limited biological research, tended to produce arbitrary generalizations of the evolutionary model, which increasingly became an abstract methodological canon, incapable of providing guidelines to research" (Pancaldi 1977, pp. 191–192).

The serious discussions of Darwin's theories by Giovanni Canestrini, the psychologist Francesco de Sarlo (1864–1937), and the philosopher Enrico Morselli (1852–1919), the editor of the influential pro-Darwinian and pro-Haeckelian *Rivista di Filosofia Scientifica* (1881–1891), failed to reach a public captured by the political and philosophical generalities of the debate over evolutionism. As far as the social uses of Darwinism and evolutionism in general were concerned, Pancaldi noted that the broad interpretations of the evolutionary model allowed opportunistic borrowing and interpretations by representatives of various and at times opposite political standpoints. In Italy, however, socialist or radical political philosophers tended to see the concept of evolution through fierce competition as the guarantee of a felicitous outcome of class struggle (Ferri 1894). It should also be pointed out that in the political as well as in the philosophical debate over evolution, from being generic, the discussion turned trivial; toward the end of the century the crisis of positivistic and fideistic evolutionism was well under way, and all too apparent. In a famous essay on socialism, and in letters to Engels, the Italian marxist philosopher Antonio Labriola (1843–1904) spoke of the sterile faith in "Madonna Evolution" and pointed out the barrenness, rhetoric, and triviality of much philosophic and sociologic evolutionism. The "Great Eunuch Spencer" could not be regarded as a reliable interpreter of Darwin's scientific ideas (Pancaldi 1977, pp. 200–201; Labriola 1949, p. 149; Labriola 1898).

The lack of a consistent group of naturalists engaged in translating the articulated Darwinian approach for the benefit of their own fields of research, and a certain leaning towards Lamarckism, prevented the professional scientists devoted to the cause of evolution from transmitting a significant legacy to the future. At the twelfth Congress of the Italian Society for Scientific Progress, Lorenzo Camerano, the veteran of Darwinian battles, made a major speech on Italian zoology in the nineteenth century, and concluded that the majority of Italian naturalists embraced Darwinism and evolutionism. Camerano also attacked the so-called reconciliations between Roman Catholic theology, spiritualistic philosophy, and generic evolutionism put forward by the liberal Catholic novelist Antonio Fogazzaro, the Jesuit Heinrich Wasmann (1859–1931), and Father Agostino Gemelli, the founder of the Catholic University of Milan (Camerano 1912, pp. 483–484). His contention that everyone was Darwinian, and that natural science had nothing to do with spiritualism or vitalism, was immediately and authoritatively rebuked by two senior colleagues, Giuseppe Cuboni (1852–1920) and Luigi Luciani (1840–1919), who argued that neo-vitalism was rampant and that every serious naturalist accepted that Darwinism had been superseded by neo-Lamarckism, which in its turn had been substantially revised in a vitalistic direction by Carl Wilhelm von Nägeli (1817–1891) (Camerano 1912, pp. 492–495). Indeed, even such an early supporter of Darwin as Federico Delpino (1833–1905), the botanist pupil of Filippo Parlatore (1816–1877), could not refrain from stressing teleology and providential supervision in nature. He firmly opposed materialism, Haeckelian monism, and socialist interpretations of evolutionism (Delpino 1867, 1868, 1895). Canestrini, who praised Delpino as "a Darwinian fully displayed", was rather embarrassed to account for the faith in vital principles displayed by his colleague in the very works in which he professed approval of crucial features of the Darwinian theory (Canestrini 1894, pp. 191–192).

Gian Battista Benasso published in 1978 the second part of his "Materials for the History of Italian Evolutionism". The first part assessed the influence of Lamarck on Italian natural sciences. He also commented upon the famous lecture delivered by De Filippi and the debate that followed. In this first part Professor Benasso appeared largely indebted to a series of important contributions to the history of Italian zoology and Lamarckism produced by Lorenzo Camerano between the late 1890s and the 1910s. Unfortunately, this important line of research opened by Camerano has not been pursued by historians of science, with the noted exception of essays by Pietro Omodeo.

The second contribution by Benasso contained a far larger amount of first-hand information and a series of portraits of naturalists who took part in the various phases of the debate over evolution from 1864 to 1900. He emphasized, as did Pancaldi, that the vehemence of the ideological and political overtones made the generic pro- or anti-evolutionist dimension of the debate prevail over the properly scientific one. Moreover, discussions and divisions within the evolutionist camp concerning Weismann's theory of heredity (rejected by many Italian evolutionists, who retained their basic Lamarckian allegiances), the significance of the paleontological record, or diverging hypotheses on the moving force of evolutionary processes, favored the penetration of neo-vitalism in Italy and the works of Hans Driesch (1867–1941) in particular (Driesch 1911).

As far as the actual impact of Darwin's theory, or of evolution in general, upon Italian naturalistic disciplines was concerned, Benasso concluded that it was minimal:

> naturalists followed a professional practice still largely empirical, and at times unconsciously linked to the fixist tradition, on which they superimposed a scaffolding of more or less advanced scientific information. Developments in modern biology were never deeply assimilated, nor did these become an integral part of the professional skill of the naturalist. (1978, p. 84)

The more biology moved toward laboratories, the more a tradition of taxonomic work in the field, or at the desk of a museum, was bound to lose touch with the major trends of European natural history.

Benasso made passing and tantalizing reference to the plurality of influences that shaped the thought of Italian naturalists who claimed to be Darwinians, or evolutionists. Thus he mentioned the impact of Oken on Paolo Mantegazza (1831–1910), the celebrated Florentine anthropologist, and the sophisticated exegesis of the *Origin* by Achille Quadri, a paleontologist who well understood Darwin's ideas but sought to improve them with the help of Gaudry, Haeckel, and a broad chain-of-being approach to taxonomic work (Benasso 1978, pp. 86–90; Canestrini 1894, pp. 179–180; Quadri 1869). In the first part of his study on Italian evolutionism, when analyzing the debate over the origin of man, Benasso reproduced without further comment the suggestion by Camerano that De Filippi had probably been exposed to Lamarckian ideas early in hs career. Unfortunately, mention and suggestion are no substitute for thorough investigation and full theoretical assessment. It will be the task of historians of science concerned with Darwinism and evolutionism in Italy to pursue the line of research opened up by past and recent commentators.

The situation is already improving. Pancaldi has completed a volume collecting case studies of Italian naturalists active in the second half of the nineteenth century. Particular emphasis is placed on the activities and ideas of Giovanni Canestrini, the first translator of the *Origin* and one of the first historians of Italian evolutionism (Pancaldi 1983). Professor Giacobini, who in 1977 published a short case study of the reactions to Darwin by naturalists active in Turin, has prepared an anthology of texts relating to the evolutionary debate from 1864 to 1900, and has written a long introduction on general features of Italian reactions to Darwin (1983).

It is appropriate to point out that the task of surveying Italian debates over evolutionism and Darwin is made particularly difficult by the remarkable polycentric nature of Italian culture. The existence, for periods of centuries, of small states and town-states, dominated by local or foreign aristocracy, and exposed to a variety of local and international cultural traditions, was responsible for many singular features of Italian social and intellectual life. The unification of Italy in 1859–1860 did not mean the end of local culture. The fragmentation of economic and political life in a plurality of centers and regional spheres of influence was also reflected in the variety of scientific traditions and institutions. Many towns of the center-north, and a few of the major towns of the south, were characterized by the presence of local natural history societies; literary, medical and scientific academies and societies informally organised by groups of amateurs and numerous universities. During the early nineteenth century, medical faculties widened the scope of their teaching in natural history. Towns like Milan, Pavia, and Turin expanded or established natural history museums, often on the Parisian model. During the second half of the century, the French model was slowly replaced by the German one. The foundation of the Zoological Station in Naples by Felix Anton Dohrn (1840–1909), officially inaugurated by the Italian Minister of Education on 11 April 1875, established lasting links between European and Italian naturalists working on marine biology (Groeben 1975; Heuss 1940).

In many cases, however, the founding of an institution did not necessarily imply commitment, and teaching rarely required research. Nevertheless, a survey of library holdings in a town like Florence reveals the sustained effort to keep pace with international developments in natural sciences during the early decades of the century. French books, dictionaries, and periodicals on natural history are particularly well represented, whereas German works are scarce. Gaps in the catalogues tended to widen with the 1840s, even though the teaching of scientific disciplines at Florence remained adequate. Landucci has written the only case study available on the impact of Darwinism and evolutionism on the culture of one provincial capital, Florence, which for a few years was the capital of the new Italian Kingdom. Although Landucci's monograph focussed on Paolo Mantegazza, the first Italian professor of anthropology at the Institute for Higher Studies, the first three chapters of the book were devoted to assessing the initial impact of evolutionary anthropology on the local scientific and philosophical community, and the appropriation of evolutionary conclusions by intellectuals engaged in the debate over the origin of language.

According to Landucci, the Florentine debate over evolutionism was concerned only indirectly with the specific proposals put forward by Darwin. A broad evolutionist interpretation of organic, human, and social life was valued by supporters of a secular view of nature. The debate on the origin of language, for instance, was more a discussion of the limits of naturalistic explanations of cultural phenomena than a specific attempt at applying Darwinian concepts and categories to linguistics (Landucci 1977, pp. 51–78; cf. Conry 1974, pp. 91–107). It was the debate over the origin of man that provoked the greatest amount of controversy in Florence, and in Italy, during the 1860s. De Filippi's 1864 lecture had already caused considerable alarm and a violent reaction from a variety of religiously oriented or philosophically more conservative sectors of contemporary culture.

De Filippi's arguments favoring descent from a common ancestor for man and monkeys were answered by, among others, the geologist, paleontologist, and botanist Giovanni Giuseppe Bianconi (1809–1878), the Director of the Natural History Museum of Bologna. Bianconi drew his counterarguments from Cuvier and Richard Owen. He maintained that the unity of plan within large groups of organic forms did not authorize phylogenetic conclusions, and he stated his belief in independent creation. Bianconi's answer to De Filippi was almost universally acknowledged by supporters of Darwin and of evolution to be moderate, civil, and technically well argued (Bianconi 1864, 1874; Canestrini 1894, pp. 196–199; Benasso 1976, pp. 92–98 finds Bianconi's answer unnecessarily technical; Martinucci 1978). In Florence the debate over the origin of mankind had a late start;

it was opened by Father Giovanni Antonelli (1818–1872), a member of the religious educational order of the Scolopi. Discussing the importance and shortcomings of the study of natural history, Father Antonelli violently attacked the doctrine of the animal origin of man, and those countrymen who were not ashamed to follow the theories of "some imbecile foreigner" (Landucci 1977, p. 83).

A later refutation of common descent by Terenzio Mamiani (1799–1885), former Minister of Education and a close friend of De Filippi, as well as the counterarguments put forward by Paolo Mantegazza, who defended evolutionary anthropology, represented the last attempt to keep the debate within the boundaries of scientific polemic. Mamiani, a spiritualist philosopher, calmly defended teleology, progressionism, and creationism. He also pointed out that Darwin himself did not appear to share the enthusiasm of his followers for the doctrine of the animal origin of man. Equally moderate was the counterattack by Mantegazza, who stressed the inductive and empirical character of science and its neutrality with respect to metaphysical and religious issues. Yet, in Landucci's words, to debate over evolutionary anthropology was already taken to imply much more than a controversy over comparative anatomy and paleontology: "Many viewed evolutionism as the faith in progress, the dismantling of prejudices, and fixism as reaction, immobility, a turning towards the past" (Landucci 1977, p. 88). As a consequence, it is difficult to find even a plain account of Darwin's works and theories in the publications relating to the contemporary Florentine debate. The consideration of the consequence of admitting evolutionism of one kind or another prevailed over the discussion of the actual articulation of Darwin's theory.

A lecture by the Russian physiologist Aleksandr Herzen (1839–1906), professor at the Florentine Institute for Higher Studies since 1867, published under the title "On the relationship between men and monkeys", created an enormous uproar. The author defended the doctrine of common descent by appealing to the views of Lamarck and Darwin. Herzen was the representative of a group of foreign naturalists invited to teach in Italian universities, in a short-lived attempt to improve the scientific culture of the country. The materialist philosopher Jacob Moleschott (1822–1893), called to Turin in 1861, and the brothers Moritz (1823–1896) and Hugo Schiff (1834–1919), invited to Florence with Herzen, were others of this group. To many Florentine intellectuals, afraid of dangerous "imbecile foreigners", the presence of three distinguished naturalists of international standing represented a threat, and Herzen's lecture in favor of evolutionary anthropology a provocation.

Locally and nationally well known intellectuals such as the pedagogist Raffaello Lambruschini (1788–1873) and the linguist and philosopher Niccoló Tommaseo (1802–1874) denounced the attempt made by materialist physiologists to undermine morality and social stability. The anti-evolutionary lecture by Bianconi and works by De Quatrefages were cavalierly quoted by supporters of the higher dignity of man, whereas Herzen and his allies denounced the perverse wish of the clergy to keep the masses in a state of permanent ignorance. There was indeed little common ground between the contenders, and, as Landucci concludes, the polemic was never a discussion, but a confrontation of monologues (Landucci 1977, pp. 92–102).

Generational factors and the wish to avoid further controversy extinguished the debate. Landucci agrees with Pancaldi's evaluation of the debates that characterized the 1870s. After the struggle of the 1860s, the diffusion of a broadly evolutionary view of nature and culture was paralleled by a process of revision and reinterpretation of Darwin's work, and of the available evolutionary mechanisms. The teaching and activities of Paolo Mantegazza, who taught at Florence from 1869 until his death in 1910, epitomized the development of evolutionary debates in Florence. A line of moderation with respect to religious and metaphysical issues was coupled with the critical evaluation of relevant features of Darwinism, Spencerism, and Haeckelism. Mantegazza attacked Haeckel and Canestrini for their ultra-Darwinism, was skeptical of the explanatory power of natural selection, and rejected sexual selection. With other Italian naturalists, he accepted Darwin's theory of pangenesis. He also put forward his doctrine of variation by saltation, which he called neogenesis, in order to explain gaps in the paleontological record, and to speed up the rate of evolutionary processes (Delpino 1868; Mantegazza 1871; Danielli 1885; Benasso 1978, pp. 92–97; Canestrini 1877, p. 131). Paolo Mantegazza was very active in promoting the study of physical and cultural anthropology. He founded the National Anthropological Museum of Florence, the Italian Society of Anthropology and Ethnology, and established the Archivi di Antropologia ed Etnologia.

In a recent study of "Science, Religion and Educational Publishing", Landucci touched upon the broader cultural dimensions of debates over evolution in Florence during the second half of the nineteenth century. Florence had traditionally been a leading center of Italian quality publishing. Various firms of the town, noted for their publication of literary, philosophical, and politico-economic writings, played an important role in nineteenth-century Italian intellectual life. Yet Florentine publishers distinctly failed to provide a platform for the vital and vocal scientific community of their town. The vast literary and educational production for numerous private religious schools and the city educational system was singularly deficient in the scientific sector. Landucci has rightly suggested that the study of textbooks of natural history disciplines, produced by several Italian publishers, especially in Rome, Turin, and Milan, would provide useful insights into the actual state of affairs in contemporary science education. Of particular relevance to the understanding of Italian natural sciences of the time is the study of curricula and courses in various Italian universities. Landucci has undertaken to publish the manuscript text of a course by the botanist

Filippo Parlatore, who discussed Darwinian and broadly evolutionary doctrines in lectures devoted to the "philosophy of botany" (Landucci 1982).

The few recent and past contributions to the history of Italian reaction to Darwin and of Italian evolutionism have not failed to mention the theological overtones of the debate — indeed, the centrality, for several commentators, of religious considerations — in the assessment of evolutionary doctrines. Yet there is no systematic study of Italian Catholic reactions to Darwin, nor case studies centered on such well-known Jesuit periodicals as the *Civiltà Cattolica.* Moreover, philosophical, theological, political, and social dimensions of the evolutionary debate were thoroughly investigated by contributors to the numerous periodicals published in Italy during the second half of the nineteenth century. In a long paper devoted to discussing "Darwinism and Nationalism", Landucci (1981) has listed a considerable number of journals active in the debate. He also pointed out that during the 1880s and the 1890s there was a proliferation of journals that had the word "evolution" in their title and enjoyed wide circulation. Thus, even though we still lack a thorough assessment of this interesting feature of Italian debates on evolution, scholars and students of the period have been alerted to the relevance of a systematic inquiry into the contemporary periodical press.

With respect to the specific Roman Catholic reaction to evolutionism, it could be argued that the debate over evolutionary anthropology, evolution in general, and Darwinism in particular, did in fact put a halt to attempts by Catholic naturalists such as the geologist Antonio Stoppani (1824–1891), or Giovan Battista Pianciani, the reviewer of the *Origin* in the *Civiltà Cattolica*, to come to terms with developments within geology, astronomy, or natural history disciplines in general. Thus Father Stoppani, who from 1871 to 1882 was at the Institute for Higher Studies of Florence, was prepared to support many of Charles Lyell's ideas, but he found evolutionism morally repulsive and socially dangerous. The view of Stoppani on the role of the clergy in directing, supervising, and sanitizing scientific development closely resembled the stand taken by such Anglican dons as Edward Copleston and Richard Whately at Oxford, who saw the establishment of scientific chairs, the geology chair in particular, as a brilliant move toward achieving control over the debates on the age of the earth, or the deluge. In terms Copleston and Whately would have subscribed to, Stoppani advocated the expansion of the scientific curriculum of the Catholic seminaries, in order to "create an army of apologists" capable of preserving the natural sciences from irresponsible deviations (Stoppani 1886, p. 219).

The spread of evolutionary ideas represented a concrete and dramatic instance of such deviations. The practitioners of natural sciences "were materialists, atheists, who conceal truth"; "socialism and nihilism are the formidable products of naturalism" (Stoppani 1886, pp. 67, 69). The opposition of so famous a geologist, and a liberal, or rather, "conservative reformer", as Stoppani caused serious worries to supporters of evolutionism and Darwinism. Camerano, in his commemoration of Michele Lessona, acknowledged that the opposition of Stoppani was much more effective than the hysteric denunciations of Niccolò Tommaseo. According to Camerano, the infighting between the followers of Antonio Rosmini-Serbati (1797–1855) — Stoppani was a leading representative of the movement — and the Church hierarchy prevented the clergy from fully appreciating the value of the suggestions put forward by the geologist. It could be added that it was only in the early decades of the twentieth century that the Church took an active role in forwarding Catholic scientific institutions and schools, or in favoring Catholic involvement in promoting sanitized interpretations of formerly "dangerous" scientific doctrines.

As far as evolutionism was concerned, the spiritualistic interpretations of evolutionary, non-Darwinian models under discussion in the last decades of the nineteenth century, and the early decades of the twentieth, favored several conciliatory attempts. It is of interest that the Church hierarchy made clear distinction among those who were active in promoting a rapprochement between evolutionism and Catholic thought, based on political and ideological grounds. Compromises put forward by Rosminians or other reformers were opposed, whereas those defended by clergymen faithful to the hierarchy, and ready to comply with dogmatic pronouncements, were tolerated and silently approved. In 1977 Rossi discussed the conciliatory model put forward by Antonio Fogazzaro, a novelist and conservative-reformer Catholic who in 1898 published a highly controversial book, *Human Ascent*. Fogazzaro, like Stoppani, was a follower of Antonio Rosmini. He embraced spiritualistic philosophy and wanted to see the Church more active in contemporary scientific, philosophic, and social debates. Defensive stands, and the stream of denunciations against every social and intellectual development of the century, tended to isolate Catholic intellectuals from the national life.

In *Human Ascent*, which collected a series of public lectures and articles, Fogazzaro tackled the issue of evolutionism. It was his view that Darwin and Haeckel were not to be regarded as the only representatives of evolutionism. Together with French and English scientists and apologists, Fogazzaro was convinced that evolution was a fact, but that the model put forward by Darwin did not represent the best explanation of it. Fogazzaro insisted on a vitalistic interpretation of evolution, teleologically oriented and supernaturally supervised. As Rossi rightly stressed, commentators have tended to judge *Human Ascent* as the idiosyncratic lucubrations of a man with no understanding of contemporary science. On the contrary, Fogazzaro was well aware of the apologetic possibilities offered by the variety of evolutionary models currently under discussion. If he misunderstood Darwin, he did it in good company. The conciliatory attempt by the novelist was bound to fail. The condemnation of modernism and of Catholic liberalism included Fogazzaro's essays on evolution.

More successful was the attempt by Father Agostino Gemelli, who in 1906 translated Heinrich Wasmann's work on modern biology, wrote a long preface to it, and published a long article on evolution. Gemelli carefully worded his approval of spiritualistic interpretations of evolution. He capitalized on the official silence of the Church on the subject. There was no encyclical or pastoral letter explicitly denouncing evolutionism. Thus Gemelli argued that a Catholic was free to embrace whatever hypothesis he liked, provided he declared his readiness to give it up as soon as the ecclesiastical authority pronounced otherwise. Gemelli was no sympathizer of Rosmini, and he firmly declared his readiness to obey his superiors. His compromise was implicitly accepted, though not without difficulty. The favor of the Church hierarchy toward Father Gemelli was shown in later years, when under his guidance the Catholic University of Milan was established.

As far as earlier reactions of Italian Catholics to evolutionism were concerned, there was, as already mentioned in the section on France, the letter to the anti-evolutionist Constantin James written by Pius IX. Yet this was probably the only official reaction of the Church authority to the new doctrines. It is appropriate to point out that the letter did not have the character of an *ex-cathedra* pronouncement, even though it well expressed the sentiments of the Pope. Vociferous anti-evolutionary Catholics, opposition by the intellectual fringe of the clergy (the Jesuits in particular), and denunciation of materialist science from the pulpit of parish churches, did not convince Church hierarchies to attack evolutionism authoritatively.

A reading of the various and frequent encyclicals and pastoral letters written by the Popes and the Church hierarchy during the nineteenth century reveals that the Catholic Church was far more concerned about the political situation in Europe, and in Italy in particular, than about evolutionism. In Italy, as well as in other European or South American countries, liberal, radical, and anti-clerical governments caused the Church to lose a considerable amount of wealth and power. Socialism, liberalism, and democracy were regarded as the greatest impending dangers. It is significant that the *Syllabus* appended to the encyclical *Quanta Cura* (8 December 1864) did not single out any scientific doctrine for particular condemnation. Science was not dangerous per se: any doctrine was condemned that was made to support materialistic philosophies, or was used to impinge upon the credibility of the Holy Scriptures. The fact that many contemporaries enthusiastically upheld a broadly evolutionary interpretation of the solar system, or of the history of life on earth, was regarded as a consequence of moral corruption, the influence of atheistic propaganda and social subversion, rather than the consequence of changing scientific, social, and philosophical values, or of the social structure of many European countries.

Natural sciences could never achieve reliable conclusions and acquire solid epistemological bases. Those who took the conclusions of scientific speculations as refutations of traditional beliefs and revealed truths, were obviously motivated by ideological or political reasons as the agents of darker forces. In Italy as well as in France, moreover, the early opponents of Darwinism and evolutionism provided the Church with ample evidence of the "unreliability" of evolutionary conclusions. During the last decades of the century, in Italy as well as in France, the spread of broad evolutionary convictions forced the Church to revise earlier hopes and standpoints, but never to change the basic claim that the final word on scientific as well as social or philosophical matters rested with the Church. When concessions to science were made, these tended to reinforce the claim that theologically oriented neo-Thomism represented the philosophical framework capable of accounting for and supervising any new development in science and philosophy. In the encyclical *Humani Generis* (12 August 1950), issued by Pope Pius XII (1876–1958), Catholic scientists were acknowledged to have the right to full freedom of research: but they were reminded that only the Church had the final word in judging whether a theory could or could not be accepted by Catholics, scientists included.

Thus, just as it would be misleading to assess the "influence" of Darwin without taking into account the actual complex articulation of national natural history traditions, it would be equally misleading to approach the issue of Roman Catholic reactions to Darwin's theory without full awareness of the philosophical, theological, and political dimensions of Roman Catholic culture at the time. Recent historiography has rightly stressed the impossibility of equating Roman Catholicism with anti-evolutionism, in the nineteenth century as well as in the twentieth (H. Paul 1979; Landucci 1982). There appears to be a risk, however, of limiting research to a listing of individual cases opposing the dated conclusions of Whiggish historiography. Moreover, it is rarely pointed out that the issue of the relationship between Roman Catholic theology, or Christian theology in general, and science or evolutionism, is marred by the essentialistic and ahistorical assumption that there is one theology and one science. As far as the debate over evolutionism was concerned, it is by now clear that there were many evolutionisms, some of which were incompatible with Darwinism, and with each other. Yet historians concerned with the relationship between science and religion seldom mention that there were as many Christian theologies, even within the same Church and sect. Because of the theocratic structure of the Roman Catholic Church, the essentialist fallacy could claim support from the absolute dogmatic authority of the Pope and of the Church. It would be wrong, however, to consider Roman Catholic theology a monolithic structure of doctrines and beliefs, free from conflicts and tensions. In Italy as well as in France, the official voice of the Church — or its silence — did not prevent individual Catholics or groups of Catholic intellectuals from holding strong views that differed from those of the keepers of the dogma: views against or in favor of evolution. But it is true that fanatical anti-evolutionists tended to be tolerated, whereas supporters of evolution, such as St. George Mivart or Fogazzaro, two sincere and devout Catholics, paid a high price for their conciliatory attempts.

The use of repressive measures, from warning to excommunication and the placing of books on the Index, were rarely the result of a careful evaluation of the theological implication of the challenge, but implied subtle consideration of wider ideological and political factors. The Church was fully aware of the problem it was confronted with. Phases of severity and impatience were followed by phases of tolerance and seeming indifference. Thus the search for evidence of obstinate Catholic opposition to science and evolutionism (or Whiggish historiography), as well as the search for favorable Catholic responses to Darwinism and evolutionism (revisionist and "justificationist" historiography), appear to be dominated by an inner tendency to misrepresent the actual historical dimension of the encounter.

In Italy as well as in Germany, the closing decades of the nineteenth century, and the early decades of the twentieth century, saw a proliferation of political and social philosophies known by the collective and rather improper label "Social Darwinism". Several commentators have touched upon this feature of the various political and philosophical movements in contemporary Italy, and have mentioned the "influence" of Darwinism and evolutionism on leading intellectuals and the public in general. There is no sympathetic study available on this subject. A broad survey of the relationship between Darwinism and nationalism in Italy recently published by Landucci has shown the pervasive use of biological metaphors in a variety of works by representatives of the most diverse political and philosophical standpoints. As Landucci has rightly stressed, opportunistic borrowing prevailed over first-hand knowledge of biological doctrines (Landucci 1981). It is important, however, to stress that the variety of evolutionary and pre-evolutionary contacts between natural and social sciences or political ideologies still requires systematic investigation. As with the other dimensions of Italian reactions to Darwinism and evolutionism reviewed here, selected case studies will help clarify this important chapter of Italian intellectual and social life.

It is appropriate to consider briefly a further important feature of Italian reactions to Darwin, to which several commentators have alluded but which has never been systematically approached. In his contribution to the history of natural sciences in Italy during the early decades of the nineteenth century, Lorenzo Camerano hinted at the persistence of Lamarckian and broadly transformist views among Italian naturalists. Giuseppe Gautieri (1769–1833), a doctor living in Pavia, published in 1805 his *Slancio sulla genealogia della terra*, in which he advocated evolutionary cosmological ideas. Gautieri's mechanism of the transformation of organisms relied on the capability of animals to acquire new habits when subjected to changing environmental conditions. The new way of life caused minor as well as major changes of anatomical and organic structure. Gautieri was familiar with analogous ideas put forward by Erasmus Darwin, but was clearly unaware of Lamarck's *Recherches sur l'organisation des corps vivans* and the *Discours d'ouverture* to the 1800 course at the Muséum. Gautieri quoted ornithological examples of transformist adaptation, which were also used by Lamarck and Erasmus Darwin, and by William Paley in his critique of the *Zoonomia*.

Unlike Gautieri, Michele Foderà (1793–1848) was familiar with Lamarck's ideas and published several memoirs in French during a long period of residence in Paris (Foderà 1826). In 1844 Francesco Constantino Marmocchi (1805–1858) published a *Prodromo della storia naturale . . . d'Italia*, in which he defended Lamarck's transformism (Omodeo 1969, Preface). Omodeo has called attention to the important collection of manuscript notes taken by Giosuè Sangiovanni at the Muséum courses given by Lamarck (Omodeo 1949a, b).

The works, ideas, and intellectual milieu of the authors mentioned above certainly deserve further investigation, in spite of the occasional nature of their contribution to natural history debates, and their isolation from the mainstream anti-speculative tradition of Italian natural sciences in the early decades of the nineteenth century. Yet it is important to stress that the study of the followers of Lamarck and of Lamarckism during the early and middle decades of the century will provide an essential contribution to the understanding of crucial features of the reaction to Darwin, Huxley, Haeckel, and Spencer by a significant section of Italian naturalists.

Michele Lessona and Lorenzo Camerano, two of the most prolific contributors to evolutionary debates, were pupils and relatives of Turin naturalists who frequently engaged in private discussion of Lamarck's views. Franco Andrea Bonelli (1784–1830) had been a pupil of Lamarck and Etienne Geoffroy Saint-Hilaire. Although he was a friend of Cuvier and Duméril, he was converted to transformism. Political caution restrained him from publishing his views in the unfavorable intellectual climate of the Restoration; an early death in 1830 prevented Bonelli from exerting a lasting influence through his writings. His efforts in the organization of the Natural History Museum of Turin, and his dedication to students and junior colleagues, did, however, produce lasting effects. During the years 1811–1830, Carlo Lessona (1784–1858), father of Michele, attended lectures given by Bonelli. He became a convinced Lamarckian, and he was keen to introduce his son to the higher theoretical dimension of zoological investigation.

Another convinced Lamarckian was Vittore Ghiliani (1812–1878), a pupil of Bonelli and teacher of Michele Lessona and Lorenzo Camerano. Ghiliani studied the entomology of Piedmont and Sardinia. He thought that the concept of adaptation through transformation of structures and habits was the key to explain the geographical distribution of animals and plants (Lessona 1884, pp. 139–258). When De Filippi, the naturalist who started the debate on Darwin in 1864, moved to Turin in 1848, he found himself surrounded by colleagues and pupils who did not make a mystery of their beliefs. Political opportunity clearly inhibited open profession of transformism, but could not prevent the young Lessona from discussing with De Filippi the ideas of Lamarck. Even though acceptance of evolutionary ideas came only in the early 1860s, De Filippi "was constantly studying the question of species", Lessona wrote when remembering the discussions he had had with his friend (Lessona 1884, pp. 193, 191–194).

De Filippi, Lessona, and Camerano read the *Origin* as the most powerful, and the best documented, advocacy of evolutionism since the days of Lamarck. Yet each of them read Darwin's work in a different way, largely determined by generational, religious, and scientific factors. The Catholic De Filippi had already revolved in his mind the question of the place of man in the evolutionary scheme. He was pleased to see that Darwin avoided the issue, but he immediately reacted against Huxley's attempt fully to include the evolution of man in the theory. The Lamarckian Lessona was prepared to go a long way with Darwin, but in later years he found Haeckel and Giard nearer to his Lamarckian upbringing and philosophical aspirations (Camerano 1896, p. 382). Camerano was taught by Lessona and Ghiliani to pay close attention to biogeography. During the 1870s and the 1880s he devoted less time to evolutionary propaganda and more to intense scientific investigation. His papers ranged from the study of insect taxonomy and the geographical distribution of insects, to the study of sexual dimorphism, polymorphism, neoteny, and mimicry. It is to be regretted that there is no study available of Camerano's career as one of the leading Italian evolutionary biologists and historians of Italian natural sciences of the nineteenth century. Thus two leading representatives of the hard core of Italian supporters of Darwin became committed to the evolutionary cause well before 1859. Their acceptance of Darwin was deeply influenced by their exposure to alternative evolutionary doctrines; as was their interpretation of the doctrines put forward in the *Origin* (Corsi 1983).

In conclusion, it is clear that in Italy as well as in France, the question of the reactions to Darwin cannot be studied as an issue in itself. If there is little doubt that the publication of the *Origin* acted as a catalyst with respect to a variety of trends in natural history and philosophic disciplines, and powerfully contributed to the reorientation of their priorities, it is equally true that close attention has to be paid to the intellectual traditions that found, in the debate on evolution and Darwinism, a new channel of expression. I have underlined the importance of the study of Lamarckism and transformism in France as well as in Italy. Attention should also be paid to the diffusion of the doctrines put forward by Lorenz Oken or Hans Christian Oersted (1777–1851), and to pre-1859 debates on idealistic philosophies of nature. It could indeed be argued that the variety of reactions to Darwin in France and Italy, as well as in Germany and elsewhere, is better understood by considering 1859 as one date — a crucial one indeed — among many in a debate that was not started by Darwin's work, and was destined to continue to the present day.

24

Darwin and Russian Evolutionary Biology

Francesco M. Scudo
and
Michele Acanfora

Introduction

As is well known, Darwin produced a large number of theories, many of which dealt explicitly with evolution or special aspects of it, and many others of which were closely related to problems in evolution, although the evolutionary aspect was only implicit. Furthermore, throughout his life Darwin varied the emphasis he placed on different processes and mechanisms, and, for some of them, he radically altered his position. In his published works, Darwin hardly ever indicated precisely how any single topic he discussed could or should be connected with other topics, in an overall interpretative framework of evolutionary phenomena. Thus there is plenty of room for speculation, even on relatively minor points such as the extent, if any, to which *The Movements and Habits of Climbing Plants* (1875) should be explicitly considered in such a framework (see Mârza and Tarnavschi 1974, and, for a more general statement of the problem, Ghiselin 1973a).

In principle the problem of interpreting Darwin's legacy should pose methodological questions in assessing its reception. In practice, however, this problem has hardly been felt in most cultures. Most debates centered at first on the acceptance of evolution as a fact, and later on the degree to which natural selection could be held as a sufficient causal mechanism for evolution. Also, in most cultures relatively little consideration was paid to the mature technical works by Darwin, that is, those that came after the *Origin*. By the 1880s, on the other hand, the problem of the mechanisms of evolution usually became polarized between two incompatible extremes, a Darwinism which could be better qualified as Weismannism, and various forms of neo-Lamarckism. In Russia, on the other hand, the unique pattern of development of evolutionary ideas poses very serious methodological problems in the interpretation of Darwin's work, ones that hardly apply to other countries. Let us briefly anticipate some of those problems, to introduce the unusual approach of our analysis.

It seems that Russian biologists reacted in a unique way to Darwin's work mainly because Russia was the only major country in which evolution, justified mostly by theories with a strong Lamarckian bent, was already rather well established among professional zoologists some time before Darwin and Wallace entered the scene, that is by the 1840s.[1] As a result, in Russia the *Origin* immediately started a technical debate concerned almost solely with mechanisms, particularly on the ways in which natural selection would be an essential complement to the already accepted ones. Unlike other countries such as France, in which a similar debate was resolved mostly against natural selection (see Conry 1974; Corsi, this volume; cf. also Roger 1979), in Russia the essential causative role of natural selection in evolution started being seriously questioned only in the 1930s, by Lysenkoism. Also uniquely, in Russia the debate continued to be centered on Darwin himself and on all his theories. Little attention was paid to various other forms of Darwinism, and Weismannism was mostly — and usually very distinctly — rejected.

The present paper aims to summarize a vast body of literature dealing with Darwin's impact on Russian evolutionary biology. Most of this literature is in Russian, and some is in German, Italian, and French. We are aware of only two works in English dealing with the core of our problem, Platonov (1955) and Vucinich (1974). The former is a scientific and political biography of Timiriazev, approached from the standpoint of orthodox Lysenkoism. Here Platonov lauds Timiriazev as having carried on the first, albeit insufficient steps to "reform" Darwin in the direction of "Creative Darwinism". At the opposite extreme, Vucinich concentrates on the reception of Darwinism in Russia, that is, on only one of the aspects we are considering. Also, his analysis suffers from the fact that it considers Russian evolutionism before Darwin hardly at all. On the other hand thorough accounts of embryology, comparative anatomy, and paleontology given by Vucinich and by Adams (1980a) exempt us from paying much attention to such key subjects in the post-Darwinian period. Also Vucinich, and even more Platonov, closely consider sociological aspects of evolution, including the attempts by the administrative and religious Russian establishments to curb the secularizing implications of evolutionary theories and their strong connections with subversive movements (on such points see also Rogers 1974a, 1974b and Tagliagambe 1983). All the other works in English known to us deal only with peripheral aspects of our inquiry, and we shall give them a disproportionate amount of attention relative to works in other languages. While this material is already known to certain specialists, it might be of interest to a broad range of biologists and historians.

Section I of our paper presents a brief overview of Russian evolutionary biology before Darwin. Section II, the core of our study, concentrates on the partial acceptance of the *Origin*; Section III deals with the generally positive reception by Russian contemporaries of Darwin's later works, that is, those making essential use of features ii) and iii) below. We shall conclude by sketching later debates on a most critical point of Darwin's theory, the role of intraspecific competition, debates that were instrumental in the rise and fall of Lysenkoism. All but the last of these developments center on the figure of Timiriazev, who dominated evolutionary biology and, later, history and philosophy of science in Russia from the 1860s to his death in 1920. Here we shall barely touch on Timiriazev, to be studied in greater detail by Acanfora and Scudo (manuscript).

In dealing with the peculiar pattern of development of evolutionary biology in Russia it is convenient from the outset to separate three main aspects or "phases" of Darwin's theories, and to distinguish these from "Weismannism":

(i) A substantial prevalence of gradualism in the evolution of any trait that would be, nearly always, adaptive — that is, it would be due mostly to natural or sexual selection. This view, associated with a multiplicity of mechanisms for speciation, is expressed more strongly in early editions of the *Origin*, and it could be compatible with moderate forms of Weismannism (cf. iv).
(ii) The view that many or most evolutionary novelties would consist, at first, of changes mainly or solely in ecology, behavior, and development (changes in conditions of life, use and disuse, correlated variations, etc.). Natural selection would come into play mainly to improve upon such changes after they became established, if, as usual, further improvements would be needed. In contemporary genetic terms, this would be equivalent to phenotypic selection (Haldane 1957), phenocopy (Piaget 1974), or genetic assimilation of acquired characters (Waddington 1975; cf. also Scudo 1976a; Rachootin and Thomson 1981).[2] Darwin's often considerable reliance on positions of this sort is related to his explicit consideration of useless characters, or characters originated as such. This is typical of the *Variation* and the *Expression*, as well as late editions of the *Origin* in which Darwin emphasized directed or spontaneous variations, produced independently of selection. Starting with Wallace, such "phenocopy-like" mechanisms of variation were often interpreted as implying inheritance of acquired characters, even in cases in which it is by no means clear whether or not Darwin meant that hereditary material entering the germ plasm from outside would play a crucial role in the process.
(iii) A great reliance on sexual selection to justify the origins of many behaviors and communicative devices, as in the *Descent* and its "Supplementary Note" (*CP* 2: 207–211, 1876). This is connected with extreme geographic mechanisms of animal speciation (that is, extreme

allopatry in the present terminology), also taking account of the effects of small population size in isolates.

(iv) Darwinism, then Weismannism, namely the view that evolution proceeds essentially, or solely, by direct selective accumulation of hereditary variations, produced at random relative to selective conditions. In this view the process in ii) is regarded as impossible or irrelevant. This view was already widespread particularly in England and Germany, long before Weismann institutionalized it around 1885. He did so through such evidence as the embryology of the germinal lines of metazoans and, perhaps more important, by minting the catchall expression of the "principle of non-inheritance of acquired characters". In this connection we are facing a semantic ambiguity that will persist throughout this paper. The term Weismannism to denote the view just stated has been used widely in Russia though used only rarely outside Russia (see Romanes 1896). Conversely "Darwinism",[3] "neo-Darwinism" and, then, "the synthesis" were usually employed to denote approvingly much the same views, while the same terms have been used only rarely in Russia and usually with negative connotations (see discussion of Pisarev in Section II).

One should also notice how, in contrast to his baffling behavior in many other cases, Darwin explicitly stated the nature of the necessary connections between (ii) and (iii) (see, for example, the Introduction and General Summary to the *Descent*).

I. Evolution in Russia Before Darwin

Three key aspects of Russian culture must be kept well in mind when one seeks to understand why the developments considered in this and the next section appear to be relatively uniform, and most of them necessary rather than fortuitous ones. In the first place the relatively late start of Russian science was rewarded by a very rapid, successful development (as for Lomonossov), that the best scientific minds of Germany and France helped to "seed". As a result, by the end of the eighteenth century Russia provided an ideal environment for innovation and synthesis in science. Also the very backwardness of the administrative and religious establishments of Russia allowed a level of philosophical freedom that was not possible in any other country, except for a short period in France before Napoleon's empire. On the other hand the Russian intelligentsia (whether of noble or plebean origin) continued to remain largely Francophone even after Napoleon's invasion. Finally, and particularly significant for the next section, by the end of the eighteenth century breeding had become a major scientific endeavor in Russia, whose harsh climate posed severe problems in plant and animal husbandry. In this specific area, however, Russia was influenced far more by English breeders, particularly Bakewell and Holling, than by the French. Already in the early decades of the nineteenth century Russia had a sizeable establishment of scientific breeders, and of academic botanists and zoologists with applied interests, who had come largely from the English tradition. On such points see, for instance, Raikov (1957) and Mikulinsky (1961).

With this background, it is not surprising that Russian professionals and intellectuals were able to embark very soon on a path toward acceptance of evolution. That this happened sooner in Russia than in any other country might also be due to the influence of K. F. Wolff. The extent to which this anti-preformist embryologist could be considered "the first truly evolutionist European scientist", as Engels maintained — rather than a "limited transformist", a naive believer in "hopeful monsters", or a "phylogenetically inclined epigeneticist", — is still open to debate (cf. Wolff 1966 and Herrlinger therein; Engels 1878 in Engels 1964, p. 243; Gott 1889; Vorlander 1907; Guyénot 1957; Lukina 1973; Roe 1981). Far less questionable is the extent to which the school Wolff helped start in St. Petersburg was instrumental in a surprisingly early spread of evolutionary ideas. Thus Gmelin had witnessed in his St. Petersburg garden the origin of a number of distinct forms from two *Delphinium* species he had transplanted from Siberia. On the basis of these observations as well as others on the same *Peloria* first discovered by Linnaeus, by the mid-eighteenth century Gmelin was making overt, general statements of an evolutionary nature.[4] He may also have helped to spread some of his own views through the last, posthumous edition of Linnaeus's *Systema Naturae* of 1788 in Lyon, about which he appears to have had a strong say as editor (see again Guyénot 1957).

As is well known, the Paris debate in 1830 between evolutionists and catastrophists was followed with great interest not just by professionals, and not just in France. The debate was clearly lost by the evolutionists on technical grounds, mainly the crushing of Geoffroy's fanciful phylogenetic reconstructions by Cuvier's precise arguments from comparative anatomy. Older accounts picture the evolutionists' cause as having been definitely buried on this occasion, and this is largely true as far as French academic science is concerned. The same reports fail to mention that ecology, ethology, and evolution continued to prosper in France outside the academic establishment, until their sudden triumph in the 1880s (with Giard's municipally endowed chair at the University of Paris; for example Jaynes 1969 and Roger 1979).

Virtually all western accounts also fail to mention the reaction elicited among Russian biologists by the Paris debate. Some brilliant young Russian biologists firmly took the losing side of the Paris debate as far as it concerned philosophy. Well aware of the technical deficiencies of this faction, these biologists proceeded at an amazing speed to correct many of them, and also to propose a number of theoretical improvements. Consequently, in Russia, both the fact of evolution and the beginnings of theories about its mechanisms became established gradually, and not as a revolution, as happened much later in England and elsewhere. Let us look briefly at three key figures in the process of maturation of evolutionary ideas in Russia.

K. F. Rul'ye was born in 1814, of a French saddle master (Roullier) and a Russian mother. He received a long education at home before undertaking formal studies of medicine and surgery at the University of Moscow. From medical research he soon switched to paleontology; subsequently his interests focussed on comparative psychology. What concerns us more directly here is that, as an influential professor at the University of Moscow, Rul'ye was able to diffuse his view that evolution in animals was mostly driven by behavior, particularly learned behavior. By the time he died in 1858 the core of Rul'ye's position did not appear to be seriously contested any longer in Russia.

A rich biographical literature on Rul'ye in Russian has been recently re-evaluated in Mikulinsky (1979). Rul'ye's role in the establishment of evolutionary biology in Russia is also sketched, in English, by Naumov (1972). He claims that a distinctive tradition, deriving from Rul'ye, persisted up to the present through an unbroken chain of master-pupil relationships (p. 17). Naumov also stresses the focus on applications, which Rul'ye gave as the ultimate purpose of all his investigations. He likewise hints how the strong anti-selection movements of the 1930s could be partly justified by a persistence of Rul'ye's ideals, particularly in the applied sectors of Russian biology.

At a modern reading, Rul'ye's *Zoopsychology* looks more interesting in its own right and for its similarity in approach to that of Darwin, than for having started the tradition leading to the objective psychology of Sechenov and Pavlov. Rul'ye's starting-point was a categorical rejection of mechanistic notions of instinct as the sole basis for animal behavior: "Either instinct does not exist, or it has a meaning" (1847a, p. 64).

At times Rul'ye combined this rejection with an equally categorical rejection of notions of maximization to account for animal behavior. For example, concerning migratory instincts he writes as follows: "What determines the irresistible upstream swimming of fishes, which causes the death of so many of them? Computing the progeny one might produce? One ought to believe that fishes carry on these calculations more precisely than God does. Is it a matter of blind instinct? Either one uses a meaningless word or, if one likes this word so much, he would have to understand where such an instinct resides, and from where it came" (1847a, p. 64). Rul'ye overcame this impasse by concluding that behavior can be understood only "on the basis of the reflexes acquired in the history of the species, as well as of each single individual" (1847b, p. 157). The relevant "history of the species" might be, in fact, a very long one, as for bird migration: "The cause of bird migration is closely related with the history of the whole earth" (p. 159).

The analogy between Rul'ye's position on behavior and Darwin's views in *Expression* is obvious. Equally obvious is the analogy between Rul'ye's analysis of animal training, such as training dancing bears (1850, p. 596), and Pavlov's analysis of conditioning (1957, 1929). For Rul'ye's zoopsychology see Mikulinsky (1979) and the specialized study by Acanfora (1980).

Many ecologists, and apparently all historians, agree on the leading role played by Rul'ye in establishing an evolutionary ecology and ethology as an academic paradigm, shared by a substantial proportion of the Russian intelligentsia. For Russia, the extent to which evolution became also a mass ideological paradigm is an altogether different problem than for England or France, for instance, and it will be considered briefly in the final remarks below.

Also G. E. Shiurovsky, known mainly for his later geological work, openly took the losing side of the Paris debate. His reservation on technical points of this side soon grew into his *Animal Organology* (1834). This is probably the first treatment of evolution explicitly presenting Lamarckian ideas in terms of internal mutations, suggested from the beginnings of cellular theories. At a somewhat different level one can recognize, at this time, what might be considered as a continuation of Wolff's tradition. Possibly its best representative is K. M. Baer, better known as K. E. von Baer in his German editions, who became widely known outside Russia mainly for *History of Animal Development* (1828-1837) and *Animal Organology* (1828). In these works he clearly shows the unity of developmental plan both within major vertebrate groups, and among vertebrates as a whole. Only in later works (for example, cf. Menzbir, ed., 1934, pp. 121 ff., especially pp. 141–142), which are not so well known abroad, had Von Baer enough observational material to deal with the relationships within and among major invertebrate taxa. Not surprisingly, Von Baer eventually rejected the Haeckelian notion of recapitulation, which must have appeared to him as a naive encroachment on the more general principles he himself had developed much earlier. Von Baerian and Haeckelian recapitulation are contrasted in Gould (1977b) and Løvtrup (1978). Von Baer's early theorizing on limited transformism did not go much beyond a teleological principle of goal reaching akin to Wolff's. Von Baer is of special interest for our purposes, being perhaps the only one among old-guard Russian evolutionists to initially reject whichever of Darwin's ideas did not overlap with his own. Only at a later stage did Von Baer partially change his mind, acknowledging a number of Darwin's contributions (cf. 1876a, e.g. p. 171, 1876b, e.g. p. 241 with 1865; 1873; see also Oppenheimer, 1959).

One can debate the extent to which the ethology and ecology typified by Rul'ye, the comparative anatomy typified by Shiurovsky, and the developmental biology typified by Von Baer can be considered as a body of scientific theories on evolution. Wholly analogous problems are posed by the contemporary attempts in France. Thus Lamarck had repeatedly claimed in his *Zoological Philosophy* that he was not at all interested in speculations about mechanisms, but only in objective descriptions of events. And yet Lamarck did produce at least one bona fide scientific theory about animal evolution: that changes in behavior should always precede changes in structures (see for example Kohn 1980 for the possible influences of this theory on Darwin). More generally, one faces the problem of the extent to which one can consider as bona fide scientific theories on evolution constructions that deal only with partial aspects of the causative agents involved, perhaps none. We shall not tackle this problem, which is still a very open one nowadays (cf. for example Cracraft and Eldredge, eds., 1979; Platnick 1980).

II. The Partial Acceptance of the *Origin*

The maturity of the Russian evolutionary establishment is more easily demonstrated by the analysis of its reaction to the publication of the *Origin* than by a speculative epistemological analysis as in Section I. Before proceeding to a dry listing of these reactions, we should like to describe briefly the ideological and sociological background from which they came (for more details, see Platonov 1955 and Vucinich 1974), and in addition their most common characteristics. In the debate started by the *Origin* in Russia a number of biologists practicing specialized disciplines, such as Beketov, Rashinsky, and Timiriazev, carried almost as much weight as people who could be better qualified as scientific philosophers or philosophers of science, such as Chernishevsky, Pisarev, Lavrov, and Kropotkin. All these people already shared materialistic philosophies. They were also committed to politics *sensu latu*, which ranged from being just liberal or progressive to being openly revolutionary. They all belonged to one or another "ism" — socialism, anarchism, nihilism — of Russian populism. Contrary to Anglo-Saxon countries, or to France, in Russia there were only few radically negative reactions to the *Origin* by the scientific establishment, and almost none at first by other social bodies (notably the Church). Ample documentation on this and the next section can be found in a number of historical works, such as Sobol (1945) and Zavadsky (1973). Last but not least, Darwin's theories had a profound impact on breeding in Russia, particularly on zootechnical theory and practice (see for example Myrzoyan 1959).

The positive reactions to the *Origin* by specialized scientists as well as philosophers were relatively uniform, and particularly uniform on three counts that have but few parallels outside Russia. First of all, even if not accepted in all its points, the principle of natural selection was considered by Russian evolutionists as an essential complement to previous interpretations of more or less direct Lamarckian origin, rather than an alternative to them. A similar position was held only by a not particularly influential minority of Anglo-Americans, such as Chauncey Wright. Emblematic of Wright's position is his statement: "It would seem, at first sight, that Mr. Darwin has won a victory, not for himself, but for Lamarck" (1871; quoted in Hull 1973b, p. 386). The same also holds for the two other main counts:

(ii) Not surprisingly, in the light of their ideological bias, many Russians tended to attach relatively little weight to competition as a selective agent in general, and to intraspecific competition in particular.
(iii) In Russia Darwin's theories were explicitly considered as distinct from various forms of Darwinism; in the extreme form of this phenomenon, Darwin's mature work could be fully accepted, while "Darwinism" was flatly rejected (see discussion of Pisarev below).

Perhaps the single older work outside Russia in which the largest number of similar ideas are concentrated is a collection of essays by Italian scientific philosophers (Morselli, ed., 1892). There Loria rejects Social Darwinism as distinct from Darwin's thinking. Cattaneo finds Darwin's theories a necessary complement to Lamarck's. He also tempers Darwin's stress on the indirect selective action of abiotic factors (through affecting the struggle for existence) at the expense of direct effects.

The promptness with which Russian scientists took account of the Darwin-Wallace papers and then of the *Origin* is typified by Kutorga's lectures in his 1860 course and by the translation of Lyell's report to the twenty-ninth meeting (1859) of the British Association for the Advancement of Science in the January 1860 issue of the Journal of the Ministry of Popular Education (that is, little over a month after the *Origin* first appeared; see Vucinich 1974).[5] Darwin's ideas were then summarized in Bogdanov's *Introductory Zoology* (1861) and in a paper by Rashinsky (1863), the same professor at the University of Moscow who presented the first Russian edition of the *Origin* in 1864. In the same year Timiriazev, a twenty-one-year-old erstwhile student[6] of natural sciences at the University of St. Petersburg, published his famous trilogy on the reactions to the *Origin*, with mostly positive comments of his own. A modified version of Timiriazev's trilogy appeared as a volume the following year (1865). A second edition of the *Origin* followed in 1865, a third in 1873, etc.[7] We spare the reader further listings of assessment of the *Origin* from 1864 onward, both in scientific journals and in popular ones such as *The Russian World*, *The Contemporary*, *Russian Thought*, etc. by people such as D. Pisarev, M. A. Metchnikov, A. O. Kovalevsky, and V. O. Kovalevsky.

Soon after the publication of the *Origin* several Russian Darwinists started exchanging correspondence with Darwin,[8] who was receiving the highest official honors in Russia, starting with his election to the Imperial Academy of Sciences in 1867. There were also close personal contacts between Darwin and Russian evolutionists. In 1865 Kovalevsky spent a period at Down, a visit he described enthusiastically in a letter to Lyell (see for example Menzbir, ed., 1907, p. 73). Timiriazev too was at Down in 1877, drawing a number of interesting insights on Darwin's personality (see Timiriazev 1949, vol. 4, particularly pp. 83 ff.). Von Baer had visited Huxley, and had continued to correspond with him. In turn Huxley corresponded with Darwin (6 August 1860) about this "new and great ally" (*LL* (NY) 2: 122). It is not clear how and precisely when Von Baer managed to communicate with Darwin, but a proper quotation of his views appears in the penultimate chapter of the *Origin* starting with its third edition (*Origin* 1959, pp. 685–686).[9] On the other hand we have no idea whether Darwin ever became aware that his plea for support from Von Baer in his 8 August 1860 letter to Huxley[10] was eventually satisfied, albeit far from fully so (cf. Sections I and III of the present paper).

Rather than continuing with such listings, let us take a closer look at the main features of the Russian reactions to the *Origin*, particularly to the first edition.[11] The notion of compatibility and complementarity between Lamarck's and Darwin's ideas has its obvious explanation in the fact that most of the people who reacted positively to the *Origin* were already Lamarckians or neo-Lamarckians of some sort. On the other hand the two connected themes of downplaying competition as a selective agent, and of rejecting most forms of Darwinism, are not so readily explained. From the outset Rashinsky (1863) had considered the Malthusian model as not essential to Darwin's theory. In its original applications to advanced human societies, the Malthusian model would be just a misleading metaphor. Criticisms such as Rashinsky's have strong analogies with the ones both Marx and Engels were developing at much the same time — that the struggle for life within human societies would essentially involve group rather than individual behavior, and that it would take place mostly through cultural rather than selective means.[12] Virtually all further reactions to the theory of natural selection up to the 1930s can be considered as variations of such themes, and the related criticism of downplaying the direct selective effects of physical factors.[13]

Take for instance Pisarev, perhaps the most influential among the scientific philosophers. He first approached Malthusianism in a chapter of his *Essay on the History of Labour* (1862), shortly after a bloody repression of popular unrest. He begins by pointing out a fallacy in much of the Malthusian reasoning, that was far from obvious at the time — how limiting effects are not imposed by the conservation of matter but rather by its rate of circulation, and this would not have any obvious upper limit. Pisarev pointed out how high living standards were generally attained by quite dense human populations, in sharp contrast with the combination of poverty and vast untapped resources that characterized the thinly populated Russia of that time. He then proceeded to reject Malthusianism in general, and its components of sexual deprivation in particular, as a dangerous weapon of class oppression. Much as it was for Rashinsky, Malthusianism for Pisarev would be just a "Victorian fashion" not at all essential in this form to Darwin's theoretical construction (1864, p. 164). Apparently lumping together Darwinism and Social Darwinism, Lavrov described Pisarev as "a follower of Darwin but an enemy of Darwinists" (as in Le Bétoyer, ed. 1983, p. 86).[14]

Timiriazev, by far the most influential Darwinian among the specialists, largely concurred with Pisarev on Malthusianism as applied to human societies and intraspecific competition. He is also noteworthy for having rejected Weismannism on counts such as the operational validity of a precise dichotomy between the acquired and the innate and the purely fortuitous, that is, random nature of variations. Much as Beketov did (1882, 1887)[15] Timiriazev too tended to question the extent to which selective effects of the physical environment would be almost solely indirect, acting through biotic mechanisms, and only exceptionally direct (that is, in very marginal ecological situations), as Darwin had maintained in the *Origin*. Timiriazev deserves a more in-depth treatment, to be given elsewhere (Acanfora and Scudo, manuscript).

III. The Reception of Darwin's Technical Works

The *Origin* is unique among Darwin's published works on a number of counts, including the conspicuous scarcity of references to other works. Particularly in the late editions, it also strives for empiricism more than other theoretical works by Darwin on evolution (on Darwin's empiricism, see Ghiselin 1969). The nine chapters (1–5, 10–13) of the first edition of the *Origin* supporting the theory, as compared with only four (6–9) dealing with difficulties and objections,[16] contrast sharply, for instance, with the *Expression*. Here the bulk of Darwin's original theories are inductively stated at the beginning or confined to the last chapter, where hardly any evidence "pro" or "con" is given. Further, as also stressed by Løvtrup (1979), the *Origin* has been subjected to constant criticism since it appeared, a surely unusual situation for any theory. By contrast Darwin's "mature, technical" works drew comparatively few open criticisms while Darwin was alive, and most later criticisms were based on a clearcut acceptance of Weismannism or on an equally extreme rejection of it.

Among Wallace's strong criticisms of a number of Darwin's mature positions, only two drew a substantial, persistent following in English-speaking cultures: on sexual selection by female choice (see Kottler 1980), and on most new behaviors originating by being first acquired, and only afterward becoming in some degree innate. In *Darwinism* Wallace also rejected Darwin's mature views of speciation as an accidental by-product of sexual selection, acting on tight, long-lasting geographic separation among populations (cf.

especially *CP* 2: 207–211, 1876). Instead Wallace proposed a mechanism of speciation based largely at first on "plastic" reactions to different habitats, and involving specific selection for the divergence of sexual signals, or infertility of hybrids, when the former mechanisms would be insufficient to prevent interbreeding (Wallace 1889, chap. 7). These views drew only a meager following, while "geographic" mechanisms of speciation akin to Wagner's have continued to remain very popular in English-speaking cultures until recently (cf. for example Mayr 1970, Grant 1971, and Scudo 1976b).

As anticipated in the introduction, possibly the most critical point in interpreting Darwin's mature works concerns the extent to which these relied on modes of variation that could be considered, in contemporary terminology, "phenotypic selection" (Haldane 1957) or "phenocopy" (cf. Scudo 1976a) rather than "direct genotypic selection". In particular Darwin's imprecise terminology and scarce cross references among his works leave room for doubt as to the extent to which such "indirect" modes of selection might also involve incorporation into germinal lines of hereditary material from the outside. Indicative of such difficulties is the fact that Baldwin (1896) proposed as a novel mechanism his "organic selection", that is, much the same precise definition of "phenocopy" in Piaget's sense that Lloyd Morgan had also independently proposed in the same year (cf. Schmalhausen 1946; Waddington 1975).

In this confusing situation only two facts stand out very clearly. One is that Darwin did make use in his mature works of two explicitly distinct selective processes, one directly on hereditary variations and the other involving such variations only indirectly, in what we might call "perfecting" or "stabilizing" variations of a "plastic" nature (cf. especially *CP* 2: 172–176, 1873). The other very clear fact is that the extent to which Weismann's principle might be violated is largely irrelevant to Baldwin's "organic selection" and to the modern theory of "phenotypic selection", "phenocopy", or "genetic assimilation of acquired characters". In fact any possible violation of Weismann's principle at the microscopic level would enter both genotypic and phenotypic selection (or any of their equivalents) exactly in the same ways, that is, as some sort of bias in rates of mutation, or as a different mode of mutation. Obviously in different terms (that is, through Galton's "stirp"), the utter irrelevance of Weismann's reasoning at the microscopic level to his conclusions at a macroscopic, selective one had already been stressed by Romanes (1896).

The largely retrospective excursion above may help understand the key reason why Darwin's mature work was read in a unique way by most Russian evolutionists, most of whom had flatly dismissed Darwinism as well as the operational validity of Weismann's dichotomy between the innate and the acquired. For example, as early as 1887 Beketov had characterized Weismann's principle as "blind internalism", incompatible with Darwin's theories. Timiriazev's position on the matter was undoubtedly very influential in the long run. He summarized it much later: "the historical process of the production of novel organic forms, discovered in nature by Darwin, is the inescapable result of the interaction of three factors, which undoubtedly act all the time on all features of organisms. The first of them, variability, provides the necessary raw material for this historical process. The second of them, heredity, fixes, integrates and elaborates this material. Finally the third one, overpopulation, gets rid of (or, to use Comte's expression, destroys) all forms that are partly or wholly unsatisfactory. Organic structures are perfected by the joint action of these three factors, a process one metaphorically calls 'natural selection' " (1892–1895, as in 1948, pp. 245–246). Timiriazev, thus, appears to go beyond Darwin in considering the direct selection of hereditary variations as a trival limit case of "genocopy", which, if at all occurring in the form claimed by Weismannists, would not suffice to explain the origin of any evolutionary novelty.

Here we cannot consider in detail the subtle changes in emphasis that Timiriazev's position on variation underwent through over half a century, both before and after the establishment of Mendelism (see again Zavadsky 1973, chap. 5, section 5). We must mention, however, how Timiriazev's habit of joking about Mendel's laws as "the laws of little peas" is often interpreted, in the west, as being derogatory to Mendel and his laws. This interpretation is a patently gratuitous one. Timiriazev indeed delighted in this joke, but mainly to poke fun at the evolutionary interpretations by early Mendelians, which he abhorred for much the same technical reasons as did Wallace (e.g., Wallace 1908b). Further, Timiriazev firmly accused the same Mendelians of pushing a grossly distorted picture of Mendel's position, much in the same vein as the more scholarly, but tentative critique by Olby (1979). On these points see particularly Timiriazev (1915), Acanfora and Scudo (manuscript), and, with due caution, Platonov (1955).

To conclude our discussion of this general problem, we should stress that the emphasis many Russian evolutionists placed on "genocopy" tended to be directly based on Darwin's examples and reasoning, rather than on some purported neo-Lamarckian alternative to Weismannism. Emblematic of this position is the anonymous entry "Darwinism" in the Russian encyclopedia of 1902. It centers on the discussion of variation on Darwin's studies of "double flowers" (cf. *CP* 1:175–177, 1843) and his theory on the acquisition and transmission of new sociosexual signals. In this respect one should also stress the unusual amount of attention Russian theorists had paid to signals, their origin and transmission. Such a level of attention goes back to Rul'ye, and persisted at least through Pavlov (1957 and 1929, for example).

Let us move from such general aspects to single, major mature works by Darwin. Perhaps the most striking feature of the Darwinian "revolution" in Russia is that the *Descent* generally met an enthusiastic reception, even by people who held strong reservations against the *Origin* (cf. Shipanov 1974, p. 371). Antonovich (1896, p. 15) is the best example of the latter, in his praise for what he perceived to be a major step in overcoming the "Malthusianism" of the *Origin*. Equally favorable, on the whole, was the reception of the *Expression*, particularly by the zoopsychologists in Rul'ye's tradition. Wallace's subsequent criticism of the *Expression* failed to draw any following. For example, Sechenov wrote: "Wallace's criticism of Darwin's theory contrasts with the experience of the naturalists who deal with the daily changes in animal activities. Nature is far less repetitive than it is often believed. Animal activities are strongly tied to the environment, since they arise through repeated series of actions and reactions. Certainly, in this way any animal acquires a precise way of reacting to the environment. However, is there any animal which would not try to actively adapt to changed conditions?" (1892, p. 8).

Equally enthusiastic was the reaction to the *Expression* by scientific philosophers such as Antonovich, who believed this would complete the process, started in the *Descent*, of "taking the moral world outside of the domain of metaphysics" (1896, p. 18). According to Timiriazev, only with the *Expression* did Darwinism become a "sociobiological science", which could deal successfully with the deepest roots of the human self. In this way, then, Darwinism would also give a deeper meaning to Marx's and Engels's theories (see Timiriazev 1892–1895; cf. again Acanfora and Scudo, manuscript). According to Severtsov, with the *Expression* Darwin ended up by fully rehabilitating the Lamarckian thesis, which he had at first discarded: "There is no contradiction between direct effects of changes in the external environment and use and disuse of organs (orthogenesis) and Darwin's natural selection. Further, psychic evolution in higher animals is closely linked to both such aspects, i.e., a Lamarckian and a Darwinian one. Changes in environmental conditions necessarily change habits and modes of action" (Severtsov 1889, p. 20).

Among the negative reactions to the *Descent*, the most illustrious came from the aging, almost blind Von Baer (Baer 1873).[17] There he finds Darwin at fault on specific points of ecology and recapitulation. Von Baer even suggests, half-jokingly, that it would be easier to interpret the evolution of men and apes as having both diverged from a common plantigrade ancestor. More than anything, however, Von Baer's paper is a tirade against the random mutation-selection scheme of Darwinists, from whom Darwin himself is studiously set apart. It is also a tirade against anti-religious usages of evolutionary theories, by a man whom the real or apparent inadequacies of such theories had forced to a dualistic or spiritualistic position (much as it was happening to Wallace).

Darwin's extreme gradualism in the *Origin*, and to a lesser degree his mature position on the divergence of species specific signals (*CP* 2:207–211, 1876), were often read to imply that all taxa would be merely nominalistic entities, from the point of view of evolution. A number of Russian Darwinists were sensitive to the problem of nominalism, and most of them rejected it as a false one, for which Darwin was hardly considered responsible. Most of the blame for this false problem went to Darwinists and Weismannists, whose extreme gradualism was very apparent. Instead some Russian Darwinists claimed the real or individual nature of species and other taxa through theoretical constructions directly based on M. Wagner, Delboeuf, Romanes or Gulick (see Zavadsky 1973, especially chap. 4, section 5, and Streltsenko 1981 for V. A. Wagner's ethological approach).[18]

To other Russian theorists nominalism and modes of speciation did not loom at all as big problems. To Timiriazev, for instance, the individuality of species would be an obvious consequence of Darwin's general theory as he was interpreting it, when operating in the context of qualitative changes in ecology or population dynamics (cf. again Acanfora and Scudo, manuscript). As a result, notions akin to "physiological segregation" — often mainly behavioral — were frequently presented as a main initial step in the formation of animal races and species. Severtsov's position is exemplary in this respect: "It is often difficult to assign precise boundaries to the distribution of species. Both in geographically neighboring areas and in distant ones there can occur populations with as large differences in behavior as to deserve being classified as different species. And yet, a detailed examination of their germinal plasm does not reveal any marked difference. It is well possible that future investigations might reveal some new features in their germ plasm. However one cannot rule out a priori that the environment might have a determining influence on habits, and on the transmission of these habits. Nowadays many Darwinists maintain that such a position is a left-over from Lamarckism. And yet, if two related species in different areas differ in behavior, and their pups are reared before being trained by their parents, they still tend to stick to parental characteristics. Only if one subjects these pups to a proper countertraining, their behaviors tend to become the same in the two cases" (1889, p. 18; cf. Shipanov 1974, p. 280).

The scatter of leading views presented in this section is indicative of a substantial, but not complete, agreement among Russian theorists of evolution. Small minorities of theorists continued to hold, in varying degrees, the "internalistic" theses of Weismann (cf. Menzbir 1893, 1900.) Not surprisingly, some of them also held views akin to western Social Darwinism (see again Rogers 1972). Perhaps the most influential among the latter, Tkachev, was holding ground in heated polemics on such matters with Engels (cf. Tkachev 1933, p. 211 with Engels 1964, p. 23). Further, as noticed in Section I, Rul'ye's empiricist tradition of Lamarckian derivation persisted, particularly among zoopsychologists and applied biologists other than specialist breeders. Until the 1930s these relatively diverse trends in evolutionary biology coexisted in Russia in generally peaceful ways, through a sort of indifference furthered by scarce contacts among schools or disciplines. No substantial change in the reception of Darwin was going to take place until the rise and fall of Lysenkoism. On all such points, and particularly on Kholokovsky's explicit "Lamarcko-Darwinism", see again Zavadsky (1973) and Acanfora and Scudo (manuscript).

Final Remarks

By having barely touched on Timiriazev in this paper, we have skipped over decades of most fertile developments in evolutionary biology up to the October Revolution. Most Russian scientists continued to hold progressive or revolutionary tendencies, and later these were mainly reflected in adherence to the Menshevik or Bolshevik movements. Also, many scientists took an active part in the October Revolution (for a case history, see Scudo and Ziegler 1976). In turn the success of this revolution gave to science a position of social pre-eminence that had hardly any parallel outside the USSR. The same revolution also ended up by rendering the works of Marx and Engels "sacred" and, as a direct consequence of their opinions, those of Darwin as well. Let us briefly look at how and why such changes soon resulted in a serious clash with views that were entrenched in the Russian evolutionary tradition since the 1860s.

Denying a major selective role, or any at all, to intraspecific competition posed serious theoretical problems to a number of Russian theorists. In particular it made it difficult to explain the origin of the cooperative features of animal and human societies, on which so much ideological emphasis was being placed. Alternative attempts to justify these features had to rely mostly on neo-Lamarckian mechanisms, whose plausibility seemed more and more at odds with the rapid development of population genetics in the USSR. Also, to deny a major selective role to intraspecific competition meant taking issue with some forms of Marxian orthodoxy, in particular with some of Engels's key points in *Anti-Dühring*.

We shall just allude here to the development of population genetics in Russia after the revolution, and in particular the culmination of knowledge attained by Chetverikov and his followers, and later on by the "evolutionary brigade" of the Kol'tsov Institute in Moscow. Many such developments are discussed in works in English, mainly Haldane (1932), Dobzhansky (1980), and Adams (1980a). It suffices to point out that in the 1920s and 1930s Chetverikov and his followers had arrived at much the same conclusions as western population geneticists (mainly Wright) on the existence of a large store of neutral or weakly selective genetic variability in natural populations. They had done so at about the same time, and independently of western population geneticists, through a quite different path of investigations (Babkov in press and in Chetverikov 1983).

The combination of this recent knowledge, and the equally recent "orthodoxy" of intraspecific competition as a selective agent, soon made relatively popular, for the first time in Russia, forms of neo-Darwinism wholly akin to those prevailing in the West. At a partial or superficial reading, Schmalhausen's mature ideas on variation (1946) might appear to have moved even further away from Darwin than neo-Darwinism, as by claiming that plastic modifications that are novel would also be indeterminate.[19] Furthermore, in the same volume Schmalhausen did not even mention the results obtained by Michurin and his school,[20] and he barely mentioned Timiriazev on a marginal point regarding selection. Such attitudes must have appeared particularly offensive to some Soviet evolutionists, since Schmalhausen, as professor of Darwinism at the University of Moscow, was expected to be Timiriazev's intellectual heir. It is more than natural, then, that Schmalhausen's position could be confused with extreme forms of Weismannism and resisted on many grounds, including that it backed up forms of Social-Darwinism such as the eugenic movement headed by Kol'tsov. The most extreme forms of such reactions, notably Prezent's "Creative Darwinism" (better known popularly as Lysenkoism), are relatively well known in the West. On Schmalhausen, see especially Rubailova (1981).

We would like to dwell, instead, on a number of reactions to "Creative Darwinism" that appear not to be widely known in the West. An influential component of such reaction came from geneticists who abhorred Weismannism as much as Lysenkoism. The same favored, instead, positions on the active interplay between the acquired and the selected analogous to Timiriazev's, that is, recasting Darwin's mature positions in Mendelian terms. Emblematic of this kind of reaction is Zavadovsky's intervention at the 1948 meeting of the Academy of Agriculture in which Lysenkoism established its relatively short-lived yet disastrous supremacy (it began to be quietly phased out in 1952 and was effectively abandoned in 1963, while its official rejection took place in connection with the centenary of Mendel's paper; see for example Lecourt 1976). Another major argument of Marxist Darwinians against "Lysenkoism", perhaps the major one, was that it went far beyond the older tradition by negating the very existence of intraspecific competition. This holds true for Russia as well as for other countries, as typified by Prenant's reaction in France (cf. again Lecourt 1976 and Prenant 1980).[21] Yet perhaps the worst theoretical deviation of Lysenkoism would have been of a historical or ideological nature — to have attempted to turn upside down Engels's reasoning on intraspecific competition in "Anti-Dühring" (cf. again Zavadovsky's intervention, which is also reprinted in its official French translation in Lecourt 1976).

In a sense, then, the debates about Lysenkoism added the last missing point in a process of acceptance of Darwin's theories in Russia, which had started so early and, for almost a century, had remained so close to completion.

At the present time a genetic-selectionism of Weismannian derivation is being seriously questioned in the West, although it survives as textbook orthodoxy. In this context the typical positions of Russian Darwinists and of the schools they started no longer appear so "exotic" or, perhaps, nonsensical. One rather wonders why such positions are not being seriously considered in the West as obvious historical antecedents of analogous positions that are now popular. One also wonders about the precise extent to which positions similar to those of Russian Darwinists have been proposed earlier in the West, and might have passed unnoticed, rather than having been overtly rejected.

It would not be difficult to extend the list of western evolutionists (cf. Section II) who maintained one or more of the typical views of their Russian colleagues on Darwin's role in evolutionary theories. The main characteristic of all the people in this list is that they were not very influential — Chauncey Wright and Giard having been among the most influential — or they were regarded as "dangerous heretics".

One also has examples of westerners who were influential on one count or another in evolutionary theory, but had little overall influence because they held one or more of the typical views of Russian Darwinists. Perhaps the best example of this sort is Teissier, who is acknowledged as a major founder both of quantitative developmental biology and of population genetics. In his historical contributions, and particularly in his last general work on evolution (1961), he clearly rejects both the inheritance of acquired characters and the validity of Weismann's principle. He finds Lamarck's theories fully compatible with Darwin's, while "neo-Darwinism" would be as incompatible with Darwin as "neo-Lamarckism" would be with Lamarck, etc.

It must be pointed out that, although Teissier had been an ardent Communist, there is no conclusive evidence that he borrowed his historical or philosophical views from Soviet colleagues. Rather there are many hints that Teissier's views matured slowly mostly on their own, as a by-product of his "amateurish" historical interests, in the elder Geoffroy in particular. These views by Teissier are not mentioned, other perhaps than through negative allusions in a specialist western work dealing with his role in evolutionary theories (Mayr and Provine, eds., 1980, chap. 10). By contrast Teissier's views were warmly acknowledged by his Soviet colleagues as, for example, Zavadsky seeking support in Teissier (1961) for his main thesis on evolution in Russia after Darwin (1973, p. 260).

We have dealt at some length with Teissier's case to partially justify our main, final contention. Up to now the typical positions of Russian Darwinists, of their followers, and analogous ones by westerners have usually appeared unacceptable to most western scientists, and inconsistent or nonsensical to most western historians. We also hope this sketch might encourage western evolutionary biologists to look at Russian ecological, ethological, developmental, and genetic approaches to evolution in their historical context, rather than as a series of miraculous successes (Pavlov, Severtsov, Chetverikov, etc.) taking place in a vacuum. Those wishing to do so will discover other success stories of Russian evolutionary biology, some of which are virtually unknown in the West (Sushkin, for example). They will also find plenty of ways to go about it — from very dense summaries such as Bielozersky and Mikulinsky (1967) to monumental, in-depth analyses such as Davitashvili (1948).

ACKNOWLEDGEMENTS

Originally we planned not to write this paper on our own, but rather to work as intermediaries for a top Soviet specialist of the subject matter, possibly K. M. Zavadsky or N. I. Rubailova. Not receiving any answer from either of them, we decided to start while still trying to get in touch. Much to our sorrow, we finally discovered that both of them had recently died, and decided to attempt our own tribute to their life-time dedication. As amateurs with little besides our private libraries as specific sources, we have largely relied from the outset on help from friends and senior colleagues from socialist countries. This was provided in the most generous manner; whatever degree of professionality this paper might have is mostly due to Academician Mârza D. Vasile, Prof. C. R. Mikulinsky and I. I. Shipanov, and Drs. V. V. Babkov and V. I. Nazarov. They labored to provide us with all kind of sources and patiently went through two drafts of our manuscript, pointing to innumerable deficiencies and mistakes. We also received much critical help on drafts from Profs. M. T. Ghiselin and S. Løvtrup and, particularly on matters of the Russian language, from our wives, Katherina (née Zimanova) and Tatiana (née Koudriavsteva). Previous missing information was provided to us by Profs. B. Hoppe, F. Vidoni, D. Kohn, P. Omodeo, and P. Corsi, who also kindly made available to us his manuscript on Italian Darwinism in this volume, for cross reference. Last but not least, our thanks to Miss Ornella Fiorani for her patient and competent typing. We have made no attempt to be exhaustive in our study; remaining deficiencies are solely our fault and may result from choices we made that were contrary to advice we were given.

Notes

1. For analogous situations in some Italian subcultures, see Corsi, this volume as well as Section II of the present paper and Pancaldi (1983).
2. Direct statements, as by Piaget (1974, p. 3), and comparison of this work with Waddington (1975) as well as with previous works by the same authors, make clear that, prior to about 1973, each of them had to some extent misinterpreted his own results and misinterpreted, or ignored, those of the other. Thus, to avoid cumbersome qualifications, we refrain from giving references to works by Piaget and Waddington prior to 1973.
3. Occasionally "Darwinism" had peculiar usages in the West also, as when Haeckel denoted by "Darwinism" what amounted to a rather extreme form of neo-Lamarckism (see Dougherty in Roger 1979).
4. Here perhaps more than elsewhere in this paper we are facing problems that are also to a large degree of a semantic nature. It is hard to qualify views about organic change through the only two relatively precise terms

now in use, "evolution" or "transformation" versus "limited transformism". Less common alternatives tend to mean precious little, such as "saltationism", which might include processes of hybridization, developmental "hopeful monsters" with any degree of genetic determinism, or, perhaps, even "non-saltations" such as Darwin's "double flowers" (*CP* 1: 175–177, 1843; cf. also *Variation*). This kind of problem becomes acute with statements such as the following one by Gmelin, in the *Sermo academicus* of 1749: "the number of plants originally made up by the Creator has doubled, tripled, or has been multiplied infinitely many times" (Guyénot 1857, p. 370).

5. Throughout, one must also keep in mind that, by the end of the nineteenth century, the Russian calendar had gotten to be ahead of the western one by about one month.
6. From 1862 to 1866 Timiriazev was not a regular student. He had been officially banished for refusing to sign a declaration of non-involvement in revolutionary movements.
7. In his preface to the sixth edition of the *Origin* Darwin mentions three Russian editions of the *Origin* instead of the two that had appeared (*Origin* 1959, p. 52 (sent. 4.6:F)). According to Shipanov (personal communication) Darwin would have been misled by the Russian title of Kovalevsky's translation of the *Variation*, which had started appearing several months prior to the English original (cf. again Vucinich 1974).
8. This correspondence can be found in Antonovich 1945 and in the two Russian editions of Darwin's Collected Works (Menzbir 1907–1909 and 1926–1934).
9. Darwin had discussed at length Von Baer's ideas (though not mentioning him by name) in the *1842 Sketch* and *1844 Essay*. However, he had apparently forgotten about them when writing the *Origin*, where at first he attributed similar ideas to Agassiz. Huxley reminded Darwin of this mistake when he was working on the second edition of the *Origin* but, meanwhile, he had lost the copy of the volume Von Baer had presented to him. Also taking account of Darwin's notorious difficulties with German, his quotations of Von Baer starting from the third edition of the *Origin* most likely came, though not so acknowledged, from the excerpts Huxley had translated (see especially Oppenheimer 1959).
10. "If you write to Von Baer, for heaven's sake tell him that we should think one nod of approbation on our side, of the greatest value; and if he does write anything, beg him to send us a copy, for I would try and get it translated and published in the *Athaeneum* and in 'Silliman' to touch up Agassiz" (*LL* (NY) 2:123). Obviously Von Baer could never get along with Darwin, after Darwin had quoted him approvingly and then proceeded on his own with much the same form of recapitulation later made popular by Haeckel. Thus when in 1876 Seidlitz, an enthusiastic supporter and popularizer of Darwinism at the University of Derpt (now Tartu), tried to present Von Baer as closer to Darwin than he actually was, Von Baer, then retired in Derpt, became indignant and set out to write a rejoinder, but death stopped him at the introduction (see Oppenheimer 1959 and Bliakher 1971).
11. As will be made clear in Section III, most later debates in Russia concentrated on subsequent works by Darwin.
12. While Marx and Engels had been in contact with a number of Russian Darwinists, Lavrov in particular, it is far from clear what reciprocal influences there might have been among the former and the latter (cf. Vidoni 1982 and Christen 1981 as somewhat extreme examples of a vast literature). Still less clear is why both Marx and Engels ignored works by Darwin after the *Origin* and relied, instead, on authors such as Tremeaux and Espinas (albeit in critical ways). On the other hand Engels's defense of Darwin's ideas against deformations by "bourgeois Darwinists" in *Anti-Dühring* bears close analogies with the position of Russian Darwinists, and it was very influential on later generations of Darwin's followers in Russia, even before the October (November) Revolution.
13. Among nineteenth-century evolutionists, Chernishevsky had taken a completely negative stand against natural selection and Darwin. He expressed his views only in the late 1880s, after a long imprisonment for subversive activities, when the debates we have considered had already "settled" (see Rogers 1974b). Among later opponents of natural selection are also Strahov, Danilevsky and Katkov (see Rubailova 1981, Tagliagambe 1983). On the other hand it is far from clear to what extent Kropotkin's analysis was meant to be mainly critical of the crude picture of the struggle for life in most forms of Darwinism (on this point, and Romanes's analogous position, see Christen 1981). At a more technical level, already since the 1870s A. S. Famintsyn had attributed a key role to symbiosis in the origin of evolutionary

novelties. Famintsyn's observational basis appears to have been much sounder than any then available for natural selection, which he considered capable only of minor modifications (see Zavadsky 1973, chap. 5, section 5).

14. Our assessment of Pisarev differs somewhat from the one in Rogers (1972), to which we refer for Darwin's impact on social sciences proper. In particular Pisarev (1862) warns against the empiricism of the *Origin*, which he regards as a genial but methodologically immature work. Notice also how a rational egotism would be the main spring of cooperation in humans, which anticipates in some ways Darwin's analysis in the *Descent*.

15. Much as Gmelin had, it appears that Beketov had at first based his analysis of plant evolution mostly on the direct effects of weather, soil, etc. (see Botnariuc 1961). After having apparently become a "pure selectionist" just after the *Origin*, Beketov began in the 1870s to question the sufficiency of natural selection, particularly in plant evolution. In the 1880s he then attempted a grand synthesis of the positions of Darwin, Lamarck, and Geoffroy Saint-Hilaire. Later on the cause of a "Lamarckian Darwinism" was furthered considerably in Russia by the translation of le Dantec's *Evolution individuelle et hérédité* and *La crise du transformisme* (see again Zavadsky 1973, chap. 5, section 5; chap. 11, section 3; chap. 13, section 5). One must be careful to take note of the different connotations of "Lamarckism", as well as of the "reform" by Geoffroy Saint-Hilaire concerning the direct effect of external conditions *on animals*. Having done this, one could argue that many themes in the opposition to Weismannism or Darwinism could as legitimately be attributed to Darwin as to Lamarck.

16. In the sixth edition of the *Origin*, Darwin added a new chapter entitled "Miscellaneous Objections to the Theory of Natural Selection".

17. An English translation appears in Hull (1973b). Hull pictures Von Baer as essentially anti-evolutionist or as a naive evolutionist *at a level of an Aristotle*(??). These comments, the most negative on Von Baer we have encountered so far, suffer from the fact that they consider only a minute part of his production (Hull, personal communication). They also appear to suffer from having been made from a strict syntheticist standpoint. Only from such a standpoint can one interpret as anti-evolutionism perplexities on the documentation of any one macro-evolutionary event, or on the powers of any one of the mechanisms proposed to justify such events. Whatever his remaining differences with Darwin might have been (this is not at all clear — cf. Oppenheimer 1959), any historian not bound to a Weismannian or synthetic viewpoint might rather recognize in Von Baer the first, and still a foremost, representative of a syndrome that has persisted to this day — scientists approaching evolution from the standpoint of embryology, who have bluntly rejected Weismannism or syntheticism, have been dissatisfied in varying degrees with the *Origin* or with Darwin in general, and much puzzled about "macro-evolution". Piaget (1974), Waddington (1975) and Teissier (mainly 1961) are among the foremost representatives of this "syndrome" among westerners with strong interests in genetics. This trend is now very much alive, and it includes extreme representatives such as Løvtrup, who openly prefers Von Baer (1978a) or Chambers (1978b) to Darwin. Indeed Løvtrup may have reasons, given the major retrogressive influence Darwin's ideas on recapitulation appear to have had (mostly through Haeckel; see Oppenheimer 1959).

18. Mechnikov stressed inadequacies in all the existing theories on evolution, in particular the inability of the theory of natural selection to account for the apparent lack of direct correlation between overpopulation and selective divergence (as in highly dense species tending to be monotypic even in wide ranges of habitats). As with Romanes who also stressed much the same deficiencies, Mechnikov has not been much attended to by historians (see however Zavadsky 1973, particularly chap. 5, section 3).

19. This is in fact Schmalhausen's starting-point, according to which morphoses would also be non-adaptive or maladaptive. The point is a delicate one, strongly dependent on what exactly is meant by novel (cf. Waddington's critique in 1975, n. 8), and it could be turned into a farce of the original argument by failing to specify that it would apply only to purely morphologico-physiological reactions (i.e., behavioral ones being generally excluded). Only much further on in his 1946 treatment (chapt. 3), did Schmalhausen consider Lloyd Morgan's and Baldwin's views as most interesting, but still largely untested (thus reflecting an epistemological position much akin to that of his mentor, Severtsov). This book is available in an almost complete English translation (Schmalhausen 1949). This is not a particularly good one, even according to Dobzhansky who, by his own admission,

largely failed in his editorial tasks (see Waddington 1975, p. 98). Here Dobzhansky enthusiastically endorsed Schmalhausen's views as completing the synthesis from a morphological standpoint. By and large Dobzhansky's syntheticist colleagues and pupils did not denounce this endorsement *ad personam*, but they did not accept it either (cf. Mayr 1970, p. 364). Dobzhansky himself continued to share most synthetic views quite regardless of his own lukewarm Baldwinism (cf. Dobzhansky 1962).

20. These results would prove that natural selection on fruit stocks grafted to different varieties was far more powerful than in non-grafted stocks. These claims clearly conflict with the view that heredity is purely nuclear, and that selection depends solely on this heredity and on selective procedures. These same views, however, are compatible with hereditary variability being substantially cytoplasmic or organellar, as it is believed to be nowadays, or with the result of selection depending also on phenotypic properties of an organism other than those due to genetic mutants.
21. After an in depth analysis also including an interview with Lysenko, Prenant ended up by dismissing "Creative Darwinism" mostly because of its position on competition, interspecific in particular. In other words, he reacted to Lysenkoism as rigorous ecologist and Darwinian scholar. His reaction was particularly effective, since it also involved his resignation from the Communist Party of France while being a member of its central committee. On the other hand Prenant was not alone in his criticism for the "synthesis" of the 1940s, nor in maintaining an open attitude towards the results of grafting experiments (which the "synthesis" dismissed as implying "inheritance of acquired characters"). Some of Prenant's reserved or overdefensive attitudes in this connection might indeed have had mainly political motivations (Prenant remained an ardent communist even after leaving the party). Further, all statements by Prenant on matters of heredity and evolution — topics he regarded as being outside his direct competence (Prenant, personal communication) — were confined to philosophical, political or popularized writings. Then, by forgetting the stand Prenant took as an ecologist, as a Darwinian scholar and as a politician, from a strict "synthetical" viewpoint Prenant's "lay" writings might be misinterpreted to the point of passing him for a "notorious Lysenkoist" (Buican 1983).

PART FOUR

Perspectives on Darwin and Darwinism

25

DARWIN'S FIVE THEORIES OF EVOLUTION

Ernst Mayr

In recent controversies on evolution one frequently finds references to "Darwin's theory of evolution", as though it were a unitary entity. In reality Darwin's "theory" of evolution was a whole bundle of theories, and it is impossible to discuss Darwin's evolutionary thought constructively if one does not distinguish the various components of which it consists. But quite aside from the fact that it helps understanding of the structure of evolutionary theory, to carry the analysis to the level of the subtheories Darwin adopted, it is important to call attention to the composite nature of the Darwinian theory for three very specific additional reasons.

The first is that this knowledge is very important for the proper understanding of the term *Darwinism.* This term has numerous meanings, depending on who has used the term and at what period. At first, in Darwin's day, Darwinism to most people simply meant a belief in evolution, and perhaps man's descent from the apes. By contrast, if a modern biologist uses the term, his emphasis is entirely on natural selection. He would define Darwinism as the theory that attributes evolutionary change to selection forces. Non-biologists, however, have often used the term in a much broader sense. I need refer only to the term *Social Darwinism*, an ideology that had more to do with Spencer than with Darwin. Indeed, in the last third of the nineteenth century the term Darwinism, often used with a distinctly derogatory connotation, was applied to a materialistic-atheistic Weltanschauung (Greene 1981b), which as a matter of fact had little to do with Darwin's own thinking. To repeat, an understanding of the meaning of the term Darwinism will be helped considerably by a discrimination among the various evolutionary theories held by Darwin.

The second reason for such discrimination is that one cannot answer the question correctly of how and when "Darwinism" was accepted in different countries through the world unless one focusses on the various Darwinian theories separately. If we look at the contributions made by several authors to the volume by Glick (1974) or the book on the fate of Darwinism in France by Conry (1974), we find that the treatment suffers from a failure to deal with the different Darwinian ideas individually. As we now know, what Darwin presented in 1859 in the *Origin* was a compound theory, and the five subtheories I shall single out had very different fates in the eighty years after Darwin.

A third reason to determine clearly and unambiguously what Darwinism is, is the frequently made recent claim that the Darwinian theory is obsolete, that neo-Darwinism has been refuted, or even that the Evolutionary Synthesis is dated if not refuted. It is quite impossible to test the validity of these claims until one has clearly determined the meaning of the terms *Darwinism* and *neo-Darwinism,* that is, has determined of what theories they consist, and how they are constructed.

I want to issue a warning at the outset. Darwin was a great pioneer, a person with an exceptionally fertile mind, but like other fertile thinkers, he had considerable trouble sticking to a consistent "party line". On almost any subject he dealt with — and this includes almost all of his own theories — he not infrequently reversed himself. For instance, he might say that only slight variations are evolutionarily important, but then on another page he might talk about rather strikingly different varieties, like the ancon sheep and the turnspit dog, both of them extremely short-legged. Darwin's pluralism has recently been emphasized by Gould and Lewontin (1979). In addition to natural selection, for instance, Darwin allowed also for use and disuse and occasionally even for a direct influence of the environment. Although Darwin at first fully supported geographic speciation, eventually he also allowed considerable scope to various forms of sympatric speciation. I could quote many more examples of his pluralism. It was this that led one of Darwin's twentieth-century critics to assert: "Darwin's hedging and self-contradiction — enabled any unscrupulous reader to choose his text from the *Origin of Species* or the *Descent of Man* with almost the same ease of accommodation to his purpose as if he had chosen from the Bible" (Barzun 1958, p. 75). Evidently, then, it is not legitimate to refute the validity of one of Darwin's multiple choices and then claim this is a refutation of Darwinism.

This flexibility of Darwin's thought forces us to raise a number of new questions. For instance, was there ever a unitary package of subtheories which, as a whole, we might consider *the* Darwinian theory? And the further question of whether one can present what was, so to speak, the end product of his often rather involved speculations, or whether one should present the detailed history of his groping for solutions? It would seem to me that to do the latter would simply duplicate what Gruber (1974), Herbert (1974, 1977), Kohn (1980), Ospovat (1981), Sulloway (1982) and Hodge (1982) have done recently in such exemplary fashion. The only places where I have made an exception are those in which Darwin was hedging or was unclear right to the end, in which a historical analysis is required in order to understand what seems to be the end product of Darwin's cogitation.

Some historians (for example, Kohn, Ospovat, Hodge) have referred to the combination of theories Darwin held at various times as his "unified theory" of that period. I will not argue against this, if this is the historians' practice. But it must not be forgotten that each of these "unified" theories consisted of a very heterogeneous set of components, each of them a full theory in its own right. There is one particularly cogent reason why Darwinism cannot be a single homogeneous theory, which is that organic evolution consists of two essentially independent processes, transformation in time and diversification in (ecological and geographical) space. The two processes require a minimum of two entirely independent and very different theories. That writers on Darwin have nevertheless almost invariably spoken of the combination of these various theories as "Darwin's theory" in the singular, is in part Darwin's own doing. He not only referred to the theory of evolution itself as "my theory", but he also called the theory of descent by natural selection "my theory" as if common descent and natural selection were a single theory.

The discrimination among his various theories was not helped by the fact that he treated speciation under natural selection in Chapter 4 of the *Origin* and that he ascribed many phenomena, particularly those of geographic distribution, to natural selection when they were really the consequences of common descent. Under the circumstances I consider it urgently necessary to dissect Darwin's conceptual framework of evolution into a number of major theories that formed the basis of his evolutionary thinking. For the sake of convenience I have partitioned Darwin's evolutionary paradigm into five theories, but of course others might prefer a different division. The selected theories are by no means all of Darwin's evolutionary theories; others were, for instance, sexual selection, pangenesis, effect of use and disuse, and character divergence. However, when later authors referred to Darwin's theory they invariably had a combination of some of the following five theories in mind. For Darwin himself these five theories were apparently much more a unity than they appear to a person who analyzes them with modern hindsight. The five theories were: (1) evolution as such, (2) common descent, (3) gradualism, (4) multiplication of species, and (5) natural selection. Someone might claim that indeed these five theories are a logically inseparable package and that Darwin was quite correct in treating them as such. This claim, however, is refuted by the fact, as I have demonstrated elsewhere (Mayr 1982b), that most evolutionists in the immediate post-1859 period - that is, authors who had accepted the first theory — rejected one or several of Darwin's other four theories. This demonstrates that the five theories are not one indivisible whole.

I. Evolution as Such

This is the theory that the world is neither constant nor perpetually cycling but rather is steadily and perhaps directionally changing, and that organisms are being transformed in time. It is difficult for a modern to visualize how widespread the belief still was in the first half of the last century, particularly in England, that the world is essentially constant and of short duration. Even the majority of those who, like Charles Lyell, were fully aware of the great age of the earth and of the steady march of extinction, refused to believe in a transformation of species.

I must stress at this point the word *transformation* as representing a process guaranteeing continuity. This was overlooked by Osborn in his *From the Greeks to Darwin* (1894), where he lists scores if not hundreds of authors whom he designated as forerunners of Darwin. To be sure, these authors proposed "new origins", that is, the production of new species or new types, but invariably by discontinuous saltations. This was inevitably so because all these authors had been essentialists, and an essence cannot evolve. Any change must be due to the production of new essences. Accordingly, even Lyell explained the steady change of faunas as due to extinction and a mysterious "introduction of new species", a discontinuous process.

When Darwin began to break away from Lyell's thinking, he did so in stages. When he discovered the second species of *Rhea* (South American "ostriches") occurring (in certain districts) side by side with the better-known large Rhea, he explained it, not as a new "introduction" (à la Lyell) to fill a vacant ecological niche in nature, but rather as derived from the older species by a saltation (*RN* 127, 130). Darwin thus adopted transmutation, but transmutation essentialistically conceived.

To the best of my knowledge Lamarck was the first author to propose a consistent theory of gradual transformation. After 1800, but before 1859, the idea of gradual evolution was accepted by a considerable number of authors on the Continent (Mayr 1982b), but none of these authors developed the idea of evolution into a consistent and well-documented theory. This is what Darwin achieved in the *Origin*.

Evolution as such is no longer a theory for a modern author. It is as much a fact as that the earth revolves around the sun rather than the reverse. The changes documented by the fossil record in precisely dated geological strata are a fact that we designate as evolution. It is the factual basis on which the other four evolutionary theories rest. For instance, all the phenomena explained by common descent would make no sense if evolution were not a fact.

II. Common Descent

The case of the species of Galapagos mockingbirds provided Darwin with an important new insight. The three species had clearly descended from a single ancestral species on the South American continent. From here it was only a small step to postulate that all mockingbirds were derived from a common ancestor — indeed, that every group of organisms descended from an ancestral species. This is Darwin's theory of common descent.

It must be emphasized that the terms *common descent* and *branching* describe exactly the same phenomenon for an evolutionist. Common descent reflects a backward-looking view and branching a forward-looking view. As an aside it might be remarked that the concept of common descent runs into difficulties in the relatively rare cases of "reticulate evolution", that is, when a phyletic lineage is the product of a merger (owing to hybridization or symbiosis) of two previously separate lineages.

With branching being inseparably connected with descent in the mind of an evolutionist, it seems at first puzzling that the term *branching* was used so widely long before 1859. Just exactly what did branching mean to an author who did not believe in evolution? Pallas expressed relationship in the form of branching trees, and Cuvier called his major phyla "embranchements". For Agassiz and Milne-Edwards, branching reflected a divergence in ontogeny, so that the adult forms were far more different than the earlier embryonic stages. From all these examples it is evident that static branching diagrams of non-evolutionists are no more indications of evolutionary thinking than branching flow charts in business or branching diagrams in administrative hierarchies. The concept of common descent was, however, not entirely original with Darwin. Buffon had already considered it for close relatives, such as horses and asses; but not accepting evolution, he had not extended this thought systematically. There are occasional suggestions of common descent in a number of other pre-Darwinian writers, but historians so far have not made a careful search for early adherents of common ancestry. It is a theory that was definitely not upheld by Lamarck, who, although he proposed the occasional splitting of "masses" (higher taxa), never thought in terms of a splitting of species and regular branching. He derived diversity from spontaneous generation and the vertical transformation of each line separately into stages of higher perfection. For him descent was linear descent within each phyletic line, and the concept of common descent was alien to him.

The concept of branching occurred to Darwin quite early in his evolutionary speculating, and the rough sketches of branching trees in his *Notebooks* have often been described (Gruber and Barrett 1974, pp. 142–143). Branching, by necessity, means divergence; and every modern student of phylogeny, regardless of which school he belongs to, takes it for granted that evolutionary lines, once they have become completely separated from each other, will steadily diverge to a greater or lesser degree. And the evidence indicates that Darwin made the same assumption in his earlier branching diagrams. Yet in the 1850s Darwin discovered his "principle of divergence", which he considered, together "with Natural Selection — [to be] the keystone of my book" (*ML* 1: 109). Ever after he discovered this principle, sometime between 1854 and 1857, he referred to it always with great excitement, as if it had been a major departure from his previous thinking (for example, *Autobiography,* pp. 120–121). His wording implies that earlier he had thought phyletic lines would remain parallel after completing the process of speciation. His earlier understanding that insular phyletic lines could diverge drastically was apparently completely forgotten. In some respects one has the impression that the principle emerged as a result of a shift in Darwin's concept of speciation from insular speciation by geographical varieties to continental speciation by ecological varieties as described by botanists. But Ospovat (1981) makes a strong case for considering the principle an outcome of Darwin's thinking about classification. Actually in the years 1844 to 1858 Darwin's concepts underwent such a strong change in several important respects that it would probably not be correct to point to a single factor as the source for the principle of divergence. It seems to me that the conceptual connections in Darwin's intellectual development between speciation, common descent, and character divergence are not yet fully understood.

None of Darwin's theories was accepted as enthusiastically as common descent; it is probably correct to say that no other of Darwin's theories had such enormous immediate explanatory powers. Everything that had seemed to be arbitrary or chaotic in natural history up to that point, now began to make sense. The archetypes of Owen and of the comparative anatomists could now be explained as the heritage from a common ancestor. The entire Linnaean hierarchy suddenly became quite logical, because it was now apparent that each higher taxon consisted of the descendants of a still more remote ancestor. Patterns of distribution that previously had seemed capricious could now be explained in terms of the dispersal of ancestors. Virtually all the proofs for evolution listed by Darwin in the *Origin* actually consist of evidence for common descent. To establish the line of descent of isolated or aberrant types, became the most popular research program of the post-*Origin* period, and has largely remained the research program of comparative anatomists and paleontologists almost up to the present day. To shed light on common ancestors also became the program of comparative embryology. Even those who did not believe in strict recapitulation often discovered similarities in embryos that were obliterated in the adults. These similarities, such as the chorda in tunicates and vertebrates, or the gill arches in fishes and terrestrial tetrapods, had been totally mystifying until they were interpreted as vestiges of a common past.

Nothing helped the rapid adoption of evolution more than the explanatory power of the theory of common descent. Soon it was demonstrated that even animals and plants, seemingly so different from each other, could be derived from a common, one-celled ancestor. This Darwin had already predicted, when he suggested that "all our plants and animals [have descended] from some one form, into which life was first breathed" (*Natural Selection,* p. 248). The studies of cytology (meiosis, chromosomal inheritance) and biochemistry fully confirmed the evidence from morphology and systematics for a common origin. It was one of the triumphs of molecular biology to be able to establish that eukaryotes and prokaryotes have the identical genetic code, thus leaving little doubt about the common origin even of these groups. Even though there are still a number of connections among higher taxa to be established, particularly among the phyla of plants and invertebrates, there is probably no biologist left today who would question that all organisms now found on the earth have descended from a single origin of life.

There was only one area in which the application of the theory of common descent encountered vigorous resistance: the inclusion of man into the total line of descent. To judge from contemporary cartoons, none of the Darwinian theories was less acceptable to the Victorians than the derivation of man from the other primates. Yet at the present time this derivation is not only remarkably well substantiated by the fossil record, but the biochemical and chromosomal similarity of man and the African apes is so great that it is quite puzzling why they are so relatively different in morphology and brain development.

III. Gradualism

Darwin's third theory was that evolutionary transformation always proceeds gradually, never in jumps. One will never understand Darwin's insistence on the gradualism of evolution, nor the strong opposition to this theory, unless one realizes that virtually everyone at that time was an essentialist. The occurrence of new species, documented by the fossil record, could take place only by new origins, that is, by the saltations. Since the new species, however, were perfectly adapted and since there was no evidence for the frequent production of maladapted species, Darwin saw only two alternatives. Either the perfect new species had been specially created by an all-powerful and all-wise Creator, or else — if such a supernatural process were unacceptable — the new species had evolved gradually from pre-existing species by a slow process, at each stage of which they maintained their adaptation. It was this second alternative that Darwin adopted.

This theory of gradualism was a drastic departure from tradition. Theories of a saltational origin of new species had existed from the pre-Socratics to Maupertuis and the progressionists among the so-called catastrophist geologists. These saltationist theories were consistent with essentialism. Perhaps one can distinguish three kinds of such saltationist theories:

(1) Extinct species are replaced by newly created ones that are more or less at the same level as those that they replace (Lyell 1830–1833).

(2) Extinct species are replaced by new creations at a higher level of organization (progressionists, such as Buckland, Sedgwick, Hugh Miller, L. Agassiz).

(3) Saltational origin of new species from pre-existing species (E. Geoffroy Saint-Hilaire, Darwin in Patagonia ("Petise"), Galton, Goldschmidt).

Darwin's totally gradualist theory of evolution — not only species but also higher taxa arise through gradual transformation — immediately encountered strong opposition. Even Darwin's closest friends were unhappy about it. T. H. Huxley wrote to Darwin on the day before the publication of the *Origin*: "You have loaded yourself with an unnecessary difficulty in adopting *Natura non facit saltum* so unreservedly. . ." (*LL* (NY) 2: 27). In spite of the urgings of Huxley, Galton, Kölliker, and other contemporaries, Darwin insisted almost obstinately on the gradualness of evolution, even though he was fully aware of the revolutionary nature of this concept. With the exceptions of Lamarck and Geoffroy almost everybody else who had ever thought about changes in the organic world had been an essentialist and had resorted to saltations.

While still on the *Beagle* Darwin had accepted Lyell's sudden introduction of new species, and even when he derived new species from pre-existing species, like the Petise "ostrich", he contended it occurred through a saltation. Even though Darwin continued to use the term *transmutation* for many more years, after March 1837 he held evolutionary change was more or less gradual, and it had become transformation instead of transmutation. Furthermore his adherence to gradualism became stronger with time; eventually (after the 1867 critique by F. Jenkin) he minimized even more the evolutionary role of "sports" (drastic variations).

The source of Darwin's strong belief in gradualism is not quite clear. The problem has not yet been analyzed adequately. Gruber (Gruber and Barrett 1974) thinks Darwin was influenced by the theologian Sumner (1824, p. 20), who suggested that sudden occurrences are indications of an intervention by the Creator and thus of supernatural origin. By stressing continuity and gradualness Darwin was stressing natural causation. Stanley (1981) thinks that Darwin applied the principle of plenitude. Most likely gradualism is the extension of Lyell's uniformitarianism from geology to the organic world. Lyell's failure to do so had rightly been criticized by Bronn. Darwin, of course, also had strictly empirical reasons for his insistence on gradualism. His work with domestic races, particularly his work with pigeons and his conversations with animal breeders, convinced him how strikingly different the end products of slow, gradual selection could be. This fitted well with his observations on the Galapagos mockingbirds and tortoises, which were best explained as the result of gradual transformation.

Finally Darwin had didactic reasons for insisting on the slow accumulation of rather small steps. He answered the argument of his opponents that one should be able to "observe" evolutionary change owing to natural selection, by saying: "As natural selection acts solely by accumulating slight successive favorable variations, it can produce no great or sudden modifications; it can act only by very short and slow steps" (*Origin*, p. 471). There is little doubt that the general emergence of population thinking in Darwin strengthened his adherence to gradualism. As soon as one adopts the concept that evolution occurs in populations and slowly transforms them — and this is what Darwin increasingly believed — one is automatically forced also to adopt gradualism. Gradualism and population thinking probably were originally independent strands in Darwin's conceptual framework, but eventually they reinforced each other powerfully.

After Darwin's death the concept of gradualism became even less popular than it had been in Darwin's own time. This began with Bateson's 1894 book and reached a climax with the mutationist theories of the Mendelians. Both Bateson and De Vries missed no opportunity to make fun of Darwin's belief in gradual evolution and upheld instead evolution by macromutations (Mayr and Provine 1980; Mayr 1982b). A mild popularity of saltationist theories continued right through the Evolutionary Synthesis (Goldschmidt 1940; Willis 1940; Schindewolf 1950).

The naturalists were the main supporters of gradual evolution, which they encountered everywhere in the form of geographic variation. Eventually geneticists arrived at the same conclusion through the discovery of ever slighter mutations, of polygeny, and of pleiotropy. The result was that gradualism was able to celebrate a complete victory during the Evolutionary Synthesis in spite of the continuing opposition by Goldschmidt and Schindewolf.

Now, some forty years later, the argument has flared up again in the wake of the theory of punctuated equilibria. But the current argument simply boils down to the question: What is gradual, and how is it to be defined? Punctuated evolution for Goldschmidt and Schindewolf, as well as for Bateson and De Vries, was the production of new species or higher taxa through the origin of a single individual that had experienced a complete genetic reorganization. Gradual evolution, as now defined, is populational evolution, during which the genetic changes are effected in the course of a series of generations. This does not preclude the possibility that the reorganization of the phenotype at the end of a series of generations may be rather dramatic, as suggested in my theory of genetic revolutions during peripatric speciation (1954a; 1982a).

Defining gradualism as populational evolution — and this is what Darwin basically had in mind — permits us to say that in spite of all the opposition to him, Darwin ultimately prevailed even with his third evolutionary theory. The only exceptions to gradualism that are clearly established are cases of stabilized hybrids that can reproduce without crossing (like allotetraploids).

Nothing is said in the theory of gradualism about the rate at which the change may occur. Darwin was aware of the fact that evolution could sometimes progress quite rapidly, but, as Andrew Huxley (1981) has recently quite rightly pointed out, it could also contain periods of complete stasis "during which these same species remained without undergoing any change".

In his well-known diagram in the *Origin* (opposite p. 117), Darwin lets one species (F) continue unchanged through 14,000 generations or even through a whole series of geological strata (p. 124). The understanding of the independence of gradualness and evolutionary rate is important for the evaluation of the theory of punctuated equilibria (Mayr 1982c).

IV. The Multiplication of Species

This theory of Darwin's dealt with the explanation of the origin of the enormous organic diversity. It is estimated that there are five to ten million species of animals and one to two million species of plants on earth. Even though in Darwin's day only a fraction of this number was known, the problem of why there are so many species and how they originated, was already present. Lamarck had ignored the possibility of a multiplication of species in his *Philosophie Zoologique* (1809). For him diversity was produced by differential adaptation. New evolutionary lines originated by spontaneous generation, he thought. In Lyell's steady-state world, species number was constant and new species were introduced to replace those that had become extinct. Any thought of the splitting of a species into several daughter species was absent among these earlier authors.

To find the solution to the problem of species diversification required an entirely new approach, and only the naturalists were in the position to find it. L. von Buch in the Canary Islands, Darwin in the Galapagos, Wagner in North Africa, and Wallace in Amazonia and the Malay Archipelago were the pioneers in this endeavor. By adding the horizontal (geography) to the vertical dimension that had previously monopolized evolutionary thought, they all were able to discover geographically representative (allopatric) species or incipient species. But more than that, these naturalists found numerous allopatric populations that were in all conceivable intermediate stages of species formation. The sharp discontinuity between species that had so impressed John Ray, Carl Linnaeus, and other students of the non-dimensional situation (the local naturalists), was now supplemented by a continuity among species owing to the incorporation of the geographical dimension.

At first the situation was not understood as clearly as it is now. This is well reflected in Wallace's statement in 1855: "Every species has come into existence coincident both in time and space with a pre-existing closely allied species", and more specifically "the most closely allied species are found in the same locality or in closely adjoining localities" (1855, pp. 6, 5). It reflects the typological species concept still prevailing at that time. There is no reference to geographically representative populations.

If one defines species simply as morphologically different types, one evades the real issue. A more realistic formulation of the problem of speciation did not occur until the development of the biological species concept (K. Jordan, Poulton, Stresemann, Mayr). Only then was it seen that the real problem is not the acquisition of difference but of "distinctness". The problem is thus the acquisition of reproductive isolation in relation to other contemporary species. Transformation of a phyletic line in the time dimension (gradual phyletic evolution, as it was later designated) sheds, of course, no light on the origin of diversity.

Darwin struggled with the problem of the multiplication of species all his life. When he discovered in Patagonia a second species of *Rhea* (which he called *ostrich*), Darwin modified only slightly Lyell's concept of a sudden introduction of new species. In contrast to Lyell, he derived it from an existing species (the northern Rhea); but he did so by an essentialistic saltation. Only after he had discovered the three new species of mockingbirds on different islands in the Galapagos did Darwin develop a fully consistent concept of geographic speciation. At once he found additional examples of apparent geographic speciation, as stated in his B *Notebook*: "Galapagos tortoises, . . . Falkland fox, Chiloe fox. — English and Irish Hare" (*B* 7). It was at this time that Darwin provided species definitions which basically agree with the modern biological species concept (*B* 24: "repugnance to intermarriage"; see also *B* 122, *B* 213, C_{ex} 161). His thinking, at that period, seems to have been derived exclusively from the zoological literature. Even though Darwin considered isolation on islands as the principal speciation mechanism, he seems to have had difficulties in explaining speciation on continents. At one time, to account for the rich species diversity in South Africa, he postulated large scale geological changes, up and down movements of the crust, during which South Africa was temporarily converted into an archipelago, setting the stage for abundant speciation.

But it was not until Darwin, with the help of Hooker, became better acquainted with the botanical literature, that he began seriously to speculate on sympatric speciation on continents. There is still some uncertainty about the development of his thoughts on this subject, in spite of extensive clarification by Kottler (1978) and Sulloway (1979). Part of Darwin's difficulty was caused by the ambiguous use of the term *variety* in the taxonomic literature. Zoologists tended to use the term for geographic races (subspecies) and these, when isolated, were incipient species. Hence, in the 1840s, varieties for Darwin were incipient species.

In the botanical literature, however, varieties more often than not were individual variants ("morphs") within a population. Applying his previously established axiom "varieties are incipient species" to such coexisting varieties required a theory of sympatric speciation. This Darwin establishes in *Natural Selection* with numerous examples, most of them actually zones of secondary hybridization (pp. 252–261). The same thesis, but in a rather abbreviated form, is presented in the *Origin* (p. 103). It was this claim of sympatric speciation that led to his controversy with M. Wagner. Evidently Darwin was not aware of some of the difficulties for sympatric speciation to which modern authors have called attention.

Although Darwin deserves credit, together with Wallace, for having posed concretely for the first time the problem of the multiplication of species, the pluralism of his proposed solution led to a history of continuous controversy that is not ended to this very day. At first, from the 1870s to the 1940s, sympatric speciation was perhaps the more popular theory of speciation, although some authors, particularly ornithologists and specialists of other groups displaying strong geographic variation, insisted on exclusive geographic speciation. The majority of entomologists, however, and likewise most botanists, even though admitting the occurrence of geographic speciation, considered sympatric speciation to be the more common and thus more important form of speciation. After 1942 allopatric speciation was more or less victorious for some twenty-five years, while now the controversy is again in full swing (M. White 1978; Mayr 1982a).

Paleontologists, on the whole, completely ignored the problem of the multiplication of species. For instance, one finds no discussion of it in the work of G. G. Simpson. Indeed, the material of the paleontologists is not suitable for an analysis of the speciation process. When paleontologists finally incorporated speciation into their theories (Eldredge and Gould 1972), their conclusions were based on the speciation research of those who study living organisms.

There are three reasons why speciation is still an open problem 125 years after the publication of the *Origin*. The first is that, as in so much of evolutionary research, the evolutionist analyzes the results of past evolutionary processes and is thus obliged to reach conclusions by inference. Consequently one encounters all the well-known difficulties met in the reconstruction of historical sequences. The second difficulty is that in spite of all the advances of genetics, we are still almost entirely ignorant as to what happens genetically during speciation. And finally, there are reasons to believe that rather different genetic mechanisms may be involved in the speciation of different kinds of organisms and under different circumstances. Yet Darwin's model of speciation, as first developed on the basis of the Galapagos mockingbirds, is still very much alive, and presumably essentially correct.

For many years I have extolled Darwin's introduction of population thinking into biology. Its importance cannot be questioned. The whole Malthus episode of 28 September 1838 would not have made sense if Darwin had not suddenly appreciated on that day that the struggle for existence was among members of the same population and that selection could operate successfully only if there were individual differences among the members of the population. Even earlier, when in March 1837 Darwin had come to the conclusion that the mockingbirds on each of three islands in the Galapagos had undergone gradual change, he could not have interpreted this in terms of essentialism; it required a populational explanation. The great stress in the B *Notebook* on generation as the source of variation is also a populational explanation. Owing to the enormous explanatory power of the populational approach, one would think that from 1838 on Darwin would have employed it for all his explanations, as the modern evolutionist does; but this is not the case. In *Natural Selection* and the *Origin* many explanations are given in a strictly typological language. No historian has yet made a detailed analysis of all of Darwin's references to variation to determine what proportion of them was based on essentialistic or on populational thinking. This much is certain, however; the development of the theory of natural selection would have been impossible without the prerequisite of populational thinking. Nor could any essentialist have come to terms with gradualism.

V. Natural Selection

Darwin's theory of natural selection was his most daring, his most novel theory. It dealt with the mechanism of evolutionary change and, more particularly, how this mechanism could account for the seeming harmony and adaptation of the organic world. It attempted to provide a natural explanation in place of the supernatural one of Natural Theology. His theory for the natural mechanism that would be able to direct evolutionary change was unique. There was nothing like it in the whole philosophical literature from the pre-Socratics to Descartes, Leibniz, or Kant. It replaced the teleology in nature with an essentially mechanical explanation.

Just how Darwin came to develop this theory, and what its components are, have been discussed so often in the recent Darwin literature that I shall not go into this in detail. Darwin based his theory on a number of concepts that he either had himself developed or at least had used in a unique way. Through the reading of Malthus, Darwin suddenly came to realize that the most intense "struggle for existence", so often referred to in the literature of the preceding 200 years, was among the individuals of a single species. From this Darwin concluded that among genetically slightly different individuals, some would have a better chance to survive in such a struggle than others. This was an insight he had presumably received from the animal breeders, although it was reinforced by the experience of the taxonomists, and indeed by Darwin's own later work on barnacles.

To judge from his writings, Darwin had a much simpler concept of natural selection than the modern evolutionist. For him there was a steady production of individuals, generation after generation, with those that were "superior" having a reproductive advantage. It seemed essentially to be a single-step process, the conveying of reproductive success. The modern evolutionist agrees with Darwin that the individual is the target of selection; but we also know, owing to a hundred years of research in cytology and genetics, that the production of a new individual is an exceedingly complex process. It begins with meiosis (including crossing-over and the separation of the paired homologous chromosomes during gamete formation), with the production of millions of gametes, with mate selection, and the random encounter of male and female gametes, resulting in the production of a new zygote. The genotypes of the preceding generation are broken up during this process, and the pieces are, so to speak, poured back into the gene pool and thoroughly mixed. From this mixture new individuals are extracted, all of them again genetically unique. This is the first step in the process of natural selection; its consequences are to make an enormous amount of genetic variation available for the exercise of natural selection in the second step. As I have already emphasized (Mayr 1962), chance reigns supreme at this first step during the entire sequence of gamete formation and fertilization that precedes the production of a new individual.

The second step in the process of natural selection is the determination of reproductive success of these individuals. Selection is not merely mortality selection as reflected in the slogan "survival of the fittest", but "success in leaving progeny", as Darwin saw quite clearly and mentioned specifically as part of natural selection (*Origin*, p. 62). This explains why modern authors usually define natural selection as differential reproduction. It is the potentially reproducing individual, and not the gene, that is the target of selection. The great emphasis on single genes in the work of the mathematical population geneticists, and their definition of evolution as a "change in gene frequencies", have led to the unfortunate conclusion by certain outsiders that neo-Darwinism means a theory of evolution in which the selection of genes is the basic thesis. In fact this misinterpretation of evolution has nothing to do with the actual meaning of neo-Darwinism. The term neo-Darwinism was coined by Romanes for theories of evolution that accept natural selection but reject any belief in an inheritance of acquired characters (1895, pp. 12–13).

Although I call the theory of natural selection Darwin's *fifth* theory, it is actually, in turn, a small package of theories. This includes the theory of the perpetual existence of a reproductive surplus (superfecundity), the theory of the heritability of individual differences, the discreteness of the determinants of heredity, and several others. Many of these were not explicitly stated by Darwin but are implicit in his model as a whole.

Evolutionists from Darwin on have always emphasized the continuity of populational evolution, in contrast to the discontinuous character of saltational evolution by way of reproductively isolated individuals. The fact is invariably ignored, however, that even continuous evolution is mildly discontinuous owing to the sequence of generations. In each generation an entirely new gene pool is reconstituted from which the new individuals are drawn that are the target of selection in that generation.

We now come to the important problem of the antecedents of the theory of natural selection. In a historical survey entitled "Natural Selection before Darwin", Zirkle (1941) enumerated a large number of reputed forerunners. When one looks more closely, one sees that virtually all of these so-called prior cases of natural selection turn out to be a rather different phenomenon, which is only superficially similar to selection. I am referring to the *elimination* of "degradations of the type". It is what we now call "stabilizing selection". Essentialism always had had great difficulty in coping with the phenomenon of variation. One of its collateral concepts was that any deviation from the type that was too drastic would be eliminated. But such a process is not natural selection in the Darwinian sense, a force that would permit directional change and an improvement of adaptation. Nevertheless, as historians have pointed out, and as Darwin himself recorded in the historical introduction added in the third edition of the *Origin*, there were a few genuine forerunners such as Wells and Matthew, even though their obscure publications were totally ignored and apparently not even quoted a single time prior to 1859. The matter is, of course, quite different with the independent discovery of natural selection by Alfred Russel Wallace, whose essay on the subject was published simultaneously with some extracts from Darwin's unpublished writings in the famous Linnean Society publication of 1858.

The steps by which Darwin pieced together his theory of natural selection have been the subject of active discussion in the recent Darwin literature. One cannot doubt that 28 September 1838, the day on which Darwin was reminded by Malthus of superfecundity, played a decisive role. It seems that it was on this day that Darwin for the first time fully realized that the struggle for existence was not so much among species as among individuals of the same species. There is little doubt that by the end of 1838 the theory was complete in Darwin's mind, as far as its major components are concerned. The full integration of the theory into Darwin's theory of evolution, however, continued in the ensuing years and was, as Ospovat (1981) has demonstrated, particularly active in 1856 and 1857. Even though Darwin modified some of his views after the publication of the *Origin* in 1859, these modifications did not affect any important component of his theory of natural selection.

The theory of natural selection was the most bitterly resisted of all of Darwin's theories. If it were true, as some sociologists have claimed, that the theory was the inevitable consequence of the Zeitgeist of early nineteenth-century Britain, of the industrial revolution, of Adam Smith and the various ideologies of the period, one would think that the theory of natural selection would have been embraced at once by almost everybody. Exactly the opposite is true: the theory was almost universally rejected. Only a few naturalists, like Wallace, Bates, Hooker, and Fritz Müller, could be called consistent selectionists in the 1860s. Lyell never had any use for natural selection, and even T. H. Huxley, defending it in public, was obviously uncomfortable with it and probably did not really believe in it (Poulton

1896; Kottler, this volume). Before 1900 not a single experimental biologist either in Britain or elsewhere adopted the theory (Weismann was basically a naturalist). Of course Darwin himself was not a total selectionist, since he always allowed for the effects of use and disuse and an occasional direct influence of the environment. The most determined resistance came from those who had been raised under the ideology of natural theology. They were quite unable to abandon the idea of a world designed by God and to accept a mechanical process instead. More importantly, a consistent application of the theory of natural selection meant a rejection of any and all cosmic teleology. Sedgwick and K. E. von Baer were particularly articulate in resisting the elimination of teleology.

Natural selection represents not only the rejection of any finalistic causes that may have a supernatural origin, but it rejects any and all determinism in the organic world. Natural selection is utterly "opportunistic", as G. G. Simpson has called it; it is a "tinkerer" (Jacob 1977). It starts, so to speak, from scratch in every generation, as I described above. Throughout the nineteenth century the physical scientists were still deterministic in their outlook, and so indeterministic a process as natural selection was simply not acceptable to them. One has only to read the critiques of the *Origin* written by some of the best-known physicists of the period (Hull 1973b) to see how strongly the physicists objected to Darwin's "law of the higgledy-piggledy" (*LL* (NY) 2: 37; Herschel 1861, p. 12). From the Greeks to the present day there has been a never-ending argument as to whether the events of nature are due to chance or due to necessity (Monod 1970). Curiously, in the controversies over natural selection, the process has been described sometimes as "pure chance" (Herschel and many other opponents of natural selection) or as a strictly deterministic optimization process. Both classes of claimants overlook the two-step nature of natural selection and the fact that in the first step, chance phenomena prevail, while the second step is decidedly of an anti-chance nature. As Sewall Wright has so correctly said: "The Darwinian process of continued interplay of a random and a selective process is not intermediate between pure chance and pure determination, but in its consequences qualitatively utterly different from either" (1967, p.117).

Even though everybody very soon accepted evolution, at first only a minority of biologists and very few non-biologists accepted consistent selectionism. This was true until the period of the Evolutionary Synthesis. Instead they adopted finalistic theories, neo-Lamarckian theories, and saltational theories. The controversy over natural selection is by no means at an end. Even today the relationship between selection and adaptation is hotly debated in the evolutionary literature, and it has been questioned whether it is legitimate to adopt an "adaptationist program", that is, to search for the adaptive significance of the various characteristics of organisms (Gould and Lewontin 1979). But the question that is really before us is not so much whether natural selection is now universally adopted by evolutionists — a question one can unhesitatingly answer affirmatively — but rather whether the modern evolutionist's concept of natural selection is still that of Darwin or is considerably modified.

When Darwin first developed his theory of natural selection, he was still inclined to think that it was able to produce near-perfect adaptation, in the spirit of natural theology (Ospovat 1981). More thinking and the realization of the numerous deficiencies in the structure and function of organisms — perhaps particularly the incompatibility of a perfection-producing mechanism with extinction — led Darwin to reduce his claims for selection, so that all he demanded in the *Origin* was that "natural selection tends only to make each organism, each organic being, as perfect as, or slightly more perfect than, the other inhabitants of the same country with which it has to struggle for existence" (p. 201). Today we are even more conscious of the numerous constraints that make it impossible for natural selection to achieve perfection, or, to state it perhaps more realistically, to come even anywhere near perfection (Gould and Lewontin 1979; Mayr 1982e).

VI. The Varying Fates of Darwin's Five Theories

We can now summarize the subsequent fate of each of the five theories of Darwin, which I have discussed above. Evolution as such, as well as the theory of common descent, were adopted very quickly, within fifteen years of the publication of the *Origin*, hardly a qualified biologist was left who had not become an evolutionist. Gradualism, by contrast, had to struggle, and populational thinking was a concept that was apparently very difficult for anyone who was not a naturalist to adopt. Even today, in the discussions of punctuated equilibria, statements are made that indicate some people still do not understand the core of population thinking. What counts is not the size of the individual mutation but only whether the introduction of evolutionary novelties proceeds through their incorporation into populations or through the productions of single new individuals that are the progenitors of new species or higher taxa.

That a theory of the multiplication of species is an essential, in fact integral, component of evolutionary theory, as first pronounced by Wallace and Darwin, is now taken for granted. How this multiplication proceeds is still controversial. That allopatric speciation, and particularly its special form of peripatric speciation (Mayr, 1954; 1982c), is the most common mode is hardly questioned. That speciation by polyploidy is common in plants is likewise admitted. How important other processes are, like sympatric and parapatric speciation, is still controversial.

Finally, the importance of natural selection, the theory that is usually meant by the modern biologist when speaking of Darwinism, is firmly accepted by nearly everyone. Rival theories — like finalistic theories, neo-Lamarckism, and saltationism — are so thoroughly refuted that they are no longer seriously discussed. Where the modern biologist perhaps differs from Darwin most is in assigning a far greater role to stochastic processes than did Darwin or the early neo-Darwinians. Chance plays a role not only during the first step of natural selection, the production of new, genetically unique individuals, but also during the probabilistic process of the determination of reproductive success of these individuals. Yet when one looks at all the modifications that have been made in the Darwinian theories between 1859 and 1984, one finds that none of these changes affects the basic structure of the Darwinian theories. There is no justification whatsoever for the claim that the Darwinian paradigm has been refuted and has to be replaced by something new. It strikes me as almost miraculous that Darwin in 1859 came so close to what would be considered valid 125 years later.

26

Darwinism as a Historical Entity: A Historiographic Proposal

David L. Hull

I. The Problem: What is Darwinism?

In a recent meeting of the American Association for the Advancement of Science, several eminent scientists addressed the question: What happened to Darwinism between the two Darwinian Centennials, 1959–1982?[1] An unanticipated problem soon arose — none of the participants could agree on what Darwinism actually was. Each speaker was sure that Darwinism has an essence, a set of tenets that all and only Darwinians hold, but no two could agree about which tenets are actually essential. Is selectionism essential? Must nearly all traits and all adaptations arise through natural selection, or does the neutralist alternative also count as part of Darwinism? Must evolution be largely or exclusively gradual to be Darwinian, or is saltative evolution merely a variant of an all-embracing Darwinism?

A sampling of the recent biological literature reveals an amazing variety of answers to these and other questions about the essence of Darwinism, neo-Darwinism, and the synthetic theory. Gould argues that the "essence of Darwinism lies in its claim that natural selection creates the fit. Variation is ubiquitous and random in direction. It supplies the raw material only" (1977a, p. 44). The saltationist Richard Goldschmidt (1933) was mistaken in thinking that his theory of "hopeful monsters" was in the least non-Darwinian. "For Goldschmidt, too, failed to heed Huxley's warning that the essence of Darwinism — the control of evolution by natural selection — does not require a belief in gradual change" (Gould 1977b, p. 24). Gould certainly has a point. T. H. Huxley disagreed with Darwin about evolution being gradual, and surely Huxley counts as a Darwinian. If Huxley's saltationism is part of Darwinism, how can we exclude Goldschmidt's saltationism?

Elsewhere Gould (1980a) presents a slightly different view. A decade ago, Eldredge and Gould (1972) produced a theory of the evolutionary process that they term "punctuated equilibria". On this view, most change occurs at speciation events and is not influenced primarily by selection. The role of selection is to fine-tune new species to their environments once they have emerged. To the question, "Is a new and general theory of evolution emerging?" Gould answers, yes (1980a). It is new, not because it is microsaltational but because of the decreased role of selection (see also Gould 1982). Eldredge argues that "neo-Darwinian theory is not at all as monolithic as we might suppose. Rather, *all* evolutionary thinking, Darwinian and non-Darwinian, pre- and post-1859, has been beset by a curious quality, which has effectively hindered a truly integrated theory of any guise from emerging" (1979, p. 7). Eldredge dubs the two poles of this duality the "taxic approach" and the "transformational approach". Traditionally, the works of G. G. Simpson (1944, 1949, 1953) and Ernst Mayr (1942, 1963) are viewed as contributing to the same theory, the synthetic theory. As Eldredge would have us parse the issues, Simpson and Mayr actually were in fundamental conflict because Simpson's approach was transformational, while Mayr's was taxic.

Lewontin agrees that Darwinism has an essence, but he disagrees with Gould and Eldredge about the precise nature of this essence (1977, p. 4). For example, in response to the "non-Darwinian" neutralist views of Kimura and Ohta (1971), Lewontin states that the "essential nature of the Darwinian revolution was neither the introduction of evolutionism as a world view (since historically that is not the case) nor the emphasis on natural selection as the main motive force in evolution (since empirically that may not be the case), but rather the replacement of a metaphysical view of variation among organisms by a materialistic view."

Andrew Huxley concurs with Eldredge and Gould that disagreements between the punctuationalists and gradualists are a "debate within the Darwinian framework" (1981, p. ii). Stebbins and Ayala second Huxley's perspective and propose to embrace the neutralists in the all-encompassing arms of the synthetic theory of evolution (1981, p. 970). "The 'selectionist' and 'neutralist' views of molecular evolution are competing hypotheses within the framework of the synthetic theory of evolution" (Stebbins and Ayala 1981, p. 967). I could continue adding more elements to this discussion; for example, the levels of selection controversy. Gene selectionists such as G. Williams (1966) and Dawkins (1979) claim that they are the true Darwinians, while group selectionists such as Wynne-Edwards (1962) are not. Darwin surely was not a group selectionist in the modern sense, but neither was he a gene selectionist (Ruse 1980b).

But, one might complain, these opinions are from scientists who are themselves engaged in the process they are evaluating. They are likely to see things in ways calculated to foster their own research programs. Scientists who are committed to a certain well-entrenched view of evolution, such as the synthetic theory, are liable to interpret any new view, once it seems to be gaining a foothold, as merely a minor modification of their own grand theory. Initially, the neutralist position was false and non-Darwinian;

now that considerable evidence exists for a large pool of neutral variation existing at the molecular level, neutralism is part of the synthetic theory. Young Turks, on the contrary, are presented with two conflicting strategies — emphasizing how different their views are from the well-entrenched views or emphasizing how similar they are. On the first strategy, the battle is harder to win, but once won, the credit accrued is much greater. On the second strategy, the assent of eminent scientists is easier to obtain but at the cost of decreasing the apparent originality of one's contributions. The choice is between a hard fight for big stakes — a new theory of evolution — and an easier contest for reduced payoff — a minor modification of received views.

I think that considerations such as these do color the ways in which scientists present their own views and react to the views of others, but it cannot be the entire story because the same multiplicity of opinion about the essence of Darwinism can be found in the writings of those who are not engaged in the intramural disputes among evolutionists but study these disputes. Ruse (1978b, pp. 409, 411) agrees with Gould. The two essential tenets of Darwinism are selection and random variation. Like Lewontin, Greene adopts a more global perspective: "*Darwinism* should be used to designate a world view" (1981b, pp. 130–131). Its components are nature as a law-bound system of matter in motion, organic evolution, change by means of competition, Lockean epistemology, and sensationalist psychology. Who was it that first combined all these components of Darwinism into a single all-embracing synthesis? Herbert Spencer, that's who.

In 1972 a group of historians gathered together to compare the reception of Darwinism around the world (Glick 1974). One commentator, after surveying the variety of beliefs that had been arrayed under the banner of "Darwinism", was appalled. Leeds's response is sufficiently instructive to quote extensively. First, Leeds sets out what he takes to be the Darwinian model and then observes:

> The social evolutionism discussed or adverted to in the papers of this volume and discussed at length below is quite different from this model in almost every respect.
>
> It is important to note that the model is not made explicit in most of the papers in this book; the writers assume not only that it is known to the readers but also that it was not merely known but *held* by the writers they discuss. This is demonstrably false for a number of cases, e.g., the Mexican one discussed by Roberto Moreno — the Mexicans *said* they were Darwinian, but any close analysis of what they were actually assuming seems clearly strongly Lamarckian (see the quotations given in Moreno's article). What appears to me striking is how few of the figures discussed in these pages — with the exception of a small number of the Spanish, the Germans, and the English — held a Darwinian view

at all. Mostly they assimilated a phrase or an aspect of Darwin's expression of his thought to their own understanding and thought, then, that they were Darwinians. The most striking case is that of the Russians, discussed in James Allen Rogers's paper, in which *not one* of the protagonists of his drama is remotely near the Darwinian model. (1974, p. 439)

II. The Filiation of Ideas

The obvious response to all this variation is that scientific theories, like biological species, evolve. Without variation, evolution is impossible, and in evolution traits that were once universally distributed in a species become replaced by other traits. Darwin himself changed his mind through the years. At one time he thought that geographic isolation was necessary for speciation, but by the time he published the *Origin*, he had changed his mind. Later, when Moritz Wagner published his migration theory (1868, 1873), Darwin treated it as in opposition to his own theory. In 1859 Darwin thought that Lamarckian inheritance played a very minor role in evolution. Later he was willing to grant it a somewhat enlarged role, "but his views are not usually considered non-Darwinian on this account" (Dobzhansky 1972, p. 161). In the hands of later Darwinians, however, such as August Weismann (1904), neo-Darwinism became totally selectionist, and any theory that included a Lamarckian mechanism was held to be in opposition. As J. W. Dawson remarked with respect to Darwinism, "it is in the nature of this protean philosophy that it should itself be in process of evolution from day to day, and thus be in so rapid motion that it changes its features momentarily while one endeavors to sketch it" (1890, p. 5).

I agree but insist that Darwinism is not the only philosophy that is "protean". Every conceptual system, to the extent that it is successful, is just as protean. The only research programs that can possibly have an essence, a set of tenets that *all* and *only* the advocates of that program hold, are those that fall stillborn from the press or degenerate into ideologies. No matter what strategy one uses to pin down conceptual systems, they always succeed in slithering off the point before one's very eyes. If Huxley was a Darwinian, if Weismann was a Darwinian, if Simpson and Mayr are Darwinians, then Darwin was no Darwinian. Aristotle, Linnaeus, Marx, Mendel, and Freud are but a few of the people credited with initiating important intellectual movements, yet present-day commentators argue with reassuring regularity that Aristotle was no Aristotelian (Balme 1980; Grene 1978; Heinaman 1979; Lennox 1980; Preus 1979), that Linnaeus was no Linnean (E. Greene 1909; Hull 1984a; Larson 1971), nor Marx a Marxist (Trigger 1979), nor Mendel a Mendelian (Brannigan 1979; Olby 1979), nor, as everyone knows by now, was Freud a Freudian (Bettelheim 1982, 1983).

In the face of comments such as these, one might be tempted to throw up one's hands in exasperation, while quoting a few lines from Whitman:

Do I contradict myself?
Very well then I contradict myself,
(I am large, I contain multitudes.)
(*Song of Myself*, §51)

The evolutionary biologists cited above understand the implications of evolutionary theory for species: if species are the things that evolve, then they need not and usually do not have any essences. Perhaps all the organisms that belong to a particular species have the same blood type, perhaps some have one blood type and others another blood type, or perhaps a dozen or so different blood types coexist in the same species in varying frequencies. It may even be the case that a blood-type that is widely distributed in a species, possibly universally distributed, may through time be eliminated totally from the species. None of the preceding states of affairs "contradict" each other. When we turn to scientific theories, research programs, and philosophies, however, we are strongly inclined to dig in our heels. It is one thing to say that Darwin held the "same" theory both when he thought geographic isolation was necessary for speciation and when he thought it was not. After all, changing one's mind is not the same thing as contradicting oneself. It is quite another thing to say that Darwin and Huxley at a particular time held the "same" theory when Darwin thought evolution was nearly always gradual and directed largely by natural selection and Huxley thought it was largely saltative and that selection played only a subsidiary role. Darwin and Huxley *were* contradicting each other. But, if conceptual systems "evolve" in anything like the way that biological species do, then it must be possible for them to include contradictory claims.

Differences of opinion about the essence of Darwinism have numerous sources. Some authors want to pin Darwinism down to a particular set of narrowly scientific views expressed explicitly by Darwin at one point in his intellectual development — for example, those views expressed in the *Origin*. Other authors want to include the scientific views of other Darwinians contemporary with Darwin, such as Wallace, Huxley, Hooker, and Gray. Still others want to include later Darwinians on this list. Not infrequently scientists attempt to gain support for their views, no matter how much they might differ from those of Darwin, by throwing the mantle of Darwin around their own shoulders. De Vries (1901–1903), Weismann (1904), G. G. Simpson (1949, 1953), Mayr (1942, 1963), and now Dawkins (1976) and Gould (1977c) claim their theories are essentially Darwinian. Still others interpret Darwinism more globally as a world view, and so it goes.

As I see it, the problem these evolutionary biologists are having with "Darwinism" is that they have failed to extend to conceptual systems the same sort of perspective they apply to species. They do not expect species to have an essence — a set of traits that all and only members of a particular species have throughout all time; but they do expect conceptual systems to have an essence — a set of tenets that all and only instances of a particular conceptual system have throughout all time. Although Dawkins is well aware that his version of evolutionary theory differs markedly from anything Darwin might have said, "Yet, in spite of all this, there is something, some essence of Darwinism, which is present in the head of every individual who understands the theory" (1976, p. 210).

The fundamental problem these authors are confronting — and it is a sufficiently fundamental problem to deserve the title "metaphysical" and not in the derogatory sense implicit in Lewontin's (1974a) reference — is not that one author or another is mistaken about the essence of Darwinism. The problem lies in the conviction that something like Darwinism can have an essence, not only that it *can* have an essence but that it *must* have an essence. But, if one views conceptual systems as the sort of thing that can change through time — "evolve" — then they need not and usually do not have essences. Hence the search for the essence of Darwinism is misdirected. If Darwinism has an essence, it is in its origin. In this paper I pursue the notion of treating Darwinism as a "lineage" in the filiation of ideas. I extend Mayr's population thinking to thinking itself. Choosing Darwinism as my expository example is fitting because it was Darwin and his fellow Darwinians who forced the scientific community as well as large sections of the public at large to view species in this way. I intend my analysis to be general, however — to apply to the evolution of all research programs, both as groups of scientists and as conceptual systems.[2]

III. Historical Entities

Traditionally philosophers have treated concepts, statements, theories, and the like as similarity classes — types with similar tokens. Anyone throughout all time who believed that new species arise from old believed in the same idea — evolution. But just as biologists distinguish between homologies and analogies, anyone interested in the filiation of ideas must make parallel distinctions for concepts (Gould 1977c). Whether for traits or concepts, in cases of homology, descent takes priority to similarity, while in cases of analogy, similarity takes priority to descent. This way of viewing conceptual development is not especially new. Nothing it seems ever is. Kierkegaard, for example, can be found saying:

> Concepts, like individuals, have their histories, and are just as incapable of withstanding the ravages of time as are individuals, but when all is said and done, they retain a sort of homesickness for their birthplaces. (1841, p. 47)

In conceptual development, unreceived messages do not count. Over and over again, the "same" idea is expressed, the "same" discovery made, but until one of these instances catches fire, these prophetic types might as well have never existed. As Emerson remarked with respect to cases of scientists working in total isolation, "one hermit finds this fact, and another finds that, and lives and dies ignorant of its value" (quoted in Morse 1876, p. 139). This list of Darwin's precursors is as endless as it is irrelevant.[3] In conceptual development, truly unappreciated precursors do not count (Sandler 1979).

Recently, however, philosophers of science have begun to take the "temporal dimension" to science more seriously. According to the "new" philosophy of science, the basic unit of appraisal is no longer an atemporal rational reconstruction of a theory but some sort of temporally extended conceptual entity. Whether these entities are termed disciplinary matrixes (Kuhn 1970), research programmes (Lakatos 1970), scientific disciplines (Toulmin 1972), theories (McMullin 1976), or research traditions (L. Laudan 1977), they all have something in common — an essence. They are historical entities. As Laudan puts this new view of conceptual development:

> Research traditions, as we have seen, are *historical* creatures. They are created and articulated within a particular intellectual milieu, they aid in the generation of specific theories — and like all other historical institutions — they wax and wane. Just as surely as research traditions are born and thrive, so they die, and cease to be seriously regarded as instruments of furthering the progress of science. (1977, p. 95)

In response to complaints about the inherent vagueness and ambiguity of his term "paradigm", Kuhn (1970) has distinguished two senses of this term which he now calls "exemplars" and "disciplinary matrixes". According to Kuhn, exemplars are concrete problem solutions, "universally recognized scientific achievements that for a time provide model problems and solutions to a community of practitioners" (1970, pp. viii, 187). Disciplinary matrixes are much more inclusive and much less definite entities. They include exemplars, symbolic generalizations, metaphysical principles, and values.

One possible function of exemplars is to help individuate disciplinary matrixes. Kuhn identifies disciplinary matrixes with scientific communities:

> A paradigm is what members of a scientific community share, *and*, conversely, a scientific community consists of men who share a paradigm. (1970, p. 176)

In spite of first appearances, Kuhn's identification is not circular because different means exist for individuating scientific communities and their conceptual correlates. To the extent that each disciplinary matrix possesses its own exemplar, that exemplar can be used to individuate its matrix. Scientific communities in turn can be distinguished sociologically in terms of attendance at conferences, distribution of draft manuscripts, and so on. If disciplinary matrixes are individuated independently from scientific communities, then the claim that the two are coextensive becomes an empirical matter that can be checked. As I shall argue shortly, when one does get around to checking, one discovers that the boundaries of scientific communities and disciplinary matrixes do not coincide perfectly. Two scientists can share the "same" paradigm and belong to different scientific communities, and conversely, two scientists can belong to the same community and disagree with each other even over fundamentals.

Toulmin (1972) has made parallel distinctions between intellectual disciplines and intellectual professions with the added twist that all the constituents of each can be exchanged while they remain the same disciplines and professions respectively. All that is necessary is that both remain internally cohesive while they change continuously through time. We are certainly used to treating groups in this way. The Republican Party still exists even though none of its founding members is still alive. Treating conceptual entities in this way is much more problematic. On Toulmin's view, later stages in an intellectual discipline might not include *any* of its original constituents, including the exemplar that initially served to individuate it, and still count as the same matrix. Every four years the Republican Party formulates a "new" platform in the sense that it contains some new planks, but it is the "same" platform in the sense that it is a modification of the previous platform (see Frankel 1979).

Rudwick (1982b) makes comparable distinctions in setting out what he terms a "processual" view of science. According to Rudwick, geology in the first half of the nineteenth century can be conceptualized as consisting of concentric zones. In the center are a half dozen or so élite geologists, surrounded by respected members of the community, surrounded in turn by amateurs. These zones are situated within the public at large, interested members of the general public nearer the circumference of the geological community as such. Rudwick then further subdivides this community in ways that cut across the concentric zones — for example, by indicating the membership of the Geological Society of London. Given this topographical map, Rudwick is then in a position to trace the trajectory of a particular scientist across its surface. For example, in 1831 Darwin entered the outermost ring of amateurs, progressed by 1835 into the respected zone, finally to be counted among the élite in 1838, only to slip back into the respected group as his research interests shifted from geology to natural history. Rudwick (1982b) differs from Kuhn and Toulmin in structuring the contours of his map simultaneously on both social and cognitive considerations — "zones are defined as much by criteria of ascribed competence in a particular field of knowledge as by criteria of social interaction." (See also Gruber's evolving systems approach (1980a).)

In the remainder of this paper, I propose to set out the constitution of both the Darwinians as a social group and Darwinism as a conceptual system borrowing from all of the works described above. Like Kuhn and Toulmin, I keep the social and conceptual mappings as separate as possible for the simple reason that social groups are much easier to individuate than are conceptual systems. If scientific communities are individuated on the basis of social criteria, then they can be used to help determine the boundaries of their conceptual systems. From Toulmin I take the notion of historical entity. I shall treat both scientific communities and conceptual systems as historical entities that remain internally cohesive as they develop indefinitely and continuously through time. I also develop Kuhn's notion of an exemplar for both scientific communities and conceptual systems in analogy to "type specimens" in biology. Just as a particular way of structuring breeding experiments can function as an exemplar for Mendelian genetics as a scientific research program, a particular scientist such as Darwin can serve as an "exemplar" for individuating a scientific community. Finally, I adopt Rudwick's notion of a single scientist's career as a trajectory, in much the same way as the trajectory of a species can be mapped on one of Wright's adaptive landscapes (1932). On a larger scale, communities and conceptual systems are themselves trajectories. As the reader has surely noticed by now, in this paper I am adopting the safe strategy mentioned previously. I am not developing a radically new view but building on the solid foundations of past achievements.

IV. The Type Specimen Method

In the early years of taxonomy, naturalists thought that the vast majority of organisms that belong to a species are essentially similar to each other. They were well aware that some variation exists within a species, but they were convinced that such variation is only accidental. The task of the systematist was to see through all this accidental variation to the essence of the species under investigation. To aid them in this process, systematists devised the type specimen method. The type specimen was an organism selected to be the name bearer for its species. As the term implies, type specimens were supposed to be typical members of their species. Finding a type specimen was thought not to be very difficult because nearly all members of a species were supposed to be essentially the same, an occasional damaged specimen or aberrant form notwithstanding. If by a stroke of bad luck, one happened to pick an atypical member as a type specimen, one could always rectify this mistake by replacing it with another more typical specimen.

As our understanding of biological species grew, however, as we came to understand the evolutionary process, systematists came to realize that the "typical" member of a species is a will-o'-the-wisp. Some species are monotypic — that is, they can be characterized with minimal falsification by means of a set of essential traits. But many are polytypic. At best they can be characterized by traits that covary only statistically.[4] But more importantly, biologists came to realize that organisms are included in a species not because they are similar to each other, but because they belong to the same chunk of the genealogical nexus (Ghiselin 1974; Hull 1976, 1978a; Mayr 1976d, 1978; Gould 1980b, 1982; Wiley 1981). The evolutionary relationships that organisms have to each other are primary, their similarities secondary. Organisms that belong to the same species tend to be similar to each other *because* they belong to the same chunk of the genealogical nexus; they do not belong to the same chunk of the genealogical nexus *because* they are similar to each other.

This change in the way evolutionary biologists conceive of species was accompanied by parallel modifications in the type specimen method. Because the notion of a "typical" member is inapplicable to many species, the codes of nomenclature were modified so that type specimens need no longer be typical. In fact, they can be monsters. The rule is, pick an organism, any organism, and use that organism to affix a name to whatever chunk of the genealogical nexus it happens to be part of. If the type specimen turns out to be aberrant, it makes no difference. It must remain the type specimen for its species. If two organisms happen to be selected as type specimens for what turns out to be the same chunk of the genealogical nexus, the earlier type specimen and name take priority. When one lineage splits into two new lineages, new species come into existence. Hence new type specimens must be selected and new names coined. If two species merge into one, producing a third hybrid species, then both a new type specimen and a new name are called for. Type specimens serve to individuate their species by being one node in the web of relations that define the species. Because these relations are primary and any similarity in make-up is secondary, the relations that the type specimen has to other organisms in its species are primary, and any similarities between it and its conspecifics are secondary. The type specimen may be similar to some of its conspecifics, quite different from others (Mayr 1969b).[5]

The preceding position is strongly counterintuitive, so much so that many biologists strenuously resist adopting it. Even so, I think that the only way that sense can be made of an evolutionary process, whether biological, sociological, or conceptual, is to view the things that are evolving in this way. But before turning to the social evolution of the Darwinians and the conceptual evolution of Darwinism, I shall make a gesture, no matter how futile, to make the preceding view of biological species seem somewhat more intuitive by discussing briefly one example of a biological species — *Homo sapiens*. So the story goes, Linnaeus was chosen; in recognition of his great contributions to the science of systematics, as the type specimen for *Homo sapiens*. In what sense was he "typical"? He was male; at least half of human beings are female. He was Caucasian; Caucasians make up a minority of the human species. He had some blood type or other, but whatever blood type he had, it could not be even the majority blood type, let alone the essential blood type. This list could be expanded indefinitely.

Even so, the belief that there must be an essence of *Homo sapiens* is all but irresistible. If all human beings do not share some characteristic or characteristics in common, how can they all belong to the same species? The answer is simple enough: to be born of human beings, to mate with human beings, and/or to parent human beings. A baby born without opposable thumbs or even the potentiality of rational thought nevertheless remains a part of the human species. More humanistically inclined commentators, such as Dobzhansky are right — "there is no single human nature common to everybody but as many variant human natures as there are men" (1973, p. 261; see also Gould 1980b).

If scientific communities are also viewed as historical entities, then the rule becomes, pick a scientist, any scientist, and follow out his social relations. In doing so, one will discover a variety of groups of varying degrees of inclusiveness and discreteness. Only some of these groups will be "scientific". The sorts of groups one individuates in this way depend on the sorts of relations one chooses. A short list of such relations includes agrees-with, owes-allegiance-to, identifies-with, writes-papers-with, refuses-to-criticize-publicly, cites-work-approvingly, and so on. It might be that certain of these relations fail to covary with each other even statistically, that is, they define quite different groups of scientists. Or it might be that certain relations will pick out roughly the same group. I think it is a fact about scientific development, a contingent fact, that the most significant innovations are invariably associated in the early stages of their development with small ephemeral groups of scientists. For these groups at least, the application of a wide variety of relational predicates results in the specification of the same group. These are the "Darwinians" in the most important sense of this term.

Similarly, if conceptual systems are to be viewed as historical entities, then the rule becomes, pick a particular instance of a concept, any instance, and follow out its conceptual relations. The important conceptual relations turn out not to be such traditional relations as is-similar-to or is-deducible-from, however, but much more problematic relations such as gave-rise-to and mutually-support. The peculiar thing about conceptual systems as historical entities is that they can contain at one stage a particular proposition, at another stage its negation. To make matters worse, at a particular point in time, a conceptual system can contain contradictory propositions — both a statement and its negation. Conversely, even though one proposition follows deductively from another in a conceptual system, it does not follow that this deductive consequence belongs in this conceptual system. If no one saw the connection, it does not.

In each case, the historical system can be picked out and individuated by the type specimen method — by selecting one element in this nexus of relations and tracing its interconnections. It is important to keep in mind that these type specimens or "exemplars" need not be especially exemplary.

The choice of a central node makes things easier, but when one first studies an historical entity, one has no way of knowing which nodes will turn out to be central and which peripheral. Whether a node is central or peripheral, however, it can still function as an exemplar. For example, if one wanted to individuate the Darwinians in 1859, Huxley would do as well as Hooker, Hooker as well as Darwin, and so on. Similarly, if one wanted to individuate Darwinism in 1859, Hooker's treatment of evolution in his *Flora of Australia* (1859a) would do as well as Darwin's treatment in the *Origin*.

One final characteristic of historical entities is that they can be recognized only in retrospect. If Eldredge and Gould (1972) are right, species are constantly throwing off peripheral isolates. Most go extinct, but every once in a while a new species becomes established. This species itself may eventually go extinct without leaving issue, or it may bud off numerous descendant species. There is no way to tell in advance. The evolutionary process is Markovian. Trajectories can be traced through a phase-space even though successive positions cannot be inferred for the distant future. Similar remarks apply to scientific communities and conceptual systems. Scientists are constantly forming alliances that last for a while and then dissipate. Only rarely do groups materialize that last long enough and are sufficiently cohesive to be noticed or worth noticing. For example, it is difficult to ignore the Darwinians, while the Quinarians that rose up around William Sharp Macleay's theory of classification have disappeared without a ripple (Ospovat 1981). At the time there is no way of predicting which of these newly emerging groups will succeed and which will fail. As Lakatos (1970) has pointed out, research programs can be appraised only retroactively, but more than this, they can be recognized only retroactively.

V. Scientific Communities and Conceptual Systems

The commonest and superficially the easiest way to individuate social groups, especially scientific communities, is in terms of shared beliefs and commitments. Unfortunately, when we turn to the Darwinians, such a procedure does not produce anything like our pre-analytic conception of the Darwinians. Huxley preferred saltative evolution and was not nearly as confident as Darwin and the early Wallace of the powers of natural selection. Gray insisted on a notion of evolution a good deal more progressive and theological than either Darwin or Huxley was able to stomach. Early in their careers, Darwin, Huxley, and Wallace insisted on a totally naturalistic view of the world. Only Darwin held fast to this commitment throughout his long life. Lyell dragged his feet on almost all counts and never could bring himself to extend evolutionary theory to include man. Initially, Wallace did and then later recanted.

Clearly, the Darwinians did not totally agree with each other, even over "essentials". Another alternative is to treat scientific communities the way some authors have suggested species be treated — as polytypic or polythetic groups.[6] All that matters is that the members of a group agree with each other on enough of the more important beliefs. As promising as this maneuver might appear, it has several drawbacks. In the area in which cluster analysis has been applied most extensively and with greatest success — biological classification — a point of diminishing returns is rapidly reached. Fairly discrete clusters can be distinguished by a variety of programs and clustering methods, but as these clusters become less discrete, no two algorithms produce the same clusters (Sneath and Sokal 1973). Current intractability does not automatically rule out a particular avenue of investigation. If it did, no progress would ever be made in science. But it does detract to some extent from its current attractiveness. From a pragmatic, operational point of view, cluster analysis is not fulfilling its early appearance of great promise.

This alternative, however, suffers from a more serious drawback with respect to scientific development. I cannot see how one can construct a cluster of beliefs that would include people like Lyell and Henslow in the early years of the group, when they were clearly members, and exclude Richard Owen (1804–1892), and St. George Jackson Mivart (1827–1900) who were clearly not members. Both Lyell and Henslow were important "Darwinians" on a host of counts. In this section I propose to delineate the Darwinians as a social group. By "social group" I mean a group of people interrelated by social ties. The social ties with which I am concerned are those that bind scientists together in cooperative ventures. Several Darwin scholars have addressed this same issue in the past — for example, Manier (1978) and his cultural circle, W. Cannon (1978) and his self-reviewing circle or network, Gruber (1980) and his evolving system, and Rudwick (1982b) and his topographical map. In each case, these authors have chosen to define their networks at least in part in terms of intellectual influence. Quite obviously, historians of science are interested in such groups as Darwin's cultural circle and the Cambridge network *because* of their intellectual influence. It is equally obvious that certain people are included in a particular network and others exlcuded at least in part *because* of intellectual considerations. For my purposes, however, holding intellectual influence to the side for the time being is extremely helpful, for the simple reason that intellectual and social considerations are so intimately connected. It is also instructive in that we as intellectual historians tend to place too much emphasis on intellectual influence. If nothing else, keeping intellectual and social considerations separate *initially* can neutralize to some extent this particular bias. Then, when the two sorts of considerations are combined, their relative importance can be assessed more consciously and explicitly.

In beginning my exposition by tracing the social relations that integrated the Darwinians into a scientific network, I do not mean to imply that I think these social relations are more fundamental than the intellectual factors that I discuss later. On the contrary, they are both important. Causal feedback loops constantly exist between the two, which is only to say that the two are part of a more inclusive sociocognitive historical entity. For example, Darwin is highly equivocal about the issue of the progressiveness of evolutionary development both in his private jottings and in his published works. His reticence in committing himself too unequivocally in print on this issue is certainly a function of his own uncertainty, but it is also a function of the firm stand that his mentor, Lyell, and his most adept disciple, Huxley, had taken on the subject. Ignoring either pole of the continuum between cognitive and social factors produces only a partial understanding of Darwin's position.

I begin my exposition with the Darwinians as a socially defined network for two reasons, one somewhat idiosyncratic, the other more central to my program. It would be self-defeating for me to ignore where my own prejudices lie and where the prejudices of most of my readers are likely to lie. By detailing the non-intellectual relations between the Darwinians first, I can to some extent neutralize these prejudices. More importantly, socially defined scientific groups are easier to delineate than conceptual systems. For example, MacLeod (1970) had no trouble discovering who the members of the X Club were. Certainly, few of the groups of scientists operative in conceptual development are as formal as the X Club,[7] but even these more informal groups of scientists are, as I have discovered in my own research, relatively easy to discern (Mullins 1968; Crane 1972; Gaston 1978; Andrews 1979). Because socially defined scientific groups are easier to discern than conceptual systems as both change through time, the former can be used to help delineate the latter. For example, once the Darwinians as a historical entity have been identified, it can be used to help individuate Darwinism as a historical entity.

Because Darwin's name appears in the term "Darwinians", it might seem only natural to use him as the type specimen in defining the Darwinian nexus. It is, but other Darwinians such as Hooker and Huxley would serve as well. If in the years preceding the publication of the *Origin*, a network of scientists gradually formed — a group that came to be known many years later as the Darwinians — then my choice of exemplar should make no difference, just so long as I pick someone who was in this network. But, one might ask, what if Darwin had not published the *Origin*? What if he had died and only his *1844 Essay* had been published posthumously? Or what if he had behaved as honorably as many critics would have had him behave and held up publishing anything on evolution until Wallace had had a chance to write and publish his own treatise, as he finally did in 1870? As I mentioned earlier, historical entities can be recognized only in retrospect. The Darwinians developed in the way they did and in no other. There is no way to predict what will happen to a historical entity, be it biological, social, or conceptual. Historical entities are not that sort of thing. They are individuals and not natural kinds (Hull 1983b).

If scientific communities and conceptual systems evolve in anything like the manner that species do, then they do not go through stages. Although early on, Darwin toyed with the idea of the life cycles of species (Kohn 1980; Ospovat 1981; Hodge 1982), by the time he published the *Origin* he had abandoned the notion. Periodically, since then, evolutionists have resurrected the idea but with little success.[8] Whether species are capable of changing indefinitely or only of making minor modifications to slight fluctuations in their environment, little evidence exists to support the view that species undergo programmed change the way that organisms do. As in the case of species, however, conceptual systems as historical entities have certain "definitional stages". One can neither develop nor disseminate views one has yet to formulate. As a mnemonic device, the "stages" in a scientific research program can be characterized as discovery, development, dissemination and dogma — as long as one keeps in mind that in science as a temporal process these stages need not occur in any prescribed order. All that matters is that sooner or later all bases are touched.

Numerous authors have pointed out the importance of Darwin's period of protracted relative isolation on the *Beagle* for the formulation of his views on species. The isolation was only relative. He was reasonably well supplied with past wisdom in the books aboard the *Beagle*. On his return to England, Darwin was no longer isolated even in this sense. Suddenly he was flooded with all the views then under dispute in the British scientific community. But in connection with his theory of evolution, Darwin maintained a one-way isolation for an unusually long time. He worked on his theory, ferreting out information that might help him, but not telling anyone else about it. He gladly received but declined to send.

Innovation is a highly psychological phenomenon. Broader sociological factors are important only to the extent that the scientist has internalized them. Who knows why Watson and Crick automatically started to build a model of DNA with its "backbone" on the inside — perhaps because of the connotations of this biochemical term, perhaps because of their own experiences as vertebrates and with vertebrates — but something led Watson to switch to an invertebrate model. The wow-feeling of discovery, of finally pulling together apparently disparate elements into a single pattern, or of putting in place the last piece of a puzzle is extremely important. Such issues have been pursued at great length by others (Gruber and Barrett 1974); I shall not treat them here. Instead I want to discuss the sociological factors that come into play in connection with development and dissemination.

Short periods of relative intellectual isolation are as helpful in the process of innovation as they are dangerous for development if they are protracted. The danger of developing one's views in isolation is that one is denied the benefits of criticism. To be sure, no new fundamental theory can withstand hostile criticism from the moment of its inception. The care that scientists exercise in sheltering their more radical views at least initially is well taken, but gentle, supportive criticism is extremely beneficial. I think that the fate of Lamarck's and Chambers's theories of evolution was due in part to the lack of supportive if critical input from other scientists during the developmental period. By the time Lamarck started work on his theory of evolution, he had so isolated himself by publishing a whole series of theories in areas in which he was far from a professional that no one was about to take another of his crackpot ideas seriously (Burkhardt 1972, 1977). Similarly, Chambers ends his *Vestiges* (1844, p. 387) with the pathetic observation that it was "composed in solitude, and almost without the cognizance of a single human being, for the sole purpose (or as nearly so as may be) of improving the knowledge of mankind and through that medium their happiness." But as published the *Vestiges* was clearly the work of an amateur. As Egerton has noted, the deficiencies in his theory and its evidential support might have been corrected, had "his general idea been sympathetically received" (1970, p. 177). It is one thing for professional scientists to help a relative amateur in the privacy of their personal communications; it is quite another to come to his rescue once he has declared himself in print.

Lamarck began his career as a legitimate scientist producing conventional work but gradually lost touch, both with the science of his day (he consistently held on to views that were being replaced) and with the accepted professional standards of expository style that were developing at the time. Chambers for his part had yet to grasp either when he foolishly published his *Vestiges*. Later editions were greatly improved, but the damage was done. The scientific community had already committed itself on the theory (Hodge 1971a, 1972).

Scientists still active in research are likely to begrudge the time spent in "socializing" at meetings, but Poulton was forced to admit that this is time well spent:

> Here, and in kindred communities, a "man sharpeneth the countenance of his friend," and there is born of the influence of mind upon mind thought which is not a mere resultant of diverse forces, but a new creation.
>
> The scientific man who shuts himself away from his fellow-men, in the belief that he is thereby obtaining conditions the most favourable for research, is grievously mistaken. Man, scientific man perhaps more inevitably than others, is a social animal, and the contrast between the lives of Darwin and Burchell [a scientific recluse] shows us that friendly sympathy with our brother naturalists is an essential element in successful and continued investigation. (1904, p. lxxxi)

The claim that science is a social activity is fast becoming a platitude.

Contrary to the fears of some philosophers, this assertion does not entail that in science "truth" is equivalent to "consensus", as if scientists determine which views are true by voting in a good democratic fashion, one scientist, one vote. The focus of this claim is quite elsewhere. It concerns objectivity rather than truth. Individual scientists are, to some extent, objective, but the source of the sort of objectivity that gives science the character it has is the scientific community and its organization. Built into the social structure of science are mechanisms to promote, though certainly not to guarantee, sufficient objectivity for science to fulfill its official goals (Hull 1978b).

VI. The Darwinians

Prior to his return from his voyage, Darwin was acquainted with only a very few people who can be termed "scientists". In Edinburgh, Darwin interacted to some extent with Robert Grant (1793-1874), the only person whom Huxley was able to recall as having a "word to say for Evolution — and his advocacy was not calculated to advance the cause" (T. H. Huxley 1887a (NY), p. 541; L. Huxley 1900, 1: 180). This is also the Grant who was known as the "British Cuvier" until Owen appropriated the title for himself. Owen also effectively blocked Grant's appointment as comparative anatomist to the Zoological Society of London. Owen's refutation of Lamarck in his 1841 address to the British Association for the Advancement of Science was actually an attack on Grant. Thirteen years later, Huxley (1854) was happy to return the compliment by attacking Owen under the guise of a review of the tenth edition of Chambers's *Vestiges* (1853). Grant also helped Darwin curate some of his collection upon his return from the *Beagle* voyage (Ashworth 1935; Jespersen 1948-1949). At the time, Darwin also met Leonard Horner (1785-1864), a professional draper and amateur geologist, who took Darwin to a meeting of the Royal Society of Edinburgh. Lyell married one of Horner's daughters.

Darwin's circle expanded rapidly when he transferred to Cambridge. Among the students at Cambridge with whom he became intimate were J. M. Herbert (1808-1882) and his second cousin W. D. Fox (1805-1880). But the man who had the greatest influence on Darwin was John Henslow (1796–1861). It was through Henslow that Darwin met such luminaries as Adam Sedgwick (1785–1873) and William Whewell (1796-1866) as well as such lesser lights as Leonard Jenyns (1880-1893), Henslow's brother-in-law. Later Henslow's eldest daughter would marry J. D. Hooker. It was also Henslow who recommended Darwin, instead of Jenyns, to accompany Captain FitzRoy on the *Beagle* after FitzRoy decided that Robert McCormick (1800-1890), the official surgeon-naturalist, was not a suitable companion. Darwin became the de facto naturalist when McCormick left the *Beagle* four months into the voyage at Rio de Janeiro. On the voyage, Darwin was limited very narrowly in his personal contacts primarily to FitzRoy himself, but Darwin was able to visit John Herschel (1792-1871) at the Cape of Good Hope in 1836.

The importance of these early relations can be seen in the people to whom Darwin wrote during his voyage about scientific matters, primarily Fox, Herbert, and Henslow. It was to Henslow that Darwin shipped his specimens and to whom he first went upon his return. It cannot be said that at this time or before, however, Darwin belonged to anything that might be construed as a scientific community or as a tightly-knit social group. This state of affairs was not to last for long. Darwin's first contacts with the scientific community at large were less than reassuring. As Darwin remarked to Henslow at the time:

> I am out of patience with the Zoologists, not because they are overworked, but for their mean, quarrelsome spirit. I went the other evening to the Zoological Society, where the speakers were snarling at each other in a manner anything but like that of gentlemen. (1967,p. 121)

Darwin was further dismayed when he discovered that Henslow was right in warning him that great men "are overwhelmed with their own business" (Darwin 1967, p. 118). The overworked zoologists were not as keen to curate Darwin's collection as he thought that they would be, but eventually Darwin was able to interest appropriate experts in most of his collection — Grant in corallines, Owen in fossil mammals, Jenyns in fish, George R. Waterhouse (1810-1888) in recent mammals, Thomas Bell (1792-1880) in reptiles, John Gould (1804-1881) and George Robert Gray (1808-1872) in birds. Darwin might have formed an alliance with any one of these men, but he did not. Instead it was to the geologist Lyell that Darwin turned. As a series of historians have amply documented, Darwin rapidly became part of the geological community in general and a Lyellian in particular (Rudwick 1972, 1982b; Bowler 1976a; Manier 1978; Ruse 1979a).

Because Lyell and his disciples wrote the early histories of geology, we tend to think that Lyell's uniformitarian research program was a good deal more successful in Great Britain than it actually was. Lyell was more than a little happy to gain a disciple who was as bright and ambitious as Darwin (Gruber and Barrett 1974, p. 90). In fact, master and disciple got along famously. As Lyell wrote to Sedgwick at the time, "It is rare even in one's own pursuits to meet with congenial souls, and Darwin is a glorious addition to my society of geologists" (Clark and Hughes 1890, 2: 484). In looking back on his early relationship with Lyell, Darwin recalled that one of Lyell's "chief characteristics was his sympathy with the work of others, and I was as much astonished as delighted at the interest which he showed when, on my return to England, I explained to him my views on coral reefs" (*Autobiography*, pp. 83-84). Lyell was enthusiastic about Darwin's early work in geology. Perhaps he was more enthusiastic about the work that seemed to support his own geological views than the work that did not, but he supported Darwin even when Darwin was forced to contradict his mentor's views (Stoddart 1976; Dean 1980). The two men stayed on good terms throughout their lives on both a personal and a professional level, even when Darwin gradually ceased being an active supporter of Lyell's research program and began to develop his own, when it became clear to Lyell that Darwin's views on species posed serious problems for his uniformitarian program in geology.

As Darwin became a professional scientist, both his alliances and allegiances shifted away from his early friends, who failed to continue their scientific pursuits, and toward his more recent colleagues, who were as serious about science as he himself was. It is sad to watch the frequency and character of Darwin's correspondence with such old friends as Herbert, Fox, and Henslow change through the years. Although Darwin remained on good terms with Henslow until Henslow's death in 1861, Darwin communicated with him less and less on scientific matters. In his *Autobiography* Darwin recalled that Henslow's "judgement was excellent, and his whole mind well balanced; but I do not suppose that any one would say that he possessed much original genius" (p. 64). Darwin informed Jenyns of his heretical views at least by 1845, and Fox by 1855, but Henslow had to wait until the appearance of the *Origin* to receive any official communication on the subject. The continuing influence of his friends such as Fox, Herbert and Henslow can be gauged by references to them in Darwin's *Natural Selection.*

Initially Darwin developed his theory of evolution in relative isolation, but at that very time he was also developing his geological theories in public. As Rudwick (1982b) has pointed out, Darwin learned how to be a scientist and gained a reputation as a scientist by working and publishing in geology. He was thereby able to transfer this ability to the development of a different subject and, when it came time for making his views public, he could trade on his reputation.

Eventually, however, Darwin began to let first one of his fellow scientists, then another into his confidence, sometimes just his belief that species change through time, sometimes his suggested mechanism of natural selection. During the years 1844-1845, Darwin confided in three of his fellow scientists — Hooker, Jenyns, and Lyell. Darwin's estimations of his contemporaries during this period can be seen by the men he listed as potential literary executors in case he died before finishing his "Big Book" on species — Lyell, Edward Forbes (1815-1854), Henslow, Hooker, and finally H. E. Strickland (1811-1853). After Strickland's name Darwin wrote but then erased "Professor Owen would be very good; but I presume he would not undertake such a work" (*LL*(NY) 1: 379).

After a ten-year hiatus, Darwin confided in two others — Huxley and Fox. Things changed drastically in 1856. On 16 April 1856 Darwin explained natural selection to Lyell at Lyell's request. On 29 April, at a meeting of the Philosophical Club, the species question was discussed by Lyell, Huxley, Hooker, John Stuart Mill (1806-1873), W. B. Carpenter (1813-1885), and George Busk (1807-1886). The very next day, Huxley, Hooker, and T.V. Wollaston (1822-1878) met with Darwin at Down to talk species. In a letter dated 18 June 1856, Darwin hinted at his views on species to Asa Gray. When Gray's response was not hostile, Darwin was even more open. Finally, on 5 September 1857, Darwin wrote Gray a letter than was eventually published along with the Darwin-Wallace papers in 1858. During this same period, Darwin told S. P. Woodward (1821-1865) and John Lubbock (1834-1913), Hooker told Hugh Falconer (1808-1865), Lyell told his brother-in-law, Charles Bunbury (1809-1886), and Darwin's heresy was hardly a secret anymore. Still Darwin did not broach the subject with Wallace when he responded in 1857 to Wallace's first letter.

Any suggestions as to the considerations that led Darwin to tell certain scientists about his heretical views on species quite early and not to tell others until the appearance of the *Origin* made the communication gratuitous are necessarily conjectural. Darwin makes at most passing remarks on the subject. As far as I can tell, he selected scientists who were likely to provide sympathetic criticisms — but criticisms nonetheless — and whose criticisms were worth having. I have already remarked on Darwin's delight in the sympathy Lyell showed in the works of others. Forbes seems to have been much the same sort of man and served for Huxley the same sort of function that Lyell served for Darwin. As Huxley wrote to his sister in 1851, Professor Forbes is "my great ally, a first-rate man, thoroughly in earnest and disinterested, and ready to give his time and influence — which is great — to help any man who is working for the cause", and concludes, "My notions are diametrically opposed to his in some matters, and he helps me to oppose him" (1900, 1: 103).

Darwin attributed these same qualities to Hooker. Darwin in his turn served much the same function for Hooker that Humboldt had served for Darwin. It was Humboldt's highly romanticized *Personal Narrative* (1814-1829) that fired the desire in Darwin, not to mention Wallace, to undertake a comparable voyage. In 1839 Lyell's father sent proof sheets of Darwin's *Journal of Researches* (1839) to the young Hooker as he prepared for his own voyage to the Antarctic on the *Erebus*. Hooker claims to have slept with these proof sheets under his pillow. Incidentally, who was the other naturalist on board the *Erebus*? None other than the ubiquitous McCormick. Hooker seems to have gotten on with the man — barely.

Hooker was the first scientist to whom Darwin divulged his heretical theory on the origin of species, giving him a copy of the *1844 Essay*. Why Hooker? One reason was that he had the Galapagos data that Darwin had failed to collect accurately and adequately on his own voyage (Sulloway 1982). When H. E. Strickland died in 1853 Darwin relied more and more heavily on Hooker for data on variation and species' distributions. The fact that Hooker was a botanist and not a zoologist had to have had an impact on Darwin's theory, for example, shifting Darwin's attention from animal varieties (geographic populations) to plant varieties (variants within populations). In 1858, however, after fourteen years of discussing the topic, Darwin was still unsure of Hooker's views on evolution. So much for how completely evolution was in the air at the time. The topic came up in connection with copies of the Darwin-Wallace papers, which Darwin was sending to Wallace. He prayed that Hooker would "not pronounce too strongly against Natural Selection, till you have read my abstract" (*LL* (NY) 1: 494). This "abstract" was, of course, the *Origin.* The next day Darwin felt guilty about doubting his friend's good will and wrote again begging his forgiveness. "I forgot for the moment that you are the one living soul from whom I have constantly received sympathy" (*LL* (NY) 1: 495). But more than just sympathy, Hooker's criticisms were useful. As Darwin noted, they "clear my mind wonderfully" (L. Huxley 1918, 1: 497).

In Darwin's correspondence, one can see Darwin feeling out first one scientist and then another to see if he might be of some use. Darwin was on good terms with both Sedgwick and Owen, and both men knew a great deal, but Darwin rapidly realized that neither man would be of any service. Sedgwick's bitter review (1845) of the *Vestiges of Creation* (1844) and his bloated *Discourse on the Studies of Cambridge* (1847) were sufficient evidence to that effect. Although Sedgwick became one of Darwin's sternest critics, the two men remained personally on good terms. Owen was quite another matter. Darwin was on as cordial terms with the man as anyone else, but Owen seems to have inspired trust in no one, at least not sufficient trust to have produced a single student or disciple. Huxley was an up-and-coming naturalist who was not above taking on the most powerful and able scientists in their own areas of expertise — chiefly Owen. He too wrote a review of Chambers's *Vestiges,* but not until 1854 after publication of the tenth edition. It was this review, a review that vied with Sedgwick's for savagery, that led Darwin to broach the subject of species with Huxley. As strange as this behavior might seem, on closer inspection of Huxley's review, some of the puzzlement disappears. The object of Huxley's review was really not the "Vestigiarian" but Owen who had written an anonymous review (1851) in the *Quarterly Review* of the eighth edition of Lyell's *Principles of Geology* (1850). And, in spite of Huxley's pugnacious reputation, he could provide careful, sympathetic criticisms of the work of others, as he did with both Darwin and Spencer.

Darwin mentions not a word to Henslow, not because he had any doubts about his consideration or sympathy but because, as Darwin reluctantly admitted, his criticisms were not likely to be of much use. Henslow would make an honourable literary executor but not a good sounding board for his research into species. Wollaston, to the contrary, was a rich source of data. He published a book on the variation of species in 1856. Later Wollaston (1860) published a review of the *Origin* that was so careful and fair that almost no one noticed it except Darwin. But one point I wish to emphasize — likely *agreement* appears *not* to have been a factor. As late as 25 January 1859, Darwin responded to a question raised by Wallace about Lyell's frame of mind on the species question: "I think he is somewhat staggered, but does not give in, and speaks of horror, often to me, of what a thing it would be, and what a job it would be for the next edition of 'The Principles,' if he were 'perverted'" (*LL* (NY) 1: 502). Lyell was unhappy enough at the prospect of having to recant his earlier views on species in public, but to rewrite long sections of his *Principles* was almost more than a body could bear.[9]

As another example, several of the men already discussed — Hooker, Lyell, Forbes, Wollaston, and Woodward — argued for the submergence of something akin to the lost continent of Atlantis in the Atlantic Ocean within the period of existing species. Darwin wrote letters of protest to Wollaston, Lyell, and Hooker. In a postscript to the letter to Hooker (18 June 1856), Darwin concludes that "I must try and cease being rabid and try to feel humble, and allow you all to make continents, as easily as a cook does pancakes" (*LL* (NY) 1:432). Wollaston responded, none too surprisingly, that "ultra-honesty" must be one of Darwin's characteristics.

Thus far I have followed Darwin as he moved from developing his theory in private to the semi-privacy of his small circle of intimates (Manier 1978; Rudwick 1982b). With the publication of the Darwin-Wallace papers in 1858 and the *Origin* in 1859, the center of gravity shifted from development to dissemination. Although Darwin was hardly finished with his theory, future development had to be carried on in public, and acceptance became an important issue. That Darwin paid at least some attention to such considerations can be seen in the list of "converts" that he made soon after the appearance of the *Origin*. He listed the names of fifteen men by their fields of specialization — Geologists, Zoologists and Palaeontologists, Physiologists, and Botanists (*LL* (NY) 2:87).[10]

Darwin was able to develop his theory with the aid of at least some sympathetic criticism, mainly from Hooker and Lyell. This emerging group of scientists was to play an even more significant role in the reception of Darwin's theory. It was Lyell and Hooker who communicated the Darwin-Wallace papers to the meeting of the Linnean Society on 1 July 1858. Hooker (1859a, 1859b, 1860a) and Huxley (1859, 1860b) wrote immediate supportive reviews of the *Origin*. Henslow was in the chair at the famous Oxford meetings of the BAAS in 1860 when Huxley and Hooker took on Bishop Wilberforce. Hooker's hand trembled enough as it was when he handed up his name to be recognized. It would have trembled even more had someone like Owen been in charge of the proceedings. When Chambers published the *Vestiges of the Natural History of Creation* (1844), not a single eminent scientist came to his defense. On the contrary, a long list of notables criticized it brutally, including Herschel, Lyell, Sedgwick, Buckland, Agassiz, Whewell, Huxley, Gray, Forbes, David Brewster (1781–1868), and Hugh Miller (1802–1859). Owen, who was later to present a theory of transmutation with a mechanism not unlike Chambers's, wisely kept his silence. He did not come out for the theory, as unlikely an eventuality as can be imagined, but at least he did not join the long list of those who publicly condemned the anonymous author of the *Vestiges* (see Brooke 1977a). The closest thing to scientific support that the *Vestiges* received was a letter from Baden Powell (1796–1860) (Chambers 1884, p. xxx). Although many eminent scientists openly criticized the *Origin*, several important voices were also raised in its defense. As Darwin remarked to Hooker in 1860, "One thing I see most plainly, that without Lyell's, your, Huxley's, and Carpenter's aid, my book would have been a mere flash in the pan"(*LL* (NY) 2:101). Almost a decade later, Darwin reiterated his conviction in a letter to Huxley, this time including Gray on his list (*ML* 1:157).

Although I have not come close in this short section to setting out the extent of the social relations that integrated the Darwinians as a social group, I do think that sufficient evidence has been presented to support two important conclusions about the Darwinians at this stage: the rapid emergence of such a group was extremely helpful to Darwin, and the members of this group did not agree with each other about evolution even on fundamentals. At the time, neither Lyell nor Henslow agreed with Darwin that species evolve, let alone that they do so by the mechanisms suggested by Darwin. Lyell for his part was as important a Darwinian as any other member of the group. His support in the absence of agreement gave Darwin courage. In 1859, Darwin wrote to Lyell, "I fully believe that I owe the comfort of the next few years of my life to your generous support, and that of a very few others. I do not think I am brave enough to have stood being odious without support" (*LL* (NY) 2:33). Henslow was certainly not a central member of the early Darwinians, but he too supported Darwin. For example, at a meeting of the Cambridge Philosophical Society in early May of 1860, Sedgwick attacked Darwin savagely. As Henslow reported to Hooker, "I got up, as Sedgwick had alluded to me, and stuck up for Darwin as well as I could, refusing to allow that he was guided by any but truthful motives" (Darwin 1967, p. 205). Darwin might be wrong about species, but no one in Henslow's presence was going to get away with impugning Darwin's character or motives.

Initially the Darwinians were a reasonably clear-cut social group. Darwin, Lyell, Huxley, and Hooker had extensive contact with each other. The contact with Henslow was much less direct and frequent. Much of it also lacked the narrowly scientific character of so much of the contact of the more central members. Gray's membership is characterized by just the opposite combination of factors. Although his presence in the United States precluded face-to-face contact, the amount of his correspondence with the other Darwinians and the influence he had on them and vice versa seem sufficient to include him among the original Darwinians as a late arrival. Rapidly, after 1859, the ranks of the Darwinians swelled to include scientists who had very little direct contact with the original Darwinians. They count as Darwinians, not because they met regularly together as did, say, the members of the X Club (MacLeod 1970), but because they took themselves to be working in the Darwinian research program, pledged allegiance to what they took to be "Darwinism", and contributed to "Darwinism". The Darwinians in this global sense verge on becoming as amorphous as the paradigm to which they pledged allegiance.

Before I turn to the topic of Darwinism as a historical entity, a few words need to be said about those contemporaries of Darwin who were *not* Darwinians. Most scientists at the time were not Darwinians, but only in the trivial sense that they had nothing to do with the events in question. They neither joined in with the Darwinians to foster their research goals, nor actively worked against them. The non-Darwinians of special interest are those who actively opposed the Darwinians — the anti-Darwinians. In most cases, these scientists in their opposition to Darwin were relatively independent of each other. They belonged to a variety of scientific communities, but with a couple of possible exceptions, they did not band together as a group to oppose the Darwinians. Chief among Darwin's opponents were first Owen and then later Mivart. All of the local Darwinians had extensive dealings with Owen, especially Huxley. Although Owen initially helped Huxley in his career, Huxley rapidly became convinced that Owen was his main enemy and set out to oppose this his most powerful opponent. Because of Owen's early opposition to the Darwinians and the animosity on both sides that resulted, one might suspect that this conflict was initiated by Darwin's theory and that Owen's reputation as being mean, petty, and self-engrossed resulted from his opponents' winning the day and being able to write the history of their dispute. Owen's reputation was well in place before 1859, however, and the animosity between Owen and the Darwinians was an outgrowth of a long-standing dispute with Huxley over "idealism" (Rudwick 1972; Ruse 1979a; Hull 1983a). Once Huxley was a Darwinian, it would have been very difficult for Owen to become a member of the group.

As W. Cannon (1976b) noted, the creationist view against which Darwin was arguing in the *Origin* was the one set out by Lyell in his *Principles of Geology* (1830–1833). Yet Lyell is never singled out by Darwin and named as an opponent. Nevertheless, Lyell was well aware of the focus of Darwin's argument. Even so, Lyell neither defended himself nor attacked Darwin in print. Owen is hardly mentioned in the *Origin*. Nor were any of the doctrines with which he identified directly opposed. Yet Owen immediately set about to oppose the Darwinians. He coached Wilberforce in preparation for his presentation at the Oxford (1860) meetings of the BAAS, he wrote one of the most damning reviews of the *Origin* and Hooker's *Flora of Australia* (Owen 1866-1868), and he kept up a steady stream of carping remarks about evolution and the Darwinians, even while he himself prepared to publish his own theory of evolution (Owen, 1866–1868). Owen's opposition to the Darwinians was not solely a function of his hatred of Huxley. It was also a function of the danger that a theory like Darwin's posed to his own philosophical research program of introducing "transcendental idealism" into Great Britain. On this score, he had good company — Whewell, Forbes, Louis Agassiz (1807–1873), and J. D. Dana (1813–1895) — but these men did not form a group. On this issue they worked in relative isolation. This conflict was so important and so fundamental that W. Cannon (1976b) has suggested that the real controversy at the time was between Whewell and Darwin over the nature of science (Rudwick 1972; Ruse 1979a; Hull 1983a).

Mivart presents an even more fascinating case than does Owen. Mivart could easily have become a Darwinian. His views about evolution differed in no important respect from several key Darwinians. Like Huxley, he thought evolution was more saltative than Darwin did. Like Gray, he thought it was directed. And like so many Darwinians, he did not think natural selection could do all that Darwin claimed of it. Yet he soon became one of the Darwinians' most effective critics (1871a, 1871b, 1872a, 1872b). According to Mivart's biographer in the *Dictionary of National Biography,* J. M. Rigg, it was Mivart's "assertion of the right of private judgment which led to an estrangement from both Darwin and Huxley" (1909, 22:1052), as if Mivart were ostracized for refusing to hue the party line. I think the extent and duration of the disagreements among the Darwinians themselves over precisely these same issues is sufficient to discount this explanation as presented. Like it or not, scientists do permit even their closest associates the right of private judgement. For example, although some of Darwin's early publications in geology supported Lyell, others did not. He was not thereby excommunicated. Similarly, Huxley was able to disagree with his mentor, Forbes, without incurring his wrath[9].

The *content* of disagreements surely matters in science, but the *style* also makes a difference. *That* one disagrees with one's colleagues is less important than *how* one disagrees. Private disagreement is always more acceptable than public disagreement. Lyell disagreed with Darwin for years over the species question, but Lyell never published a blistering attack on Darwin. Instead he confined his observations to his private notebooks (Lyell 1970). Similar observations hold for other Darwinians. As time went by, they raised their differences of opinion in print, but very respectfully and very gently. Huxley was one of the most skilled polemicists science has ever produced. He seemed to thrive on vitriol. But when he addressed the work of his scientific allies, he retracted his talons. He never ridiculed Darwin's belief in gradual evolution the way he ridiculed Owen's belief in a progressive order in the fossil record. Huxley frequently stepped over the boundary between proper and improper scientific behavior, but never in connection with one of his allies. Either Mivart was unaware of this boundary or else he lacked the talent for stepping over it selectively. In response to the entry on Mivart in the *Dictionary of National Biography*, Hooker acknowledged that the Darwinians did resent Mivart, not for his disagreeing with them, but rather for the way in which this disagreement was conducted: "True that they resented, and Mivart privately apologized for, the personalities of his *Quarterly* article; the breach took place three years later owing to a repetition of the offense in a peculiarly hurtful form" (L. Huxley 1918, 2:128). So instrumental, it seems, in this breach was Mivart's commenting on the immoral implications of statements made on eugenics by one of Darwin's sons, George Darwin (J. Gruber 1960).

Before leaving the topic of the amount and sort of agreement that exists among scientists working together in research groups, I must make one final observation that is extremely important but is of such a character that evidence for it is very difficult to obtain. Although the members of a group like the early Darwinians need not agree even over fundamentals, they must firmly believe that such a consensus exists. My evidence for this claim rests, not with my study of the Darwinians but with my research into present-day scientific groups. Over and over again, I have discovered that scientists believe that there is an essence to their research program and that everyone working in it agrees over these essentials. Any attempt to make explicit the fundamental disagreements that actually divide them is met with extreme hostility. A belief in consensus seems to be necessary even when it is illusory (Hull 1984b).

As far as I have been able to discern, the only social group of scientists of the X Club sort that formed to oppose Darwin had as its sociological exemplar Lord Kelvin. The number of Darwin's critics that had close ties to Lord Kelvin is so high that it is difficult not to suspect some sort of conscious intent on their part. Among these Kelvinians were Bowen (1860), Haughton (1860), Hopkins (1860), Jenkin (1867), and Tait (1869). Proving that a Kelvin conspiracy actually existed requires the same sort of labor that has been lavished on discerning the make-up of the Darwinians. Brock and MacLeod (1976) have charted the opposition to the wave of naturalism that both Darwin's *Origin* and *Essays and Reviews* epitomized as it was reflected in a famous "Declaration" of scientists that no conflict could possibly exist between true science and true religion and its institutionalization in the Victoria Institute.

I mentioned above that scientific communities of the sort that are instrumental in the development and dissemination of new ideas are relatively easy to individuate. One reason for this state of affairs is that science, like the fine arts, is extremely élitist. No matter how science is investigated, the same conclusion is forced on the investigator — scientists act as if certain scientists are much more important than others. They are the ones whose views are cited in approbation, whose views are worth attacking, and whose good will is worth having. As Laudan observes, scientific revolutions "*can be*, and often have been, *achieved by a relatively small proportion of scientists in any particular field*" (1977, p. 137), and one of his examples is the Darwinian revolution. Late in life, Darwin wrote to William Graham (1839–1911) complaining of the "enormous importance" which he attributed to "our greatest men" (*LL* (NY) 1: 285) in his *Creed of Science* (1881). Darwin, on the contrary, thought "second, third, and fourth rate men of very high importance, at least in the case of Science."

Perhaps Darwin is thinking of the extensive use he made of the data gathered by all sorts of amateur scientists and non-scientists alike while he was working up his several theories, but at a crucial time in the history of his theory of evolution, when the issue of dissemination was forced upon him, Darwin thought differently. In a letter to Hooker (23 October 1859), Darwin remembers thinking about a "year ago, that if ever I lived to see Lyell, yourself, and Huxley come round, partly by my book, and partly by their own reflections, I should feel that the subject is safe, and all the world might rail, but that ultimately the theory of Natural Selection (though no doubt, imperfect in its present condition, and embracing many errors) would prevail" (*LL* (NY) 1:529). A month later, he reiterated his conviction in a letter to Huxley, observing that "fifteen months ago, when I put pen to paper for this volume, I had awful misgivings; and thought perhaps I had deluded myself, like so many have done, and I then fixed in my mind three judges, on whose decision, I determined to abide. The judges were Lyell, Hooker, and yourself" (*LL* (NY) 2:28).

Great Men social and political histories are nowadays definitely out of fashion. But in the sciences (not to mention the fine arts) the disproportionate impact of a very few figures is difficult to gainsay. Even those commentators who argue most strongly for the importance of socioeconomic factors in the development of science, after a few words of apology, concentrate almost exclusively on a very few workers. In this section on the Darwinians, I have concentrated too exclusively on Darwin. Although Darwin was an extremely important Darwinian, he was not as important as the emphasis in this paper would rightly lead a reader to infer. In order to get a more balanced view, the story needs to be retold using first Hooker as the point of entry into the Darwinians, then Huxley, then Lyell, etc. By the time the structure of the Darwinians as a social group has been delineated, with three or four different figures as the focus for the relevant social relations, the outlines of the Darwinians as a historical entity are clear enough. Three or four Darwinians will do, however. One need not sample hundreds of Victorian scientists at random. As much as one might abhor the élitist nature of science in Victorian Great Britain (not to mention today) from a moral or a political perspective, it certainly makes the delineation of scientific groups like the Darwinians a good deal more tractable.

In conclusion, just as the organisms that belong together in the same species are more similar to each other on average than they are to organisms belonging to different species, the scientists who belong to the same scientific community agree more with each other on average than they do with scientists who belong to other research groups, especially competing groups. But in both cases the correlations are not primary. In species, cohesion and continuity in development are what count. The same can be said for scientific communities. I agree with Ruse that factors other than acceptance of ideas are also operative in sustaining research groups, factors such as identification. For Ruse, a "Darwinian" is "someone who identifies with Darwin, but not necessarily someone who accepted all of Darwin's ideas" (1979a, p. 203). In this sense the Darwinians persist down to the present. Darwin indicated a profound understanding of the nature of science when he commented to Hooker in 1859 that he was fully convinced that the "future progress (which is the really important point) of the subject will have depended on really good and well-known workers, like yourself, Lyell, and Huxley, having taken up the subject, than on my own work" (*LL* (NY) 2: 47). Darwin was a major contributor to "Darwinism" as a research program, possibly the most important contributor; but he was not the only scientist to leave his mark on Darwinism.

VII. Darwinism

The day in which a chronological list of similar ideas can count as intellectual history is past, and Darwin scholars have been extremely influential in bringing about this change. Possibly because so much attention has been paid to Darwin's precursors, Darwin scholars have been firm in insisting that reference to unappreciated precursors has no place in the history of Darwinism. Actual influence is necessary in the filiation of ideas. As Ruse has argued (1979a), the real "precursors" of Darwin were not authors who may or may not have held ideas similar to those enunciated by Darwin, but those workers who actually influenced him even if they themselves believed that species were immutable. In fact, in the development of Darwinism as a historical entity, the work of early evolutionists was much less influential than the writings of such immutabilists as Lyell, Herschel, and Whewell.

In this section my goal is to set out the general characteristics of conceptual historical entities with particular reference to Darwinism. The historical research necessary to trace conceptual historical entities has been carried out for only a very few scientists. Darwin is among them. Thanks to Darwin's extensive correspondence and habit of keeping notebooks, Darwin scholars have been able to chronicle Darwin's intellectual development year by year, month by month, almost day by day. In several papers in this volume, the process continues. In this section I cannot hope to treat Darwinism in its entirety. Instead I intend to follow two strands — Lamarckism and gradualism — to illustrate the considerations that are relevant in the individuation of conceptual historical entities.

In the preceding section I emphasized the social relationships that integrated the Darwinians into a socially cohesive group. In this section I concentrate on conceptual cohesion and continuity — the ideas that were included under the term "Darwinism" at any one time. The activity is largely descriptive. In no instance can I say what *must* happen or *could not* happen. In point of fact, the Lamarckian element in Darwinism was gradually eliminated until anyone who thought that Lamarckian inheritance played a role in evolution was considered a non-Darwinian. Could anyone have predicted this course of events? No, historical entities are not that sort of thing. Trajectories resulting from Markovian processes can be plotted, their next position in phase space can be inferred with reasonable certainty, but their overall path cannot be predicted. Intellectual justice is also not relevant. Even though Lamarck did not claim that an organism's "wishful thinking" could produce heritable change in its physical make-up, that was how later workers interpreted him. What did Lamarck *really* say? As H. G. Cannon discovered (1957, 1959), no one but historians care.[11]

Manier (1978), in his investigation of Darwin's cultural circle between 1837 and 1844, has undertaken the sort of study necessary to trace actual conceptual influence. He takes as a rough indicator of influence explicit reference in one of Darwin's early manuscripts, notebooks, or autobiography, or Darwin's having owned a work by a particular author at the time. Men whom we would class today as scientists appear prominently on Manier's list — Lyell, Owen, Gould, and Lamarck; but so do several authors whom we would consider more philosophers than scientists — James Mackintosh (1765–1832), Whewell, John MacCulloch (1773–1835), and David Hume (1711–1776). As Manier is concerned to argue, philosophical ideas were as influential in the development of Darwinism as were more narrowly scientific ideas.[12]

As Manier emphasizes, however, counting references is not enough. Herschel, for example, was much more important in Darwin's intellectual development than the four references in Darwin's notebooks and early manuscripts would imply. Manier extends this line of reasoning to include James Ferrier (1808–1864), even though Darwin never refers to him by name and mentions only a single paper by Ferrier, which he may or may not have read (1978, p. 57). Manier's justification for including Ferrier in Darwin's cultural circle is that "it is apparent that Darwin did *not* understand the central thesis of his articles on the philosophy of consciousness when they appeared in *Blackwood's Magazine*" (p. 57).

I agree with Manier that, if Darwin read Ferrier's papers, he would not have understood his "idealist" conception of mind. Further, if Darwin had understood it, he would have rejected it (Hull 1983c). I also agree with Manier (1978, p. 69) that influence can appear in all possible degrees from positive to negative — the incorporation of a view as is, incorporation of a view but in a changed form, or outright rejection. Opposing views are as important in the shaping of a conceptual historical entity as those that are part of it. But if Ferrier's views are considered relevant in the individuation of Darwin's particular version of Darwinism, it is difficult to see how any views can be excluded. Strangely, Manier also includes a discussion of William Kirby (1759–1850), although he fulfills *none* of his stated criteria (Manier 1978, p. 69). Darwin may have read some of Kirby's works or may have been influenced by them indirectly through others, but until some evidence of such connections is discovered, Kirby is not entitled to be included in Darwin's cultural circle. As it turns out, Darwin refers to an introductory entomology text by Kirby and William Spence in his *Natural Selection*.

Lamarck figures in discussions of Darwinism, but usually for all the wrong reasons — does Lamarck deserve to be termed the Father of Evolution? The relevant issue is the extent to which Lamarck's views, whether interpreted correctly or not, influenced later evolutionists, including Darwin (Corsi, 1978). Darwin had clearly heard of Lamarck before Grant mentioned him, but the extent to which such common knowledge influenced Darwin only a psychoanalyst could estimate. Darwin owned an 1830 printing of the first volume of Lamarck's *Philosophie Zoologique* (1809), which he heavily annotated, and he mentions the second volume as well. As Kohn argues, Darwin's opinion of Lamarck was at its highest during the early part of 1838 prior to his stumbling on natural selection (1980, p. 131). Just as Lyell provided substance and proof for Hutton's vague intimations in geology, Darwin would do the same for Lamarck in natural history. Darwin's opinion of Lamarck was to sour, however. Instead of viewing Lamarck as a fellow evolutionist, an early victim of the "blindness of preconceived opinion" (*Origin*, p. 483), Darwin came to consider Lamarck, not to mention Chambers, as negative influences in the reception of his ideas. Darwin attempted to distance himself from Lamarck and Chambers, not for fear of compromising his own claims to originality but for fear that the reputations of these theories as being unscientific would tar his theory as well. Darwin's fears were certainly realized on the appearance of the *Origin*. A common ploy of those opposed to Darwin's theory was to class it with the empty speculations of Lamarck and the Vestigiarian. Lyell did not help by constantly referring to Darwin's theory as the theory of Lamarck as modified by Darwin. Darwin was irate:

> Lastly, you refer repeatedly to my view as a modification of Lamarck's doctrine of development and progression. If this is your deliberate opinion

> there is nothing to be said, but it does not seem so to me. Plato, Buffon, my grandfather before Lamarck, and others, propounded the *obvious* views that if species were not created separately they must have descended from other species, and I can see nothing else in common between the 'Origin' and Lamarck. I believe this way of putting the case is very injurious to its acceptance, as it implies necessary progression, and closely connects Wallace's and my views with what I consider, after two deliberate readings, as a wretched book, and one from which (I well remember my surprise) I gained nothing. But I know you rank it higher, which is curious, as it did not in the least shake your belief. But enough, and more than enough. Please remember you have brought it all down on yourself!! (*LL* (NY) 2: 198–199)

Darwin found himself in opposition to Lyell in defining the contours of Darwinism. If Lyell had his way, Darwinism would be viewed as a natural outgrowth of Lamarckism because, for him, it was. Lamarck played an important role in Lyell's conceptual development. Lamarck was as much a "uniformitarian" as Lyell — more so, in fact (Hodge 1971a). He was willing to extend Lyell's "presentism" to species as well. For Lamarck, spontaneous generation was occurring *now*, organisms were gradually being impelled up the great escalator of being *now* by the same processes that *always* produced these effects. For Lamarck, species are as eternal and immutable as for Lyell. Whenever the right circumstances occur, organisms of the appropriate species would develop to fill the appropriate niche (Hull 1984a). The difference between the two is that Lamarck suggested a naturalistic mechanism for this process while Lyell did not, and Lamarck's mechanism was progressive in opposition to Lyell's steady-state predilections (Hodge 1971a; R. Burkhardt 1977).

Although direct, positive impact of Lamarck on Darwin is doubtful, his indirect influence is clear. Darwin first confronted a detailed explication of the species problem in the context of Lyell's refutation of Lamarck in his *Principles of Geology*. As if this were not enough to guarantee the importance of Lamarck for the development of Darwinism, others were influenced both directly and indirectly by Lamarck. People like Grant had been converted to evolutionism by reading Lamarck; others like Spencer and Chambers were converted by reading Lyell's refutation of Lamarck; still others like Wallace and Powell were led to entertain the possibility of evolution by reading Chambers, James Croll (1821–1913) by reading Spencer, and so on. As Egerton has remarked, "It seems very likely that Darwin learned more from Lyell's attack on Lamarck than he learned from Lamarck, and that he learned more from Sedgwick's attack on Chambers than he learned from Chambers" (1970, p. 179). It should also be kept in mind that Darwin was not the only person who contributed to Darwinism.

Once Darwin and Wallace went public, opponents of Darwinism played an increasingly important role in shaping it. The critics of a new view do not necessarily attack those parts of a conceptual system that the proponents think are most important. Instead they concentrate their attacks on the parts that *they* think are both most important and most vulnerable. If advocates of a new view can be made to defend a minor part of their system at some length, it automatically becomes raised in importance. In this way, a minor theme can be converted to a central tenet. In the face of opposition, advocates of a new conceptual system can harden their position, making it even more extreme, or co-opt the objection by incorporating it into their own position. Both of these reactions can be seen in the development of the Lamarckian thread in the Darwinian tapestry.

In early versions of Darwin's theorizing on evolution, "Lamarckian" mechanisms were more prominent than in the version that appeared in the *Origin*. By the time that Darwin went public, however, use and disuse, the inheritance of acquired characteristics, and the direct effect of the environment had become of only minor importance when compared with natural selection. By the time he wrote the *Descent* Darwin was willing to grant "Lamarckian" mechanisms somewhat greater effect. Darwin's critics jumped on such admissions as a retreat. For example, Mivart's usual mode of attack was to attribute a categorical belief to Darwin and then present counterexamples, as if Darwin believed that *all* characteristics of organisms arise *solely* through the action of natural selection or that *all* differences between the sexes arise *solely* through the action of sexual selection (1871a, 1871b, 1872a, 1872b).

In the face of such objections, Darwin remain pluralistic. Weismann (1904) did not. In the latter decade of the nineteenth century and the first decade of the twentieth, Weismann and the biometricians succeeded in getting themselves recognized as Darwin's true descendants. Thanks to Weismann, "Darwinian inheritance" has come to mean "non-Lamarckian inheritance" even though in later life Weismann himself softened on the topic. Neo-Lamarckism might have emerged as a theme within Darwinism, instead of being in direct opposition. As things turned out, the neo-Lamarckian attacks hardened the Darwinians on this score. In the middle of this century, H. Graham Cannon (1958) once again raised the Lamarckian standard against the Darwinian orthodoxy with the usual effect.

The story continues to the present. E. J. Steele (1981) presents his theory of adaptive evolution through somatic selection as being opposed to the Darwinian orthodoxy and adopts Lamarck as his martyred patron saint. Needless to say, Steele takes considerable pains to rectify past injustices to the reputation of the man who "originated some 170 years ago" the "idea that environmentally induced characteristics are inherited" (1981, p. 7). Historical accuracy to one side, it is highly unlikely that Steele's reading of the *Philosophie Zoologique* had much to do with the generation of his ideas on the inheritance of immunological systems. That a particular scientist identifies with an earlier scientist is liable to tell us more about the intellectual climate of the time than about any actual intellectual connections. Giving the screw yet another turn, Dawkins (1982) argues that Steele's neo-Lamarckism is well within the Darwinian paradigm. Although changes in somatic cells are being transmitted to the germ cells, the original changes are in the genetic material of the somatic cells.

As much as Darwinism has changed through the years, it forms a continuously developing historical entity. In a review of Richard Burkhardt's *The Spirit of System* (1977), Camille Limoges argues that Lamarckism does not form such an entity:

> As Burkhardt points out, the idea of the inheritance of acquired characteristics had a long history before Lamarck's time and moreover never was central to his thought. As for neo-Lamarckism (in contrast to neo-Darwinism in relation to Darwin), be it American or French, it bore no continuity in tradition with Lamarck. Indeed, there is a considerable body of evidence to show that the so-called neo-Lamarckian conceptions initially emerged in reaction against Darwin's theory and some of its 19th-century reshapings and not as the product of a research program under development from Lamarck's time on. (1978, p. 1427)

A comparable story can be told for the gradualistic component in Darwinism. It had its immediate origins in Lyell's uniformitarianism. Because Lyell (1830–1833) portrayed his catastrophist opponents in geology as advocating supernatural miracles, Darwin tended to identify "gradual" with "natural". Darwin was well aware that "sports" periodically appear in nature, but he discounted them as a likely source of evolutionary change, both because they were so rare and because the difficulties involved with their mating with their untransformed congeners were even greater than the difficulties associated with slightly modified forms mating with their untransformed congeners. As is well known, Huxley thought Darwin loaded himself down unnecessarily by his opposition to evolution by *saltus*. Even so, Huxley did not openly attack Darwin on the issue. It continued, however, to be a bone of contention among the Darwinians and emerged as a major objection to Darwinism in the hands of such critics as Jenkin (1867) and Mivart (1871a, 1871b, 1872a, 1872b). Under the continuing pressure of its critics, Darwin's gradual evolution by means of small imperceptible steps evolved into the continuous variation of the biometricians.[13]

By the turn of the century, the general consensus was that the variations operative in Darwinian evolution were both non-Lamarckian and continuous — Darwin's own views on the subject notwithstanding. As a result, early Mendelians found themselves in opposition to the Darwinians. From the present-day perspective, the dozen or so years spent in haggling over the apparent conflict between Mendelian genetics and evolutionary theory is an inexplicable embarrassment. The chief source for this period of opposition is to be found in the person of William Bateson (1861–1926). Prior to the "rediscovery" of Mendel's laws, Bateson (1894) was arguing for discontinuous variation as the source of evolutionary change, not the continuous variation of the biometricians. The constantly differentiating characters of Mendelian genetics were just the sort of discontinuous variation that Bateson needed. That Yule (1906) showed that the two theories are perfectly compatible, and that one of the "rediscoverers" of Mendel's laws, Hugo De Vries (1901–1903; 1904), claimed that his theory of intracellular pangenesis and mutation theory were lineal descendants of Darwin's, did no good. These conceptual systems developed initially in opposition to each other. As a result, they had to merge in the first stage of the evolutionary synthesis (Provine 1971; Mayr and Provine 1980).

At the risk of belaboring the obvious, I must repeat that in referring to "merging", "opposition", and the like, I do not intend to imply strictly logical notions. Two research programs can be in opposition even though they are conceptually all but indistinguishable from each other. Conversely, both sides of a contradiction can be included in the same conceptual system. If one takes deductive logic as a model for good reasoning, most of the inferences that scientists make are fallacious. For example, every time a scientist reasons from the successful results of an experiment to the truth of the law being tested, he is committing the fallacy of affirming the consequent. But scientists also frequently reason in ways that currently cannot be accommodated in any system of informal logic either. The fault lies, at least in some instances, with current formalizations of informal logic, not the scientists, and as I have remarked elsewhere today's ad hoc hypothesis may well be tomorrow's law of nature (Hull 1978b, p. 146).

Certainly fallacious reasoning should not be lauded, but an inference that may appear superficially to be fallacious might play an important and positive role in science. For our purposes, the important feature of scientific conceptual systems as they develop is that little of the internal cohesiveness results from deductive inferences or even inductive inferences of the sort currently treated successfully by logicians. For example, if the elements in a research program must be integrated deductively, then the evolutionary synthesis is a myth. Very few of the elements of the synthetic theory of evolution are connected deductively. At a minimum the founders of the synthetic theory showed that the conclusions of various relevant disciplines such as paleontology, genetics, and natural history were *not contradictory*. As Stebbins and Ayala have remarked recently, "The theory of population genetics is compatible with both punctuationalism and gradualism. Logically, therefore, it does not entail either" (1981, p. 970). Not all non-contradictory propositions belong in the same research program; conversely, sometimes contradictory propositions can be found coexisting innocuously in the same research program. In short, simple inference is not sufficient for analyzing the sort of relations that make conceptual systems cohesive.

In my earlier discussion, I placed the term "rediscoverers" in quotation marks because I find it not very appropriate for what happened at the turn of the century in connection with Mendel's laws. Neither De Vries (1900), Correns (1900), nor Tschermak (1900) rediscovered Mendel's laws for a variety of reasons, not the least of which is the disparity between what each man took to be "Mendel's laws" and anything that Mendel himself may have thought or said (Brannigan 1979; Olby 1979). People were not interested in what Mendel really said, but in the use they could make of him. De Vries may or may not have come across Mendel's paper in time for it to have affected his own formulations (Darden 1977; Kottler 1979; M. Campbell 1980); but whether he did or not, De Vries had his own research program, and it was not coincident with that of Bateson and the Mendelians. For De Vries, Mendel's laws were only a minor element in his Darwinian mutation theory.

The various elements in the synthetic theory had partially independent origins. They became the "same" in the relevant sense only later when they merged into a single conceptual system. As Lakatos emphasized, "Some alleged simultaneous discoveries or novel programmes are seen as having been simultaneous only with false hindsight: in fact they are *different* discoveries, merged only later into a single one" (1970, p. 103). Those of us who comment on science have spent so much time investing past episodes with our own moral convictions that the real significance of such contingencies as Darwin and Wallace arriving at the "same" theory has too often been missed. One can follow the development of Darwin's views on evolution. One can also follow with equal interest and justification Wallace's conceptual development. It had many of the same origins as Darwin's views. He too read Humboldt's *Personal Narrative* (1814–1829), Lyell's *Principles of Geology* (1830–1833), and Malthus's *Essay on Population* (1826) with much the same effect. But Wallace was also influenced by reading Darwin's *Journal of Researches* (1839), and his reading of Chambers's *Vestiges* (1844) came at a very different time in his conceptual development than it did for Darwin. From the point of view of conceptual development, Darwin and Wallace produced the "same" theory, not because the two had similar content but because they had some of the same roots and more importantly because the two theories merged into one beginning in 1858. Once again, by "merging", I do not mean to imply that Darwin and Wallace were ever in total agreement or whatever agreement existed between them remained unchanged through the years. Just the opposite is the case. As Kottler has detailed at some length, Darwin and Wallace disagreed on numerous counts, including the "role of natural selection in the origin of (1) man, (2) cross- and hybrid-sterility, and (3) sexual dimorphism" (Kottler 1980, p. 203, and this volume; see also McKinney 1972).

In the midst of the modern synthesis, Richard Goldschmidt objected both to the Mendelian theory of the gene and to gradualistic versions of evolutionary theory (1933, 1938, 1940). According to Goldschmidt, each genome is a well-organized system that cannot be modified piecemeal. Instead, a minor change produces a drastic reorganization of the genome resulting in a genuinely new sort of organism, a sort of "hopeful monster". Both Simpson (1944, 1953) and Mayr (1942) rejected Goldschmidt's views in no uncertain terms, ridiculing his idea of "hopeless monsters". As Gould has argued recently, Goldschmidt's views were not quite as outlandish as they were made to appear, nor was the received view all that gradualistic (1977b, 1980a). Simpson (1944) argued for quantum evolution, while Mayr (1954a) argued for the efficacy of small founder populations. More recently, Eldredge and Gould (1972) have argued for at least a microsaltational view reminiscent in many ways of Mayr's views. In contrast to Goldschmidt, however, Simpson and Mayr's views are populational, not embryological, while Eldredge and Gould's model is both populational and embryological.

Given the "protean" nature of conceptual historical entities, the problem of identifying and individuating them is of more than minor interest. Previously I have suggested that a method akin to the type specimen method in biological systematics might be of some help for both social and conceptual systems. For biological species, the type specimen is one node in the reproductive nexus, while in social systems it is one node in a nexus of social relations. One way of individuating conceptual systems is to pick one element at one moment in its development as a type specimen or exemplar for that system. For Darwinism, one might pick Darwin's metaphor of a phylogenetic "tree". Darwin's earliest discussion of his tree metaphor occurs in 1837 in the midst of his monad theory (Gruber and Barrett 1974, pp. 141–144):

> Organized beings represent a tree, *irregularly branched*; some branches far more branched. — hence genera. — As many terminal buds dying, as new ones generated. There is nothing stranger in death of species, than individuals. (*B* 21–22)

Immediately Darwin realizes how inappropriate the tree metaphor is. The base of a tree is much broader than any of its subsequent branches and twigs. In evolution all branches are the same size: they are all equally species. In a tree, the trunk, branches, and twigs are all cotemporal. In evolution only the terminal twigs still exist. The contours of the phylogenetic tree are represented now only by the remnants of ancestors long dead. Thus, Darwin quickly adds:

> The tree of life should perhaps be called the coral of life, base of branches dead, so that passages cannot be seen. (*B* 25)

Unfortunately, Darwin immediately returns to the tree metaphor and provides a sketch that begs to be misinterpreted, as if the base of the tree represented a higher taxon such as a family, the next branches genera, and the terminal twigs species. Once again, in April 1868, Darwin represents a section of the phylogenetic tree as if it were a taxonomic tree (Gruber and Barrett 1974, pp. 143, 197). But the diagram that Darwin published in the *Origin* (p. 117) and his discussions of this diagram (pp. 331, 412, 420) are unequivocal. As accurate as the scheme is, it is not visually very appealing. Subsequent representations of phylogeny tend to resemble Haeckel's (1866) famous tree of life. In Haeckel's tree, the Monera trunk gives rise to Plantae, Protista, and Animalia. Each of these branches gives rise to taxa at lower taxonomic levels, and so on. As attractive as such trees are, they are a discordant mix of phylogenetic and taxonomic considerations. (For an excellent history of branching diagrams, from Theophrastus, Aristotle, and Porphyry to Simpson, Sokal and Sneath, and Hennig, see Nelson and Platnick 1981, pp. 63–168.)

I agree with Kohn that the "tree, which is simpler to draw than coral, became Darwin's standard emblem, not only of phylogeny, but of his entire theory" (1980, p. 97), as long as one realizes that this emblem was many things to many people. As far as cognitive implications are concerned, Darwin's tree was exemplary. Many of its descendants were much less so. Even if Darwin had published one of his more misleading diagrams, however it could serve as well as his "standard emblem". Exemplars need not be exemplary to serve the function of individuating the historical entities of which they are part. Both Lamarck (1809) and Chambers (1844) published "trees" of sorts. Although neither of these representations was adequate for Darwin's purposes (Hodge 1971a, 1972), they were no more misleading than many later representations stemming from Darwin. What matters is that later representations be lineal descendants (no matter how misconstrued) from Darwin and not from Lamarck or Chambers. One advantage of using Darwin as a focus for integrating the Darwinians into a social historical entity, and of using some element of Darwin's work — such as his tree metaphor — as the focus for integrating all the diverse elements of Darwinism into a conceptual historical entity, is that these two historical entities themselves are thereby interconnected. One has a point of entry for each of these historical entities and for moving from one to the other.

Conclusion

If the historiographic proposal that I have set out in this paper is adequate, it should be able to resolve the terminological chaos sketched in the first section of this paper. The first distinction that must be made is between the Darwinians as a social group and Darwinism as a conceptual system. A scientist can be a Darwinian without accepting all or even a large proportion of the elements of Darwinism. Conversely, a scientist can by and large accept the tenets of Darwinism without being a Darwinian. Furthermore, one must distinguish between the evolution of scientific theories, narrowly defined, and the evolution of the research program of which they are part. A single research program inevitably contains numerous versions of the same theory as well as different theories. Lewontin and Greene use the term "Darwinism" to refer to those more inclusive historical entities. The other authors mentioned restrict the term more narrowly to a family of theories.

Historical entities are essentially non-essentialistic. At any one time in a conceptual historical entity, no one tenet may be essential. As they are followed through time, the importance of various tenets changes. Early on, gradualism was central to the Darwinian research program, primarily because of Darwin's strong partiality to this view. As a result of attacks, first by the Mendelians and later by the anti-Mendelian Goldschmidt, the gradualistic position hardened even while Simpson and Mayr were presenting less than perfectly gradualistic models. Later, when Eldredge and Gould presented their punctuational model, they vacillated on whether it was supposed to be a modification of earlier Darwinian models such as Mayr's founder principle or a non-Darwinian alternative. The fate of the Eldredge and Gould model has yet to be determined. It seems likely at this stage that the model will be considered a form of Darwinism whether it is accepted in its present form, or in a highly modified form, or whether it is rejected. Although a few gestures have been made toward rehabilitating Goldschmidt, the current trend is to subdivide the Simpson-Mayr school into two different strands and trace more saltative views back to Mayr (Bush 1975; Gould 1980a, 1980b, 1982; Stanley 1979). Simpson has been selected as the arch gradualist while Mayr has been cast as the prophetic type, no matter that Mayr was one of Goldschmidt's sternest critics.

All the disputes in the recent literature about the "essence of Darwinism" are not misdirected antiquarianism but the latest efforts of scientists to establish the boundaries of their research program, both for the recent past and for the immediate future. Because different workers hold different views and have developed these views along historically different trajectories, they see the issues differently. The individuation of research programs is a creative process: scientists are doing it as they proceed. That scientists see their own development the way that they do, mistaken or not, influences that very development. The scientists engaged in the ongoing process are not entirely unbiased in their perceptions. Scientific evolution is hardly less opportunistic than biological evolution (C. D. Darlington 1959). The task of the intellectual historian is to discern as accurately as possible the actual constitution of conceptual historical entities.

Notes

1. The participants in the Darwinism symposium at the 1982 meetings of the AAAS in Washington, D.C. were William Provine, Ernst Mayr, Walter Fitch, Stephen Jay Gould, and G. Ledyard Stebbins.
2. Because I have argued so consistently against the inappropriate attribution of essences (Hull 1965), I have gained the reputation of opposing essences *tout court*. On the contrary, I think that genuine natural kinds might well have essences (Hull 1983b). The species category might well have an essence without individual species also having essences. The general position is that the notion of a historical entity might be a natural kind characterizable in terms of essential traits without particular historical entities having such traits.
3. It is only fair to note that immediately after saying that concepts are historical entities, Kierkegaard adds: "As philosophy cannot be indifferent to the subsequent history of this concept, so neither can it content itself with the history of its origin, though it be ever so complete and interesting a history as such. Philosophy always requires something more, requires the eternal, the true, in contrast to which even the fullest existence as such is but a happy moment" (1841, p. 47).
4. Among those alleged to be Darwin's precursors with respect to the evolution of species are Aristotle (384–322 B.C.), Theophrastus (370–285 B.C.), Francis Bacon (1561–1626), Benoit Demaillet (1657–1738), Charles Louis de Secondat Montesquieu (1689–1755), Carl Linnaeus (1707–1778), George Louis Leclerc, Comte de Buffon (1707–1788), Denis Diderot (1713–1802), Erasmus Darwin (1731–1802), Jean Baptiste Pierre Antoine Monet de Lamarck (1744–1829), Bernard-Germain-Etienne Lacépède (1756–1825), Thomas Robert Malthus (1766–1834), Baron L. von Buch (1774–1853), William Herbert (1778–1869), William Lawrence (1783–1867), Robert Grant (1793–1874), Baden Powell (1796–1860), Etienne Geoffroy Saint-Hilaire (1799–1853), Robert Chambers (1802–1871), Richard Owen (1804–1892), and Herbert Spencer (1820–1903). Those authors to whom natural selection is attributed include Charles William Wells (1757–1817), Patrick Matthew (1790–1864), Edward Blyth (1810–1873), and Charles Naudin (1815–1899).
5. According to two of the currently prominent schools of taxonomy, the pheneticists (Sneath and Sokal 1973) and the evolutionists (Simpson 1961a; Mayr 1969b), taxa are typically polythetic. According to a third school, the cladists (Nelson and Platnick 1981), the apparent polythetic character of taxa results from an improper identification of traits as evolutionary novelties. Manier discusses nineteenth-century anticipations of the present-day notion of family resemblances and their connection to essential attribution (1978, p. 38).
6. Both the strengths and the weaknesses of the type specimen method can be seen clearly in Sulloway's (1982a, b) discussion of the terrible mess Darwin made of the Galapagos finches and the subsequent efforts of later taxonomists to straighten it out.
7. The members of the X Club were Huxley, Hooker, Spencer, Lubbock, Busk, John Tyndall (1820–1893), Thomas Hirst (1830–1892), William Spottiswoode (1825–1883), and Edward Frankland (1825–1899). All members of this club had come to accept evolution at least by 1864 (Hull, Tessner and Diamond 1978).
8. As long as species are viewed as natural kinds, they are not the sorts of things that can have life cycles. If one construes species to be individuals (historical entities), however, then they *are* the sorts of things that can have life cycles. It is merely a contingent fact that they do not. The closest anyone now comes to treating species as having life cycles is to claim that they are programmed for no change whatsoever. That is a "life-cycle" of sorts.
9. Darwin's frustration over Lyell's reticence about evolutionary theory can be seen in the following quotations concerning Lyell's views on the immutability of species in successive editions of the *Origin* (Darwin 1959, p. 519):
   "But I have reason to believe that one great authority, Sir Charles Lyell, from further reflection entertains grave doubts on this subject" (1st ed. 1859, sent. 245).
   "But it is evident from the recent works of Sir Charles Lyell that he now almost gives up his view; and some other great geologists and palaeontologists are much shaken in their confidence" (4th ed. 1866, sent. 245:d).
   "But Sir Charles Lyell now gives the support of his high authority to the opposite side; and most other geologists and palaeontologists are much shaken in their former belief" (5th ed. 1869, sent. 245:e).
10. Under Geologists, Darwin lists Lyell, Andrew Ramsey (1814–1891), J. B. Jukes (1811–1869),

H. D. Rogers (1809–1866); under Zoologists and Palaeontologists, Huxley, Lubbock, Jenyns (to large extent), and Searles Wood (1798–1880); under Physiologists, Carpenter and H. Holland (1788–1873) (to large extent), and among Botanists, Hooker, H. C. Watson (1804–1881), Gray (to some extent), F. Boott (1792–1863) (to large extent), and G. H. K. Thwaites (1811–1882). For 3 March 1860, Darwin's list of converts is not especially impressive, especially since Darwin was a bit overly optimistic about several of them. Incidentally, I use the term "convert", with its religious overtones, because this is the term that Darwin used. By it I do not intend that those who came to accept various parts of Darwin's theory did so by means of some conversion experience. They might easily have been "convinced".

1. The story of how Lamarck's views on evolution were caricatured under the encouragement of Cuvier has been told often and well. Even those who knew better found it difficult not to attribute to Lamarck the view that "wishful thinking" played a role in Lamarck's theory. Although Lyell's (1830–1833) discussion is by and large fair, he cannot resist an occasional reference to the causal efficacy of the desires of organisms. Chambers feels obligated to dissociate himself from Lamarck's "hypothesis of organic progress which deservedly incurred much ridicule, although it contained a glimmer of the truth" (1844, p. 230). Although he was perfectly fluent in French, Owen was not above perpetuating the traditional parody of Lamarck (Owen 1866–1868, 2: 801). Even the saintly Darwin, after admitting to Hooker (11 January 1844) that he doubted the immutability of species, felt forced to add: "Heaven forfend me from Lamarck nonsense of a 'tendency to progression,' 'adaptations from the slow willing of animals,' etc.! (*LL* (NY) 1: 384).

12. I realize that in using terms like "philosopher" and "scientist" in this way, I run the risk of superimposing a present-day conceptual system on the past, but I have no choice but to speak to present-day readers in the only language at my disposal. Besides, declining pedantically to use the term "biology" until Lamarck introduced the term "biologie" or refusing to term someone a "scientist" until Whewell coined that term does not begin to resolve the problem.

13. For what it is worth, I have yet to find a single instance of Darwin's using "continuous" to modify "evolution" or "variation" the way that the biometricians did. Instead, he used this term in conjunction with references to land masses and geographic distributions.

27
Darwinism Today (Commentary)

Jacques Roger

If I borrow my title from Kellogg (1907), it is because it aptly defines, I think, the real meaning of our colloquium and perhaps the present state of affairs about evolutionary theory. After 1909 and 1959, this is the third Darwinian centenary year. On the situation in 1909 we have a valuable testimony, the book published by the Cambridge University Press "at the suggestion of the Cambridge Philosophical Society" with the title *Darwin and Modern Science*. In all the essays that composed the book, Darwin's genius and outstanding role in biology, natural history, and allied sciences were unanimously praised, but there was clearly no agreement over the actual scientific value of his ideas about the mechanism of evolution. What was clearly realized was the fact that Darwin had revolutionized the whole field of natural history, created new links between the sciences of nature and the sciences of man, and introduced a new era in the western thought. What was not at all clear was the true nature of mutations or the actual role of natural selection, that is, the most precise and scientific tenets of the Darwinian creed. On the contrary, in the many meetings and colloquia that took place in 1959, Darwin was unanimously declared to have been right in his explanation of evolution. With the new science of genetics explaining away most of the early difficulties of Darwinism, the then prevailing synthetic theory of evolution essentially was, or was considered to be, the direct development of Darwin's ideas. "Back to Darwin" was the word of the day or, as H. J. Muller put it: "One hundred years without Darwin are enough" (1959).

Things are more complicated today. The last ten years have witnessed the emergence of new theories — "non-Darwinist" evolution and "punctuated equilibria" — that more or less question some tenets of the synthetic theory, which itself evolved significantly in the last twenty years. It is no longer certain that "Darwin was right"; or at least we feel that it is more necessary than ever to go back to "what Darwin really said" and thought, to put the synthetic theory in historical perspective, and to examine the extent to which it gave a distorted idea of Darwin himself. Briefly speaking and paradoxically enough, we are in a more historical mood than in 1959. Our colloquium at least is definitely and purposedly historical.

But this raises a new problem: how to deal with Darwin and Darwinism *historically* when many biologists are still speaking of Darwinism as a living entity and define themselves as Darwinians, including those biologists who are introducing new theories. This question in turn necessarily raises the problem of the very nature of Darwinism, a problem that is central to Hull's paper and ought to be central to our reflection. It is also a problem that cannot be addressed and solved without having recourse to various historical methods. What makes Darwin and Darwinism such a difficult and inexhaustible subject-matter for historians and philosophers is not, or not only, the complexity of Darwin's human personality or the complexity of his theory of evolution, but rather the fact that Darwinism is a social and cultural phenomenon as well as a scientific one. We must also deal with the intellectual complexity, now better recognized, and the philosophical richness of a thought that contributed a great deal to the shaping of our own way of thinking. Thus it is not a matter of chance that the problem of Darwinism itself finally emerged in this last day of our colloquium, a day devoted to "Darwin and Victorian culture". Darwinism is much more than a scientific theory. This perhaps is one of the reasons why we still are speaking of Darwinism today. There are no such things as Maxwellism or Einsteinianism. Only historians speak of Copernicanism or Newtonianism. But there is a Darwinism, in the same way that there is Freudianism or Marxism. We are therefore obliged to take the historical phenomenon of Darwinism in its entirety, without neglecting either its socio-cultural or its intellectual dimensions. Hence the diversity of the papers in this volume, to the richness of which I am afraid I cannot do full justice in the following remarks.

Starting with James Moore's penetrating description of Darwin's ways of dealing with the affairs of the parish of Down, we apparently are very far from the great problems I have just alluded to. Actually, however, several difficult questions are already at stake with Darwin's daily behavior as a traditional country squire or even as a "squarson". Not only the question of Darwin's personal religious feelings, which still is a controversial one, but also that of his political and social attitudes. Moore seems to be surprised by the contradiction between Darwin's conservative social behavior in his parish and his allegedly irreligious ideas on man and nature, and to think that such an apparent contradiction requires a more precise analysis and a better explanation than those usually accepted. As he puts it: "One who served as a country magistrate, founded a Friendly Club and became treasurer of a parish Sunday school was an agent of the very mechanisms of subordination that were tended by country parsons in mid-Victorian England." Which is probably true. But, tempted as we are to identify a man with his thought, we must not forget that there is no necessary link between political and scientific ideas, at least at the level of clear consciousness. After all, Copernicus was a very traditional canon of the Roman Catholic Church and Buffon a typical French bourgeois of the Old Régime. And yet, their scientific thought was "revolutionary".

But Darwin's social behavior, and his choosing to live in a small village rather than in the noisy and feverish London, are not, or not only, a matter of personal taste and conservative temper. It seems strange that one of the greatest representatives of the intellectual life in Victorian England refused to live in the very center of that intellectual life. But was Darwin really a Victorian? To be sure, he was one of the luminaries of Victorian England, but he was born and grew to manhood in the Georgian era. He was older by ten to sixteen years than most of the great Victorians, George Eliot, Herbert Spencer, Alfred Russel Wallace, or Thomas Huxley. He had been educated as a country gentleman, in a world more similar to that of Jane Austen than to that of Charles Dickens. Had it not been for some peculiarities of his mind and some strokes of good luck in his life, he could have been, like Edward Ferrars in *Sense and Sensibility,* "entered at Oxford and have been properly idle ever since", waiting to be ordained and become the parson of a country parish somewhere in Shropshire. A project that does not look so strange if we consider how religious feelings and preoccupations are completely foreign to Jane Austen's world, even for the young men who, like the same Edward Ferrars or Edmund Bertram in *Mansfield Park,* intend to perform seriously their duties as country clergymen. It seems that religion had a purely social function in that world, which may account for the fact that Charles Darwin, as a layman and an agnostic one at that, eventually could play the part of a parson *honoris causa.*

As Edward Manier has pointed out (1978, p. 151), the British cultural environment no longer was in the 1860s what it had been during Darwin's youth and creative years. Nor was the social environment. Apparently Darwin could adapt himself less easily to the new society than to the new ways of thinking, which is not at all surprising. It would be interesting, however, to know more precisely how Darwin personally reacted to the social and political turmoils that shook England in the early Victorian era, and if he did not keep a somewhat nostalgic memory of a time when a strict social order was associated with a relative freedom of thought, at least for those rich enough to afford it. On the other hand, there were in the middle of the nineteenth century many bourgeois intellectuals who were free-thinkers in religious matters and conservative about social problems. Things become very complicated when individual psychology and historical movements interact, as they always do. Therefore, if we want to understand the relationship between Darwin's thought and its cultural environment, we must be careful not to overlook the chronology nor to mix up the different aspects of intellectual life.

The same thing could be said of Darwin's literary readings as studied by Gillian Beer. Since I am not an expert in English literature I do not intend to dwell on that part of her paper, interesting as it is, except to say that Darwin's extensive reading of literature may be considered evidence for the interest he had in *human* affairs. It might well be that the problem of man was central to Darwin's reflections, as it was for many other evolutionists before and after Darwin. Implicitly Gillian Beer reinforces the conclusions of Howard Gruber (Gruber and Barrett 1974) and the general idea that evolutionary theory primarily is an answer to the questions raised by the new idea of man that emerged in the eighteenth century, a man who no longer was created in the image of God and had to establish new relationships with nature in general and his fellow living beings in particular. But I prefer here to dwell a little longer on Beer's literary analysis of Darwin's writings, especially the *Origin*, because I found it extremely interesting and able to lead us to some general reflections on Darwin's enterprise.

My first remark is about Darwin's relatively *unscientific* style, especially in the first edition of the *Origin*. Darwin consciously tries to express his ideas "in the commonest language that occurs to (him)". Since every writer addresses himself, consciously or not, to a particular category of readers, this clearly indicates that the kind of readers Darwin wanted to convince was the general reader, cultivated, interested in science, but not a specialist. Incidentally, it would be interesting to compare Darwin's style in the *Origin* and in the long manuscript of *Natural Selection*. I have not made the comparison, but I have the feeling that the differences are not so great. We know how timid and cautious Darwin was when he wrote to his fellow scientists, how afraid he was of their possible hostile reactions. He seems not to have considered himself as really belonging to the scientific community, relegated as he was afraid to be because of his unorthodox ideas. He wrote for a reader whose mind would not be prejudiced against his theory by the creationist paradigm then prevailing among specialists and professional scientists, and he was eventually justified by the success of the book. But this also tells us something about the strength and prestige of official science in Britain at that time, and also perhaps about its relative estrangement from public opinion, which finally ensured Darwin's success. It might be said that Galileo and, to some extent, Buffon, had been in the same situation and used the same strategy with the same stylistic consequences. But it is doubtful if a physicist, a chemist, or even a physiologist would have succeeded, in the 1860s, in having recourse to the help of the general reader against prejudiced specialists. This again points to the very peculiar nature of evolutionary theory in general and Darwinism in particular.

But if we follow Beer's paper, we are brought very soon into more subtle difficulties. A well-known feature of common language is the polysemic value of the words. If, according to a famous definition, science is "a well made language", it is precisely because, at least theoretically, one word codes for one concept and only one. Now, Darwin not only accepts what Beer calls the "multivocality" of common language, but he seems to enjoy it and to use it for his own purposes. Is it, as Beer suggests, because multivocality better answers the very nature of a theory that emphasizes "relations and transformations" rather than essences and stability? But the same question may be asked about the Darwinian use of metaphors, whose heuristic value is now well known. There are many metaphors in Darwin's writings; we must be careful not to forget the original metaphoric value of some Darwinian phrases like "natural selection" itself, which have now lost that metaphoric value and have become well defined concepts. Indeed it could be said that the very history of Darwinism, from the *Origin* to the synthetic theory, is best exemplified by the transformation of the Darwinian metaphors of "natural selection" or "struggle for existence" into precise concepts of population genetics.

If this be true, metaphors and polysemic words should be considered as incipient concepts, some bound to extinction, some succeeding in establishing themselves, at least temporarily. If so, Darwin's way of writing does not only answer a peculiar view of nature, according to which diversity, polymorphic interrelations, and evolving equilibria are of higher value and significance than clear-cut distinctions, well-defined entities and static one-to-one relationships. It also witnesses an infant-stage of a theory struggling for existence, that is, trying to pass from the state of a powerful and blurred vision of nature into a consistent and articulate scientific theory. How much of the work was done by Darwin himself, from the first drafts he wrote down to the time of his death, and how much remained to be done by others, is open to discussion. But how much of the original complexity and polysemic value of Darwin's metaphors has been lost in the process is also open to discussion. I am not at all sure, for example, that our modern translation in terms of population genetics of the Darwinian phrase "struggle for existence" really conveys the full meaning of Darwin's consciously and carefully chosen metaphor. What has been lost perhaps in that particular case is a sense of reality, a perception of the concrete environmental conditions under which living things, and not genotypes only, have to struggle to survive and reproduce.

Another interesting remark in Beer's paper underlines the role of narrative in the *Origin*, a narrative whose organization "emphasizes variability rather than development" and whose time "is not one that begins at the beginning but rather in the moment of observation". Samuel Butler was right when he complained that, in the *Origin*, the origin was "cut out", but he showed at the same time that he did not understand what "origin" means in Darwin's title. The subject-matter of Darwin's book is not "the origin" of living things or life itself, but, as Beer points out, "the origin of species by means of . . . ," that is, the *process* of speciation, not as a historical and unique phenomenon having occurred in the past, but as a perpetual process, obeying a law as immutable as those that regulate physical processes. Darwin's science is not historical — here I disagree slightly with Beer's conclusion — and by this I mean that, like physics or chemistry, it intends to deal with current and general processes, not with unique events that never are to occur again. Hence the often misunderstood meaning of the famous diagram of Chapter 4 of *Origin*, a diagram that is *not* a phylogenetic tree, and the necessity of a careful reading of the last paragraph of the same chapter, where Darwin himself seems to accept the metaphor of the tree: Darwin's tree has several "limbs divided into great branches, and these into lesser and lesser branches", but *it has no trunk*. At the beginning, it was "a mere bush". It may represent "the affinities of all the beings of the same class", but not the history of life. Haeckel's use of the image of the phylogenetic tree probably was more Lamarckian than Darwinian, although in Lamarck himself the meaning of the metaphor is not always clear. Another stylistic consequence of the non-historical character of Darwin's theory, and of its epistemological status at the same time, is the frequent use of the conditional mood *at the present time,* especially in the "imaginary instances" that Darwin offers as "illustrations of the action of the Natural Selection" in the same chapter (*Origin*, pp. 90–95).

There are many other interesting points in Beer's paper, which demonstrates how useful a careful literary analysis may be for the history of science. But perhaps I am too much of a literary historian myself not to be prejudiced in that matter.

I now come to David Hull's particularly rich, complex, and controversial paper, and I am all the more embarrassed that, to put it very briefly, I agree with most of Hull's particular conclusions, I admire the acuteness of his analyses, I am ready to accept almost all his remarks, and yet I disagree with some essential aspects of his paper. To summarize my own position, I would say:

First, that I fully accept the concept of "historical entity" and find it a useful tool to deal with the particular objects with which, as historians of ideas or science, we have to deal in our daily practice. The phrase is self-contradictory and this is precisely why it is useful: we are dealing with theories, opinions, creeds, ideas, social groups, and institutions that have a real existence and can be described more or less accurately at a particular moment of history. But at the same time, those entities are perpetually evolving. They have no stable essence; they are not Platonic ideas; they are historically situated and can be understood in history only.

But second, I am very reluctant to introduce into cultural history models borrowed from the history of nature. Not because cultural phenomena have a special status or a special dignity or a transcendental character. I do believe that, if we want to understand cultural phenomena, we must treat them as "natural" phenomena and try to account for them by "natural" processes. This is, so to speak, an essential condition for any scientific work and, on that particular point, there is no difference between the sciences of man and the sciences of nature. But cultural processes are different and perhaps more complicated than the processes that govern the evolution of living things.

To begin with, for all its genetical variability, a natural species in its present state may be generally identified and sometimes very easily. There are some "essential features" and we are not to mistake a cat for a dog, whereas we could very easily mistake Haeckel for a Lamarckian, which perhaps would not even be a mistake. I know that, for many species, identification is not as easy as for cats and dogs. In such cases it is necessary to study the filiation; the possibility, impossibility, or difficulty of interbreeding; and so on. Hull exemplifies that difficulty by saying that the only scientific definition of man is: "to be born of human beings, to mate with human beings, and/or to parent human beings." But here precisely there is a great difference between cultural and natural history. It is often possible to experiment on living things and to discover their links of filiation, whereas we have no such possibility in cultural history, where the processes of inheritance are awfully obscure. I am ready to accept Hull's remark that an unknown precursor is no precursor at all. But the links of filiation between a *known* precursor and the scientist we are studying are not that clear. If I were to indulge in rhetorical questions, I would ask: who was Darwin's father? or did he have any father? or how many fathers did he have? Or, to take another example, is Stephen Gould the legitimate and unfaithful son of George Gaylord Simpson or the natural child of Ernst Mayr? Nobody breeds true in the cultural world, which is full of hybrids, interspecific, intergeneric, inter-what-you-like hybrids, hopeful and hopeless monsters, not to speak of chimaeras. There are no interspecific barriers nor Mendelian laws in cultural genetics.

If the interplay between cultural entities and their historical environment is of a Lamarckian or a Darwinian type may be open to discussion: there is obviously a direct pressure of the environment that induces "adaptive" alterations of the theories, but one could also say that new theories are screened by a kind of "natural selection" that allows some of them to survive the competition with the others and to "reproduce" themselves in the next generation of scientists or thinkers. But there is no question about the cultural inheritance of acquired characters nor about the role of use and disuse. Scientists may change their minds during their lives because of the pressure of the environment, and this was precisely the case with Darwin. Thus he will successively beget true disciples very different from each other. I know that societies have powerful institutions that may be analogically described as reproductive organs of cultural life. Sociologists study those institutions and the constraints they put on those who enter the cultural world. This is what Thomas Kuhn has described as the introduction of students to "normal science". But for all those constraints, such institutions, be they the workshops of the Italian painters of the Renaissance or our modern universities, never were able to impose a mere replication process and produce students really similar to their masters. Of course, the same could be said of the reproduction of natural species, but again, if from general analogies we go down to historical details, we discover very soon that the reproduction processes are different, especially because cultural hybridization has no limits and acquired characters are as easily inherited as the innate ones, which anyway would be extremely difficult to define in any individual.

The pace of cultural evolution is much too fast to allow us to use lasting labels. If we were to borrow a model from natural sciences in order to study the evolution of a cultural entity, I would not borrow it from genetics, but rather from traditional paleontology. When paleontologists follow a phyletic line they wisely change the specific name as soon as they think it suitable to the morphological alterations they have observed in the series of fossil remains. Like theirs, our filiations are always constructed: we can only describe and compare the "morphology" of textual "remains" historically dated. If we use the same name for a cultural entity evolving through time we risk being confronted with endless and perhaps unnecessary difficulties. Personally I would prefer to use the label "Darwinism" only for the thought of Darwin himself, which is not so easy to describe; to speak of "neo-Darwinism" for Weismann, and of "the synthetic theory" for that which prevailed in the 1950s, although that theory itself was and still is incessantly evolving. This is not a perfectly satisfactory solution, but, in any case, no label can be substituted for a careful description of a theory, with all its nuances and even its internal contradictions.

Until now I have spoken of intellectual history only, but an important aspect of Hull's paper is that it is not limited to conceptual entities. It applies the concept of historical entities to social groups as well. As a historian I cannot but agree with such a move. Like Hull I believe the study of social groups may be of great use, not only for a complete description of cultural history but even for our understanding of intellectual history. Social groups are also historical entities. Like natural species, they appear under certain conditions but never out of nothing. They evolve and sometimes become extinct. The relationships between social and intellectual history are not always clear, however, and there is a great deal of disagreement among historians about them. Here I would just say that some historians of ideas have fallen into what I would call the "group fallacy", that is, they have more or less assumed that people who met in some formal or informal groups should have many ideas, aims, or enterprises in common. This is sometimes the case, but sometimes it is not. Hull does not fall into that trap: when he speaks of "Darwinians" as a social group he is very careful to emphasize that they did not necessarily think the same way, even about evolution itself. What he actually describes is the network of personal relationships that Darwin built in about the first half of his life. The main result of this careful and illuminating analysis is to show that scientists who knew Darwin personally, were interested in his work, supported his efforts, respected his scientific character, and occasionally defended him against unjustified criticism, did not necessarily accept his views on evolution. This gives us an interesting insight into the way things are really going in a scientific community and should prevent us from too easy oversimplifications. The only thing I am tempted to object to, is the use of the word "Darwinians" to identify such people. If we were to extend the list to the second half of Darwin's life, we could easily present some unexpected Continental candidates to that honorable title of "Darwinians": old Quatrefages, for example, who was as much of a fixist as old Henslow, but had the highest regard for Darwin's work and did his best to have him elected to the Paris Academy of Sciences. Maybe it would be better to speak of "Darwin's network" when, at the beginning, Darwin was nothing more than a modest member of the group, of "Darwin's circle" when Darwin slowly emerged as the most original figure of the network, and of "Darwinians" for those who understood — or believed they understood — and accepted Darwin's ideas, even when those ideas were self-contradictory. Even in this restricted use, I think that the word "Darwinian" would need qualification, because I find it difficult to put in the same category two scientists like De Saporta and Fritz Müller; Saporta who proudly claimed he was Darwin's disciple but did not understand Darwin, as Professor Conry has convincingly shown, and Fritz Müller who, according to Francis Darwin, was "of all his unseen friends the one for whom he [Charles Darwin] had the strongest regard", the one also who probably was the first to conceive and follow a truly Darwinian research program. Here, we are slowly going from "Darwinians" to "Darwinists," a shift that is aptly described by Hull when he speaks of "the Darwinians in this global sense [who] verge on becoming as amorphous as the paradigm to which they pledged allegiance."

But how can such an "amorphous paradigm" be at the same time a "research program" and how can such an "amorphous" group be considered as a "historical entity"? If we want to describe and understand the evolution of social and conceptual entities, we need some tentative definition, at least for practical reasons, and we must, I think, introduce some hierarchic order and distinguish different levels of integration. Darwinism itself is possibly the best example for that necessity. As far as social groups are concerned, it is clear that they are of different kinds. Networks of personal relationships, political parties, religious groups and churches, national academies, international communities defined by the study of a scientific discipline, and university departments do not evolve the same way or according to the same laws. In each group, the rules of admission, the relative importance of social and/or intellectual conditions, the strength of the personal feeling toward the community, may be different; each of these factors, and probably many others, influence the historical fate of the group. I do not know if a typology of social groups is possible, but I do believe that, for each group, we must study carefully the rules that govern its life before trying to understand that life itself.

As for conceptual entities, that necessity of distinguishing different levels seems to me still more obvious, given the chaotic state of affairs that we now observe and that Hull describes very well at the beginning of his paper. Taking Darwinism as an example, we may distinguish a lowest level, that of the hard facts Darwin borrowed from almost everybody, from breeders to scientists of whatever discipline or opinions, but to which he gave a new meaning by introducing them into his theory. Then comes the level of particular theories, sometimes borrowed from others, like that of the division of labor, sometimes more original, like the theory of variation, the theory of divergence, the theory of migration, and so on. In my opinion such particular theories are not research programs, nor is Darwinism itself. They may inspire different research programs in various disciplines, each program being adapted to the particular requirements of the discipline itself. Then come the theory of natural selection, the theory of evolution as a whole and, as a necessary corollary of the theory, the development of a Darwinian theory of history (Gould 1982, n. 1), a Darwinian method (Ghiselin 1969), and a Darwinian philosophy of science. This is of course not a historical order: philosophical considerations probably came first for Darwin himself, and then the concept of natural selection was formulated, around which all the particular theories were organized. As a tool for an analysis of a complex system of thought like Darwinism, such distinctions may be useful, provided we do not consider them as ontological categories and are careful enough not to forget their constant interplay and unceasing evolution.

With such a tool we could perhaps see a little more clearly what happened to Darwinism, how it slowly constituted itself, how it evolved, and why it still is evolving. For example, we could perhaps better understand how natural selection, the central concept of Darwinian evolution, was at the same time adopted and misunderstood by Lamarckians who tried to introduce that extraneous element into a system of thought completely foreign to the Darwinian one. Or how Mendelian genetics, first introduced into Darwinism in order to replace an obsolete theory of heredity and variation, finally and surreptitiously, so to speak, transformed the whole theory, including the very concept of natural selection, putting in the forefront some neglected aspects of Darwin's ideas and shaping the so-called synthetic theory. In any system of thought, every element influences the whole system and is influenced by it.

But it is perhaps at the highest level of generality and integration that we may hope to find some essential features of Darwinism and understand why we still are speaking of Darwinism today when we speak of the modern evolutionary theory. For that extraordinary survival there are some good reasons, more sociological than intellectual, that Hull has analyzed very well at the beginning of his paper when he explains, for example, how neutralism, first rejected as "non-Darwinian", became an accepted part of the synthetic theory when it was clear that it was supported by many indisputable facts. But beyond the strategic choices offered to Young Turks or to older members of the scientific community who want to acquire or to preserve a respectable status, the survival of Darwinism may be accounted for by its being much more than a scientific theory. About many problems, like the relative importance of the various disciplines interested in evolution, even about speciation or natural selection, there is still a great deal of disagreement among scientists, or even a great deal of ignorance. About the general framework of the Darwinian view of nature's laws; about the place of man among the other living species and not above them; about the necessity to study living things not as isolated individuals but as members of complex networks, there is little discussion today. All those ideas were not original with Darwin or were not explicitly formulated by Darwin himself, but derived almost necessarily from his theory and his method, and were more easily recognized when the revolution of twentieth-century physics helped scientists to accept a new idea of science. That Darwin belonged to Victorian culture is a historical fact, but he was much more than a Victorian. He was one of the first and perhaps the first of the scientists and philosophers who shaped the thought of the twentieth century. This explains why we still are "Darwinians". The mere historical fact of the lasting life of Darwinism as a historical entity obliges us to recognize the full dimension of the Darwinian revolution.

28

Adaptation and Mechanisms of Evolution after Darwin: A Study in Persistent Controversies

William B. Provine

Introduction

Do the primary mechanisms of evolution in nature lead to adaptation? This has remained a persistently controversial question from the appearance of Charles Darwin's *Origin* in 1859 until the present. The main reason for this persistent controversy is that evolutionists have disagreed about whether or not the observed differences between closely related taxa (especially at the species level) are adaptive, and disagreed about the prevailing mechanisms of microevolution (evolution up to the level of geographical races or subspecies), speciation, and macroevolution (evolution above the species level). Although the question of adaptation in relation to mechanisms of evolution became more narrowly focussed during the period of the Evolutionary Synthesis of the 1930s and 1940s, the controversy remained sufficiently intense to fuel disagreement and stimulate a high percentage of both theory and field research in evolutionary biology. Controversy over the question continues today. Through historical analysis, I attempt in this essay to elucidate the reasons for the persistent controversy and to evaluate the current standing of the question.

The question "What is the mechanism of evolution?" must be clarified at the outset. The question is imprecise because the answer depends upon the level of evolution addressed (Lewontin 1970). For Darwin, individual natural selection was the primary mechanism by which both geographical races and species evolved. For De Vries, in contrast, individual natural selection might explain the evolution of geographical races, but not the evolution of species, which required macromutations followed by "selection between elementary species", a wholly different mechanism (De Vries 1904, pp. 92–120). Using the same mechanism for all levels of the evolutionary process is beset with difficulties, but this is equalled and probably exceeded by the difficulty of discovering distinct mechanisms of evolution for some levels or each level of the process. Historically, most of the controversy has been associated with moving from the levels of evolution in populations or geographical races to the level of speciation. Mechanisms of macroevolution have also generated many controversies, but most of these controversies reflected the views participants held about evolution at or below the species level. This is understandable because the geological record is consistent with a wide variety of highly divergent views of the mechanisms of speciation or subspeciation.

Most of the current controversies about mechanisms of evolution center upon proposed revisions of the "neo-Darwinian" or "synthetic" views developed in the 1930s and 1940s and expressed most clearly in the host of publications at about the time of the Darwin Centennial of 1959. Darwin's own views actually differed substantively from those of neo-Darwinians in 1959.

I. Darwin's Mechanisms of Evolution

In this section I have used the sixth edition of the *Origin*, published with additions and corrections in 1872. This was the edition most persons read. John Murray printed 9750 copies of the first through fifth editions between 1859 and 1869; over 100,000 copies of the sixth edition came from Murray between 1872 and 1929. Appleton printed tens of thousands of copies of the sixth edition in New York (R. Freeman 1977). The great emphasis now upon the first edition of 1859 reflects the mood a century later of neo-Darwinians who found that it fit their ideas of Darwinism better than did the sixth edition. Ernst Mayr (Harvard University Press) and John Burrow (Penguin) are the editors of the two most widely selling reprints of the first edition. The sixth edition of the *Origin* is the one of greatest interest to those studying Darwin's influence after 1872, by which time the first edition had become scarce.

Darwin's conception of the basic mechanism of evolution is too well known to require exposition here. He summarized the view as follows:

> Hence I look at individual differences, though of small interest to the systematist, as of the highest importance for us, as being the first steps towards such slight varieties as are barely thought worth recording in works on natural history. And I look at varieties which are in any degree more distinct and permanent, as steps towards more strongly-marked and permanent varieties; and at the latter, as leading to sub-species, and then to species. The passage from one stage of difference to another may, in many cases, be the simple result of the nature of the organism and of the different physical conditions to which it has long been exposed;

> but with respect to the more important and adaptive characters, the passage from one stage of difference to another, may be safely attributed to the cumulative action of natural selection, hereafter to be explained, and to the effects of the increased use or disuse of parts. A well-marked variety may therefore be called an incipient species. (*Origin* 1872, pp. 41–42)

When Darwin's evidence of all degrees of sterility between species is added to this quote, his general position is clear. Natural selection of individual differences (those small ubiquitous variations found in every population) was the primary mechanism of evolution at every level of the evolutionary process. Certainly this is the dominant view neo-Darwinians today attribute to Darwin.

Darwin constantly reminds the reader that natural selection does not operate to make all parts of an organism exquisitely adapted to its surroundings, as the natural theologians would have one believe:

> Natural selection tends only to make each organic being as perfect as, or slightly more perfect than, the other inhabitants of the same country with which it comes into competition. And we see that this is the standard of perfection attained under nature. (*Origin* 1872, p. 163)

Thus Darwin explained we should expect to see the bizarre contraptions possessed by organisms, and to observe an introduced species outcompete a native species even though the native species appeared well adapted to its surroundings. Natural selection was not a mechanism of perfection, but rather of adaptation of organisms to their immediate environments.

All this sounded very familiar to neo-Darwinians of 1959. A note of disquiet is, however, already present in the first quotation above. Darwin plainly stated that use and disuse of parts was a substantial mechanism of evolution, right along with natural selection. As so many opponents of neo-Darwinism in the late nineteenth and early twentieth century pointed out, Darwin's introduction to the *Origin* ends with the statement, "I am convinced that Natural Selection has been the most important, but not the exclusive, means of modification" (*Origin* 1872, p. 4).

What indeed were Darwin's mechanisms of evolution in addition to natural selection, and what were their relationships to adaptive evolution? The list is longer, more substantial, and with greater implications for non-adaptive evolution and classification than most neo-Darwinians after Alfred Russel Wallace have realized.

1. Use and disuse of parts. Next to natural selection, Darwin considered this to be the most important mechanism of adaptive evolution. The *Origin* is filled with references to the evolutionary effects of use and disuse, and Darwin provided a biological justification with his provisional hypothesis of pangenesis in the *Variation*.

2. Sexual selection. This mechanism was less rigorous than natural

selection, and was basically non-adaptive or maladaptive. Darwin in the *Descent* frequently provided examples of sexual selection leading to characters disadvantageous in the general struggle for existence. Natural selection prevented sexual selection from leading to highly disadvantageous characters, however.

3. Directed variation:

> Certain rather strongly marked variations, which no one would rank as mere individual differences, frequently recur owing to a similar organisation being similarly acted on, — of which fact numerous instances could be given with our domestic productions. In such cases, if the varying individual did not actually transmit to its offspring its newly-acquired character, it would undoubtedly transmit to them, as long as the existing conditions remained the same, a still stronger tendency to vary in the same manner. There can also be little doubt that the tendency to vary in the same manner has often been so strong that all the individuals of the same species have been similarly modified without the aid of any form of selection (*Origin*, 1872, p. 72)

Directed variation was predominantly non- or maladaptive, since it did not result from selection of any kind.

4. Correlated variation. This was one of Darwin's favorite ways of explaining non- or maladaptive features. The argument was that the maladaptive character was correlated with another of adaptive value sufficiently high that their combination had positive adaptive value. Adaptationists since Darwin have loved this argument and used it frequently. The hypothesized linkage is in most cases very difficult to prove or disprove.

5. Spontaneous variations. These variations simply appeared spontaneously and then were passed on by heredity. Spontaneous variations were not induced by changed conditions, Darwin's favorite cause for the appearance of new increased variability. Darwin offered the "appearance of a moss-rose on a common rose, or of a nectarine on a peach tree" as "good instances of spontaneous variations".

> In the earlier editions of this work I under-rated, as it now seems probable, the frequency and importance of modifications due to spontaneous variability. But it is impossible to attribute to this cause the innumerable structures which are so well adapted to the habits of life of each species. (*Origin* 1872, p. 171)

Spontaneous variations passed on by heredity accounted for many of the non-adaptive characters of organisms.

6. Family selection. Darwin invented this mechanism to explain the evolution of altruistic social behavior and neuter castes, apparently antithetical to individual natural selection. The difficulties of these examples, "though appearing insuperable, is lessened, or as I believe, disappears, when it is

remembered that selection may be applied to the family, as well as to the individual, and may thus gain the desired end" (*Origin* 1872, p. 230). Many modern biologists have suggested that Darwin would have been pleased by W. D. Hamilton's calculus of inclusive fitness to explain such anomalous cases by individual selection rather than familial, and I see no reason to doubt the suggestion. Darwin's familial selection was adaptive.

Darwin's mechanisms 2, 3, 4, and 5 above could all lead to non-adaptive differentiation in local populations, and by heredity after that to the levels of subspecies and species, and to even higher taxa. But did Darwin think that animals and plants in nature really exhibited a considerable number of non-adaptive characters?

The answer is unquestionably affirmative. Chapter 6 contains an important section titled "Organs of little apparent Importance, as affected by Natural Selection". Such organs, Darwin said, presented for the theory of evolution by natural selection great difficulties, "almost as great, though of a very different kind, as in the case of the most perfect and complex organs" (*Origin* 1872, p. 157). Darwin suggested in the first place that these apparently unimportant characters might have hidden adaptive value: "we are much too ignorant in regard to the whole economy of any one organic being, to say what slight modifications would be of importance or not" (p. 157). He further suggested that "Organs now of trifling importance have probably in some cases been of high importance to an early progenitor, and, after having been slowly perfected at a former period, have been transmitted to existing species in nearly the same state, although now of very slight use; but any actually injurious deviations in their structure would of course have been checked by natural selection" (p. 57). But having stated the caveats for natural selection, Darwin added:

> In the second place, we may easily err in attributing importance to characters, and in believing that they have been developed through natural selection. We must by no means overlook the effects of the definite action of changed conditions of life, — of so-called spontaneous variations, which seem to depend in a quite subordinate degree on the nature of the conditions, — of the tendency to reversion to long-lost characters, — of the complex laws of growth, such as of correlation, compensation, of the pressure of one part on another, &c., — and finally of sexual selection, by which characters of use to one sex are often gained and then transmitted more or less perfectly to the other sex, though of no use to this sex. But structures thus indirectly gained, although at first of no advantage to a species, may subsequently have been taken advantage of by its modified descendants, under new conditions of life and newly acquired habits. (*Origin* 1872, pp. 157–158)

A primary example of non-adaptive variation, Darwin thought, could be seen in polymorphic genera and species:

> There is one point connected with individual differences, which is extremely perplexing: I refer to those genera which have been called "protean" or "polymorphic," in which the species present an inordinate amount of variation. With respect to many of these forms, hardly two naturalists agree whether to rank them as species or as varieties. We may instance Rubus, Rosa, and Hieracium amongst plants, several genera of insects and of Brachiopod shells. In most polymorphic genera some of the species have fixed and definite characters. Genera which are polymorphic in one country seem to be, with a few exceptions, polymorphic in other countries, and like-wise, judging from Brachiopod shells, at former periods of time. These facts are very perplexing, for they seem to show that this kind of variability is independent of the conditions of life. I am inclined to suspect that we see, at least in some of these polymorphic genera, variations which are of no service or disservice to the species, and which consequently have not been seized on and rendered definite by natural selection. (*Origin* 1872, p. 35)

After the R. A. Fisher/E. B. Ford collaboration on polymorphism, Darwin's statement here appears most un-Darwinian.

In Chapter 7 Darwin gives a substantial list of non-adaptive characters of animals and plants. Why, one might ask, would someone clearly promoting the idea of natural selection as the primary mechanism of evolution spend considerable energy documenting the existence of non-adaptive variation that did not evolve by natural selection, at least in Darwin's opinion? The answer is that Darwin held allegiance not only to natural selection, but to the idea of evolution by descent, and to the implications of evolution by descent for the problem of classification:

> From the fact of the above characters being unimportant for the welfare of the species, any slight variations which occurred in them would not have been accumulated and augmented through natural selection. A structure which has been developed through long-continued selection, when it ceases to be of service to a species, generally becomes variable, as we see with rudimentary organs; for it will no longer be regulated by this same power of selection. But when, from the nature of the organism and of the conditions, modifications have been induced which are unimportant for the welfare of the species, they may be, and apparently often have been transmitted in nearly the same state to numerous, otherwise modified, descendants. It cannot have been of much importance to the greater number of mammals, birds, or reptiles, whether they were clothed with hair, feathers, or scales; yet hair has been transmitted to almost all mammals, feathers to all birds, and scales to all true reptiles. A structure, whatever it may be, which is common to many allied forms, is ranked by us as of high systematic importance, and consequently is often assumed to be of high vital importance to the species. Thus, as I am inclined

> to believe, morphological differences, which we consider as important — such as the arrangement of the leaves, the divisions of the flower or of the ovarium, the position of the ovules, &c. — first appeared in many cases as fluctuating variations, which sooner or later became constant through the nature of the organism and of the surrounding conditions, as well as through the intercrossing of distinct individuals, but not through natural selection; for as these morphological characters do not affect the welfare of the species, any slight deviations in them could not have been governed or accumulated through this latter agency. It is a strange result which we thus arrive at, namely that characters of slight vital importance to the species, are the most important to the systematists; but, as we shall hereafter see when we treat of the genetic principle of classification, this is by no means so paradoxical as it may first appear. (*Origin* 1872, pp. 175–176)

Turning, as Darwin suggests, to his chapter on classification, the importance of the non-adaptive characters documented above becomes obvious. The purpose of classification to an evolutionist is to reveal community of descent; the true system of classification is based upon the evolutionary tree. Should not the classification be based upon those distinctive characters most clearly adapting the organism to its environment?

> It might have been thought . . . that those parts of the structure which determined the habits of life, and the general place of each being in the economy of nature, would be of very high importance in classification. Nothing can be more false. No one regards the external similarity of a mouse to a shrew, of a dugong to a whale, of a whale to a fish, as of any importance. (*Origin* 1872, p. 365)

And here, of course, is where the importance of non-adaptive characters enters, as revealed by the following quotes:

> In formerly discussing certain morphological characters which are not functionally important, we have seen that they are often of the highest service in classification. This depends on their constancy throughout many allied groups; and their constancy chiefly depends on any slight deviations not having been preserved and accumulated by natural selection, which acts only on serviceable characters.
>
> That the mere physiological importance of an organ does not determine its classificatory value, is almost proved by the fact, that in allied groups, in which the same organ, as we have every reason to suppose, has nearly the same physiological value, its classificatory value is widely different.

> No one will say that rudimentary or atrophied organs are of high physiological or vital importance; yet, undoubtedly, organs in this condition are often of much value in classification.

> Numerous instances could be given the characters derived from parts which must be considered of very trifling physiological importance, but which are universally admitted as highly serviceable in the definition of whole groups.
>
> On the view of characters being of real importance for classification, only in so far as they reveal descent, we can clearly understand why analogical or adaptive characters, although of the utmost importance to the welfare of the being, are almost valueless to the systematist. For animals, belonging to two most distinct lines of descent, may have become adapted to similar conditions, and thus have assumed a close external resemblance; but such resemblances will not reveal — will rather tend to conceal their blood-relationship. (*Origin* 1872, pp. 366–367, 374)

So Darwin is actually concerned to show that natural selection does not determine all species characters; in that case, the natural system of classification by descent would be impossible to establish. Non-adaptive characters were, to Darwin, the essential keys to accurate systematics.

Darwin frequently expressed in correspondence in his later years his belief in the importance of non-adaptive variation in natural populations. Thus in a letter to Moritz Wagner dated 13 October 1876, in which Darwin agreed with Wagner as to the necessity of isolation for the splitting of species, he confessed:

> In my opinion the greatest error which I have committed, has not been allowing sufficient weight to the direct action of the environment, i.e. food, climate, etc., independently of natural selection. Modifications thus caused, which are neither of advantage nor disadvantage to the modified organism, would be especially favored, as I can now see chiefly through your observations, by isolation in a small area, where only a few individuals lived under nearly uniform conditions. (*LL* 3: 159)

From this quote, Darwin might be expected to have agreed with J. T. Gulick about non-adaptive differentiation in Hawaiian snails (Achatinellidae).

But Darwin was a complex man facing overwhelmingly complex data. Alfred Russel Wallace, for reasons I shall clarify below, much preferred to quote Darwin's letter of 30 November 1878, to Karl Semper:

> As our knowledge advances, very slight differences, considered by systematists as of no importance in structure, are continually found to be functionally important; and I have been especially struck with this fact in the case of plants to which my observations have of late years been confined. Therefore it seems to me rather rash to consider the slight differences between representative species, for instance those inhabiting the different islands of the same archipelago, as of no functional importance, and as not in any way due to natural selection. (*LL* 3:61; quoted in Wallace 1889, p. 142)

In the light of subsequent history, this quote, rather than the previous one, appears to have been the clarion call to those who have shouldered the mantle of "neo-Darwinism".

II. Neo-Darwinian Adaptationism and its Detractors in the Late Nineteenth and Early Twentieth Centuries

Alfred Russel Wallace, co-inventor of the concept of natural selection, became the pre-eminent Darwinian after Darwin's death in 1882. Wallace had far more direct acquaintance than Darwin with field research on natural populations (although Darwin certainly held the edge on knowledge of domestic populations). The overwhelming impression that Wallace gained from his work on natural populations was that Darwin had underestimated the effectiveness, rapidity, and importance of natural selection. In his 1889 summary of evolutionary views, *Darwinism*, Wallace argued that since the sixth edition of the *Origin* in 1872, much evidence had accumulated for greater variability in natural populations than Darwin had imagined. The availability of this variability insured the effectiveness of natural selection. Except in the case of man (see Turner 1974a; Kottler 1974 and this volume), Wallace was a far more thoroughgoing selectionist than Darwin. The differences in view between Wallace and Darwin emerge clearly on the issue of whether the characters used as taxonomic markers are adaptive or not.

Wallace's answer to G. J. Romanes's theory of physiological selection is helpful in this analysis. Romanes, who had worked closely with Darwin on mental evolution, argued in his essay "Physiological Selection" that natural selection was "not, strictly speaking, a theory of the origin of *species*: it is a theory of the origin — or rather of the cumulative development — of *adaptations*" (1886, p. 345; see Lesch 1975). Natural selection did not, he claimed, directly account for the mutual sterility of closely related species, nor for non-adaptive taxonomic markers:

> The features, even other than sterility *inter se*, which serve to distinguish allied species, are frequently, if not usually, of a kind with which natural selection can have had nothing whatever to do; for distinctions of specific value frequently have reference to structures which are without any utilitarian significance. It is not until we advance to the more important distinctions between genera, families, and orders that we begin to find, on any large or general scale, unmistakeable evidence of utilitarian meaning. (Romanes 1886, p. 338)

> The only answer which Mr. Darwin makes to this difficulty is, that structures and instincts which appear to us useless may nevertheless be useful. But this seems to me a wholly inadequate answer. Although in many cases it may be true, as indeed it is shown to be by a number of selected illustrations furnished by Mr. Darwin, still it is impossible to believe that it is always, or even generally so. In other words, it is impossible to believe that in all, or even in most, cases where minute specific differences of structure or of instinct are to all appearance useless, they are nevertheless useful . . . it surely becomes the reverse of reasonable so to pin our faith to natural selection as to conclude that all these peculiarities must be useful, whether or not we can perceive their utility. For by doing this we are but reasoning in a circle . . . . But I need not argue this point, because in the later editions of his works Mr. Darwin freely acknowledges that a large proportion of specific distinctions must be conceded to be useless to the species presenting them; and, therefore, that they resemble the great and general distinction of mutual sterility in not admitting of any explanation by the theory of natural selection. (Romanes 1886, pp. 344–345)

Wallace vigorously denied that Darwin ever stated that the particular characters used by systematists to distinguish one species from another "are ever useless, much less that a 'large proportion of them' are so, as Mr. Romanes makes him 'freely acknowledge' " (1889, p. 132). Wallace then proceeded to give several pages of recent evidence that taxonomic characters, formerly supposed to be non-adaptive, were strictly adaptive. He concluded:

> On the whole, then, I submit, not only has it not been proved that an "enormous number of specific peculiarities" are useless, and that, as a logical result, natural selection is "not a theory of the origin of species," but only of the origin of adaptations which are usually common to many species, or, more commonly, to genera and families; but, I urge further, it has not even been proved that any truly "specific" characters — those which either singly or in combination distinguish each species from its nearest allies — are entirely unadaptive, useless, and meaningless; while a great body of facts on the one hand, and some weighty arguments on the other, alike prove that specific characters have been and could only have been, developed and fixed by natural selection because of their utility. We may admit, that among the great number of variations and sports which continually arise many are altogether useless without being hurtful; but no cause or influence has been adduced adequate to render such characters fixed and constant throughout the vast number of individuals which constitute any of the more dominant species. (1889, pp. 141–142)

Wallace and Romanes were basically arguing about the mechanism of speciation. Wallace said it was natural selection; Romanes said it had to be a mechanism other than natural selection, and proposed his theory of "physiological selection" (selection for sterility factors) as an alternative. If the differences between species were non-adaptive, then natural selection could not be the primary mechanism. Darwin, of course, had in part already finessed this argument with his concept of "correlated" non-adaptive characters. In this instance natural selection could be the primary determinant of speciation; it was just that the adaptive characters by which the species had evolved were not as useful as the non-adaptive correlated characters for purposes of classification by descent.

Other neo-Darwinians in England followed Wallace's strong selectionist, adaptationist view. Prominent among them were E. Ray Lankester, Raphael Meldola, and E. B. Poulton. Wallace himself lived until 1913 and Poulton until 1943, by which time Julian Huxley, R. A. Fisher, E. B. Ford, Ernst Mayr, and others had firmly established modern neo-Darwinism. The selectionist, adaptationist view of Wallace has had a continuous existence since 1889. Just how strongly the view was held by some at the turn of the century is well exemplified in Poulton's 1894 essay "Theories of Evolution", which he delivered in Boston in an attempt to impress Darwin's idea of natural selection upon American neo-Lamarckians.

> The more we study the characters of animals in general, even though we at first can see no utility, the more we come to admit this principle, and to believe that either now or in some past time, the characters have been useful. I can certainly say of many characters which I have studied in some of my investigations, that at first they seemed to be meaningless, but afterwards appeared to be of much importance in the struggle for existence. I think we may safely assume with regards to many characters of which we can now see no explanation that ultimately the explanation will be forthcoming.
>
> Being unable to prove utility does not invalidate Natural Selection. If inutility could be proved for any large class of characters, the theory would certainly be destroyed as a wide-reaching and significant process. I do not think, however, that any such evidence has been forthcoming. (Poulton 1908, pp. 106–107)

If only one grants Poulton's presumption of utility and adaptation, then his position becomes almost unassailable. He grants that if inutility could be *proved* for a character, then natural selection as an explanation for the character would be destroyed. But how would the proof proceed? How many hypotheses can an inventive person dream up for the possible adaptive value of an apparently useless character; or its net adaptive value when correlated with other characters; or its adaptive value at an earlier time in evolutionary history? Very many, indeed an almost unlimited number. And each one would have to be refuted. Granting the presumption of utility gives the adaptationist an insurmountable advantage; indeed, as Poulton says, inability to prove utility (so *much* easier than proving inutility) does not invalidate natural selection. William Bateson nicely summarized Poulton's kind of argument in the same year as being equivalent to this: " 'If,' say we with much circumlocution, 'the course of Nature followed the line we have suggested, then, in short, it did.' That is the sum of the argument" (W. Bateson 1894, p. v).

If Wallace's selectionist view enjoyed a degree of continuity from 1889 on, it certainly did not gain continuous approval from other biologists. William Bateson had come to dislike the Darwinian view that evolution proceeded very gradually by natural selection working upon small individual differences, and advocated instead the view of Galton and T. H. Huxley that evolution proceeded by large discontinuous variations, of which his 1894 book *Materials for the Study of Variation* was a catalogue. In the introduction Bateson vigorously attacked the adaptationist view that natural selection was the mechanism of speciation:

> The Study of Adaptation ceases to help us at the exact point at which help is most needed. We are seeking for the cause of the differences between species and species, and it is precisely on the utility of Specific Differences that the students of Adaptation are silent. For, as Darwin and many others have often pointed out, the characters which visibly differentiate species are not as a rule capital facts in the constitution of vital organs, but more often they are just those features which seem to us useless and trivial, such as the patterns of scales, the details of sculpture on chitin or shells, differences in number of hairs or spines, differences between the sexual prehensile organs, and so forth. These differences are often complex and are strikingly constant, but their utility is in almost every case problematical. (1894, p. 11)

An even deeper objection, Bateson said, was that even if a character could be shown to be useful in some way, this fact was useless "unless we know also the degree to which its presence is harmful; unless, in fact, we know how its presence affects the profit and loss account of the organism" (1894, p. 12).

Bateson therefore absolutely refused, in his huge catalogue of 886 discontinuous variations, to speculate on the usefulness or harmfulness of the variations.

> Such speculation, whether applied to normal structures or to Variation, is barren and profitless. If any one is curious on these questions of adaptation, he may easily thus exercise his imagination. In any case of Variation there are a hundred ways in which it may be beneficial, or detrimental. For instance, if the "hairy" variety of the moor-hen became established on an island, as many strange varieties have been, I do not doubt that

> ingenious persons would invite us to see how the hairiness fitted the bird in some special way for life on that island in particular. Their contention would be hard to deny, for on this class of speculation the only limitations are those of the ingenuity of the author. While the only test of utility is the success of the organism, even this does not indicate the utility of one part of the economy, but rather the net fitness of the whole. (1894, pp. 79–80)

In 1894, it was Bateson rather than Poulton who sounded the dominant note of the succeeding two decades. Bateson's fellow experimental biologists would have little respect for the "just so" stories of Poulton or Wallace, and the neo-Darwinian selectionist-adaptationist view would suffer its deepest decline in the entire time between the first publication of the *Origin* and the present.

The most prominent and influential critic of the neo-Darwinian emphasis upon the natural selection of small differences as the mechanism of speciation was Hugo de Vries. He was influential for several reasons. He had clear experimental evidence for the sudden appearance of new true breeding varieties of *Oenothera* (the evening primrose) at a time when such experimental evidence was greatly admired; he published voluminously and traveled widely; he was one of the rediscoverers of Mendelian heredity; and his arguments just made good sense to experimental breeders and laboratory scientists, as well as to paleontologists looking at a discontinuous geological record.

Perhaps the best place to turn to in De Vries's published work for his view of the mechanism of evolution is his 1904 book *Species and Varieties: Their Origin by Mutation*, comprised of lectures originally delivered at the University of California (Berkeley) in the Summer of 1904. Here De Vries argued that Darwin was precisely right in making the analogy between artificial and natural selection; the problem was that Darwin had misunderstood what the most scientific breeders were really doing, and therefore had made incorrect inferences about the action of natural selection in nature.

Darwin thought breeders did their work primarily through individual selection. Giving evidence from many scientific breeders, De Vries challenged this view. The most successful breeders, he said, selected not individuals but "elementary species". Choosing the right variety as the foundation stock was the key to rapid success. True, breeders used individual selection once the elementary species was established, and it was important for fine-tuning of the population according to desire; but individual selection alone could not create the varieties produced by breeders.

The situation in nature was exactly analogous. Natural selection of individuals merely adapted local populations to the local conditions of their environment. Thus natural selection of individuals

> produces the local races, the marks of which disappear as soon as the

> special external conditions cease to act. It is responsible only for the smallest lateral branches of the pedigree, but has nothing in common with the evolution of the main stems. It is of very subordinate importance. (De Vries 1904, p. 802)

Corresponding to the "variety testing" or selection of elementary species practiced by breeders was what De Vries called "survival of species" or "selection between species" in nature:

> The fact that recent types show large numbers, and in some instances even hundreds of minor constant forms, while the older genera are considerably reduced in this respect, is commonly explained by the assumption of extinction of species on a correspondingly large scale. This extinction is considered to affect the unfit in a higher measure than the fit. Consequently the former vanish, often without leaving any trace of their existence, and only those that prove to be adapted to the surrounding external conditions, resist and survive. (1904, p. 799)

Microevolution and speciation both depended upon natural selection. Selection acted upon individual differences to produce geographical races; but to produce new species, natural selection acted upon new "elementary species" originating by large mutations. Using Darwin's analogy for the similarity of artificial and natural selection, De Vries deduced a mechanism of speciation that Darwin had pointedly denied.

Here again is the view that the *production* of new species can be basically a non-adaptive or maladaptive process, yet by the genus level adaptation was the rule. Thus a selective process had to occur at the species level. This process could not be the usual individual selection of small heritable differences, which was incapable of producing new species. In contrast, the basic position of neo-Darwinians such as Wallace and Poulton was that individual selection produced new varieties, species, genera, etc. But was there really a "species selection" in nature corresponding to breeders consciously testing varieties to see which ones would be foundation stocks?

In the early twentieth century, when De Vries was so popular and influential, the concept of a "species selection" tied to the mutation theory was attractive to many biologists, especially the geneticists and experimental biologists. On the surface, however, this seems inconsistent. Experimentalists emphasized tangible, decisive experiments; yet who of them had ever observed species selection in action? I think that De Vries's concept of species selection was not so much attractive per se as it was a conclusion to which experimentalists felt driven by the state of available knowledge in heredity and evolution.

One way to view the issue is to examine carefully T. H. Morgan's 1903 book, *Evolution and Adaptation*. This book contains the most systematic and careful critique of the Darwinian adaptationist view that I know from the early twentieth century (and has been carefully analyzed by G. Allen 1968; 1978, pp. 108–125). As a militant experimentalist, Morgan was antagonistic to adaptationist rhetoric, just as Bateson was. Morgan flatly rejected the inheritance of acquired characters, and raised what he considered insuperable objections to gradual natural selection of small differences as the prevailing mechanism of evolution. Indeed, it cannot be emphasized enough that natural selection of individual differences did not appear intuitively to have sufficient power to create new species; this was a major objection to Darwinism raised by critics everywhere. Where then could Morgan turn for a mechanism of evolution? The only available alternative was De Vries's theory, which was not only "experimental" but also fit the paleontological record much better than did gradual evolution.

Morgan observed that taxonomists used non-adaptive characters for classifying species, so clearly these species did not arise by natural selection:

> Animals and plants are not changed in this or that part in order to become better adjusted to a given environment, as the Darwinian theory postulates. Species exist that are in some respects very poorly adapted to the environment in which they must live. If competition were as severe as the selection theory assumes, this imperfection would not exist.
>
> In other cases a structure may be more perfect than the requirements of selection demand. We must admit, therefore, that we cannot measure the organic world by the measure of utility alone. If it be granted that selection is not a moulding force in the organic world, we can more easily understand how both less perfection and greater perfection may be present than the demands of survival require. (1903, p. 464)

But despite this statement and his extensive critique of adaptationist rhetoric, Morgan believed that above the species level most taxonomic features were adaptive, and indeed that adaptation was a fundamental aspect of animals and plants. If selection were not the key to the production of species ("Nature does not remodel old forms through a process of individual selection") it still had to play a significant role at the next level. Here Morgan's reasoning and conclusions are worth quoting at some length:

> We find that the great majority of animals and plants show distinct evidence of being suited or adapted to live in a special environment, i.e. their structure and their responses are such that they can live and leave descendants behind them. I can see but two ways in which to account for this condition, either (1) teleologically, by assuming that only adaptive variations arise, or (2) by the survival of only those mutations that are sufficiently adapted to get a foothold. Against the former view is to be urged that the evidence shows quite clearly that variations (mutations) arise that are not adaptive. On the latter view the dual nature of the problem that we have to deal with becomes evident, for we assume

> that, while the origin of the adaptive structures must be due to purely physical principles in the widest sense, yet whether an organism that arises in this way shall persist depends on whether it can find a suitable environment. This latter is in one sense selection, although the word has come to have a different significance, and, therefore, I prefer to use the term *survival of species*.
>
> The origin of a new form and its survival after it has appeared have been often confused by the Darwinian school and have given the critics of this school a fair chance for ridiculing the selection theory. The Darwinian school has supposed that it could explain the origin of adaptations on the basis of their usefulness. In this it seems to me they are wrong. Their opponents, on the other hand, have, I believe, gone too far when they state that the present condition of animals and plants can be explained without applying the test of survival, or in a broad sense the principle of selection amongst species.
>
> It will be clear, therefore, in spite of the criticism that I have not hesitated to apply to many of the phases of the selection theory, especially in relation to the selection of the individuals of a species, that I am not unappreciative of the great value of that part of Darwin's idea which claims that the *condition* of the organic world, as we find it, cannot be accounted for entirely without applying the principle of selection in one form or another. This idea will remain, I think, a most important contribution to the theory of evolution. (Morgan 1903, pp. 462–464)

Morgan had never seen his "principle of selection amongst species" in action in nature, any more than the neo-Darwinians had ever seen natural selection in action in nature. So it is curious to see this avowed experimentalist turn on the last two pages of his book to a mechanism of evolution that he not only had never seen, but for the existence of which he proposed no possible experiments. I detect no enthusiasm from Morgan for the idea of species selection. The evidence before him drove him reluctantly to this idea. De Vries was delighted (De Vries 1904, p. 9).

Enthusiasm for De Vries's mutation theory was high among early geneticists. C. B. Davenport, E. M. East, W. E. Castle, Raymond Pearl, G. H. Shull, Liberty Hyde Bailey, E. G. Conklin, and many others declared their admiration for the theory. Davenport organized a session devoted to the mutation theory at the annual meeting of the American Society of Naturalists in December 1904. Here Castle argued, with Darwin, that

> there is no essential difference between breeds and species, and if we can ascertain how breeds originate we can infer much as to the origin of species. On the whole, it appears that the formation of new breeds begins with the discovery of an exceptional individual, or with the production of such an individual by means of cross-breeding. Such exceptional individuals are mutations. (Castle 1905, pp. 522, 524)

Castle's basic assumption, in agreement with De Vries, Bateson, Galton, and Huxley, was that selection of small differences was ineffective in changing substantially the genetic constitution of a population; for that, large mutations were required. Reasoning from Wilhelm Johannsen's pure line theory and his experimental work, geneticists opted for the mutation theory. Yet in pursuing their experimental work geneticists soon began to produce incontrovertible evidence that selection of small continuous variations, far from being ineffective as predicted, actually could change a population far beyond its original range of variability, the new strains thus produced not regressing significantly when selection was relaxed. The most influential example was Castle's selection experiment with hooded rats. After 1907, Castle never again advocated the mutation theory of evolution, instead arguing for Darwinian selection as the primary mechanism of evolutionary change. The arguments for and against the selection theory in the period 1900–1918 I have discussed in detail elsewhere (Provine 1971, pp. 90–129).

By 1918 most prominent geneticists had accepted Mendelism and Darwinism as complementary (including substantial changes of mind by Morgan, Castle, Jennings, G. H. Shull, East, Baur, and many others), and believed natural selection of small Mendelian differences was the mechanism of evolution. To be sure, some geneticists, including Pearl, Punnett, Gates, and Bateson, still advocated discontinuous evolution. Their position, however, eroded seriously after a series of successful selection experiments and the demonstration that De Vries's "mutants" were actually balanced-lethal hybrids.

If the experimental biologists found the mutation theory convincing in the early twentieth century, the naturalists who knew natural populations well did not. They found very few large mutations in natural populations; those they did find were almost invariably reproductive misfits. Such mutants were not the source of evolutionary change. Yet the naturalists, other than the neo-Darwinians who were a minority, also were convinced by arguments like those so ably advanced by Morgan in 1903 against the neo-Darwinian adaptationist/selectionist view. Systematists generally believed that the inheritance of the effects of the direct action of the environment upon organisms was the primary mechanism of differentiation of varieties and species in nature, although most of them also believed natural selection was a significant force. Different environments produced different effects upon two populations derived from an original one. F. B. Sumner, Ernst Mayr, and Bernhard Rensch all believed in the direct action of the environment as the primary mechanism causing the differentiation of geographical races or *Rassenkreise*. All three of course became neo-Darwinians between 1925 and 1932.

I would emphasize that belief in the direct action of the environment did not necessarily lead to an adaptationist view. Sumner, for example, believed strongly up until 1925 that the differences between geographical races of his deer mouse *Peromyscus* were wholly non-adaptive. Not until he moved to a selectionist view (based upon his evidence of multiple-factor Mendelian differences between geographical races) did Sumner gravitate toward an adaptationist interpretation of racial differences (Provine 1979). Rensch, on the other hand, was always strongly adaptationist in his neo-Lamarckian views.

Some of the best-known and most spectacular taxonomic work before the evolutionary synthesis was on land snails. The Rev. John T. Gulick had examined the many species of the genus *Achatinella* in the Hawaiian Islands, and H. E. Crampton the species of *Partula* in Polynesia, particularly on Tahiti and Moorea. Both Gulick and Crampton came firmly to the view that natural selection had no significant role in the differentiation of races of snails, and that distinguishing taxonomic characters had no possible adaptational value. Instead, racial differentiation appeared to result from the chance isolation from the original population of a few members with a genetic complement different on average from that of the original population.

Although by the mid-1920s geneticists had generally come to the view that Mendelism and gradual selection were complementary, they had little knowledge of the systematics of natural populations or of paleontology. The systematists worked in isolation from the experimental geneticists and paleontologists. There is little need to review here the details of this situation, so thoroughly discussed recently by Ernst Mayr (Mayr and Provine 1980, pp. 1–48). I shall summarize the situation by saying that no group of biologists had a sophisticated and consistent view of the mechanisms of evolution in nature, and there existed fundamental disagreement on mechanisms of evolution within many fields as well as between all of them. Darwin originally had a synthetic vision of evolution, but in the mid-1920s, the synthesis appeared far away.

III. Mechanisms of Evolution Early in the Evolutionary Synthesis: Fisher and Wright

The quantitative models held by the theoretical population geneticists Fisher, Haldane, and Wright strongly influenced evolutionary thinking in the synthesis period in at least four ways (Provine 1978). First, the models demonstrated that Mendelism and natural selection, in combination with known processes in natural populations, were sufficient to account for observed monophyletic evolution in nature. Darwin's belief that a very small selection rate could alter the hereditary constitution of a population, a far from intuitively obvious proposition, was verified by the models. Second, the models indicated that some earlier views were untenable. One popular conception of laboratory geneticists (but not systematists) had been that mutation pressure was the dominant factor in evolutionary change. By elucidating the relationships between mutation rates, selection pressures, and changes of gene frequencies, the quantitative models showed unmistakably that selection was vastly more effective than mutation as an agent of evolutionary change. Third, the models clarified and complemented field researches already completed or in progress, thus giving the field research greater significance. One prominent example was Haldane's use in 1924 of available data on the frequency of melanic and non-melanic forms of the moth *Biston betularia* in the area of Manchester, England. Haldane calculated that the melanic form was twice as likely to survive as the previously prevalent non-melanic form. Fourth, the models stimulated and provided the intellectual framework for later field research. The most impressive examples of this were the great influence of Sewall Wright's models on Theodosius Dobzhansky's monumental series of forty-three papers on field researches with *Drosophila pseudoobscura* and relatives under the title "The Genetics of Natural Populations" (Lewontin et al. 1981), and the important influence of R. A. Fisher's work upon the field researches of E. B. Ford and his associates.

R. A. Fisher and Sewall Wright were the two most influential evolutionary theorists of the evolutionary synthesis. I shall show in this section that despite their great influences, neither of them thought clearly about the relation of microevolution to mechanisms of speciation, and that their views on speciation were simplistic, contradictory, or ambiguous.

R. A. Fisher chose as one of his two prizes awarded at graduation from Harrow in the Spring of 1909 the collected works of Charles Darwin (the other was a collection of Greek plays in translation). That Fall he entered Gonville and Caius College of Cambridge University in time to witness some of the celebrations of the centenary of Darwin's birth and the half-century of the *Origin*. The centenary itself, as a whole, did not turn Fisher into a Darwinian because, as Fisher himself later recalled, much of the outpouring of literature on Darwinism in 1909 was antagonistic to the idea of natural selection as the dominant mechanism of evolution (Box 1978, pp. 17, 23). Fisher did, however, become a staunch believer in natural selection at about this time. By 1911, at age twenty-one, he delivered a paper to the Cambridge University Eugenics Society, arguing, as Yule had before him, that Mendelism, biometry, and selection together provided a synthetic view of the mechanism of evolution in nature and eugenical improvement in man.

Mathematics, physics, and astronomy were among Fisher's early loves. He especially liked the way simple quantitative laws had deep explanatory power. Newton, for example, had deduced the motions of the planets and comets, the behavior of falling bodies on the surface of the earth, and the action of the tides all from the inverse-square law of attraction. Boyle's gas laws and the second law of thermodynamics appealed to Fisher for the same reasons. In turning his attention to evolution, Fisher clearly wished to find the simple quantitative law that would allow evolutionary phenomena to all fall in place.

Fisher was no blind follower of Darwin's ideas. He adopted Mendelism with the explicit belief that Darwin had misunderstood heredity and had therefore come to many unnecessary or wrong conclusions about the mechanisms of evolution in nature. Darwin, in Fisher's view, had discovered in natural selection the primary determinant of evolution; but then Darwin had spoiled the power of his mechanism by overlaying it with other hypotheses. It wasn't Darwin's fault — he couldn't be expected to invent Mendelism also. Fisher wanted to reinterpret natural selection in terms of the new genetics and to sweep away the unnecessary hypotheses.

His first order of business was to attack Karl Pearson's assertion that Mendelian heredity was inconsistent with observed correlations between relatives (Pearson and Lee 1903). Fisher demonstrated the consistency in 1918 by utilizing the analysis of variance (squared standard deviation) and by taking dominance into account, something Pearson had not done (Fisher 1918). The larger picture of Fisher's view of evolution appeared in his 1922 paper, "On the Dominance Ratio". In physics, it was sometimes possible to analyze an apparently complex process into a relatively few variables, the interaction of which yielded the observed process. Frequently the behavior of molecules could be formulated into a simple stochastic distribution, thus obviating the necessity for tracking the path of each molecule, as in Boyle's gas laws. Perhaps Fisher's greatest contribution to evolutionary theory was his insight that an equation representing the stochastic distribution of Mendelian determinants in a population over time was the key to an accurate and quantitative understanding of monophyletic evolution in that population. So far as I am aware, his 1922 paper represents the first attempt in this direction.

In the paper Fisher examined the influence of selection, dominance, mutation rate, random extinction of genes, and assortative mating upon the statistical distribution of genes in the population. Among many simplifying assumptions, Fisher assumed his population was extremely large and consequently had high storage of genetic variability. In such a population, his stochastic distribution led to the certain conclusion that selection, acting upon single genes, was the supreme determinant of the evolutionary process. A mutation rate far higher than any observed in nature could be balanced by a minuscule selection rate against it.

Furthermore, Fisher assumed that the larger the effects of a mutation, the more likely that its effects would be deleterious. Thus natural selection of single genes of small effect was the key, and the only one, to evolution in natural populations. Natural selection was slow but certain in its effects. Fisher likened the stochastic distribution of genes, dominated by natural selection, to the general laws of the behavior of gases:

> The investigation of natural selection may be compared to the analytic treatment of the Theory of Gases, in which it is possible to make the most varied assumptions as to the accidental circumstances, and even the essential nature of the individual molecules, and yet to develop the general laws as to the behavior of gases, leaving but a few fundamental constants to be determined by experiment. (1922, pp. 321–322)

Effects of genetic interaction and random genetic drift were two of the accidental circumstances irrelevant in the evolutionary process, Fisher argued.

In 1927 and 1928, Fisher published major papers using the deterministic effect of gradual selection acting upon small modifiers to explain the evolution of mimicry and dominance (1927, 1928a). In the paper on mimicry, Fisher tried to demolish the argument advanced by Punnett (1915) that mimicry patterns determined by sharply discontinuous Mendelian factors must have evolved by correspondingly discontinuous leaps. Fisher argued that the sharply discontinuous patterns had evolved through the gradual accumulation by natural selection of small modifiers of the Mendelian factors determining the mimicry patterns. Similarly, he argued (1928a) that natural selection accumulated small modifiers of dominance. Aboriginal mutations were not recessive, like those seen in most organisms, where dominance had already evolved. Fisher's hypothesized selection rates were tiny, on the order of mutation rates (one per million), in the evolution of dominance. But given his assumptions of populations of effectively infinite size, at least in theory these tiny rates of selection could cause the evolution of dominance.

The publication of Fisher's theory of the evolution of dominance (1928a, 1928b) stirred Sewall Wright into action. Like Fisher, Wright had read Darwin's *Origin* early in life; but later influences led him to develop a view of the mechanism of evolution that differed substantively from that held by neo-Darwinians. Four major research projects were most influential in shaping Wright's theory of evolution in nature: (1) Castle's selection experiment with hooded rats, (2) Wright's thesis research on interaction effects of Mendelian color factors, (3) inbreeding, outbreeding, and selection in guinea pigs, and (4) analysis of the transformation of the Shorthorn breed of cattle over time (Wright 1978).

From Castle's selection experiment on hooded rats Wright learned two crucial points: that mass selection (meaning selection on a random breeding population) of a merely quantitatively varying character could substantially and permanently change the expression of the character, and that this selection process had built-in limitations. The limitations stemmed primarily from a truth long observed by professional animal and plant breeders. Severe mass selection might indeed change a population rather rapidly, but at the cost of deleterious side effects, mostly expressed as loss of fecundity. In the breeding of large animals such as cattle, mass selection was a slow and tedious process, particularly when the character or characters being selected were not highly heritable. Many geneticists who in the 1930s and 1940s were influenced by the growing wave of neo-Darwinism tended to think of natural selection as only mass selection, as Darwin himself apparently did. Wright, while understanding the power of mass selection, was from the beginning of his work on evolution in nature also keenly aware of the limitations of mass selection in animal breeding and probably therefore also in nature.

His thesis research (Harvard 1912–1915) upon interaction effects in color characters in guinea pigs taught Wright clearly that organisms were built up of complex interaction systems rather than being, as Wright frequently said, mere mosaics of unit characters, each determined by a single gene. The same color gene might be expressed very differently in different genetic combinations; it followed that each gene had many multiple, if indirect, effects. To the animal breeder this meant that selection would be most effective by operating upon whole interaction systems rather than upon single genes. But in a large random breeding population, distinctive interaction systems of genes are rarely clearly expressed and therefore cannot be seized upon by the selective process. Thus in a large random breeding population the basic process of selection is limited to mass selection.

The start of a solution to this dilemma came from Wright's work with the highly inbred strains of guinea pigs at the United States Department of Agriculture between 1915 and 1925. Because of the random fixation of genes caused by the many generations of intense inbreeding, each strain became fixed with a mostly homozygous genetic complement, so that particular interaction systems were clearly expressed. Each inbred strain was easily distinguished from the others, and the wide range of variation in all characters between strains was striking. The inbreeding process had revealed the interaction systems so well hidden in the original random breeding population, making them available for the selection process. In actual animal breeding operations, intermediate rather than such intense inbreeding should be practiced to avoid the general decline in vigor and fecundity. Wright was, of course, aware of the use of inbreeding in hybrid corn production.

Finally, from his analysis of the breeding history of Shorthorn cattle, Wright found that a major breed had indeed experienced rather intense inbreeding during its foundation period. Selection had accompanied the differentiation from the inbreeding, and diffusion from the selected few herds had then made over the entire breed. Mass selection had played a relatively minor role. By 1923 Wright had a comprehensive view of what he thought was the best process of animal breeding.

To get from his theory of animal breeding to his theory of evolution in nature, Wright proceeded upon the plausible but wholly unproved assumption that evolution in nature occurred primarily by the mechanisms utilized by the best animal breeders. Because he had ample evidence from animal breeders who found that mass selection was frequently a slow, unsure, or even ineffective process, Wright decided that evolution in nature must proceed from a more efficient and effective process than mere mass selection. Judging from animal breeding, he thought that natural populations must be subdivided into partially isolated subgroups small enough to cause a kaleidoscopic random drifting of genes, but large enough that the random drifting did not lead directly to fixation of genes, for this was the road to degeneration and extinction. Selective diffusion from subgroups with successful genetic combinations was the step required for transformation of the whole species.

Population structure was the essential key. A breeder who practiced mass selection upon a very large, randomly breeding herd of cattle made very slow progress toward the desired type; but by artificially changing this population into inbred subgroups the breeder could soon reveal hidden variability, and use this as a basis for selection. In nature, there was no breeder to artificially alter population structure. Wright's belief, based upon very little evidence, was that Fisher's assumption of large random breeding populations in nature was unwarranted; instead, populations were probably more or less subdivided into partially isolated subpopulations in which some random drifting of gene frequencies occurred. Indeed, Wright's theory of evolution in nature was impossible if natural populations were large and random breeding. By the same reasoning, the view championed by R. A. Fisher during the late 1920s and early 1930s, that evolution proceeded by mass selection of single genes in large random breeding populations, was impossible if natural populations were subdivided in the way Wright thought. Unsurprisingly, so far as experimental evidence in natural populations for their differing views is concerned, the disagreements between Fisher and Wright hinged upon analyses of population structure in nature.

Fisher and Wright met in 1924, and soon after Wright read Fisher's paper "On the Dominance Ratio" (1922). Wright was greatly impressed by Fisher's attempt to derive a stochastic distribution of gene frequencies in populations; and using different quantitative methods, and a different qualitative view of evolution, Wright wrote a long manuscript on evolution in 1925. He discovered that he and Fisher disagreed on some quantitative points, and he did not revise the paper for publication until 1929 (it was published in 1931). After Fisher's papers on the evolution of dominance appeared in 1928, Wright published a reply (Wright 1929a, 1929b). Fisher then initiated a correspondence that ended in 1931. They were corresponding, therefore, as Wright was revising his "Evolution in Mendelian Populations" (1931) and as Fisher was writing his *Genetical Theory of Natural Selection* (1930).

The correspondence is revealing for many reasons, only two of which I shall mention here. First, neither Fisher nor Wright knew much about the genetics or systematics of natural populations. Fisher had used some of E. B. Ford's data from natural populations of Lepidoptera and Wright had collected and identified spiders, but both admitted to a severe lack of knowledge of natural populations. Second, Fisher was truly naive about effective population sizes and mechanisms of speciation. Until writing *The Genetical Theory of Natural Selection*, Fisher appears never to have even thought about speciation as a problem, focussing only upon changes of gene frequencies in idealized populations. On 13 August 1929, Fisher wrote to tell Wright that the effective population size of a species "must usually be the total population on the planet, enumerated at sexual maturity, and at the minimum of the annual or other periodic fluctuation. For birds twice the number of nests would be good." In other words, an entire species for the purpose of the statistical analysis of evolution was a random breeding Mendelian population. It was enough to make a person wonder how, given such a view, speciation could possibly occur.

When Fisher came to consider speciation, he realized that it required a break in effective population size and that "a gene frequency gradient is maintained by selection between different parts of a species' range. So that well marked local variations may or may not be incipient species, according as real fission, cessation of diffusion, ultimately supervenes" (Fisher to Wright, 9 September 1929). To Fisher, mechanisms of speciation were an afterthought, a simple extension of his theory of natural selection.

With *The Genetical Theory of Natural Selection* Fisher emerged as a highly influential evolutionary theorist, and as the most radical neo-Darwinist, totally emphasizing natural selection and adaptation as the only means of evolutionary change. In the first chapter, Fisher argued that Darwin's mistake was in accepting blending inheritance; but upon Mendelism, "a rational theory of Natural Selection can be based". Under Mendelism, none of the mechanisms other than natural selection were required to explain evolution. Thus Fisher criticized Weismann, who had a reputation as an arch-selectionist neo-Darwinian, for hypothesizing germinal selection and other schemes to assist natural selection, which Fisher said needed no help. He concluded:

> The tacit assumption of the blending theory of inheritance led Darwin, by a perfectly cogent argument, into a series of speculations, respecting the causes of variations, and the possible evolutionary effects of these causes . . . . The whole group of theories which ascribe to hypothetical physiological mechanisms, controlling the influence of mutations, a power of directing the course of evolution, must be set aside, once the blending theory of inheritance is abandoned. The sole surviving theory is that of Natural Selection, and it would appear impossible to avoid the conclusion that if any evolutionary phenomenon appears to be inexplicable on this theory, it must be accepted at present merely as one of the facts which in the present state of knowledge seems inexplicable. (1930, pp. 20–21)

So natural selection directed even the tiniest evolutionary change. Fisher derived in the book his "fundamental theorem of natural selection", which he thought must be the grand law governing all evolutionary change: "The rate of increase in fitness of any organism at any time is equal to its genetic variance in fitness at that time" (1930, p. 35). Fisher likened his fundamental theorem to the second law of thermodynamics: "Professor Eddington has recently remarked that 'The law that entropy always increases — the second law of thermodynamics — holds, I think, the supreme position among the laws of nature.' It is not a little instructive that so similar a law should hold the supreme position among the biological sciences" (1930, pp. 36–37).

Sewall Wright's theory of evolution was far more strongly tied to biology than was Fisher's. Wright's theory stemmed directly from laboratory genetics and the experience of breeders. Contrasting with Fisher's view that natural populations were extremely large and panmictic, Wright started from the assumption that natural populations were subdivided into partially isolated local populations. Fisher's fundamental theorem of natural selection, Wright thought, was a mathematical abstraction from the world of biology. He suggested that Fisher restate it to read: "The rate of increase in fitness of any population at any time is equal to its genetic variance in fitness at that time, except as affected by mutation, migration, change of environment, and effects of random sampling" (Wright to Fisher, 3 February 1931). In his review of Fisher's *Genetical Theory of Natural Selection*, Wright suggested alternative views of evolution in nature, ending with his own.

> If the population is not too large, the effects of random sampling of gametes in each generation brings about a random drifting of the gene frequencies about their mean positions of equilibrium. In such a population we can not speak of single equilibrium values but of probability arrays for each gene, even under constant external conditions. If the population is too small, this random drifting about leads inevitably to fixation of one or the other allelomorph, loss of variance, and degeneration. At a certain intermediate size of population, however (relative to prevailing mutation and selection rates), there will be a continuous kaleidoscopic shifting of the prevailing gene combinations, not adaptive itself, but providing an opportunity for the occasional appearance of new adaptive combinations of types which would never be reached by a direct selection process. There would follow thorough-going changes in the system of selection coefficients, changes in the probability arrays themselves of the various genes and in the long run an essentially irreversible adaptive advance of the species. It has seemed to me that the conditions for evolution would be more favorable here than in the indefinitely large population of Dr. Fisher's scheme. It would, however, be very slow, even in terms of geologic time, since it can be shown to be limited by mutation rate. A much more favorable condition would be that of a large population, broken up into imperfectly isolated local strains . . . . The rate of

> evolutionary change depends primarily on the balance between the effective size of population in the local strain (N) and the amount of interchange of individuals with the species as a whole (m) and is therefore not limited by mutation rates. The consequence would seem to be a rapid differentiation of local strains, in itself non-adaptive, but permitting selective increase or decrease of the numbers in different strains and thus leading to relatively rapid adaptive advance of the species as a whole. Thus I would hold that a condition of subdivision of the species is important in evolution not merely as an occasional precursor of fission, but also as an essential factor in its evolution as a single group. (Wright 1930, pp. 354–355)

Wright's shifting balance theory of evolution was certainly an alternative to Fisher's conception of natural selection as the mechanism of evolution. In domestic populations, the shifting balance process was highly "adaptive", that is, it conformed to the breeders' ideals at the level of the breed. Indeed, as Wright pointed out in the passage above, he expected the shifting balance process to lead to more rapid adaptation at the species level than did Fisher's conception of mass selection. Wright has claimed for years that his conception of random drift was misunderstood by many who assumed that he advocated non-adaptive differentiation at the level of species by means of random drift, a position he now vigorously denies: "I emphasize here that while I have attributed great importance to random drift in small local populations as providing material for natural selection among interaction systems, I have never attributed importance to nonadaptive differentiation of species" (1982, p. 12). This may be Wright's firm position now, but I think his position was far less clear in the period 1929–1932.

A major reason for his lack of clarity can be traced to a view held by many prominent systematists at that time. Like G. Robson, these systematists held that the taxonomic differences at the lowest levels, through the species and even up to the genus level, were basically non-adaptive (I shall develop this point further in the next section). Wright then had to fit his shifting balance theory, which supposedly led to adaptive advance above the demic level, with a science of systematics that told him adaptation was not the rule until well above the species level. Wright's response was confusing, as illustrated by following ten direct quotes from his published papers in the years 1929–1932:

> 1. [Assessing the influence of random drift in isolated populations] The non-adaptive nature of the differences which usually seem to characterize local races, subspecies, and even species of the same genus indicates that this factor of isolation is in fact of first importance in the evolutionary origin of such groups, a point on which field naturalists (e.g. Wagner, Gulick, Jordan, Osborn, and Crampton) have long insisted. (1929b, pp. 560–561)
> 2. The actual differences among natural geographical races and subspecies

are to a large extent of the nonadaptive sort expected from random drifting apart. (1931, p. 127)

3. [Fisher's] theory is one of complete and direct control by natural selection while I attribute greatest immediate importance to the effects of incomplete isolation. (1931, p. 149, fn.)

4. The direction of evolution of the species as a whole will be closely responsive to the prevailing conditions, orthogenetic as long as these are constant, but changing with sufficiently long continued environmental change. (1931, p. 151)

5. Adaptive orthogenetic advances for moderate periods of geological time, a winding course in the long run, nonadaptive branching following isolation as the usual mode of origin of subspecies, species, perhaps even genera, adaptive branching giving rise occasionally to species which may originate new families, orders, etc . . . . are all in harmony with this interpretation. (1931, p. 153)

6. [The shifting balance process] originates new species differing for the most part in nonadaptive respects but is capable of initiating an adaptive radiation as well as of parallel orthogenetic lines, in accordance with the conditions. (1931, p. 158)

7. [Under the shifting balance theory] complete isolation of a portion of a species should result relatively rapidly in specific differentiation, and one that is not necessarily adaptive. The effective intergroup competition leading to adaptive advance may be between species rather than races. Such isolation is doubtless usually geographic in character at the outset but may be clinched by the development of hybrid sterility. (1932, p. 363)

8. That evolution involves nonadaptive differentiation to a large extent at the subspecies and even the species level is indicated by the kinds of differences by which such groups are actually distinguished by systematists. It is only at the subfamily and family levels that clear-cut adaptive differences become the rule (Robson 1928; Jacot 1932). The principal evolutionary mechanism in the origin of species must then be an essentially nonadaptive one. (1932, pp. 363–364)

9. Subdivision into numerous local races whose differences are largely nonadaptive has been recorded in other organisms wherever a sufficiently detailed study has been made. [There follows citation of the work of Gulick, Crampton, David Starr Jordan, Ruthven, Kellogg, Osgood, Kinsey, Osborn, Rensch, Schmidt, David Thompson, and Sumner.] (1932, pp. 364–365)

And finally, Alfred C. Kinsey, who was a taxonomist specializing in the gall wasps of the genus *Cynips* before turning his attention to human sexuality, wrote to Wright for his 1931 paper, and enclosed his own monograph on

Cynips. Wright, looking for corroboration of his theory, hoped that Kinsey's wasps might show random differentiation:

> 10. I am especially interested in the question as to how far there is subdivision of species into small local strains differentiated in the *random* fashion expected of inbreeding (instead of in adaptive ways by natural selection). My results seem to indicate that such a condition is the most favorable for progressive evolution of the species as a single group. (Wright to Kinsey, 14 April 1931)

Kinsey replied, incidentally, that his team routinely searched over a wide area "in order to avoid such local strains and to obtain a more complete idea of the species as a whole" (Kinsey to Wright, 22 April 1931). Thus Kinsey's field work could not furnish the precise corroboration Wright desired.

Viewed all together at one time, these ten citations shed some light upon the question of why Wright was so much misunderstood during the 1930s and later. The statement that "the direction of evolution of the species as a whole will be closely responsive to the prevailing conditions" (citation #4) is inconsistent with the statement that shifting balance leads to "nonadaptive branching following isolation as the usual mode of origin of subspecies, species, and even genera" (citation #5), or "the principal evolutionary mechanism in the origin of species must thus be an essentially nonadaptive one" (citation #8). The careful reader in 1932 would almost certainly conclude that Wright believed non-adaptive random drift was a primary mechanism in the origin of races, subspecies, species, and perhaps genera. Wright's more recent view that the shifting balance theory should lead to adaptive responses at least by the subspecies level is found nowhere in the 1931 and 1932 papers, or in the letter to Kinsey. In any case, with these citations in mind one can easily understand why some biologists understood Wright to be saying that random drift played the dominant role in the origin of subspecies and species. Certainly this is what R. A. Fisher, E. B. Ford, Julian Huxley, and Theodosius Dobzhansky took Wright to be saying.

The essential element of confusion in Wright's shifting balance theory concerned the problem of levels of selection in relation to adaptation. The shifting balance theory requires at the very least "interdemic selection". But "interdemic selection" is not itself a mechanism, but a description of the interaction of individual selection in combination with population structure and migration. Wright fully admits that the cutting edge of interdemic selection is individual selection. Then what should one make of Wright's remark in quote #7 that "the effective intergroup competition leading to adaptive advance may be between species rather than races?" I don't know what to make of it except to conclude that Wright was confused about mechanisms of speciation.

Fisher and Wright left an influential legacy to evolutionary biology after the early 1930s. Each had a coherent mechanism of monophyletic microevolution. Together with J. B. S. Haldane and a few others, the mathematical population geneticists provided a framework in which the evolutionary synthesis could develop. The tension between the two mechanisms of microevolution proposed by Fisher and Wright was enormously creative. The only possible way to really understand the work on the genetics of natural populations done by Dobzhansky and his followers or by E. B. Ford and his followers is in terms of the tensions between Fisherean and Wrightean viewpoints, as I shall show in the next section.

But if a clear difference between the views of Fisher and Wright could be detected at the level of microevolution, the same was not true regarding the population genetic mechanisms of speciation. Here Fisher and Wright left a legacy of theoretical confusion that continues to exist today. Goldschmidt, whether or not one accepts his hopeful monsters as the key to speciation, accurately pinpointed the legacy of Fisher and Wright in 1948, when the evolutionary synthesis had already occurred: "In spite of many assertions to the contrary, [population genetics] has failed to throw any light upon evolution above the subspecific and subsubspecific level of the investigations" (Goldschmidt 1948, p. 14). Goldschmidt was right. As one moved from the level of evolution in populations to the level of speciation, the foundation stone of population genetics turned into sand.

IV. The Adaptationist Program in the Evolutionary Synthesis

Gould has recently argued that the adaptationist program grew substantially in strength during the period of the evolutionary synthesis (1982, pp. 381–382). He used as primary evidence the work of George Gaylord Simpson and also Malcolm Kottler's (as yet unpublished) analysis of David Lack's changing interpretations of adaptation in Darwin's finches. Gould further suggests that this "hardening of the synthesis" cries for explanation. I think Gould's thesis is basically accurate, although the development is very complex; and I shall offer here only a few suggestions for explanation in relation to the question of mechanisms of evolution.

The question of adaptation was closely tied to the situation in systematics. Ernst Mayr has described this situation on the eve of the synthesis in the 1920s as extremely complex (Mayr and Provine 1980, chap. 4). He suggests, however, that most systematists believed that geographic variation was adaptive and that Robson, Crampton, Richards, Kinsey and others who stressed non-adaptive differentiation were "very much in the minority". The evidence strongly indicates to me that, on the contrary, systematists who advocated non-adaptive differentiation were at least as numerous as those emphasizing adaptation. The strong influence of the non-adaptationists, especially upon geneticists, is undeniable.

The most ardent neo-Darwinians in the 1920s, including Poulton, Fisher, Ford, Goodrich, Julian Huxley, and others, were in England. Yet at the same time there flourished in England an influential group of neo-Lamarckians (R. Burkhardt 1980) and a non-adaptationist trend in systematics. The two most visible systematists with this view were G. C. Robson and O. W. Richards, who began publishing together in the mid-1920s. They strongly advocated the view that the taxonomic characters used to distinguish closely allied species were generally non-adaptive, and they convinced others.

A good example of their influence can be seen in the work of Charles Elton, a pioneer in the field of ecology during the synthesis period. Elton was one of the outstanding students, along with E. B. Ford, who studied with Julian Huxley at Oxford in the early 1920s. Huxley believed that ecology was destined to have a great future and encouraged Elton to go into the subject. Elton's first book, *Animal Ecology* (1927), appeared in a series edited by Huxley, and began with a glowing introduction written by Huxley. In the chapter on ecology and evolution, Elton summarized and agreed with the conclusions of a recent article by Richards and Robson (1926):

> The gist of their conclusions is that very closely allied species practically never differ in characters which can by any stretch of the imagination be called adaptive. If natural selection exercises any important influence upon the divergence of species, we should expect to find that the characters separating species would in many cases be of obvious survival value. But the odd thing is that although the characters which distinguish genera or distantly allied species from one another are often obviously adaptive, those separating closely allied species are nearly always quite trivial and apparently meaningless.
>
> It seems probable that the process of evolution may take place along these lines: genotypic variations arise in one or a few individuals in the population of any species and spread by some means that is not natural selection; this process, combined with various factors which lead to the isolation of different sections of the population from one another, results in the establishment of varieties and species which differ in comparatively trivial and unimportant characters. Later on, natural selection is ultimately effective, probably acting rather on populations than on individuals. (Elton 1927, pp. 184–185)

Elton, who had already begun to study the subject of animal numbers that would make him a famous ecologist, offered a suggestion about the mechanism leading to non-adaptive differentiation:

> Many animals periodically undergo rapid increase with practically no

> checks at all. In fact, the struggle for existence sometimes tends to disappear almost entirely. During the expansion in numbers from a minimum, almost every animal survives, or at any rate a very high proportion of them do so, and an immeasurably larger number survives than when the population remains constant. If therefore a heritable variation were to occur in the small nucleus of animals left at a minimum of numbers, it would spread very quickly and automatically, so that a very large proportion of numbers of individuals would possess it when the species had regained its normal numbers. In this way it would be possible for non-adaptive (indifferent) characters to spread in the population, and we should have partial explanation of the puzzling facts about closely allied species, and of the existence of so many apparently non-adaptive characters in animals. (1927, p. 187)

I have cited Elton at some length here for two reasons. The first is to show that he, like Wright, was influenced by the views of Robson and Richards. The two most widely read and cited books on systematics (in English) were Robson's *The Species Problem* (1928) and *The Variation of Animals in Nature* (1936, but written by 1933) by Robson and Richards. Not until *The New Systematics* edited by J. Huxley (1940) was a strongly adaptationist single general work on systematics available in the English language. And in that book, Huxley himself in the long introduction left plenty of room for non-adaptive differentiation, citing Wright, Diver, and others in the volume as representing that point of view. The other reason for citing Elton is to demonstrate that he and Wright felt compelled to devise mechanisms of evolution that would lead to non-adaptive speciation. Only in this light is it possible to understand why Wright published the suggestion that random drift generally led to non-adaptive speciation in the years 1929–1932.

Many systematists adopted an adaptationist view before the appearance of the evolutionary work of the mathematical population geneticists. An excellent example is F. B. Sumner. From the time he began full-time research on *Peromyscus* in 1913 until 1925 he strongly rejected the concept of Mendelian multiple factors and the protective coloration hypothesis and advocated the neo-Lamarckian view that humidity in the environment caused non-adaptive differentiation of coat colors. In 1925, Sumner's own crosses indicated Mendelian multi-factorial differences between geographical races of *Peromyscus* and he began to move toward the view that gradual selection of small heritable differences was the key to evolution in nature. By the time Sumner published his influential monograph on *Peromyscus* in 1932, he was strongly adaptationist and neo-Darwinian in his thinking (Sumner 1932; for an account of Sumner's development see Provine 1979). In 1932, Wright was still citing Sumner's work on *Peromyscus* as indicating non-adaptive differentiation between geographical races.

In botany, by 1922 Turesson had developed his very influential concept of "ecotypes" (1922). He argued that ecological factors working by natural selection always molded organisms to their immediate environments, thus creating subspecies highly adapted to their ecological settings. But Turesson was quick to point out that he meant genotypic not morphological adaptation; morphological differences between geographical races were inadequate guides to adaptational differences.

Generally speaking, as systematists moved closer to an adaptationist view during the synthesis period, the need for models to explain non-adaptive differentiation declined. One reason, however, why the situation is so complex is that systematists were affected by the developments in genetics and by the work of the mathematical population geneticists. A major reason why Mayr and Rensch became neo-Darwinians was their increasing knowledge of Mendelian multifactorial inheritance and the realization, rendered inescapable by simple quantitative models, that even small selection rates could drastically alter gene frequencies in a population in surprisingly few generations. The move toward a more selectionist and adaptationist outlook thus involved in part an interrelationship between systematists and geneticists.

The most enlightening way I have found to investigate the move toward a more selectionist/adaptationist view during the synthesis is to examine the consequences of the tension clearly visible between the general evolutionary views of Fisher and Wright.

Fisher was unquestionably the "Hyper-Darwinian" described by Goldschmidt (1955). Fisher was more thoroughly a selectionist/adaptationist than any other evolutionist before him, and perhaps any after him. His great quantitative abilities impressed geneticists and gave credence to his models. Fisher believed in his fundamental theorem of natural selection and thought that selection so dominated evolution that non-adaptive characters were virtually non-existent, except for some secondary sexual characteristics produced by sexual selection.

E. B. Ford, who came from the neo-Darwinian naturalist tradition of Lankester and Poulton (see Mayr and Provine 1980, pp. 329–342), was the ideal person to complement Fisher's abstract models with his own very meticulous field research on natural populations. Together, Fisher and Ford constituted a powerful team pushing always toward the selectionist/adaptationist view.

Fisher and Ford found each other in 1923 and were actively working and publishing together by the late 1920s. Wright strongly felt a need for someone with a greater knowledge of natural populations, but not until 1936 did he find the right person. Then Dobzhansky came seeking Wright's aid for his projected series on the genetics of natural populations of *Drosophila pseudoobscura*, and there ensued a very fruitful and influential collaboration. It began with Dobzhansky finding different frequencies and arrangements of inversions in local populations of *D. pseudoobscura*, assuming that such differences should be expected on Wright's shifting balance theory. From 1937 until 1942 (that is, after the second edition of Dobzhansky's *Genetics and the Origin of Species* in 1941), Dobzhansky and Wright advocated the view that the inversions were non-adaptive and differed in frequency in local populations from the effects of random drift. In England, Cyril Diver was at the same time following the tradition of Gulick and Crampton and finding non-adaptive differences between local populations of snails of the genus *Cepaea*, declaring the cause to be random differentiation in small partially isolated populations (see Diver 1940 for summary).

During the 1940s and 1950s Fisher, Ford, and Ford's students vigorously pursued the adaptationist program. The notable successes of the adaptationists did much to further their cause. In 1940, three of the foremost examples of non-adaptive differentiation used by Wright were inversions in *D. pseudoobscura*, bloodgroups, and differences in local populations of snails. By the mid-1950s all three had been shown (or at least persuasively argued) to be subject to considerable selection pressures; furthermore, the adaptationists cogently argued that systematists such as Gulick or Crampton had not employed proper experimental procedure to detect the selective value of the apparently non-adaptive characters they observed in snails. Strong support for the adaptationist view came from David Lack's conversion from a non-adaptationist to an adaptationist view between the first (1945, but submitted in 1940) and second (1947, but submitted in 1944) of his monographs on Darwin's finches. E. B. Ford's writings, beginning with the first edition of his *Mendelism and Evolution* (1931), celebrated the findings that apparently non-adaptive characters really had adaptive value after all. The books of Cain (1954), Sheppard (1958), Kettlewell (1973), and the succeeding editions of Ford's *Mendelism and Evolution* and his *Ecological Genetics* (first published 1964) amply document the successes of the adaptationist view. To fully document the thesis of this paragraph would require a substantial monograph; I can only refer the reader to the works cited.

The influence of the adaptationist view of Fisher and Ford can be traced in a particularly nice example in the writings of Julian Huxley, who prided himself on the scope of his evolutionary views. His 1942 book, *Evolution: The Modern Synthesis*, relied heavily and explicitly upon Fisher and Wright. At this time Huxley believed that the selectionist viewpoint, after suffering during the anti-selectionist interlude dominated by Bateson and De Vries, was making a strong revival. Yet he also trusted the systematists who argued that many of the differences between the lower taxonomic groups were non-adaptive, and he thought Wright's idea of random genetic drift in small partially isolated populations was the causal agent. So in 1942 Huxley wavered back and forth between Fisher and Wright, citing Fisher thirty-four times and Wright thirty-seven times, trying to make room for both of their mechanisms of evolution.

By 1959 Huxley was extolling "the omnipresence of detailed adaptation, or biological fitness, as some modern workers prefer to call it" (J. Huxley 1960a, p. 11); and in his new introduction to the 1963 photographic reprint of *Evolution: The Modern Synthesis*, he specifically retreated from his earlier non-adaptationist position, warning the reader that on the authority of Ford and Cain, local differentiation in the snail genera *Partula* and *Cepaea* was adaptive rather than non-adaptive as his book stated.

After the appearance of his book *Systematics and the Origin of Species* (1942), Ernst Mayr became a major force in promoting an adaptationist view. Mayr kept insisting in private, and in print (1945, pp. 73–76), that the inversions of *Drosophila pseudoobscura* were subject to substantial selection coefficients. But Mayr was not an arch-selectionist comparable to Fisher and Ford. For example, in *Systematics and the Origin of Species* Mayr stated clearly:

> It should not be assumed that all the differences between populations and species are purely adaptational and that they owe their existence to their superior selective qualities. We have already pointed out the fallacy of such a point of view in the discussion of neutral polymorphism. Many combinations of color patterns, sports, and bands, as well as extra bristles and wing veins, are probably largely accidental. This is particularly true in regions with many stationary, small, and well-isolated populations, such as we find commonly in tropical and insular species . . . . We must stress the point that not all geographic variation is adaptive (1942, p. 86)

By 1963 Mayr had moved closer to a fully adaptationist position, while still leaving room for some non-adaptive geographic variation:

> Each local population is the product of a continuing selection process. By definition, then, the genotype of each local population has been selected for the production of a well-adapted phenotype. It does not follow from this conclusion, however, that every detail of the phenotype is maximally adaptive. If a given subspecies of ladybird beetles has more spots on the elytra than another subspecies, it does not necessarily mean that the extra spots are essential for survival in the range of that subspecies. It merely means that the genotype that has evolved in this area as the result of selection develops additional spots on the elytra. When studying geographic variation in the voice of birds, in the plumes of birds of paradise, or in the color patterns of parrots and pigeons, one must never ignore the possibility that some of the phenotype is merely the incidental by-product of the pleiotropic action of genes selected for other contributions to the viability of the phenotype. Yet close analysis often reveals unsuspected adaptive qualities even in minute details of the phenotype, for instance, in the body proportions of island lizards. (1963, p. 311)

Although Mayr in 1942 was well short of the adaptationist views of Fisher and Ford, his views (which he had held in 1942 for a decade) nevertheless represented a significant shift toward adaptationism from the position delineated in the Robson and Richards book appearing only six years earlier in 1936. Mayr's book became the new major book in systematics, and it unquestionably drew young systematists away from the non-adaptationist views held by many systematists up into the 1930s.

Focussing as it does upon species and speciation, Mayr's *Systematics and the Origin of Species* might be expected to offer more than Fisher or Wright did on the problem of extending microevolution to speciation and the evolution of higher taxa. It does so by presenting the "founder" principle:

> The reduced variability of small populations is not always due to accidental gene loss, but sometimes to the fact that the entire population was started by a single pair or by a single fertilized female. These "founders" of the population carried with them only a very small proportion of the variability of the parent population. This "founder" principle sometimes explains even the uniformity of rather large populations, particularly if they are well isolated and near the borders of the range of the species. (Mayr 1942, p. 237)

Mayr claimed no originality for the concept of "founder" principle, clearly outlined before him by Wagner, Gulick, Rensch, Crampton, and others; but he did invent a highly appropriate term for the principle, and emphasized it more than his predecessors.

The "founder" principle appears in the long chapter of the 1942 book on the biology of speciation (Chapter 9), in a section entitled "Population Size and Variability". The section was explicitly based upon Wright's theoretical work as well as on the observations of naturalists. Mayr's central conclusion was "that evolution should proceed more rapidly in small populations than in large ones, and this is exactly what we find" (1942, p. 236). The problem was that the founder principle did not by itself explain speciation because recently isolated groups were still fully fertile with, and in the usual range of variability of, the mother population. Also required were biological isolating mechanisms, and Mayr later in the chapter devoted twelve pages to detailing them.

Still missing was an understanding of the genetic basis, and acquisition, of the biological isolating mechanisms, and indeed of the evolution of the morphological differences used by taxonomists to distinguish species. In other words, the founder principle is not by itself even a mechanism of microevolution, much less of speciation. For Mayr, after initial isolation, selection became the primary determinant. In the introduction to the section entitled "Selective Factors and Species Formation", Mayr pinpointed the difficulty I have stressed throughout this paper, addressing

> the question of how selective factors influence the establishment of discontinuities and what selective factors tend to enlarge the gaps between

> incipient species. Competition and predation are generally listed as the two most important factors to be considered in this connection. A survey of this field indicates, unfortunately, that our knowledge of the actual influence of these factors on the speciation process is still very slight. In fact, it is surprising how badly ecologists have neglected these questions. (1942, p. 271)

Mayr's theory of geographic speciation in 1942 had crucial gaps; at best, his evidence pertained directly to subspeciation, not speciation, and he purposely avoided discussion of the genetics of speciation, referring the reader to the second edition of Dobzhansky's *Genetics and the Origin of Species* (1941). Despite its title, that book was through and through dealing with the genetics of taxonomic groups *below* the rank of species. Yet neither Dobzhansky nor Mayr had any doubt, despite the lack of evidence, that speciation and macroevolution were extrapolations of microevolution. Mayr's *Systematics and the Origin of Species* ended:

> In conclusion we may say that all the available evidence indicates that the origin of the higher categories is a process which is nothing but an extrapolation of speciation. All the processes and phenomena of macroevolution and of the origin of the higher categories can be traced back to intraspecific variation, even though the first steps of such processes are usually very minute. (1942, p. 298)

In 1954, Mayr buttressed his founder principle with the concept of "genetic revolutions" occurring in small isolated populations.

> *The mere change of the genetic environment may change the selective value of a gene very considerably.* Isolating a few individuals (the "founders") from a variable population which is situated in the midst of the stream of genes which flows ceaselessly through every widespread species will produce a sudden change of the genetic environment of most loci. This change, in fact, is the most drastic genetic change (except for polyploidy and hybridization) which may occur in a natural population, since it may affect all loci at once. Indeed, it may have the character of a veritable "genetic revolution." Furthermore, this "genetic revolution," released by the isolation of the founder population, may well have the character of a chain reaction. Changes in any locus will in turn affect the selective values at many other loci, until finally the system has reached a new state of equilibrium. (1954a, pp. 169–170)

Sewall Wright had long emphasized the revolutionary effects (not merely of random drift alone) of isolation of small groups of individuals, and Mayr's use of the idea of genetic revolution was important to better explain speciation. The limitation in 1954 was that Mayr simply stated the idea, and it was too general to elucidate precise genetic mechanisms of speciation.

Botanists in general tended to be less adaptationist than geologists during the synthesis period, and of course emphasized more the possibilities of discontinuous evolution from chromosomal rearrangements and polyploidy. Ledyard Stebbins reflected this view in his *Variation and Evolution in Plants* (1950). He left considerable room for non-adaptive differentiation, and argued that "random fixation", "drift", or "the Sewall Wright effect" was "undoubtedly the chief source of differences between populations, races, and species in non-adaptive characteristics" (1950, p. 145). Stebbins did not pretend to know what the balance was in nature between selection pressures, heritability, and effective population sizes: "At present, far too little is known about any of these factors" (p. 145). Yet I think it would be fair to say that, as in the case of Mayr's *Systematics and the Origin of Species*, Stebbins in 1950 represented a shift among botanists toward a more adaptationist viewpoint. Certainly his book became the standard source for the evolutionary synthesis in botany.

If Stebbins was unsure of the exact mechanisms of microevolution in particular cases, he had no doubt that these mechanisms led directly to speciation and macroevolution.

> Individual variation, in the form of mutation (in the broadest sense) and gene recombination, exists in all populations; and the molding of this raw material into variation on the level of populations, by means of natural selection, fluctuation of population size, random fixation, and isolation is sufficient to account for all the differences, both adaptive and nonadaptive, which exist between related races and species. In other words, we do not need to seek unknown causes or motivating agencies for the evolution going on at present. And the differences between genera, families, orders, and higher groups of organisms . . . are similar enough to interspecific differences so that we need only to project the action of these same known processes into long periods of time to account for all of evolution. (1950, p. 152)

The evidence I see during the synthesis period yields the conclusion that a move toward a stronger adaptationist view did occur, although by no means did all evolutionists move to the very extreme view held by Fisher, Ford, and their followers. The effect can be seen upon Sewall Wright. Although he said in print that random drift accounted for non-adaptive differentiation of races and species — and this is the view other evolutionists took from him — by 1949 Wright was asserting that his shifting balance process was highly adaptive at least by the subspecies level and certainly by the species level; and furthermore he asserted that he had always held this view (Wright 1948).

Along with the move toward greater adaptationism came the conclusion so clearly reached by Mayr and Stebbins — that speciation and macroevolution proceeded from long continued microevolution. I cannot help sympathizing with Goldschmidt's frustration as he saw this conclusion become dominant. He saw clearly that the evidence available did not drive one to accept the view that microevolution led to speciation and macroevolution. Goldschmidt's hopeful monsters, however, were no match for the adaptationism of the evolutionary synthesis. The evolutionary synthesis occurred without resolution of the Wright/Fisher tension. The primary mechanism of microevolution was still undecided; but whatever it was, it led in time (in conjunction with geographic isolation) directly to speciation and macroevolution. Elucidation of the genetic mechanisms of speciation is not one of the great triumphs of the evolutionary synthesis.

V. Beyond the Evolutionary Synthesis: Mechanism of Evolution and Adaptation Today

The period of the evolutionary synthesis began in the 1930s with conflicting theoretical views of microevolution and ended in 1959 with the conflict unresolved. On certain quantitative issues, however, agreement is universal. From Fisher, Haldane, and Wright to Lewontin, Kimura, and Karlin, population geneticists have agreed that evolution in nature is strongly dependent upon population structure (effective population size in particular), migration, and rates of selection. The essential problem is that the magnitudes and effects of these crucial variables in nature are poorly documented, and no one will truly know how microevolution proceeds in nature until they are well documented.

Population structure in nature is very difficult to analyze to the degree necessary for precise understanding of microevolution. If a species is subdivided the way Wright envisions in his shifting balance theory as expressed in volume three of his *Evolution and the Genetics of Populations* (1977), then microevolution proceeds differently from the way it proceeds under Fisher's view of rather large panmictic populations. Effective population sizes, however, are highly sensitive to migration — very small migration rates, determined only with the greatest difficulty, dramatically expand effective population sizes. On the other hand, periodic population bottlenecks recurring at intervals sufficiently far apart to make continuous monitoring difficult or impossible can greatly reduce effective population sizes. An added problem is that population structure changes, often drastically, over just a few generations. Determination of effective population sizes in relation to migration is one of the most essential tasks of the future study of natural populations.

Fisher argued that genic selection rates of one in a million were sufficient to cause observed evolutionary change. Wright said such small selection coefficients would be overcome by other factors, random drift in particular, affecting the frequency of the same genes. Unfortunately, selection rates far larger than those Fisher envisioned are still undetectable in the study of natural populations. Biologists do study cases in which large selection rates can be determined; but no way exists to assess how much of evolution in nature is caused by selection rates comparable to those that can be measured, and how much by selection rates too small to be measured. If Fisher's general view were correct, then the selective processes that cause evolution could never be measured in natural populations.

If effective population sizes, migration rates, and selection rates are insufficiently known, the inevitable conclusion is that we cannot know whether microevolution is primarily adaptive or non-adaptive. Moreover, we cannot evaluate the claim that speciation results (or does not result) from a continuation of the mechanisms of microevolution in relation to geographic isolation. It is possible, as Wright's shifting balance theory suggests, for microevolution to be largely non-adaptive at the level of local populations, but highly adaptive by the species level through the mechanism of interdemic selection (that is, individual selection in combination with population structure and migrational diffusion).

Motivated primarily by the wish to temper the highly adaptationist claims of some sociobiologists, Gould and Lewontin have in recent years re-examined the adaptationist program (Lewontin's term that I have used freely in this paper; Lewontin 1979; Gould and Lewontin 1979). They present a list of non-adaptive mechanisms of evolution, and ask for a return to Charles Darwin's "pluralistic" approach to mechanisms of evolution (Gould and Lewontin 1979, pp. 590–593). The mechanisms they propose are plausible enough. They are, however, primarily (though not entirely) mechanisms of evolution at the level of populations, and cannot necessarily be extrapolated to the level of speciation. The other problem is familiar — no one knows how much weight to place upon these non-adaptive mechanisms, because too little is known of microevolution.

Can we not simply turn to working systematists, who have had the benefits of the "new systematics" of the evolutionary synthesis, to discover whether the differences between closely allied species are adaptive or non-adaptive? My own informal (should I say unsystematic?) survey suggests the negative answer. A taxonomist of ants declares that yes, of course the closely allied species differ by morphological characters of significant selective value, including those characters used for classification. A taxonomist of beetles, on the contrary, asserts that probably most of the characters he uses to distinguish closely allied species have no conceivable adaptive value. The crucial variable in this great difference of view has little to do with differences between ants and beetles, but is instead a function of where and with whom the systematist took his graduate training. Whether the observed differences between closely allied species are adaptive or non-adaptive is directly pertinent to mechanisms of microevolution and speciation, and a critical survey of current knowledge in relation to this question would be an important addition to evolutionary thought.

To examine ideas about mechanisms of speciation since the synthesis period, I shall take three representative works, separated by about ten years: Mayr's *Animal Species and Evolution* (1963), Lewontin's *Genetic Basis of Evolutionary Change* (1974a), and A. Templeton's "Mechanisms of Speciation — A Population Genetic Approach" (1981). Since 1959 Mayr has frequently stated that population genetics has contributed little to the understanding of speciation. I think Mayr is basically correct in this assertion (one of the rather few points on which Goldschmidt, Waddington, and Mayr could all agree!) To account for speciation, Mayr emphasized his concept of "founder" populations with their concomitant "genetic revolutions", and sufficient periods of geographic isolation for at least the initial events of speciation. Mayr's concept of speciation has been highly influential, but it shares a problem of the sort that accompanies the idea of natural selection. The concept is too general to give a very accurate idea of what actually happens genetically in founder populations, nor does the concept explain why so many isolate populations do not undergo genetic revolutions and lead to speciation (Lande 1980; Charlesworth, Lande, and Slatkin 1982). In short, what Mayr's concept of speciation required for concreteness was population genetics theory and greater knowledge of the genetics of natural populations.

Lewontin began his chapter "The Genetics of Species Formation" with the statement that

> It is an irony of evolutionary genetics that, although it is a fusion of Mendelism and Darwinism, it has made no direct contribution to what Darwin obviously saw as the fundamental problem: the origin of species. (1974, p. 159)

So Lewontin and Mayr agreed on this point, and Lewontin further emphasized that "*we know virtually nothing about the genetic changes that occur in species formation*" (1974a, p. 159)

A significant advance, however, had occurred since Mayr's 1963 book. The only way to study sterility factors and genetic differences before the advent of electrophoretic allozyme studies was by breeding experiments that were possible only with species producing fertile hybrids. But after the work of Hubby, Throckmorton, and Lewontin in the late 1960s and early 1970s it was possible to document at least certain kinds of genetic differences between species without crossbreeding experiments. The evidence in 1972 was very sparse, particularly on the early stages of speciation, but Lewontin emphasized that many more cases should be studied: "the tools are readily available."

The large amount of genetic evidence on species differences that has accumulated since Lewontin wrote his book in the early 1970s is apparent in Alan Templeton's review article (1981). Part of this genetic evidence came from karyotyping and protein or DNA analyses rather than crossbreeding experiments. The implications of this evidence for the understanding of speciation were disappointing. Lewontin and many others had already pointed out that although the initial speciation events required some genetic change, measurements of genetic species differences had to cover both those required for initial speciation and those acquired since. According to Templeton,

> It is virtually impossible to sort out what differences are actually associated with the process of speciation and which are consequences of evolution subsequent to speciation. This is a real problem because many species differences contribute little or nothing to reproductive isolation. In the light of this perplexity, hybridization experiments, although of limited applicability, provide the best tool for distinguishing between the genetics of speciation and the genetics of species differences. (1981, p. 25)

Templeton's assessment, with which I certainly agree, is that we do not yet have much understanding of genetic mechanisms of speciation.

Working from observed punctuated equilibria in the geological record, and faced with mechanisms of speciation of uncertain explanatory value, Stanley (1979) and Gould (1982) have revived the old concept of "species selection". I sympathize with this attempt to find mechanisms other than the straightforward "extension of microevolution" to explain patterns of speciation and evolution in the higher taxa. But the idea of species selection does not appear to be a mechanism, but rather a term describing patterns of speciation. Hierarchical levels of selection corresponding to genes, individuals, demes, groups, populations, species, genera, etc. break down early in the series. Interdemic selection is nothing more than individual selection interacting with population structure and migrational diffusion. Group selection may be a reality (supposing that someone can finally turn up a convincing example of it in nature), but I agree with David Sloan Wilson that group selection cannot be logically extended to a concept of species selection. If there is species selection, is there also genus or family selection? "Selection" at these levels can be deduced from patterns of evolutionary descent, but again selection at these levels appears to be a name for the pattern and not a biological mechanism. Here I agree with Templeton:

> The predictions of macroevolutionary theories depend critically upon the assumed speciation mechanisms and upon a proper understanding of microevolutionary processes. Hence, a comprehensive theory of evolution that integrates macro- and microevolution can be formed only after the integration of population genetics and speciation theory. (Templeton 1981, p. 40)

The prospects of knowing more about the mechanisms of evolution at all levels are, however, much brighter now than during the period of evolutionary synthesis of the 1930s and 1940s. For microevolution, systematic compilation of already extant data on population structure, effective population size, and migration rates might be very helpful, especially if coupled with new critical field work. Templeton (1980) has himself pointed to ways of integrating population genetics theory with qualitative mechanisms of speciation by his reinterpretation of Mayr's founder principle, and by his later more detailed attempt to relate most modes of speciation to population genetics theory (Templeton 1981, see especially his table on p. 27). Walter Fitch has recently been devising methods for using the new technologies of molecular biology for constructing phylogenies of the genetic changes involved in the beginnings of species formation. I predict that during the next decade the cutting edge of the analysis of the genetic mechanisms of speciation will focus upon the very initial stages of speciation, where reproductive isolation is not only incomplete, but barely detectable. In this case both crossbreeding experiments and analysis by the techniques of molecular biology are possible, and the differences are fewer and less complex. The near future holds the possibility of much deeper understanding of the mechanisms of evolution in nature.

29

Darwin on Natural Selection: A Philosophical Perspective

Elliott Sober

Introduction

Whig history is full of threats and promises. Interpreting the past in terms of the present has its dangers; since the present did not cause the past, one can be misled in the search for explanation. But when the question we put to the past concerns its meaning, matters change; seeing the significance of the past may well essentially involve seeing it in terms of the present.

To discern Darwin's achievement — identifying both what he saw and what he failed to see — is in part to locate his ideas in conceptual space. We now understand evolution and natural selection in a much more systematic and thorough way than Darwin did. Because of this, it is remarkable that so many of his ideas have remained canonical. He speaks to us as a contemporary; his concepts are ours. Yet in other respects, Darwin's ideas have been supplanted and survive now only as special cases of our more general outlook. Perhaps it is not too misleading to compare this aspect of his relationship to the present with the relationship of Newtonian physics to relativity theory. From the vantage-point of contemporary theory, we can see the older theory as holding approximately within a certain range of parameter values. In the physical case, it is well known what these parameter values are; for low velocities, classical dynamics approximates relativistic dynamics. Stray from these values, and the fit deteriorates. One task of this paper is to establish that a similar relationship obtains in the evolutionary case.

Whig history falls prey to fallacy when the significance of a past event is conflated with its cause. We understand the meaning of Darwin's theory by locating it in our more general conception of evolution. But it would be misleading to assume automatically that Darwin constructed his theory by discerning that general structure and then imposing special assumptions. The present conception is often just what Darwin lacked, so this explanation of his achievement is precluded. Still, a Whig reconstruction of the significance of the theory allows one to pose a historical question: where Darwin's ideas are special cases of modern theory, why did he arrive at the special conception he did? The familiar dichotomy between internalist and externalist historical explanation is relevant here. Perhaps the special conception was the first one because the general conception was not directly accessible; one needs to climb the ladder in order to then kick it away. Or alternatively, it may be that social conditions made the special case especially salient, and the general conception became available only because advancing inquiry was able to peel away undefended presuppositions and reveal the rational core.

The next three sections of this paper attempt to identify fundamental features of Darwin's understanding of natural selection that have stood the test of time. Both Darwin and modern biology conceive of natural selection as a *force* of evolution. This "Newtonian" interpretation allows us to understand how the theory of evolution is constructed, and also to defuse an objection that has been raised against the theory, namely that it is unexplanatory because one of its central postulates — the so-called survival of the fittest — is a tautology. I next examine Ernst Mayr's important observations that Darwin's view of evolution involved a rejection of typological or essentialist modes of thought in favor of population thinking (1963, 1976b). Discussing this insight will allow us to gain an additional appreciation of the explanatory structure of Darwin's theory. A related distinction is the one that Lewontin (1981, 1983) has drawn between selectional and developmental theories. Darwin's explanation of evolution was novel, in large measure, because it departed from the developmental paradigm followed by other evolutionists (for example, Lamarck).

After discussing these issues in which our understanding of natural selection has not changed materially from Darwin's, I shall focus on instances in which Darwin's perspective is a special case of the more general outlook of contemporary biology. Darwin thought of natural selection as involving the competition of some organisms with others within a single population; yet the modern conception allows for the possibility that natural selection acts on different units of organization in the hierarchy from single "selfish" genes through whole organisms, to groups, species, and communities. Darwin assumed that natural selection was an improver — within a stable environment, the level of adaptedness in a population would increase as a result of natural selection; yet the modern idea of selection guarantees no such optimization. The point here is not to deny that individual selection and increasing adaptation are facts of life. Rather, the goal is to show that these ideas proceed from contingent assumptions about populations and not from the principle of natural selection alone. An internalist might explain these contrasts by claiming that the general conception was not available to Darwin, any more than relativistic dynamics was available to Newton. An externalist might explain them as an artifact of Darwin's social milieu — as a result of reading his view of the social world into nature at large. Without wishing to dismiss internalist explanations of these phenomena,

I shall make some externalist remarks on the sources of Darwin's special conceptions. The Scottish economists — pre-eminently Adam Smith — loomed large on Darwin's intellectual horizon (Schweber 1977, 1980). They understood the market in terms of individuals competing with each other. It is perhaps no accident that for Darwin the individual organism was the pre-eminent unit of selection. The Scottish economists also saw the *laissez-faire* market as having a benign effect on the collective welfare. It is no surprise that Darwin thought of selection as an improver.

Externalist explanation is not limited to Darwin's shortcomings. Darwin was profoundly *lucky* to live in a society that afforded so rich a source of ideas about nature. The Scottish economists not only suggested the special conceptions that have been superseded; they also gave Darwin a set of concepts that have remained canonical. Adam Smith saw social harmony evolving as an unintended consequence of individual competition (as if by an "invisible hand"). Individuals acted according to their own free will; from these "random" interactions, stable population properties emerged. Darwin's strategy of explaining design in nature was the same: the adaptedness of a population of individuals was not to be explained by a developmental theory that showed how each organism constructed its own advantageous phenotype. Rather, the invisible hand explanation held that random variations were sifted, and the well designed survived. The idea that macroregularities were to be explained by microfluctuations was a powerful ingredient in Darwin's "population thinking" and played an important role in other nineteenth-century sciences as well (Schweber 1982).

I. Natural Selection as a Newtonian Force

Darwin's reading of Herschel and Whewell in the formative years in which he was working towards his theory of natural selection played a significant role in shaping his view of what an adequate scientific theory should be (Hull 1973b; Schweber 1977; Ruse 1979a). Historians have stressed the impact on Darwin of the idea that a theory should be quantitative and predictive. Additionally, Darwin absorbed from physics the distinction between the initial conditions governing a system and the laws that determine its change in state. And Darwin also thought of the theory of gravitation as a paradigm case of explanation by appeal to *vera causa*.

Yet, when we look at Darwin's theory, it is not entirely clear to what degree these Newtonian standards are met. Darwin was attracted to Malthus's views on population in part because they allowed a quantitative formulation to be given of the impact of natural selection (Ruse 1979a; Schweber 1977). But the principle of natural selection, which is the core of Darwin's theory, is not especially mathematical. The ease with which it can be expressed in ordinary English makes a mathematical statement of it seem wholly gratuitous: if organisms exhibit variation, and if some variants are better able to survive and reproduce than others, and if this advantage is a characteristic transmitted from parent to offspring, then the composition of the population will change. As we say today, heritable variation in fitness leads to evolution by natural selection (unless some contrary force cancels its effects, of course).

Nor is the schematic statement especially predictive. It says nothing about which biological characteristics are apt to promote survival and reproduction in which environments. If this were all we knew about fitness, we might be in the embarrassing position of Molière's good doctor, who explained the fact that opium put people to sleep by saying that it possessed a dormative virtue. Fitness is the property that accounts for the fact that some organisms outsurvive and outreproduce others. It threatens to be the dormative virtue of evolutionary theory.

Yet Darwin devoted a great deal of time in his "one long argument" to describing fitness and its multifold manifestations in different organisms. Indeed, most of the *Origin* is devoted to showing how the simple theoretical schema just stated can be applied. But let us be clear about what this detailed argumentation provided. Its main point was to show that life on earth exhibited two properties: (1) it evolved, and (2) its evolution was primarily guided by natural selection. Indeed, this is how Darwin himself described his efforts (*Autobiography*). Notice that both of these claims are historical statements about what has happened in a certain part of the universe. If we carefully distinguish the *history* of evolution from the *theory* of evolution — and think of the latter as involving the laws governing evolutionary processes — it seems clear that most of Darwin's energy went into defending a certain view of the *history* of evolution. He had comparatively little to say about the *laws*. It is startling how much mileage he was able to get out of so simple a theoretical conception.

I say this not to chide Darwin. It would be silly to question the fundamental nature of his insights. Yet it would be a distortion of his achievements to think of him the way we think of Newton — as laying down a rich mathematical framework of laws. Still the comparison with Newton is a useful one in another respect, when it comes to refuting a standard criticism that has been made against Darwin's idea of natural selection. This is the charge that the principle of the survival of the fittest is an empty tautology, since the fit organisms are, by definition, those organisms that happen to survive.

Critics have intended this charge as marking a difference between evolutionary theory and a "real" science, like Newtonian mechanics. Defenders have sometimes risen to the bait and cast about to find empirical elements in Darwin's conception. Ghiselin, for example, has noted that it is an empirical matter that the organisms in a given population exhibit heritable variation in fitness. It is a priori possible that they be clones of each other and therefore show no variation. It is a priori possible that what variation there is confers no selective advantage (no variation in fitness). And it is a priori possible that the variation in fitness not be transmitted from parent to offspring (no heritability). So the principle of natural selection is empirical (falsifiable) after all (1969, pp. 64–66). But one should note that this fails to vindicate Darwin's *theory* from the intended charge. It is, of course, an empirical question which populations happen to fall within the scope of the principle of natural selection. But the question is whether the principle itself is empirical. Laws of nature, so the tradition says, must be empirical, and not definitional. One does not show that "bachelors are unmarried" is empirical by remarking that it is an empirical question whether someone is a bachelor.

Another line of counterattack has been to point out that evolutionary theory does not say that the fit always outsurvive and outreproduce the less fit. The alleged tautology is in fact a falsehood. Scriven (1959), for example, remarked that if one of two identical twins is struck by lightning, we do not explain the resulting difference in survival and reproduction by invoking selection. We say that the difference is "due to chance". This is correct, both at the level of informal discussion of evolutionary theory and at the level of contemporary and mathematical formulations of the subject. "Chance" is given a formal representation under the rubric of "random drift"; selection has a deterministic effect on the frequencies of traits in a population only on the assumption of "infinite population size" (that is, zero drift).

But this reply, useful though it is in circumscribing what is and is not part of the theory, also fails to dispatch the objection completely. The charge of tautology can be reformulated. Isn't it a matter of definition that fitter organisms have a higher *probability* of survival and reproduction? There is no guarantee that such individuals *will* be more successful, but doesn't the definition of fitness guarantee that they have a better chance?

The right reply to this charge is, I think: perhaps-but-so-what?[1] And here is where the Newtonian analogy becomes apt. The same sort of question has been put to Newton's conception of force. There is a long history of scientists and philosophers asking whether "$F = ma$" is a definition of force. The law says that there is a net force acting on an object (with fixed mass) just in those cases when it accelerates. A component force need not produce an acceleration, if it is cancelled by another that is equal and opposite. But the law does describe the existence of a force in terms of the effects it will have if no other force acts. The force is characterized in terms of its *ceteris paribus* effects (or better yet, *ceteris absentibus*, as my colleague Geoffrey Joseph has remarked).

Contemporary population genetics has a law that plays the same explanatory role. The Hardy-Weinberg Law describes how the frequencies of genotypes can be predicted from the frequencies of alleles, if no evolutionary forces (mutation, migration, selection, random drift, and some others) act on the population. In Darwin's own theory, a representation of the mechanism of inheritance was not incorporated.[2] But the principle that Darwin appealed to — namely that heritable variation in fitness would, if nothing else interfered, lead to a change in the composition of the population — characterizes selection in terms of its *ceteris paribus* effects. Fitness, like Newtonian force, is understood in terms of the changes it can produce.

If this were all there were to fitness — or to force — we might wonder how the concepts could function in explanations. Indeed, the concepts would be dormative virtues — items that are accessible to us only via the effects that are mentioned in their definitions. But Newtonian theory does more for the concept of force than merely provide the mechanical law "$F=ma$". In addition, it provides some *source laws*.

A source law specifies what physical arrangements will produce what kinds of forces. The law of gravitation, for example, says that a pair of objects will generate a gravitational force equal to the products of their masses divided by the square of the distance between them (times a constant). Coulomb's law of electric force plays the same role. Source laws give us a way of finding out about the presence or absence of forces *besides* looking at the effect the forces will have on the motions of bodies. A mechanical law will tell us about the *effect* of a force; a source law will tell us about its *cause*.

Did Darwin provide us with an additional kind of access to the fitness of organisms besides the effects that selection can have in survival and reproduction? Does contemporary evolutionary theory do this? Certainly Darwin and contemporary biology both provide a great deal of anecdotal information about which traits have been selectively advantageous in which environments. Often this kind of information is derived from a design analysis of the organism and the environment it is in (Gould 1977a). To take a familiar example, a biologist might recognize the selective advantage of melanic coloration in *Biston betularia* before observing the reduced mortality of melanic moths. The inference that this trait represents a selective advantage is, of course, fraught with dangers. One might be wrong, for any number of reasons. But this is no more than to say that the hypothesis is an empirical one. One may have to look at other, similar populations. One may have to examine carefully the search strategies of predators, the pleiotropic effects of melanism, and so on. But, in principle, it seems to be possible to confirm hypotheses about design advantages of traits in a population without actually observing the survival and reproduction of the organisms in that population.

Initially, biology might have been quite inaccurate in such conjectures. Arguably, it still is. But a bootstrapping operation seems to be available for improving one's track record. One formulates a design hypothesis, then checks it against the data of actual mortality and reproduction. One then revises the design hypothesis and uses it on a new population to try to predict fitness differences. This can then be checked in turn. By a process of this kind, contemporary biology has accumulated a systematic lore that allows biologists to make claims about fitness independently of the data about the effects of selection. Fortunately, adaptations are not totally unique. Darwin pursued this methodology as well.

Although Darwin and contemporary biology have aimed at characterizing source laws for selection, it would be a mistake to think that either has provided a complete edifice. Richard Lewontin has described population genetics as the auto mechanics of evolution. I would shorten his description simply to *mechanics*, in the Newtonian sense. Population genetics describes the *consequences* of selection, migration, mutation, and so on. It does not describe what ecological conditions generate what kinds of evolutionary forces. This is a task for theoretical ecology. Unfortunately, systematic laws about the kinds of environments that will generate certain kinds of selective advantages are only in the fledgling stage. But there is nothing in principle that bars evolutionary theory from producing such source laws.

So natural selection, in both its Darwinian and contemporary settings, is a Newtonian force. It is understood in two ways — by its (*ceteris paribus*) effects and by its causes. Source laws in evolutionary theory may turn out to be empirical, just as their counterparts in physics are. A priori reflection alone does not guarantee that the masses of two separated objects produce a force (the force due to gravitational attraction), nor that the force will have the mathematical character that Newton laid down. Similar results may perhaps obtain in evolutionary theory, if a *science* of design, and not just an *ad hoc* collection of anecdotal descriptions of designs, can be produced.

But this still leaves the *effect* laws of the two sciences to be addressed. There are two reasons to think that "$F=ma$" is an empirical claim. These reasons seem to have no analogues in the case of "the probable survival of the fitter."[3] First, pre-Newtonian physics had other ideas about what an object would do were it subject to no force at all. Aristotle thought the object would stop moving. If we assume that he and Newton were talking about the same thing — an assumption that is not only not obvious, but that we currently lack the theoretical equipment to discuss seriously — then the law of motion is empirical. Second, the potential interaction of source laws and effect laws allows us to foresee how "$F=ma$" might conceivably be abandoned. If an object subject to no known force is observed to accelerate, there are two possible explanations. It may be that our present catalogue of forces is incomplete; a new source law is required. Or it is conceivable that, having failed to uncover any new force, we should decide to revise the effect law itself. The latter is certainly not the first revision we would propose, but it is not ruled out a priori. An overall more coherent theory may result from modifying the effect law.

Notice that in considering the question whether "$F=ma$" is empirical, we have had to imagine theoretical settings other than the Newtonian one. If we were to look at that theory alone, we would find "$F=ma$" duly recorded as truth. It would be listed alongside the law of gravitation. Laws do not wear their epistemological status on their sleeves. We must look at alternative theoretical contexts to see if the proposition is empirically revisable.

What makes it hard to approach the effect law of natural selection in the same way is the unavailability of alternative theories of fitness. Aristotle and Newton may both have talked about *force*, but Aristotle and Darwin did not both talk about evolutionary *fitness*.[4] To see whether empirical considerations could displace it from its present role as a truism of evolutionary theory, we must see whether there are alternative theories of natural selection in which the principle of natural selection is supplanted by some competing conception.

It is very hard to justify negative answers to such questions. The fact that there is no obvious experiment that one can dream up that would tell against the principle is not decisive. Before the development of non-Euclidean geometries philosophers might have had a similar problem. We now see that this was due to their limited imaginative powers, not a reflection of the fact that Euclidean geometry is a priori true. Although the notion of fitness has undergone several transformations in the biology of the last twenty years, I don't think any of these changes has placed the principle in jeopardy.[5] What the future will bring I am not going to predict.

The apparent a priori character of the principle of natural selection has always seemed a problem, because it has been assumed that a priori principles cannot be explanatory. Science is an empirical activity, and so, it is claimed, the explanatory principles of science must be empirical. It is just this inference that strikes me as a fallacy. There is plenty of room for a science to be empirical, even when it exploits a priori principles in its explanations.

If fitness means probability of survival and reproductive success, then it may seem rather thin to explain why one organism outsurvived and outreproduced another by saying that the former was fitter.[6] This would be similar to explaining why one coin came up heads more frequently than another in a run of tosses by saying that the former had a higher probability of heads. Notice that neither of these explanations is trivially true: fitter organisms may be less reproductively successful, and coins biased towards heads may come up heads less frequently in a run of tosses. What has gone wrong is that the explanatory information provided, though non-trivial, is rather sparse. It amounts to little more than the claim that what happened happened because it was more probable than the alternative.

I have already noted one way to make the explanation more interesting. A design analysis of the organisms may say *why* the first organism was fitter than the second. And this analysis may even provide empirical laws relating general design features to patterns of fitness difference. Here the explanation would be enriched by importing empirical generalizations. This perhaps fosters the impression that the initial explanation was impoverished precisely because its only generalization was a priori true.

But there is another way to make the story more interesting. Not all mathematical principles are unsurprising. Harnessing non-trivial ones to the empirical world can be a genuine source of illumination. This avenue of explanation was not explored by Darwin, but has been exploited since the rise of population biology in the Evolutionary Synthesis (see Crow and Kimura 1970; Roughgarden 1979).

A priori propositions can be known to be true without empirical evidence. This does not mean that cursory inspection shows them to be obviously true, or that they are "unsurprising" or "uninformative". Philosophers like Wittgenstein and Carnap used these quoted labels to describe the whole of mathematics. Probably they did not have the usual meaning of those terms in mind; for it is hard to see how anyone could hold that there is no such thing as a surprising piece of mathematical information. This philosophical confusion persists in the idea that an explanation must include an *empirical* law of nature (Hempel 1965), and is reflected in the worry that evolutionary theory is in trouble, if its laws are non-empirical.

Physics worship also has contributed to the confusion. Newton's laws are empirical. Hence, each law in evolutionary theory must be so as well. But the fact of the matter is that much of population genetics simply traces out the algebraic consequences of algebraic assumptions. The Hardy-Weinberg Law is indistinguishable from an elementary consequence of probability theory: if I have a large number of coins, each with the same probability of landing heads when tossed, then forming them into pairs after all have been tossed can be expected to yield certain frequencies of the three types HH, HT, and TT. Population genetics is rarely as simple as this, but it is in the same spirit.

Another confusion has turned a priori truths into a bogey man — the idea that an explanatory story will be non-empirical if the law it appeals to is. But even the simplest example of explanation by appeal to natural selection shows that this is a mistake. It is hardly an a priori truth that the dark moths are fitter than the light ones in a particular region. Still less is it an a priori truth that they enjoy their fitness advantage because melanism conceals them from predators. Each of these singular claims is empirical; but by virtue of being singular, neither qualifies as a law. Even when a proposition of the form "Each A is (probably) B" is a priori, which objects have property A will still be an empirical matter.

So I conclude that the "principle of natural selection" may well be an a priori truth, although I must remind the reader of the problems noted above in establishing a claim like this. Nevertheless a priori truths may be explanatory. And even when a generalization is a priori, which systems in nature the generalization applies to will only be settled empirically.[7]

II. Population versus Typological Thinking

Darwin was not the first biologist to argue for the fact of evolution. Nor was the idea of a selection process modifying the composition of a population entirely novel with him.[8] But the combination of these two ideas — the idea that evolution, including speciation, is principally propelled by the force of natural selection — was his.

Ernst Mayr has argued that Darwin's idea was not just a new theory to account for the diversity of life. It was a new *kind* of theory, displacing traditional typological or essentialist modes of thought in favor of a new way of conceptualizing nature — population thinking (1963, 1976b). According to Mayr, the essentialist doctrines that Darwin emancipated himself from held that

> [t]here are a limited number of fixed, unchangeable "ideas" underlying the observed variability, with the *eidos* (idea) being the only thing that is fixed and real, while the observed variability has no more reality than the shadows of an object on a cave wall . . . . [In contrast], the populationist stresses the uniqueness of everything in the organic world. . . . All organisms and organic phenomena are composed of unique features and can be described collectively only in statistical terms. Individuals, or any kind of organic entities, form populations of which we can determine the arithmetic mean and the statistics of variation. Averages are merely statistical abstractions, only the individuals of which the population are composed have reality. The ultimate conclusions of the population thinker and of the typologist are precisely the opposite. For the typologist the type (average) is an abstraction and only the variation is real. No two ways of looking at nature could be more different. (Mayr 1976b, pp. 27–28)

Although Mayr's distinction cuts very deep, it raises a number of puzzles (Sober 1980). How was it ever possible for scientists to indulge in typological thinking? Mayr's characterization of essentialism suggests that it is essentially anti-scientific; the essentialist appears to ignore the fact of variability in nature and invent some altogether mysterious and unverifiable subject matter to describe. But if essentialism does involve a principled failure to attend to the evidence, it is hard to see how the *details* of any scientific theory could refute it. Mayr's description suggests that what essentialists needed was not a new explanation of old phenomena, but for someone to rub their noses in the data.

Other perplexities arise when we try to understand modern population biology as embodying "population thinking". If "only the individuals of which the populations are composed have reality", it seems that much of modern science has its head in the clouds. Much of evolutionary theory describes the dynamics of *populations*, with little or no case by case attention to individual organisms. Statistical properties of populations are the inputs and outputs of the equations of population genetics and theoretical ecology. Here the population biologist and the typologist seem to have something in common: both place their fundamental emphasis on descriptors that do *not* apply in the first instance to individuals.

Another issue is that of how we are to understand the distinction between "reality" and "abstraction". One natural way is simply to interpret "reality" as meaning *existence.* But presumably no population thinker will deny that there are such things as averages. If there are groups of individuals, then there are numerous properties that those groups possess. The *average* fecundity in a population is no more a property that we invent by "mere abstraction" than is the fecundity of individual organisms. Individual and group properties are equally "out there". Similarly, it is unclear how one could say that typologists held that variability is unreal; surely the historical record shows that they realized that differences between individuals *exist*. How then are we to understand the difference between typological and population thinking in terms of what each holds to be "real"?

These interpretive questions can be answered, I think, by attending to the kind of explanation that typological thinkers offered of the observed variation in nature. Aristotle typifies typological thinking in his formulation of The Natural State Model of variation. This model played a canonical role in biology for a very long time; it has played, *and continues to play,* a fundamental part in physics (Lewontin 1974a, p. 5). Darwin and the evolutionary biology he gave rise to provide specifically biological reasons for rejecting essentialism in biology; but essentialism is alive and well in other sciences.

The Natural State Model partitions the forces that can act on a set of objects into two kinds. First, there are the forces that underlie the *natural tendency* of the kind of object considered. Second, there are the *interfering forces* that can prevent the objects in question from arriving at their *natural state.* Aristotle applied this model in both his physics and his biology. In the sublunar sphere, heavy objects have the natural tendency to move to the center of the earth. But, of course, many heavy objects fail to be there. The source of this divergence from what is natural is that these objects are acted on by interfering forces that frustrate their natural tendency. Variability within nature is explained by deviation from what is natural; were there no interfering forces, all heavy objects would be located in the same place (Lloyd 1968).

This kind of model survives in Newtonian and even in relativistic physics. Even though the terms "natural" and "interfering" do not occur in these theories, one can still discern a distinction between two kinds of states that physical objects can occupy. *Rest or uniform motion* is, in Newtonian mechanics, the state an object will be in if it is not acted on by a force. And in general relativity, the geometry of space-time specifies a set of geodesics along which an object will move if it is not subjected to a force. If no force acts on a body, a fortiori, no interfering force acts on it either. And in the absence of interfering forces, all objects will have the same dynamic property.

It is striking that although the Natural State Model is familiar in physics, Aristotle's deployment of it in biology strikes us as utterly alien. This is a sign of the extent to which Darwinism has permeated both science and common sense. Aristotle's theory of reproduction was built around identifying what was natural:

> [for] any living thing that has reached its normal development and which is unmutilated, and whose mode of generation is not spontaneous, the most natural act is the production of another like itself, an animal producing an animal, a plant a plant. . . . (*De Anima*, 415a26)

Spontaneous generation, it should be noted, is a form of reproduction in which it is *not* natural for like to produce like. But aside from this exception, the natural state has been specified.

What counts as an interfering force? Aristotle thought that in sexual reproduction, the male semen provided a set of instructions that dictated how the female matter was to be shaped into an organism. Interference might arise when the form failed to completely master the matter. This might happen, for example, when one or both parents were abnormal, or when the parents were from different species, or when there was trauma during foetal development. Such interferences were anything but rare, according to Aristotle. Mules — sterile hybrids — counted as deviations from the natural state (*Generation of Animals*, ii, 8). In fact, the females of a species did too, even though they were necessary for the species to reproduce itself (*Generation of Animals*, ii, 732a; ii, 3, 737a27; iv, 3, 767b8; iv, 6, 775a15)! In fact, reproduction that was completely free of interference would result in an offspring that exactly resembled the father. Deviations from type, whether mild or extreme, Aristotle labeled "*terata*" — monsters. They were the result of interfering forces (*biaion*) deflecting reproduction from its natural path (Furth 1975; Sober 1980).

I won't take the space to describe how the Natural State Model continued to be used in biology. I will mention, however, that the emergence of statistical concepts in the nineteenth century adopted, but then broke with, this mode of thought. What we now call the *Normal* Law was called the Law of *Errors*. Statisticians like Laplace and Quetelet thought that variation was caused by interfering forces deflecting individuals from their natural tendencies. Quetelet, for example, thought that variation in the girths of Scottish soldiers was to be explained by various interfering forces modifying a central tendency, represented by the average man (Hilts 1973; Sober 1980).

Darwinian biology has discredited the Natural State Model in a number of ways. To begin with, we now know that this model is simply mistaken in its within-population explanation of how variation arises. Averages in a population are standardly *effects*, not *causes*. Individuals do not aim in their development at some median value, only to be deflected from that goal in different directions. Nor is it even true of individuals that they have one possible developmental outcome as their natural state, with which accidents of development can interfere. At the level of individual development, the guiding evolutionary concept is the *norm of reaction*. The norm of reaction of a genotype describes what phenotype will be produced in what environment; it might describe, for example, how a corn plant with a given genotype will grow taller or shorter, depending on how much moisture there is in the soil. There is no such thing as a "natural phenotype" for a given genotype to display (Lewontin 1977b). It isn't just that the Natural State Model was wrong in its particular guess about what every individual in a population naturally tended to be like; rather, the mistake was the more fundamental one of thinking that there was any such thing as a natural tendency of this kind.

The Natural State Model saw uniformity as the natural state of a population and variation as the result of interference. If we look at current evolutionary theory and ask whether it has anything like Newton's first law of motion — a law that says what will happen to a population when no evolutionary forces are at work — we come up with the Hardy-Weinberg Law. And this law says that genetic variation will be *preserved* in the absence of evolutionary forces. What is more, the pre-eminent Darwinian evolutionary force — natural selection — has the effect of *destroying variation* (in fitness). So at the level of population dynamics, when no forces act, we get variation. And when natural selection acts, variation is eliminated. This is Aristotle stood on his head.

Darwin, of course, did not know the Hardy-Weinberg Law. He did realize, however, that selection acts on a pre-existing variation, and serves to destroy it. Because of this, it is perhaps not too great an anachronism to view him as reversing the typological pattern of explanation: Aristotle explained variation as arising from uniformity (by the effects of interfering forces); Darwin explained uniformity as arising from variation (by selection).[9]

This inversion of the explanatory pattern is not simply a relationship that obtains between Darwin and an ancient predecessor. Bowler (1977a) has pointed out how Darwin differed from a number of so-called idealist biologists in the status he accorded to the concept of unity of type. Biologists like Chambers, Owen, Mivart, and Cope thought of types as exercising a powerful constraint on the similarities we observe within and between species. Darwin, on the other hand, viewed unity of type as a *consequence* of common descent. It is interesting that this controversy lives on today, without its theological trappings, in the question about the power of natural selection as a shaper of evolution versus the role of architectural constraints — *baupläne* — in limiting the diversity that can exist within species (Gould and Lewontin 1979; Kauffman 1983). Darwin, as well as the natural theologians who wrote the *Bridgewater Treatises*, tended to describe species in terms of accumulations of adaptations, with natural selection or Divine benevolence as the shaper of a completely plastic organic matter. The idealist tradition that Bowler describes, however, envisioned a powerful material constraint on the efficacy of natural selection. Recent defenders of this point of view have dropped the "idealism", in at least one sense of this word, but have retained the idea of constraints.

One detail of Mayr's characterization of population thinking should be liberalized, I think, in view of a line of thinking that Darwin considered and rejected, but other evolutionary biologists have taken more seriously. Darwin's *individual* selectionism (to be discussed in Section V) led him to view population averages as consequences rather than causes. For him, it was differences between organisms, not between groups, that propelled the process of selection. On the other hand, biologists who have accorded more weight to the idea of *group* selection — like Alfred Russel Wallace and some theorists today — see population properties, such as averages, as causally efficacious, and not just as the effects of processes occurring at lower levels of organization. For example, one group selectionist account of reduced virulence among parasites is that a population of parasites may enjoy a selection advantage if it has a lower than average degree of virulence (Lewontin 1970; Sober 1982b). I will look in more detail at what this issue amounts to in Section V. My point here is just to note that group selection hypotheses are examples of population thinking *par excellence*. The difference between population and typological thinking, therefore, is to be sought elsewhere.

The heart of population thinking, I suggest, consists in the idea that theories may be stated relating the interactions of population properties and magnitudes. Aristotle's was a theory of *individual* development. Population properties — like a bell curve distribution — are explained as arising out of this individualistic basis. But models in population biology treat population phenomena autonomously, as subject to their own laws.

Not that population thinking has achieved complete hegemony in current evolutionary thought. Although participants in the units of selection debate who reject group selection often wrap themselves in the mantle of Darwin's own individualist orientation, one of their arguments in favor of lower-level "genic" selection involves precisely the mistake typologists make when they view an *average* as a causally efficacious property. If we define evolution, as is customary now, as change in gene frequencies, it follows trivially that when evolution by natural selection occurs, there must be differences in the fitness values attaching to individual genes. The fitness values that are assignable to genes will be averages over the different genetic contexts in which the genes occur. A given gene may be advantageous in one context and deleterious in another; but the genic fitness value will simply combine these two values, weighted by their frequencies of occurrence in the population, into a single number. So much is mathematically uncontroversial. The problem arises when it is further asserted (as do G. Williams 1966 and Dawkins 1976) that all natural selection is really selection for or against single genes. This thesis amounts to the claim that the *cause* of evolution under natural selection is the differences in genic fitness values. But this is frequently to mistake effect for cause. Genic fitness values are artifacts — they represent average effects — that are no more causally efficacious than the average girth of Quetelet's soldiers. Although Darwin discredited typological modes of thought, the central problem — of distinguishing cause from effect in evolution — is with us still (Sober and Lewontin 1982).

III. Selectional versus Developmental Theories

Another way of grasping how Darwin's explanation of evolution by natural selection was novel is to compare it with the picture of evolution that was prevalent in the eighteenth century and culminated in the work of Lamarck. Lovejoy (1936) described how the Enlightenment "temporalized" the great chain of being. Lamarck's evolutionary theory was in this mold (Mayr 1976a). Lamarck conceived of a fixed sequence of phylogenetic stages through which living forms moved. Each lineage began with simple life forms like amoebas, and gradually gained in complexity. Our lineage has gone the farthest, since it has reached the human state of development. All the other populations observed in the present are undergoing the same process, but had started later. Since the early stages were constantly being vacated in this upward march, Lamarck had a mechanism (spontaneous generation) for introducing rudimentary living forms at the bottom of the ladder.

As might be expected in a Natural State Model, the central tendency of ascending the phylogenetic ladder is complicated by "forces of circumstances" that cause diversity among populations at the same stage of development. The net result of these two sorts of forces is a branching tree. Darwin's theory predicted the same shape for organic diversity, but from an entirely different picture of the underlying mechanism.

Lamarck's theory was a developmental theory; it specified a sequence of steps that all things of a given kind had to move through. This kind of theory is standard fare in the social sciences. Piaget's theory of cognitive development in children, Marx's theory of the economic transformations that human societies undergo, and Chomsky's conception of language acquisition are all stage theories. A *selectional* theory, as Lewontin has observed (1981, 1983), is a different kind of theory entirely. Let us try to characterize the structural differences that Lewontin has drawn our attention to.[10]

A simple example: Suppose you observe a room full of school children, all of whom have a certain level of reading ability. You would like to know why all the children in the room have the level of reading skill they do. A developmental theory might describe the ontogenetic trajectory that each child has moved through. Jimmy can read at the third-grade level because last year he was reading at the second-grade level, and he learned and matured in certain ways. Ideally, a description of these earlier states would be inputs into a developmental theory that predicts his present reading level. And what we say about Jimmy we can repeat for Sally and Tommy. We explain why all the children read at the third-grade level by saying how each developed.

An alternative explanation would describe the population configuration — all the children being at the third-grade reading level — as the result of a selection process. Perhaps there was an admission test, and no one at a different reading level was permitted to enter the room. Here one is in the peculiar position of knowing why all the children in the room are at the third-grade reading level, even though one does not know why Jimmy is, or why Sally is, or why Tommy is. In this selectional account, one has no idea why the individuals are as they are, even though one does know why the population has the configuration it does.

A developmental theory can *aggregate* its explanations; it can account for why all (or most, or n%) of the individuals in the population have some characteristic by taking up this problem for each individual, and then aggregating the separate stories. But selectional stories often do not explain population phenomena by aggregating individual scenarios. This is perhaps another sense in which a selection theory embodies a kind of population thinking.

Explanation is a contrastive phenomenon (Dretske 1973; Garfinkel 1981). To explain why something is true is to explain why it, *rather than some contrasting alternative*, is true. Vary the contrasting alternative, and you have changed the explanatory problem. Garfinkel's story about the bank robber Willie Sutton illustrates the idea nicely. A priest once asked Sutton why he robbed banks. Sutton replied that he did so because that is where the money is. Explaining why you rob banks rather than not robbing anything is one thing; explaining why you rob banks rather than robbing candy stores is quite another (Garfinkel 1981).

The developmental explanation of why the children all read at the third-grade level says of each child why he or she reads at the third-grade level rather than reading at some other grade level. For each individual x, the story explains why the first and not the second of the following contrasting alternatives is true: (x reads at level 3, x does not read at level 3). The aggregation of these individual stories explains why all the children in the room read at the third-grade level by saying why the first and not the second of the following contrasting alternatives is true: (they read at level 3, they do not read at level 3).

The selectional explanation of why the children all read at the third-grade level focusses on a different contrast. It does not offer a different explanation of the same fact addressed by the developmental story; it explains a different fact. The selection story says why the room is filled with children reading at the third-grade level rather than with children having different reading levels. In the developmental story, both contrasting alternatives are about the same children; the question is why Tommy, Sally, Jimmy, and so on have one property rather than another. In the selectional story, the domain of individuals is not held constant between the two contrasting alternatives; the question is why the domain is filled with individuals of one type rather than with (possibly different) individuals of another type.

Science sometimes progresses by finding new explanations for old phenomena. And sometimes it progresses by discovering new phenomena and then providing them with explanations. A novel phenomenon may be brought into focus by the invention of a new descriptive vocabulary. The concept of *variance*, for example, was introduced into science at a certain historical juncture and thereafter could serve as a device for posing explanatory problems. But the sort of conceptual innovation we are now considering in the contrast between selectional and developmental theories is different. Selectional theories differ from developmental theories over what the object of explanation is. But the novelty of the selectional theory's object of explanation does not consist in any new concept being used. Both the explanations we have considered use the concept of reading at the third-grade level and the idea of generality. The selectional theory creates a new object of explanation by placing the populational fact to be explained in a new contrastive context: the contrast posed by the developmental explanation holds the domain of objects fixed and varies the properties, whereas the selectional explanation varies both the domain and the properties at once.

Given this preliminary clarification of the difference between selectional and developmental explanations, we now need to focus more carefully on the explanatory structure of Darwin's theory. An account of evolution by natural selection crucially involves assumptions about the hereditary mechanism. Darwin recognized this in his discussion of the problem of blending inheritance. And the modern Synthetic Theory of Evolution has so intertwined the mechanism of Mendelism with the force of natural selection that it is difficult to discern the explanatory contribution of each. But we must try to separate assumptions about the transmission of characters from the hypothesis of selection if we are to grasp their respective roles in our present view of evolution and also the historical novelty of Darwin's idea. Selection and heredity are partners in a division of explanatory labor. The success of the partnership is remarkable in view of the partners' very different forms of explanation. Natural selection, purged of assumptions about transmission, is, as one would expect, a concept that figures in selectional explanations. But laws of heredity, whether they are Mendelian or something closer to the ideas deployed by Darwin (the theory of pangenesis) provide developmental theories. It is remarkable that these two kinds of ideas can be synthesized so successfully into a single theory.

I have an opposable thumb because my parents did and they passed this trait along to me. The same can be said of practically all members of our species who have opposable thumbs. The exceptions are those individuals whose parents gave a trait to their children because their gametes contained mutations of the appropriate kind. Heredity explains by tracing traits backwards from offspring to parent.

Students of human evolution tell another, complementary, story about the opposable thumb. Although its selective importance is by no means clear, the opposable thumb is often described as having conferred a selective advantage owing to the fact that it facilitated tool use. Selection for this trait resulted in its fixation in the population. We observe now that virtually all human beings have opposable thumbs and explain this by picturing a distant past in which some individuals had opposable thumbs while others did not. Selectional explanations of the frequency of a trait in a population describe some earlier frequency distribution and the selective forces that drive the population to its equilibrium configuration.

Fitness is selection by another name. Its twin components are viability and fertility. Viability selection explains why some individuals but not others survive from one stage of the life cycle to another. Fertility selection explains why some individuals have more offspring than others. Neither of these two modes of explanation says anything about the characteristics of individuals *after* selection. To do this, some further assumptions about transmission must be added. This is clearer in the case of fertility selection. Perhaps dark moths outreproduce light ones. But what impact does this have on the resulting composition of the population? If the trait has zero heritability, then selection will leave the population unchanged; change the assumptions about inheritance and you get different predictions about the effect of selection. The same is true in viability selection, except here the automatic assumption is that the traits that organisms have before selection *persist*. But some transmission assumption is essential if anything is to be said about the characteristics that individuals have after selection.

Selection, stripped of assumptions about transmission, explains only two properties of individuals — survival and reproduction. It does not explain why I have an opposable thumb. In conjunction with transmission assumptions, natural selection can explain the frequency of traits in a population. Heredity, once it is separated from assumptions about selection, explains why individuals have the traits they do. It says why I have an opposable thumb. But in a certain sense, it is silent on the question of why virtually all human beings now have opposable thumbs. The explanatory labor is divided between levels of organization: natural selection focusses on population patterns, while heredity homes in on individual characteristics.

Lamarckian biology treated evolution and individual development within a single conceptual framework. The origin of species, like the onset of puberty, was the result of a transformational process. Darwin split these domains apart by substituting a selectional theory of evolution. The individualistic explanations provided by heredity and development aggregate; heredity and development may explain why all (or most, or some) of the individuals in a population have opposable thumbs by telling a separate story about each individual. But this is no substitute for a selectional explanation of why human beings have opposable thumbs. The selectional story is not obtainable by aggregating explanations about why the individuals in the population are as they are. Saying why Jimmy, Sally, and Tom have third-grade reading levels does not explain why they were admitted to the room, or why other individuals were not admitted. The concept of natural selection required a novel kind of population thinking.[11]

Developmental theories and selectional theories are different kinds of theories. Specific forms of each can be incompatible. One can explain a population configuration by postulating that the objects in it were changed (developed) in certain ways from their earlier states. Or one can explain that configuration by seeing the objects as static entities that were selected from some larger population. One can opt for the dynamic developmental story or the static selectional one, but one cannot do both. This does not, however, preclude the possibility that a selectional theory at one level may be compatible with a developmental theory at a higher level. Biologists, from the time of Darwin to the present, have tried to describe lawful patterns that govern the history of life. Even if natural selection at the individual level governs the origin of species, it still may be true that something like Cope's rule is true — that species tend to increase in size. A selectional process at one level may provide the material basis for a developmental process at a higher level.

To claim that the history of evolution obeys some developmental laws is not simply to say that certain regularities are exhibited. Laws, to be laws, cannot be accidental generalizations. Rather, the developmental laws must assert that there is something inherent in the evolutionary process that endows the branching tree of life with a certain shape. Bowler (1977a) has described the theistic sources of this idea during the time Darwin wrote. Darwin, of course, opposed this kind of idea. But biology today allows us to discard the theological problematic that once surrounded this question. And, in a way, the theology is incidental; biologists like Vavilov proposed developmental laws about the pattern of evolution, but not for any theistic motives (see Gould 1983). The real opposition to the possibility of there being developmental laws comes from the idea that natural selection is limitlessly *opportunistic* — that there are no endogenous constraints on the successive forms that a lineage can exhibit. If organisms are mere putty in the hands of nature, there will be no laws of development. What regularities we observe will simply be fortuitous reflections of the slow, consistent transitions made by the environment. A developmental theory of ontogeny is made possible by the existence of endogenous constraints; the possibility of a developmental theory of phylogeny turns on the same question. Darwin reversed the theistic idea that environments are made to accommodate organisms; but the reversal frequently took the form of thinking of the environment as a total shaper of the organism. A more dialectical conception in which each element in the relation obeys certain endogenous constraints and also constrains the other, may make more room for finding developmental laws about the course of evolution (Lewontin and Levins 1980).

IV. Selection and Progress

Darwin shared with his fellow "transmutationists" an interest in discovering the principles of progress that underlie the sequence of events that compromise the history of life (Ospovat 1981). Although he was much attracted to the idea that evolution by natural selection tended to be "progressive", Darwin realized that this "tendency" was difficult to characterize and would in any case be probabilistic in character rather than universal and necessary.

Darwin is famous for the reminder he gave himself to "never use the words 'higher' and 'lower' ". Gould has interpreted this remark as indicating that natural selection is "opportunistic" (1980a). Natural selection will favor any characteristic that is advantageous — whether it happens to coincide with some preconceived picture of what is "progressive" or not. Ospovat (1981), on the other hand, interprets the remark as a warning that Darwin gave to himself of how difficult it was to use these notions precisely. But, says Ospovat, this was a warning that Darwin consistently refused to heed.

The idea that progress might consist in increasing specialization and complexity had considerable plausibility for Darwin. But he realized that barnacles probably represented a counterexample. Modern evolutionists concur in the same critical point. Obligate parasites are a favorite example; they often simplify their organization in the process of adapting.[12]

It isn't just the failure of simple suggestions like "complexity and specialization" that makes the idea of defining a notion of progress look hopeless. There is, in addition, the deeper fact that natural selection adapts organisms to their *local* environments. What signals progress in one environment may be disadvantageous in another, and simply inapplicable in a third. As Ghiselin has pointed out, Darwin replaced the linear ordering of the great chain of being with a tree of life (1969, pp. 70–71). Although the former arrangement facilitates comparison between pairs of organisms, the latter does not.

Ospovat suggested that theological motives may have prompted Darwin to cling to the idea that evolution meant progress, in spite of the difficulties that picture presented. Darwin had discovered that the natural world was a world of struggle and disharmony. How to reconcile so much suffering with a beneficent deity, if that struggle did not produce a higher good?

But there is another reason to seek a principle of progress. It is intellectually satisfying to be able to describe a changing system as increasing in some particular quantity. If that quantity can be seen as morally or aesthetically desirable, so much the better. But even without this additional pay-off, the intellectual benefits are considerable.

In the twentieth century, R. A. Fisher (1930) achieved this goal, to a limited degree at least, in what he called *The Fundamental Theorem of Natural Selection*. He was able to prove that under certain less-than-universal conditions, natural selection will improve the average level of adaptedness found in a population. It isn't that any particular organism is better off after a selection process is over than it was before. When the process runs over many generations, there will be no such single organism still around to point to as the beneficiary. Rather, there is a measure of the population — the average level of fitness of the organisms found in it — that natural selection will have raised.

The sort of progress that Fisher's theorem proclaims is less concrete than that which Darwin and his contemporaries wished to identify. Increase in fitness *may* mean increasing specialization and complexity, but it need not. It may mean enhanced protective coloration in one species, fast locomotion in another, or the ability to exploit some new food source in a third. Fisher thought of increasing fitness as playing the role in evolution by natural selection that increasing entropy plays in thermodynamics. But he noted that the two properties differed in at least one way: "Fitness, although measured by a uniform method, is qualitatively different for every different organism, whereas entropy, like temperature, is taken to have the same meaning for all physical systems" (1930, pp. 39–40).[13]

Let us be more precise about what this principle of progress asserts. As noted earlier, the laws governing evolution do not guarantee that evolution will occur; they merely describe what will happen *if* certain initial conditions are satisfied. Nor do the laws guarantee that if there is evolution, then natural selection will be the pre-eminent force that determines its trajectory. But if one assumes, as Darwin and R. A. Fisher both did, that natural selection is the most important force shaping the history of life, then it becomes possible to describe that process in terms of the increase in a certain quantity. It is here that the idea of progress gets its foothold in evolutionary theory.

For natural selection to produce progress, it isn't necessary for it to produce perfection. The best alternative found in a population does not have to be the best conceivable engineering solution to a design problem. For one thing, there may be trade-offs: a characteristic may be advantageous in some respects and deleterious in others. Darwin believed that this was the situation in cases of sexual selection (*Descent*). The females of a population may prefer mates who have a characteristic that is in fact disadvantageous when it comes to avoiding predators. Here viability and fertility are at odds with each other in the context of male display.

Darwin was no vulgar optimizer. He didn't imagine that traits exist in a vacuum and can be perfected one by one, thereby arriving at a perfect organism. He understood the idea of design constraints. A characteristic can have numerous different effects on fitness. If natural selection plays the pre-eminent role in determining the course of evolution, one should not expect that *each* effect of a trait is advantageous, but only that the *average* effect is. If large horns place a stag at greater risk from predators, but represent an advantage in mating, then having the horns will be selected for only if benefits outweigh costs (*ceteris paribus*, of course). Cost/benefit analysis is quite different from the naive idea that there are no costs.[14]

Yet, in an important sense, Darwin thought of natural selection as an improver. Granted, there are trade-offs of the kinds just mentioned; and the level of adaptedness present in a population is constrained by the raw materials thrown up at random "by mutation"; and, if the environment changes faster than the population can evolve, the general level of well-being may decline. But if the population can track the environment, and if the population exhibits any heritable variation in fitness, the Darwinian assumption was that the level of fitness in the population would increase under the guidance of natural selection.

It is of some interest that this structural assumption about natural selection is not in general justified. We can see one source of the limited validity of this idea by considering a graph as shown in Figure 1. This graph represents the fitness of two types of organism — *A* and *B* — as a function of their frequency in the population. The graph tells us what will happen if we introduce an *A* mutant into a population of *B* individuals, or a *B* mutant into a population of *A* individuals. So as to avoid the complications introduced by transmission assumptions (which characteristic does the offspring of an *A* mother and a *B* father have?), we will assume that the individuals involved reproduce asexually and like always produces like. This graph then allows us to see how the frequencies of the two types in the population are modified by selection. Fitnesses are represented in terms of numbers of offspring. Dropping an *A* type into a population of *B* will eventually result in the fixation of *A* and the disappearance of *B*.

Notice what happens in this selection process to the average fitness of the population. The per capita output before the selection process began was 3. As selection proceeds, the average fitness steadily rises, until, at its completion, the average fitness is 5. Here we see how selection can be an improver — fitness increases under the guidance of natural selection.

The fitnesses of types *A* and *B* are constant on the above graph; they are unaffected by the shifting frequencies of the two types. But frequency dependent selection is a common phenomenon; indeed some biologists now

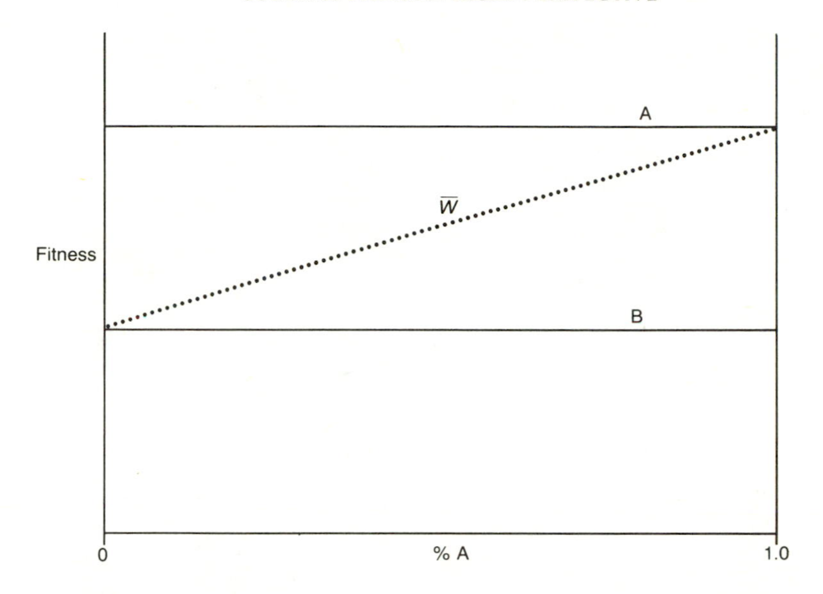

Figure 1. Frequency independent selection leads *A* to replace *B* and the average fitness of organisms ($\overline{W}$) in the population increases.

think it is the rule rather than the exception (for example, Lewontin 1974). A classic example is the evolution of mimicry. The monarch butterfly *Danaus plexippus* tastes terrible to blue jays. Another butterfly species, *Limenitis archippus*, has evolved the characteristic appearance of the monarch, but lacks its bad flavor. The selective advantage of this form of mimicry depends on the frequency of the mimics relative to the models. If the unpalatable monarchs (the models) predominate, mimicry will be advantageous to *Limenitis archippus*, since the blue jays will be fooled. But if the tasty mimics predominate, the blue jays will learn how nice they are to eat. So the fitness of mimicry declines as it becomes more common (Brower 1969; Brower, Pough, and Meck 1970).

We might picture the fitnesses of mimics (*M*) and non-mimics (*N*) within a population of *Limenitis* as having the shape shown in Figure 2. The exact character of this fitness function will depend on the number of *Limenitis* mimics relative to monarch models, but we can take Figure 2 as an example of what can happen. Here the fitness of the mimic is frequency dependent, although the absolute fitness of the non-mimic is not. Notice that selection will push mimicry (*M*) to 100 percent, and that the average fitness of organisms in the population will increase.

But suppose we represent the fitness relations as shown in Figure 3. In Figure 3 *M'* will also go to 100 percent, but the population will be no better off after the process than it was before. It isn't that the abiotic environment has changed. Rather the process of adapting steadily destroys

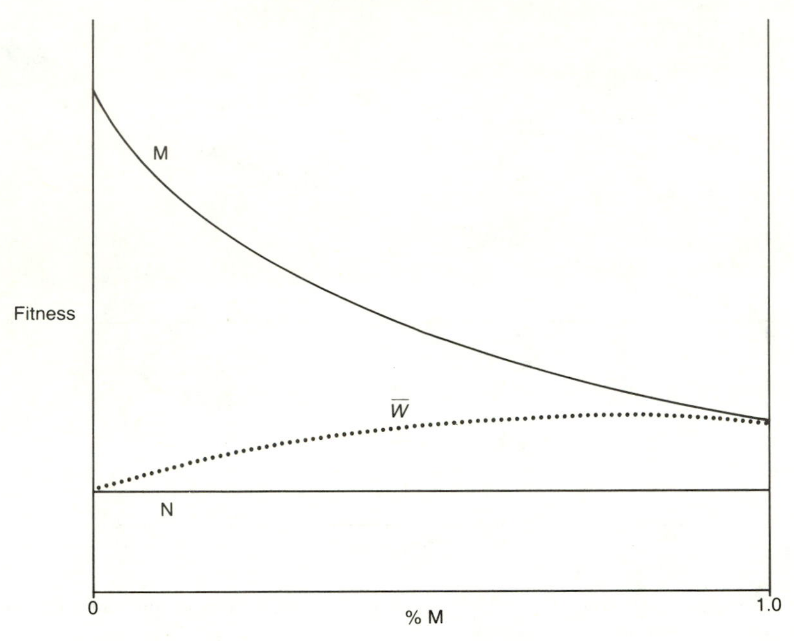

Figure 2. Frequency dependent selection, as might be expected to occur in the case of mimicry, leads *M* to replace *N*, and the average fitness of organisms in the population ($\overline{W}$) increases.

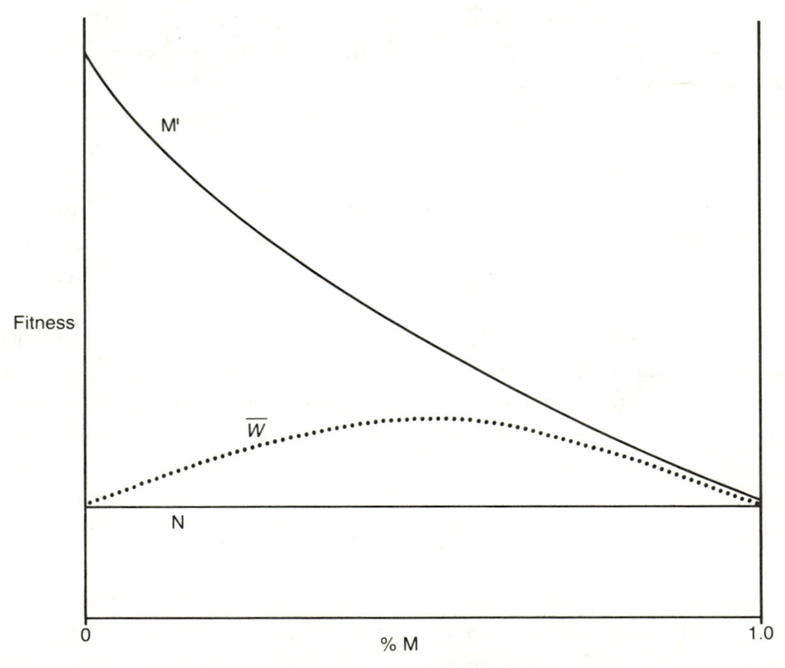

Figure 3. Frequency dependent selection leads *M'* to replace *N*, but the average fitness of organisms in the population ($\overline{W}$) remains unchanged.

the condition that makes mimicry advantageous. Here we have selection without improvement.

An even more depressing fitness function is represented by the pair of traits shown in Figure 4. *S* is a "spoiler". When introduced into a population of *R* individuals, it is at a reproductive advantage, and so increases in frequency. Indeed, at every point in the evolution of the population, *S* is fitter than *R*, and the selection process takes *S* to 100 percent. But as the process continues, both *S* and *R* decline in their absolute fitnesses. The net result is that the average fitness in the population *declines* as a result of selection.

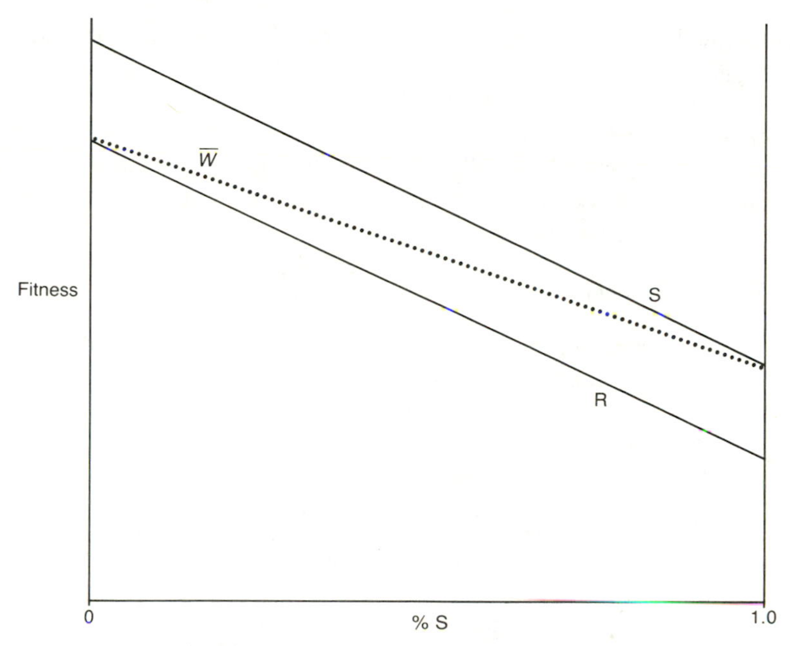

Figure 4. Frequency dependent selection leads *S* to replace *R*, and the average fitness of organisms in the population ($\overline{W}$) declines.

The method of representing fitnesses as a function of frequencies is now a familiar one in population genetics. And the argument that selection need not be an improver is just an informal presentation of the idea that Fisher's (1930) Fundamental Theorem of Natural Selection fails to hold if fitnesses are frequency dependent. But this borrowing from population genetics carries no assumptions about Mendelism. The characteristics described in these graphs may be phenotypes, genes, anything you like. The point here is not to chide Darwin for failing to have made the acquaintance of Gregor Mendel.

The fitness functions just discussed are hypothetical. There is no a priori reason to think that nature often approximates one of them any more than any other. There is no a priori reason to expect selection to lead to improvement, nor any to think that selection will grind a population into the dust. So why did Darwin think that selection is an improver? Perhaps the conceptual tools needed to represent the idea of frequency dependent selection were not available to him, and this may provide a straightforward internalist explanation. But another, externalist perspective is available.

Schweber (1980) has described Darwin's readings in the political economists. They held very definite opinions on the effects that unrestrained competition would have on the collective welfare. Schweber describes how Darwin read McCulloch's *Principles of Political Economy* in 1840, and found in it a view characteristic of British economic thought since Adam Smith:

> every individual is constantly exerting himself to find out the most advantageous methods of employing his capital and labour. It is true, that it is his own advantage, and not that of society, which he has in view; but a society being nothing more than a collection of individuals, it is plain that each, in steadily pursuing his own aggrandisement, is following that precise line of conduct which is most for the public advantage. (Quoted in Schweber 1980, p. 268)

McCulloch's book concluded with the following observation:

> The true line of policy is to leave individuals to pursue their own interest in their own way, and never to lose sight of the maxim *pas trop gouverner.* It is by the spontaneous and unconstrained. . . efforts of individuals to improve their conditions. . . and by them only, that nations become rich and powerful. (Quoted in Schweber 1980, p. 268)

Darwin's reading was not completely confined to this optimistic view of the consequences of competition. Jean Charles Sismondi, an eminent Swiss economist who had married Emma Darwin's favourite aunt, died in 1842, and the *Quarterly Review* in 1843 published a long review of Sismondi's views. The article described Sismondi's differences with the British school:

> Division of labour, according to Adam Smith, is the great source of national [wealth], of "general plenty, diffusing itself through all the different ranks of society." Sismondi says, "No". . . . Unlimited competition according to the popular theory, is the great source of national riches. Sismondi says "no" — Unlimited competition renders the whole system of commerce a vast game of "beggar-my-neighbor". . . . "Permit each person" — quoth the political economist call him Adam Smith, call him McCulloch, call him Chalmers, it is all the same — "to seek his own interest in the way which suits him best, and you must be, since society consists only of individuals, promoting the general interest of society." Sismondi contradicts this doctrine. . . . (Quoted in Schweber 1980, p. 269)

Darwin went on in 1847 to read Sismondi's *Political Economy*, and labeled the work "poor" in his reading notebook (Schweber 1980, p. 270).

We now find it quite natural to think about the deleterious consequences of unrestrained competition. A standard way of understanding how competition can make everyone worse off is in terms of the metaphor of the "tragedy of the commons". Suppose that we are dairy farmers who share a field for grazing. We all sell our milk and butter, and each earns $95 annually. Each has prospered and is considering whether or not to add another cow to his herd. You realize that if everyone adds a new cow, the grass supply will deteriorate, and so everyone's profit will decline to $90 a year. On the other hand, if you add a new cow and no one else does, your income will rise to $100. But if everyone else adds a cow and you don't, you will only make $85. Your decision problem can be represented as follows:

		Possible States of the World	
		Others Add	Others Do Not Add
Possible Actions	You add	$90	$100
	You do not add	$85	$95

Decision theory counsels that you should add a cow — it is a "best action". Regardless of what the others do, you are better off adding than you would be if you didn't (formulation from Giere 1979). But everyone else is in the same situation. So everyone adds a cow and *per capita* income declines to $90. This outcome is not, as the saying goes, Pareto optimal, in that everyone would be better off by shifting to the outcome in which no one adds a cow.

There is an intuitive way to avoid tragedies of the common, made precise in economics under the rubric of Coase's Theorem (Coase 1960; discussed in Hirshleifer 1980). The farmers could jointly agree that only some of them will introduce a cow and that the ones who do not will be paid compensation. Then everyone may be better off. Another possibility is that the state intervene and plan the economy in the hope of destroying the conditions of competition that set tragedies of the common in motion. One way or another, human beings can (at least in principle) transform the structure of their relationships when that structure makes everyone worse off. Organisms undergoing natural selection are usually not so lucky; there is generally no way to opt out of a selection process in which selection fails to improve the average fitness of individuals in a population.[15]

The tragedy of the commons shows how a process of competition in which individuals act on their own behalf can lead to a situation that is not optimal. Given our historical experience it is not uncommon for us to see economic life as often involving this sort of dilemma. Darwin, from his different historical perspective, apparently did not, and the suggestion of this possibility that he found in the work of Sismondi he rated "Poor". Perhaps it is not just the conceptual difficulties of the idea that competition can fail to improve that made the idea unappealing to Darwin; his social views may have provided a powerful filter as well.

V. Individualism

Darwin's conception of the process of natural selection can be represented in terms of a set of abstract conditions that leave open exactly what the objects are on which selection acts (Lewontin 1970). If the objects are different from each other, and if those differences allow some to survive and reproduce more successfully than others, and if the advantages of parents are transmitted to offspring, then the ensemble of objects will evolve by natural selection (assuming, of course, that no other force counters the effects of selection). In short, evolution in which natural selection is the only cause requires and is entailed by heritable variation in fitness.

These abstract conditions may apply to many different kinds of things. In the last thirty years, a theory of the business firm has been developed according to which businesses are subject to a selection process (Alchian 1950; Winter 1964; reviewed in Hirshleifer 1977). It also has been suggested that this model might be applied to cultural evolution, so that some ideas find their way into more heads more successfully than others, and so the frequency of beliefs in a population changes with time (Dawkins 1976; Richerson and Boyd 1978; Cavalli-Sforza and Feldman 1981). But Darwin's application of this model was not in the sphere of cultural evolution, but to the case of biological traits whose mode of transmission is by biological heredity (genetic, as we now say). And, in particular, his fundamental outlook was that *organisms* compete with other organisms in a selection process.

That Darwin chose this position after consciously thinking about alternatives is quite clear from his consideration of the evolution of hybrid sterility, as illustrated by his correspondence on the subject with Wallace (see Kottler, this volume). His only clear departure from this monolithic point of view occurred in the discussion of human morality in the *Descent*. Given that the issue engaged him so directly, one cannot attribute his specific position to the unavailability of alternatives. On the contrary, one of Darwin's achievements — and arguably one of the lessons he extracted from reading Malthus — was to stop thinking about the good of the species and focus exclusively on the good of individual organisms (Herbert 1971). It is difficult to discern any terribly good reason that Darwin had for rejecting hypotheses of group selection. He and Wallace went back and forth without being able to make much headway. An externalist explanation may therefore have some plausibility. The British economists were individualists rather than holists in their explanations. Interactions of individuals explain social facts; social facts do not explain what happens to individuals. Given this format, it is no surprise that Darwin was able to discern numerous cases in which population properties were consequences of individual selection, but with one exception, never claimed that individual properties were the consequences of group selection.

When Darwin took up the question of the existence of sterile castes in the social insects in the *Origin* he came close to invoking a group selectionist account:

> How the workers had been rendered sterile is a difficulty; but not much greater than that of any striking modification of structure; for it can be shown that some insects and other articulate animals in a state of nature occasionally become sterile; and if such insects had been social, and it had been profitable *to the community* that a number should have been annually born capable of work, but incapable of procreation, I can see no very great difficulty in this being effected by natural selection. (p. 236; my emphasis)

Darwin's subsequent discussion makes clear, however, that he did not think of himself as offering a non-individualist model.

Parents might be benefitted, he said, by producing offspring some of whom were sterile, if these sterile offspring helped their sibs. He compared this process with "family selection" practiced by plant and animal breeders: parents were selected for further breeding by looking at the characteristics of offspring they had already produced. That these offspring were killed before they could reproduce in no way implied that the process departed from the tenets of individual selection. What was curious about this sort of process, Darwin thought, was that (parental) organisms were selected for the traits they had, but the traits involved a relationship to other organisms. The phenotype was extended beyond the organism's own body (cf. Dawkins 1982).

Darwin's way of finessing the question of group selection has retained its attractions. He described sterile castes as emerging by a process of individual selection: parents that produced offspring, some of whom were sterile, would be more reproductively successful than parents that produced offspring who were all fertile. But what is to prevent some group selectionist from redescribing this phenomenon in terms of *kin groups* being favored if they included sterile members? Here the unit of selection has suddenly shifted from an individual organism to a group of (related) organisms. Is there a real difference here, or only a terminological one? This very question has recently exercised biologists, with no clear consensus in sight. The problem of what makes for real differences in units of selection is still not clearly understood (Sober 1982b; Sober and Lewontin 1982).

In the fourth edition of the *Origin*, Darwin considered, and rejected, another hypothesis of group selection. Hybrid sterility would be to a species' advantage, in that it would prevent the species from losing its identity by blending. But Darwin could see no selective reason for expecting natural selection to produce sterility (*Origin* 1959, p. 446).

Granted, individual selection might lead organisms to avoid mating with members of other species, if there would be no offspring. But Darwin could not see how individual selection might prefer sterile hybrids to fertile ones (Ruse 1980b). And since individual selection could not have this effect, he viewed the phenomenon as an artifact of the speciation process — not as an adaptation at all.

Not that Wallace did not try to persuade him. In correspondence, Wallace proposed the following hypothesis:

> It appears to me that, given a differentiation of a species into two forms, each of which was adapted to a special sphere of existence, every slight degree of sterility would be a positive advantage, not to the individuals who were sterile, but to each form. If you work it out, and suppose the two incipient species a. . .b to be divided into two groups, one of which contains those which are fertile when the two are crossed, the other being slightly sterile, you will find that the latter will certainly supplant the former in the struggle for existence; remembering that you have shown that in such a cross the offspring would be more vigorous than the pure breed, and therefore would certainly soon supplant them, and as these would not be so well adapted to any special sphere of existence as the pure species *a* and *b*, they would certainly in their turn give way to *a* and *b*. (*ML* 1: 288)

In most of the letters that Darwin and Wallace exchanged on "this terrible problem" (*ML* 1: 288–297), false steps and non sequiturs abounded. How could hybrids simultaneously be more vigorous and also less adapted to their conditions of life? Still, Wallace had in mind a scenario in which hybrid sterility would be promoted by selection acting above the level of the individual: A pair of species (or incipient species) that produced *fertile* hybrids of greater vigor would eventually disappear, leaving only the hybrid form; a pair of species producing *sterile* hybrids would not. So species selection would favor hybrid sterility, even though individual selection would favor hybrid fertility. The result of these opposing forces could not be calculated a priori.

Darwin remained unconvinced. Wallace remarked that perhaps it was he, Wallace, who was mistaken, but later his confidence returned and he produced the same argument in his book *Darwinism* (1889).

Wallace reasoned that since natural selection was the pre-eminent force of natural selection, hybrid sterility must have emerged in consequence of it. Seeing no individual advantage in the production of sterile hybrids, he invoked a higher level selection hypothesis. Darwin agreed that natural selection was the principal force, and agreed that the account of hybrid sterility had to be tied to it. But instead of ascending to a hypothesis of species selection, he appealed to *pleiotropy*: a trait may emerge as a consequence of a selection process without there being selection *for it*. For example, a neutral phenotypic trait may be correlated with an advantageous one (perhaps because one gene complex controls them both) and thereby be a "free-rider". This contrast between Wallace and Darwin surfaced in other contexts. Wallace despaired of explaining human intelligence by natural selection in part because he could not see how there could be direct utility in the multitude of intellectual activities of which we are capable. Darwin, on the other hand, was much more willing to appeal to pleiotropy (Gould 1980b; see Kottler, this volume). This explanatory concept permitted him to resist the Siren songs of group selectionism and divine intervention. Modern Darwinians have followed suit. G. Williams (1966) and Dawkins (1976) will postulate the existence of group selection only if they are forced to; individual selection hypotheses are, for them, "more parsimonious". This is "Darwinism" properly so-called, in that Darwin had the same cast of mind. Pleiotropy was and is a safety valve for those who do not wish to be pushed.

Darwin was not entirely single minded. In the *Descent*, he argued that altruism in human moral behavior was to be accounted for in terms of group selection. First he described the problem that self-sacrifice poses:

> It is extremely doubtful whether the offspring of the more sympathetic and benevolent parents, or of those which were the most faithful to their comrades, would be reared in greater number than the children of selfish and treacherous parents of the same tribe. He who was ready to sacrifice his life, as many a savage has been, rather than betray his comrades, would often leave no offspring to inherit his noble nature. The bravest men, who were always willing to come to the front in war, and who freely risked their lives for others, would on average perish in larger numbers than other men. (1: 163)

Then Darwin gave his selectionist explanation:

> It must not be forgotten that although a high standard of morality gives but a slight or no advantage to each individual man and his children over the other men of the same tribe, yet that an advancement in the standard of morality and an increase in the number of well-endowed men will certainly give an immense advantage to one tribe over another. (1: 166)

Darwin went on to note that members of tribes may have been relatives, so that he could have treated the phenomenon in the way he did sterile castes in the social insects — as a case of individualistic kin (or family) selection. Also, Darwin supplemented his group selectionist story with an individualistic one: once individuals had developed their powers of reasoning and foresight, they might see that it was in their self-interest to aid others, in the expectation that the help would be returned ("reciprocal altruism").

With the exception of this one flirtation with the idea of group selection, Darwin was able to remain consistently an individual selectionist. Besides grandeur, there is a certain simplicity in this view of life (G. Williams 1966; Dawkins 1976). Postulating fewer processes is more parsimonious than postulating more. Although the majority opinion among biologists now appears to side with Darwin on the question of group selection, I venture to say that we have yet to definitively establish that this more parsimonious theory has the additional virtue of being true.

ACKNOWLEDGEMENTS

My thanks to Jack Hirshleifer, Richard Lewontin, and Ernst Mayr for useful discussion. I also am grateful to the National Science Foundation and the University of Wisconsin-Madison Graduate School for financial support.

Notes

1. Here I dissent from Mary Williams's reply that fitness is an undefined primitive concept in evolutionary theory, and since every theory must have undefined terms, there can be no objection to evolutionary theory on this score (1973). My disagreement begins with the fact that fitness *is* definable in terms of the probabilities of survival and reproduction. And there is more to fitness even than this.
2. Darwin, of course, developed his theory of pangenesis. But this was presented as an additional theory. Darwin never attempted a "synthetic theory" of the kind given by Wright, Fisher, and Haldane, in which the mechanism of Mendelism is used to give a canonical formulation of the effects of all other evolutionary forces.
3. The similarities between "the principle of natural selection" and "$F = ma$" have been noted in Ruse (1973a), Hull (1974a); and Bradie and Gromko (1981).
4. Don't be misled by the fact that Aristotle discussed the harmonious "fit" of organisms to their environments. The point is that the force for evolutionary change that we now call "fitness" was not something that Aristotle had much to say about.
5. Hamilton's idea of inclusive fitness may superficially appear to upset the apple cart here (1964). Sterile workers in the social insects may have non-zero inclusive fitness, even though their chances of reproduction are nil. But I see no reason to interpret inclusive fitness in this way. Why not say that it simply requires an expanded picture of what *reproduction* involves? Sterile workers "reproduce" in this expanded sense, in that they may cause their genes to pass into future generations by helping their sibs to reproduce (in the narrow, more traditional, sense).
6. Shifting the question to explaining why one group of organisms — say, those sharing a particular genotype — does better than another doesn't improve matters.
7. The position taken here has affinities with the one adopted in Brandon (1978). It also has something in common with the so-called "semantic view of theories" (Beatty 1980), except that I regard that view as overgeneral. I have suggested that *some* laws are a priori whereas others are not. This should be evident from the contrast with Newtonian theory.
8. As mentioned earlier, the idea of a selection process transforming the composition of a population is to be found in the Scottish economists and in Hobbes before them.
9. Lamarck's theory of evolution also incorporated a version of the Natural State Model

in its explanation of differences between species that are in the same genus or family. A central force drove populations up the ladder of life while a secondary category of forces — namely forces of circumstances — led to diversity between species at the same level. The result was a branching tree, not a one-dimensional line. This will be discussed in Section III.

10. Lewontin (1981, 1983) labels the alternatives *developmental* and *variational*. This latter term is well chosen, in that explanations appealing to random drift, as well as ones citing natural selection, have the property of interest here. I have chosen the narrower term "selectional explanation" for its aptness in characterizing Darwin's use of the idea.
11. As Lewontin (1981) points out, a fundamental question for sociobiology that goes beyond the question of nature versus nurture concerns whether cultural change is best conceptualized from the point of view of a developmental or a selectional model. Traditional theories in the social sciences have often been of the former kind.
12. G. Williams (1966) provides an interesting discussion of the problems involved in defining progress as an increase in the information content of DNA in an organism.
13. Philosophers discussing the fact that fitness is not a single physical property include Rosenberg (1978), Brandon (1978), and Mills and Beatty (1979).
14. The dispute between Darwin and Wallace over human intelligence, which will be briefly discussed in Section V, is another instance in which Darwin appealed to the fact that a trait might be favored by natural selection even when many of its consequences have no selective importance whatever.
15. See Hirshleifer (1982) for further discussion of the parallel analyses of competition in evolutionary theory and in economics.

30

Images of Darwin: A Historiographic Overview

Antonello La Vergata

To the memory of my father

Introduction

The members of any community sooner or later begin to reflect on their past with an eye to their future. Darwin scholars are no exception. They have increasingly found themselves discussing methodological problems and more general "philosophical" questions, such as their relationship to other areas of the history of science and to studies on the nature of science.

What, it is now time to ask, have been the issues most commonly dealt with in Darwin historiography? What have been the dominant interpretative approaches to Darwin problems? Which ones have proved most fruitful? Which deserve greater attention in the future? These and other questions have already been raised, although not systematically discussed. But there is room for a further, deeper level of inquiry, for patterns of historiography are often (if not always) conditioned by assumptions about the nature of science and its history. It is, then, fair to ask if the unsatisfactory solution or the non-perception of some problems in Darwin studies might be due to the intrinsic limitations of the approaches that have been used. An attempt at answering this question is required in order to place the progress of Darwin studies in a broader context. In addition, such a discussion could be of interest to people not belonging to the circle of Darwin scholars.

Although this paper is not meant to present a complete history or panorama,[1] it is a historiographic critique of Darwin research over the last thirty years. It aims to discuss the ways in which the major problems concerning Darwin's achievement have been treated.

I have confined my study to the last thirty or thirty-five years for three main reasons: 1) Darwin studies benefitted from the "evolutionary synthesis" of the 1940s. After that synthesis, Darwin's stock rose; he was seen in a new light by scientists and his work was taken more and more seriously as a subject worthy of historical research. Moreover, the architects of the synthesis themselves began to rationalize the path that had led from Darwin to the synthesis. 2) Late in the 1940s and early in the 1950s a number of books were published that, although none was decisive in itself, can be used to mark, with their cumulative effect, a renewal of interest in Darwin.[2] 3) Prior to the synthesis, the most significant works on Darwin and the history of evolutionism had been, with few exceptions (for instance Lovejoy 1904, 1909, and other papers collected in Glass, Temkin, and Straus eds. 1959; Loewenberg 1933, 1935, 1941; West 1938), either reconstructions of the historical background to contemporary debates or personal views of a theory that to many still seemed a bold adventure of reason. Writing the history, and unearthing the antecedents, of Darwin's theory often was a way to declare for or against it, as well as this or that variety of Darwinism. For instance Butler (1879), Wigand (1874–1877), Osborn (1894), Quatrefages (1870, 1892, 1894), Perrier (1884), Poulton (1896, 1908, 1909), Dacqué (1903), Kellogg (1907), Delage and Goldsmith (1912), Ungerer (1923), Tschülock (1936), all more or less conformed to the same basic pattern. They were half histories, half discussions of problems in contemporary evolutionary theory (a pattern still conformed to, for instance, by Fothergill 1952 and Jacob 1970, both being, significantly, scientist-historians). In the 1950s historians progressively took the upper hand.

Since most of the papers in the present volume also contain historiographic surveys of their topics, I have not organized mine into sections corresponding to specific aspects of Darwin's achievement. Nor shall I discuss what has been written on issues such as Darwin's illness, his impact on literature and philosophy, his contacts with Marx, the "Darwinian" model for the individual and social growth of knowledge (what Rescher styled "methodological Darwinism" (1977) and Bloor "social Darwinism in the field of science" (1976, p. 62)), the implications of Darwinism for ethics and religion, religious anti-Darwinism, etc.[3] I have also not dealt with such philosophical discussions as those on the alleged tautological character of the theory of natural selection,[4] on whether it fits a reductionist research program, or on whether it supplies the basis for a new conception of teleology (or teleonomy). As a result, works in the philosophy of biology by writers such as Beckner, Goudge, Canguilhem, Grene, Mayr, Hull, Ruse, Manier, Nowinski, and Ayala have not been taken into account.[5] I have concentrated on writings that help us to understand the historical Darwin in his context, and on the images of Darwin that emerge from, or are implicit in, such works.

However, the fact of confining oneself, with few exceptions, to people writing as historians entails a major limitation. It is as though one were to write, say, the history of the idea of progress, and decided to take into account only leading figures who made explicit declarations on progress, and to neglect images of progress hidden in a number of other writings not directly bearing on the subject. Implicit images are not less pervasive by that. To avoid this historiographic fallacy, and to see Darwin research in a broader context, I have tried to give attention to the images of Darwin implicit in, or supposed by, the writings I discuss.

The images people have of Darwin can disclose their ideas of science, man and values. They can offer important insight into aspects of our culture. For this purpose, implicit or "submerged" images are more telling than explicitly stated ones. Seen in this perspective, professional Darwin students are only a category, although a very important one, of people interested in, and by, the Darwinian revolution. And this paper can be seen as part of a wider historical study of Darwin's place in our culture.

It would be rewarding to discuss the more or less distinct images of Darwin in semi-popular and popular writings by Ardrey, Barash, G. Bateson, Dawkins, Eibl-Eibesfeldt, Lorenz, Midgley, Morris, E. O. Wilson, and, in general in the debates over ethology and sociobiology, those implicit in literary critics, historians of the Victorian period, and social historians. It would be interesting to inquire what lies behind Popper's caricature of Darwin (see criticism in Ruse 1981, pp. 65–84), why I. Lakatos once called Darwin "a lousy philosopher" (Grene 1976, p. 211), and the reasons for the general neglect of Darwin by philosophers of science. (I do not think I am too *naïf* if I wonder how books bearing such titles as *The Structure of Science, The Scientific Inference, Logic of Scientific Discovery, The Structure of Scientific Revolutions, The Structure of Scientific Theories, Scientific Explanation, Philosophy of the Natural Sciences*, etc. can give Darwin's name only a couple of passing mentions on aggregate).

It would be interesting to disentangle the mixed image of Darwin and neo-Darwinism that characterizes the writings, especially the more "philosophical" ones, of leading biologists such as Bertalanffy, Dobzhansky, Grassé, J. S. Huxley, Jacob, Jacquard, Monod, Rensch, Ruffié, and Simpson. All the more so because scientists obviously influence historians, if only by presenting them with images of what could be considered the end result of the process Darwin started and historians have to reconstruct. It would be interesting to study the images of Darwin in the debates (and trials) raised by creationists and the religious opponents of Darwin. And would a historian of culture not be intrigued by titles such as *Evolution and Darwinism Foreshadowed in the Apocalypse, Darwin: the Evil Genius of Science and His Nordic Religion*, or *Darwin Is Not for Children*?[6]

This paper consists of four sections. The first covers Darwin research before the publication of the *Notebooks on Transmutation of Species* (1960). This date is a watershed between two phases of Darwin scholarship, for it marks the beginning of what has been called the "Darwin industry". The first section also deals with attempts to evaluate Darwin's place in western intellectual history. Section II deals with the problems raised by the publication of the *Notebooks*, and in particular with the genesis of the theory of natural selection. In Section III I discuss the literature on Darwin's method, "philosophy", strategy, language, and frame of mind. Section IV concerns

Darwin's development after 1859 and his relationship to his professional, intellectual, and social contexts.

I. Before the *Notebooks*

DARWIN'S PLACE IN INTELLECTUAL HISTORY

J. C. Greene is not only the author of the best history of evolutionary ideas, and of one of the best works of intellectual history in the field (Greene 1959a); he is also the author of pertinent criticisms of Darwin studies by intellectual historians (Greene 1975). His characterization of the merits and limits of works by Lovejoy, Barzun, Loewenberg, Himmelfarb, Eiseley and himself is concise and accurate, so I shall refer the reader to it, use his comments as a starting point for my own, and keep his historical work in the forefront in the following discussion.

As Greene remarks, the above-mentioned writers

> have been less concerned with the description and evaluation of the specific researches of Darwin and his predecessors and contemporaries than with the general ideas that have informed their work [. . .] For these historians Darwin was primarily a thinker. His ideas were to be studied, and evaluated, not in reference to his empirical investigations, but in reference to the ideas of his predecessors, contemporaries and successors. (1975, pp. 250–251)

These historians presented Darwin's ideas as the culmination of some major trends in western thought, not merely as an event in the internal history of biology. It may be pertinent here to recall Sears's 1950 title, *Charles Darwin: The Naturalist as a Cultural Force.* Intellectual historians, then, deserve high credit for distinctly broadening the perspective of Darwin studies. What was thus gained in breadth of view was, however, lost when attempts were made (in particular by Barzun and Himmelfarb) to evaluate Darwin's scientific achievement (as was pointed out by Mayr 1959d, De Beer 1959c, and Ellegård 1960–1961 in their reviews of Himmelfarb 1959). Similar criticisms were leveled by Kuhn in 1968 and 1971 (now in Kuhn 1977, pp. 111–112, 139 n.), who remarked that Darwin scholars remained at "a level of philosophical generality" and neglected internal scientific issues.

It is my contention that this criticism holds true not only for writers who are intellectual historians by training, but often also for scientist-historians, as I hope I shall be able to demonstrate in this and the following sections. Indeed, what I discuss in this section under the heading "intellectual history" is not the work of a category of professionals, but a *pattern of approach* to Darwin. The "Darwin's-place-in-history approach" was dominant before 1960, but has been followed by a great variety of writers even since then. To my mind, it has not harmonized with the approach implicit in the more narrowly focussed studies that grew up as a result of two decisive events in Darwin scholarship: the publication of Darwin's *Notebooks* and the rise of a community of professional Darwin students. The title of this section might legitimately have been "Before and without the *Notebooks*". In what follows I shall list and comment on some features of the intellectual history approach in the wide sense indicated above. Obviously, I shall be obliged to focus on a few representative works.

Darwin's way of looking at nature. According to Greene, "the starting point of the Darwinian approach to nature was the mechanical philosophy of the seventeenth century, the idea of nature as a law-bound system of matter in motion" (1981, p. 89). To this extent he follows a historiographic "topos" that goes back as far as T. H. Huxley (cf. for instance 1893b, pp. 206–207, 213). But to Greene Darwinism is the "logical outcome" of a combination of this Cartesian philosophy of nature with other elements: the eighteenth-century corrosion of religion, the notion of progress (that is, of a direction in change in society), British political economy and competitive ethos, and a type of deism verging on agnosticism. Kant and Laplace in cosmology; Hutton and Lyell in geology; Buffon, Erasmus Darwin, and Lamarck in biology developed the implications of Cartesian mechanical philosophy along three lines that eventually converged in Charles Darwin. But Darwin was not alone in this, for it would be more correct to speak of "a more general Spencerian-Darwinian world view in which Spencer, Darwin, Wallace, and Huxley converged about 1860, only to diverge again before the century had run its course" (Greene 1981a, p. 7). One is not surprised, then, to find Greene saying that progress was the leitmotiv of Darwinism and that English Darwinism might better be called Spencerianism (pp. 128–155). Characteristically, when Greene wants to investigate the Darwinian concept of order, he takes Spencer's and Huxley's writings in conjunction with Darwin's own (Greene 1968, p. 89). But the very category of "Darwinism" has subsequently proved on close scrutiny to be enormously complex, if not misleading. "Darwinism" was "many things to many people" (Hull 1974b, p. 388), as happened with other categories such as "Newtonianism", "Positivism", and so on (see also Roger 1976).

According to Gillispie, Darwin's theory belongs to the mechanical and mathematical world-picture built up by Galileo, Newton, and modern science in general. Being "quantitative in method and matter of thought", though not in numerical expression, "the theory of natural selection is what turned the study of all living nature into an objective science" (Gillispie 1960, p. 261). This is evident, for instance, in Darwin's treating variation as a phenomenological fact: he assumed that variations occurred, without feeling it necessary to set himself the task of seeking out their causes, and he distinguished between the origin of variations and their preservation (Gillispie neglects Darwin's life-long interest in the origin of variation, which led eventually to his theory of pangenesis, and dismisses this latter as speculation).

Compare this, Gillispie suggests, with Galileo's attitude towards atoms. Again, in the phenomena of cross-fertilization and hybridization of plants, Darwin might demonstrate the exchange of old species for new ones "as explicitly as in Lavoisier's combinatorial analysis-synthesis applied to chemical species (or reagents)" (p. 312). And were not also Mendel's particulate view of inheritance and his experiments of recombinations an application and confirmation of an atomistic mathematical approach? (pp. 337–339). Though an old-fashioned naturalist, like Wallace and Mendel, Darwin was the "Newton of the grass blade" whom Kant hoped for. The reason for this lies not only in the nature of the empirical work of those three men, but more specifically "in the nature of their reasoning, which was concerned with quantity and circumstance [. . .] They [. . .] liberated biology from its limiting dependence on classification and dissection", that is, from its focussing on mere form (p. 340). By bringing atomism into biology — through the application to it of the individualistic assumptions of classical political economy — Darwin defeated an age-old tendency to stress unity and continuity throughout nature that extended back through Hegel, Lamarck, Goethe, and Diderot to the Stoics and Heraclitus. Biological time was no longer the all-embracing dimension symbolizing the continuity of nature, no longer "a refuge of becoming or a locus of flux"; it was a dimensional coordinate of events that were as physical as any others. Gillispie's interpretation stands against those that characterize Darwin's achievement as the final act of the introduction of time into our view of nature. Far from bringing a Hegelian sense of becoming within the pale of science, "the Darwinian theory of evolution by natural selection turned the problem of becoming into a problem of being, and permitted the eventual mathematicization of that vast area of living nature which until Darwin had been protected from logos in the wrapping of process" (p. 342). Gillispie believes that all advances in science are moves toward particulate thinking.[7]

Whatever one may think of Greene's and Gillispie's evaluations of Darwin's achievement, it must be noted that they represent a tendency among intellectual historians to lay emphasis on Darwin's world view, and to discuss it in comparison with broad historical trends in western thought. This proved a useful corrective to the so-called "positivistic" image of Darwin as a "pure scientist" and a "fact-gatherer"; ideas, not bare facts, were important in Darwin's achievement.

Ingredients and combinations. Greene's main aim was to trace "the various currents of Western thought that came together to form what I call Darwinism and [. . .] the way in which these currents were blended in the thought of each of the four men [Darwin, Spencer, Huxley, and Wallace]." He attempted to outline the origins and development of these elements up to the point at which they "lay waiting for the architect who could combine them into a single all-embracing synthesis" (1981b, pp. 130, 133). But sometimes this approach degenerates into what might be called a "combination-of-ingredients" historiography, as exemplified by the following passage:

> If Wells had been a zoologist as well as a physician, Charles Darwin's theory of the origin of species might have been anticipated by almost fifty years. All the elements of the theory were present in the scientific world by 1818. Buffon, Kant, and Laplace had derived the origin of the solar system from the operation of a universal system of laws, elements, and forces. Hutton had conceived the surface of the earth as a system of matter in motion millions of years old. Cuvier had applied the resources of comparative anatomy to the reconstruction of extinct species and, with William Smith, had discovered how to read the fossil record embedded in the globe's crust. Buffon had suggested the variability of organic forms, and Lamarck had postulated their gradual evolution from monad to man. Buffon had seen that the extinction of species was related to the struggle for survival among the various creatures produced by nature's endless combinations. Maupertuis, Prichard, and Wells had sensed the possibility that new types might be formed from chance variations thrown up in the course of procreation, and Wells had used the notion of natural selection to explain the origin of the Negro race. Even Malthus' *Essay on the Principle of Population*, the book which Darwin said gave him the clue to the origin of species, was available.
>
> But although the elements of Darwin's theory lay at hand, they were not embraced in one powerful and well-informed mind. Moreover, the traditional view of nature, though greatly weakened by these developments, still exerted a powerful influence on scientific thought. (Greene 1959a, p. 245)

Or consider the following statements by Eiseley:

> Many of the ideas Darwin was later to use come from the researches of these very men [his forerunners]. Lamarck, for example, observed the struggle for existence and recognized the significance of vestigial organs before Darwin [. . .] He [Darwin] took the providential 'localizing principle' of a neozoologist like Blyth and added to it the infinity of geological chance. (1959b, p. 111)

R. M. Young has rightly remarked that "the argument of Eiseley's *Darwin's Century* is laid out like a detective story or a jigsaw puzzle, in which the clues or pieces are seen exclusively in the light of their contribution to the 'solution,' the picture on the cover of the box" (1973a, p. 365). This criticism I think holds true for much that has been written by intellectual historians about Darwin and his antecedents. The idea that these antecedents were to become "elements" of Darwin's thought pervaded much of the literature. Just think of the phrases that were, and still are, commonly used to describe Darwin's alleged debt to Malthus on the score of the idea of the struggle for existence (or of natural selection): "he took it from Malthus", "he stumbled on it when reading Malthus", "he borrowed it from Malthus", "he derived it from Malthus", "he found it in Malthus", and so on, as though what were to become elements of Darwin's thought were just there, simply waiting for Darwin to grasp them.[8] Furthermore, what are considered to be the intellectual features and results of Darwin's achievement are transformed by the historian into contributory factors. It is as though one first abstracted some general ideas from a set of intellectual phenomena over a certain time span, and then used them as leading ideas not only to characterize a broad context or a period in which to locate individual achievements, but also to account for the very intellectual phenomena from which these ideas were abstracted. The interpretation of an individual's work in terms of the results he, together with others, produced is misleading. It is like trying to explain Darwin by some variety of subsequent Darwinism. Did Darwinism mean the triumph of secularism, naturalism, empiricism, and anti-teleologism? If so, secularism, naturalism, empiricism, and anti-teleologism must have been key factors in the origin and development of Darwin's theory itself.

Overconcern with general patterns of ideas. According to Greene, "the primary function of the intellectual historian is to delineate the presuppositions of thought in given historical epochs and to explain the changes that these presuppositions undergo from epoch to epoch" (1981a, p. 10).[9] But this emphasis on general ideas and long-term trends disregards — or misconceives — the individual thinker's own development. It is no accident that up to the late 1960s many attempts were made to capture the essence of Darwinism, but none was devoted to the evolution of Darwin's own thinking; that biographies of Darwin, including the quasi-official one by De Beer (1963), failed to account for Darwin's struggle in constructing his theory; and that Greene (1959a), in his chapter on Darwin, makes no use of the *Origin*, as if there were no difference between it and the *1844 Essay*, on which he relies almost exclusively. Furthermore, what about "individual variations" among scientists who shared these common assumptions? Here again, the ambiguity of the category "Darwinism" is a case in point. The fact is that under such treatment individuals seem to receive attention mostly as bearers or representatives of ideas: they are receptacles for the combination and development of ideas. Scientists, like other men, do not have ideas: ideas have men.[10] Look at the very phrasing:

> In the writings of Erasmus Darwin and Lamarck, Western thought adumbrated the second component of what was to be known as Darwinism. (Greene 1981b, p. 131)

This tendency to present great scientists as spokesmen of the great concepts, or conceptual assemblages, that inform the thought of their age is evident in Jacob (1973), a fascinating narrative, which, however, exemplifies many of the shortcomings of the intellectual history approach to Darwin.[11]

In writers less aware of the polycentric character of the process said to have culminated in Darwin, this cast of mind produced a mere harping on stale commonplaces, such as: "evolution was in the air"; or "the time was ripe for Darwin's theory" because "the scientific and philosophical background for it were there"; and Darwin's age was able "to recognize itself in him" (Carter 1957, p. 45; Barzun 1958, p. 80).

On a different plane, this reminds me of the striking analogy between what Mandelbaum (1965) called Greene's "cultural monism" and the cultural determinism in support of which the anthropologist L. A. White instanced Darwin. According to White, exceptional individuals are the result of the convergence of various cultural currents into a synthesis, "the neural locus of an historical fact in cultural development". The theory of evolution did not originate with Darwin, for we find it in one form or another in the "neural reactions" of Buffon, Lamarck, Erasmus Darwin, and others. As a matter of fact, White says, virtually all the ideas which together we call Darwinism are to be found in the writings of J. C. Prichard. These various ideas were interacting with each other and with current theological beliefs for decades, until

> the time finally came, i.e., the stage of development was reached, where the theological systems broke down and the rising tide of scientific interpretation inundated the land. Here again the new synthesis of concepts found expression simultaneously in the nervous systems of two men working independently of each other: A. R. Wallace and Charles Darwin. The event had to take place when it did. If Darwin had died in infancy, the culture process would have found another neural medium of expression [. . .] At the time he read Malthus, Darwin's mind was filled with various ideas, (i.e., he had been moulded, shaped, animated and equipped by the cultural milieu into which he happened to have been born and reared) [. . .] These ideas reacted upon one another, competing, eliminating, strengthening, combining. Into this situation was introduced, *by chance*, a peculiar combination of cultural elements (ideas) which bears the name of Malthus. Instantly a reaction took place, a new synthesis was formed [. . .] Darwin's nervous system was merely the place where these cultural elements came together and formed a new synthesis. It was something that *happened* to Darwin rather than something he *did*. (L. White 1969, pp. 294–295, White's italics)

The tendency to see in great men the culmination of the development of great ideas explains still another feature of much that was written on Darwin and his surroundings from the point of view of intellectual history: the frequency with which comparisons were drawn between Darwin and this or that thinker. These were comparisons between eminent individuals embodying different world views. This is evident in Gillispie's (1958, 1960) classic comparison between Darwin the progressive, quantitative scientist and Lamarck the spokesman of "the contracting and self-defeating history of subjective science", to which belong "Diderot's organismic and metamorphic philosophy of nature", German "biological romanticism" — which "gave neo-Lamarckism a hospitable climate" — Samuel Butler's and G. B. Shaw's venturings into biology, Driesch's and Bergson's vitalism, the moralists' resentment of "the survival of rascals", Lysenko's demagogy, the sentimental-Rousseauist hostility to mathematics and blind stochastic processes, and all manner of what Lovejoy styled "romantic evolutionism" (Gillispie 1958, pp. 276; 1960, pp. 261–262, 322, 344–347). Other instances of this comparison approach are Grassé (1960), Vandel (1960), Willey (1960), Simpson (1961b, also 1964, chap. 3), Wilkie (1959), Rousseau (1969), and B. Coleman (1974). But perhaps the pervasiveness and persistence of this approach is best illustrated by the studies on the relationship between Lyell and Darwin. In fact, these studies have consisted mostly of a comparison between essences, that is between the spirit of Lyell's uniformitarianism and that of Darwin's evolutionism. Lovejoy (1959b, pp. 366–373), Gillispie (1959*, p. 131; 1960, p. 30), Eiseley (1961, p. 100), Irvine (1955, p. 58), and Greene (1959a, pp. 246–253) have all followed T. H. Huxley in claiming that uniformitarianism in geology almost cries out for, if it does not logically involve, evolutionism in biology. And even those who have criticized this view have argued along the same methodological lines. Hooykaas (1959, 1965, 1966, 1970a, b), W. Cannon (1960a, b, 1961b, c, 1969, 1976a), Rudwick (1962, 1970, 1971, 1972), L. G. Wilson (1967, 1969, 1971a, 1972, 1973), Limoges (1970c, pp. 10–16), and Mayr (1972a, pp. 983–987) have all maintained that Lyell's view of the history of the earth was intrinsically anti-evolutionary. Whatever the answer, the argumentative method was the same: a comparison of two distinct world pictures.

Logical implications. The attitude of many intellectual historians toward the relationship between Lyell and Darwin disclosed another negative consequence of their concentration on relations between general ideas: intellectual development is often presented as a "drawing out of implications" from pre-existing ideas. Thus, for instance, Greene says that "mutability, not stability was the logical outcome [of the] law-bound system of matter in motion" (1981a, p. 14).[12] It should be borne in mind that presenting Darwin's achievement as the logical culmination of some long-term processes could be a powerful weapon in the battle for Darwinism, provide fuel for myth-making, and foster ideological self-satisfaction: Darwin's theory was the necessary result of the winning, really "scientific" stream in western thought. T. H. Huxley and especially Haeckel are cases in point. But another facet of this approach was the tendency to present Darwin as the powerful synthesizer who drew the inevitable *theoretical* consequences from the *empirical* premises represented by the data amassed by preceding observers. As instances of slight variations on this theme, we may adduce passages from Sachs, Pantin, and Eiseley.[13] Obviously those who stressed Darwin's role as a synthesizer and the author of a cumulative argument based on cumulative evidence often tended to belittle the novelty of his theory. A good instance of this is Fothergill (1952, particularly pp. 117, 143).

Precursors. The forerunner hunt started in the midst of the Darwinian controversies. Talk about forerunners was to some a way of belittling the importance of Darwin's achievement by showing that it was not that new (Butler 1879 is a good instance of this, but the same applies to many French critics of Darwin). To others, however, unearthing forerunners served to show that Darwin was the legitimate heir to the mainstream of scientific thought. Between these two extremes the creativity of the forerunner-dealers has often been given full rein. From the time of Matthew Arnold's "it is all in Lucretius" onward, the number of Darwin's alleged precursors has been legion, and includes such people as Empedocles, Theognis, Alberuni, Confucius, Lavater, and Franklin.[14] How many know that Dante recognized natural selection as acting between ecotypes (Haldane 1959, p. 102)? But here it is sufficient to refer to Zirkle (1941, 1946), the champion of precursor-listing and the author of an exemplary caricature of Lovejoy's method of tracing the history of "unit-ideas".

Although the notion of "forerunner", a notion belonging "to the prehistory of the history of science" (Koyré), was not always taken literally by the best historians, it was still very pervasive in its many guises. For instance, the 1974 supplement of the *Isis* Cumulative Bibliography still had a subsection on the "forerunners" of Darwin. One of the best products of the 1959–1960 flood of works on Darwin bore the very title *Forerunners of Darwin* (Glass, Temkin, and Straus Jr. eds. 1959). It was "devoted to the forerunners who made Charles Darwin's achievement possible." Glass, who wrote the preface, was well aware that "certain of them were hardly evolutionists; others, in their own eyes, not evolutionists at all. Some, who lived in the period after 1859, even hated the Darwinian teaching and fought it vehemently. Yet one and all they formed the great, steadily enlarging current of biological thought which eventuated in Darwin." Glass also apologized for not including Malthus in "the gallery of precursors" (p. vi). This gives an idea of the gulf that divides a large part of pre-*Notebook* Darwin scholarship from post-*Notebook* work, for today the most radically externalist historian of ideas would hardly dare refer to Malthus as a forerunner of Darwin.

As late as 1959 Wilkie felt it necessary to vindicate Darwin's originality against the "Empedoclean view" of the history of evolutionism that, for instance, a bit of Maupertuis plus a bit of Buffon produce something similar to Darwin. Interestingly, however, Wilkie was still somewhat a victim of the very approach whose results he was criticizing. For instance, "to make the deficiencies [note the term] of Maupertuis's theory the more apparent", he compared it with "a full anticipation of the idea of evolution by selection of chance variations", that is, that of W. C. Wells. Wilkie's conclusion was puzzling:

> Although Darwin was little conscious of the preliminary [note the term] work done by Buffon and by Lamarck, he must have profited by it, if only indirectly [. . .] Yet it was Darwin who completed the edifice. (1959, p. 307)

Darwin's "forerunners" were variously presented: now as contributors to a common cause, now as those who brought forth the elements that converged in the Darwinian synthesis, now as the authors of "not completely scientific" theories (Ellegård 1960–1961), or "glimpses", "hints", "guesses", "wild speculations", "anticipations", "preconditions," or "raw facts" waiting for a theory. However they were judged, they were the protagonists of much historiography until well into the 1960s. Eiseley's (1959b, 1979) well-known fixation about Blyth is only one exaggerated aspect of a widespread tendency. For instance, did C. D. Darlington (1959a, 1959b) not insinuate that Darwin fell short of the standards of scientific rectitude in not acknowledging his debt to Lawrence and other forerunners? Did he not reveal in passing, among other things, that Lucretius was a precursor of Mendel? Did he not go so far as to say that, whatever the causes of Darwin's failing to give due credit to his forerunners, the latter paved the way for the acceptance of evolution? It is no accident that the book that inaugurated a new era in Darwin studies, Limoges's *La sélection naturelle* (1970c), devoted almost a whole chapter ("Vraie et fausses identitées") to rejecting the claim that Wells, Lawrence, Blyth, and Matthew had anticipated Darwin. It must also be remembered that Darwin studies benefitted greatly, if indirectly, from the publication of J. Roger's classic *Les sciences de la vie dans la pensée française au XVIIIe siècle* (1963). This book helped shift the focus from a study of eighteenth-century evolutionists as forerunners of Darwin to a study of the same figures per se and in their own context.

Finally, overconcern with precursors was often accompanied by a tendency to play down the scope of Darwin's revolution, a tendency that Ellegård (1960–1961, p. 71) considered very common among historians. This attitude is exemplified in Himmelfarb (1959), who did a good deal by describing Darwin's family and social background and overhauling the plaster-of-Paris image of Darwin the Scientist (for instance, it opposed his "peculiarly imaginative, inventive mode of reasoning" to the fact-gathering-plus-deduction image of Darwin). Himmelfarb also wrote, however, that "not only the raw material, but even the very terms of his theory were common to the entire scientific community" (1959, p. 177). Darwin's was a "conservative revolution" both in its scientific and cultural aspects (chap. 20). The theory of natural selection itself, "posing as a massive deduction from the evidence, [. . .] ends up as an ingenious argument from ignorance" (p. 276).

Many other authors were liable to the criticism Mandelbaum addressed to Lovejoy, namely that stressing the continuity of the unit-ideas that enter a particular theory or system usually fails to account for the motivating force or the formative influences that help determine the patterns into which these elements fit (1965, p. 37). All this is evident in Lovejoy's revaluation of "poor Chambers" and discussion of the argument for organic evolution before 1859 (1959b).

Easy formulas. From the preceding pages I think it emerges that what was written from the point of view of intellectual history, mainly before 1970, was often spoiled by a tendency to trace anticipatory ideas and their purely logical implications. As a result, the coexistence of ideas was often interpreted as causal connections, and similarities as real influences. These excesses were criticized by Hull (1975) and Ghiselin (1976). Indeed, "influence", "impact", and the like were keywords in much literature. Many writings dealt with the impact of Darwin's ideas, especially in philosophy (Höffding 1909; Lovejoy 1909; Dewey 1910; Fisch 1947; Wiener 1949; Passmore 1957, 1959; Collins 1959; Fulton 1959; Randall, Jr 1961; Marnell 1966). The tendency to attribute too much to Darwin's direct influence was particularly strong in the literature on the history of human sciences. It was criticized by Bock (1955) and Burrow (1966a).

But nowhere is the tendency to resort to easy formulas more evident than in the frequency with which Darwin's view is linked to the British competitive ideology. For instance Greene writes:

> British political economy, based on the idea of the fittest in the market-place, and the British competitive ethos generally, predisposed Britons to think in terms of competitive struggle in theorizing about plants and animals as well as man. (1981a, p. 7)

Statements such as this may be found in any section of the literature on Darwin and are too numerous and too well known to need stressing here. It was confidently, if vaguely, believed that "in order to invent the theory of descent with modification by natural selection, it was necessary to be English in the middle of the nineteenth century" (W. Cannon 1968, p. 164). Intellectual history dealing only with general ideas and their spontaneous dialectics, and transforming, as pointed out above, coexistences into causal connections, is very likely to become external history. Ideas philosophical, political, biological, artistic, etc. communicate with each other without limit through the medium of so-called world views, or other versions of the Zeitgeist.

When discussion reaches a too generic level, almost anything can be said or refuted, and there is hardly any limit to fanciful comparisons. However, few reached the level of Zirkle (1959a) (nothing less than a cold war pamphlet), or of Northrop (1950). The latter contrasted the "Darwinian Anglo-American" (and to a lesser extent French) concept of evolution and the "Hegelian-Germanic-Russian" (and Marxian) one, and went on to state that, if one does not understand this contrast, "the ideological conflicts between the communistic Russians and the Anglo-American democracies at the present moment cannot be fully appreciated" (Northrop 1950, p. 45).

SOME FEATURES OF THE FLOOD

On the occasion of, and immediately after, the centenary of the *Origin*, there was a flood of writings on Darwin and his surroundings (for surveys see Loewenberg 1959a, 1965; Ellegård 1960–1961; Fleming 1959). Allowing for some simplification, these writings can be divided into two main categories, which corresponded to a division of labor between scholars: 1) works by intellectual historians concerned with Darwin's place in intellectual history; 2) evaluations of aspects of Darwin's achievement in the light of modern knowledge (according to a pattern already exemplified by Muller 1949; a late flowering of this approach is P. Ekman ed. 1973). Two other subordinate categories were: 3) writings by intellectual historians on Darwin's intellectual antecedents; 4) writings by specialists about Darwin's impact on their disciplines. This division is particularly evident in symposia volumes, in special issues of periodicals, and in collections of essays published in the wake of the centennial celebrations.[15] The number of writings falling under headings 2) and 4) was greater than that of analyses of single aspects of Darwin's work per se. For instance, scientist-historians seemed to prefer to discuss post-Darwinian developments rather than Darwin's own views (a partial exception is G. G. Simpson 1959). But focussing on how subsequent developments in science have clarified, integrated or corrected Darwin's views on various problems is scarcely the *via regia* to the historical reconstruction of Darwin's own work. As I shall argue in Section IV, Darwin studies still suffer from the consequences of this. The celebrative and laudatory tone of historiographic overviews such as Loewenberg (1965) is now somewhat disconcerting.

VARIOUS IMAGES

The following is a list of some images of Darwin in addition to those discussed in the foregoing pages. They are drawn, somewhat at random, from books before and after 1959 that aim to assess Darwin's place in a wide intellectual context or to capture the "essence" of his theory. The list shows the multiplicity of these images, and also the contradiction between some of them.

Darwin the scientist who asserted the primacy of historical knowledge in biology, and who marked the irruption into it of nineteenth-century

Historismus — that is, the idea of universal development that had triumphed in metaphysics with Hegel (Cassirer 1950).[16]

Darwin the thinker who introduced a sense of meaninglessness, pessimism, discomfort, if not tragedy into our world view (Loewenberg 1934, 1941, 1957, 1962; Irvine 1955, 1959; Hyman 1962; Worster 1977).

Darwin the scientist who introduced an ecological dimension and a philosophy of ecological precariousness into our view of nature's economy (Limoges 1970c; Conry 1974; La Vergata 1979; Worster 1977).

Darwin a major figure in European Positivism (Kolakowski 1966).

Darwin the "evolutionary deist" (Greene 1975; 1981, chap. 6).

Darwin the individualist, who, accordingly, established the importance of the individual in biology (Ghiselin 1971).

Darwin the haunted dreamer (Eiseley 1979).

Darwin the "fragmentary man" bequeathing to the world a fragmentary, partial truth, "seeing all in acquisitive terms, subordinating the whole to the part, making the quantitative aspects his total consideration till 'everything about him dries up' and everything about him becomes a wilderness, all life dries, all value disappears'." This man compensates with personal courage for the inevitable shortcomings of any theory such as his (West — quoting Sombart — 1938, p. 337).

Darwin the scientist who tried (but failed) to reduce anything that is human to biological laws, for he could not grasp the difference between a biological and a human fact (Greene 1961a; Farrington 1966). Such criticism was leveled in the name of humanism. To Montagu, Darwin freed man from one prison only to place him in another (1952a, pp. 99–100). To Lovejoy (1961), Darwinian evolution was a trap, mechanistic and cruel. Darwin was unable to be a consistent mechanist, as his science required (Willey 1960). To J. S. Huxley Darwin failed to realize "man's truly unique and most important characteristic — cumulative tradition" (1960a, p. 17).

Darwin the author of a genuine materialist revolution (Hollitscher 1964; Lewontin 1974b).

Darwin the rigid mechanist (Arber 1950).

Darwin the materialist utilitarian, opposing Romanticism in science (W. Cannon 1976b).

Darwin the scientist and thinker who worked consistently in the reductionist and deterministic tradition of Descartes (Ghiselin 1975, p. 47; cf. 1974a; Montalenti 1982).

Darwin not a positivist-mechanist. To interpret him thus means to accept uncritically a nineteenth-century stereotype (Schneer 1968, pp. 140–141).

Darwin the proponent of a "compositionist" approach to the problems of life as an alternative to the "Cartesian or reductionist" (Dobzhansky 1968; Dobzhansky and Boesiger 1968, chap. 1; Mayr 1972a).

Darwin the pure scientist, whose name, in spite of his lack of interest on ontological questions, was abused to support a philosophy and a metaphysics. Spencer and Huxley were responsible for the "imbroglio remarquable d'où est sort le mythe de l'évolutionnisme darwinien" (Gilson 1971, p. 121. Other Catholic writers share this view).

Darwin the man who taught philosophers to be "disposed to abandon system building and synoptic truth for the piecemeal study of the basic concepts, procedures, and language of the sciences" (Wiener 1949, p. 193). According to this view, the pragmatists were "Darwin's spiritual offspring in the philosophical fields".

Darwin the bad logician and "rag-and-bone" rhetorician (W. Cannon 1968).

Darwin the fine rhetorician (J. A. Campbell 1968, 1970, 1975).

Darwin the author of a tragedy written in symbolic and ritual language and entitled *On the Origin of Species* (Hyman 1959, 1962).

Darwin the author of a comedy entitled *On the Origin of Species* (Culler 1968).

Darwin the author of a book entitled *On the Origin of Species*, which contains "epic traces" (Scheick 1973).

II. The *Notebooks*

MANUSCRIPTS

The importance of the publication of Darwin's *Notebooks on Transmutation of Species* (De Beer ed. 1960, De Beer and Rowlands eds. 1961; De Beer, Rowlands and Skramowsky eds. 1967; also see Barrett ed. 1960) could scarcely be overestimated. Certainly many a Darwin student would have preferred a more accurate edition, and some pedantic spirit might also wonder what use an edition is that so often obliges one to go to the originals (a new edition is now being prepared). But the publication of the *Notebooks* was De Beer's best service to Darwin scholarship, and a watershed. A whole species of *historici Darwiniani* has been feeding and growing on them since. And nowadays the genuine Darwin student is recognized by the fact that at least once in his or her life he or she went to Mecca (the Manuscripts Room of the Cambridge University Library) to try Mr. P. J. Gautrey's and Dr. S. Smith's fathomless patience, unrivalled kindness, and sovereign mastery of the Darwin documents, or resorted to that modern substitute for scholarly pilgrimage: massive microfilm order.[17]

The *Notebooks* have played the leading role in a very large part of what has been written on Darwin since 1960. They have found their way into popular books as well: for instance Brent (1981) devotes almost a whole chapter to them (and thinks he sees in them clues to Darwin's sex life). But they also started a rush to other unpublished material: his early pangenesis manuscript, the privately circulated "Queries about Expression" and the "Questions about the Breeding of Animals", what Vorzimmer wrongly identified as the "outline and draft of 1839",[18] the notebooks on "Man,

Mind and Materialism", the "Old and Useless Notes", the "Big Book" of 1856–1858 (*Natural Selection*), the "Red Notebook" (*RN*), the papers contained until recently in what was called the "Black Box".[19] A complete edition of the letters to and from Darwin is being prepared under the editorship of F. Burkhardt and S. Smith. Also Darwin's marginalia, the importance of which was stressed by S. Smith (1959a, b, 1960b, and 1965), have enjoyed an ever-growing fame, so that M. Di Gregorio (pers. comm.) is preparing an edition of those in the books of Darwin's library, which he hopes will be followed by an edition of the marginalia in Darwin's offprint collection.

Now, what has been the impact of the publication of the *Notebooks* and other manuscript material? On the one hand, they have opened a number of directions for research; on the other hand, they have closed an epoch in Darwin studies. By facilitating our access to Darwin's mind at work, to its wanderings, waverings, and strivings toward a final goal, they have not only made it possible to study Darwin's actual process of discovery; they have also revealed its affective dimension and his philosophical reflections on nature, God, mind, and man. Not only have they presented historians with the new task of accounting for the relationship between Darwin's private reflections and his argumentative strategy and pattern of publication; they have also opened a wide gap between the previously prevailing historiographic approaches and the new perspectives. Indeed, when compared with the works of the *nouvelle vague* of professional Darwin scholars, much of what was previously written from the point of view of intellectual history now seems almost *naïf*, if not superficial, in that it was said at a too general level of historiographic abstraction. For instance, the alleged influence of various people on Darwin can now be studied concretely, without relying on mere external similarities. The *Notebooks* have laid to rest some established historiographic commonplaces: that the analogy between artificial and natural selection guided Darwin's quest for the mechanism of evolution from the beginning; that Darwin long avoided tackling the problem of man; that he really worked according to a "Baconian-inductive" method; that the argumentative structure of the *Origin* reflected Darwin's path to the discovery of natural selection. Indeed, the *Notebooks* have shown that the very word "discovery" must be replaced by "construction". In general, Darwin historians have been progressively attracted more by the phenomenological study of Darwin's development than by the anatomy of the theory in its final shape (if there is any), more by the logic (or logics) of discovery than by the logic of the discovered. This seems to be part of an ever more widespread interest among historians of science, and of biology in particular, in the *genesis* of scientific theories. It has to be seen as the result of the progressive emancipation of the history of biology as an autonomous discipline, and also of a dissatisfaction with too hasty generalizations by some philosophers of science. Awareness has been growing that, to use Grmek's analogy, "descriptive embryology" must precede "causal-dynamic embryology" (1973,

p. 4). This seemed to call for a series of narrowly focussed studies to serve as bases for a general reinterpretation of the history of Darwin's theory (for a recent statement to this effect see Herbert 1977, p. 155). Indeed, this is what happened, mainly through the *Journal of the History of Biology*. This new tendency in Darwin studies was started by a book, one of the rare French studies in *histoire épistémologique* in which the perspicuity of the historical reconstruction is not endangered by brilliant epistemological tricks.

THE CONSTITUTION OF A CONCEPT

Limoges's *La sélection naturelle* (1970c) marked the beginning of a new phase in Darwin studies, although it was more seminal than influential; it was comparatively little quoted, and still less discussed, at least until Kohn (1980) (for instance, neither Young 1973a, Gruber and Barrett 1974, nor Manier 1978 mention it). The promised English translation never appeared. But a mere list of its features will give an idea of its importance.

It was the first broadly based and systematic study focussed on Darwin's actual process of discovering natural selection. It dismissed once and for all the claim that Blyth, Lawrence, Prichard, Wells, or Matthew was a forerunner or anticipator of Darwin. It asserted peremptorily the revolutionary character of Darwin's accomplishment: "the presence of randomness at the very heart of the process of transformation of species [. . .] broke violently with the whole tradition of natural history" (Limoges 1970c, p. 151).[20] It emphasized that Darwin disrupted the traditional concept of adaptation, and thereby the natural theology image of nature. It showed beyond any doubt how decisive the study of biogeography was in leading Darwin to evolution and natural selection. It denied the traditional view that the analogy between artificial and natural selection played a heuristic role in Darwin's development. It refuted the idea that Darwin's development from the Galapagos to the *Origin* was continuous, and attacked the "retrospective illusions" by which some, like Francis Darwin, had antedated Darwin's abandonment of fixism (Limoges 1970c, pp. 7–25). It tried to rescue Darwin from the charge of nominalism levelled against him by Mayr (1947, 1954b, 1957, 1959c), and it described Darwin's passage from a view of speciation through geographic isolation to a view in which both phyletic evolution and speciation were due to ecological-reproductive forms of isolation. (Limoges was the first who faced up to Darwin's concept of species from a purely historical point of view.) Finally, it gave a well-balanced account of Wallace's path to natural selection and its similarity to Darwin's.

But it was Limoges's approach that was new. And that was due to the philosophical options that inspired him and led him to attack the *naïf* empiricist prejudice on which what I have called "ingredients historiography" was based. Limoges set out to study the concept of natural selection "in its constitution, that is in the very act that brings it into being" (1970c,

p. 5). For "what is to be pursued is not the mere gathering of those elements the sum of which 'would be equal to' a transformist pronouncement; instead, we have to establish where these heretofore separate elements were first connected into a system of relations that was necessarily to be imposed from outside" (p. 20). It was only after Darwin had become an evolutionist that these elements could be given an evolutionary meaning they could not have had previously, either separately or in connection with each other. So the historian, according to Limoges, must take Darwin's "conversion" to evolution for granted, however it occurred, and start from there to inquire into the ways Darwin reached a first point of arrival: the *1842 Sketch*. The question Limoges tries to answer is then: "What happened to Darwin between 1837 and 1842?" The influence of Bachelard's idea that the history of science is mainly concerned with showing the paths that led to the achievements of rationality is evident here (Bachelard 1951a, pp. 25–27, 1951b; Lecourt 1969, 1972; Canguilhem 1975, pp. 1–23, 173–186). Nor does Limoges conceal his epistemological engagement. He declares from the beginning that his is not only a historical study, "but also a critique referring one back to the epistemology on which it is founded, and which might give conceptual history its guarantee of rigor" (1970c, p. 5). But the influence of Bachelard put some constraints on Limoges's work. For instance, he was perhaps too much concerned with the point of arrival of the process he was reconstructing. As he describes it, Darwin's development throughout the *Notebooks* seems to be rather unilinear. Maybe it was Limoges's very approach that prevented him from giving due emphasis to the phases, the alternative explanations, or the provisional theories Darwin worked his way through, or constructed and then dismissed, before "discovering" natural selection (see criticism of this in Kohn 1980, passim). The *Notebooks* themselves had their own story, which had significant events in it. As Limoges describes the process, after revising the traditional conception of adaptation, Darwin was already on the road to natural selection (with Malthus merely precipitating a process already underway).

Bachelard's influence is also disclosed by Limoges's conclusion:

> It was not necessary that *Darwin* or *Wallace* should work out their theory for it to appear; what was necessary was a modification of the conceptual shape of contemporary natural history, since works in biogeography were bringing about results which, to minds formed in the tradition of English natural theology, could not be reconciled with the accepted view of the adaptation of organisms to their environment. The so-called Darwinian theory emerged when the theoretical framework peculiar to the scientific practice of English naturalists in the first half of the nineteenth century was faced with new research in biogeography. The scientific act that produced that theory has not its *raison d'être* in any form of subjectivity. (1970c, p. 152)

Limoges says that Darwin only appears to be the real center of the book. He does so because his works are, owing to purely accidental circumstances, the best place to study the conceptual development that took place in British natural history. According to Limoges, historians should be concerned not with individuals, but with the formation and transformation of concepts, theories, and research methods. When he says that "in the history of science proper nouns should only serve as indexes", Limoges discloses the influence of the strong idealist strand in Bachelard's thought, which makes *Rationalité* the real protagonist of the history of science.[21]

Still more evident is Limoges's debt to Bachelard when he discusses Darwin's "stubborn use" of the anthropomorphic image of selection. The demiurgical metaphor, which presents nature as a sublime workman, revives, according to Limoges, a Paleyan model, that very "scheme of artifact" Darwin had dismantled when it took the form of belief in perfect adaptation. Likewise, the persistence of the selection image marks a "theoretical gap", an "error", an "epistemological obstacle" to the right use of the central concept of the theory (Limoges 1970a, pp. 370–374). Wallace, on the contrary, avoided this trap. "Were it not for the factual (*événementielle*) importance of Darwin's contribution, the theory could as well have been given Wallace's name", for he sensed the difficulties of that teleology-laden concept well before population genetics purged the theory of it (Limoges 1970c, pp. 149–150). Yet, one could ask, how come it was Wallace, and not Darwin, who eventually fell for spiritualism? Had not Wallace's very hyperselectionism something to do with this, as Gould has most acutely remarked (Gould 1980a, pp. 54, 58)? But that is another story.

DISCOVERY

From what I have said, it appears that Darwin scholars have played their part in the general reaction to the neo-positivist view of science, which, by being concerned almost exclusively with justification, had neglected discovery (Nickles 1980a, b). As a consequence of this, discovery had been either left to some variety of the mystique of the creative act, or explained away by such commonplace notions as "borrowing of concepts", "influence", dependence on "intellectual climate", "ideas in the air", "psychological peculiarities of the individual," and the like. A large part of what Darwin scholars have written in the last fifteen years, however, would be sneered at by such philosophers as Popper as exercises in psychologism. Furthermore, although most Darwin students did not intend it to be so, focus on discovery was a way to escape from the never-dead alternatives of externalism and internalism. For instance, discussion of the relation between science and ideologies would be greatly improved by ascertaining whether factors external to science played a role not only in the debate raised by a scientific theory but also in the very genesis of that theory. In other words, the study of the genesis of a theory may cast light on whether such extra-scientific factors shaped only the form of a theory or its content as well. To this end one should evaluate the relative strength of the variables of which a scientist's mind at work is a function. Provided, of course, some of these variables are not discarded a priori. Is there such a risk in the attempts that have been made to reconstruct Darwin's process of discovery? Since the various interpretations of this that have been brought out[22] are dealt with by other contributors to this volume, I shall not discuss them here. I shall only say that a general agreement is forming, and that Kohn (1980) stands out as almost definitive on many points. Some general remarks on it, however, are in order.

Kohn has brought order into the tangle of the *Notebooks* by not attempting to account for all the multiple lines (and segments) of thoughts they exhibit and by leaving in the dark many aspects that could not be related to Darwin's quest for the mechanism of evolution. It seems as though there was no other way of getting out of the maze. By deciding not to refer to the debates around Darwin, Kohn gives the illusion that he was a more solitary mind than Kohn himself would admit. In his reconstruction Darwin's context resolves into one major interlocutor — Lyell. Darwin's path to natural selection is dominated by Lyell's shadow. It is a subterranean dialogue with, and it results in a rebellion against, him. On the other hand, Herbert (1974, 1977) has paid more attention to the family and professional influences on Darwin, but has presented a less complex view of the phases of his early development. Manier (1978) has not dealt with the process of discovery itself and has neglected naturalists almost completely in his study of the young Darwin's cultural circle. Schweber (1977, 1980) has called attention to the decisive influence of philosophers and social thinkers in leading Darwin to natural selection, but has presented Darwin's progress as almost straightforward, and has not given equal prominence to other naturalists and to the "technical" problems in natural history that Darwin was faced with.

Coming back to Kohn, his paper should be influential in shifting the attention of Darwin scholars to other problems, and should serve as a model of rigorous analysis. Indeed, emphasis on discovery has tended to monopolize the interest of professional Darwin scholars to the disadvantage of other, equally important aspects of Darwin's achievement. As has happened with other scientists (like Newton, Faraday, and Claude Bernard), extensive use of manuscript sources has made research procedures more rigorous and has made possible a better understanding of how those scientific minds really worked. But there is still a gap between manuscript-focussed studies and reconstructions wider in scope and ambition (for instance, Ghiselin 1969; Young 1971a; Gillespie 1979). Is the stress on "internal" problems of Darwin's development as documented in manuscript sources the best guarantee of rigor? On the other hand, a book like Manier (1978) (which I think is very important: see below, Section III, "Images and Metaphors") makes me wonder whether this emphasis on manuscripts and on the young Darwin may lend itself to operations similar to those that took place on the young Marx of the 1844 manuscripts.* This is probably only a personal hallucination, but it is a matter of fact that post-1839 Darwin is less studied than 1837–1839 Darwin, and post-1859 Darwin still less so. The best recent studies give us images, not of Darwin, but of the young Darwin.

III. Darwin's Method, Darwin's Philosophy, Darwin's Mind

NAÏVETÉ

Until recently the idea was not rare among some commentators that Darwin was "*naïf*". De Beer, for instance, wrote that "the keynote of Darwin's character was simplicity amounting to naïvety" (1963, p. 252). Didn't he speak of himself as an "overgrown child"? He was capable of asking basic, direct, "*naïf*" questions, of not recoiling from making a jump in the dark. To many, this "*naïveté*" was merely another name for originality, spontaneity, and intellectual courage. But to others this image of Darwin had a bit more to it than simply that. It was the idea that, being free from philosophical sophistication, Darwin was capable of going straight to things themselves; that he was, after all, lucky in not falling prey from the beginning to a rigid and highly formalized education; that his contact with, and interest in, natural populations and the richness and intricacies of nature, rather than in the sophisticated anatomical types of the morphologists, secured him from becoming an erudite Platonist, as Forbes or Agassiz did. Hull wrote of Darwin's "philosophical naïveté", attributing to it his originality, and added that Darwin's ignorance of certain basic metaphysical issues was an advantage in formulating the theory, even though a disadvantage in defending it (1967b, pp. 334–336). (But, one might ask, is professional philosophy so completely separated from the conceptual tools of creative individuals as Hull implies?) W. Cannon suggested that Darwin and Wallace were lucky in being citizens of a country where a real educational system virtually did not exist (1961b, p. 112). Mayr has argued forcefully that only naturalists unencumbered with the intricacies of a tardy scholastic logic could embrace population thinking (1976b, 1980).[23] Gruber's study of Darwin (see below) has pointed out the necessity for the creative mind of adding the adult factors of courage and tenacity to the infantile factors of curiosity and creativity (Gruber 1973). To some popularizers, Darwin's *naïveté* enabled him to have access to great truths without the tortuous labors of common men. Chancellor (1973, p. 213) suggested that he was like the child portrayed

* This reflection was stimulated by a discussion with Sebastiano Timpanaro, whom I thank.

by Wordsworth: "Mighty Prophet, Seer blest!/On whom those truths do rest / Which we are toiling all our lives to find [. . .]" He too said that Darwin was "unencumbered with sophisticated knowledge" and had a "simpler, more direct approach to complicated subjects than his more learned friends Hooker, Lyell, and Huxley". Thus some have contrasted Darwin with Huxley, "who was at least half a philosopher [. . .] more brilliant and quick-thinking" than Darwin the practical biologist (Lack 1957, p. 81). It took William Irvine a joint biography of Darwin and Huxley to persuade himself that, paradoxical though it might seem, it was Darwin, and not Huxley, who discovered natural selection (1955, pp. 55–56).

So Darwin did not get enmeshed in philosophical niceties, and this was a good thing, it seems. What is interesting is the extent to which the idea of Darwin's *naïveté* derived from nineteenth-century stereotypes. It was encouraged by some of the very protagonists of the Darwinian debate. T. H. Huxley presented Darwin as the prototype of the scientist, who must "sit before a fact as a little child" and be as ingenuous before Nature as a Cinderella (1893c, 9:146; L. Huxley ed. 1900, 1:235). But we might also remember Wallace, who, at the Darwin-Wallace celebration of 1908, made his contribution to mythology by saying:

> It is this superficial and almost child-like interest in the outward form of living things which, though often despised as unscientific, happened to be *the only one* which would lead us [Darwin and Wallace] towards a solution of the problem of species. (Linnean Society of London 1908, p. 8)

If there is any *naïveté* here, it is the *naïveté* with which scientists often speak of themselves and of their colleagues. I don't think it is any longer possible to speak of Darwin's *naïveté* while there is ever-growing evidence that he was anything but misinformed in matters philosophical and methodological. And maybe we have flown from one extreme to the other. For instance, Schweber presents the "polymathic Darwin" as "thoroughly acquainted" with all the sciences, culture and socio-political events of the day (1978, p. 321).

METHOD

Cassirer considered Darwin's theory "a model of pure inductive science and pure inductive demonstration" (1950, book 2, chap. 4). J. Barzun went so far as to say that Darwin was "a great assembler of facts and a poor joiner of ideas" (1958, p. 74).[24] Both these judgements unconsciously echoed a traditional, "positivist" image of Darwin as the scientist who used to pile up facts until they almost spoke for themselves. Among other things, Huxley's propaganda mentioned above must have been an influential cause of this persisting image of Darwin. Darwin himself was another source of it. Occasionally, as is well known, he presented himself as working inductively. His descriptions of his methods, his ways of working, and his "philosophy of science" were not always clear, and sometimes they were even contradictory. Both inductivists and Popperians have found in his writings and letters statements that support their own views. Indeed, trying to answer the question "What was Darwin's method?" by merely collecting and comparing Darwin's statements on the subject would prove depressing and useless. If anything exists that could be labeled unequivocally "Darwin's method", it should be analyzed in relation to its historical sources, to Darwin's grappling with concrete problems, and to the scientific and philosophical debates in which Darwin got involved, as well as those that were taking place around him throughout his whole career. From the 1830s things went on changing both inside Darwin and outside him. Influences on him did not stop when he received Wallace's famous letter and manuscript in 1858. More important, Darwin, as is normal with any creative mind, went on influencing himself, so to speak. A study of "Darwin's method", accounting for the interactions of all these variables throughout Darwin's lifetime, has still to be made.

Perhaps the most commonly agreed-upon result of twenty years of Darwin studies has been the establishing of Darwin's strength as a theorist and as an imaginative thinker, who was continuously readjusting, or rather constructing, his bold theoretical framework and struggling with hard facts. This complex image of Darwin is the result more of the general progress of Darwin scholarship than of studies tackling directly the problem of Darwin's method.

J. S. Huxley characterized Darwinism as a "blend of induction and deduction", that is, as a theory based on three facts (geometrical increase of all populations, constancy in the number of their individuals, universal variation) and two deductions (struggle and selection) (1939; 1942, pp. 13–14; 1963, pp. 36–37, 40). As time went by, statements that Darwin's theory has a deductive structure became ever more frequent. Earlier statements by Pantin (1953, pp. 129ff) and Feibleman (1959) were followed by Crombie's claim that Darwin used the hypothetico-deductive or "retrodictive" method (A. C. Crombie 1960, p. 360) and by A. Flew's pointing out the existence of a "deductive core" to Darwin's theory (1959, p. 28; 1967, p. 8). Flew later altered the "deductive core" into a "deductive skeleton" that was "the uniting framework of the *Origin of Species*" (1978, p. 13). According to him, Darwin put together various elements into a form and deployed an enormous mass of empirical material under that scheme; "by itself," Flew added, "the deductive scheme proves little" (1978, p. 18). So Darwinism conforms to a model evident in Malthus's essay: deductive arguments with contingent, empirical premises and conclusions (Flew 1963). Other writers have since argued that Darwin's theory fits more or less closely the paradigm of nomological-deductive theory (M. B. Williams 1971; Mayr 1977; Caplan 1979). Williams in particular has disguised Darwin's theory under a highly sophisticated mathematical reformulation. Other commentators, however, while admitting that there is something deductive in Darwin's argument, have felt it necessary to add a number of qualifications. D. L. Hull, for instance, wrote:

> Although there is, broadly speaking, a deductive core to the *Origin*, by and large it is one long, involved inductive argument conducted in the midst of a mass of very concrete facts. Darwin's argument as presented in the *Origin* is a genuinely inductive argument, not just a deduction set up on end. (1967b, p. 335; see also 1973b, pp. 3–36)

An evaluation of the place of deduction in Darwin's theory in many aspects similar to this may be found in Ayer (1959). Ruse went further on this road, pointing out that the overall structure of Darwin's theory presents, to be sure, deductive parts, but that they are "inevitably in a sketchy form", that the "semideductive" inferences are more numerous, and that the role played by analogical and inductive reasoning is still more important, the result being a "very fine network, where many different threads mesh together to make the whole" (Ruse 1975b, p. 241, where a diagram is given). More recently Løvtrup has reasserted that Darwin's theory is not amenable to the model of theory structure as a set of hierarchically organized nomological premises and initial conditions (1976, 1977). An articulated view of the structure of the Darwin-Wallace theory is given in Oldroyd (1980). The issue of the inductive or deductive character of Darwin's theory is important for many reasons: first, because of the frequent charge that the concept of natural selection is tautological; second, because of its bearing on the related charge that Darwin's theory is not falsifiable; third, because it is its deductive structure that seemed to many to be the foundation of Darwin's revolution, which consisted in giving a causal explanation of what looked purposeful and only attributable to God's caprice (see for instance Grene 1974, p. 190).[25] Some features common to most of the studies just mentioned appear, however, in a sense, to be limiting factors. First, there studies deal almost uniquely with the *Origin*, which is the heart of Darwin's achievement but is not all of it (and which edition of the *Origin* must be taken as the truest expression of Darwin's theory?). Furthermore, I don't think it legitimate to exclude any temporal dimension from the study of the logic of Darwin's theory. Second, over all the writings mentioned above there hangs the ghost of the logical positivist model of science and scientific explanation; after all, it is not so much per se as in relation to this model that, implicitly or explicitly, they analyze Darwin's theory, whatever the results of this analysis are. Third, and more important, what is analyzed in these writings is the logical structure of the end result of Darwin's reflections on the species question. By viewing the *Origin* exclusively as a completed work, without regard to the long development which preceded it, and that which followed, these writings implicitly accept the strong neopositivist distinction between "discovery" and "justification". Now, what relation is there between, say, the first edition of the *Origin* and the method, or methods, actually deployed in the process by which Darwin struggled his way toward that result? Ruse is right when he argues that philosophers should be aware of processes of discovery, for they may throw new light on the structure of theories itself, and also overthrow some oldfashioned interpretations of them (1980a). This is the point where historians and philosophers of science should cooperate, something that very rarely happens.

A major discussion of Darwin's method was Ghiselin's book *The Triumph of the Darwinian Method* (1969). Ghiselin himself usually presents it as the book that "stirred things up" at a moment when Darwin's works were not fully understood (interview in *La Repubblica*, 31 January 1982; preface to the Italian edition). As a matter of fact, it had the unique merit of covering a wider range of Darwin's production than any other work had, and of drawing attention to areas that had been neglected. It also showed that virtually all that Darwin touched in natural history was part of his evolutionary theorizing and part of a great revolutionary synthesis. But perhaps its best feature was the attempt to discuss formal similarities in Darwin's procedures and views in different fields. Darwin's success was due mainly to his "ability to transfer methodologies and theoretical points of view across disciplinary boundaries" (Ghiselin 1969, p. 16). He pursued a unitary, slow, positional, but bold strategy. As Ghiselin put it, "Darwin was a master of the scientific method [. . .] a theoretician of the first rank, a thinker of both originality and rigor, and a philosopher of no mean competence [. . .] He was a speculator who tended to formulate intricate and subtle hypothetical systems. He was a great methodologist — indeed, one of the best [. . .]" (1969, pp. 4, 15). He had a tremendous "grasp of the intricacies of applied logic" (p. 42).

Unfortunately, Ghiselin's approach is largely unhistorical (see Greene 1975). It tends to reduce historiography to an exercise in logic and methodology. For instance, although he sharply distinguishes the context of discovery from that of justification, he declares that "an understanding of the logic of the theories casts a flood of light upon the history of their discovery" (p. 7).[26] To Ghiselin there is only one way of being scientific, so there is no difference, from the point of view of the method, between Darwin the discoverer and Darwin the "logician" arguing cogently in his published works. Indeed, the title of the book is misleading: it should have been *The Triumph of the Scientific Method in Darwin*.

Ghiselin is not content with saying that we would not understand a single thing in Darwin unless we bore in mind "that Darwin applied, rigorously and consistently, the modern hypothetico-deductive scientific method" (1969, p. 4). He goes so far as to say that "understanding Darwin becomes for us largely a question of methodology" (p. 7). "The structure of Darwin's system explains his successes and failures alike" (p. 75), but the latter are "errors which can be explained as instances of the same mistake in formal logic" (p. 7). For instance, it was a "particular kind of fallacious reasoning" that led Darwin to endorse blending inheritance, just as it was an error in logic that led Aristotle to confusion "over the whole structure of the physical universe" (p. 138).

Ghiselin's is not a historical book. It does not even hint at any sort of development in Darwin's works and ideas. Darwin's mind emerges as something static and pre-ordained, as a *machine à penser*. Darwin's method has no source anywhere, nor development during and after his working life. This is a Darwin in a vacuum. His contemporaries are mentioned, if ever, only to show that they were wrong. To discuss Darwin's theories, Ghiselin often integrates them with post-Darwinian knowledge and often substitutes himself for Darwin to provide the link that is missing in an argument. He sometimes presents ideas that he claims derived from Darwin's achievement without quoting Darwin. He hardly adduces a single Darwin text when he discusses Darwin's population thinking. The *Origin* is referred to only twice in the seventy-one notes to a chapter entitled "Natural Selection".

Ghiselin does not discuss Darwin's context; rather, he attacks his own foes and some straw men: "professional philosophers" (1969, p. 237), "biological Platonists generally" (p. 263, n. 14), "metaphysicians", "essentialists", "formalists" (in morphology), "certain philosophical physicists" (p. 65), and especially Plato and Aristotle, who really seem to deserve to have their ears boxed. Darwin's critics were simply prejudiced and wrong; they just did not understand; they were blinded by essentialism. Progress is due to science, error to metaphysics (p. 127) or predominance of language over things (p. 88).[27] "The fact that Darwin's accomplishment has been controversial merely reflects the degree to which it has been misunderstood" (p. 232).

Darwin's method as portrayed by Ghiselin is both too rigid to do justice to the richness of Darwin's thought[28] and too generic, for it boils down to the mere testing of hypotheses.[29] Ghiselin succumbs to the temptation "to emphasize the importance of method as being largely independent of subject matter [. . .] Darwin could and did employ several methodologies simultaneously, as they were appropriate" (Egerton 1971a, p. 284; cf. Kleiner 1979, p. 309, and Gaukroger 1976, p. 199).

THEORY AND PHILOSOPHY

The philosophical importance of Darwin's accomplishment had been stressed by other writers before Ghiselin. Fulton's (1959) view that within fifty years of the publication of the *Origin* evolution had been absorbed into common sense and evolutionary philosophy had gone out of fashion, was not shared by Wiener (1949), Collins (1959), and Randall, Jr. (1961). These discussions mostly concerned the impact of Darwinism on professional philosophers (see also Passmore 1959 and Marnell 1966). Others stressed the conceptual novelty implicit in Darwin's achievements: Peirce, Dewey, and, more recently, Mayr, Hull, and Jacob.[30] Although Mayr claimed that "no other work advertised to the world the emancipation of science from philosophy as did Darwin's *Origin*", and added that to many philosophers it is still "unphilosophical", he also stated clearly that "Darwin's conceptual framework is, indeed, a new philosophical system" (1964, pp. xii, viii). Hull echoed this judgement: "Darwin was breaking new ground both biologically and philosophically" (1967b, pp. 334–335). In subsequent studies Hull has argued that most of the nineteenth-century philosophies of science (mainly those of Herschel, Whewell, J. S. Mill, and also Peirce himself) were inadequate to Darwinism and unable to come to terms with it (1973b, 1974a). He has followed Mayr in making a sharp contrast between population thinking and essentialism, having already pointed out the deleterious consequences of essentialism in taxonomy (Hull 1964–1965). His survey of a number of important scientific and philosophical reactions to Darwin has stressed the novelty of Darwin's accomplishment (1973b). Hull seems to me to go too far, however, when he says:

> There is nothing so well calculated to turn a man into a neoplatonist than to put him in a storeroom full of fossil remains and set him the task of reconstructing the original organisms. The idea that there are series of basic plans of organization with numerous variations emerges quite forcibly. (1967b, p. 326)

This seems to me a too crudely deterministic view. In such drastic contrapositions as that between population thinking and typological thinking, what is gained in philosophical insight may be lost in historical accuracy. For instance, one feels there is something unsatisfactory in saying that, on the score of the species concept, "even Locke is akin to Plato", or in presenting Buffon as now a nominalist, now a Platonist (Mayr 1959a, now in Mayr 1976a, p. 257; Hull 1967b, pp. 322, 324–325). One feels as though Mayr's and Hull's emphasis on Darwin's conceptual revolution, though perceptive, was founded on a too sketchy historical analysis, if not, to a certain extent, on the imposition of a posteriori philosophical analysis on historical complexity. Some aspects of Whewell's or Herschel's philosophies of science may really seem to be "schizophrenic", as Hull puts it, but such a judgement, besides looking a bit Whiggish, might be legitimate only after careful and detailed analysis, say, of the relationship between "natural laws" and "divine intervention" in Whewell's thought.

However, Hull has the merit of having stressed the importance of the "metaphysical" aspects of the impact of Darwin's theory on British nineteenth-century philosophies of science. He has also improved on Ellegård's (1957, 1958) view of that debate as a conflict between empiricist and idealist philosophies of science. Ellegård's interpretation owed too much to Huxley's biased presentation of the whole affair.

Mayr has analyzed perceptively some crucial aspects of Darwin's theory. Looking at him from the vantage point of modern knowledge, Mayr is oriented toward ascertaining what Darwin lacked and why.[31] This may give the historian tremendously useful insights into Darwin's science and conceptual presuppositions and provide him with *aperçus* of wide scope, but these must be handled with care. It is easy to slip from a concern for why scientists did something the way they did to a concern for why they failed to do something else that science has done since. Excessive reliance on general interpretative categories like "essentialism" or "population thinking" may prove detrimental. They might become a sort of *deus ex machina*, or useful devices for filling the gaps in detailed historical reconstruction, or — worse — dispensing with it altogether. The idea may creep in that detailed reconstruction is merely useful in confirming and implementing general insights reached by logic, through mere comparison of texts and ideas. Commenting on the attitude of "well-informed and broadly educated lay people" who embraced evolution more readily than did professional scientists, Mayr says: "A view from the distance is sometimes more revealing, for the understanding of broad issues, than the myopic scrutiny of the specialist" (1972a, p. 982). This I think characterizes most aptly not only the perceptive layman's "holistic" way of looking at the problem, but also Mayr's way of looking at the Darwinian revolution. Those historians who plod on among the intricacies of debates that are often anything but clear-cut receive little help when they read that "Lyell showed no understanding of the nature of genetic variation" and learn about the "total absence in his arguments of any thinking in terms of populations" (p. 984). They feel that these clear-cut judgements, perspicacious as they may be, are of little or no help in casting light on Lyell's real motives for rejecting evolution for so long before his "conversion with reservations". They will be perplexed when they read that Lyell eventually adopted population thinking; for they know that he was never converted wholeheartedly to Darwin's idea of natural selection, let alone population thinking, which, as Mayr himself has convincingly argued several times over, was not at all synonymous with belief in evolution and was loathed by many an evolutionist well into this century (Mayr 1980). Other historians will be left wondering at Mayr's peremptory statement that "progressionism [. . .] was intellectually a backward step from the widespread 18th-century belief that the running of the universe required only occasional, but definitely not incessant, active intervention by the Creator" (1972a, p. 985). Also what Mayr says of catastrophism and uniformitarianism seems to be grounded more on a highly personal view of the logic of these categories than on how things really happened. For instance, he speaks of "assumptions" that one had to make for the progression in fossil faunas to become "automatically [. . .] evidence in favor of evolution". Mayr's approach leads him to embrace Lovejoy's opinion that abundant evidence in favor of evolution existed by

1830, and, accordingly, to stress the "power of retarding concepts" such as creationism and essentialism (p. 982; Lovejoy 1959). Mayr quite rightly shows that Darwin introduced some profound methodological innovations. But isn't it too much to say that the "demand for conclusive proof", the tendency to deal in rigid pairs of alternatives, and the incapacity to make important distinctions, such as that between the reality and the fixity of species, were "important weaknesses in the scientific methodology of the period", "violations of sound scientific method" and of "scientific logic"? If carried to its extreme, this approach produces Ghiselin's logico-formal analyses.

If Ghiselin did not even attempt to suggest possible sources for Darwin's method, other scholars have pointed out what they thought were decisive influences not only on Darwin's methodology but also on his philosophy, and, by no means the least important, on his self-image as a scientist and as an intellectual.

Gruber made a somewhat vague attempt to describe Darwin's "family *Weltanschauung*" (Gruber and Barrett 1974, chaps. 3–4; see Corsi's criticisms, 1975). Indeed, his very approach (see below) seems to prevent him from being interested in strictly philosophical and methodological issues. Cannon, Hodge, Rudwick, Ruse, Schweber, and Manier, on the other hand, have devoted much attention to them. Ruse has emphasized the influence of Herschel and Whewell in providing Darwin with epistemological criteria and standards, and in shaping some important aspects of both his style of argument and his techniques of research. Not only were principles like those of *vera causa* and "consilience of inductions" familiar to Darwin, but he very frequently modeled his research and arguments on them (Ruse 1975c, 1978a; Thagard 1977).[32] Hodge (1977) has argued convincingly that the structure, distribution, and strategy of Darwin's expositions of his theory, from the *1842 Sketch* through the *Origin*, conformed to the requirements of the *vera causa* principle and way of argument as exemplified by Herschel and Lyell. But it should be borne in mind that *vera causa* did not mean the same thing to everyone (Kavaloski 1974), and that the methodological and epistemological debate was intertwined with the theological debate (Brooke 1977a, 1977b; Corsi 1980a). Careful historical research in a detailed case study has not only confirmed Ruse's views, but also set on a better historical footing than Ghiselin had Ghiselin's own idea of formal similarities between various aspects of Darwin's thought and work. By studying precisely what Darwin himself called his "gigantic blunder" over the Parallel Roads of Glen Roy, Rudwick has shown that there were links of theoretical content between Darwin's tectonic theory and his speculations about the species question. These were parts of "a broader unified research programme" (1974a, p. 164). The logical structure of Darwin's Glen Roy paper shows not only that Darwin "incorporated conscious methodological themes derived from

Herschel and Whewell" (p. 169), but also that his paper was a "trial run" for a particular style of scientific argument (p. 175).

To Herschel and Whewell, and to the already well-known Malthus, Schweber (1977, 1980) has added Adam Smith, Comte (through Brewster's famous review of the *Cours*), Quetelet, and Dugald Stewart. The pervasive influence of Stewart and other Scottish philosophers even among naturalists should not surprise anyone, were it not for the lack of previous detailed study and a certain tendency toward "internal" history in Darwin scholarship in spite of R. M. Young's provocative papers (see below, Section IV, "Naturalism"). The presence of Hume in Darwin's circle had been pointed out, though in a generic way, in Huntley (1972). The impact of Hume and the sensationalists on Darwin has been stressed by Richards (1979). Lastly, Manier has reinforced the claim for more attention to what he has called Darwin's "cultural circle". Manier has done what common sense should have prompted Darwin students to do earlier; he has tried to trace Darwin's "metaphysical" discussions in the *Notebooks* back to their sources. So, then, here come a number of new characters in the panorama of Darwin studies, and, again and with more force, the new revelation, Dugald Stewart. According to Manier (but see criticism of this in Corsi 1980a), he provided Darwin with a theory of language and metaphorical expression that was an antidote to the Cartesian and Herschelian request for a neutral, objective, technical and quasi-mathematical language of science. To Ruse's case, Manier has added an analysis of Darwin's reflections on such concepts as "law", "prediction", "cause", "chance", which he discussed in a sort of dialogue with these philosophers. Now more than ever it is evident that Darwin did not operate in a vacuum. One can object to Manier's thesis that Darwin's thinking was animated by a tension between anthropomorphism and positivism that was reflected in his style of research, language, and logic. But it is no longer possible to consider Darwin's method ahistorically. In short, to the question "Was Darwin a philosopher?", we should answer, according to Manier: "Yes, because he took his science seriously."

STRATEGY, UNITY, AND PROLIFERATION

Ghiselin wrote: "The entire corpus of Darwinian writings constitutes a unitary system of interconnected ideas. It strives, with astonishing success, to encompass all organic phenomena within the structure of one comprehensive theory" (1969, p. 12). This statement reflects an ever more widespread consensus among Darwin scholars. For, if we accept Isaiah Berlin's division of philosophers into "foxes" and "hedgehogs", the first being those who "know many things", and whose thinking is centrifugal, diffuse, multifarious, and lacking a unifying focal point, and the latter being those who "know just one big thing", and are centripetal in thinking, single-minded, and aspiring to a unitary vision, then there is little doubt that most Darwin scholars would agree with Lerner that Darwin "stands supreme as a hedgehog"

(1959, p. 173). Provided, of course, that this judgement is accompanied by some qualifying formula like "in the sense that Darwin's thinking consists of the dynamic correlation and unification of a whole set of interdependent enterprises". And certainly, also, without the slightly pejorative flavor of Himmelfarb's characterization of Darwin as "a single-minded, hard-working naturalist" (1959, p. 357).

From the scholarship of the last twenty years Darwin has emerged as an interdisciplinary thinker who could move from field to field with unusual ease (Ghiselin 1969; Rudwick 1974a; Herbert 1974, 1977),[33] a pluralist (Gould 1980a, p. 54; 1981, p. 83), a tenacious and even stubborn man steadfastly sticking to certain principles, but also imaginative, versatile, capable of entertaining a multiplicity of hypotheses at the same time, readier to adapt than to give up a theory, a highly strategical mind, often rigid but sometimes ready to rethink and revise a good many ideas after some crucial influence has modified the whole context (Browne 1980; Kohn 1980; Richards 1981, p. 208), and a scientist by no means alienated, at least in his most creative period, from the object of his researches (Manier 1978). S. Herbert has very recently written that "the tentative and empirical nature of Darwin's (early geological) inquiries is paramount"; but she has also stressed "Darwin's enthusiasm for large theoretical issues" (Herbert, ed., 1980, pp. 8, 15). Eng (1977) has presented Darwin's theoretical work as a "confrontation between reason and imagination". Such a man could not "discover" his theory of natural selection, nor collect "elements" of it until the puzzle was solved. Nor, indeed, could anyone have done so. He had to construct it, and get entirely absorbed in the search, cerebrally and emotionally.

According to Kohn, Darwin's path to natural selection shows five phases or "episodes", each of them being comparatively autonomous with respect to the others. Each is characterized by a theoretical framework that suggests later steps in the inquiry, establishes the criteria against which explanations must be tested, pushes research in some directions while blocking others, with momentous consequences. Darwin's development up to the moment he read Malthus is not "an unfolding drama of gradual and progressive intellectual development [. . .] There is highly ordered activity in Darwin's preselectionist intellectual world", but nothing like "an epigenetic sequence". His sexual theory, for instance, underwent little development. Significant changes did occur, but mostly as extensions of the dominant general framework, which was gradually embracing additional areas of natural history until the moment came when a rethinking of the fundamental assumptions was required (Kohn 1980, pp. 114, 153). "The transition between the early B Notebook and the discovery of natural selection in D is a movement between two stable states, punctuated by occasional rejection of destabilizing alternatives" (p. 140). Nevertheless, the impact of Malthus was "dramatic and sustained", a "sudden revelation". "Slow change, preparation, stasis, and sudden reorganizations are all familiar facets of developmental processes.

They are not mutually exclusive and we should not be surprised that Darwin's case exhibits them all" (p. 148). "The content of these 'theories to work by' is not as important as the conclusion that, at each stage, what Darwin could and could not perceive of the transmutation problem was conditioned by the way he explained transmutation. Darwin's insight, like that of all men, was both clarified and constrained by what he believed at any point in time. Perhaps the essence of his creativity was not that he could formulate a theory, but that he continued to formulate theories until he reached one which he felt was correct" (p. 150). In the way in which "Darwin practised the art of theory making" there was no room for scruples in resorting to ad hoc hypotheses. Any theory was better than no theory. "Whatever peculiar form his explanations of transformism took, the sheer fact that he was able to continually come up with new explanations confirmed his faith in transformism. By believing that he was solving the species problem, he kept himself at work on the problem without having to face the brutal fact that he had no idea how evolution worked. Thus, Darwin's capacity to creatively delude himself by concocting *ad hoc* hypotheses probably played a very positive role in his career" (p. 152). The construction aspect of the growth of a scientific theory upon itself, this "process of constructing theory after theory" (p. 154), could not be better emphasized.

From the studies on the construction of various aspects of Darwin's theories we get an enormously complex image of a great mind at work, a mind whose strength consisted in pursuing an incredible number of interdependent lines of research. Early intuitions of this complexity have been confirmed and qualified. If Loewenberg (1959b) spoke of the "mosaic of Darwin's thought", scholars now prefer using phrases that, while conveying the idea of the richness of Darwin's mind, do not portray it as static. In a discussion of Darwin's theories of instinct, we find an allusion to the "sedimented character of Darwin's thinking", meaning that "older ideas were often preserved and put to new uses" (Richards 1981, p. 208). Likewise, Ospovat has shown that the "facts" Darwin searched the literature for were, more often than not, ideas. These "facts" often reacted on the voracious mind that was absorbing them; and reading, and keeping abreast with, literature was Darwin's way of discussing, second only to his letter-writing (Ospovat 1981, pp. 90–114). Conversely, there has been a growing awareness of the action certain aspects of Darwin's thought or some of his general assumptions may have exerted in retarding or blocking some developments in Darwin's thinking. Here again Kohn provides us with a concise and penetrating discussion.

There is a central paradox in Darwin's intellectual development up to the reading of Malthus. Kohn has called it the "paradox of realization": why was Darwin both so close to and so far from natural selection for the whole period July 1837–October 1838? There were "forces internal to Darwin's consciousness" that prevented him from realizing the evolutionary importance of struggle. For instance, even when he became an evolutionist, he kept some "metaphysical" assumptions from the earliest framework he received from natural theology and Lyell, for instance the idea that there is harmony in nature. These remnants constrained his further reasoning. In addition, the power and coherence of his non-selectionist theory, which simply did not require struggle, "built a wall" against the realization of its importance (Kohn 1980, pp. 152–153). For Darwin to recognize the importance of struggle, he had to be stimulated from outside. "He had to be jolted from the harmonious interpretation of nature that he had inherited from the orthodox tradition and upon which he had constructed his preselectionist theories" (p. 153).

Ospovat has expanded Kohn's remarks and argued, in disagreement with him, that Darwin persisted in believing in perfect adaptation well after Malthus and up to the 1850s. This persistence directed Darwin's speculations along certain lines instead of others (for instance, leaving virtually undisturbed the idea that variation was rare in nature, arising mainly in response to environmental change). Darwin persisted in this direction until he found the solution to the problem of divergence (Ospovat 1979; 1980; 1981, chaps. 3, 6–8).

Richards (1981) and R. Richardson (1981), too, have called in question the idea that Malthus's influence was definitive in shaping Darwin's theory of natural selection. They have argued that, after arriving at the formulation of the principle, Darwin did *not* discard all his pre-Malthusian mechanisms. The delay in publishing the theory, Richards suggests, was due not so much to the fear of being accused of materialism as to "the several conceptual obstacles he had to overcome if his theory of evolution by natural selection were to be made scientifically acceptable" (Richards 1981, p. 229). Richards adds that in the case of Darwin's studies of behavior, "the inertia of his older ideas about instinct" is particularly evident until as late as 1848. This description confirms in many ways Rudwick's observation that Darwin's attitude to the Glen Roy problem might better be described as due to what Lakatos called the "co-presence of proliferation [of hypotheses] and tenacity" (Rudwick 1974a, p. 178; Lakatos 1970).

AN EVOLVING SYSTEM

In spite of minor divergences, Darwin students today generally agree that Darwin's theory was constructed, not discovered, and that it was the result of the evolution of a creative system: Darwin's mind. This convergence is indirect confirmation of the pioneering character of H. Gruber's work. It has not had a commensurate influence, however, because the image of Darwin's mind at work that I have been discussing is rather a side-effect of the research done by careful historians than the direct result of the impact of Gruber's work. Neither have all of Gruber's theses concerning particular aspects of Darwin's development and theories had a lasting influence on

Darwin studies.[34] Gruber's papers state his point of view on this issue more concisely and penetratingly than his 1974 book, so I shall concentrate on them.

According to Gruber, "scientific work is not a single process but a complex group of activities organized and orchestrated toward certain ends." Each of these enterprises is itself a group of tasks, a "branching network of enterprises" (1981b, p. 311). Scientific work is then to be conceptualized as protracted, continuous, patient, coherent, purposeful, constructive, pluralistic, multifaceted, often interdisciplinary, composed of indissociable, interacting substructures, animated by tensions and a strong emotional commitment, self-regulating, dynamic, evolving, perpetually reorganizing itself. Each phase in the work of a scientific mind is the result of a dialectic of conservation and innovation. Gruber pleads for what he calls a "*systemic view* of the whole thinking person engaged in scientific work", or "the constructivist view of intellectual growth", or, more often, "the *evolving system* approach to the study of creative thinking" (1980b, p. 114, Gruber's italics).

The fact that thinking is a pluralistic enterprise necessitates a "pluralistic approach" (Gruber 1981b, p. 314). The historian, therefore, must not be afraid of complexity. His task is not to simplify, but "to describe organized complexity" (p. 320). Implicitly criticizing what is happening in some quarters of Darwin studies, Gruber mistrusts short-term, "hit-and-run" case studies, as well as other approaches that are almost exclusively concerned with contents, not processes, of thought. They often lead to treating particular achievements or sets of ideas in isolation from each other. For instance, Limoges does not take up the interplay of the contents of the transmutation notebooks with the notebooks on "Man, mind, and materialism" (p. 296). (And one feels that, from Gruber's point of view, the same criticism could be extended to Kohn (1980), although Kohn's stress on the interdependence of many issues in Darwin's development and "the paradox of realization" enables him to meet some of Gruber's requirements). Gruber also criticizes the tendency to examine single topics, for instance geographic distribution, variation, etc. According to him, this leads to neglecting the place they occupy in the evolving structure of Darwin's research, and to missing some of the real causes of Darwin's changes on those very issues. This approach, in short, fragments the unity of Darwin's thought.[35] Purely biographical treatments are also criticized, and so are attempts at psychohistory like that of Greenacre (1963), which do not explain why the creative mind "went from one idea to another" or are belied by the fact that Darwin and Wallace "discovered" natural selection independently of each other. Even if Gruber does not say so explicitly, his approach points out a way to go beyond the dichotomy between "internal" and "external" history of science. To be influential, events in the historical context must be related to the living core of the mental processes the individual is engaged in.

Historical sketches of the period in which the problem under consideration is set often prove to be mere appendages to, or embellishments of, the main narrative. Arguing against the two extremes of psychological determinism on the one hand and the belief in some form of Zeitgeist on the other, Gruber demands that intellectual history and history of science "ought to be closely related to psychology, especially cognitive psychology, since they deal with the way human beings get ideas and elaborate them" (1977, p. 240). Finally, he argues that it is useless to try to characterize Darwin's thought with one "image of the world" ascribed to him, for each individual has many "modalities of representation" (1981b, p. 315). Gruber's study of Darwin's "images of wide scope", which I shall discuss later, destroys the foundations of much of what intellectual historians have written on Darwin.

In Gruber's perspective, Darwin emerges as a great strategist, "a past master of the separation of issues when this would serve his purpose" (p. 314). For scientific behavior, like any behavior, is choice, and a creative mind is faced with "a bewildering assortment of possibilities". "To be effective, out of its manifold possibilities, out of its *pluralism*, it [the cognitive structure] must perpetually re-organize and regulate itself in order to produce singular outcomes. This is why the organization of knowledge and the organization of purpose are ineradicably intertwined" (p. 309, Gruber's italics). Darwin was thus able to study barnacles, leaving other concerns aside for the time being, but without forgetting his overall aim; he was able to separate the study of the causes of variation from the study of the role of variation in evolution; he was able to exclude man from treatment in the *Origin*, etc. "Darwin's clear grasp of his own network of enterprises permitted him to plan his work more purposefully, to concentrate his thinking on different subjects in a flexible and adaptive way, and to time his publications strategically." But protracting parallel "trains of thought" (to use Darwin's phrase) caused what Gruber calls in Piaget's language a "mutual assimilation of schemes". These enrich each other — for instance, when work in one direction is resumed after a certain time and is considered from the vantage point of the new achievements made in the meantime. Techniques or styles of thought may be extended from one context to another.

Gruber stresses duration, continuity, and interaction in the development of Darwin's thought. Changes in this process are systematic and are brought about under the fairly steady pressure of new insights. But even the larger changes lead to thought-complexes that preserve unvaried certain structures of previous stages. Gruber deflates the importance of "eureka" or "aha" experiences (1981d). There are, to be sure, "qualitative leaps in a series of structural transformations", but they are "expressions of the relatively stable functioning of a system, rather than of its overthrow, [. . . .] a series of relatively small transformations representing both purposeful growth and structural changes". There are thousands of insights in a creative lifetime

(1980a, pp. 127–128). Thus, for instance, Darwin's reading of Malthus should not be overrated (Gruber and Barrett 1974, pp. 161–174). "Even when a theory has reached maturity and has been thoroughly examined and well formulated, no one can think about it all at once. Why should we suppose that the inventor can?" (1980a, p. 128; contrast S. Smith's argument (1959a, b; 1960) that Darwin's theory was "intuitively complete" by March–July 1837).

It seems that Gruber conceives of his approach as capable of reconciling in a synthesis the different interpretations that have been offered of Darwin's path to natural selection. By stressing the changing and multifarious character of Darwin's thought, one can see these interpretations as due to focussing on different phases or aspects of the same growth structure. To me this is also the limit of Gruber's contribution to Darwin studies: his approach can explain too much, can absorb too much. Gruber's stress is more on the dynamic of scientific creative thinking as exemplified in Darwin than on Darwin himself. Darwin is sometimes not so much explained by Gruber as "used" to explain Gruber's stimulating and enlightening general view of cognitive processes. Moreover, Darwin's mind as described by Gruber seems to evolve only by an internal dialectic, the influence of other people on him being somehow put in the shade. Certainly, Gruber is right in saying that the traditional account of "influence" describes the individual far too passively (1980a, p. 117). For instance, much of what has been written on the Malthus-Darwin relation was inspired by a one-to-one, cause-and-effect view of "influence" (see below, Section IV, "Malthus"). But Gruber's reconstruction of Darwin's development pays little attention to the information or attitudes Darwin acquired, or "inherited", from others. For instance, in *Darwin on Man* Gruber scarcely analyzes the ideas on moral sense of the Scottish philosophers Darwin read and discussed; and, after all, this book is intended not only as a tract on scientific creativity, but also as a reconstruction of Darwin's ideas on man in the crucial years 1837–1839. It is not only the "moves" in an overall strategy that are selected by the creative scientist, but also information, methodological rules, images. What is "social" and what is "individual", what purposeful and what casual, or mechanical, in this selection or filtering? How does the "evolving system" approach to Darwin's development tackle these problems?

Gruber's Darwin is too active and original. This criticism may seem absurd, but I cannot resist the impression that the Darwin Gruber portrays is, in a sense, too conscious, or rather too strategic. He appears to be a master chess player, or a rich sage who is continually investing a certain amount of money in some enterprise and waiting patiently to see the results and reinvest the profits. Darwin's work appears to have been regulated mainly by an inner watch. I think Herbert is nearer the mark when she says that sometimes "Darwin's entrance into new fields altered his theory in a direction and to an extent which he could not have imagined beforehand"

(Herbert 1977, p. 216). There were more "unanticipated effects" in Darwin's development than Gruber assumes. There may seem to be merely a shade of difference in emphasis between Herbert and Gruber. But look at how they view Malthus's influence on Darwin. To Herbert, it was the "enormous" unanticipated result of a direction of inquiry Darwin adopted both deliberately and because he was influenced by his family's philosophical leanings (pp. 211–213, 216); to Gruber, Malthus's influence was no sudden revelation, but a relation that must be analyzed in terms of "increasing the importance of a variable" (superfecundity) and "transforming the importance of a process" (natural selection). Gruber does not seem to be interested in what in Darwin's overall strategy led him to read Malthus (Gruber and Barrett 1974, pp. 161, 163–164 n., 165–174). The difference between Gruber and Herbert is due to their different approaches, not to their referring to different internal evidence. Herbert tries to see Darwin in a social dimension Gruber does not take into account precisely because of his "evolving system" approach.

IMAGES AND METAPHORS

The social dimension of Darwin's language, logic, and imagery has been studied by Manier. He has reached some conclusions very similar to Gruber's, but has investigated aspects that Gruber overlooked. Gruber rightly called attention to the "complex and lively interaction between different levels of experience, such as the conceptual and the imaginal" (1977, p. 233). Darwin's metaphors ("artificial selection", "chance", "struggle", "the branching tree (or "the coral") of life", the "hundred thousand wedges", the "entangled bank", etc.) were not merely didactic or communicative devices. They seem to have played a role in the actual generation of central parts of the theory of natural selection. "Even when expressed in very general form (vague, intuitive, poetic) such images have generative and regulatory power, both governing the search for more explicit formulations and giving rise to them" (p. 234). These "images of wide scope" are flexible, regulative, constructive, ready to hand and easy to use, but also relatively independent of the metaphors they become part of. They are capable of modifying the perception of the subject of the metaphor; they are instructive precisely because they do not fit snugly to the subject and may be coupled to them only loosely. For instance, the analogy between artificial and natural selection is useful in that it points out the differences as well as the similarities between them. Gruber also had the merit of showing how the scientific study of nature can be influenced by the predominance of an "aesthetic of disorder" over one of order, that is by opposing the orderly, harmonious world view with one of nature as "irregular, non-repeating, unpredictable, incomplete, indeterminate, complex, open-ended, and inventive" (1977, pp. 233, 238; 1978).

But Gruber fell short of evaluating the polysemy and multiple functions of Darwin's metaphors. On the contrary, Manier discussed five functions of them: critical-persuasive, heuristic, semantic, explanatory, affective (1978, p. 182). Darwin, says Manier, was opening new paths in science and tackling problems that not only were not covered by a generally admitted theory, but did not even exist as a unified field of inquiry. Therefore Darwin had to mold new conceptual tools, untranslatable in contemporary scientific language; but he also had to create his public, and to come to terms with many different (too many and too different) demands by influential philosophers and scientists. He reacted to this challenge by inventing a flexible, colorful, colloquial, and metaphorical language, not immune from ambiguity but capable of resolving many tensions, intellectual, moral, and strategical (p. 150). Accordingly, his metaphors aimed more at persuading than at demonstrating. They were not — indeed, could not be — rigorous arguments meeting the standards of mathematical precision, since the theory Darwin was struggling toward was not empirically founded in all its passages. Rather, Darwin's metaphors were "plausible stories" for a certain public (p. 151). However, they were not so much didactic as *auto*didactic devices. By them, Darwin was able to cross disciplinary and methodological boundaries, to free himself from traditional perspectives and the need to conform to certain standards. Consider, for instance, the demiurgical metaphor of the "Being infinitely more sagacious than man". He is not an omnipotent Creator, he does not create variations; he is obliged to wait for useful variations to appear, and only then can he select among them; he "cannot unite the causes of organic variability and the adaptive requirements of reproductive efficiency"; in short, he is "unable to improve on the elementary strategy of trial-and-error" (pp. 154, 173–175). This metaphor shows that it is possible to conceive evolution and go a long way with it even if only some of the envisaged processes are known in detail. According to Manier, this shows that Darwin's theory, at least prior to 1859, does not fit the Cartesian-reductionist ideal.[36] Whatever one thinks of Descartes (and of the "fable" he tells in *Le monde*), Manier's discussion suggests one major reason why Darwin's theory was so sharply contested by many nineteenth-century scientists as empirically unsound, or incomplete insofar as it did not provide an explanation of the origin of variations.

But metaphors sometimes were also useful to disguise, as it were, Darwin's thought. Take, for instance, the concept of "chance". It implies, among other things, that many different causal chains cannot be reduced to one first cause, and thereby undermines the idea that there is only one end towards which all of nature tends. Now, this revolutionary concept is both enriched and disguised, that is softened, by the concept of selection, which stresses the non-random character of the accumulation of variations (Manier 1978, p. 186). Metaphors, therefore, can react upon one another. They are themselves intersections of many chains of ideas. Thus, for instance, "struggle" has many meanings. It cannot always be resolved into "war", "conflict", or "battle". Its polysemy, as Manier shows successfully, has not been fully recognized and appreciated by Darwin scholars. Some have interpreted it as merely a colloquial, unnecessary, confusing shorthand for "differential reproduction" and tried to strip it of its gladiatorial overtones, which were responsible for so many distortions (G. G. Simpson 1949, pp. 95–96; 1967, pp. 221–222; Dobzhansky 1955, p. 112); others have distinguished between its "purely biological meaning" and unnecessary "metaphorical concepts from Malthus and Spencer" (Rogers 1972, pp. 268–269); still others have stated that the literal and the metaphorical meanings of "struggle" refer to "just about opposite" situations (Gale 1972, p. 323).[37] Complaining of Darwin's ambiguity, Ruse has tried to translate "struggle for life" into a formula capable of covering all the senses in which Darwin used the phrase; the result is impressive: "organisms (or groups of organisms) respond in certain kinds of ways and they must so respond in order to survive long enough to reproduce — in so doing they usually bring about the deaths of other organisms, or at least, the failures of others to reproduce, and if they fail to do so, then their own deaths probably follow" (1971b, pp. 317–318).

Referring to M. Black's (1962) and M. B. Hesse's (1966) studies of models and metaphors in science, Manier agrees that many important aspects of Darwin's theory could not have been expressed without metaphors. These, therefore, could not have been replaced by literal descriptions of the conceptual systems connected by the metaphors and transformed by this very connection. This is an important point for both epistemology and history. For it can throw light on some debated issues in theory reduction. Moreover, many evolutionary biologists and Darwin scholars are sure that recent progress in genetics and evolutionary biology has made it possible to get rid of all the Victorian slag from Darwin's concepts, and to translate them into purely rigorous biological terms (for a clear expression of this idea, see H. M. Peters 1972, p. 347). This attitude has obvious implications for the evaluation of Darwin's "debt" to social and political thought (see Section IV, "Malthus", "Social Darwinism").

Darwin's metaphors had still another function. They were "not only non-mechanist and non-reductionist, they were thoroughly anthropomorphic and moralistic" (Manier 1978, p. 159). Their cognitive and affective dimensions were inseparable. They were also ways of resolving strong speculative and emotional tensions in the young Darwin. He was anything but the prototype of the scientist detached from the object of his research. His science was looking for the laws of life and did not shrink from discussing the great themes of chance, suffering, hope, and love (pp. 95–96, 167–168). The young Darwin felt something grand in his view of nature, life, man, and science. But it was not the cold grandeur of deism or of a reassuring theodicy. The young Darwin had a strong aesthetic sensitivity. His model was

Wordsworth's Wanderer, who went on looking for the meaning of life in nature and human experience, not theology.

The Darwin Manier presents is not easy to label. He was neither a teleologist nor a positivist; he did not hesitate to biologize human behavior and humanize animal behavior; he was not a mechanist-reductionist, but certainly not a dualist; he was not a deist, nor a theist, nor an agnostic, still less an atheist; he was anything but an individualist or a utilitarian in ethics (contrast Ghiselin 1971 and Schweber 1980); he was a realist with shades of romanticism (contrast W. Cannon 1976b); he was a materialist in that, in his integrally biological perspective, he tended to soften the distinction between the physical and the psychical; he was influenced by Herschel and Whewell and by Comte (indirectly), but did not conform to all their methodological requirements. But what happened to the young Darwin as time went by? Why did that enthusiastic young biologist turn into the "anaesthetic man" (Fleming 1961)[38] who once described himself as a "machine for grinding theories"? What modifications did Darwin's self-image undergo in the course of time? And under what influences? And, since Manier has shown that matters of language shade into matters of content, was his way of practicing science influenced by such changes?

IV. Darwin and Other People

DARWIN AFTER DARWINISM

Darwin did not retire after 1859. Yet, taken as a whole, Darwin studies compel one to think the contrary. For what he did after the *Origin* has often been either simply ignored or treated as of lesser importance. This latter attitude, represented for instance by Himmelfarb's definition of Darwin's later works as "a footnote to the *Origin*" (1959, p. 359), has been opposed only by general cursory statements that "the post-*Origin* Darwin had shifted tactics from the pre-*Origin* Darwin" (Ruse 1978a, p. 329) or that the context in which Darwin published the *Origin* was very different from the one in which he had conceived the theory (Manier 1980a, pp. 4, 11–12). Histories of biology devote only a few lines in passing to Darwin's later writings. So do biographies. One of the few exceptions, De Beer (1963), when treating the post-*Origin* works, is even less revealing than usual. We shall, then, not be surprised to find that good recent syntheses like those of Ruse (1979a) and Oldroyd (1980) pay comparatively little attention to those works. On the whole, one is given the impression that the essential Darwin is contained in the first four chapters and the conclusion of the *Origin* (plus some pages of the *Notebooks*, of course). This tendency reached one of its peaks with Vorzimmer (1970). He portrayed Darwin as a stubborn man who, as time went on, loaded his over-cherished theory with so many ad hoc arguments or detours to meet criticisms as to render it almost unrecognizable and create "knotty confusions and inconsistencies [. . .] in his theoretical framework" that he was not able to disentangle.[39] This, however, did not save Darwin, for criticisms (especially those of Mivart) drove him into a corner and badgered him "into the state of frustrating confusion which marked him on the eve of his retirement" (p. 251). Vorzimmer's argument has some truth in it, but is too narrowly focussed; hence it is misleading. There *were* changes in Darwin's mind and theory after 1859, and Darwin was receptive to, and confused by, many criticisms. But it is only by ignoring the great books he wrote in the second half of his career and the multiplicity of his interests that one can present him as a scientist fading away after the *floruit* of 1859. Here again lack of contextualization, which amounts to mere narrow-minded internalism, has proved fatal.

It is Ghiselin's great merit that he has treated the post-*Origin* works as not anti-climactic. But he, too, thinks that Darwin "had most of his ideas when a young man, and his later years were spent expounding and elaborating them, without much substantive change" (1975, p. 55; cf. 1973a, p. 964). This is true in the sense that all Darwin's works led to one large synthesis and were elaborations upon the dominant theme of the transformation of our entire world view (Ghiselin 1973b, pp. 163–164). But a one-sided emphasis on the unity of Darwin's work may hinder our understanding of the changes his mind and strategy underwent.

For changes there were. For instance, Darwin did not always give natural selection the same emphasis, status, and role in the explanation of evolutionary processes. There has been much talk of his alleged shift toward Lamarckism,[40] but it has at last been recognized that his subsequent greater emphasis on the direct action of the environment was "only a matter of degree" (Mayr 1971, p. 278; cf. 1964, pp. xxiv–xxvi. See also Vorzimmer 1970, chap. 5). To take another instance, the opinions on man and society Darwin expressed in the *Descent of Man* were only partly an application of the theories of the *Origin*. They were also the result of: 1) a resumption of ideas worked out long before, in the *Notebooks*; 2) the adaptation of some of these ideas to the new developments of Darwin's thought; 3) new ideas that had arisen in the light of new problems in a changed cultural context requiring new strategies (see below, "Social Darwinism"). And the same considerations hold true of Darwin's work on the expression of the emotions.

Yet studies of the various disciplines covered by Darwin exhibit little concern for the *development* of Darwin's thought on these topics. This is one of the reasons why we still lack comprehensive studies of Darwin's views on psychology, anthropology and "sociology", botany, embryology, and taxonomy.[41] The case of pangenesis is particularly revealing of some negative tendencies in Darwin scholarship. Many accounts of Darwin's life and work almost suppressed this infamous error (for instance, R. Moore

1957; Himmelfarb 1959; Wichler 1961). Others presented it as more or less an ad hoc Lamarckian device (Zirkle 1946; Keith 1955; C. D. Darlington 1953, 1959a; Eiseley 1958), a "retreat" (C. U. M. Smith 1976, p. 255), or a series of "fumbling efforts" (Fleming 1959, p. 444). Heslop-Harrison (1958, p. 289) said that "it is doubtful whether he [Darwin] ever took it with any very great seriousness". Gillispie dismissed it as "vague and contradictory speculations" (1960, p. 322). De Beer (1963) seemed almost to suggest that it was something not really worthy of Darwin. Neither Dunn (1965) nor Carlson (1966) took much trouble over it. And in a recent, authoritative textbook we still read that it was "an unfortunate anomaly. It was almost his [Darwin's] only venture into the field of pure speculation" (G. L. Stebbins in Dobzhansky et al. 1977, pp. 14–15).[42] On the other hand, some authors have taken it seriously. They have pointed out its connection with Darwin's ideas on sexual and asexual reproduction (Olby 1966a) and embryology (Ghiselin 1975), described its development (Geison 1969a), shown its historical importance (Darden 1976), and argued that it was as ambitious a theory, and as worthy of respect, as natural selection itself (Ghiselin 1975).[43] However, pangenesis has still to be connected in a detailed way with Darwin's ideas on variation and his early interest in invertebrate zoology. Kohn's (1980) stress on the importance of the subject of "propagation" in Darwin should provide a decisive stimulus.

One reason for the lack of extensive studies of post-*Origin* Darwin is the fact that his work in various disciplines has been examined for the most part by specialists in those disciplines, who were chiefly concerned with evaluating his contribution to modern knowledge in the field. A second reason is the heritage of the intellectual history approach, which leads one to focus on the more eye-catching conceptual innovations in (the first four chapters of) the *Origin* rather than the more technical subjects of climbing plants, orchids, cross- and self-fertilization, variation and pangenesis, and also coral reefs, cirripedes, etc. Most of Darwin's post-*Origin* works have simply followed the historiographical fate of the more technical sections of the *Origin* (like those on instincts in neuter insects, geographical distribution, embryology, or taxonomy). After what Browne (1980) and Ospovat (1981) have shown of Darwin's tabulations concerning the numerical relations of species to genera, intellectual historians should realize that the study of these "technicalities" may disclose the intellectual presuppositions that inform the scientist's research: this may be more fascinating and revealing than the usual celebrations of the "triumph of chance and change", the "discovery of time", or the attack on essentialism.

DARWINIAN PARTY OR DARWIN'S MOONS? DARWIN'S CIRCLE — OR CIRCLES

The lack of studies on the post-*Origin* Darwin is also due to the lack of extensive studies on major problems or figures of his context that may have influenced, or been influenced by, him.[44] Most figures and problems in Darwin's context seem to have been studied much less for their own sake than in Darwin's shadow. We are still in need of comprehensive studies of, say, Owen, Hooker, T. H. Huxley, Romanes, and the biology of Spencer. Lyell and Wallace have fared better. But few new studies have been added to Millhauser (1959) on Chambers, Mayr (1959a) and Lurie (1959) and (1960) on Agassiz, Dupree (1959) on Asa Gray, and J. W. Gruber (1960) on Mivart.[45] We have almost no study of how widely Lamarck, Cuvier, E. Geoffroy Saint-Hilaire, or other continental scientists were read in Britain, or how great their influence was, before Darwin (exceptions are Ospovat 1976, 1981; Corsi 1978, 1980b). Nor have we any account of the debates on methods and systems in pre-Darwinian British taxonomy. However, S. Smith (1965), Winsor (1976), Heilbroner (1976), La Vergata (1981, forthcoming), Knight (1981), and Ospovat (1981) have shown the importance of MacLeay and the quinarians, whom Streseman (1975), and especially De Beer, (ed., 1960) presented almost as mystical fools.[46] As to the debates on science and its methods, in which Herschel and Whewell were the leading figures, those who have studied them in relation to Darwin have given little attention to their underlying philosophical and theological aspects. Ruse, in particular, has discussed Darwin's debt to Whewell and especially Herschel mainly as a more or less faithful conforming to methodological principles (1975c, 1978a). On the other hand, Manier has argued that Darwin's language, logic, and imagery cannot be separated from philosophical, epistemological, and methodological issues, nor from ethical or aesthetic ones. It is Darwin's rhetoric that best reveals some fundamental aspects of his image of science and his self-perception as a member of a scientific community. This "sociological" dimension of Darwin's science as evinced by his "scientific rhetoric" influenced the very content of Darwin's theory, for his language was not merely a contingent envelope in which his theory happened to be wrapped. This variety of sociological approach to science is, then, far from merely providing footnotes to rational reconstructions or supporting far-fetched "externalist" ones (Manier 1978, 1980a, b).[47]

On a different plane, Herbert (1974) has emphasized the professional character of Darwin's early activities. His first transmutationist musings were connected with his occupation as a voyaging naturalist, which required reflection on the geographical distribution of species and were conducive to reflection on their origin. But we must also keep in mind the audience Darwin was writing for. His pattern of publication and division of labor in note-taking reflected his perception of his audience and professional boundaries. This sensitivity led to successful strategy, but also caused ambiguity and consequent confusion to the reader. In Britain "there did not exist a role for the theorist in science generally" (Herbert 1977, p. 189; but cf. Rudwick 1982b). This prevented Darwin from publishing his theoretical work of 1837–1839 as a single piece, although it was a coherent whole. Herbert has also called attention to Darwin's family background. From it he derived assurance, self-awareness, and "a satisfactory and supporting cosmology", a "family taste for empiricism in philosophy", which, together with the Unitarianism and "rather phlegmatic rationalism" of the Wedgwoods, acted as an instrument of selection in his readings and, for instance, guided him towards reading political economy (pp. 211–213).

As to the latter, Schweber (1977, and especially 1980) has argued that Darwin's commitment to some of its basic tenets was decisive in leading him to Malthus and natural selection and to the principle of divergence. The Benthamite optimalization principle and deductive approach to husbandry and political economy, Adam Smith's division of labor, a strong commitment to individualism and to a progressive view of history, a quantitative-statistic way of looking at nature derived from political economy — all these were transformed into biological principles by Darwin and gave his theory its "uniquely British character" (Schweber 1980, p. 198). "Darwin was aware that he was 'biologizing' the explanations political economy gave for the dynamics of the wealth of nations" (p. 212). But how did these external factors act on the content of Darwin's theory? Schweber talks of "interaction of external and internal factors" and of "heuristical transfers" of explanatory principles and models from one field to another. This was all the more easy because to Darwin political economy, morals, and psychology were biological problems and had to become branches of evolutionary biology (pp. 212, 276). I think Schweber has succeeded in showing a multiplicity of strands and a coexistence of interests in Darwin rather than a real intertwining and interaction of external and internal factors. For the borrowing and extension of models from one field to another is one thing; the perception of various sets of problems as belonging to the same fundamental area is another. Transfers can take place only between fields that are distinct, at least in principle. What, then, about borderline problems, such as man's moral sense? Presenting the influence of political economy on Darwin merely as an offer of useful explanatory models (as did Canguilhem 1975, pp. 108–110) is oversimplifying, and it also attributes to Darwin a too neatly modern epistemology. Did he perceive as distinct the fields whose boundaries he so easily crossed? Or does he seem so interdisciplinary a thinker precisely because he did not see so many distinctions as we do? To answer these questions neither old "externalist" commonplaces on Darwinism as a reflection of a socioeconomic situation (see below, pp. 955–959), nor the too generic formulas dear to traditional history of ideas (see above, pp. 913–914) is adequate. What is really important in discussing the role of "external" factors in the development of Darwin's *science* is to show how some highly mediated implicit assumptions become part of Darwin's language and logic, and how they conditioned the very way Darwin perceived some biological problems, "invented" some problems, and devised certain explanations. These mediations may be at work even when no bodily transfer from one field to another takes place. Mine is not a plea against taking into account "external" factors; it is a plea for detecting less visible mediations and distrusting too easy ones. Some of the analogies Schweber points out between Darwin and political economists are too generic for us to get a clear idea of how they may have influenced Darwin in his struggle with hard technical problems. For instance, I wonder what an "individualistic mechanism of inheritance" is (Schweber 1980, p. 109).[48]

Whatever his results, Schweber has tried to link two lines of research that, with very few exceptions (for instance Pancaldi 1977), had been pursued separately: the study of the development of Darwin's theory and the discussion of aspects of its broader intellectual context. In short, Schweber has the merit of reformulating some tenets of the externalist approach in the light of the results of recent, more rigorous scholarship. This attempt is important for inquiries into the nature of science. For, whatever its sources, Darwin's theory is a scientific one, or, to put it another way, it raises fewer suspicions that it has been conditioned by ideologies than does, for instance, Malthus's population theory. As Pancaldi writes me:

> Darwin produced good science even using some materials that were scientifically spurious or ideology-laden (like Malthus's theory, the idea of "struggle", that of progress, some aspects of medical and philosophical culture, Scottish moral philosophy) [. . .] These elements are "scientifically spurious" in that they do not conform to subsequent standards of rigour in biology [. . .] Some of them appeared to be such to Darwin himself: this may be why he did not treat man or free will in the *Origin*. On the other hand, such was not Malthus's case: his theory appeared to Darwin and to many of his contemporaries to be scientifically acceptable. However, those "spurious elements" provided invaluable intellectual stimuli and even some key concepts to what we now consider a legitimate scientific theory. (personal communication)

This raises grave philosophical problems. The fact that Darwin's theory was a scientific one is obviously not due to these "spurious elements" (and of course Pancaldi does not claim that it *is*). Pancaldi (1977) also implies that it is the fact that Darwin worked on technical biological problems that rescued him from ideology and merely speculative evolutionism. But does this not boil down to saying that "external factors" merely provided favorable personal accidents in a "purely" scientific undertaking? These are matters that cannot be left to philosophers of science arguing, without historical support, about a priori criteria of demarcation. Darwin scholars *must* concern themselves with the relationship between science and ideology. Yet there still are few signs that such is going to be the case, even in discussions of the *locus classicus*, Malthus's influence (see below, "Malthus").

Discussions of the intellectual and cultural circles in which Darwin moved will not be improved until another major gap in Darwin scholarship is filled. For, if studies on social and institutional aspects of nineteenth-century British science are being cultivated more than before, we still lack comprehensive studies, of, say, how natural history information was collected, filtered, evaluated, and systematized. Apart from the excellent study by D. E. Allen (1976), we do not have comprehensive studies of what being a naturalist meant and what were the methods, the ideals, the images, and the place of natural history among the life sciences and with respect to the physical sciences. Narrow-focussed case studies are also lacking, and the few exceptions (such as J. W. Gruber 1969, Burstyn 1975, and K. S. Thomson 1979) show how interesting clues can be gained to the general period by pursuing this line of research. Such studies would contribute decisively to charting a territory in which to locate internal analyses. What sense is there in talking about Darwin's interdisciplinarity, or his being a new (or an oldfashioned) biologist, if one does not really know what the relations were between natural history and other disciplines? Even a book as packed with information as that of Morrell and Thackray (1981) devotes incredibly little space to the life sciences. So does W. Cannon (1978), although he discusses Darwin's position as a leading exponent of what is called "Humboldtian science" and as a disrupter of the physics-based "Truth-complex".[49]

In conclusion, the backward state of studies on Darwin's intellectual context does not authorize such blunt generic statements as the following one in Ruse's otherwise excellent book:

> Driven by their internal forces and prodded by external influences such as Continental science and speculators on the fringe of science, like Chambers, the community [that is the scientific group of which Darwin was a member] *produced* Darwin's *Origin* and accepted it to the degree we have seen. (1979a, p. 266, my italics; see a concise criticism in Gruber 1981e, p. 326)[50]

One final gap to be pointed out is that of a full account of the debates and conflicts in natural theology in the first half of the nineteenth century. On considering this, one wonders how even such a well-documented book as Moore (1979) can launch on a reconstruction of the post-Darwinian religious controversies without connecting them to the pre-Darwinian ones. This consideration also holds good for most of the existing studies on the reception of Darwinism, which look somehow premature and one-sided (the big exception being the monumental Conry 1974). For there is an intrinsic limitation in studying the reaction of something that is not well known in its condition before the reaction.[51] Moreover, most of these studies deal exclusively with the reception of the *Origin* or of the *Descent*, and neglect the other works.

RELIGION

Today few would describe Darwin's impact on culture as simply a struggle between positive science and theological chatter doomed to be swept away by the Darwinian apocalypse. The idea of a "Darwin vs Paley & Co." match is deceptively simple, for natural theology was not a monolithic body of trivialities on design in nature. Well before 1859 natural theology was torn by inner tensions and conflicting responses to scientific issues (Brooke 1977a, b; Corsi 1980b).[52] Accordingly, the traditional accounts of the pre-Darwinian argument from design (for instance Ellegård 1956; Gillispie 1959*) have given way to attempts to show that there was not only one version of it (Bowler 1977a; Ospovat 1978, 1979, 1980; Yeo 1979). Taking this revaluation as a starting point, Ospovat has reformulated Darwin's position with respect to both natural theology and natural history. According to him, Darwin stood with those scientists and intellectuals who adopted "a doctrine of perfection limited by general laws". Like Owen, Carpenter, Chambers, or Baden Powell, and unlike Lyell, Buckland, or Sedgwick, Darwin abandoned the Paleyan interpretation of perfect adaptation, but "continued to share with that school a general view of the world, that it is a harmonious system, the creation of a benevolent God" (Ospovat 1979, p. 215). As a consequence, "for a number of years, Darwin conceived of natural selection as operating in an essentially natural-theological framework [. . .] The idea of harmony furnished not only the final cause of transmutation, but also the framework and constraints within which Darwin's mechanisms operated" (pp. 212, 217). Darwin was a theist, believed in design and progress, and had a theodicy that coincided with that of Malthus (p. 220).

What is important in Ospovat's thesis is the attempt to show how religious and philosophical assumptions helped shape, and acted as structural constraints on, Darwin's theory: "Perfect adaptation," he says, "was not a neutral or theoretically unimportant assumption [. . . .] The assumption of perfect adaptation played a regulative role in his theory. Through it, his theism shaped in subtle ways his understanding of the mechanism, as well as the products, of evolutionary change" (Ospovat 1980, pp. 191–192). Nature was planned to achieve certain general ends, for instance to maximize utility. Ospovat does not go so far, but the following seems to be a legitimate inference from his thesis: to a large extent, the "adaptationist programme", which Lewontin (1978) calls a "caricature" of Darwin's theory, a metaphysical postulate that cannot be refuted, was, at least up to the mid-1850s, ingrained in Darwin's theory and view of nature. But Ospovat also described the process by which Darwin gradually rejected this harmonious view. This process, he argued, was a social one. Almost all attempts to relate Darwin's theories to their intellectual (and ideological) context have focussed on the relationship between those theories and social-political thought. Some have also suggested that social forces and interests shaped Darwin's theory and that ideological concepts were transferred bodily from those fields into biology. Ospovat, more subtly, argued that Darwin's study of nature was "mediated by assumptions and ways of perceiving nature" that he derived from other naturalists and various quarters of the culture around him. Some of these basic assumptions were "ideologically loaded", but economic and social relations acted on Darwin's science through "labyrinthine and obscure pathways", not in a reflexive way (1981, pp. 230–233). Darwin's development was a social process in yet another sense. Aiming at restructuring the whole of contemporary natural history in evolutionary terms, Darwin had to reckon with the generalizations of his leading colleagues. This both opened new problems and led him to incorporate some of their concepts, which, in turn, altered his original theory. For instance, he absorbed developmentalism into his theory, which was not at first a theory of development, but of the process of adaptation to environmental change. The principle of divergence, too, emerged out of Darwin's "complex creative response" to the thought of some leading naturalists of his day. It embodied their belief in the orderly, upward character of natural processes, but it produced Darwin's concept of relative adaptation, which marked the exclusion of natural theological assumptions from his view of evolution.

Some details of Ospovat's argument puzzle me, namely his presentation of Von Baer, Cuvier, and the French post-Cuvierian biology; and the ease with which he uses such categories as "developmentalism" and "the branching conception in natural history". One may also wonder to what extent the "idealist" position on teleological explanation was seen to be compatible with religious orthodoxy. But, whatever one thinks on these comparatively minor points, his case must be taken very seriously. Before Ospovat (1981), with the partial exceptions of Manier and Schweber, the influence of contextual elements on Darwin's theory had been claimed rather than studied in detail. The "social context" had remained an empty slogan good only for making the most simple-minded internalists angry. In the case of religion, it had been debated to what extent Darwin's own opinions on the subject were influenced by the implications of his scientific views (Mandelbaum 1958); it had been argued that these implications did not succeed in destroying completely all forms of natural theology (Greene 1959c, 1961a); Darwin's ambiguities on the issue of theism had been pointed out (Young 1971a; N. C. Gillespie 1979); and it had been said that Darwin's world was Paley's turned on its head, while others saw in Darwinism a "vanishing" or "decapitated" teleology, and still others derived some satisfaction from pointing out aspects of his language that seemed to disclose a cryptoteleology or to cry out for an explicit one (a recent instance is Kass 1978). Among philosophers, as is well known, Darwin has been criticized for being a disguised teleologist, for being a rank teleologist, for being a crass anti-teleologist, and also for being a contradictory mixture of all these things (see for instance Bertalanffy 1932, 1933, 1937, 1952, 1975; Grene 1966, 1974b). Among historians, W. F. Cannon maintained that "the triumph of Darwinism is the triumph of a Christian way of picturing the world over the other ways available to scientists" (1959a, p. 110). But no one, Cannon included, elaborated in a serious way on this provocative *boutade*.

Ospovat's works are the best products of the recent revival of interest in the relationship between Darwin and theology. It is no accident that the other two representatives of this tendency, N. C. Gillespie and J. Moore, are, like him, intellectual historians. Only Ospovat, however, has measured his strength against some of the most technical aspects of Darwin's theory. And this was enough to secure him from historiographic simplifications. While Ospovat aimed at an integrated image of Darwin, Gillespie points out the "impressive mental ambivalence" of Darwin and his colleagues, whom he labels "theist scientists" (Gillespie 1979, pp. 6, 18, 86). Faced with the complexity of Darwin's mind, Gillespie solves any problem concerning the interaction between its various facets by presenting them as a series of intermediate nuances between two extremities or opposing "epistemes": "creationism" and "positivism". Or rather he presents "two Darwins", complementary but in tension with each other and varying inversely. One was a positivist simply because he was a scientist; the other was a man who felt that there must be some foundation for morals, rationality, and science, deeper than science itself. And here comes the crunch, for Gillespie says:

> Though a positivist in science, and despite his insistence on the autonomy of science, Darwin was not able to jettison the idea of God. He needed it to underwrite the possibility of science, to guarantee its rationality, that is, the correspondence of the scientist's activity with truth, and to preserve his optimistic view of the evolutionary process. (p. 125)

Theology, then, had "an evident integral function in the *Origin*" (p. 134).

Moore (1979) went perhaps even further in suggesting that religious orthodoxy was the foundation of scientific rationality in Darwin and the Darwinians. Apologists of various tendencies have been trying many ways of reconciling Darwinism with some version of Christianity, and some intellectual historians have followed them (for instance, Willey 1960). Moore, who sometimes seems to be writing like an apologist for some view of Christianity, beats them all with a brilliant strategy. He first launches a well-supported offensive against the traditional, warlike interpretation of the relationship between science and religion. Then he charges with a straightforward and paradoxical thesis: Darwin's theory was accepted more by orthodox Christians than by either conservative or liberal Christians, who adopted, or rejected, some form of "Darwinisticism", to use Peckham's (1959) concept. Since Darwin's theory was shaped by an orthodox theology of nature, Christian anti-Darwinians misunderstood Darwin and should not even be called Christians. Darwin derived his universe from "orthodox natural theologians", mainly from Paley and Malthus and, although "Darwin's theology" eventually declined, his universe remained the same to the end of his life. Now, leaving aside what Moore means by "orthodoxy" (one may, for instance, doubt that Paley was a representative of it), it is easy to note that he is indebted to Chadwick's (1966–1970, 1975) and W. Cannon's (1961b) arguments, of which he often offers only an aggressive paraphrase. A discussion of the details of Moore's impressively annotated book would be out of place here, so I shall express my opinion bluntly. In spite of the bulk of the volume, many of the arguments on which he bases his belligerent thesis are as innocent of demonstration as battle orders. The image of Darwin that emerges is, to say the least, partial. Moore spins his specious argument *around* Darwin's science. Although he devotes the second part of the book to it, he nowhere discusses the interaction between the contents of Darwin's theory and his theological views.

It is not to be supposed that Moore would hail Gillespie as an ally in his battle. For Gillespie believes that religion and science are ultimately in tension, a notion Moore cannot swallow, at least as far as Darwin is concerned. He says that Gillespie's book "both typifies and consummates the positivist historiography in Darwinian scholarship" (Moore 1981b, p. 179). He is perfectly right. And so he is in calling urgently for a new, integrated portrait of Darwin. But the premises he broaches are rather ominous. For there seems to have been a shift in Moore's approach. Not only has he contributed an admirably written paper to Chapman (ed. 1982) in the spirit of R. M. Young, but, judging from his review of Gillespie's book, he seems to have passed, by an easy transition, from his ecumenical view of the relationship between religion and science to a view in which echoes of Burtt's (1924) approach to the "metaphysical foundations of science" are mingled with aspects of a worn out, 1930s Marxism.

NATURALISM

It has become increasingly frequent to place the debates on Darwinism in the context of Victorian scientific naturalism.[53] Accordingly, attempts have been made to relate the ideological aspects of this movement to the scientific issues of the debate on evolution. The strongest case in this direction has been made by R. M. Young in a number of papers, some of them still unpublished. Young claims that "the fine texture of the debate directly involves theological and philosophical issues. These were constitutive, not contextual" (1971a, p. 444). "One simply cannot demarcate the biological from the social aspects of the debate" (1971b, p. 221). Young argues cogently that, whatever Darwin's ambiguities and oscillations on metaphysical, methodological, and more technically scientific issues, his theory contributed greatly to the mainstream of scientific naturalism. By eliciting faith in the principle of the uniformity of nature, Darwin's theory helped the reconciliation of evolution with a theistic view of nature. Once arbitrary interventions had been banished, God's government of nature became easily and increasingly identified with the uniformity of nature. The popular version of Darwinism pointed to "a grander view" of the Deity, and was as reassuring as the natural theology it was replacing. No matter how misunderstood, Darwin's anthropomorphic language, and especially the metaphor of selection, softened the blow of evolutionism and gained adherents to *some* evolutionary doctrine. Darwin gave a decisive contribution to a general, gradual process, which culminated in the naturalisation of man and the substitution of a secular, naturalistic world view for the old natural theology of perfect adaptation and divine contrivance. The associationists, the utilitarian philosophers, the radicals, Bentham, Malthus, and the two Mills were followed by Darwin, Wallace, Huxley, Tyndall, and the *Essays and Reviews*. All were part of a larger movement embracing a number of naturalistic approaches to earth, life, and man, and including phrenology, geological uniformitarianism, and psychology. Darwinism was perhaps more an effect than a cause in this movement toward a new theodicy. This rising tide substituted law and progress for God, but "the evolutionary debate produced an adjustment within a basically theistic view of nature rather than a rejection of theism"; moreover, "the evolutionary debate was seen by its participants as occurring within natural theology" (Young 1970b, pp. 27–30; cf. 1969, p. 111).

Prior to Young, others had occasionally said that there was a continuity between the ideologies of natural theology and Darwinism, or aspects of it.[54] Young's interpretation is part of a wider commitment, however. It supports a plea for a militant "radical historiography":

> To sequester the social and political debate from the scientific one is to falsify the texture of the nineteenth-century debate and to mystify oppression in the form of science. (1973a, p. 373)

Discussing Marxist views of science, society, and history, Young argued for "the need for a subtle and complex theory of mediations [between social, political, cultural, and scientific levels]" (1973a, p. 384). His aim was to free the historiography of science from the "internalist" approach and the "relatively isolationist view of science" exemplified in De Beer (1963) (Young 1971a, p. 453). This led Young to discuss the debates and their context, rather than the theories that were debated. He does not argue about the relationship between more "technical" debates and this "larger" ideological debate. Nor does he mention the *degrees* of influence of "non-scientific" factors on the various levels of the debate. He focusses less on the content of the theories — that is, the explanatory and directive principles — than on their philosophical, and ideological, implications and overtones. Young is quite right that studying the context of scientific theories throws light on the "nature of science" (1971a, p. 500). But why sequester the study of the technical core of theories from the study of their context? As Kuhn pertinently remarks, analyses of the intellectual milieu that neglect the technical problems that theories like Darwin's are credited with solving, restore a historiographic tradition that preserves the very separation between contexts Young deplores (Kuhn 1977, p. 139 n.). For Young's provocation stops before the scientific core of Darwin's theory. As Shapin and Barnes (1979, p. 128) remark, Young, who seems to go very far in an externalist approach, does not hesitate to use an equivocal concept such as "purity" to characterize Darwin's scientific work. He wrote, for instance, that "[Lyell and Darwin] are, relatively speaking, the purest of the scientists in the Victorian debate and as such are nearer to the positions of physicists, chemists, and mathematicians" (Young 1973a, pp. 386–387). Now, granted that "the role of social and political factors in the work of both Lyell and especially Darwin is a highly mediated one" (p. 384), why does Young use elsewhere such blunt expressions as "the movement of thought which *produced* the Darwin-Wallace theory of natural selection" (1970b, p. 31; my italics), or "[social factors] were *determinant* of the biological theories themselves" (1971b, p. 221; Young's italics)? And what about the statement that "Darwinism was an extension of *laissez-faire* economics from social science to biology" (1970b, p. 15)? This seems to dispense with *any* theory of mediation, let alone a "subtle and complex one".

MALTHUS

Young argued that the debates on evolution were grafted onto the debates on Malthus: this was a good instance of what he called "the common context of biological and social theory" (1969). Pushing his argument further, he claimed that "the link from Malthus to Darwin and on to the so-called 'Social Darwinism' is unbroken and continues to the recent writings on biology and society of, for example, Morris, Ardrey, and [C. D.] Darlington" (Young 1973a, p. 372; cf. 1972b). Functionalism, too, was part of this intellectual continuum.[55]

The relationship between Darwin and Malthus is perhaps the issue in Darwin studies on which there has been most discussion and disagreement. To some, Darwin's debt to Malthus was a clear instance of the extra-scientific factors that strongly conditioned the very essence of the theory of natural selection. To others, Malthus was just one factor among others in the development of Darwin's theory, which was purely scientific and by no means reflected its ideological context. Between these two extremes there have been almost all possible shades of opinion. Furthermore, interpretations intersect each other, so much so that charting them in a short discussion is impossible. Finally, the Darwin-Malthus issue has frequently been merged into more general problems, such as that of the relationship between the content of the theory and the language it was expressed in, between Darwin and scientific naturalism, between Darwin and Social Darwinism, or simply between science and ideology. For instance, in Rogers (1972) and D. Freeman (1974), minimizing the importance of Malthus was connected to an attempt to dissociate Darwin from Social Darwinism and

Spencerism. On the other hand, Harris (1968) stressed Darwin's debt to Malthus in order to show that Darwin shared with him and others an "intellectual matrix" that, according to Harris, might properly be called "Spencerianism".

If some argued that Malthus's influence had been overrated,[56] a host of others wrote that Darwin apprehended something vital for his theory through his reading of Malthus.[57] As I argued in Section I, to say that Darwin "took" something from Malthus has been very common among authors writing from the point of view of intellectual history or influenced by it. But, there has been no agreement as to what Malthus gave Darwin that was so important. Was it "un modèle explicatif fondamental" (Canguilhem),[58] an argumentative method, or the missing element in the final synthesis? And, in this latter case, was it the struggle for existence, or elements of it, or natural selection, or the idea of constant population pressure, or the concept of *intraspecific* competition (Herbert 1971), or the analogy prompting him to move from artificial to natural selection (Bowler 1976b)? In short, did Malthus act as a contributor or simply as a catalyst (Limoges 1970c)?[59] Vorzimmer (1969a, p. 541) argued that he was both things, and even more, since he provided Darwin with an "all-encompassing context through which Darwin would relate a large number of previously unrelated ideas". According to Ghiselin, Malthus stimulated Darwin to conceive of species in terms of populations, thus contributing to the evolution in metaphysics marked by the introduction of population thinking (cf. Jacob 1970). Malthus's doctrine was an expression of that individualism and belief in competition as the basic feature of life that Ghiselin presents as a major tenet of Darwin's and his own world view (1969, pp. 49–77; 1971, 1974a). But Malthus gave only a "heuristic aid", and "provided only a conceptual system or model, not the argument for an empirical proposition, and his contribution is mainly of psychological or historical interest" (Ghiselin 1969, p. 60).

Nor does there seem to be much agreement as to how intense the effects of Darwin's reading of Malthus were. To Kohn, it produced a "sudden revelation" (1980). To Gruber, who does not like flashes of insight, quite the contrary was the case (Gruber and Barrett 1974, pp. 173–174). To Herbert it was the highest point of a climax and was followed by a "period of detachment" (1977, p. 221). Schweber used the word "intoxication" to describe Darwin's state of mind after the event (1980, p. 225). Ospovat, as we saw earlier, denied that reading Malthus was decisive in leading Darwin to revise his concept of adaptation (1979, 1981).

As Manier has shown, "struggle" in Darwin is no single concept with one dominant meaning, but is a bundle of interrelated concepts (1978, pp. 177–181; cf. Gale 1972). It is their interaction that gives Darwin's notion of struggle its unique character. This notion cannot be reduced to the sum of its components; therefore its novelty cannot reside in only one of them.

Moreover, the notion makes sense only within Darwin's view of nature and theory of natural selection, and thus cannot be equated with that of any other pre-Darwinian writer (La Vergata 1977). So Darwin could not take it over from Malthus, and his relation to him cannot be described as the bodily "transference of a concept" (as Cowles 1937 and many others have argued). Darwin found in Malthus an impressive demonstration of superfecundity, limited food supply, and positive checks to population, but he transformed these ideas by inserting them into a new context, which, in turn, they helped shape.

But why did Malthus interest Darwin? Historians have challenged Darwin's own statement that he happened to read him "for amusement". To Schweber (1977, 1980), Malthus's was part of a more general influence that philosophers, moralists, and economists exerted on Darwin as a member of an intelligentsia who had a deep interest in individualism, empiricism, utilitarianism, *laissez-faire*, and all sorts of related issues. To Herbert it was Darwin's interest in human behavior as an evolutionary problem that led him to read Malthus in the context of extensive readings in philosophy and political economy (1977, pp. 216, 277). On another plane, Ruse argued that Darwin and Wallace reacted favorably to Malthus because he pointed the way toward a biological equivalent of Newtonian astronomy (1975c, pp. 171–173; 1979a, p. 179). He showed Darwin how to place struggle and selection in a hypothetico-deductive network of quantitative laws, fitting the criteria set by Herschel and Whewell. Others maintained that it was Malthus's quantitative, mathematical way of arguing that particularly impressed Darwin (McKinney 1972; Limoges 1970c; Conry 1974; Gruber and Barrett 1974). Or was Malthus influential because he touched the chord of Darwin's Victorian sensitivity to the "eat-or-be-eaten" credo? Here we come to a major problem. Was Malthus's influence different in kind from Malthus? I think it was partly the desire to remove any suspicion of contamination with ideology that led authors such as de Beer to downgrade Malthus's importance.[60] Actually, while some recited the usual litany of Darwin's theory being a reflection of British society and economic thought,[61] others simply tried to explain away the problem. One favorite solution was to confine Malthus's influence to the wrapping of Darwin's theory. To Gillispie, the Malthusian jargon simply happened to be lying ready to hand; if confusion arose, that was just too bad. "In the relatively inexact state of biology, he [Darwin] borrowed from common language [. . .] Nor could Darwin stop at every paragraph to explain himself. When he did, it was evident that he was capable of thinking clearly even in loose language" (1960, p. 343): in sum, objective truth in Victorian smog.[62] Likewise, Herbert does not find it necessary to expand critically on her statement that "because of the enormous effect of Malthus on Darwin's work, biology remains perpetually indebted to the field of political economy" (1977, p. 216). Others have solved the problem by implying that Malthus deserved the title of scientist as much as Darwin himself: no ideological taint, then, in their relationship (Himmelfarb 1959, pp. 132–139).[63] Really the whole matter seems to be one of "purity versus history", as Shapin and Barnes have put it (1979).

Marxists have been more sensitive to the problem, but as a rule they have tended, as Young has pointed out (1969, pp. 138–140), to evaluate Malthus and Darwin separately. From Marx and Engels on, Malthus was parson Malthus the plagiarist, the bought advocate, and the enemy of the working class. He was an ideologist, in that he proposed a view of man, nature, and society that was not only historically conditioned, but also false and mystifying. Darwin, on the other hand, was a contributor to the establishment of a materialistic, scientific, and objective view of reality (except when, discussing man and society, he fell victim to what Marx called "abstract materialism moulded hastily on natural sciences"). The attempt to separate genuine Darwin from the transient Malthusian element was characteristic of Lysenko.[64] I end this paragraph by quoting S. J. Gould. What he says on Darwin's "creative transfer to biology of Adam Smith's basic argument for a rational economy" seems to me as rash as the facile slogans I have been criticizing so far. But, unlike many others, he does recognize the problem, and offers a clear-cut view of it. His solution can be useful as a "hypothesis to work by" to historians not dodging the philosophical problems their job raises.

> I believe that the theory of natural selection should be viewed as an extended analogy — whether conscious or unconscious on Darwin's part I do not know — to the laissez-faire economics of Adam Smith [. . .] But the source of an idea is one thing; its truth or fruitfulness is another. The psychology and utility of discovery are very different subjects indeed. Darwin may have cribbed the idea of natural selection from economics, but it may still be right. (Gould 1980a, pp. 67–68)

Be that as it may, the Darwin-Malthus problem calls for a broadening of perspective; that is, it demands to be treated with an eye to broad issues concerning the very nature of science. I think the foregoing discussion confirms that decision on how to deal with a historiographic problem is a function of the historian's options on matters of science and ideology.

Clues to broadening our perspective have been offered by Egerton and Pancaldi. Egerton set Malthus in the tradition of the studies of animal populations from Leeuwenhoek to Darwin via A. von Humboldt. Writing as an intellectual historian, Pancaldi extended the focus to the physiocrats and the tradition of the "economy of nature". According to him, Malthus's influence on Darwin was an episode in the continual interaction between natural history and the social sciences. In particular, "Malthus's work marked the end of confidence in the natural and spontaneous character of social equilibrium. A dynamic element was imperceptibly gaining ground within the 'organic economy' tradition [. . . .] If one has to study Malthus's relation to Darwin, one must not ignore that many other naturalists had been interested in the problems of 'natural economy' and were well known to Darwin before he read Malthus" (Pancaldi 1977, p. 59). What Darwin found in Malthus's law was a confirmation of the basic precariousness of the "economy of nature" (pp. 78–79). Pancaldi here takes the same position as Young, according to whom "the Malthusian spectre was a direct challenge to the harmonious view of nature", so much so that Paley and Chambers felt it necessary to absorb and to neutralize Malthus's gloomy law into a reassuring theodicy (Young 1969, pp. 111–114; 1970b; 1971a, b). This view of Malthus is shared also by Bowler (1976b) and Kohn (1980, pp. 144–145).[65] It must not be forgotten, however, that Malthus *was* a natural theologian. One of his main aims was to show that the difficulties of life were means to exert man's faculties, rescue him from sluggishness and foster the "creation of mind" (see LeMahieu 1979). Certainly, Malthus's essay underwent great alterations from the first, more theological, edition to the sixth, which is the one Darwin read. Still, it is legitimate to ask how such a thinker could destroy the belief in harmony.

Here I may be allowed some general considerations. Many commentators have sought the key to understanding Malthus's influence on Darwin in some aspects of the former's doctrine. Now, what Malthus actually thought is not necessarily what Darwin "took" from him or saw in him. Therefore, I find some flaws in the arguments of those who, like Himmelfarb, Limoges, or Bowler, tried to assess the Darwin-Malthus relationship by pointing out that this or that aspect of Darwin's theory was intrinsically anti-Malthusian in character. Comparing the "spirit" of two well-defined bodies of ideas is not necessarily a way to understanding their position in a debate, or the "influence" one exerted upon the other. In other words, to understand Malthus's place in the history of ideas is not automatically to understand Malthus's place in Darwin's development. These two historiographic levels must be kept distinct. What was happening in Darwin's mind when he read political economists was not a miniature version of what was going on in the debates of political economy. Furthermore, it is possible that what Darwin really found in Malthus was different from what he thought he had taken from Malthus. I think it is only partially true that by using Malthus's principle Darwin inserted his own theory into the Malthusian debate. Darwin's position in the larger debate was determined by a much more complex network of mediations than his mere "debt" to Malthus. These mediations shaped Darwin's language, logic, concepts, and, as it were, mind. Young's claim that we ought to consider a broader intellectual context is absolutely justified; but in a sense this task can be performed only at the price of deflating the importance of the Darwin-Malthus relationship as a historiographic problem. Conversely, I think that it is not by simply digging more deeply into the Malthusian debate that we can assess Malthus's influence on Darwin. We also need, for instance, to rethink the concept of influence. An important step in this direction has been made by Manier, who has analyzed the Darwin-Malthus relationship as an instance of the "multi-dimensional structure of influence" (1978, pp. 75–85, 190–192). To sum up my contention, Darwin's contacts with ideologies cannot be reduced to the direct influence of this or that leading figure. Darwin's debt to his intellectual (and ideological) context must be traced to the formation of a language, in the broadest sense of the term, that could condition the perception of the contents of scientific inquiry. This approach can connect studies in the history, the logic, and the sociology of science.

SOCIAL DARWINISM

Social Darwinism, everybody agrees, is a bad thing. Yet, we have no commonly accepted definition of it. In a sense, to define it as an extension of Darwin's theories to society begs one of the questions it poses. For it has been maintained that Darwin had made a similar extension the other way round. In addition, "Social Darwinism" covers too many different things. To many it was merely a reactionary phenomenon, closely allied to racism, and leading to Nazism (Prenant 1938; Michalova 1958; Lukàcs 1959, chap. 7; Gottschalk 1959; Harris, 1968). To others it was mainly an apology of *laissez-faire*, "Robber Barons", and political skullduggery (Hofstadter 1944; Persons ed. 1950b; Young 1969, 1973a).[66] Some saw in it an episode in the general phenomenon of "biologism" (Di Siena 1969, 1976). Others have restricted its meaning to more particular phenomena, such as eugenics (Conrad Martius 1955; Halliday 1971). It has rightly been maintained that there were many kinds of Social Darwinism (Himmelfarb 1968; G. Jones 1980), and that many of them were hardly Darwinian (Stocking 1962, 1968). Burrow (1966a), Harris (1968), H. M. Peters (1972), and Freeman (1974) have suggested "Spencerianism" as a better label. But not all of the so-called Social Darwinists liked Spencer, and not all of Spencer fits the stereotype of Social Darwinism (Peel 1971; La Vergata 1980a). There was a socialist Social Darwinism (Bulferetti 1951; Semmel 1958; Benton 1982) and an anarchist one (think of Kropotkin). And certainly Spencer, Bagehot, Galton, W. G. Sumner, Kidd, Kropotkin, Kautsky, and Hitler were not members of the same team. Some historians have attempted to draw distinctions between "reform" and "reaction" Social Darwinism (Loewenberg 1957), or between "internal" and "external" Social Darwinism (Semmel 1960). But others have lost their patience and have proposed — rightly, I should add — to abolish the very term (La Vergata 1982a).

From a different point of view, Shapin and Barnes (1979) have argued that we should stop making distinctions inappropriate to the context in which Darwin worked: Social Darwinism, as a problem, might then disappear.

This sounds a bit optimistic, but I am not here to discuss Social Darwinism per se.[67] Rather, let me ask: where does Darwin stand in all this turmoil? The debate between Grace, Stern, Montagu, and Haldane in *Science and Society* in 1941–1942 set the pattern for much of the ensuing discussion of the issue. Basically, it consisted in comparing Darwin's declarations on man and society with one image of Social Darwinism, and emitting a verdict. McConnaughey (1950), Rogers (1972), D. Freeman (1974 and the participants in the discussion on Freeman's paper) conform to this method, and are good evidence of the contradictory results it leads to. It can degenerate into a "phrase hunt", at the end of which you can have Darwin say almost everything, all the more so because of his changing and sometimes seemingly ambiguous or contradictory statements. His ideas on man and society evolved through time, from the *Notebooks* to the *Expression*. The same basic pattern of evaluation was followed both by those who thought Social Darwinists had misunderstood and arbitrarily carried over Darwin's biological concepts into the study of society (for instance, Rogers 1972, but see also Becker and Barnes 1961, 2: 701–702, 704), and by those who, on the contrary, thought the Social Darwinists were in a sense authorized in bringing Darwin's theory back to its sources, that is, society and economics. Thus Harris (1968) characterized Malthus's, Darwin's, Spencer's, and the Social Darwinists' theories alike as the result of a simple addition (progress + struggle + racism) and insisted on Spencer's priority. Neither party showed interest in Darwin's own development. Neither gave him the human right to change his own mind, and to hesitate and waver when faced with enormous problems. And then there was the practice of trying to assess Darwin's relation to Social Darwinism by focussing exclusively on his relation with the ubiquitous Spencer. There is an inherent limit to such comparisons (for instance, by D. Freeman 1974 and Ruse 1982), since Social Darwinism is not completely amenable to Spencerianism.

This narrowing of perspective was also due to the fact that historians of the human sciences contributed little to Darwin studies. A survey shows, with very few exceptions, confusion, generic formulas, and an excessive reliance on the methods of the history of ideas, at the expense of detailed historical reconstruction.[68] Evolutionism and Darwinism were often taken as synonyms (Goldman 1959, who scarcely mentions Darwin's name). Too much was attributed to Darwin's direct influence (Becker and Barnes 1961, 2: 747; Kroeber 1960, p. 10). The alleged Darwin-Morgan relationship monopolized the attention of many anthropologists interested in the history of their discipline. Reaction by Bock (1955, 1964), Burrow (1966a), and Stocking (1968) led to the recognition that the rise of Victorian anthropology and social theory was not due to Darwin but to a preceding tradition.[69] Burrow went too far, however, in dissociating Darwinism from the tradition to which Victorian anthropology was heir. He did not discuss Social Darwinism. As a result, it was considered as "standing somewhat outside the mainstream of anthropological development" (Malefijt 1974, pp. 122–123). On the contrary, British anthropology was interested in next-to-biological issues. It was based on "the assumption of a homology between nature and culture" (Weber 1974, pp. 281–282). Victorian attitudes to the relationship between culture and biology have yet to be the subject of a detailed study, which would cast a flood of light on Darwin himself.

The same applies, obviously, to Darwin's own activity in the field of what we now term the human sciences. Little has been written on these matters (see note 41). For instance, historians of psychology, classical and recent, either offer commonplaces or are singularly defective (Boring 1950a, b; Hilgard 1960; R. Thomson 1968; Mueller 1976; see discussion in Young 1966).

Today, however, few could skim over Darwin's interest in social evolution as Ghiselin (1969) did. And few could dispute Greene's (1975) thesis that Darwin was a social evolutionist who saw in natural selection a powerful means of interpreting human and social evolution as well. Manier has claimed that the young Darwin hardly fits the common image of individualistic Social Darwinism (1978, p. 138–146). Greene and Manier refer to different evidence and phases in Darwin's development. They agree, however, that often Darwin did not hesitate to biologize human and social behavior. One could say that Darwin was one of the very few Social Darwinists who was really a Darwinian.

At any rate, he was a fluctuating Social Darwinist, and the causes of this are many and varied. As Herbert convincingly put it, the confusion about what he really thought on human society was prompted by his "extreme caution in choosing and addressing his audience. His inaction left the field to others." The cost ran high. By the time he did speak on man publicly, "his emotional ties to the subject were gone" (1977, pp. 194–195). Other causes of change have been discussed, though very briefly, by G. Jones (1978), to the best of my knowledge the only attempt to connect the development of Darwin's views on man with changes in his context. Jones has interpreted Darwin's relation to associationism and "developmentalism" as an alliance with strong intellectual currents that could, and did, secure the success of both his theory and his attempt at an overall explanation of life. Darwin was concerned not only with developing and implementing his theory, but also with creating an "ideological space" in which he could operate. Jones follows Canguilhem's (1977, pp. 33–45) and Lecourt's (1972, p. 106) concept of "*idéologie scientifique*" as "a doctrine which [. . .] lends support to a new scientific concept in order to extend this concept outside the domain of its validation". Scientific ideologies are part of epistemological obstacles (which scientific discoveries expel from a particular field), but also a condition of the possibility of science: they rationalize experience, supply order, and, by being recreated as science advances, provide access to totality and restore the unity of science in the face of its divisions and contradictions. Now, Darwin was trying to give his theory "the character of a world view, he was in search of totality". By allying himself to the associationist-developmentalist current, he borrowed from it. For instance, contrary to what he had done in the *Notebooks*, he now subscribed to an anthropology of faculties that was teleological in so far as it saw evolution as bringing into existence the pre-defined characteristics of these faculties (G. Jones 1978, p. 20). But, in his "attempt to disguise the regional character of natural selection", Darwin entered "an area at which scientific ideology shades off into ideology with political and social determinants" (p. 4).

> Entering into the realms of social thought was, in a sense, a reaction to the demands placed upon him by a particular characterisation of scientific practice. Darwin, it could be argued, was influenced by a theory of knowledge in two ways, in the construction of the evolution of Man and in his acceptance of the applicability of concepts across the particular region in which they are formed. (G. Jones 1978, p. 20)

However, because of the fluidity of this alliance and the inevitable consequences of transferring a set of problems from one area into others, Darwin was exposed to waverings and also contradictions, which were amplified as they reverberated in other contexts. His theory opened up many combinations of ideas. Consequently, "there can be no inner political character attributed to the theory of natural selection itself" (p. 19).[70]

The implications of this view are the very opposite of those of Bannister (1979), who believes there *is* one privileged moral message to be drawn from Darwin's theory. I cannot discuss here his attempt to revise a whole historiographic tradition that culminated in the classical Hofstadter (1944). Bannister's provocative thesis, that Social Darwinism was merely a polemical stereotype, has many merits but is connected with an utterly oversimplified and misleading image of Darwin.[71] For instance, he thinks that Darwin and the early Darwinians were not Social Darwinists (Bannister 1979, p. 16). He is right that Darwin's views were "moderate", but not that Darwinism "opened a gap between society and nature" (pp. 31–33). He arbitrarily likens Darwin, Huxley, and Wallace to each other and says that "they rejected the idea that sound social policy was the result of allowing free play to the automatic operation of natural law". According to Bannister, the more Darwin stressed natural selection through struggle among plants and animals, "the more it appeared that society operated on different principles. In extending Malthus, Darwin removed the one prop that made society possible: Malthus's assumption that the struggle for existence was a collective one against nature. To reapply the extended version to society had implications Darwin and his contemporaries found unacceptable" (pp. 31–33). To Bannister, Darwinism fostered "the idea that men must transcend nature rather than follow her dictates" (p. 10; cf. Malefijt 1974, pp. 273–274). So he thinks it warranted to conclude that "the principal legacy of the *Origin of Species* [. . .] was the reform Darwinism that flourished in various forms from the 1880s onward" (p. 11). This presentation of Darwin's message to ethics, politics, and sociology is blatantly partial, to say the least. Bannister simply does not consider the aspects of Darwin's thinking that do not fit his scheme.

Luckily for Bannister, his image of Darwin has little bearing on his main thesis. But this says a great deal about the lack of communication between historians of science and historians of the social sciences. And on this sad note I close this section.

V. Final Considerations

The foregoing overview shows, I think, that in the field of Darwin studies there is need for a revision of some traditional historiographic categories. Let me give a few examples. Darwin and Darwinism cannot be used as synonyms. There were many varieties of Darwinism, and not all of them bore the same relation to Darwin (although one should not go so far as Lorenz (1965, p. 6), who says that to speak of Darwinism is to calumniate Darwin). Likewise, it has been recognized that "materialism" is a label that must be used with extreme caution in relation to Darwin and his context (Mandelbaum 1971; R. Smith 1973; Corsi 1975, 1980a; Manier 1978). New, subtler tools are necessary to discuss the relationship between "internal" and "external" factors in the genesis and development of Darwin's theory. There seems to be a growing sense of distrust in monocausal explanations. Thus the fact, now ascertained, that Darwin was sensitive to his audience, forces us to give up univocal interpretations of the relationship between him and the recipients of his theories, and to look for more flexible models. Darwin's evolving thought and strategy should be viewed, at least in part, as functions of a changing intellectual context. Also attempts at capturing *the* logical structure and *the* method of Darwin's theory have not been very successful. Darwin escapes from the cages in which some have tried to enclose him; and it is no accident that philosophers of science have, on the whole, contributed very little to Darwin studies.

Cooperation, pluralism, and especially philosophical sensitivity are necessary to connect the different levels of historical research. This is all the more necessary because there continues to be a tension between works aiming at synthetic vistas (either written from the point of view of general intellectual history or dealing with trends in nineteenth-century biology) on the one hand, and works on "local" problems in Darwin and Darwinism on the other. To my mind, there are two main ways to avoid the negative while leaving the positive aspects of this tension: 1) intellectual history should turn into the study of assumptions informing actual scientific practice, both routine and innovative; 2) historians should give their attention not only to national contexts, paradigms, archives, and research programs, but also to the "dictionaries" of the authors they study (Bellone 1976), to "themata" in scientific thought (Holton 1973, 1975), and especially to disciplines, or better "fields", and traditions or styles of research. Although difficult to define, many of these existed independently of Darwin. His theory absorbed some of them (for instance parts of contemporary taxonomy, comparative anatomy, and embryology); others it penetrated with great difficulty. It is well known, for instance, that physiologists gave a very cold reception to it (Mendelsohn 1964; Schiller 1965, 1980; La Vergata 1982b), so much so that the fact that late nineteenth-century English physiology was in some way "evolutionary" is enough to denote a "national style" (Geison 1978, p. 335; cf. French 1970a). Now, both what was absorbing and what was being absorbed were altered in the process. Studies on particular research fields, such as Winsor (1976), may throw light on Darwin by enlightening the interplay between aspects of his work as a naturalist and established bodies of facts, rules, and problems.

So far, however, we have only some general evaluations of the place of Darwin's theory in nineteenth-century biology (for instance J. W. Wilson 1959 (*naïf*), Coleman 1971a, and La Vergata 1982d, plus the relevant sections in the various histories of biology). Canguilhem (1977, pp. 101–119) gives interesting but rapid insights into how different biological disciplines reacted to Darwin.

Recently, there has been discussion about the conflict between "naturalists" and "experimentalists" which took place in the late nineteenth and early twentieth centuries (G. E. Allen 1975, 1979; Mayr 1980).[72] It has been argued that natural selection was supported by the former and opposed by the latter; Mayr has rightly written that "naïve physicalism" retarded the acceptance of Darwin's theory (1973, p. 133). But no extensive historical study exists that faces the problem: What is the relation of Darwin's theory to the mechanist-reductionist ideal in late nineteenth-century biology? Partial exceptions are Conry (1974), La Vergata (1982b), and Roger (1982).

And what about morphology itself? In spite of Gould (1977b) and Balan (1979), the best book on Darwin's place in the development of morphology is still that of the anti-Darwinian E. S. Russell (1916). Paleontology has fared a little better, thanks to Rudwick (1972) and Bowler (1976a) (the latter being concerned with a typical intellectual-history problem: the history of the concept of progression of living forms; see also Bartholomew 1976).

It is to be hoped that current debates on evolution, particularly the critiques of gradualism and of the so-called adaptationist program will stimulate new historical research. There have already been instances of this. The historical part of Gould (1977b) grew out of his concern for the role of developmental constraints in evolution and his revision of aspects of the neo-Darwinian synthesis (see also Gould 1982 and Gould and Lewontin 1979). On a different plane, Mayr's unique position as an architect of modern evolutionary biology and a leading systematist has led him to challenge the traditional interpretation of the role genetics and systematics played in the path from Darwin to the evolutionary synthesis of the 1940s (Mayr 1980; 1982b). This revision cannot help influencing Darwin studies. The debates on sociobiology, too, have given spur to some historical work (for instance, Ruse 1980b; Durant 1980, 1981a, b). These are instances that Darwin studies can also contribute to current debates by clarifying some key concepts around which polemics, fueled by confusion, rages. And maybe the renewal of interest in Darwin and religion is connected with the recrudescence of religious attacks on Darwin?

Conclusions

My conclusion can be simply stated. Like other species, that of Darwin scholars has evolved. And it has evolved according to a truly Darwinian pattern: divergence from common ancestors, division of labor, increasing specialization, piecemeal variation, competition, extinction of some varieties, promotion of others to the rank of species, isolating barriers, both ecological and reproductive.

On the whole, Darwin students, or most of them, today seem to be a much more compact family than heretofore. However, I am afraid the danger is always there that increasing specialization leads to a fragmentation of the image of Darwin. I think the lack of a good up-to-date biography is less innocent than it might appear. It may reveal the lack of communication between some quarters of Darwin studies, but it certainly also reveals a state of ferment, which is the least suited for a synthesis, even for a provisional one.

If left to themselves, intellectual historians may tend to the superficial, philosophers abstract and unhistorical, scientists *naïf* and Whiggish, professional Darwin students erudite and narrowminded. Therefore it is a matter of cooperating. Historiography of science is the result of many histories, a collective undertaking with a plurality of approaches. Darwin studies are no exception. They are de facto an interdisciplinary activity and a pluralistic affair. But what "invisible hand" harmonizes these different approaches? Maybe an ideal reader capable of reading everything on the subject? To avoid the dangers just pointed out, it is to be hoped that each Darwin scholar will strive to become as pluralist as possible, that is to see from his or her own point of view as many images of Darwin as possible. This is clearly unrealizable now. Nevertheless, we must try. To be pluralist we have an excellent model: Charles Darwin.

NOTE

All translations are mine. All personal communications are quoted with their authors' permission.

ACKNOWLEDGEMENTS

In preparing this paper I have become indebted to many people. In particular I wish to thank Giulio Barsanti, Roger Chapman, Yvette Conry, Pietro Corsi, Mario A. Di Gregorio, Dietrich von Engelhardt, Donatella Ermini, Bernardino Fantini, Marta Fekete, Peter J. Gautrey, Barbara Gerard, Nick Gill, Jane Ireland, Béla Köpeczi, Maria Landolfi, Giovanni Landucci, Giuliano Pancaldi, Stefano Poggi, Hans Querner, Jacques Roger, László Vekerdi. Very special thanks are due to Virginia Browne-Wilkinson, Maurizio Bossi, David Kohn, Dorian Kottler, and Paolo Rossi.

Notes

1. For more bibliographical information and/or discussion, consult Fleming 1959; Loewenberg 1959a, 1965; Ellegård 1960–1961; Mendelsohn 1964; Young 1873a; Smit 1974; Guédès 1974–1975; Greene 1975; Moore 1979; the Isis Critical Bibliographies and the Isis Cumulative Bibliography.
2. Barzun 1958; Cassirer 1950; Ostoya 1951; Persons ed. 1950; Sears 1950; Stresemann 1975; Fothergill 1952; Montagu 1952a; Keith 1955; Irvine 1955; Gillispie 1951.
3. On Darwin's illness, see Adler 1959; Foster 1965; Brussel 1966; Roberts 1966; Woodruff 1968; Winslow 1971; Pickering 1974; Colp 1977; Bean 1978; on "methodological Darwinism," see Campbell 1974, which has a rich bibliography; on the implications of Darwinism for ethics, see Ganz 1939; Quillian 1950; Daiches Raphael 1959; Marnell 1966; Flew 1967, 1978; on religion, see Lack 1957; Dillenberger 1961; Fothergill 1961; Nogar 1961, 1963, 1966; Centore 1969, 1971; Deely 1969; Hegenbarth 1972; Deely and Nogar 1973; Spilsbury 1974. On the relationship between Darwin and Marx, see note 64.
4. See Lerner 1958, p. 10; Popper 1959, 1972, 1976, 1978; Waddington in Tax ed. 1960, 1: 385; Smart 1963, p. 59; Simpson 1964, p. 79; Manser 1965; Barker 1969; Grene 1974; R. H. Peters 1976; Medawar and Medawar 1977. Reaction to this charge included J. S. Huxley 1938; Weizsäcker 1951; Goudge 1961; Flew 1966; Ruse 1971a, b; M. Williams 1973; G. L. Stebbins 1977; Caplan 1977. On the problem of the falsifiability of the theory of natural selection, see Popper 1959, 1972, 1976, 1978; Lee 1969; Løvtrup 1976; Olding 1978; Platnick 1978; Wassermann 1978; Ruse 1981. On the explanatory structure of evolutionary theory, see also Scriven 1959; Bunge 1961; Mayr 1961; Lehman 1966; Ruse 1981, pp. 1–27. General philosophical evaluations are Meyer-Abich 1964; Cimutta 1969; Goll 1972.
5. These writers, like those mentioned in note 4, tend to discuss problems in neo-Darwinism rather than in Darwin's own theory.
6. I have not been able to trace the author of the first book; it was published in 1913 at Eastbourne. The other two are Reinheimer (1915) and Barclay (1950).
7. For a criticism of this image of science with respect to physics, see Capek (1961).
8. A variety of the "ingredients approach" is evident in the step-by-step reconstruction of Darwin's own development, as exemplified in the following statement by De Beer: "It may well be asked what he [Darwin] had to go upon in 1844. The answer is his own observations made during the voyage of the *Beagle*, Lyell's principles of geology, von Baer's laws of embryonic resemblance, Malthus's *Essay on Population*, a few fossils such as *Mylodon*, *Macrauchenia*, *Palaeotherium*, and *Mastodon*, and an English country gentleman's knowledge of domestic plants and animals and their breeding" (1958a, pp. 3–4). To some extent, the same historiographic scheme underlies the following list of "fundamental dispositions, faculties and characteristics which contributed to Darwin's great achievements" by Wichler: his love of and leaning for science manifested in his intense effort to pose questions and to answer them; his powers of observation; his collaboration with other scientists; his clarity; his non-dogmatic way of thinking; his lack of any tendency to philosophize; his singlemindedness; his equanimity; his favorable financial situation; a harmonious marriage and a happy family life; his circle of friends (1961, pp. 173–192).
9. Here is a list of other terms Greene uses to denote those "most general ideas or patterns

of ideas which inform the thought of an age": "general conceptions of nature"; "basic patterns"; "pattern of thought, or dominant view of nature"; "presuppositions of thought" that "do not become less real when they are left unexpressed"; "implicit major premises"; "habit of thought", "basic presuppositions", "lowest common denominator" of various works in various fields; "cast of mind"; "climate of opinion"; "state of mind" (cf. Lovejoy 1936, chap. 1). These are not Lovejoy's "unit-ideas". Greene acknowledges that they are analogous to Kuhn's paradigms, but, unlike these, do not concern scientists only. They may be found in the writings of avant-garde thinkers in any field as well as in the sermon of a country parson (Greene 1981, pp. 3–4, 7, 10–11, 13, 23, 49, 52, 57).

10. The tendency to see the individual (the "first-degree reality" in Gruber's phrase) in terms of the broader intellectual trends of the period (a "second-degree reality") has been criticized by S. Drake 1970, pp. 3–5; Grmek in Grmek, Cohen and Cimino eds. 1981, pp. 17–18; Gruber 1981b.
11. Jacob presents the history of biology as the *logical* development, both through continuity and through breaks, of a few major concepts. These are presented as embodied in great books, not in actual debates. Not for nothing, as Jacob's narrative arrives at recent developments, does it become a sort of nameless, logico-intellectual autobiography. A perceptive criticism of Jacob is Holmes (1977), who also points out striking similarities between Jacob and Foucault.
12. See also the following passage: "Geological uniformitarianism and its *corollary* of indefinite mutability in the organic world were *implied* in the Cartesian program of deriving the present structures of nature from a simpler, more homogeneous state of the system of matter in motion by the operation of the laws of nature. *The drawing out of this implication* by Buffon, Lamarck, Lyell, and others sprang more from the appeal of this vision of nature and natural science to imaginative minds than it did from factual discoveries, which could always be interpreted differently by less imaginative observers" (Greene 1981, pp. 51–52; my italics).
13. "It was reserved for Darwin's wonderful talent for combination to sum up the product of the investigation of a hundred years . . . . Here [in plant physiology], as in morphology and systematic botany, Darwin found the premisses given and drew the conclusion from them; here too the certainty of his theory rests on the results of the best observers, on investigations which find in that theory their necessary logical and historical consummation" (Sachs 1890, p. 431). Pantin claimed that the Darwin-Wallace theory was a "deductive argument based on commonly accepted ideas", their merit being that before them "no one had shown precisely what these ideas implied" (1958, p. 221). Eiseley described Darwin's "creative synthesis" thus: "Darwin's solution, in essence, was merely another way of looking at the world from the same set of data, but it was the dispassionate observation of a man on a height to which no else had climbed . . . . Sir James Paget once remarked that Darwin's volumes exemplified in a most remarkable manner Darwin's power of utilizing the waste material of other men's laboratories. One might venture the observation . . . that he was equally adept in the utilization of those stray sparks which fell from other men's minds, but which in his own head underwent a marvelous transformation" (1959b, pp. 111–112, 114).
14. See for instance J. A. Thomson 1909; Lanessan 1914; Kunkle 1917; Tischner 1928; Tze Tuan 1929; Zirkle 1941, 1946, 1957; Shryock 1944; Montagu 1947; Dufrenoy and Dufrenoy 1954; Doeschate 1959; Wilczynski 1959; Rostand 1960; Kanaev 1962; Heim ed. 1963.
15. See for instance American Philosophical Society 1959, Rice Institute Pamphlets 1959, Antioch Review 1959, Victorian Studies 1959, Glass, Temkin, and Straus Jr. eds. 1959, Barnett ed. 1958a, Bell ed. 1959a, Meggers ed. 1959, Tax ed. 1960, Heberer and Schwanitz eds. 1960, Mason ed. 1960, Banton ed. 1961.
16. To some extent, the roots of this interpretation could be traced back to Kuno Fischer's *Hegels Leben, Werke und Lehre* (Heidelberg 1901), vol. viii of his *Geschichte der Neueren Philosophie.* Compare Cassirer's view with that of Gillispie discussed above. On Darwin's place in the so-called "discovery of time", see Haber 1958, 1959, 1971, 1972; Buckley 1967; Toulmin and Goodfield 1965; von Engelhardt 1977, 1979; Balan 1979.
17. If he or she has done both things he or she is a serious candidate for a paper in *The Darwinian Heritage.*
18. D. Kohn, S. Smith, and R. C. Stauffer (1982) prove the "Outline and draft of 1839" is bogus.
19. Olby 1963; Freeman and Gautrey 1969, 1972; Vorzimmer 1969b, 1975, 1977; Gruber and Barrett 1974; Stauffer ed. 1975; Herbert ed. 1980a; Ospovat 1981. For a list of Darwin manuscript material published by De Beer,

Barlow, Barrett, Conry, Stecher, and Trenn, see Greene 1975, pp. 244–247. To these there should be added Baehni 1955; De Beer ed. 1958d, 1959b, 1968; Stecher 1969; Fischer 1970; van der Pas 1970; Barrett and Corcos 1972; Schwartz 1980.

20. Other French writers have emphasized the radical novelty of Darwin's achievement both as a biologist and as a thinker. They have done so with impressive style, panache and virtuosity, and sometimes also with verbal exuberance (see Canguilhem 1965, pp. 135–137; Dagognet 1970, pp. 173–187; Jacob 1970, passim). It may seem a bit strange that this has happened in a country where no variety of Darwinism may be truly said to be an accepted orthodoxy.
21. On this point, Limoges shows striking similarities to some features of intellectual history I discussed above, Section I, "Before the Notebooks".
22. For instance by Gruber and Gruber 1962; Vorzimmer 1969a; Limoges 1970c; Gruber and Barrett 1974; Herbert 1974, 1977; Grinnell 1974; Pancaldi 1977; Kohn 1980; Schweber 1977, 1978, 1980; Richardson 1981.
23. "In retrospect, it is quite evident how exceedingly difficult it would be for one steeped in the tradition of Plato's philosophy to accept the idea of 'common descent'. This is the reason the great German zoologists of the first half of the nineteenth century failed so completely to solve the problem of evolution. They had been thoroughly indoctrinated in the concepts of idealistic philosophy, while the two 'dile • ıntes', Darwin and Wallace in England, had spent their time watching birds, collecting insects, and reading Malthus and *Vestiges of the Natural History of Creation*, thus happily remaining unaffected by the lofty fallacies of idealistic philosophy" (Mayr 1976, pp. 257–258).
24. Irvine presented Darwin as a plodder who "was frequently to discover kingdoms while searching for asses", and therefore "muddled into genius and greatness like a true Englishman". He attributed to Darwin what Bagehot called the "alluvial mind", that is the mind "in which an idea develops so slowly that it hardly seems to have been there at all until it seems to have been there always" (1956, pp. 34, 37).
25. See notes 4 and 5. Following Hartmann (1950) and Topitsch (1958), H. W. Peters has characterized Darwin's use of "technomorphic" and "sociomorphic" explanatory models (selections and competition) as conforming to the *als–ob* (as–if) pattern of explanation (1972, p. 333; see also H. W. Peters 1960, 1965; and Uschmann 1968).
26. Compare Medawar's (1963) paradox that the scientific paper is a fraud, because it does not tell the whole story of discovery. For this reason, and from internal evidence, the following statement by Young seems highly objectionable: "The path by which Darwin arrived at the mechanism of natural selection was also the one which he chose to follow in setting out his argument" (1971a, p. 454).
27. A further illustration of Ghiselin's mythological and Manichaean view of history is his attempt to exhume poor Haeckel, whom he reveres as a fighter against Prussian despotism, from the heap of lies under which "reactionaries" buried him. His "battle" with Virchow was "an episode in the long conflict between Platonism, aristocratic privileges, and ecclesiastical tyranny on the one hand, and empiricism, popular democracy, and freedom of conscience on the other" (1969, p. 123).
28. What does not fit Ghiselin's method of discussing Darwin's method is simply explained away or omitted. He excludes from treatment Darwin's Glen Roy theory and his papers on glacial phenomena as having "little direct bearing on the problems under consideration" (1969, p. 31). He does not discuss Social Darwinism, because "Darwin embraced no such notions" (p. 70). This may be disputed but, still, is a motivation. What is puzzling is the exclusion of any mention of Darwin's ideas on social evolution and physical and cultural anthropology. Darwin's pronouncements on induction are at one point explained as lip service to philosophical orthodoxy and labeled "hypocrisy" (p. 35); but later, when Darwin is reported as saying something that squares with Ghiselin's reconstruction, we read that "there is no reason whatever to treat his perfectly ingenuous accounts of the discovery as mistaken, contradictory, or hypocritical" (p. 75).
29. Ghiselin accepts Popper's theory of falsification, but does not say that Popper himself has leveled against Darwin's theory the very charge of not being refutable that Ghiselin attributes to Von Bertalanffy alone and reacts to indignantly (1969, p. 63). He distinguishes between discovery and justification, but seems to suggest that testing takes place through induction (or through the reciprocal testing of joint implications, which is something quite different). This amounts to inverting the neo-positivist characterization of the two processes. Ghiselin says that Darwin *refuted*

Lamarckism and creationism (pp. 62–63). He also thinks that scientific theories can be *verified* and *confirmed* (p. 43). All this would be branded as impossible by Sir Karl. Finally, I think Ghiselin is wrong and inconsistent in saying that "Darwin's philosophy was akin to that of a pragmatist or a logical positivist" (p. 5).

30. Peirce 1877, 1935, 6: 297; Dewey 1910; Mayr 1959a, b, d, 1961, 1964, 1971, 1972a, 1976a, 1977; Hull 1964–1965, 1967b, 1973a, b; Jacob 1970.

31. I must mention some aspects of Mayr's interpretation of Darwin very schematically. He has repeatedly remarked that Darwin failed on five points: (1) he still had a morphological, not a biological, species concept; (2) he did not distinguish between phyletic evolution and speciation; (3) he identified the reality with the fixity of species; (4) he underestimated the role of isolation as an evolutionary factor; (5) he did not distinguish clearly between varieties and individual variations (Mayr 1949, 1957, 1959b, c, 1963, 1976a). As I cannot discuss these crucial points here, I will only mention the bearing of some recent studies on them. Point 1 has been denied by Kottler (1978). Sulloway (1979) has discussed Darwin's shift from an early biological to a later morphological species concept. (Ghiselin 1969 points out that Darwin occasionally lapsed into a morphological habit of thought). Point 2 should be partly revised in the light of Sulloway (1979) and recent studies of Darwin's principle of divergence (Browne 1980, Schweber 1980, Ospovat 1981). Point 3 has been challenged convincingly, by Limoges (1970c). Ghiselin (1969, pp. 82, 85, 89–102), followed by Kottler (1978, pp. 291–294), has attempted a reconciliation of Darwin's seemingly contradictory views on species. Point 4 has been qualified by Kottler (1978) and Sulloway (1979). Sulloway has adduced various plausible reasons for Darwin's abandoning his early emphasis on geographic isolation. In addition, he has shown that Moritz Wagner, whom Mayr had presented as far-sighted about the importance of isolation, was far from being all that modern, had many Lamarckian aspects, and failed to grasp the complexity of Darwin's concept of isolation. Mayr himself however has somewhat revised his own judgement of Wagner (compare Mayr 1959c, p. 224; 1963, pp. 6, 516; 1976a, p. 124). Point 5 is where Mayr's critique holds up best; but his interpretation can now be integrated in the light of recent studies on Darwin's conception of variation by Bowler 1974a, 1976c, and, partly, Gaissinovitch 1970 (see also Kottler 1978, p. 289 and Browne 1980, pp. 72–73). On all these points, however, Mayr's opinions have undergone some changes (see, for instance, Mayr 1976a, pp. 117–118 on isolation). Moreover, most of his views were expressed before the publication of the *Notebooks* and the rise of Darwin scholarship to a highly specialized activity.

32. Ruse (1975a, c) has argued that Herschel's and Whewell's canons also prompted Darwin to investigate the artificial selection practiced by breeders and horticulturists.

33. Herbert has also stressed that Darwin's being a theorist put him in a very particular position with respect to both his fellow naturalists and the contemporary standard image of what natural science should be. This influenced Darwin's strategy considerably. Similar remarks have been made on Lyell by Rudwick (1974b, 1977, 1979). Secord provides good instances of Darwin's strategical ability in dealing with breeders and the problems of breeding (1981; this volume). Grinnell has turned Darwin's strength as a theorist into a mysterious tendency to arbitrary choice of explanatory models, and to the invention of ad hoc hypotheses. Darwin used to "superimpose" a "prior point of view" on to the data and their anomalies. He was strongly committed to a "philosophy of nature" and then "inclined to transmutation theories for reasons that transcended the empirical data with which he originally worked" (1974, p. 273). To him, therefore, empirical data seem to have counted for almost nothing. And so does internal evidence for Grinnell, who fails to establish his case.

34. On Gruber's reconstruction of Darwin's family background, see Corsi (1975), and compare Gruber's discussion of "materialism" with Mandelbaum 1971; R. Smith 1970, 1973; and Manier 1978. Gruber's reconstruction of the genesis of Darwin's theory — particularly his emphasis on what he calls the "monad theory" — is criticized in Kohn (1980).

35. A propos of Gruber's plea for greater recognition of unity and complexity in thought processes (and as a confirmation that students of the history of science often *are* guided by assumptions concerning the nature of science and culture), let me quote the following passage: "It is perhaps not too much to say that the scientific culture that oversimplifies Darwin is part of a larger civilization that has elevated fragmentation and simplification to high principles for the conduct of life. Is a job interesting and

complex, placing a demand on the intellect and character of a person? Break it up into many jobs that will make no such demands! Is some nuance of nature unnecessary to the life of this society of simplified human beings? Uproot the tree, fill in the marsh, cover the earth with cement! Nor is it too much to say that in the struggle toward something better for our descendants we need a theory of intellectual functioning that enjoys and does justice to human complexity" (Gruber 1981b, p. 320).

36. Cf. Young (1971a, pp. 461–470). Similar conclusions to those of Manier on metaphors have been reached by Fellmann (1977), although from a different point of view. Rudwick has discussed Lyell's creative use of metaphors and analogies from non-geological fields (1974b, 1977, 1979).

37. Many commentators had argued, or implied, that since Darwin's ideas could be given today a rigorous, mathematical formulation, they had only a contingent link with their social and extra-biological context. For instance, Gillispie (comment to D. Freeman 1974, p. 224) sees "classical political economy as the environment rather than the motivation" of Darwin's theory, and adds that "theory is not the less scientific for that since it has repeatedly proved its strength when expressed in other terms." On the relation between Darwin and the social sciences see below, Section IV, "Darwinian Party or Darwin's Moons?", "Naturalism", "Malthus" and "Social Darwinism".

38. A partial revision of Fleming's judgement is J. A. Campbell (1974).

39. "His genius was to construct, not to render consistent . . . . Perhaps this is one of the facets of a genius: an ability to weather all crisis, those induced by criticism from outside, as well as those springing from an inner lack of consistency. Such a man never accepts defeat; he is therefore never defeated" (Vorzimmer 1970, p. 271).

40. Fothergill 1952, p. 132; Hardin 1959, pp. 116–120; Eiseley 1961, pp. 240–247; De Beer 1963, p. 175. Also consider Young: "When it was pointed out that . . . his theory was insufficient, Darwin went into a dignified retreat and was left with the very sort of mixed bag of factors which he had rejected at the outset of his studies . . . . It is a useful exaggeration to say that by the sixth edition the book was mistitled and should have read *On the Origin of Species by Means of Natural Selection and All Sorts of Other Things*" (1971a, pp. 470, 497).

41. On Darwin's psychology and its context see Marler 1959; Twiesselmann 1959–1960; Canguilhem 1975 pp. 112–125; Barnett 1958b; Swisher 1967; Ghiselin 1969, 1973a; Young 1967a, 1970a, 1973b; R. Smith 1970, 1973, 1977; Gruber and Barrett 1974; Ekman ed. 1973; C.U.M. Smith 1978; Richards 1977, 1979, 1981; Gilman 1979. On his anthropology and "sociology": Stewart 1959; Meggers ed. 1959; Mason ed. 1960; Banton ed. 1961; Rubailova 1973. On botany: Whitehouse 1959; Heslop-Harrison 1958; Brabec 1960; Schwanitz 1960; Wichler 1960; Haustein 1960; Basalla 1964; Baker 1965; Baillaud 1966; Marza and Tarnavschi 1967; Ghiselin 1969; Kilburn 1969; Allan 1977; Browne 1978a, 1980. On geology and paleontology: Challinor 1959; Romer 1962; Yonge 1962; Andrée 1960; H. Schmidt 1960; Gould 1968; Hattiangadi 1971; Barrett 1973, 1974; Rudwick 1972, 1974a, 1976; Burchfield 1974, 1975; Bowler 1976a. On cirripedes: S. Smith 1965; Ghiselin and Jaffee 1973; Trenn 1974; Gunther 1979. On ecology: Stauffer 1957, 1960; Vorzimmer 1965; Glacken 1967; Worster 1977. On embryology and morphology: De Beer 1958c; Lovejoy 1959a; Oppenheimer 1967; Coleman 1976; Gould 1977b; Balan 1979; Ospovat 1981. On taxonomy: Gilmour 1951; Cain 1959a, b; Mertens 1960; Crowson 1958; Nelson 1974; Winsor 1976; Kottler 1978; Sulloway 1979. On geographic distribution: de Lattin 1960; Nelson 1978; Sulloway 1979; Richardson 1981. On sex: Ghiselin 1974a; Kottler 1980. On mimicry: Evans 1965.

42. This judgement is also an effect of the rough-and-ready philosophy of science that peeps out in the book.

43. Ghiselin also says that, though anachronistic and based on insufficient empirical foundations, pangenesis was "the price to pay" for Darwin's very propensity to work on a grand scale and to think big that produced the theory of natural selection (1975, pp. 55–56). On pangenesis see also Roberts 1929, Schierbeek 1943, and Stubbe 1965.

44. Among the few notable exceptions are Sanford 1965; Geison 1969a, 1978; French 1970a; Santucci 1971; Conry 1972a, 1974; Cowan 1972a, b, 1976; Churchill 1968; Ospovat 1976; R. W. Burkhardt, Jr. 1979; Sulloway 1979.

45. On Owen we have only McLeod 1965; Rudwick 1972; Bowler 1976a; Ospovat 1976, 1978, 1981; and Desmond 1979; on Hooker, Turrill 1953, 1963; Allan 1967; Browne 1978a, b; on Romanes, Turner 1974a and Lesch 1975. Bibby 1959a, b, 1972, Stanley 1957, Eisen 1964, Ashforth 1969, Helfand 1977, Paradis 1978,

Gilbert 1979 are almost silent on Huxley's scientific work. On specific aspects of it see Geison 1969b; Blinderman 1971; Friday 1974; Bartholomew 1975; Winsor 1976; Eng 1978; Querner 1978; Di Gregorio 1982b, c. Di Gregorio has prepared a comprehensive study of Huxley's science: his treatment, as he himself is proud to admit, is "strictly internalist" (I am grateful to him for allowing me to read his typescript and discussing my criticism). Aspects of Spencer's biology have been discussed in Burrow 1966a; Medawar 1967; Young 1968b, 1970a; Peel 1971 (the best comprehensive work on Spencer); Wiltshire 1978 (less original); Sharlin 1976; McQuire 1977; Francis 1978; Kennedy 1978; Toscano 1980; and especially Plochmann 1959; Bliakher 1973; Churchill 1978; Burkhardt, Jr. 1979; La Vergata 1980a. Spencer's biology suffers from being considered mainly in conjunction with social Darwinism or as a sort of pseudo-scientific antithesis to Darwin's biology (as in D. Freeman 1974). I am preparing a study of Spencer's biology per se. On Lyell see the books mentioned in Section I, p. 910; here I will add only Rudwick 1974b, 1975a, 1977, 1979a; Corsi 1978; Bartholomew 1973. Considering Wallace's importance, he has been rather neglected. See Wichler 1938; Mayr 1954b; Eiseley 1958, 1959a; McKinney 1966; Beddall 1968, 1972; Brooks 1969, 1972; Vorzimmer 1970; Bowler 1976c; Fichman 1977, 1981; Nelson 1978. The only comprehensive study, George 1964, is almost purely descriptive and underestimates Wallace's philosophical, moral and social commitment. So does another book, McKinney 1972, which stops at exactly the start of the process that led Wallace to become "Darwin's moon" (Williams-Ellis 1966). Nor are attempts such as Brackman's (1980) likely to rescue him from his role (see Kohn 1981 for stringent criticism). Kottler 1980 is excellent on Wallace's views on sexual dimorphism. R. Smith 1972, Turner 1974a, Kottler 1974, and Durant 1979 have discussed aspects of Wallace's concern with man, society, and scientific naturalism. But this line of research has not been followed in conjunction with that dealing with scientific issues. We are therefore left with a confirmation of the coexistence of many interests in Wallace, rather than with the demonstration of a real interaction between scientific and "non-scientific" aspects. Among recent studies on other figures in Darwin's context are Beddall ed. 1969; Woodcock 1969; Hodge 1972; Brooke 1977a; Russell-Gebbett 1977; N. C. Gillespie 1977; V. D. Hall 1979. On Agassiz see Baron 1956; Carozzi 1966, 1973; Weir 1968; D. E. Pfeifer 1970; Balmer 1974; Thuillier 1974; Gould 1979; V. D. Hall 1979; Winsor 1979.

46. Now see also Di Gregorio (1982a).
47. Manier's approach to Darwin's language and logic is connected with his opinion that "it is time for philosophers of biology to expand their horizons". They must go beyond exclusive concern with matters of the logic of demonstration and empirical evidence, and with traditional issues like reduction, functional explanation, historical explanation, etc. They should open a sociological perspective in their work and come to deal with those non-mathematical or non-deductive forms of communication that are the most common in the life species (1980b, p. 305).
48. The same remarks apply to Secord, who writes: "Just as Adam Smith's invisible hand of the economic realm brought the actions of competitive individuals into a functioning whole, so unconscious selection resulted from the actions of thousands of individual fanciers, thinking only of their own ends in pursuing their hobby" (1981b, p. 183).
49. By this Cannon means the alliance between science and religion that issued from Newton's achievement and that assured science its exalted role. Darwin disintegrated this intellectual totality, which consisted of a hierarchy of disciplines under a universal norm (Cannon 1978, pp. 3, 56, 63, 86–105, 263–287). This thesis has been followed, less wittily, by Garland (1980). Elsewhere Cannon (1976b) portrayed Darwin as the opponent of the Christian Romantic position in science represented by Sedgwick and Whewell, and the exponent of a utilitarian-materialistic theory in the wake of Locke, Hartley and eighteenth-century materialism.
50. Ruse is right, however, in saying that Darwin and the *Origin* were not "the natural culmination of a long line of evolutionists", but an expression of "the scientific group from which he came" (1979a, p. 200). This point had already been made most forcefully by Pancaldi (1977).
51. Professor Jacques Roger writes me: "A research group is currently working in Paris on the condition of natural and biological sciences in Europe in 1859. The work is based on the perusal of a number of periodicals in order to evaluate the comparative weight of the various disciplines. In addition, a reading of the main books and articles published during that year enables us to study the issues that most interested naturalists and biologists." Here I may add a list of studies on the reception of Darwin's theories in different

countries: Ellegård 1958; Rubailova 1971; Hull 1973b; F. Burkhardt 1974; Hodge 1974a, b; Hull, Tessner and Diamond 1978 (England); Loewenberg 1933, 1934, 1935, 1941, 1957; Hofstadter 1944; Persons ed. 1950b; Curti 1951; Kultgen ed. 1959; Dupree 1959; Lurie 1960; R. J. Wilson ed. 1967; Daniels ed. 1968; Pfeifer 1965, 1974; Aldrich 1974; G. E. Allen 1975; Russett 1976 (United States); Mullen 1969; Rajkov 1969; Querner 1972, 1975; Montgomery 1974a, b; Kelly 1981 (Germany); L. L. Clark 1968; R. Stebbins 1965, 1974; Seidler 1969; Conry 1971, 1972a, b, 1974, 1982; H. W. Paul 1979 (France); Savorelli 1974, 1977; Benasso 1976, 1978; Pancaldi 1977, 1983; Giacobini 1977; Landucci 1977, 1981; Martucci 1978, 1980, 1982; Giacobini and Panattoni eds. 1983 (Italy); Rajkov 1951; Rubailova 1968; Gaissinovitch 1973; Rogers 1960, 1963, 1973, 1974a, b; Vucinich 1974 (Russia); Glick 1974a; Nuñez ed. 1977 (Spain); Rapaics 1952; Réti 1958a, b, 1962a, b, 1964; Boros 1959a, b; Êhikné Süle 1959; Székely 1959; Bartucz 1964 (Hungary); Lappalainen 1956, 1967; Leikola 1981, 1982; Vepsäläinen 1982, Voipio 1982 (Finland); Simonsson 1958; Danielsson 1963–1964; 1965–1966; Schopf and Bassett 1973 (Sweden); Bulhof 1974 (Netherlands); Calcoen 1960 (Belgium); Watanabe 1971, Shimao 1981 (Japan); Marchant 1957, 1959; Goodwin 1964; Mozley 1967 (Australia); Bezirgan 1974 (the Islamic world).

52. Corsi points out strongly that Paley was far from representing orthodox natural theology and that different incompatible traditions were conflicting in it. I am indebted to Pietro Corsi for much information on this point.
53. Burrow 1966a; Young 1967a, 1969, 1970a, b, 1971a, b, 1972a, b, 1973a, b; Mandelbaum 1971; R. Smith 1972, 1977; Turner 1974a; Durant 1977, 1979, 1982. Weber refers to the "moralizing naturalism which characterized the dominant Darwinian tradition from the 1870s on" (1974, p. 280). Scientific naturalism has been pointed out as the supporting structure of Lamarck's overall ideological program (Jordanova 1976, 1981; cf. Barsanti 1979). On scientific naturalism in Germany, see Gregory (1977a, b). The very term "scientific naturalism" was given currency as early as the end of the nineteenth century in the works of such writers as T. H. Huxley, J. Ward, A. J. Balfour, and C. Lloyd Morgan.
54. For instance, Geddes wrote that Darwinism had replaced the anthropomorphism of the eighteenth century with that of the nineteenth (which took the form of the extension to nature of the belief in the severity of industrial competition): "the place vacated by Paley's theological and metaphysical explanation has simply been occupied by that suggested to Darwin and Wallace by Malthus" (1882, p. 116). Likewise, J. S. Huxley spoke of late nineteenth-century Darwinism as "Paley redivivus" (1942, p. 23).
55. This view has been challenged by Bowler. He argued that Malthus was not the source of the view of nature that led to Social Darwinism, because "there is a real conceptual gulf between Darwin's struggle for existence and the *laissez-faire* philosophy, particularly as represented by Malthus" (1976b, p. 636).
56. For instance Himmelfarb 1959; Eiseley 1958; and especially De Beer 1961, 1963, 1964b, 1969a, 1970; De Beer ed. 1960a; De Beer, Rowlands, and Skramowski eds. 1967.
57. For instance, before 1970, Ràdl 1905–1909, 1930; Nordenskiöld 1928; West 1938; Montagu 1952a; Irvine 1955; Butterfield 1957, chap. 12; Carter 1957; R. A. Fisher 1958; Sirks and Zirkle 1964; Canguilhem 1975, pp. 108–110.
58. On the model-analogy issue see discussion in Ruse (1973b, c).
59. Limoges (1970c, pp. 79–80) argued that Malthus merely reinforced upon Darwin the idea of the intensity of the struggle for existence. His contribution to Darwin's theory, therefore, was not indispensable and consisted in "crystallizing" it, without affecting in any way "its decidedly anti-Malthusian character". For Malthus, Limoges rightly said, admitted only a blind, non-selective elimination of excess population. Here Limoges followed Himmelfarb (1959, pp. 132–139) and was followed by Conry: "S'il est indéniable que le recours à Malthus a constitué, sinon un moment, du moins un élément d'intégration, on ne saurait sans abus théorique le convertir en nécessité ou en modèle" (1974, p. 397). According to her, Darwin could borrow only the idea of a "pressure" from Malthusian ideology; he had to superimpose an ecology on a mere mathematical reckoning. Moreover Malthus's system was still an expression of the traditional concept of natural economy. Conry showed that, in many French authors, seeing Darwin in close connection with Malthus and *laissez-faire* economists was the premise for a reading that "denatured" Darwin's theory.
60. However, De Beer (1961a) made some concession to it.
61. Merz 1896–1914, 2: 395–396; Ràdl 1930, pp. 18, 25–31; Nordenskiöld 1928, p. 470; Prenant, 1935, 1938; Sandow 1938; Montagu 1952a, pp. 28–32; Irvine 1955, p. 76; S. F. Mason 1962; Bernal 1965, 2: 662; 4: 1233; Ben-David

1971; Hall and Hall 1964; Mulkay 1979.

62. A similar view is expressed by Rogers (1972): Darwin's "Malthusian" and "Spencerian" metaphors were inessential and caused pernicious misunderstandings. They were responsible for much Social Darwinism. It is clear that Manier's reinterpretation of Darwin's metaphors can also shed new light on the relationship between Darwin and Social Darwinism.
63. Curiously enough, Young, too, wrote that Malthus "was a biologist, a human ecologist" (1969, p. 111).
64. On the relationship between Marx, Engels, Darwin, Darwinism and Marxism see Prenant 1935, 1938; Rostand 1947; Schneider 1951; Meek ed. 1954; Preti 1955; Selsam 1959; Runkle 1961; A. Schmidt 1962; Hollitscher 1964; Lucas 1964; Avineri 1967; Mikulak 1970; Timpanaro 1970; Zavadskii, Georgievskii, and Mozelov 1971; Gerratana 1972; Hirst 1976; Lecourt 1976, pp. 119–121; Yokoyama 1978; Pancaldi 1977, pp. 141–160; Bethell 1978; Ball 1979; Benton 1979; Naccache 1980; Christen 1981; Vidoni and Guerraggio 1982. On the contacts between Darwin and Marx (including the "Darwin-Marx correspondence") see also H. Gruber 1961; Colp 1974, 1976; Colp and Fay 1979; Feuer 1975, 1976; Carroll 1976b; Gaissinovitch 1977; Berlin 1978; Fay 1979; Tee 1979. A recent overview is Müller 1983.
65. Young has argued that Darwin synthesized Paley and Malthus by showing that "struggle both *explains and produces* adaptation" (1969, p. 118; Young's italics).
66. Hofstadter's view has been challenged by Corwin 1950; Wyllie 1959; R. J. Wilson 1967; Bannister 1979.
67. I can only list some works on the topic in addition to those mentioned in the text: Faris 1950; Barié 1953; Curti 1951; Nachtwey 1959; G. E. Simpson 1959; Bogardus 1960; Lenz 1960; Murphree 1961; McRae 1958; Zmarzlik 1963, 1969, 1972; Leibowitz 1969; Bannister 1970, 1979; Breck 1972; Williams 1973; Nichols 1974; Koch 1973; Russet 1976; Mackenzie 1978, 1981; Szacky 1979; Jones 1980; Santucci 1982. We lack an extensive series of case studies. Exceptions are: Persons ed. 1963; Marchant 1957, 1959; Goodwin 1964; Burton 1965; Mozley 1967; Clark 1968; Gasmann 1971; Conry 1974; Wall 1976; R. Smith 1972; Durant 1977, 1979, 1981a, b, 1982; D. C. Bell 1979; Schungel 1980; Crossley 1981; Weindling 1981; Landucci 1981; La Vergata 1982c; Benton 1982.
68. On this point I am very much indebted to Maurizio Bossi, who assisted me with his knowledge of the literature on the history of anthropology and discussed with me works by Godelier, Harris, Hays, Hirst, Kardiner and Preble, Krader, Legros, Llobera, Mafeje, Makarius, Malefijt, Meillassoux, Mercier, Murphree, Stocking, Terray, and White.
69. Mandelbaum (1971) stressed the methodological and conceptual parallelism between Darwin's biology and anthropo-sociology, particularly with respect to the use of comparative method and to the basic and pervasive belief in progress. Mandelbaum pointed out a parallelism due to common heritage where others had previously seen a direct reciprocal influence.
70. Unfortunately, Jones does not fully develop these insights in her recent book on Social Darwinism (1980). Its merits remain more in the preliminary, programmatic parts (for instance in the chapter on "The moral economy of nature") than in the treatment itself, which is sometimes sketchy and poorly organized.
71. Bannister sees Hofstadter's interpretation of Social Darwinism as heir to the democratic and humanitarian currents that coined the very term "Social Darwinism" as a polemical stereotype. According to Bannister, proponents of social control — from the new liberals of the 1880s to the varieties of "reform Darwinism", from the American "psychological school of sociology" (as opposed to the "biological" school of Spencer) to the New Deal intelligentsia — used the myth of Social Darwinism in their battle against *laissez-faire* and utilitarianism. To them the label Social Darwinism meant "the charge, often unsubstantiated or quite out of proportion to the evidence, that Darwinism was widely and wantonly abused by the forces of reaction" (Bannister 1977, p. 9). On the urge toward social control and prediction in relation to the demise of Social Darwinism and a more experimental approach to social science, see Cravens (1978).
72. Allen's view of early twentieth-century American morphology has been challenged (Maienschein, Benson, and Rainger 1981; Maienschein 1981; Benson 1981; Rainger 1981; G. E. Allen 1981; Churchill 1981).

31

THE *BEAGLE* COLLECTOR AND HIS COLLECTIONS

Duncan M. Porter

Collections to be made by the naturalist who was to accompany Captain Robert FitzRoy (1805–1865) as a companion on the second surveying voyage of H.M.S. *Beagle* to South America were discussed even before Charles Darwin was recruited. In a note added to his original letter to the Reverend John Stevens Henslow (1796–1861, Professor of Botany in the University of Cambridge), the Reverend George Peacock (1791–1858, Tutor in Mathematics in Trinity College) stated, "What a glorious opportunity this would be for forming collections for our museums" (Darwin 1967, p. 29).[1]

Peacock was approached by his friend Captain Francis Beaufort (1774–1857, Hydrographer to the Navy) to nominate someone as naturalist. He in turn wrote to Henslow, asking him if there might be someone he could recommend strongly for the position. The letters between Henslow, Peacock, and Darwin have been printed several times, as have those between Darwin and his father (Dr. Robert Waring Darwin, 1766–1848, physician of Shrewsbury) and uncle (Josiah Wedgwood, Jr., 1769–1843, potter of Maer Hall), in which Darwin first declined and then accepted the offer to accompany the voyage (for example, *LL,* Darwin 1967).

It is well known that Henslow recommended Darwin to Peacock in August 1831 "as the best qualified person I know of who is likely to undertake such a situation — I state this not on the supposition of yr. being a *finished* Naturalist, but as amply qualified for collecting, observing, & noting anything new to be noted in Natural History" (Darwin 1967, p. 30).[2] Henslow had had plenty of opportunity to observe Darwin's prowess in natural history while he was a student at Cambridge in 1828–1831. During this time, Darwin came more and more under the influence of Henslow. Their relationship has been well documented by Barlow (in Darwin 1967) and Allan (1977).

I. The Young Collector

One of Charles Darwin's recollections of his childhood shows that by the time he began to attend the Reverend George Case's day-school in Shrewsbury at age eight he had already begun to follow the path toward natural history.

> By the time I went to this day-school my taste for natural history, and more especially for collecting, was well developed. I tried to make out the names of plants, and collected all sorts of things, shells, seals, franks, coins, and minerals. The passion for collecting, which leads a man to be a systematic naturalist, a virtuoso or a miser, was very strong in me, and was clearly innate, as none of my sisters or brother ever had this taste. (*Autobiography*, pp. 22–23)[3]

Lucky we are that Darwin kept up this collecting throughout his life. If he had not collected and saved specimens, notes, manuscripts, letters, and annotated books, our understanding of his life and motivations would be close to nil. Darwin scholars, sifting through his accumulated mass of collected materials at Cambridge, Downe, and elsewhere, are doing much to illuminate the processes by which this extraordinary man accomplished his works and to set right many of the myths surrounding him that have accumulated over the years in the writings of the uninformed.

Darwin's autobiography, written in 1876, with later additions, contains several recollections of collecting animals, plants, and minerals while he was a youth in Shrewsbury. That his interests were not purely those of a collector is revealed by such entries as this one:

> From reading White's *Selborne* I took much pleasure in watching the habits of birds, and even made notes on the subject. In my simplicity I remember wondering why every gentleman did not become an ornithologist. (*Autobiography*, p. 45)

Interest in natural history continued for Darwin when he entered Edinburgh University as a medical student in 1825, joining his older brother Erasmus Alvey Darwin (1804–1881). One of those in Edinburgh who took a fancy to young Charles, certainly because of mutual interests, was the zoologist Dr. Robert Edmond Grant (1793–1874). Darwin stated that he often accompanied Grant

> to collect animals in the tidal pools, which I dissected as well as I could. I also became friends with some of the Newhaven fishermen, and sometimes accompanied them when they trawled for oysters, and thus got many specimens. But from not having had any regular practice in dissection, and from possessing only a wretched microscope my attempts were very poor. (*Autobiography*, pp. 49–50)

In spite of this self-effacing comment, Darwin was proficient enough with his "wretched microscope" to discover that the supposed "ova" of the bryozoan *Flustra carbasea* were in fact its larvae and that the "ova" of the brown alga *Fucus loreus* were the egg cases of the leech *Pontobdella muricata*. These observations were reported by him to the Plinian Natural History

Society in March 1827. They and others occur in a notebook (DAR 118) that was published in part by Ashworth (1935) and in full by Barrett (*CP* 2:285–291). The notes and their accompanying drawings are strikingly similar to those he was to make in his Zoology Diary (DAR 30, 31) while on the *Beagle*.

In 1827 Darwin left Edinburgh, switching careers from physician to clergyman, and entered Christ's College, University of Cambridge. Natural history continued to play a role in his life, indeed a greater role than before. Whereas at Edinburgh he had been befriended by Dr. Grant and the naturalist William Macgillivray (1796–1852), who gave him some rare marine mollusk shells, he was repelled by most of the learned Scottish professors.

On the other hand, at Cambridge his interests were fostered at every turn by such eminent scientists as Henslow, the Reverend William Whewell (1794–1866, Professor of Mineralogy and later Master of Trinity College), and the Reverend Adam Sedgwick (1785–1873, Woodwardian Professor of Geology). Henslow was the greatest influence on the young Darwin, beginning as his tutor and soon becoming his mentor.

In 1822, during Erasmus's first year as an undergraduate student at Cambridge, he wrote to thirteen-year-old Charles his good impressions of professors Henslow and Sedgwick (Brent 1981, p. 31–32). Following his own arrival in Cambridge, Charles was befriended by these scientific worthies, leading members of what W. Cannon (1964d, 1978) termed the "Cambridge Network". Their influence on and in behalf of Darwin extended far beyond the *Beagle* voyage.

Knowledgeable in both botany and entomology, Henslow encouraged his students to collect and observe. He previously had been Professor of Mineralogy and was later to interest Darwin in geology as well. This in spite of Robert Jameson's (1774–1854, Professor of Natural History at Edinburgh) dry lectures having previously killed any earlier interest in the field.

But Darwin's passion during most of his time at Cambridge was entomology: "no pursuit at Cambridge was followed with nearly so much eagerness or gave me so much pleasure as collecting beetles" (*Autobiography*, p. 62). His autobiographical recollections for this time are full of entomological comments. "No poet ever felt more delight at seeing his first poem published than I did at seeing in Stephen's *Illustrations of British Insects* the magic words, 'Captured by C. Darwin, Esq.' " (p. 63).

Darwin met James Francis Stephens (1792–1852) in 1829; "his cabinet is more magnificent than the most zealous entomologist could dream of" (*LL* (NY) 1: 150–151). Another well-known entomologist who befriended him was the Reverend Frederick William Hope (1797–1862): "his collection is most magnificent, and he himself is the most generous of entomologists; he has given me about 160 new species, and actually wanted to give me the rarest insects of which he had only two specimens" (*LL* (NY) 1: 150).

Both these comments are from a letter of 26 February 1829 to William Darwin Fox (1805–1860), Charles's second cousin, also a student at Christ's College (and later Vicar of Delamere, Cheshire), who introduced him to beetle trapping.

Some of the others who engaged in this activity with Darwin were John Maurice Herbert (1808–1882, later a County Court Judge in Wales) of St. John's College; Harry Stephen Meysey Thompson (1809–1874, later a Member of Parliament and Baronet); and Albert Way (1805–1874, to become a well-known antiquary), both of Trinity College. These relationships led Darwin in old age to write, "It seems therefore that a taste for collecting beetles is some indication of future success in life!" (*Autobiography*, p. 63).

Like his cousin Fox, Darwin also proselytized for the field. In March 1830 he wrote to Fox, "I have two very promising pupils in Entomology, and we will make regular campaigns into the Fens. Heaven protect the beetles and Mr. Jenyns, for we won't leave him a pair in the whole country" (*LL* (NY) 1: 156). The Reverend Leonard Jenyns (1800–1893, Vicar of Swaffham Bulbeck, Henslow's brother-in-law, and friend of Darwin) was adding much to the knowledge of the natural history of Cambridgeshire.

Like many serious naturalists of the time, Darwin invested in a cabinet in which to display his entomological finds. In his letter of March 1830 to Fox, he also wrote: "My new cabinet is come down, and a gay little affair it is" (*LL* (NY) 1: 156). Thirteen years later (28 March 1843), he wrote again to Fox: "I was looking over my arranged cabinet (the only remnant I have preserved of all my English insects), and was admiring *Panagaeus Crux-major* [a carabid beetle]: it is curious the vivid manner in which this insect calls up in mind your appearance, with little Fan [Fox's dog] trotting after, when I was first introduced to you" (*LL* (NY) 1: 291).

Darwin was not only influenced by others while at Cambridge, he had a certain amount of influence himself. After his death, some of his university contemporaries wrote reminiscences of Charles Darwin to his son Francis Darwin (1848–1925, Reader in Botany at Cambridge). Several mentioned beetles, as might be imagined, but the Reverend Thomas Butler (1806–1886, late Canon of Lincoln) wrote that, "He inoculated me with a taste for Botany which has stuck by me all my life" (*LL* (NY) 1: 144).

Except for his remark that he attended Henslow's botany lectures, there is little in Darwin's autobiography of this time on plants, but there is plenty on beetles. Of Reverend Frederick Watkins (1808–1888, Archdeacon of York), Francis Darwin wrote:

> another old college friend of my father's, remembers him unearthing beetles in the willows between Cambridge and Grantchester, and speaks of a certain beetle the rememberance of whose name is 'Crux major.' [*Panagaeus Crux-major*] How enthusiastically must my father have exulted over this beetle to have impressed its name on a companion so that he remembers it after half a century. (*LL* (NY) 1: 144)

This enthusiasm runs through Darwin's letters to Fox and others during these Cambridge days.

Fired with reading the German naturalist Baron Friedrich Heinrich Alexander von Humboldt's (1769–1859) *Personal Narrative* (von Humboldt 1814–1829), Darwin hatched a plan to visit the Canary Islands with some of his entomologizing friends. He went so far as to begin to learn Spanish in preparation for the trip, which was planned for June 1832. Such preparation stood him in good stead when he visited South America on the *Beagle*. Henslow was involved with this scheme, which is another reason why he immediately thought of Darwin when approached by Peacock to recommend someone as the *Beagle's* naturalist.

Many people have the impression that Darwin was prepared for the *Beagle* voyage only through his geologizing in North Wales with Professor Sedgwick in the late Summer of 1831. Certainly he was filled with enthusiasm for this newly acquired interest, and doubly so after reading the first volume of Charles Lyell's (1797–1875, Professor of Geology at King's College, London) *Principles of Geology* (Lyell 1830), given to him on departure by Captain FitzRoy. I have, however, attempted to show in the foregoing discussion that, in spite of his later protestations to the contrary, Charles Darwin was as prepared to be the naturalist of the *Beagle* as any contemporary university graduate could have been.

In his first letter to Darwin discussing the possibility of Charles's accompanying the *Beagle* (24 August 1831), Henslow added, "Capt. F. wants a man (I understand) more as a companion than a mere collector & would not take anyone however good a Naturalist who was not recommended to him likewise as a *gentleman*" (Darwin 1967, p. 30). Darwin qualified on both grounds, collector and gentleman. Peacock, on the other hand, appears to have been most concerned with specimens, writing in his first letter to Darwin, "I look forward with great interest to the benefit which our collections of natural history may receive from your labours" (Darwin 1967, p. 31). Thus the importance of his collections was impressed on Darwin from the beginning.

This led Darwin, not surprisingly, to give thought to where his collections might best be deposited upon his return. He wrote to Henslow on 9 September 1831:

> but about my collections. Cap. Beaufort said his first impression was, that they ought to be given to British Museum? but I think I convinced ["him" added in brackets by Barlow] of the impropriety of this & he finished by saying he thought I should have no difficulty so that I presented them to some public body, as Zoological & Geological[4] etc. — But I do not think the Admiralty would approve of my sending them to a Country collection, let it be ever so good. — & really I doubt myself, whether it is not more for the advancement of Nat. Hist. that new things

should be presented to the largest and most central collection. (1967, pp. 39–40).

On the 17th, having visited the *Beagle* in Plymouth, he wrote again to Henslow: "My Cabin is more comfortable than I expected; & my only difficulty is about the disposal of my collection when I come back" (1967, p. 42). As we shall see, this difficulty was still plaguing Darwin on his return from the voyage five years later.

At the same time that Darwin was worrying about which institutions should receive his collections, he was also concerned with to whom they should be sent for safekeeping from the *Beagle*. Apparently after making several inquiries, he wrote to Henslow on 18 October 1831:

> I seize the opportunity of writing to you on the subject of consignment. — I have talked to everybody: & you are my only resourse [sic]; if you will take charge, it will be doing me the greatest kindness. — The land carriage to Cambridge, will be as nothing compared to having some safe place to stow them; & what is more having somebody to see that they are safe. — I suppose plants & Birdsskins are the only things that will give trouble: but I know you will do what is proper for them. . . . About paying for them, I should think the best plan will be, after the arrival of one or two cases, to write to my Father, & he will place the sum to your account at any bank in Cambridge you may choose: — I will write to him on the subject: . . . (1967, p. 43)

So, as was to prove true for all of Charles's other expenses on the voyage, payment for the shipment of his specimens from their points of arrival in England to Cambridge was borne by Dr. Darwin. Sea carriage was on "His Majesty's Service".

Shipments of books and materials in the other direction were arranged by Charles's brother Erasmus, whose good services were mentioned by Darwin in many of his letters home to his sisters. Most of these were printed by Barlow who stated, "In the letters home all the arrangements about purchases and despatching books to Charles, and the plans for the reception of his specimens in England were placed in Erasmus' hands" (Darwin 1945, p. 13). It is true that Erasmus played a role in making sure that specimens arrived where they were intended, but for the most part the intended was Henslow.

After Henslow agreed to receive the specimens, Darwin wrote to him from Devonport on 30 October 1831:

> I am very much obliged for your direction about consingment [sic], — I believe most of the things will first go to Falmouth (where I must get an agent) & then to Cambridge. — I will tell my Father that you will send him a note with an account of what you pay for me. — and I do not think you will find him as careless as I am. — I hope to be

> able to assist the Philosoph. Society[5] when I come back. — but from all I hear, I suppose I shall be in honor bound to give largely to British Museum, — Everything here goes on very prosperously. (1967, p. 46)

In the same letter, Darwin wrote: "What an important Epoch 1831 will be in my life. — taking ones degree & starting for Patagonia are each in their respective ways memorable events" (p. 47). But in his next letter (15 November 1831) he added, "Yet I should not call it one of the very best opportunities for Nat. Hist. that has ever occurred. — the absolute want of room is an evil, that nothing can surmount" (1967, pp. 48–49). Henslow responded on the 20th with these prophetic words: "With a little self denial on your part I am quite satisfied you must reap an abundant harvest of future satisfaction" (Darwin 1967, p. 50).

II. On the *Beagle*

Darwin kept several sets of notebooks while on the voyage, each devoted to certain aspects of his interests. The most general were the fifteen small (from 6½ by 4 inches to 3¾ by 3 inches in size) Pocket Notebooks, which he took into the field in order to record observations, impressions, and thoughts as they came to him. They are often telegraphic in style and also contain such things as lists of purchases to make. The Pocket Notebooks are now at Down House, along with many other mementos of the voyage. Excerpts from them were published by Darwin (1945), and they are currently being edited for publication in full by Dr. Gordon Chancellor of the Oxford University Museum.

Also at Down House are six Specimen Notebooks, identical in size to the largest Pocket Notebooks. These are of two series, three being devoted to dried biological collections, three to those preserved in spirits. They are numbered independently of each other. The Specimen Notebooks are discussed in more detail below.

Four Geological Specimen Notebooks, which enumerate Darwin's mineralogical and paleontological collections, were recently conveyed from the Department of Earth Sciences to the Cambridge University Library by Dr. Sandra Herbert. The Geological Specimen Notebooks originally were deposited in Cambridge's Department of Mineralogy and Petrology by Darwin's sons Francis and George (1845–1912, Professor of Astronomy and Experimental Philosophy at Cambridge) in 1897 (Porter 1982). They are identical to the other Specimen Notebooks, and are numbered in the same series as the dried biological collections. Thus Darwin used two series of numbers for his collections. Dried materials were given colored paper tags with their numbers printed on them, while specimens in spirits were given metal tags with stamped numbers.

Darwin paid close attention to the tagging of specimens, being well aware of the worthlessness of specimens lacking numbers and therefore their locality data. Fearing that William Clift (1775–1849, Conservator of the Royal College of Surgeons Museum) had removed the numbered tags from some of his fossil bones, Darwin wrote his sister Caroline (1800–1888) in August 1834 to ask Erasmus to call on Clift and express his concern. He also wrote Henslow of this concern, but this did not deter the latter from removing most of Darwin's tags from his plant specimens. Near the end of the first edition of the *Journal of Researches*, several pages are devoted to "advice to collectors". These describe Darwin's methods of preparing specimens for shipment, including tagging and the keeping of several series of notebooks.

Darwin also kept a personal diary, which was used as the basis for the *Journal of Researches*. The manuscript, written on 9 by 11 inch paper, is deposited at Down House. It was published unabridged by Barlow (*Diary*). In addition, Darwin kept a Zoology Diary (DAR 30–31), which consists mainly of detailed observations on certain of his animal and plant collections; a Geology Diary (DAR 32–33); and Geology Notes (DAR 34–38). They are mostly on 9 by 11 inch sheets also. The latter were used as the sources for his three books on the geological findings of the voyage, *Coral Reefs* (1842), *Geological Observations on Volcanic Islands* (1844c), and *Geological Observations on South America* (1846a).

As soon as he got over his initial seasickness after leaving England on 27 December 1831, Darwin began to collect and observe. The first entry in the Zoology Diary was made on 6 January 1832, the day that Darwin wrote to his father that he "now first felt even moderately well" (1945, p. 53). It describes a "Luminous Sea" observed in the bay of Santa Cruz, Tenerife, Canary Islands:

> The sea was luminous in specks & in the wake of the vessel, of an uniform slightly milky colour. — When the water was put into a bottle, it gave out sparks for some minutes after having been drawn up. — When examined both at night & next morning. it was found full of numerous small (but many bits visible to naked eye) irregular pieces of (a gelatinous?) matter. — The sea next morning was in the same place equally impure. (DAR 30.1: 1)

He soon put a net over the side and began to collect the marine fauna familiar to him since his days at Edinburgh. In his first letter home, sent to his father from Bahia, Brazil in March 1832 (but begun at sea on 8 February), Darwin described his initial collecting between the Canary and Cape Verde Islands:

> From Teneriffe to St. Jago the voyage was extremely pleasant. — I had a net astern the vessel which caught great numbers of curious animals,

> and fully occupied my time in my cabin, and on deck the weather was so delightful and clear, that the sky and water together made a picture. (1945, p. 53)

In his *Diary* for this initial period of the voyage, remarks on his collections begin to appear, which continue throughout the *Diary* and figure prominently in his letters home to his family and to Henslow. The *Diary* was sent periodically to his family as well. In his entry for 10 January 1832, Darwin wrote:

> I proved to day the utility of a contrivance which will afford me many hours of amusement & work, it is a bag four feet deep, made of bunting, & attached to ["a" added in brackets by Barlow] semicircular bow: this by lines is kept upright, & dragged behind the vessel. This evening it brought up a mass of small animals & tomorrow I look forward to a greater harvest. (*Diary* p. 23)

Darwin's Zoology Diary for 10 January reads, "Lat. 21. Sea very luminous. chiefly from a crustacean animal. which gave a very green light retaining for some time after having been taken out of water" (DAR 30.1: 1). On the back of the page is given the first of many morphological descriptions of the organisms collected.

Darwin wrote his observations on the right-hand pages of the Zoology Diary and added any additional comments on the facing left-hand pages. These comments are keyed to the organisms in question by letters, like footnotes. Here also were placed the specimen numbers of each. He was later to write of these notes in his *Autobiography*:

> Another of my occupations was collecting animals of all classes, briefly describing and roughly dissecting many of the marine ones; but from not being able to draw and from not having sufficient anatomical knowledge a great pile of MS. which I made during the voyage has proved almost useless. I thus lost much time, with the exception of that spent acquiring some knowledge of the Crustaceans, as this was of servitude when in after years I undertook a monograph of the Cirripedia. (pp. 77–78)

Darwin's estimation of the worth of his Zoology Diary would have been higher had he received more encouragement from zoologists regarding the real value of his collections, particularly of the marine invertebrates. His exasperation with their lack of interest in examining his specimens led him to write to Henslow soon after his return from the voyage (30 October 1836), "I only wish I had known the Botanists cared so much for specimens & the Zoologists so little; the proportional number of specimens in the two branches should have ["worn *del*" added in brackets by Barlow] had a very different appearance" (1967, p. 121). This was written while he was desperately searching for specialists to identify the specimens, particularly the invertebrates.

In spite of Darwin's feeling as to the usefulness of his Zoology Diary, the notes are uniformly of high quality. They should be examined by marine invertebrate zoologists especially, who will find much of interest in them, including unpublished observations on a number of organisms. I have examined them in detail and have found the information on plants included in them quite helpful in my studies on Darwin's Plant Notes (Porter 1981, 1982). Darwin himself found some of his observations and descriptions of use, utilizing them for several of the papers he wrote following his return to England (for example, *CP* 1: 177–182, *CP* 1: 182–193).

In his *Diary* for 11 January 1832, Darwin continued:

> I am quite tired having worked all day at the produce of my net. The number of animals that the net collects is very great & fully explains the manner so many animals of a large size live so far from land. Many of these creatures, so low in the scale of nature, are most exquisite in their forms & rich colours. It creates a feeling of wonder that so much beauty should be apparently created for such little purpose. (p. 23)

Darwin returned to the theme of creation several times in his various series of notes made while on the *Beagle*. Certainly it played on his mind during much of the voyage. For example, in writing of the trip up the Rio Santa Cruz in southern Argentina, made in April and May 1834, he recorded in the Zoology Diary:

> I suspect Patagonia has but few productions of its own. — is the Botany sufficiently known. to tell. — The extreme infertility, even close to running water, has often much surprised me. — At different times I have attributed this general sterility, to the salt [to *del*] contained in the sandy clay. — the extreme dryness of the climate, (which is an undoubted fact). — the poorness of the soil of the gravel beds. — & to no creation having taken place. since this country was elevated (I yet think this applies to the Northern parts): I am now most inclined to attribute it all to the poorness of the soil. — Yet in the Lava country, where there was water, it was but little better!. (DAR 31.1: 260, 260 *verso*)

There is no mention of creation in his observations of Galapagos Islands organisms in the Zoology Diary, but in his *Diary* Darwin entered the following for 26–27 September 1835 while on James Island:

> I industriously collected all the animals, plants, insects & reptiles from this island. It will be very interesting to find from future comparison to what district or 'centre of creation' the organized beings of this archipelago must be attached. (p. 337)

Thus began one of the lines of evidence that led Darwin to evolution by natural selection.

That this line of evidence was not pursued until the end of the voyage or after return to England, however, is perhaps shown by his remark in a letter to Henslow, sent from the island of St. Helena on 9 July 1836, four months before his return: "It seems strange, that this little centre of a distinct creation should, as is asserted, bear marks of recent elevation" (1967, p. 115). Had Darwin already become a believer in evolutionary change, there would have been no reason to make this statement. Indeed, Herbert's analysis (1980) of the pocket notebook known as the *Red Notebook*, which she dates as begun in late May 1836, shows that Darwin did not mention what he came to call "transmutation" until soon after the voyage was over: "if one species does change into another it must be per saltum" (*RN* 130). This statement appears to have been written in March 1837 (cf. Sulloway 1982c).

To return to the beginning of the voyage, Darwin also began to collect geological specimens before the first landfall was made by the *Beagle*. The first was dust that fell on the ship while it was sailing to the Cape Verde Islands. An entry in the Zoology Diary for 16 January 1832 begins: "At 8 oclock this morning. the vane was taken down from the mast head & found on the under side to be covered with a very impalpable soft yellow-brown dust" (DAR 30.1: 3). This dust was the subject of one of Darwin's later papers resulting from the voyage (*CP* 1: 199–203, 1846).

The first entry in the Geological Specimen Notebooks, however, is for specimens number 12 through 15, collected in the Cape Verde Islands: "Jan. 17 1832 St Jago Quail Isld 12 The following specimens were collected at Quail Island. Jan. 17th near Porto Praya. St. Jago. Feldspathic rocks forming a horizontal cap for island 13 do [ditto] (aluminium smell) 14. do 15 do" (DAR deposit, MS p. 1). All are marked in the margin "Poor Specimens," and each number has been changed from the one higher, that is the series was originally 13 through 16.

In May 1832, Darwin wrote his first letter to Henslow, from Rio de Janeiro. In it, he reported:

> In the one thing collecting, I cannot go wrong. — St Jago is singularly barren & produces few plants or insects. — so that my hammer was my usual companion & in its company most delightful hours I spent. — On the coast I collected many marine animals chiefly gasteropodous (I think ["many *del*" in brackets by Barlow] some new). (1967, p. 53)

Thus, although his scientific background was mainly in zoology, Darwin's interest in geology, newly awakened by Sedgwick and Lyell, was to profoundly affect him on the voyage from its beginning. In the same letter to Henslow, he added prophetically: "Geology & the invertebrate animals will be my chief object of pursuit through the whole voyage" (1967,

p. 54). A look through the Zoology Diary, Geology Diary, and Geology Notes shows that this became true (see Sulloway, Sloan, this volume).

The first to point out the relative importance of geology, zoology, and botany to Darwin on the voyage were Gruber and Gruber (1962). They showed that he wrote 1383 pages of geological notes, 368 pages of notes on zoology, and no separate series of plant notes, although they felt that the latter may not have yet been discovered. This does not include the Specimen Notebooks, the Geological Specimen notebooks, the Pocket Notebooks, or the 779 pages of his personal *Diary*. A close examination of the Zoology Diary, however, reveals that about 20 percent of the pages are devoted to notes on his plant specimens or to general notes on vegetation, so that geology predominates over zoology to an even larger extent than Gruber and Gruber reported. Except for 1832, as the voyage progressed Darwin wrote from twice to 11½ times as many pages of geological notes per year as biological notes.

It is no wonder that most of the papers and books authored by Darwin as a result of his *Beagle* experience were geological in nature. Most of the uninformed today think of Darwin as a biologist only, certainly because of what he published in the *Origin* and later. It is useful, however, for biologists to remember that the only time he referred to himself in print as a scientist, it was as a geologist (Darwin 1855). As he wrote to his sister Catherine (Emily Catherine Darwin 1810–1866) on 6 April 1834: "There is nothing like Geology; the pleasure of the first day's partridge shooting or the first day's hunting cannot be compared to finding a fine group of fossil bones, which tell their story of former times with almost a living tongue" (1945, p. 96). A surprising admission from one who before the voyage had counted partridge shooting among life's primary pleasures (*Autobiography*, p. 71).

Darwin's various *Beagle* notes and letters contain many observations on collecting and his collections. Those given above must suffice for the present paper. To do them complete justice would require either a much longer paper or a book.

Although Darwin was the *Beagle's* official naturalist, he was not the only one to collect specimens on the voyage. In mid-1833, Syms Covington (ca. 1816–1861), "Fiddler and Boy to the Poop Cabin", became his servant. As Charles wrote to his sister Catherine (22 May and 6 July 1833, both in the same letter):[6] "I have taught him to shoot & skin birds, so that in my main object he is very useful. . . . I shall now make a fine collection in birds & quadrupeds, which before took up far too much time" (1945, pp. 85, 88). The evidence available indicates that Covington collected only vertebrates, while Darwin attended to geology, invertebrates, and plants.

Covington's collections were entered into the Specimen Notebooks under Darwin's collecting numbers, for the most part. But Sulloway (1982b) shows that at least six Covington bird specimens were not. Others known to have collected birds were Captain FitzRoy, Harry Fuller (dates unknown, FitzRoy's personal steward), Benjamin Bynoe (ca. 1804–1865, Acting Surgeon), and Edward Hellyer (?-1833, Ship's Clerk), who drowned in the Falkland Islands in March 1833 attempting to retrieve a bird he had shot. Their specimens are discussed by Sulloway (1982b).

FitzRoy is also known to have collected other vertebrates, and perhaps Fuller, Bynoe, and Hellyer collected them on his behalf as well, as they did birds. The only other crew member who is known to have collected biological specimens was the Second Lieutenant, Bartholomew James Sulivan (1810–1890), who gathered plants on the voyage. These were not given to FitzRoy or to Darwin, however. Sulivan sent them to his father, who in turn sent them for identification to Dr. John Lindley (1799–1865, Professor of Botany at University College, London) (Sulivan 1896).

There is one other piece of evidence that indicates others may have given specimens to Darwin. After the return of the *Beagle*, Darwin sent to FitzRoy the introduction to what in 1839 became his *Journal of Researches*. Part of FitzRoy's reply to this draft reads:

> I was also astonished at the total omission of any notice of the officers, either particular or general. My memory is rather tenacious respecting a variety of transactions in which you were concerned with them and others in the *Beagle*. Perhaps you are not aware that the ship which carried us safely was first employed in exploring and surveying, whose officers were not ordered to collect and were therefore at liberty to keep the best of all, nay all, their specimens for themselves. To their honour, they gave you the preference. (Stanbury 1977, p. 20)

Darwin's acknowledgement to the crew of the *Beagle*, however, reads only, "Both to Captain FitzRoy and to all the Officers of the Beagle, I shall ever feel most thankful for the undeviating kindness with which I was treated, during our long voyage" (*Journal of Researches*, pp. vii–viii).

This is rather surprising, since a reading of the Geological Specimen Notebooks shows that the above members of the *Beagle's* crew were not the only ones to favor Darwin with their specimens. Geological specimens were obtained from Covington and FitzRoy, as might be expected, and also from Edward Main Chaffers (Master of the *Beagle*), William Kent (Assistant Surgeon), John Lort Stokes (1812–1885, Mate and Assistant Surveyor), and John Clements Wickham (1798–1864, First Lieutenant). It is no wonder that FitzRoy was provoked with his old shipmate. In addition, specimens were received from John Augustus Lloyd (1800–1884, Surveyor General of Mauritius), from a Mr. Fox in Argentina, and a Dr. Smith at the Cape of Good Hope. All the foregoing were entered into Darwin's series of collecting numbers, a common practice then and now.

Darwin had a limited amount of space available to him on the *Beagle* for storing his collections. In the 1880s, Sulivan (now Admiral Sir James

Sulivan) wrote to Francis Darwin that, "For specimens he had a very small cabin under the forecastle" (*LL* (NY) 1: 192). Probably for this reason, he shipped them back to Henslow whenever a sizeable number had accumulated. By the time the *Beagle* left South America in early September 1835, Darwin had forwarded eight consignments of specimens to Henslow in Cambridge. The following record of shipments was gleaned from Darwin's letters to Henslow and to his family:

August 1832: a box from Monte Video, Uruguay to Falmouth via H.M.S. *Emulous*.
November 1832: three casks and a box from Monte Video to Falmouth via H.M.S. *Duke of York*.
July 1833: four barrels from Rio de la Plata, Argentina via an unnamed packet.
November 1833: two boxes and one cask from Monte Video to Portsmouth via H.M.S. *Samarang*.

Also a box of fossil bones was sent to Dr. Robert Armstrong (dates unknown), Physician at the Royal Naval Hospital in Plymouth and Inspector of Fleets. As Darwin wrote to Henslow, "I do this to avoid the long land-carriage [that is, Portsmouth to Cambridge, to be paid by his father — now the Navy would bear the expense!]: & as they do not want any care it does not much signify where kept" (1967, p. 81).

March 1834: seeds in a letter from the Falkland Islands to Captain Beaufort in London.
May 1834: a box from Buenos Aires, Argentina to Liverpool via H.M.S. *Basenthwaite*. This was forwarded by Edward Lumb (dates unknown), an English merchant resident in Argentina.
January 1835: two boxes from Valparaiso, Chile via H.M.S. *Challenger*.
June 1835: two boxes from Valparaiso via H.M.S. *Conway*.

The boxes, casks, and barrels presumably were constructed by Jonathan May, the *Beagle's* carpenter, and James Lester, her cooper.

Upon arrival in the Galapagos Islands in mid-September, Darwin again began to collect with fervor. These specimens remained on the *Beagle*, however, until she returned to England in October 1836, as did all others collected following leave of Valparaiso. Conditions were crowded, and few collections were made save in the Galapagos and Cocos-Keeling Islands and King George Sound, Australia. As Darwin wrote to Henslow in January 1836 from Sydney:

> In our passage across the Pacifick, we only touched at Tahiti & New Zealand: at neither of these places, or at sea had I much opportunity of working. . . . During the remainder of our voyage, we shall only visit places generally acknowledged as civilized & nearly all under the

> British Flag. There will be a poor field for Nat: History & without it, I have lately discovered that the pleasure of seeing new places is as nothing. (1967, p. 114)

Thus the emphasis on geological observation and reduced specimen collecting on the last leg of the voyage is not surprising.

Much of the latter part of the voyage was spent by Darwin poring over his geological notes. He wrote to Caroline on 29 April 1836 from the Indian Ocean island of Mauritius, "Whilst we are at sea & the weather is fine, my time passes smoothly because I am very busy. My occupation consists in rearranging old geological notes: the rearranging generally consists in totally rewriting them" (1945, p. 138). Presumably, much of this was concerned with coral reefs, first examined closely earlier that month in the Cocos-Keeling Islands.

While Darwin was thus engaged, Covington was preparing lists of specimens for the taxonomists whom Darwin hoped might identify them. The dating of these lists is discussed by Sulloway (1982b). Darwin took the Specimen Notebooks and marked them in pencil (the original entries are in ink) for Covington. Inside the first page of the second Specimen Notebook is a column in pencil that reads:

a animal [that is, mammal]
B bird
I insect
S Shell
P Plant

In the Catalogue for Animals in Spirits of Wine, he added R (reptiles and amphibians), C (crustaceans), and F (fishes). These notations in pencil were added for all collections in the six Specimen Notebooks. In addition, Darwin added an occasional "Copy" in pencil in the margin, or a reference to a specific page in the Zoology Diary where the collection in question is described or is discussed in more detail. Covington and Darwin then sat down with the Specimen Notebooks and the Zoology Diary and produced the lists now known as the Ornithological Notes (Darwin 1963), Plant Notes (Porter 1981), and Insect Notes (Porter 1983b).

There appear to be thirteen of these lists in all. In the Cambridge University Library are found lists titled "Animals" (that is, mammals, DAR 29.1: 32 pp; only about half of the first page is in Covington's handwriting, the rest is by Darwin); "Fish in Spirits of Wine" (DAR 29.1: 20 pp.); "Shells in Spirits of Wine" (DAR 29.1: 8 pp.); "Birds" (the Ornithological Notes, DAR 29.2: 85 pp.; the first page is by Covington, the following by Darwin); "Shells" (DAR 29.3: 8 pp.); "Insects in Spirits of Wine" (DAR 29.3: 1 p.; in Darwin's handwriting); "Mammalia in Spirits of Wine" (DAR 29.3: 1 p.; in Darwin's hand); "Birds &c &c in Spirits of Wine" (DAR 29.3:

1 p.; also in Darwin's hand); and "Plants" (the Plant Notes, DAR deposit, 11 pp.).

Two lists are found in the British Museum (Natural History). "Reptiles in Spirits of Wine" (26 pp.; most in Covington's handwriting, but most of pp. 4, 5, and 6 by Darwin) is in the General Library. While "Insects" (the Insect Notes, 26 pp.; in Covington's hand) is in the Library of the Department of Entomology. Also in the British Museum (Natural History) is a one-page list of Darwin's corals collected in the Cocos-Keeling Islands. Unlike the other lists, it does not enumerate all specimens of its group collected on the voyage, so it is not counted as one of the thirteen. It is in Darwin's handwriting, and like many notes in DAR 29 may have been made for his own use.

Three separate pages at the Cambridge University Library actually are parts of two of the lists in DAR 29. They are in Covington's handwriting. "Diodon" (DAR 29.1: 49) is part of the Animal Notes, while "Insecta. June." and "Pediculus. Chiloe. July" (DAR 29.3: 2 pp.) are part of the Insects in Spirits of Wine Notes.

Still missing are the lists for Plants in Spirits of Wine and Crustaceans in Spirits of Wine. For reasons explained in the next section of this paper, the former probably is at Trinity College, Dublin, and the latter perhaps at the Oxford University Museum.

III. After the *Beagle*

Reports of Darwin's scientific findings on the *Beagle* began to appear even before his return in October 1836. The first of these was the exhibition of the head of a giant fossil ground sloth he had collected in Argentina in 1832, at the British Association for the Advancement of Science meeting at Cambridge in the Summer of 1833.

Geological and zoological comments from his letters were read by Henslow to a meeting of the Cambridge Philosophical Society on 16 November 1835, and published the next month (*CP* 1: 3–16, 1835). On 18 November, Sedgwick read some geological extracts from the same letters to a meeting of the Geological Society of London (*CP* 1: 16–19, 1838).[7] On 14 December 1835, Henslow read a "Communication on viviparous lizards, and on red snow" to the Cambridge Philosophical Society. This was not published separately, but the information was included in the *Journal of Researches* (pp. 394–395). Further notice of Darwin's geological studies in South America was made by Charles Lyell in his presidential address to the Geological Society of London on 19 February 1836 (Lyell 1837).

Following his return, Darwin and others made a number of presentations to scientific bodies on his collections and observations. Those of which I am aware follow:

2 January 1837: Entomological Society of London; George R. Waterhouse (1810–1888, Honorary Curator of the Society and Curator of the Zoological Society of London) read a paper on some of Darwin's beetles from New South Wales (Waterhouse 1838b).

4 January 1837: Geological Society of London; Darwin read a paper on the elevation of the coast of Chile (*CP* 1: 41–43, 1838).

10 January 1837: Zoological Society of London; papers on Darwin's collections were read by William Charles Linnaeus Martin (1798–1864, Superintendent of the Society's Museum) on South American wildcats (Martin 1837a), by James Reid on an opossum and viscacha (Reid 1837), and by John Gould (1804–1881, Taxidermist to the Society) on Galapagos Islands finches (Gould 1837a).

24 January 1837: Zoological Society; papers on raptorial birds by Gould (1837b), the Chiloe Island fox by Martin (1837b), and a new opossum by Martin (1837c).

14 February 1837: Zoological Society; papers by Waterhouse on new species of mice (Waterhouse 1837a) and by Gould on the fissirostral birds (Gould 1837c).

27 February 1837: Cambridge Philosophical Society; a paper was read by Darwin on fused sand tubes caused by lightning (published in *Journal of Researches*).

28 February 1837: Zoological Society; papers by Gould on Australian and Galapagos Islands birds (Gould 1837d) and by Waterhouse on the small rodents (Waterhouse 1837a, 1837b).

14 March 1837: Zoological Society; papers by Gould on Darwin's rhea (Gould 1837e) and by Darwin on Patagonian rheas (*CP* 1: 38–40, 1837).

3 April 1837: Entomological Society; Darwin exhibited specimens of southern South American carabid beetles, later published on by Hope (1838).

19 April 1837: Geological Society; a paper was read by Richard Owen (1804–1892, Conservator and Hunterian Professor, Royal College of Surgeons) on the fossil South American mammal *Toxodon* (Owen 1838a, 1838b).

1 May 1837: Entomological Society; a paper by Hope on Darwin's carabid beetles (Hope 1838).

3 May 1837: Geological Society; Darwin read a paper on fossil mammals from Argentina (*CP* 1: 44–45; 1837, 1838).

10 May 1837: Zoological Society; Darwin exhibited and remarked on his Galapagos Islands finches (*CP* 1: 40, 1837).

31 May 1837: Geological Society; Darwin read a paper on elevation, subsidence, and coral reefs (*CP* 1: 46–49, 1838c).

25 July 1837: Zoological Society; an exhibition and talk on more of Darwin's birds by Gould (1837f).

9 January 1838: Zoological Society; a continuation of the exhibition and talk by Gould (1838).
16 February 1838: Geological Society; in his presidential address to the Society, Whewell appraised Darwin's geological work while on the *Beagle* (Whewell 1839).
28 February 1838: Zoological Society; a paper on FitzRoy's dolphin by Waterhouse (1838a).
7 March 1838: Geological Society; Darwin read a paper on volcanoes and earthquakes in South America (*CP* 1: 53–86, 1840).
12 March 1838: Cambridge Philosophical Society; a paper on the plants from the Cocos-Keeling Islands was read by Henslow (1838).
16 March 1841: Linnean Society of London; Rev. Miles Joseph Berkeley (1803–1879, Curate of Apthorne and Wood Newton) read a paper on an edible fungus from southern South America (Berkeley 1845).
14 April 1841: Geological Society; a paper on the distribution of erratic boulders in South America was read (*CP* 1: 145–163, 1842).
14 December 1841: Zoological Society; a paper on Darwin's beetles from southern South America by Waterhouse (1841c).
4 March, 6 May and 18 December 1845; Linnean Society; Dr. Joseph Dalton Hooker (1817–1911, Assistant at the Royal Botanic Gardens, Kew) read installments of a paper on Darwin's Galapagos Islands plant collections (J. Hooker 1847a).
4 June 1845: Geological Society; a paper was read on volcanic dust (*CP* 1: 199–203, 1846).
1 and 15 December 1846: Linnean Society; Hooker read installments of a paper on the relationships of the plants of the Galapagos Islands, based primarily on Darwin's specimens (J. Hooker 1847b).

When Charles Darwin returned home to England on 2 October 1836, he left the *Beagle* in Falmouth as soon as possible, making for Shrewsbury, where he arrived on the 4th to be reunited with his family after a hiatus of over five years. He was, however, soon (6 October) writing to Henslow that "it will be necessary in four or five days to return to London to get my goods & chattels out of the Beagle" (1967, p. 117). These, of course, included his post-Valparaiso specimens.

The *Beagle* did not reach Greenwich until 28 October, however, and Darwin visited the ship on the 29th. He spent the interim in Shrewsbury, Maer, Cambridge, and London, visiting family and friends and making new friends among the scientific community of London. On the 29th he retrieved his Galapagos Islands plants, which greatly interested him and about which he queried Henslow until the publication of the *Journal of Researches*. Unfortunately, Henslow was unable to answer most of his questions, and the first edition contains much less information on plants than Darwin wished to include (Porter 1980a).

The Galapagos plants were boxed, and they and four other boxes of specimens were shipped to Henslow at Cambridge, to join those sent from South America. On 13 December 1836, after several days at the Henslows, Darwin and Covington moved into a row house on Fitzwilliam Street in Cambridge and began to sort through the five years' worth of collections and separate them for identification. As we shall see, some experts were quite interested in examining the specimens of certain groups, some were not, and some who were did not.

Darwin wrote to his Cambridge friend Charles Whitley (1808–1895) on 24 October 1836, "I am at present at an utter loss to know how to begin the arrangements of specimens and observations collected during the five long years. All I know is, that I must work far harder than poor shoulders have ever been accustomed to do" (De Beer 1958d, p. 111). He first met Charles Lyell about this time, certainly after this letter to Whitley was penned.[8] Lyell took him under his wing and advised Darwin as to who might be interested in identifying certain of his collections.

Following a tea party at the Lyells' on 29 October, Darwin on the 30th sent his mentor Henslow a long, chatty letter from London full of plans for the future. It mentioned the names of a number of scientific worthies whom Darwin had talked to regarding the identification of his specimens, or whom he planned to talk to. He also questioned Henslow about several of them. Darwin's comments and queries to Henslow in this and a few subsequent letters are quoted below in the sections devoted to the various groups collected.

In this letter, Darwin expressed some anxiety, also felt before the *Beagle* set sail, as to where the specimens should be deposited:

> I see it is quite unreasonable to hope for a minute, that any man will undertake the examination of an whole order. — It is clear the collectors so much outnumber the real naturalists, the latter have no time to spare. — I do not even find that the collections care for receiving the unnamed specimens. — The Zoological Museum is nearly full & upward of a thousand specimens remain unmounted. I daresay the British Museum would receive them, but I cannot feel, from all I hear, any great respect even for the present state of that establishment.

In spite of this, the British Museum and the Museum of the Zoological Society of London became two of the primary depositories for his animals. He continued:

> Your plan will be not only the best, but the only one, namely to come down to Cambridge, arrange & group together the different families & then wait till ["any one *del*" added in brackets by Barlow] people, who are already working in different branches may want specimens. — But it appears to me, to do this, it will be almost necessary to reside

> in London. — As far as I can yet see, my best plan will be to spend ["some *del*" added in brackets by Barlow] several months in Cambridge, & then, when by your assistance, I know on what grounds I stand, to emigrate to London, when I can complete my geology, & try *to push on the Zoology*. (1967, p. 119)

This, in fact, is how it happened.

Darwin and Covington remained in Cambridge for three months, sorting the collections into the various groups represented by the lists of specimens they had prepared. Also, "I began preparing my Journal of travels, which was not hard work, as my MS. Journal had been written with care, and my chief labour was making an abstract of my more interesting scientific results" (*Autobiography*, p. 83). Much of his interest in the identification of specific collections at this time was directed toward including them in the *Journal of Researches*.

In March 1837, after the *Beagle* collections had been put in order, Darwin and Covington moved to London, where Darwin remained until moving to Downe in 1842, Covington leaving his service in 1839. Darwin continued preparing his *Journal of Researches*, completing the manuscript in the fall of 1837. In March he wrote to his cousin Fox:

> In your last letter you urge me to get ready *the* book. I am now hard at work and give up everything else for it. Our plan is as follows: Captain Fitz-Roy writes two volumes out of the materials collected during the last voyage under Capt. King to Tierra del Fuego, and during our circumnavigation. I am to have the third volume, in which I intend giving a kind of journal of a naturalist, not following, however, always the order of time, but rather the order of position.[9] The habits of animals will occupy a large portion, sketches of the geology, the appearance of the country, and personal details will make the hodge-podge complete. Afterwards I shall write an account of the geology in detail, and draw up some zoological papers. So that I have plenty of work for the next year or two, and till that is finished I will have no holidays. (*LL* (NY) 1: 250)

Here was a man who knew what he wanted to accomplish and how he was going to accomplish it. Besides writing his *Journal of Researches*, excellently outlined above, Darwin in 1837 worked on several geological papers and began in earnest to gather information on the transmutation of species. There has been a tendency on the parts of some contemporary writers to paint Darwin at this time as an ailing recluse, too ill to attend to his science except between bouts of sickness. Eiseley, for example, wrote:

> When Darwin reached home after the voyage of the *Beagle*, he was an ailing man, and he remained so to the end of his life. . . . For twenty-two years after the *Beagle's* return he published not one word beyond

> the bare journal of his trip (later titled *A Naturalist's Voyage around the World*) and technical monographs on his observations. (1979, pp. 9, 10)

Both of these comments shade the truth. According to his journal (Darwin 1959a), Darwin's first bout of illness came upon him in May 1838, a year and a half after his return. It is well known that he suffered periodic bouts of illness throughout the rest of his life, yet he accomplished a great deal in spite of them. Even in later years, when he is popularly conceived to have spent much of his time bed-ridden, he spent significant periods of time away from home. Between 1845 and 1854, he spent sixty weeks away from Downe, an average of five weeks per year. Atkins (1974) lists these and later trips, many made for reasons of health, it is true; but the majority can be considered family vacations, and a few were made for scientific purposes.

Eiseley's second comment is far more damning to Eiseley himself than to Darwin. The years between 1837 and 1859 were the most productive of his life for Darwin. He published or edited thirteen books (not counting second editions) and published fifty-one scientific papers and notes and one book review during this time. In spite of recent speculation to the contrary, the four books and twelve papers published during his first five years back from the voyage certainly would earn Darwin tenure in any university in the United States today.

Even while preparing his *Journal of Researches*, Darwin was giving thought to the publication of the zoological results of the voyage. On 10 April 1837 he wrote to Leonard Jenyns from London,

> During the last week several of the zoologists of this place have been urging me to consider the possibility of publishing the "Zoology of the *Beagle's* Voyage" on some uniform plan. Mr Macleay[10] has taken a great deal of interest in the subject, and maintains that such a publication is very desirable, because it keeps together a series of observations made respecting animals inhabiting the same part of the world, and allows any future traveller taking them with him. How far this facility of reference is of any consequence I am very doubtful; but if such is the case, it would be more satisfactory to myself to see the gleanings of my hands, after having passed through the brains of other naturalists, collected together in one work. . . . I apprehend the whole will be impracticable, without Government will aid in engraving the plates, and this I fear is a mere chance, only I think I can put in a strong claim, and get myself well backed by the naturalists of this place, who nearly all take a good deal of interest in my collections. (*LL* (NY) 1: 252–253)

With the aid of such naturalists as Edward Adolphus Seymour (1775–1855, Duke of Somerset and President of the Linnean Society), Edward Smith Stanley (1775–1851, Earl of Derby and past-President of the Linnean Society), and William Whewell (President of the Geological Society), in August 1837 a grant of £1000 was made by the Treasury to publish the *Zoology*.

Early in 1838, Darwin began to play a role in the administration of two of the scientific societies to which he had been elected. In spite of protestations to Henslow that he was poorly qualified and that it would take up too much of his time, he was appointed Secretary of the Geological Society, serving from 16 February 1838 through 19 February 1841. It is not generally known that Darwin was also elected to the council of the Entomological Society on 22 January 1838, and on 5 February was appointed to act as one of the four Vice Presidents for the year (Anonymous 1840). Thus Darwin soon became part of the scientific establishment.

IV. The Collections

In this section, I have attempted to chronicle the subsequent histories of the various groups of materials collected by Darwin on the *Beagle* and the publications that resulted. Only those papers that deal primarily with Darwin specimens are noted, however; those in which a specimen is cited only in passing generally are not. Thus the nineteenth-century literature is particularly well represented, while that of the twentieth century is included only if truly relevant. It is also evident that much remains to be discovered regarding where the specimens are and how they arrived.

GEOLOGY

Charles Lyell and William Lonsdale (1794–1871, Curator and Librarian of the Geological Society) provided Darwin with his entrée into the London scientific establishment. As he wrote to Henslow on 30 October 1836, "If I was not much more inclined for geology, than the other branches of Natural History, I am sure Mr. Lyell's & Lonsdale ["*s*" added in brackets by Barlow] kindness ought to fix me" (1967, p. 122). In July 1837 he wrote to Fox:

> I have read some short papers to the Geological Society, and they were favourably received by the great guns, and this gives me much confidence, and I hope not a very great deal of vanity, though I confess I feel too often like a peacock admiring his tail. I never expected that my Geology would ever have been worth the consideration of such men as Lyell, who has been to me, since my return, a most active friend. (*LL* (NY) 1: 251)

The publication of his great mass of geological notes weighed heavily on Darwin at first. In a letter to Henslow of 28 March 1837 he queried:

> Have you ever had an opportunity of sounding any of the great Cambridge Dons about the publication of my geology. I hope they will prove gracious

> for it would be a great bore to be half killed with seasickness, and then in reward half starved with poverty. (1967, p. 126)

He need not have worried.

Darwin's Geology Diary (DAR 32–33) and Geology Notes (DAR 34–38) were the major sources for the three books he published subtitled "the Geology of the Voyage of the Beagle" (*Coral Reefs*; Darwin 1844a; 1846) They also produced a number of papers over the years (*CP* 1: 44–45, 1837; 41–43, 1838; 46–49, 1838; 44–45, 1838; 137–139, 1839; 53–86, 1840; 139–142, 1841; 145–163, 1842; 203–212, 1846; 212–213, 1846; 214, 1847; 2: 74–77, 1863), plus brief notices in other of his geological papers. In addition, his experiences as a field geologist while on the *Beagle* led to a chapter on geology in the Navy's *A Manual of Scientific Inquiry* (Herschel 1849). Darwin's geological collections are mentioned in a number of these publications, but they are not treated systematically by him in any of them.

One basically geological paper (*CP* 1: 199–203, 1846) resulted from notes in the Zoology Diary (DAR 31.1). The volcanic dust described in this paper was examined by the Prussian protozoologist Christian Gottfried Ehrenberg (1795–1876, Professor of Zoology at Berlin University), who reported on its organic constituents (Ehrenberg 1844, 1845b). He also described infusoria from Darwin's collections of Fuegean Indian body paint and from volcanic ash tuff from Ascension Island and the mountains of Patagonia (Ehrenberg 1845a, 1845c, 1845d). Presumably, these specimens, if extant, are still in Berlin.

Minerals

In his first letter to Henslow following the return of the *Beagle* (6 October 1836), Darwin wrote: "I want your advice on many points, indeed I am in the clouds & neither know what to do, or where to go. My chief puzzle is about the geological specimens, who will have the charity to help me in describing their mineralogical nature?" (1967, p. 117). On the 30th he wrote, "I am anxious to know, whether Prof. Sedgwick recommends any particular nomenclature for the rocks" (1967, p. 121).

The Professor of Mineralogy at Cambridge, William Hallowes Miller (1801–1880), and not Sedgwick, aided with the examination of the mineralogical specimens while Darwin was in residence at Cambridge. Notes in the Geological Specimen Notebooks added after his return to England show that Henslow, the ex-Professor of Mineralogy, assisted Darwin as well. After moving to London, Darwin wrote to Henslow (28 March 1837): "When you next meet Prof. Miller, pray ["tel *del*" added in brackets by Barlow] remember me to him and tell him I shall not look at any more geological specimens for a few months, so that there is not the slightest hurry about the specimens which he has of mine" (1967, p. 126). By 4 November, however, he was writing: "I left with Miller last winter some geological specimens. — I should be very much obliged if he would make

["SOON *added*" added in brackets by Barlow] a list of the numbers (*specifying the colour of the paper*), for otherwise I might be hunting in vain for hours" (1967, p. 141).

Darwin was concerned about the color of his labels because the color indicated to which thousand the printed number belonged. Before leaving on the *Beagle*, he had 4,000 labels printed, each thousand a different color, and each printed from 1 through 1,000. The system is explained on the inside front covers of several of his series of notes. For example, the cover of the Shell Notes is annotated by him: "Red=1000 / Green=2000 / Yellow=3000 / For instance the number 242 printed on yellow paper has the value of 2000+242 or 2242 [changed from 2442]." White labels indicated specimens numbered from 1 through 1,000. Darwin never claimed to be much of a mathematician. This is well illustrated in the above example, as the correct number is 3,242!

There is no indication that Miller ever published on Darwin's rocks. Darwin's geological books are full of information on them, but the only ones examined systematically were those of the Cape Verde Islands (Harker 1907) and the Galapagos Islands (C. Richardson 1933). About 2,000 mineralogical specimens now reside in the Mineralogy and Petrology Museum, apparently having remained at Cambridge since 1836. "A few are in the Geological Survey Museum, London" (Richardson 1933, p. 45). Although Darwin's Geological Specimen Notebooks are now in the University Library, a Catalogue of the *Beagle* Collection of Rocks, copied from them, is in the Museum.

Fossil Invertebrates

Darwin paid particular attention to fossils, collecting them whenever he could, even at a height of 12,000 feet in the Andes of South America. He wrote to his sister Susan Elizabeth Darwin (1803–1866) in April 1835 that, "I think an examination of these will give an approximate age to these mountains as compared to the strata of Europe" (1945, p. 117).

Darwin inquired of Henslow in his letter of 30 October 1836 about the naturalist George Brettingham Sowerby (1778–1854), "Also about fossil shells. Is Sowerby a good man? I understand his assistance can be purchased" (1967, p. 120). Whether his assistance was purchased or not is unknown, but Sowerby did describe Darwin's fossil shells from the Cape Verde Islands, St. Helena, and Tasmania (Sowerby 1844) and some from South America (Sowerby 1846). Edward Forbes (1815–1854, Professor of Botany at King's College, London) also described some of the South American fossil shells (Forbes 1846), while William Lonsdale described six fossil corals from Tasmania (Lonsdale 1844), and John Morris (1810–1886, later Professor of Geology at University College, London) and Daniel Sharpe (1806–1856, a businessman and amateur geologist) described fossil brachiopods from the Falkland Islands (Morris and Sharpe 1846). The invertebrate fossils appear to be in Cambridge at the Sedgwick Museum and in London at the British Museum (Natural History).

Fossil Vertebrates

The vertebrate fossils were the first of the collections to catch the eye of the scientific public. In Punta Alta, Argentina in October 1832, Darwin entered into his first Geological Specimen Notebook the following collection: "821 Great head: (Megalonyx?) it was found in horizontal position in the cemented gravel; the upper jaw & molars exposed" (DAR deposit). He noted on the previous page that, "The anterior part is broken into 3 pieces: they can be joined by the shape of curious anterior cavity:" This skull of an extinct ground sloth was sent to Henslow in the shipment of November 1832. Recognizing its importance, Henslow sent it to William Clift of the Hunterian museum of the Royal College of Surgeons. Of its impact, in one of his few letters to Darwin on the *Beagle*, Henslow wrote (31 August 1833):

> The fossil portions of the Megatherium turned out to be extremely interesting as serving to illustrate certain parts of the animal which the specimens formerly received in this country & in France had failed to do — Buckland & Clift exhibited them at the Geological Section[11] (what this means you will learn from the report I send you) — & I have just received a letter from Clift requesting me to forward the whole to him, that he may pick them out carefully repair them, get them figured, & return them to me with a description of what they are & how far they serve to illustrate the osteology of the Great Beast — This I shall do in another week when I return to Cambridge. . . . (1967, pp. 77–78)

The Reverend William Buckland (1784–1856) was Professor of Mineralogy at the University of Oxford. He published a short note (Buckland 1837) on Darwin's fossils. Later in the same letter, Henslow added: "Send home every scrap of Megatherium skull you can set your eyes upon — & *all* fossils" (1967, p. 79).

In September 1832, Darwin also collected at Punta Alta "735. Pentagonal open plates in [th *del*] an earthy intervening bed" and "736 : 737 : 738. — Fragments of the latter: Is it a sort of hide?" (DAR deposit). Indeed, it was. The next month he wrote to Caroline that,

> I have been wonderfully lucky with fossil bones. Some of the animals must have been of great dimensions: I am almost sure that many of them are quite new: this is always pleasant, but with the antediluvian animals it is doubly so. I found parts of the curious osseous coat which is attributed to the Megatherium: as the only specimens in Europe are at Madrid (originally in 1798 from Buenos Aires) this alone is enough to repay some wearisome minutes. (1945, p. 76)

One of his sisters must have commented on these specimens in a letter to him now lost, as he wrote to Catherine in June 1833, "I am quite delighted to find the hide of the Megatherium has given you all some little interest in my employments. These fragments are not however by any means the most valuable of the geological relics" (1945, p. 86). In spite of this, they remain some of the most intriguing of his fossils.

All the mammalian fossils, apparently, were sent by Henslow to Clift as they were unpacked after arrival. Henslow in another letter (22 July 1834) wrote after Darwin's shipment of specimens to Dr. Armstrong in Portsmouth had arrived:

> He tells me however that everything is safe, & that he had used the precaution of opening the cases & airing everything for you — I recommended the fossils to be all sent to Mr. Clift at Surgeon's Hall[12] who has kindly undertaken to repair them & prepare them so that they shall be preserved without injury — Judging from what you sent before I did not hesitate to do this as they will be well worth the carriage to London, & could not possibly be in better hands than Clift's. (1967, p. 89)

In a subsequent letter (9 August 1834) to his sister Caroline, however, Darwin cautioned,

> Another point must clearly be explained to Mr. Clift; it is with reference to the Coll. of Surgeons paying the expence of the carriage. The ultimum destinatum of *all* my collections will of course be to wherever they may be of most service to Natural History. But *ceteris paribus*[13] the British Museum has the first claims, owing to my being on board a King's Ship. Mr. Clift must understand that *at present* I cannot say that any of the fossil Bones shall go to any particular Museum. As you may well believe I am quite delighted that I should have had the good fortune (in spite of sundry sneers about Seal & Whale bones) to have found fossil remains which can interest people such as Mr. Clift. (1945, p. 105)

Clift presumably was in contact with the family through Erasmus. From the foregoing, it would appear that Clift had requested that the mammal bones be given to his museum.

In his letter to Henslow of 30 October 1836, after he had returned, Darwin reported that, "Mr. Clift says he will ask Prof. Buckland to look at the bones; I should think he would rather like it, as Mr. Clift says some belong to forms ["which *added*" added in brackets by Barlow] he ["himself *added*" added in brackets by Barlow] does not at all know" (1967, p. 120). Neither Clift nor Buckland, however, described the fossils. This was to be done by Clift's former assistant, Richard Owen.

Owen and Darwin apparently first met at the Lyells' home in London on 29 October 1836, although in his letter of invitation to Owen, Lyell wrote: "Among others you will meet Mr. Charles Darwin, whom I believe you have seen, just returned from South America, where he has laboured for zoologists as well as for hammer-bearers" (R. S. Owen 1894, p. 102). They hit it off immediately, and Darwin soon resolved to have Owen describe his fossil mammals. He wrote to Owen on 19 December 1836, "I have scarcely begun to unpack my cases; in the course of a week I shall have everything open, and I already Know of one very large bone (of a Mastodon??) which I will forward to the College" (De Beer 1959b, p. 49). By 23 January 1837, Owen sent Lyell a list of his identifications of these fossils (Wilson 1972). Owen published several papers on Darwin's fossil *Toxodon* from Uruguay (1838a, 1838b, 1838c), plus the first part of the *Zoology*. In spite of later differences, Owen was at this time of great help to Darwin with his *Beagle* vertebrates, both fossil and modern.

Only the fossil mammals, which made up almost the entirety of the vertebrate animals, were worked up. According to Francis Darwin (*LL* (NY) 1: 247), the following was written to Henslow, but it does not appear in Darwin 1967: "I ["have" added in brackets by F. Darwin] disposed of the most important part ["of" added in brackets by F. Darwin] my collections, by giving all the fossil bones to the College of Surgeons, casts of them will be distributed, and descriptions published." ". . . the remnants came to the British Museum (Natural History) in 1946" (De Beer 1959b, p. 49). The few other vertebrate fossils are here also or at the Sedgwick Museum in Cambridge. Drawings of the fossil mammals, made for or by Owen, are in the British Museum (Natural History) (Ingles and Sawyer 1979).

Plant Fossils

One of the members of London's scientific establishment whom Darwin met before he embarked on the *Beagle* was the botanist Robert Brown (1773–1858, Keeper of Botany at the British Museum). Brown had sailed on an expedition to Australia in 1802 and collected plants. Darwin sought him out for advice on scientific equipment, particularly what kind of microscope he should take with him. He wrote to Henslow on 18 October 1831: "Mr. Brown has been of great use to me, & most exceedingly pleasant & goodnatured" (1967, p. 44).

Upon Darwin's return, Brown showed an interest in Darwin's plants, both the fossils and, as we shall see, the recent specimens. Darwin wrote to Leonard Jenyns on 10 April 1837, to "Tell Henslow, I think my silicified wood has unflinted Mr. Brown's heart, for he was very gracious to me, and talked about the Galapagos plants; but before he never would say a word" (*LL* (NY) 1: 253–254).

In a letter to his sister Susan from Valparaiso (23 April 1835), Charles described some fossil trees which he had come upon in the Chilean Andes:

> In these same beds (& close to a Gold mine) I found a clump of petrified trees, standing upright, with the layers of fine Sandstone deposited round

> them, bearing the impression of their bark. These trees are covered by other sandstones & streams of Lava to the thickness of several thousand feet. These rocks have been deposited beneath water, yet it is clear the spot where the trees grew, must once have been above the level of the sea, so that it is certain the land must have been depressed by at least as many thousand feet, as the superincumbent subaqueous deposits are thick. (1945, pp. 117–118)

There are several pages devoted to them in *Geological Observations on South America* (1846).

Darwin wrote to Henslow in May 1837 that

> Mr. Brown has been taking a good deal of interest in my affairs & in a most kind manner. I want therefore to oblige him any way I can. — He was much pleased with the fossil woods & has gone to the expence of having several of them cut & ground. — The clump of trees which were growing vertically ["were fine *del*, are *added*" added in brackets by Barlow] allied to Araucaria, but in some respects resembling yews. (1967, p. 127)

Near the end of the fourth Geological Specimen Notebook are six pages of notes on the fossil plants, including the araucaria-like wood discussed above. It is obvious that, although Brown never published on Darwin's plant fossils, he provided the information regarding them that Darwin needed for his various geological publications. Presumably, they are at the British Museum (Natural History), although a few are with the other fossils in the Sedgwick Museum at Cambridge.

ZOOLOGY

Darwin's introduction to the London zoologists, whom he had counted on to help him with the identification of his animals, was less than auspicious. He wrote to Henslow in his letter of 30 October 1836:

> I am out of patience with the Zoologists, not because they are overworked, but for their mean quarrelsome spirit. I went the other evening to the Zoological Soc. where the speakers were snarling at each other, in a manner anything but like ["that of *added*" added in brackets by Barlow] gentlemen. (1967, p. 121)

In spite of this, he was soon able to enlist a number of specialists who described many of the groups of animals he collected. By the next week (6 November 1836), he was able to write Henslow that, "All my affairs, indeed, are most prosperous; I find there are plenty who will undertake the description of whole tribes of animals, of which I know nothing" (F. Darwin 1892, p. 150).[14] As we shall see, however, many of the arthropods, except for the insects, remain unidentified.

Mammals

As was true for the fossil mammals, Richard Owen also showed an interest in the recent mammals. Darwin informed Henslow (30 October 1836) that "Mr. Owen seems anxious to dissect some of the animals in spirit. . . ." (1967, pp. 118–119). He wrote to Owen less than two months later (19 December): "When separating out the animals in Spirit I will put by any I think will interest you" (De Beer 1969b, p. 49).

The Mammal Notes and Birds in Spirits of Wine Notes at the Cambridge University Library (DAR 29.3) are both stamped "Coll. Sherborn / ex litt. Ricardi Owen. / don. R. S. Owen". They are in Darwin's hand, and both are begun by him in a latter annotation: "Numbers with + refer to additional information on the back of Page." His comments on a Chilean armadillo show that he did more than merely collect and preserve his specimens, he often also dissected them: "*1038*. Very common Valparaiso. uses its tail. but little; — in stomach larva of beetles" (DAR 29.3: 1 *verso*). In spite of Owen's early interest in the mammals, however, he appears not to have published on any other than the fossils.

The bulk of the collection was described in the *Zoology* by George Robert Waterhouse (1810–1888, Keeper of Mineralogy and Geology at the British Museum) as Part II, *Living Mammalia*, Owen having described Part I, *Fossil Mammalia*. There are several pages of notes by Waterhouse on the mammals in DAR 29.1. He also worked up a number of Darwin's insects.

Prior to the publication of the *Zoology*, several papers on a few of the mammals were published. The first of these were on some South American wildcats (Martin 1837a), an opossum and a viscacha, a chinchilla-like rodent from Argentina (Reid 1837), a fox from Chiloe Island, Chile (Martin 1837b), and an armadillo (Martin 1837c). Of the yaguarundi from Buenos Aires described by Martin (1837a), it was written that "in the event of its ultimately being considered distinct, he proposed that it should be called *Felis darwinii*". It is a synonym of *F. yaguarundi* Desmarest (R. Freeman 1978). A footnote to Reid's publication of a new species of opossum reads: "The characters of species newly described which have not yet been furnished by the respective authors, and are therefore necessarily omitted, will be inserted, if subsequently sent in, at the termination of the volume" (Reid 1837, p. 4). This was not done by either Reid or Martin.

Waterhouse published three papers on the voyage's mammals before his contributions to the *Zoology* appeared. The first of these described nineteen new species of mice:

> At the request of the Chairman, Mr. Waterhouse brought under the notice of the Meeting numerous species of the genus *Mus*, forming part of the collection presented to this Society[15] by Charles Darwin, Esq., a Corresponding Member. . . . Most of these numerous species were considered by Mr. Waterhouse as hitherto undescribed, and drawings

> were exhibited by him illustrative of the modifications observable in their dentition. (Waterhouse 1837a, p. 15)

The second paper (Waterhouse 1837b) described two new genera and four new species of South American rodents. The third (Waterhouse 1838a) described a new species of dolphin, named for Captain FitzRoy (*Delphinus fitzroyi*): "The figure which illustrates this description agrees with the dimensions, which were carefully taken by Mr. Darwin immediately after the animal was captured, and hence is correct" (Waterhouse 1838a, p. 23).

Darwin's mammals and birds were presented by him to the Zoological Society of London on 4 January 1837 (Sulloway 1982c). They were transferred to the British Museum in 1855 when the Museum of the Zoological Society was dispersed. De Beer stated that

> A number of specimens collected by Darwin and described in the *Zoology of the voyage of the Beagle* were presented to the British Museum in 1837 by Sir William Burnett (Physician-General of the Navy) and Captain FitzRoy. Some of Darwin's specimens were given to the Zoological Society's Museum, from which they were transferred to the British Museum in 1855. (1959b, p. 49)

So far as is known, however, all were given to the Zoological Society, none going directly to the British Museum. Those forwarded by Burnett and FitzRoy were the latter's collections, which, unlike Darwin's, were the property of the Navy and under Burnett's control (Sulloway 1982c).

Birds

There has probably been more published on Darwin's ornithological collections than on all the other groups combined. In spite of this, there are still a number of misconceptions regarding them, particularly having to do with the influence of the Galapagos Islands finches on Darwin's formulation of his theory of evolution by natural selection. These misconceptions are well discussed in Sulloway (1982a).

Darwin took copious notes on the birds he collected. His Ornithological Notes run for eighty-six numbered pages, plus ten others inserted but unnumbered. They are well over twice as long as the Mammal Notes, the second longest list. They were published, with annotations, by Barlow (Darwin 1963). As we have seen, the Birds in Spirits of Wine Notes were given to Owen, who did not publish on the birds. But some of his comments to Darwin on them were included in the *Zoology*.

Like the mammals, the birds attracted much interest following Darwin's return to England, and their skins were exhibited at the Zoological Society of London over a series of meetings in 1837 and 1838. They were presented and discussed by the ornithologist John Gould, to whom Darwin had entrusted their skins, and by Darwin himself. The first of these exhibitions was of the remarkable Galapagos Islands finches, which Gould characterized as "so peculiar in form that he was induced to regard them as constituting an entirely new group, containing 14 species, and appearing to be strictly confined to the Galapagos Islands" (1837a, p. 4). The paper ended: "Mr. Gould deferred entering into any further details respecting the species under consideration until Mr. Darwin had furnished him with some information relating to their habits and manners" (Gould 1837a, p. 7). They were discussed at a second meeting by Darwin (*CP* 1: 40, 1837).

Next, "Mr. Gould exhibited the Raptorial Birds included in the collection recently presented to the Society by Charles Darwin, Esq. . . . ." (Gould 1837b, p. 9). This was followed by exhibitions and notes on the swallows and swifts (Gould 1837c), on some Australian and Galapagos Islands specimens (Gould 1837d), and a new species of rhea from Patagonia:

> Mr. Gould, in conclusion, adverted to the important accessions to science resulting from the exertions of Mr. Darwin, and to his liberality in presenting the Society with his valuable Zoological Collection; to commemorate which he proposed to designate this interesting species by the name of *Rhea darwini*. (Gould 1837e, p. 35)

"Mr. Darwin then read some notes upon the *Rhea americana*, and upon the newly described species, but principally returning to the former" (*CP* 1: 39). Two final exhibitions (Gould 1837f, 1838) ended the series.

The birds for the *Zoology* were described and classified primarily by Gould. He, however, left England for Australia in 1838, and additional descriptions were added by George Robert Gray (1808–1872, Assistant in the Natural History Department of the British Museum). Gould's departure also caused Darwin to add much to the manuscript, and Barlow states that "in his Preface Darwin has greatly underrated his own share, consisting of the habits and ranges" (1963, p. 206). This is true to a lesser extent for the other numbers of the *Zoology* as well. In addition, an anatomical appendix was added by Darwin's Cambridge classmate and ornithologist Thomas Campbell Eyton (1809–1880) (*Zoology* 3: 147–156).

Not only did Gould's departure cause problems with the completion of the birds for the *Zoology*. Gray appears to have been slow in completing his part of the manuscript as well. Darwin wrote to him in 1840,

> I trust now you have completed this work,[16] you will oblige me by kindly finishing the remaining MS. for the Birds of the Beagles Voyage. — I had hoped to have finished the part, but I have of late been so frequently unwell that all my plans have disarranged.
>
> I shall esteem it a great favour should you be able to finish at once Gould's MS. (De Beer 1958, p. 91)

Over thirty years later, Darwin published another paper, which included some of his field observations. This was on the Pampas woodpecker, and was written in answer to allegations of the Anglo-Argentine naturalist

William Henry Hudson (1841–1922) regarding Darwin's observations on the bird (*CP* 2: 161–162, 1870).

As was true for the mammals, Darwin's bird specimens were presented to the Zoological Society of London on 4 January 1837 (Sulloway 1982c). According to Sulloway, when the Zoological Society Museum was closed in December 1855, the British Museum was given first choice of the specimens, but they did not acquire all of Darwin's birds. Thus, some of them have disappeared; either they were destroyed or their whereabouts are now unknown.

Darwin kept at least a few specimens, since he presented some to the British Museum in 1856. Gould and Gray also had some in their possessions, Gould selling some to the British Museum in 1857, and Gray acquiring some types when the museum of the Zoological Society was liquidated. Gould's collection eventually was sold to the British Museum in 1881 after his death. Thus most of the Darwin *Beagle* birds extant are now in the British Museum (Natural History) at Tring. However, Sulloway (1982c) presents evidence that Darwin specimens also may be contained in the Netherlands' Leiden Rijksmuseum van Natuurlijke Historie.

Reptiles and Amphibians

Regarding this group of animals, Darwin wrote to Henslow in his 30 October 1836 letter: "Mr. Bell I hear is so much occupied that there is no chance of his wishing for specimens of reptiles" (1967, p. 122). But after attending a meeting of the Linnean Society on 2 November, he wrote again, "I became acquainted with Mr. Bell, who, to my surprise, expressed a good deal of interest about my crustaceae & reptiles & seems willing to work at them" (1967, p. 123).

Thomas Bell (1792–1880) was Professor of Zoology, King's College, London. He provided the manuscript of the lizards and frogs for the *Zoology*. Although expressing an interest in the crustaceans, he took the specimens but never worked them up. His two parts on the *Reptiles* for the *Zoology* were the last to be published. According to R. Freeman, he "delayed completion for nearly 2 years through procrastination and ill-health" (1978, p. 34).

The Reptiles in Spirits of Wine Notes are in the General Library of the British Museum (Natural History), having been deposited there by Bell in 1845. Most of the twenty-six pages are by Covington, but there are large additions by Darwin, and notes by Bell and by John Edward Gray (1800–1875, Keeper of Zoology at the British Museum). Because of this, it has been assumed that all the reptiles and amphibians are at the British Museum (Natural History). In his preface, however, Bell states,

> The Ophidians have been placed in the hands of Mons. Bibron, who is at the present time engaged in completing his admirable history of Reptiles, by the publication of those volumes which are devoted to this

> order; and it must be considered a fortunate circumstance that the delay which has taken place in the appearance of that portion of his labours, has thus afforded the opportunity of embodying in so perfect a work, the numerous discoveries of Mr. Darwin in this particular department of Erpetology. (*Zoology* 5: vi)

Alas, this was not to be, for Professor Bibron was too ill to finish his herpetological masterpiece. According to Donoso-Barros:

> Bell mentions that the snakes were given to Bibron for a separate account, but Bibron died from tuberculosis in 1848 before the snake volumes of *Erpétologie Générale* appeared, and no mention of Darwin's snakes is given there. Previously, however, J. E. Gray had mentioned some of Darwin's material and A. Günther described *Dromicus biserialis* based on a specimen collected by Darwin in the Galapagos Islands. Darwin mentions certain turtles and other herpetological specimens in his narrative, but nothing is known about the disposition of the remaining material. (1975, p. iii)

These specimens should be searched for at the British Museum (Natural History). Presumably the specimens in Bell's possession were deposited in 1845 along with the Reptiles in Spirits of Wine Notes. The publications of Gray and Günther cited by Donoso-Barros have not been seen. But Günther (1877) does cite some Darwin Galapagos Islands reptiles.

Fish

The fish were all accounted for in the *Zoology*, no separate papers on them being published. They were authored by the Reverend Leonard Jenyns, another of Darwin's friends from Cambridge days. The Fish in Spirits of Wine Notes (DAR 29.1: 20 pp.) are in Covington's handwriting, with some corrections and additions added by Darwin. On the left-hand pages are many determinations by Jenyns. The inside page of the front cover has the statement, "+ Signifies the fish the names of which I am anxious to know. —" It is certainly a note from Darwin to Jenyns, seeking information to be used in the first edition of the *Journal of Researches*. Many numbers are marked with a plus.

An 1838 article by Jenyns on the history of the Museum of the Cambridge Philosophical Society was in part quoted by J. W. Clark as follows:

> The foreign department of the Museum is not extensive, consisting for the most part of single specimens which have been presented at different times by different individuals. . . . It is also rich in Icthyological specimens . . . more recently, with the entire collection of Fish brought here from South America and some other portions of the globe by C. Darwin, Esq., of Christ's College, and accompanying Naturalist in the late voyage of the Beagle, under the command of Captain FitzRoy. The whole of the fish above alluded to, as well as those belonging to the British collection

> are preserved in spirits. They amount to several hundred species; and many of those comprised in the Darwin collection are entirely new. (1890, p. xv)

Presumably the dried fish, mentioned in Darwin's letter of 28 March 1837 to Henslow, also went to the Cambridge Philosophical Society.

At least some of the fish, like some birds, were examined anatomically by Thomas Eyton, as Darwin wrote to Henslow on 4 November 1837, "My message to L. Jenyns is simply that I expect T. Eyton to pay me a visit before long, when he comes up to town, & that the fish had better be sent soon by waggon to 36 Great Marlborough St" (1967, p. 141). This was Darwin's address in London, where he was at the time working up the scheme for the *Zoology*.

In 1865, the collections of the Cambridge Philosophical Society were given to the University:

> Apart from the several hundred species of fish presented by Darwin, however, the natural history museum never really attained scientific importance. . . . Accordingly, when the University resolved in 1865 to build some new museum and lecture rooms the Society agreed (27 February) to ask for accommodation for its Library in the new buildings, and somewhat later offered the University as a gift the whole of its own Natural History collections, which were incorporated into the Museum of the new Department of Zoology and Comparative Anatomy. (Hall 1969, pp. 26, 28)

A few of them are now on public display in the Museum.

Corals

Darwin was particularly keen on collecting corals. The Zoology Diary has many descriptions of corals, and there are a number of unpublished drawings of these organisms in DAR 29.3, which are keyed to the descriptions. He wrote to Catherine from Chile on 20 July 1834, "Amongst Animals, on principle I have lately determined to work chiefly amongst the Zoophites or Coralls; it is an enormous branch of the organized world, very little known or arranged, and abounding with most curious yet simple forms of structures" (1945, p. 101).

Upon his return, Darwin found that his old Edinburgh professor Robert Grant (now Professor of Zoology and Comparative Anatomy at University College, London) was interested in seeing them. He wrote to Henslow on 30 October 1836, "I have scarcely met anyone who seems to wish to possess any of my specimens. — I must except Dr. Grant, who is willing to examine some of the corallines" (1967, p. 119).

What came of this is unknown. A number of the specimens discussed in the Zoology Diary actually are coralline algae, and this may have cooled Grant's interest. However that may be, I suspect that it was Darwin's feeling of frustration regarding these specimens in particular that led him to downgrade the importance of his Zoology Diary, as discussed earlier in this paper. A number of comments on the corals from the Zoology Diary do appear in *Coral Reefs*. But they were never treated systematically.

Around 1869, Darwin wrote to the invertebrate paleontologist Peter Martin Duncan (1824–1891) that he would send him his Cocos-Keeling Islands coral specimens (Carroll 1976c). This is perhaps the source of the one-page list of Cocos-Keeling corals at the British Museum (Natural History). The corals are there as well, several being illustrated by Whitehead and Keates (1981, p. 23). In 1876 he again wrote to Martin to advise him that he would send a fossil coral and a related manuscript by William Lonsdale (Carroll 1976c).

More recently, Hickson (1921) described two of Darwin's sea pens from the Galapagos Islands, found at Cambridge. A few years later, he added "Some years ago, I discovered in the cellars of the Cambridge Museum of Zoology a specimen of the sea-pen *Cavernularia* with the label 'Voyage of the Beagle. C. Darwin. Galapagos Is' " (Hickson 1936, p. 909). Perhaps following further scrutiny these cellars will yield up more invertebrate specimens of Darwin.

Flatworms

One of the groups that Darwin himself published was the planarians (*CP* 1: 182–193, 1844). He took a great deal of interest in them and made extensive notes on their morphology and behavior in his Zoology Diary. The material in his paper is taken almost verbatim from the Zoology Diary. Several new species and a new genus, *Diplanaria*, the only genus erected by him, are described.

Henslow wrote to him on 15 January 1833,

> L. Jenyns does not know what to make of your land Planariae. Do you mistake for such the curious genus, "Oncidium" allied to slug, of which a fig. is given in Linn. Transact. & one not the marine species also *mollusca,* perhaps Doris & other genera. — Specimens & observations upon these wd. be highly interesting. (Darwin 1967, pp. 66—67)

This was in response to Darwin's comments to Henslow of 15 August 1832 from Monte Video, Uruguay:

> Amongst the lower animals, nothing has so much interested me as finding 2 species of elegantly coloured true Planariae, inhabiting the dry forest! The false relation they bear to Snails is the most extraordinary thing of the kind I have ever seen. — In the same genus (or more truly family) some of the marine species possess an organization so marvellous — that I can scarcely credit my eyesight. (1967, pp. 58–59)

Here was one case where the observations of the student were correct over the skepticism of the teacher. Darwin spent a great deal of time and effort on his collections of invertebrates, and he was keenly interested in describing them. Regarding his plans for the immediate future, he informed Henslow in October 1837:

> I have had hopes by giving up society & not wasting an hour, that I should be able to finish my geology in a year & a half, by which time the descriptions of the higher animals by others would be completed & my whole time would then necessarily be required to complete myself the description of the invertebrate ones. (1967, p. 139)

Darwin did publish two short systematic papers (*CP* 1: 177–182, 1844; 182–193, 1844) and parts of two books (1851, 1854) on his invertebrates. But prior to the present, the only group of invertebrates that has received much attention from others is the insects. The whereabouts of the planaria and most other groups of invertebrates are unknown, but they should be sought at the British Museum (Natural History) and the Museum of Zoology at Cambridge University.

Crustaceans

We have seen that Thomas Bell expressed an interest not only in the reptiles, but in Darwin's crustacea as well. The latter also were turned over to Bell, but he appears to have done even less with them than he did with the reptiles and amphibians. In spite of Darwin's hope that Bell would identify and publish a monograph on them, this was never done.

Darwin's old entomological friend the Reverend Frederick William Hope founded the Hope Chair of Zoology at Oxford University in 1862. The history of the zoological collections at the Oxford University Museum has recently been published (Davies and Hull 1976). There is much information in this history that bears on the specimens Darwin gave to Bell:

> A large part of Bell's collections of crustacea and reptiles was purchased in 1862 for the Hope Collections by J. O. Westwood, the first Hope Professor of Zoology: the bulk of this material was transferred to the zoological collections in 1889, a further transfer of assorted spirit material took place in 1949, and in 1962 the bulk of the spirit collections were transferred from the Hope Department to the Zoological Collections. More recently, in 1975, the large collection of dried crustacea was taken over by the Zoological Collections. (Davies and Hull 1976, p. 78)

Darwin wrote to John Obadiah Westwood (1805–1893) in 1860 that he agreed with Bell in the sending of his crustacea to Oxford. I have examined some of Darwin's dried crustacea in the collection and found that most have metal tags attached, indicating that they were originally preserved in spirits. It is safe to infer that they all were so prepared.

The curation of the specimens still in spirits has been undertaken since

1962 by Mr. Jimmy Hull, who has found specimens of other invertebrate phyla included as well. The dried crustacea have been under study since 1976 by Dr. Angelo Di Mauro (see Di Mauro 1982). These investigators have found a number of Darwin specimens, but the Crustacea in Spirits of Wine Notes have not yet been discovered. An exhaustive review of the collection will soon be published (Di Mauro, King, and Chancellor, in press). According to these authors, most of the material at Oxford that has been found so far is of the higher crustacea (Malacostraca). The whereabouts of the lower crustacea (Entomostraca), except for the barnacles, are unknown. Some of Darwin's crustacea are now on display at the Oxford University Museum.

Barnacles

The best known systematic work of Darwin is that on barnacles, which started with his curiosity over a specimen collected on the *Beagle* in 1835. In his *Autobiography* he states:

> In October, 1846, I began to work on Cirripedia. When on the coast of Chile, I found a most curious form, which burrowed into the shells of Concholepas,[17] and which differed so much from all other Cirripedes that I had to form a new sub-order for its sole reception. . . . To understand the structure of my new Cirripede I had to examine and dissect many of the common forms: and this gradually led me to take up the whole group. I worked steadily on the subject for the next eight years, and ultimately published two thick volumes, describing all the known living species, and two thin quartos on the extinct species. (p. 117)

Darwin's *Beagle* barnacles were cited in these "two thick volumes" (1851, 1854). They remain the standard references for barnacle taxonomy. The specimens were deposited by him in the British Museum. There is a possibility that some *Beagle* specimens are in the Department of Invertebrate Zoology of the Merseyside County Museums. According to Francis Darwin, the "duplicate type-specimens" were deposited in the "Liverpool Free Public Museum" (1892, p. 173).

Insects

This is the largest group of invertebrates collected by Darwin, which is not surprising given his long-time interest in them, especially the beetles. They figure prominently in his various notes kept while on the *Beagle*. He paid particular attention to them, writing to Henslow in March 1834, "I have forgotten to mention, that for some time past & for the future, I will put a pencil cross on the pill-boxes containing insects, as these alone will require being kept particularly dry, it may perhaps save you some trouble" (1967, pp. 86–87).

In spite of his keen interest in insects, Darwin never published separately on his *Beagle* collections. The first paper on them was by his friend Hope

(1838), describing some new ground beetles. They are cited as being "In Museo Dom. Darwin.", indicating that they had not yet been given to a museum. The description of *Carabus darwini* Hope ends, "This beautiful insect I have named in honour of my friend Charles Darwin, Esq., a zealous Entomologist. His exertions in advancing the progress of Zoology in general entitle him to the thanks of the scientific world" (Hope 1838, p. 129).

Most of the papers on Darwin's insects, however, were authored by the man who also worked up his recent mammals, George Waterhouse. The Insect Notes, which are in the Department of Entomology Library, British Museum (Natural History), are in Covington's hand, with Darwin's additions and corrections (Porter, 1983b). There are a few additions of names by Waterhouse, and a map of southern South America giving some of Darwin's localities in Waterhouse's handwriting.

Waterhouse's first paper on Darwin's beetles described some from Australia (Waterhouse 1838b). The rest, however, were from South America (Waterhouse 1840a, 1840b, 1840c, 1841a, 1841b, 1841c, 1842a, 1842b, 1843, 1845). The specimens were deposited in the Museum of the Entomological Society of London, eventually finding their way into the British Museum (Natural History), where his insect collections now reside.

Darwin took advantage of collecting whenever he could. This could not be better shown than by quoting Waterhouse on *Cardiophthalmus clivinoides* Curtis: "This specimen was 'found dead in the sea, 40 miles off the Straits of Magellan.' — Mr. Darwin's Notes" (1840c, p. 254).

The other group of Darwin's insects that received much early attention was the chalcid wasps. The entomologist Francis Walker (1809–1874, Assistant at the British Museum) produced two volumes (Walker 1839) and a series of papers on them (Walker 1838, 1842a, 1842b, 1843a, 1843b, 1843c, 1843d, 1843e). Unfortunately, few specimens are cited. Walker described many new species and genera, which bear scrutiny by present-day hymenopterists, as "Walker's name has come to be a by-word amongst insect taxonomists for his inaccuracy and superficiality" (Doncaster 1961, p. v).

Some of Darwin's Galapagos Islands insects were later cited (in Günther 1877), including the types of several new species and genera. The beetles were treated by Charles Owen Waterhouse (1843–1917, George's son), the wasps and flies by Frederick Smith (1805–1879, entomologist at the British Museum), and the butterflies, grasshoppers, and bugs by Arthur Gardiner Butler (1844–1925). The specimens had been recently deposited in the museum by George Waterhouse.

The Insects in Spirits of Wine Notes actually enumerate parasitic arthropods. These were reported on by the parasitologist Thomas Spencer Cobbold (1828–1886) (1885). A note on lice in Chiloe Island, Chile (DAR 29.3: 1 p.) belongs with these notes.

Two major papers on the *Beagle* beetles from South America and Australia were published in the twentieth century (Champion 1918, Lea 1926). Both describe numerous new species, and the first several new genera. Champion wrote in his introductory remarks:

> Darwin, as is well known, was a keen Coleopterist, as shown by the representative collection made by him of our British forms, still preserved in the University Museum at Cambridge. During the voyage of the 'Beagle', 1832–1836, he captured beetles at every opportunity, and frequently mentions them in his published Journal. These insects were sent direct to specialists for determination, and most of them subsequently passed into the British Museum, the last instalment of his unnamed collections having been presented to that Institution by Mr. C. O. Waterhouse in 1885. The conspicuous South American *Carabidae*, *Dysticidae*, *Tenebrionidae*, etc. were named or described long ago by Babington, G. R. Waterhouse, and others, but the rest of the American beetles have remained untouched to this day amongst the 'Accessions' in the Museum. (1918, p. 43)

George Waterhouse appears to have retained insect specimens until they could be identified, or, as in this case, until he was no longer able to work on them.

In 1841, the printer and naturalist Edward Newman (1801–1876) published a drawing of one of Darwin's long-horned beetles from Chile (Newman, 1841), and Charles Cardale Babington (1808–1895, Professor of Botany at Cambridge following Henslow) described and illustrated some of his predacious diving beetles (Babington 1841). In a letter to Darwin of 1 July 1837, Babington wrote,

> I returned here yesterday evening & found your letter lying upon my table. Will you tell Hope that I have only one insect from Australia, that is K. Georges Sound. It is an Hydroporus allied to 12 — punctatus but smaller & less marked with yellow. I will endeavour to complete a description of it during the following week & send it to you for Hope. (DAR 29.3)

Babington then describes several other specimens from New Zealand and Tierra del Fuego, continuing:

> I an sorry to say that I have been prevented from examining the insects with care, but propose doing so & drawing up the descriptions after the long vacation. . . . I will however endeavour to complete the K.G.S. species for Hope & leave it with you or Waterhouse as I pass through town. I propose sending my complete account of your Insects to the Entom. Society when finished, but will not do so if you wish any other plan to be adopted. Still I think that that *is* the best place for them.

The remaining Australian specimens were described by Lea, who wrote:

> The British Museum having at various times sent to me for identification

> specimens taken by Darwin, mixed with others, I suggested to Mr. Arrow[18] that it appeared to be desirable to identify all the remaining Australian beetles taken by the great naturalist and deposited in that institution. . . . The specimens sent were all small, and in fact Mr. Arrow wrote: — 'Darwin did not give his collection to the Museum, but allowed different individuals to take particular groups which interested them, and the unsorted mass of minute specimens was given to G. R. Waterhouse, only coming here in 1887.' (1926, p. 279)

It is more likely that this "unsorted mass" represented what Waterhouse had not been able to identify. Be that as it may, there are plenty more Darwin insects at the British Museum (Natural History) that remain to be identified.

Spiders

On 18 May 1832, Darwin wrote to Henslow from Rio de Janeiro that, "I am at present red-hot with Spiders, they are very interesting & if I am not mistaken I have already taken some new genera" (1967, p. 55). The Zoology diary has several long entries on the morphology and behavior of his specimens.

The spiders were given for identification to Adam White (1817–1879, Assistant in the Zoology Department of the British Museum). He wrote of them:

> Having been favoured by Mr. Darwin with the whole of the extensive collection of Arachnida, made by him on the voyage of H.M.S. Beagle, I intend describing them occasionally in this journal. . . . They are all preserved in spirits of wine, as spiders should always be if possible, and, to some of Mr. Darwin's notes are occasionally added, which I have that gentleman's permission to extract from his copious manuscript journal.

This is footnoted:

> These notes, there is no use saying, were always made amid the hurry and bustle of a campaign in which annulose animals found but a small part of the subjects of research. I prefer giving them as I found them, as there is a *freshness* about them which would be *rubbed off* were I to attempt to improve them. (White 1841, p. 471)

In spite of White's intent to publish a series of papers on Darwin's "extensive collections of Arachnida", this was the only one that appeared. White continued:

> I may add, that specimens of all the species here described, unless otherwise intimated, will be found in the collection of the British Museum, and that I have made figures of most of them, which I intend to publish hereafter.

Alas, this was not done. White had great hopes for publishing on the spiders, which never came to fruition.

In a later letter (25 February 1877) to the British Museum zoologist Albert Karl Ludwig Gotthilf Günther (1830–1914), probably in relation to Günther (1877), Darwin wrote:

> Your note has led me to discover to my dismay that my catalogue of specimens, which I lent several years ago to the Museum at Cambridge, has never been returned to me. I have written about it, & if it has not been lost will hereafter answer your query. I shall be very sorry if it is lost, but it will not signify much with respect to the spiders, as the labels have been detached. That poor mad creature Adam White no doubt was the sinner. It was too bad of him, for I told him that I had notes about the habits of some of the species. (De Beer 1958d, p. 96)

The "catalogue of specimens" must have been the Specimens in Spirits of Wine Notebooks. A letter from Darwin of 3 May 1877 answers several of Günther's questions about spiders, indicating that the notebooks had been returned to him. The spiders, presumably, are in the British Museum (Natural History).

Mollusks

"I also heard, that Mr. Broderip would be glad to look over the S. American shells. — So that things flourish well with me," wrote Darwin to Henslow on 2 November 1836 (1967, p. 123). However, there is no further evidence that the conchologist William John Broderip (1789–1859) worked on Darwin's mollusks. A large number were collected, and the Shell Notes (DAR 29.3) are eight pages long, as are the Shells in Spirits of Wine Notes (DAR 29.1). Both are in Covington's hand, with additions and corrections by Darwin.

Included with the Shell Notes in DAR 29.3 are a one-page "list of Cape de Verd Shells" and a seven-page list with a cover reading "Mr. Darwin's Shells". The latter, perhaps in Covington's handwriting, lists the specimens and their localities. Darwin has written in pencil across the top of the first page: "N.B. The shells which I want out are marked with a cross // about 100 //." It is not clear to whom the note is written, nor for what purpose the shells were used. There is little information on recent mollusks in Darwin's published writings, and where the bulk now reside is not known, but a few are to be found in the Zoological Museum at Cambridge.

Arrowworms

Of the two papers that Darwin published on his collections, the first was a report of his observations on the marine genus *Sagitta* (*CP* 1: 177–182, 1844). These are taken from the Zoology Diary. I have found no reference to *Sagitta* in any of his other writings.

BOTANY

It should be clear by now that Darwin's primary interest while on the voyage did not lie with plants. Yet his friendship with Henslow had instilled in him at least an interest in collecting them for others. This eventually led to an interest in their geographical distributions, and played a role in providing evidence for evolution by natural selection.

In his second letter to Henslow, mailed from Monte Video on 15 August 1832, he wrote:

> It is positively distressing to walk in the glorious forest, amidst such treasures, & feel they are all thrown away upon one. — My collection from the Abrolhos is interesting as I suspect it nearly contains the whole flowering Vegetation, & indeed from extreme sterility the same may almost be said of St. Jago. (1967, p. 58)

Indeed, throughout the trip, when Darwin was in close proximity to the *Beagle*, and specimens could be conveniently dried, he collected a surprisingly large number of plants.

Henslow offered good advice and encouragement, writing on 15 January 1833,

> Your account of the Tropical forest is delightful, I can't help envying you. So far from being disappointed with the Box — I think you have done wonders — as I know you do not confine yourself to collecting, but are careful to describe. Most of the plants are very desirable to *me*. (1967, p. 66)

In spite of his interest in Darwin's plants, Henslow eventually was unable to identify more than a few (Porter 1980a).

After his return, Darwin placed his plants at Henslow's disposal. When he moved to London early in 1837, however, he found that Robert Brown also was interested in them. Brown had in his possession at the British Museum the plant collections from the 1826–1830 *Beagle* voyage. But Darwin was not too keen in having Brown examine his collections. He wrote to Henslow on 28 March 1837:

> I met Mr. Brown a few days after you had called on him, he asked me in a rather ominous manner what I meant to do with my plants. — In the course of conversation Mr. Broderip who was present remarked to him 'you forget how long it is since Capt. King's expedition.' He answered, 'Indeed I have something, in the shape of Capt. King's undescribed plants to make me recollect it.' Could a better reason be given ["if I had been asked *added*" added in brackets by Barlow] by me for not giving the plants to the Brit. Museum. (1967, p. 125)

Captain King's plants never were identified by Brown.

Fungi and Lichens

The fungi were described by the Reverend Miles Joseph Berkeley, to whom they were sent by Henslow. Three papers on them were offered by Berkeley (1839, 1842, 1845), although their first mention was in Henslow's (1838) paper on the flora of the Cocos-Keeling Islands. J. D. Hooker (1844–1847, 1847a) also listed several of Darwin's fungi, while Henslow (1844) published Darwin's notes on wheat rust in Patagonia.[19]

Robert Brown expressed a great deal of interest in one of Darwin's fungi, an edible species from Tierra del Fuego. In May 1837, Darwin informed Henslow,

> Mr. Brown is very curious about the fungi from the beech trees in T. del Fuego. — He has some specimens, but is very curious to see mine, but I do not know whether he wants to describe them: as your hands are so full, would you object to send them to me, & allow Mr. Brown to do what he likes with them. — If you particularly care about them, of course do not send them, but otherwise I should be glad to oblige Mr. Brown (1967, p. 127)

Henslow's answering letter is not extant, but Darwin responded to it on 28 May:

> I fear by your letter you cared more about the edible Fungi than I thought. — I took them to Mr. Brown, who said he had never seen anything of the sort before, & appeared interested on the subject, but whether he means to describe ["to *del*" added in brackets by Barlow] them, & for what he wants them, — I have not a guess, — at some future time, if I can summon courage, I will ask him, but I stand in great awe of Robertus Brown. (1967, p. 130)

On 14 July Darwin again wrote regarding this fungus:

> You can tell me what genus of fungi the edible one from T. del Fuego comes nearest to; Mr. Brown of course has not only never looked at it a second time, but cannot even lay his hand on the specimens. — I fear I must trouble you to send me one more *good* dried specimen, for I am thinking of having a wood cut. (1967, p. 132)

A woodcut was made, but as *Cyttaria darwinii* was not described by Berkeley until 1845, it was included in the second edition of the *Journal of Researches*, not the first.

Berkeley's type specimen of *Cyttaria darwinii* is on display at the Botany School, University of Cambridge, but most of the other Darwin fungi are at the Royal Botanic Gardens, Kew. On the other hand, the lichens that I have found are at the British Museum (Natural History). A few of the lichens were listed by Berkeley (1842) and by J. D. Hooker (1844–1847). But there are a number of fungi and lichens still unaccounted for as well.

Since Berkeley's herbarium was given to Kew, and Hooker worked there, and the Kew lichens are now at the British Museum (Natural History), this is where they should be sought. Some have been found recently by Dr. David Galloway in boxes stored elsewhere in the Museum from the Cryptogamic Herbarium.

Algae

Darwin's Plant Notes contain several references to "Conferva" and "Fucus", but the only group of his algae that appears to have been systematically identified is the corallines. Most were treated by Darwin as corals in the Zoology Diary, but upon his return they were separated out and sent to the Irish botanist Dr. William Henry Harvey (1811–1866, Curator of the Herbarium, Trinity College, Dublin). They are in the Dublin University Herbarium, but the other groups of algae are not. Only a few are marked "P" in the Specimens in Spirits of Wine Notebooks, so the Plants in Spirits of Wine Notes perhaps would not have been sent to Harvey with these specimens.

In the preface to his work that includes Darwin's corallines, Harvey, among others, acknowledged: "Charles Darwin, Esq., for the liberal donation to our Herbarium of all those which he collected while accompanying H.M.S. 'Beagle' in her voyage round the world, and for liberty to make the freest use of his manuscript notes respecting them" (1847, pp. vii–viii). Those "manuscript notes" which he quotes are from the Zoology Diary, indicating that a separate set of Coralline Algae Notes may have been prepared for him.

Several other algae were included by J. D. Hooker (1844–1847) in the *Flora Antarctica*. They presumably are in the British Museum (Natural History), as this is where the Kew cryptogamic herbarium now resides. The whereabouts of the other algae listed in the Specimens in Spirits of Wine Notebook are unknown. Unlike the bulk of the other plants, they are not at Cambridge.

Mosses and Liverworts

The first bryophyte of Darwin's *Beagle* specimens to be reported in the literature was also the first to be illustrated. *Polytrichium dendroides* was pictured in November 1836 (W. J. Hooker 1836c), soon after Darwin's return. William Jackson Hooker (1785–1865, Regius Professor of Botany at Glasgow University) was an expert on mosses and assisted Henslow with both them and vascular plants. Many were sent to him by Henslow even before Darwin returned to England.

William's son Joseph (1844–1847) described several new bryophytes based on Darwin's collections, as did he and the barrister William Wilson (J. D. Hooker 1847a). These specimens are now in the British Museum (Natural History). A number of unidentified mosses and liverworts are in the Herbarium of the Botany School, University of Cambridge. Presumably these are duplicates of those originally sent to William Hooker in Glasgow. There also are three Darwin *Beagle* mosses in the Herbarium of the Manchester Museum.

Vascular Plants

In spite of the fact that the majority of Darwin's vascular plants of the voyage were not identified until quite recently, they received a surprising amount of attention in the 1830s and 1840s. The first mention of his plant specimens in print, indeed the first printed notice of any of his collections, was published on 1 September 1836 (W. J. Hooker and Arnott 1836). This resulted from Henslow's sending Darwin's specimens to William Hooker before the *Beagle* returned (Porter 1980a).

Starting in 1833, William Hooker and George Walker Arnott (1799–1868, a Scots botanist) began publishing a series of "Contributions towards a flora of South America and the islands of the Pacific". These resulted from the large number of specimens sent to Britain resulting from increased exploration and trade. It was logical for Henslow to appeal to William Hooker for help in identifying Darwin's plants. The series eventually ran to ten numbers, Darwin's specimens being cited in many (W. J. Hooker and Arnott 1836, 1837, 1840, 1841).

The Darwin collections were introduced as follows:

> In addition to the collections of *extratropical South American plants* mentioned at p. 234 of our first volume, as having been lately received by us, we have now the pleasure to announce another, which we owe to the kindness of the Rev. Professor Henslow. It was formed by C. Darwin, Esq., of H.M.S. Beagle, in various countries between Maldonado, in the North, and Terra del Fuego, in the South, including the Falkland Islands, and hence, as may be supposed, it has afforded several new plants and new localities for some rarities which had been described before. (W. J. Hooker and Arnott 1836, pp. 41–42)

Indeed, I would roughly estimate that 10 percent of Darwin's flowering plants were used as the type specimens for new species by W. J. Hooker, Arnott, and others. In addition to the above cited papers, Hooker and Arnott (1842) described and illustrated an aster based on a Darwin specimen, as William Hooker did for several other species (1836a, 1836b, 1842, 1844a, 1844b).

Henslow himself, with William Hooker's assistance, published two papers on Darwin's plants, on a prickly pear cactus (Henslow 1837b) and on the flora of the Cocos-Keeling Islands (Henslow 1838). In the latter, Henslow wrote:

> Mr. Darwin, who accompanied the Beagle in her late voyage round the world, visited these islands in 1836, and is about to give an account of their geological conditions, as well as of the scanty zoology which

> they furnish. As he obligingly presented me with the plants which he collected, together with his memoranda respecting them, I have thought that a list of the species, accompanied by a few remarks, might be of interest; and chiefly as serving to point out a set of plants whose seeds must be provided in a very eminent degree with the means of resisting the influence of sea water. (1838, p. 337)

Given Darwin's later interest in the oceanic dispersal of plants, one cannot but wonder whether here Darwin is influencing Henslow, or vice versa.

The "memoranda" referred to by Henslow are the Plant Notes, found in the Cambridge Herbarium in 1980 (Porter 1981, 1982). Found with them were two notebooks on Darwin's plants compiled by Henslow. It appears that Henslow did more with the plants than he has hitherto been given credit for.

A few Darwin plants are cited here and there in nineteenth-century taxonomic books and papers, but there is no need to list these here. I have now identified almost all of his vascular plants and will soon publish a paper citing all the relevant literature. Publications in which Darwin plant specimens play a major role were written by Joseph Hooker (1844–1847, 1846, 1847a, 1847b) and by the British botanist Philip Barker Webb (1793–1855), who published a flora of the Cape Verde Islands (Webb 1849), citing a number of Darwin specimens. More recently, I have discussed his Galapagos Islands plants (Porter 1980b).

When William Hooker moved from Glasgow University to Kew in 1841 to become first Director of the Royal Botanic Gardens, presumably his herbarium went with him (for example, Turrill 1963). This presumption has not proved to be entirely correct, however. Although the majority of Darwin's (mostly) duplicate plant specimens are now at Kew, a few ferns remain at Glasgow University. A number are at the Royal Botanic Garden, Edinburgh, whence Glasgow's foreign flowering plants were sent in 1965. The Manchester Museum, Oxford University, the Gray Herbarium of Harvard University, and the Missouri Botanical Garden also have a few duplicates. In addition, I have recently discovered that the Webb Herbarium at the University of Florence has some of Darwin's Cape Verde Island collections (Porter 1983a). In spite of a number of specimens sent to William Hooker that were never returned, however, the largest set of Darwin's vascular plants remains at his alma mater, Cambridge University.

ACKNOWLEDGEMENTS

Too many people have provided me with information or assistance for me to acknowledge them individually, so I hereby thank them collectively. However, I should like to especially mention Mr. Peter Sell, Dr. Sydney Smith, Mr. Peter Gautrey, and the late Miss Mea Allan for their initial and continued encouragement of my Darwin research, Dr. Gordon Chancellor for valuable information, and Dr. David Kohn for suggesting the present project and for his patience. Much of the research for this paper was undertaken while I was a Visiting Fellow at Clare Hall, University of Cambridge, and was funded by a grant from the National Geographic Society.

Notes

1. Most of the correspondence cited in this chapter was examined by the author. However, if it has been published, reference is made to the place of publication, not to the archive in which it is to be found.
2. Much of Darwin's correspondence was published by Francis Darwin (*LL*; 1892) and Nora Barlow (Darwin 1945, 1967). Lady Barlow's versions of the letters are cited, as they are less tampered with editorially than those of Sir Francis.
3. Likewise, Francis Darwin (*LL*; 1892) published his father's autobiography in part, but a complete, unedited edition did not appear until that of Barlow (*Autobiography*).
4. That is, the Zoological Society of London or the Geological Society of London.
5. Darwin here probably was referring to the Cambridge Philosophical Society.
6. Many of Darwin's letters home to his family or to Henslow were written over a period of several days or weeks.
7. Author and date of publication for these reports are given here and in the following list.
8. Although Wilson (1972) indicates that Darwin and Lyell may have met at Henslow's home in Cambridge in May 1831.
9. Darwin did author the third volume, and FitzRoy the second and an appendix volume, but the first was by Captain Philip Parker King (1793–1856), captain of the *Adventure* on the 1826–1830 voyage.
10. William Sharp MacLeay (1792–1865), Secretary of the Linnean Society of London.
11. "Of the British Association's third meeting held under the presidency of Professor Adam Sedgwick" (Darwin 1967, p. 77 fn.). The report that Henslow mentions is untraced.
12. That is, the Hunterian Museum of the Royal College of Surgeons.
13. Other things being equal.
14. This letter is not included in Darwin (1967).
15. The Zoological Society of London.
16. *A List of the Genera of Birds*, London, 1840.
17. Another genus of barnacles.
18. Gilbert John Arrow (1873–1948), Deputy Keeper of Entomology at the British Museum (Natural History).
19. This paper was kindly brought to my attention by Dr. James Secord. It is not included in *Collected Papers*.

BIBLIOGRAPHY

Short Titles

Autobiography — Barlow, N., ed. 1958. *The autobiography of Charles Darwin 1809–1882, with the original omissions restored,* edited with appendix and notes by his granddaughter Nora Barlow. London: Collins.

CP — Barrett, P. H., ed. 1977. *The collected papers of Charles Darwin,* with a foreword by Theodosius Dobzhansky. 2 vols. Chicago: University of Chicago Press.

Coral Reefs — Darwin, C. R. 1842. *The structure and distribution of coral reefs, being the first part of the geology of the voyage of the Beagle, under the command of Capt. FitzRoy, R. N. during the years 1832 to 1836.* London: Smith Elder. Reprint 1984. Introduction by M. T. Ghiselin.

1889. 3rd ed. Appendix by T. G. Bonney. London: John Murray.

Descent — Darwin, C. R. 1871. *The descent of man, and selection in relation to sex.* 2 vols. London: John Murray. Reprint 1981. Introduction by J. T. Bonner and R. May. Princeton: Princeton University Press.

1874. 2d ed. London: John Murray.

Diary — Barlow, N., ed. 1933. *Charles Darwin's diary of the voyage of H.M.S. Beagle.* Cambridge: Cambridge University Press.

1842 Sketch — Darwin, C. R. 1958. *Charles Darwin's Sketch of 1842.* In C. Darwin and A. R. Wallace, *Evolution by natural selection,* with a foreword by Sir Gavin de Beer, pp. 39–88. Cambridge: Cambridge University Press.

1844 Essay — Darwin, C. R. 1958. *Charles Darwin's Essay of 1844.* In C. Darwin and A. R. Wallace, *Evolution by natural selection,* with a foreword by Sir Gavin de Beer, pp. 89–254. Cambridge: Cambridge University Press.

Expression — Darwin, C. R. 1872. *The expression of the emotions in man and animals.* London: John Murray. Reprint 1965. Preface by K. Lorenz. Chicago: Chicago University Press.

Journal of Researches — Darwin, C. R. 1839. *Journal of researches into the geology and natural history of the various countries visited by H.M.S. Beagle, under the command of Captain FitzRoy, R. N. from 1832 to 1836.* London: Henry Colburn.

1845. *Journal of researches into the natural history and geology of the countries visited during the voyage of H.M.S. Beagle round the world, under the command of Capt. FitzRoy, R. N.* 2d ed. London: John Murray.

1889. *A naturalist's voyage; journal of researches into the natural history and geology of the countries visited during the voyage of H.M.S. Beagle round the world, under the command of Capt. FitzRoy, R. N.* London: John Murray.

LL — Darwin, F., ed. 1887. *The life and letters of Charles Darwin, including an* *autobiographical chapter.* 3 vols. London: John Murray.

LL (NY) — Darwin, F., ed. 1887. *The life and letters of Charles Darwin, including an autobiographical chapter.* 2 vols. New York: Appleton.

ML — Darwin, F., and Seward, A. C., eds. 1903. *More letters of Charles Darwin; a record of his work in a series of hitherto unpublished letters.* 2 vols. London: John Murray.

Natural Selection — Stauffer, R. C., ed. 1975. *Charles Darwin's Natural Selection; being the second part of his big species book written from 1856 to 1858.* London: Cambridge University Press.

Notebooks

B, C, D, E — De Beer, G.,; Rowlands, M. J.; and Skramovsky, B. M. (eds.) 1960–1967. Darwin's notebooks on transmutation of species. *Bull. Brit. Mus. (Nat. Hist.) Hist. ser.* 2, nos. 2–6: 23–183; 3, no. 5: 129–176.

M, N, OUN — Barrett, P. H., ed. 1974. Darwin's early and unpublished notebooks. In H. E. Gruber and P. H. Barrett, *Darwin on man, a psychological study of scientific creativity; together with Darwin's early and unpublished notebooks,* transcribed and annotated by P. H. Barrett, pp. 266–305; 329–360; 382–413.

Barrett, P. H., 1980. *Metaphysics, materialism and the evolution of the mind. Early writings of Charles Darwin.* Chicago: University of Chicago Press.

RN — Herbert, S. (ed.) 1980. The Red Notebook of Charles Darwin. *Bull. Brit. Mus. (Nat. Hist.) Hist. ser.* 7. Also Ithaca: Cornell University Press.

Orchids — Darwin, C. R. 1862. *On the various contrivances by which British and foreign orchids are fertilised by insects, and on the good effects of intercrossing.* London: John Murray. Reprint 1984. Introduction by M. T. Ghiselin. Chicago: University of Chicago Press.

1877. *The various contrivances by which orchids are fertilised by insects.* 2d ed. London: John Murray.

Origin — Darwin, C. R. 1859. *On the origin of species by means of natural selection, or the preservation of favoured races in the struggle for life.* London: John Murray. Reprint 1975. Introduction by E. Mayr. Cambridge: Harvard University Press.

1872. 6th ed. London: John Murray.

1959. *The origin of species by Charles Darwin; a variorum text,* ed. by M. Peckham. Philadelphia: University of Pennsylvania Press.

Variation — Darwin, C. R. 1868. *The variation of animals and plants under domestication.* 2 vols. London: John Murray.

1875. 2d ed. London: John Murray.

Zoology — Darwin, C. R., ed. 1839–1843. *The zoology of the voyage of H.M.S. Beagle, under the command of Captain FitzRoy, during the years 1832 to 1836.* London: Smith Elder.
5 parts
Part 1. Owen, R. *Fossil mammalia.*
Part 2. Waterhouse, G. R. *Mammalia.*
Part 3. Gould, J. *Birds.*
Part 4. Jenyns, L. *Fish.*
Part 5. Bell, T. *Reptiles.*

Bibliography

Abbott, L. F. 1927. Charles R. Darwin, the saint. In *Twelve great modernists*, pp. 225–251. New York: Doubleday, Page.

Acanfora, M. 1980. La zoopsicologia di K. F. Rulè. *Med. secoli* 17: 225–236.

--- **and Scudo, F. M.** Klim Arkadevich Timiriazev and evolutionary theory. Unpublished manuscript.

Ackermann, R. 1800. *Le Brun traversed, or caricatures of the passions*. London: Ackermann.

Adams, M.B. 1980a. Severtsov and Schmalhausen: Russian morphology and the evolutionary synthesis. In *The evolutionary synthesis: Perspectives on the unification of biology*, eds. E. Mayr and W. B. Provine, pp. 193–225, Cambridge: Harvard University Press.

--- 1980b. Sergei Chetverikov, the Kol'tsov Institute and the evolutionary synthesis. In *The evolutionary synthesis: perspectives on the unification of biology*, eds. E. Mayr and W. B. Provine, pp. 242–278. Cambridge: Harvard University Press.

Addison, W. 1947. *The English country parson*. London: J. M. Dent.

Adler, S. 1959. Darwin's illness. *Nature* 184: 1102–1103.

Agassiz, L. 1859. *An essay on classification*. London: Longmans, Green. 1962. Reprinted, ed. E. Lurie. Cambridge: Harvard University Press.

--- 1860. *Contributions to the natural history of the United States of America*. Vol. 3. Boston: Little, Brown.

--- 1874. Evolution and permanence of type. *Atlantic monthly* 33: 92–101.

Alchian, A. 1950. Uncertainty, evolution, and economic theory, *J. polit. econ.* 58: 211–222.

Aldred, G. A. [1942.] *The devil's chaplain: the story of the Rev. Robert Taylor, M.A., M.R.C.S. (1784–1844)*. Glasgow: Strickland Press.

Aldrich, M. 1974. United States: bibliographical essay. In *The comparative reception of Darwinism*, ed. T. F. Glick, pp. 207–226. Austin: University of Texas Press.

Allan, M. 1967. *The Hookers of Kew, 1785–1911*. London: Michael Joseph.

--- 1977. *Darwin and his flowers: the key to natural selection*. London: Faber and Faber.

Allen, D. E. 1976. *The naturalist in Britain: a social history*. London: Allen and Unwin.

--- 1979. The lost limb: geology and natural history. In *Images of the earth*, eds. L. J. Jordanova and R. Porter, pp. 200–212. Chalfont St. Giles, Bucks: British Society for the History of Science.

Allen, G. E. 1968. Thomas Hunt Morgan and the problem of natural selection. *J. hist. biol.* 1: 113–139.

--- 1969. Hugo de Vries and the reception of the mutation theory. *J. hist. biol.* 2: 55–87.

--- 1975. *Life science in the twentieth century*. New York: John Wiley.

--- 1978. *Thomas Hunt Morgan: the man and his science*. Princeton: Princeton University Press.

--- 1979. Naturalists and experimentalists: the genotype and the phenotype. *Stud. hist. biol.* 3: 179–209.

--- 1981. Morphology and twentieth-century biology: a response. *J. hist. biol.* 14: 159–176.

Altholz, J. L. 1980. The Huxley-Wilberforce debate revisited. *J. hist. med.* 35: 313–316.

Altner, G. 1966. *Charles Darwin und Ernst Haeckel: ein Vergleich nach theologischen Aspekten*. Zurich: EVZ-Verlag.

--- 1981. *Der Darwinismus: die Geschichte einer Theorie*. Darmstadt: Wissenschaftliche Buchgesellschaft.

American Philosophical Society. 1959. Commemoration of the centennial of the publication of the *Origin of Species* by Charles Darwin. *Proc. Am. Phil. Soc.* 103: 159–319, 609–644, 716–725.

Anderson, P. 1966. Socialism and pseudo-empiricism. *New left rev.* 35: 2–42.

Andrée, K. 1960. Charles Darwin als Geologer. In *Hundert Jahre Evolutionsforschung*, eds. G. Herberer and F. Schwanitz, pp. 277–289. Stuttgart: Gustav Fischer.

Andrews, F. M., ed. 1979. *Scientific productivity*. New York: Cambridge University Press.

Annan, N. G. 1955. The intellectual aristocracy. In *Studies in social history: a tribute to G. M. Trevelyan*, ed. J. H. Plumb, pp. 241–287. London: Longmans, Green.

Anon. 1840. Journal of proceedings. *Trans. Entomol. Soc. Lond.* 2: xli, lxxii–lxxiii.

--- 1858. *Cottage gardener* 19: 256.

--- 1860. Sulla origine delle specie e la variazione delle razze, di Carlo Darwin. *Il politecnico* 9: 110–112.

--- 1871. Artistic feeling of the lower animals. *The spectator.* (11 March 1871): pp. 281, 319–320.

--- 1888. Darwin in Edinburgh. *St. James gaz.* 16: part 1, pp. 5–6; part 2, pp. 7–8.

--- 1978. A genetic defense of the free market. *Business week*, 10 April, pp. 103–104.

--- 1980. Editorial. *Computer weekly* 24 January.

Antioch Review. 1959. *"The origin of species" — 100 years later.* Special issue, 19 (Spring): 5–68.

Antonovich, M. A. 1896. *Charl'za Darwina i ego teoria* (Charles Darwin and his Theory), brochure, St. Petersburg.

Appel, T. A. 1980. Henri de Blainville and the animal series. *J. hist. biol.* 13: 291–319.

Appleby, J. O. 1978. *Economic thought and ideology in seventeenth-century England.* Princeton: Princeton University Press.

Appleman, P. 1959a. The logic of evolution: some reconsiderations. *Victorian stud.* 3: 115–125.

---, ed. 1959b. *1859: entering an age of crisis.* Bloomington: Indiana University Press.

---, ed. 1970. *Darwin: a Norton critical edition.* New York: Norton.

Arber, A. 1950. *The natural philosophy of plant form.* Cambridge: Cambridge University Press.

Argyll, G. D. Campbell, 8th Duke of, 1867. *The reign of law.* London: Alexander Strahan.

--- 1884. *The unity of nature.* London: Alexander Strahan.

--- 1885. What is science? *Good words* 26: 236–245.

--- 1888. Prophetic germs. *Nature* 38: 564.

--- 1898. *Organic evolution cross-examined.* London: John Murray.

Aristotle. 1963. *De anima.* Ed. W. D. Ross. Oxford: Oxford University Press.

--- 1972. *De partibus animalium I and De generatione animalium I.* Ed. D. Balme. Oxford: Oxford University Press.

Arrow, K. J. 1979. *The division of labor in the economy, the polity and society.* In *Adam Smith and modern political economy*, ed. C. P. O'Driscoll Jr. Ames : Iowa State University Press.

Ashforth, A. 1969. *Thomas Henry Huxley.* New York: Twayne.

Ashton, R. D. 1980. *The German Idea: Four English writers and the reception of German thought, 1800–1860.* Cambridge: Cambridge University Press.

Ashworth, J. H. 1935. Charles Darwin as a student in Edinburgh, 1825–1827. *Proc. Roy. Soc. Edin.* 55: 97–113.

Asimov, I. 1964. *A short history of biology.* New York: Doubleday.

Atkins, H. 1974. *Down, the home of the Darwins: the story of a house and the people who lived there.* [rev. ed. 1976] London: published under the auspices of the Royal College of Surgeons of England, Lincoln's Inn Fields.

Audouin, J. V. et al., eds. 1822–1831. *Dictionnaire classique d'histoire naturelle.* 17 vols. Paris: Rey and Gravier.

Aveling, E. B. 1882a. Charles Darwin on religion. *National reformer* N.S. 40: 273.

--- 1882b. A visit to Charles Darwin. *National reformer* N.S. 40: 273–274, 291–293.

--- 1883. *The religious views of Charles Darwin.* London: Free Thought Publishing Co.

Avineri, S. 1967. From hoax to dogma: a footnote on Marx and Darwin. *Encounter* 28: 30–32.

Ayala, F. J. 1977. Philosophical issues. In *Evolution*, eds. T. Dobzhansky, F. Ayala, G. L. Stebbins, and J. W. Valentine, pp. 474–516. San Francisco: W. H. Freeman.

Ayer, A. J. 1959. The structure of Darwinism. *New biol.* 22: 25–44.

Babbage, C. 1832. *On the economy of machinery and manufactures.* London: John Murray.

--- 1838. *The ninth Bridgewater treatise: a fragment.* 2d ed. London: John Murray.

Babington, C. C. 1841. Dytiscidae darwinianae; or, descriptions of the species of Dytiscidae collected by Charles Darwin, Esq., M. A. Sec. G. S. &c., in South America and Australia, during his voyage in H.M.S. Beagle. *Trans. Entomol. Soc. Lond.* 3: 1–17.

Babkov, V. V. forthcoming. *Moscowskaia Shkola Evoliuzionnoi genetiki* (The Moscow school of evolutionary genetics) Moscow: Nauka.

Bachelard, G. 1951a. *L'activité rationaliste de la physique contemporaine.* Paris: Presses Universitaires de France.

--- 1951b. *L'actualité de l'histoire des sciences.* Paris: Editions du Palais de la Découverte.

Baehni, C. 1955. Correspondance de Charles Darwin et d'Alphonse de Candolle. *Gesnerus*

12: 109–156.

Baer, K. E. von (Baer, K. M.) 1828. *Organologia zhivotnykh* (Animal organology), brochure, University of St. Petersburg (also Menzbir, ed. 1938).

--- 1828–1837. *Über Entwickelungsgeschichte der Thiere.* Königsberg: Bornträger.

--- 1865. *Nachrichten über Leben und Schriften des Geheimaraths Dr. Karl Ernst von Baer, mittheilt von ihm selbst.* St. Petersburg: Kaiserliche Akad. der wissenschaften. 1886. 2d ed. Braunschweig: Vieweg und Sohn. Russian translation (Moscow, 1950).

--- 1873. Zum Streit über den Darwinismus. *Allg. Zeit.* 130: 1986–1988.

--- 1876a. *Razvitie zhivotnykh i razvitie cheloveka* (The development of animals and the development of humans, lecture), brochure, St. Petersburg.

--- 1876b. *Über Darwins Lehre.* brochure, St. Petersburg.

Bailey, E. 1962. *Charles Lyell.* London: Nelson.

Baillaud, L. 1966. Le mémoire de Charles Darwin sur les plantes grimpantes. *Arch. int. hist. sci.* 19: 235–246.

Bain, A. 1864. *Senses and the intellect.* 2d ed. London: Longmans.

--- 1865. *Emotions and the will.* 2d ed. London: Longmans.

--- 1873. *Senses and the intellect.* 3rd ed. London: Longmans.

--- 1904. *Autobiography.* London: Longmans.

Baker, H. G. 1965. Charles Darwin and the perennial flax — a controversy and its implications. *Huntia* 2: 141–161.

Baker, W. 1977. Herbert Spencer and "evolution" — a further note. *J. hist. ideas* 38: 476.

Balan, B. 1979. *L'ordre et le temps: l'anatomie comparée et l'histoire des vivants au XIXe siècle.* Paris: Vrin.

Baldwin, J. M. 1896. A new factor in evolution. *Am. nat.* 30: 536–553.

Ball, T. 1979. Marx and Darwin: a reconsideration. *Polit. theory* 7: 469–483.

Balme, D. 1980. Aristotle's biology was not essentialist. *Arch. für Geschichte der Philosophie* 62: 1–12.

Balmer, H. 1974. Louis Agassiz, 1807–1873. *Gesnerus* 31: 1–18.

Bannister, R. C. 1970. "The survival of the fittest is our doctrine": history or histrionics? *J. hist. ideas* 31: 377–397.

--- 1979. *Social Darwinism: science and myth in Anglo-American social thought.* Philadelphia: Temple University Press.

Banton, M., ed. 1961. *Darwinism and the study of society. A centenary symposium.* London: Tavistock; Chicago: Quadrangle Books.

Barasch, M. 1976. *Gestures of despair in Medieval and early Renaissance art.* New York: N5ew York University Press.

Barber, B. 1961. Resistance by scientists to scientific discovery. *Science* 134: 596–602.

Barber, L. 1980. *The heyday of natural history, 1820–1870.* London: Jonathan Cape.

Barclay, V. C. 1950. *Darwin is not for children.* London: Herbert Jenkins.

Barié, O. 1953. *Idee e dottrine imperialistiche nell'Inghilterra vittoriana.* Bari: Laterza.

Barker, A. D. 1969. An approach to the theory of natural selection. *Philosophy* 44: 271–290.

Barnes, B. 1974. *Scientific knowledge and sociological theory.* London: Routledge and Kegan Paul.

--- **and Shapin, S.**, eds. 1979. *Natural order: historical studies of scientific culture.* Beverly Hills and London: Sage Publications.

Barnes, H. E., ed. 1948. *An introduction to the history of sociology.* Chicago: University of Chicago Press.

Barnes, J. A. 1960. Anthropology in Britain before and after Darwin. *Mankind* 5: 369–385.

Barnett, S. A., ed. 1958a. *A century of Darwin.* Cambridge: Harvard University Press.

--- 1958b. The expression of the emotions. In *A century of Darwin,* ed. S. A. Barnett, pp. 206–230. Cambridge: Harvard University Press.

Baron, W. 1956. Zu L. Agassiz's Beurteilung des Darwinismus. *Sudhoffs Arch.* 40: 259–277.

--- 1968. Wissenschaftliche Analyse der Begriffe Entwicklung, Abstammung und Entstehung im 19. Jahrhundert. *Technikgeschichte* 35: 68–79.

Barrett, P. H., ed. 1960. A transcription of Darwin's first notebook on "Transmutation of species." *Bull. Mus. Comp. Zool. Harvard* 122: 245–296.

--- 1972. *Computerized print-out: Darwin's transmutation notebooks — B, C, D, E.* Lansing: Michigan State University.

--- 1973. Darwin's "gigantic blunder". *J. geol. educ.* 21: 19–28.

--- 1974. The Sedgwick-Darwin geologic tour of North Wales. *Proc. Am. Phil. Soc.* 118: 146–164.

---, ed. 1977. *The collected papers of Charles*

Darwin: 2 vols. Chicago: University of Chicago Press.

--- **and Corcos, A.** 1972. A letter from Alexander Humboldt to Charles Darwin. *J. hist. Med.* 27: 159–172.

---; **Weinshank, D. J.; and Gottleber, T. T.** 1981. *A concordance to Darwin's* Origin of Species, *first edition*. Ithaca: Cornell University Press.

Barsanti, G. 1979. *Dalla storia naturale alla storia della natura. Saggio su Lamarck*. Milano: Feltrinelli.

Barthes, R. 1973. *Le plaisir du texte*. Paris: Editions du Seuil.

Bartholomew, M. 1973. Lyell and evolution: an account of Lyell's response to the prospect of an evolutionary ancestry for man. *Brit. j. hist. sci.* 6: 261–303.

--- 1975. Huxley's defence of Darwin. *Ann. sci.* 32: 525–535.

--- 1976. The non-progress of non-progression: two responses to Lyell's doctrine. *Brit. j. hist. sci.* 9: 166–174.

--- 1979. The singularity of Lyell. *Hist. sci.* 7: 276–293.

Bartov, H. 1977. A fortiori arguments in the Bible, in Paley's writings and in the "Origin of Species". *Janus* 64: 131–145.

Bartucz, L. 1964. Világnézeti viták és egyéb tényezök a budapesti embertani tanszék felállításának hatterében 85 év elött. [Ideological debates and other factors in the establishment of the Chair of Anthropology at the University of Budapest 85 years ago.] *Anthropológiai közlemények* nos. 3–4: 51–68.

Barzun, J. 1958. *Darwin, Marx and Wagner: critique of a heritage*. 2d ed. Garden City: Anchor.

Basalla, G. 1964. Darwin's orchid book. *Proc. 10th int. cong. hist. sci.* (Ithaca 1962) Paris: Hermann.

Basalla, G.; Coleman, W; and Kargon, R. H., eds. 1970. *Victorian science, a self-portrait from the presidential addresses to the British Association for the Advancement of Science*. New York: Doubleday.

Bates, H. W. 1862. Contributions to an insect fauna of the Amazon valley. Lepidoptera: Heliconidae. *Trans. Linn. Soc. Lond.* 23: 495–566.

Bates, M., and Humphrey, P. S. 1956. *The Darwin reader*. New York and London: Macmillan.

Bateson, G. 1980. *Mind and nature: a necessary unity*. Glasgow: Fontana/Collins.

Bateson, W. 1894. *Materials for the study of variation: treated with especial regard to discontinuity in the origin of species*. London: Macmillan.

--- 1928. *The scientific papers of William Bateson*. Ed. R. C. Punnett. 2 vols. Cambridge: Cambridge University Press.

Baumel, H. 1978. *Biology: its historical development*. New York: Philosophical Library.

Baynes, T. S. 1873. Darwin on expression. *Edin. rev.* 137: 492–528.

Bean, W. B. 1978. The illness of Charles Darwin. *Am. j. med.* 65: 572–574.

Beatty, J. 1980. Optimal-design models and the strategy of model building in evolutionary biology. *Phil. sci.* 47: 532–562.

--- 1982. What's in a word: Coming to terms in the Darwinian revolution. *J. hist. biol.* 15: 215–239.

Bechstein, J. M. 1789–1795. *Gemeinnützige Naturgeschichte Deutschlands nach allen drey Reichen*. 4 vols. Leipzig: S.L. Crusius.

Becker, H., and Barnes, H. E. 1961. *Social thought from lore to science*. 3rd ed. New York: Dover.

Beckner, M. 1959. *The biological way of thought*. Berkeley: University of California Press.

--- 1967. Darwinism. In *The encyclopaedia of philosophy*, ed. P. Edwards, 2: 296–306. New York: Macmillan.

Beddall, B. G. 1968. Wallace, Darwin, and the theory of natural selection, a study in the development of ideas and attitudes. *J. hist. biol.* 1: 261–323.

---, ed. 1969. *Wallace and Bates in the tropics: an introduction to the theory of natural selection*. New York: Macmillan.

--- 1972. Wallace, Darwin, and Edward Blyth: further notes on the development of evolutionary theory. *J. hist. biol.* 5: 153–158.

--- 1973. "Notes for Mr. Darwin": Letters to Charles Darwin from Edward Blyth at Calcutta: a study in the process of discovery *J. hist. biol.* 6: 69–95.

--- 1975. "Un naturalista original": Don Felix de Azara, 1746–1821. *J. hist. biol.* 8: 15–66.

Beer, G. 1983a. *Darwin's plots: evolutionary narrative in Darwin, George Eliot and nineteenth-century fiction*. London: Routledge.

--- 1983b. Anxiety and interchange: *Daniel Deronda* and the implications of Darwin's writing. *J. hist. behav. sci.* 19.

--- forthcoming. The face of nature: anthropomorphic elements in Darwin's style. In *The language of nature*, ed. L. Jordanova.

Beishon, J. and Peters, G., eds. 1972. *Systems behaviour*. New York: Harper and Row.

Beketov, A. N. 1882. *Darvinisms tochki zrenia obshchefizicheskikh nauk* (Darwinism from the point of view of general physical sciences.) Obsc. estestvoispit., St. Petersburg.

--- 1887. Izuchenie krabov zhivushchikh v vodakh raznoi stepeni salionosti (A study on crabs living in waters of different salinities), *Zool. Inst.*, Odessa.

Bell, C. 1806. *Essays on the anatomy of expression in painting*. London: Longman.

--- 1821. *On the nerves; giving an account of some experiments on their structure and functions, which lead to a new arrangement of the system*. London: Bulmer.

--- 1822. *Of the nerves which associate the muscles of the chest in the actions of breathing, speaking, and expression: being a continuation of the paper on the structure and functions of the nerves*. London: Nichol.

--- 1824. *Essays on the anatomy and philosophy of expression*. 2d ed. London: John Murray.

--- 1833. *The hand: its mechanism and vital endowments as evincing design*. London: W. Pickering.

--- 1844. *The anatomy and philosophy of expression as connected with the fine arts*. 3rd ed. London: John Murray.

--- 1872. *The anatomy and philosophy of expression as connected with the fine arts*. 6th ed. London: Henry G. Bohn.

Bell, D. C. 1979. William Graham Sumner as an antisocial Darwinist. *Pacific sociol. rev.* 22: 309–331.

Bell, P. R., ed. 1959a. *Darwin's biological work: some aspects reconsidered*. Cambridge: Cambridge University Press.

--- 1959b. The movement of plants in response to light. In *Darwin's biological work*, ed. P. R. Bell, pp.1–49. Cambridge: Cambridge University Press.

Bell, S. 1981. George Henry Lewes: a man of his time. *J. hist. biol.* 14: 277–298.

Bellone, E. 1976. *Il mondo di carta*. Milano: Mondadori.

Belt, T. 1874. *The naturalist in Nicaragua*. London: John Murray.

Benasso, G. 1976. Materiali per una storia dell'evoluzionismo italiano: da Bonelli a De Filippi (1811–1864). *Atti dell'Accademia Roveretana degli Agiati* 14–15: 5–106.

--- 1978. Materiali per una storia dell'evoluzionismo Italiano: un approccio al darwinismo, 1864–1900. *Atti della Accademia Roveretana degli Agiati. Classe di scienze matematiche, fisiche e naturali* 16–17: 73–152.

Ben-David, J. 1971. *The scientist's role in society: a comparative study*. Englewood Cliffs: Prentice Hall.

Benedek, I. 1961. *A darwinizmus kibontakozása* [The development of Darwinism]. Budapest: Tankönyvkiadó.

Benson, K. 1981. Problems of individual development: descriptive morphology in America at the turn of the century. *J. hist. biol.* 14: 115–128.

Benton, T. 1979. Natural science and cultural struggle: Engels on philosophy and the natural sciences. In *Issues in Marxist philosophy, vol. 2: materialism*, eds. J. Mepham and D. H. Ruben, pp. 101–142. Brighton: Harvester Press.

--- 1982. Social Darwinism and socialist Darwinism in Germany, 1860 to 1900. *Rivista di filosofia* 22–23: 79–121.

Berg, M. 1980. *The machinery question and the making of political economy, 1815–1848*. Cambridge: Cambridge University Press.

Berkeley, M. J. 1839. Notice of some fungi collected by C. Darwin, Esq., during the expedition of H. M. Ship *Beagle*. *Ann. nat. hist.* 4: 291–293.

--- 1842. Notice of some fungi collected by C. Darwin, Esq., in South America and the islands of the Pacific. *Ann. mag. nat. hist.* 9: 443–448.

--- 1845. On an edible fungus from Tierra del Fuego, and an allied Chilean species. *Trans. Linn. Soc. Lond.* 19: 37–43.

Berlin, I. 1978. Marx's *Kapital* and Darwin. *J. hist. ideas* 39: 519.

Bernal, J. 1965. *Science in history*. 3rd ed. London: Watts.

Bertalanffy, L. von. 1932. *Theoretische Biologie*: Berlin: Borntraeger.

--- 1933. *Modern theories of development*. London: Oxford University Press.

--- 1937. *Das Gefüge des Lebens*. Leipzig und Berlin: B. G. Teubner.

--- 1952. *Problems of life*. London: Watts.

--- 1969. *General system theory: foundations, development, applications*. New York: Braziller.

--- 1975. *Perspectives on general system theory:*

scientific-philosophical studies. Ed. E. Taschdjian. New York: Braziller.

Bethell, T. 1976. Darwin's mistake. *Harper's mag.* 252: 70–75.

--- 1978. Burning Darwin to save Marx. *Harper's mag.* 254: 31–38, 91–92.

Bettelheim, B. 1982. Freud and the soul. *The New Yorker* 1 March, pp. 52–93.

--- 1983. *Freud and man's soul*. New York: Knopf.

Bezirgan, N. A. 1974. The Islamic world. In *The comparative reception of Darwinism*, ed. T. F. Glick pp. 357–387. Austin: University of Texas Press.

Bianconi, G. G. 1864. *La teoria dell'uomo scimmia esaminata sotto il rapporto dell'organizzazione*. Bologna: Gamberini e Parmeggiani.

--- 1874. *La teoria Darwiniana e la creazione detta indipendente*. Bologna: Zanichelli.

Bibby, C. 1959a. *Thomas Henry Huxley: scientist, humanist and educator*. London: Watts.

--- 1959b. Huxley and the reception of the "Origin". *Victorian stud.* 3: 76–86.

--- 1972. *Scientist extraordinary: the life and work of Thomas Henry Huxley, 1825–1895*. Oxford: Pergamon Press.

Bielozersky, A. N., and Mikulinsky, S. R. 1967. *Uspekci sovietskoi biologii* (The success of Soviet biology.) Moscow: Snanie.

Biermann, K. R. 1972. Friedrich Wilhelm Heinrich Alexander von Humboldt. In *Dictionary of scientific biography*, vol. 6, 549–555. New York: Charles Scribner's Sons.

Bikaplan, E. 1948. Duchenne of Boulogne and the *Physiologie des mouvements*. In *Victor Robinson memorial volume*, ed. S. R. Kagan, pp. 172–192. New York: Froben.

Birch, L. C. 1957. The meanings of competition. *Am. nat.* 91: 5–18.

Black, M. 1962. *Models and metaphors: studies in language and philosophy*. Ithaca: Cornell University Press.

Blainville, H. 1816–1830. *Planches: vers et zoophytes* and *conchologie et malacologie*. In *Dictionnaire des sciences naturelles: supplément*. Paris: Levrault.

Blake, W. H. 1933. *A preliminary study of the interpretation of bodily expression*. New York: Columbia University, Teacher's College.

Blanckaert, C. 1979. Edmond Perrier et l'étiologie du "Polyzoïsome organique". *Revue de synthèse* 100, nos. 95–96: 353–376.

--- 1981. *Monogénisme et Polygénisme en France de Buffon à P.Broca (1749–1880)*. 3 vols. Thèse de Doctorat de 3e cycle. Paris: Université de Paris I. Panthéon-Sorbonne.

Bliakher, L. I. 1971. Georg Seidlitz und seine Vorlesungsreihe Über Darwinismus an der Universität Dorpat in den 60er–70er Jahren des 19. Jahrhunderts. In *Colloquium: Evolutionstheorie und Genetik*, Moscow: Nauka.

--- 1973. Die Diskussion zwischen Spencer und Weismann über die Bedeutung der natürlichen Zuchtwahl und der direkten Anpassung für Evolution. *Zeitschrift für Geschichte der Naturwissenschaften, Technik, und Medizin* 10: 50–58.

Blinderman, C. S. 1971. The great bone case. *Perspect. biol. med.* 14: 370–393.

Blomefield, L. [formerly Jenyns]. 1889. *Chapters in my life, with appendix containing special notices of particular incidents and persons; also, thoughts on certain subjects*. Rev. ed. Bath: printed for private circulation.

Bloor, D. 1976. *Knowledge and social imagery*. London: Routledge and Kegan Paul.

Bock, K. E. 1955. Darwin and social theory. *Phil. sci.* 22: 122–134.

--- 1964. Theories of progress and evolution. In *Sociology and History*, eds. W. J. Cahman and A. Boskoff, pp. 21–41. Glencoe: Free Press.

--- 1980. *Human nature and history: a response to sociobiology*. New York: Columbia University Press.

Bodenheimer, F. S. 1958. *A history of biology: an introduction*. London: W. Dawson.

Boelsche, W. 1906. *Haeckel: his life and work*. London: T. Fisher Unwin.

Boesiger, E. 1967. La signification évolutive de la sélection sexuelle chez les animaux. *Scientia* 102: 207–223.

Bogardus, E. S. 1960. *The development of social thought*. New York: Longmans and Green.

Bogdanov, A. P. 1861. *Zoologicheskaia khrestomatia* (An introduction to zoology). University of St. Petersburg.

Bolle, F. 1981. Monistische Mauerei. *Medizinhistorisches j.* 16: 280–301.

Boring, E. 1950a. *A history of experimental psychology*. 2d ed. New York: Appleton-Century-Crofts.

--- 1950b. The influence of evolutionary thought upon American psychological thought. In *Evolutionary thought in America*, ed. S. Persons, pp. 268–298. New Haven: Yale University Press.

Boros, I. 1959a. A 100 esztendös darwinizmus magyarországi pályafutása [The centennial career of Darwinism in Hungary]. *Elővilág* 4: 3–7, 20–25.

--- 1959b. Darwin és a darwinizmus [Darwin and Darwinism]. *Élet és tudomány* 51: 1603–1607.

Bory de Saint-Vincent, J.-B.-G.-M., ed. 1822–1831. *Dictionnaire classique des sciences naturelles*. 17 vols. Paris: Rey and Gravier.

Bostetter, E. E. 1970. Coleridge's manuscript essay *On the Passions*. *J. hist. ideas* 31: 99–108.

Botnariuc, N. 1961. *Din Historia Biologhiei Generale*. Bucharest: Editura Siintifica.

Bouanchaud, D. H. 1976. *Charles Darwin et le transformisme*. Paris: Payot.

Bourdier, F. 1960. Trois siècles d'hypothèses sur l'origine et la transformation des êtres vivants (1550–1859). *Rev. d'hist. sci.* 13: 1–44.

Bowen, F. 1860. Darwin on the origin of species. *North Am. rev.* 90: 474–506.

Bowler, P. J. 1974a. Darwin's concepts of variation. *J. hist. med.* 29: 196–212.

--- 1974b. Evolutionism in the enlightenment. *Hist. sci.* 12: 159–183.

--- 1975a. The changing meaning of "evolution". *J. hist. ideas* 36: 95–114.

--- 1975b. Herbert Spencer and "evolution" — an additional note. *J. hist. ideas* 35: 367.

--- 1976a. *Fossils and progress: paleontology and the idea of progressive evolution in the nineteenth century*. New York: Science History Publications.

--- 1976b. Malthus, Darwin, and the concept of struggle. *J. hist. ideas* 37: 631–650.

--- 1976c. Alfred Russel Wallace's concepts of variation. *J. hist. med.* 31: 17–29.

--- 1977a. Darwinism and the argument from design: suggestions for a re-evaluation. *J. hist. biol.* 10: 29–43.

--- 1977b. Edward Drinker Cope and the changing structure of evolution theory. *Isis* 68: 249–265.

--- 1978. Hugo de Vries and Thomas Hunt Morgan: the mutation theory and the spirit of Darwinism. *Ann. sci.* 35: 55–73.

--- 1979. Theodor Eimer and orthogenesis: evolution by "definitely directed variation". *J. hist. med.* 34: 40–73.

--- 1983. *The eclipse of Darwinism: anti-Darwinian evolution theories in the decades around 1900.* Baltimore: Johns Hopkins University Press.

--- 1984. *Evolution: the history of an idea.* Berkeley: University of California Press.

Box, J. F. 1978. *R. A. Fisher: the life of a scientist.* New York: Wiley.

Brabec, F. 1960. Darwins Genetik im Lichte der modernen Vererbungslehre erläutert an botanischen Beispielen. In *Hundert Jahre Evolutionsforschung*, eds. G. Heberer and F. Schwanitz, pp. 99–122. Stuttgart: Gustav Fischer.

Brackman, A. C. 1980. *A delicate arrangement: the strange case of Charles Darwin and Alfred Russel Wallace*. New York: Times Books.

Bradie, M., and Gromko, M. 1981. The status of the principle of natural selection. *Nature and system* 3: 3–12.

Braithwaite, R. B. 1953. *Scientific explanation: a study of the function of theory, probability and law in science*. Cambridge: Cambridge University Press.

Brandon, R. 1978. Adaptation and evolutionary theory. *Stud. hist. phil. sci.* 9: 181–206. 1983. Reprinted in *Conceptual issues in evolutionary biology*, ed. E. Sober, pp. 58–82. Cambridge: M.I.T. Press.

Brannigan, A. 1979. The reification of Mendel. *Soc. stud. sci.* 9: 423–454.

--- 1981. *The social basis of scientific discoveries.* Cambridge: Cambridge University Press.

Bratchell, D. F. 1981. *The impact of Darwinism: texts and commentary illlustrating nineteenth century religious, scientific and literary attitudes.* London: Avebury.

Breck, A. D. 1972. The use of biological concepts in the writing of history. In *Biology, history, and natural philosophy*, eds. A. D. Breck and W. Yourgrau, pp. 217–232. New York: Plenum Press.

--- **and Yourgrau, W.,** eds. 1972. *Biology, history, and natural philosophy*. New York: Plenum Press.

Breitenbach, W. 1913. *Die Gründung und erste Entwicklung des deutschen Monistenbundes.* Brackwede: W. Breitenbach.

Brent, P. 1981. *Charles Darwin: "A man of enlarged curiosity".* London: Heinemann.

Bresson, F., and de Montmollin, M., eds. 1966. *Psychologie et épistémologie: thèmes piagetiens*. Paris: Dunod.

Brian, A. D., and Crane, E. E. 1959. Charles Darwin and bees. *Bee world* 40: 297–303.

Briegel, M. 1963. *Evolution: Geschichte eines*

Fremdworts in Deutschen. Diss. Freiburg i. B.

Brien, P. 1960a. *Bryozoaires.* In *Traite de zoologie,* ed. P. Grassé, vol. 5. Paris: Masson.

--- 1960b. À l'occasion d'un glorieux centenaire: "l'origine des espèces" de Charles Darwin et le problème de l'évolution. *Scientia* 95: 156–160.

Brock, W. H. 1979. Chemical geology or geological chemistry? In *Images of the earth,* eds. L. J. Jordanova and R. Porter, pp. 147–170. Chalfont St. Giles, Bucks: British Society for the History of Science.

--- **and MacLeod, R. M.** 1976. The scientists' declaration: reflexions on science and belief in the wake of "Essays and reviews", 1864–1865. *Brit. j. hist. sci.* 11: 39–66.

Brockway, L. H. 1979. *Science and colonial expansion: the role of the British Royal Botanic Gardens.* New York: Academic Press.

Brodie, B. C. 1854. *Psychological inquiries in a series of essays, intended to illustrate the mutual relations of the physical organisation and the mental faculties.* London: Longman, Brown, Green, and Longmans.

Bronn, H. G. 1858. *Morphologische Studien über die Gestaltungs-Gesetze der Naturkörper überhaupt, und der organischen insbesondere.* Leipzig and Heidelberg: C. F. Winter.

Brooke, J. H. 1977a. Richard Owen, William Whewell, and the "Vestiges". *Brit. j. hist. sci.* 35: 132–145.

--- 1977b. Natural theology and the plurality of worlds: observations on the Brewster-Whewell debate. *Ann. sci.* 34: 221–286.

--- **and Richardson A.** 1974. *The crisis of evolution.* Milton Keynes: Open University Press.

---; **Hooykaas, R.; and Lawless, C.** 1974. *New interactions between theology and natural science.* Milton Keynes: Open University Press.

Brooks, J. L. 1969. Re-assessment of A. R. Wallace's contribution to the theory of organic evolution. *Year Book 1968, Am. Phil. Soc.*, 534–535.

--- 1972. Extinction and the origin of organic diversity. *Trans. Conn. Acad. Arts Sci.* 44: 19–56.

Brougham, H., Lord. 1839. *Dissertations on subjects of science connected with natural theology.* 2 vols. London: C. Knight.

--- 1845. *Dialogues on instinct; with analytical view of the researches on fossil osteology.* Philadelphia: E. Ferrett.

Brower, L. 1969. Ecological chemistry, *Sci. Amer.* 220 (Feb.): 22–29.

---; **Pough, F. H.; and Meck, H. R.** 1970. Theoretical investigations of automimicry I. *P.N.A.S.* 66: 1059–1066.

Brown, R. 1828. A brief account of microscopical observations made in the months of June, July and August 1827, on the particles contained in the pollen of plants, and on the general existence of active molecules in organic and inorganic bodies. *Edin. new phil. j.* 5: 358–371.

Browne, C. A. 1944. *A source book of agricultural chemistry.* Chronica Botanica, vol. 8. Waltham: Chronica Botanica.

Browne, E. J. 1978a. *C. R. Darwin and J. D. Hooker: episodes in the history of plant geography.* Ph.D. thesis. London University — Imperial College.

--- 1978b. The Charles Darwin — Joseph Hooker correspondence: an analysis of manuscript resources and their use in biography. *J. Soc. Bibl. Nat. Hist.* 8: 352–366.

--- 1980. Darwin's botanical arithmetic and the principle of divergence, 1854–1858. *J. hist. biol.* 13: 53–89.

--- 1985. Darwin and the face of madness. In *The anatomy of madness: essays in the history of psychiatry.* eds. W. F. Bynum and R. S. Porter. London: Tavistock.

Browne, J. C. 1871. Memoir on the vapour of the nitrite of amyl. *West Riding Lunatic Asylum med. rept.*: 95–98.

Browne, T. 1964. *Religio Medici and other works.* Ed. L. C. Martin. Oxford: Clarendon Press.

Brown-Séquard, C. E. 1860. Hereditary transmission of an epileptiform affection, accidentally produced. *Proc. Roy. Soc. Lond.* 10: 297–298.

Browning, R. 1940. *Poetical works.* London: Oxford University Press.

Brush, S. G. 1978. Planetary science: from underground to underdog. *Scientia* 113: 771–787.

--- 1979. Nineteenth-century debates about the inside of the earth: solid, liquid or gas? *Ann. sci.* 36: 225–254.

Brussel, J. A. 1966. The nature of the naturalist's unnatural illness: a study of Charles Robert Darwin. *Psychiat. q. supp.* 40: 315–331.

Buckland, W. 1823. *Reliquiae diluvianae; or, observations on the organic remains contained in caves, fissures, and diluvial gravel, and on other geological phenomena, attesting the action of an universal deluge*. London: John Murray.

--- 1837. *Geology and mineralogy, considered with reference to natural theology: with supplementary notes*. 2d ed., vol. 1. London: Pickering.

Buckley, J. H. 1967. *The triumph of time*. Cambridge: Harvard University Press.

Bucknill, J. C., and Tuke, D. H. 1858. *A manual of psychological medicine*. London: J. Churchill.

Buffon, G. L. 1749. *Histoire naturelle*. Vol. 2. Paris: Imprimerie Royale.

--- 1753. *Histoire naturelle, générale et particulière*. Vol. 4. Paris: Imprimerie Royale.

Buican, D. 1983. *Histoire de la Génetique et de l'Evolutionnisme en France*. Paris: Presses Universitaires de France.

Bulferetti, L. 1951. *Le ideologie socialistiche in Italia nell'età del positivismo evoluzionistico (1870–1892)*. Firenze: Le Monnier.

Bulgarini, G. B. 1887–1888. La filosofia monistica in Italia. *Il rosmini* 2: 531–541; 3: 293–303, 556–567, 726–739.

Bulhof, I. 1974. The Netherlands. In *The comparative reception of Darwinism*, ed. T. F. Glick, pp. 269–306. Austin: University of Texas Press.

Bunge, M. 1961. The weight of simplicity in the construction and assaying of scientific theories. *Phil. sci.* 28: 120–149.

Bunting, J. 1974. *Charles Darwin*. Folkestone: Bailey and Swinfen.

Burchfield, J. D. 1974. Darwin and the dilemma of geological time. *Isis* 65: 301–321.

--- 1975. *Lord Kelvin and the age of the earth*. New York: Science History Publications.

Burgess, S. W. 1825. *Historical illustrations of the origin and progress of the passions, and their influence on the conduct of mankind, with some subordinate sketches of human nature and human life*. London: Longman.

Burgess, T. H. 1839. *The physiology or mechanism of blushing; illustrative of the influence of mental emotion on the capillary circulation; with a general view of the sympathies, and the organic relations of those structures with which they seem to be connected*. London: Churchill.

Burkhardt, F. 1974. England and Scotland: the learned societies. In *The comparative reception of Darwinism*, ed. T. F. Glick, pp. 32–74. Austin: University of Texas Press.

Burkhardt, R. W., Jr. 1972. The inspiration of Lamarck's belief in evolution. *J. hist. biol.* 5: 413–438.

--- 1976. Review of Y. Conry, *L'introduction du darwinisme en France au XIXe siècle*. *Isis* 67: 494–496.

--- 1977. *The spirit of system: Lamarck and evolutionary biology*. Cambridge: Harvard University Press.

--- 1979. Closing the door on Lord Morton's mare: the rise and fall of telegony. *Stud. hist. biol.* 3: 1–21.

--- 1980. Lamarckism in Britain and the United States. In *The evolutionary synthesis: perspectives on the unification of biology*, eds. E. Mayr and W. B. Provine, pp. 343–352. Cambridge: Harvard University Press.

--- 1981. Lamarck's understanding of animal behavior. In *Lamarck et son temps; Lamarck et notre temps*, pp. 11–28. Paris: Vrin.

--- 1983. The development of an evolutionary ethology. In *Evolution from molecules to men*, ed. D. S. Bendall, pp. 429–444. Cambridge: Cambridge University Press.

Burrow, J. W. 1966a. *Evolution and society: a study in Victorian social theory*. Cambridge: Cambridge University Press.

--- 1966b. Charles Darwin. *Horizon* 8: 41–47.

Burstyn, H. L. 1975. If Darwin wasn't the "Beagle" 's naturalist, why was he on board? *Brit. j. hist. sci.* 8: 62–69.

Burton, D. H. 1965. Theodore Roosevelt's social Darwinism and views on imperialism. *J. hist. ideas* 26: 103–118.

Burtt, E. A. 1924. *The metaphysical foundations of modern physical science*. London: Routledge and Kegan Paul.

Bush, G. L. 1975. Modes of animal speciation. *Ann. rev. ecol. syst.* 6: 339–364.

Butler, S. 1877. *Life and habit*. London: Fifield.

--- 1879. *Evolution, old and new: or the theories of Buffon, Dr. Erasmus Darwin, and Lamarck, as compared with those of Mr. Charles Darwin*. London: Hardwick and Bogue.

--- 1880. *Unconscious memory*. London: David Bogue.

--- 1887. *Luck, or cunning, as the main means of organic modification?* London: Trübner.

--- 1908. *Essays on life, art, and science*. London: Fifield.

--- 1920. *Unconscious memory*. 3rd ed. London:

Fifield.
Butterfield, H. 1957. *The origins of modern science.* London: G. Bell & Sons.
Bynum, W. F. 1974a. *Time's noblest offspring:the problem of man in the British natural historical sciences, 1800–1863.* Ph.D. thesis. University of Cambridge.
--- 1974b. Rationales for therapy in British psychiatry: 1780–1835. *Med. hist.* 18: 317–334.
--- 1975. The great chain of being after forty years: an appraisal. *Hist. Sci.* 13: 1–28.

Cadinouche, H. 1929. *La médicine dans l'oeuvre de Gericault.* Paris: Vigne.
Cadman, S. P. 1911. *Charles Darwin and other English thinkers with reference to their religious and ethical value.* London: James Clarke.
Cain, A. J. 1954. *Animal species and their evolution.* London: Hutchinson.
--- 1959a. Deductive and inductive methods in post-Linnaean taxonomy. *Proc. Linn. Soc. Lond.* 170: 185–217.
--- 1959b. The post-Linnaean development of taxonomy. *Proc. Linn. Soc. Lond.* 170: 234–244.
Calcoen, R. 1960. Darwin en België. *Soc. Standpunkten* 7: 178–183.
Camerano, L. 1877. Polimorfismo nella femmina dell'*Hydrophilius piceus. Atti della Reale Accademia delle Scienze di Torino* 13: 79–96.
--- 1880. Dell'equilibrio dei viventi mercè la reciproca distruzione. *Atti della Reale Accademia delle Scienze di Torino* 15: 445–450.
--- 1883. Intorno alla neotenia ed allo sviluppo degli anfibi. *Atti della Reale Accademia delle Scienze di Torino* 19: 84–93.
--- 1884. Intorno alla distribuzione dei colori nel regno animale. *Memorie della Reale Accademia delle Scienze di Torino* 36: 329–360.
--- 1890. Osservazioni intorno al dimorfismo sessuale degli Echinodermi. *Bollettino dei Musei di Zoologia e di Anatomia Comparata della Reale Università di Torino* 5: n. 91.
--- 1896. La vita scientifica di Michele Lessona. *Memorie della Reale Accademia delle Scienze di Torino* 45: 331–388.
--- 1902. Contributi alla storia delle teorie Lamarckiane in Italia. Il corso di zoologia di Franco Andrea Bonelli. *Atti della Reale Accademia delle Scienze di Torino* 37: 455–464.
--- 1904. I manoscritti di Franco Andrea Bonelli. Contributo alla storia delle teorie Lamarckiane in Italia nel principio del secolo xix. *Atti del congresso internazionale di scienze storiche (Roma, 1–9 Aprile 1903)* 12: 203–209.
--- 1905–1909. Materiali per la storia della zoologia in Italia nella prima metà del secolo xix. *Bollettino dei Musei di Zoologia ed Anatomia Comparata della Regia Università di Torino* vols. 20–24, nos. 486, 526, 535, 536, 579, 586, 591, 601, 606.
--- 1912. Gli studi zoologici in Italia nel primo cinquantennio di vita nazionale. *Atti della Societa' Italiana per il Progresso delle Scienze. Quinta Riunione. Roma, Ottobre 1911* 5: 463–495.
Campbell, B., ed. 1972. *Sexual selection and the descent of man, 1871–1971.* Chicago: Aldine.
Campbell, D. T. 1960. Blind variation and selective retention in creative thought as in other knowledge processes. *Psych. rev.* 67: 380–400.
--- 1974. Evolutionary epistemology. In *The philosophy of K. R. Popper*, ed. P. A. Schilpp, vol. 1, pp. 413–463. La Salle: Open Court.
Campbell, J. A. 1968. *A rhetorical analysis of* The Origin of Species *and of American Christianity's response to Darwinism.* Ph.D. thesis. University of Pittsburgh.
--- 1970. Darwin and "The origin of species": the rhetorical ancestry of an idea. *Speech monog.* 37: 1–14.
--- 1974. Nature, religion, and emotional response: a reconsideration of Darwin's affective decline. *Victorian stud.* 18: 159–174.
--- 1975. Charles Darwin and the crisis of ecology: a rhetorical perspective. *Q. j. speech* 60: 442–449.
Campbell, L., and Garnett, W. 1882. *The life of James Clerk Maxwell.* London: Macmillan.
Campbell, M. 1980. Did de Vries discover the law of segregation independently? *Ann. sci.* 37: 639–655.
Candolle, A. P. 1813. *Théorie élémentaire de la botanique.* Paris: Deterville.
--- **and Candolle, A.** 1824–1873. *Prodromus systematis naturalis regni vegetabilis . . . .* 17 vols. Paris: Treuttel et Würtz (vols. 1–7); Fortin, Masson (vols. 8–9); V. Masson (vols. 10–14); Masson et Filii (vols. 15–16); G. Masson (vol. 17).
--- **and Sprengel, K.** 1821. *Elements of the philosophy of plants: containing the principles of scientific botany.* Edinburgh: Blackwood.
Candolle, A. 1882. *Darwin considéré au point de*

vue des causes de son succès. Geneva: H. Georg.

Canestrini, G. 1877. *La teoria dell'evoluzione, esposta nei suoi fondamenti, come introduzione alla lettura delle opere del Darwin*. Torino: Unione Tipografica Editrice.

--- 1879. *Sulla produzione dei sessi: considerazioni*. Padova: Prosperini.

--- 1880. *La teoria di Darwin criticamente esposta*. Milano: Dumolard.

--- 1894. *Per l'evoluzione: recensioni e studi*. Torino: Unione Tipografica Editrice.

Canguilhem, G. 1960. L'homme et l'animal du point de vue psychologique selon Charles Darwin. *Rev. d'hist. sci.* 13: 81–94.

--- 1965. *La connaissance de la vie*. 2d ed. Paris: Vrin.

--- 1975. *Études d'histoire et de philosophie des sciences*. 3rd ed. Paris: Vrin.

--- 1977. *Idéologie et rationalité dans l'histoire des sciences de la vie*. Paris: Vrin.

--- et al. 1962. Du développement à l'évolution au XIXe siècle. *Thalès* 11, année 1960. Paris: Presses Universitaires de France.

Cannon, H. G. 1957. What Lamarck really said. *Proc. Linn. Soc. Lond.* 168: 70–85.

--- 1958. *The evolution of living things*. Manchester: Manchester University Press.

--- 1959. *Lamarck and modern genetics*. Springfield: Charles C. Thomas.

Cannon, W. F. (S. F.) 1960a. The problem of miracles in the 1830's. *Victorian stud.* 4: 5–32.

--- 1960b. The uniformitarian-catastrophist debate. *Isis* 51: 38–55.

--- 1961a. John Herschel and the idea of science. *J. hist. ideas* 22: 215–239.

--- 1961b. The bases of Darwin's achievement: a revaluation. *Victorian stud.* 5: 109–134.

--- 1961c. The impact of uniformitarianism: two letters from John Herschel to Charles Lyell, 1836–1837. *Proc. Am. Phil. Soc.* 105: 301–314.

--- 1964a. History in depth: the early Victorian period. *Hist. sci.* 3: 20–38.

--- 1964b. Scientist and broad churchmen: an early Victorian intellectual network. *J. Brit. stud.* 4: 65–88.

--- 1964c. The normative role of science in early Victorian thought. *J. hist. ideas* 25: 487–502.

--- 1964d. The role of the Cambridge movement in early 19th century science. *Proc. 10th Int. cong. hist. sci.*, pp. 317–320.

--- 1968. Darwin's vision in *On the origin of species*. In *The art of Victorian prose*, eds. G. Levine and W. Madden, pp. 154–176. New York, Oxford: Oxford University Press.

--- 1969. Charles Lyell is permitted to speak for himself: an abstract. In *Toward a history of geology*, ed. C. J. Schneer, pp. 78–79. Cambridge: M.I.T. Press.

--- 1976a. Charles Lyell, radical actualism, and theory. *Brit. j. hist. sci.* 9: 104–120.

--- 1976b. The Whewell-Darwin controversy. *J. Geol. Soc. Lond.* 132: 377–384.

--- 1978. *Science in culture: the early Victorian period*. New York: Dawson and Science History Publications.

Cantor, G. N. 1975a. The Edinburgh phrenology debate: 1803–1828. *Ann. sci.* 32: 195–218.

--- 1975b. A critique of Shapin's social interpretation of the Edinburgh phrenology debate. *Ann. sci.* 32: 245–256.

Čapek, M. 1961. *Philosophical impacts of contemporary physics*. Princeton: Van Nostrand.

Caplan, A. L. 1977. Tautology, circularity, and biological theory. *Am. nat.* 111: 390–393.

---, ed. 1978. *The sociobiology debate: readings on ethical and scientific issues*. New York: Harper and Row.

--- 1979. Darwinism and deductivist models of theory structure. *Stud. hist. phil. sci.* 10: 341–353.

Cappelletti, V. 1965. *Entelechia: saggi sulle dottrine biologiche del secolo decimonono*. Firenze: Sansoni.

Carlès, J. 1952. *Le transformisme*. Paris: Presses Universitaires de France.

Carlson, E. A. 1966. *The gene: a critical history*. Philadelphia: W. B. Saunders.

Carlson, E. T. and Dain, N. 1962. The meaning of moral insanity. *Bull. hist. med.* 36: 130–140.

Carozzi, A. V. 1966. Agassiz's amazing geological speculation: the ice-age. *Stud. in romanticism* 5: 57–83.

--- 1973. Agassiz's influence on geological thinking in the Americas. *Arch. sci.* 27: 5–38.

Carpenter, W. B. 1873. On the hereditary transmission of acquired psychical habits. *Contemporary rev.* 21: 295–314, 779–795, 867–885.

--- 1888. *Nature and man: essays scientific and*

philosophical. London: K. Paul, Trench.

Carroll, P. T. 1976a. On the utility of collating the Darwin correspondence. *Ann. sci.* 33: 384–385.

--- 1976b. Further evidence that Karl Marx was not the recipient of Charles Darwin's letter dated 13 October 1880. *Ann. sci.* 33: 385–386.

---, 1976c. *An annotated calendar of the letters of Charles Darwin in the Library of the American Philosophical Society*. Wilmington, Del.: Scholarly Resources.

Carter, G. S. 1957. *A hundred years of evolution*. New York: Macmillan.

Cassirer, E. 1925. *Sprache und Mythos, ein Beitrag zum Problem der Götternamen*. Leipzig: B. G. Teubner.

--- 1944. *An essay on man: an introduction to the philosophy of human culture*. New Haven: Yale University Press.

--- 1950. *The problem of knowledge: philosophy, science and history since Hegel*. New Haven: Yale University Press.

Castle, W. E. 1905. The mutation theory of organic evolution, from the standpoint of animal breeding. *Science* N.S. 21: 521–525.

Cattaneo, G. 1882. Le colonie lineari e la morfologia dei molluschi. *Zoologischer Anzeiger* 5: 682–685.

--- 1886. *Lamarck e Darwin*. Milano: Dumolard.

Caullery, M. 1966. *A history of biology*. New York: Walker.

Cavalli-Sforza, L., and Feldman, M. 1981. *Cultural transmission and evolution*. Princeton: Princeton University Press.

Caverni, R. 1877. *L'antichità dell'uomo*. Firenze: G. Carnesecchi.

Cawood, J. 1977. Terrestrial magnetism and the development of international collaboration in the early nineteenth century. *Ann. sci.* 34: 551–587.

Centore, F. F. 1969. Darwin on evolution: a re-estimation. *Thomist* 33: 456–496.

--- 1971. Neo-Darwinian reactions to the social consequences of Darwin's nominalism. *Thomist* 35: 113–142.

Centre de Recherche sur l'Histoire des Ideés de l'Université de Picardie 1981. *Lamarck et son temps; Lamarck et notre temps*. Colloque international dans le cadre du Centre d'Études et de Recherches interdisciplinaires de Chantilly (C.E.R.I.C.). Paris: Vrin.

Cesca, G. 1902. *Il monismo di Ernesto Haeckel*. Pavia: Bizzoni.

Chadwick, O. 1966–1970. *The Victorian church*. 2 vols. London: Black.

--- 1975. *The secularization of the European mind in the nineteenth century*. Cambridge: Cambridge University Press.

Challinor, J. 1959. Paleontology and evolution. In *Darwin's biological work*, ed. P. R. Bell, pp. 50–100. Cambridge: Cambridge University Press.

--- 1970. The progress of British geology during the early part of the nineteenth century. *Ann. sci.* 26: 177–234.

Chamberlin, J. E. 1981. An anatomy of cultural melancholy. *J. hist. ideas* 42: 691–705.

Chambers, R. 1844. *Vestiges of the natural history of creation*. London: Churchill. 1969. Reprinted with an introduction by Sir G. de Beer. Leicester: Leicester University Press.

--- 1846. *Vestiges of the natural history of creation*. 5th ed. London: Churchill.

--- 1853. *Vestiges of the natural history of creation*. 10th ed. London: Churchill.

--- 1860. *Vestiges of the natural history of creation*. 11th ed. London: Churchill.

--- 1884. *Vestiges of the natural history of creation*. 12th ed. London: W. & R. Chambers.

Champion, G. C. 1918. Notes on various South American Coleoptera collected by Charles Darwin during the voyage of the *Beagle*, with descriptions of new genera and species. *Entomol. monthly mag.* 54: 43–55.

Chancellor, J. 1973. *Charles Darwin*. London: Weidenfeld and Nicolson.

Chapeville, F. et al. 1979. *Le darwinisme aujourd'hui*. Paris: Editions du Seuil.

Chapman, R. G., ed. 1982. *Charles Darwin 1809–1882: a centennial commemorative*. Wellington: Nova Pacifica.

Charlesworth, B.; Lande, R.; and Slatkin, M. 1982. A neo-Darwinian commentary on macroevolution. *Evolution* 36: 474–498.

Chase, A. 1977. *The legacy of Malthus: the social costs of the new scientific racism*. New York: Knopf.

Chetverikov, S. S. 1983. (Z. S. Nikoro, ed.) *Problemi Obscheii Biologii i Genetiki*. (Problems in General Biology and Genetics) Novosibirsk: Nauka.

Christen, Y. 1981. *Le grand affrontement: Marx et Darwin*. Paris: Albin Michel.

Churchill, F. B. 1968. August Weismann and a break from tradition. *J. hist. biol.* 1: 91–112.

--- 1970. Hertwig, Weismann, and the meaning of reduction division *circa* 1890. *Isis* 61: 429–457.

--- 1978. The Weismann-Spencer controversy over the inheritance of acquired characters. *Proc. 15th int. cong. hist. sci.* (Edinburgh), pp. 451–468.

--- 1981. In search of a new biology: an epilogue. *J. hist. biol.* 14: 177–191.

Cimutta, J. 1969. Die Dialektik von Zufall und Notwendigkeit in Evolutionsgeschehen. *Deutsche Zeitschrift für Philosophie* 17: 967–984.

Clagett, M., ed. 1959. *Critical problems in the history of science*. Madison: University of Wisconsin Press.

Claparède, E. 1861. M. Darwin et sa théorie de la formation des espèces. *Revue germanique* 16: 523–569.

Clark, G. K. 1973. *Churchmen and the condition of England, 1832–1885: a study in the development of social ideas and practice from the old regime to the modern state*. London: Methuen.

Clark, J. W. 1890. The foundation and early years of the society. *Proc. Cambridge Phil. Soc.* 7:i–l.

--- **and Hughes, T. M.** 1890. *The life and letters of the Reverend Adam Sedgwick*. 2 vols. London: Cambridge University Press.

Clark, L. L. 1968. *Social Darwinism and French intellectuals, 1860–1915*. Ph. D. thesis. University of North Carolina.

Clark, R. E. D. 1948. *Darwin: before and after: the story of evolution*. London: Paternoster Press.

Clarke, S. 1977. Marxism, sociology and Poulantzas' theory of the state. *Capital and class* 2: 1–31.

The Clergy List. 1841ff. London: C. Cox.

Clifford, J. 1898. Charles Darwin; or, evolution and Christianity. In *Typical Christian leaders*, pp. 213–236. London: Horace Marshall.

Coase, R. H. 1960. The problem of social cost. *J. law and econ.* 3: 1–45.

Cobbold, T. S. 1885. *Linn. Soc. Lond. (Zool.)* 19: 174–178.

Cock, A. G. 1973. William Bateson, Mendelism and biometry. *J. hist. biol.* 6: 1–36.

Cohen, I.B. 1980. *The Newtonian revolution: with illustrations of the transformation of scientific ideas.* New York: Cambridge University Press.

Coleman, B. 1974. Butler, Darwin, and Darwinism. *J. Soc. Bibl. Nat. Hist.* 7: 93–105.

Coleman, W. 1962. Lyell and the reality of species. *Isis* 53: 325–338.

--- 1964. *Georges Cuvier zoologist: a study in the history of evolution theory*. Cambridge: Harvard University Press.

--- 1965. Cell, nucleus and inheritance: an historical study. *Proc. Am. Phil. Soc.* 109: 124–158.

--- 1971a. *Biology in the nineteenth century: problems of form, function, and transmutation*. New York: John Wiley. [Rev. ed. 1977, C.U.P.]

--- 1971b. Commentary on the paper of John C. Greene [Greene 1971]. In *Perspectives in the history of science and technology*, ed. D. H. D. Roller, pp. 26–30. Norman: Oklahoma University Press.

--- 1976. Morphology between type concept and descent theory. *J. hist. med.* 31: 149–175.

Collins, H. M. 1981. The place of the "core-set" in modern science: social contingency with methodological propriety in science. *Hist. sci.* 19: 6–19.

Collins, J. 1959. Darwin's impact on philosophy. *Thought* 34: 185–248.

Colloms, B. 1977. Victorian country parsons. London: Constable.

Colp, R. 1972. Charles Darwin and Mrs. Whitby. *Bull. acad. med.* 2nd. ser. 48: 870–876.

--- 1974. The contacts between Karl Marx and Charles Darwin. *J. hist. ideas* 35: 329–338.

--- 1976. The contacts of Charles Darwin with Edward Aveling and Karl Marx. *Ann. sci.* 33: 387–394.

--- 1977a. *To be an invalid: the illness of Charles Darwin*. Chicago: University of Chicago Press.

--- 1977b. Charles Darwin and the Galápagos. *New York State j. med.* 77: 262–267.

--- 1978. Charles Darwin: slavery and the American civil war. *Harvard Libr. bull.* 26: 471–489.

--- 1979. Charles Darwin's vision of organic nature. *New York State j. med.* 79: 1622–1629.

--- 1980. "I was born a naturalist": Charles Darwin's 1838 notes about himself. *J. hist. med.* 35: 8–39.

--- **and Fay, M. A.** 1979. Multiple independent discovery: the Darwin-Marx letter. *J. hist.*

ideas 40: 479.

Comte, A. 1830–1842, *Cours de philosophie positive*. Paris: Bachelier.

Conolly, J. 1858–1859. On the physiognomy of insanity. *Med. times and gaz.* N. S. 16, 17, and 18.

Conrad Martius, H. 1955. *Utopien der Menschenzüchtung. Der Sozialdarwinismus und seine Folgen*. München: Kösel.

Conry, Y. 1971. Broca et Darwin. *Actes du XIIe Congrès* international d'histoire des sciences (Paris 1968), 8: 25–29. Paris: Blanchard.

--- 1972a. *Correspondence entre Charles Darwin et Gaston de Saporta précédée de l'histoire de la paléobotanique en France au XIXe siècle*. Paris: Presses Universitaires de France.

--- 1972b. Darwin et Mendel dans la biologie contemporaine. *Revue de l'enseignement philosophique* 23, no. 1: 1–16.

--- 1974. *L'introduction du darwinisme en France au XIXe siècle*. Paris: Vrin.

--- 1982. Le darwinisme en France. *Rivista di filosofia* 22–23: 53–78.

Cope, E. D. 1868. On the origin of genera. *Proc. Acad. Nat. Sci. Phila.* 20: 242–300.

--- 1886. The phylogeny of the Camelidae. *Am. nat.* 20: 611–624.

--- 1887. *The origin of the fittest: essays in evolution*. New York: Macmillan.

--- 1891. The Litopterna. *Am. nat.* 25: 685–693.

--- 1896. *The primary factors of organic evolution*. Chicago: Open Court.

Cornell, J. F. 1984. Analogy and technology in Darwin's vision of nature. *J. hist. biol* 17: 303–344.

Correns, C. 1900. G. Mendel's Regel über das verhalten der nachkommenschaft der rassenbastarde. *Berichte der deutschen botanischen Gesellschaft* 18: 158–168.

Corsi, P. 1975. Essay review of H. E. Gruber, *Darwin on man: a psychological study of scientific creativity*. *Ann. sci.* 32: 583–586.

--- 1978. The importance of French transformist ideas for the second volume of Lyell's *Principles of geology*. *Brit. j. hist. sci.* 11: 221–244.

--- 1980a. Essay review of E. Manier, *The young Darwin and his cultural circle*. *Ann. sci.* 37: 673–678.

--- 1980b. *Natural theology, the methodology of science, and the question of species in the works of the Reverend Baden Powell*. D. Phil. thesis. University of Oxford.

--- 1983. "Lamarckiens" et "Darwiniens" à Tourin (1812–1894). In *De Darwin au Darwinisme: science et idéologie*, ed. Y. Conry. Paris: Vrin.

--- 1983. *Ultre il mito. Lamarck e le Scienza naturale nel suo tempo*. Bologna: Il Mulino.

--- forthcoming. *Science and religion: Baden Powell and the Anglican debate*, 1820–1860. Cambridge: Cambridge University Press.

Corwin, E. S. 1950. The impact of the idea of evolution on the American political constitutional tradition. In *Evolutionary thought in America*, ed. S. Persons, pp. 182–199. New Haven: Yale University Press.

Cowan, R. S. 1972a. Francis Galton's contribution to genetics. *J. hist. biol.* 5: 389–412.

--- 1972b. Francis Galton's statistical ideas. *Isis* 63: 509–528.

--- 1977. Nature and nurture: the interplay of biology and politics in the work of Francis Galton. *Stud. hist. biol.* 1: 133–208.

Cowles, T. C. 1937. Malthus, Darwin, and Bagehot: a study in the transference of a concept. *Isis* 26: 341–348.

Cracraft, J., and Eldredge, N., eds. 1979. *Phylogenetic analysis and paleontology*. New York: Columbia University Press.

Crane, D. 1972. *Invisible colleges: diffusion of knowledge in scientific communities*. Chicago: University of Chicago Press.

Cranefield, P. F. 1974. *The way in and the way out: Francois Magendie, Charles Bell and the roots of the spinal nerves*. New York: Futura.

Cravens, H. 1978. *The triumph of evolution*. Philadelphia: University of Pennsylvania Press.

--- **and Burnham, J.** 1971. Psychology and evolutionary naturalism in American thought, 1890–1940. *Am. q.* 23: 635–657.

Crellin, K. 1968. *Darwin and evolution*. London: Cape.

Crockford's Clerical Directory 1858ff. London: John Crockford/Horace Cox et al.

Crombie, A. C. 1960. Darwin's scientific method. *Actes du IXe cong. int. d'hist. sci.* (Barcelona), 1: 354–362. Paris: Blanchard.

Crombie, D. L. 1967. Back to Darwin. *J. Roy. Coll. Gen. Practitioners* 13: 22–29.

Crossley, C. 1981. *La Création* d'Edgar Quinet et le darwinisme social. *Romantisme* 11: 65–73.

Crow, J., and Kimura, M. 1970. *An introduction to population genetic theory.* Minneapolis: Burgess.

Crowson, R. A. 1958. Darwin and classification. In *A century of Darwin,* ed. S. A. Barnett, pp. 102–129. Cambridge: Harvard University Press.

Crowther, J. G. 1972. *Charles Darwin.* London: Methuen.

Cuboni, G. 1910. *L'opera di Carlo Darwin e la critica moderna: discorso.* Pavia: Fusi.

Culler, A. D. 1968. The Darwinian revolution and literary form. In *The Art of Victorian Prose,* eds. G. Levine and W. Madden, pp. 224–246. New York: Oxford University Press.

Cunningham, J. T. 1895. The origin and evolution of flatfishes. *Natural sci.* 6: 169–177, 233–239.

Curti, M. 1951. *The growth of American thought.* New York: Harper.

Cuvier, G. 1813. *Essay on the theory of the earth, with mineralogical notes and an account of Cuvier's geological discoveries* by Professor Jameson. Trans. R. Kerr. New York: Kirk and Mercein.

--- 1829–1830. *Le règne animale.* 2d ed. 5 vols. Paris: Deterville.

--- 1831. *The animal kingdom arranged in conformity with its organization.* Trans. M. M'Murtie. 4 vols. New York: G. & C. & H. Carvill.

Dacqué, E. 1903. *Der Deszendenzgedanke und seine Geschichte vom Altertum bis zur Neuzeit.* München: Reinhardt.

Dagognet, F. 1970. *Le catalogue de la vie: étude méthodologique sur la taxonomie.* Paris: Presses Universitaires de France.

Dally, E. 1868. L'ordre des primats et le transformisme. *Bulletin de la Société d'Anthropologie* 3: 673–712.

Danielli, I. 1885. *Alcuni fatti spiegabili colla pangenesi di Darwin.* Pisa.

Daniels, G. H., ed. 1968. *Darwinism comes to America.* Waltham: Blaisdell.

Danielsson, U. 1963–1964, 1965–1966. Darwinismens inträngande i Sverige [The introduction of Darwinism in Sweden]. *Lychnos* 1963–1964 (publ. 1965): 157–210; 1965–1966 (publ. 1967): 261–334.

Darden, L. 1976. Reasoning in scientific change: Charles Darwin, Hugo de Vries, and the discovery of segregation. *Stud. hist. phil. sci.* 7: 127–169.

--- 1977. William Bateson and the promise of Mendelism. *J. hist. biol.* 10: 87–106.

Darlington, C. D. 1937. The early hybridizers and the origins of genetics. *Herbertia* 4: 63–69.

--- 1953. Purpose and particles in the study of heredity. In *Science, medicine, and history: essays on the evolution of scientific thought and medical practice . . . in honour of Sir Charles Singer,* ed. E. A. Underwood, vol. 2, pp. 472–481. London and Oxford: Oxford University Press.

--- 1959a. *Darwin's place in history.* Oxford: Basil Blackwell.

--- 1959b. The origin of Darwinism. *Sci. Am.* (May) 200: 60–66.

--- 1969a. *The evolution of man and society.* London: Allen and Unwin.

--- 1969b. The genetics of society. *Past and Present* 43: 3–33.

Darlington, P. J., Jr. 1959. Darwin and zoogeography. *Proc. Am. Phil. Soc.* 103: 307–319.

Darnell, R. 1977. History of anthropology in historical perspective. *Ann. rev. anthro.* 6: 399–417.

Darwin. 1888. *British weekly* 4: 233.

Darwin, C. R. 1844a. *Geological observations on the volcanic islands visited during the voyage of H.M.S. Beagle, together with some brief notices of the geology of Australia and the Cape of Good Hope. Being the second part of the geology of the voyage of the Beagle, under the command of Captain Fitzroy, R. N. during the years 1832 to 1836.* London: Smith Elder.

--- 1844b. *Charles Darwin's naturwissenschaftliche Reisen.* Trans. E. Dieffenbach. Braunschweig: F. Vieweg und Sohn.

--- 1846. *Geological observations on South America. Being the third part of the geology of the voyage of the Beagle, under the command of Capt. Fitzroy, R. N. during the years 1832 to 1836.* London: Smith Elder.

--- 1851a. *A monograph of the sub-class Cirripedia, with figures of all the species. The Lepadidae; or, pedunculated cirripedes.* London: The Ray Society.

--- 1851b. *A monograph of the Fossil Lepadidae, or pedunculated cirripedes of Great Britain.* London: Palaeontographical Society.

--- 1854a. *A monograph of the sub-class Cirripedia, with figures of all the species. The Balanidae, (or sessile cirripedes); the Verrucidae, etc., etc., etc.*

London: The Ray Society.
--- 1854b. *A monograph of the Fossil Balanidae and Verrucidae of Great Britain*. London: Palaeontographical Society.
--- 1860. *Über die Entstehung der Arten im Thierund Pflanzen-Reich durch natürliche Züchtung*. Trans. H. G. Bronn. Stuttgart: E. Schweizerbart.
--- 1862a. Do bees vary in different parts of Great Britain? *Journal of Horticulture and Cottage Gardener*. N. S. 3: 207 [published 10 June 1862].
--- 1862b. Findet bei den Bienen in den verschiedenen Theilen Deutschlands ein Unterschied statt? *Bienenzeitung* 18: 145 [dated Bromley, Kent, England 18 June 1862; published 20 August 1862].
--- 1862c. *De l'origine des espèces, ou des Lois du progrès chez les êtres organisés* par Charles Darwin, traduit en français sur la 3e édition . . . . par M.lle Clémence-Auguste Royer, avec une preface et des notes du traducteur. Paris : Guillaumin.
--- 1863. *Über die Entstehung der Arten im Thier- und Pflanzen-Reich durch natürliche Züchtung* . . . Nach der 3. englischen Auflage 2 verb. und sehr verm. Aufl. Trans. H. G. Bronn. Stuttgart: E. Schweizerbart.
--- 1864a. *Sull'origine delle specie per elezione naturale ovvero conservazione delle razze perfezionate nella lotta per l'esistenza*. Prima traduzione italiana col consenso dell'autore per cura di G. Canestrini e L. Salimbeni. First installment, Chapters 1–3. Modena : Zanichelli.
--- 1864b. *Sull'origine delle specie per elezione naturale ovvero conservazione delle razze perfezionate nella lotta per l'esistenza*. Prima traduzione italiana col consenso dell'autore per cura di G. Canestrini e L. Salimbeni. Modena : Zanichelli.
--- 1866. *De l'origine des espèces, ou des Lois de tranformation des êtres organisés* . . . 2d ed. Paris: V. Masson.
--- 1867. *Über die Entstehung der Arten durch natürliche Züchtwahl; oder, die Erhaltung der begünstigsten Rassen im Kampfe um's Dasein*. Trans. J. V. Carus. Stuttgart: E. Schweizerbart.
--- 1870. *De l'origine des espèces, ou des Lois de transformation des êtres organisés*. 3rd ed. Paris: V. Masson.
--- 1873. *L'origine des espèces au moyen de la sélection naturelle, ou la lutte pour l'existance dans la nature*, par Charles Darwin. Traduit . . . sur la 5me et 6me éditions anglaises par J.-J. Moulinié. Paris : C. Reinwald.
--- 1875. *On the movements and habits of climbing plants*. 2d ed. rev. London: John Murray.
--- 1876. *The effects of cross and self-fertilisation in the vegetable kingdom*. London: John Murray.
--- 1877. *The different forms of flowers on plants of the same species*. London: John Murray.
--- 1878. *The effects of cross and self fertilisation in the vegetable kingdom*. 2d ed. London: John Murray.
--- 1881. *The formation of vegetable mould, through the action of worms, with observations on their habits*. London: John Murray.
--- 1882. A hitherto unpublished letter of Charles Darwin. *National reformer* 40: 235.
--- 1888. Darwin. *Brit. Weekly*. 4:233.
--- 1945. *Charles Darwin and the voyage of the Beagle*. Ed. N. Barlow. London: Pilot Press.
--- 1959. *Darwin's Journal*. Ed. G. de Beer. *Bull. Brit. Mus. (Nat. Hist.) Hist. ser.* 2, no. 1: 1–21.
--- 1963. *Darwin's ornithological notes*. Ed. N. Barlow. *Bull. Brit. Mus. (Nat. Hist.) Hist. ser.* 2: 203–278.
--- 1967. *Darwin and Henslow: the growth of an idea. Letters 1831–1860*. Ed. N. Barlow. Berkeley: University of California Press.
--- 1968. *Questions about the breeding of animals*. Ed. G. de Beer. Sherborn Fund Facsimile No. 3. London: Society for the Bibliography of Natural History.
--- **and Wallace, A.R.** 1958. *Evolution by natural selection*. With a foreword by Sir Gavin de Beer. Cambridge: Cambridge University Press.
Darwin, E. 1794–1796. *Zoonomia; or, the laws of organic life*. 2 vols. London: J. Johnson.
--- 1800. *Phytologia; or, the philosophy of agriculture and gardening*. London: J. Johnson.
--- 1803. *The temple of nature, or the origin of society*. London: J. Johnson.
Darwin, F., ed. 1892. *Charles Darwin: his life told in an autobiographical chapter, and in a selected series of his published letters*. London: John Murray.
--- 1908. President's address. *Rept. Brit. Assoc. Adv. Sci.* 3–27.
---, ed. 1909. *The foundations of the Origin of Species: two essays written in 1842 and 1844 by Charles Darwin*. Cambridge: Cambridge University Press.
--- 1920. *Springtime and other essays*. London: John

Murray.

Daudin, H. 1926a. *Les classes zoologiques et l'idée de série animale en France à l'époque de Lamarck et Cuvier (1790–1830)*. Paris: Alcan.

--- 1926b. *Les méthodes de la classification et l'idée de série en botanique et en zoologie de Linnée à Lamarck (1740–1790)*. Paris: Alcan.

Davie, D. 1961. *The heyday of Sir Walter Scott*. London: Routledge & Paul.

--- 1963. *The language of science and the language of literature 1700–1740*. London.

Davies, G. L. 1969. *The earth in decay: a history of British geomorphology, 1578–1878*. New York: American Elsevier.

Davies, K. C., and Hull, J. 1976. *The zoological collections of the University of Oxford: a historical review and general account, with comprehensive donor index to the year 1975*. Oxford: University Museum.

Davitashvili, L. Sh. 1948. *Istoria evoliuzionnoi paleontologii ot Darwina do nashikh dnei*. (History of evolutionary paleontology from Darwin to present times) Moscow: Akad. Nauk.

Davy, H. 1813. *The elements of agricultural chemistry in a course of lectures for the Board of Agriculture*. London: Longman, Hurst, Rees, Orme, and Brown.

--- 1839–1840. *The collected works of Humphrey Davy*. 9 vols. London: Smith, Elder & Co. Vol. 1: *Memoirs of the life of Sir Humphrey Davy*. Vols. 7–8: *Discourses delivered before the Royal Society; Elements of agricultural chemistry*.

Dawkins, R. 1976. *The selfish gene*. Oxford: Oxford University Press.

--- 1979. Defining sociobiology. *Nature* 280: 427–428.

--- 1982. *The extended phenotype*. San Francisco: W. H. Freeman.

Dawson, J. W. 1890. *Modern ideas of evolution*. London: Religious Tract Society.

Deacon, M. 1971. *Scientists and the sea, 1650–1900: a study of marine science*. New York: Academic Press.

Dean, D. 1980. Graham Island, Charles Lyell, and the craters of elevation. *Isis* 71: 571–588.

De Beer, G. R. 1958a. Foreword to C. Darwin and A. R. Wallace, *Evolution by natural selection*. Cambridge: Cambridge University Press.

--- 1958b. The Darwin-Wallace centenary. *Endeavour* 17: 61–76.

--- 1958c. Darwin and embryology. In *A century of Darwin*, ed. S. A. Barnett pp. 153–172. Cambridge: Harvard University Press.

---, ed. 1958d. Further unpublished letters of Charles Darwin. *Ann. sci.* 14: 83–115.

---, ed. 1959a. Darwin's journal. *Bull. Brit. Mus. (Nat. Hist.) Hist. ser.* 2, no. 1: 1–21.

---, ed. 1959b. Some unpublished letters of Charles Darwin. *Notes rec. Roy. Soc. Lond.* 14: 12–66.

--- 1959c. "Darwin without modern science" [review of Himmelfarb 1959. *Darwin and the Darwinian revolution.*]. *Nature* 184: 385–388.

---, ed. 1960. Darwin's notebooks on transmutation of species. *Bull. Brit. Mus. (Nat. Hist.) Hist. ser.* 2, nos. 2–5: 23–183.

--- 1961. The origins of Darwin's ideas on evolution and natural selection. *Proc. Roy. Soc. Lond.* 155: 321–338.

--- 1962. *Reflections of a Darwinian: essays and addresses*. London: Nelson.

--- 1963. *Charles Darwin: evolution by natural selection*. London: Nelson.

--- 1964a. Mendel, Darwin, and Fisher (1865–1965). *Notes rec. Roy. Soc. Lond.* 19: 192–226.

--- 1964b. Other men's shoulders. *Ann. sci.* 20: 303–322.

--- 1966. Mendel, Darwin, and Fisher: addendum. *Notes rec. Roy. Soc. Lond.* 20: 64–71.

---, ed. 1968. The Darwin letters at Shrewsbury School. *Notes rec. Roy. Soc. Lond.* 23: 68–85.

--- 1969a. *Streams of culture*. Philadelphia: Lippincott.

--- 1969b. Introduction to R. Chambers, *Vestiges of the natural history of creation*. Leicester: Leicester University Press.

--- 1970. The evolution of Charles Darwin. *New York rev. of books*, December 17.

--- **and Rowlands, M. J.**, eds. 1961. Darwin's notebooks on transmutation of species: addenda et corrigenda. *Bull. Brit. Mus. (Nat. Hist.) Hist. ser.* 2, no. 6: 185–200.

---; **Rowlands, M. J.; and Skramovsky, B. M.**, eds. 1967. Darwin's notebooks on the transmutation of species: pages excised by Darwin. *Bull. Brit. Mus. (Nat. Hist.) Hist. ser.* 3, no. 5: 129–176.

Deely, J. N. 1969. The philosophical dimensions of the *Origin of species*. *Thomist* 33: 75–147, 251–335.

--- **and Nogar, R. J.**, eds. 1973. *The problem of evolution: a study of the philosophical repercussions of evolutionary science*. New York:

Appleton-Century-Croft.

De Filippi, F. 1864. L'uomo e le scimie. *Il politecnico* 21: 5–32.

DeGrood, D. H. 1965. *Haeckel's theory of the unity of nature: a monograph in the history of philosophy*. Boston: Christopher Publishing House.

De la Beche, H. T. 1828. *A tabular and proportional view of the superior, supermedial and medial rocks*. 2d ed. London: Treuttel.

Delage, Y., and Goldsmith, M. 1912. *Theories of evolution*. Trans. A. Tridon. New York: B. W. Huebsch.

De Lattin, G. 1960. Darwin als Klassiker der Tiergeographie. In *Hundert Jahren Evolutionsforschung*, eds G. Heberer and F. Schwanitz, pp. 203–233. Stuttgart: Gustav Fischer.

Delpino, F. 1867. Pensieri sulla biologia vegetale, sulla tassonomia, sul valore tassonomico dei caratteri biologici, e proposta di un genere nuovo della famiglia delle labiate. *Il nuovo cimento* 25: 283–304, 321–398.

--- 1868. *Sulla darwiniana teoria della pangenesi*. Torino : Negro.

--- 1888–1896. *Applicazione di nuovi criteri per la classificazione delle piante*. Bologna : Gamberini e Parmeggiani.

--- 1895. *Socialismo e storia naturale: discorso*. Napoli: Tipografia della Università.

De Marrais, R. 1974. The double-edged effect of Sir Francis Galton: a search for the motives in the biometrician-Mendelian debate. *J. hist. biol.* 7: 141–174.

De Meis, A. C. 1872–1875. *I tipi animali*. Bologna.

--- 1886. *Darwin e la scienza moderna*. Bologna: Monti.

Depéret, C. 1909. *The transformations of the animal world*. London: Kegan Paul.

Derrida, J. 1970. Structure, sign, and play. In *The Languages of Criticism and the Science of Man*, eds. R. Macksey and E. Donato. Baltimore and London.

--- 1976. *Of grammatology*. Trans. G. K. Spivak. Baltimore: Johns Hopkins University Press.

De Sarlo, F. 1887. *Studi sul Darwinismo*. Napoli: Stabilimento Tipografico A. Tocco.

Desmond, A. J. 1979. Designing the dinosaur: Richard Owen's response to Robert Edmund Grant. *Isis* 70: 224–234.

--- 1982. *Archetypes and ancestors: palaeontology in Victorian London, 1850–1875*. London: Blond and Briggs.

--- 1984. Robert E. Grant: the social predicament of a pre-Darwinian transmutationist. *J. hist. biol.* 17: 189-223.

--- 1985. The making of institutional zoology in London, 1822-1836. *Hist. sci.* 23: 153–185, 223–250.

De Stefano, G. 1907. *Il pensiero filosofico di un evoluzionista (Alb. Gaudry)*. Roma : Unione Tipografica Cooperativa.

De Vries, H. 1889. *Intracellulàre pangenesis*. Jena: Gustav Fischer.

--- 1900. Das Spaltungsgesetz der Bastarde: vorlaussige Mitteilung. *Berichte der deutschen botanischen Gesellschaft*. 18: 83–90.

--- 1901–1903. *Die mutationstheorie*. 2 vols. Leipzig: Veit.

--- 1904. *Species and varieties: their origin by mutation*. Chicago: Open Court.

--- 1910. *Intracellular pangenesis*. Trans. C. S. Gager. Chicago: Open Court.

Dewey, J. 1910. *The influence of Darwin on philosophy and other essays in contemporary thought*. New York: Holt.

Dexter, R. W. 1965. The "Salem secession" of Agassiz zoologists. *Essex Inst. hist. coll.* 101: 27–39.

--- 1979. The impact of evolutionary theories on the Salem group of Agassiz zoologists. *Essex Inst. hist. coll.* 115: 144–171.

Diara, A. 1981. Sens et définition du mot "espèce" dans l'oeuvre biologique de Félix Le Dantec. *Rev. de synthèse* 101–102: 73–86.

Dibner, B. 1960. *Darwin of the "Beagle"*. Norwalk : Burndy Library.

Dickinson, A. 1971. *Charles Darwin and natural selection*. London : Watts.

Di Gregorio, M. 1979. *On the side of the apes: the scientific and philosophical work of T. H. Huxley*. Ph. D. thesis. London University.

--- 1982a. In search of the natural system: zoological systematics in Victorian England. Paper read at the 3rd Course of the International School of the History of Biological Sciences, Napoli-Ischia, Stazione Zoologica, 4–14 July, 1982.

--- 1982b. The Dinosaur connection: a reinterpretation of T. H. Huxley's evolutionary view. *J. hist. biol.* 15: 397–418.

--- 1982c. Order or process of nature: Huxley's and Darwin's different approaches to natural sciences. *Hist. phil. life sci.* 3: 217–241.

--- 1984. *T. H. Huxley's place in natural science*.

New Haven: Yale University Press.

Dillenberger, J. 1961. *Protestant thought and natural science: a historical interpretation*. London: Collins.

Di Mauro, A. A. 1982. Rediscovery of Professor Thomas Bell's type Crustacea (Brachyura) in the dry crustacean Collection of the Zoological Collections, University Museum, Oxford. *Zool. j. Linn. Soc. Lond.* 76: 155–182.

---; **King, G. and Chancellor, G.** Forthcoming. Charles Darwin's zoological collections in the Oxford University Museum. *Biol. Linn. Soc. Lond.*

Di Siena, G. 1969. Ideologie del biologismo. *Ideologie* 9–10: 69–138.

--- 1976. Biologia, darwinismo sociale e marxismo. *Critica marxista* 6: 241–253.

Diver, C. 1940. The problem of closely related species living in the same area. In *The new systematics*, ed. J. S. Huxley, pp. 303–328. Oxford: Oxford University Press.

[Dixon, E. S.] 1851. Poultry literature. *Q. rev.* 88: 317–351.

Dobzhansky, T. 1937. *Genetics and the origin of species*. New York: Columbia University Press.

--- 1941. *Genetics and the origin of species*. 2d ed. New York: Columbia University Press.

--- 1951. *Genetics and the origin of species*. 3rd ed. New York: Columbia University Press.

--- 1955. *Evolution, genetics and man*. New York: J. Wiley.

--- 1959. Blyth, Darwin, and natural selection. *Am. nat.* 93: 204–205.

--- 1962. *Mankind evolving: the evolution of the human being*. New Haven: Yale University Press.

--- 1968. On some fundamental concepts of Darwinian biology. In *Evolutionary biology*, eds. T. Dobzhansky, M. K. Hecht, and W. C. Steere, vol. 2, pp. 1–34. Amsterdam: North Holland Publishing Company.

--- 1972. Darwinian evolution and the problem of extraterrestrial life. *Perspect. biol. and med.* 15: 157–175.

--- 1973. Ethics and values in biological and cultural evolution. *Zygon* 8: 261–281.

--- 1980. The birth of the genetic theory of evolution in the Soviet Union in the 1920s. In *The evolutionary synthesis: perspectives on the unification of biology*, eds. E. Mayr and W. B. Provine, pp. 229–242. Cambridge: Harvard University Press.

---; **Ayala, F.; Stebbins, G. L.; and Valentine, J. W.** 1977. *Evolution*. San Francisco: W. H. Freeman.

--- **and Boesiger, E.** 1968. *Essai sur l'evolution*. Paris: Masson.

Doesschate, G. 1959. Donders als voorloper van Darwin ten aanzien van de evolutieleer. *Bijdr. Gesch. Geneesk.* 39: 56.

Dolby, R. G. A. 1977. The transmission of science. *Hist. Sci.* 15: 1–43.

Doncaster, J. P. 1961. *Francis Walker's aphids*. London: British Museum (Natural History).

Donoso-Barros, R. 1975. Introduction to *The zoology of the voyage of H.M.S. 'Beagle.' Part V: Reptiles, by Thomas Bell*. Lawrence, Kansas: Society for the Study of Amphibians and Reptiles.

Dorber, H., and Plesse, W. 1968. Zur philosophischen und politischen Position des von Ernst Haeckel begründeten Monismus. *Deutsche Zeitschrift für Philosophie*. 16: 1325–1339.

Dörpinghaus, H. J. 1969. *Darwins Theorie und der deutsche Vulgärmaterialismus im Urteil deutscher katholischer Zeitschriften zwischen 1854 und 1914*. Diss. Freiburg i. B.

Dougherty, F. W. P. 1979. Les fondements scientifiques et métaphysiques du monisme Haeckelien. *Rev. de synthèse* 100, nos. 95–96 : 311–336.

Drachman, J. M. 1930. *Studies in the literature of natural science*. New York: Macmillan.

Drake, S. 1970. *Galileo studies: personality, tradition, and revolution*. Ann Arbor: University of Michigan Press.

Drapiez, P.-A.-J., ed. 1853. *Dictionnaire classique des sciences naturelles*. 10 vols. Bruxelles : Meline.

Dretske, F. 1973. Contrastive statements. *Phil. rev.* 81: 411–437.

Driesch, H. 1911. *Il vitalismo: storia e dottrina*. Trans. M. Stenta. Palermo : Sandron.

Duchenne, G. B. 1862. *Mécanisme de la physionomie humaine ou analyse electro-physiologique de l'expression des passions*. Paris: Renouard.

Duffin, K. E. 1980. Arthur O. Lovejoy and the emergence of novelty. *J. hist. ideas* 41: 267–281.

Dufrenoy, M. L., and Dufrenoy, J. 1954. Benôit de Maillet as precursor to the theory of evolution. *Arch. int. hist. sci.* 7: 161–167.

Dunn, L. C. 1965. *A short history of genetics*. New York: McGraw Hill.

Dunphy, D. C. 1966. Social change in self-analytic groups. In *The general inquirer: a computer approach to content analysis*, eds. J. Stone, D. C. Dunphy, M. S. Smith, and D. M. Ogilvie, pp. 287–340. Cambridge: M.I.T. Press.

Dupree, A. H. 1959. *Asa Gray, 1810–1888*. Cambridge : Harvard University Press.

Durant, J. 1977. *The meaning of evolution: post-Darwinian debates on the significance for man of the theory of evolution, 1858–1908*. Ph.D. thesis. University of Cambridge.

--- 1979. Scientific naturalism and social reform in the thought of Alfred Russel Wallace. *Brit. j. hist. sci.* 12: 31–58.

--- 1980. Exploring the roots of sociobiology. *Brit. j. hist. sci.* 13: 55–60.

--- 1981a. Innate character in animals and man: a perspective on the origins of ethology. In *Biology, medicine and society 1840–1940*, ed. C. Webster, pp. 157–192. Cambridge: Cambridge University Press.

--- 1981b. The beast in man. Perspective on the biology of human aggression. In *The biology of aggression*, eds. P. F. Brain and D. Benton, pp. 17–46.

--- 1982. The great debate: evolution and society in the nineteenth century. In *Charles Darwin 1809-1882: a centennial commemorative*, ed. R. Chapman. Wellington: Nova Pacifica.

Duval, M. 1886. *Le Darwinisme*. Paris: A. Delahaye and E. Lecrosnier.

Dzierzon, J. 1862. Do bees vary? — Dzierzon's opinion on the point. *Horticulture and Cottage Gardener*. N.S. 3: 463–464.

Egerton, F. N. 1968. Studies of animal populations from Lamarck to Darwin. *J. hist. biol.* 1: 225–259.

--- 1970. Humboldt, Darwin, and population. *J hist. biol.* 3: 325–360.

--- 1970–1971. Refutation and conjecture: Darwin's response to Sedgwick's attack on Chambers. *Stud. hist. phil. sci.* 1: 176–183.

--- 1971a. Darwin's method or methods? *Stud. hist. phil. sci.* 2: 281–286.

--- 1971b. The concept of competition in nature before Darwin. *Actes du XIIe cong. int. d'hist. sci.* 8: 41–46. Paris: Blanchard.

--- 1973. Changing concepts of the balance of nature. *Q. rev. biol.* 48: 322–350.

--- 1976. Darwin's early reading of Lamarck. *Isis* 67: 452–456.

--- 1977. A bibliographical guide to the history of general ecology and population ecology. *Hist. sci.* 15: 189–215.

Éhikné Süle, M. 1959. A darwinizmus biológiaoktatásunkban [Darwinism in the teaching of biology in Hungary]. *A természettudományok tanítása* no. 3: 55–59.

Ehrenberg, C. G. 1844. Über einen die ganze Luft längere Zeit trübenden Staubregen im hohen atlantischen Ocean, in 70° 43′ N.B. 26 W.L., und dessen Mischung aus zahlreichen Kieselthieren. *Ber. Bekanntm. Verh. Königl. Preuss. Akad. Wiss. Berlin* 1844: 194–207.

--- 1845a. Über eine aus feinstem Kieselmehl von Infusorien bestehende Schminke der Feuerländer. *Ber. Bekanntm. Verh. Königl. Preuss. Akad. Wiss. Berlin* 1845: 63–64.

--- 1845b. Weitere Untersuchungen des atmosphärischen Staubes aus dem atlantischen Ocean an den Capverdischen Inseln. *Ber. Bekanntm. Verh. Königl. Preuss. Akad. Wiss. Berlin* 1845: 64–66, 85–87.

--- 1845c. Über einen bedeutenden Infusorien haltenden vulkanischen Aschen-Tuff (Pyrobiolith) auf den Insel Ascension. *Ber. Bekanntm. Verh. Königl. Preuss. Akad. Wiss. Berlin* 1845: 140–142.

--- 1845d. Über einen See-Infusorien haltenden weissen vulkanischen Aschen-Tuff (Pyrobiolith) als sehr grosse Gebirgmasse in Patagonien. *Ber. Bekanntm. Verh. Königl. Preuss. Akad. Wiss. Berlin* 1845: 143–157.

Eibl-Eibesfeldt, I. 1970. *Liebe und Hass: Zur Naturgeschichte elementarer Verhaltensweisen*. München: R. Piper.

Eimer, G. H. T. 1890. *Organic evolution as the result of the inheritance of acquired characters according to the laws of organic growth*. Trans. J. T. Cunningham. London: Macmillan.

Eiseley, L. 1956. Charles Darwin. *Sci. Am.* 194 (Feb.): 62–72.

--- 1958. *Darwin's century: evolution and the men who discovered it*. Garden City: Doubleday.

--- 1959a. Alfred Russel Wallace. *Sci. Am.* 200: 70–84.

--- 1959b. Charles Darwin, Edward Blyth, and the theory of natural selection. *Proc. Am. Phil. Soc.* 103: 94–158.

--- 1959c. Charles Lyell. *Sci. Am.* 201 (Aug.): 98–106.

--- 1961. *Darwin's century*. Paperback ed. Garden City, New York: Doubleday Anchor Paperback.

--- 1965. Darwin, Coleridge, and the theory of unconscious creation. *Daedalus* 94: 588–602.

--- 1972. The intellectual antecedents of *The descent of man*. In *Sexual selection and the descent of man*, ed. B. Campbell, pp. 1–16. Chicago: Aldine.

--- 1979. *Darwin and the mysterious Mr. X. New light on the evolutionists*. New York: E. P. Dutton.

Eisen, S. 1964. Huxley and the positivists. *Victorian stud.* 7: 337–358.

Eisenberg, L. 1972. The human nature of human nature. *Science* 176: 123–128.

Ekman, P., ed. 1973. *Darwin and facial expression: a century of research in review*. New York: Academic Press.

--- **Friesen, W., and Ellsworth, P.** 1972. *Emotion in the human face*. Oxford: Pergamon Press.

Ekman, S. 1953. *Zoogeography of the sea*. Trans. E. Palmer. London: Sidgwick and Jackson.

Eldredge, N. 1979. Alternative approaches to evolutionary theory. *Bull. Carnegie Mus. Nat. Hist.* 13: 7–19.

--- **and Gould, S. J.** 1972. Punctuated equilibria: an alternative to phyletic gradualism. In *Models in paleobiology*, ed. T. J. M. Schopf, pp. 82–115. San Francisco: Freeman.

Eliot, George. 1860. *The Mill on the Floss*. 1980. Ed. G. S. Haight. Oxford: Clarendon Press.

--- 1954–1978. *The George Eliot Letters*. Ed. G. S. Haight, 9 vols. New Haven: Yale University Press.

--- 1963. *Essays*. Ed. T. Pinney. New York: Columbia University Press.

Elkana, Y., ed. 1974. *The interaction between science and philosophy* (The Van Leer Jerusalem Foundation Series). Atlantic Highlands: Humanities Press.

Ellegård, A. 1956. The Darwinian theory and the argument from design. *Lychnos* 173–192.

--- 1957. The Darwinian theory and nineteenth-century philosophies of science. *J. hist. ideas* 18: 362–393. Also in *Roots of scientific thought: a cultural perspective*, eds. P. P. Wiener and A. Noland, pp. 537–568. New York: Basic Books.

--- 1958. *Darwin and the general reader: the reception of Darwin's theory of evolution in the British periodical press, 1859–1872*. Gothenburg Studies in English, no. 8. Stockholm: Almqvist and Wiksell.

--- 1960–1961. The Darwinian revolution: a review article. *Lychnos* 55–85.

Elton, C. 1927. *Animal ecology*. London: Macmillan.

Emery, F. E. 1969. *Systems thinking: selected readings*. Harmondsworth: Penguin.

Eng, E. 1977. The confrontation between reason and imagination: the example of Darwin. *Diogenes* 95: 58–67.

--- 1978. Thomas Henry Huxley's understanding of evolution. *Hist. sci.* 16: 291–303.

Engel, H., and Engel, M. S. J. 1960. Charles Robert Darwin. *Janus* 49: 53–66.

Engelhardt, D. von. 1977. Naturwissenschaft und Geschichtlichkeit in der Neuzeit. *Fortschritte der Medizin* 95: 2203–2205.

--- 1979. *Historische Bewusstsein in der Naturwissenschaft von der Aufklärung bis zum Positivismus*. Freiburg, München: Karl Alber.

Engels, F. 1964. *Selected works*. Moscow: Progress.

Evans, E. C. 1969. Physiognomics in the ancient world. *Trans. Am. Phil. Soc.*, N.S. 59, no. 5.

Evans, L. T. 1984. Darwin's use of analogy between artificial and natural selection. *J. hist biol.* 17: 113–140.

Evans, M. A. 1965. Mimicry and the Darwinian heritage. *J. hist. ideas* 26: 211–220.

Farber, P. L. 1982. Discussion paper: the transformation of natural history in the nineteenth century. *J. hist. biol.* 15: 145–152.

Faris, R. E. L. 1950. Evolution and American sociology. In *Evolutionary thought in America*, ed. S. Persons, pp. 160–180. New Haven: Yale University Press.

Farley, J. 1972. The spontaneous generation controversy (1859–1880): British and German reaction to the problem of abiogenesis. *J. hist. biol.* 5: 285–319.

--- 1974. The initial reactions of French biologists to Darwin's *Origin of species*. *J. hist. biol.* 7: 275–300.

--- 1977. *The spontaneous generation controversy from Descartes to Oparin*. Baltimore: Johns Hopkins University Press.

--- 1982. *Gametes and spores: ideas about sexual reproduction, 1750–1914*. Baltimore: Johns

Hopkins University Press.

--- **and Geison, G.** 1974. Science, politics and spontaneous generation in 19th century France: the Pasteur-Pouchet debate. *Bull. hist. med.* 48: 161–198.

Farrall, L. A. 1979. The history of eugenics: a bibliographical review. *Ann. sci.* 36: 111–123.

Farrington, B. 1966. *What Darwin really said.* London: Macdonald.

Fay, M. A. 1979. Did Marx offer to dedicate *Capital* to Darwin? *J. hist. ideas* 39: 133–146.

Febvre, L. 1927. Un chapitre d'histoire de l'esprit humain: les sciences naturelles de Linné à Lamarck et à Georges Cuvier. *Rev. de synthèse* 43: 37–60.

Feibleman, J. 1959. Darwin and scientific method. *Tulane stud. phil.* 8: 3–14.

Feigl, H. 1956. Some major issues in the philosophy of science of logical positivism. In *The Foundations of science and the concepts of psychology and psychoanalysis*, eds. H. Feigl and M. Scriven, pp. 3–37. Minneapolis: University of Minnesota Press.

Fellmann, F. 1977. Darwins Metaphern. *Arch. für Begriffsgeschichte* 21: 285–297.

Fenizia, C. 1901. *Storia della evoluzione, con un breve saggio di bibliografia evoluzionista.* Milano: Hoepli.

--- 1905. *L'evoluzione biologica e le sue prove di fatto.* Palermo : Sandron.

Ferri, E. 1894. *Socialismo e scienza positiva (Darwin, Spencer, Marx).* Roma : Casa Editrice Italiana.

Feuer, L. S. 1975. Is the "Darwin-Marx correspondence" authentic? *Ann. sci.* 32: 1–12.

--- 1976. The "Darwin-Marx correspondence": a correction and revision. *Ann. sci.* 33: 383–384.

Feyerabend, P. 1970a. Consolations for the specialist. In *Criticism and the growth of knowledge*, eds. I. Lakatos and A. Musgrave, pp. 197–230. Cambridge : Cambridge University Press.

--- 1970b. Problems of empiricism: part II. In *The nature and function of scientific theories*. Ed. R. Colodny, pp. 275–353. Pittsburgh.

--- 1975. *Against method.* London: NLB.

Fichman, M. 1977. Wallace: zoogeography and the problem of land bridges. *J. hist. biol.* 10: 45–63.

--- 1981. *Alfred Russel Wallace.* Boston: Twayne.

Figlio, K. 1975. Theories of perception and the physiology of mind in the late 18th century. *Hist. sci.* 13: 177–212.

Finlayson, C. P. 1958. Records of scientific and medical societies preserved in the university library, Edinburgh, *The Bibliothek* 1: 14–19.

Fisch, M. H. 1947. Evolution in American philosophy. *Phil. rev.* 56: 357–373.

Fischer, J.-L. 1970. Lettre inédite de Charles Darwin à Dareste. *Arch. int. hist. sci.* 23: 81–86.

--- 1973. *La vie et la carrière d'un biologiste du XIX siècle : Camille Dareste, 1822–1899, fondateur de la tératologie experimentale.* 3 vols. Thèse de 3e cycle. Paris : École Pratique des Hautes Études, VI Section.

--- 1979. Yves Delages (1854–1920) : L'épigènese néo-Lamarckienne contre la prédétermination Weismanienne. *Rev. de synthèse* 100, nos. 95–96 : 443–461.

--- 1981. L'hybridologie et la zootaxie du siècle des Lumières à l'*Origine des espèces. Rev. de synthèse* 102: 47–72.

Fisher, R. A. 1918. The correlation between relatives on the supposition of mendelian inheritance. *Trans. Roy. Soc. Edin.* 52: 399–433.

--- 1922. On the dominance ratio. *Proc. Roy. Soc. Edin.* 42: 321–341.

--- 1927. On some objections to mimicry theory — statistical and genetic. *Trans. Roy. Entomol. Soc. Lond.* 75: 269–278.

--- 1928a. The possible modification of the response of the wild type to recurrent mutations. *Am. nat.* 62: 115–126.

--- 1928b. Two further notes on the origin of dominance. *Am. nat.* 62: 571–574.

--- 1930. *The genetical theory of natural selection.* Oxford: Oxford University Press.

--- 1932. The bearing of genetics on theories of evolution. *Sci. progress* 27: 273–287.

--- 1958. *The genetical theory of natural selection.* Rev. and enl. ed. New York: Dover.

Fitton, W. 1828. Presidential address. *Proc. Geol. Soc. Lond.* 1: 50–62.

--- 1829. Presidential address. *Proc. Geol. Soc. Lond.* 1: 112–134.

FitzRoy, R. 1839. *Narrative of the surveying voyages of His Majesty's ships Adventure and Beagle, between the years 1826 and 1836, describing their examination of the southern shores of South America, and the Beagle's circumnavigation of the globe. Vol. 2: proceedings of the second expedition,*

1831–1836, under the command of Captain Robert Fitz-Roy, R.N. With *appendix*. London: Henry Colburn.

Fleming, D. 1959. The centenary of the *Origin of species*. *J. hist. ideas* 20: 437–446.

--- 1961. Charles Darwin, the anaesthetic man. *Victorian stud.* 4: 219–236.

Fletcher, H. R. 1969. *Story of the Royal Horticultural Society 1804–1968*. London: Oxford University Press for the Royal Horticultural Society.

Flew, A. G. N. 1959. The structure of Darwinism. *New biol.* 28: 25–44.

--- 1963. The structure of Malthus' population theory. In *Philosophy of science. The Delaware seminar*, ed. B. Baumrin, vol. 1, pp. 283–307. New York: Interscience.

--- 1966. The concept of evolution: a comment. *Philosophy* 41: 70–75.

--- 1967. *Evolutionary ethics*. London: Macmillan.

--- 1978. *A rational animal, and other philosophical essays on the nature of man*. Oxford: Clarendon Press.

Flourens, P. 1864. *Examen du livre de M. Darwin sur l'origine des espèces*. Paris : Garnier.

Foderà, M. 1826. *Discours sur la biologie ou science de la vie; suivi d'un tableau des connaissances naturelles envisagés d'après la nature de leur filiation*. Paris : J. B. Baillière.

Fogazzaro, A. 1898. *Ascensioni umane*. Milano: Baldini e Castoldi.

Foote, G. W. 1889. *Darwin on God*. London: Progressive Publishing Co.

Foote, S. c. 1750. *A treatise on the passions, so far as they regard the stage*. London.

Forbes, E. 1846. Descriptions of Secondary fossil shells from South America. In C. Darwin, *Geological observations on South America*, pp. 265–268. London: Smith Elder.

Forbes, T. R. 1962. William Yarrell, British naturalist. *Proc. Am. Phil. Soc.* 106: 505–515.

Ford, E. B. 1931. *Mendelism and evolution*. London: Methuen.

--- 1964. *Ecological genetics*. London: Methuen.

Fordyce, J. 1883. *Aspects of scepticism, with special reference to the present time*. London: E. Stock.

Foresti, L. 1863. *Una lezione del Prof. Cav. Giovanni Capellini sull' antichitá* dell'uomo. Bologna : Vitali.

Forman, P. 1971. Weimar culture, causality and quantum theory, 1918–1927: adaptation by German physics and mathematics to a hostile intellectual environment. *Hist. stud. phys. sci.* 3: 1–115.

Forrest, D. W. 1974. *Francis Galton: the life and work of a Victorian genius*. London: Paul Elek.

Forti, U. 1969. *Storia della scienza nei suoi rapporti con la filosofia, le religioni, la societa, Milano: Dall 'Oglio.*

Foster, J., ed. 1888. *Alumni Oxonienses . . . 1715–1886* . . . 4 vols. Oxford: Joseph Foster and Parker.

---, ed. 1890. *Index ecclesiasticus; or, alphabetical lists of all ecclesiastical dignitaries in England and Wales since the Reformation, containing 150,000 hitherto unpublished entries from the bishops' certificates of institutions to livings, etc. now deposited in the Public Record Office*. Oxford: Parker.

Foster, J. E. 1885. *The art of expression, a book for clergymen, barristers, vocalists, actors, and for all persons of culture*. London: Simpkin and Marshall.

Foster, W. D. 1965. A contribution to the problem of Darwin's ill-health. *Bull. hist. med.* 39: 476–478.

Fothergill, P. G. 1952. *Historical aspects of organic evolution*. London: Hollis and Carter.

--- 1961. *Evolution and Christians*. London: Longmans.

Foucault, M. 1965. *Madness and civilization: a history of insanity in the age of reason*. Trans. R. Howard. New York: Pantheon.

--- 1970. *The order of things: an archaeology of the human sciences*. London: Tavistock; New York: Random House.

Fournié, E. 1882. *Ch. Darwin, étude critique*. Paris: Imprimerie Chaix.

Fox, R., ed. 1976. Lyell centenary issue: papers delivered at the Charles Lyell Centenary Symposium, London, 1975. *Brit. j. hist. sci.* 9 (part 2): 91–242.

Fox, S. B. 1862. Do bees vary. *J. horticulture and cottage gardener*. N.S. 3: 284.

Francis, M. 1978. Herbert Spencer and the myth of laissez-faire. *J. hist. ideas* 39: 317–328.

Frankel, M. 1979. The career of continental drift theory, an application of Imre Lakatos' analysis of scientific growth to the rise of drift theory. *Stud. hist. phil. sci.* 10: 21–66.

Franz, V., ed. 1941. *Ernst Haeckel: sein Leben, Denken und Wirken*. Vol. 1. Jena and Leipzig: Wilhelm Gronau, W. Agricola.

---, ed. 1944. *Ernst Haeckel: eine Schriftenfolge zur Pflege seines geistigen Erbes*. Vol. 2. Jena and

Leipzig: W. Gronau, W. Agricola.

Freeman, D. 1974. The evolutionary theories of Charles Darwin and Herbert Spencer. *Current anthro.* 15: 211–237.

Freeman, R. B. 1968. Charles Darwin on the routes of male humble bees. *Bull. Brit. Mus. (Nat. Hist.), Hist. ser.* 3: 177–189.

--- 1977. *The works of Charles Darwin: an annotated bibliographical handlist*. 2d ed. Folkestone: Dawson.

--- 1978. *Charles Darwin : a companion*. Folkestone: Dawson.

--- **and Gautrey, P. J.** 1969. Darwin's *Questions about the breeding of animals*, with a note on *Queries about expression. J. Soc. Bibl. Nat. Hist.* 5: 220–225.

--- **and Gautrey, P. J.** 1972. Charles Darwin's *Queries about expression. Bull. Brit. Mus. (Nat. Hist.). Hist. ser.* 4: 207–219.

French, R. D. 1970a. Darwin and the physiologists, or the medusa and modern cardiology. *J. hist. biol.* 3: 253–274.

--- 1970b. Some concepts of nerve structure and function in Britain 1875–1885: background to Sir Charles Sherrington and the synapse concept. *Med. hist.* 14: 154–165.

Friday, J. 1974. A microscopic incident in a monumental struggle: Huxley and antibiosis in 1875. *Brit. j. hist. sci.* 7: 61–71.

Froggatt, P., and Nevin, N. C. 1971a. Galton's law of ancestral heredity: its influence on the early development of human genetics. *Hist. sci.* 10: 1–27.

--- 1971b. The "law of ancestral heredity" and the Mendelian-ancestrian controversy in England, 1889–1900. *J. med. genet.* 8: 1–36.

Fullerton, W. Y. [1930]. *J. W. C. Fegan: a tribute*. London: Marshall, Morgan, & Scott.

Fullinwider, S. P. 1975. Insanity as the loss of self: the moral insanity controversy revisited. *Bull. hist. med.* 49: 87–101.

Fulton, J. S. 1959. Philosophical adventures of the idea of evolution, 1859–1959. *Rice Institute pamphlet* 46: 1–31.

Furth, M. 1975. *Essence and individual: reconstruction of an Aristotelian metaphysics*, Chapter 11, unpublished mimeo, UCLA.

Gaissinovitch, A. E. 1967. *Zarozhdenie genetiki.* [Origins of genetics] Moscow: Isdatel'stvo Nauka.

--- 1970. Vzgliady Ch. Darvina na izmenchivost'i nasledstvennost' [Darwin's views on variability and heredity]. *Iz istorii biologii* 2: 33–59.

--- 1973. Problems of variation and heredity in Russian biology in the late nineteenth century. *J. hist. biol.* 6: 97–123.

--- 1977. Novye dannye o pis'makh Darvina Marksu [New facts on the Darwin-Marx correspondence]. *Priroda* 2: 92–97.

Gale, B. G. 1972. Darwin and the concept of a struggle for existence: a study in the extrascientific origins of scientific ideas. *Isis* 63: 321–344.

Gallesio, G. 1811. *Traité du citrus*. Paris: L. Fantin.

--- 1816. *Teoria della riproduzione vegetale*. Pisa: Presso N. Capurro.

Galton, F. 1869. *Hereditary genius: an inquiry into its laws and consequences*. London: Macmillan.

--- 1873. Hereditary improvement. *Fraser's mag.* 7: 116–130.

--- 1874. *English men of science: their nature and nurture*. London: Macmillan.

--- 1883. *Inquiries into human faculty and its development*. London: Macmillan.

--- 1889. *Natural inheritance*. London: Macmillan.

Ganz, K. F. 1939. The beginnings of Darwinian ethics. *Univ. Texas Publ. Stud. in English* 3939: 180–209.

Garfinkel, A. 1981. *Forms of explanation: rethinking the questions of social theory*. New Haven: Yale University Press.

Garin, E. 1962. *La cultura italiana tra '800 e '900*. Bari : Laterza.

Garland, M. M. 1980. *Cambridge before Darwin: the ideal of a liberal education, 1800–1860*. Cambridge: Cambridge University Press.

Gasman, D. 1971. *The scientific origins of national-socialism: social Darwinism in Ernst Haeckel and the German Monist League*. London: Macdonald; New York: American Elsevier.

Gaston, J. 1978. *The reward system in British and American science*. New York: Wiley.

Gaudant, M., and Gaudant, J. 1971. *Théories classiques de l'evolution*. Paris: Dunod.

Gaukroger, S. W. 1976. Bachelard and the problem of epistemological analysis. *Stud. hist. phil. sci.* 7: 189–244.

Gaull, M. 1979. From Wordsworth to Darwin. *The Wordsworth Circle* 10: 33–48.

Gautieri, G. 1805. *Slancio sulla genealogia della terra e sulla costituzione dinamica dell'organizzazione, seguito da una ricerca sull'origine dei vermi abitanti le interiora degli animali*. Jena [Pavia].

Geddes, P. 1882. Biology. In *Chambers' encyclopaedia*, vol. 4. Edinburgh: Chambers.
Gegenbaur, C. 1912. Ueber die Entwicklung der Sagitta. In *Gesammelte Abhandlungen von Carl Gegenbaur*, ed. M. Fürbringer and H. Bluntschli. Leipzig: W. Engelmann.
Geikie, A. 1909. *Charles Darwin as geologist.* Cambridge: Cambridge University Press.
--- 1924. *A long life's work.* London: Macmillan.
Geison, G. L. 1969a. Darwin and heredity: the evolution of his hypothesis of pangenesis. *J. hist. med.* 24: 375–411.
--- 1969b. The protoplasmic theory of life and the vitalist-mechanist debate. *Isis* 60: 273–292.
--- 1975. Schultze, Max Johann Sigismund. In *Dictionary of scientific biography*, 12: 230–233.
--- 1978. *Michael Foster and the Cambridge school of physiology: the scientific enterprise in late Victorian society.* Princeton: Princeton University Press.
Gemelli, A. 1906. *Per l'evoluzione.* Pavia : Fusi.
--- 1912. *Recenti scoperte e recenti teorie nello studio dell'origine dell'uomo: conferenza.* Firenze : Libreria Editrice Fiorentina.
--- 1914. *L'enigma della vita e i nuovi orizzonti della biologia: introduzione allo studio delle scienze biologiche.* Firenze : Libreria Editrice Fiorentina.
Gentile, G. 1922. *Gino Capponi e la cultura toscana nel secolo decimonono.* Firenze : Vallecchi.
Geoffroy Saint-Hilaire, I. 1854–1862. *Histoire naturelle générale des règnes organiques principalement etudiée chez l'homme et les animaux.* 3 books in 4 vols. Paris : V. Masson.
George, W. 1964. *Biologist philosopher: a study of the life and writings of Alfred Russel Wallace.* London: Abelard Schuman.
Gerratana, V. 1972. *Ricerche di storia del marxismo.* Roma: Editori Riuniti.
--- 1973. Marx and Darwin (Engl. trans. of above, chap. 2). *New left rev.* 82: 60–82.
Geymonat, L. 1957. *Galileo Galilei.* Torino : Einaudi.
--- 1965. *Galileo Galilei: a biography and inquiry into his philosophy of science.* New York : McGraw Hill.
Ghiliani, V. 1854. Materiali per servire alla compilazione della Fauna Entomologica Italiana, ossia elenco delle specie di Lepidotteri riconosciuti esistenti negli Stati Sardi. *Memorie della Reale Accademia delle Scienze di Torino* 14: 131–248.
Ghiselin, M. T. 1966. On psychologism in the logic of taxonomic principles. *Syst. zool.* 15: 207–215.
--- 1969. *The triumph of the Darwinian method.* Berkeley: University of California Press.
--- 1971. The individual in the Darwinian revolution. *New literary hist.* 3: 113–134.
--- 1973a. Darwin and evolutionary psychology. *Science* 179: 964–968.
--- 1973b. Essay review: Mr. Darwin's critics, old and new. *J. hist. biol.* 6: 155–165.
--- 1974a. *The economy of nature and the evolution of sex.* Berkeley: University of California Press.
--- 1974b. A radical solution to the species problem. *Syst. zool.* 23: 536–544.
--- 1975. The rationale of pangenesis. *Genetics* 79 (Supp.): 47–57.
--- 1976. Two Darwins: history versus criticism. *J. hist. biol.* 9: 121–132.
--- 1980. Review of I. Stone, *The Origin. California monthly* 91, no. 2: 12–13.
---; **and Jaffe, L.** 1973. Phylogenetic classification in Darwin's *Monograph of the Sub-Class Cirripedia. Syst. zool.* 22: 132–140.
Giacobini, G. 1977. Il problema dell'origine dell'uomo e la critica postdarwiniana a Torino, 1864–1900. *Studi piemontesi* 6: 75–81.
---; **and Panattoni, G. L.,** eds. 1983. *Il Darwinismo in Italia: testi di Filippo de Filippi, Michele Lessona, Paolo Mantegazza, Giovanni Canestrini.* Torino: Unione Tipografico Editrice Torinese.
Giere, R. N. 1979. *Understanding scientific reasoning.* New York: Holt, Rinehart and Winston.
--- **and Westfall, R. S.,** eds. 1973. *Foundations of scientific method: the nineteenth century.* Bloomington: Indiana University Press.
Gilbert, S. F. 1979. Altruism and other unnatural acts: T. H. Huxley on nature, man and society. *Perspect. biol. med.* 22: 346–358.
Gillespie, N. C. 1977. The Duke of Argyll, evolutionary anthropology, and the art of scientific controversy. *Isis* 68: 40–54.
--- 1979. *Charles Darwin and the problem of creation.* Chicago: University of Chicago Press.
Gillispie, C. C. 1951. *Genesis and geology: a study in the relations of scientific thought, natural theology, and social opinion in Great Britain, 1790–1850.* Cambridge: Harvard University Press.
--- 1958. Lamarck and Darwin in the history of science. *Am. sci.* 46: 388–409.

--- 1959. Lamarck and Darwin in the history of science. In *Forerunners of Darwin, 1745–1859*, eds. B. Glass, O. Temkin, and W. L. Straus Jr., pp. 265–291. Baltimore: Johns Hopkins University Press.

--- 1959*. *Genesis and geology*. 2d ed. Harper Torchbooks.

--- 1960. *The edge of objectivity: an essay in the history of scientific ideas*. Princeton: Princeton University Press.

--- 1974. Commentary on D. Freeman, The evolutionary theories of Charles Darwin and Herbert Spencer. [D. Freeman 1974] *Current anthro.* 15: 224.

Gillmor, C. S. 1975. The place of the geophysical sciences in nineteenth century natural philosophy. *Eos* 56: 4–7.

Gilman, S. L. 1976. *The face of madness*. New York: Brunner/Mazel.

--- 1979. Darwin sees the insane. *J. hist. behav. sci.* 15: 253–262.

Gilmour, J. S. L. 1951. The development of taxonomic theory since 1851. *Nature* 168: 400–402.

Gilson, E. 1971. *D'Aristote à Darwin et retour*. Paris: Vrin:

--- 1979. *Le Darwinisme aujourd'hui*. Conférence à Radio France. Paris: Le Seuil.

Giustino, D. de. 1975. *Conquest of mind: phrenology and Victorian social thought*. London: Croom Helm.

Glacken, C. J. 1967. *Traces on the Rhodian shore: nature and culture in western thought from ancient times to the end of the eighteenth century*. Berkeley: University of California Press.

Glass, B.; Temkin, O.; and Straus, W. L., Jr., eds. 1959. *Forerunners of Darwin, 1745–1859*. Baltimore: Johns Hopkins University Press.

Glick, T. F. 1974a. Spain. In *The comparative reception of Darwinism*, ed. T. F. Glick pp. 307–345. Austin: University of Texas Press.

---, ed. 1974b. *The comparative reception of Darwinism*. Austin: University of Texas Press.

Gliserman, S. 1975. Early Victorian science writers and Tennyson's *In memoriam*. *Victorian stud.* 18: 277–308, 437–459.

Godelier, M. 1973. *Horizon, trajets marxistes en anthropologie*. Paris: Maspero.

Godron, D. A. 1859. *De l'espèce et des races dans les êtres organisés et spécialement de l'unité de l'espèce humaine*. 2 vols. Paris: J.-B. Baillière.

Gohau, G. 1978. *Biologie et biologistes*. Paris: Magnard.

--- 1979. Alfred Giard. *Rev. de synthèse* 100, nos. 95–96: 393–406.

Goldman, I. 1959. Evolution and anthropology. *Victorian stud.* 3: 55–75.

Goldschmidt, R. 1933. Some aspects of evolution. *Science* 78: 539–547.

--- 1938. The theory of the gene. *Sci. monthly* 46: 268–273.

--- 1940. *The material basis of evolution*. New Haven: Yale University Press.

--- 1948. Ecotype, ecospecies, and macroevolution. *Experientia* 4: 1–22.

--- 1955. Different philosophies of genetics. *Science* 119: 703–710.

Goldstein, P. 1965. *Triumphs of biology*. New York: Garden City.

Goll, R. 1972. *Der Evolutionismus: Analyse eines Grundbegriffs neuzeitlichen Denkens*. München: C. H. Beck.

Gombrich, E. 1960. *Art and illusion: a study in the psychology of pictorial representation*. London: Phaidon.

--- 1963. *Meditations on a hobby horse and other essays on the theory of art*. London: Phaidon.

--- 1978. *The story of art*. 13th ed. London: Phaidon.

Goodman, D. C., and Brooke, J. H. 1974. Towards a mechanistic philosophy. (*Science and belief: from Copernicus to Darwin*, Block II, Units 4–5) Milton Keynes: The Open University Press.

Goodwin, C. 1964. Evolutionary theory in Australian social thought. *J. hist. ideas* 25: 393–416.

Gosden, P. H. J. H. 1961. *The friendly societies in England, 1815–1875*. Manchester: Manchester University Press.

--- 1973. *Self-help: voluntary associations in the 19th century*. London: B. T. Batsford.

Gosse, P. H. 1851. *A naturalist's sojourn in Jamaica*. London: Longmans.

Gott, M. F. 1889. *K. F. Wolff: eine biographische Skizze*. St. Petersburg: Salenskji.

Gottschalk, R. 1959. Darwin und der Sozialdarwinismus. *Deutsche Zeitschrift für Philosophie* 4: 521–539.

Goudge, T. A. 1954. The concept of evolution. *Mind* 63: 16–25.

--- 1961. *The ascent of life: a philosophical study of the theory of evolution*. London: Allen &

Unwin.
--- 1967. Darwin. In *The encyclopaedia of philosophy*, ed. P. Edwards, vol. 2, pp. 294–295. London: Macmillan.
--- 1973. Evolutionism. In *Dictionary of the history of ideas*, ed. P. P. Wiener et al., vol. 2, pp. 174–189. New York: Charles Scribner's Sons.
Gould, J. 1837a. Remarks on a group of ground finches from Mr. Darwin's collection, with characters of the new species. *Proc. Zool. Soc. Lond.* 5: 4–7.
--- 1837b. Observations on the raptorial birds in Mr. Darwin's collection, with characters of the new species. *Proc. Zool. Soc. Lond.* 5: 9–11.
--- 1837c. Exhibition of the fissirostral birds from Mr. Darwin's collection, and characters of the new species. *Proc. Zool. Soc. Lond.* 5: 22.
--- 1837d. Three species of the genus *Orpheus*, from the Galapagos, in the collection of Mr. Darwin. *Proc. Zool. Soc. Lond.* 5: 27.
--- 1837e. On a new *Rhea* (*Rhea darwinii*) from Mr. Darwin's collection. *Proc. Zool. Soc. Lond.* 5: 35–36.
--- 1837f. Exhibition of Mr. Darwin's birds, and description of a new species of wagtail (*Motacilla leucopsis*) from India. *Proc. Zool. Soc. Lond.* 5: 77–78.
--- 1838. [Exhibition of "another portion of the birds collected by Charles Darwin, Esq."] *Proc. Zool. Soc. Lond.* 6: 4.
Gould, S. J. 1968. *Trigonia* and the Origin of species. *J. hist. biol.* 1: 41–56.
--- 1977a. *Ever since Darwin: reflections in natural history*. New York: Norton.
--- 1977b. *Ontogeny and phylogeny*. Cambridge: Harvard University Press.
--- 1977c. Eternal metaphors of palaeontology. In *Patterns of evolution as illustrated by the fossil record*, ed. A. Hallam, pp. 1–26. New York: Elsevier.
--- 1977d. Darwin's dilemma: the odyssey of evolution. In *Ever Since Darwin*, pp. 34–38. New York: W. W. Norton.
--- 1977e. Darwin's untimely burial. In *Ever since Darwin*, pp. 39–48. New York: W. W. Norton.
--- 1977f. The return of hopeful monsters. *Nat. hist.* 86: 22–30. 1980, Reprinted in *The panda's thumb*, pp. 186–193. New York: Norton.
--- 1979. Agassiz's marginalia in Lyell's *Principles*, or the perils of uniformity and the ambiguity of heroes. *Stud. hist. biol.* 3: 119–138.
--- 1980a. *The panda's thumb: more reflections in natural history*. New York: Norton.
--- 1980b. Natural selection and the human brain: Darwin *vs.* Wallace. In *The panda's thumb*, pp. 47–58. New York: W. W. Norton.
--- 1980c. Is a new and general theory of evolution emerging? *Paleobiology* 6: 119–130.
--- 1980d. The promise of paleobiology as a nomothetic science. *Paleobiology* 6: 96–118.
--- 1981. The rise of neo-Lamarckism in America. In *Lamarck et son temps. Lamarck et notre temps*. Colloque international organisé par le Centre de Recherche sur l'Histoire des Idêes de l'Université de Picardie, pp. 81–91. Paris: Vrin.
--- 1982. Darwinism and the expansion of evolutionary theory. *Science* 216: 380–387.
--- 1983. A hearing for Vavilov. In *Hen's teeth and horse's toes*, pp. 134–146. New York: W. W. Norton.
--- **and Eldredge, N.** 1977. Punctuated equilibria: the tempo and mode of evolution reconsidered. *Paleobiology* 3: 115–151.
--- **and Lewontin, R. C.** 1979. The spandrels of San Marco and the Panglossian paradigm: a critique of the adaptationist programme. *Proc. Roy. Soc. Lond.* 205B: 581–598.
--- **and Vrba, E. S.** 1982. Exaptation — a missing term in the science of form. *Paleobiology* 8: 4–15.
Grace, E. 1942. More on Social Darwinism. *Science and society* 6: 71–74.
Graham, J. 1961. Lavater's *Physiognomy* in England. *J. hist. ideas* 22: 561–572.
Graham, W. 1881. *Creed of science: religious, moral and social*. London: C. K. Paul.
Grant, R. E. 1825–1826. Observations and experiments on the structure and functions of the sponge. *Edin. phil. j.* part 1, 13: 94–107; 343–346; part 2, 14: 113–124; 343–346.
--- 1826a. Notice of a new zoophyte (Cliona celata, Gr.) from the Firth of Forth. *Edin. new phil. j.* 1: 78–81.
--- 1826b. Observations on the spontaneous motions of the ova of the *Campanularia dichomota, Gorgonia verrucosa, Caryophyllea calycularis, Spongia tomentosa,* and *Plumularia falcata. Edin. new phil. j.* 1: 150–156.
--- 1826c. Observations on the structure of some silicious sponges. *Edin. new phil. j.* 1: 341–351.

--- 1826d. Observations on the nature and importance of geology. *Edin. new phil. j.* 1: 293–302.

--- 1827a. Notice regarding the structure and mode of generation of the *Virgularia mirabilis* and *Pennatula phosphorea. Edin. j. sci.* 7: 330–334.

--- 1827b. Observations on the structure and nature of the Flustrae. *Edin. new phil. j.* 3: 107–118; 337–342.

--- 1828a. Observations on the generation of the *Lobularia digitata,* Lam. (*Alcyonium lobatum,* Pall.). *Edin. j. sci.* 8: 109–110.

--- 1828b. *An essay on the study of the animal kingdom: being an introductory lecture delivered in the University of London on the 23rd of October, 1828*. London: Taylor.

Grant, V. 1966. The selective origin of incompatibility barriers in the plant genus *Gilia. Am. nat.* 100: 99–118.

--- 1971. *Plant speciation*. New York: Columbia University Press.

--- 1981. *Plant speciation*. 2d ed. New York: Columbia University Press.

Grant Duff, Mrs. Adrian [Ursula], ed. 1924. *The life-work of Lord Avebury (Sir John Lubbock), 1834–1913* . . . London: Watts & Co.

Grant Watson, E. L. 1946. *But to what purpose? the autobiography of a contemporary*. London: Cresset Press.

Grassé, P.-P. 1960. Lamarck, Wallace, and Darwin. *Rev. d'hist. sci.* 13: 73–79.

--- 1973. *L'évolution du vivant: matériaux pour une nouvelle théorie tranformiste*. Paris: Albin Michel.

--- 1977. *Evolution of living organisms: evidence for a new theory of transformation*. New York: Academic Press.

Gratiolet, P. 1865. *De la physionomie et des mouvements d'expression*. Paris: Bibliothèque D'Education.

Gray, A. 1876. *Darwiniana: essays and reviews pertaining to Darwinism*. New York: Appleton. 1963. Reprinted, ed. A. H. Dupree. Cambridge: Harvard University Press.

Gray, R. 1977. Bourgeois hegemony in Victorian Britain. In *The communist university of London: papers on class, hegemony, and party,* ed. J. Bloomfield, pp. 73–93. London: Lawrence & Wishart.

Great Britain. 1837. *Evidence, oral and documentary taken and received by the commissioners appointed . . . for visiting the universities of Scotland*. Vol. 1. London: Clowes and sons.

--- 1852. *Report of Her Majesty's commissioners appointed to inquire in the state, discipline, studies and revenues of the University and Colleges of Cambridge*. London: Clowes and sons.

Greenacre, P. 1963. *The quest for the father: a study of the Darwin-Butler controversy, as a contribution to the understanding of the creative individual*. New York: International Universities Press.

Greene, E. L. 1909. Linnaeus as an evolutionist. *Proc. Wash. Acad. Sci.* 11: 17–26.

Greene, J. C. 1954a. Some early speculations on the origin of human races. *Am. anthropologist* 56: 31–41.

--- 1954b. The American debate on the Negro's place in nature, 1780–1815. *J. hist. ideas* 15: 384–396.

--- 1959a. *The death of Adam. Evolution and its impact on Western thought*. Ames: Iowa State University Press.

--- 1959b. Biology and social theory in the nineteenth century: Auguste Comte and Herbert Spencer. In *Critical problems in the history of science*, ed. M. Clagett, pp. 419–446. Madison: University of Wisconsin Press.

--- 1959c. Darwin and religion. *Proc. Am. Phil. Soc.* 103: 716–725.

--- 1961a. *Darwin and the modern world view*. Baton Rouge: Louisiana State University Press.

--- 1961b. *The death of Adam. Evolution and its impact on Western thought*. New York: Mentor Books.

--- 1968. The concept of order in Darwinism. In *The concept of order*, ed. P. G. Kuntz, pp. 89–103. Seattle: University of Washington Press.

--- 1971. The Kuhnian paradigm and the Darwinian revolution. In *Perspectives in the history of science and technology*, ed. D. H. D. Roller, pp. 3–25. Norman: University of Oklahoma Press.

--- 1975. Reflections on the progress of Darwin studies. *J. hist. biol.* 8: 243–273.

--- 1977. Darwin as a social evolutionist. *J. hist. biol.* 10: 1–27.

--- 1981a. *Science, ideology, and world view: essays in the history of evolutionary ideas*. Berkeley: University of California Press.

--- 1981b. Darwinism as a world-view. In J. C. Greene, *Science, ideology and world-view: essays in the history of evolutionary ideas*, pp. 128–157. Berkeley: University of California

Press.

Greene, M. T. 1982. *Geology in the nineteenth century. Changing view of a changing world.* Ithaca: Cornell University Press.

Greenough, G. 1834. Presidential address. *Proc. Geol. Soc. Lond.* 2: 42–70.

--- 1835. Presidential address. *Proc. Geol. Soc. Lond.* 2: 145–175.

Greg, W. R. 1853. *Essays on political and social science, contributed chiefly to the Edinburgh Review.* 2 vols. London: Longman, Brown, Green, and Longmans.

--- 1868. On the failure of "natural selection" in the case of man. *Fraser's mag.* 78: 353–362.

Gregory, F. 1977a. Scientific and dialectical materialism. *Isis* 68: 206–223.

--- 1977b. *Scientific materialism in nineteenth century Germany.* Dordrecht: Reidel.

Grene, M. 1958. Two evolutionary theories. *Brit. j. phil. sci.* 9: 110–127, 185–193.

--- 1966. *The knower and the known.* London: Faber and Faber.

--- 1974. The understanding of nature. In *Essays in the philosophy of biology.* Boston Studies in the Philosophy of Science, vol. 23. Dordrecht, Boston: Reidel.

--- 1975. Darwin and philosophy. In *Connaissance scientifique et philosophie*, pp. 133–145. Bruxelles: Palais des Académies.

--- 1976. Imre Lakatos: some recollections. In *Essays in memory of Imre Lakatos*, eds. R. S. Cohen, P. K. Feyerabend, and M. W. Wartofsky, pp. 209–212. Boston Studies in the Philosophy of Science, vol. 29. Dordrecht, Boston: Reidel.

--- 1978. Individuals and their kinds: Aristotelian foundations of biology. In *Organism, medicine, and metaphysics*, ed. S. F. Spicker, pp. 121–136. Dordrecht: Reidel.

--- **and Mendelsohn, E.,** eds. 1976. *Topics in the philosophy of biology.* Boston Studies in the Philosophy of Science, vol. 27. Dordrecht, Boston: Reidel.

Grinnell, G. 1974. The rise and fall of Darwin's first theory of transmutation. *J. hist. biol.* 7: 259–273.

Griscom, J. 1824. *A year in Europe.* 2d ed. 2 vols. New York: Collins and Hannay.

Grmek, M. D. 1973. *Raisonnement expérimental et recherches toxicologiques chez Claude Bernard.* Genève: Droz.

---; **Cohen, R. S.; and Cimino, G.,** eds. 1981. *On scientific discovery.* Boston Studies in the Philosophy of Science, vol. 34. Dordrecht, Boston: Reidel.

Groeben, C. 1975. *The Naples Zoological station at the time of Anton Dohrn.* Cava dei Tirreni: E. Di Mauro.

Grose, F. 1788. *Rules for drawing caricatures, with an essay on comic painting.* London.

Gross, M. G. 1972. *Oceanography.* Englewood Cliffs: Prentice Hall.

Gruber, H. E. 1961. Darwin and *Das Kapital. Isis* 52: 582–583.

--- 1966. Pensée créatrice et vitesse du changement adaptif: le développement de la pensée de Darwin. In *Psychologie et épistémologie: thèmes piagetiens*, eds. F. Bresson et M. de Montmollin. Paris: Dunod.

--- 1973. Courage and cognitive growth in children and scientists. In *Piaget in the classroom*, eds. M. Schwebel and J. Raph, pp. 73–105. New York: Basic Books.

--- 1976–1977. Créativité et fonction constructive de la répétition. *Bull. de psychologie de l'Univ. de Paris*: 30: 235–239.

--- 1977. The fortunes of a basic Darwinian idea: chance. In *The roots of American psychology: historical influences and implications for the future*, eds. R. W. Rieber and K. Salzinger. *Ann. N. Y. Acad. Sci.* 291: 233–245.

--- 1978. Darwin's "Tree of nature" and other images of wide scope. In *Aesthetics in science*, ed. J. Wechsler, pp. 121–140. Cambridge: M.I.T. Press.

--- 1980a. The evolving systems approach to creative scientific work. In *Scientific discovery: case studies*, ed. T. Nickles, pp. 113–130. Dordrecht: Reidel.

--- 1980b. And the bush was not consumed: the evolving systems approach to creativity. In *Toward a theory of psychological development*, eds. S. Modgil and C. Modgil, pp. 269–299. Windsor: NFER Publishers.

--- 1981a. Appendix. The many voyages of the *Beagle*. In *Darwin on man: a psychological study of scientific creativity*. 2d. ed., pp. 259–299. Chicago: University of Chicago Press.

--- 1981b. Cognitive psychology, scientific creativity, and the case study method. In *On scientific discovery*, eds. M. D. Grmek, R. S. Cohen, and G. Cimino, pp. 295–322. Dordrecht, Boston: Reidel.

--- 1981c. *Darwin on man: a psychological study*

of scientific creativity. 2d ed. Chicago: University of Chicago Press.

--- 1981d. On the relation between 'aha experiences' and the construction of ideas. *Hist. sci.* 19: 41–59.

--- 1981e. Review of M. Ruse, *The Darwinian revolution. Isis* 72: 326–327.

--- 1982. On the hypothesized relation between giftedness and creativity. In *Developmental approaches to giftedness and creativity*, ed. D. H. Feldman. San Francisco: Jossey-Bass, 1982.

--- **and Barrett, P. H.** 1974. *Darwin on man: a psychological study of scientific creativity*. Together with *Darwin's early and unpublished notebooks*. Transcribed and annotated by P. H. Barrett. Foreword by J. Piaget. New York: E. P. Dutton.

--- **and Gruber, V.** 1962. The eye of reason: Darwin's development during the *Beagle* voyage. *Isis* 53: 186–200.

Gruber, J. W. 1960. *A conscience in conflict: the life of St. George Jackson Mivart*. New York: published for Temple University by Columbia University Press.

--- 1964. Darwinism and its critics. *Hist. sci.* 3: 115–123.

--- 1969. Who was the *Beagle's* naturalist? *Brit. j. hist. sci.* 4: 266–282.

Guédès, M. 1974–1975. Évolutionnisme. *Histoire et nature* 3–4: 160–169.

Guilhot, J. 1976. Review of Y. Conry, *L'introduction du darwinisme en France au XIXe siècle. Rev. d'hist. sci.* 29: 282–285.

Guimond, A. A. 1972. William Herbert. In *Dictionary of scientific biography*; 5: 295–297. New York: Charles Scribner's Sons.

Gunther, A. E. 1979. J. E. Gray, Charles Darwin, and Cirripedes. *Notes rec. Roy. Soc. Lond.* 34: 53–63.

Günther, A. K. L. G., ed. 1877. Account of the zoological collection made during the visit of H.M.S. "Peterel" to the Galapagos Islands. *Proc. Zool. Soc. Lond.*

Gursch, R. 1980. *Die Auseinandersetzungen um Ernst Haeckels Abbildungen*. Diss. Marburg.

Guyénot, E. 1957. *Les sciences de la vie aux XVIIe et XVIIIe siècles; l'idée d'evolution*. Paris: Albin Michel.

Haber, F. C. 1959a. *The age of the world: Moses to Darwin*. Baltimore: Johns Hopkins University Press.

--- 1959b. Fossils and the idea of a process of time in natural history. In *Forerunners of Darwin*, eds. B. Glass, O. Temkin, and W. L. Straus, Jr., pp. 222–264. Baltimore: Johns Hopkins University Press.

--- 1971. The Darwinian revolution in the concept of time. *Studium generale* 24: 289–307.

--- 1972. The Darwinian revolution in the concept of time. In *The study of time*, eds. J. T. Fraser, F. C. Haber, and G. H. Müller, pp. 382–401. New York: Springer.

Haeckel, E. 1862. *Die Radiolarien. (Rhizopoda radiaria)*. 2 vols. Berlin: Georg Reimer.

--- 1863. Ueber die Entwickelungstheorie Darwin's. *Amtliche Bericht über die acht und dreissigste Versammlung Deutscher Naturforsche und Äerzte in Stettin*, pp. 17–30. Stettin: F. Hessenland's Buchdruckerei.

--- 1866. *Generelle Morphologie der Organismen: allgemeine Grundzüge der organischen Formen-Wissenschaft mechanisch begründet durch die von Charles Darwin reformirte Descendenz-Theorie*. 2 vols. Berlin: Georg Reimer.

--- 1868. *Natürliche Schöpfungsgeschichte*. Berlin.

--- 1869a. *Zur Entwickelungsgeschichte der Siphonophoren*. Utrecht: C. van der Post.

--- 1869b. *Über Arbeitstheilung in Natur und Menschenleben. Vortrag gehalten in Saale des Berliner Handwerker-Vereins am 17. Dezember 1868*. Berlin: C. G. Lüdentz'sche Verlagsbuchhandlung.

--- 1874. The Gastraea-theory, the phylogenetic classification of the animal kingdom and the homology of the germ-lamellae. *Q. j. micro. sci.* N.S. 14: 142–165, 223–247.

--- 1876. *The history of creation, or the development of the earth and its inhabitants by the action of natural causes: a popular exposition of the doctrine of evolution in general and of that of Darwin, Lamarck and Goethe in particular*. 2 vols. New York: Appleton.

--- 1877. Ueber die heutige Entwickelungslehre im Verhältnisse zur Gesammtwissenschaft. *Amtlicher Bericht der 50. Versammlung Deutscher Naturforscher und Aerzte in München*, pp. 14–22. München: F. Straub.

--- 1878. *Freie Wissenschaft und Freie Lehre*. Stuttgart: Schweizerbart.

--- 1892. *Storia della creazione: conferenze scientifico-popolari sulla teoria dell'evoluzione e specialmente su quella di Darwin, Goethe e Lamarck*. Trad. D. Rosa. Torino: Unione Tipografica Editrice.

--- 1895. *Antropogenia e storia dell'evoluzione umana.*

Trad. D. Rosa. Torino: Unione Tipografica Editrice.

--- 1899a. *Die Welträthsel: gemeinverständliche Studien über Monistische Philosophie*. Stuttgart: Alfred Kröner.

--- 1899b. *The history of creation*. 4th ed. From 8th German ed. The translation revised by E. R. Lankester. 2 vols. London: Kegan Paul.

--- 1905. *Über die Biologie in Jena während des 19.Jahrhunderts*. Jena: Gustav Fischer.

--- 1914. *Il Monismo: professione di fede di un naturalista*. Milano: Università Popolare.

--- 1921a. *Entwicklungsgeschichte einer Jugend: Briefe an die Eltern 1852–1856*. Leipzig: K. F. Koehler.

--- 1921b. *Italienfahrt: Briefe an die Braut 1859–1860*. Leipzig: K. F. Koehler.

--- 1923. *Zelseelen und Seelenzellen: Vortrag gehalten am 22.März 1878 in der "Concardia" zu Wien*. 2d ed. Leipzig: Alfred Kröner.

Haldane, J. B. S. 1932. *The causes of evolution*. London: Longmans.

--- 1941. Concerning social Darwinism. *Science and society* 5: 373–374.

--- 1957. The cost of natural selection. *J. genet.* 55: 511–524.

---1959. Natural selection. In *Darwin's biological work. Some aspects reconsidered*, ed. P. R. Bell, pp. 101–149. Cambridge: Cambridge University Press.

Halévy, E. 1924–1926. *A history of the English people in the nineteenth century*. 6 vols. London: Ernest Benn.

Hall, A. R. 1969. *The Cambridge Philosophical Society: a history, 1819–1969*. Cambridge: Cambridge Philosophical Society.

--- **and Hall, M. B.** 1964. *A brief history of science*. New York: Signet Science Library.

Hall, V. D. 1979. The contribution of the physiologist, William Benjamin Carpenter (1813–1885), to the development of the principles of the correlation of forces and the conservation of energy. *Med. hist.* 23: 129–155.

Haller, J. S., Jr. 1971. *Outcasts from evolution: scientific attitudes of racial inferiority, 1859–1900*. Urbana: University of Illinois Press.

Halliday, R. J. 1971. Social Darwinism. *Victorian stud.* 14: 389–405.

Hamilton, G. 1812. *The elements of drawing, in its various branches*. London.

Hamilton, W. D. 1964. The genetical evolution of social behavior. *J. theor. biol.* 7: 1–16; 17–52.

Hankins, T. L. 1980. *Sir William Rowan Hamilton*. Baltimore: Johns Hopkins University Press.

Hansen, B. 1970. The early history of glacial theory in British geology. *J. glaciol.* 9: 135–141.

Hanson, N. R. 1965. *Patterns of discovery*. Cambridge: Cambridge University Press.

Haraway, D. 1981–1982. The high cost of information in post-World War II evolutionary biology: ergonomics, semiotics, and the sociobiology of communications systems. *Phil. forum* 13: 244–278.

Hardin, G. 1959. *Nature and man's fate*. New York: Rinehart.

Harker, A. 1907. Notes on the rocks of the "Beagle" collection. — I. *Geol. mag.*, N.S., Decade V, 4: 100–106.

Harmer, S. F. 1910. Polyzoa. In *The Cambridge natural history*, vol 2. London: Macmillan.

Harris, M. 1968. *The rise of anthropological theory: a history of theories of culture*. New York: Thomas Y. Crowell.

Harrison, J. 1971. Tennyson and evolution. *Durham Univ. j.* 64: 26–31.

Hart, A. T. 1959. *The country priest in English history*. London: Phoenix House.

--- **and Carpenter, E.** 1954. *The nineteenth century country parson (circa 1832–1900)*. Shrewsbury: Wilding & Son.

Hart, J. 1977. Religion and social control in the mid-nineteenth century. In *Social control in nineteenth century Britain*, ed. A. P. Donajgrodzki, pp. 108–137. London: Croom Helm.

Hartmann, N. 1950. *Philosophie der Natur*. Berlin: De Gruyter.

Harvey, W. H. 1847. *Nereis australis, or algae of the Southern Ocean: being figures and descriptions of marine plants, collected on the shores of the Cape of Good Hope, the extra-tropical Australian Colonies, Tasmania, New Zealand, and the Antarctic regions; deposited in the Herbarium of the Dublin University*. London: Reeve Bros.

Hattiangadi, J. N. 1971. Alternatives and incommensurables: the case of Darwin and Kelvin. *Phil. sci.* 38: 502–507.

Haughton, S. 1860. Bióyévéis *Nat. hist. rev.* 7: 23–32. 1973. Reprinted in D. Hull, *Darwin and his critics*, pp. 217–227. Cambridge: Harvard University Press.

--- 1863. *Ann. mag. nat. hist.* 11: 420–429.

Haustein, E. 1960. Darwin als Botaniker. In

Hundert Jahre Evolutionsforschung, eds. G. Heberer and F. Schwanitz, pp. 169–185. Stuttgart: Gustav Fischer.

Hays, H. R. 1958. *From ape to angel.* New York: Knopf.

Heberer, G. 1960a. Darwins Bild der abstammungsgeschichtlichen Herkunft des Menschen und die modern Forschung. In *Hundert Jahre Evolutionsforschung*, eds. G. Heberer and F. Schwanitz, pp. 397–448.

--- 1960b. *Was heisst heute Darwinismus?* 2. neubearbeitete Auflage aus Anlass der 100. Wiederkehr der Begründung der Abstammungslehre durch Charles Darwin. Berlin, Frankfurt, Göttingen: Musterschmidt.

--- **and Schwanitz, F.**, eds. 1960. *Hundert Jahre Evolutionsforschung*. Stuttgart: Gustav Fischer.

Heeney, B. 1976. *A different kind of gentleman: parish clergy as professional men in early and mid-Victorian England.* Hamden: Archon Books for The Conference on British Studies and Wittenburg University.

Hegenbarth, H. 1972. *Darwin, die Bibel und die Tatsache*. Graz: Steiermärkische Landregierung.

Heilbroner, P. L. 1976. *Circular logic: the context, structure, and influence of William Sharp MacLeay's quinarism*. Senior honors thesis. Harvard University.

Heim, R. ed. 1963. *Précurseurs et fondateurs de l'évolutionnisme. Texte des allocutions prononcées le 5 juin 1959 au Muséum National d'Histoire Naturelle*. Paris: Editions du Muséum.

Heinaman, R. 1979. Aristotle's tenth aporia. *Arch. für Geschichte der Philosophie* 61: 249–270.

Helfand, M. S. 1977. T. H. Huxley's *Evolution and ethics:* the politics of evolution and the evolution of politics. *Victorian stud.* 20: 159–177.

Hemleben, J. 1968. *Charles Darwin in Selbstzeugnissen und Bilddokumenten*. Reinbeck bei Hamburg: Rowohlt.

Hempel, C. G. 1965. *Aspects of scientific explanation*. New York: Free Press.

Henderson, G. 1958. *Alfred Russel Wallace: his role and influence in nineteenth century evolutionary thought*. Ph.D. thesis. University of Pennsylvania.

Henderson, L. J. 1970. *On the social system: selected writings*. Chicago: University of Chicago Press.

Henkin, L. J. 1940. *Darwinism in the English novel. The impact of evolution on Victorian fiction.* New York: Corporate Press.

Henry, W. 1819. *The elements of experimental chemistry*. 1st Am. ed. from 8th London ed. Philadelphia: Robert Desilver.

--- 1823. *The elements of experimental chemistry.* 9th ed. London: Baldwin, Cradock and Joy.

Henslow, G. 1888. *The origin of floral structures through insect and other agencies*. London: Kegan Paul.

--- 1895. *The origin of plant structures by self-adaptation to the environment*. London: Kegan Paul.

--- 1898. Scientific proofs vs. "*a priori*" assumptions. *Natural sci.* 13: 103–108.

Henslow, J. S. 1828. *Syllabus of a course of lectures on botany*. Cambridge: Privately printed.

--- 1833. *Syllabus of a course of lectures on botany.* Cambridge: Privately printed.

--- 1836. *Descriptive and physiological botany*. In *The Cabinet cyclopedia*, ed. D. Lardner. London: Longman, Rees, Orme, Brown, Green and Longman.

--- 1837a. *Descriptive and physiological botany*, 2d ed. London: Longman, Rees, Orme, Brown, Green and Longman.

--- 1837b. Description of two new species of *Opuntia*; with remarks on the structure of the fruit of *Rhipsalis*. *Mag. zool. bot.* 1: 466–469.

--- 1838. Flora keelingensis: an account of the native plants of the Keeling Islands. *Ann. nat. hist.* 1: 337–347.

--- 1844. On the registration of facts tending to illustrate questions of scientific interest. *Gardeners' chron.* 659.

Herbert, S. 1968. *The logic of Darwin's discovery.* Ph.D. thesis. Brandeis University.

--- 1971. Darwin, Malthus, and selection. *J. hist. biol.* 4: 209–217.

--- 1974. The place of man in the development of Darwin's theory of transmutation. Part I. *J. hist. biol.* 7: 217–258.

--- 1976. Review of T. F. Glick, ed. *The comparative reception of Darwinism. Isis* 67: 497–499.

--- 1977. The place of man in the development of Darwin's theory of transmutation. Part II. *J. hist. biol.* 10: 155–227.

--- 1979. Darwin and philosophers. [Review of E. Manier, *The young Darwin and his cultural circle.*] *Science* 204: 726–727.

--- 1980. Introduction. In *The Red Notebook of*

Charles Darwin, ed. S. Herbert, pp. 5–29. Ithaca: Cornell University Press. Also *Bull. Brit. Mus. (Nat. Hist.) Hist. ser.* 7.

--- 1982. Remembering Charles Darwin as a geologist. In *Charles Darwin 1809–1882: a centennial commemorative* ed. R. G. Chapman. Wellington, Nova Pacifica.

--- 1983. Les divergences entre Darwin et Lyell sur quelques questions géologiques. In *De Darwin au Darwinisme: science et idéologie* ed. Y. Conry. Paris: J. Vrin.

Herbert, W. 1837. *Amaryllidaceae: followed by a treatise on cross-bred vegetables*. London: James Ridgway.

Herschel, J. F. W. 1830. *Preliminary discourse on the study of natural philosophy*. London: Longmans, Rees, Orme, Brown and Green.

--- 1831. Field Notebook. Listed in Newsletter Hist. Geol. Div., Geol. Soc. Am., vol. 5, no. 2.

---, ed. 1849. *A manual of scientific inquiry; prepared for the use of Her Majesty's Navy: and adapted for travellers in general*. London: John Murray.

--- 1861. *Physical geography of the globe*. London: Longmans, Green..

Hertwig, O. 1916. *Das Werden der Organismen*. Jena: Gustav Fischer.

Herzen, A. 1869. *Sulla parentela fra l'uomo e le scimie: lettura del Dott. Alessandro Herzen, fatta a Firenze nel R. Museo di Storia Naturale il 21 Marzo 1869, seconda edizione dell'articolo del Sen. Lambruschini e la risposta del Dott. Herzen*. Firenze: A. Bettini.

Heslop-Harrison, J. 1958. Darwin as a botanist. In *A century of Darwin*, ed. S. A. Barnett, pp. 267–295. Cambridge: Harvard University Press.

--- 1962. In 2d ed. of *A century of Darwin*. London: Mercury Books.

Hesse, M. B. 1966. *Models and analogies in science*. Notre Dame: University of Notre Dame Press.

--- 1974. *The structure of scientific inference*. London: Macmillan.

Heuss, T. 1940. Anton Dohrn in Neapel. Berlin: Atlantisverlag.

Heyl, B. 1968. The Harvard "Pareto Circle". *J. hist. behav. sci.* 4: 316–334. 1980. Reprinted in *Darwin to Einstein: historical studies on science and belief*, eds. C. Chant and J. Fauvel, pp. 134–155. New York and London: Longmans.

Hickson, S. J. 1921. On some Alcyonaria in the Cambridge Museum. *Proc. Cambridge Phil. Soc.* 20: 366–373.

--- 1936. Darwin's *Cavernularia*. *Nature* 137: 909.

Hilgard, E. R. 1960. Psychology after Darwin. In *Evolution after Darwin*, ed. S. Tax, vol. 2, pp. 269–287. Chicago: University of Chicago Press.

Hilts, V. L. 1973. Statistics and social science. In *Foundations of scientific method in the nineteenth century*, eds. R. Giere and R. Westfall, pp. 206–233. Bloomington: Indiana University Press.

--- 1975. A guide to Francis Galton's "English men of science". *Trans. Am. Phil. Soc.* N.S. 65: pt. 5.

Himmelfarb, G. 1959. *Darwin and the Darwinian revolution*. Garden City: Doubleday;London: Chatto and Windus.

--- 1968. Varieties of social Darwinism. In *Victorian minds*, pp. 314–332. New York: Knopf.

Hinds, J. I. D. 1900 [1889]. *Charles Darwin: a sketch of his life, writings, theory, character, mental characteristics and religious views*. Rev. ed. Nashville: Cumberland Presbyterian Publishing House.

Hinkle, R. C. 1979. *Founding theory of American sociology, 1881–1915*. London: Routledge and Kegan Paul.

Hintikka, J. 1981. Gaps in the great chain of being: an exercise in the methodology of the history of ideas. In *Reforging the Great Chain of Being: studies of the history of modal theories*, ed. S. Knuuttila, pp. 1–17. Synthese Historical Library, vol. 30. Dordrecht, Boston, London: Reidel.

Hirshleifer, J. 1977. Economics from a biological point of view. *J. law and econ.* 20: 1–52.

--- 1980. *Price theory and applications*. Englewood Cliffs: Prentice-Hall.

--- 1982. Evolutionary models in economics and law: cooperation and conflict strategies. *Res. law and econ.* 4: 1–60.

Hirst, P. Q. 1976. *Social evolution and sociological categories*. London: Allen & Unwin.

Hodge, M. J. S. 1971a. Lamarck's science of living bodies. *Brit. j. hist. sci.* 5: 323–352.

--- 1971b. On the origins of Darwinism in Lyellian historical geography. Paper read to the British Society for the History of Science, 10 July 1971.

--- 1972. The universal gestation of nature: Chambers' *Vestiges* and *Explanations*. *J. hist.*

biol. 5: 127–152.

--- 1974a. England. In *The comparative reception of Darwinism*, ed. T. F. Glick, pp. 3–31. Austin: University of Texas Press.

--- 1974b. England: bibliographical essay. In *The comparative reception of Darwinism*, ed. T. F. Glick, pp. 75–80. Austin: University of Texas Press.

--- 1977. The structure and strategy of Darwin's "long argument". *Brit. j. hist. sci.* 10: 237–245.

--- 1982. Darwin and the laws of the animate part of the terrestrial system (1835–1837): on the Lyellian origins of his zoonomical explanatory program. *Stud. hist. biol.* 7: 1–106.

--- 1983. The development of Darwin's general biological theorizing. In *Evolution from molecules to men*, ed. D. S. Bendall, pp. 43–62. Cambridge: Cambridge University Press.

--- forthcoming. *Darwin and the theory of natural selection: roles for epistemological and methodological ideals in a scientific innovation.*

Hodgkin, L. 1976. Politics and physical sciences. *Radical sci. j.* 4: 29–60.

Höffding, H. 1909. The influence of the conception of evolution on modern philosophy. In *Darwin and modern science*, ed. A. C. Seward, pp. 446–464. Cambridge: Cambridge University Press.

Hofstadter, R. 1944. *Social Darwinism in American thought.* Philadelphia: University of Pennsylvania Press.

--- 1955. *Social Darwinism in American thought.* 2d ed. Boston: Beacon.

Hogarth, W. 1753. *The analysis of beauty written with a view of fixing the fluctuating ideas of taste.* London: Reeves.

Hogben, L. 1960. *Man, race, and Darwin.* London: Oxford University Press.

Hollitscher, W. 1964. Die Natur im Weltbild der Wissenschaft. 2d ed. Wien.

Holmes, F. L. 1977. Conceptual history: a review of François Jacob, *La logique du vivant* — The logic of life. *Stud. hist. biol.* 1: 209–218.

Holt, N. R. 1971. Ernst Haeckel's monistic religion. *J. hist. ideas.* 32: 265–280.

Holton, G. 1973. *Thematic origins of scientific thought: Kepler to Einstein.* Cambridge: Harvard University Press.

--- 1975. On the role of themata in scientific thought. *Science* 188: 328–334.

Home, R. W. 1970. Electricity and the nervous fluid. *J. hist. biol.* 3: 235–251.

Hood, T. L. 1933. Letters of Robert Browning. London.

Hooker, J. D. 1844–1847. *The botany of the Antarctic voyage of H.M. discovery ships Erebus and Terror in the years 1839–1843: under the command of Captain Sir James Clark Ross.* Part 1. (2 vols.) *Flora Antarctica.* London: Reeve.

--- 1846. Description of *Pleuropetalum*, a new genus of Portulacaceae from the Galapagos Islands. *Lond. j. bot.* 5: 108–109.

--- 1847a. An enumeration of the plants of the Galapagos Archipelago; with descriptions of those which are new. *Trans. Linn. Soc. Lond.* (*Bot.*) 20: 163–233.

--- 1847b. On the vegetation of the Galapagos Archipelago as compared with that of some other tropical islands and of the continent of America. *Trans. Linn. Soc. Lond.* (*Bot.*) 20: 235–262.

--- 1853–1855. *Flora Nova-Zelandiae: part 2 of the Botany of the Antarctic voyage of H.M. discovery ships* Erebus *and* Terror. . . . 2 vols. London: Reeve.

--- 1854. *Himalayan journals; or, notes of a naturalist in Bengal, the Sikkim and Nepal Himalayas.* . . . 2 vols. London: John Murray.

--- 1859a. *On the flora of Australia, its origin, affinities and distribution: being an introductory essay to the flora of Tasmania.* London: Lovell Reest.

--- 1859b. Review of the "Origin of species". *Gardeners' chron.* 31 Dec., pp. 1051–1053. 1973. Reprinted in D. Hull, *Darwin and his critics*, pp. 81–85. Cambridge: Harvard University Press.

--- 1860a. Review of the "Origin of species". *Gardeners' chron.* 7 Jan., pp. 3–4.

--- 1860b. On the origination and distribution of vegetable species; introductory essay to the Flora of Tasmania. *Am. j. sci.* 2d series, 29: 1–25, 305–326.

Hooker, W. J. 1836a. *Pernettia pumila. Hook. Ic. Pl.* t. IX.

--- 1836b. *Donatia magellanica. Hook. Ic. Pl.* t. XVI.

--- 1836c. *Polytrichum dendroides. Hook. Ic. Pl.* t. XXV.

--- 1842. *Homoianthus echinatulus. Cass. Hook. Ic. Pl.* t. CDXCI.

--- 1844a. *Berberis darwinii. Hook. Hook. Ic. Pl.* t. DCLXXII.

--- 1844b. *Callixene polyphylla. Hook. Hook. Ic. Pl.* t. DCLXXIV.

--- **and Arnott, G. A. W.** 1836. Contributions towards a flora of South America and the islands of the Pacific. *Comp. bot. mag.* 2: 41–52.

--- **and** --- 1837. Contributions towards a flora of South America and the islands of the Pacific. *Comp. bot. mag.* 2: 250–254.

--- **and** --- 1840. Contributions towards a flora of South America and the islands of the Pacific. *Hook. j. bot.* 3: 19–47.

--- **and** --- 1841. Contributions towards a flora of South America and the islands of the Pacific. *Hook. j. bot.* 3: 310–348.

--- **and** --- 1842. *Aster vahlii. Hook. et Arn. Hook. Ic. Pl.* t. CDLXXXVI.

Hooykaas, R. 1959. *Natural law and divine miracle: historical-critical study of the principle of uniformity in geology, biology, and theology.* Leiden: Brill.

--- 1965. Der Aktualismus in Natur und Geschichte. *Gesnerus* 22: 1–16.

--- 1966. Geological uniformitarianism and evolution. *Arch. int. hist. sci.* 19: 3–19.

--- 1970a. *Catastrophism in geology, its scientific character in relation to actualism and uniformitarianism.* Amsterdam and London: North-Holland Publishing Co.

--- 1970b. *Continuité et discontinuité en géologie et en biologie.* Paris: Le Seuil.

Hope, F. W. 1838. Descriptions of some species of Carabidae, collected by Charles Darwin, Esq., in his late voyage. *Trans. Entomol. Soc. Lond.* 2: 128–131.

Hopkins, W. 1860. Physical theories of the phenomena of life. *Fraser's mag.* 61: 739–752; 62: 74–90. 1973. Reprinted in D. Hull, *Darwin and his critics*, pp. 229–272. Cambridge: Harvard University Press.

Hoppe, B. 1971. Die Geschichtlichkeit der Natur und des Menschen: die Entwicklungstheorie Alexander Brauns. In *Medizingeschichte in unserer Zeit: Festgabe für Edith Heischkel-Artelt und Walter Artelt*, eds. H.-H. Eulner et al., pp. 393–421. Stuttgart: F. Enke.

--- 1972. Die Entwicklung der biologischen Fächer an der Universität München im 19. Jahrhundert unter Berücksichtigung des Unterrichts. In *Die Ludwig-Maximilians-Universität in ihren Fakultäten*, eds. L. Boehm and J. Spörl, vol. 1 pp. 354–389. Berlin: Duncker und Humblot.

Howard, D. T. 1927. The influence of evolutionary theory on psychology. *Psych. rev.* 34: 305–312.

Howarth, O. J. R., and Howarth, E. K. [1933]. *A history of Darwin's parish, Downe, Kent.* Southampton: Russell & Co. (Southern Counties).

Hudson, K. 1972. *Patriotism with profit: British agricultural societies in the eighteenth and nineteenth centuries.* London: Evelyn.

Huet, B. 1975. L'école d'anthropologie de Paris. In *100e Congrès national des Sociétés Savantes. Paris, 21–25 Mars 1975*, p. 90.

Hughes, A. 1959. *A history of cytology.* London and New York: Abelard-Schuman.

Hull, D. L. 1965. The effect of essentialism in taxonomy — Two thousand years of stasis. *Brit. j. phil. sci.* 15: 314–326, 16: 1–18.

--- 1967a. Certainty and circularity in evolutionary taxonomy. *Evolution* 21: 174–189.

--- 1967b. The metaphysics of evolution. *Brit. j. hist. sci.* 3: 309–337.

--- 1969. What philosophy of biology is not. *Synthese* 20: 157–184 (also in *J. hist. biol.* 2: 120–149).

--- 1973a. Charles Darwin and nineteenth-century philosophies of science. In *Foundations of scientific method: the nineteenth century*, eds. R. N. Giere and R. S. Westfall, pp. 115–132. Bloomington: Indiana University Press.

--- 1973b. *Darwin and his critics. The reception of Darwin's theory of evolution by the scientific community.* Cambridge: Harvard University Press.

--- 1974a. *Philosophy of biological science.* Englewood Cliffs: Prentice Hall.

--- 1974b. Darwinism and historiography. In *The comparative reception of Darwinism,* ed. T. F. Glick, pp. 388–402. Austin: University of Texas Press.

--- 1975. Central subjects and historical narratives. *History and theory* 14: 253–274.

--- 1976. Are species really individuals? *Syst. zool.* 25: 174–191.

--- 1978a. A matter of individuality. *Phil. sci.* 45: 335–360.

--- 1978b. Sociobiology: scientific bandwagon or traveling medicine show? In *Sociobiology and human nature*, eds. M. S. Gregory, A. Silvers, and D. Sutch, pp. 136–163. San

Francisco: Josey-Bass.
--- 1980. Individuality and selection. *Ann. rev. ecol. syst.* 11: 311–332.
--- 1983a. Darwin and the nature of science. In *Evolution from molecules to man*, ed. D. S. Bendall, pp. 63–80. Cambridge: Cambridge University Press.
--- 1983b. Karl Popper and Plato's metaphor. In *Advances in cladistics*, vol. 2, eds. N. Platnick and V. A. Funk, pp. 177–189. New York: Columbia University Press.
--- 1984a. Linné as an Aristotelian. In *Contemporary perspectives on Linnaeus*, ed. J. M. Weinstock. Baltimore: University Press of America.
--- 1984b. Cladistic theory: hypotheses that blur and grow. In *Cladistic perspectives on the reconstruction of evolutionary history*, eds. T. Steussy and T. Duncan, pp. 5–23. New York: Columbia University Press.
--- forthcoming. Why species cannot evolve.
---; **Tessner, P. D.; and Diamond, A. M.** 1978. Planck's principle: do younger scientists accept new scientific ideas with greater alacrity than older scientists? *Science* 202: 717–723.

Humboldt, A. von, and Bonpland, A. 1814–1829. *Personal narrative of travels to the equinoctial regions of the new continent during the years 1799–1804*. 7 vols. in 9. Trans. H. M. Williams. London: Longmans, Hurst, Rees, Orme, and Brown.

Hunter, R. A., and Macalpine, I. 1963. *Three hundred years of psychiatry 1535–1860: a history presented in selected English texts*. London: Oxford University Press.

Huntley, W. B. 1972. David Hume and Charles Darwin. *J. hist. ideas* 33: 457–470.

Hutton, R. H. 1885. The Metaphysical Society: a reminiscence. *Nineteenth century* 18: 177–196.

Huxley, A. 1981. Anniversary address of the president. Supplement to *Royal Society News*, Issue 12, pp. i–vii.

Huxley, J. S. 1938. Darwin's theory of sexual selection. *Am. nat.* 72: 416–433.
--- 1939. *The living thoughts of Darwin*. London: Cassell.
--- ed. 1940. *The new systematics*. Oxford: Oxford University Press.
--- 1942. *Evolution: the modern synthesis*. London: Allen & Unwin.
--- 1953. *Evolution in action*. New York: Harper.
--- 1960a. The emergence of Darwinism. In *Evolution after Darwin*, ed. S. Tax, vol. 1, pp. 1–21. Chicago: University of Chicago Press.
--- 1960b. Darwin und der Gedanke der Evolution. In *Hundert Jahre Evolutionsforschung*, eds. G. Heberer and F. Schwanitz, pp. 1–10. Stuttgart: Gustav Fischer.
--- 1963. *Evolution in action*. Harmondsworth: Penguin.
--- **and Kettlewell, H. B. D.** 1965. *Charles Darwin and his world*. New York: Viking Press; London: Thames and Hudson.
---; **Hardy, A. C.; and Ford, E. B.**, eds. 1954. *Evolution as a process*. London: Allen & Unwin.

Huxley, L., ed. 1900. *The life and letters of Thomas Henry Huxley*. 2 vols. London: Macmillan.
--- 1918. *Life and letters of Sir Joseph Dalton Hooker O.M., G.C.S.I.: based on materials collected and arranged by Lady Hooker*. 2 vols. London: John Murray.
--- 1921. *Charles Darwin*. London: Life-stories of famous men.

Huxley, T. H. 1854. Review of *Vestiges of the natural history of creation. Brit. foreign medico-chir. rev.* 19: 425–439.
--- 1859. The Darwinian hypothesis. 1893. Reprinted in *Darwiniana*, pp. 1–21. New York: Appleton.
--- 1860a. On species and races and their origin. 1899. Reprinted in *The Scientific memoirs of Thomas Henry Huxley*, eds. M. Foster and E. R. Lankester, 5 vols., vol. 2, pp. 388–394. London: Macmillan.
--- 1860b. The origin of species. 1893. Reprinted in *Darwiniana*, pp. 22–79. New York: Appleton.
--- 1861. On a new species of *Macrauchenia (M. Boliviensis). Quart. J. Geol. Soc., Lond.* 17: 73–84. 1899. Reprinted in *The Scientific Memoirs of Thomas Henry Huxley*, eds. M. Foster and E. R. Lankester, 5 vols, vol. 2, pp. 403–416. London: Macmillan.
--- 1863a. *Evidence as to man's place in nature*. London: Williams & Norgate.
--- 1863b. *On our knowledge of the causes of the phenomena of organic nature*. London: Robert Hardwicke.
--- 1865. Emancipation — black and white. *The Reader* 20 May 1865. 1903. Reprinted in *Lay Sermons, addresses and reviews*. 3rd ed., pp. 17–23. London: Macmillan.
--- 1869. *Prove di fatto intorno al posto che l'uomo*

tiene nella natura. Milano : Treves.

--- 1870. *La place de l'homme dans la nature.* Paris: J. B. Baillière.

--- 1871. Mr. Darwin's critics. 1873. Reprinted in *Critiques and addresses*, pp. 218–269. New York: Appleton.

--- 1877. *American addresses: with a lecture on the study of biology*. New York: Appleton.

--- 1887a. On the reception of the "Origin of species". In *The life and letters of Charles Darwin*, ed. F. Darwin, vol. 1, pp. 533–558. New York: Appleton.

--- 1887b. On the reception of the "Origin of Species". In *The life and letters of Charles Darwin,* ed. F. Darwin, vol. 2, pp. 179–204. London: John Murray.

--- 1893a. *Methods and results*. London: Macmillan.

--- 1893b. *Darwiniana*. London: Macmillan.

--- 1893c. *Evolution and ethics, and other essays*. London: Macmillan.

--- 1899–1903. *The Scientific Memoirs of Thomas Henry Huxley*. Eds. M. Foster and E. R. Lankester. 5 vols. London: Macmillan.

--- 1935. *Diary of the voyage of H. M. S. Rattlesnake*. Ed. J. Huxley. London: Chatto and Windus.

Hyatt, A. 1866. On the parallelism between the different stages of life in the individual and in the entire group of the molluscous order Tetrabranchiata. *Mem. Boston Soc. Nat. Hist.* 1: 193–209.

--- 1889. *Genesis of the Arietidae*. Smithsonian Contributions to Knowledge, no. 673. Washington: Smithsonian Institution.

Hyman, L. H. 1940. *Protozoa through Ctenophora. The invertebrates*, vol. 1. New York: McGraw-Hill.

Hyman, S. E. 1959. Darwin the dramatist. *Centennial rev. arts sci.* 3: 364–375.

--- 1962. *The tangled bank: Darwin, Marx, Frazer and Freud as imaginative writers*. New York: Atheneum.

Ingles, J. M., and Sawyer, F. C. 1979. A catalogue of the Richard Owen collection of palaeontological and zoological drawings in the British Museum. *Bull. Brit. Mus. (Nat. Hist.) Hist. ser.* 6: 109–197.

Irvine, W. 1955 *Apes, angels, and Victorians: a joint biography of Darwin and Huxley*. London: Weidenfeld and Nicolson; Toronto: McGraw-Hill.

--- 1959. The influence of Darwin on literature. *Proc. Am. Phil. Soc.* 103: 616–628.

Iser, W. 1978. *The act of reading: a theory of aesthetic response*. Baltimore: Johns Hopkins University Press.

Issel, A. 1865. *Sulla variabilita' nelle specie : breve cenno sulla teoria di Darwin*. Genova : Tipografia Sordomuti.

Jackson, S. W. 1970. Force and kindred notions in 18th century neurophysiology and medical psychology. *Bull. hist. med.* 44: 397–410, 539–554.

Jacob, F. 1973. *The logic of life: a history of heredity*. New York: Pantheon.

--- 1977. Evolution and tinkering. *Science* 196: 1161–1166.

--- 1981. *Le jeu des possibles: essai sur la diversité du vivant*. Paris: Fayard. *The possible and the actual.* Seattle: University of Washington Press.

Jacot, A. P. 1932. The status of the species and the genus. *Am. nat.* 66: 346–364.

Jacyna, L. S. 1980a. Science and social order in the thought of A. J. Balfour. *Isis* 71: 11–34.

--- 1980b. *Scientific naturalism in Victorian Britain: an essay in the social history of ideas*. Ph.D. thesis. University of Edinburgh.

--- 1981. The physiology of mind, the unity of nature, and the moral order in Victorian thought. *Brit. j. hist. sci.*. 14: 109–132.

Jahn, I.; Löther, R.; and Senglaub, K., eds. 1982. *Geschichte der Biologie: Theorien, Methoden, Institutionen, Kurzbiographien*. Jena: Gustav Fischer.

James, C. 1877. *Du Darwinisme; ou, l'homme singe*. Paris : E. Plon.

--- 1882. *Mes entretiens avec S. M. l'Empereur don Pedro sur le darwinisme*. Paris.

James, P. 1979. *Population Malthus: his life and times*. London: Routledge and Kegan Paul.

Jay, W. 1980. Charles Darwin: photography and everything else. *Brit. j. photography* 7 November: 1116–1118.

Jaynes, J. 1969. The historical origins of "ethology" and "comparative psychology". *Animal behav.* 17: 601–606.

Jeffrey, F. 1806. Essays on the anatomy of expression in painting. *Edin. rev.* 8 (16): 365–378.

Jenkin, F. 1867. The origin of species. *North Brit. rev.* (Am. ed.) 46: 149–171. 1973.

Reprinted in D. Hull, *Darwin and his critics*, pp. 303–344. Cambridge : Harvard University Press.

Jenyns, L. 1862. *Memoir of the Rev. John Stevens Henslow*. London: van Voorst.

Jenyns, L. *See* Blomefield, L. [formerly Jenyns].

Jepsen, G. L. 1949. Selection, "orthogenesis", and the fossil record. *Proc. Am. Phil. Soc.* 93: 479–500.

Jespersen, P. H. 1948–1949. Charles Darwin and Dr. Grant. *Lychnos* 1: 159–167.

Jodl, M., ed. 1922. *Bartholomäus von Carneri's Briefwechsel mit Ernst Haeckel und Friedrich Jodl*. Leipzig: K. F. Koehler.

Johnson, S. 1826. *A dictionary of the English language in which the words are deduced from their originals, explained in their different meanings, and authorized by the names of the writers in whose books they are found.* Abridged from the Rev. H. J. Todd's corrected and enlarged quarto edition by Alexander Chalmers, F.S.A. London: C. and J. Rivington.

Johnston, R. 1976. Contextual knowledge: a model for the overthrow of the internal-external dichotomy in science. *Australian and New Zealand j. sociol.* 12: 193–203.

Jones, A. E. 1950. *Darwinism and its relationship to realism and naturalism in American fiction*, 1860–1900. Madison, N. J.: Drew University.

Jones, E. Y. 1973. *Father of art photography: O. G. Rejlander 1813–1875*. Newton Abbot: David and Charles.

Jones, G. 1978. The social history of Darwin's *Descent of man. Economy and society* 7: 1–23.

--- 1980. *Social Darwinism and English thought: the interaction between biological and social theory*. Brighton: Harvester Press.

Jones, H. F. 1911. *Charles Darwin and Samuel Butler: a step toward reconciliation*. London: Fifield.

Jordan, K. 1905. Der Gegensatz zwischen geographischer und nicht-geographischer Variationen. *Zeitschrift für wissenschaftliche Zoologie* 83: 151–210.

Jordan, W. D. 1968. *White over black: American attitudes toward the Negro, 1550–1812*. Chapel Hill: University of North Carolina Press.

Jordanova, L. J. 1976. *The natural philosophy of Lamarck in its historical context*. Ph. D. thesis. University of Cambridge.

--- 1981. La psychologie naturaliste et le "problème des niveaux": la notion du sentiment intérieur chez Lamarck. In *Lamarck et son temps. Lamarck et notre temps.* Colloque international organisé par le Centre de Recherche sur l'Histoire des Idées de l'Université de Picardie, pp. 69–80. Paris: Vrin.

--- **and Porter, R.**, eds. 1979. *Images of the earth*. Chalfont St. Giles, Bucks : British Society for the History of Science.

Judd, J. W. 1909. Darwin and geology. In *Darwin and modern science*, ed. A. C. Seward, pp. 337–384. Cambridge: Cambridge University Press.

Kalikow, T. J. 1978. Konrad Lorenz's "brown past": a reply to Alec Nisbett. *J. hist. behav. sci.* 14: 173–180.

Kanaev, I. I. 1962. Mopertiui kak predshestvennik Darvina [Maupertuis as a precursor of Darwin]. *Tr. Inst. Ist. Est. Tekh.* 41: 29–43.

Kardiner, A., and Preble, E. 1961. *They studied man*. Cleveland: World Pub. & Co.

Kass, R. L. 1978. Teleology and Darwin's *The origin of species*: beyond chance and necessity? In *Organism, medicine, and metaphysics: essays in honor of Hans Jonas on his 75th birthday*, ed. S. F. Spicker, pp. 97–120. Dordrecht: Reidel.

Kauffman, R. 1983. Filling some epistemological gaps: new patterns of inference in evolutionary theory. In *PSA 1983*, eds. P. Asquith and R. Giere, vol. 2. E. Lansing: Philosophy of Science Association.

Kavaloski, V. K. 1974. *The* Vera Causa *principle: an historico-philosophical study of a metatheoretical concept from Newton through Darwin*. Ph. D. thesis. University of Chicago.

Keegan, R. T., and Gruber, H. E. 1983. Love, death and continuity in Darwin's Thinking. *J. hist. behav. sci.* 19: 15–30.

Keith, A. 1955. *Darwin revalued*. London: Watts.

Kellogg, V. L. 1907. *Darwinism today*. New York: Henry Holt.

Kelly, A. 1981. *The descent of Darwin: the popularization of Darwinism in Germany, 1860–1914*. Chapel Hill: University of North Carolina Press.

Kelvin, W. Thomson, Baron, 1863. On the secular cooling of the earth. *Phil. mag.* 4th series, 25: 1–14.

--- 1871. On geological time. *Trans. Glasgow Geol. Soc.* 3: 1–28.

--- 1891–1894. *Popular lectures and addresses*. 3 vols. London: Macmillan.

Kennedy, J. G. 1978. Herbert Spencer. Boston: Twayne.

[Keppel, T. E.] 1887. The country parson as he was, and as he is. *Blackwood's mag.* 142: 317–328.

Kettlewell, H. B. 1973. *The evolution of melanism*. Oxford: Oxford University Press.

Keynes, R. D. 1979. *The Beagle record: selections from the original pictorial records and written accounts of the voyage of H.M.S. Beagle*. Cambridge: Cambridge University Press.

Kierkegaard, S. 1841. *The concept of irony*. New York: Harper and Row.

Kilburn, P. D. 1969. Plants of the Galápagos. *Isis* 60: 386–388.

Kimura, M., and Ohta, T. 1971. *Theoretical aspects of population genetics*. Princeton: Princeton University Press.

King-Hele, D. G. 1963a. Dr. Erasmus Darwin and the theory of evolution. *Nature* 200: 304–306.

--- 1963b. *Erasmus Darwin*. London: Macmillan.

---, ed. 1968. *The essential writings of Erasmus Darwin*. London: MacGibbon & Kee.

--- 1977. *Doctor of revolution: the life and genius of Erasmus Darwin*. London: Faber and Faber.

Kingsland, S. 1978. Abbott Thayer and the protective coloration debate. *J. hist. biol.* 11: 223–244.

Kirby, W., and Spence, W. 1818–1826. *An introduction to entomology: or elements of the natural history of insects*. 4 vols. London: Longman, Hurst, Rees, Orme, and Brown.

Kitcher, P. 1979. Theories, theorists, and theory change. *Philosophical rev.* 87: 519–547.

Klatt, B. 1960. Darwin und die Haustierforschung. In *Hundert Jahre Evolutionsforschung*, eds. G. Heberer and F. Schwanitz, pp. 149–168. Stuttgart: Gustav Fischer.

Kleine, G. 1862. Do bees vary in the different parts of Germany? *J. horticulture and Cottage gardener*. N. S. 3: 642.

Kleiner, S. A. 1979. Feyerabend, Galileo and Darwin: how to make the best out of what you have — or think you can get. *Stud. hist. phil. sci.* 10: 285–309.

Knapp, G. 1973. *Der antimetaphysiche Mensch: Darwin, Marx, Freud*. Stuttgart: Klett.

Knight, D. 1976. *The nature of science: the history of science in western culture since 1600*. London: André Deutsch.

--- 1981. *Ordering the world: a history of classifying man*. London: Burnett Books in association with André Deutsch.

Koch, H. J. W. 1973. *Der Sozialdarwinismus: Seine Genese und sein Einfluss auf das imperialistische Denken*. München: C. H. Beck.

Koehler, O. 1960. Darwin und wir. In *Hundert Jahre Evolutionsforschung*, eds. G. Heberer and F. Schwanitz, pp. 11–31. Stuttgart: Gustav Fischer.

Koestler, A. 1965. An exercise in analogy: biological and mental evolution. *Nature* 208: 1033–1036.

--- 1967. *The act of creation*. New York: Laurel.

--- 1971. *The case of the midwife toad*. London: Hutchinson.

Kohler, R. E. 1981. Discipline history. In *Dictionary of the history of science*, eds. W. F. Bynum, E. J. Browne, and R. Porter, p. 104. London: Macmillan.

Kohn, D. 1980. Theories to work by: rejected theories, reproduction, and Darwin's path to natural selection. *Stud. hist. biol.* 4: 67–170.

--- 1981. On the origin of the principle of divergence [Review of A. C. Brackman, *A delicate arrangement*]. *Science* 213: 1105–1108.

---; **Smith, S. and Stauffer, R. C.** 1982. New light on *The Foundations of the Origin of Species*: a reconstruction of the archival record. *J. hist. biol.* 15: 419–442.

Kolakowski, L. 1966. *Filozofia pozytywistyczna (Od Hume'a do Kola Wiedenskiego)* [Positivist philosophy]. Warsaw: Panstwowe Wydawnictwo Naukowe.

Komarov, V. L. 1935. Marx and Engels on biology. In *Marxism in modern thought*, ed. N. S. Bukharin, pp. 190–234. New York: Harcourt.

Koop, R. 1941. *Haeckel und Allmers*. Bremen: Arthur Geist.

Korsunskaia, V. M. 1969. *Charles Darwin*. [In Russian] Moscow: Prosveshchenie.

Kottler, M. J. 1974. Alfred Russel Wallace, the origin of man, and spiritualism. *Isis* 65: 145–192.

--- 1976. *Isolation and speciation, 1837–1900*. Ph.D. thesis. Yale University.

--- 1978. Charles Darwin's biological species concept and theory of geographic speciation: the transmutation notebooks. *Ann. sci.* 35: 275–297.

--- 1979. Hugo de Vries and the rediscovery of Mendel's laws. *Ann. sci.* 36: 517–538.

--- 1980. Darwin, Wallace, and the origin of

sexual dimorphism. *Proc. Am. Phil. Soc.* 124: 203–226.

Krader, L. 1972. Introduction to *The ethnological notebooks of Karl Marx*. Assen: Van Gorcum.

--- 1976. The dialectical critique of the nature of human nature. *Critique of anthro.* 2, no. 6: 4–22.

Kraus, G. 1973. *Homo Sapiens in decline: a reappraisal of natural selection*. Sandy, Bedfordshire: New Diffusionist Press.

Krebs, H. 1971. Some facts of life: biology and politics. *Proc. Roy. Inst. of Great Britain* 44: 169–184. 1972. Reprinted in *Perspect. in biol. and med.* 15: 491–506.

Kroeber, A. L. 1960. Evolution, history, and culture. In *Evolution after Darwin*, ed. S. Tax, vol. 2, pp. 1–16. Chicago: University of Chicago Press.

Krohn, W., and Schäfner, W. 1976. The origin and structure of agricultural chemistry. In *Perspectives on the Emergence of Scientific Disciplines,* eds. G. Lemaine et al., pp. 27–52. The Hague, Paris: Mouton.

Kuhl, J. 1977. *Darwin und die Sprachwissenschaft*. Leipzig, Mainz: Lesimple.

Kuhn, T. S. 1962a. Historical structure of scientific discovery. *Science* 136: 760–764.

--- 1962b. *The structure of scientific revolutions*. Chicago: University of Chicago Press.

--- 1970. *The structure of scientific revolutions*. 2d ed. Chicago: University of Chicago Press.

--- 1976. Mathematical vs. experimental traditions in the development of physical science. *J. interdisciplinary hist.* 7: 1–31.

--- 1977. *The essential tension: selected studies in scientific tradition and change*. Chicago: University of Chicago Press.

Kultgen, J. H., ed. 1959. *The impact of Darwinian thought on American life and culture: Papers read at the Fourth Annual Meeting of the American Studies Association of Texas at Houston, Texas*. Austin: University of Texas Press.

Kunkle, B. W. 1917. Benjamin Franklin and the struggle for existence. *Science* 46: 437.

Labriola, A. 1898. *Discorrendo di socialismo e di filosofia: lettere a G. Sorel.* Roma: Loescher.

--- 1949. *Lettere a Engels*. Roma: Rinascita.

Lack, D. 1945. The Galapagos Finches (Geospizinae): a study in variation. *Occ. papers Calif. Acad. Sci.* no. 21.

--- 1947. *Darwin's finches*. Cambridge: Cambridge University Press.

--- 1957. *Evolutionary theory and Christian belief: the unresolved conflict*. London: Methuen.

--- 1968. *Ecological adaptations for breeding in birds*. London: Methuen.

Lagarde, A. 1979. Jean de Lanessan (1843–1919). Analyse d'un transformisme. *Rev. de synthèse* 100, nos. 95–96: 337–351.

Laird, D. 1955. *Charles Darwin, naturalist*. London, Glasgow: Blackie & Son.

Lakatos, I. 1970. Falsification and the methodology of scientific research programmes. In *Criticism and the growth of knowledge*, eds. I. Lakatos, and A. Musgrave, pp. 91–195. Cambridge: Cambridge University Press.

--- **and Musgrave, A.,** eds. 1970. *Criticism and the growth of knowledge*. Cambridge: Cambridge University Press.

Lamarck, J. B. 1809. *Philosophie zoologique*, 2 vols. Paris: Savy. 1914. Trans. H. Elliot. London: Macmillan.

--- 1815–1822. *Histoire naturelle des animaux sans vertèbres*. 7 vols. Paris.

--- 1835–1845. *Histoire naturelle des animaux sans vertèbres*. 2d ed. 11 vols. Paris: Baillière.

Lamouroux, J. V. F. 1821. *Exposition méthodique des genres de l'ordre des polypiers*. Paris: Mme. veure Agasse.

Lande, R. 1980. Genetic variation and phenotypic evolution during allopatric speciation. *Am. nat.* 116: 463–479.

Landrieu, M. 1909. *Lamarck le fondateur du transformisme: sa vie, son oeuvre*. Paris: Société Zoologique de France.

Landucci, G. 1977. *Darwinismo a Firenze: tra scienza e ideologia (1860–1900)*. Firenze: Olschki.

--- 1981. Darwinismo e nazionalismo. In *La cultura italiana tra '800 e '900 e le origini del nazionalismo*, pp. 103–187. Firenze: Olschki.

--- 1982. Scienza, religione ed editoria scolastica (1850–1900). *Ricerche bollettino degli scolopi*: n. 1.

Lanessan, J.-L. de 1914. L'attitude de Darwin a l'égard de ses précurseurs au sujet de l'origine des espèce. *Rev. anthropologique* 24: 33–45.

Lanham, U. N. 1968. *Origins of modern biology*. New York: Columbia University Press.

Lankester, E. R. 1876. An account of Professor Haeckel's recent additions to the Gastraea-theory. *Q. j. micro. sci.* N.S. 16: 51–66.

--- 1888–1889. Inheritance of acquired characters. *Nature* 39: 485.

Lappalainen, P. 1965. A. J. Melan darwinisten "herääminen". *Luonnon Tutkija* 60: 97–112.

--- 1967. Darwinin teorioiden ja darwinistisen maailmanselityksen tulo Suomeen. Historica II. *Studia Historica Jyväskyläensis* 5: 130–166.

Larson, J. L. 1971. *Reason and experience: the representation of natural order in the work of Carl von Linné.* Berkeley: University of California Press.

Laudan, L. 1968. Theories of scientific method from Plato to Mach: a bibliographical review. *Hist. sci.* 7: 1–63.

--- 1971. William Whewell on the consilience of inductions. *Monist* 55: 368–391.

--- 1977. *Progress and its problems: towards a theory of scientific growth.* Berkeley: University of California Press.

Laudan, R. 1976. William Smith: stratigraphy without palaeontology. *Centaurus* 20: 210–226.

--- 1982. Tensions in the concept of geology: natural history or natural philosophy? *Earth sci. hist.* 1: 7–13.

Laurent, G. 1980. Gaudry et la nomenclature. *Rev. de synthèse* 99–100: 297–312.

La Vergata, A. 1977. *La "lotta per l'esistenza" nell'eta' di Darwin.* Ph.D. thesis. University of Florence.

--- 1979. *L'evoluzione biologica: da Linneo a Darwin 1735–1871.* Torino: Loescher.

--- 1980a. H. Spencer: sopravvivenza del più adatto ed evoluzione cosmica. In *Scienza e Storia. Analisi critica e problemi attuali*, eds. S. Tagliagambe and A. Di Meo, Quaderni di Critica marxista, no. 2, pp. 205–232. Roma: Editori Riuniti.

--- 1980b. L' "ipotesi di derivazione delle specie": l'"evoluzionismo' di Charles Lyell. Due sue lettere, del 1859 e del 1862, a T. H. Huxley. In *Annali dell'Istituto di Filosofia dell'Università di Firenze, Facoltà di Lettere e Filosofia* 2: 289–310. Firenze: Olschki.

--- 1981. Lamarck, MacLeay, and Darwin. Paper read at the international conference *Development, institutionalization, and methods in the natural sciences: 1780–1840*, Florence, Florence Centre for the History and Philosophy of Science, 22–26 September, 1982.

--- 1982a. Biologia, scienze umane, e "darwinismo sociale": riflessioni contro una categoria storiografica dannosa. *Intersezioni* 2: 77–97.

--- 1982b. Claude Bernard, "punti di vista fisiologici" ed economia della natura. In *Claude Bernard: scienza, filosofia, letteratura*, ed. M. Di Giandomenico, pp. 285–313. Verona: Bertani. (also *Rivista di Filosofia* 24: 361–395).

--- 1982c. "Fisica", politica e natura umana: Walter Bagehot. *Il pensiero politico* 15: 248–256.

--- 1982d. La teoria di Darwin e la biologia dell'Ottocento. In *Scienza e filosofia nella cultura positivistica*, ed. A. Santucci, pp. 289–312. Milano: Feltrinelli.

--- 1982e. Une nouveauté dans le domaine des études darwiniennes. Revue critique de E. Manier, *The young Darwin and his cultural circle. Rev. de Synthèse* 102: 433–446.

--- forthcoming. Dichotomists vs. quinarians: a debate on taxonomic systems and methods just before Darwin. Paper read at the 3rd Course of the International School of the History of Biological Sciences. Napoli-Ischia, Stazione Zoologica, 1982. July 4–14.

Lawrence, P. 1977. Heaven and earth — the relation of the nebular hypothesis to geology. In *Cosmology, history, and theology*, eds. W. Yourgrau and A. D. Breck, pp. 253–281. New York and London: Plenum Press.

--- 1978. Charles Lyell versus the theory of central heat: a reappraisal of Lyell's place in the history of geology. *J. hist. biol.* 11: 101–128.

Lea, A. M. 1926.

Leatherdale, W. H. 1974. *The role of analogy, model and metaphor in science.* Amsterdam, Oxford, New York: New Holland.

Le Bétoyer, F. S. ed. 1936. La Pensée Russe. Paris: Aurore.

Le Brun, C. 1974. *A method to learn to design the passions.* 1980. Facs. ed. introduced by A. T. McKenzie. Los Angeles: William Andrews Clark Memorial Library, University of California, Augustan Reprint Society.

Lecourt, D. 1969. *L'épistémologie historique de Gaston Bachelard.* Paris: Vrin.

--- 1972. *Pour une critique de l'épistémologie: Bachelard, Canguilhem, Foucault.* Paris: Maspéro.

--- 1976. *Lyssenko. Histoire réelle d'une "science prolétarienne".* Paris: Maspéro. 1977. *Proletarian science?: the case of Lysenko.* London: NLB.

Le Dantec, F. 1899. *Lamarckiens et Darwiniens, discussion de quelques théories sur la formation des* *espèces*. Paris: F. Alcan.

Lee, K. K. 1969. Popper's falsifiability and Darwin's natural selection, *Philosophy* 44: 291–302.

Leeds, A. 1974. Darwinian and "Darwinian" evolutionism in the study of society and culture. In *The comparative reception of Darwinism*, ed. T. F. Glick, pp. 437–485. Austin: University of Texas Press.

Legros, D. 1977. Chance, necessity, and mode of production: a Marxist critique of cultural evolutionism. *Am. anthropologist* 79: 26–41.

Lehman, H. 1966. On the form of explanation in evolutionary theory. *Theoria* 32: 14–24.

Leibowitz, L. 1969. Dilemma for social evolution: the impact of Darwin. *J. theor. biol.* 25: 255–275.

Leikola, A. 1981. Miten Darwin löysi tiensä Suomeen. *Savon Luonto* 13: 61–84.

--- 1982. J. A. Palmén, the Darwinist reformer of zoology in Finland. *Eidema* 1: 206–220.

Leith, J. A. 1965. *The idea of art as propaganda in France 1750–1799*. Toronto: University of Toronto Press.

LeMahieu, D. L. 1979. Malthus and the theology of scarcity. *J. hist. ideas* 40: 467–474.

Lemaine, G.; Macleod, R.; Mulkay, M.; and Weingart, P., eds. 1976. *Perspectives on the emergence of scientific disciplines*. The Hague and Paris: Mouton and Maison des Sciences de l'Homme.

Le Monnier, G. 1882. *Darwin, sa vie, son oeuvre*. Nancy: Imprimerie Berger-Levrault.

Lenin, V. I. 1948. *Materialism and empirio-criticism: critical comments on a reactionary philosophy*. London: Lawrence and Wishart.

Lennox, J. G. 1980. Aristotle on general species and "the more and the less". *J. hist. biol.* 13: 321–346.

Lenz, L. 1960. Die soziologische Bedeutung der Selektion. In *Hundert Jahre Evolutionsforschung*, eds. G. Heberer and F. Schwanitz, pp. 368–396. Stuttgart: Gustav Fischer.

Leopoldina. 1982. *Neujahrgrüsse*. Halle a.S. Deutsche Akademie der Naturfreunde der Leopoldina.

Lerner, I. M. 1958. *The genetic basis of selection*. New York: John Wiley.

--- 1959. The concept of natural selection: a centennial view. *Proc. Am. Phil. Soc.* 103: 173–182.

Leroy, J. F. 1966. *Charles Darwin et la théorie moderne de l'evolution*. Paris: Seghers.

Lesch, J. E. 1975. The role of isolation in evolution: George J. Romanes and John T. Gulick. *Isis* 66: 483–503.

Lessona, M. 1883. *Carlo Darwin*. Roma: Casa Editrice A. Sommaruga.

--- 1884. *Naturalisti Italiani*. Roma: Casa Editrice A. Sommaruga.

--- 1897. *Pagine autobiografiche di Michele Lessona, per Antonio di Nino*. Roma: Forzani.

Levidow, L., and Young, R. M., eds. 1981. *Science, technology and the labour process: Marxist studies*, vol. 1. London: CSE Books.

Levin, D. 1959. *History as romantic art: Bancroft, Prescott, Motley and Parkman*. (Stanford Studies in Language and Literature, 20), Stanford: Stanford University Press.

Levin, S. M. 1966. Malthus and the idea of progress. *J. hist. ideas* 27: 92–108.

Levine, G. 1980. George Eliot's hypothesis of reality. *Nineteenth century fiction* 35: 1–28.

--- **and Madden, W.** eds. 1968. *The art of Victorian prose*. New York: Oxford University Press.

Lewins, R. 1882. Mr. Darwin and professor Haeckel. *J. sci.*, 3d ser. 4: 751–752.

Lewontin, R. C. 1970. The units of selection. *Ann. rev. ecol. syst.* 1: 1–18.

--- 1974a. *The genetic basis of evolutionary change*. New York: Columbia University Press.

--- 1974b. Darwin and Mendel — The materialistic revolution. In *The heritage of Copernicus: theories "pleasing to mind"*, ed. J. Neyman, pp. 166–183. Cambridge: M.I.T. Press.

--- 1977a. Sociobiology. A caricature of Darwinism. In *PSA 1976*, eds. F. Suppe and P. Asquith, vol. 2, pp. 22–31. E. Lansing, Mich.: Philosophy of Science Association.

--- 1977b. Biological determinism as a social weapon. In Ann Arbor Science for The People Editorial Collective: *Biology as a social weapon*, pp. 6–20. Minneapolis: Burgess.

--- 1978. Adaptation. *Sci. Am.* 1978 (Sept.): 212–230.

--- 1979. Sociobiology as an adaptationist program. *Behav. sci.* 24: 5–14.

--- 1981. Laws of biology and laws of social science, unpublished mimeo, Museum of Comparative Zoology, Harvard University.

--- 1983. Darwin's revolution. *New York review of books* 30: 21–27.

--- **and Levins, R.** 1980. Dialectics and reductionism in ecology. *Synthese* 43: 47–78.

---; **Moore, J. A.; Provine, W. B.; and Wallace, B.,** eds. 1981. *Dobzhansky's Genetics of natural populations I — XLIII.* New York: Columbia University Press.

Liebig, J. 1840. *Chemistry in its application to agriculture and physiology.* Ed. L. Playfair. London: Taylor and Walton.

--- 1842. *Animal chemistry, or organic chemistry in its application to Physiology and pathology.* Ed. W. Gregory. London: Taylor and Walton. Reprinted in 1964, with a new introduction by F. L. Holmes. New York: Johnson Reprint Corp.

--- 1843. *Familiar letters on chemistry and its relation to commerce, physiology, and agriculture.* Ed. J. Gardner. London: Taylor and Walton.

Limoges, C. 1969. Une lecture nouvelle de Darwin. *Sciences* 58–59: 70–73.

--- 1970a. Darwinisme et adaptation. *Rev. questions scientifiques* 141: 353–374.

--- 1970b. L'économie naturelle et le principe de corrélation chez Cuvier et Darwin. *Rev. d'hist. sci.* 23: 35–48.

--- 1970c. *La sélection naturelle: étude sur la première constitution d'un concept (1837–1859).* Paris: Presses Universitaires de France.

--- 1971. Darwin, Milne-Edwards et le principe de divergence. *Actes du XIIe cong. int. d'hist. sci.* (Paris 1968) 8e: 111–115. Paris: Blanchard.

--- 1978. Lamarck and his milieu. *Science* 199: 1427–1428.

Lindroth, S. 1946. *Charles Darwin.* Stockholm: Lindfors.

Linnean Society of London 1908. *The Darwin-Wallace celebration held on Thursday 1st July, 1908, by the Linnean Society of London.* London: Printed for the Linnean Society.

Lister, J. J. 1834. Some observations on the structure and function of tubular and cellular polypi, and of Ascidiae. *Phil. trans. Roy. Soc. Lond.* 124: 365–388.

[Litchfield (*née* Darwin), H. E.] 1910. *Richard Buckley Litchfield: a memoir written for his friends.* Cambridge: privately printed at Cambridge University Press.

Litchfield [*née* Darwin], H. E. ed. 1904. *Emma Darwin, wife of Charles Darwin: a century of family letters.* 2 vols. Cambridge: privately printed at Cambridge University Press.

Littlejohn, M. J. 1981. Reproductive isolation: a critical review. In *Evolution and speciation: essays in honor of M. J. D. White,* eds. W. R. Atchley and D. S. Woodruff, pp. 298–334. Cambridge: Cambridge University Press.

Livingstone, D. 1857. *Missionary travels and researches in South Africa.* London: John Murray.

Llobera, J. R. 1976. The history of anthropology as a problem. *Critique of Anthro.* 2: 17–42.

Lloyd, G. E. R. 1968. *Aristotle: the growth and structure of his thought.* Cambridge: Cambridge University Press.

Loades, A. L. 1979. Analogy, and the indictment of the deity: some interrelated themes. *Studia theologica* 33: 25–43.

Lodge, D. J. 1977. *The modes of modern writing; metaphor, metonymy, and the typology of modern writing.* Ithaca: Cornell University Press.

Loewenberg, B. J. 1933. The reaction of American scientists to Darwinism. *Am. hist. rev.* 38: 687–701.

--- 1934. The impact of the doctrine of evolution on American thought. Ph.D. thesis. Harvard University.

--- 1935. The controversy over evolution in New England, 1859–1873. *New England q.* 8: 232–257.

--- 1941. Darwinism comes to America, 1859–1900. *Mississippi Valley hist. rev.* 28: 339–368.

--- 1957. *Darwinism: reaction or reform?* New York: Holt, Rinehart & Winston.

--- 1959a. Darwin scholarship in the Darwin year. *Am. Q.* 11: 526–533.

--- 1959b. The mosaic of Darwinian thought. *Victorian stud.* 3: 3–18.

--- 1959c. *Darwin, Wallace, and the theory of natural selection, including the Linnean Society Papers.* Cambridge: Arlington Books.

---, ed. 1959d. *Charles Darwin: evolution and natural selection.* Boston: Beacon.

--- 1962. Darwin and the tragic vision. In *Am. Q.* 14: 618–622.

--- 1965. Darwin and Darwin studies, 1959–1963. *Hist. sci.* 4: 15–54.

Lombroso, C. 1871. *L'uomo bianco e l'uomo di colore: letture sull'origine e le varietà delle razze umane.* Padova: F. Sacchetto.

Lonsdale, W. 1844. Description of six species of corals from the Palaeozoic Formation of Van Diemen's Land. In C. Darwin, *Geological observations on the volcanic islands visited during the voyage of H.M.S. Beagle, together with some brief notices of the geology of Australia and the*

Cape of Good Hope, pp. 161–169. London: Smith Elder.

Lorenz, K. 1935. Der Kumpan in der Umwelt des Vogels. *J. ornithol.* 83: 137–215, 289–413. 1970. Companions as factors in the bird's environment. In *Studies in animal and human behaviour*, vol. 1, pp. 101–258. Cambridge: Harvard University Press.

--- 1965. *Darwin hat recht gesehen*. Pfullingen: Günther Neske.

Loria, A. 1884. *Carlo Darwin e l'economia politica*. Milano : Dumolard.

Lovejoy, A. O. 1904. Some eighteenth century evolutionists. *Pop. sci. monthly* 65: 238–251, 323–340.

--- 1909. Some aspects of Darwin's influence upon modern thought. *Bull. Washington Univ.*, April: 85–99.

--- 1936. *The great chain of being: a study of the history of an idea*. Cambridge: Harvard University Press.

--- 1959a. Recent criticism of Darwin's theory of recapitulation: its grounds and its initiator. In *Forerunners of Darwin*, eds. B. Glass, O. Temkin, and W. L. Straus, Jr. pp. 438–458. Baltimore: Johns Hopkins University Press.

--- 1959b. The argument for organic evolution before the *Origin of Species*, 1830–1858. In *Forerunners of Darwin 1745–1859*, eds. B. Glass, O. Temkin, and W. L. Straus Jr., pp. 356–414. Baltimore: Johns Hopkins University Press.

--- 1959c. Buffon and the problem of species. In *Forerunners of Darwin*, eds. B. Glass, O. Temkin, and W. L. Straus, Jr., pp. 84–113. Baltimore: Johns Hopkins University Press.

--- 1961. *The reason, the understanding, and time*. Baltimore: Johns Hopkins University Press.

Løvtrup, S. 1976. On the falsifiability of neo-Darwinism. *Evol. theory* 1: 267–283.

--- 1977. Variation, selection, isolation, environment: an analysis of Darwin's theory. *Theory* 43: 65–83.

--- 1978a. On von Baerian and Haeckelian recapitulation. *Syst. zool.* 27: 348–352.

--- 1978b. Robert Chambers, Charles Darwin and the mechanism of evolution. *Bull. biol. France Belg.* 62: 113–128.

--- 1979. Semantics, logic and vulgate neo-Darwinism. *Evol. theory*, 4: 157–172.

Lowe, J. 1860. A peculiar variety of the honey bee. *Cottage gardener* 24: 110.

--- 1862. Do bees vary in different parts of Great Britain? *J. horticulture and cottage gardener*. N.S. 3: 242–243.

Lucas, E. 1964. Marx' und Engels' Auseinandersetzung mit Darwin: zur Differenz zwischen Marx und Engels. *Int. rev. soc. hist.* 9: 433–469.

Lucas, J. R. 1979. Wilberforce and Huxley: a legendary encounter. *The hist. j.* 32: 313–330.

Luciani, L. 1901–1911. *Fisiologia dell'uomo: trattato didattico*. 2 vols. Milano : Società Editrice Libraria.

Lücker, R. R. 1977. *Max J. S. Schultze (1885–1874) und die Zellenlehre des 19. Jahrhunderts*. Diss. Bonn.

Ludwig, W. 1960. Die heutige Gestalt der Selectionstheorie. In *Hundert Jahre Evolutionsforschung*, eds. G. Heberer and F. Schwanitz, pp. 45–80. Stuttgart: Gustav Fischer.

Lukács, G. 1959. *Die Zerstörung der Vernunft*. Berlin: Aufbau-Verlag.

Lukina, T. A. 1973. Caspar Friedrich Wolff und die Petersburger Akademie der Wissenschaften. *Acta hist. Leopoldina 9*: 411–425.

Lurie, E. 1959. Louis Agassiz and the idea of evolution. *Victorian stud.* 3: 87–108.

--- 1960. *Louis Agassiz: a life in science*. Chicago: University of Chicago Press.

--- 1974. *Nature and the American mind: Louis Agassiz and the culture of science*. New York: Science History Publications.

Luzzatti, L. 1901. Sulle idee filosofiche e religiose di Darwin, sotto l'influenza delle sue dottrine naturali. *Atti della Reale Accademia dei Lincei* (5th ser., Rendiconti classe di scienze fisiche, matematiche, e naturali) 10: 60–72.

Lyell, C. 1830–1833. *Principles of geology, being an attempt to explain the former changes of the earth's surface, by reference to causes now in operation*. 3 vols. London: John Murray.

--- 1835. *Principles of geology*. 4th ed. 4 vols. London: John Murray.

--- 1836. Presidential address. *Proc. Geol. Soc. Lond.* 2: 357–390.

--- 1837. Presidential address. *Proc. Geol. Soc. Lond.* 2: 479–523.

--- 1838. *Elements of geology*. London: John Murray.

--- 1863. *The geological evidences of the antiquity of man, with remarks on theories of the origin of species by variation*. London: John Murray.

--- 1868. *Principles of geology: or the modern changes of the earth and its inhabitants considered as illustrative of geology*. 10th ed. 2 vols. London: John Murray.

--- 1970. *Sir Charles Lyell's scientific journals on the species question*. Ed. L. G. Wilson. New Haven: Yale University Press.

Lyell, K. M., ed. 1881. *The life, letters, and journals of Sir Charles Lyell*. 2 vols. London: John Murray.

Lyons, J. 1972. Immediate reactions to Darwin: the English Catholic press' first reviews of the *Origin of the* [sic] *Species*. *Church hist.* 41: 78–93.

McAtee, W. L. 1947. The cats-to-clover chain. *Scientific monthly* 65: 241–242.

Macbeth, N. 1971. *Darwin retried*. Boston: Gambit.

McConnaughey, G. 1950. Darwin and Social Darwinism. *Osiris* 9: 397–412.

McCrie, G. M. 1891. Mr. Darwin and professor Haeckel. In *Further reliques of Constance Naden: being essays and tracts for our times*, ed. G. M. McCrie, pp. 258–259. London: Bickers and Son.

MacKenzie, D. 1978. *Biological ideas in politics*. Harmondsworth: Pelican Books.

--- 1981. Sociobiologies in competition: the biometrician-Mendelian debate. In *Biology, medicine and society 1840–1940*, ed. C. Webster, pp. 243–288. Cambridge: Cambridge University Press.

McKinney, H. L. 1966. Alfred Russel Wallace and the discovery of natural selection. *J. hist. med.* 21: 333–357.

---, ed. 1971. *Lamarck to Darwin: contributions to evolutionary biology 1809–1859*. Lawrence: Coronado Press.

--- 1972. *Wallace and natural selection*. New Haven: Yale University Press.

McLachlan, H. 1934. *The Unitarian movement in the religious life of England. I. Its contribution to thought and learning, 1700–1900*. London: George Allen & Unwin.

MacLeay, W. 1819–1821. *Horae entomologicae: or essays on the annulose animals*. London: S. Bagster.

MacLeod, R. M. 1965. Evolutionism and Richard Owen, 1830–1868: an episode in Darwin's century. *Isis* 66: 259–280.

--- 1969. The genesis of *Nature*. *Nature* 224: 423–461.

--- 1970. The X-club: a scientific network in late Victorian England. *Notes rec. Roy. Soc. Lond.* 24: 305–322.

McMillan, N. F. 1979. William Swainson in New Zealand, with notes on his drawings held in New Zealand. *J. Soc. Bibl. Nat. Hist.* 9: 161–169.

McMullin, E. 1976. The fertility of theory and the unit for appraisal of science. In *Essays in memory of Imre Lakatos*, eds. R. S. Cohen, P. K. Feyerabend, and M. W. Wartofsky, pp. 395–432. Dordrecht: Reidel.

McPherson, T. 1972. *The argument from design*. London: Macmillan.

McQuire, D. 1977. Herbert Spencer's factors in social evolution. *Sociol. anal. theory* 7: 99–115.

Mac Rae, D. G. 1958. Darwinism and the social sciences. In *A century of Darwin*, ed. S. A. Barnett, pp. 296–312. Cambridge: Harvard University Press.

Mafeje, A. 1976. The problem of anthropology in historical perspective: an inquiry into the growth on the social sciences. *Rev. Canadienne des Études Africaines/Canadian j. African stud.* 10, no. 2: 307–333.

Magner, L. N. 1980. *A history of the life science*. New York, Basel: Marcel Dekker.

Magoun, H. W. 1960. Evolutionary concepts of brain function following Darwin and Spencer. In *Evolution after Darwin*, ed. S. Tax, vol. 2, pp. 187–209. Chicago: University of Chicago Press.

--- 1961. Darwin and concepts of brain function. In *Brain mechanism and learning*, ed. J. F. Delafresnaye, pp. 1–20. Oxford: Blackwell.

Maienschein, J. 1981. Shifting assumptions in American biology: embryology, 1890–1910. *J. hist. biol.* 14: 89–113.

---; **Rainger, R.; and Benson, K.** 1981. Introduction: were American morphologists in revolt? *J. hist. biol.* 14: 83–87.

Malefijt, A. 1974. *Images of man: a history of anthropological thought*. New York: Knopf.

Malthus, T. R. 1798. *An essay on the principle of population, as it affects the future improvement of society, with remarks on the speculations of Mr. Goodwin, M. Condorcet, and other writers*. London: J. Johnson.

--- 1826. *An essay on the principle of population; or, a view of its past and present effects on human happiness; with an inquiry into our prospects respecting the future removal or mitigation of the evils which it occasions*. 6th ed. 2 vols. London: John Murray.

Mamiani, T. 1868. Nuove considerazioni intorno al sistema di Darwin. *Nuova antologia* 8: 472–503.

Mandelbaum, M. 1948. Arthur O. Lovejoy and the theory of historiography. *J. hist. ideas* 9: 412–423.

--- 1957. The scientific background of evolutionary theory in biology. *J. hist. ideas* 18: 342–361. Also in *Roots of scientific thought*, eds. P. P. Wiener and A. Noland, pp. 517–536. New York: Basic Books.

--- 1958. Darwin's religious views. *J. hist. ideas* 19: 363–378.

--- 1965. The history of ideas, intellectual history, and the history of philosophy. *History and theory* Beiheft 5: 33–66.

--- 1971. *History, man, and reason: a study in nineteenth-century thought*. Baltimore: Johns Hopkins University Press.

Manier, E. 1965. The theory of evolution as personal knowledge. *Phil. sci.* 32: 244–252.

--- 1969. "Fitness" and some explanatory patterns in biology. *Synthese* 20: 206–218.

--- 1978. *The young Darwin and his cultural circle: a study of influences which helped shape the language and logic of the first drafts of the theory of natural selection*. Dordrecht: Reidel.

--- 1980a. History, philosophy and sociology of biology: a family romance. *Stud. hist. phil. sci.* 11: 1–24.

--- 1980b. Darwin's language and logic. *Stud. hist. phil. sci.* 11: 305–323.

Mann, G., ed. 1973a. *Biologismus im 19. Jahrhundert*. Stuttgart: Ferdinand Enke.

--- 1973b. Rassehygiene — Sozialdarwinismus. In *Biologismus in 19. Jahrhundert*, ed. G. Mann, pp. 73–93. Stuttgart: Ferdinand Enke.

--- 1975. Biologie und Geschichte: Ansätze und Versuche zur biologistischen Theorie der Geschichte im 19. und beginnenden 20. Jahrhundert. *Medizinhistorische Journal* 10: 281–306.

--- 1980. Ernst Haeckel und der Darwinismus: Popularisierung, Propaganda und Ideologisierung. *Medizinhistorisches J.* 15: 269–283.

Manser, A. R. 1965. The concept of evolution. *Philosophy* 40: 18–34.

Mantegazza, P. 1871. L'elezione sessuale e la neogenesi. Lettera del Prof. P. Mantegazza a Carlo Darwin. *Archivio per l'antropologia e l'etnologia* 1: 306–325.

--- 1882. *Commemorazione di Carlo Darwin: discorso*. Firenze : Tipografia dell'Arte della Stampa.

Manuel, F. 1956. From equality to organicism. *J. hist. ideas*. 17: 54–69.

--- 1972. Henri Saint-Simon on the role of the scientist. In *Freedom from history and other essays*, pp. 205–218. London: University of London Press.

Marchant, P. D. 1957. Social Darwinism. *Australian j. polit. and hist.* 3: 46–59.

--- 1959. Darwin and the social theory. *Australian j. polit. and hist.* 5: 213–217.

Marler, P. 1959. Developments in the study of animal communication. In *Darwin's biological work*, ed. P. R. Bell, pp. 150–206. Cambridge: Cambridge University Press.

Marmocchi, F. C. 1844. *Prodromo della storia naturale, generale e comparata d'Italia*. 2 vols. Firenze : Società Editrice Fiorentina.

Marnell, W. H. 1966. *Man-made morals: four philosophies that shaped America*. New York: Doubleday.

Marquez-Breton, B. 1977. Review of Y. Conry, *L'introduction du darwinisme en France au XIXe siècle*. *Rev. de synthèse* 85–86 : 154–156.

Marsh, H., and Langenheim, J. 1961. Natural selection as an ecological concept. *Ecology* 42: 158–165.

Marsh, O. C. 1878. *Introduction and succession of vertebrate life in North America*. New York: Appleton.

--- 1880. *Odontornithes: a monograph on the extinct toothed birds of North America*. Report of the Geological Survey of the Fortieth Parallel, vol. 7. Washington: Government Printing Office.

Marshall, A. J. 1971. *Darwin and Huxley in Australia*. Sydney: Verry.

Marshall, P. T. 1969. *The development of modern biology*. Oxford: Pergamon Press.

Martin, R. N. D. 1982. Darwin and Duhem. *Hist. sci.* 20: 64–74.

Martin, W. 1837a. Observations on three specimens of the genus *Felis* presented to the Society by Charles Darwin, Esq., Corr. Memb. Z. S. *Proc. Zool. Soc. Lond.* 5: 3–4.

--- 1837b. Observations upon a new fox from Mr. Darwin's collection (*Vulpes fulvipes*). *Proc. Zool. Soc. Lond.* 5: 11–12.

--- 1837c. Observations on a specimen of *Dasypus hybridus*, Desm., from Mr. Darwin's collection. *Proc. Zool. Soc. Lond.* 5: 13–14.

Martucci, V. 1978. Un interlocutore italiano di Darwin: Giuseppe Bianconi e la "creazione indipendente". *Physis* 20: 349–355.

--- 1980. Sviluppo ontogenetico e citologia nell'ologenesi di Daniele Rosa. *Physis* 22: 73–84.

--- 1982. Il discusso fascino di una ipotesi: la pangenesi di Darwin e gli studiosi italiani (1868–1888). *Hist. phil. Life sci.* 3: 243–257.

Marx, K. 1961. *Economic and philosophic manuscripts of 1844*. Moscow: Foreign Languages Publishing House.

--- **and Engels, F.** 1965. *Selected correspondence.* 2d ed. Moscow: Progress.

--- **and Engels, F.** 1968. *The German ideology.* Moscow: Progress.

Mârza, V. D., and Tarnavschi, I. T. 1967. The problem of the fertilization and evolution of phanerogams in Darwin's work: a critical study. *Indian j. hist. sci.* 2: 71–104.

--- **and Tarnavschi, I. T.** 1974. Darwin's theory of unity of reacting mechanisms in plants and animals: its present day importance. *Ind. j. hist. sci.*, 9: 185–220.

Mason, P., ed. 1960. *Man, race, and Darwin.* London: Oxford University Press.

Mason, S. F. 1962. *A history of the sciences*. Rev. ed. New York: Collier.

Matthew, P. 1831. *On naval timber and aboriculture.* London: Longmans.

Mayr, E. 1942. *Systematics and the origin of species from the viewpoint of a zoologist.* New York: Columbia University Press.

--- 1945. Symposium on age of the distribution pattern of the gene arrangement in *Drosophila pseudoobscura.* Introduction and some evidence in favor of a recent date. *Lloydia* 8: 70–83.

--- 1947. Ecological factors in speciation. *Evolution* 1: 263–288.

--- 1949. Speciation and selection. *Proc. Am. Phil. Soc.* 93: 514–519.

--- 1954a. Change of genetic environment and evolution. In *Evolution as a process*, eds. J. Huxley, A. C. Hardy, and E. B. Ford, pp. 157–180. London: Allen and Unwin.

--- 1954b. Wallace's line in the light of recent zoogeographical studies. *Q. rev. biol.* 29: 1–14.

--- 1955. Karl Jordan's contributions to current concepts in systematics and evolution. *Trans. Roy. Entomol. Soc. Lond.* 107: 45–66.

--- 1957. Species concepts and definitions, the species problem. In *The species problem,* ed. E. Mayr, pp. 1–22. Washington: American Association for the Advancement of Science.

--- 1959a. Agassiz, Darwin, and evolution. *Harvard Libr. bull.* 13: 165–194.

--- 1959b. Darwin and the evolutionary theory in biology. In *Evolution and anthropology: a centennial appraisal*, ed. B. J. Meggers, pp. 1–10. Washington: Anthropological Society of Washington.

--- 1959c. Isolation as an evolutionary factor. *Proc. Am. Phil. Soc.* 103: 221–230.

--- 1959d. Review of G. Himmelfarb, *Darwin and the Darwinian revolution. Sci. Am.* 201 (Nov.): 209–216.

--- 1960. The emergence of evolutionary novelties. In *Evolution after Darwin*, ed. S. Tax, vol. 1, pp. 349–380. Chicago: University of Chicago Press.

--- 1961. Cause and effect in biology. *Science* 134: 1501–1506.

--- 1962. Accident or design, the paradox of evolution. In *The evolution of living organisms,* ed. G. W. Leeper, pp. 1–14. Melbourne: Melbourne University Press.

--- 1963. *Animal species and evolution.* Cambridge: Harvard University Press.

--- 1964. Introduction to *C. Darwin "On the origin of species". A facsimile of the first edition.* Cambridge: Harvard University Press.

--- 1965. Selektion und gerichtete Evolution. *Naturwissenschaften* 52: 173–180.

--- 1969a. Grundgedanken der Evolutionsbiologie. *Naturwissenschaften* 56: 392–397.

--- 1969b. *Principles of systematic zoology.* New York: McGraw-Hill.

--- 1970. *Populations, species, and evolution.* Cambridge: Harvard University Press.

--- 1971. Open problems of Darwin research. *Stud. hist. phil. sci.* 2: 273–280.

--- 1972a. The nature of the Darwinian revolution. *Science* 176: 981–989.

--- 1972b. Sexual selection and natural selection. In *Sexual selection and the descent of man,* ed. B. Campbell, pp. 87–104. Chicago: Aldine.

--- 1973. Essay review: the recent historiography of genetics. *J. hist. biol.* 6: 125–154.

--- 1976a. *Evolution and the diversity of life: selected essays.* Cambridge Harvard University Press.

--- 1976b. Typological versus population thinking. In *Evolution and the diversity of life,* pp. 26–29. Cambridge: Harvard University Press.

--- 1976c: Lamarck revisited. In *Evolution and the diversity of life*, pp. 222–250. Cambridge: Harvard University Press.

--- 1976d. Is the species a class or an individual? *Syst. zool.* 25:192.

--- 1977. Darwin and natural selection: how Darwin may have discovered his highly unconventional theory. *Am. sci.* 65: 321–327.

--- 1978. Evolution. *Sci. Am.* 239 (Sept.): 46–55.

--- 1980. Prologue. In *The evolutionary synthesis*, eds. Mayr and Provine, pp. 1–48. Cambridge: Harvard University Press.

--- 1982a. Patterns of speciation in animals. In *Mechanisms of speciation*, ed. C. Barigozzi, pp. 1–19. Rome: Academia dei Lincei.

--- 1982b. *The growth of biological thought*. Cambridge: Harvard University Press.

--- 1982c. Speciation and macroevolution. *Evolution* 36: 1119–1132.

--- 1982d. Adaptation and selection. *Biologisches Zentralblatt* 101: 161–174.

--- 1982e. How to carry out the adaptationist program? *Am. nat.* 121: 324–334.

--- **and Provine, W. B.**, eds. 1980. *The evolutionary synthesis: perspectives on the unification of biology*. Cambridge: Harvard University Press.

Medawar, P. B. 1963. Is the scientific paper a fraud? *The listener*, September 12, 1963.

--- 1967. *The art of the soluble*. London: Methuen.

--- **and Medawar, J. S.** 1977. *The life science: current ideas of biology*. London: Wildwood House.

Meek, R. L., ed. 1954. *Marx and Engels on Malthus: selections from the writings of Marx and Engels dealing with the theories of Thomas Robert Malthus*. New York: International.

Meggers, B. J., ed. 1959. *Evolution and anthropology: a centennial appraisal*. Washington: Anthropological Society of Washington.

Mellersch, H. E. L. 1964. *Charles Darwin: pioneer of the theory of evolution*. London: Barker.

Mendelsohn, E. 1964. The biological sciences in the nineteenth century: some problems and sources. *Hist. sci.* 3: 39–59.

Menzbir, M. A. 1893. Opyt teorii nasledstvennosti (Problems in the theory of heredity). *Russk. mysl*, no. 10: 219–

--- 1900. Glaveneshie predstaviteli darwinisma v Zapadnoj Evrope (The main representatives of Darwinism in Western Europe) *Russk. mysl*, no. 6: 12–

---, ed. 1907. *Pisma Darwina*. (Darwin's letters) Moscow: Darwin Museum.

--- 1926–1934. *Polnoe sobranie sochinenii Charl'za Darwina* (Ch. Darwin's complete works) Moscow: Darwin Museum.

--- 1934. *Ber i organologie zhivotnykh* (Baer and animal organology) Moscow: Akad. Nauk.

Mercier, P. 1966. *Histoire de l'anthropologie*. Paris: Presses Universitaires de France.

Mertens, R. 1960. Von der statischen zur dynamischen Systematik in der Zoologie. In *Hundert Jahre Evolutionsforschung*, eds. G. Heberer and F. Schwanitz, pp. 186–202. Stuttgart: Gustav Fischer.

Merton, R. K. 1962. Priorities in scientific discovery: a chapter in the sociology of science. In *The sociology of science*, eds. B. Barber and W. Hirsch, pp. 447–485. Glencoe: Free Press.

--- 1973. *The sociology of science: theoretical and empirical investigations*. Chicago: University of Chicago Press.

Merz, J. T. 1896–1914. *A history of European thought in the nineteenth century*. 4 vols. Edinburgh: W. Blackwood and Sons.

Meyer-Abich, A. 1964. The historico-philosophical background of the modern evolution biology. *Acta biotheoretica* Supp. 2.

Michalova, C. 1958. Sozialdarwinismus und Rassentheorie. *Urania* 21: 88–91.

Midgley, M. 1978. *Beast and Man: the roots of human nature.* Ithaca: Cornell University Press.

Mikulak, M. W. 1970. Darwinism, Soviet genetics, and Marxism-Leninism. *J. hist. ideas* 31: 359–376.

Mikulinsky, S. R. 1961. *Problem obshchei biologii v Rossii, Pervaia polovina XIX veka* (The development of a general biological problematic in Russia in the first half of the 19th century). Moscow: Akad. Nauk.

--- 1979. *Karl Franzovich Rul'ye, uchenyi, chelovek i uchitel* (K. F. R., scholar, man and Teacher), *1814–1858*. Moscow: Nauka.

Mill, J. S. 1874. *A system of logic.* 8th ed. London: Longmans, Green.

Miller, J., ed. 1974. *The limits of human nature.* New York: Dutton.

Miller, J. H. 1976. *The disappearance of God: five nineteenth century writers.* Cambridge: Harvard University Press.

Miller, M. 1940–1941. Géricault's paintings of the insane. *J. Warburg and Courtauld Inst.* 4: 151–163.

Millhauser, M. 1959. *Just before Darwin: Robert Chambers and Vestiges*. Middletown: Wesleyan University Press.

Mills, E. 1984. A view of Edward Forbes, naturalist. *Arch. nat. hist.* 11: 365–393.

Mills, S., and Beatty, J. 1979. The propensity interpretation of fitness. *Phil. Sci.* 46: 263–286. 1983. Reprinted in *Conceptual issues in evolutionary biology*, ed. E. Sober, pp. 36–57. Cambridge: M.I.T. Press.

Milne-Edwards, H. 1834. *Élémens de zoologie: lecons sur l'anatomie, la physiologie, la classification des moeurs des animaux.* Paris: Crochard.

--- 1844. Considérations sur quelques principes relatifs à la classification naturelle des animaux. *Ann. des sci. nat.* 3rd ser., 1: 65–99.

Milne Edwards, E. 1867. *Rapport sur les progrès récents des sciences zoologiques en France*. Paris: Imprimerie Impériale.

Mischel, T. 1966. "Emotion" and "motivation" in the development of English psychology: D. Hartley, James Mill, and A. Bain. *J. hist. behav. sci.* 2: 123–144.

Mitchell, I. 1978. Marxism and German scientific materialism. *Ann. sci.* 35: 379–400.

Mitchell, P. C. 1900. *Thomas Henry Huxley: a sketch of his life and work*. New York: G. P. Putnam.

Mivart, St. G. J. 1871a. *On the genesis of species*. London: Macmillan.

--- 1871b. Darwin's descent of man. *Q. rev.* 131: 47–90.

--- 1872a. Specific genesis. *North Am. rev.* 114: 451–468.

--- 1872b. Evolution and its consequences: a reply to Professor Huxley. *Contemporary rev.* 19: 168–197.

--- 1884. On the development of the individual and of the species. *Proc. Zool. Soc. Lond.* 462–474.

Mocek, R. 1974. *Wilhelm Roux — Hans Driesch*. Jena: VEB Gustav Fischer.

--- 1982. Das Lamarck-Darwinische Prinzip des Gebrauchs und Nichgebrauchs der Organe und seine Rolle in der entwicklungsbiologische Diskussion. In *Darwin 1809–1882*, ed. S. Kirschke. Halle a. S.: Martin-Luther-Universität Halle-Wittenberg.

Mondella, F. 1971. La teoria dell'evoluzione e l'opera di Charles Darwin. In *Storia del pensiero filosofico e scientifico*, ed. L. Geymonat, vol. 5, pp. 247–293. Milano: Garzanti.

Monod, J. 1970. *Le hasard et la nécessité*. Paris: Éditions du Seuil. 1971. *Chance and necessity: an essay on the natural philosophy of modern biology*. New York: Knopf.

Montagu, M. F. A. 1942. Further comment. *Science and society* 6: 74–75.

--- 1947. Theognis, Darwin, and social selection. *Isis* 37: 24–26.

--- 1952a. *Darwin: competition and cooperation*. New York: Henry Schuman.

--- 1952b. *Man's most dangerous myth. The fallacy of race*. 3rd ed. New York: Harper.

Montague, J. 1959. *Charles Le Brun's Conference sur l'expression générale et particulière* [1668]. Ph.D. thesis. University of London.

Montalenti, G. 1958. *L'evoluzione*. Torino : Edizioni Radio Italiana.

--- 1982. *Carlo Darwin*. Roma: Editori Riuniti.

Montgomery, W. M. 1974a. *Evolution and Darwinism in German biology, 1800–1883*. Ph. D. thesis. University of Texas.

--- 1974b. Germany. In *The comparative reception of Darwinism*, ed. T. F. Glick, pp. 81–116. Austin: University of Texas Press.

Moody, J. W. T. 1971. The reading of the Darwin and Wallace papers: an historical "non-event". *J. Soc. Bibl. Nat. Hist.* 5: 474–476.

Moore, J. R. 1977a. Could Darwinism be introduced in France? *Brit. j. hist. sci.* 35: 246–251.

--- 1977b. On the education of Darwin's sons: the correspondence between Charles Darwin and the Reverend G. V. Reed, 1857–1864. *Notes rec. Roy. Soc. Lond.* 32: 51–70.

--- 1979. *The post-Darwinian controversies: a study of the Protestant struggle to come to terms with Darwin in Great Britain and America*. Cambridge: Cambridge University Press.

--- 1980a. Review of M. Ruse, *The Darwinian revolution: science red in tooth and claw*. *Am. hist. rev.* 85: 860–861.

--- 1980b. Preface. In *Darwinian impacts: an introduction to the Darwinian revolution* by D. R. Oldroyd, pp. ix–x. Milton Keynes, Bucks: Open University Press.

--- 1981a. *Beliefs in science: an introduction*. Milton Keynes: Open University Press.

--- 1981b. Essay review of N. C. Gillespie, *Charles Darwin and the problem of creation*. *Brit.*

j. hist. sci. 14: 189–200.

--- 1982a. 1859 and all that: re-making the story of evolution-and-religion. In *Charles Darwin 1809-1882: a centennial commemorative*, ed. R. Chapman, pp. 167–194. Wellington: Nova Pacifica.

--- 1982b. Charles Darwin lies in Westminster Abbey. *Biol. j. Linn. Soc.* 17: 97–113.

--- Forthcoming. Darwinism *is* social [essay review of G. Jones, *Social Darwinism in English thought: the interaction between biological and social theory*]. *Radical sci. j.*

Moore, R. 1957. *Charles Darwin.* London: Hutchinson.

Moore, W. E. 1979. Functionalism. In *A history of sociological analysis*, ed. T. Bottomore, pp. 321–361. London: Heinemann.

Moorehead, A. 1969. *Darwin and the Beagle.* New York: Harper and Row.

Moorhead, P. S., and Kaplan, M. M., eds. 1967. *Mathematical challenges to the neo-Darwinian interpretation of evolution.* Philadelphia: Wistar Institute Press.

Moreno, R. 1974. Mexico. In *The comparative reception of Darwinism*, ed. T. F. Glick, pp. 346–374. Austin: University of Texas Press.

Morgan, T. H. 1903. *Evolution and adaptation.* New York: Macmillan.

Morison, A. 1840. The physiognomy of mental diseases. London: Longmans.

Morrell, J. B. 1971. Individualism and the structure of British science in 1830. *Hist. stud. phys. sci.* 3: 183–204.

--- **and Thackray, A.** 1981. *Gentlemen of science: early years of the British Association for the Advancement of Science.* Oxford: Clarendon Press.

Morris, J., and Sharpe, D. 1846. Description of Eight Species of Brachiopodus Shells from the Palaeozoic Rocks of the Falkland Islands *Proc. Geol. Soc. London*, pt. 1, 2: 267–274.

Morse, E. 1876. Address on the contributions of American zoologists to the Darwinian theory of evolution. *Proc. Am. Assoc. Adv. Sci.* 25: 137–176.

Morselli, E., ed. 1881–1891. *Rivista di filosofia scientifica.* Milano : Dumolard.

--- 1887. La filosofia monistica in Italia. *Rivista di filosofia scientifica* 6 : 1–36.

--- 1892. *Carlo Darwin e il Darwinismo nelle scienze biologiche e sociali.* Milano : Dumolard.

Mozley, A. 1967. Evolution and the climate of opinion in Australia, 1840–1876. *Victorian stud.* 10: 411–430.

Mudford, P. G. 1968. William Lawrence and *The natural history of man. J. hist. ideas* 29: 430–436.

Mueller, F.-L. 1976. *Histoire de la psychologie.* 4ᵉ éd. entièrement revue et augmentée. Paris: Payot.

Mulkay, M. 1979. *Science and the sociology of knowledge.* London: Allen and Unwin.

Mullen, P. C. 1964. *The preconditions and reception of Darwinian biology in Germany, 1800–1870.* Ph.D. thesis. University of California, Berkeley.

Müller, F. 1864. *Für Darwin.* Leipzig: W. Engelmann.

Müller, G. 1983. Darwin, Marx, Aveling — Briefe and Spekulationen: eine bibliographische Betrachtung. *Dialektik* 6: 149–159.

Muller, H. J. 1949. The Darwinian and modern conceptions of natural selection. *Proc. Am. Phil. Soc.* 93: 459–470.

--- 1959. One hundred years without Darwinism are enough. *School sci. and math.* 59: 304–316.

Müller, J. 1838–1842. *Elements of physiology.* Trans. W. Baly. 2 vols. London: Taylor and Walton.

Mullins, N. C. 1968. The distribution of social and cultural properties in informal communications networks among biological scientists. *Amer. sociol. rev.* 3: 786–797.

Müntzing, A. 1959a. Darwin's views on variation under domestication in the light of present-day knowledge. *Proc. Am. Phil. Soc.* 103: 190–220.

--- 1959b. Darwin's views on variation under domestication. *Am. sci.* 47: 314–325.

Murchison, R. 1832. Presidential address. *Proc. Geol. Soc. Lond.* 1: 362–386.

--- 1839. *The Silurian system.* London: John Murray.

--- 1842. Presidential address. *Proc. Geol. Soc. Lond.* 3: 637–687.

--- 1843. Presidential address. *Proc. Geol. Soc. Lond.* 4: 65–151.

Murphree, I. L. 1961. The evolutionary anthropologists: the progress of mankind. The concepts of progress and culture in the thought of John Lubbock, Edward B. Tylor, and Lewis H. Morgan. *Proc. Am. Phil. Soc.* 105: 265–300.

Murray, R. 1925. *Science and scientists in the nineteenth century.* London: Sheldon Press.

Myers, F. W. H. 1888. Charles Darwin and agnosticism. 1893. In *Science and a future life, with other essays*, pp. 51–75. London: Macmillan.

Myrzoyan, E. N. 1959. Ch. Darwin i razvitie zootekhnicheskoii mysli v Rossii v 60–80 gody veka (Darwin and the development of zootechnical thought in Russia in the 1860s to 1880s). *Ann. biol.* 2: 91–110.

Naccache, B. 1980. *Marx critique de Darwin*. Paris: Vrin.

Nachtwey, R. 1959. *Der Irrweg des Darwinismus*. Berlin: Morus.

Nagel, E. 1961. *The structure of science: problems in the logic of scientific explanation*. London: Routledge and Kegan Paul; New York: Harcourt, Brace and World.

Nash, J. V. 1928. The religious evolution of Darwin. *Open court* 42: 449–463.

Nasmyth, G. 1916. *Social progress and the Darwinian theory*. New York: Putnam.

Naumov, N. P. 1972. *The ecology of animals*. Urbana: University of Illinois Press.

Nelson, G. J. 1974. Darwin-Hennig classification: reply to E. Mayr. *Syst. zool.* 23: 452–458.

--- 1978. From Candolle to Croizat: comments on the history of biogeography. *J. hist. biol.* 11: 269–305.

--- **and Platnick, N.** 1981. *Systematics and biogeography: cladistics and vicariance*. New York: Columbia University Press.

Newell, A., and Simon, H. A. 1972. In *Human problem solving*, pp. 163–259. Englewood Cliffs: Prentice-Hall.

Newman, E. 1841. Entomological notes. *Entomologist* 1: 7–13.

Newman, H. W. 1862. Do bees vary in different parts of Great Britain? *J. horticulture and cottage gardener*. N.S. 3: 225.

Newton, A. 1888a. Early days of Darwinism. *Macmillan's mag.* 57: 241–249.

--- 1888b. [Address of the president of section D — biology]. In *Rept. 57th meeting B.A.A.S.* [Manchester 1887], pp. 726–733. London: John Murray.

--- 1888c. [Review of *Life and letters of Charles Darwin*] *Q. rev.* 145: 1–30.

Nichols, C. 1974. Darwinism and the social sciences. *Phil. soc. sci.* 4: 255–277.

Nicholson, A. J. 1960. The role of population dynamics in natural selection. In *Evolution after Darwin*, ed. S. Tax, vol. 1, pp. 477–522. Chicago: University of Chicago Press.

Nickles, T., ed. 1980a. *Scientific discovery, logic and rationality*. Boston Studies in the Philosophy of Science, vol. 59. Dordrecht, Boston, London: Reidel.

---, ed. 1980b. *Scientific discovery: case studies*. Boston Studies in the Philosophy of Science, vol. 60. Dordrecht, Boston, London: Reidel.

Noble, D. 1976. George Henry Lewes, George Eliot, and the physiological society. *J. physiol.* 263: 45–54.

Nogar, R. J. 1961. From the fact of evolution to the philosophy of evolutionism. *The Thomist* 24.

--- 1963. *The wisdom of evolution*. New York: Doubleday.

--- 1966. *The Lord of the absurd*. New York: Herder.

Nordenskiöld, E. 1928. *The history of biology: a survey*. New York: Knopf.

North, J. 1969. Charles Darwin. In *Mid-Nineteenth Century*, ed. J. North, pp. 35–71. Oxford: Pergamon Press.

Northrop, F. S. C. 1950. Evolution in its relation to the philosophy of nature and the philosophy of culture. In *Evolutionary thought in America*, ed. S. Persons, pp. 44–84. New Haven: Yale University Press.

Norton, B. J. 1973. The biometric defense of Darwin. *J. hist. biol.* 6: 283–316.

--- 1975a. Metaphysics and population genetics: Karl Pearson and the background to Fisher's multi-factorial theory of inheritance. *Ann. sci.* 32: 537–553.

--- 1975b. Biology and philosophy: the methodological foundations of biometry. *J. hist. biol.* 8: 85–93.

Nowiński, C. 1967. Biologie, théorie du développement et dialectique. In *Logique et connaissance scientifique*, ed. J. Piaget, Encyclopédie de la Pléiade, vol. 22. Paris: Gallimard.

--- 1974a. Pojecie doboru naturalnego. [The concept of natural selection]. In *Ewolucja biologiczna*, ed. C. Nowiński, pp. 39–122. Warsaw: Ossolineum.

---, ed. 1974b. *Ewolucja biologiczna. Szkice teoretyczne i metodologiczne*. [Biological evolution. Theoretical and methodological sketches]. Warsaw: Ossolineum.

Nuñez, D., ed. 1977. *El darwinisimo en España*. Madrid: Editorial Castalia.

O'Brien, C. F. 1971. *Sir William Dawson: a life in science and religion. Mem. Am. Phil. Soc.*, vol. 84. Philadelphia: American Philosophical Society.

O'Brien, J. A. 1932. *Evolution and religion: a study of the bearing of evolution upon the philosophy of religion.* New York: Century.

Oken, L. 1840. *Idee sulla classificazione filosofica dei tre regni della natura esposte dal Professor Oken alla riunione dei naturalisti italiani in Pisa nell' Ottobre del 1839.* Milano : Giacomo Pirola.

Olby, R. 1963. Charles Darwin's manuscript of *Pangenesis. Brit. j. hist. sci.* 1: 251–263.

--- 1966a. *Origins of Mendelism.* London: Constable.

--- 1966b. Joseph Koelreuter, 1733–1806. In *Late eighteenth century European scientists*, ed. R. C. Olby. Oxford: Pergamon Press.

--- 1967. *Charles Darwin.* London: Oxford University Press.

--- 1979. Mendel no Mendelian? *Hist. sci.* 17: 53–72.

Olding, A. 1978. A defence of evolutionary laws. *Brit. phil. sci.* 29: 131–143.

Oldrini, G. 1973. *La cultura filosofica napoletana dell'Ottocento.* Bari: Laterza.

Oldroyd, D. R. 1979. Historicism and the rise of historical geology, part 2. *Hist. sci.* 17: 227–257.

--- 1980. *Darwinian impacts: an introduction to the Darwinian revolution.* Milton Keynes: Open University Press.

Olmsted, J. M. D. 1946. *Charles Eduard Brown-Séquard: a nineteenth century neurologist and endocrinologist.* Baltimore: Johns Hopkins University Press.

Omalius D'Halloy, J.–B.–J. d' 1831. Éléments de géologie. Paris: F. G. Levrault.

--- 1855. *Abrégé de géologie.* Bruxelles. A. Schnée.

--- 1868. Correspondance générale. 3me série. 1862–1871. A. Gaudry to D'Omalius D'Halloy, 12 January 1868, ff. 727–728. Bruxelles : Académie Royale des Sciences.

Omodeo, P. 1949a. Documenti per la storia delle scienze naturali al principio del XIX secolo. 1. La vita e le opere di Giosuè Sangiovanni. *Bollettino di zoologia* 16: 107–117.

--- 1949b. Documenti per la storia delle scienze naturali al principio del XIX secolo.

--- 1959. Centocinquant'anni di evoluzionismo. *Società* 15: 833–883.

--- 1960. Darwin e l'ereditarietà dei caratteri acquisiti. *Scientia* 95: 22–31.

--- 1969. *Opere di Jean-Baptiste de Lamarck.* Torino: Unione Tipografica Editrice.

Ong, W. J., ed. 1960. *Darwin's vision and Christian perspectives.* New York: Macmillan.

--- 1966. Evolution, myth and poetic vision. *Comp. lit. stud.*, 3: 1–20.

Oppenheimer, J. M. 1940. The non-specificity of the germ-layers. *Q. rev. biol.* 15: 1–27.

--- 1959. An embryological enigma in the *Origin of species.* In *Forerunners of Darwin: 1745–1859*, eds. B. Glass, O. Temkin, and W. L. Straus Jr., pp. 292–322. Baltimore: Johns Hopkins University Press.

--- 1967. *Essays in the history of embryology and biology.* Cambridge: M.I.T. Press.

Orbigny, A. D. 1850–1852. *Prodrome de paléontologie stratigraphique universelle des animaux mollusques & rayonnés.* 3 vols. Paris: V. Masson.

Orwin, S. C., and Whetham, E. H. 1964. *History of British agriculture, 1846–1914.* London: Longmans.

Osborn, H. F. 1894. *From the Greeks to Darwin: an outline of the development of the evolution idea.* New York: Columbia University Press.

--- 1922. William Bateson and Darwinism. *Science* 55: 194.

--- 1929. *The Titanotheres of ancient Wyoming, Dakota and Nebraska.* 2 vols. U. S. Geological Survey Monographs, no. 55. Washington.

Ospovat, D. 1976. The influence of Karl Ernst von Baer's embryology, 1828–1859: a reappraisal in light of Richard Owen's and William B. Carpenter's "Palaeontological application of 'von Baer's law' ". *J. hist. biol.* 9: 1–28.

--- 1977. Lyell's theory of climate. *J. hist. biol.* 10: 317–339.

--- 1978. Perfect adaptation and teleological explanation: approaches to the problem of the history of life in the mid-nineteenth century. *Stud. hist. biol.* 2: 33–56.

--- 1979. Darwin after Malthus. *J. hist. biol.* 12: 211–230.

--- 1980. God and natural selection: the Darwinian idea of design. *J. hist. biol.* 13: 169–194.

--- 1981. *The development of Darwin's theory: natural history, natural theology, and natural selection, 1838–1859.* Cambridge: Cambridge University Press.

Ostoya, P. 1951. *Les théories de l'évolution: origines et histoire du transformisme et des idées qui s'y rattachent.* Paris: Payot.

Owen, R. 1837. Description of the cranium of the *Toxodon platensis*. *Proc. Geol. Soc. Lond.* 2: 541–542.

--- 1838a. A description of the cranium of the *Toxodon platensis*, a gigantic extinct mammiferous species, referrible by its dentition to the *Rodentia*, but with affinities to the *Pachydermata* and the *Herbivorous Cetacea*. *Proc. Geol. Soc. Lond.* 2: 541–542.

--- 1838b. Description du crâne du *Toxodon platensis*, grand Mammifère perdu que l'on doit rapporter à l'ordre des Pachydermes, mais qui offre en même temps des affinités avec les Rongeurs, les Edentés et les Cetacés herbivores. *Ann. sci. nat. zool.*, ser. 2, 9: 25–45.

--- 1838c. Description d'une mâchoire inférieure et de dents de *Toxodon* trouvées à Bahia-Blancha, à 39° de latitude sur le côte Est de L'Amérique méridionale. *Ann. sci. nat. zool.* 9: 45–54.

--- 1840–1845. *Odontography; or, a treatise on the comparative anatomy of the teeth, their physical relations, mode of development, and microscopic structure, in the vertebrate animals*. London: H. Baillière.

--- 1846. Report on the archetype and homologies of the vertebrate skeleton. *Rept. 16th Meeting B.A.A.S.*, pp. 169–340. London: John Murray.

--- 1848. *On the archetype and homologies of the vertebrate skeleton*. London: van Voorst.

--- 1849a. *On the nature of limbs*. London: van Voorst.

--- 1849b. *On parthenogenesis, or the successive production of procreating individuals from a single ovum*. London: van Voorst.

--- 1851. Review of "Principles of geology" by Sir Charles Lyell. *Q. rev.* 89: 412–451.

--- 1853a. *Descriptive catalogue of the osteological series contained in the museum of the Royal College of Surgeons of England: vol. II. Mammalia Placentalia*. London: Taylor and Francis.

--- 1853b. Description of some species of the extinct genus *Nesodon*, with remarks on the primary group (Toxodontia) of hoofed quadrupeds. *Phil. Trans. Roy. Soc. Lond.* 143: 291–310.

--- 1855. *Lectures on the comparative anatomy and physiology of the invertebrate animals*. 2d ed. London: Longman, Brown, Green, and Longmans.

--- 1858. Address. In *Rept. 28th meeting B.A.A.S.*, pp. xlix–cx. London: John Murray.

--- 1860. Darwin on the origin of species. *Edin. rev.* 111: 487–532. 1973. Reprinted in D. Hull, *Darwin and his critics*, pp. 175–213. Cambridge: Harvard University Press.

--- 1866–1868. *On the anatomy of vertebrates*. 3 vols. London: Longmans, Green.

Owen, R. S., ed. 1894. *The Life of Richard Owen*. 2 vols. London: John Murray.

Packard, A. S., Jr. 1870. On the embryology of *Limulus polyphemus*. *Proc. Am. Assoc. Adv. Sci.* 18: 247–255.

--- 1872. On the development of *Limulus polyphemus*. *Mem. Boston Soc. Nat. Hist.* 2, part 2: 155–201.

--- 1901. *Lamarck, the founder of evolution: his life and work*. New York: Longmans, Green.

Pagliaini, A. 1903–1928. *Catalogo generale della libreria Italiana dall'anno 1847 a tutto il 1899*. 2 vols. Milano : Associazione Tipografico-Libraria Italiana. *Primo Supplemento. Dal 1900 al 1910*. 2 vols. Milano : Associazione Tipografico-Libraria Italiana. *Secondo Supplemento. Dal 1911 al 1920*. Milano: Associazione Editoriale-Libraria Italiana.

Paley, W. 1816. *Natural theology; or, evidences of the existence and attributes of the Deity, collected from the appearances of nature*. new ed. London: Baynes.

Pancaldi, G. 1973. Spazio e tempo nella teoria darwiniana. *Rivista di filosofia* 64: 3–17.

--- 1975. L' "economia della natura" da Cuvier a Darwin. *Rivista di filosofia* 66: 77–111.

---, ed. 1976. *Evoluzione: biologia e scienze umane*. Bologna: Il Mulino.

--- 1977. *Charles Darwin: "storia" ed "economia" della natura*. Firenze: La Nuova Italia.

--- 1981. Scientific internationalism and the British Association. In *The parliament of science: the British Association for the Advancement of Science*, 1831–1981, eds. R. MacLeod and P. Collins, pp. 145–169. Norwood, Middlesex: Science Reviews.

--- 1982. Charles Darwin e il pensiero sociale. Alcune prospettive storiografiche recenti. *Il pensiero politico* 15: 222–230.

--- 1983. *Darwin in Italia. Impresa scientifica e Frontiere culturali*. Bologna: Il Mulino.

Pantin, C. F. A. 1953. *History of science*. London: Cohen and West.

--- 1958. President's address at the unveiling of the Darwin-Wallace memorial plaque. *Proc. Linn. Soc. Lond.* 170: 219–226.

--- 1968. *The relations between the sciences*. Eds. A. M. Pantin and W. H. Thorpe. Cambridge: Cambridge University Press.

Papert, S. 1980. *Mindstorms: children, computers, and powerful ideas*. New York: Basic Books.

Paradis, J. G. 1978. *T. H. Huxley: man's place in nature*. Lincoln: University of Nebraska Press.

--- **and Postlewait, T.**, eds. 1981. Victorian science and Victorian values: literary perspectives. *Ann. N. Y. Acad. Sci.* vol. 360.

Parkes, S. 1807. *Manufacturing chymist: A chymical catechism or the application of chymistry to the arts for the use of young people, artists, tradesmen and the amusement of leisure hours*. Philadelphia: Humphreys.

--- 1818. *The chemical catechism with notes, illustrations and experiments*. From the 8th London ed. NY: Collins.

Pas, P. van der 1970. The correspondence of Hugo de Vries and Charles Darwin. *Janus* 57: 173–215.

Passmore, J. 1957. *A hundred years of philosophy*. London: Duckworth.

--- 1959. Darwin's impact on British metaphysics. *Victorian stud.* 3: 41–54.

Paul, D. B. 1979. Marxism, Darwinism, and the theory of two sciences. *Marxist perspectives* (spring): 116–143.

Paul, H. W. 1971. Science and the Catholic institutes in nineteenth century France. *Societas. A rev. soc. hist.* 1: 271–285.

--- 1972. The crucifix and the crucible: Catholic scientists in the Third Republic. *Catholic hist. rev.* 58: 195–219.

--- 1974. Religion and Darwinism: varieties of Catholic reaction. In *The comparative reception of Darwinism*, ed. T. F. Glick, pp. 403–436. Austin: University of Texas Press.

--- 1979. *The edge of contingency: French Catholic reaction to scientific change from Darwin to Duhem*. Gainesville: University Presses of Florida.

Pavlov, I. P. 1929. *Leçons sur l'activité du cortex cérébral*. Paris: Legrand.

--- 1957. *Experimental psychology and other essays*. New York: Philosophical Library.

Peach, L. du Garde 1973. *Charles Darwin*. Loughborough: Ladybird Books.

Pearson, K., and Lee, A. 1903. On the laws of inheritance in man: 1. inheritance of physical characters. *Biometrika* 2: 357–462.

Peckham, M. 1959. Darwin and Darwinisticism. *Victorian stud.* 3: 19–40.

--- 1965. *Man's rage for chaos: biology, behavior and the arts*. Philadelphia: Chilton.

Peel, J. D. Y. 1971. *Herbert Spencer: the evolution of a sociologist*. London: Heinemann.

Peirce, C. S. 1877. The fixation of belief. *Pop. sci. monthly* 12: 1–15.

--- 1935. Scientific metaphysics. In *Collected papers of Charles Sanders Peirce*, eds. C. Hartshorne and P. Weiss, vol. 6. Cambridge: Harvard University Press.

Perrier, E. 1884. *La philosophie zoologique avant Darwin*. Paris: Alcan.

--- 1889. *Le transformisme*. Paris: J. B. Baillière.

Persons, S. 1950a. Evolution and theology in America. In *Evolutionary thought in America*, ed. S. Persons, pp. 422–453. New Haven: Yale University Press.

---, ed. 1950b. *Evolutionary thought in America*. New Haven: Yale University Press.

---, ed. 1963. *Social Darwinism: selected essays of William Graham Sumner*. Englewood Cliffs: Prentice Hall.

Peters, D. S. 1980. Das Biogenetische Grundgesetz — Vorgeschichte und Folgerungen. *Medizinhistorische J.* 15: 57–69.

Peters, H. M. 1960. Soziomorphe Modelle in der Biologie. *Ratio* 1: 22–37.

--- 1965. Modellbeispiele aus der Geschichte der Biologie. *Stadium Generale* 18: 298–305.

--- 1972. Historische, soziologische und erkenntniskritische Aspeckte der Lehre Darwins. In *Neue Anthropologie*, eds. H.-G. Gadamer and P. Vogler, vol. 1, pp. 326–352. München: Deutscher Taschenbuch.

Peters, R. H. 1976. Tautology in evolution and ecology. *Am. nat.* 110: 1–12.

Peterson, H. 1932. *Huxley: prophet of science*. New York: Longmans, Green.

Pfeifer, D. E. 1970. Louis Agassiz and the origin of species. In *Studies in philosophy and the history of science: Essays in honor of Max Fisch*, ed. R. Tursman, pp. 87–105. Lawrence: Coronado.

Pfeifer, E. J. 1965. The genesis of American neo-Lamarckism. *Isis* 56: 156–167.

--- 1974. America. In *The comparative reception of Darwinism*, ed. T. F. Glick, pp. 168–206. Austin: University of Texas Press.

Piaget, J. 1974. *Adaptation vitale et psychologie de l'intelligence* (*Selection organique et phénocopie*). Paris: Hermann. 1980. *Adaptation and intelligence: organic selection and phenocopy*. Trans. S. Eames. Chicago: University of Chicago

Press.

Pianciani, G. B. 1860a. Cosmogonia: distruzione e creazione delle specie. *Civiltà Cattolica* 5: 55–76.

--- 1860b. Cosmogonia: della origine delle specie organizzate. *Civiltà Cattolica* 7: 164–179, 272–283.

--- 1862. *Cosmogonia naturale comparata col Genesi.* Roma: Morini.

Pickering, G. 1974. *Creative malady: illness in the lives and minds of Charles Darwin, Florence Nightingale, Mary Baker Eddy, Sigmund Freud, Marcel Proust, Elisabeth Barrett Browning.* London: Allen and Unwin.

Pictet, F.-J. 1860. Sur l'origine des espèces. *Bibliothèque universelle. Archives des sciences physiques et naturelles.* 7: 233–255. 1973. In D. Hull, *Darwin and his critics*, pp. 142–152. Cambridge: Harvard University Press.

Pisarev, D. I. 1862. Ocherk istorii truda (Essay on the history of labor) *Russkoe slovo* Jan. pp. 2–38, Feb., pp. 15–60, March, pp. 20–52, April, pp. 1–130, June, pp. 2–40.

--- 1864. Progress v mire zhivotnykh i rastenii. (Progress in the animal and vegetable world) *Russk slovo*, Apr., pp. 1–52, May, pp. 43–70, June, pp. 233–274, July, pp. 1–46, Sept., pp. 1–40.

Pitoni, R. 1862. *Un tentativo di darwinismo cattolico.* Roma: Crivelli.

Platnick, N. 1978. Evolutionary biology: a Popperian perspective. *Syst. zool.* 27: 137–141.

--- 1980. Philosophy and the transformation of cladistics. *Syst. zool.* 28: 537–546.

Platonov, C. I. 1955. *Klim Arkadevic Timiryazeff.* Moscow: Foreign Languages Publishing House.

Platt, R. 1959. Darwin, Mendel, and Galton. *Med. hist.* 3: 87–99.

Plochmann, G. K. 1959. Darwin or Spencer? Why has Darwin's reputation risen, while that of Herbert Spencer has declined? *Science* 130: 1452–1456.

Popper, K. R. 1959. *The logic of scientific discovery.* London: Hutchinson. New York: Basic Books.

--- 1972. *Objective knowledge: an evolutionary approach.* Oxford: Clarendon Press.

--- 1976. *The unended quest.* Glasgow: Fontana/Collins.

--- 1978. Natural selection and the emergence of mind. *Dialectica* 32: 339–355.

--- **and Eccles, J. C.** 1977. *The self and its brain.* Berlin: Springer.

Porter, D. M. 1980a. Charles Darwin's plant collections from the voyage of the *Beagle. Jour. Soc. Bibl. Nat. Hist.* 9: 515–525.

--- 1980b. The vascular plants of Joseph Dalton Hooker's *An enumeration of the plants of the Galapagos Archipelago; with descriptions of those which are new. Bot. j. Linn. Soc. Lond.* 81: 79–134.

--- 1981. Darwin's missing notebooks come to light, *Nature* 291: 13.

--- 1982. Charles Darwin's notes on plants of the *Beagle* voyage. *Taxon* 31: 503–506.

--- 1983b. More Darwin *Beagle* notes resurface. *Arch. nat. hist.* 11: 315–316.

--- 1983a. Charles Darwin, the Cape Verde Islands and the Herbarium Webb. *Webbia*, 36: 225–228.

Porter, R. 1973a. The history of palaeontology. *Hist. sci.* 11: 130–138.

--- 1973b. The industrial revolution and the rise of the science of geology. In *Changing perspectives in the history of science*, eds. M. Teich and R. Young, pp. 320–343. London: Heinemann.

--- 1977. *The making of geology: earth science in Britain, 1660–1815.* Cambridge: Cambridge University Press.

--- 1978. Gentlemen and geology: the emergence of a scientific career, 1660–1920. *The hist. j.* 21: 809–836.

--- 1982. The descent of genius: Charles Darwin's brilliant career. *Hist. today*, July, pp. 16–22.

Poulton, E. B. 1890. *The colours of animals, their meaning and use, especially considered in the case of insects.* London: K. Paul, Trench, Trübner.

--- 1896. *Charles Darwin and the theory of natural selection.* London: Cassell.

--- 1898. Physiological selection. *Nature* 59: 121–122.

--- 1904. What is a species? *Proc. Roy. Soc. Lond.* 44: lxxvii–cxvi. 1908. Reprinted in *Essays on evolution*, pp. 46–94. Oxford: Clarendon Press.

--- 1908. *Essays on evolution 1889–1907.* Oxford: Clarendon Press.

--- 1909. *Charles Darwin and the origin of species.* London: Longmans.

--- 1924. Alfred Russel Wallace, 1823–1913. *Proc. Roy. Soc. Lond.* 95B: i–xxxv.

Poynter, F. N. L., ed. 1958. *The history and*

philosophy of knowledge of the brain and its functions. Oxford: Oxford University Press.

Prenant, M. 1935. *Biologie et Marxisme*. Paris: Editions Sociales Internationales. 1938. *Biology and Marxism*. New York: International.

--- 1938. *Darwin*. Paris: Éditions Sociales Internationales.

--- 1980. *Toute une vie à gauche*. Paris: Encre.

Prescott, W. H. 1843. *History of the conquest of Mexico with a preliminary view of the ancient Mexican civilization and the life of conqueror Hernando Cortes*. 3 vols. New York: Harper and brothers.

Preti, G. 1955. Materialismo storico e teoria dell'evoluzione. *Rivista di filosofia* 46: 18–47.

Preus, A. 1979. *Eidos* as norm in Aristotle's biology. *Nature and system* 1: 79–101.

Prichard, J. C. 1813. *Researches into the physical history of man*. London: J. & A. Arch.

Provine, W. B. 1971. *The origins of theoretical population genetics*. Chicago: University of Chicago Press.

--- 1978. The role of mathematical population geneticists in the evolutionary synthesis of the 1930s and 1940s. *Stud. hist. biol.* 2: 167–192.

--- 1979. Francis B. Sumner and the evolutionary synthesis. *Stud. hist. biol.* 3: 211–240.

--- 1980. Epilogue. In *The evolutionary synthesis*, eds. E. Mayr and W. B. Provine, pp. 399–411. Cambridge: Harvard University Press.

Punnett, R. C. 1915. *Mimicry in butterflies*. Cambridge: Cambridge University Press.

Pusey, B. 1840. On the present state of the science of agriculture in England. *J. Roy. Agric. Soc. England* I: 1–21.

Quadri, A. 1869. *Note alla teoria darwiniana*. Bologna: Vitali.

Quatrefages, J.-L.-A. de 1860–1861. Histoire naturelle de l'homme. *Rev. des deux mondes* 30: 807–833; 31: 155–175, 412–435, 635–671, 938–969; 32: 145–177, 436–464, 635–671.

--- 1870. *Charles Darwin et ses précurseurs français*. Paris: Baillière.

--- 1892. *Darwin et ses précurseurs français*. 2d ed. Paris: Alcan.

--- 1894. *Les émules de Darwin*. Paris: Alcan.

Querner, H. 1971. Ideologisch-weltanschauliche Konsequenzen der Lehre Darwins. *Studium Generale* 24: 231–245.

--- 1972. Probleme der Biologie um 1900 auf den Versammlungen der Deutschen Naturforscher und Ärzte. In *Wege der Naturforschung 1822–1972 im Spiegel der Versammlungen Deutscher naturforscher und Ärzte*, eds. H. Querner and H. Schipperges, pp. 186–202. Berlin, Heidelberg, New York: Springer.

--- 1973. Darwin, sein Werk und der Darwinismus. In *Biologismus in 19 Jahrhundert*, ed. G. Mann, pp. 10–29. Stuttgart: Ferdinand Enke.

--- 1975. Darwins Deszendenz und Selektionslehre auf den Deutschen Naturforscher-Versammlungen. In *Beiträge zur Geschichte der Naturwissenschaften und der Medizin. Festschrift für Georg Uschmann*, eds. K. Mothes and J. H. Scharf, pp. 439–456. *Acta Historica Leopoldina*, vol. 9.

--- 1978. Karl Ernst von Baer und Thomas Henry Huxley: Unveröffentlichte Briefe aus den Jahren 1860–1868. *Sudhoffs Archiv* 62: 131–147.

Quillian, W. F., Jr. 1950. Evolution and moral theory. In *Evolutionary thought in America*, ed. S. Persons, pp. 398–419. New Haven: Yale University Press.

Rabelais, F. 1533. *Gargantua and Pantagruel*. Trans. T. Urquhart and P. A. Motteux. Chicago.

Rachootin, S. P., and Thomson, K. S. 1981. Epigenetics, paleontology and evolution. *Evolution Today*, eds. G.C.E. Scudder and J. L. Reveal, pp. 181–193. *Proc. 2nd int. cong. syst. and evol. biol.*

Radcliffe-Brown, A. R. 1935. On the concept of function in social science. 1952. Reprinted in *Structure and function in primitive society: essays and addresses*. pp. 178–187. London: Cohen and West.

Radical Science Journal Collective 1981. Science, technology, medicine, and the socialist movement. *Radical sci. j.* 11: 3–70.

Ràdl, E. 1905–1909. *Geschichte der biologischen Theorien*. Leipzig: Englemann.

--- 1930. *The history of biological theories*. Trans. E. J. Hatfield. London: H. Milford, Oxford University Press.

Raikov, B. E. 1951. *Russkie biologi-evolyutsionisty do Darwina. Materialy k istorii evolyutsionnoj idei v Rossii* [Russian evolutionary biologists before Darwin]. Moscow: AN SSSR.

--- ed. 1955. *Ruskii biologi-evoluzionisti do Darwina*, Vol. III, Moscow: Akad. Nauk.

--- 1957. Iz istorii darwinizma v Rossii (From

the history of Darwinism in Russia). *Tr. Inst. Ist. Estes. i Techn*, 31: 17–81.

--- 1969. *Germanskie biologi-evolyutsionisty do Darwina* [German evolutionary biologists before Darwin]. Leningrad: Izdat. Nauka.

Rainger, R. 1981. The continuation of the morphological tradition: American paleontology, 1880–1910. *J. hist. biol.* 14: 129–158.

Randall, J. H., Jr. 1961. The changing impact of Darwin in philosophy. *J. hist. ideas* 22: 435–462.

Rapaics, R. 1952. A darwinizmus Magyarországon [Darwinism in Hungary]. In *Darwin*, pp. 27–46. Budapest: Müvelt Nép.

Raphael, D. D. 1958. Darwinism and ethics. In *A century of Darwin*, ed. S. A. Barnett, pp. 334–359. Cambridge: Harvard University Press.

Rashinsky, S. A. 1863. *Kurs biologii* (Biology lectures, lithograph) University of St. Petersburg.

[Raspe, R. E.] 1786. *Baron Munchausen's narrative of his marvellous travels and campaigns in Russia: humbly dedicated and recommended to country gentlemen; and, if they please, to be repeated as their Own. . . .* Oxford: Printed for the Editor [R. E. Raspe].

Rather, L. J. 1959. *Disease, life and man: essays by Rudolf Virchow*. Stanford: Stanford University Press.

--- 1965. Old and new views of the emotions and bodily changes: Wright and Harvey versus Descartes, James and Cannon. *Clio medica* 1: 1–25.

Raverat, G. 1952. *Period piece: a Cambridge childhood*. London: Faber and Faber.

Reed, E. S. 1981. The lawfulness of natural selection. *Am. nat.* 118: 61–71.

Reid, J. 1837. Notes on several quadrupeds in Mr. Darwin's collection. *Proc. Zool. Soc. Lond.* 5: 4.

Reinheimer, H. 1915. *Darwin: the evil genius of science and his nordic religion*. Surbiton: Grevett.

Rescher, N. 1977. *Methodological pragmatism: a systems-theoretic approach to the theory of knowledge*. Oxford: Blackwell.

Réti, E. 1958a. A darwinizmus utja hazánkban [The career of Darwinism in Hungary]. *Természettudományi közlöny* no. 2: 68–71.

--- 1958b. Darwinizmus és antidarwinizmus hazánkban [Darwinism and anti-Darwinism in Hungary]. *Természettudományi Közlöny* no. 3: 100–104.

--- 1962a. Darwinista humanizmus Apáthy és Lenhossék szemléletében [Darwinian humanism in Apáthy's and Lenhossék's conception]. *Elövilág* no. 6: 51–53.

--- 1962b. Darwinisták és antidarwinisták Magyarországon [Darwinists and anti-Darwinists in Hungary]. *Világosság* nos. 7-8: 61–65.

--- 1964. Magyar darwinista orvosok 1945–ig. [Darwinian physicians in Hungary up to 1945]. *Comm. Bibl. Hist. Med. Hung.* 31: 117–133.

--- A darwinizmus fogalmának fejlödése [The evolution of the notion of Darwinism]. *Elövilág* 2: 45–52.

Revue de Synthèse 1979. *Les néo-Lamarckiens français* [special issue], 100, nos. 95–96.

Reynolds, J. 1778. *Seven discourses delivered in the Royal Academy by the President*. London: T. Cadell.

Richards, O. W., and Robson, G. C. 1926. The species problem and evolution. *Nature* 177: 345, 382.

Richards, R. J. 1977. Lloyd Morgan's theory of instinct: from Darwinism to neo-Darwinism. *J. hist. behav. sci.* 13: 12–32.

--- 1979. Influence of sensationalist tradition on early theories of the evolution of behavior. *J. hist. ideas.* 40: 85–105.

--- 1981. Instinct and intelligence in British natural theology: some contributions to Darwin's theory of the evolution of behavior. *J. hist. biol.* 14: 193–230.

--- 1982. The emergence of evolutionary biology of behavior in the early nineteenth century. *Brit. j. hist. sci.* 15: 241–280.

Richardson, C. 1933. Petrology of the Galapagos Islands. *Bishop Mus. bull.* 110: 45–67.

Richardson, E. W. 1916. *A Veteran Naturalist: Being the Life and Work of W. B. Tegetmeier*. London: Witherby.

Richardson, R. A. 1981. Biogeography and the genesis of Darwin's ideas of transmutation. *J. hist. biol.* 14: 1–41.

Richerand, A. 1802. *Nouveaux élémens de physiologie*. Vol. 1., Paris: Crapart, Caille and Ravier.

Richerson, P., and Boyd, R. 1978. A dual inheritance model of the human evolutionary process I: basic postulates and a simple model. *J. social and biol. structure* 1: 127–154.

Ricks, C. 1976. *Keats and embarrassment*. Oxford: Oxford University Press.

Ricoeur, P. 1976. *Interpretation theory: discourse and the surplus of meaning*. Fort Worth: Texas Christian University Press.

Ridley, M. 1982. Coadaptation and the inadequacy of natural selection. *Brit. j. hist. sci.* 15: 45–68.

Rinard, R. 1981. The problem of the organic individual: Ernst Haeckel and the development of the biogenetic law. *J. hist. biol.* 14: 249–275.

Ritter, W. E. 1954. *Charles Darwin and the golden rule*. Washington: Science Service. New York: Storm Publishers.

Ritvo, L. B. 1965. Darwin as the source of Freud's neo-Lamarckism. *J. Am. Psychiat. Assoc.* 13: 499–517.

--- 1972. Carl Claus as Freud's professor of the new Darwinian biology. *Int. j. of psychoanalysis* 53: 277–283.

--- 1974. The impact of Darwin on Freud. *Psychiat. Q.* 43: 177–192.

Roberts, H. F. 1929. *Plant hybridization before Mendel*. Princeton: Princeton University Press.

Roberts, H. J. 1966. Reflections on Darwin's illness. *J. chronic dis.* 19: 723–725.

Robson, G. C. 1928. The species problem. London: Oliver and Boyd.

--- **and Richards, O. W.** 1936. *The variation of animals in nature*. London: Longmans, Green.

Roe, S. A. 1981. *Matter, life and generation*. Cambridge: Cambridge University Press.

Roger, J. 1963. *Les sciences de la vie dans la pensée française au XVIIIe siècle*. Paris: Vrin.

--- 1972. Les conditions intellectuelles de l'apparition du transformisme. In *Épistémologie et marxisme*, pp. 99–114. Paris: Union Général d'Éditions.

--- 1976. Darwin en France. Essay review of Y. Conry, *L'introduction du darwinisme en France au XIXe siècle*. *Ann. sci.* 33: 481–484.

--- 1979. Presentation. Les néo-Lamarckiens français. *Rev. de synthèse* 100, nos. 95–96: 79–82.

--- 1982. La Filosofia dei neolamarckiani francesi. In *Scienza e filosofia nella cultura positivistica*, ed. A. Santucci, pp. 234–244. Milano: Feltrinelli.

Rogers, J. A. 1960. Darwinism, scientism and nihilism. *Russ. rev.* 19: 10–23.

--- 1963. The Russian populists' response to Darwin. *Slavic rev.* 22: 456–468.

--- 1972. Darwinism and social Darwinism. *J. hist. ideas* 33: 265–280.

--- 1973. The reception of Darwin's *Origin of species* by Russian scientists. *Isis* 64: 484–503.

--- 1974a. Russia: social sciences. In *The comparative reception of Darwinism,* ed. T. F. Glick, pp. 256–268. Austin: University of Texas Press.

--- 1974b. Russian opposition to Darwinism in the nineteenth century. *Isis* 65: 487–505.

Roget, P. M. 1840. *Animal and vegetable physiology considered with reference to natural theology*. 2 vols. 3d ed. London: William Pickering.

Rokitansky, C. 1869. *Die Solidarität alles Thierlebens*. Vienna: k.k. Hof- und Staatsdruckerei.

Romanes, G. J. 1883. *Mental evolution in animals, with a posthumous essay on instinct by Charles Darwin*. London: Kegan Paul, Trench.

--- 1886. Physiological selection; an additional suggestion on the origin of species. *J. Linn. Soc. Lond.* 19: 337–411.

--- 1893. *An examination of Weismannism*. London: Longmans.

--- 1895. *Darwin and after Darwin. Post-Darwinian questions: heredity and utility*. vol. 2. Chicago: Open Court.

--- 1896. *The life and letters of George John Romanes*, ed. E. Romanes. London: Longmans, Green.

Romer, A. S. 1958. Darwin and the fossil record. In *A century of Darwin*, ed. S. A. Barnett, pp. 130–152. Cambridge: Harvard University Press.

Roppen, G. 1956. *Evolution and poetic belief, a study in some Victorian and modern writers*. Oslo: Oslo University Press.

Rosa, D. 1891. La "Zoogenia" di F. C. Marmocchi. *Bollettino dei Musei di Zoologia e di Anatomia Comparata della Reale Università di Torino* 6: no. 95.

--- 1918. *Ologenesi: teoria dell'evoluzione e della distribuzione geografica dei viventi*. Firenze: Bemporad.

Rosenberg, A. 1978. The supervenience of biological concepts. *Phil. sci.* 45: 368–386.

Rossi, P. 1957. *Francesco Bacone, dalla magia alla scienza*. Bari: Laterza. 1968. *Francis Bacon from Magic to Science*. Chicago: University of Chicago Press.

--- 1977a. *Immagini della scienza*. Roma: Editori Riuniti.

---, ed. 1977b. Prefazione. In A. Fogazzaro, *Ascensioni Umane*. Milano: Longanesi.

--- 1979. *I segni del tempo: storia della terra e storia delle nazioni da Hooker a Vico*. Milano: Feltrinelli.

Rossiter, M. W. 1975. *The emergence of agricultural science: Justus Liebig and the Americans, 1840–1880*. New Haven: Yale University Press.

Rostand, J. 1932. *L'evolution des espèces: histoire des idées tranformistes*. Paris: Hachette.

--- 1947. *Charles Darwin*.. Paris: Gallimard.

--- 1960. Les précurseurs français de Charles Darwin. *Rev. hist. sci.* 13: 45–58.

Roughgarden, G. 1979. *Theory of population genetics and evolutionary ecology*. New York: Macmillan.

Rousseau, G. 1969. Lamarck et Darwin. *Bull. Mus. Nat. Hist.* 41: 1029–1041.

Royden, A. M. 1924. A modern prophet: Charles Darwin. In *The friendship of God*, pp. 43–60. London: G. P. Putnam's Sons.

Royle, E. 1974. *Victorian infidels: the origins of the British Secularist movement, 1791–1866*. Manchester: Manchester University Press.

Rubailova, N. G. 1963. Stoletnii iubilei evoliutsionnoi teorii Ch. Darvina. [The 100th anniversary of Charles Darwin's theory of evolution]. *Vop. Ist. Est. Tekh.* 14: 87–91.

--- 1968. The study of Charles Darwin's works in the USSR. *Actes XIe cong. int. hist. sci. (1965)* 5: 93–95.

--- 1971. O reakcii anglijskogo obscestva na teorju Čarlza Darwina [The reaction of English society to Darwin's theory]. In *Naucnoe oktrytie i ego vosprijatie* [Scientific discovery and its reception], eds. S. R. Mikulinskij and M. G. Jarosevskij, pp. 168–177. Moscow: Nauka.

--- 1973. Ch. Darvin i problema proiskhozhdeniia cheloveka. [C. Darwin and the problem of man's origin]. *Iz. Istor. Biol.* 4: 89–111.

--- 1981. (A. A. Jablokov, ed.) *Formirovanie i Razvitie Teorii Estestvennogo Otbora*. (The Origins and Developments of Theories on Natural Selection) Moscow: Nauka.

Rudwick, M. J. S. 1962. The principle of uniformity. A review of R. Hooykaas, *Natural law and divine miracle*. *Hist. sci.* 1: 82–86.

--- 1963. The foundation of the Geological Society of London: its scheme for co-operative research and its struggle for independence. *Brit. j. hist. sci.* 1: 325–355.

--- 1970. The strategy of Lyell's *Principles of geology*. *Isis* 61: 5–33.

--- 1971. Uniformity and progression: reflections on the structure of geological theory in the age of Lyell. In *Perspectives in the history of science and technology*, ed. D. H . D. Roller, pp. 209–227. Norman: University of Oklahoma Press.

--- 1972. *The meaning of fossils: episodes in the history of paleontology*. London: Macdonald [rev. ed. 1976].

--- 1974a. Darwin and Glen Roy: a "great failure" in scientific method? *Stud. hist. phil. sci.* 5: 97–185.

--- 1974b. Poulett Scrope on the volcanoes of Auvergne: Lyellian time and political economy. *Brit. j. hist. sci.* 7: 205–242.

--- 1975a. Charles Lyell, F. R. S. (1797–1875) and his London lectures on geology. 1832–1833. *Notes rec. Roy. Soc. Lond.* 29: 231–263.

--- 1975b. Caricature as a source for the history of science: De la Beche's anti-Lyellian sketches of 1831. *Isis* 66: 534–560.

--- 1976. The emergency of a visual language for geology, 1760–1840. *Hist. sci.* 14: 149–195.

--- 1977. Historical analogies in the geological work of Charles Lyell. *Janus* 64: 89–107.

--- 1978. Charles Lyell's dream of a statistical palaeontology. *Palaeontology* 21: 225–244.

--- 1979a. Transposed concepts from the human sciences in the early work of Charles Lyell. In *Images of the earth. Essays in the history of the environmental sciences*, eds. L. J. Jordanova and R. S. Porter, pp. 67–83. Chalfont St. Giles: British Society for the History of Science.

--- 1979b. The Devonian: a system born from conflict. In *The Devonian system, special papers in palaeontology*, vol. 23. London: The Palaeontological Association.

--- 1982a. Cognitive styles in geology. In *Essays in the sociology of perception*, ed. Mary Douglas, pp. 219–241. London: Routledge and Kegan Paul.

--- 1982b. Charles Darwin in London: the integration of public and private science. *Isis* 73: 186–206.

--- 1985. *The great Devonian controversy. The shaping of scientific knowledge among gentlemanly specialists*. Chicago: University of Chicago

Press.

Ruffié, J. 1976. *De la biologie à la culture*. Paris: Flammarion.

Rul'ye, K. F. 1846. Khod riby protiv techenia vody (Upstream swimming of fishes). 1856. In Aksakov, S. T. *Zapisky ob uchenii ryby* (Notes on ichthyology). Moscow: IZD.Z.

--- 1847a. Rybo-iashteritsa (Ichtiozavar). In *Zhivopisnaia entsiklopedia*, pp. 60–64. Moscow.

--- 1847b. Ptitsy (The birds). In *Zhiropisnaia entsiklopedia*, pp. 154–159. Moscow.

--- 1850. *Obshchaia Zoologia* (General zoology, lithographed lectures of the 2nd free course for students of the department of mathematics and natural sciences); see also *Nauchn. nasled.* 2: 594–640, the Appendix to Raikov, ed. 1955 and Rul'ye 1954.

--- 1954. *Izbrannye biologicheski e proizvedenia* (Selected biological works). Moscow: Isb. biol. prosvid.

Runkle, G. 1961. Marxism and Charles Darwin. *J. politics* 23: 108–126.

Ruse, M. 1969. Confirmation and falsification of theories of evolution. *Scientia* 104: 329–357.

--- 1970. The revolution in biology. *Theoria* 36: 1–22.

--- 1971a. Is the theory of evolution different? I: The central core of the theory. II: The structure of the entire theory. *Scientia* 106: 765–783, 1069–1093.

--- 1971b. Natural selection in the *Origin of species*. *Stud. hist. phil. sci.* 1: 311–351.

--- 1971c. Two biological revolutions. *Dialectica* 25: 17–38.

--- 1973a. *Philosophy of biology*. London: Hutchinson's University Library.

--- 1973b. The debate about models: formal versus material analogy. *Phil. soc. sci.* 3: 63–80.

--- 1973c. The value of analogical models in science. *Dialogue* 12: 246–253.

--- 1974. The Darwin industry. A critical evaluation. *Hist. sci.* 12: 43–58.

--- 1975a. Charles Darwin and artificial selection. *J. hist. ideas* 36: 339–350.

--- 1975b. Charles Darwin's theory of evolution: an analysis. *J. hist. biol.* 8: 219–241.

--- 1975c. Darwin's debt to philosophy: an examination of the influence of the philosophical ideas of John F. W. Herschel and William Whewell on the development of Charles Darwin's theory of evolution. *Stud. hist. phil. sci.* 6: 159–181.

--- 1975d. The relationship between science and religion in Britain, 1830–1870. *Church rev.* 44: 505–522.

--- 1976a. The scientific methodology of William Whewell. *Centaurus* 20: 227–257.

--- 1976b. Charles Lyell and the philosophers of science. *Brit. j. hist. sci.* 9: 121–131.

--- 1978a. Darwin and Herschel. *Stud. hist. phil. sci.* 9: 323–331.

--- 1978b. Rev. of Harré, "Problems of scientific revolutions." *Erkenntnis* 13: 407–416.

--- 1979a. *The Darwinian revolution: science red in tooth and claw*. Chicago: University of Chicago Press.

--- 1979b. *Sociobiology: sense or nonsense?* Dordrecht, Boston, London: Reidel.

--- 1979c. Philosophical factors in the Darwinian revolution. In *Pragmatism and purpose*, eds. L. W. Sumner, J. G. Slater, and F. Wilson, pp. 220–235. Toronto: University of Toronto Press.

--- 1980a. Ought philosophers consider scientific creativity? A Darwinian case-study. In *Scientific discovery: case studies*, ed. T. Nickles, pp. 131–149. Dordrecht, Boston, London: Reidel.

--- 1980b. Charles Darwin and group selection. *Ann. sci.* 37: 615–630.

--- 1981. *Is science sexist? And other problems in the biomedical sciences*. Dordrecht, Boston, London: Reidel.

--- 1982. Social Darwinism: the two sources. *Rivista di filosofia* 22–23: 36–52.

Russell, A. J. 1934. Darwin. In *Their religion*, pp. 273–297. London: Hodder & Stoughton.

Russell, E. J. 1966. *A history of agricultural science in Great Britain, 1620–1954*. London: George Allen and Unwin.

Russell, E. S. 1916. *Form and function: a contribution to the history of animal morphology*. London: John Murray.

Russell-Gebbett, J. 1977. *Henslow of Hitcham: botanist, educationalist and clergyman*. Lavenham, Suffolk: Terence Dalton.

Russett, C. E. 1966 *The concept of equilibrium in American social thought*. New Haven: Yale University Press.

--- 1976. *Darwin in America: the intellectual response 1816–1912*. San Francisco: W. H. Freeman.

Rylands, J. S. 1970. *Bryozoans*. London: Hutchinson.

Sachs, J. von 1890. *History of botany (1530–1860).* Trans. H. E. F. Garnsey. Rev. B. Balfour. Oxford: Clarendon Press.

Sahlins, M. D. 1977. *The use and abuse of biology: an anthropological critique of sociobiology.* London: Tavistock.

--- **and Service, E. R.,** eds. 1960. *Evolution and culture.* Ann Arbor: University of Michigan Press.

Said, E. W. 1975. *Beginnings: intention and method.* New York: Basic Books.

Salvadori, G. 1900. *Herbert Spencer e l'opera sua.* Firenze: Lumachi.

Sandler, I. 1979. Some reflections on the protean nature of the scientific precursor. *Hist. sci.* 17: 170–190.

Sandow, A. 1938. Social factors in the origin of Darwinism. *Q. rev. biol.* 13: 315–326.

Sanford, W. F, Jr. 1965. Dana and Darwinism. *J. hist. ideas* 26: 531–546.

Santucci, A. 1971. Scienza ed evoluzione in Chauncey Wright. *Rivista di filosofia* 62: 3–43.

--- 1982. Darwin, Spencer e la filosofia sociale americana nella seconda metà dell'Ottocento. *Il pensiero politico* 15: 48–73.

Savile, B. W. 1882a. The late Mr. Darwin [correspondence]. *The record* (supp.) N.S. 1: 149.

--- 1882b. The late Mr. Darwin [correspondence]. *The record* (supp.) N.S. 1: 177.

Savorelli, A. 1974. Un frammento inedito di Bertrando Spaventa su Vico e Darwin. *Bollettino del Centro Studi Vichiani* 4: 171–175.

--- 1977. Da Darwin a Vaihinger: scienza e filosofia nell'ultimo Spaventa. *Atti dell'Accademia di Scienze Morali e Politiche* 88: 57–80.

Scheele, I. 1981. *Von Lüben bis Schmeil: die Entwicklung von der Schulnaturgeschichte zum Biologieunterricht zwischen 1830 und 1833.* (West-) Berlin: Dietrich Reimer.

Scheick, W. J. 1973. Epic traces in Darwin's *Origin of species. S. Atlant. Q.* 72: 270–279.

Schierbeek, A. 1943. De pangenesistheorie van Darwin. *Bijdr. Gesch. Geneesk.* 23: 29–35.

--- 1958. *Darwin's werk en persoonlijkheid.* Amsterdam: Wereldbiblioteek.

Schiller, F. 1979. *Paul Broca: founder of anthropology, explorer of the brain.* Berkeley: University of California Press.

Schiller, J. 1965. Claude Bernard et Darwin. *Physis* 7: 476–488.

--- 1978. *La notion d'organisation dans l'histoire de la biologie.* Paris: Maloine.

--- 1980. *Physiology and classification: historical relations.* Paris: Maloine.

Schilpp, P. A. ed. 1974. *The philosophy of Karl Popper.* 2 vols. La Salle: Open Court.

Schindewolf, O. H. 1950. *Grundfragen der Paläontologie.* Stuttgart: Schweizerbart.

Schmalhausen, I. I. 1946. *Faktory evoliutsii; teoriia stabilitsiruiushchego otbora* (Factors of evolution; the theory of stabilizing selection). Moscow: Akad. Nauk.

--- 1949. *Factors of evolution: the theory of stabilizing selection.* Philadelphia: Blakiston.

Schmidt, A. 1962. *Der Begriff der Natur in der Lehre von Marx.* Frankfurt a.M.: Europäische Verlagsanstalt.

Schmidt, G. 1974. *Die literarische Rezeption des Darwinismus: das Problem der Vererbung bei Emile Zola und im Drama des deutschen Naturalismus.* Berlin: Akademie-Verlag.

Schmidt, H. 1926. *Ernst Haeckel: Leben und Werke.* Berlin: Deutsche Buchgemeinschaft.

--- 1960. Darwins Erbe in der Paläontologie. In *Hundert Jahre Evolutionsforschung,* eds. G. Heberer and F. Schwanitz, pp. 234–276. Stuttgart: Gustav Fischer.

Schneer, C. J. 1968. Science and history. In *The concept of order,* ed. P. G. Kuntz, pp. 122–149. Seattle: University of Washington Press.

---, ed. 1969. *Toward a history of geology.* Cambridge: M.I.T. Press.

Schneider, G. 1951. *Die Evolutionstheorie, das Grundproblem der modernen biologie.* 2d ed. Berlin: Deutscher Bauernverlag.

Schoenherr, C. J. 1849. *Genera et species curculionidum. Catalogues.* Paris: auctor.

Schofield, R. E. 1963. *The Lunar Society of Birmingham: a social history of provincial science and industry in 18th century England.* Oxford: Clarendon Press.

Schon, D. 1967. *Invention and the evolution of ideas.* London.

Schopf, T. J. M., and Bassett, E. L. 1973. F. A. Smitt (1839–1904), marine Bryozoa, and the introduction of Darwin into Sweden. *Trans. Am. Phil. Soc.* 63: 1–30.

Schultze, M. J. S. 1861. Ueber Muskelkörperchen und das, was man eine Zelle zu nennen habe. *Arch. für Anatomie, Physiologie und wissenschaftliche Medicin*: 1–27.

--- 1865a. Prospectus. *Arch. für mikroskopische Anatomie.* 1:v–vi.

--- 1865b. Ein heizbarer Objecttisch und seine

Verwendung bei untersuchungen des Blutes. *Arch. für mikroskopische Anatomie.* 1: 1–41.

Schungel, W. 1980. *Alexander Tille (1866–1912): Leben und Ideen eines Sozialdarwinisten*. Husum: Matthiesen.

Schwanitz, F. 1960. Darwin und die Evolution der Kulturpflanzen. In *Hundert Jahren Evolutionsforschung*, eds. G. Heberer and F. Schwanitz, pp. 123–148. Stuttgart: Gustav Fischer.

Schwartz, J. S. 1974. Charles Darwin's debt to Malthus and Edward Blyth. *J. hist. biol.* 7: 301–318.

--- 1980. Three unpublished letters to Charles Darwin: the solution to a geometrico-geological problem. *Ann. sci.* 37: 631–637.

Schweber, S. S. 1977. The origin of the *Origin* revisited. *J. hist. biol.* 10: 229–316.

--- 1978. The genesis of natural selection — 1838: some further insights. *BioScience* 28: 321–326.

---1979. The young Darwin. *J. hist. biol.* 12: 175–192.

--- 1980. Darwin and the political economists: divergence of character. *J. hist. biol.* 13: 195–289.

--- 1982. Demons, angels, and probability: some aspects of British science in the nineteenth century, unpub. mimeo, Brandeis Univ.

--- 1983. Ideological and intellectual factors in the genesis of the theory of natural selection. In *Ideologie et la Revolution Darwiniere*, ed. Y. Conry. Paris: Vrin.

Schweigger, A. 1826. Observations on the anatomy of the *Corallina opuntia*, and some other species of corallines. *Edin. new phil. j.* 1: 220–224.

Scoon, R. 1950. The rise and impact of evolutionary ideas. In *Evolutionary thought in America*, ed. S. Persons, pp. 4–42. New Haven: Yale University Press.

Scott, G. H. 1963. Uniformitarianism, the uniformity of nature, and paleoecology. *New Zealand j. geol. geoph.* 6: 510–527.

Scott, J. 1864. Observations on the functions and structure of the reproductive organs in the *Primulaceae*. *J. Proc. Linn. Soc. Lond.* (Bot.) 8: 78–126.

Scott, W. 1814. *Waverley; or tis sixty years since.* 3 vols. Edinburgh: Archibald Constable.

Scott, W. B. 1929. *A history of land mammals in the Western hemisphere*. New York: Macmillan.

Scriven, M. 1959. Explanation and prediction in evolutionary theory. *Science* 130: 477–482.

Scudo, F. M. 1976a. The role of phenocopy in evolution. *Atti A.G.I.*, 21: 196–201.

--- 1976b. "Imprinting", speciation and avoidance of inbreeding. In *Evolutionary Biology*, eds. V. J. A. Novâk and B. Palctová. Czech. Biol. Soc., Praha.

---, ed. *Biological associations, ecology and evolution.* Unpublished manuscript.

--- **and Ziegler, J. T.** 1976. Vladimir Aleksandrovich Kostitzin and theoretical ecology. *Theor. pop. biol.* 10: 395–412.

Scull, A. T. 1979. *Museums of madness: the social organisation of insanity in 19th century England.* London: Allen Lane.

Sears, P. B. 1950. *Charles Darwin: the naturalist as a cultural force*. London: Scribner.

Sebright, J. 1836. *Observations upon the instinct of animals*. London: Gossling & Egley.

Sechenov, M. I. 1892. Cited in N. Z. Tscholodkovskii. Biologceski ocerki, p. 355. In *Foprosi Biologii*. E. N. Pavlovskii. Petrograd.

Secord, J. A. 1981a. *Cambria/Siluria: the anatomy of a Victorian geological debate*. Ph.D. thesis. Princeton University.

--- 1981b. Nature's fancy; Charles Darwin and the breeding of pigeons. *Isis* 72: 163–186.

--- 1985. Natural history in depth. *Soc. stud. sci.* 15: 181-200.

--- 1986. *Controversy in Victorian geology: the Cambrian — Silurian dispute*. Princeton: Princeton University Press.

Sedgwick, A. 1845. Review of "Vestiges of the natural history of creation." *Edin. rev.* 82: 1–45.

--- 1850. *Discourse on the studies of the University of Cambridge*. 5th ed. London: J. W. Parker.

Seidler, E. 1969. Evolutionismus in Frankreich. *Sudhoffs Arch.* 53: 362–377.

Seidlitz, G. 1871. *Die Darwin'sche Theorie.* Dorpat: C. Mattiesen.

Selsam, H. 1959. Charles Darwin and Karl Marx. *Mainstream* 12: 23–36.

Semmel, B. 1958. Karl Pearson: socialist and Darwinist. *Brit. j. sociol.* 9: 111–125.

--- 1960. *Imperialism and social reform. English social-imperial thought 1895–1914*. London: Allen & Unwin.

Sercjeski, M. H. 1970. Charles Darwin's views on history. *Scientia* 105: 757–761.

Sergi, G. 1914. *L'evoluzione organica e le origini umane: induzioni paleontologiche*. Torino: Bocca.

Serres, M. 1975. *Feux et Signaux de brume : Zola.* Paris.

Severtsov, A. N. 1889. Neskol'ko slov ob aklimatizazii zhivotnykh (A few words on animal acclimatization) *Bull. obsest. ispitat. prid.* 22: 13–24.

Seward, A. C., ed. 1909. *Darwin and modern science.* Cambridge: Cambridge University Press.

Shapin, S. 1974. The audience for science in eighteenth century Edinburgh. *Hist. sci.* 12: 95–121.

--- 1975. Phrenological knowledge and the social structure of early nineteenth-century Edinburgh. *Ann. sci.* 32: 219–243.

--- 1982. History of science and its sociological reconstructions. *Hist. sci.* 20: 157-211.

--- and Barnes, B. 1979. Darwinism and social Darwinism: purity and history. In *Natural order: historical studies of scientific culture,* eds. B. Barnes and S. Shapin, pp. 125–142. Beverly Hills and London : Sage Publications.

Sharlin, H. I. 1976. Herbert Spencer and scientism. *Ann. sci.* 33: 457–465.

Shaw, A. 1839. *Narrative of the discoveries of Sir Charles Bell in the nervous system.* London: Longmans.

--- 1860. *An account of Sir Charles Bell's discoveries in the nervous system.* London: Longmans.

--- 1868. *Reprint of Idea of a new anatomy of the brain, together with extracts of letters showing the origin and progress of his discoveries in the nervous system.* London: Macmillan.

Sheets-Pyenson, S. 1981. Darwin's data: his reading of natural history journals, 1837–1842. *J. hist. biol.* 14: 231–248.

Sheppard, P. M. 1958. *Natural selection and heredity.* London: Hutchinson.

Shibles, W. A. 1971. *Metaphor: an annotated bibliography and history.* Whitewater, Wis: Language Press.

Shimao, E. 1981. Darwinism in Japan, 1877–1927. *Ann. sci.* 38: 93–102.

Shipanov, I. I. 1974. *Istoria filosofii narodov S.S.S.R.* (History of philosophy in the U.S.S.R.). Vol. 3, Moscow: Nauka.

Shiurovsky, G. E. 1834. *Organologia zhivotnikh.* University of St. Petersburg.

Shryock, R. H. 1944. The strange case of Wells' theory of natural selection (1813): some comments on the dissemination of scientific ideas. In *Studies and essays offered to George Sarton,* ed. A. Montagu, pp. 195–207. New York: Schuman.

Siciliani, P. 1885 *La nuova biologia: saggio storico-critico in servigio delle scienze antropologiche.* Milano : Dumolard.

Siddons, H. 1822. *Practical illustrations of rhetorical gesture and action; adapted to the English drama from the work on the subject by M. Engel.* London.

Sillard, J.-C. 1979. Quatrefages et le transformisme. *Revue de synthèse* 100, nos. 95–96: 283–295.

Simonsson, T. 1958. *Face to face with Darwin: a critical analysis of the Christian front in Swedish discussion of the later nineteenth century.* Lund: CWK Glerup.

Simpson, G. E. 1959. Darwin and social Darwinism. *Antioch rev.* 19: 33–45.

Simpson, G. G. 1944. *Tempo and mode in evolution.* New York: Columbia University Press.

--- 1949. *The meaning of evolution.* New Haven: Yale University Press.

--- 1951. The species concept. *Evolution* 5: 285–298.

--- 1953. *The major features of evolution.* New York: Columbia University Press.

--- 1959. Anatomy and morphology: classification and evolution 1859 and 1959. *Proc. Am. Phil. Soc.* 103: 286–306.

--- 1960. The world into which Darwin led us. *Science* 131: 966–974.

--- 1961a. *Principles of animal taxonomy.* New York: Columbia University Press.

--- 1961b. Lamarck, Darwin and Butler: three approaches to evolution. *Am. Scholar* 30: 238–249.

--- 1964. *This view of life.* New York: Harcourt, Brace and World.

--- 1967. *The meaning of evolution.* Rev. ed. New Haven: Yale University Press.

--- 1980. *Splendid isolation.* New Haven: Yale University Press.

Singer, C. 1959. *A short history of scientific ideas to 1900.* Oxford: Oxford University Press.

Sirks, M. S., and Zirkle, C. 1964. *The evolution of biology.* New York: Ronald.

Skultans, V. 1975. *Madness and morals: ideas on insanity in the 19th century.* London: Routledge and Kegan Paul.

--- 1977. Bodily madness and the spread of the blush. In *The anthropology of the body,* ed. J. Blacking, pp. 145–160. London: Academic Press.

Sloan, P. 1979. Buffon, German biology, and the historical interpretation of species. *Brit. j. hist. sci.* 12: 109–153.

--- forthcoming. Darwin, vital matter and the unity of nature.

Smart, J. J. C. 1963. *Philosophy and scientific realism*. London: Routledge and Kegan Paul.

Smit, P. 1974. *History of the life sciences: an annotated bibliography*. Amsterdam: A. Asher.

Smith, A. 1937. *An inquiry into the nature and causes of the wealth of nations*. Ed. E. Cannan. New York: Modern Library.

Smith, C. U. M. 1976. *The problem of life: an essay in the origin of biological thought*. London: Macmillan.

--- 1978. Charles Darwin, the origin of consciousness, and panpsychism. *J. hist. biol.*. 11: 245–267.

Smith, J. M. 1958. *The theory of evolution*. Harmondsworth: Pelican Books.

Smith, R. 1970. *Physiological psychology and the philosophy of nature in mid-nineteenth century Britain*. Ph. D. thesis. University of Cambridge.

--- 1972. Alfred Russel Wallace: philosophy of nature and man. *Brit. j. hist. sci.* 6: 177–199.

--- 1973. The background of physiological psychology in natural philosophy. *Hist. sci.* 11: 75–123.

--- 1974. Essay review of D. Hull, *Darwin and his critics*. *Brit. j. hist. sci.* 7: 278–285.

--- 1977. The human significance of biology: Carpenter, Darwin, and the *vera causa*. In *Nature and Victorian Imagination*, eds. U. C. Knoepfmacher and G. B. Tennyson, pp. 216–230. Berkeley: University of California Press.

Smith, S. 1959a. Evolution: two books and some Darwin marginalia. *Victorian stud.* 3: 109–114.

--- 1959b. The origin of the *Origin*. *Impulse* 11: 2–4.

--- 1960. The origin of the *Origin*, as discerned from Charles Darwin's notebooks and his annotations in the books he read between 1837 and 1842. *Advancement sci.* 16: 391–401.

--- 1965. The Darwin collection at Cambridge, with one example of its use: Charles Darwin and Cirripedes. *Actes du XIe cong. int. d'hist. sci.* 5; 96–100.

Sneath, P. H. A., and Sokal, R. R. 1973. *Numerical taxonomy*. San Francisco: Freeman.

Sober, E. 1980. Evolution, Population Thinking, and Essentialism. *Phil. sci.* 47: 350–383.

--- 1982a. Dispositions and subjunctive conditionals; or, dormative virtues are no laughing matter. *Phil. rev.* 91: 591–596.

--- 1982b. Holism, individualism, and the units of selection. In *PSA 1980*, eds. R. Giere and P. Asquith, vol. 2, pp. 93–121. E. Lansing: Philosophy of Science Association.

---, ed. 1983. *Conceptual issues in evolutionary biology: an anthology*. Cambridge: Bradford/ M.I.T. Press.

--- **and Lewontin, R. C.** 1982. Artifact, Cause and Genetic Selection, *Phil. sci.* 49: 157–180.

Sobol, S. L. 1945. Piervye reaktsii na teorii Darwina v Russkoi pechati (First reactions to Darwin's theories in the Russian press) *Bull. Mosc. Obsc. Ispitat. Priod.*

--- 1960. Recent literature on the life and work of Charles Darwin. [In Russian]. *Vop. Ist. Est. Tekh.* 9: 171–175.

Somenzi, V. 1981. Scientific discovery from the viewpoint of evolutionary epistemology. In *On scientific discovery*, eds. M. D. Grmek, R. S. Cohen, and G. Cimino, pp. 167–177. Dordrecht, Boston: Reidel.

Somkin, F. 1962. The contributions of Sir John Lubbock, F.R.S., to the *Origin of species*: some annotations to Darwin. *Not. rec. Roy. Soc. Lond.* 17: 183–191.

South American Missionary Society. 1882. *The record* N.S. 1: 134–135.

Sowerby, G. B. 1844. Description of fossil shells. In C. Darwin, *Geological observations on the volcanic islands visited during the voyage of H.M.S. Beagle, together with some brief notices of the geology of Australia and the Cape of Good Hope*, pp. 153–160. London: Smith Elder.

--- 1846. Descriptions of Tertiary fossil shells from South America. In C. Darwin, *Geological observations on South America*, pp. 249–264. London: Smith Elder.

Spencer, H. 1851. *Social statics: or, the conditions essential to human happiness specified, and the first of them developed*. London: Chapman.

--- 1860. The social organism. In Essays: scientific, political, and speculative, vol. 1. London: Williams and Norgate.

--- 1864–1867. *Principles of biology*. 2 vols. London: Williams and Norgate.

--- 1868. *Essays: scientific, political, and speculative*. 2 vols. London: Williams and Norgate.

--- 1875. *Grundlagen der Philosophie*. Autorisierte deutsche Ausgabe. Nach der vierten englischen Auflage übersetzt von B. Vetter. Stuttgart: E. Schweizerbart.

--- 1876, 1877. *Die Principien der Biologie*. Autorisierte deutsche Ausgabe. Nach der zweiten englischen Auflage übersetzt von B. Vetter. 2 vols. Stuttgart: E. Schweizerbart (E. Koch).

--- 1877–1897. *Die Principien der Sociologie*. Autorisierte deutsche Ausgabe von B. Vetter [fortgesetzt von Prof. J. V. Carus]. 4 vols. Stuttgart: E. Schweizerbart.

--- 1879–1895. *Die Principien der Ethik . . . Aus dem Englischen von B. Vetter, fortgesetzt von J. V. Carus*. 2 vols. Stuttgart: E. Schweizerbart.

--- 1882–1886. *Die Principien der Psychologie*. 2 vols. Stuttgart: E. Schweizerbart.

--- 1887. *The factors of organic evolution*. London: Williams and Norgate.

--- 1893. The inadequacy of natural selection. *Contemporary rev*. 63: 153–166, 439–456.

--- 1901. Progress: its law and cause. In *Essays: scientific, political, and speculative*, vol. 1, pp. 8–62. London: Williams and Norgate.

Spencer, T. J. B. 1959. *From Gibbon to Darwin*. Birmingham: University Press.

Spengel, J. W. 1872. *Die Darwinsche theorie: Verzeichnis der über dieselbe in Deutschland, England, Amerika, Frankreich, Italien, Holland, Belgien und den Skandinavien Reichen erschienen schriften und aufsätze*. 2d ed. Berlin: Wiegandt und Hempel.

Spengler, J. J. 1950. Evolutionism in American economics. In *Evolutionary thought in America*, ed. S. Persons, pp. 202–266. New Haven: Yale University Press.

Spiller, G. 1914. *Darwinism and sociology*. London: Sherratt and Hughes.

Spilsbury, R. 1974. *Providence lost: a critique of Darwinism*. London: Oxford University Press.

Spurzheim, J. G. 1815. *The physiognomical system of Drs. Gall and Spurzheim; founded on an anatomical and physiological examination of the nervous system in general and of the brain in particular; and indicating the dispositions and manifestations of the mind*. London: Baldwin.

Stanbury, D. ed. 1977. *A narrative of the voyage of H.M.S.* Beagle. London: Folio Society.

Stanley, O. 1957. Thomas Henry Huxley's treatment of nature. *J. hist. ideas* 18: 120–127.

Stanley, S. M. 1979. *Macroevolution: pattern and process*. San Francisco: Freeman.

--- 1981. *The new evolutionary time table*. New York: Basic Books.

Stannard, D. E. 1980. *Shrinking history: On Freud and the failure of psychohistory*. New York: Oxford University Press.

Stark, W. 1961. Natural and social selection. In *Darwinism and the study of society: a centennial symposium*, ed. M. Banton, pp. 49–61. London: Tavistock; Chicago: Quadrangle Books.

Stauffer, R. C. 1957. Haeckel, Darwin and ecology. *Q. rev. biol.* 32: 138–144.

--- 1959. *On the Origin of Species*: an unpublished version. *Science* 130: 1449–1452.

--- 1960. Ecology in the long manuscript version of Darwin's *Origin of species* and Linnaeus's *Economy of nature*. *Proc. Am. Phil. Soc.* 104: 235–241.

---, ed. 1975. *Charles Darwin's* Natural Selection, *being the second part of his big species book written from 1856 to 1858*. London and New York: Cambridge University Press.

Stearn, W. T. 1981. *The Natural History Museum at South Kensington: a history of the British Museum (Natural History) 1753–1980*. London: Heinemann.

Stebbins, G. L. 1950. *Variation and evolution in plants*. New York: Columbia University Press.

--- 1977. In defence of evolution: tautology or theory? *Am. nat.* 111: 386–390.

--- **and Ayala, F. J.** 1981. Is a new evolutionary synthesis necessary? *Science* 213: 967–971.

Stebbins, R. E. 1965. *French reactions to Darwin. 1859–1882*. Ph.D. thesis. University of Minnesota.

--- 1974a. France. In *The comparative reception of Darwinism*, ed. T. F. Glick, pp. 117–163. Austin: University of Texas Press.

--- 1974b. France: bibliographical essay. In *The comparative reception of Darwinism*, ed. T. F. Glick, pp. 164–167. Austin: University of Texas Press.

Stecher, R. M. 1961. The Darwin-Innes letters: the correspondence of an evolutionist with his vicar, 1848–1884. *Ann. sci.* 17: 201–258.

--- 1969. The Darwin-Bates letters — Correspondence between two nineteenth-century travellers and naturalists: parts I and II. *Ann. sci.* 25: 1–47, 95–125.

Steele, E. J. 1981. *Somatic selection and adaptive evolution.* 2d ed. Chicago: University of Chicago Press.

Steenstrup, J. J. S. 1845. *On the alternation of generations; or, the propagation and development of animals through alternate generations.* Trans. G. Busk. London: Ray Society.

Stephens, J. F. 1829. *Illustrations of British entomology; or, a synopsis of indigenous insects* . . . Vol. 2. London: Baldwin and Cradock.

Stern, B. J. 1941a. Recent literature of race and culture contacts. *Science and society* 5: 173–188.

--- 1941b. Reply to J. B. S. Haldane. *Science and society* 5: 374–375.

--- 1942. Reply to M. F. A. Montagu. *Science and society* 6: 75–78.

Steuart, J. 1966. *An inquiry into the principles of political economy.* Ed. with an introduction by A. Skinner, 2 vols. Chicago: University of Chicago Press.

Stevenson, L. 1932. *Darwin among the poets.* Chicago: University of Chicago Press.

Stewart, D. 1855–1856. *The collected works of Dugald Stewart.* Ed. W. Hamilton 11 vols. Vols. 8–9: *Lectures on political economy.* Edinburgh: Thomas Constable.

Stewart, T. D. 1959. The effect of Darwin's theory of evolution on physical anthropology. In *Evolution and anthropology*, ed. B. J. Meggers, pp. 11–25. Washington: Anthropological Society of Washington.

Stocking, G. W., Jr. 1962. Lamarckianism in American social science: 1890–1915. *J. hist. ideas* 23: 239–256.

--- 1968. *Race, culture and evolution: essays in the history of anthropology.* New York: Free Press.

Stoddart, D. R. 1976. Darwin, Lyell, and the geological significance of coral reefs. *Brit. j. hist. sci.* 11: 199–218.

Stone, I. 1980. *The "Origin": a biographical novel of Charles Darwin.* London: Cassell.

Stone, P. J. et al., eds. 1966. *The general inquirer: A computer approach to content analysis.* Cambridge: M.I.T. Press.

Stoppani, A. 1871–1873. *Corso di Geologia.* 3 vols. Milano : G. Bernardini and G. Brigola.

--- 1884. *Il dogma e le scienze positive; ossia, La missione apologetica del clero nel moderno conflitto tra la regione e la fede.* [2d ed. 1886] Milano: Dumolard.

Street, B. V. 1975. *The savage in literature: representatives of "primitive" society in English fiction, 1858–1920.* London: Routledge and Kegan Paul.

Strel'tsenko, V. I. 1981. Behavior as a factor of evolution in the Wagner's Concept. In *Evolution and environment*, eds. V. J. A. Novák and J. Milkovsky. Czech. Acad. of Sciences, Praha.

Stresemann, E. 1975. *Ornithology from Aristotle to the present.* Trans. H. J. and C. Epstein. Ed. G. W. Cottrell. With a foreword and an epilogue on American ornithology by E. Mayr. Cambridge: Harvard University Press.

Stubbe, H. 1965. *Kurze Geschichte der Genetik bis zur Wiederentdeckung der Vererbungsregeln Gregor Mendels.* 2nd ed. Jena: Gustav Fischer. 1972. *History of genetics, from prehistoric times to the rediscovery of Mendel's laws.* Trans. T. R. W. Water. Cambridge: M.I.T. Press.

Sturtevant, A. H. 1965. *A history of genetics.* New York: Harper and Row.

Sulivan, H. N. ed. 1896. *Life and letters of the late Admiral Sir Bartholomew James Sulivan, K.C.B. 1810–1890.* London: John Murray.

Sulloway, F. J. 1969. Charles Darwin and the voyage of the *Beagle*. Senior honors thesis. Harvard University.

--- 1979a. *Freud, biologist of the mind: beyond the psychoanalytic legend.* New York: Basic Books.

--- 1979b. Geographic isolation in Darwin's thinking: the vicissitudes of a crucial idea. *Stud. hist. biol.* 3: 23–65.

--- 1982a. Darwin and his finches: the evolution of a legend. *J. hist. biol.* 15: 1–53.

--- 1982b. The *Beagle* collections of Darwin's finches (Geospizinae). *Bull. Brit. Mus. (Nat. Hist.) Zool. ser.*, 43, no. 2.

--- 1982c. Darwin's conversion: the *Beagle* voyage and its aftermath. *J. hist. biol.* 15: 327–398.

--- 1983. Further remarks on Darwin's voyage spelling habits. *J. hist. biol.* 16: 361–390.

--- 1984. Darwin and the Galapagos. *Biol. J. Linn. Soc.* 21: 29–59; and in *Evolution in the Galapagos*, ed. R. J. Berry. London: Academic Press.

Sumner, F. B. 1932. Genetic, distributional, and evolutionary studies of the subspecies of deer-mice (*Peromyscus*). *Bibliographia genetica* 9: 1–106.

Sumner, J. B. 1824. *The evidence of Christianity derived from its nature and reception.* London: Hatchford.

Suppe, F. ed. 1979. *The structure of scientific theories*. 2d ed. Urbana: University of Illinois Press.

"Surrey Highlander". 1862. Do bees vary in different parts of Great Britain? *J. horticulture and cottage gardener*. N.S. 3: 242.

Swisher, C. 1967. Charles Darwin on the origins of behavior. *Bull. hist. med.* 41: 24–43.

Symonds, J. A. 1893. Darwin's thoughts about God. In *Essays speculative and suggestive*, pp. 425–428. New ed. London: Chapman and Hall.

Szacki, J. 1979. *History of sociological thought*. London: Aldwych Press.

Szarsky, H. 1968. The concept of progress in evolution. *Scientia* 103: 152–160.

Székely, S. 1959. Hermann Ottó és a darwinizmus [Otto Hermann and Darwinism]. *Élővilág* no. 1: 44–45.

Tagliagambe, S. 1983. Darwin in Russia and Galileo's trial. In *Scientia* 118: 247–265.

Tait, P. 1869. Geological time. *North Brit. rev.* 50: 215–233.

Tax, S. ed. 1960. *Evolution after Darwin*. 3 vols. Chicago: University of Chicago Press.

Taylor, G. R. 1963. *The science of life*. London: Thames and Hudson.

Tee, G. J. 1979. Another link between Marx and Darwin. *Ann. sci.* 36: 176.

Teich, M., and Young, R. M., eds. 1973. *Changing perspectives in the history of science*. London: Heinemann.

Teidman, S. J. 1963. Darwin's reverend friend. *Modern churchman* N.S. 6: 286–290.

Teissier, G. 1961. Transformisme d'aujourd-'hui. In *Deuxième des conférences consacrées a l'évolution*. Stat. Biol. Roscoff (also 1962, *Ann. Biol*, (4e serie), *1:* 359–375).

Temkin, O. 1959. The idea of descent in post-Romantic German biology: 1848–1858. In *Forerunners of Darwin*, eds. B. Glass, O. Temkin and W. L. Straus, Jr., pp. 323–355. Baltimore: Johns Hopkins University Press.

Templeton, A. R. 1980. The theory of speciation *via* the founder principle. *Genetics* 94: 1011–1038.

--- 1981. Mechanisms of speciation — a population genetic approach. *Ann. rev. ecol. syst.* 12: 23–48.

Terray, E. 1969. *Le marxisme devant les sociétés "primitives": deux études*. Paris: Maspéro.

Tetry, A. 1974. Deux grandes biologistes français: A. Giard et L. Cuénot. *Actes XIIIe cong. int. d'hist. sci.* (Moscou 1971) 9: 77–81. Moscow: Nauka.

Thackray, A. 1974. Natural knowledge in cultural context: the Manchester model. *Am. hist. rev.* 79: 672–709.

Thackray, J. C. 1977. T. T. Lewis and Murchison's Silurian system. *Trans. Woolhope Naturalists' Field Club* 42: 186–193.

Thagard, P. R. 1977. Darwin and Whewell. *Stud. hist. phil. sci.* 8: 353–356.

Théodoridès, J. 1976. La correspondance scientifique d'Henri de Lacaze-Duthiers (1821–1901). *Rev. de Synthèse* 81–82: 147–148.

Thompson, E. P. 1978. *The poverty of theory and other essays*. London: Merlin.

Thompson, F. M. L. 1963. *English landed society in the nineteenth century*. London: Routledge and Kegan Paul.

Thompson, H. S. 1864. Agricultural progress and the Royal Agricultural Society. *J. Roy. Agric. Soc. England* 25: 1–52.

Thompson, W. P., ed. 1960. *Evolution: its science and doctrine*. Toronto: University of Toronto Press.

Thomson, D. 1977. Social and political thought. In *The new Cambridge modern history*, vol. 9: *Material progress and world-wide problems*, ed. F. H. Hinsley., pp. 101–120. Cambridge University Press.

Thomson, J. A. 1909. Darwin's predecessors. In *Darwin and modern science*, ed. A. C. Seward. pp. 3–17. Cambridge: Cambridge University Press.

--- 1912. *Heredity*. 2d ed. London: John Murray.

Thomson, K. S. 1979. Why was Charles Darwin on board the HMS *Beagle*? *Discovery* (Peabody Museum) 14: 2–10.

Thomson, R. 1968. *The pelican history of psychology*. Harmondsworth: Penguin Books.

Thomson, T. 1802. *A system of chemistry*. 4 vols. Edinburgh: Bell and Bradfute.

--- 1807. *A system of chemistry*. 5 vols. 3rd ed. Edinburgh: Bell and Bradfute.

--- 1810. *A system of chemistry*. 5 vols. 4th ed. Edinburgh: Bell and Bradfute.

--- 1818. *A system of chemistry*. 4 vols. 5th London ed. with Notes by T. Cooper. Philadelphia: Abraham Small.

--- 1825. *An attempt to establish the first principles of chemistry*. 2 volumes. London: Baldwin, Gradock and Joy.

--- 1838. *Chemistry of organic bodies: vegetables*. London: J. B. Baillière.

Thorpe, W. H. 1979. *The origin and rise of ethology*. London: Heinemann.

Thuillier, P. 1974. Un anti-évolutionniste exemplaire: Louis Agassiz. Il y a cent ans mourait Louis Agassiz. *Rev. Questions Scientifiques* 145: 195–215, 405–424.

--- 1979. Les ruses de Darwin. *La recherche* 102: 794–798.

--- 1981. *Darwin & Cie*. Bruxelles: Éditions Complexe.

Tiedemann, F. 1834. *A systematic treatise on comparative physiology, introductory to the physiology of man*. Trans. J. M. Gully and J. H. Lane. London: Churchill.

Timiriazev, K. A. 1865. *Kratkii ocherk teorii Darwina* (A short essay on Darwin's theory). St. Petersburg.

--- 1892–1895. Istoricheskii metod v biologii (The historical method in biology). Installments in *Russia Mysl.*; reprinted in Timiriazev 1937–1940; 1948.

--- 1915. Mendel. In *Entsiklopedskii slovar Granat*, 7th ed., vol. 25.

--- 1937–1940. *Sobrania sochinenii* (Collected works) Moscow: O.G.I.Z.

--- 1948. *El metodo historico en la biologia*. Moscow: Progress.

--- 1949. (V. L. Komarov and T. D. Lysenko, eds.) *Izbrannie Sochinenii*. (Selected works) 4 vols. Moscow: O.G.I.Z.

Timpanaro, S. 1970. *Sul materialismo*. Pisa: Nistri-Lischi.

Tinbergen, N. 1952. "Derived" activities: their causation, biological significance, origin, and emancipation during evolution. *Q. rev. biol.* 27: 1–32.

Tischner, R. 1928. Franz Anton Mesmer als vorläufer Darwins. *Arch. für Geschichte der Mathematik, der Naturwissenschaften und der Technik* 10: 476–477.

Tkachev, P. N. 1933. Izbrannye sochinenia na sozial'nopoliticheskie temi (Selected works on socio-political issues). Moscow: Gospolitizdat.

Tobach, E.; Gianutsos, J.; Topoff, H.; and Gross, C., eds. 1974. *The four horsemen: racism, sexism, militarism, and social Darwinism*. New York: Behavioral Publications.

Tobien, H. 1974. Alcide Charles Victor Dessalines d'Orbigny. In *Dictionary of Scientific biography*, vol. 10, pp. 221–222. New York: Charles Scribner's Sons.

Tommaseo, N. 1869. *L'uomo e la scimmia: lettere dieci con un discorso sugli urli datici per origine delle lingue*. Milano: G. Agnelli.

Topitsch, S. 1958. *Über Darwins Selektionslehre: eine Studie zur Weltanschauungskritik*. Wien: Spring.

Toscano, M. A. 1980. *Malgrado la storia: per una lettura critica di Herbert Spencer*. Milano: Feltrinelli.

Toulmin, S. 1972. *Human understanding, vol. 1. Concepts: their collective use and evolution*. Princeton: Princeton University Press.

--- **and Goodfield, J.** 1965. *The ancestry of science: the discovery of time*. London: Hutchinson.

Towers, R. 1842. Thoughts on botany. *Gardeners' chron.* 37.

Trenn, T. J. 1974. Charles Darwin, fossil Cirripedes, and Robert Fitch: presenting sixteen hitherto unpublished Darwin letters of 1849 to 1851. *Proc. Am. Phil. Soc.* 108: 471–491.

Treviranus, G. R. 1803. *Biologie oder Philosophie der lebenden Natur*. Vol. 2. Göttingen: Rower.

Trigger, B. G. 1979. The materialist strategy. *Science* 205: 890–891.

Tristram, H. B. 1859. On the ornithology of Northern Africa. Part III. The Sahara continued. *The Ibis* 1: 415–435.

Trow-Smith, R. 1959. *A history of British livestock husbandry, 1700–1900*. London: Routledge and Kegan Paul.

Tschermak, E. 1900. Ueber künstliche Kreuzung bei Pisum sativum. *Berichte der deutschen botanischen Gesellschaft*. 18: 232–239.

Tschülock, S. 1936. Über Darwins Selektionslehre. Historisch-Kritische Betracht ungen. *Vierteljahrsschr. Naturforsch. Ges. Zürich* 81, Supp. 26: 1–68.

Turbayne, C. 1970. *The myth of metaphor*. Rev. ed. Columbia: University of South Carolina Press.

Turesson, G. 1922. The species and the variety as ecological units. *Hereditas* 3: 100–113.

Turner, F. M. 1974a. *Between science and religion: the reaction to scientific naturalism in late Victorian England*. New Haven: Yale University Press.

--- 1974b. Rainfall, plagues, and the Prince of Wales: a chapter in the conflict of religion and science. *J. Brit. stud.* 13: 46–65.

--- 1974–1975. Victorian scientific naturalism and Thomas Carlyle. *Victorian Stud.* 18: 325–

343.

--- 1978. The Victorian conflict between science and religion: a professional dimension. *Isis* 69: 356–376.

--- 1981. John Tyndall and Victorian scientific naturalism. In *John Tyndall: essays on a natural philosopher*, eds. W. H. Brock, N. D. McMillan, and R. C. Mollan, pp. 169–191. *Roy. Dublin Soc. Hist. Stud. in Irish Science and Technology*, no. 3. Dublin: Royal Dublin Society.

Turrill, W. B. 1953. *Pioneer of plant geography: the phytogeographical researches of Sir Joseph Dalton Hooker*. The Hague: M. Nijhoff.

--- 1963. *Joseph Dalton Hooker. Botanist, explorer and administrator*. London: Nelson.

Twiesselmann, F. 1959–1960. Darwin et les causes de l'évolution de l'homme. *Ann. Soc. Zool. Belg.* 90: 27–35.

Tytler, G. 1982. *Physiognomy in the European novel: faces and fortunes*. Princeton: Princeton University Press.

Tze Tuan, C. 1929. Twenty-five centuries before Charles Darwin. *Sci. monthly* 29: 49–52.

Ungerer, E. 1923. *Lamarck-Darwin: die Entwicklung des Lebens*. Stuttgart: Frommann.

--- 1941. *Die Erkenntnisgrundlagen der Biologie: ihre Geschichte und ihr gegenwärtiger Stand*. Kostanz: Akademische Verlagsgesellschaft Athenaion.

Urbanek, A. 1974. Hystoryczne aspekty nauk biologicznych. [Historical aspects of biological sciences]. In *Ewolucja biologiczna*, ed. C. Nowiński pp. 139–156. Warsaw: Ossolineum.

Ure, A. 1823. *A dictionary of chemistry*. 2d ed. London: Tegg.

--- 1835. *The philosophy of manufactures or, an exposition of the scientific, moral and commercial economy of the factory system of Great Britain*. London: Charles Knight.

Uschmann, G., ed. 1954. *Ernst Haeckel: Forscher, Künstler, Mensch* Jena: Urania.

--- 1959. *Geschichte der Zoologie und der zoologischen Anstalten in Jena 1779–1919*. Jena: VEB Gustav Fischer.

--- 1968. Die Naturgeschichte des biologischen Modells. *Nova Acta Leopold.*, N.S. 33: 43–64.

--- **and Hassenstein, B.** 1965. Der Briefwechsel zwischen Ernst Haeckel und August Weismann. In *Kleine Festgabe aus Anlass der hundertjährigen Wiederkehr der Gründung des zoologischen Institutes der Friedrich-Schiller-Universität Jena im Jahre 1865 durch Ernst Haeckel*, ed. M. Gersch, pp. 7–68. Jena: Friedrich-Schiller-Universität.

--- **and Jahn, I.** 1959–1960. Der Briefwechsel zwischen Thomas Henry Huxley und Ernst Haeckel: ein Beitrag zum Darwin-Jahr. *Wissenschaftliche Zeitschrift der Friedrich-Schiller-Universität Jena. Mathematisch-naturwissenschaftliche Reihe* 9: 7–33.

Van Cittert, P. H. 1934. *Descriptive catalogue of the collection of microscopes in charge of the Utrecht university museum, with an introductory historical survey of the resolving power of the microscope*. Gröningen: Noordhoff.

Vandel, A. 1960. Lamarck et Darwin. *Rev. d'hist. sci.* 13: 59–72.

Venn, J. A., ed. 1940–1954. Alumni Cantabrigienses . . . to 1900. 6 vols. Part 2. Cambridge: Cambridge University Press.

Vepsäläinen, K. 1982. Darwinism in Finland in the 20th century. *Eidema* 1: 140–155.

Verrall, R. 1979. Sociobiology: the instincts in our genes. *Spearhead* (March): 10–11.

Vianna de Lima, A. 1885. *Exposé des doctrines transformistes (Lamarck, Darwin, Haeckel)*. Paris: C. Delagrave.

Victorian Studies. 1959. Darwin anniversary issue. 3: 1–128.

Vidoni, F. 1982. Ruolo delle scienze naturali nel pensiero marxiano. In *Nel Laboratorio di Marx: scienze naturali e matematica*, ed. A. Guerraggio and F. Vidoni. Milano: Franco Angeli.

--- **and Guerraggio, A.** 1982. *Nel laboratorio di Marx: scienze naturali e matematiche*. Milano: Franco Angeli.

Virchow, R. 1860. *Cellular pathology*. Trans. F. Chance. London: Churchill.

--- 1864. Ueber den vermeintlichen Materialismus der heutigen Naturwissenschaft. *Amtliche Bericht über die acht und dreissigste Versammlung Deutscher Naturforscher und Ärzte in Stettin*, pp. 35–42, 74–76. Stettin: F. Hessenland.

--- 1877. *Amtlicher Bericht der 50. Versammlung Deutscher Naturforscher und Aerzte in München*, pp. 65–77. München: F. Straub.

Viré, M. 1979. La création de la chaire de "L'evolution des êtres organisés" à la Sorbonne en 1888. *Rev. de synthèse* 100, nos. 95–96, 377–391.

Virey, J. J. 1801. *Histoire naturelle du genre humain ou Recherches sur ses principaux fondemens physiques et moraux.* 2 vols. [New ed. 1824, 1834.] Paris: F. Dufart.

---, ed. 1803–1804. *Nouveau dictionnaire d'histoire naturelle.* 24 vols. Paris: Deterville.

---, ed 1816–1819. *Nouveau dictionnaire d'histoire naturelle.* 2d ed. 36 vols. Paris: Deterville.

--- 1822. *Histoire des moeurs et de l'instinct des animaux, avec les distributions méthodiques et naturelles de toutes leurs classes.* 2 vols. Paris: Deterville.

--- 1841. *Des causes physiologiques de la sociabilité chez les animaux, et de la civilization dans l'homme.* Paris: J. B. Baillière.

Vogt, C. 1863. *Lectures on man: his place in creation, and in the history of the earth.* Ed. J. Hunt. London: Anthropological Society.

Voipio, P. 1982. Darwinismi suomessa 1800-luvulla. (Darwinism in Finland in the 19th century.) *Eidema* 1: 124–139 (with a summary in English).

Vorlander, G. 1907. *K. F. Wolff.* Leipzig: Teubner.

Vorzimmer, P. J. 1963a. Charles Darwin and blending inheritance. *Isis* 54: 371–390.

--- 1963b. *The development of Darwin's evolutionary thought after 1859.* Ph. D. thesis. University of Cambridge.

--- 1965. Darwin's ecology and its influence upon his theory. *Isis* 56: 148–155.

--- 1968. Darwin and Mendel: the historical connection. *Isis* 59: 77–82.

--- 1969a. Darwin, Malthus, and the theory of natural selection. *J. hist. ideas.* 30: 527–542.

--- 1969b. Darwin's *Questions about the breeding of animals* (1839). *J. hist. biol.* 2: 269–281.

--- 1969–1970. Darwin's "Lamarckism" and the "flat-fish controversy": 1863–1871. *Lychnos*: 121–170.

--- 1970. *Charles Darwin: the years of controversy. The "Origin of Species" and its critics, 1859–1882.* Philadelphia: Temple University Press.

--- 1975. An early Darwin manuscript: the "Outline and Draft of 1839". *J. hist. biol.* 8: 191–217.

--- 1977. The Darwin reading notebooks (1838–1860). *J. hist. biol.* 10: 107–153.

Vucinich, A. 1974. Russia: biological sciences. In *The comparative reception of Darwinism*, ed. T. F. Glick, pp. 227–255. Austin: University of Texas Press.

Waddington, C. H. 1958. Theories of evolution. In *A century of Darwin*, ed. S. A. Barnett, pp. 1–18. Cambridge: Harvard University Press.

--- 1975. *The evolution of an evolutionist.* Edinburgh: Edinburgh University Press.

Wagar, W. W. 1972. *Good tidings: the belief in progress from Darwin to Marcuse.* Bloomington: Indiana University Press.

Wagner, M. 1868. *Die Darwinsche Theorie und das Migrationgesetz der Organismen.* Leipzig: Dunker und Humblot.

--- 1873. *The Darwinian theory and the law of the migration of organisms.* Trans. J. L. Laird. London: E. Stanford.

Walker, F. 1838. Descriptions of some chalcidites discovered by C. Darwin, Esq. *Entomol. mag.* 5: 469–477.

--- 1839. *Monographia chalciditum, species collected by C. Darwin, Esq.* 2 vols. London: Baillière.

--- 1842a. Descriptions of *Chalcidites* discovered by C. Darwin, Esq., near Valparaiso. *Ann. mag. nat. hist.* 10: 113–117.

--- 1842b. Descriptions of *Chalcidites* discovered in Valdivia by C. Darwin, Esq. *Ann. mag. nat. hist.* 10: 271–274.

--- 1843a. Descriptions of *Chalcidites* discovered near Conception, in South America, by C. Darwin, Esq. *Ann. mag. nat. hist.* 11: 30–32.

--- 1843b. Descriptions of *Chalcidites* found near Lima by C. Darwin, Esq. *Ann. mag. nat. hist.* 11: 115–117.

--- 1843c. Descriptions of *Chalcidites* discovered in the Isle of Chonos by C. Darwin, Esq. *Ann. mag. nat. hist.* 11: 184–185.

--- 1843d. Descriptions of *Chalcidites* discovered in Coquimbo by C. Darwin, Esq. *Ann. mag. nat. hist.* 11: 185–188.

--- 1843e. Descriptions of *Chalcidites* discovered by C. Darwin, Esq. *Ann. mag. nat. hist.* 12: 45–46.

Wall, J. F. 1976. Social Darwinism and constitutional law with special reference to *Lochner v. New York. Ann. sci.* 33: 465–476.

Wallace, A. R. 1855. On the law which has regulated the introduction of new species. 1891. Reprinted in *Natural selection and tropical nature*, pp. 3–19. London: Macmillan.

--- 1856a. On the habits of the orang-utan of Borneo. *Ann. mag. nat. hist.* II. 18: 26–32.

--- 1856b. Attempts at a natural arrangement of birds. *Ann. mag. nat. hist.* II. 18: 193–216.

--- 1858a. Note on the theory of permanent

and geographical varieties. *The Zoologist* 16: 5887–5888.
--- 1858b. On the tendency of varieties to depart indefinitely from the original type. In C. Darwin and A. R. Wallace, *Evolution by natural selection*, ed. G. de Beer, pp. 268–279. Cambridge: Cambridge University Press, 1958.
--- 1864. The origin of human races and the antiquity of man deduced from the theory of natural selection. *J. Anthropol. Soc. Lond.* 2: clvii–clxxxvii.
--- 1865. On the phenomena of variation and geographical distribution as illustrated by the Papilionidae of the Malayan region. *Trans. Linn. Soc. Lond.* 25: 1–71.
--- 1867. Mimicry and other protective resemblances among animals. *Westminster rev.* [American ed.] 88: 1–20.
--- 1869. Sir Charles Lyell on geological climates and the origin of species. *Q. rev.* 126: 359–394.
--- 1870a. *Contributions to the theory of natural selection.* London: Macmillan.
--- 1870b. The limits of natural selection as applied to man. In *Contributions to the theory of natural selection*, pp. 332–371. London: Macmillan.
--- 1871. [Review of C. Darwin,] The descent of man and selection in relation to sex. *The academy* 2: 177–183.
--- 1873. Review of C. R. Darwin, *Expression of the emotions. Quart. j. sci.* N.S. 3: 113–118.
--- 1876. *The geographical distribution of animals.* 2 vols. London: Macmillan.
--- 1880. *Island life: or the phenomena and causes of insular faunas.* London: Macmillan.
--- 1889. *Darwinism: an exposition of the theory of natural selection with some of its applications.* London: Macmillan.
--- 1891. *Natural selection and tropical nature.* London: Macmillan.
--- 1896. The problem of utility: are specific characters always or generally useful? 1900. Reprinted in *Studies scientific and social*, 2 vols., vol. 1, pp. 378–398. London: Macmillan.
--- 1905. *My life; a record of events and opinions.* 2 vols. London: Chapman and Hall.
--- 1908a. Notes on the passages of Malthus's "Principles of population" which suggested the idea of natural selection to Darwin and myself. In the Darwin-Wallace celebration held on Thursday, 1st July, 1908, by the Linnean Society of London, pp. 111–118.
--- 1908b. The present position of Darwinism. *Contemporary rev.* 94: 129–141.
--- 1916. *Alfred Russel Wallace; letters and reminiscences.* Ed. J. Marchant. New York: Harper.
Wallace, B. 1968. *Topics in population genetics.* New York: W. W. Norton.
--- 1975. Hard and soft selection revisited. *Evolution* 29: 465–474.
Wallich, G. C. 1870. *Eminent men of the day: photographed by G. C. Wallich.* London.
Walters, S. M. 1981. *The shaping of Cambridge botany.* Cambridge: Cambridge University Press.
Warfield, B. B. 1889. Darwin's arguments against Christianity and against religion. *Homiletic rev.* 17: 9–16.
--- 1932 [1888]. Charles Darwin's religious life: a sketch in spiritual biography. In *Studies in theology*, pp. 541–582. New York: Oxford University Press.
Wasmann, E. 1906. *La biologia moderna e la teoria dell'evoluzione.* Trad. A. Gemelli. Firenze : Libreria Editrice Fiorentina.
--- 1908. *Istinto e intelligenza nel regno animale, contributo critico alla zoopsicologia moderna.* Preface by A. Gemelli. Firenze : Libreria Editrice Fiorentina.
Wassermann, G. 1978. Testability and the role of natural selection within theories of population genetics and evolution. *Brit. j. phil. sci.* 29: 223–242.
Watanabe, M. 1971. Darwinism in Japan in the late 19th century. *Actes du XIIe cong. int. d'hist. sci.* 11: 149–154. Paris: Blanchard.
Waterhouse, G. R. 1837a. Characters of new species of the genus *Mus*, from the collection of Mr. Darwin. *Proc. Zool. Soc. Lond.* 5: 15–21, 27–29.
--- 1837b. Characters of two genera of Rodentia (*Reithrodon* and *Abrocoma*), from Mr. Darwin's collection. *Proc. Zool. Soc. Lond.* 5: 29–32.
--- 1838a. On a new species of the genus *Delphinus. Proc. Zool. Soc. Lond.* 6: 23–24.
--- 1838b. Descriptions of some of the insects brought to this country by C. Darwin, Esq. *Trans. Entomol. Soc. Lond.* 2: 131–135.
--- 1840a. Description of a new species of the genus *Lophotus*, from the collection of Charles Darwin, Esq. *Ann. nat. hist.* 5: 329–332.
--- 1840b. Descriptions of some new species

of carabideous insects, from the collection made by C. Darwin, Esq., in the southern parts of S. America. *Mag. nat. hist.* 4: 354–362.

——— 1840c. Carabideous insects collected by Mr. Darwin during the voyage of Her Majesty's Ship Beagle. *Ann. mag. nat. hist.* 6: 254–257.

——— 1841a. Carabideous insects followed by Charles Darwin during the voyage of Her Majesty's Ship Beagle. *Ann. mag. nat. hist.* 6: 351–355.

——— 1841b. Carabideous insects collected by Charles Darwin during the voyage of Her Majesty's Ship Beagle. *Ann. mag. nat. hist.* 7: 120–129.

——— 1841c. Descriptions of some new coleopterous insects from the southern parts of S. America, collected by C. Darwin, Esq. and T. Bridges, Esq. *Proc. Zool. Soc. Lond.* 9: 105–128.

——— 1842a. Carabideous insects followed by M. Charles Darwin during the voyage of Her Majesty's Ship Beagle. *Ann. mag. nat. hist.* 9: 134–139.

——— 1842b. Description of a new species of lamellicorn beetle brought from Valdivia by C. Darwin, Esq. *Entomologist* 1: 281–283.

——— 1843. Description of a new genus of carabideous insects brought from the Falkland Islands by Charles Darwin, Esq. *Ann. mag. nat. hist.* 11: 281–283.

——— 1845. Descriptions of coleopterous insects collected by Charles Darwin, Esq., in the Galapagos Islands. *Ann. mag. nat. hist.* 16: 19–41.

Watson, H. C. 1843. Remarks on the distinction of species in nature, and in books; preliminary to the notice of some variations and transitions of character, observed in the native plants of Britain. *Lond. j. bot.* 2: 613–622.

——— 1845a. On the theory of "progressive development," applied in explanation of the origin and transmutation of species. *Phytologist* 2: 108–113, 140–147.

——— 1845b. Report of an experiment which bears upon the specific identity of the cowslip and primrose. *Phytologist* 2: 217–219.

——— 1859. *Cybele Britannica, or, British plants and their geographical relations*. Vol. 4. London: Longman.

Watson, J. A. S. 1939. *The history of the Royal Agricultural Society of England, 1839–1939*. London: Royal Agricultural Society.

Watt, I. 1979. *Conrad in the nineteenth century*. Berkeley: University of California Press.

Watt, M. H. 1943. *The history of the parson's wife*. London: Faber & Faber.

Webb, P. B. 1849. Spicilegia gorgonia; or a catalogue of all the plants as yet discovered in the Cape de Verd Islands, from the collections of J. D. Hooker, Esq. M.D. R.N., Dr. T. Vogel, & other travellers. In *Niger flora: or, an enumeration of the plants of western tropical Africa, collected by the late Dr. Theodore Vogel, botanist to the voyage of the expedition sent by Her Britannic Majesty to the River Niger in 1841, under the command of Capt. H. D. Trotter, R. N., etc.* ed. W. J. Hooker, pp. 89–197. London: Baillière.

Webb, S. D. 1965. The osteology of *Camelops*. *Bull. Los Angeles Co. Mus. Sci.* 1: 1–54.

Weber, G. 1974. Science and society in nineteenth century anthropology. *Hist. sci.* 12: 260–283.

Webster, C., ed. 1981. *Biology, medicine and society 1840–1940*. Cambridge: Cambridge University Press.

Wedgwood, H. 1859–1865. *A dictionary of English etymology*. 3 vols. in 4. London: Trübner.

Weindling, P. J. 1981. Theories of the cell state in imperial Germany. In *Biology, medicine and society 1840–1940*, ed. C. Webster, pp. 99–155. Cambridge: Cambridge University Press.

——— 1982. *Cell biology and Darwinism in imperial Germany: the contribution of Oscar Hertwig (1849–1922)*. Ph.D. thesis. University College London.

Weir, J. A. 1968. Agassiz, Mendel, and heredity. *J. hist. biol.* 1: 179–204.

Weismann, A. 1891–1892. *Essays upon heredity and kindred biological subjects*. 2 vols. Eds. E. B. Poulton, S. Schönland, and A. E. Shipley Oxford: Oxford University Press.

——— 1893a. *The germ plasm: a theory of heredity*. Trans. W. N. Parker and H. Rönnfeldt. London: Scott.

——— 1893b. The all-sufficiency of natural selection. *Contemporary rev.* 64: 309–338, 596–610.

——— 1904. The evolution theory, 2 vols. Trans. J. A. Thomson and M. R. Thomson. London: E. Arnold.

Weizsäcker, C. F. von 1951. *The history of nature*.

London: Routledge and Kegan Paul.

Weldon, W. F. R. 1894–1895. An attempt to measure the death-rate due to the selective destruction of *Carcinas moenas* with respect to particular dimensions. *Proc. Roy. Soc. Lond.* 57: 360–379.

--- 1898. President's address, zoology section. *Rept. Brit. Assoc. Adv. Sci.*. 887–902.

--- 1901. A further study of natural selection in *Clausilla laminata. Biometrika* 1: 109–124.

Welleman, K. 1979. Félix Le Dantec et le néo-Lamarckisme français. *Rev. de synthèse* 100, nos. 95–96 : 296–336.

Wells, K. D. 1971. Sir William Lawrence (1783–1867). A study of pre-Darwinian ideas on heredity and variation. *J. hist. biol.* 4: 319–361.

--- 1973. The historical context of natural selection: the case of Patrick Matthew. *J. hist. biol.* 6: 225–258.

Werner, J., ed. 1930. *The love letters of Ernst Haeckel written between 1898 and 1903*. London: Methuen.

West, G. 1938. *Charles Darwin: a portrait*. New Haven: Yale University Press.

Weyl, H. 1952. *Symmetry*. Princeton: Princeton University Press.

Wheeler, W. M. 1939. *Essays in philosophical biology*. Cambridge: Harvard University Press.

Whewell, W. 1834. *Astronomy and general physics considered with reference to natural theology*. London: William Pickering.

--- 1837. *History of the inductive sciences*. 3 vols. London: Parker.

--- 1838. Presidential address. *Proc. Geol. Soc. Lond.* 2: 624–649.

--- 1839. Presidential address. *Proc. Geol. Soc. Lond.* 3: 61–98.

White, A. 1841. Descriptions of new or little known Arachnida. *Ann. mag. nat. hist.* 7: 471–477.

White, G. 1974. *The natural history of Selborne*. London: Oxford University Press.

White, H. V. 1973. *Metahistory: the historical imagination in nineteenth-century Europe*. Baltimore: Johns Hopkins University Press.

--- 1976. The fictions of factual representation, In *The literature of fact*, ed. J. S. Fletcher, pp. 21–44. New York: Columbia University Press.

White, L. A. 1944. Morgan's attitude toward religion and science. *Am. anthropologist* 46: 218–230.

--- 1947. Evolutionism in cultural anthropology: a rejoinder. *Am. anthropologist* 49: 400–411.

--- 1969. *The science of culture. A study of man and civilization*. 2d ed. New York: Farrar, Strauss & Giroux.

White, M. J. D. 1978. *Modes of speciation*. San Francisco: W. H. Freeman.

Whitehead, P., and Keates, C. 1981. *The British Museum (Natural History)*. London: Scala/ Philip Wilson.

Whitehouse, H. L. K. 1959. Cross- and self fertilization in plants. In *Darwin's biological work*, ed. P. R. Bell, pp. 207–261. Cambridge: Cambridge University Press.

Wichler, G. 1938. Alfred Russel Wallace (1823–1913), sein Leben, seine Arbeiten, sein Wesen, Zugleich ein Beitrag zu dem Verhältnis von Wallace zu Darwin. *Sudhoffs Arch.*: 30: 364–400.

--- 1960. Darwin als botaniker. *Sudhoffs Arch.* 44: 289–313.

--- 1961. *Charles Darwin: the founder of the theory of evolution and natural selection*. New York: Pergamon Press.

Wiener, P. P. 1949. *Evolution and the founders of pragmatism*. Cambridge: Harvard University Press.

--- **and Noland, A.**, eds. 1957. *Roots of scientific thought: a cultural perspective*. New York: Basic Books.

Wigand, A. 1874–1877. *Der Darwinismus und die Naturforschung Newtons und Cuviers*. 3 vols. Braunschweig: Vieweg und Sohn.

Wilberforce, S. 1860. [Review of *Origin of species*] *Q. rev.* 108 : 225–264.

--- 1874. Essays contributed to the "Quarterly review". 2 vols. London: John Murray.

Wilczynski, J. Z. 1959. On the presumed Darwinism of Alberuni eight hundred years before Darwin. *Isis* 50: 459–466.

Wiley, E. O. 1981. *Phylogenetics, the theory and practice of phylogenetic systematics*. New York: Wiley.

Wilkie, J. S. 1959. Buffon, Lamarck and Darwin: the originality of Darwin's theory of evolution. In *Darwin's biological work. Some aspects reconsidered*, ed. P. R. Bell, pp. 262–307. Cambridge: Cambridge University Press.

Willey, B. 1960. *Darwin and Butler. Two versions of evolution*. London: Chatto and Windus.

Williams, G. C. 1966. *Adaptation and natural selection.* Princeton: Princeton University Press.

Williams, M. B. 1971. Deducing the consequences of evolution: a mathematical model. *J. theor. biol.* 28: 343–385.

--- 1973. The logical status of the theory of natural selection and other evolutionary controversies: resolution by an axiomatization. In *The methodological unity of science*, ed. M. Bunge, pp. 84–102. Dordrecht: Reidel.

Williams, R. 1973. Social Darwinism. In *The limits of human nature*, ed. J. Benthall, pp. 115–130. London: Allen Lane.

Williams-Ellis, A. 1966. *Darwin's moon: a biography of Alfred Russel Wallace.* London, Glasgow: Blackie.

Willis, J. C. 1940. *The course of evolution by differentiation or divergent evolution rather than by selection.* Cambridge: Cambridge University Press.

Wilson, E. B. 1900. *The cell in development and inheritance.* 2d ed. New York: Macmillan.

Wilson, E. O. 1975. *Sociobiology: the new synthesis.* Cambridge: Harvard University Press.

--- 1978. *On human nature.* Cambridge: Harvard University Press.

Wilson, J. B. 1965. Darwin and the transcendentalists. *J. hist. ideas* 26: 286–290.

Wilson, J. W. 1959. Biology attains maturity in the nineteenth century. In *Critical problems in the history of science*, ed. M. Clagett, pp. 401–418. Madison: University of Wisconsin Press.

Wilson, L. G. 1960. Review of G. Himmelfarb, *Darwin and the Darwinian revolution. Arch. int. hist. sci.* 13: 343–351.

--- 1964. The development of the concept of uniformitarianism in the mind of Charles Lyell. *Actes du Xe cong. int. d'hist. sci.* (Ithaca 1962) 11: 993–996. Paris: Blanchard.

--- 1967. The origins of Charles Lyell's uniformitarianism. *Geol. Soc. Am.* 89: 35–62.

--- 1969. The intellectual background to Charles Lyell's *Principles of Geology*, 1830–1833. In *Toward a history of geology*, ed. C. L. Schneer, pp. 426–433. Cambridge: M.I.T. Press.

--- 1971a. Sir Charles Lyell and the species question. *Am. sci.* 59: 43–55.

--- 1971b. Commentary on the paper of John C. Greene. [Greene 1971] In *Perspectives in the history of science and technology*, ed. D. H. D. Roller, pp. 31–37. Norman: Oklahoma University Press.

--- 1972. *Charles Lyell. The years to 1841: the revolution in geology.* New Haven: Yale University Press.

--- 1973. Uniformitarianism and catastrophism. In *Dictionary of the history of ideas*, ed. P. Wiener, pp. 417–423. New York: Charles Scribner's Sons.

Wilson, R. J. 1967. *Darwinism and the American intellectual.* Homewood: Dorset Press.

Wiltshire, D. 1978. *The social and political thought of Herbert Spencer.* Oxford: Oxford University Press.

Winau, R. 1981. Ernst Haeckels Vorstellungen von Wert und Werden menschlicher Rassen und Kulturen. *Medizinhistorisches j.* 16: 270–279.

Winslow, J. H. 1971. *Darwin's Victorian malady: evidence for its medically induced origin.* Philadelphia: American Philosophical Society.

Winsor, M. P. 1969. Barnacle larvae in the nineteenth century: a case study in taxonomic theory. *J. hist. med.* 29: 294–309.

--- 1976. *Starfish, jellyfish and the order of life: issues in nineteenth-century science.* New Haven: Yale University Press.

--- 1979. Louis Agassiz and the species question. *Stud. hist. biol.* 3: 89–117.

Winter, S. G. 1964. Economic "natural selection" and the theory of the firm. *Yale Econ. Essays* 4: 225–272.

Withers, G. R. A. 1971. *Charles Darwin and the theory of evolution.* London: Edward Arnold.

Wolff, K. F. 1966. *Theorie von der Generation.* Intro. by R. Herrlinger. Hildesheim: George Olms.

Wollaston, A. F. R. 1921. *Life of Albert Newton.* London: John Murray.

Wollaston, T. V. 1856. *On the variation of species.* London: Van Voorst.

--- 1860. Bibliographical notice. *Ann. and mag. nat. hist.* 5: 132–143. 1973. Reprinted in D. Hull, *Darwin and his critics*, pp. 127–140. Cambridge: Harvard University Press.

Wood, H. G. 1955. *Belief and unbelief since 1850.* Cambridge: Cambridge University Press.

[Woodbury]. 1862. Are Ligurians larger than common bees? *J. horticulture and cottage gardener.* N.S. 3: 225.

[Woodbury]. 1863. Exotic honey-bees — size

of their cells. *J. horticulture and cottage gardener.* N.S. 4: 252.

Woodcock, G. 1969. *Henry Walter Bates: naturalist of the Amazons*. London: Faber and Faber.

Woodruff, A. W. 1968. The impact of Darwin's voyage to South America on his work and health. *Bull. N.Y. Acad. Med.* 44: 661–672.

Woodward, H. B. 1907. *The history of the Geological Society of London*. London: Geological Society.

Worboys, M. 1981. The British Association and empire: science and social imperialism, 1880–1940. In *The parliament of science: the British Association for the Advancement of Science, 1831–1981*, eds. R. MacLeod and P. Collins, pp. 170–187. Norwood, Middlesex: Science Reviews.

Wordsworth, W. 1798–1805. *The lyrical ballads.* 1940. Ed. G. Sampson. London: Methuen.

--- 1949. *The poetical works*. Eds. E. de Selincourt and H. Darbishire. Oxford : Clarendon Press.

Worster, D. 1977. *Nature's economy: the roots of ecology*. San Francisco: Sierra Club Books.

Wright, C. 1870. Limits of natural selection. 1877. Reprinted in *Philosophical discussions*, pp. 97–125. New York: Henry Holt.

--- 1871. The genesis of species. *North Am. rev.* 113: 63–103.

Wright, S. 1929a. Fisher's theory of dominance. *Am. nat.* 63: 274–279.

--- 1929b. The evolution of dominance: comment on Dr. Fisher's reply. *Am. nat.* 63: 556–561.

--- 1930. The genetical theory of natural selection. *J. hered.* 21: 349–356.

--- 1931. Evolution in Mendelian populations. *Genetics* 16: 97–159.

--- 1932. The roles of mutation, inbreeding, crossbreeding, and selection in evolution. *Proc. 6th int. cong. genet.* 1: 356–366.

--- 1948. On the roles of directed and random changes in gene frequency in the genetics of populations. *Evolution* 2: 279–294.

--- 1967. Comments. In *Mathematical challenges to the neo-Darwinian interpretation of evolution*, eds. P. S. Moorhead and M. M. Kaplan, pp. 117–120. Philadelphia: Wistar Institute Press.

--- 1968–1978. *Evolution and the genetics of populations*. 4 vols. Chicago: University of Chicago Press.

--- 1978. The relation of livestock breeding to theories of evolution. *J. animal sci.* 46: 1192–1200.

--- 1982. Character change, speciation, and the higher taxa. *Evolution* 36: 427–443.

Wright-St. Clair, R. E. 1956. David Monro's lecture on the expression of passion, 1840. *Bull. hist. med.* 30: 450–464.

Wyllie, I. G. 1959. Social Darwinism and the businessman. *Proc. Am. Phil. Soc.* 103: 629–635.

Wynne-Edwards, V. C. 1962. *Animal dispersion in relation to social behavior*. Edinburgh: Oliver and Boyd.

Wyss, W. von 1959. *Charles Darwin, ein Forscherleben*. Zürich: Artemis-Verlag.

Yeo, R. 1979. William Whewell, natural theology and philosophy of science in mid-nineteenth century Britain. *Ann. sci.* 36: 493–516.

Yokoyama, T. 1971. The influence of theological thought on Charles Darwin. Consideration of the relation between William Paley and Charles Darwin [in Japanese]. *Kagakusi Kenkyu* 10: 49–59.

--- 1978. Comparative philosophy of *The origin of species* and *Das Kapital*: an essay on the unitary understanding of natural history and social history [in Japanese]. *Kagakusi Kenkyu* 17: 31–40, 81–89.

Yonge, C. M. 1958. Darwin and coral reefs. In *A century of Darwin*, ed. S. A. Barnett, pp. 245–266. Cambridge: Harvard University Press.

Youatt, W. 1831. *The horse; with a treatise on draught*. London: Baldwin and Cradock.

--- 1834. *Cattle; their breeds, management, and diseases*. London: R. Baldwin.

--- 1837. *Sheep; their breeds, management, and diseases*. London: Baldwin and Cradock.

--- 1847. *The pig: a treatise on the breeds, management, feeding, and medical treatment, of swine; with directions for salting pork, and curing bacon and ham*. London: Cradock.

Young, A. 1774. *Political arithmetic*. London: W. Nicoll.

--- 1808. 1792–1794. *Travels in France and Italy during the years 1787, 1788, and 1789*. 2d ed. 4 vols. Edmunds: J. Rackham.

Young, G. M. 1977. *Portrait of an age: Victorian England*. Ed. G. K. Clark. London: Oxford University Press.

Young, R. M. 1966. Scholarship and the history of the behavioural sciences. *Hist. sci.* 5: 1–51.

--- 1967a. Animal soul. In *The encylopaedia of philosophy*, ed. P. Edwards, vol. 1, pp. 122–127. New York: Macmillan.

--- 1967b. Philosophy of mind and related issues. *Brit. j. phil. sci.* 18: 325–330.

--- 1968a. The functions of the brain: Gall to Ferrier (1808–1886). *Isis* 59: 251–268.

--- 1968b. The development of Herbert Spencer's concept of evolution. *Actes du XIe cong. int. d'hist. sci.* 2: 273–278. Paris: Blanchard.

--- 1969. Malthus and the evolutionists: the common context of biological and social theory. *Past and present* 43: 109–145.

--- 1970a. *Mind, brain and adaptation in the nineteenth century: cerebral localization and its biological context from Gall to Ferrier*. Oxford: Clarendon Press.

--- 1970b. The impact of Darwin on conventional thought. In *The Victorian crisis of faith*, ed. A. Simondson, pp. 13–35. London: Society for Promoting Christian Knowledge.

--- 1971a. Darwin's metaphor: does nature select? *Monist* 55: 442–503.

--- 1971b. Non-scientific factors in the Darwinian debate. *Actes du XIIe cong. int. d'hist. sci.* (Paris 1968) 8: 221–226. Paris: Blanchard.

--- 1972a. Darwinism and the division of labour. *The listener* 2264 (17 August 1972): 202–205.

--- 1972b. Evolutionary biology and ideology: then and now. In *The biological revolution*, ed. W. Fuller, pp. 241–282. Garden City: Doubleday.

--- 1972c. Franz Joseph Gall. In *Dictionary of scientific biography*, vol. 5, pp. 250–256. New York: Charles Scribner's Sons.

--- 1973a. The historiographic and ideological contexts of the nineteenth century debate on man's place in nature. In *Changing perspectives in the history of science: essays in honour of Joseph Needham*, eds. M. Teich and R. M. Young, pp. 344–438. London: Heinemann.

--- 1973b. The role of psychology in the nineteenth-century evolutionary debate. In *Historical conceptions of psychology*, eds. M. Henle, J. Jaynes, and J. J. Sullivan, pp. 180–204. New York: Springer.

--- 1973c. Association of ideas. In *Dictionary of the history of ideas*, vol. 1, pp. 111–118, New York: Charles Scribner's Sons.

--- 1973d. The human limits of nature. In *The limits of human nature*, ed. J. Benthall, pp. 235–274. London: Allen Lane.

--- 1977. Science *is* social relations. *Radical sci. j.* 5: 65–129.

--- 1979. Science is a labour process. *Science for people* 43/44: 31–37.

--- 1980. Natural theology, Victorian periodicals and the fragmentation of a common context. In *Darwin to Einstein: historical studies on science and belief*, eds. C. Chant and J. Fauvel, pp. 69–107. London: Longman and Open University Press.

--- 1981. The naturalization of value systems in the human sciences. In M. Bartholomew et al., *Problems in the Biological and Human Sciences*, pp. 63–110. Milton Keynes: Open University Press.

--- 1982a. How societies constitute their knowledge: prolegomena to a labour process perspective. Unpublished manuscript.

--- 1982b. The Darwin debate. *Marxism today* 26(4): 20–22.

--- 1982c. Is nature a labour process? Unpublished manuscript.

Yule, G. U. 1906. On the theory of inheritance of quantitative compound characters on the basis of Mendel's laws — a preliminary note. *Rept. 3rd int. conf. genet.* pp. 140–142.

Zangerl, C. H. E. 1971–1972. The social composition of the country magistracy in England and Wales, 1831–1887. *J. Brit. stud.* 11: 113–125.

Zavadskii, K. M. 1973. *Razvitie evolutsionnoi teorii posle Darwina* (The development of evolutionary theories after Darwin) Moscow: Nauka.

---; **Georgievskii, A. B.; and Mozelov, A. P.** 1971. Engels and Darwinism. *Soviet Stud.-Philos.* 10: 63–80.

Zimmerman, W. 1953. *Evolution: die Geschichte ihrer Probleme und Erkenntnisse*. München: Freiburg.

--- 1960. Die Auseinandersetzung mit den Ideen Darwins. Der "Darwinismus" als ideengeschichtliches Phänomen. In *Hundert Jahre Evolutionsforschung*, eds. G. Heberer and F. Schwanitz, pp. 290–354. Stuttgart: Gustav

Fischer.
--- 1968. *Evolution und Naturphilosophie*. Berlin: Dunker und Humblot.
Zirkle, C. 1941. Natural selection before the "*Origin of Species*". *Proc. Am. Phil. Soc.* 84: 71–123.
--- 1946. The early history of the idea of the inheritance of acquired characters and of pangenesis. *Trans. Am. Phil. Soc.* 35: 91–151.
--- 1957. Benjamin Franklin, Thomas Malthus and the United States census. *Isis* 48: 58–62.
--- 1959a. *Evolution, Marxian biology and the social scene*. Philadelphia: University of Pennsylvania Press.
--- 1959b. Species before Darwin. *Proc. Am. Phil. Soc.* 103: 636–644.
--- 1964. Some oddities in the delayed discovery of Mendelism. *J. hered.* 55: 65–72.
Zirnstein, G. 1974. Charles Darwin. Leipzig: Teubner.
--- 1977. Zu Charles Darwin's Briefwechsel mit seinem deutschen Übersetzer Victor Carus. *NTM* 14: 59–73.
Zittel, K. A. von 1891–1893. *Handbuch der Palaeontologie*. Vol. 4. Munich and Leipzig: R. Oldenbourg.
Zmarzlik, H. G. 1963. Der Sozialdarwinismus in Deutschland als geschichtliches Problem. *Vjh. Zeitgesch.* 11: 246–273.
--- 1969. Der Sozialdarwinismus in Deutschland. In *Kreatur Mensch*, ed. G. Altner, pp. 147–156. München: Heinz Moos. 1976. In *The nature of human behaviour*, ed. G. Altner, pp. 346–377. London: Allen and Unwin.
--- 1972. Social Darwinism in Germany, seen as an historical problem. In *Republic to Reich: the making of the Nazi revolution*, ed. H. Holborn, pp. 435–474. New York.
Zuckerman, S. 1976. *The Zoological Society of London, 1826–1976 and beyond*. London: Academic Press.

INDEX

Page numbers in italics refer to chapter end notes.

Library of Congress Cataloging in Publication Data
Main entry under title:

The Darwinian heritage.

Bibliography: p.
Includes index.
1. Darwin, Charles, 1809-1882 -- Congresses.
2. Naturalists -- England -- Biography -- Congresses.
I. Kohn, David, 1941–
II. Charles Darwin Centenary Conference (1982 :
Florence Center for the History and Philosophy of
Science)
QH31.D2D385 1985 575′.0092′4 84-13386
ISBN 0-691-08356-8 (alk. paper)